12TH INTERNATIONAL CONFERENCE ON MESON-NUCLEON PHYSICS AND THE STRUCTURE OF THE NUCLEON (MENU 2010)

12TH INTERNATIONAL CONFERENCE ON MESON-NUCLEON PHYSICS AND THE STRUCTURE OF THE NUCLEON (MENU 2010)

Williamsburg, Virginia, USA   31 May – 4 June 2010

EDITORS

David Armstrong
College of William and Mary, Williamsburg, VA, USA

Volker Burkert
Jian-Ping Chen
Jefferson Lab, Newport News, VA, USA

Will Detmold
College of William and Mary, Williamsburg, VA, USA

Jo Dudek
Old Dominion University, Norfolk, VA, USA

Wally Melnitchouk
David Richards
Jefferson Lab, Newport News, VA, USA

SPONSORING ORGANIZATIONS
The College of William & Mary
Jefferson Lab
Jefferson Science Associates
Forschungszentrum Juelich
IUPAP (International Union of Pure and Applied Physics)

AIP
American Institute of Physics

Melville, New York, 2011
AIP | CONFERENCE PROCEEDINGS ■ 1374

Editors

David Armstrong
Physics Department
P.O. Box 8795
College of William & Mary
Williamsburg, Virginia 3187-8795, USA

Volker Burkert
Jian-Ping Chen

Jefferson Laboratory
12000 Jefferson Avenue
Suite #1
Newport News, VA 23606, USA

E-mail: jpchen@jlab.org

Will Detmold
Physics Department
P.O. Box 8795
College of William & Mary
Williamsburg, Virginia 3187-8795, USA

Jo Dudek
Department of Physics
Old Dominion University
OCNPS Bldg., Room 306
4600 Elkhorn Ave.
Norfolk, VA 23529, USA

Wally Melnitchouk
David Richards

Jefferson Laboratory
12000 Jefferson Avenue
Suite #1
Newport News, VA 23606, USA

E-mail: dgr@jlab.org

L.C. Catalog Card No. 2011908783
ISBN 978-0-7354-0934-7
ISSN 0094-243X
Printed in the United States of America

AIP Conference Proceedings, **Volume 1374**

12th International Conference on Meson-Nucleon Physics and the Structure of the Nucleon (MENU 2010)
UCM-NPSN 2010

Table of Contents

PARALLEL SESSION CONTRIBUTIONS
Session A: Few Body

Session B: Strange Hadrons

Session C: Spin Structure

Session D: Meson Spectroscopy

Session G: Photoproduction

Session H: Transition Form Factors

Session I: Structure Functions

Session M: Nuclear Structure Functions

Session N: Meson-Baryon Interactions

Session O: Form Factors and Two-Boson Exchange

Session P: Hard Exclusive Reactions

Session Q: Heavy Hadrons

Preface: 12th International Conference on Meson-Nucleon Physics and the Structure of the Nucleon (MENU 2010)

It was our great pleasure to host the 12th International Conference on Meson-Nucleon Physics and the Structure of the Nucleon (MENU 2010). The conference continues a series held previously in Karlsruhe (1983), Los Alamos (1987), Gatchina/Leningrad (1989), Bad Honnef (1991), Boulder (1993), Blaubeuren (1995), Vancouver (1997), Zuoz (1999), Washington DC (2001), Beiiing (2004) and Juelich (2007). Having the meeting co-hosted by the Thomas Jefferson National Accelerator Facility (Jefferson Lab) was particularly timely. Jefferson Lab plays a major role as a center of research on hadronic physics, and is poised to continue and expand on that role, as the energy-doubling 12 GeV upgrade to the facility is just around the corner – the major shutdown for the accelerator upgrade will begin in May of 2012.

We would like to thank the International Advisory Committee for guiding us in defining an exciting program, covering a broad range of topics in hadronic physics, as well as suggesting many key speakers. The program consisted of invited plenary talks each morning, with parallel sessions in the afternoon, which featured invited parallel talks as well as contributed talks. There was also a poster session the first evening of the meeting, held in conjunction with an opening reception. All presenters, both speakers and poster presenters, were invited to contribute to the present proceedings. Copies of almost all the talks are available at the conference website http://conferences.jlab.org/menu10/.

We also offer our thanks and appreciation to our sponsors, whose support was essential to the success of the conference. These include the International Union of Pure and Applied Physics (IUPAP), Jefferson Science Associates, Jefferson Lab, Forschungzentrum Juelich, and the College of William & Mary. In particular, this support allowed us to offer reduced registration fees for graduate students, and we were delighted by the number of presentations by students and other young physicists at the meeting.

The meeting was held on the campus of the College of William & Mary, in Williamsburg, Virginia, some 20 miles from Jefferson Lab. There were over 200 participants at the conference. We were fortunate to have great weather for the meeting, which made for a pleasant setting for discussions between sessions. In addition to the scientific program, the meeting featured an excursion during one afternoon, which featured a tour of Historic Jamestown and the Jamestown Settlement. Jamestown, Virginia is the site of the first permanent English settlement in North America, and features a museum, an archaeological site, as well as a "living history" recreation of the original settlement and an aboriginal village. In a similar historical vein, the College, being the second oldest university in the USA (founded in 1693), is next door to Colonial Williamsburg, which was the Colonial Capitol of Virginia (1699-1780). Colonial Williamsburg features a restored 122 hectare "Living History Museum", which conference participants were free to explore. Following the excursion, there was a well-attended banquet on campus in the evening.

We wish to thank support staff at both Jefferson Lab and at William & Mary for the excellent efforts in making the meeting a success. In particular, we thank Ms. Mary Fox and Ms. Ruth Bizot of Jefferson Lab, and Ms. Liz Kiewiet of Conference Services at William & Mary for their efficient, tireless and good-natured work on many aspects of the conference, to David Chopard for his excellent work developing and maintaining the conference web pages, to Joanna Griffin for her design of the poster, and again to Mary Fox for her diligent work helping to put these Proceedings together.

The next conference in the MENU series is planned to be held in Italy in 2013, hosted by INFN/Rome and INFN/Genoa; the contact person is Prof. Annalisa D'Angelo (INFN/Rome).

David Armstrong (co-chair)
Volker Burkert
Jian-Ping Chen
Will Detmold
Jo Dudek
Wally Melnitchouk (co-chair)
David Richards

12th International Conference on Meson-Nucleon Physics and the Structure of the Nucleon (MENU 2010)
AIP Conf. Proc. 1374, 1-1 (2011); doi: 10.1063/1.3647091

International Advisory Committee

Stan Brodsky (SLAC)
Larry Cardman (Jefferson Lab)
Annalisa D'Angelo (INFN, Rome)
Simon Eidelman (Novosibirsk)
Haiyan Gao (Duke)
Eberhard Klempt (Bonn)
Siggy Krewald (Juelich)
Harry Lee (ANL)
Hartmut Machner (Juelich)
Takashi Nakano (Osaka)
Ben Nefkens (UCLA)
Klaus Peters (GSI)
Marco Ripani (Genoa)
Craig Roberts (ANL)
Andreas Schaefer (Regensburg)
Madeleine Soyeur (Saclay)
Hans Stroeher (Juelich)
Anthony Thomas (Adelaide)
Marc Vanderhaeghen (Mainz)
Wim van Oers (TRIUMF/Manitoba)
Fan Wang (Nanjing)
Shin-Nan Yang (Taiwan)
Bingsong Zou (IHEP, Beijing)

Local Organizing Committee

David Armstrong (William and Mary) co-chair
Volker Burkert (JLab)
Jian-Ping Chen (JLab)
Will Detmold (William and Mary/JLab)
Jo Dudek (Old Dominion/JLab)
Wally Melnitchouk (JLab) co-chair
David Richards (JLab)

Name	Institution	Email
Accardi, Alberto	Hampton University & Jefferson Laboratory	accardi@jlab.org
Afonin, Sergey	St. Petersburg State University	afonin24@mail.ru
Anderson, Bryon	Kent State University	bdanders@kent.edu
Arends, Juergen	Mainz University	arends@kph.uni-mainz.de
Armstrong, David	College of William & Mary	dsarms@wm.edu
Atashbar Tehrani, Shahin	Institute for Studies in Theoretical Physics and Mathematics	atashbar@ipm.ir
Aubin, Christopher	Fordham University	aubin@fordham.edu
Avagyan, Harut	Jefferson Lab	avakian@jlab.org
Averett, Todd	College of William and Mary	tdaver@wm.edu
Bacchetta, Alessandro	University of Pavia	alessandro.bacchetta@unipv.it
Beck, Reinhard	University Bonn	beck@hiskp.uni-bonn.de
Bennett, Robert	Old Dominion University	rbennett@odu.edu
Berger, Niklaus	Institute of High Energy Physics	nberger@ihep.ac.cn
Bhatt, Himani	Indian Institute of Technology Bombay	himani2711@gmail.com
Biselli, Angela	Fairfield University	biselli@jlab.org
Bizzeti, Andrea	INFN, Sezione di Firenze and University of Modena and Reggio Emilia	andrea.bizzeti@fi.infn.it
Blunden, Peter	University of Manitoba	blunden@physics.umanitoba.ca
Braaten, Eric	Ohio State University	braaten@mps.ohio-state.edu
Brash, Edward	Christopher Newport University	brash@jlab.org
Burkardt, Matthias	New Mexico State University	burkardt@nmsu.edu
Burkert, Volker	Jefferson Lab	burkert@jlab.org
Burtin, Etienne	CEA- Centre de Saclay	eburtin@cea.fr
Calancha-Paredes, Constantino	CIEMAT	calancha@fnal.gov
Calle Cordon, Alvaro	University of Granada	alvarocalle@ugr.es
Cardman, Lawrence	Jefferson Lab	cardman@jlab.org

Cargnelli, Michael	Austriam Academy of Sciences, Stefan Meyer Institute	michael.cargnelli@oeaw.ac.at
Carlson, Carl	College of William & Mary	carlson@physics.wm.edu
Carroll, Jonathan	University of Adelaide	jonathan.carroll@adelaide.edu.au
Chang, Wen-Chen	Institute of Physics, Acdemia Sinica	changwc@phys.sinica.edu.tw
Chen, Jian-Ping	Jefferson Lab	jpchen@jlab.org
Choi, Seonho	Seoul National University	choi@phya.snu.ac.kr
Cloet, Ian	University of Washington	icloet@uw.edu
Cole, Philip	Idaho State University	colephil@isu.edu
Crabb, Donald	University of Virginia	dgc3q@virginia.edu
D'Angelo, Annalisa	University of Rome Tor Vergata and INFN Rome Tor Vergata	annalisa.dangelo@roma2.infn.it
Daniel, Aji	Ohio University	adaniel@jlab.org
De La Cruz Burelo, Eduard	CINVESTAV-IPN Mexico	eduard@fnal.gov
De Vita, Raffaella	INFN - Genova	devita@ge.infn.it
Deconinck, Wouter	Massachusetts Institute of Technology	wdconinc@mit.edu
Deng, Xiaoyan	University of Virginia	deng@jlab.org
Detmold, William	College of William & Mary/Jefferson Laboratory	wdetmold@jlab.org
Dickson, Ryan	Carnegie Mellon	dickson.66@gmail.com
Djalali, Chaden	University of South Carolina	Djalali@sc.edu
Doering, Michael	Fz-Jeulich	m.doering@fz-juelich.de
Doi, Takumi	University of Tsukuba	doi@ribf.riken.jp
Dudek, Jozef	Jefferson Laboratory / Old Dominion University	dudek@jlab.org
Dugger, Michael	Arizona State University	dugger@jlab.org
Edwards, Robert	Jefferson Laboratory	edwards@jlab.org
Eichmann, Gernot	Darmstadt University of Technology	gernot.eichmann@physik.tu-darmstadt.de
Elster, Charlotte	Ohio University	elster@ohiou.edu
Erlich, Josh	College of William and Mary	erlich@physics.wm.edu

Filikhin, Igor	North Carolina Central University	ifilikhin@nccu.edu
Fischer, Christian	TU Darmstadt/GSI	christian.fischer@physik.tu-darmstadt.de
Fradi, Ahmed	IPN-ORSAY	fradi@ipno.in2p3.fr
Fuchey, Eric	Université Blaise Pascal	efuchey@jlab.org
Gamberg, Leonard	Penn State University	lpg10@psu.edu
Garzon, Javier	University of Valencia	garzon@ific.uv.es
Gasparyan, Ashot	GSI	a.gasparyan@gsi.de
Giovannella, Simona	INFN Frascati	simona.giovannella@lnf.infn.it
Girod, Francois-Xavier	JLab/CLAS	fxgirod@jlab.org
Gothe, Ralf W.	University of South Carolina	gothe@sc.edu
Guo, Feng-Kun	IKP, Forschungszentrum Juelich	f.k.guo@fz-juelich.de
Haas, Florian	Technische Universität München	florian.haas@ph.tum.de
Haegler, Philipp	TU Munich	haegler@p2h.de
Haidenbauer, Johann	Forschungszentrum Juelich	j.haidenbauer@fz-juelich.de
Hansen, Jens-Ole	Jefferson Lab	ole@jlab.org
Hayano, Ryugo	Department of Physics, The University of Tokyo	hayano@phys.s.u-tokyo.ac.jp
Hicks, Kenneth	Ohio University	hicks@ohio.edu
Higinbotham, Douglas	Jefferson Lab	doug@jlab.org
Hinterberger, Frank	University of Bonn	fh@hiskp.uni-bonn.de
Hoop, Bernard	Walden University	bhoop@waldenu.edu
Horn, Tanja	Catholic University of America	hornt@jlab.org
Huang, Fei	University of Georgia	huang@physast.uga.edu
Huang, Jin	Massachusetts Institute of Technology	jinhuang@mit.edu
Jaegle, Igal	Department of Physics	igal.jaegle@unibas.ch
Jeschonnek, Sabine	Ohio State	jeschonnek.1@osu.edu
Ji, Chueng-Ryong	North Carolina State University	ji@ncsu.edu
Jin, Ge	University of Virginia	gjin@jlab.org
Jiang, Xiaodong	Los Alamos National Laboratory	jiang@jlab.org

Julia-Diaz, Bruno	U. Barcelona	bjulia@gmail.com
Kamano, Hiroyuki	Jefferson Lab	hkamano@jlab.org
Kao, Chung-Wen	Chung Yuan Christian University	cwkao@phys.cycu.edu.tw
Khrykin, Anatoly	JINR	khrykin@nusun.jinr.ru
Kohl, Michael	Hampton University	kohlm@jlab.org
Krewald, Siegfried	Forschungszentrum Juelich	s.krewald@fz-juelich.de
Kroll, Peter	Universitaet Wuppertal	kroll@physik.uni-wuppertal.de
Kubarovsky, Valery	Jefferson Lab	vpk@jlab.org
Kupsc, Andrzej	Uppsala University	Andrzej.Kupsc@physics.uu.se
Lalwani, Kavita	Indian Institute of Technology Bombay	pankavi@gmail.com
Lange, Jens Soeren	II. Physikalisches Institut, University Giessen	soeren.lange@exp2.physik.uni-giessen.de
Ledwig, Tim	University Mainz	ledwig@kph.uni-mainz.de
Lewis, Randy	York University	randy.lewis@yorku.ca
Liuti, Simonetta	University of Virginia	sl4y@virginia.edu
Liu, Fang	IHEP,CAS,CHINA	liuf@mail.ihep.ac.cn
Liu, Ming	Los Alamos National Lab	ming@bnl.gov
Luo, Wei	Lanzhou University	weiluo@jlab.org
Machner, Hartmut	FZ Juelich	h.machner@fz-juelich.de
Mai, Maxim	Uni-Bonn	mai@hiskp.uni-bonn.de
Malace, Simona	University of South Carolina	simona@jlab.org
Matevosyan, Hrayr	CSSM, University of Adelaide	hrayr.matevosyan@adelaide.edu.a
Mathur, Nilmani	Tata Institute, India	nilmani@theory.tifr.res.in
Mattione, Paul	Rice University	pmatt@jlab.org
McKeown, Robert	Jefferson Lab	bmck@jlab.org
Melnitchouk, Wally	Jefferson Laboratory	wmelnitc@jlab.org
Mestayer, Mac	Jefferson Lab	mestayer@jlab.org
Meziane, Mehdi	College of William and Mary	mezianem@jlab.org
Mihovilovic, Miha	Jožef Stefan Institute	miham@jlab.org
Mitchell, Ryan	Indiana University	remitche@indiana.edu

Mokeev, Victor	Jefferson Laboratory	mokeev@jlab.org
Moriya, Kei	Carnegie Mellon University	kmoriya@andrew.cmu.edu
Moutarde, Hervé	CEA-Saclay	herve.moutarde@cea.fr
Musch, Bernhard	Jefferson Lab	bmusch@jlab.org
Muto, Takumi	Chiba Institute of Technology	takumi.muto@it-chiba.ac.jp
Nadel-Turonski, Pawel	Jefferson Lab	
Nakamura, Satoshi	Jefferson Laboratory	sxnakamura@gmail.com
Nakano, Takashi	RCNP, Osaka University	nakano@rcnp.osaka-ua.cjp
Nakayama, Kanzo	University of Georgia	nakayama@uga.edu
Nasseripour, Rakhsha	George Washington University	rakhsha@jlab.org
Nebreda, Jenifer	Universidad Complutense de Madrid	jenifer.nebreda@fis.ucm.es
Nefkens, Ben	University of California, Los Angeles	nefkens@physics.ucla.edu
Neubert, Sebastian	Technische Universität München	sneubert@e18.physik.tu-muenchen.de
Oh, Yongseok	Kyungpook National University	yohphy@knu.ac.kr
Oset, Eulogio	University of Valencia	oset@ific.uv.es
Paris, Mark	George Washington University	mparis@gwu.edu
Parno, Diana	Carnegie Mellon University	dparno@cmu.edu
Pascalutsa, Vladimir	Johannes Gutenberg-Universität Mainz	vladipas@kph.uni-mainz.de
Paschke, Kent	University of Virginia	paschke@virginia.edu
Perazzini, Stefano	INFN & Università di Bologna	stefano.perazzini@bo.infn.it
Perdrisat, Charles	College of William and Mary	perdrisa@jlab.org
Petratos, Gerassimos	Kent State University	gpetrato@kent.edu
Phillips, Daniel	Ohio University	phillips@phy.ohiou.edu
Pitt, Mark	Virginia Tech	pitt@vt.edu
Prokudin, Alexei	JLab	prokudin@jlab.org
Punjabi, Vina	Norfolk State University	punjai@jlab.org
Qiang, Yi	Jefferson Lab	yqiang@jlab.org
Ramalho, Gilberto	Center for Theoretical Particle Physics	ramalho@jlab.org

Reis, Alberto	Centro Brasileiro de Pesquisas Fisicas	alberto@cbpf.br
Richards, David	Jefferson Laboratory	dgr@jlab.org
Riedl, Caroline	DESY	criedl@mail.desy.de
Rios, Guillermo	Universidad Comlputense de Madrid	griosmar@fis.ucm.es
Ritchie, Barry	Arizona State University	Barry.Ritchie@ASU.edu
Rossi, Patrizia	Laboratori Nazionali di Frascati - INFN	rossi@lnf.infn.it
Rozpedzik, Dagmara	Jagiellonian University	dagmara.rozpedzik@uj.edu.pl
Rubin, Philip	George Mason University	prubin@gmu.edu
Sabintsev, Arthur	The Geroge Washington University	arthur@Jlab.org
Sadler, Michael	Abilene Christian University	sadler@physics.acu.edu
Saghai, Bijan	CEA-Saclay	bijan.saghai@cea.fr
Saha, Arun	Jefferson Lab	saha@jlab.org
Salme`, Giovanni	INFN	salmeg@roma1.infn.it
Sarantsev, Andrei	HISKP Bonn-Uni	andsar@hiskp.uni-bonn.de
Sargsian, Misak	Florida International University	sargsian@fiu.edu
Sato, Masaharu	University of Tokyo	m-sato@nucl.phys.s.u-tokyo.ac.jp
Savage, Martin	university of washington	savage@phys.washington.edu
Sawada, Shinya	KEK	shinya.sawada@kek.jp
Schat, Carlos	Ohio University and University of Buenos Aires	carlos.schat@gmail.com
Schumacher, Reinhard	Carnegie Mellon University	schumacher@cmu.edu
Segovia, Jorge	University of Salamanca	segonza@usal.es
Shah, Neha	Indian Institute of Technology Bombay	neau2802@gmail.com
Sirca, Simon	University of Ljubljana	simon.sirca@fmf.uni-lj.si
Sokhan, Daria	IPN Orsay	daria@jlab.org
Somov, Alexander	Jefferson Laboratory	somov@jlab.org
Strakovsky, Igor	The George Washington University	igor@gwu.edu
Subedi, Ramesh	George Washington University	rsubedi@jlab.org
Sudol, Malgorzata	IPNO/CNRS	sudol@ipno.in2p3.fr

Sulkosky, Vincent	Massachusetts Institute of Technology	vasulk@jlab.org
Suslov, Vladimir	North Carolina Central University	vsuslov@nccu.edu
Suzuki, Takatoshi	The University of Tokyo	tsuzuki@phys.s.u-tokyo.ac.jp
Syritsyn, Sergey	Massachusetts Inst. of Tech	syritsyn@mit.edu
Tandy, Peter	Kent State University	tandy@kent.edu
Thomas, Anthony	University of Adelaide	anthony.thomas@adelaide.edu.au
Thomas, Christopher	Jefferson Laboratory	thomasc@jlab.org
Tiator, Lothar	Institute for Kernphysik	tiator@kph.uni-mainz.de
Torok, Aaron	Indiana University	amtorok@indiana.edu
Ungaro, Maurizio	UCONN/JLAB	ungaro@jlab.org
Valdau, Yury	Forschungszentrum Jeulich	y.valdau@fz-juelich.de
van Beveren, Eef	University Coimbra	eef@teor.fis.uc.pt
Van Oers, Willem	University of Manitoba/TRIUMF	vanoers@triumf.ca
Vancraeyveld, Pieter	Ghent University	pieter.vancraeyveld@ugent.be
Vlahovic, Branislav	North Carolina Central University	vlahovic@nccu.edu
Walker-Loud, Andre	College of William & Mary	
Wang, Diancheng	University of Virginia	dwang@jlab.org
Wang, Fan	Nanjing University, China	wangfan19340729@yahoo.com.cn
Wiedner, Ulrich	Ruhr-Universität Bochum	
Weiss, Christian	Jefferson Lab	weiss@jlab.org
Weller, Henry	Duke University and TUNL	weller@tunl.duke.edu
Williams, Richard	TU Darmstadt	richard.williams@physik.tu-darmstadt.de
Wronska, Aleksandra	Jagiellonian University, Cracow, Poland	aleksandra.wronska@uj.edu.pl
Wu, Jia-Jun	Institute of High Energy Physics	wujiajun@ihep.ac.cn
Xu, Guofa	Institute of High Energy Physics	xugf@ihep.ac.cn
Yamazaki, Takeshi	CCS, Univ. of Tsukuba	yamazaki@ccs.tsukuba.ac.jp
Young, Glenn	Jefferson Laboratory	gyoung@jlab.org
Young, Ross	University of Adelaide	ross.young@adelaide.edu.au

Zhan, Xiaohui	Argonne	zhanxh@jlab.org
Zhang, Jixie	ODU	jixie@jlab.org
Zhao, Qiang	Institute of High Energy Physics	zhaoq@ihep.ac.cn
Ziegler, Veronique	SLAC	vziegler@slac.stanford.edu
Zou, Bing-Song	IHEP, CAS	zoubs@ihep.ac.cn

Sponsors

The College of William & Mary,
Jefferson Lab,
Jefferson Science Associates,
Forschungszentrum Juelich,
IUPAP (International Union of Pure and Applied Physics)

PLENARY SESSION PRESENTATIONS

Complete Photo-production Experiments

A. D'Angelo[*,†], K. Ardashev[**,‡], C. Bade[§], O. Bartalini[*,†], V. Bellini[¶,‖], M. Blecher[††], J. -P. Bocquet[‡‡], M. Capogni[*,§§], A. Caracappa[¶¶], L. E. Casano[†], M. Castoldi[***], R. Di Salvo[†], A. Fantini[*,†], D. Franco[*,†††], G. Gervino[‡‡‡], F. Ghio[§§§], G. Giardina[a,¶¶¶], C. Gibson[‡], B. Girolami[§§§], A. Giusa[a,¶], H. Glükler[b], K. Hicks[§], S. Hoblit[**,¶¶], A. Honig[c], T. Kageya[d,¶¶], M. Khandaker[d,e], O. C. Kistner[¶¶], S. Kizilgul[§], S. Kucuker[**], A. Lapik[f], A. Lehmann[‡], P. Levi Sandri[g], A. Lleres[‡‡], M. Lowry[d,¶¶], M. Lucas[§], J. Mahon[§], F. Mammoliti[a,¶], G. Mandaglio[a,¶¶¶], M. Manganaro[a,¶¶¶], L. Miceli[¶¶], D. Moricciani[†], A. Mushkarenkov[f], V. Nedorezov[f], B. Norum[**], M. Pap[b], B. Preedom[‡], H. Seyfarth[b], C. Randieri[a,¶], D. Rebreyend[‡‡], N. Rudnev[f], G. Russo[a,¶], A. Sandorfi[d,¶¶], C. Schaerf[*,†], M. -L. Sperduto[a,¶], H. Ströher[b], M. -C. Sutera[a,¶], C. E. Thorn[¶¶], A. Turinge[f], V. Vegna[*,†], C. S. Whisnant[h], K. Wang[**] and X. Wei[d,¶¶]

[*]*Dipartimento di Fisica, Universitá di Roma "Tor Vergata", Via della Ricerca Scientifica 1, I-00133 Roma, Italy*
[†]*INFN Sezione di Roma Tor Vergata, Via della Ricerca Scientifica 1, I-00133 Roma, Italy*
[**]*Department of Physics, University of Virginia, Charlottesville, Virginia 22901, USA*
[‡]*Department of Physics, University of South Carolina, Columbia, South Carolina 29208, USA*
[§]*Department of Physics, Ohio University, Athens Ohio 45701, USA*
[¶]*Dipartimento di Fisica, Universitá di Catania , Via Santa Sofia 44, I-95123 Catania, Italy*
[‖]*INFN Laboratori Nazionali del Sud, Via Santa Sofia 44, I-95123 Catania, Italy*
[††]*Physics Department, Virginia Polytechnic Institute and State University, Blacksburg, Virginia 24061, USA*
[‡‡]*IN2P3, Laboratoire de Physique Subatomique et de Cosmologie, 38026 Grenoble, France*
[§§]*Present affiliation: ENEA - C.R. Casaccia, via Anguillarese 301, I-00060 Roma, Italy*
[¶¶]*Physics Department, Brookhaven National Laboratory, Upton, New York 11973, USA*
[***]*Dipartimento di Fisica, Università degli Studi di Genova, via Dodecaneso 33, I-16146 Genova, Italy*
[†††]*Present affiliation: Albert Einstein Center for Fundamental Physics - LHEP - University of Bern, Switzerland*
[‡‡‡]*Dipartimento di Fisica Sperimentale, Università degli Studi di Torino and INFN Sezione di Torino, via Pietro Giuria 1, I-10125 Torino, Italy*
[§§§]*Istituto Superiore di Sanità, viale Regina Elena 299, I-00161 Roma, Italy and INFN Sezione di Roma, piazzale Aldo Moro 2, I-00185 Roma, Italy*
[¶¶¶]*Dipartimento di Fisica, Università di Messina, salita Sperone 31, I-98166 Messina, Italy*
[a]*INFN Sezione di Catania, via Santa Sofia 64, I-95123 Catania, Italy*
[b]*Forschungszentrum Jülich GmbH, D-52425 Jülich, Germany*
[c]*Department of Physics, Syracuse University, Syracuse, New York 13210, USA*
[d]*Present affiliation:Thomas Jefferson National Accelerator Facility, Newport News, Virginia 23606, USA*
[e]*Norfolk State University, Norfolk, Virginia 23606, USA*
[f]*Institute for Nuclear Research, 60-letiya Oktyabrya prospekt 7a, 117312 Moscow, Russia*
[g]*INFN Laboratori Nazionali di Frascati, via E. Fermi 40, I-00044 Frascati, Italy*
[h]*James Madison University, Harrisonburg, Virginia 22807, USA*

Abstract. The extraction of resonance parameters from meson photo-reaction data is a challenging effort, that would greatly benefit from the availability of several polarization observables, measured for each reaction channel on both proton and neutron targets. In the aim of obtaining such *complete* experiments, polarized photon beams and targets have been developed at facilities, worldwide. We report on the latest results from the LEGS and GRAAL collaborations, providing single and double polarization measurements on pseudo-scalar meson photo-production from the nucleon.

Keywords: Photoreactions, polarized photon beams, polarized targets.

12th International Conference on Meson-Nucleon Physics and the Structure of the Nucleon (MENU 2010)
AIP Conf. Proc. 1374, 17-22 (2011); doi: 10.1063/1.3647092

PACS: 25.20-x,24.70+s

1. INTRODUCTION

One of the most significant challenges of hadron physics is the quantitative description of the properties of hadrons from QCD. Recently, both numerical simulations of unquenched lattice QCD (LQCD)[1] and solutions of the Dyson-Swinger equation (DSE) for the quark propagator[2] have found that the current-quarks of perturbative QCD evolve into constituent quarks at low momenta; that is to say that the constituent quark masses arise from low momentum gluons attaching themselves to current quarks.

This finding provides a theoretical justification for QCD -inspired Constituent Quark Models (CQM) that have been used in the last decades to predict the resonant structure of the nucleon[3, 4, 5]. In these models the asymmetry of the baryon wave function is guaranteed by color, but color degrees of freedom are integrated out and play no dynamical role. Excited baryon states (N^*) are classified by isospin, parity and spin within each oscillator band. Comparison with experimental data shows, however, that only the lowest few resonances in each band have been seen in πN reactions. Moreover the $g(\pi N)$ couplings are predicted to decrease rapidly with increasing mass in each oscillator band and higher levels are expected to have larger couplings to $K\Lambda$, $K\Sigma, \pi\pi N, \eta N$ and ωN channels. Therefore, unseen or *missing* resonances may show their presence in other reaction channels such as K, η, ω and multi-pion photo-production on the nucleons.

An alternative explanation comes from the *di-quark* model in which two quarks in nucleon are assumed to be quasi-bound in a color isotriplet; the baryon states are made of diquark-quark combinations that are net color iso-singlet. If all possible internal di-quark excitations are considered the full spectrum of CQM is obtained. But if internal di-quark excitations are frozen out (having both spin and and isospin equal to zero) a large reduction in the number of baryon degrees of freedom is predicted, providing about the same number of N^* states, seen in πN reactions.

Unravelling the N^* spectrum is still a challenge. Photo-nuclear reactions on the proton, in the resonance energy region, have been deeply investigated in recent years and detailed knowledge of differential and total cross sections is presently available for most of the possible reaction channels. Extraction of resonance parameters from experiments is not straightforward. The idealized path to search for N^* and Δ^* states via meson photo-production would be to determine the production amplitude from experiments and to search for resonant structure in a model independent way (using Argand circles, phase motion, speed plots, etc.). Then models would be needed to separate resonance from background components and determine resonant γN^* and decay couplings, which could be compared with LQCD, DSE and other hadron models predictions.

After fifty years of photo-nuclear experiments, however, amplitude extraction from data has never been accomplished and models have conjectured resonances and adjusted couplings to compare with limited data. The next experimental challenge is to provide measurements for the highest number of spin and isospin dependent observables.

In the case of pseudo-scalar meson photo-production on the nucleon, there are eight possible spin states, corresponding to four independent complex amplitudes, which describe the transition matrix.

From four complex amplitudes it is possible to construct fifteen polarization observables, in addition to the unpolarized differential cross section. While seven carefully chosen polarization observables could be enough to completely determine the amplitudes from the data, it has been found that an higher number of observables are needed to compensate for limited experimental precision[6]. Moreover each reaction amplitude has two components resulting from coupling of the $I = 1/2$ nucleon isospin with the isoscalar and the isovector components of the photon. These two isospin-dependent contributions appear in linear combinations in the final reaction amplitude and they may be disentangled only using measurements on both the neutron and the proton targets.

An experiment capable of providing enough single and double polarization observable on both proton and neutron targets, so that a model independent extraction of all spin- and isospin- dependent amplitudes may be performed from the data, is called *complete* experiment[7, 6]. It requires linearly and circularly polarized photon beams and longitudinally and transversely polarized targets. A large effort is going on in worldwide facilities to realize this goal. The most recent results from the LEGS and the GRAAL experiments are here shown, providing new single and double polarization measurements on single pion and kaon photo-production channels.

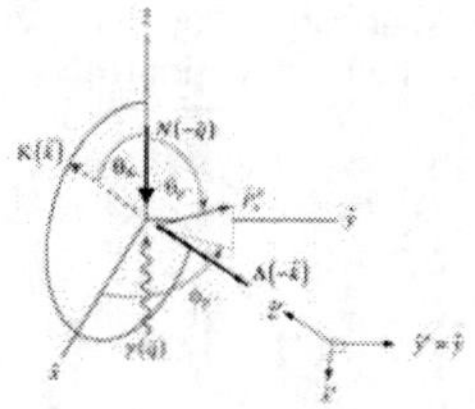

FIGURE 1. Reference frames used to define the beam, target and recoil polarization directions.

2. THE LEGS AND GRAAL EXPERIMENTS

Both LEGS and GRAAL experiments used Compton backscattered photon beams, produced by the interaction of an Ion-Argon Laser with the electrons circulating inside a storage ring: the National Synchrotron Radiation Facility at the Brookhaven National Laboratory (USA), in the case of LEGS, and the European Synchrotron Radiation Facility in Grenoble (France), in the case of GRAAL. The two tagged photon beams had complementary energy ranges: (220-480) MeV, covering the first nucleon resonance region, in the case of LEGS; (550-1480)MeV, corresponding to the second and third nucleon resonance region, in the case of GRAAL. Since relativistic electrons conserve helicity in their scattering process with the incoming Laser photon beam, also backscattered photons retain their initial degree of polarization. Therefore if the Laser light is linearly or circularly polarized, the Compton backscattered photon beam is also linearly or circularly polarized to a high degree. The polarization degree slowly decreases as a function of the electron scattering angle and scattered photon energy, but it may be higher than 90% at LEGS and 60% at GRAAL, over the entire tagged energy range, by changing the Ion-Argon Laser line. Both experiments were equipped with a large solid angle detector. In the case of the LEGS experiment a NaI crystal box was used to detect photons and charged particles for polar angles between 20° and 160°. It was alternatively equipped with an internal neutron barrel, made of plastic scintillators, optimized for neutron detection or with a Time Projection Chamber (TPC), for charged particle tracking. At forward angles two walls, made of plastic scintillators and of Lead Glass crystals, respectively, allowed to detect both charged and neutral particles. A polarized frozen-spin HD target completed the experimental set-up. Both proton and deuterons could be independently polarized, with polarization degrees up to 60% and 35% for protons and deuterons, respectively.

In the case of the GRAAL set-up the LAGRANγE detector was made of a BGO ball covering polar angles between 25° and 155°, internally equipped with a plastic scintillator barrel and a double layer of cylindrical Multi Wire Proportional Chambers (MWPC). The detector was optimized for photon detection up to 1.5 GeV and for proton detection up to 300 MeV. At forward angles a double wall of planar MWPC was backed by a double wall of plastic scintillators for charged particles detection and a shower wall made of sandwiches of plastic scintillators and lead, for neutral particles detection.

Both experiments were characterized by very low levels of backgrounds, due to unpolarized or untagged photons impinging the target. Since only H or D targets were used, very low backgrounds coming from heavier nuclear target components (such as Carbon in buthanol targets or Nitrogen in ammonia targets) contaminated the data acquisition, allowing for excellent signal/background separation.

3. POLARIZED OBSERVABLES

For a polarized photon beam impinging on a polarized target and assuming the spins of the outgoing baryons are measured, the differential cross section for a pseudo-scalar meson photo-production reaction on a nucleon, may be expressed in terms of polarization observables, to leading order in polarization, as follows:

$$\begin{aligned}
d\sigma &= \frac{1}{2}\{d\sigma_0 + \hat{\Sigma}[-P_L^\gamma\, cos(2\phi_\gamma)] + \hat{T}[P_y^T] + \hat{P}[P_{y'}^R] \\
&+ \hat{E}[-P_c^\gamma P_z^T] + \hat{G}[P_L^\gamma P_z^T\, sin(2\phi_\gamma)] + \hat{F}[P_c^\gamma P_x^T] + \hat{H}[P_L^\gamma P_x^T\, sin(2\phi_\gamma)] \\
&+ \hat{C}_{x'}[P_C^\gamma P_{x'}^R] + \hat{C}_{z'}[P_C^\gamma P_{z'}^R] + \hat{O}_{x'}[P_L^\gamma P_{x'}^R sin(2\phi_\gamma)] + \hat{O}_{z'}[P_L^\gamma P_{z'}^R sin(2\phi_\gamma)] \\
&+ \hat{L}_{x'}[P_z^T P_{x'}^R] + \hat{L}_{z'}[P_z^T P_{z'}^R] + \hat{T}_{x'}[P_x^T P_{x'}^R] + \hat{T}_{z'}[P_x^T P_{z'}^R]\}
\end{aligned}$$

TABLE 1. Polarization observables for pseudoscalar meson photoproduction[6].

Simbol	Required Polarization Beam,Target,Recoil	Type
$\frac{d\sigma}{dt}_0$	-;-;-	Single
$\hat{\Sigma} = \Sigma\frac{d\sigma}{dt}_0$	L($\frac{\pi}{2}$,0);-;-	
$\hat{T} = T\frac{d\sigma}{dt}_0$	-;y;-	
$\hat{P} = T\frac{d\sigma}{dt}_0$	-;-;y	
$\hat{G} = G\frac{d\sigma}{dt}_0$	L($\frac{\pi}{4}$,0);z;-	Beam
$\hat{H} = H\frac{d\sigma}{dt}_0$	L($\frac{\pi}{4}$,0);x;-	and
$\hat{E} = E\frac{d\sigma}{dt}_0$	C;z;-	Target
$\hat{F} = F\frac{d\sigma}{dt}_0$	C;x;-	
$\hat{O}_x = O_x\frac{d\sigma}{dt}_0$	L($\frac{\pi}{4}$,0);-;x'	Beam
$\hat{O}_z = O_z\frac{d\sigma}{dt}_0$	L($\frac{\pi}{4}$,0);-;z'	and
$\hat{C}_x = C_x\frac{d\sigma}{dt}_0$	C;-;x'	Recoil
$\hat{C}_z = C_z\frac{d\sigma}{dt}_0$	C;-;z'	
$\hat{T}_x = T_x\frac{d\sigma}{dt}_0$	-;x;x'	Target
$\hat{T}_z = T_z\frac{d\sigma}{dt}_0$	-;x;z'	and
$\hat{L}_x = L_x\frac{d\sigma}{dt}_0$	-;z;x'	Recoil
$\hat{L}_z = L_z\frac{d\sigma}{dt}_0$	-;z;z'	

where $d\sigma_0$ is the unpolarized differential cross section, $P_L^\gamma(P_C^\gamma)$ is the degree of linear (circular) beam polarization, P_x^T (P_y^T or P_z^T) is the degree of target polarization along the x (y or z) direction (see Fig. 1), $P_{x'}^R$ ($P_{y'}^R$ or $P_{z'}^R$) is the degree of recoil polarization along the x' (y' or z') directions (see Fig. 1) and ϕ_γ is the relative angle between the linear beam polarization direction and the scattering plane. The single ($\hat{\Sigma}, \hat{T}, \hat{P}$) and double ($\hat{G}, \hat{H}, \hat{E}, \hat{F}, \hat{O}_x, \hat{O}_z, \hat{C}_x, \hat{C}_z, \hat{T}_x, \hat{T}_z, \hat{L}_x, \hat{L}_z$) polarization observables are shown schematically in Table 1 and defined explicitly in [6]. The sixteen observables are not independent and are related by several relations, called *Fierz* identities. These relations may be used to check consistency among different measured observables.

4. RESULTS FROM THE LEGS AND GRAAL EXPERIMENTS

Circularly polarized photon beams have been used on a longitudinally polarized HD target to measure the $\hat{E}$ observable for both $\vec{\gamma}\vec{p} \to \pi^+ n$ and $\vec{\gamma}\vec{p} \to \pi^0 p$ reactions, in the energy range between 220 and 480 MeV, by the LEGS experiment. The $\hat{E}$ observable may be directly related to the difference of cross sections for parallel and anti-parallel polarizations of beam and target, as it appears in the GDH spin sum rule. The results are shown in Fig. 2 for a few energy bins, together with predictions of the SAID[FA07k] solution of partial wave analysis[8]. The curves slightly overestimate the experimental results and LEGS measurements allowed for a new evaluation of the GDH sum rule[9], closer to the theoretical prediction.

Using linearly polarized photons on a longitudinally polarized HD target it was also possible to measure the $\hat{G}$ observable for the same pion photo-production channels on the proton. Some preliminary results are shown in Fig. 3 together with previous data from [10] and predictions form the SAID[FA07k] solution. The new data show that while a reasonable agreement exists with both existing results and predictions for the charged pion production, opposite sign and higher strength than predicted are found for the neutral pion channel. This effect is still under investigation and it may indicate that a larger D-wave component of the $\Delta(1232)$ than currently established is necessary to describe the data.

Σ beam asymmetries of pseudo-scalar meson photo-production reactions on the nucleons have been systematically measured by the GRAAL collaboration[11]. Some of latest results from π^- photo-production on the neutron[12] are shown in Fig.4 together with existing data and partial wave analysis predictions from SAID and MAID[13]. The new precise GRAAL data required a revision of the SAID solution leading to several changes from the previous [SP09]

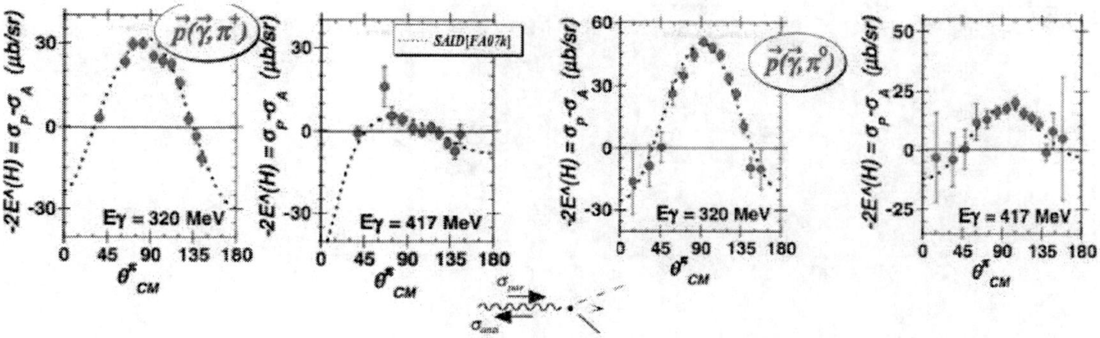

FIGURE 2. Measurements of the $\hat{E}$ polarization observable for the $\vec{\gamma}\vec{p} \to \pi^+ n$ (left two panels), and $\vec{\gamma}\vec{p} \to \pi^0 p$ (right two panels) at LEGS.

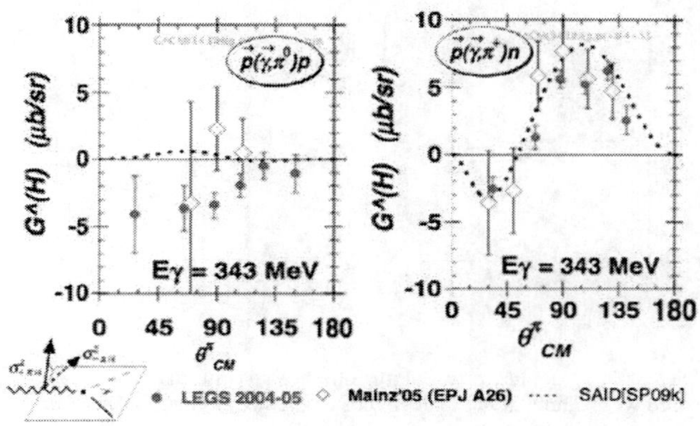

FIGURE 3. Measurements of the $\hat{G}$ polarization observable for the $\vec{\gamma}\vec{p} \to \pi^+ n$ (left panel), and $\vec{\gamma}\vec{p} \to \pi^0 p$ (right panel) at LEGS. Results are compared with previous measures at Mainz and with SAID[SP09k] partial wave analysis predictions.

version to the updated [MA09].

The study of the $\vec{\gamma} + p \to K^+ + \Lambda$ reaction is particularly interesting because the weak decay of the Λ recoil allows for the measurement of the P recoil asymmetry, from the $\vec{\Lambda} \to \pi^- + p$ decay distribution. If a linearly polarized photon beam is used, the O_x and the O_z double polarization asymmetries may be measured. A sample of the results obtained by the GRAAL collaboration on this observables[14] are shown in Fig.5, together with Ghent Isobar (Regge plus resonance) model (dashed curve)[15] and the Bonn-Gatchina model (soulid curves)[16]. A general good agreement is found between data and the Bonn-Gatchina Model, which require the introduction of new or poorly known resonances in the 1900 MeV mass region (P and/or D) to correctly reproduce the data.

5. CONCLUSIONS

The latest results from the LEGS and GRAAL collaboration have been reported in the aim of enlarging the experimental data-set on single and double polarization observables for meson photo-production reactions, in the nucleon resonance energy region. Similar efforts are currently made by other collaborations such as CLAS at the Thomas Jefferson National Accelerator Facility, the CB-ELSA collaboration at the Bonn University, Mami at Mainz and LEPS at Spring8 in Japan. A very attractive reaction candidate is the $\vec{\gamma}\vec{N} \to K + \vec{\Lambda}$ reaction, because of the "self-analyzing" properties of the Λ recoil. These reactions are being measured by the CLAS collaboration on both buthanol and $\vec{H}\vec{D}$ polarized targets with polarized photon beams. The goal of performing the first *complete* experiment is very close to being achieved.

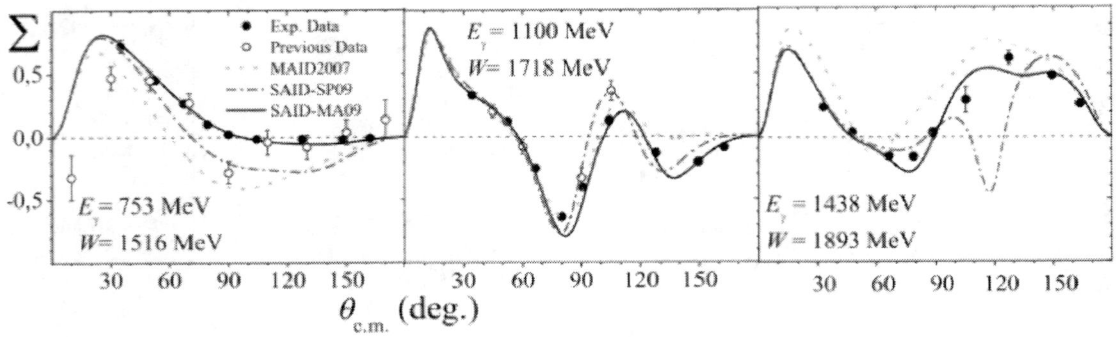

FIGURE 4. Results on the Σ beam asymmetry of the π^- photo-production on a quasi-free neutron in a deuteron target from the GRAAL collaboration. Results are compared with existing data and partial wave analysis predictions.

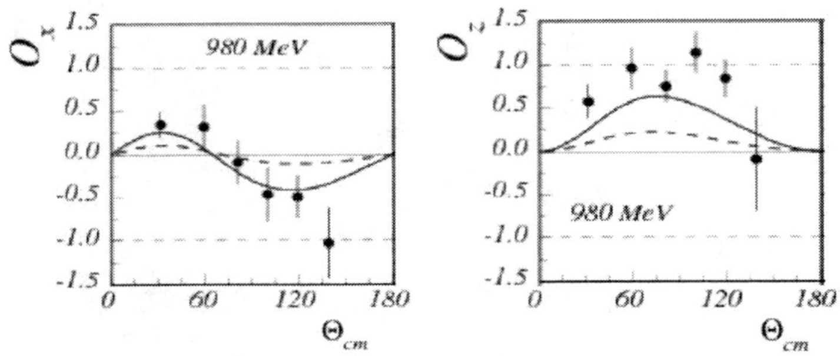

FIGURE 5. Results on the O_x and O_z double polarization observables for the $\vec{\gamma}+p\rightarrow K^+\vec{\Lambda}$ reaction from the GRAAL collaboration. Results are compared with Ghent Isobar (Regge plus resonance) model (dashed curve) and the Bonn-Gatchina model (soulid curves).

REFERENCES

1. W. Kamleh, P.O. Bowman, D.B. Leinweber, A.G. Williams and J. ZhanM. P. Brown, and K. Austin, Phys. Rev, **D 76**, 094501 (2007).
2. M.S. Bhagwat,Ph.D Thesis,Kent State Univ.2005;P. C. Tandy, arXiv:1011.5250v1[nucl-th]2010;S.J. Brodsky, C.D. Roberts, R. Shrock and P.C. Tandy, Phys. Rev **C 82**, 022201(R) (2010).
3. S. Capstick and W. Roberts, Phys. Rev. D **49**, 4570 (1994); **47** 1994 (1993); S. Capstick Phys. Rev D **46**, 2864 (1992); S. Capstick and N. Isgur, Phys. Rev D **34**, 289 (1986).
4. D.O. Riska and G.E. Brown, Nucl. Phys. **A679**, 577 (2001).
5. Koniuk and Isgur, Phys. Rev. Lett. **44** , 845(1980) .
6. A.M. Sandorfi , S. Hoblit , H. Kamano, T.-S.H. Lee, ,J.Phys. **G 38**, 053001 (2011).
7. I.S. Barker, A. Donnachie, J.K. Storrow, Nucl.Phys. **B 95**, 347 (1975).
8. http://gwdac.phys.gwu.edu/
9. S. Hoblit. et al. LSC Collaboration, Phys.Rev.Lett. **102**, 172002 (2009).
10. J. Ahrens, et al., Eur.Phys.J. **A26**, 135-140 (2005).
11. R.Di Salvo et al.,Eur.Phys.J. **A42**, 151-157 (2009); A. Fantini et al., Phys.Rev. **C78**, 015203 (2008); O. Bartalini et al., Phys.Lett. **B544**, 113-120 (2002); J. Ajaka et al.,Phys.Lett. **B475**, 372-377 (2000); J. Ajaka et al.,Phys.Rev.Lett. **81**, 1797-1800 (1998).
12. G. Mandaglio et al., Phys.Rev. **C82**, 045209 (2010).
13. http://wwwkph.kph.uni-mainz.de/MAID//
14. A. Lleres et al., Eur.Phys.J. **A39**, 149-161 (2009).
15. T. Corthals, J. Ryckebusch and T. Van Cauteren, Phys. Rev. **C 73**, 045207 (2006);T. Corthals et al., Phys. Lett. **B 656**, 186 (2007).
16. A.V. Anisovich et al., Eur. Phys. J. **A. 25**, 427 (2005); A.V. Sarantsev et al., Eur. Phys. J. **A. 25**, 441 (2005); A.V. Anisovich et al., Eur. Phys. J **A 34**, 243 (2007);V.A. Nikonov et al., arXiv:hep-ph/0707.3600 (2007).

Recent results from MAMI

Hans-Jürgen Arends

Institut für Kernphysik, Johannes Gutenberg-Universität Mainz, J.-J.-Becher-Weg 45
D 55099 Mainz
arends@kph.uni-mainz.de

Abstract. The Mainz Microtron MAMI is an ideal facility to study the hadron structure with the electromagnetic probe. With the new accelerator stage (HDSM), which went into operation in 2007, high-intensity polarized electron and photon beams with energies up to 1.6 GeV are delivered to the experiments. Polarized targets and recoil polarimeters in combination with dedicated detectors are available for precision experiments in hadron physics. In this article, an overview over selected recent results is given.

Keywords: Electromagnetic form factors, Baryon resonances, Meson production
PACS: 13.40.Gp, 14.20.Gk, 13.60.Le

MAMI-C AND EXPERIMENTAL SETUPS

The Mainz microtron MAMI is a unique facility in Europe to study the hadron structure with the electromagnetic probe at small momentum transfers. It consists of three race track microtrons and delivers high quality polarized electron and photon beams with energies up to 885 MeV into the experimental halls, see Fig. 1. A fourth stage, the Harmonic Double-Sided Microtron (HDSM)[1], upgrading the maximum energy to 1.5 GeV, has gone into routine operation in February 2007 and delivers high-quality beams for the experiments. Recently, a maximum energy of 1.6 GeV was achieved.

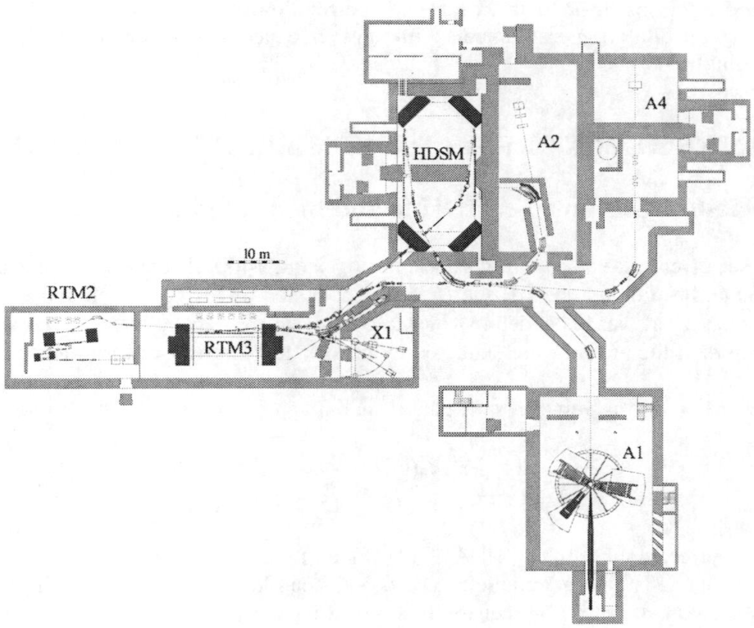

FIGURE 1. Floor plan of the Mainz microtron MAMI-C, including the experimental areas A1, A2, A4, X1 and the new stage HDSM (harmonic double-sided microtron).

12th International Conference on Meson-Nucleon Physics and the Structure of the Nucleon (MENU 2010)
AIP Conf. Proc. 1374, 23-28 (2011); doi: 10.1063/1.3647093

The largest experimental hall of the MAMI accelerator complex houses the spectrometer setup for electron scattering experiments operated by the A1 collaboration. The working horses of the setup are three high-resolution spectrometers with $\delta p/p \leq 10^{-4}$, a large solid angle acceptance of up to 28 msr, and a momentum acceptance of up to 25%. All three high-resolution spectrometers can be used simultaneously for triple coincidence experiments. One of the spectrometers can be tilted up to an out-of-plane angle of 10 degrees. A proton recoil polarimeter gives, in combination with the polarized MAMI beam and a polarized Helium gas target, access to a broad variety of polarization observables. A short-orbit spectrometer is used to detect low-energetic pions. A new kaon-spectrometer (KAOS/A1) covering high momenta with a moderate path length has recently been commissioned. This equipment is well suited to explore fundamental properties of the nucleon, for example, distributions of charge and magnetization by elastic form factor measurements and generalized polarizabilities by virtual Compton scattering. Meson production gives access to the resonance structure of the nucleon.

Reactions induced by circularly and linearly polarized, energy-tagged, real photons with energies up to 1.5 GeV are being studied by the A2 collaboration at MAMI. Since 2004, the Crystal Ball calorimeter, which was moved from BNL/ AGS to Mainz, is the central part of a hermetic detector system for present and future experiments. The Crystal Ball consists of 672 NaI(Tl) crystals covering 93% of the full solid angle with an energy resolution of 1.7% for electromagnetic showers at 1 GeV. For charged particle tracking and identification two layers of coaxial multi-wire proportional chambers and a barrel of 24 scintillation counters surrounding the target were installed inside the Crystal Ball sphere. The forward angular range is covered by the TAPS calorimeter consisting of 384 BaF2 detectors and a Cherenkov detector. The Crystal Ball/TAPS is an ideal instrument for multi-photon final states.

As spin degrees of freedom are essential to understand nucleon excitations, the experimental setup is being complemented by polarized targets and a recoil polarimeter. In addition to the study of single and double meson production processes the high intensity photon beam in combination with a hermetic detector allows us to investigate also hadronic and even rare meson decays. Especially, decays of the isoscalar η and η' mesons offer unique possibilities to study symmetries and symmetry breaking patterns in QCD. Several experiments aiming at a precise determination of neutral decay parameters as well as the mass of the η meson have already been performed. The magnetic moment of the $\Delta(1232)$ resonance is being studied via radiative π^0 production.

The A4 collaboration is using a setup of 1022 PbF_2 crystals that can be arranged to cover a solid angle of 0.6 sr at forward or backward angles. The A4 collaboration is measuring parity violating cross-section asymmetries using longitudinally polarized electrons in order to extract the contribution of the strange quarks to the electromagnetic form factors G_E and G_M. In addition, measurements with transverse electron spin determine the imaginary part of the two-photon exchange amplitude.

SELECTED RECENT RESULTS FROM ELECTRON SCATTERING

High-precision ep-scattering and the proton charge radius

A substantial data set of elastic ep-scattering consisting of about 1400 different kinematical settings covering a momentum transfer range from 0.004 up to 1 $(\mathrm{GeV/c})^2$ was collected by the A1 collaboration. The statistical errors are below 0.2 % and great effort was expanded in order to control the systematic errors to the 1 % level. The electric and magnetic form factors of the proton were extracted by fits of a large variety of form factor models directly to the cross sections, see Fig. 2.

From these analyses the following values for the electric and magnetic rms radii of the proton result [2]:

$$< r_E^2 >^{1/2} = 0.875(5)_{stat}(4)_{syst}(2)_{model}(4)_{group} \text{ fm}, \tag{1}$$

$$< r_M^2 >^{1/2} = 0.777(13)_{stat}(9)_{syst}(5)_{model}(2)_{group} \text{fm}. \tag{2}$$

The electric radius is in agreement with the CODATA [3] value of 0.8768(69) fm which is dominated by Lamb shift measurements on hydrogen. It is also in complete accord with the old Mainz result [4] when Coulomb corrections are applied. However, the very recent high-precision Lamb shift data from muonic hydrogen [5] are about 5 standard deviations smaller. This difference is still unexplained. The magnetic radius is smaller than results from previous fits, however, is in good agreement with results from hyperfine splitting in hydrogen.

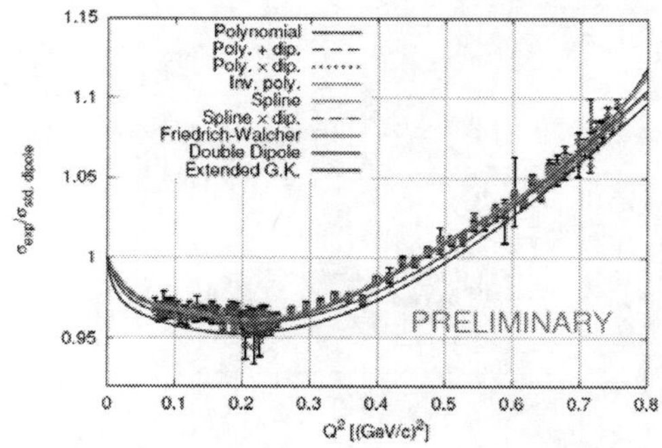

FIGURE 2. Cross section normalized to the standard dipole.

Recoil polarization in η-electroproduction

The first experiment carried out with MAMI-C was η electroproduction in the $S_{11}(1535)$ region [9] in order to help clarifying an open issue near η threshold, where some contribution of the $D_{13}(1520)$ resonance is expected. The polarized target asymmetry measured at ELSA [6] showed a surprising angular structure which cannot be described by the standard phenomenological models like η-MAID [7]. However, one possibility to describe these data was found [8] by including a strong phase shift between the S- and D-waves, basically giving up the standard Breit-Wigner phase assignment. The resulting data of the A1 experiment [9] is consistent with the polarized target data from ELSA and is well described by the model of Tiator et al.[8] including the strong phase shift. More data at other kinematics has been taken. A second preliminary data point at W = 1500 MeV (see Fig. 3) is also consistent with this model.

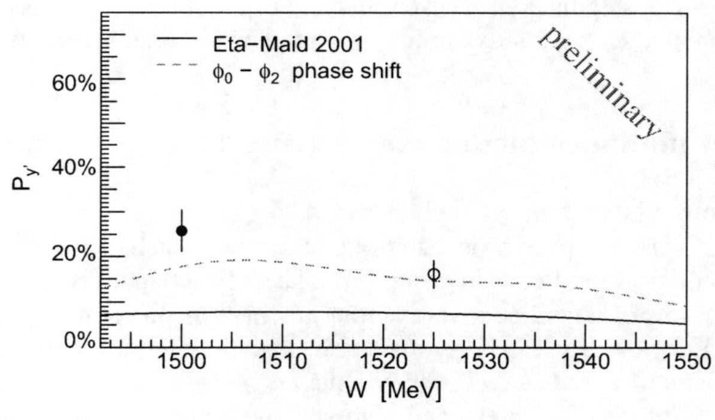

FIGURE 3. Recoil polarization for η electroproduction at $\Theta_\eta = 120^o$. Open point from [9], full point: new preliminary data [10].

SELECTED RECENT RESULTS WITH TAGGED PHOTONS

π^0 photoproduction near threshold

The pion production threshold region is an important field for testing low-energy theorems (LET). In previous experiments, total and differential cross sections as well as the photon asymmetry Σ has been measured [11]. With this data set, the S-wave multipole E_{0+} and three P-wave combinations P_1, P_2, P_3 could be separated and compared to results from chiral perturbation theory (ChPT). In order to extend the asymmetry data to a wider energy range and to improve the statistical accuracy, a new measurement was carried out [12].

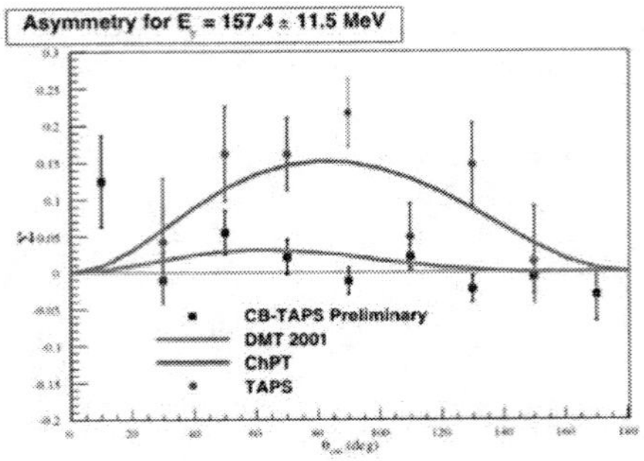

FIGURE 4. Preliminary data of the photon asymmetry Σ for π^0 photoproduction near threshold. Dots: data from [11], squares: new preliminary data [12]. Curves: DMT from [13], ChPT from [14].

The new preliminary photon asymmetry data are shown in Fig. 4 together with the previous data. The new asymmetries turned out significantly smaller than the previous ones. The reason for this discrepancy could be traced down to empty-target contributions, mainly coherent π^0 production from carbon in the target foils. The new preliminary data are in good accordance with the DMT model [13]. The ChPT curve [14] had been fitted to the old data by adjusting the low-energy constant b_P. The effect of the new data on the P-wave combinations is currently under investigation.

High-precision η-photoproduction

The $\gamma p \rightarrow \eta p$ differential cross sections have been measured using the Crystal Ball and TAPS spectrometers [15]. The data span the photon energy range from 707 to 1402 MeV and the full angular range in the c.m. frame. The accumulation of 3×10^6 events via η-decay into 3 π^0 allows a fine binning in energy (4 MeV) and angle (9 deg.). The present data agree well with previous equivalent measurements, but are markedly superior in terms of precision and energy resolution. This will enable the reaction dynamics to be studied in greater detail than previously possible.

$\pi^0\eta$ photoproduction and the $D_{33}(1700)$ resonance

The $\gamma p \rightarrow \pi^0\eta$ reaction seems to be dominated in the threshold region by one single resonance, the $D_{33}(1700)$. The major part of the D_{33} decay into $\pi\eta N$ seems to proceed through the $\eta\Delta$ channel. This observation is in agreement with predictions of the dynamical model of the Valencia group [16]. The πS_{11} channel is another possible mechanism, but may be interpreted entirely in terms of a final-state interaction in which the nucleon appearing after Δ decay interacts with the η meson via excitation of the $S_{11}(1535)$ resonance. In this model, the production of πS_{11} is a higher order process in comparison to $\eta\Delta$, which is produced directly via the $D_{33} \rightarrow \eta\Delta$ decay.

Using the Crystal Ball/ TAPS setup, a very clean and high-statistics data set could be assembled [17]. In Fig. 5 the total cross section is shown together with other available data from GRAAL [18], CB-ELSA [19], and LNS [20].

Up to $E_\gamma = 1.3$ GeV the cross section exhibits a smooth rise, reproducing the energy dependence of a three-body phase space (solid line in Fig. 5). This suggests that in the low-energy region the reaction mainly proceeds via formation of S-waves in the final system.

The dominance of the D_{33} partial-wave amplitude in the energy range $E_\gamma < 1.4$ GeV is confirmed by the angular distributions of the pions in the πp rest frame. In order to describe the final-state kinematics, the canonical (K) and helicity (H) reference frames are used. Within the single resonance ansatz [21] the angular distributions are determined (apart from the resonance quantum numbers J^P) by the ratio of the partial decay widths, r, and the squared ratio of the helicity amplitudes, $a(W)$. A fit with these two quantities used as adjustable parameters gives a remarkably good description of the angular distributions.

In the meantime, first precise beam helicity asymmetry measurements have also been carried out [22], see Fig. 6. The simple model described above also reproduces the gross features of this observable. Nevertheless, systematic deviations at higher photon energies point to the necessity for further improvement of the theory [23].

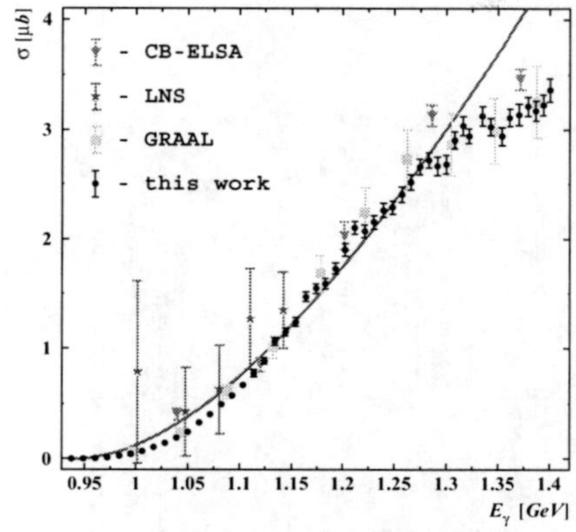

FIGURE 5. Total cross sections for $\gamma p \rightarrow \pi^0 \eta p$ reaction. The LNS [20] and GRAAL [18] data points include both statistical and systematic uncertainties. For CB-ELSA [19] and our points [17] only statistical error bars are given. The curve shows the energy dependence of the reaction phase space with arbitrary normalization.

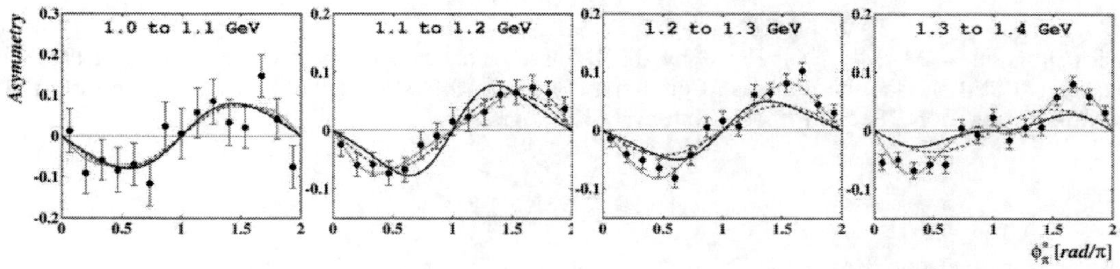

FIGURE 6. Angular distributions [22] of the cross-section asymmetry for the reaction $\vec{\gamma} p \rightarrow \pi^0 \eta p$. Dotted curves: our Fourier fit with three terms, solid curves: full isobar model with 6 resonances [23], dotted curves: similar model with only the D_{33} resonance.

K-photoproduction

While the KAOS spectrometer is dedicated to K-electroproduction and hypernuclear physics, photoproduction of K^+ mesons can be studied from threshold up to 1.4 GeV with the Crystal Ball/TAPS setup. In this energy range the most recent cross section measurements at JLab [24] and ELSA [25] contain significant discrepancies. The technique is based on a stopped K^+ signal in the Crystal Ball and a delayed signal from the weak decays via $K^+ \rightarrow \mu^+ \nu_\mu$ or $K^+ \rightarrow \pi^+ \pi^0$ with branching ratios of 64% and 36% respectively and a mean lifetime of 12 ns. The time difference spectrum between the first and the second cluster shows the expected exponential shape, the energy spectrum allows to distinguish between the two decay modes. Preliminary cross sections for $\gamma(p, K^+)\Lambda$ with high statistics and unprecedented beam energy resolution were obtained [26]. With the ongoing analysis the double polarization observables C_X and C_Z can also be extracted.

Double polarization observables in π^0 production

The new Mainz/Dubna polarized frozen-spin target with longitudinal and transverse polarization is fully operational since the beginning of 2010. It provides proton polarization degrees of 90% and relaxation times of about 1000 hours. A series of data taking periods for Compton scattering, π^0, and η production has been carried out. An example for first preliminary results (from an 80 h test run with limited acceptance) of the double polarization observable F with circularly polarized photons and transversely polarized protons for π^0 production in the energy range of 300 to 600

MeV is shown in Fig. 7.

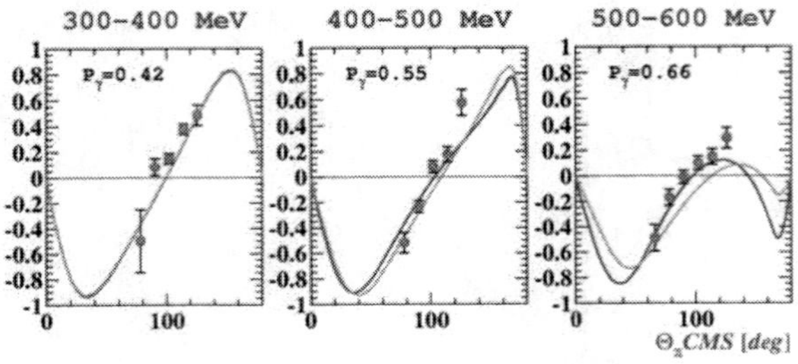

FIGURE 7. First preliminary data of the double polarization observable F in π^0 production in the energy range of 300 to 600 MeV. Blue curve: MAID[27], green curve: SAID[28].

Using the polarimeter in Crystal Ball, the polarization transfer from circularly polarized photons to the recoil proton, C_X, could also be measured.

ACKNOWLEDGMENTS

The physics program at MAMI is supported by the Deutsche Forschungsgemeinschaft (DFG) via the Sonderforschungsbereich SFB443 and the European Community-Research Infrastructure Activity under the FP6 (HadronPhysics, RII3-CT-2004-506078) and FP7 (HadronPhysicsII) programs.

REFERENCES

1. K.-H. Kaiser et al., *Nucl. Instr. Meth. A* **593** (2008) 159.
2. J.C. Bernauer et al., *Phys. Rev. Lett.* submitted for publication, *arXiv:1007.5076v2 [nucl-ex] 30 Jul 2010.*
3. P.J. Mohr et al., *Rev. Mod. Phys.* **80**, 633 (2008).
4. G.G. Simon et al., *Nucl. Phys. A* **333**, 381 (1980).
5. R. Pohl et al., *Nature* **466**, 213 (2010).
6. A. Bock et al., *Phys. Rev. Lett.* **81** (1998) 534.
7. W. T. Chiang et al., *Nucl. Phys. A* **700** (2002) 429.
8. L. Tiator et al., *Phys. Rev. C* **60** (1999) 035210.
9. H. Merkel et al., *Phys. Rev. Lett.* **99** (2007) 132301.
10. K. Griessinger et al., to be published.
11. A. Schmidt et al., *Phys. Rev. Lett.* **87**, 232501 (2001).
12. D. Hornidge et al., publication in preparation.
13. S. Kamalov et al., *Phys. Rev. C*, 03220 (2001).
14. V. Bernard et al., *Eur. J. Phys. A* **11**, 209 (2001).
15. E.F. McNicoll et al., *Phys. Rev. C* **82**, 035208 (2010).
16. M. Döring, E. Oset, and D. Strottman, *Phys. Rev. C* **73**, 045209 (2006).
17. V. Kashevarov et al., *Eur. Phys. J. A* **42**, 141 (2009).
18. J. Ajaka et al., *Phys. Rev. Lett.* **100**, 052003 (2008).
19. I. Horn et al., *Eur. Phys. J. A* **38**, 173 (2008).
20. T. Nakabayashi et al., *Phys. Rev. C* **74**, 035202 (2006).
21. A. Fix, M. Ostrick, and L. Tiator, *Eur. Phys. J. A* **36**, 61 (2008).
22. V. Kashevarov et al., *Phys. Lett. B*, accepted for publication.
23. A. Fix et al., *Phys. Rev. C* **82**, 035207 (2010).
24. R. Bradford et al., *Phys. Rev. C* **73**, 035202 (2006).
25. K.H. Glander et al., *Eur. Phys. J. A* **19**, 251 (2004).
26. T. Jude et al., *Chinese Physics C* **33**, (2009).
27. D. Drechsel et al., *Nucl. Phys. A* **645**, 145 (1999) and *Eur. Phys. J. A* **34**, 69 (2007).
28. R.A. Arndt et al., *Phys. Rev. C* **66**, 055213 (2002).

Transverse-momentum-dependent parton distributions (TMDs)

Alessandro Bacchetta

Dip. di Fisica Nucleare e Teorica, Università di Pavia, and INFN, Sez. di Pavia, via Bassi 6, I-27100 Pavia, Italy

Abstract. Transverse-momentum-dependent parton distributions (TMDs) provide three-dimensional images of the partonic structure of the nucleon in momentum space. We made impressive progress in understanding TMDs, both from the theoretical and experimental point of view. This brief overview on TMDs is divided in two parts: in the first, an essential list of achievements is presented. In the second, a selection of open questions is discussed.

Keywords: parton distribution functions, semi-inclusive DIS, transverse momentum
PACS: 12.38.-t, 13.60.-r, 13.88.+e

Providing detailed maps of the nucleon partonic structure is one of the main goals of subnuclear physics. Such maps are the starting steps toward the comprehension of confined QCD dynamics. They are essential for the interpretation of any high-energy process involving hadrons.

After about forty years of study, we have reached a good knowledge of (some) collinear parton distribution functions (PDFs). They map the partonic structure in a monodimensional momentum space, i.e., as a function of partonic longitudinal momentum. "Longitudinal" is defined as parallel to the hard probe, which must be always present in hard processes where PDFs are observed (e.g., in DIS the photon with virtuality Q^2).

TMDs (an acronym for Transverse Momentum Distributions or Transverse Momentum Dependent parton distribution functions) are natural extensions of standard PDFs: they are three-dimensional maps of the partonic structure in momentum space. They depend not only on the longitudinal momentum, but also on the transverse momentum.

Ultimately, the knowledge of TMDs will allow us to build tomographic images of the inner structure of the nucleon in momentum space, complementary to the impact-parameter space tomography that can be achieved by studying generalized parton distribution functions (GPDs). An example of tomographic images of the nucleon based on a model calculation of TMDs [1] is presented in Fig. 1.

The information contained in TMDs is as fundamental as that contained in standard PDFs. They reveal crucial aspects of the dynamics of confined partons, they can be extracted from experimental data, and allow us to make prediction for hard-scattering experiments involving nucleons.

Our understanding of TMDs and their extraction from data has made giant steps in the last years, thanks to new theoretical ideas and experimental measurements (see Ref. [2] for a recent review). In the near future, more experimental data are expected from HERMES, COMPASS, BELLE and JLab. A high-energy, high-luminosity, polarized electron-proton collider, such as the envisaged EIC project, would undoubtedly be a precision machine for the study of TMDs.

WHAT WE KNOW ABOUT TMDS

Similarly to standard collinear PDFs, it is essential to define TMDs in a formally clear way, through the proof of factorization theorems. TMDs appear when factorizing semi-inclusive processes. For instance, while totally inclusive DIS can be described introducing collinear PDFs, TMDs appear in semi-inclusive DIS if the transverse momentum of one outgoing hadron, $P_{h\perp}$, is measured.

Dealing with semi-inclusive processes pushes the difficulty of proving factorization theorems to a higher level of complications. TMD factorization is in fact a challenging arena where many of the simplifications used in collinear factorization cannot be applied. Nevertheless, factorization for semi-inclusive DIS has been worked out explicitly at leading twist (twist 2) and one-loop order [3]. For instance, the structure function $F_{UU,T}$ in the region $P_{h\perp}^2 \ll Q^2$ can

12th International Conference on Meson-Nucleon Physics and the Structure of the Nucleon (MENU 2010)
AIP Conf. Proc. 1374, 29-34 (2011); doi: 10.1063/1.3647094

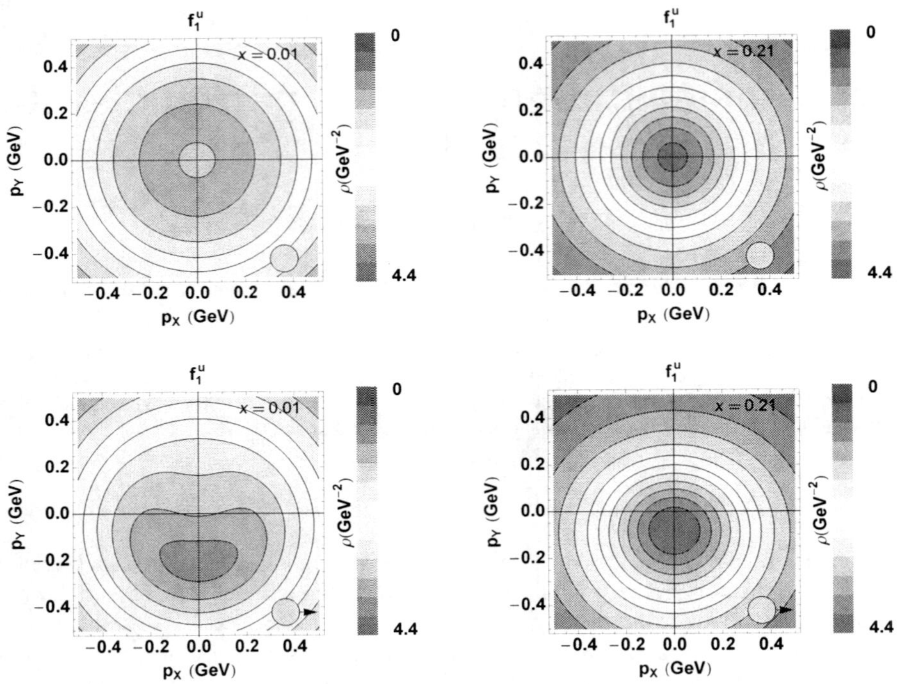

FIGURE 1. Momentum-space tomographic "images" of the up quarks in a nucleon obtained from a model calculation of TMDs [1]. The nucleon is assumed to move into the page. The circle with the arrow indicates the nucleon and its spin orientation. The distortion in the lower panels is due to the Sivers function.

be expressed as

$$F_{UU,T} = \left|H\left(x\zeta^{1/2}, z^{-1}\zeta_h^{1/2}, \mu_F\right)\right|^2 \sum_a x e_a^2 \int d^2p_T\, d^2k_T\, d^2l_T \\ \times \delta^{(2)}\left(p_T - k_T + l_T - P_{h\perp}/z\right) f_1^a(x, p_T^2; \zeta, \mu_F)\, D_1^a(z, k_T^2; \zeta_h, \mu_F)\, U(l_T^2; \mu_F)\,. \quad (1)$$

The formula contains the (calculable) hard scattering factor H, the transverse-momentum-dependent PDFs and fragmentation functions, and the soft factor U, a nonperturbative, process-independent object. However, it should be also possible to include the soft factor into the PDFs and fragmentation functions by redefining them in a suitable way [4].

A schematic definition of quark TMDs is (taking, e.g., the unpolarized distribution of a quark with flavor a)

$$f_1^a(x, p_T^2; \zeta, \mu_F) = \int \frac{d\xi^- d^2\xi_T}{(2\pi)^3}\, e^{ip\cdot\xi}\, \langle P|\bar{\psi}^a(0)\,\mathscr{L}^{v\dagger}_{(\pm\infty,0)}\,\gamma^+\,\mathscr{L}^{v}_{(\pm\infty,\xi)}\,\psi^a(\xi)|P\rangle\bigg|_{\xi^+=0}. \quad (2)$$

The Wilson lines, $\mathscr{L}$, guarantee the gauge invariance of the TMDs. They depend on the gauge vector v and contain also components at infinity running in the transverse direction [5]. Beyond tree level, there are several subtleties related to the treatment of soft and rapidity divergences. Various ways to deal with these divergences have been proposed [3, 6]. In general, they require the introduction of a rapidity cutoff ζ, on which the TMDs depend.

A remarkable property of TMDs is that the detailed shape of the Wilson line is process-dependent. This immediately leads to the conclusion that TMDs are not universal. However, for transverse-momentum-dependent fragmentation functions, the shape of the Wilson line appears to have no influence on physical observables [7]. In SIDIS and Drell–Yan, the difference between the Wilson line consists in a simple direction reversal and leads to calculable effects, namely a simple sign reversal of all T-odd TMDs [8].

In general and in simplified terms, the study of TMD factorization requires a deeper understanding of what happens when a quark is hit inside a nucleon, with a particular attention to the infinitely many gluons that surround the quark, beyond the simple case when its transverse momentum is integrated over.

nucleon pol. \ quark pol.	U	L	T	
U	$\boldsymbol{f_1}$		$h_1^{\perp}$	
L		$\boldsymbol{g_1}$	$h_{1L}^{\perp}$	
T	$f_{1T}^{\perp}$	g_{1T}	$\boldsymbol{h_1}$	$h_{1T}^{\perp}$

(a)

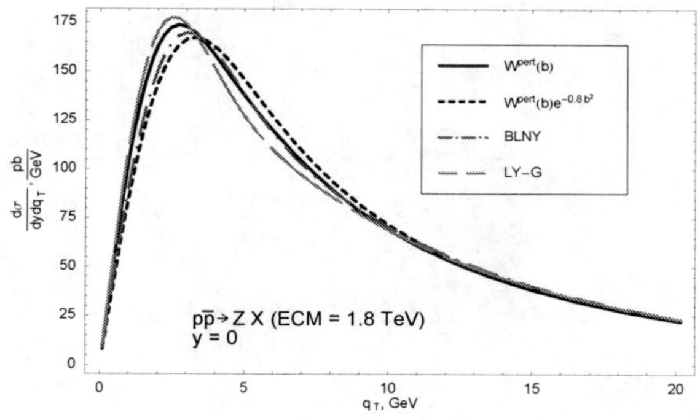

(b)

FIGURE 2. (a) Twist-2 transverse-momentum-dependent distribution functions. The U,L,T correspond to unpolarized, longitudinally polarized and transversely polarized nucleons (rows) and quarks (columns). Functions in boldface survive transverse momentum integration. Functions in gray cells are T-odd. (b) The cross sections of Z boson production at the Tevatron, computed in the CSS formalism. The difference between the curves shows the impact of choosing different nonperturbative components for the TMDs. See Ref. [12] for details.

At present, especially for azimuthally-dependent structure functions, phenomenological analyses are often carried out using the tree-level approximated expression (which corresponds to using parton-model formulas for collinear PDFs and neglecting scaling violations).

$$F_{UU,T} = \sum_a x e_a^2 \int d^2 p_T \, d^2 k_T \, \delta^{(2)}\big(p_T - k_T - P_{h\perp}/z\big) \, f_1^a(x, p_T^2) \, D_1^a(z, k_T^2) \,. \tag{3}$$

The transverse-momentum dependence of the partonic functions is usually assumed to be a flavor-independent Gaussian [9]. The tree-level approximation and the Gaussian assumption are known to be inadequate at $P_{h\perp}^2 \gg M^2$, but they could effectively describe the physics at $P_{h\perp}^2 \approx M^2$. Especially for low-energy experiments, this is where the bulk of the data is.

There is an extensive literature where the analysis is carried out to a higher level of complication, but only for the specific case of unpolarized observables integrated over the azimuthal angle of the measured transverse momentum. The analysis is usually performed in the space of the Fourier-conjugate to $P_{h\perp}$ (b-space) in the Collins–Soper–Sterman (CSS) framework [10]. The region of $P_{h\perp}^2 \gg M^2$, or $b^2 \ll 1/M^2$, can be calculated perturbatively, but when $P_{h\perp}^2 \approx M^2$ a nonperturbative component has to be introduced and its parameters must be fitted to experimental data. This component is usually assumed to be a flavor-independent Gaussian [11].

At present, we can make the conservative statement that unpolarized quark TMDs seems to be well described by flavor-independent Gaussians with $\sqrt{\langle p_T^2 \rangle} \approx 0.4 - 0.8$ GeV, depending on the kinematics.

The knowledge of the details of the unpolarized TMDs has an impact also on high-energy physics. In Fig. 2b, the cross section for Z boson production at the Tevatron is plotted [12]. The difference between the curves originates from different models and fits of the nonperturbative component of the TMDs. Apart from the details, the plot shows that the knowledge of TMDs is essential for precision studies at the Tevatron.

Even the determination of a fundamental parameter of the Standard Model, the mass of the W boson, is affected by the uncertainties of the knowledge of unpolarized TMDs. In Ref. [13], the CDF collaboration discussed several ways to fit the W mass. According to the analysis, TMDs uncertainties generate an error of 3.9 MeV on the W mass determination (the total systematic error is about 34 MeV).

When the spin of the nucleon and that of the quark are taken into account, eight twist-2 TMDs can be introduced. They are listed in Fig. 2a. As with collinear PDFs, extracting TMDs calls for global fits to semi-inclusive DIS, Drell–Yan, and e^+e^--annihilation data.

From the theoretical point of view, we know positivity bounds for TMDs [14], behavior at high transverse momentum [15], behavior at high x [16].

Apart from the f_1 (unpolarized function) and the p_T-integral of g_1 (helicity distribution), we have at the moment some extractions of h_1 (transversity distribution), $f_{1T}^{\perp}$ (Sivers function) [17, 18] and $h_1^{\perp}$ (Boer–Mulders function) [19].

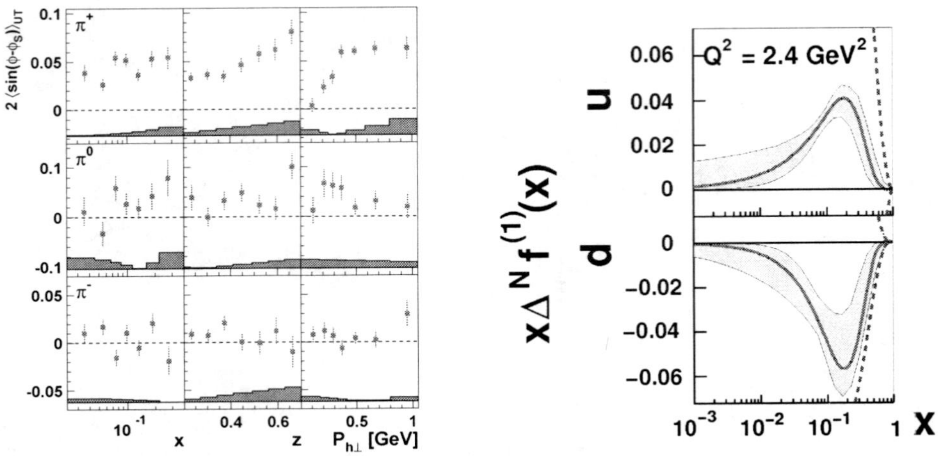

(a) (b)

FIGURE 3. (a) Example of experimental data on the Sivers asymmetries [20]. (b) Sivers distribution for up and down quark from a recent extraction [17].

As illustrated in Fig. 1, a nonzero Sivers function means that the distribution of quarks in transverse momentum is affected by the direction of the nucleon's spin. Thanks to the experimental measurements of the HERMES [20] and COMPASS [21] collaborations, it has become possible to extract for the first time the Sivers distribution function. Fig. 3 shows a representative subset of the data and one of the most recent parametrizations [17, 18]).

The conclusions we can draw from these first studies are already interesting: the Sivers function is nonzero, at least for up and down quarks; the Sivers function for up quark is negative, while for down quark is positive and could be even larger than that of up quarks. These findings are in agreement with the idea suggested in Ref. [22] of a link between the anomalous magnetic moment, κ, and the Sivers function. Assuming isospin symmetry and no strange quark contribution, we know that $\kappa_u \approx 1.673$ and $\kappa_d \approx -2.033$. Since quarks are subject to attractive final-state color interactions, this indeed translates in a negative up and a large, positive down Sivers, in qualitative agreement with some models [23].

WHAT WE STILL DON'T KNOW ABOUT TMDS

There are many essential questions that have not been addressed sufficiently in the past.

Concerning the formalism, there are a few alternative definitions of TMDs, based on different ways to handle soft and rapidity divergences. It is at present not clear if the different definitions are all formally acceptable, if they can be seen as different "schemes" to define TMDs, and if some of them are more convenient than others.

It is extremely important to clarify if and how TMDs can be studied in hadron-hadron collisions to hadrons. The treatment of Wilson lines in this case is more intricate and recent studies lead to the conclusion that TMD factorization does not work [24]. Efforts should be made to look for experimental evidence of factorization-breaking effects, and to find ways to avoid problems [].

For what concerns our phenomenological knowledge of TMDs, we do not know whether the transverse momentum distribution is different for different quark flavors and for gluons. We know that collinear PDFs are flavor dependent: not only their normalization is different, but also their shape. Their transverse spatial distributions, as reconstructed from form factor measurements, are also sharply different (see, e.g., [25]). Some models predict different shapes [1, 26]. There is no fundamental reason to believe that their transverse momentum distributions should be equal. However, up to now they have been assumed to be exactly the same in all phenomenological analyses. There are in fact already some indications that this might not be a good assumption, both from experiments [27] and lattice-QCD calculations [28] (see Fig. 4a).

We don't know enough about the detailed shape of TMDs. In all studies, it is assumed to be Gaussian. However, this is a choice based on convenience and tradition, but not on a fundamental reason. There is no model calculation of TMDs that predicts a pure Gaussian shape. The presence of nonzero orbital angular momentum actually implies that

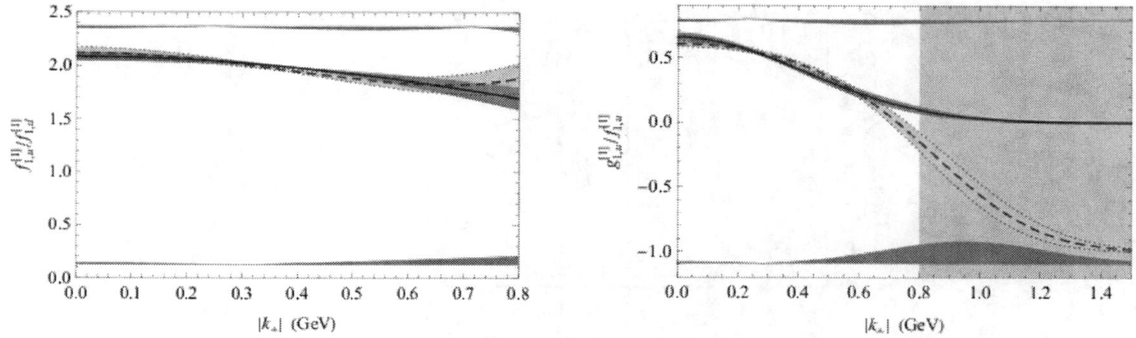

(a) (b)

FIGURE 4. First hints on TMD behavior based on lattice QCD computations [28]. (a) Ratio between the TMD of up and down quarks from two different fits to lattice QCD data: the non-constant ratio suggests that up and down quark have different transverse momentum distributions. (b) Ratio between the helicity distribution and the unpolarized distribution for up quarks: the non-constant ratio suggests that quarks with different spin orientation have different transverse momentum distributions.

the shape cannot be a simple Gaussian. In nonrelativistic quantum mechanics, it is well known that wave-functions with orbital angular momentum vanish at zero momentum. This feature is reflected also in TMDs: contributions from partons with nonzero angular momentum have to vanish at zero transverse momentum and therefore cannot be described by a simple Gaussian. We can assume that the wave-function of the proton contains quarks with $L_z = 0$ (s wave) and $L_z = 1$ (p wave), where L_z is defined in the Jaffe–Manohar way. Then we know that

$$\lim_{p_T \to 0} |\psi_{p-\text{wave}}|^2 \propto p_T^2/M^4 \tag{4}$$

The full TMD will be the sum of the squares of different components of the wave-function, making it difficult to identify the contributions with nonzero L. In any case, a downturn of a TMD at small transverse momentum may signal the presence of nonzero orbital angular momentum. While this effects could barely be visible in unpolarized TMDs, certain combinations of polarized TMDs could filter out more clearly the configurations with nonzero orbital angular momentum.

Apart from the details of their shape, all the TMDs that are not boldface in Fig. 2a vanish in the absence of orbital angular momentum due to angular momentum conservation. Measuring any one of them to be nonzero is already a indisputable indication of the presence of partonic orbital angular momentum. We know already from other sources (e.g., the measurement of nucleons' anomalous magnetic moments) that partonic orbital angular momentum is not zero, however TMDs have the advantage that they can be flavor-separated and that they are x dependent. Thus, they allow us to say if orbital angular momentum is present for each quark flavor and for gluons, and at each value of x.

If stating that a fraction of partons have nonzero orbital angular momentum is relatively simple, it is not easy to make a quantitative estimate of the net partonic orbital angular momentum using TMDs. Any statement in this direction is bound to be model-dependent. Generally speaking, TMDs have to be computed in a model and the parameters of the model have to be fixed to reproduce the TMDs extracted from data. Then, the total orbital angular momentum can be computed in the model. Unfortunately, it is possible that two models describe the data equally well, but give two different values for the total orbital angular momentum.

Another aspect that requires further investigations is to which extent transverse momentum distribution is influenced by the spin of the quarks and of the nucleon. For instance, it may be possible that the average transverse momentum of quarks with spin antiparallel to the nucleon is larger than that of quarks with spin parallel to the nucleon. Also in this case there is already some evidence from experiments [29] (see Fig. 5) and lattice-QCD calculations [28] (see Fig. 4b).

Finally, it is worth mentioning that during this conference the first measurement connected to the so-called "worm-gear" TMD g_{1T} has been reported [30] and seems to confirm the sign of the distribution predicted by lattice QCD studies [31].

In summary, TMDs open new dimensions in the exploration of the partonic structure of the nucleon. They require challenging extensions of the standard formalism used for collinear parton distribution functions, leading us to a deeper understanding of QCD. Among other things, they give evidence of the presence of partonic orbital angular momentum and, with model assumptions, they can help constraining its size.

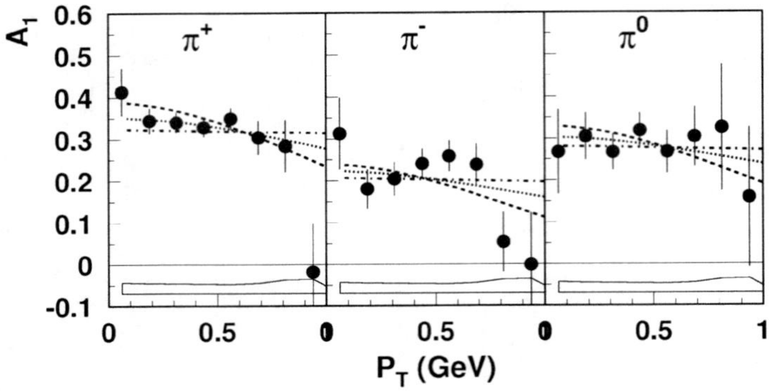

FIGURE 5. Data from the CLAS collaboration showing the transverse-momentum dependence of the double longitudinal spin asymmetry A_1 [29]. The dependence would be flat if TMDs for quarks with opposite helicity were the same.

REFERENCES

1. A. Bacchetta, F. Conti, and M. Radici, *Phys. Rev.* **D78**, 074010 (2008); A. Bacchetta, M. Radici, F. Conti, and M. Guagnelli, *Eur. Phys. J.* **A45**, 373–388 (2010).
2. V. Barone, F. Bradamante, and A. Martin, *Prog. Part. Nucl. Phys.* **65**, 267–333 (2010).
3. J. C. Collins, and D. E. Soper, *Nucl. Phys.* **B193**, 381 (1981); X. Ji, J.-P. Ma, and F. Yuan, *Phys. Rev.* **D71**, 034005 (2005).
4. J. C. Collins, and F. Hautmann, *JHEP* **03**, 016 (2001).
5. X. Ji, and F. Yuan, *Phys. Lett.* **B543**, 66–72 (2002).
6. J. C. Collins, *Acta Phys. Polon.* **B34**, 3103 (2003); I. O. Cherednikov, and N. G. Stefanis, *Phys. Rev.* **D80**, 054008 (2009).
7. J. C. Collins, and A. Metz, *Phys. Rev. Lett.* **93**, 252001 (2004); F. Yuan, *Phys. Rev.* **D77**, 074019 (2008); L. P. Gamberg, A. Mukherjee, and P. J. Mulders, *Phys. Rev.* **D77**, 114026 (2008); S. Meissner, and A. Metz, *Phys. Rev. Lett.* **102**, 172003 (2009).
8. J. C. Collins, *Phys. Lett.* **B536**, 43–48 (2002).
9. U. D'Alesio, and F. Murgia, *Phys. Rev.* **D70**, 074009 (2004); P. Schweitzer, T. Teckentrup, and A. Metz, *Phys. Rev.* **D81**, 094019 (2010).
10. J. C. Collins, D. E. Soper, and G. Sterman, *Nucl. Phys.* **B250**, 199 (1985).
11. F. Landry, R. Brock, P. M. Nadolsky, and C. P. Yuan, *Phys. Rev.* **D67**, 073016 (2003).
12. P. M. Nadolsky, *AIP Conf. Proc.* **753**, 158–170 (2005).
13. T. Aaltonen, et al., *Phys. Rev.* **D77**, 112001 (2008).
14. A. Bacchetta, M. Boglione, A. Henneman, and P. J. Mulders, *Phys. Rev. Lett.* **85**, 712–715 (2000).
15. X. Ji, J.-W. Qiu, W. Vogelsang, and F. Yuan, *Phys. Rev. Lett.* **97**, 082002 (2006); A. Bacchetta, D. Boer, M. Diehl, and P. J. Mulders, *JHEP* **08**, 023 (2008); J. Zhou, F. Yuan, and Z.-T. Liang, arXiv:0909.2238[hep-ph].
16. S. J. Brodsky, and F. Yuan, *Phys. Rev.* **D74**, 094018 (2006).
17. M. Anselmino, et al., *Eur. Phys. J.* **A39**, 89–100 (2009).
18. S. Arnold, A. V. Efremov, K. Goeke, M. Schlegel, and P. Schweitzer, arXiv:0805.2137[hep-ph].
19. Z. Lu, and I. Schmidt, *Phys. Rev.* **D81**, 034023 (2010); V. Barone, S. Melis, and A. Prokudin, *Phys. Rev.* **D81**, 114026 (2010).
20. A. Airapetian, et al., *Phys. Rev. Lett.* **103**, 152002 (2009).
21. M. Alekseev, et al., *Phys. Lett.* **B673**, 127–135 (2009).
22. M. Burkardt, *Phys. Rev.* **D66**, 114005 (2002).
23. A. Courtoy, F. Fratini, S. Scopetta, and V. Vento, *Phys. Rev.* **D78**, 034002 (2008); A. Courtoy, S. Scopetta, and V. Vento, *Phys. Rev.* **D79**, 074001 (2009); B. Pasquini, and F. Yuan, *Phys. Rev.* **D81**, 114013 (2010).
24. T. C. Rogers, and P. J. Mulders, *Phys. Rev.* **D81**, 094006 (2010).
25. G. A. Miller, *Phys. Rev. Lett.* **99**, 112001 (2007).
26. M. Wakamatsu, *Phys. Rev.* **D79**, 094028 (2009).
27. H. Mkrtchyan, et al., *Phys. Lett.* **B665**, 20–25 (2008).
28. B. U. Musch, P. Hagler, J. W. Negele, and A. Schafer, arXiv:1011.1213[hep-lat].
29. H. Avakian, et al., arXiv:1003.4549[hep-ex].
30. J. Huang, and X. Jiang, these proceedings; JLab experiment E06-010/E06-011, J.-P. Chen, E. Cisbani, H. Gao, X. Jiang, J.-C. Peng, spokespersons (2011).
31. P. Hagler, B. U. Musch, J. W. Negele, and A. Schafer, *Europhys. Lett.* **88**, 61001 (2009).

Results from the Beijing Spectrometer III

Niklaus Berger for the BES III Collaboration

IHEP, Beijing

Abstract. The new BES III detector at BEPCII collected 106 million ψ' and about 226 million J/ψ events in 2009. Results based on these samples are reported, including branching ratios for $\chi_{cJ} \to \pi^0\pi^0$, $\eta\eta$, $4\pi^0$, γV and VV, where V is a ρ, ω or ϕ as well as first measurements of the branching ratios for $\psi' \to \pi^0 h_c$ and $h_c \to \gamma\eta_c$ and measurements of the mass and width of the h_c. Results in hadron spectroscopy include the confirmation of the $p\bar{p}$ threshold enhancement in $J/\psi \to \gamma p\bar{p}$ and confirmation of the $X(1835)$ in $J/\psi \to \gamma\eta'\pi^+\pi^-$.

Keywords: Charmonium, Hadron spectroscopy, BES III
PACS: 13.25.Gv, 13.25.Jx, 13.66.Bc, 13.66.Jn, 14.40.Pq, 14.40.Rt

INTRODUCTION

The Beijing Electron-Positron Collider has been upgraded to a two-ring machine (BEPC II) with a design luminosity of $1 \times 10^{33}\mathrm{cm}^{-2}\,\mathrm{s}^{-1}$ at a centre-of-mass energy of 3.78 GeV. It will operate between 2 and 4.6 GeV, enabling precision studies of charmonium (J/ψ, ψ', $\psi(3770)$, η_c, χ_{cJ}, and h_c), open charm (D and D_s meson pairs), the τ lepton mass and the hadronic cross section (R) in this energy region. BEPC II and the BES III detector are described in detail in [1], whilst the physics programme is outlined in [2].

During 2009, BES III recorded a sample of 106 million ψ' and 226 million J/ψ events, about a factor of four improvement in statistics over the previous samples at CLEOc and BES II respectively. This paper reports results based on these samples.

CHARMONIUM PHYSICS

χ_{cJ} decays

The large ψ' sample combined with the large $\psi' \to \gamma\chi_{cJ}$ branching fractions make BES III an excellent laboratory to study inclusive and exclusive χ_{cJ} decays; as these decays proceed via two gluons, they are an excellent place to search for gluonium [3]. Also, colour octet contributions are important in these decays [4] and precise measurements allow tests of theoretical models.

BES III has studied $\psi' \to \gamma\chi_{cJ}, \chi_{cJ} \to \pi^0\pi^0$ and $\eta\eta$, where π^0 and η decay to two photons [5]. The results are given in table 1, where they are compared to CLEOc results [6] and the Particle Data Group (PDG) averages [7]. The recoil photon energy spectra are shown in figure 1.

BES III has preliminary results of a study of $\psi' \to \gamma\chi_{cJ}, \chi_{cJ} \to 4\pi^0$, where both branching fractions excluding intermediate K_s (see table 2) and the branching fractions for $\chi_{cJ} \to K_sK_s$ (table 3) are derived.

BES III is also studying $\psi' \to \gamma\chi_{cJ}, \chi_{cJ} \to \gamma V$, where V is $\phi \to K^+K^-$, $\rho \to \pi^+\pi^-$, or $\omega \to \pi^+\pi^-\pi^0$. Invariant mass distributions of the high energy photon and V are shown in figure 2, where clear χ_{c1} signals are visible. This is the first observation of $\chi_{c1} \to \gamma\phi$. Preliminary branching fractions are given in table 4 and compared to CLEOc results [8] and perturbative QCD predictions [9].

The decays $\chi_{cJ} \to \phi\phi$, $\omega\omega$ and $\omega\phi$ are also under study at BES III. This has led to the first observation of the doubly OZI-suppressed decay $\chi_{cJ} \to \omega\phi$ and the first observation of $\chi_{c1} \to \phi\phi$ and $\chi_{cJ} \to omega\omega$. The recoil photon mass spectra are shown in figure 3, preliminary branching fractions for the decay to two ϕs can be found in table 5.

12th International Conference on Meson-Nucleon Physics and the Structure of the Nucleon (MENU 2010)
AIP Conf. Proc. 1374, 35-40 (2011); doi: 10.1063/1.3647095

TABLE 1. Branching fraction results for $\chi_{cJ} \to \pi^0\pi^0$ and $\eta\eta$. The last errors for BES III and CLEOc are the branching fraction uncertainties for $\psi' \to \gamma\chi_{cJ}$. Note that CLEOc used their own $\psi' \to \gamma\chi_{cJ}$ branching fractions for their results.

Decay		χ_{c0} (10^{-3})	χ_{c2} (10^{-3})
$\pi^0\pi^0$	BES III [5]	$3.23 \pm 0.03 \pm 0.23 \pm 0.14$	$0.88 \pm 0.02 \pm 0.06 \pm 0.04$
	CLEOc [6]	$2.94 \pm 0.07 \pm 0.32 \pm 0.15$	$0.68 \pm 0.03 \pm 0.07 \pm 0.04$
	PDG08 [7]	2.43 ± 0.20	0.71 ± 0.08
	Theory [4]	2.3	0.95
$\eta\eta$	BES III [5]	$3.44 \pm 0.10 \pm 0.24 \pm 0.20$	$0.65 \pm 0.04 \pm 0.05 \pm 0.03$
	CLEOc [6]	$3.18 \pm 0.13 \pm 0.31 \pm 0.16$	$0.51 \pm 0.05 \pm 0.05 \pm 0.03$
	PDG08 [7]	2.4 ± 0.4	< 0.5
	Theory [4]	3.2	1.3

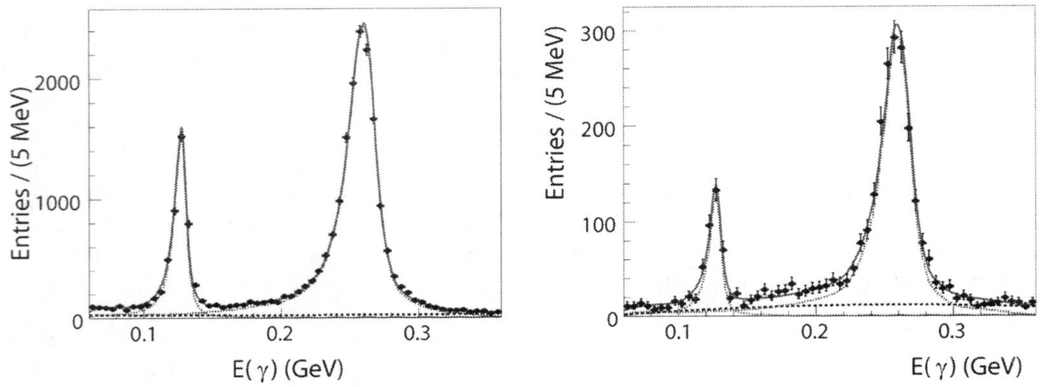

FIGURE 1. Recoil photon energy spectra for $\psi' \to \gamma\chi_{cJ}, \chi_{cJ} \to \pi^0\pi^0$ (left) and $\eta\eta$ (right).

h_c studies

In 2005, CLEO [10] reported an observation of the $h_c(^1P_1)$ in $e^+e^- \to \psi' \to \pi^0 h_c$, $h_c \to \gamma\eta_c$, using both inclusive and exclusive η_c decays. The analysis was repeated in 2008 with the full sample of 25 million ψ' events [11], resulting in a combined mass measurement of $m(h_c)_{AVG} = 3525.2 \pm 0.18 \pm 0.12$ MeV/c^2. A precise determination of the mass is important to learn about the hyperfine (spin-spin) interaction of P wave states. Using the spin weighted centroid of the ^3P_J states, $< m(^3P_J) >$, to represent $m(^3P_J)$, they obtain $\Delta m_{hf}(1P) = \; < m(^3P_J) > - m(^1P_1) = +0.08 \pm 0.18 \pm 0.12$ MeV/c^2. This is consistent with the lowest order expectation of zero.

BES III has measured $m(h_c)$ from the distribution of the invariant mass of the system recoiling against the π^0, both in the inclusive $\psi' \to \pi^0 h_c$ sample and a sample tagged by the $E1$ photon from the $h_c \to \gamma\eta_c$ transition. This allows for the first measurements of the branching fractions $B(\psi' \to \pi^0 h_c)$ and $B(h_c \to \gamma\eta_c)$. BES III also measured for the first time the width of the h_c, $\Gamma(h_c)$. Results and a comparison with the CLEO measurements and theoretical predictions are given in table 6; a detailed description of the measurement is given in [12]. BES III also measured for the first time the width of the h_c, $\Gamma(h_c)$. Results and a comparison with the CLEO measurements and theoretical predictions

TABLE 2. Preliminary branching fraction results for $\chi_{cJ} \to 4\pi^0$ excluding $K_s \to \pi^0\pi^0$.

Decay	Branching fraction excluding $K_s \to \pi^0\pi^0$ (10^{-3})
$\chi_{c0} \to 4\pi^0$	$3.42 \pm 0.07 \pm 0.45$
$\chi_{c1} \to 4\pi^0$	$0.60 \pm 0.03 \pm 0.09$
$\chi_{c2} \to 4\pi^0$	$1.13 \pm 0.04 \pm 0.15$

TABLE 3. Branching fraction results for $\chi_{cJ} \to K_s K_s$. The error for BES III is statistical only. The last error for CLEOc is the branching fraction uncertainty for $\psi' \to \gamma\chi_{cJ}$. Note that the CLEOc result is based on the charged decay of the K_s.

	χ_{c0} (10^{-3})	χ_{c2} (10^{-3})
BES III preliminary	4.1 ± 0.4	0.6 ± 0.2
CLEOc [6]	$3.49 \pm 0.08 \pm 0.17 \pm 0.17$	$0.53 \pm 0.03 \pm 0.03 \pm 0.03$
PDG08 [7]	2.82 ± 0.28	0.68 ± 0.11

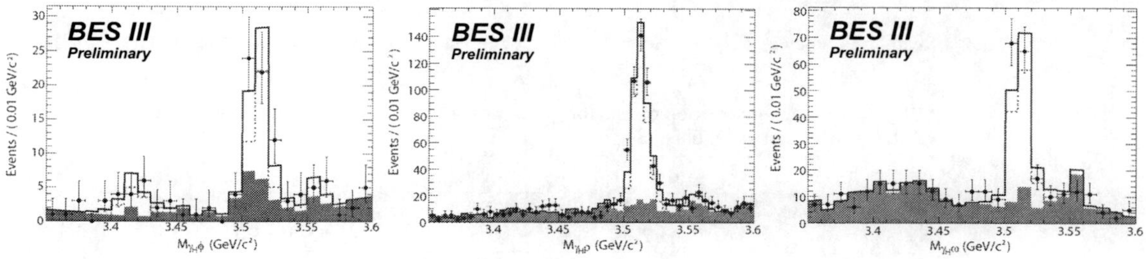

FIGURE 2. Invariant mass distributions of the high energy gamma and V from $\psi' \to \gamma\chi_{cJ}, \chi_{cJ} \to \gamma V$, where V is ϕ, ρ, or ω. Dots with error bars are data, the shaded histogram is background estimated from Monte Carlo simulation, and the open black histogram is the fit. Clear signals for $\chi_{c1} \to \gamma V$ are observed.

are given in table 6; a detailed description of the measurement is given in [12].

LIGHT HADRON SPECTROSCOPY

X(1860)

BES II observed a $p\bar{p}$ mass threshold enhancement in $J/\psi \to \gamma p\bar{p}$ [16], a fit to an S-wave resonance gave a mass below the $p\bar{p}$ threshold of $M(p\bar{p}) = 1859^{+3+5}_{-10-25}$ MeV/c^2. BES III has confirmed this observation [17] in $\psi' \to \pi^+\pi^- J/\psi, J/\psi \to \gamma p\bar{p}$ with a fitted mass of $M(p\bar{p}) = 1861^{+6+7}_{-13-26}$ MeV/c^2 and directly in $J/\psi \to \gamma p\bar{p}$ with a fitted mass of $M(p\bar{p}) = 1859 \pm 0.8$ (preliminary and statistical error only), shown in figure 5. CLEOc also observes a $p\bar{p}$ threshold enhancement in $\psi' \to \pi^+\pi^- J/\psi, J/\psi \to \gamma p\bar{p}$ and obtains a consistent fitted mass [18]. No significant enhancement is found by either BES III or CLEOc in $\psi' \to \gamma p\bar{p}$, thus disfavouring the interpretation of the the X(1860) arising purely due to final state interaction.

TABLE 4. Branching fraction results for $\chi_{cJ} \to \gamma V$, where V is a ϕ, ρ, or ω. For BES III, results are preliminary, and errors are statistical only.

Decay	BES III (10^{-6})	CLEOc [8] (10^{-6})	pQCD [9] (10^{-6})
$\chi_{c0} \to \gamma\phi$	< 14.8	< 6.4	0.46
$\chi_{c1} \to \gamma\phi$	27.3 ± 5.5	< 26	3.6
$\chi_{c2} \to \gamma\phi$	< 7.8	< 13	1.1
$\chi_{c0} \to \gamma\rho^0$	< 9.5	< 9.6	1.2
$\chi_{c1} \to \gamma\rho^0$	241 ± 14	$243 \pm 19 \pm 22$	14
$\chi_{c2} \to \gamma\rho^0$	< 19.7	< 50	4.4
$\chi_{c0} \to \gamma\omega$	< 11.7	< 8.8	0.13
$\chi_{c1} \to \gamma\omega$	73.5 ± 7.6	$83 \pm 15 \pm 12$	1.6
$\chi_{c2} \to \gamma\omega$	< 5.8	< 7.0	0.5

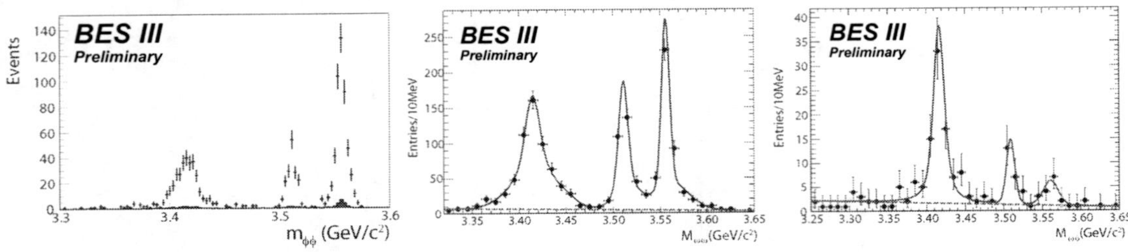

FIGURE 3. Recoil photon energy spectra for $\psi' \to \gamma\chi_{cJ}, \chi_{cJ} \to \phi\phi$ (left) and $\omega\omega$ (centre) and $\omega\phi$ (right).

TABLE 5. Branching fraction results for $\chi_{cJ} \to \phi\phi$. The error for the preliminary BES III results is statistical only.

	$\chi_{c0} \to \phi\phi$ (10^{-4})	$\chi_{c1} \to \phi\phi$ (10^{-4})	$\chi_{c2} \to \phi\phi$ (10^{-4})
BES III preliminary	8.0 ± 0.4	4.2 ± 0.3	11.3 ± 0.4
PDG08 [7]	9.3 ± 2.0	—	15.4 ± 3.0

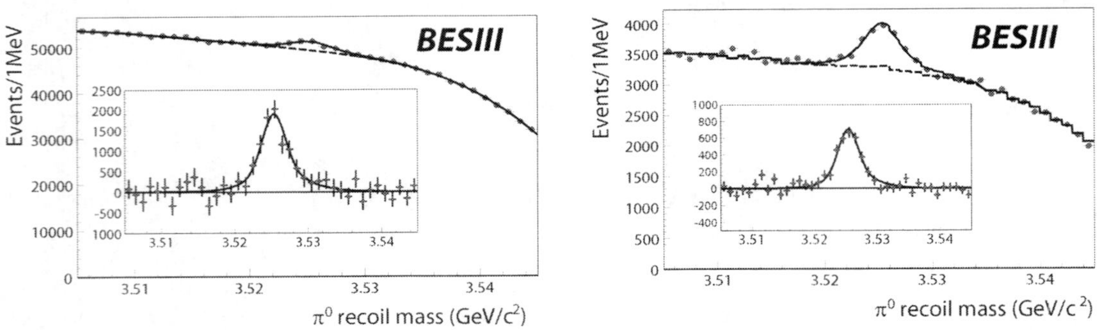

FIGURE 4. The π^0 recoil-mass spectrum and fit for the inclusive analysis of $\psi' \to \pi^0 h_c$ (left) and recoil-mass spectrum and fit for the E1-tagged analysis of $\psi' \to \pi^0 h_c$, $h_c \to \gamma\eta_c$ (right). Fits are shown as solid lines, background as dashed lines. The insets show the background-subtracted spectra.

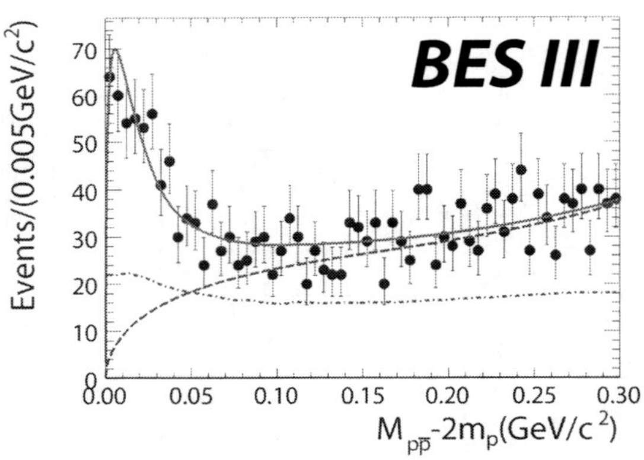

FIGURE 5. The $p\bar{p}$ invariant mass spectrum in $\psi' \to \pi^+\pi^- J/\psi, J/\psi \to \gamma p\bar{p}$

TABLE 6. BES III h_c results compared with CLEOc and theory. Note that the lattice prediction was for the partial width; the large uncertainty stems mostly from the conversion to a branching fraction using the BES III measurement of the (total) width with its large error.

	BES III	CLEOc [11]	Theory
$B(\psi' \to \pi^0 h_c)$ $\times B(h_c \to \gamma\eta_c)$ $[10^{-4}]$	$4.58 \pm 0.40 \pm 0.50$	$4.16 \pm 0.30 \pm 0.37$	
$B(\psi' \to \pi^0 h_c)$ $[10^{-4}]$	$8.4 \pm 1.3 \pm 1.0$		$(4-13)$ [13]
$B(h_c \to \gamma\eta_c)$ [%]	$54.3 \pm 6.7 \pm 5.2$		41 (NRQCD) [13] 88 (PQCD) [13] 38 [14] 51-100 (Lattice) [15]
$m(h_c)$ [MeV/c^2]	$3525.40 \pm 0.13 \pm 0.18$	$3525.20 \pm 0.18 \pm 0.12$	
$\Gamma(h_c)$ [MeV/c^2]	$0.73 \pm 0.45 \pm 0.28$ < 1.44 @ 90% C.L.		1.1 (NRQCD) [13] 0.51 (PQCD) [13]
$\Delta M_{hf}(1P)$ [MeV/c^2]	$0.10 \pm 0.13 \pm 0.18$	$0.08 \pm 0.18 \pm 0.12$	

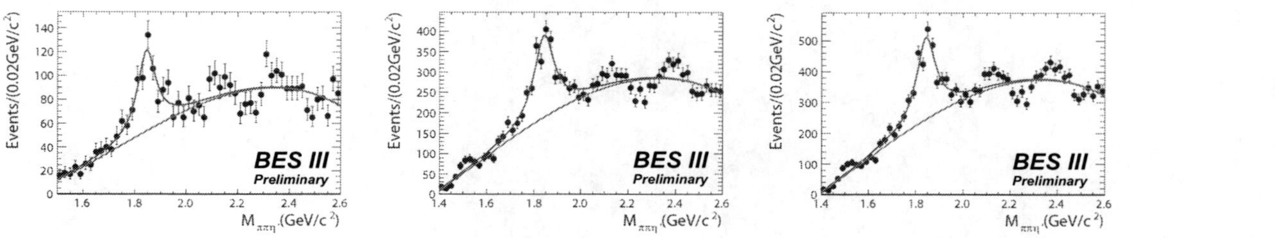

FIGURE 6. The $\eta'\pi^+\pi^-$ mass spectra in $J/\psi \to \gamma\eta'\pi^+\pi^-$. Left for $\eta' \to \pi^+\pi^-\eta$, centre for $\eta' \to \gamma\rho$ and right for the two channels combined.

X(1835)

BESII observed a resonance at $M(\eta'\pi^+\pi^-) = 1833.7 \pm 6.1 \pm 2.7$ MeV/c^2 in $J/\psi \to \gamma\eta'\pi^+\pi^-$ where $\eta' \to \pi^+\pi^-\eta$ and $\gamma\rho$ [19]. BES III also confirms this observation (in both η' decay channels) with a mass of $M(\eta'\pi^+\pi^-) = 1842.4 \pm 2.8$ (preliminary and statistical error only), as shown in Fig. 6.

CONCLUSION AND OUTLOOK

Results, both published and preliminary have been presented based on the first BES III samples of 106 million ψ' and about 226 million J/ψ events. More preliminary results have been released since the MENU 2010 conference (see e.g. [20]) and many more are to be expected in the near future. This includes results from partial wave analyses in various channels; refer to the *Partial Wave Analysis at BES III harnessing the power of GPUs* contribution in these proceedings for details on the techniques and challenges in these analyses. In 2010, BES III acquired nearly $1 fb^{-1}$ of data at the $\psi(3770)$ resonance, including approximately $75 pb^{-1}$ of scan data around the $\psi(3770)$ peak. Decays of the $\psi(3770)$ produce quantum correlated *DD* pairs, which are ideal for mixing and CP violation studies, as well measurements of absolute branching fractions and studies of semi-leptonic decays, thus launching the charm physics program at BES III.

REFERENCES

1. M. Ablikim, et al., *Nucl. Instr. Meth.* **A614**, 345–399 (2010), `0911.4960`.
2. D. M. Asner, et al., *Physics at BES-III* (2008), `0809.1869`.

3. C. Amsler, and F. E. Close, *Phys. Rev.* **D53**, 295–311 (1996), hep-ph/9507326.
4. J. Bolz, P. Kroll, and G. A. Schuler, *Eur. Phys. J.* **C2**, 705–719 (1998), hep-ph/9704378.
5. M. Ablikim, et al., *Phys. Rev.* **D81**, 052005 (2010), 1001.5360.
6. D. M. Asner, et al., *Phys. Rev.* **D79**, 072007 (2009), 0811.0586.
7. C. Amsler, et al., *Phys. Lett.* **B667**, 1 (2008).
8. J. V. Bennett, et al., *Phys. Rev. Lett.* **101**, 151801 (2008), 0807.3718.
9. Y.-J. Gao, Y.-J. Zhang, and K.-T. Chao (2007), hep-ph/0701009.
10. J. L. Rosner, et al., *Phys. Rev. Lett.* **95**, 102003 (2005), hep-ex/0505073.
11. S. Dobbs, et al., *Phys. Rev. Lett.* **101**, 182003 (2008), 0805.4599.
12. M. Ablikim, et al., *Phys. Rev. Lett.* **104**, 132002 (2010), 1002.0501.
13. Y.-P. Kuang, *Phys. Rev.* **D65**, 094024 (2002), hep-ph/0201210.
14. S. Godfrey, and J. L. Rosner, *Phys. Rev.* **D66**, 014012 (2002), hep-ph/0205255.
15. J. J. Dudek, R. G. Edwards, and D. G. Richards, *Phys. Rev.* **D73**, 074507 (2006), hep-ph/0601137.
16. J. Z. Bai, et al., *Phys. Rev. Lett.* **91**, 022001 (2003), hep-ex/0303006.
17. M. Ablikim, et al., *Chinese Physics* **C34**, 421–426 (2010), 1001.5328.
18. J. P. Alexander, et al. (2010), 1007.2886.
19. M. Ablikim, et al., *Phys. Rev. Lett.* **95**, 262001 (2005), hep-ex/0508025.
20. F. A. H. f. t. B. Collaboration (2010), 1008.3569.

Quark and Gluon Angular Momentum in the Nucleon

Matthias Burkardt, Hikmat BC and Abdullah Jarrah

New Mexico State University

Abstract. Parton distributions in impact parameter space, which are obtained by Fourier transforming GPDs, are exhibit a significant deviation from axial symmetry when the target and/or quark is transversely polarized. From this deformation, we present an intuitive derivation of the Ji relation. In a scalar diquark model and in QED, we compare the Ji and Jaffe-Manohar decompositions of the nucleon spin. Using the MIT bag model, we estimate spectator effects through the presence of the gluon vector potential in the definitions of the quark orbital angular momentum.

DISTRIBUTION OF QUARKS IN THE TRANSVERSE PLANE

In the case of transversely polarized quarks and/or nucleons, parton distributions in impact parameter space [1] show a significant transverse deformation. In the case of unpolarized quarks in a nucleon polarized in the $+\hat{x}$ direction, this deformation is described by the $\perp$ gradient of the Fourier transform of the GPD E^q [2]

$$q_{q/p\uparrow}(x,\mathbf{b}_\perp) = \int \frac{d^2\mathbf{x}_\perp}{(2\pi)^2} e^{-i\mathbf{b}_\perp\cdot\Delta_\perp} H^q(x,0,-\Delta_\perp^2) - \frac{1}{2M}\partial_y \int \frac{d^2\mathbf{x}_\perp}{(2\pi)^2} e^{-i\mathbf{b}_\perp\cdot\Delta_\perp} E^q(x,0,-\Delta_\perp^2) \tag{1}$$

for quarks of flavor q. Since $E^q(x,0,t)$ also arises in the decomposition of the Pauli form factor $F_2^q = \int_{-1}^{1} dx E^q(x,0,t)$ for quarks with flavor q (here it is always understood that charge factors have been taken out) w.r.t. x, this allows to relate the $\perp$ flavor dipole moment to the contribution from quarks with flavor q to the nucleon anomalous magnetic moment (here it is always understood that charge factors have been taken out)

$$d^q \equiv \int d^2\mathbf{b}_\perp q_{+\hat{x}}(x,\mathbf{b}_\perp) b_y = \frac{1}{2M} F_2^q(0) = \frac{1}{2M}\kappa_{q/p}. \tag{2}$$

Here $e_q\kappa_{q/p}$ is the contribution from flavor q to the anomalous magnetic moment of the proton. Neglecting the contribution from heavier quarks to the nucleon anomalous magnetic moment, one can use the proton and neutron anomalous magnetic moment to solve for the contributions from $q = u,d$, yielding $\kappa_{u/p} \approx 1.67$ and $\kappa_{d/p} \approx -2.03$. The resulting deformation ($|d_q| \sim 0.1\text{fm}$) of impact parameter dependent PDFs in the transverse direction (fig. 1) is rather significant and it is in opposite directions for u and d quarks. The sideways displacement of the center of momentum for each quark flavor from the origin provides a very intuitive derivation of the Ji-relation [3] for the contribution from quarks with flavor q to the nucleon angular momentum: Consider nucleon state that is an eigenstate under rotation about the $\hat{x}$-axis (e.g. a wave packet

12th International Conference on Meson-Nucleon Physics and the Structure of the Nucleon (MENU 2010)
AIP Conf. Proc. 1374, 41-48 (2011); doi: 10.1063/1.3647096

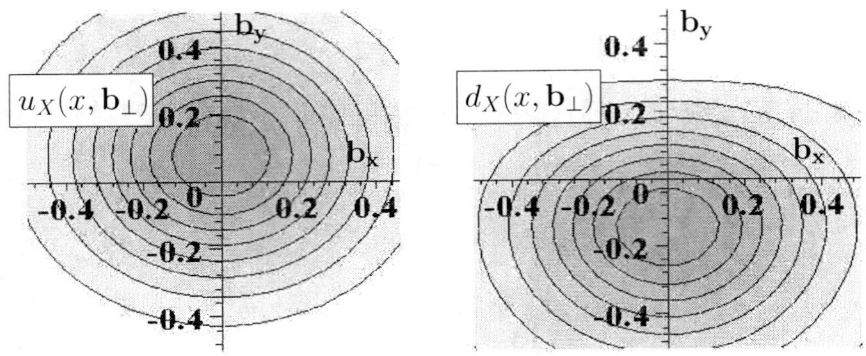

FIGURE 1. Distribution of the j^+ density for u and d quarks in the $\perp$ plane ($x = 0.3$ is fixed) for a proton that is polarized in the x direction in the model from Ref. [2]. For other values of x the distortion looks similar.

describing a nucleon polarized in the $\hat{x}$ direction with $\vec{p} = 0$). For such a state, $\langle T_q^{00} y\rangle = 0 = \langle T_q^{zz} y\rangle$ and $\langle T_q^{0y} z\rangle = -\langle T_q^{0z} y\rangle$, and therefore $\langle T_q^{++} y\rangle = \langle T_q^{0y} z - T_q^{0z} y\rangle = \langle J_q^x\rangle$. This result allows to relate the $\perp$ shift of the center of momentum for quark flavor q to the angular momentum J_q^x carried by that quark flavor. The displacement of the $\perp$ center of momentum is a sum of two effects:

- as discussed above, $\langle T_q^{++} y\rangle$ for a quark relative to the center of momentum of a transversely polarized nucleon (Fig. 1)
- however, already for a point-like transversely polarized spin $\frac{1}{2}$ particle, the $\perp$ center of momentum $\langle T^{++} y\rangle$ is shifted by $\frac{1}{2}$ a Compton wavelength away from the origin (the center of the wave packet in the rest frame)

In order to understand the 2nd effect, i.e. the $\perp$ shift of the center of momentum for a $\perp$ polarized spin $\frac{1}{2}$ particle, let us consider 'bag model' [5] type wave functions[1] for the wave packet of the target, but with a 'bag radius' R that will be sent to ∞ at the end

$$\psi = \begin{pmatrix} f(r) \\ \frac{\vec{\sigma}\cdot\vec{p}}{E+m} f(r) \end{pmatrix} \chi \qquad \text{with} \qquad \chi = \frac{1}{\sqrt{2}} \begin{pmatrix} 1 \\ 1 \end{pmatrix}. \tag{3}$$

Since $\psi^\dagger \partial_z \psi$ is even under $y \to -y$, $i\bar{q}\gamma^0 \partial^z q$ does not contribute to $\langle T^{0z} y\rangle = \langle i\bar{q}\left(\gamma^0 \partial^z + \gamma^z \partial^0\right) q\rangle$. Using $i\partial_0 \psi = E\psi$, one thus finds

$$\begin{aligned} \langle T^{0z} b_y\rangle &= E\int d^3r \psi^\dagger \gamma^0 \gamma^z \psi y = E\int d^3r \psi^\dagger \begin{pmatrix} 0 & \sigma^z \\ \sigma^z & 0 \end{pmatrix} \psi y \\ &= \frac{2E}{E+M}\int d^3r \chi^\dagger \sigma^z \sigma^y \chi f(r)(-i)\partial^y f(r) y = \frac{E}{E+M}\int d^3r f^2(r). \end{aligned} \tag{4}$$

[1] The precise shape of the wave packet does not matter as long as its size is sent to ∞.

In the limit when the bag radius goes to infinity $E = M$ and $\int d^3r f^2(r) = 1$ and therefore the 2nd moment of $\perp$ flavor dipole moment for this wave packet reads

$$\langle T_q^{++} y \rangle = \langle T^{0z} b_y \rangle = \frac{1}{2}, \tag{5}$$

i.e. for a a spherically symmetric, delocalized (Dirac-)wave packet with $J_x = \frac{1}{2}$ centered around the origin the $\perp$ center of momentum $\frac{1}{M}\langle T_q^{++} b_y \rangle$ is not at origin, but at $\frac{1}{2M}$!

This 'overall shift' of the nucleon $\perp$ center of momentum implies a contribution $\frac{1}{2}\langle x_q \rangle$ to $\langle T^{++} y \rangle$ from quarks carrying momentum fraction $\frac{1}{2}\langle x_q \rangle$, which gives rise to the first term in the Ji relation. In addition, Eq. (1) implies a shift of the center of momentum for quark flavor q relative to that of the nucleon, by $\frac{1}{2M}\int dx x E(x,0,0)$. Combining these two effects yields the Ji relation [3]

$$J_q = \langle T_q^{++} b_y \rangle = \frac{1}{2}\int dx x \left[q(x) + E_q(x,0,0)\right] \tag{6}$$

Since the famous EMC experiments revealed that only a small fraction of the nucleon spin is due to quark spins [4], there has been a great interest in 'solving the spin puzzle', i.e. in decomposing the nucleon spin into contributions from quark/gluon spin and orbital degrees of freedom. In this effort, the Ji decomposition [3]

$$\frac{1}{2} = \frac{1}{2}\sum_q \Delta q + \sum_q L_q^z + J_g^z \tag{7}$$

appears to be very useful, as not only the quark spin contributions Δq but also the quark total angular momenta $J_q \equiv \frac{1}{2}\Delta q + L_q^z$ (and by subtracting the spin piece also the the quark orbital angular momenta L_q^z) entering this decomposition can be accessed experimentally through GPDs. The terms in (7) are defined as expectation values of the corresponding terms in the angular momentum tensor

$$M^{0xy} = \sum_q \frac{1}{2} q^\dagger \Sigma^z q + \sum_q q^\dagger \left(\vec{r} \times i\vec{D}\right)^z q + \left[\vec{r} \times \left(\vec{E} \times \vec{B}\right)\right]^z \tag{8}$$

in an appropriate nucleon wave packet. Here $i\vec{D} = i\vec{\partial} - g\vec{A}$ is the gauge-covariant derivative. The main advantages of this decomposition are that each term can be expressed as the expectation value of a manifestly gauge invariant local operator and that $J_q^z = \frac{1}{2}\Delta q + L_q^z$ can be related to GPDs (6) and is thus accessible in deeply virtual Compton scattering and meson production and can also be calculated in lattice QCD. However, due to the presence of interactions through the vector potential in the gauge covariant derivative L_q^z does not have a parton interpretation.

Recent lattice calculations of GPDs surprised in several ways [6]. First, the light quark orbital angular momentum (OAM) is consistent with $L_u \approx -L_d$, i.e. $L_u + L_d \approx 0$, which would imply that $J_g \approx \frac{1}{2} \cdot 0.7$ represents the largest piece in the nucleon spin decomposition. Secondly, $L_u \approx -0.15$ and $L_u \approx +0.15$ in these calculations, i.e. the opposite signs from what one would expect from many quark models with relativistic effects, as we will also illustrate in the following section. While the inclusion of still-omitted disconnected diagrams may change the sum $L_u + L_d$, it does not affect the difference $L_u - L_d$. In Ref. [7], it was pointed out that evolution from a quark model scale of few hundred MeV to the lattice scale of few GeV might explain the difference.

Jaffe and Manohar have proposed an alternative decomposition of the nucleon spin, which does have a partonic interpretation [8]

$$\frac{1}{2} = \frac{1}{2}\sum_q \Delta q + \sum_q \mathcal{L}_q^z + \frac{1}{2}\Delta G + \mathcal{L}_g^z, \tag{9}$$

and whose terms are defined as matrix elements of the corresponding terms in the $+12$ component of the angular momentum tensor

$$M^{+12} = \frac{1}{2}\sum_q q_+^\dagger \gamma_5 q_+ + \sum_q q_+^\dagger \left(\vec{r}\times i\vec{\partial}\right)^z q_+ + \varepsilon^{+-ij}\mathrm{Tr}F^{+i}A^j + 2\mathrm{Tr}F^{+j}\left(\vec{r}\times i\vec{\partial}\right)^z A^j. \tag{10}$$

The first and third term in (9),(10) are the 'intrinsic' contributions (no factor of $\vec{r}\times$) to the nucleon's angular momentum $J^z = +\frac{1}{2}$ and have a physical interpretation as quark and gluon spin respectively, while the second and fourth term can be identified with the quark/gluon OAM. Here $q_+ \equiv \frac{1}{2}\gamma^-\gamma^+ q$ is the dynamical component of the quark field operators, and light-cone gauge $A^+ \equiv A^0 + A^z = 0$ is implied. The residual gauge invariance can be fixed by imposing anti-periodic boundary conditions $\mathbf{A}_\perp(\mathbf{x}_\perp, \infty^-) = -\mathbf{A}_\perp(\mathbf{x}_\perp, -\infty^-)$ on the transverse components of the vector potential.

Only the Δq are common to both decompositions. While for a nucleon at rest the difference in the Dirac structure between L_q^z and $\mathcal{L}_q^z$ plays no role [9], the appearance of the gluon vector potential in the operator defining L_q^z implies that in general $\mathcal{L}_q^z \neq L_q^z$. Other nucleon spin decompositions have been proposed in Refs. [10, 11].

QUARK OAM IN THE SCALAR DIQUARK MODEL

In the scalar diquark model [12], the light-cone wave function for a 'nucleon' with spin $\uparrow$ and the quark spin aligned and anti-aligned respectively reads [13]

$$\psi^\uparrow_{+\frac{1}{2}}(x,\mathbf{k}_\perp) = \left(M + \frac{m}{x}\right)\phi(x,\mathbf{k}_\perp^2) \qquad \psi^\uparrow_{-\frac{1}{2}}(x,\mathbf{k}_\perp) = -\frac{k^1 + ik^2}{x}\phi(x,\mathbf{k}_\perp^2) \tag{11}$$

with $\phi = \frac{g/\sqrt{1-x}}{M^2 - \frac{\mathbf{k}_\perp^2 + m^2}{x} - \frac{\mathbf{k}_\perp^2 + \lambda^2}{1-x}}$. Here g is the Yukawa coupling and $M/m/\lambda$ are the masses of the 'nucleon'/'quark'/diquark respectively. Furthermore x is the momentum fraction carried by the quark and $\mathbf{k}_\perp \equiv \mathbf{k}_{\perp e} - \mathbf{k}_{\perp\gamma}$ represents the relative $\perp$ momentum. Using these light-cone wave functions it is straightforward to calculate a variety of observables that appear in the context of nucleon spin physics. For example, since only $\psi^\uparrow_{-\frac{1}{2}}$ carries (one positive) unit of OAM, the quark OAM according to JM is obtained as

$$\mathcal{L}_q = \int_0^1 dx \int \frac{d^2\mathbf{k}_\perp}{16\pi^3}(1-x)\left|\psi^\uparrow_{-\frac{1}{2}}\right|^2, \tag{12}$$

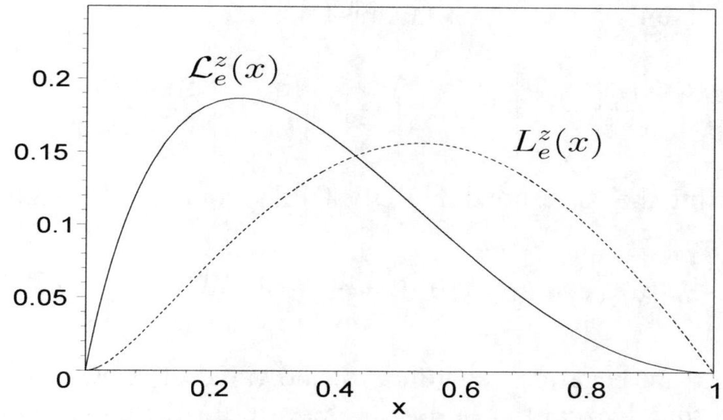

FIGURE 2. x distribution of the 'quark' OAM $\mathcal{L}_q^z(x)$ (full) compared to $L_q^z(x)$ from the unintegrated Ji relation (dotted) in the scalar diquark model for parameters $\Lambda^2 = 10m^2 = 10\lambda^2$. Both in units of $\frac{g^2}{16\pi^2}$.

where the factor $(1-x)$ is the fraction of OAM carried by the quark with momentum fraction x in a two-body system with one unit of OAM [9]. One can also use the above light-cone wave functions to calculate the GPDs that enter the Ji relation, to calculate

$$L_q \equiv \frac{1}{2}\int_0^1 dx\,[xq(x) + xE(x,0,0) - \Delta q(x)]. \tag{13}$$

Using a Lorentz invariant regulator, such as a Pauli-Villars regulator, one finds that the two definitions for quark OAM agree, i.e. $L_q = \mathcal{L}_q$ in the scalar diquark model, as was expected since L_q^z in the scalar diquark model does not contain a gauge field term.

However, despite $L_q = \mathcal{L}_q$, no such equality holds for the corresponding unintegrated quantities. If one defines $L_q(x)$ by Eq. (13) without the x integral [14], one does not obtain the OAM distribution that would be obtained from (12) without x integral (Fig.2), which renders the interpretation of $L_q(x)$ as the quark OAM distribution questionable.

ELECTRON OAM IN QED

In QED, there are four polarization states in the $e\gamma$ Fock component. To lowest order, the respective Fock space amplitudes for a dressed electron with $J^z = +\frac{1}{2}$ read

$$\Psi^{\uparrow}_{+\frac{1}{2}+1}(x,\mathbf{k}_\perp) = \frac{k^1 - ik^2}{x(1-x)}\phi(x,\mathbf{k}_\perp^2) \qquad \Psi^{\uparrow}_{+\frac{1}{2}-1}(x,\mathbf{k}_\perp) = -\frac{k^1 + ik^2}{1-x}\phi(x,\mathbf{k}_\perp^2)$$
$$\Psi^{\uparrow}_{-\frac{1}{2}+1}(x,\mathbf{k}_\perp) = \left(\frac{m}{x} - m\right)\phi(x,\mathbf{k}_\perp^2) \qquad \Psi^{\uparrow}_{-\frac{1}{2}-1}(x,\mathbf{k}_\perp) = 0 \tag{14}$$

with $\phi(x,\mathbf{k}_\perp^2) = \frac{\sqrt{2}}{\sqrt{1-x}}\frac{e}{M^2 - \frac{\mathbf{k}_\perp^2 + m^2}{x} - \frac{\mathbf{k}_\perp^2 + \lambda^2}{1-x}}$. The label $\pm\frac{1}{2}$ represents the spin of the electron in the $e\gamma$ Fock component and ± 1 that of the γ.

Using these light-cone wave functions, it is again straightforward to calculate the OAM of the electron in the JM [8] decomposition. Including a Pauli-Villars subtraction with a heavy 'photon' with mass Λ one thus finds

$$\mathcal{L}_e^z = \int_0^1 dx \int \frac{d^2\mathbf{k}_\perp}{16\pi^3}(1-x)\left[\left|\Psi^{\uparrow}_{+\frac{1}{2}-1}\right|^2 - \left|\Psi^{\uparrow}_{+\frac{1}{2}+1}\right|^2\right] \overset{\Lambda\to\infty}{\underset{\lambda\to 0}{\longrightarrow}} -\frac{\alpha}{4\pi}\left[\frac{4}{3}\log\frac{\Lambda^2}{m^2} - \frac{2}{9}\right] \quad (15)$$

Likewise, inserting the corresponding PDFs/GPDs from Ref. [13] into (13) yields

$$L_e^z = \frac{1}{2}\int_0^1 dx\,[xq_e(x) + xE_e(x,0,0) - \Delta q_e(x)] \overset{\Lambda\to\infty}{\underset{\lambda\to 0}{\longrightarrow}} -\frac{\alpha}{4\pi}\left[\frac{4}{3}\log\frac{\Lambda^2}{m^2} + \frac{7}{9}\right]. \quad (16)$$

Both $\mathcal{L}_e^z$ and L_e^z are negative, regardless of the value of Λ^2 (as long as $\Lambda^2 > \lambda^2$). In the case of $\mathcal{L}_e^z$ the physical reason is a preference for the emission of photons/gluons with the spin parallel (as compared to anti-parallel) to the original quark spin [15] — resulting more likely in a state with negative OAM. In fact, when $x \to 0$, this preference reflects the more general principle of helicity retention [16], which favors the lead parton (i.e. the parton carrying most of the momentum) to carry a spin as close as possible to that of the parent. It is also encoded in the evolution equations derived in Ref. [14]. The divergent parts of $\mathcal{L}_e^z$ and L_e^z are the same so that their difference is UV finite

$$\mathcal{L}_e^z - L_e^z \overset{\Lambda\to\infty}{\underset{\lambda\to 0}{\longrightarrow}} \frac{\alpha}{4\pi}. \quad (17)$$

Applying these results to a (massive) quark with $J^z = +\frac{1}{2}$ yields to $\mathcal{O}(\alpha_s)$

$$\mathcal{L}_q^z - L_q^z = \frac{\alpha_s}{3\pi}. \quad (18)$$

This result may have important phenomenological consequences. Recent lattice QCD calculations for GPDs yielded $L_u^z < 0$ and $L_d^z > 0$, which is opposite to the sign obtained in typical relativistic quark models — such as the MIT bag model.

In QCD, the gluon spin is experimentally accessible, but the gluon OAM $\mathcal{L}_g^z$ is not. On the other hand, the gluon (total) angular momentum J_g^z appearing in the Ji decomposition is accessible, either indirectly (by subtraction, using quark GPDs), or directly, using gluon GPDs from lattice and/or deeply virtual J/ψ production. Even though $\frac{1}{2}\Delta G$ and J_g^z belong to two incommensurable decompositions of the nucleon spin, it is thus tempting to consider the difference between these two quantities, hoping to learn something about gluon OAM. Subtracting (9) from (7), it is straightforward to convince oneself that $J_g^z - \frac{1}{2}\Delta G = \mathcal{L}_g^z + \sum_q \left(\mathcal{L}_q^z - L_q^z\right)$, i.e. numerically $J_g^z - \frac{1}{2}\Delta G$ differs from $\mathcal{L}_g^z$ by the same amount that $\sum_q \mathcal{L}_q^z$ differs from $\sum_q L_q^z$. In our QED example, with

$$\Delta\gamma = \int_0^1 dx \int \frac{d^2\mathbf{k}_\perp}{16\pi^3}\left[\left|\psi^{\uparrow}_{+\frac{1}{2},+1}\right|^2 - \left|\psi^{\uparrow}_{+\frac{1}{2},-1}\right|^2 + \left|\psi^{\uparrow}_{-\frac{1}{2},+1}\right|^2 - \left|\psi^{\uparrow}_{-\frac{1}{2},-1}\right|^2\right] \quad (19)$$

being the photon spin contribution, one thus finds (for $\lambda \to 0$, $\Lambda \to \infty$)

$$J_\gamma^z - \frac{1}{2}\Delta\gamma = \mathcal{L}_\gamma^z + \frac{\alpha}{4\pi}. \quad (20)$$

As was the case in (17), $\frac{\alpha}{4\pi}$ appears to be a small correction, but one needs to keep in mind that for an electron J_γ^z, $\Delta\gamma$, and $\mathscr{L}_\gamma^z$ are also only of order α.

QUARK OAM IN THE BAG MODEL

Since quark models have, to lowest order in α_s, no vector potential, it makes perhaps more sense to identify the quark OAM from these models with $\mathscr{L}_q^z$ rather than with the GPD-based L_q^z. In Ref. [9] the difference $\Delta_e^z \equiv \mathscr{L}_e^z - L_e^z$ was calculated to order α for a single electron in QED and the result then also applied to a single quark in QCD. However, in QCD quarks are never alone and the question arises regarding the effects from 'spectator currents' on the orbital angular momentum of each quark.

In order to address this issue, we will in the following use the MIT bag model [5] to estimate $\mathscr{O}(\alpha_s)$ corrections to the difference

$$\Delta_q^z \equiv L_q^z - \mathscr{L}_q^z = \langle q^\dagger \left(\vec{r} \times g\vec{A}\right)^z q\rangle. \tag{21}$$

The vector potential in (21) is calculated from the spectator currents, which are obtained by taking matrix elements in the corresponding ground state bag model wave functions. The vector potential resulting from these static currents is obtained by solving

$$\vec{\nabla}^2 \vec{A}^a(\vec{r}) = -\vec{j}^a(\vec{r}) = -\sum_{s'} g\psi_{s'}^\dagger(\vec{r})\vec{\alpha}\frac{\lambda^a}{2}\psi_{s'}(\vec{r}) \tag{22}$$

for each color component a and where the summation is over the spectators (here we pick the gauge $\vec{\nabla}\cdot\vec{A}^a = 0$, but to $\mathscr{O}(\alpha_s)$ the result is actually gauge invariant — at least in the subclass of all gauges where all color components are treated (globally) SU(3)-symmetrically. I such gauges, matrix elements of operators of the type $q^\dagger\lambda^a\Gamma q A^a$, where Γ is some Dirac matrix, and A^a is calculated to $\mathscr{O}(\alpha_s)$, are proportional to the matrix elements of the corresponding abelian operators. It is thus sufficient to establish gauge invariance of $q^\dagger\vec{r}\times\vec{A}q$ for abelian fields. The key observation is that the bag model wave functions contain no correlations between the positions of the quarks. Therefore, after eliminating the color in this calculation and introducing abelian currents, $\Delta_{s'}^z$ factorizes into the density of the active quark $\psi_s^\dagger(\vec{r})\psi_s(\vec{r})$ times $(\vec{r}\times\vec{A})_z = rA_\phi$. Writing the volume integral $\int d^3r$ in cylindrical coordinates, one can isolate the only ϕ-dependent term $r\int_0^{2\pi} d\phi A_\phi = \oint d\vec{r}\cdot\vec{A}(\vec{r})$ as a closed loop integral with fixed r and z. The closed loop integral is gauge invariant (its numerical value represents the color-magnetic flux through a circle with radius r) and so is the volume integral in which it enters.

The contribution from a spectator with $j^z = s'$ to $\langle q(\vec{r}\times\vec{A})^z q\rangle$ thus reads

$$\Delta_{s'}^z = -\frac{2}{3}\frac{g^2}{4\pi}\int d^3r d^3r' \psi_s^\dagger(\vec{r})\psi_s(\vec{r})\psi_{s'}^\dagger(\vec{r}')\frac{(\vec{r}\times\vec{\alpha})^z}{|\vec{r}-\vec{r}'|}\psi_{s'}(\vec{r}'). \tag{23}$$

Note that $\vec{\Delta}_{s'}$ is independent of the angular momentum $j^z = s$ of the 'active quark', since $\psi_{\frac{1}{2}}^\dagger(\vec{r})\psi_{\frac{1}{2}}(\vec{r}) = \psi_{-\frac{1}{2}}^\dagger(\vec{r})\psi_{-\frac{1}{2}}(\vec{r})$. However, it depends on the spin of the spectator since the orientation of the vector potential entering (21) depends on the latter. For example, for $s' = +\frac{1}{2}$, one finds

$$\psi^\dagger_{s'}(\vec{r}')(\vec{r}\times\vec{\alpha})^z\psi_{s'}(\vec{r}') = |\mathcal{N}|^2 j_0(kr')j_1(kr')2\frac{xx'+yy'}{r'} \tag{24}$$

and hence independent of the bag radius [17]

$$\Delta^z_{\pm\frac{1}{2}} = \mp\frac{2}{3}\alpha_s|\mathcal{N}|^4\int_{r<R}d^3r\int_{r'<R}d^3r'\left[j_0^2(kr)+j_1^2(kr)\right]2\frac{xx'+yy'}{|\vec{r}-\vec{r}'|}\frac{j_0(kr')j_1(kr')}{r'} = \mp 0.78\alpha_s. \tag{25}$$

When the active quark has s aligned with that of the proton, the two spectators must have opposite s' and their contribution to $\vec{\Delta}$ cancels, i.e. $\vec{\Delta}$ is nonzero only in those wave function components where the active quark has $s = -\frac{1}{2}$, in which case both spectators have $s' = +\frac{1}{2}$. As a result, $\Delta^z_{q/p}$ is equal to twice $\Delta^z_{+\frac{1}{2}}$ times the probability to find that quark flavor with $s = -\frac{1}{2}$ (which is $\frac{1}{3}$ for $q = u$ and $\frac{2}{3}$ for $q = d$, and hence

$$\Delta^z_{u/p} = \frac{2}{3}\Delta^z_{+\frac{1}{2}} = -0.052\alpha_s \qquad \Delta^z_{d/p} = \frac{4}{3}\Delta^z_{+\frac{1}{2}} = -0.104\alpha_s. \tag{26}$$

ACKNOWLEDGMENTS

This work was supported by the DOE under grant number DE-FG03-95ER40965.

REFERENCES

1. M.Burkardt, Phys. Rev. D **62**, 071503 (2000), Erratum-ibid. D **66**, 119903 (2002); M. Diehl, Eur. Phys. J. C **25**, 223 (2002); J.P. Ralston and B. Pire, Phys. Rev. D **66**, 111501 (2002).
2. M. Burkardt, Int. J. Mod. Phys. A **18**, 173 (2003).
3. X. Ji, Phys. Rev. Lett. **78**, 610 (1997).
4. J. Ashman et al (EMC), Phys. Lett. B **206**, 364 (1988); Nucl. Phys. B **328**, 1 (1989).
5. A. Chodos et al., Phys. Rev. D**9**, 3471 (1974); T.A. DeGrand et al., *ibid.* **12** 2060 (1975).
6. Ph. Hägler *et al.* (LHPC Collaboration), Phys. Rev. D **77**, 094502 (2008).
7. A.W. Thomas, and F. Myhrer, Phys. Lett. B **663**, 302 (2008); A.W. Thomas, Phys. Rev. Lett. **101**, 102003 (2008).
8. R.L. Jaffe and A. Manohar, Nucl. Phys. **B337**, 509 (1990).
9. M. Burkardt and H. BC, Phys. Rev. D**79**, 071501 (2009).
10. X.S. Chen et al., Phys. Rev. Lett. **100**, 232002 (2008); **103**, 062001 (2009).
11. M. Wakamatsu, Phys. Rev. D **81**, 114010 (2010); Eur. Phys. J. A **44**, 297 (2010).
12. S.J. Brodsky, D.S. Hwang, and I. Schmidt, Nucl. Phys. B **642**, 344 (2002); M. Burkardt and D.S. Hwang, Phys. Rev. **D69**, 074032 (2004); L.P. Gamberg et al., Phys. Rev. D **67**, 071504 (2003); D. Boer, S.J. Brodsky, and D.S. Hwang, Phys. Rev. D **67**, 054003 (2003); A. Bacchetta et al., Phys. Lett **B578**, 109 (2004); M. Radici et al., hep-ph/0708.0232; L.P. Gamberg et al. hep-ph/0708.0324; D. Boer et al., Phys. Rev. D **67**, 054003 (2003); L.P. Gamberg et al., Phys. Rev. D **67** (2003) 071504.
13. S.J. Brodsky and S. Drell, Phys. Rev. D**22**, 2236 (1980); S.J. Brodsky, D.S. Hwang, B.-Q. Ma and I. Schmidt, Nucl. Phys. B **593**, 311 (2001).
14. P. Hoodbhoy, X. Ji, and W. Lu, Phys. Rev. D**59**, 014013 (1999).
15. P.G. Ratcliffe, Phys. Lett. **B192**, 180 (1987).
16. S.J. Brodsky, M. Burkardt, and I. Schmidt, Nucl. Phys. **B441**, 197 (1995).
17. M. Burkardt and A. Jarrah, arXiv:1011.1041

Proposal for GPD studies at COMPASS

E. Burtin, on behalf of the COMPASS collaboration

CEA - Centre de Saclay
Irfu, Service de Physique Nucléaire
F-91191 Gif-sur Yvette Cedex, France
email=eburtin@cea.fr

Abstract. The study of nucleon structure through Generalised Parton Distributions (GPD) is one major part of the future COMPASS-II physics program [1] and can be performed using exclusive reactions like Deeply Virtual Compton Scattering (DVCS) and Meson Production. The high energy of the muon beam allows to measure the x_B-dependence of the t-slope of the DVCS cross section. The use of positive and negative polarised muon beams allows to determine the Beam Charge and Spin Difference of the DVCS cross sections to access the real part of the Compton form factor related to the dominant GPD H. The sensitivity of both measurements is examined and confronted to existing models or global fits of the data. Preliminary beam test data were analyzed and demonstrated the feasibility of the identification of the DVCS reaction using the COMPASS experimental set-up.

Keywords: Nucleon Structure, Generalised Parton Distributions, Deeply Virtual Compton Scattering, Exclusive reactions
PACS: -

KINEMATIC DOMAIN ACCESSIBLE AT COMPASS

The COMPASS experiment is located on the unique high-energy (200 GeV) and highly polarized $\mu^{\pm}$ beam line of the CERN SPS and uses a high resolution forward spectrometer in conjunction with a fixed target (unpolarized or longitudinally or transversely polarized). By installing a recoil proton detector around the target to ensure exclusivity of Deeply Virtual Compton Scattering (DVCS) and Deeply Virtual Meson Production (DVMP) events, it can be converted into a facility measuring exclusive reactions within a kinematic subspace ranging from $x \sim 0.005$ to ~ 0.1, which cannot be explored by any other existing or planned facility in the near future. Figure 1 displays the kinematic domains of the fixed-target experiments COMPASS, HERMES and JLab. These domains are indicated between the dotted lines defined by the maximum energy of the facility and the curve W > 2 GeV, i.e. above the resonance domain. COMPASS will thus explore the uncharted x domain between those of H1 and ZEUS at the HERA collider and of fixed-target experiments as HERMES and the planned 12 GeV extension of the JLab accelerator.

DVCS MEASUREMENT WITH POLARIZED μ^+ AND μ^- BEAMS

DVCS is considered to be the theoretically cleanest of the experimentally accessible processes to measure GPDs because effects of next-to-leading order and subleading twist are under theoretical control [2]. The competing Bethe-Heitler (BH) process which is elastic lepton-nucleon scattering with a hard photon emitted by either the incoming or outgoing lepton, has a final state identical to that of DVCS so that both processes interfere at the level of amplitudes $\mathscr{A}$:

$$d\sigma(\mu\mathrm{N} \to \mu\mathrm{N}\gamma) \propto |\mathscr{A}_{BH}|^2 + |\mathscr{A}_{DVCS}|^2 + \underbrace{\mathscr{A}_{BH}\mathscr{A}^*_{DVCS} + \mathscr{A}^*_{BH}\mathscr{A}_{DVCS}}_{I}. \tag{1}$$

The Bethe-Heitler amplitude can be calculated using QED and elastic form-factor measurements. The collection of almost pure BH events at small x allows one to get an excellent reference yield and to control accurately the efficiency of the detection. In contrast the collection of almost pure DVCS events at larger x will allow for the measurement of the x-dependence of the t-slope of the cross section which is related to the tomographic partonic image of the nucleon. In the intermediate domain, the DVCS contribution will be enhanced by the BH process through their interference. COMPASS is presently the only facility to provide polarized leptons with either charge: polarized μ^+ and μ^- beams. As the BH is independent of charge and polarization, this contribution can be removed by subtracting 2 separate

12th International Conference on Meson-Nucleon Physics and the Structure of the Nucleon (MENU 2010)
AIP Conf. Proc. 1374, 49-54 (2011); doi: 10.1063/1.3647097

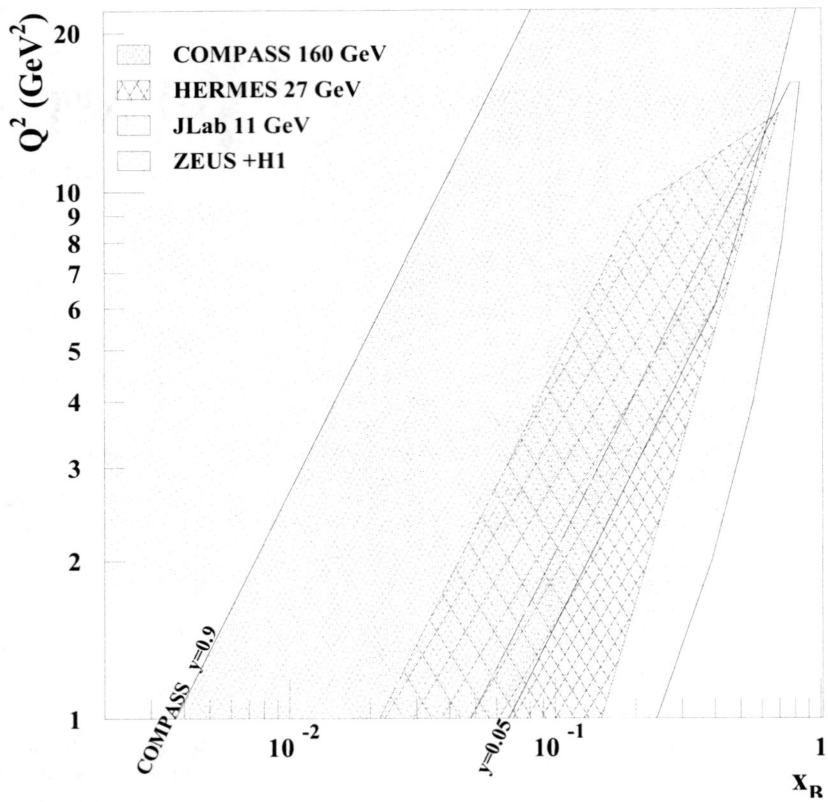

FIGURE 1. Kinematic domains for measurements of hard exclusive processes shown for the COMPASS (green area enclosed by the lines $y = 0.9$ and $y = 0.05$), HERMES and JLab fixed-target experiments and the HERA collider experiments H1 and ZEUS.

measurements obtained for the two beam charges. Moreover the *natural* polarization of the muon beam produced from pion decay changes sign when the beam charge is reversed and the different topologies of μ^+ and μ^-, polarized with opposite direction, allow one to select only the real part or the imaginary part of the complex amplitude of DVCS. COMPASS offers the advantage to provide various kinematic domains where either BH or DVCS dominate (see Fig. 2).

For the muo-production of real photons off an *unpolarized proton* target, the differential cross section can be written as :

$$\frac{d^4\sigma(\mu p \to \mu p \gamma)}{dx_B dQ^2 dt d\phi} = d\sigma^{BH} + [d\sigma^{DVCS}_{unpol} + P_\mu \, d\sigma^{DVSC}_{pol}] + e_\mu \left[\mathrm{Re}\, I + P_\mu \, \mathrm{Im}\, I\right], \qquad (2)$$

where I is the interference term of Eq. (1), P_μ is the beam polarization and e_μ its charge in units of the elementary charge. The DVCS amplitude can be expanded in $1/Q$ beyond leading twist-2 including all twist-3 contributions [2]. The dependence on ϕ, the azimuthal angle between lepton scattering plane and photon production plane, is a characteristic feature of the cross section. Integration over ϕ and/or analysis of the angular dependence in ϕ allows us to isolate specific contributions that are sensitive to different combinations of quark GPDs. Gluon GPDs enter in DVCS only beyond leading order in α_s (LO), analogous to DIS.

At the CERN SPS M2 beamline, the 'natural' polarization of the muon beam produced from pion decay changes sign when the beam charge is reversed. Hence with the *same* apparatus the COMPASS experiment can perform separate measurements for the two beam charge/polarization states $\overset{+}{\leftarrow}$ and $\overset{-}{\rightarrow}$, which can be used to calculate:
the 'Beam Charge (C) and Spin (S) Difference' (for Unpolarized (U) proton target)

$$\mathscr{D}_{U,CS} \equiv d\sigma^{\overset{+}{\leftarrow}} - d\sigma^{\overset{-}{\rightarrow}} = 2[P_\mu d\sigma^{DVCS}_{pol} + e_\mu \mathrm{Re}\, I] \qquad (3)$$

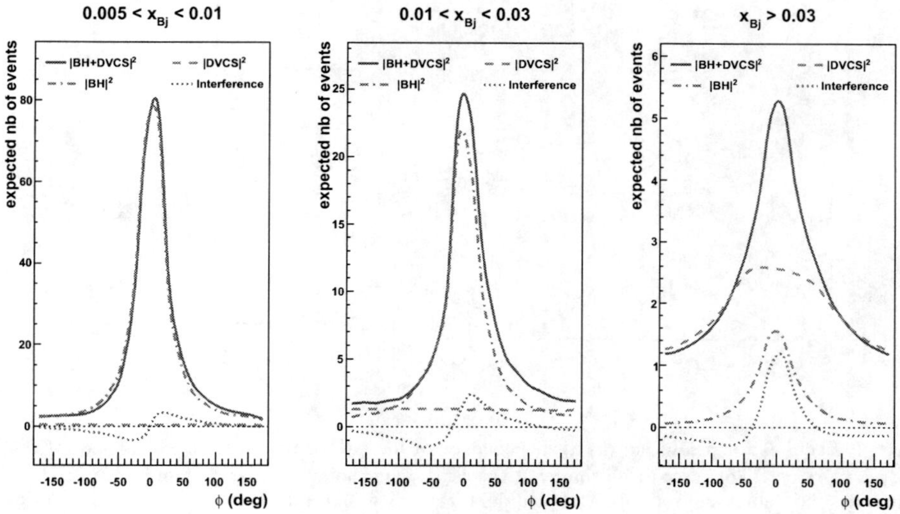

FIGURE 2. Monte Carlo simulation of the exclusive process $\mu^+ p \to \mu'^+ p\gamma$ for $Q^2 > 1$ GeV^2, showing the ϕ angle distribution for three bins in x_B: $0.005 < x_B < 0.01$ (left), $0.01 < x_B < 0.03$ (middle) and $0.03 < x_B$ (right). The event yield shown is normalised to the integrated luminosity of the 2009 DVCS test run. It is based on the acceptance of the presently existing COMPASS set-up.

$$\propto \quad \left(\{s_1^{DVCS}\sin\phi\}\right) + \left(c_0^I + c_1^I\cos\phi + \{c_2^I\cos 2\phi + c_3^I\cos 3\phi\}\right)$$

in which the pure BH contribution *cancels out*. The coefficients $c_i^{I,DVCS}$ and $s_i^{I,DVCS}$ are related to certain combinations of Compton Form Factors (CFFs). A CFF $\mathscr{F}$ is a sum over flavors f, of convolutions of the respective GPDs F^f with a perturbatively calculable kernel describing the hard $\gamma^* q$ interaction. Note that the contributions shown between a pair of braces corresponds to higher-twist or higher-order effects. The analysis of the ϕ-dependence of the beam charge and spin difference $\mathscr{D}_{U,CS}$ will provide via the term Re I the two leading twist-2 expansion coefficients c_0^I and c_1^I, the dominant contribution to which is related to the *real* part of the Compton form factor $\mathscr{H}$ that is in LO given by a flavor sum of convolutions involving the GPDs H^f.

The 'Beam Charge and Spin Sum' of cross sections can also be evaluated:

$$\begin{aligned} \mathscr{S}_{U,CS} &\equiv d\sigma^{\overset{+}{\leftarrow}} + d\sigma^{\overset{-}{\rightarrow}} = 2[d\sigma^{BH} + d\sigma^{DVCS}_{unpol} + e_\mu P_\mu \mathrm{Im}\, I] \qquad (4)\\ &\propto 2[d\sigma^{BH}] + \left(c_0^{DVCS} + \{c_1^{DVCS}\cos\phi + c_2^{DVCS}\cos 2\phi\}\right) + \left(s_1^I\sin\phi + \{s_2^I\sin 2\phi\}\right) \end{aligned}$$

in which the BH contribution *does not cancel out.* 1) The analysis of the ϕ-dependence of $\mathscr{S}_{U,CS}$ will provide via the term Im I the leading twist-2 quantity s_1^I. Its dominant contribution is related to the *imaginary* part of the Compton form factor $\mathscr{H}$. 2) A parallel analysis can be performed subtracting the BH contribution when it is not too large, and integrating over ϕ to get rid of the complete interference term and of the ϕ-dependent terms of the DVCS contribution. Thus the DVCS leading twist-2 quantity c_0^{DVCS} can be isolated and its characteristic t-slope can be determined as a function of x, from which conclusions can be drawn on the transverse size of the nucleon over the x-range accessible to COMPASS in what is know as 'nucleon tomography'.

Shrinkage of the nucleon size

Using the ϕ-integrated beam charge and spin sum after BH subtraction, figure 3 (left) shows the projected statistical accuracy for a measurement at COMPASS of the x-dependence of the t-slope parameter $B(x)$ of the DVCS cross section $\frac{d\sigma}{dt}(x) \propto \exp(-B(x)\,\mathrm{t})$. In the simple ansatz $B(x) = B_0 + 2\,\alpha' \log(\frac{x_0}{x})$, the shrinkage parameter α' is known

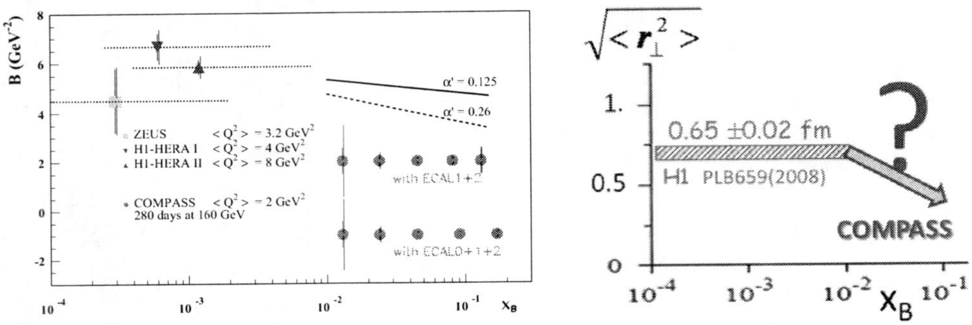

FIGURE 3. **Left:** Projections for measuring the x_B dependence of the t-slope parameter $B(x_B)$ of the DVCS cross section, calculated for $1 < Q^2 < 8$ GeV^2. For comparison some HERA results with similar $\langle Q^2 \rangle$ are shown [6, 9]. The left vertical bar on each data point indicates the statistical error only while the right one includes also the systematic uncertainty, using only ECAL1 and ECAL2 (first row) and also ECAL0 (second row). Two different parameterisations are shown using $\alpha' = 0.125$ GeV^{-2} and 0.26 GeV^{-2}. **Right:** Transverse proton radius as a function of x_B. New and significant information will be obtained in the uncharted x_B region, further elucidating the issue of "nucleon tomography".

for a long time to describe the decrease in nucleon size with increasing x. More recently, this ansatz was also used for the 'reggeized' description of a correlated (x,t) dependence of GPDs. The t-slope of the GPD H^f at a given x was shown [4] to be related to the average impact parameter $\langle (b_\perp^f)^2 \rangle$ in the distribution of partons of flavor f carrying the longitudinal momentum fraction x at a given t: $B^f(x) \sim 1/2\,\langle (b_\perp^f)^2 \rangle(x)$.

Data on B exist only for the HERA collider x-range from 10^{-4} to 0.01 [6, 9], below the COMPASS range $0.01 < x < 0.1$. In Fig. 3 only HERA results are reported for which the mean value $\langle Q^2 \rangle$ is in the investigated domain by COMPASS. In the valence region, where no experimental determinations of B exist, some information comes from fits adjusted to form factor data which give $\alpha' \simeq 1$ GeV^2 [10, 11]. For the low-x sector, H1 results on α' from exclusive J/ψ production [7], which involves the generalized gluon distribution, are smaller by two standard deviations in the total experimental uncertainty than the corresponding value $\alpha' = 0.25$ for Pomeron exchange in soft scattering processes. For the simulation shown in figure 3 we chose the value $\alpha' = 0.125$.

This measurement will yield new and significant information in the context of 'nucleon tomography' as it is expected in chiral-dynamics approach [8]. In this approach, the gluon density is generated by the 'pion cloud' of the nucleon so that a significant increase in the transverse size of the nucleon is predicted for x below the ratio of pion and proton masses, $m_\pi/m_p \approx 0.15$ which lies in the COMPASS domain.

Beam Charge and Spin Asymmetry and sensitivity to models

Using the beam charge and spin asymmetry $\mathscr{D}_{U,CS}/\mathscr{S}_{U,CS}$, figure 4 shows the projected statistical accuracy in a particular (x, Q^2) bin, for a measurement of its ϕ-dependence. Two of the curves are calculated using the 'VGG' GPD model [3]. As this model is meant to be applied mostly in the valence region, typically the value $\alpha' = 0.8$ is used in the 'reggeized' parameterization of the correlated (x,t) dependence of GPDs. For comparison, also the model result for the 'factorized' x,t dependence is shown, which corresponds to $\alpha' \approx 0.1$ in the 'reggeized' ansatz.

A recent theoretical development exploits dispersion relations for Compton form factors. In this context, the additional curve is the result of a fitting procedure [5] including next-to-next-to leading order (NNLO) corrections which was developed and successfully applied to describe DVCS observables from very small values of x for the HERA collider to large x for HERMES and JLab.

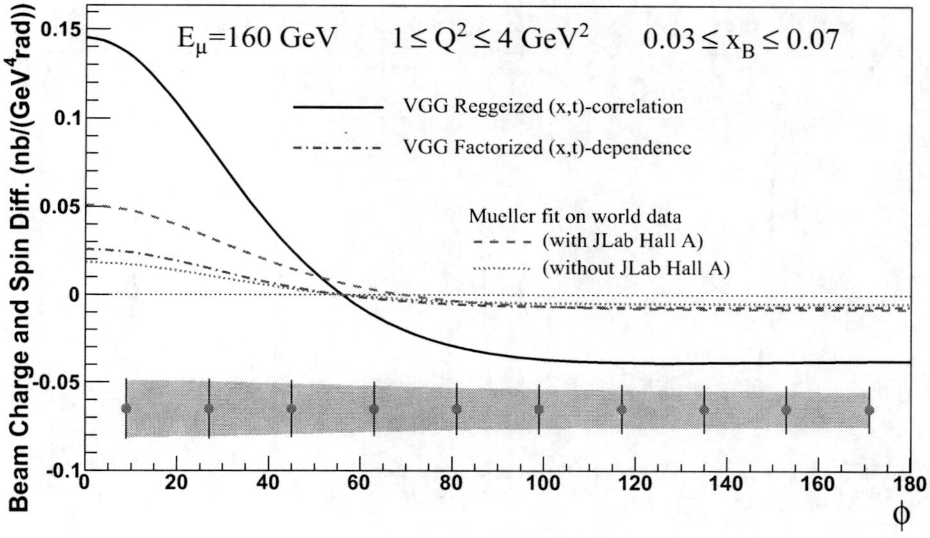

FIGURE 4. Projected statistical (error bars) and systematic (grey band) accuracy for a measurement of the ϕ dependence of the Beam Charge and Spin Difference at COMPASS for $0.03 \leq x \leq 0.07$ and $1 \leq Q^2 \leq 4$ GeV^2 and for 280 days of running time with a 2.5m LH_2 target, an intensity of $4.6 \times 10^8 \mu$ in a 48 s SPS spill period and an overall *global efficiency* $\varepsilon_{global} = 0.1$. The solid and dash-dotted curves correspond to different variants of the VGG model [3] while the dashed and dotted curves show predictions based on first fits on world data [5].

OBSERVATION OF EXCLUSIVE PHOTON PRODUCTION IN 2008 AND 2009 TEST BEAM DATA

A preliminary study of the feasibility of DVCS measurements was performed in 2008 and 2009 with minimum disturbance of the ongoing hadron spectroscopy measurements. The experimental set-up makes use of the 40 cm long LH_2 target, surrounded by a Recoil Proton Detector (RPD). It includes all detectors presently available including the ECAL1 and ECAL2 electromagnetic calorimeters for photon detection. Apart from the shorter target and the corresponding RPD lengths, this set-up has the principal features of the one foreseen for the proposed future GPD program. DVCS test measurements were performed using both μ^+ and μ^- beams of 160 GeV energy, with an intensity of about 1/3 of the maximum in 2008 and close to the maximum intensity in 2009. The 2008 DVCS test data allowed one to clearly identify exclusive photon production and to obtain a first direct determination of the global overall efficiency. The latter, including also SPS beam and spectrometer availabilities, trigger and DAQ efficiencies, was found to be close to the value of 10% assumed for the projections. The 2009 DVCS test data provided a first direct measurement of the relative contributions of the BH and DVCS in the x_B domain available at COMPASS (Fig. 5). Note that the predicted ϕ dependence of the DVCS cross section is not flat in ϕ due to acceptance effects in the 2009 set-up as there is no electromagnetic calorimeter at large photon angles. This will be improved by the addition of the new calorimeter ECAL0 already mentioned above.

REFERENCES

1. COMPASS-II proposal, SPSC-P-340, http://cdsweb.cern.ch/record/1265628/files/SPSC-P-340.pdf, May 17, 2010
2. A.V. Belitsky, D. Müller and A. Kirchner, Nucl. Phys. B **629** (2002) 323.
3. M. Vanderhaeghen, P.A.M. Guichon and M. Guidal, Phys. Rev. Lett. **80** (1998) 5064; Phys. Rev. D **60** (1999) 094017; K. Goeke, M.V. Polyakov and M. Vanderhaeghen, Prog. Part. Nucl. Phys. **47** (2001) 401.
4. M. Burkardt, Phys. Rev. D **62** (2000) 071503; erratum-ibid. d **66** (2002) 119903;
5. K. Kumericki,D. Mueller and K. Passek-Kumericki, Nucl. Phys. **B 794** (2008) 244, K. Kumericki and D. Mueller, arXiv 0904.0458[hep-ph]

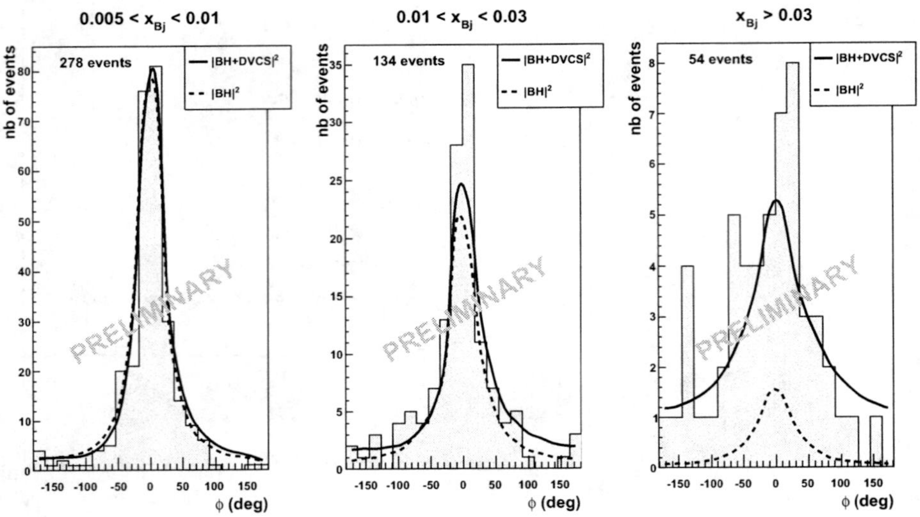

FIGURE 5. Distribution in the azimuthal angle ϕ for measured exclusive single-photon events, $\mu p \rightarrow \mu' p \gamma$ with $Q^2 > 1$ GeV^2, in the same three x_B bins as in Fig. 1. Shown here is a Monte Carlo simulation of only the BH process and of both the BH and DVCS processes.

6. H1, A. Aktas *et al.*, Eur. Phys. J.C **44** (2005) 1, F.D. Aaron *et al.*, Phys. Lett. B **659** (2008) 796.
7. H1, A. Aktas *et al.*, Eur. Phys. J.C **46** (2006) 585.
8. M. Strikman and C. Weiss, Phys. Rev. **D69** (2004) 054012.
9. ZEUS, S. Chekanov *et al.*, DESY-08-178, arXiv:hep-exp:0812.2517v3.
10. M. Diehl, Th. Feldmann, R. Jakob and P. Kroll, Eur. Phys. J. C **39** (2005) 1.
11. M. Guidal, M.V. Polyakov, A.V. Radyushkin and M. Vanderhaeghen, Phys. Rev. D **72** (2005) 054013.

Meson Spectroscopy at CLAS and CLAS12: the present and the future

R. De Vita

Istituto Nazionale di Fisica Nucleare, Sezione di Genova, Via Dodecaneso 33, 16146 Genova, Italy
for the CLAS Collaboration

Abstract. Mesons are the simplest quark bound system, being made by a quark and an anti-quark pair. Studying their structure and properties is a fundamental step to reach a deep understanding of QCD. For this purpose both the precise determination of the meson spectrum for conventional states and the search for states beyond the simple $q\bar{q}$ configurations, as hybrids (qqg) or glueballs, are needed. Finding evidence for these unconventional states would help in understanding some of the open issues in hadronic physics, as how the quarks are confined within hadrons and what is the role of gluons. These topics are presently studied with the CLAS detector at Jefferson Lab and will be studied with the novel CLAS12 experiment after the 12 GeV upgrade of the facility. In my talk I will present the physics program that is presently in progress and the future perspectives.

Keywords: Mesons, Spectroscopy, Detectors
PACS: 13.60.Le, 14.40.Rt, 11.80.Et, 29.40.-n

INTRODUCTION

Understanding the hadron spectrum is one of the fundamental issue in modern particle physics. We know that existing hadron configurations include baryons, made of three quarks, and mesons, made of a quark-antiquark pairs. However most of the mass of the hadrons is not due to the mass of these elementary constituents but to the force that binds them. Studying the hadron spectrum is therefore a tool to explore one of the fundamental forces in nature, i.e. the strong force, and Quantum Chromo Dynamics (QCD), i.e. the theory that describes it. This investigation can provide an answer to fundamental questions as what is the origin of the mass of hadrons, what is the origin of quark confinement, what are the relevant degrees of freedom to describe these complex systems and how the transition between the elementary constituents, quarks and gluons, and baryons and mesons occurs.

In this field a key tool is given by meson spectroscopy. Meson, being made by a quark and an anti-quark, are the simplest quark bound system and therefore the ideal benchmark to study the interaction between quarks and understand what the role of gluons is. In this investigation, it is fundamental to determine precisely the spectrum and properties of mesons but also to search for possible unconventional states beyond the $q\bar{q}$ configuration as tetraquarks ($qqqq$), hybrids (qqg) and glueballs. These unusual states can be distinguished unambiguously from regular mesons when they have exotic quantum numbers, i.e. combinations of total angular momentum, spin and parity that are not allowed for $q\bar{q}$ states. These are called *exotic* quantum numbers and the corresponding states are referred to as *exotics*.

So far, most experiments that have studied the light-quark meson spectrum used hadronic probes and in particular pion beams. A powerful alternative is given by electromagnetic probes and, in particular, photon beams. In fact photons, having spin equal to 1, provide complementary information to pions that have spin equal to 0. The full determination of the initial state that is obtained if the energy of the photon is known allows us to extract information about the meson production mechanism, while the measurement of the meson decay products and in particular of their angular distribution allows us to determine the quantum numbers of the state.

Existing data in photoproduction show that the production cross sections are sizable allowing for a detail study of meson states. In addition it has been suggested that the probability of producing exotics may be higher with photon beams than with pion beams since the presence of a $S = 1$ particle in the initial state may favor the production of quark pairs with aligned spins, which is more likely to result into an exotic configuration. The lowest mass exotics are expected to have masses of about 2 GeV according to lattice QCD calculations as for example the one of Ref. [1]. The existing estimates for exotic meson production also show that the cross sections are expected to be comparable to regular mesons [2].

A thorough study of the meson spectrum and an effective search for exotics states require high intensity beams and large acceptance detectors in order to produce abundantly the resonances to be studied and to measure their decays.

12th International Conference on Meson-Nucleon Physics and the Structure of the Nucleon (MENU 2010)
AIP Conf. Proc. 1374, 55-60 (2011); doi: 10.1063/1.3647098

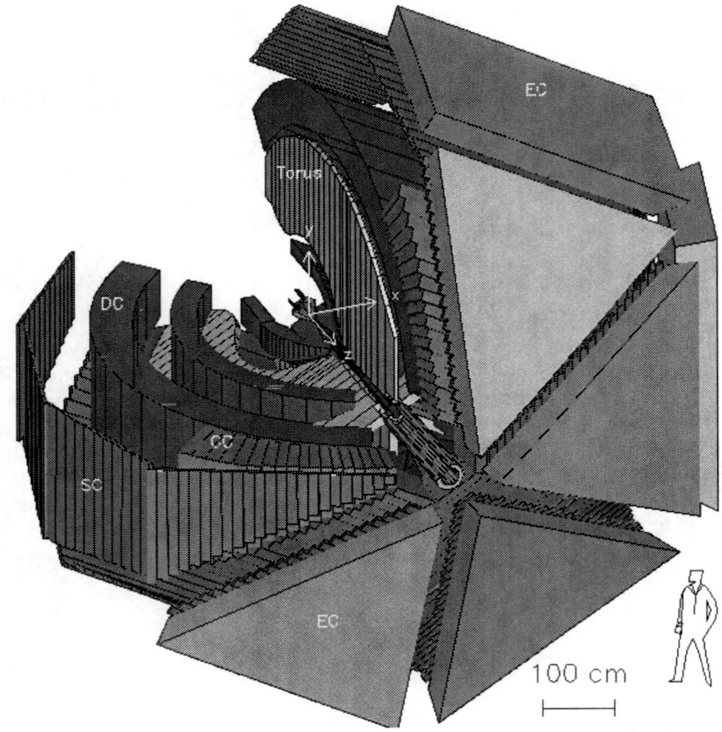

FIGURE 1. Schematic of the CLAS detector. The toroidal magnet, shown in yellow, divides the spectrometer into 6 identical sector, equipped with Drift Chambers (DC), Cerenkov Counters (CC), Scintillator time of flight (SC) and Electromagnetic Calorimeter (EC).

Both this elements are available at Jefferson Laboratory and have motivated the broad physics program that will be described in the next sections.

THE CLAS DETECTOR AT JEFFERSON LABORATORY

The core of Jefferson Lab is the Continuous Electron Beam Accelerator Facility (CEBAF), which delivers simultaneously a low emittance, 100% duty-cycle electron beam to the three different experimental Halls (A, B and C). The maximum energy is presently 6 GeV (with 80% polarization available) with a maximum current of 180 μA. The experimental Hall B is mainly devoted to measurements that require the detection of several particles in the final state, as baryon and meson spectroscopy. It hosts the CEBAF Large Acceptance Spectrometer (CLAS) [3], which is a large acceptance detector capable of measuring multi-particle final states with both charge and neutral particles over a large portion of the solid angle. A schematic of the detector is shown in Figure 1. The spectrometer is made of six identical sectors, placed in between the six superconducting coils of a toroidal magnet. Each sector is equipped with three layers of drift chambers (DC) for track reconstruction, one layer of scintillators (SC) for time-of-flight measurements and hadron identification, forward Cerenkov counters (CC) for electron-pion discrimination, and electromagnetic calorimeters (EC) to identify electrons and neutral particles.

Hall B is also equipped with a Bremsstrahlung photon tagger. The electron beam scatters over a gold radiator producing photons by Bremsstrahlung. The scattered electron is deflected by a dipole magnet toward a scintillator hodoscope, which determines the electron energy and therefore the corresponding photon energy on an event-by-event basis. The produced photon beam is therefore *tagged*. The energy and time resolution of the photon tagger are of the order of 10^{-3} and 100 ps respectively. The maximum photon energy achievable for 6 GeV electrons is of about 5.7 GeV, corresponding to a maximum invariant mass on proton target of $W \sim 3.4$ GeV and meson resonance mass of about 2.5 GeV.

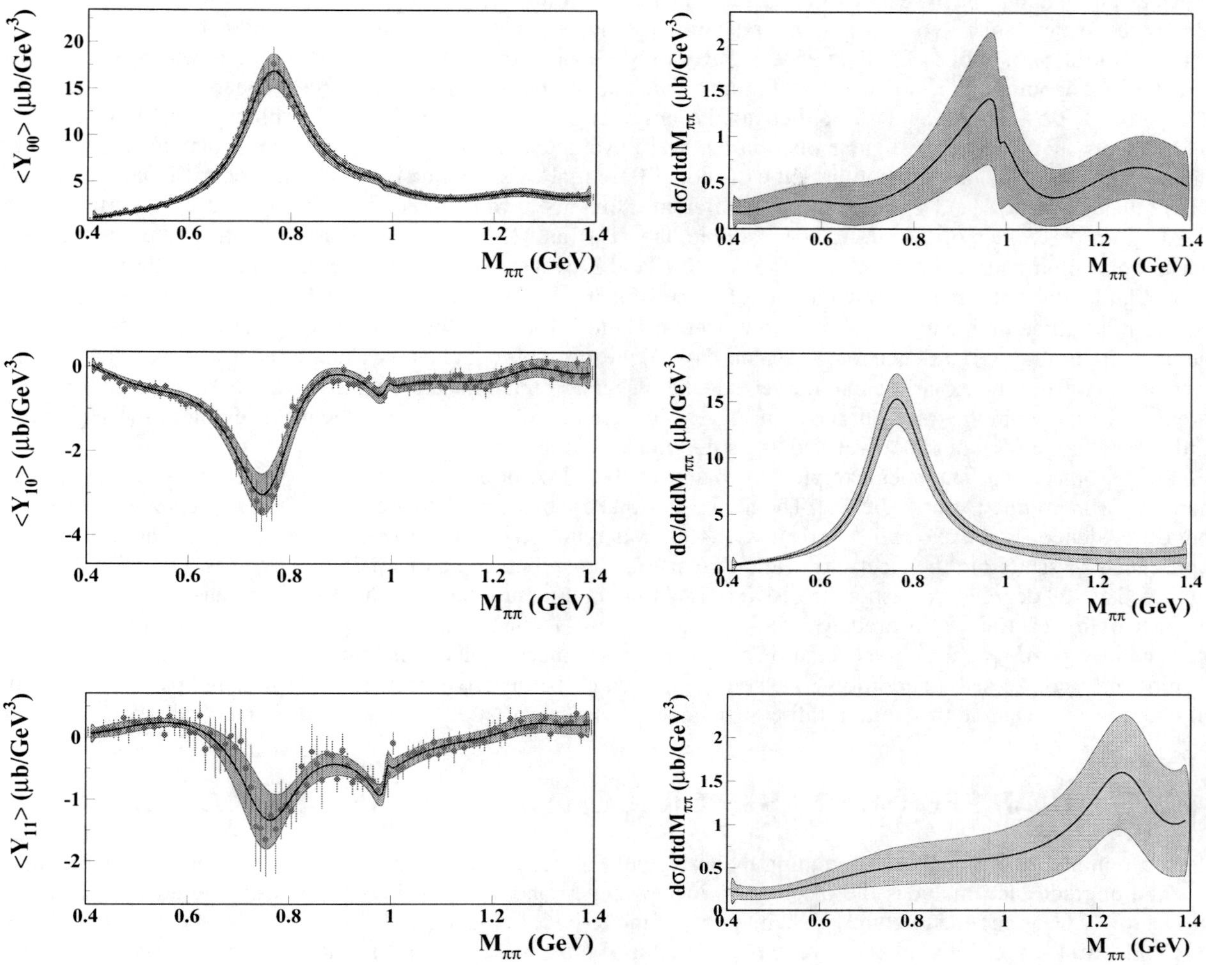

FIGURE 2. Left: Y_{00}, Y_{10}, Y_{11} moments of the di-pion decay angular distribution in the photon energy range $E_\gamma = 3.2 - 3.4$GeV and momentum transfer range $-t = 0.5 - 0.6$ GeV^2 as a function of the di-pion mass; all moments are dominated by the contribution of the $\rho(770)$ meson in the $p-$wave, while a structure associated to the $f_0(980)$ becomes evident in the Y_{10}, Y_{11} because of the interference between the *S* and *P* waves. Right: differential cross section associated to the *S* (top), *P* (middle) and *D* (bottom) waves.

MESON SPECTROSCOPY AT CLAS

Meson spectroscopy represents a significant part of the physics program of the CLAS Collaboration. Several exclusive final states have been measured looking for contribution of specific resonances. Since meson resonances are numerous, broad and often overlapping, the extraction of a single state requires sophisticated Partial Wave Analysis (PWA) techniques. Two different approaches have been applied to the analysis of the CLAS data. The first is based on the so-called *isobar model*, where the amplitudes that are fitted to the data are built assuming the decay of the resonance to the multi-particle final state occurs through a series of subsequent two-body decays. This approach has been widely adopted in meson spectroscopy studies as it allows to construct amplitudes for a number of complex final states starting from well known two-body decays. The second approach is based on the extraction of angular moments that are then analyzed in terms of partial waves using dispersion relations.

The isobar model approach was applied to CLAS data of three pion photoproduction on proton target [4]. This reaction channel is particularly interesting for the search for exotics. Evidence for the exotic $\pi_1(1600)$ was in fact reported from the analysis of three pion data recorded by the E852 Collaboration at BNL [5]. This first result was not confirmed by a later re-analysis of the same data with a PWA that includes a larger number of of waves [6], leaving an open question about the existence of this state. The results of the analysis of the CLAS data showed clear evidence for know resonances as the $a_2(1320)$ but no indication for the exotic π_1, setting an upper limit on the production cross section for the state of about 13.5 nb at 95% CL. Recently, new evidence for the same exotic state was reported by the COMPASS Collaboration [7], opening once more the debate on the real existence of this resonance.

The second PWA approach was applied to the analysis of the photoproduction of pion pairs [9, 10]. In this case moments of the di-pion angular distribution were evaluated from the data using an unbinned likelihood fit. Angular moments $\langle Y_{LM}\rangle$ are projection of the di-pion differential cross section onto specific spherical harmonics with total angular moment L and z-projection M. Different moments have different sensitivity to specific partial waves, depending on the values of L and M. For example, the $\langle Y_{10}\rangle$ and $\langle Y_{11}\rangle$ are particularly sensitive to the waves with lower angular momentum and specifically to the S, P and D waves. As an example, the first three moments of the di-pion angular distribution are shown in the left panels of Fig. 2. The dominant contribution of the $\rho(770)$ is clearly observed in all three moments, while a narrow structure below 1 GeV is found in the $\langle Y_{10}\rangle$ and $\langle Y_{11}\rangle$ moments. This structure is interpreted as evidence for the scaler $f_0(980)$ meson. This conclusion is supported by the partial waves that were evaluated from the experimental moments using dispersion relations and in particular by the structure observed in the S-wave shown in the top-right panel of Fig. 2. The strength of the S-wave in the mass region of the $f_0(1980)$ is found to be a factor of about 50 lower than the strength of the dominant P-wave.

Presently, other large statistics samples are being analyzed. Among these, data are being analyzed looking for coherent meson production on ^4He [11]. The final states that are being studied are the $\pi\eta$ and $\pi\eta'$ channels, where the strongest evidence for the exotic $J^{PC}=1^{-+}$ $\pi_1(1400)$ was reported by the E852 Collaboration [8]. These final states are produced by scattering the electrons of the primary Jefferson Lab beam on an 7 atm. ^4He gas target. The electron is scattered at "0" degrees, corresponding to very low four-momentum transfer and therefore quasi-real photons. The coherence of the reactions is ensured by detecting the recoiling ^4He in a radial TPC, which surrounds the high-pressure target. In this type of reactions, the excitation of nucleon resonances, which constitutes a major source of background, is significantly suppressed. In addition ^4He, being a spin and isospin zero object, acts as a spin and parity filter for the final state mesons. Data in this configuration were taken in the fall of 2009 and results are expected in the near future.

THE JEFFERSON LAB UPGRADE AND THE CLAS12 DETECTOR

Jefferson Lab is in the process of upgrading the energy of the accelerator from the 6 GeV, presently available, to 12 GeV. The upgrade plan includes the installation of new superconductive cavities to increase the energy of the two LINACs to 1.1 GeV and the upgrade of the magnets of the re-circulation arcs. This will lead the accelerator to deliver 11 GeV electron beam to the three existing experimental Halls and 12 GeV beam the new Hall D that is being built as part of the upgrade scope. The equipment of the Halls A,B and C will also be upgraded as part of the project. Hall B will hosts the newly designed CLAS12 detector. This spectrometer will retain the most significant features of the existing CLAS detector with enhanced capabilities in terms of kinematic coverage, resolution, particle identification and operating luminosity (10^{35}cm^{-2}s^{-1}). CLAS12 will consist of a forward spectrometer and a central spectrometer, based respectively on a toroidal and solenoidal magnet. The forward spectrometer, including new drift chambers, time of flight scintillators, Cerenkov counter and electromagnetic calorimeter, will provide the detection capability for the scattered electron and forward-going hadrons. The central spectrometer is designed to detect large-angle recoiling hadrons, using a silicon tracker, time of flight detector and a neutron detector. Most of these elements are part of the basic equipment that is part of the Jefferson Lab upgrade project, funded by the US Department of Energy for a total of 310 million dollars.

While the construction of the new detector components has already started to be ready for the beginning of operation planned for 2013-2014, upgrades to the approved equipment have already been proposed by several participating institutions and are being designed. These include a RICH detector to extend the particle identification in the kaon sector, the above-mentioned neutron detector, the micromegas central tracker which will complement the silicon tracker and the *forward photon tagger* that will allow to continue the photoproduction program at 12 GeV.

QUASI-REAL PHOTOPRODUCTION WITH CLAS12

The existing Hall B photon tagger will not be upgraded to operate at 12 GeV, both because of cost consideration and space constraints. An alternative possibility to continue measuring photoproduction reactions at higher energy is to use

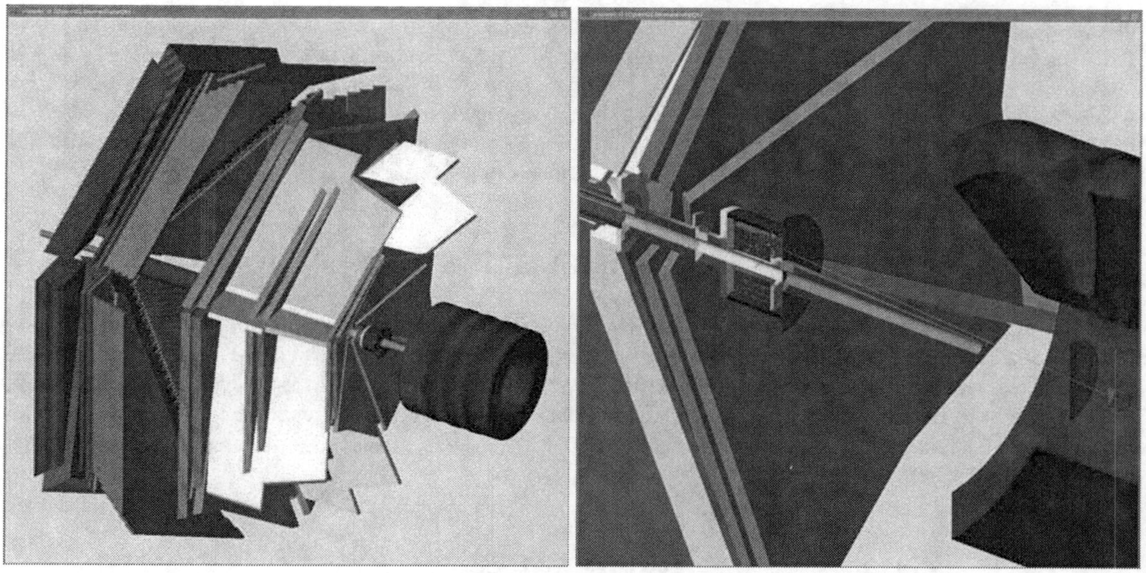

FIGURE 3. GEANT4 visualization of CLAS12 and of the low Q^2 photon tagging facility. The system consist of an electromagnetic calorimeter, shown in green in the left panel, a veto counter and a tracker, shown in magenta, located between the forward and central CLAS12 spectrometer. A tungsten cone, shown in orange, shield the calorimeter by Möller electrons. In the right panel, a longitudinal section of the CLAS12 setup is shown for an event with 2 GeV, 3° electron impinging on the tagging facility.

electro-production at very small electron scattering angles, of the order of few degrees or smaller. In this kinematics, the four-momentum transfer, Q^2, associated to the reaction is very small and the virtual photon which is produced can be considered as quasi-real. This technique has been widely used in high energy experiments as H1 and ZEUS at DESY or COMPASS at CERN and has several advantages with respect to the traditional Bremmstrahlung photon beams. The virtual photons associated with small-angle electrons have an intrinsic and sizeable degree of linear polarization with values that depend on the photon energy. The linear polarization can be established on an-event-by-event basis by measuring the electron energy and scattering plane. This information is crucial to understand the meson production mechanism and provides important constraints in the PWA, reducing therefore ambiguities in the physics analysis. In addition, the usage of electron beams gives the possibility of running with high luminosity on thin targets as for example, the ^{4}He gas target mentioned before.

To exploit this possibility, a low-Q^2 tagging facility or *forward tagger* has been proposed for CLAS12. The facility will be used to detect electrons between 2.5 and 4.5 degrees in the energy range between 500 MeV and 4.5 GeV. The corresponding virtual photons will be in the energy range 6.5-10.5 GeV, which is adequate for the production of mesons with masses around 2 GeV. The Q^2 will be between 0.01 and 0.3 GeV^2 and the degree of linear polarization will vary between 10 and 65%. The forward tagger will consist of an electromagnetic calorimeter to identify the electrons and measure their energy, a veto counter to distinguish electrons from photons and a tracker to measure the electron angles precisely and therefore determine the photon linar polarization. The facility will be located between the CLAS12 forward and central spectrometers and will be designed to be compatible with the detector base equipment.

The use of the *forward tagger* to measure low-Q^2 electrons in conjunction with the detection of multiparticle final states in CLAS12 will open new possibilities for high quality physics beyond the already approved program. The large acceptance for charged and neutral particles, the excellent momentum resolution and the particle-identification capabilities will allow the study of a broad range of exclusive reactions and continue the study of meson and baryon spectroscopy at 12 GeV.

SUMMARY

Meson spectroscopy is a fundamental tool to deepen our understanding of strong interaction. The investigation of meson production and the search for exotics quark-gluon configurations is presently an important part of the physics program that is being carried on with CLAS at 6 GeV. First results from PWA analysis of multiparticle final states have been published, demonstrating the feasibility of these complex analyses in CLAS and contributing to the worldwide effort in this field. New high statistics data sets are already available and are now being analyzed.

The extension of this physics program at 12 GeV is being proposed based on quasi-real photoproduction. For this purpose a new detector component, called *forward tagger*, is being designed to measure electrons scattered at very small angles and therefore tag the corresponding virtual photons. This facility will allow to run high intensity beam on proton and light-nuclei targets, giving access to a rich physics program.

ACKNOWLEDGMENTS

We would like to acknowledge the outstanding efforts of the staff of the Accelerator and the Physics Divisions at Jefferson Lab that made this experiment possible. This work was supported in part by the Italian Istituto Nazionale di Fisica Nucleare, the French Centre National de la Recherche Scientifique and Commissariat à l'Energie Atomique, the U.S. Department of Energy and National Science Foundation, and the Korea Science and Engineering Foundation. Jefferson Science Associates, LLC, operates Jefferson Lab for the United States Department of Energy under U.S. DOE contract DE-AC05- 060R23177.

REFERENCES

1. J. J. Dudek, R. G. Edwards, M. J. Peardon, D. G. Richards and C. E. Thomas, Phys. Rev. D **82**, 034508 (2010).
2. A. Szczepaniak and M. Swat, Phys. Lett. B516 (2001) 72.
3. B. A. Mecking *et al.* [CLAS Collaboration], Nucl. Instrum. Meth. A **503**, 513 (2003).
4. M. Nozar *et al.* [CLAS Collaboration], Phys. Rev. Lett. **102**, 102002 (2009).
5. G. S. Adams *et al.* [E852 Collaboration], Phys. Rev. Lett. **81**, 5760 (1998).
6. A. R. Dzierba *et al.*, Phys. Rev. D **73**, 072001 (2006).
7. M. Alekseev *et al.* [COMPASS Collaboration], Phys. Rev. Lett. **104**, 241803 (2010).
8. D.R. Thompson *et al.* [E852 Collaboration], Phys. Rev. Lett. **79**, 1630 (1997).
9. M. Battaglieri *et al.* [CLAS Collaboration], Phys. Rev. Lett. **102**, 102001 (2009).
10. M. Battaglieri *et al.* [CLAS Collaboration], Phys. Rev. D **80**, 072005 (2009).
11. S. Stepanyan *et al.*, E-07-009: *Meson spectroscopy in the Coherent Production on* ^{4}*He with CLAS*, http://www.jlab.org/exp_prog/proposals/07/PR-07-009.pdf. K. Hafidi *et al.*, E-08-024: *Deeply Virtual Compton Scattering off* ^{4}*He*, http://www.jlab.org/exp_prog/proposals/08/PR-08-024.pdf.
12. *Hadron Spectroscopy with low-*Q^2 *electro-scattering at CLAS12*, JLAB PAC35 Letter-of-Intent, Contact Person: M. Battaglieri.

Vector Meson Photoproduction on Nuclei

Chaden Djalali

Department of Physics and Astronomy, University of South Carolina, 712 Main Street, Columbia, SC 29208, USA

Abstract. Chiral symmetry, which is spontaneously broken in vacuum, is predicted to be partially restored in ordinary nuclear matter. The properties of vector mesons, such as their masses and/or widths are expected to change in the medium. The photoproduction reaction off nuclei is a very clean way of producing the ρ, ω and φ mesons, and detect them via their hadronic or leptonic decays. The leptonic decay to e^+e^- has a small branching ratio but has the advantage of being free from final state interactions. One critical aspect in all these experiments is the correct determination of the shape and magnitude of the combinatorial background. The in-medium mass distributions and yields are compared to those measured in vacuum. No significant shift is observed in the masses of the mesons, however substantial increase in their widths is reported.

Keywords: Medium modifications, vector mesons, di-lepton decay.
PACS: 11.15.Ex, 13.20.-v, 13.40.Gp, 13.60.Le, 14.40.-n, 25.20.Lj.

INTRODUCTION

Quantum Chromodynamics (QCD) has by now been firmly established as the theory of the strong interaction. It is remarkably successful in describing high-energy and short-distance-scale regime involving quarks and gluons. However it has been a major challenge to apply QCD in the low energies and large-distance-scale regime. Lattice QCD, which is the proper way of treating the non-perturbative regime of QCD, has recently made tremendous progress[1] but is still far from fully explaining all the experimental observations. Meanwhile, effective theories based on the fundamental symmetries of the QCD lagrangian are used to understand the low energy regime. One of these symmetries is Chiral Symmetry, which is spontaneously broken in the vacuum. Several QCD-inspired models predict a partial restoration of Chiral Symmetry in nuclear matter. The properties of hadrons in the medium are expected to change from their free-space values. The predicted changes in the quark-gluon sector are related to observables in the hadronic sector using QCD sum rules.[2] The in medium spectral functions of hadrons are predicted to change, showing mass shifts and/or in-medium broadening. For the light vector mesons such as the ρ, ω, and φ, additional structures appear in the spectral function due to the coupling of these mesons to nucleon resonances.[3] Many different experiments using a variety of probes have been studying the production of the light vector mesons in the medium in order to test these predictions. Heavy-ion reactions have the advantage of probing media with high densities and high temperatures, however the interpretation of the data is complicated by the fact that the evolution of the system is extremely complex and far from equilibrium. The experiments using elementary probes, such as γ, π and p, have the advantage of leaving the medium in equilibrium and the predicted medium modifications for the vector mesons nuclei are large enough to be detected.

VECTOR MESONS IN THE MEDIUM

The existence of the light vector mesons was predicted in the late fifties, early sixties. Sakurai[4] using gauge field theories and applying conserved current to the strong interactions predicted the existence of vector mesons. The interpretation of the first electromagnetic form factors of the nucleons measured in electron scattering led Nambu[5] to give an indirect evidence for the existence of an isoscalar vector meson decaying into three pions ($\omega \rightarrow 3\pi$). Frazer and Fulco[6] proposed the existence of an isovector meson decaying into two pions ($\rho \rightarrow 2\pi$). Shortly after, the ρ was discovered as a I=1, J=1 π–π resonance at the BNL-Cosmotron[7], and the ω as a T=0 3π resonance in anti-proton

12th International Conference on Meson-Nucleon Physics and the Structure of the Nucleon (MENU 2010)
AIP Conf. Proc. 1374, 61-65 (2011); doi: 10.1063/1.3647099

proton annihilation at the LBL 72" H-bubble chamber.[8] The following year, the φ meson was discovered at the BNL 20" H-bubble chamber.[9] These vector mesons have the same quantum numbers as the photon. The analogies between photon- and hadron- induced processes such as a total cross section varying very slowly with energy, a diffractive pattern at small t and shadowing, lead to the Vector Meson Dominance Model (VMD or VDM). The photon fluctuates between a "bare photon" and its hadronic component, which undergoes conventional hadronic interactions. The photo production of vector mesons on the nucleon has been extensively studied at different Eγ and t. A coherent picture of the reaction mechanism has emerged.[10,11,12]

The nucleus can be used as a "laboratory" to study the effect of the nuclear medium (a strongly interacting medium in equilibrium at T=0 and known density) on the properties of the vector mesons. There are many models predicting significant medium modifications of the vector mesons. Two recent comprehensive reviews[13,14] cover all the different models and compare their predictions with the latest experimental results. Hatsuda and Lee[15] have predicted that the mass of the vector mesons drops linearly with the density. Brown and Rho[16] conjectured that the masses of light vector mesons (ρ, ω) scale universally as a function of density and/or temperature.

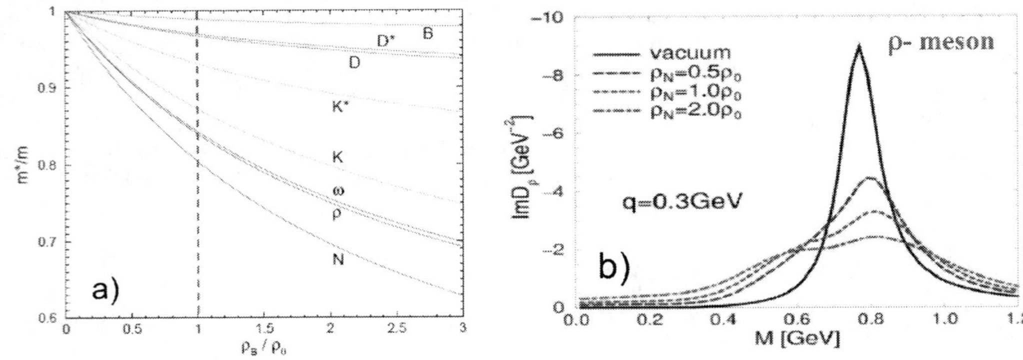

Figure 1. Predicted medium effects: (a) QMC mass modifications of hadrons, (b) ρ-meson spectral function, as a function of nuclear density.

The Quark-meson coupling model (QMC), is a phenomenological theory in which quarks and gluons are confined in a "bag" inside non-perturbative QCD vacuum.[17] It is interesting to note that at normal nuclear density (ρ_0), the ρ and ω -masses have dropped by ≈ 15%; the nucleon-mass has dropped by ≈ 20% and the D-meson mass has dropped by ≈ 3% (Fig.1-a). Hadronic models use a purely hadronic description of the mesons in the medium. The in medium self energy of the meson receives contributions from the low-energy particle-hole (p-h) excitations and the high energy nucleon–antinucleon excitations. These models provide much "richer" information about the in-medium properties of the mesons. The spectral functions are modified in non-trivial manners such as spectral shifts, spectral broadening and new spectral peaks (Fig.1-b).[18] Experimentally, one needs to measure and compare the properties of the hadrons in the vacuum and in different media (T and/or $\rho \neq 0$). Table 1 shows the properties of the vector mesons in the vacuum.

TABLE (1). Properties of the light vector mesons. Jπ = 1-

Mesons	**Mass (MeV/c^2)**	**Γ (MeV/c^2)**	**cτ (fm)**
ρ	775.49±0.34	149.1±0.8	1.3
ω	782.65±0.12	8.49±0.08	23.2
φ	1019.45±0.02	4.26±0.04	46.2

The ρ meson has a very short lifetime leading to the largest probability of decaying in the medium, while the ω and φ mesons will mostly decay outside the medium in which they are produced. In order to study their in-medium properties, it is crucial to choose a proper reaction allowing the observation of slow moving ω and φ mesons. When produced with large momenta, the ω and φ mesons (due to their longer lifetimes) will decay outside the nucleus and therefore be detected with properties of their free-space values. However, information on the in-medium width of these mesons can be extracted from their interactions with the nucleons as the mesons escape the nucleus using the transparency ratio T defined (for a vector meson V) as $T_A = (\sigma_{\gamma A\rightarrow VX} / A\cdot \sigma_{\gamma N\rightarrow VX})$. $\sigma_{\gamma A\rightarrow VX}$ is the inclusive nuclear vector-meson (V) photo-production cross section and $\sigma_{\gamma N\rightarrow VX}$ is cross section on a free nucleon. The transparency ratio T is a measure of the loss of vector-meson flux via inelastic processes in the nucleus, and is related to the absorptive part of the meson-nucleus potential. Extracting the in-medium meson width from the A dependence of the ratio T requires comparison with theoretical calculations.[19]

PHOTOPRODUCTION EXPERIMENTS

Table 2 list some experiments which have studied the vector mesons by photoproducing them on different nuclei. Contrary to hadronic experiments where most of the reactions occur at the surface of the nucleus, in photoproduction experiments, the reactions occur throughout the whole nucleus leading to larger effective densities being probed. The experiments having hadrons in the final state must carefully treat the final state interactions. The e^+e^- decay channel has the advantage of being free from final state interactions but has a very small branching ratio ($\sim 10^{-5}$).

TABLE (2). Phoproduction experiments

Experiment	Reactions	Decay Channel
TAGX [20,21]	γ + H, ^{3}He, C $\rightarrow \rho$ + X	$\rho \rightarrow \pi^+\pi^-$
SPring-8 [22]	γ + Li, C, Al, Cu $\rightarrow \phi$ + X	$\phi \rightarrow K^+K^-$
CBELSA-TAPS [23,24,25]	γ + H, C, Ca, Nb, Pb $\rightarrow \omega$ + X	$\omega \rightarrow \pi^0\gamma \rightarrow \gamma\gamma\gamma$
JLab-g7 [26,27,28]	γ + D, C, Ti, Fe, Pb $\rightarrow (\rho, \omega, \phi)$ + X	$\rho, \omega, \phi \rightarrow e^+e^-$

All experiments have to deal with large combinatorial backgrounds. The shape and magnitude of this background have to be carefully determined and subtracted from the raw yields.

TAGX Experiment

One of the first experiments that attempted to study the medium modification of the ρ-meson with elementary reactions, was carried out by the TAGX collaboration. This experiment at the 1.3-GeV INS Electron Synchrotron (Institute for Nuclear Study, Tokyo University) used a tagged photon beam in the energy range of 600-1120 MeV to study the $\gamma A \rightarrow \pi^+\pi^- X$ reaction on H, ^{3}He and ^{12}C targets. They originally reported a large decrease of the ρ-meson mass (~20%) in ^{3}He where the pion pairs result from sub-threshold ρ-meson production and decay.[20] However these results were met with skepticism due to the final-state interactions on the pion pairs and the large mass shift observed on such a light nucleus as ^{3}He. A second analysis[21] based on the longitudinal polarization of the ρ mesons reported a smaller but still significant decrease in the mass of the meson in ^{3}He. Surprisingly no effects are observed in ^{12}C.

CBELSA-TAPS Experiment

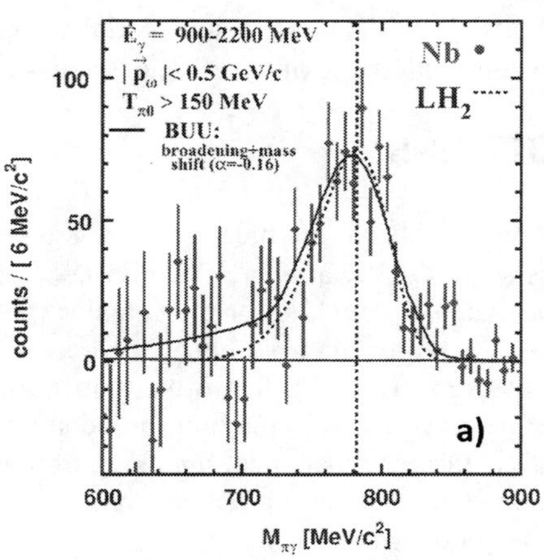

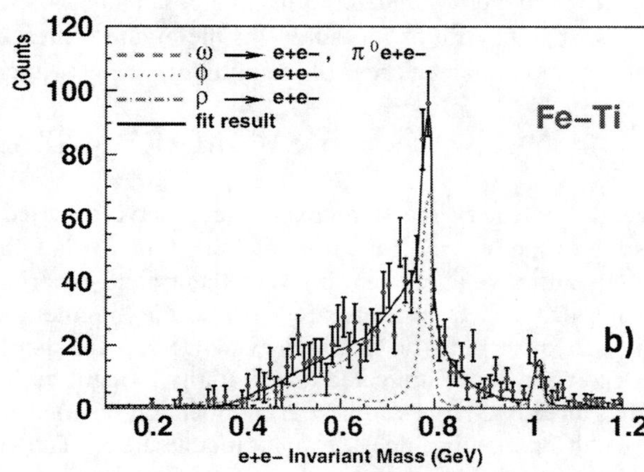

Figure 2. Medium modification of vector mesons: (a) CBELSA/TAPS mass spectrum for the ω, (b) JLab(g7) mass spectrum for ρ, ω and φ.

The CBELSA/TAPS collaboration at the electron stretcher accelerator (ELSA) in Bonn used the $\gamma A \rightarrow \pi^0\gamma X$ reaction to study the ω meson in-medium behavior in LH_2 and Nb. The branching ratio for $\omega \rightarrow \pi^0\gamma$ decay is the order of 9%. This channel provides a clean and exclusive mode to study the ω in-medium properties. The ρ is highly suppressed since the $\rho \rightarrow \pi^0\gamma$ branching ratio is $< 10^{-3}$. However, a serious disadvantage is the possible strong final-stage interactions of the π^0 meson within the nucleus. In order to maximize the in nucleus decay probability, slow-moving ω-mesons with $|p_\omega| < 0.5$ GeV/c were selected. In a first analysis, a shoulder on the low-mass side of the ω peak was found in the Nb target and was interpreted as medium modified ω-mesons with a downward mass shift of 14%.[23,24] In a reanalysis of the data,[25] an improved background determination and subtraction method is applied. The earlier claim of an in medium lowering of the *ω* mass is not confirmed. The authors claim that the strong broadening of the *ω* meson in the nuclear medium due to inelastic processes suppresses the contributions to the observed *ω* signal from the interior of the nucleus. The line-shape analysis is therefore less sensitive to a direct observation of in-medium modifications (Fig. 2-a). Data with much higher statistics will be needed to gain further insight. A corresponding experiment has been performed at the Mainz Microtron MAMI-C electron accelerator using the Crystal Ball/TAPS detector setup.

SPring-8 (LEPS)

The photo-production of φ mesons from Li, C, Al and Cu targets has been measured at Eγ = 1.5 – 2.4 GeV at SPring-8 (LEPS), in the $\gamma A \rightarrow K^+K^-X$ channel.[22] Using a Glauber-type model calculation, the analysis of the measured transparency ratios, leads to an in-medium φ-nucleon cross section of the order of $\sigma_{\varphi N}$ =37[+17,-11] mb. This is much larger than the free-space value used as an input in most transport model calculations. The Spring-8 results are compatible within statistical uncertainties with the JLab results**Error! Bookmark not defined.** and seem to indicate an in-medium width for the ϕ as large as Γϕ~ 70 MeV.

JLAB Experiment

The J-Lab E01-112 experiment was conducted in Hall B at the Thomas Jefferson National Accelerator Facility (TJNAF). The vector mesons were photo-produced on LD_2, C, Fe, Ti and Pb. The outgoing vector mesons were reconstructed from their decay to e^+e^- pairs. The CLAS detector[29] is ideal for this experiment owing to its ability to detect multi-particle final states and its high e/π rejection factor. Effective discrimination between lepton and pion pairs to the level of 10^{-7} is easily achieved. The same-charge pairs method was successfully used to determine the combinatorial background. Once the combinatorial background subtracted, the spectrum mainly consists of the ρ, ω and ϕ mesons (Fig.2-b). The fits use the spectral shape calculated by the semi-classical Boltzmann-Uehling-Uhlenbeck (BUU) transport model developed by the Giessen group.[30] The fits describe the data very well and are consistent with the PDG values (no mass shift) and the widths are consistent with the collisional broadening (ΔΓ~ +70 MeV in Fe) [26,27]. Recent calculations by the Texas A&M group[31] nicely reproduce g7 data. Confirms no major medium effect (beyond standard collisional broadening) expected for momenta Pρ > 1 GeV.

The transparency ratios measured for the ω and ϕ mesons**Error! Bookmark not defined.** show a substantial widening in the medium (more than an order of magnitude). These results are consistent with those of TAPS and Spring-8.

SUMMARY AND CONCLUSIONS

Several elementary reactions experiments have reported medium modifications for the ρ, ω and ϕ mesons. A broadening of the widths of the mesons in the nuclear medium has been reported by almost all experiments. For the ρ meson quantitative agreement between theory and experiment has been achieved while for the ω and φ meson the broadening deduced from transparency-ratio measurements is a factor 2-3 larger than predicted theoretically. The majority of experiments do not find evidence for mass shifts in the medium. Looking for in-medium mass shifts turns out to be more complicated than initially thought because of the observed large in-medium broadening. The ρ meson has already a large width in vacuum and in the medium its width is further broadened. Speaking of mass shift versus width broadening doesn't make much sense. The ω and φ meson are narrow in vacuum but are found to widen by a factor of the order of 15 in nuclear matter. This drastically reduces the branching ratio for in-medium decays into the channels of interest (di-leptons, πγ), making these experiments less sensitive to a direct observation of medium modifications. The latest models predict that most of the medium modifications happen when the meson have momenta less than 0.7 GeV/c with respect to the medium. Next generation experiments with much higher statistics and improved acceptance for low momenta mesons are needed to obtain reliable results. Several experiments are planned at JPARC, JLab, RHIC, FAIR and CERN. During the next decade, substantial theoretical and experimental efforts will continue to be being carried out in this very active field.

ACKNOWLEDGMENTS

The author would like to thank his close collaborators D. Weygand, M. Wood, R. Nasseripour and M. Paolone. Special thanks go to S. Strauch, U. Mosel, E. Oset, M. Soyeur, V. Metag and R. Hayano for extensive discussions on issues related to medium modifications of hadrons. This work was supported in part by the National Science Foundation under the award NSF-0555604.

REFERENCES

1. S. Durr *et al.*, "Ab Initio Determination of Light Hadron Masses" in *Science* **322**, 1224 (2008).
2. Y. Kwon et al, *Phys. Rev. C* **78**, 055203 (2008).
3. R. Rapp, J. Wambach, and H. van Hees, "The Chiral Restoration Transition of QCD and Low Mass Dileptons" in *Landolt-Bo rnstein New Series I-23/A* (Springer, Berlin, 2010), p. 4-38.
4. J. J. Sakhurai, *Annals of Phys.* **11**, 1 (1960).
5. Y. Nambu, *Phys. Rev.* **106**, 1366 (1957).
6. W. Frazer and J. Fulco, *Phys. Rev. Lett.* **2**, 365 (1959).
7. A. R. Erwin et al., *Phys. Rev. Lett.* **6**, 628 (1961).
8. B. C. Maglic et al., *Phys. Rev. Lett.* **7**, 178 (1961).
9. L. Bertanza et al., *Phys. Rev. Lett.* **9**, 180 (1962).
10. M. Battaglieri et al., *Phys. Rev. Lett.* **87**, 172002 (2001).
11. M. Battaglieri et al., *Phys. Rev. Lett.* **90**, 022002 (2003).
12. E. Anciant et al., *Phys. Rev. Lett.* **85**, 4682 (2000).
13. R. S. Hayano and T. Hatsuda, *arXiv:nucl-ex/***0812.1702v2** (2009).
14. S. Leupold, V. Metag and U. Mosel, *arXiv:*nucl-th/**0907.2388v1** (2009).
15. T. Hatsuda and S. H. Lee, *Nucl. Phys. B* **394**, 221 (1993).
16. G. E. Brown and M. Rho, *Phys. Rev. Lett.* **66**, 2720 (1991).
17. K. Saito et al., *Prog. Part. Nucl. Phys.* **58**,1 (2007).
18. R. Rapp et al., *Eur. Phys. Jour. A* **6**, 415 (1999).
19. M. Kaskulov et al., *Phys. Rev. C* **75**, 064616 (2007);
    P. Muehlich and U. Mosel, *Nucl. Phys. A* **765**,188 (2006).
20. G. J. Lolos *et al., Phys. Rev. Lett.* **80**, 241 (1998).
21. G. M. Huber *et al., Phys. Rev. C* **68**, 065202 (2003).
22. T. Ishikawa et al., *Phys. Lett. B* **608**, 215 (2005).
23. D. Trnka *et al., Phys. Rev. Lett.* **94**, 192303 (2005).
24. M. Kotulla et al., *Phys. Rev. Lett.* **100**, 192302 (2008).
25. M. Nanova et al., *Phys. Rev. C* **82**, 035209 (2010).
26. R. Nasseripour *et al., Phys. Rev. Lett.* **99**, 262302 (2007).
27. M. H. Wood et al., *Phys. Rev. C* **78**, 015201 (2008).
28. B. A. Mecking et al., *Nucl. Instrum. Methods Phys. Res., Sect. A* **503**, 513 (2003).
29. P Muehlich et al., *arXiv:nucl-th/***0210079** (2002).
30. F. Riek et al., *Phys Let B* **677**, 116 (2009);
    F. Riek et al., arXiv:**1003.0910v1** (2010)

Hadrons and Holography

Joshua Erlich

College of William and Mary, P.O. Box 8795, Williamsburg, VA 23187-8795, USA

Abstract. Holographic QCD is an extra-dimensional approach to modeling QCD resonances and their interactions. Holographic models encode information about chiral symmetry breaking, Weinberg sum rules, vector meson dominance, and other phenomenological features of QCD. There are two complementary approaches to holographic model building: a top-down approach which begins with string-theory brane configurations, and a bottom-up approach which is more phenomenological. In this talk I will describe the techniques used to build holographic models of QCD and the tools used to calculate observables in those models.

Keywords: Holographic QCD, AdS/QCD

WHAT IS HOLOGRAPHIC QCD?

Holographic QCD, also known as AdS/QCD, is an extra-dimensional approach to modeling the light hadronic resonances in QCD, motivated by the AdS/CFT correspondence in string theory [1]. Holographic QCD models combine several features of previous approaches to modeling the spectrum and interactions of light hadrons, including chiral symmetry breaking, hidden local symmetry [2], large-N, and the Weinberg sum rules.

There are two complementary classes of AdS/QCD models: top-down models rooted in string theory, and bottom-up models which are more phenomenological. The benefit of top-down models is that both sides of the AdS/CFT duality are often well understood. The benefit of bottom-up models is that there is more freedom to build in properties of QCD.

Top-Down AdS/QCD

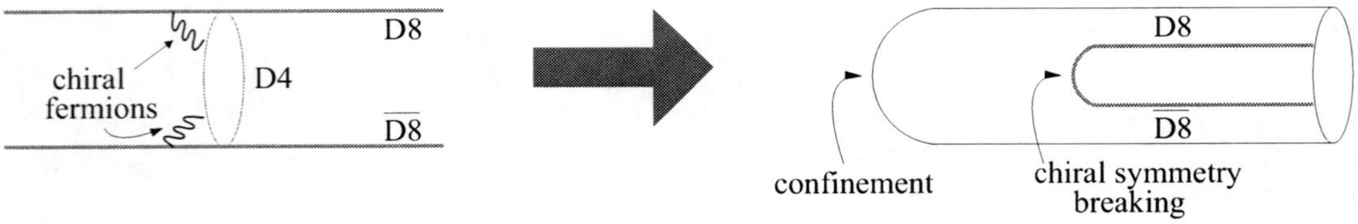

FIGURE 1. Brane configuration in the Sakai-Sugimoto model. The D4-branes create a spacetime horizon, forcing the D8 and $\overline{\text{D8}}$-branes to intersect.

In the top-down approach, a brane configuration in string theory is engineered whose low-energy spectrum of open-string fluctuations has a known field-theoretic interpretation. According to Maldacena's AdS/CFT correspondence, for certain brane configurations describing large-N gauge theories with large 't Hooft coupling g^2N, a dual description exists in terms of supergravity on a fixed spacetime background [1]. The basic dictionary between the dual theories was mapped out independently by Witten [3] and by Gubser, Klebanov and Polyakov [4]. There now exists a large number of examples of field theories with gravity duals. Despite the fact that prototypical examples of the AdS/CFT correspondence describe supersymmetric conformal theories ("CFT" in AdS/CFT is short for conformal field theory), neither conformal invariance nor supersymmetry is essential for the existence of a gravitational dual. The field theory can be confining with chiral symmetry breaking, which in those respects is similar to QCD. Some examples of

12th International Conference on Meson-Nucleon Physics and the Structure of the Nucleon (MENU 2010)
AIP Conf. Proc. 1374, 66-72 (2011); doi: 10.1063/1.3647100

confining theories with known supergravity duals are the $\mathcal{N}=1^*$ theory of Polchinski and Strassler [5], the Klebanov-Strassler cascading gauge theory [6], the D3-D7 system of Kruczenski *et al.* [7], and the D4-D8 system of Sakai and Sugimoto [8].

The Sakai-Sugimoto model is so far the example of the AdS/CFT correspondence most closely related to QCD. Ignoring the gravitational backreaction, the system includes a stack of N D4-branes (Dp-branes extend in p spatial dimensions) wrapped on a circle on which fermions satisfy antiperiodic boundary conditions which break supersymmetry. From an effective 3+1 dimensional point of view, the massless spectrum includes SU(N) gauge fields, but the fermions and scalars are massive. This theory was analyzed by Witten shortly after the initial AdS/CFT conjecture [9]. Sakai and Sugimoto added N_f D8-branes and N_f $\overline{\text{D8}}$-branes transverse to the circle. The D8 and $\overline{\text{D8}}$-branes intersect the D4-branes on 3+1 dimensional manifolds at definite positions along the circle, as in Fig. 1. The massless fluctuations of open strings connecting the D4 and D8 or $\overline{\text{D8}}$-branes at their intersections describe 3+1 dimensional chiral fermions, with opposite chirality at the D8 and $\overline{\text{D8}}$-branes. This is the Sakai-Sugimoto model [8].

In the supergravity limit ($N \to \infty$, $g^2N \gg 1$) with $N_f \ll N$, the D4-branes generate a horizon which effectively cuts off the spacetime geometry. The location of this horizon sets the mass scale of certain fluctuations in the 3+1 dimensional theory, including glueballs and some of their excitations. Hence, the location of the horizon is related to the confinement scale in the model. For the geometry to be smooth the location of the horizon is correlated with the size of the compact circle on which the D4-branes wrap [7, 9]. This has the implication that Kaluza-Klein modes around the circle, which do not correspond to QCD states, have masses comparable to the hadrons which we would like to model. Hence, the five-dimensional nature of the effective theory on the D4-branes becomes apparent at the same scale as the hadron masses we are interested in. This is an important distinction between this theory and QCD, and the extraneous states are normally ignored when predicting QCD observables from the model.

Because of the horizon, the D8 and $\overline{\text{D8}}$-branes intersect, which reflects the breaking of the SU(N_f)$\times$SU(N_f) chiral symmetry in the theory. The perturbative massless spectrum is that of SU(N) QCD with N_f flavors of quarks. If we ignore the Kaluza-Klein modes around the circle, the fluctuations of 4+1 dimensional SU(N_f)$\times$SU(N_f) gauge fields on the D8 and $\overline{\text{D8}}$-branes are identified with vector mesons, axial-vector mesons, and pions. The quantum numbers of the corresponding states can be identified with symmetries of the D-brane system. 3+1 dimensional parity, for example, is identified with a 4+1 dimensional parity, which also exchanges the two sets of SU(N_f) gauge fields. Then the effective 3+1 dimensional action on the D8-branes describes the effective action for the light mesons, and easily allows for the calculation of decay constants (f_π, F_ρ, *etc.*) and couplings (*e.g.* $g_{\rho\pi\pi}$). Most results agree relatively well with experimental data (at around the 10-20% level). The light baryons in AdS/QCD have been identified with solitonic configurations of the 4+1 dimensional fields, which are closely analogous to the baryons of the Skyrme model [8].

Bottom-Up AdS/QCD

In the bottom-up approach, we begin with the observation that the Kaluza-Klein modes of fields in an extra dimension might be identified with the radial excitations of hadrons in a confining gauge theory. As in the top-down approach, the quantum numbers of those excitations depend on the transformations of the Kaluza-Klein modes under corresponding 5D symmetries. In the hard-wall model [10, 11], we begin with a 5D SU(2)$\times$SU(2) gauge theory. To reproduce the discrete spectrum of hadronic excitations, the spacetime geometry must be such that the spectrum of Kaluza-Klein gauge fields is discrete. The metric of the five-dimensional Anti-de Sitter spacetime (AdS$_5$) can be written,

$$ds^2 = \frac{R^2}{z^2}\left(dx_\mu \eta^{\mu\nu} dx_\nu - dz^2\right), \tag{1}$$

where $\eta_{\mu\nu}$ is the 3+1 dimensional Minkowski metric with components diag(1,-1,-1,-1); R is the AdS curvature, which we will normalize to 1. The isometries of AdS$_5$ were identified with the conformal symmetry of $\mathcal{N}=4$ supersymmetric Yang-Mills theory in the original AdS/CFT correspondence [1]. For simplicity, we take as the spacetime geometry a slice of AdS$_5$ between an ultraviolet cutoff scale $z = \varepsilon$ and an infrared scale $z = z_m$ related to Λ_{QCD}. Other choices for the geometry may better match various aspects of QCD (such as running of the QCD coupling and the Regge spectrum [12]). In addition to its simplicity, the AdS spacetime has been motivated by appealing to the conformal invariance of the asymptotically free theory in the ultraviolet (UV), and also to hints of conformal behavior at low energies (*e.g.* [13]), based on measurements of the running QCD coupling at JLab [14].

The SU(2)×SU(2) gauge invariance is related to the approximate SU(2)×SU(2) chiral symmetry of the up and down quarks. (This may be extended to SU(3)×SU(3) in order to include the strange quark.) In order to break the chiral symmetry we introduce a scalar field that transforms in the bifundamental representation of the chiral symmetry, in analogy with the operator $\bar{q}q$. If this field is arranged to have a nonvanishing background profile, the gauge invariance is spontaneously broken. The 4+1 dimensional action takes the form,

$$S = \int d^5x\sqrt{-g}\left(-\frac{1}{2g_5^2}\mathrm{Tr}\left(L_{MN}L^{MN} + R_{MN}R^{MN}\right) + \mathrm{Tr}\left(|D_MX|^2 + m_X^2|X|^2\right)\right), \tag{2}$$

where L_{MN} and R_{MN} are the field strengths of the two sets of SU(2) gauge fields and m_X^2 is the squared mass of the bifundamental field X; contractions of indices are by the AdS$_5$ metric.

The scaling dimension of the operator $\bar{q}q$ in the ultraviolet is three. By the AdS/CFT dictionary the mass of a field X with scaling dimension Δ is given by $m_X^2 = \Delta(\Delta-4)$ in units of the AdS curvature, or in our case $m_X^2 = -3$. As a result of the curvature of Anti-de Sitter space, the negative squared mass does not lead to instability [15]. It is important to note that the choice $m_X^2 = -3$ is made here for definiteness, but it is not necessary to fix m_X^2 in this way. The AdS/CFT correspondence in the classical limit is not valid for QCD with finite N. Furthermore, reference to the scaling dimension in the ultraviolet ignores the effects of renormalization. However, for definiteness we may still choose to apply the AdS/CFT correspondence to fix parameters in the model.

The equations of motion for the z-dependent scalar field background, with the gauge fields turned off, are,

$$\partial_z\left(\frac{1}{z^3}\partial_z X(z)\right) + \frac{3}{z^5}X(z) = 0. \tag{3}$$

The solutions for the scalar field background are,

$$X(z) = \left(\frac{m_q}{2}z + \frac{\sigma}{2}z^3\right), \tag{4}$$

where m_q and σ are arbitrary. By the AdS/CFT correspondence, the coefficient of the solution with divergent action has the interpretation of the source for the corresponding operator [3, 4]. The quark mass is a source for the operator $\bar{q}q$, so m_q has the interpretation of the quark mass (which for simplicity is isospin-preserving in the model). Similarly, the coefficient of the finite-action (*i.e.* normalizable) solution has the interpretation of the expectation value of the corresponding operator, so σ has the interpretation of the chiral condensate $\langle\bar{q}q\rangle$. The factors of $1/2$ in (4) are chosen such that the product of coefficients, $m_q\sigma/4$, is properly normalized. A more accurate scaling of the quark mass and chiral condensate with N can be incorporated by rescaling the model parameters m_q and σ appropriately [16]. In any case, since the hard-wall model is phenomenological and does not follow from a precise AdS/CFT correspondence, the physical identification of the parameters m_q and σ is not precise.

If we want, we can fix g_5 by comparison with perturbative QCD at large $-q^2$, although the hard-wall model is not expected to be valid at high energies. The vector current two-point function at one-loop in SU(N) QCD with two flavors, valid for large $-q^2$, is

$$i\int d^4x\, e^{iq\cdot x}\langle J_\mu^a(x)J_\nu^b(0)\rangle = \left(q_\mu q_\nu - g_{\mu\nu}q^2\right)\delta^{ab}\frac{N}{24\pi^2}\log(-q^2). \tag{5}$$

The AdS/QCD prediction of the vector current two-point function is precisely of the perturbative form of Eq. (5) if we set $g_5^2 = \frac{24\pi^2}{N} = 8\pi^2$ when $N = 3$ [10, 11]. It is not necessary to fix g_5 by matching to the ultraviolet, just as it was not necessary to fix the mass of the field X by matching to the conformal dimension of the operator $\bar{q}q$. These choices are made for definiteness, but they may be relaxed. If we do fix m_X^2 and g_5 this way, the model has three remaining parameters. The masses of the Kaluza-Klein modes are identified with the masses of the hadrons with the same quantum numbers, and are determined by solutions to the eigenvalue problem specified by the linearized equations of motion with boundary conditions on the 5D fields and their derivatives. Decay constants are determined by the residues of poles in current-current two-point functions, which are determined by derivatives of the solutions to the linearized equations of motion [10, 11, 19]. Hadronic couplings are determined by the effective 4D theory after integrating the action over the extra dimension. A root-mean-squared fit of the remaining parameters to the central values of experimental and lattice data for seven observables gives $z_m = 1/(346\text{ MeV})$, $\sigma = (308\text{ MeV})^3$, $m_q = 2.3$ MeV [10]. The hard-wall AdS/QCD predictions with these values for the parameters are given in Table 1.

TABLE 1. RMS fit of hard-wall model to central values of several observables, from Refs. [10] and [17].

Observable	Measured (Central Value in MeV)	Model (MeV)
m_π	140	141
m_ρ	776	832
m_{a_1}	1230	1220
f_π	92.4	84.0
$F_\rho^{1/2}$	345	353
$F_{a_1}^{1/2}$	433	440
$g_{\rho\pi\pi}$	6.03	5.29
m_{f_2}	1275	1236

TABLE 2. Predictions of hard-wall model with three quark flavors, from Ref. [18].

Observable	Measured (Central Value - MeV)	Model (MeV)
m_π	140	134
f_π	92.4	86.6
m_K	496	514
f_K	113	101
$m_{K_0^*}$	672	697
$f_{K_0^*}$		36
m_ρ	776	789
$F_\rho^{1/2}$	345	335
m_{K^*}	894	821
$F_{K^*}^{1/2}$		337
m_{a_1}	1230	1270
$F_{a_1}^{1/2}$	433	453
m_{K_1}	1270	1400
F_{K_1}		488

A number of additional predictions in an extension of the model which includes a strange quark mass parameter are given in Table 2, from an analysis by Abidin and Carlson [18].

A relation between observables and the parameters m_q and σ can be derived which resembles the Gell-Mann–Oakes–Renner (GOR) relation [10],

$$m_\pi^2 f_\pi^2 = 2 m_q \sigma. \tag{6}$$

The SU(3) extensions of the GOR relation (6) are satisfied in appropriate extensions of the 5D model [18].

Soft-Wall AdS/QCD

The AdS/QCD models described above are not expected to be valid much above the scale of the lightest vector resonances. For heavy resonances, the vector and axial-vector masses in the hard-wall model scale as $m_n^2 \sim n^2$. However, experimental data confirm the Regge behavior $m_n^2 \sim n$. Misha Shifman and others have stressed this difference between AdS/QCD and QCD [20]. It is possible to reproduce the Regge spectrum by effectively modifying the AdS_5 geometry [12]. One way to do this is to couple the fields in the AdS/QCD model to a background dilaton:

$$S = \int d^5x \sqrt{-g} e^{-\Phi(x,z)} \mathscr{L}, \tag{7}$$

where the appropriate background for the dilaton is $\Phi_0(z) \sim z^2$. Low-energy predictions are comparable to, but not the same as, those of the hard-wall model.

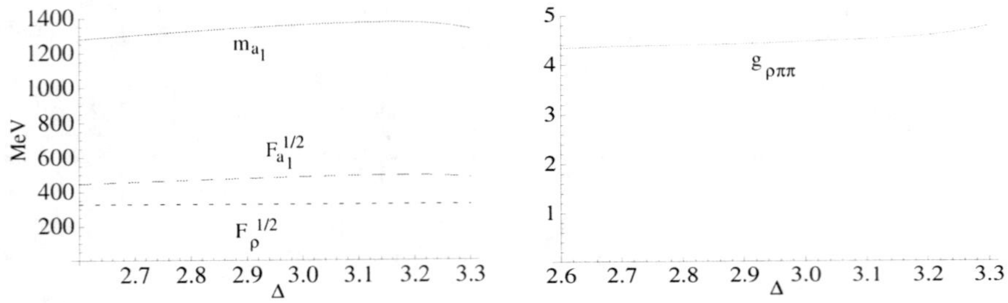

FIGURE 2. Dependence of observables on $m_X^2 = \Delta(\Delta - 4)$ in hard-wall model, holding fixed f_π, m_π, and m_ρ. Plot from Ref. [25].

Form Factors

It is straightforward to calculate form factors in AdS/QCD models, as the AdS/CFT correspondence teaches us that the contribution of a tower of resonances can be summed by use of the bulk-to-boundary propagator (see also Ref. [21]). Several authors have calculated various meson form factors. From these form factors were deduced moments of generalized parton distributions, charge radii and gravitational radii. For example, in the hard wall model with a particular choice of parameters [22, 23, 24]:

$$\langle r_\pi^2 \rangle_{charge} = 0.33 \text{ fm}^2, \quad \langle r_\pi^2 \rangle_{grav} = 0.13 \text{ fm}^2$$

$$\langle r_\rho^2 \rangle_{charge} = 0.53 \text{ fm}^2, \quad \langle r_\rho^2 \rangle_{grav} = 0.21 \text{ fm}^2$$

$$\langle r_{a_1}^2 \rangle_{charge} = 0.39 \text{ fm}^2, \quad \langle r_{a_1}^2 \rangle_{grav} = 0.15 \text{ fm}^2$$

Baryons

If one naively truncates an AdS/QCD model to the lightest modes, the 5D gauge kinetic terms include the 4D Skyrme term with coefficient that depends on the 5D dimensional gauge coupling and the spacetime geometry [8]. The light baryons have therefore been identified as Skyrmions in AdS/QCD. There has been some debate as to the stability of predictions for baryons in this approach, as in the original Skyrme model, based on the relative importance of higher-dimension operators that are neglected in this approach. A description of baryons in terms of 4+1 dimensional solitons, as opposed to the Skyrmions in the effective 3+1 dimensional theory, has been discussed by several authors, for example in Refs. [26, 27]. A comparison of these approaches appears in Ref. [28]. An alternative description of baryons modeled as fundamental fermions in the extra dimension has also been considered, and accurately reproduces the spectrum of excited Delta and nucleon resonances (see, for example, Refs. [29, 30]).

Light-front wave functions

Stan Brodsky and Guy de Teramond have noticed an intriguing relationship between the equations of motion for Kaluza-Klein modes of fields with general spin and scaling dimension in AdS/QCD models, and equations describing light-front wavefunctions of hadrons with general spin and orbital angular momentum (*e.g.* Ref. [31]). This is an interesting observation, and relates the radial direction of Anti-de Sitter space with partonic momenta inside the hadrons.

UNIVERSALITY IN ADS/QCD

Certain observables in AdS/CFT models at finite temperature have been found to be completely independent of the details of the model. A famous example is the ratio of shear viscosity η to entropy density s, which in natural units is predicted to be [32],

$$\frac{\eta}{s} = \frac{1}{4\pi}. \tag{8}$$

A more recent example is the ratio of electrical conductivity σ to charge susceptibility χ, which depends only on the temperature and number of spacetime dimensions [33]. Univeral predictions provide the strongest test of the AdS/CFT correspondence as applied to QCD. Despite the absence of complete universality of most AdS/QCD predictions, it is interesting that the various AdS/QCD models make comparable predictions for low-energy observables. It is worthwhile to better understand which observables are approximately independent of the details of the AdS/QCD model, and which details of the model are unimportant. For example, upon varying the mass of the field X by $\pm 20\%$, as long as the remaining parameters are chosen so as to correctly reproduce a small number of observables (m_π, f_π and m_ρ) the remaining low-energy observables of Table 1 were found to vary by only a few percent, as in Fig. 2 [25]. Perhaps the surprising success of AdS/QCD models at low energies is the result of such universality in its predictions.

ADDITIONAL APPLICATIONS OF ADS/CFT MODELS

Models similar in spirit to AdS/QCD have been applied to other strongly-interacting dynamical systems. Holographic technicolor models [34] are based on AdS/QCD, and AdS/CFT methods allow for calculation of precision electroweak observables in these models. These models are examples of Higgsless models of electroweak symmetry breaking, in which unitarity of longitudinal W boson scattering is the result of interactions involving the massive Kaluza-Klein excitations of the electroweak gauge bosons.

An exciting recent application of the AdS/CFT correspondence is to condensed matter systems. An important development in this direction was the construction of a spacetime geometry whose isometries are the same as the nonrelativistic conformal group [35].There are intriguing similarities of some models to high-temperature superconductors and systems of cold atoms [36].

SUMMARY

AdS/QCD models share the features of a number of earlier approaches to modeling QCD at low energies. AdS/QCD models generally predict low-energy observables at the 10-20% level, but do not fare as well at high energies. It is not yet clear why these models work as well as they do, but some predictions have been found to be universal as details of the model are varied.

ACKNOWLEDGMENTS

The author thanks Zainul Abidin, Stan Brodsky, Carl Carlson, Guy de Teramond, Hovhannes Grigoryan, Emanuel Katz, Herry Kwee, Richard Lebed, and Anatoly Radyushkin for discussions and for use of some of their results in this talk. The author's work is supported by NSF grant PHY-0757481.

REFERENCES

1. J. M. Maldacena, "The large N limit of superconformal field theories and supergravity," Adv. Theor. Math. Phys. **2**, 231 (1998) [Int. J. Theor. Phys. **38**, 1113 (1999)] [arXiv:hep-th/9711200].
2. M. Bando, T. Kugo, S. Uehara, K. Yamawaki and T. Yanagida, "Is Rho Meson A Dynamical Gauge Boson Of Hidden Local Symmetry?," Phys. Rev. Lett. **54**, 1215 (1985).
3. E. Witten, "Anti-de Sitter space and holography," Adv. Theor. Math. Phys. **2**, 253 (1998) [arXiv:hep-th/9802150].
4. S. S. Gubser, I. R. Klebanov and A. M. Polyakov, "Gauge theory correlators from non-critical string theory," Phys. Lett. B **428**, 105 (1998) [arXiv:hep-th/9802109].
5. J. Polchinski and M. J. Strassler, "The string dual of a confining four-dimensional gauge theory," arXiv:hep-th/0003136.
6. I. R. Klebanov and M. J. Strassler, "Supergravity and a confining gauge theory: Duality cascades and chiSB-resolution of naked singularities," JHEP **0008**, 052 (2000) [arXiv:hep-th/0007191].
7. M. Kruczenski, D. Mateos, R. C. Myers and D. J. Winters, "Towards a holographic dual of large-N(c) QCD," JHEP **0405**, 041 (2004) [arXiv:hep-th/0311270].

8. T. Sakai and S. Sugimoto, "Low energy hadron physics in holographic QCD," Prog. Theor. Phys. **113**, 843 (2005) [arXiv:hep-th/0412141]; "More on a holographic dual of QCD," Prog. Theor. Phys. **114**, 1083 (2005) [arXiv:hep-th/0507073].
9. E. Witten, "Anti-de Sitter space, thermal phase transition, and confinement in gauge theories," Adv. Theor. Math. Phys. **2**, 505 (1998) [arXiv:hep-th/9803131].
10. J. Erlich, E. Katz, D. T. Son and M. A. Stephanov, "QCD and a Holographic Model of Hadrons," Phys. Rev. Lett. **95**, 261602 (2005) [arXiv:hep-ph/0501128].
11. L. Da Rold and A. Pomarol, "Chiral symmetry breaking from five dimensional spaces," Nucl. Phys. B **721**, 79 (2005) [arXiv:hep-ph/0501218].
12. A. Karch, E. Katz, D. T. Son and M. A. Stephanov, "Linear Confinement and AdS/QCD," Phys. Rev. D **74**, 015005 (2006) [arXiv:hep-ph/0602229].
13. S. J. Brodsky and R. Shrock, "Maximum Wavelength of Confined Quarks and Gluons and Properties of Quantum Chromodynamics," Phys. Lett. B **666**, 95 (2008) [arXiv:0806.1535 [hep-th]].
14. A. Deur, V. Burkert, J. P. Chen and W. Korsch, "Determination of the effective strong coupling constant $\alpha_{s,g_1}(Q^2)$ from CLAS spin structure function data," Phys. Lett. B **665**, 349 (2008) [arXiv:0803.4119 [hep-ph]].
15. P. Breitenlohner and D. Z. Freedman, "Positive Energy In Anti-De Sitter Backgrounds And Gauged Extended Supergravity," Phys. Lett. B **115**, 197 (1982).
16. A. Cherman, T. D. Cohen and E. S. Werbos, "The chiral condensate in holographic models of QCD," Phys. Rev. C **79**, 045203 (2009) [arXiv:0804.1096 [hep-ph]].
17. E. Katz, A. Lewandowski and M. D. Schwartz, "Tensor mesons in AdS/QCD," Phys. Rev. D **74**, 086004 (2006) [arXiv:hep-ph/0510388].
18. Z. Abidin and C. E. Carlson, "Strange hadrons and kaon-to-pion transition form factors from holography," Phys. Rev. D **80**, 115010 (2009) [arXiv:0908.2452 [hep-ph]].
19. S. Hong, S. Yoon, M. J. Strassler, "On the couplings of vector mesons in AdS / QCD," JHEP **0604**, 003 (2006). [hep-th/0409118].
20. M. Shifman, "Highly excited hadrons in QCD and beyond," arXiv:hep-ph/0507246.
21. J. Hirn and V. Sanz, "(Not) summing over Kaluza-Kleins," Phys. Rev. D **76**, 044022 (2007) [arXiv:hep-ph/0702005].
22. H. J. Kwee and R. F. Lebed, "Pion Form Factors in Holographic QCD," JHEP **0801**, 027 (2008) [arXiv:0708.4054 [hep-ph]].
23. H. R. Grigoryan and A. V. Radyushkin, "Form Factors and Wave Functions of Vector Mesons in Holographic QCD," Phys. Lett. B **650**, 421 (2007) [arXiv:hep-ph/0703069].
24. Z. Abidin and C. E. Carlson, "Hadronic Momentum Densities in the Transverse Plane," Phys. Rev. D **78**, 071502 (2008) [arXiv:0808.3097 [hep-ph]].
25. J. Erlich and C. Westenberger, Phys. Rev. D **79**, 066014 (2009) [arXiv:0812.5105 [hep-ph]].
26. K. Nawa, H. Suganuma and T. Kojo, "Baryons in Holographic QCD," Phys. Rev. D **75**, 086003 (2007) [arXiv:hep-th/0612187].
27. A. Pomarol and A. Wulzer, "Baryon Physics in Holographic QCD," Nucl. Phys. B **809**, 347 (2009) [arXiv:0807.0316 [hep-ph]].
28. K. Nawa, H. Suganuma and T. Kojo, "Brane-induced Skyrmion on S^3: baryonic matter in holographic QCD," Phys. Rev. D **79**, 026005 (2009) [arXiv:0810.1005 [hep-th]].
29. G. F. de Teramond and S. J. Brodsky, "Baryonic states in QCD from gauge / string duality at large N(c)," arXiv:hep-th/0409074.
30. D. K. Hong, T. Inami and H. U. Yee, "Baryons in AdS/QCD," Phys. Lett. B **646**, 165 (2007) [arXiv:hep-ph/0609270].
31. S. J. Brodsky and G. F. de Teramond, "Hadronic spectra and light-front wavefunctions in holographic QCD," Phys. Rev. Lett. **96**, 201601 (2006) [arXiv:hep-ph/0602252].
32. P. Kovtun, D. T. Son and A. O. Starinets, "Viscosity in strongly interacting quantum field theories from black hole physics," Phys. Rev. Lett. **94**, 111601 (2005) [arXiv:hep-th/0405231].
33. P. Kovtun and A. Ritz, "Universal conductivity and central charges," arXiv:0806.0110 [hep-th].
34. J. Hirn and V. Sanz, "A negative S parameter from holographic technicolor," Phys. Rev. Lett. **97**, 121803 (2006) [arXiv:hep-ph/0606086]; C. D. Carone, J. Erlich and J. A. Tan, "Holographic Bosonic Technicolor," Phys. Rev. D **75**, 075005 (2007) [arXiv:hep-ph/0612242]; C. D. Carone, J. Erlich and M. Sher, "Holographic Electroweak Symmetry Breaking from D-branes," Phys. Rev. D **76**, 015015 (2007) [arXiv:0704.3084 [hep-th]]; D. K. Hong and H. U. Yee, "Holographic estimate of oblique corrections for technicolor," Phys. Rev. D **74**, 015011 (2006) [arXiv:hep-ph/0602177].
35. D. T. Son, "Toward an AdS/cold atoms correspondence: a geometric realization of the Schroedinger symmetry," Phys. Rev. D **78**, 046003 (2008) [arXiv:0804.3972 [hep-th]]; A. Adams, K. Balasubramanian and J. McGreevy, "Hot Spacetimes for Cold Atoms," JHEP **0811**, 059 (2008) [arXiv:0807.1111 [hep-th]].
36. S. S. Gubser and S. S. Pufu, "The gravity dual of a p-wave superconductor," JHEP **0811**, 033 (2008) [arXiv:0805.2960 [hep-th]]; M. M. Roberts and S. A. Hartnoll, "Pseudogap and time reversal breaking in a holographic superconductor," JHEP **0808**, 035 (2008) [arXiv:0805.3898 [hep-th]].

Physics with KLOE and KLOE-2 Experiments at DAΦNE

S. Giovannella

Laboratori Nazionali di Frascati dell'INFN, via Enrico Fermi 40, 00044 Roma, Italy

on behalf of the KLOE and KLOE-2 Collaborations

Abstract. The KLOE experiment has collected 2.5 fb^{-1} of e^+e^- collisions at the ϕ peak and about 250 pb^{-1} in the center of mass energy 1 GeV. Data taken on peak are used to study properties of light scalar and pseudoscalar mesons, produced through ϕ radiative decays. Off-peak data are used to perform preliminary analyses of the reactions $e^+e^- \to e^+e^-\pi^0\pi^0$ and $e^+e^- \to e^+e^-\eta$. All these studies will be extended with the coming KLOE-2 run.

Keywords: e^+e^- collisions, light meson spectroscopy, $\gamma\gamma$ collisions
PACS: 13.66.Bc, 14.40.Be

FROM KLOE TO KLOE-2

From 2000 to 2006, the KLOE experiment collected 2.5 fb^{-1} of e^+e^- collisions at the ϕ peak and about 240 pb^{-1} below the ϕ resonance ($\sqrt{s} = 1$ GeV). The ϕ meson predominantly decays into charged and neutral kaons, thus allowing KLOE to make precision studies in the fields of flavor physics, low energy QCD and test of discrete symmetries. At the same time, properties of pseudoscalar and scalar mesons have been extensively studied using high statistic samples of these particles produced through ϕ radiative decays and $\gamma\gamma$ interactions.

A new beam crossing scheme allowing for a reduced beam size and increased luminosity is operating at DAΦNE [1]. The KLOE-2 detector was successfully rolled in in this new interaction region and has been upgraded with small angle tagging devices to detect both high and low energy electrons or positrons in $e^+e^- \to e^+e^-X$ events. About 5 fb^{-1} are expected in the first year of running. An inner tracker and small angle calorimeters are scheduled to be installed in a subsequent step, providing larger acceptance both for charged particles and photons. A detailed description of the KLOE-2 physics program can be found in Ref. [2]

SCALARS BELOW 1 GEV

The still unresolved structure of light scalar mesons ($S = f_0(980)$, $a_0(980)$, σ) motivated a detailed studies on these decays using electric dipole transitions $\phi \to S\gamma$ [3, 4, 5, 6, 7, 8] and searching for the process $\phi \to [a_0(980) + f_0(980)]\gamma \to K\bar{K}\gamma$ [9].

Both branching ratios and spectral functions have been studied for the decays $\phi \to S\gamma \to \pi\pi\gamma/\pi^0\eta\gamma$, obtaining the coupling of scalar mesons to the $\pi\pi$ and $\eta\pi^0$ pairs. While in the latter only the $a_0(980)$ can contribute, in the $\pi\pi$ channel both the $f_0(980)$ and the σ, whose existence has been debated since a long time ago, are expected. KLOE has fitted the $\pi^0\pi^0$ invariant with the $f_0(980)$ ans σ contributions, taking mass and width of σ from other measurements and phenomenological fits. Data from $\pi^+\pi^-$ phase shift were also taken into account. For the charged channel, due to the large background, no sensitivity to the σ meson is shown by the fit. The present analysis of KLOE data is based on $\sim$500 pb^{-1}. KLOE-2 will have a 20 times larger sample, so that the σ could also show up in the $\pi^+\pi^-$ channel. Moreover, the $\pi^+\pi^-$ forward-backward asymmetry (A_{FB}) is expected to be affected by the presence of the σ. This asymmetry is generated by the different behaviour under C-conjugation of the $\pi^+\pi^-$ amplitude for initial state radiation (ISR) with respect to the other contributions (final state radiation and $e^+e^- \to \phi \to [f_0(980) + \sigma]\ \gamma$). In Fig. 1, the distribution of A_{FB} is shown for two sets of KLOE data, collected at $\sqrt{s} = M_\phi$ and 1 GeV. The Monte Carlo (MC) shape, simulated with PHOKARA generator with scalar and VMD contribution extracted from our $\pi^0\pi^0\gamma$ analysis [10], nicely reproduce data.

In the $\phi \to K\bar{K}\gamma$ process, never been observed, the $K\bar{K}$ system has scalar quantum numbers, therefore it is expected to proceeds through the $\phi \to [a_0(980) + f_0(980)]\gamma \to K\bar{K}\gamma$ decay chain. The selected channel for this search is

12th International Conference on Meson-Nucleon Physics and the Structure of the Nucleon (MENU 2010)
AIP Conf. Proc. 1374, 73-78 (2011); doi: 10.1063/1.3647101

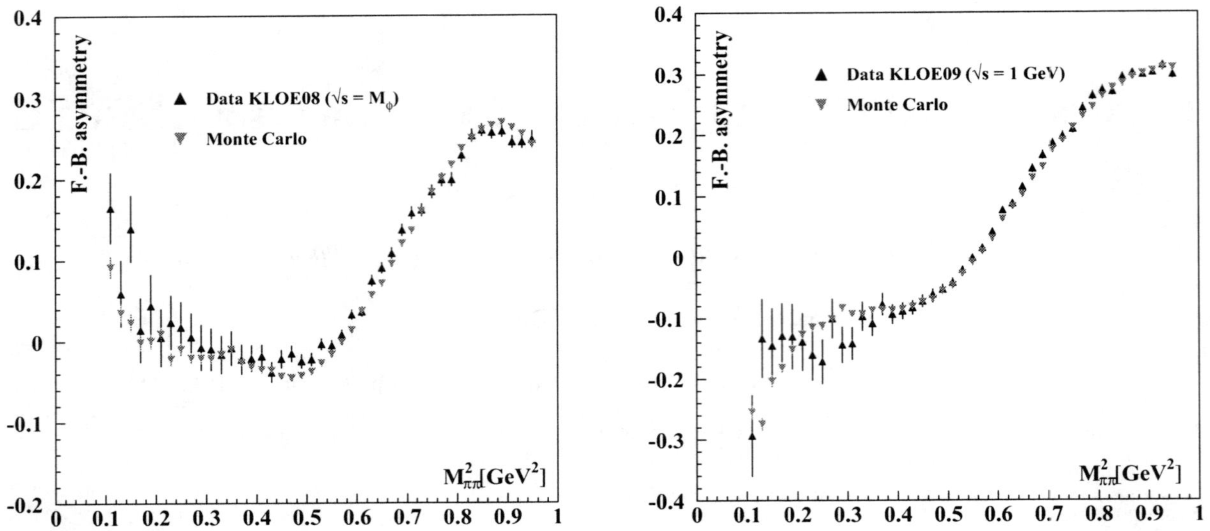

FIGURE 1. Forward-backward asymmetry as a function of $M^2_{\pi\pi}$ in $e^+e^- \rightarrow \pi^+\pi^-\gamma$ events for data at $\sqrt{s} = M_\phi$ (left) and at $\sqrt{s} = 1$ GeV (right).

$\phi \rightarrow K_S K_S \gamma \rightarrow \pi^+\pi^-\pi^+\pi^-\gamma$. The main backgrounds are the resonant $e^+e^- \rightarrow \phi\gamma \rightarrow K_S K_L \gamma$ with an ISR photon and the continuum $e^+e^- \rightarrow \pi^+\pi^-\pi^+\pi^-\gamma$ processes. MC signal has been simulated according to the phase space and the radiative decay dynamics. Selection cuts have been optimized on MC. From an integrated luminosity of 2.2 fb^{-1}, 5 candidate events are found in data, while 3.2 ± 0.7 background events are expected from MC . This leads to BR($\phi \rightarrow K\bar{K}\gamma$) $< 1.9 \times 10^{-8}$ at the 90% CL. This upper limit is consistent with the BR($\phi \rightarrow K\bar{K}\gamma$) prediction computed with $f_0(980)$ and $a_0(980)$ couplings measured by KLOE [5, 6, 8]. Scaling these numbers to the KLOE-2 statistics a sensitivity of $BR(\phi \rightarrow K^0\bar{K}^0\gamma) < 1 \times 10^{-8}$ is expected. The inner tracker will provide in the second phase of the experiment three times better vertex resolution, which is beneficial for the rejection capability of the background. This, together with the expected 20 fb^{-1} of integrated luminosity, could lead to the first observation of the decay.

$\eta/\eta' \rightarrow \pi^+\pi^-\gamma$

The decays $\eta/\eta' \rightarrow \gamma\gamma$ proceed through the triangle anomaly and are accounted for by the Wess-Zumino-Witten (WZW) term into the ChPT Lagrangian. The box anomaly is a higher order term of WZW describing the direct coupling of three pseudoscalar mesons with the photon. The $\eta/\eta' \rightarrow \pi^+\pi^-\gamma$ decays provide a good tool to investigate the box anomaly, which describes the not–resonant part of the coupling. The invariant mass of the pions is a good observable to disentangle this contribution from other possible resonant ones, e.g. from the ρ-meson. Several theoretical approaches have been developed to treat the contributions of the anomalies of these decays [11, 12, 13].

The $\eta \rightarrow \pi^+\pi^-\gamma$ decay has been measured in the 1970s with data samples of the order of 10^4 events [14] [15]. However the theoretical papers which tried to combine the two data sets reported discrepancy in data treatment and, as a consequence, difficulty in obtaining consistent results. Therefore, to clarify the situation, results from experiments with large statistics are really needed. Recently the CLEO collaboration published their results on the ratio $\Gamma(\eta \rightarrow \pi^+\pi^-\gamma)/\Gamma(\eta \rightarrow \pi^+\pi^-\pi^0)$, which differ more than 3 σ's from old results [16]. From the analysis of 1.2 fb^{-1}, the preliminary KLOE measurement of the same ratio is:

$$\frac{\Gamma(\eta \rightarrow \pi^+\pi^-\gamma)}{\Gamma(\eta \rightarrow \pi^+\pi^-\pi^0)} = 0.2014 \pm 0.0004_{stat} \pm 0.0060_{stat}. \tag{1}$$

The final systematic error, under evaluation, is expected to be less than 1%. Our number is in agreement with the old results from Thaler and Gormley and differs significantly from the recent CLEO results.

The relevance of the box anomaly asks for more data. In particular, with the first KLOE-2 run, a simultaneous analysis of the η and η' decays can be performed. With an estimated analysis efficiency of 20%, it is possible to obtain with the first KLOE-2 run $\sim 10^5$ $\eta' \to \pi^+\pi^-\gamma$ decays.

$\eta \to \pi^+\pi^-e^+e^-/e^+e^-e^+e^-$

The $\eta \to \pi^+\pi^-e^+e^-$ decay allows to probe the structure of the η meson [17], to compare the predictions of the branching ratio value based on Vector Meson Dominance model and Chiral Perturbation Theory [18, 19, 20, 21] and to study CP violation beyond the prediction of the Standard Model by measuring the angular asymmetry between the pion and the electron decay planes [22].

The analysis has been performed on 1.73 fb^{-1} [23]. The events are required to have at least four tracks coming from a cylinder around the Interaction Point. A cluster not associated to any track, having a time compatible with the photon time of flight, an energy of at least 250 MeV and in the polar-angle range ($23° - 157°$), is also required.

Background sources can be grouped into ϕ-decays and events in the continuum. The former is mainly due to $\phi \to \pi^+\pi^-\pi^0$ events (with π^0 Dalitz decay) and to $\phi \to \eta\gamma$ events either with $\eta \to \pi^+\pi^-\pi^0$ (with π^0 Dalitz decay) or with $\eta \to \pi^+\pi^-\gamma$ (with photon conversion on the beam pipe). The latter comes mainly from $e^+e^- \to e^+e^-(\gamma)$ events. Because of poor MC statistics, they have been studied using off-peak data taken at $\sqrt{s} = 1$ GeV, where ϕ decays are negligible. Backgrounds are reduced by applying cuts on track momenta and rejecting photon conversions on beam pipe. Residual background contamination is evaluated by performing a fit on the sidebands of the $\pi^+\pi^-e^+e^-$ invariant mass after the cuts on track momenta with background shapes. For the signal estimate the η mass region ($[535, 555]$ MeV) is considered, performing the event counting after background subtraction. The resulting number of signal events, $N_{\pi\pi ee} = 1555 \pm 52$, is used to extract the branching ratio:

$$BR(\eta \to \pi^+\pi^-e^+e^-\gamma) = (26.8 \pm 0.9_{\text{stat}} \pm 0.7_{\text{syst}}) \times 10^{-5}. \tag{2}$$

The decay plane asymmetry A_ϕ is defined as the sign asymmetry of the quantity $\sin\phi \cdot \cos\phi$, ϕ being the angle between the pion and the electron planes in the η rest frame. It has been evaluated for the events in the signal region after background subtraction. The value obtained is:

$$A_\phi = (-0.6 \pm 2.5_{\text{stat}} \pm 1.8_{\text{syst}}) \times 10^{-2} \tag{3}$$

which is the first measurement of this asymmetry. The distribution of the $\sin\phi\cos\phi$ variable is shown in the left panel of Fig. 2.

With the same data sample, the $\eta \to e^+e^-e^+e^-$ decay has been studied. Events with four electrons are selected using time of flight information provided by the calorimeter. Backgrounds from $\eta \to \gamma\gamma/e^+e^-\gamma$ with photon conversion are rejected by reconstructing the conversion point on the beam pipe or drift chamber walls. Most of the residual background comes from continuum events and a small contribution is due to ϕ decays. The latter is subtracted from data using the MC spectrum. The number of signal events is obtained by fitting the data distribution of the four electron invariant mass, M_{eeee}, with signal and background shapes (Fig. 2-right). From the fit we extract $N_{\eta \to e^+e^-e^+e^-} = 413 \pm 31$. This constitutes the first observation of this decay.

SEARCH FOR GLUONIUM IN η'

The η' meson, being almost a pure SU(3) singlet, is a good candidate for a sizeable gluonium content. In this hypothesis, the η and η' wave functions can be written in terms of the u,d quark wave function ($|q\bar{q}> = \frac{1}{\sqrt{2}}(|u\bar{u}> + |d\bar{d}>)$), the strange component ($|s\bar{s}>$) and the gluonium ($|GG>$) [24]:

$$|\eta'> = \cos\varphi_G \sin\varphi_P |q\bar{q}> + \cos\varphi_G \cos\varphi_P |s\bar{s}> + \sin\varphi_G |GG> \tag{4}$$

$$|\eta> = \cos\varphi_P |q\bar{q}> - \sin\varphi_P |s\bar{s}> \tag{5}$$

where φ_P is the η-η' mixing angle and $Z_G^2 = \sin^2\varphi_G$ is the gluonium fraction in the η' meson.

From the study of $\phi \to \eta'\gamma \to \pi^+\pi^-\eta\gamma$, with $\eta \to \pi^0\pi^0\pi^0$, and $\phi \to \eta\gamma \to \pi^0\pi^0\pi^0\gamma$'s decays, the ratio $R_\phi = BR(\phi \to \eta'\gamma)/BR(\phi \to \eta\gamma)$ has been measured [25]: $R_\phi = (4.77 \pm 0.09_{\text{stat}} \pm 0.19_{\text{syst}}) \times 10^{-3}$. Using the approach of

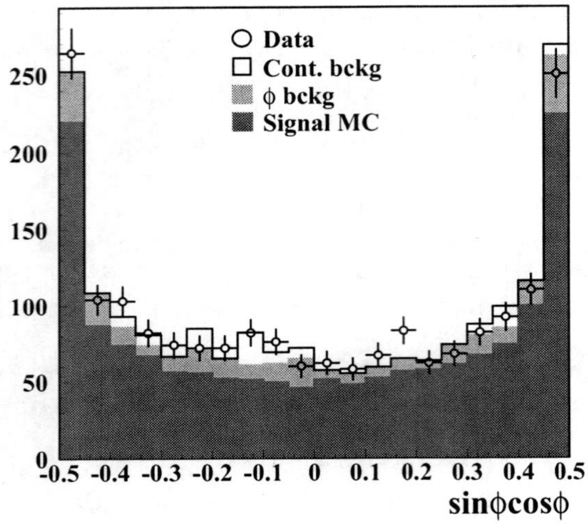

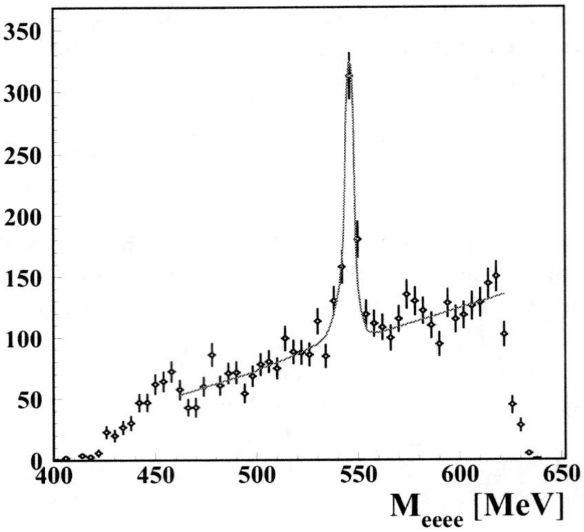

FIGURE 2. Left: angular asymmetry between pions and electrons decay planes for $\eta \to \pi^+\pi^-e^+e^-$ events. Data (dots) are compared with the expected distribution in case of null A_ϕ value for the signal. Right: fit to the four electron invariant mass for events $\eta \to e^+e^-e^+e^-$.

TABLE 1. Fit results for the evaluation of the η' gluonium content. In the first column, only the first two parameters are left free.

Parameter	KLOE '07	KLOE '09	KLOE '09 (no glue)
$Z^2_{\eta'}$	0.14 ± 0.04	0.12 ± 0.04	0 (fixed)
φ_P	$(39.7 \pm 0.7)^\circ$	$(40.4 \pm 0.6)^\circ$	$(41.4 \pm 0.5)^\circ$
Z_{NS}	0.91 ± 0.05 (fixed)	0.94 ± 0.03	0.93 ± 0.02
Z_S	0.89 ± 0.07 (fixed)	0.83 ± 0.05	0.82 ± 0.05
φ_V	3.2° (fixed)	$(3.32 \pm 0.09)^\circ$	$(3.34 \pm 0.09)^\circ$
m_s/m	1.24 ± 0.07 (fixed)	1.24 ± 0.07	1.24 ± 0.07
$P(\chi^2)$	49%	20.5%	0.5%

Refs. [26, 27], where the SU(3) breaking is taken into account via the constituent quark mass ratio $m_s/\bar{m}$, R_ϕ can be parametrized as

$$R_\phi = \cot^2 \varphi_P \cos^2 \varphi_G \left(1 - \frac{m_s}{\bar{m}} \frac{C_{NS}}{C_S} \frac{\tan \varphi_V}{\sin 2\varphi_P}\right)^2 \left(\frac{p_{\eta'}}{p_\eta}\right)^3, \tag{6}$$

where $p_{\eta'}$ and p_η are the momenta of the η' and η meson respectively, φ_V is the ϕ-ω mixing angle and C_{NS}, C_S take into account the vector and pseudoscalar wave-function overlap. Combining our R_ϕ result with other experimental constraints and using the corresponding SU(3) relations between decay modes [27, 28], we found a $3\,\sigma$ evidence for gluonium content in η' (Tab. 1, left column). Since the parameters C_{NS}, C_S, φ_V and $m_s/\bar{m}$ were taken from [26], obtained in the hypothesis of no η' gluonium content, we repeat the fit adding other SU(3) relations and enlarging the number of free parameters [29]. As shown in Tab. 1, the new result is in good agreement for all parameters, confirming the $3\,\sigma$ evidence for gluonium in η'. The quality of the fit get worse when $Z^2_{\eta'}$ is fixed to zero.

The $BR(\eta' \to \gamma\gamma)$ limit the accuracy on the η' decay width while the $\eta' \to \pi^+\pi^-\eta/\pi^0\pi^0\eta$ branching ratios dominate the systematic error on the R_ϕ measurement, as the η' is identified through these decays. Assuming that the branching fractions and the correlation matrix of the measurements used in the fit remains the same, a $4\,\sigma$'s evidence for a gluonium content in η' could be obtained also independently from the $\Gamma(\eta' \to \gamma\gamma)/\Gamma(\pi^0 \to \gamma\gamma)$ ratio if the fractional accuracy on the η' partial widths is brought to 1% level. Therefore is very important to measure accurately the main decay channels of η' at KLOE-2. With the KLOE-2 data-taking above the ϕ peak, e.g. at $\sqrt{s} \sim 1.2$ GeV, it is possible to measure the η' decay width $\Gamma(\eta' \to \gamma\gamma)$ through $\sigma(e^+e^- \to e^+e^-\gamma^*\gamma^* \to e^+e^-\eta')$ as discussed in next section. The measurement to 1% accuracy of both the cross section and the $BR(\eta' \to \gamma\gamma)$ would bring the fractional

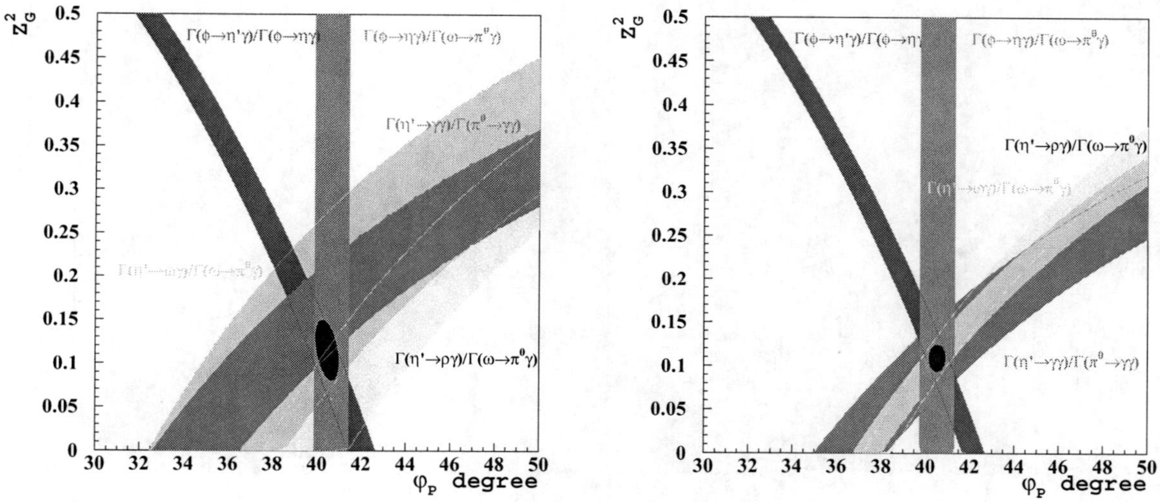

FIGURE 3. The 68% C.L. region in the ψ_P, Z_G^2 plane. Left: current KLOE measurement. Right: expectations with η' branching ratios and η' total width improved to 1% and 1.4% precision respectively.

error on the η' total width to 1.4%. In Fig. 3 the 68% C.L. region in the φ_P, Z_G^2 plane obtained with the improvements discussed above is compared with the present KLOE measurement.

$\gamma\gamma$ PHYSICS

The study of the reaction $e^+e^- \to e^+e^-\gamma^*\gamma^* \to e^+e^-X$, where X is some arbitrary final state resulting from fusion of two photons, gives access to states with $J^{PC} = 0^{\pm+}, 2^{\pm+}$. In the low-energy region covered by the KLOE detector, $m_{\pi^0} < W_{\gamma\gamma} < 700$ MeV, existing measurements [30, 31] are affected by large statistical and systematic uncertainties. At KLOE, where there is no tagging of the outgoing e^+e^-, $\gamma\gamma$ interactions have been studied using off-peak data (240 pb^{-1} collected at $\sqrt{s} = 1$ GeV), as the main source of background comes from ϕ decays.

The study of the $e^+e^- \to e^+e^-\pi^0\pi^0$ process is interesting because of a possible contribution from the controversial σ scalar meson. In Fig. 4-right, the preliminary spectrum of the four-photon invariant mass, $M_{4\gamma}$, is compared with accountable background yields. An excess of about 4000 events is evident at low $M_{4\gamma}$ values, where the σ contribution is expected. Detailed studies of the background contamination and the extraction of the $\sigma \to \pi^0\pi^0$ contribution are in progress.

The η radiative width, $\Gamma_{\eta\gamma\gamma}$, is usually extracted from the measurement of the $e^+e^- \to e^+e^-\eta$ cross section. The KLOE analysis is performed using the $\eta \to \pi^+\pi^-\pi^0$ decay channel. After background rejection, distributions of longitudinal momentum (p_L) and recoil missing mass (M^2_{miss}) are independently fitted with the superposition of MC shapes for signal and background (Fig. 4-right). More than 600 $e^+e^- \to e^+e^-\eta \to e^+e^-\pi^+\pi^-\pi^0$ events are obtained. Determination of $\sigma(e^+e^- \to e^+e^-\eta)$ and $\Gamma_{\eta\gamma\gamma}$ are under way.

The upgrade of the KLOE detector with the installation of four stations [32] to tag electrons and positrons from the reaction $e^+e^- \to e^+e^-\gamma^*\gamma^* \to e^+e^-X$, will give the opportunity to investigate $\gamma\gamma$ physics also at the ϕ resonance for the reactions $\gamma\gamma \to \pi^0/\pi\pi/\eta/\eta\pi$. Running DA$\Phi$NE at center-of-mass energies up to 1.4 GeV can eventually complete the low energy spectroscopy studies with the detection of the heavier η', $f_0(980)$, $a_0(980)$ and $K\bar{K}$ final states. Single pseudoscalar production would improve the determination of the two–photon decay widths of these mesons, relevant for the measurement of the pseudoscalar mixing angle and the measurement of the valence gluon content in the η' wavefunction. Moreover, it would be possible to measure the transition form factors $\mathscr{F}_{X\gamma^*\gamma^*}(q_1^2, q_2^2)$ as a function of the momentum of the virtual photons, q_1^2 and q_2^2. Such form factors are crucial for the theoretical evaluation of the hadronic light-by-light contribution to the muon magnetic anomaly.

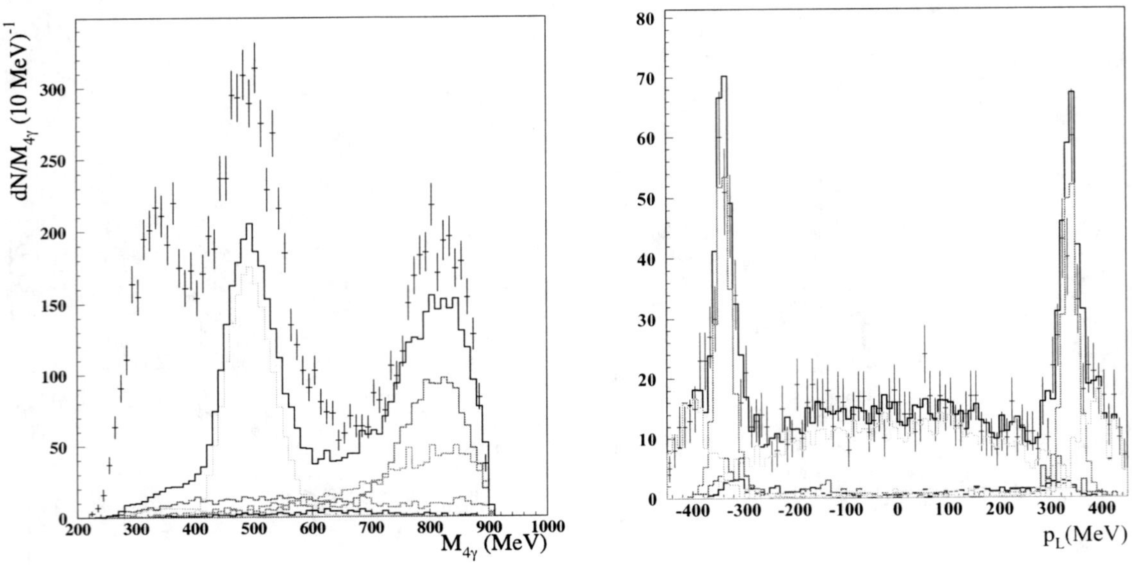

FIGURE 4. $\gamma\gamma$ physics studies at KLOE with off-peak data. Left: spectrum of the four-photon invariant mass (crosses) compared with the sum of the expected backgrounds (black solid line) for $\gamma\gamma \to \pi^0\pi^0$ events. The main background contributions are shown in colours. Right: fit to the p_L distribution for $\gamma\gamma \to \eta$ events. $\phi \to \eta\gamma$ and radiative Bhabha backgrounds are peaked in the $|p_L| > 300$ MeV regions while signal is concentrated in the center of the histogram.

REFERENCES

1. C. Milardi *et al.*, ICFA Beam Dyn. Newslett. 48 (2009) 23.
2. G. Amelino-Camelia *et al.*, Eur. Phys. J. C 68 (2010) 619.
3. A. Aloisio *et al.* [KLOE Collaboration], Phys. Lett. B 536 (2002) 209.
4. A. Aloisio *et al.* [KLOE Collaboration], Phys. Lett. B 537 (2002) 21.
5. F. Ambrosino *et al.* [KLOE Collaboration], Phys. Lett. B 634 (2006) 148.
6. F. Ambrosino *et al.* [KLOE Collaboration], Eur. Phys. J. C 49 (2007) 473.
7. F. Ambrosino *et al.* [KLOE Collaboration], Nucl. Phys. B (Proc. Suppl.) 186 (2009) 290.
8. F. Ambrosino *et al.* [KLOE Collaboration], Phys. Lett. B 681 (2009) 5.
9. F. Ambrosino *et al.* [KLOE Collaboration], Phys. Lett. B 679 (2009) 10.
10. O. Shekhovtsova, G. Venanzoni and G. Pancheri, arXiv:0901.4440 [hep-ph].
11. B. R. Holstein, Phys. Scripta T 99 (2002) 55.
12. M. Benayoun *et al*, Eur. Phys. J. C 31 (2003) 525.
13. B. Borasoy and R. Nissler, Nucl. Phys. A 740 (2004) 362.
14. P. Gormley, Phys. Rev. D 2 (1970) 501.
15. J. Layter, Phys. Rev. D 2 7 (1973) 2565.
16. A. Lopez *et al* [CLEO Collaboration], Phys. Rev. Lett. 99 (2007) 122001.
17. L. G. Landsberg, Phys. Rept. 128 (1985) 301.
18. C. Jarlskog and H. Pilkuhn, Nucl. Phys. B 1 (1967) 264.
19. A. Faessler, C. Fuchs and M. I. Krivoruchenko, Phys. Rev. C 61 (2000) 035206.
20. C. Picciotto and S. Richardson, Phys. Rev. D 48 (1993) 3395.
21. B. Borasoy and R. Nissler, Eur. Phys. J. A 33 (2007) 95.
22. D. N. Gao, Mod. Phys. Lett. A 17 (2002) 1583.
23. F. Ambrosino *et al.* [KLOE Collaboration], Phys. Lett. B 675 (2009) 283.
24. J. L. Rosner, Phys. Rev. D 27 (1983) 1101.
25. F. Ambrosino *et al.* [KLOE Collaboration], Phys. Lett. B 648 (2007) 267.
26. A. Bramon, R. Escribano and M. D. Scadron, Phys. Lett. B 503 (2001) 271.
27. R. Escribano and J. Nadal, JHEP 05 (2007) 6.
28. E. Kuo, Phys. Rev. D 63 (2001) 054027.
29. F. Ambrosino *et al.* [KLOE Collaboration], JHEP 07 (2009) 105.
30. H. Marsiske *et al.* [Crystal Ball Coll.], Phys. Rev. D 41 (1990) 3324.
31. T. Oest *et al.* [JADE COll.], Z. Phys. C 47 (1990) 343.
32. D. Babusci *et al.*, arXiv:0906.0875 (2009).

Nucleon Structure from Lattice QCD

Philipp Hägler

Institut für Theoretische Physik T39, TU München, James-Franck-Strasse, D-85747 Garching, Germany
and
Institut für Theoretische Physik Universität Regensburg, D-93040 Regensburg, Germany
E-mail: haegler@p2h.de

Abstract. Hadron structure calculations in lattice QCD have seen substantial progress during recent years. We illustrate the achievements that have been made by discussing latest lattice results for a limited number of important observables related to nucleon form factors and generalized parton distributions. A particular focus is placed on the decomposition of the nucleon spin $1/2$ in terms of quark spin and orbital angular momentum contributions. Results and limitations of the necessary chiral extrapolations based on ChPT will be briefly discussed.

Keywords: lattice QCD, nucleon structure, GPDs
PACS: 12.38.Gc, 14.20.Dh

INTRODUCTION

During the last decades, an impressive number of theoretical, phenomenological and experimental studies have provided remarkable insights into the inner structure of hadrons. Complementary to these efforts, ab-initio calculations in the framework of lattice QCD represent a highly promising approach from first principles to a wide range of fundamental hadron properties. Detailed accounts of the advances that have been made on the lattice, in particular with respect to the quark and gluon structure of the pion and the nucleon, can be found in [1, 2, 3, 4, 5, 6].

In this contribution, we concentrate on a small number of results from recent dynamical lattice QCD studies. We begin with an overview of lattice results for the anomalous magnetic moment and the Pauli mean square radius, including a brief discussion of the corresponding chiral extrapolations in the framework of heavy baryon chiral perturbation theory (HBChPT). This will be followed by a presentation of moments of generalized parton distributions (GPDs) obtained by the LHP collaboration. The impact of these lattice results on the decomposition of the nucleon spin in terms of spin- and orbital angular momentum contributions is discussed in the final section, where lattice values for the up- and down-quark angular momentum will also be compared to other experimental, phenomenological and model studies. Due to space limitations, we point the interested reader to Refs. [7, 8] for a first exploratory lattice study of transverse momentum dependent distributions (TMDs) that already provided interesting insights into non-trivial correlations between spin- and momentum degrees of freedom for polarized quarks in a polarized nucleon.

NUCLEON FORM FACTORS

Significant progress has been made during the recent years in dynamical lattice QCD calculations of the Dirac and Pauli nucleon form factors, $F_1(Q^2)$ and $F_2(Q^2)$, by a number of collaborations [9, 10, 11, 12, 13, 14, 15], with pion masses as low as ≈ 300 MeV. Very recently, QCDSF/UKQCD presented result for $N_f = 2$ Wilson fermions reaching down to $m_\pi \approx 180$ MeV [16] (albeit with larger statistical uncertainties at the lowest pion mass). Fundamental observables derived from the form factors are the mean square (ms) radii, $\langle r^2\rangle_{1,2} \propto -(dF_{1,2}(Q^2)/dQ^2)_{Q^2=0}$, as well as the anomalous magnetic moment, $\kappa = F_2(0)$. An overview of results for the latter in the isovector channel is displayed on the left in Figure 1 as a function of the pion mass. It is encouraging to see that the data points obtained for the different lattice actions and for $N_f = 2$ and $N_f = 2+1$ flavors are overall in good agreement within the statistical errors. Despite the steady increase towards smaller pion masses, the lattice results are still about 30% below the experimental value even at pion masses of ≈ 250 MeV. In a first attempt to understand this difference, we have fitted the data below $m_\pi = 300$ MeV, including the experimental value, using results from HBChPT in the small scale expansion (SSE) approach [17, 18], as illustrated by the shaded band on the left in Fig. 1. Fixing the axial nucleon-Δ

12th International Conference on Meson-Nucleon Physics and the Structure of the Nucleon (MENU 2010)
AIP Conf. Proc. 1374, 79-84 (2011); doi: 10.1063/1.3647102

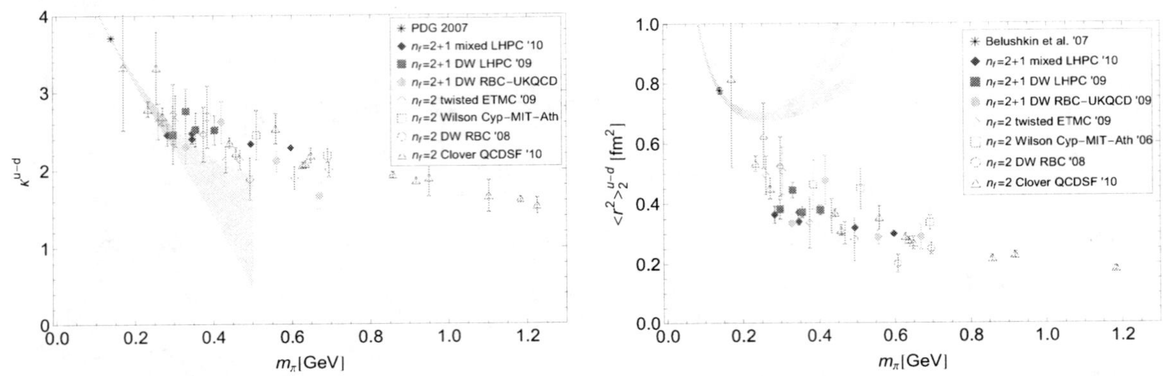

FIGURE 1. Overviews of results for the isovector anomalous magnetic moment (left) and the isovector Pauli mean square radius (right) obtained by different lattice collaborations. The shaded bands are the result of SSE-HBChPT fits including the experimental values, as described in the text. References are provided in the text.

coupling to $c_A = 1.5$, and varying the isovector nucleon-Δ coupling between $c_V = -1.5,\ldots,-3.5\,\mathrm{GeV}^{-1}$, we find a good description of the data points with a comparatively large value for the isovector anomalous magnetic moment in the chiral limit, $\kappa^{0,u-d} \sim 5.2,\ldots,5.4$.

Corresponding results for the isovector Pauli mean square radius are shown on the right in Fig. 1. Although in particular the most recent data points from QCDSF/UKQCD and ETMC show an upward trend, the lattice values in the range of $m_\pi \sim 200\ldots300$ MeV are remarkable 30-50% below the phenomenological number of Ref. [19]. Qualitatively, this doesn't come as a surprise, since ChPT predicts a linearly diverging isovector Pauli radius in the chiral limit, $\langle r^2\rangle_2 \propto m_\pi^{-1}$ as $m_\pi \to 0$. To see if the results on the right in Fig. 1 can be understood more quantitatively, we have employed the SSE-predictions of Refs. [17, 18] together with the low energy constants obtained from the fit to κ on the left in Fig. 1 to extrapolate the phenomenological value for $\langle r^2\rangle_2$ from the physical point to *larger* pion masses. The immediate change of curvature and strong upwards bending towards larger pion masses, illustrated by the shaded band on the right in Fig. 1, indicates that the SSE HBChPT expansion for $\langle r^2\rangle_2$ breaks down at the latest shortly above m_π^{phys}. This suggest that the range of applicability in m_π of ChPT for radii is rather limited (at least in this particular approach, to the given order), in particular in comparison to the anomalous magnetic moment discussed above. As before, this is not unexpected since radii are given by derivatives with respect to a small expansion parameter (Q^2) and therefore correspond to a lower order in the chiral expansion relative to, e.g., $\kappa = F_2(0)$.

GPDS AND THE SPIN STRUCTURE OF THE NUCLEON

Generalized parton distributions (GPDs)[1] provide a particularly successful framework for the investigation of hadron structure, that not only encompasses the well known hadron form factors and the PDFs, but that also provides a solid basis for the decomposition of the nucleon spin in terms of spin and (orbital) angular momentum contributions of quarks and gluons. According to [23], the nucleon spin can be given in terms of x-moments of the GPDs $H(x,\xi,t)$ and $E(x,\xi,t)$,

$$\frac{1}{2} = \sum_{q,g} J_{q,g} = \sum_{q,g} \frac{1}{2}\int_{-1}^{1} dx x \{H^{q,g}(x,\xi,t) + E^{q,g}(x,\xi,t)\}_{t=0} = \sum_{q,g}\frac{1}{2}\left(A_{20}^{q,g}(t) + B_{20}^{q,g}(t)\right)_{t=0}, \tag{1}$$

where A_{20} and B_{20} are generalized form factors (GFFs) that will be discussed in greater detail below. It is important to keep in mind that the total angular momentum of quarks can be further decomposed in a manifestly gauge-invariant manner in terms of quark spin and orbital angular momentum contributions, $J_q = \Delta\Sigma_q/2 + L_q$ [23].

[1] for reviews, we refer to Refs. [20, 21, 22]

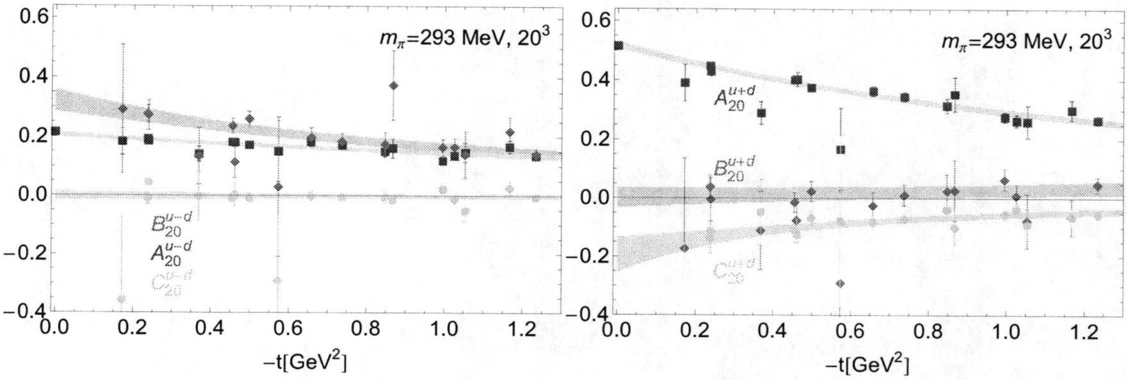

FIGURE 2. Generalized form factors A, B, C as functions of the squared momentum transfer t, in the $\overline{\text{MS}}$ scheme at a scale of $\mu = 2$ GeV, for a pion mass of $m_\pi \approx 300$ MeV [27]. The t-dependence can be parametrized by, e.g., a dipole ansatz, as illustrated by the shaded error bands.

GPDs can be measured for example in deeply virtual Compton scattering employing QCD factorization, however their extraction from experimental data turns out to be very challenging. This holds in particular for their dependence on x at fixed ξ, and therefore also for the x-moments in the decomposition in Eq. (1). As it turns out, this is different in the framework of lattice QCD, which represents at least in principle a straightforward (but in practice also challenging) approach from first principles to the lowest x-moments of PDFs and GPDs.

Significant advances in particular with respect to moments of GPDs have been made since the pioneering calculations by the LHPC and QCDSF collaborations in 2003 [24, 25]. A comprehensive lattice study of GPDs by LHPC using a mixed action approach was presented in Ref. [26]. The discussions in the following sections will be based on a recent update of this study [27], which includes an additional ensemble at a lower pion mass of $m_\pi \approx 300$ MeV, a factor of 8 increased statistics, and a strongly improved statistical analysis.

The x-moments of generalized parton distributions can be written in terms of polynomials in powers of ξ with the GFFs as coefficients. For the $n = 2$-(x-)moments in the unpolarized case, the relevant GFFs A_{20}, B_{20}, C_{20} parametrize the off-forward matrix element of the (symmetric, traceless) energy momentum tensor, e.g. for quarks

$$\langle P'|\overline{q}\gamma^{\{\mu}D^{\nu\}}q|P\rangle = \langle P'|T_q^{\mu\nu}|P\rangle = \overline{U}(P')\left\{\gamma^{\{\mu}\overline{P}^{\nu\}}A_{20}(t) - \frac{i\Delta_\rho\sigma^{\rho\{\mu}\overline{P}^{\nu\}}}{2m_N}B_{20}(t) + \frac{\Delta^\mu\Delta^\nu}{m_N}C_{20}(t)\right\}U(P), \tag{2}$$

where $\overline{P} = (P' - P)/2$, $\Delta = P' - P$, and $t = \Delta^2$. Examples of lattice results from LHPC [27] for the t-dependence of these GFFs in the isovector and isosinglet channels, are displayed in Fig. 2 for a comparatively low pion mass of ≈ 300 MeV. We note at least two characteristic features: First, while the B_{20}-GFF is dominant in the isovector case, it is very small and fully compatible with zero within the uncertainties for all accessible values of t in the isosinglet channel. Conversely, the GFF C_{20} (which is directly related to the $n = 2$-moment of the so-called D-term [28]) is zero within relatively small errors in the isovector channel, but clearly non-zero and negative for $u+d$-quarks. From the sum of A_{20} and B_{20} at zero momentum transfer, we can compute the quark angular momentum J_q, cf. Eq. (1). The above findings for B_{20} obviously have a significant impact on the decomposition of the nucleon spin in terms of quark and gluon degrees of freedom.

DECOMPOSITION OF THE NUCLEON SPIN

Instead of calculating matrix elements of the quark OAM operator directly on the lattice, we can employ the relation $L_q = J_q - \Delta\Sigma_q/2$, together with corresponding results for the quark spin fraction $\Delta\Sigma_q$, to compute the OAM carried by the quarks. The numerical results for $u+d$-quarks are shown on the left in Fig. 3 as a function of the squared pion mass. For the extrapolation of $\Delta\Sigma_{u+d}$ to the physical pion mass, we employed the heavy baryon ChPT results of Ref. [29]. Although HBChPT at the given order is most probably not applicable at the accessible pion masses, the two-parameter

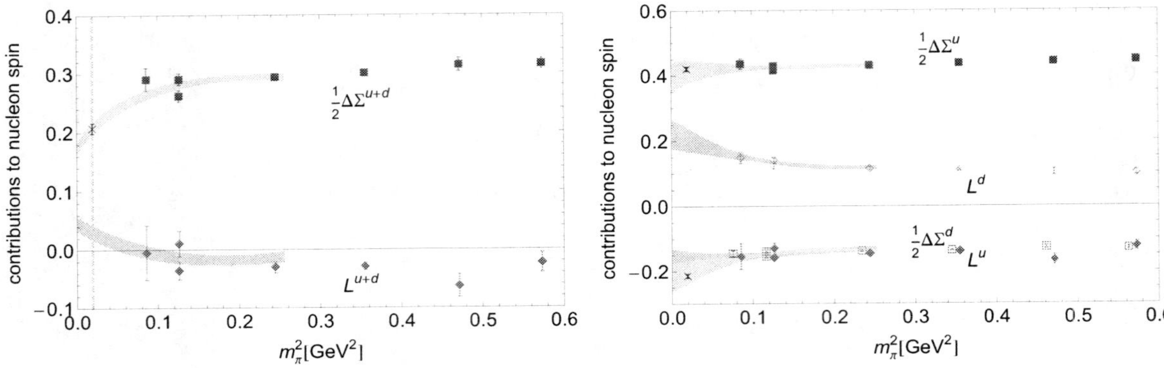

FIGURE 3. Contributions to the nucleon spin in the $\overline{\text{MS}}$ scheme at a scale of $\mu = 2$ GeV [27].

fit to the lattice data points, indicated by the error band, shows an excellent (but likely accidental) agreement with the value from HERMES [30], indicated by the cross, at m_π^{phys}.

We used the results of a covariant baryon chiral perturbation theory calculation [31] for the simultaneous extrapolation of the GFFs A_{20}, B_{20}, C_{20} in m_π and in t, in particular to obtain $B_{20}(t=0, m_\pi^{\text{phys}})$. The results of the fits to the lattice data were subsequently used to compute $L_q = (A+B)_{20}^q(t=0))/2 - \Delta\Sigma_q/2$, which is shown on the left in Fig. 3 for $q = u+d$ by the diamonds and the error band. Most remarkably, following this procedure, we find a very small OAM contribution of only $L_{u+d} \approx (6 \pm 3)\%$ of $1/2$ at the physical pion mass. Note that we obtain a slightly larger value, $L_{u+d} \approx (11 \pm 3)\%$, on the basis of a HBChPT extrapolation of J_{u+d} employing the results of Ref. [32]. These results may be surprising and at first sight even appear to be in clear contradiction to expectations from relativistic quark models, where the quark OAM contributes about $\approx 30-40\%$ to the total nucleon spin. Moreover, it has been frequently pointed out in the past that non-zero quark orbital angular momentum is strictly required with respect to certain non-vanishing single spin asymmetries related to, e.g., the Sivers effect, as well as for the Pauli form factor $F_2(Q^2)$ to be non-zero [33, 34, 35, 36].

Concerning the latter, it is important to be precise in the use of the term *quark orbital angular momentum*: The lattice results on the one hand correspond to the proton matrix element of the manifestly gauge invariant quark OAM operator as given in, e.g., Ref. [23], which is part of Jis nucleon spin sumrule, see Eq. (1) and the adjacent discussion. On the other hand, what is required with respect to a non-vanishing $F_2(Q^2)$ and the Sivers effect are light cone wave functions with non-zero orbital angular momentum in an overlap representation of the corresponding matrix elements. Also, the more heuristic approach to explain single spin asymmetries, e.g. the Sivers effect, proposed in Ref. [35], requires a generically non-zero *quark orbital motion*. The notions of OAM and orbital motion in these contexts are, however, not necessarily in one-to-one correspondence with the gauge invariant L_q discussed before.

Furthermore, we note that L_q is in general not positive definite. To fully appreciate the potential consequences of this basic observation, it is useful to study the individual up and down quark OAM contributions to the nucleon spin, which are shown on the right in Fig. 3 as functions of m_π^2, together with $\Delta\Sigma_{u,d}/2$ and the corresponding chiral extrapolations (given by the error bands). It indeed turns out that the orbital angular momentum of up and of down quarks is in each case *large* and of similar magnitude, but opposite in sign, and therefore nearly cancels in the sum over a wide range of pion masses. From the chiral extrapolations, we find that the individual quark OAM contributions are also substantial at the physical pion mass, amounting to $|L_{u,d}| \approx 33\%$. We also note that cancellations of this sort in the isosinglet channel are not unusual. A well known example is the anomalous magnetic moment, which is large both for the proton and for the neutron, $\kappa_p = F_2^p(0) \sim 1.79$ and $\kappa_n = F_2^n(0) \sim -1.91$, respectively. Using isospin symmetry, however, one finds that the sizeable, individual contributions from up and down quarks to κ_p largely cancel out in the sum, $\kappa_p^{u+d} \approx -0.36$. Keeping in mind that $F_2(Q^2 = -t)$ correspond to the first, and the GFF $B_{20}(t)$ to the second x-moment of the GPD $E(x, \xi = 0, t)$, we also find this to be perfectly consistent with the results in Fig. 2, where in particular $B_{20}^{u+d}(t) \approx 0$, while $B_{20}^{u-d}(t)$ is large and positive.

We now briefly comment on the apparent discrepancy between the common expectations from relativistic quark model calculations that $L_{u+d} \approx 30-40\%$, and the lattice QCD result of a nearly vanishing total quark OAM, $L_{u+d} \approx 0$. As has been pointed already out some time ago [37, 38], the model calculations generically correspond to a low

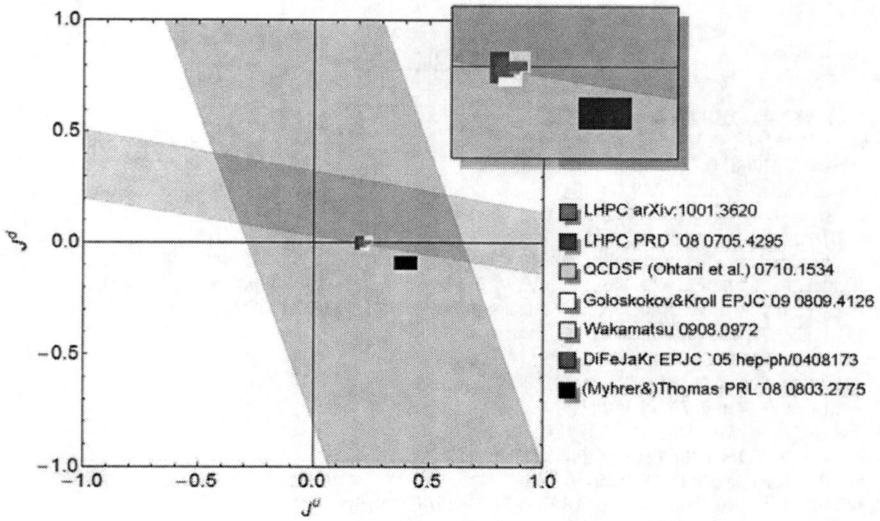

FIGURE 4. Comparison of quark angular momentum contributions to the nucleon spin, obtained in different lattice (for systematic uncertainties, see footnote [3]), model and phenomenological studies [11, 26, 41, 42, 43, 38], displayed in the J_u, J_d-plane in the style of Refs. [44, 45], for the $\overline{\text{MS}}$ scheme at a scale of $\mu = 2$ GeV. Note that the shaded bands are mainly given for illustration purposes and only approximately represent the constraints obtained in Refs. [44, 45], which are still strongly model dependent.

hadronic scale $\ll 1$ GeV, while a typical renormalization scale for which the lattice results are given is $\mu \sim 2$ GeV (e.g. in the $\overline{\text{MS}}$-scheme). Since OAM is a scale and scheme dependent quantity [39], a naive, direct comparison of lattice and model values at the respective scales is therefore in general meaningless. In practice, based on LO and NLO QCD evolution, it indeed turns out that the scale dependence of quark OAM can be very strong at low scales[2], and even lead to a change in sign for, e.g., L_{u-d}, as one goes down from larger scales to low hadronic scales [37, 38, 27]. A more detailed comparison of improved model calculations with results from lattice QCD will be given in [40].

Finally, we show a comparison of the results for the total quark angular momentum $J_q = (A^q_{20} + B^q_{20})_{t=0}/2$, as obtained from the chiral extrapolations of A_{20} and B_{20} mentioned above, with earlier lattice as well as model- and phenomenological calculations in Fig. 4 in the J_u, J_d-plane. It turns out that the numerical values [27], $J_u = 0.236(6)$ and $J_d = 0.0018(37)$, are in surprisingly good agreement not only with the previous lattice computations, but also with most of the phenomenological and model results, at least when they are evolved to the common renormalization point of 4 GeV^2. We note in passing that in the lattice calculations, the smallness of J_d can be traced back to a cancellation of the separately sizeable spin- and OAM contributions of down quarks, as can be inferred directly from the plot on the right in Fig. 3. It will be very interesting to see if the slight discrepancy to the cloudy bag model (CBM) study of Ref. [38] is due to systematic effects, missing contributions or similar on the side of the lattice[3] and the other approaches, or if it points towards the need for a further improvement of the CBM calculation.

ACKNOWLEDGMENTS

It is a pleasure to thank M. Altenbuchinger, B. Musch, W. Weise and the colleagues from QCDSF/UKQCD and LHPC for successful collaboration. I gratefully acknowledge the support by the DFG Emmy-Noether program, cluster of

[2] keeping in mind that the evolution equations in QCD perturbation theory might not be quantitatively applicable at the low scales

[3] most notable issues are: discretization and finite volume effects, missing disconnected contributions, and the so far neglected mixing of quark and gluon operators under renormalization and evolution in the singlet sector

excellence "Origin and Structure of the Universe", and Sonderforschungsbereich/Transregio 55.

REFERENCES

1. K. Orginos, *PoS* **LAT2006**, 018 (2006).
2. P. Hägler, *PoS* **LAT2007**, 013 (2007), 0711.0819.
3. J. M. Zanotti, *PoS* **LAT2008**, 007 (2008), 0812.3845.
4. A. Schäfer, *Nucl. Phys.* **A805**, 230–239 (2008).
5. P. Hägler, *Phys. Rept.* **490**, 49–175 (2010), 0912.5483.
6. D. B. Renner (2010), 1002.0925.
7. P. Hägler, B. U. Musch, J. W. Negele, and A. Schäfer, *Europhys. Lett.* **88**, 61001 (2009), 0908.1283.
8. B. Musch, P. Hägler, A. Schäfer, and J. Negele, *MENU 2010 proceedings (this conference)* (2010).
9. C. Alexandrou, G. Koutsou, J. W. Negele, and A. Tsapalis, *Phys. Rev.* **D74**, 034508 (2006), hep-lat/0605017.
10. M. Göckeler, et al. (2007), 0709.3370.
11. J. D. Bratt, et al., *PoS* **LATTICE2008**, 141 (2008), 0810.1933.
12. H.-W. Lin, et al., *Phys. Rev.* **D78**, 014505 (2008), 0802.0863.
13. T. Yamazaki, et al., *Phys. Rev.* **D79**, 114505 (2009), 0904.2039.
14. S. N. Syritsyn, et al., *Phys. Rev.* **D81**, 034507 (2010), 0907.4194.
15. C. Alexandrou, et al., *PoS* **LAT2009**, 145 (2009), 0910.3309.
16. D. Pleiter, et al., *PoS* **Lattice 2010**, 153 (2010).
17. T. R. Hemmert, and W. Weise, *Eur. Phys. J.* **A15**, 487–504 (2002), hep-lat/0204005.
18. M. Göckeler, et al., *Phys. Rev.* **D71**, 034508 (2005), hep-lat/0303019.
19. M. A. Belushkin, H. W. Hammer, and U. G. Meissner, *Phys. Rev.* **C75**, 035202 (2007), hep-ph/0608337.
20. K. Goeke, M. V. Polyakov, and M. Vanderhaeghen, *Prog. Part. Nucl. Phys.* **47**, 401–515 (2001), hep-ph/0106012.
21. M. Diehl, *Phys. Rept.* **388**, 41–277 (2003), hep-ph/0307382.
22. A. V. Belitsky, and A. V. Radyushkin, *Phys. Rept.* **418**, 1–387 (2005), hep-ph/0504030.
23. X.-D. Ji, *Phys. Rev. Lett.* **78**, 610–613 (1997), hep-ph/9603249.
24. P. Hägler, et al., *Phys. Rev.* **D68**, 034505 (2003), hep-lat/0304018.
25. M. Göckeler, et al., *Phys. Rev. Lett.* **92**, 042002 (2004), hep-ph/0304249.
26. P. Hägler, et al., *Phys. Rev.* **D77**, 094502 (2008), 0705.4295.
27. J. D. Bratt, et al., *Phys. Rev.* (2010), 1001.3620.
28. M. V. Polyakov, and C. Weiss, *Phys. Rev.* **D60**, 114017 (1999), hep-ph/9902451.
29. M. Diehl, A. Manashov, and A. Schäfer, *Eur. Phys. J.* **A31**, 335–355 (2007), hep-ph/0611101.
30. A. Airapetian, et al., *Phys. Rev.* **D75**, 012007 (2007).
31. M. Dorati, T. A. Gail, and T. R. Hemmert, *Nucl. Phys.* **A798**, 96–131 (2008), nucl-th/0703073.
32. J.-W. Chen, and X.-D. Ji, *Phys. Rev. Lett.* **88**, 052003 (2002), hep-ph/0111048.
33. S. J. Brodsky, D. S. Hwang, B.-Q. Ma, and I. Schmidt, *Nucl. Phys.* **B593**, 311–335 (2001), hep-th/0003082.
34. S. J. Brodsky, D. S. Hwang, and I. Schmidt, *Phys. Lett.* **B530**, 99–107 (2002), hep-ph/0201296.
35. M. Burkardt, *Phys. Rev.* **D66**, 114005 (2002), hep-ph/0209179.
36. M. Burkardt, and D. S. Hwang, *Phys. Rev.* **D69**, 074032 (2004), hep-ph/0309072.
37. M. Wakamatsu, and Y. Nakakoji, *Phys. Rev.* **D77**, 074011 (2008), 0712.2079.
38. A. W. Thomas, *Phys. Rev. Lett.* **101**, 102003 (2008), 0803.2775.
39. X.-D. Ji, J. Tang, and P. Hoodbhoy, *Phys. Rev. Lett.* **76**, 740–743 (1996), hep-ph/9510304.
40. M. Altenbuchinger, P. Hägler, and W. Weise, in preparation (2010).
41. D. Brömmel, et al., *PoS* **LAT2007**, 158 (2007), 0710.1534.
42. M. Wakamatsu, *Eur. Phys. J.* **A44**, 297–303 (2010), 0908.0972.
43. M. Diehl, T. Feldmann, R. Jakob, and P. Kroll, *Eur. Phys. J.* **C39**, 1–39 (2005), hep-ph/0408173.
44. A. Airapetian, et al., *JHEP* **06**, 066 (2008), 0802.2499.
45. M. Mazouz, et al., *Phys. Rev. Lett.* **99**, 242501 (2007), 0709.0450.

Short-Range Nucleon-Nucleon Correlations

Douglas W. Higinbotham

Thomas Jefferson National Accelerator Facility, Newport News, VA 23601, USA

Abstract. Valence-shell nucleon knock-out experiments, such as ^{12}C(e,e'p)^{11}B, measure less strength then is predicted by independent particle shell model calculations. The theoretical solution to this problem is to include the correlations between the nucleons in the nucleus in the calculations. Motivated by these results, many electron scattering experiments have tried to isolate the signal from these correlations in order to gain new insight into the short-range part of the nucleon-nucleon potential. Unfortunately, many competing mechanisms can cause the same observable final-state as an initial-state correlation, making truly isolating the signal extremely challenging. This paper reviews the recent experimental evidence for short-range correlations, as well as explores the possibility that such correlations are responsible for the EMC effect in the $0.3 < x_B < 0.7$ deep inelastic scattering ratios.

Keywords: Correlations, Nucleon-Nucleon Potential
PACS: 13.60.Hb,21.30.-x,21.45.Bc

INTRODUCTION

The independent particle shell model enjoyed many successes, but failed to predict the strength of valance-shell nucleon knock-out experiments [1, 2]. The theoretical solution to this problem was simply to include the correlations that must exist between the nucleons in the nucleus. When compared to an early independent particle shell model calculations, these correlations move strength away from the bound-states and into a continuum of energies and momentums. This effect is illustrated in the work of Benhar *et al.* where it is shown how the ^{12}C momentum distribution from an independent particle model calculation is reduced at low momentum and enhanced at momentums above $\sim$250 MeV/c by the addition of nucleon-nucleon correlations [3]. An *ab-initio* calculation of the depletion of the nuclear Fermi sea by correlations can be found in Ref. [4].

If one could cleanly isolate the high momentum part of the momentum distribution, it would allow electron scattering to provide new information about the short-range part of the nucleon-nucleon potential. This is shown diagrammatically in Fig. 1 where a) shows simple single nucleon knock-out and b) shows the electron knocking-out a correlated pair of nucleons.. Unfortunately, it has proven experimentally challenging to cleanly isolate the effects of short-range correlations from other effects such as meson-exchange currents, isobar-configurations, and final-state interactions. Thus, while the 1981 D(e,e'p)n momentum distribution extraction up to $\approx$250 MeV/c has stood the test of time [5], that same paper's prediction that one could probe the short-range part of the nucleon-nucleon potential by simply extending the range of kinematics has proven extremely challenging.

ELECTRON SCATTERING KINEMATICS

The story of the search for nucleon-nucleon correlations is best told with the Brojken x_B variable:

$$x_B = \frac{Q^2}{2m\omega} \tag{1}$$

where m is the mass of the proton, ω is the energy transfer given by the difference between the beam energy and the scattered electron energy, Q^2 is the four-momentum transfer given by $\mathbf{q}^2$ - ω^2 where $\mathbf{q}$ is the three-momentum transfer vector given by the difference between the incident and scattered electron momentum vectors. While this kinematic variable is typically used for deep inelastic physics, it turns out to be a convent variable for counting how many nucleons must be involved in a knock-out reaction. For example, while elastic scattering on a proton has a x_B = 1, quasi-elastic scattering of a nucleon from a nucleus can kinematically extended into the $x_B > 1$ region due to the presence of the second nucleon. For the A(e,e') reaction, the x_B and Q^2 variables can be determined by knowing the properties of the incident beam and measuring the energy and momentum of the scattered electron.

12th International Conference on Meson-Nucleon Physics and the Structure of the Nucleon (MENU 2010)
AIP Conf. Proc. 1374, 85-90 (2011); doi: 10.1063/1.3647103

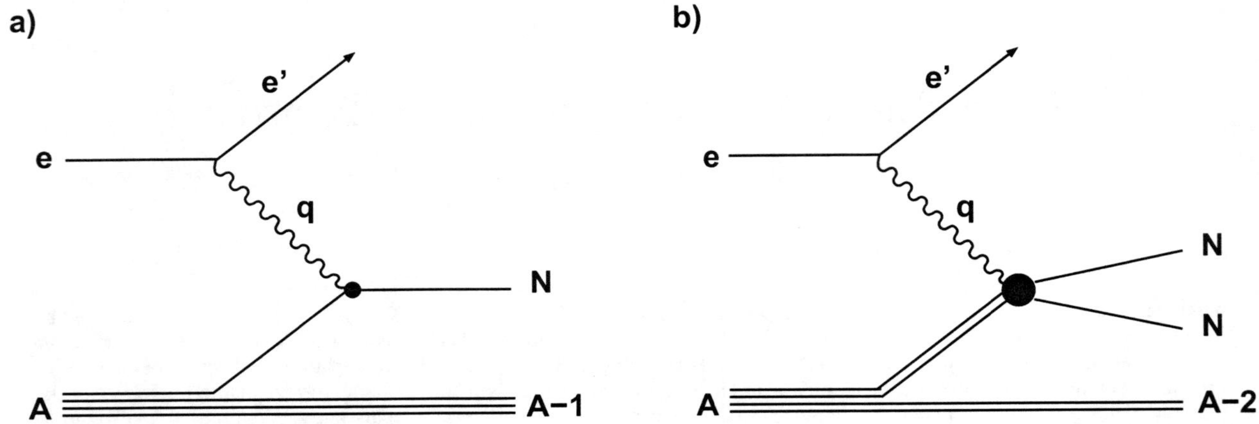

FIGURE 1. The simple goal of short-range nucleon-nucleon correlation studies is to cleanly isolate diagram b) from a). Unfortunately, there are many other diagrams, including those with final-state interactions, that can produce the same final state as the diagram scientists would like to isolate. If one could find kinematics that were dominated by diagram b) it would finally allow electron scattering to provide new insights into the short-range part of the nucleon-nucleon potential.

For A(e,e'p) reactions, one can determine not only the energy and moment transferred, but also the energy and momentum of the knocked-out nucleon. The difference between the transferred and detected energy and momentum is referred to as the missing energy, E_{miss} and missing momentum, p_{miss}, respectively. From the theoretical works on how short-range nucleon-nucleon correlations effects the momentum distribution of nucleons in the nucleus [6], it is clear one must probe beyond the simple particle in an average potential motion of the nucleon in the nucleus of approximately 250 MeV/c in order to observe the effects of correlations.

With the construction of the Jefferson Lab Continuous Electron Beam Facility (CEBAF) [7], it was possible to do high-luminosity knock-out reactions in *ideal* quasi-elastic kinematics into the $p_{miss} > 250$ MeV/c region. In the early Jefferson Lab knock-out reaction proposals, such as E89-044 ^{3}He(e,e'p)pn and ^{3}He(e,e'p)d, these kinematics were argued as the key to cleanly observe the effects of short-range correlations. And while final results of the experiments were clearly effected by the presence of correlations, the magnitude of the cross sections in the high missing momentum region was dominated by final-state interaction effects [8, 9]. Equally striking was the D(e,e'p)n data from CLAS taken at $Q^2 > 5$ [GeV/c]^2 in $x_B < 1$ kinematics [10]. Here it was shown that meson-exchange currents, final-state interaction, and delta-isobar configurations mask cleanly probing nucleon-nucleons even at extremely high Q^2 in $x_B < 1$ kinematics.

NUCLEAR SCALING

With both the $x_B < 1$ and $x_B = 1$ kinematics practically ruled out for ever being able to cleanly probe short-range correlations; there is only one region left to explore: $x_B > 1$. This is a special region, since it is kinematically forbidden for a free nucleon, and thus seems to be a natural place to observe effects of multi-nucleon interactions. These kinematics were probed with limited statistics at SLAC [11] and the plateaus in the per nucleon ratios, r(A/d), were claimed at to be evidence for short-range correlations [12].

In 2003, CLAS published high statics data in the same kinematic region. The results clearly showed that the plateaus could only be seen for $Q^2 > 1$ [GeV/c]^2 and $x_B > 1$ kinematics [13] as predicted by Frankfurt and Strikman [14]. But plateaus alone are not evidence for correlations, just evidence that the functional form of the cross section is the same for the two nuclei; so data was taken the $x_B > 2$ region. By logic, if $1 < x_B < 2$ is a region of two-nucleon correlations, then the $x_B > 2$ region should be dominated by three-nucleon correlations. The CLAS $Q^2 > 1$ and $x_B > 2$ experiment reported observing a second scaling plateau as shown in Fig. 2 [15]. Preliminary results of Hall C high precision data have shown roughly the same magnitude for these plateaus as CLAS and shown that there is no Q^2 dependence in the $2 < Q^2 < 4$ [GeV/c]^2 range [16, 17].

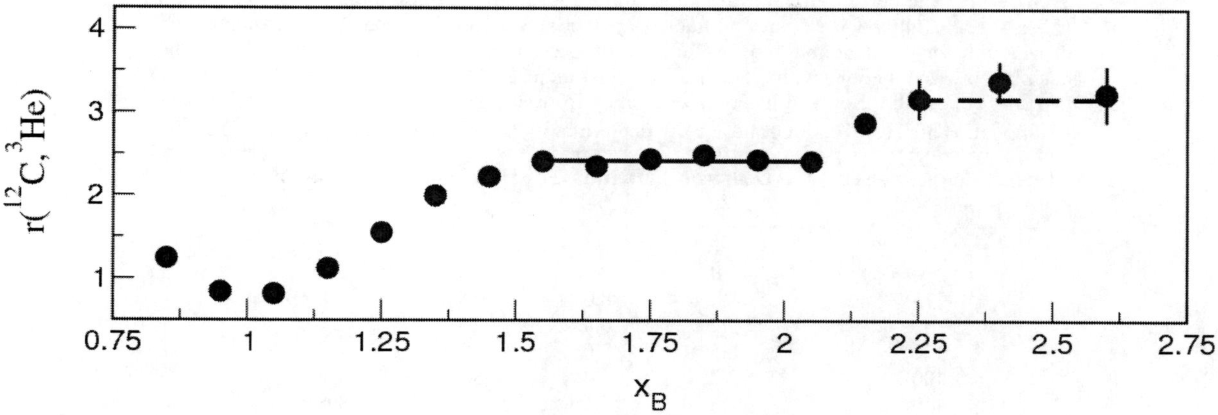

FIGURE 2. Shown is the per nucleon ratio of ^{12}C(e,e') to ^{3}He(e,e') as a function of x_B from CLAS [13, 15]. The data show two scaling plateaus. The first, in the $x_B > 1$ region, is claimed to be due to two-nucleon correlations; while the second, in the $x_B > 2$ region, is claimed to be due to three-nucleon correlations.

EXCLUSIVE REACTIONS

While the inclusive results alone are rather convincing, one would like to observe that the final-state is truly dominated by a two-nucleon knock-out process. This can be done by measuring two high-momentum nucleons in the final-state and showing that the rest of the nuclear system is left nearly at rest. Two independent experiments have been carried out that have done this, one at Brookhaven with a proton beam and another at Jefferson Lab with an electron beam.

The Brookhaven experiment made use of a proton beam on a carbon target and detected two forward going protons. Summing the momentum of the two protons and subtracting the incident proton's momentum, allowed them to determined the missing momentum. They then looked for recoiling neutrons with a large acceptance detector. For missing momentum below 250 MeV/c, an isotropic distribution of neutrons were observed; but above 250 MeV/c the signal was completely dominated by neutrons traveling in the direction of the missing momentum vector [18, 19]. Analysis of this result showed the reaction was dominated by initial-state neutron-proton pairs [20].

The Jefferson Lab experiment made use of the two Hall A high resolution spectrometers [21] to detect the electron and proton from the ^{12}C(e,e'p) reaction with $Q^2 = 2$ [GeV/c]^2 and $x_B > 1$. In the direction of the missing momentum vector, a proton detector known as BigBite and the Hall A Neutron Detector, HAND, were placed in such a way that both recoiling protons and neutron could be detected in nearly the same solid angle acceptance. While the experiment saw the signal of correlated proton-proton pairs [22]; what they discovered was that an overwhelming faction of the events were coming from initial-state neutron-proton pairs [23]. The dominance of neutron-proton paris over proton-proton pair was explained by several groups [24, 25, 26, 27] as being due to short-range tensor correlations. In 2011, the experiment will be repeated, but will go to even larger missing momentum. The new experiment expects to finally get into exclusive kinematics where the short-range part of the nucleon-nucleon potential is dominating the reaction.

STRENGTH OF CORRELATIONS IN NUCLEI

With the exclusive data showing that the inclusive scaling is dominated by initial-state quasi-deuteron neutron-proton correlations, one can confidently use the inclusive scaling results to calculate the strength of the high-momentum part of the momentum distribution for various nuclei. This is done by using a realistic nucleon-nucleon potential [28] to calculate the deuteron's nucleon momentum distribution, $n_D(p)$, where the function was normalized such that

$$1 = 4\pi \int_0^\infty p^2 \times n_D(p) \times dp \tag{2}$$

TABLE 1. The percent initial-state momentums above different minimum initial momentum values.. The Argonne v-18 nucleon-nucleon potential was used to generate the momentum distribution for the deuteron and the function properly normalized. For the other nuclei, the calculation was only done with momentums greater then 250 MeV/c where the relation $n(p)_A = n(p)_d \times a_{2N}$ can be used and where a_{2N} is taken from experimental $x_B > 1$ inclusive nuclear scaling ratios of ^{3}He, ^{4}He, ^{12}C, and ^{56}Fe to deuterium of 2, 4, 4.8 and 5.7, respectively [12, 13].

Integration Lower Limit [MeV/c]	**Deuteron** [%]	**Helium-3** [%]	**Helium-4** [%]	**Carbon-12** [%]	**Iron-56** [%]
0	100*	-†	-	-	
250	5.0	10	20	24	28
275	4.0	8.0	16	19	22
300	3.2	6.4	13	15	18
370	2.5	5.0	10	12	15
400	2.2	4.4	8.8	10	13
440	1.8	3.6	7.2	8.6	10
500	1.3	2.6	5.2	6.2	7.4

* Check of proper normalization of the deuteron momentum distribution.
† Many body systems are only calculated above 250 MeV/c where Eq. 3 should be valid.

. Since the plateaus in the inclusive scaling indicate that the functional forms of the distribution must all have the same shape in this region, the relation for the high momentum part of the distribution can be written

$$n_A(p) = n_D(p) \times a_{2N} \tag{3}$$

where in this case a_{2N} is simply the magnitude of the scaling plateau taken from data. To calculate the fraction of the initial-state above a certain momentum for any nuclei where a_{2N} has been measured, one simply calculates

$$4\pi \int_{p_{min}}^{\infty} p^2 \times n_A(p) \times dp. \tag{4}$$

The integrations of $n_D(p)$ and $n_A(p)$ starting from various minimum momentums, p_{min}, are shown in Table 1.

THE NUCLEAR EMC EFFECT

From the results in Table 1, it is clear the magnitude of the high-momentum tail is non-negligible fraction of the initial-state. If this is truly an initial-state effect, then it must appear globally in nuclear physics data. One place to look to see if these numbers provide a solution is to the long standing puzzle of the EMC effect.

Data from the EMC collaboration showed that if one takes ratios of deep-inelastic scattering data from various nuclei one finds structure in the ratios [29]. This is very different from a simple minded picture where one would expect the partons in nucleons to all act the same and the per nucleon ratio to simply be one. Many possible explanations have been put forward to explain this phenomenon [30], yet to this day there is no generally accepted solution.

It is interesting, is to plot EMC type data with a different nuclei in both the numerator and denominator as shown in Fig. 3. There even though the ratio of the number of nuclei is bigger for ^{12}C/^{4}He and ^{4}He/D, 3:1 vs 2:1, the EMC is dramatically bigger for the ^{4}He/D ratio. This makes one immediately ask what is nearly the same between ^{12}C and ^{4}He, but very different for ^{4}He and D. One obvious answer is the average binding energy; but another answer, related to the first, is the strength of short-range correlations. In fact, as pointed out in Refs. [31, 32], the magnitude of the EMC effect seems to follow the magnitudes of the nuclear scaling plateaus. This phenomenological correlation between the EMC effect and the scaling plateaus is shown in Fig. 4. Assuming this correlation is showing the reason for the EMC effect, it is even possible to use the magnitudes of the scaling plateaus to correct the deuteron back to the case of a free proton and neutron allowing for a phenomenological extraction of the F2n structure function [32].

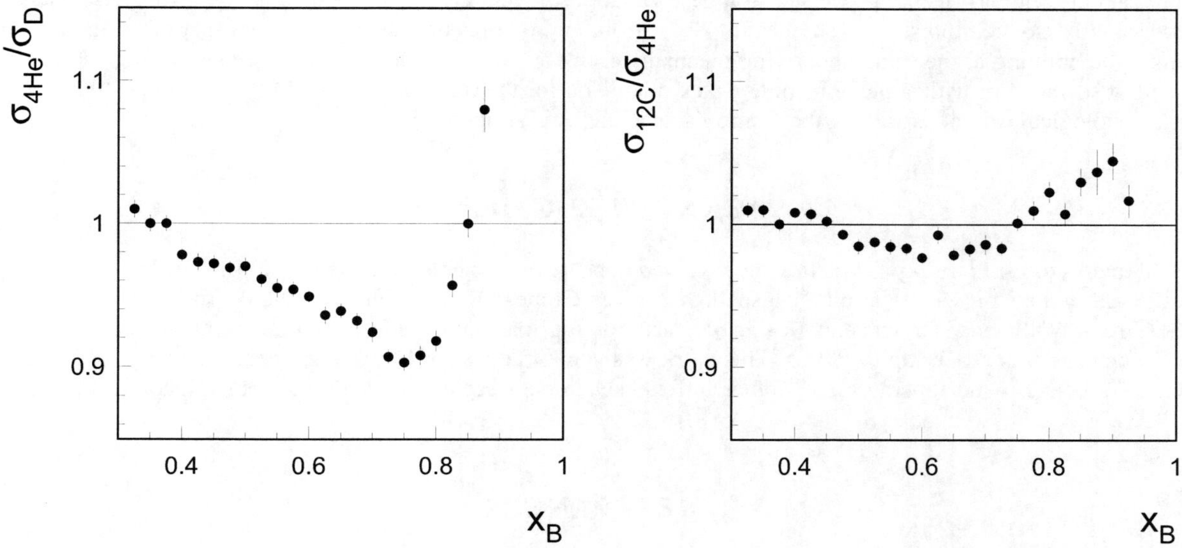

FIGURE 3. Shown on the left is the Hall C deep inelastic per nucleon cross section ratio of ^{4}He to D with its characteristic dip around x_B of 0.75 [33]. On the other hand, the ratio of ^{12}C to ^{4}He the effect is relatively small, which is consistent with the claim EMC is a local density effect.

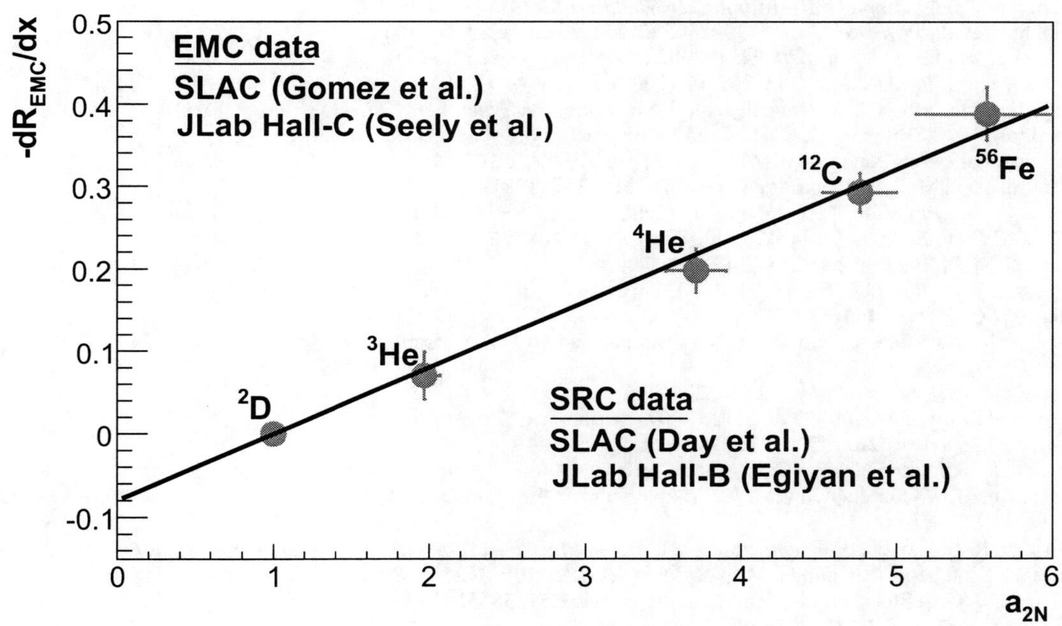

FIGURE 4. Shown is the $0.3 < x_B < 0.7$ EMC slopes, dR/dx, for various nuclei from Seely and Gomez [33, 34] plotted vs the magnitude of the $x_B > 1$ nuclear scaling ratio, a_{2N}, for various nuclei from Day and Egiyan [11, 13]. The linear correction between these two seemly disconnected regions is already striking and soon to be published data will further enhance the results [16].

SUMMARY

After years of trying to cleanly probe short-range correlations, recent $Q^2 > 1$ [GeV/c]^2 and $x_B > 1$ experiments seem to have finally isolated the signal. The beauty is not from any one result, but from the fact that all of the new data seems to be pointing at the same underlying mechanism. In fact, with the the magnitude of short-range correlations in hand, it seems to provide a phenomenological explanation for the $0.3 < x_B < 0.7$ EMC effect slope and provide a phenomenological way of extracting the neutron's deep inelastic structure function.

ACKNOWLEDGMENTS

Special thanks to Eli Piatetzky for getting me involved in short-range nucleon-nucleon studies and also for suggesting to the organizers I present this talk. I also thank Javier Gomez, Larry Weinstein, Mark Strikman, Jerry Miller, Franz Gross, Wally van Orden, and Misak Sargsian for the many interesting discussions about what these new experimental results are trying to tell us. This work was supported by the U.S. Department of Energy and Jefferson Science Associates which operates the Thomas Jefferson National Accelerator Facility under DOE contract DE-AC05-06OR23177.

REFERENCES

1. L. Lapikas, *Nucl. Phys.* **A553**, 297–308 (1993).
2. J. J. Kelly, *Adv. Nucl. Phys.* **23**, 75–294 (1996).
3. O. Benhar, D. day, and I. Sick, *Rev. Mod. Phys.* **80**, 189–224 (2008), nucl-ex/0603029.
4. A. Rios, A. Polls, and W. H. Dickhoff, *Phys. Rev.* **C79**, 064308 (2009), arXiv:0904.2183.
5. M. Bernheim, et al., *Nucl. Phys.* **A365**, 349–370 (1981).
6. C. Ciofi degli Atti, and S. Simula, *Phys. Rev.* **C53**, 1689 (1996), nucl-th/9507024.
7. C. Leemann, D. Douglas, and G. Krafft, *Ann.Rev.Nucl.Part.Sci.* **51**, 413–450 (2001).
8. F. Benmokhtar, et al., *Phys. Rev. Lett.* **94**, 082305 (2005), nucl-ex/0408015.
9. M. M. Rvachev, et al., *Phys. Rev. Lett.* **94**, 192302 (2005), nucl-ex/0409005.
10. K. S. Egiyan, et al., *Phys. Rev. Lett.* **98**, 262502 (2007), nucl-ex/0701013.
11. D. B. Day, J. S. McCarthy, T. W. Donnelly, and I. Sick, *Ann. Rev. Nucl. Part. Sci.* **40**, 357–410 (1990).
12. L. L. Frankfurt, M. I. Strikman, D. B. Day, and M. Sargsian, *Phys. Rev.* **C48**, 2451–2461 (1993).
13. K. S. Egiyan, et al., *Phys. Rev.* **C68**, 014313 (2003), nucl-ex/0301008.
14. L. L. Frankfurt, and M. I. Strikman, *Phys. Rept.* **76**, 215–347 (1981).
15. K. S. Egiyan, et al., *Phys. Rev. Lett.* **96**, 082501 (2006), nucl-ex/0508026.
16. N. Fomin, *AIP Conf. Proc.* **947**, 174–177 (2007), arXiv:0808.2625.
17. N. Fomin (2008), Ph.D.Thesis, arXiv:0812.2144.
18. A. Tang, et al., *Phys. Rev. Lett.* **90**, 042301 (2003), nucl-ex/0206003.
19. J. Watson, *AIP Conf.Proc.* **1056**, 295–300 (2008).
20. E. Piasetzky, M. Sargsian, L. Frankfurt, M. Strikman, and J. W. Watson, *Phys. Rev. Lett.* **97**, 162504 (2006), nucl-th/0604012.
21. J. Alcorn, et al., *Nucl. Instrum. Meth.* **A522**, 294–346 (2004).
22. R. Shneor, et al., *Phys. Rev. Lett.* **99**, 072501 (2007), nucl-ex/0703023.
23. R. Subedi, et al., *Science* **320**, 1476–1478 (2008), arXiv:0908.1514.
24. R. Schiavilla, R. B. Wiringa, S. C. Pieper, and J. Carlson, *Phys. Rev. Lett.* **98**, 132501 (2007), nucl-th/0611037.
25. M. M. Sargsian, T. V. Abrahamyan, M. I. Strikman, and L. L. Frankfurt, *Phys. Rev.* **C71**, 044615 (2005), nucl-th/0501018.
26. R. B. Wiringa, R. Schiavilla, S. C. Pieper, and J. Carlson, *Phys. Rev.* **C78**, 021001 (2008), arXiv:0806.1718.
27. M. Alvioli, C. Ciofi degli Atti, and H. Morita, *Phys. Rev. Lett.* **100**, 162503 (2008).
28. R. B. Wiringa, V. G. J. Stoks, and R. Schiavilla, *Phys. Rev.* **C51**, 38–51 (1995), nucl-th/9408016.
29. J. Aubert, et al., *Phys.Lett.* **B123**, 275 (1983).
30. D. F. Geesaman, K. Saito, and A. W. Thomas, *Ann. Rev. Nucl. Part. Sci.* **45**, 337–390 (1995).
31. D. W. Higinbotham, J. Gomez, and E. Piasetzky, *submitted to Phys. Lett. B* (2010), arXiv:1003.4497.
32. L. B. Weinstein, et al., *Phys. Rev. Lett.* **106**, 052301 (2011), arXiv:1009.5666.
33. J. Seely, et al., *Phys. Rev. Lett.* **103**, 202301 (2009), arXiv:0904.4448.
34. J. Gomez, R. Arnold, P. E. Bosted, C. Chang, A. Katramatou, et al., *Phys.Rev.* **D49**, 4348–4372 (1994).

Strangeness Production at COSY

Frank Hinterberger*, Hartmut Machner† and Regina Siudak**

*HISKP, Universität Bonn, Nussallee 14-16, D 53115 Bonn, Germany
†Fakultät für Physik, Universität Duisburg-Essen, Lotharstr. 1, D 47048 Duisburg, Germany
**Inst. Nucl. Physics, PAN, ul. Radzikowskiego 152, 31-342 Kraków, Poland

Abstract. The paper gives an overview of strangeness-production experiments at the Cooler Synchrotron COSY. Results on kaon-pair and ϕ meson production in pp, pd and dd collisions, hyperon-production experiments and Λp final-state interaction studies are presented.

Keywords: Associated strangeness production, Final state interactions, Hyperon-nucleon interactions
PACS: 13.75.Cs, 13.75.-n, 14.20Jn, 14.40.Df, 14.20.Jn 25.40.Ve

INTRODUCTION

The cooler synchrotron COSY [1] at the Forschungszentrum Jülich in Germany can accelerate protons and deuterons up to about 3.7 GeV/c. Both, unpolarized and polarized beams are available. Excellent beam quality can be achieved using electron- and/or stochastic cooling. COSY can be used as an accelerator for external target experiments and as storage ring for internal target experiments. The strangeness production experiments have been performed at the internal spectrometer ANKE by the COSY-ANKE collaboration, at the internal COSY-11 spectrometer by the COSY-11 collaboration, at the external TOF facility by the COSY-TOF collaboration and at the external BIG KARL spectrometer by the COSY-MOMO and COSY-HIRES collaborations.

KAON-PAIR PRODUCTION EXPERIMENTS

Extensive measurements of kaon-pair production have been performed at several COSY facilities. Total and differential cross sections are now available for a variety of reactions. The world data set of total cross sections for kaon pair production is shown in Fig. 1, left panel, as a function of the excess energy ε. The $pp \rightarrow ppK^+K^-$ reaction (black symbols) has been studied by the COSY-11 [2, 3, 4] and COSY-ANKE [5, 6] collaborations at excitation energies between 3 and 108 MeV. The closely related reaction $pp \rightarrow dK^+\bar{K}^0$ (green) has also been studied by COSY-ANKE [7, 8]. In addition the reaction $pn \rightarrow dK^+K^-$ (red) has been investigated by COSY-ANKE [9, 10] using a deuterium cluster-jet as a neutron target. The momentum of the non-observed proton spectator and the excess energy ε have been reconstructed from the four-momenta of the detected deuteron and kaon pair. The $pd \rightarrow {}^3HeK^+K^-$ (pink) reaction has been measured by the COSY-MOMO collaboration [11]. The $dd \rightarrow {}^4HeK^+K^-$ reaction studied by COSY-ANKE is an ideal isospin zero filter. It could be sensitive to the production of the scalar meson $f_0(980)$. But the total cross section (blue point) amounts only to 5 pb [12]. The high energy $pp \rightarrow ppK^+K^-$ result (open circle) has been measured at SATURNE [13].

The invariant mass distribution of K^+K^- pairs has been measured by COSY-ANKE [6] at three excess energies $\varepsilon = 51, 67$ and 108 MeV. The internal ANKE spectrometer detects simultaneously slow positive and negative particles in the side detectors and fast positive particles in the forward detector. The spectrum taken at $\varepsilon = 51$ MeV ($T_p =$ 2.65 GeV) is shown in Fig. 1, right panel. All three spectra show a strong peak which is due to the production of the $\phi(1020)$ meson which decays with about 50 % into K^+K^-. The prompt K^+K^- distribution can be described by a 4-body phase-space distribution modified by the pp and K^-p final state interaction (FSI). All three distributions show a low mass enhancement over the fit curves. This enhancement has been discussed as a coupled channel effect due to the $K^0\bar{K}^0$ channel whose threshold (dashed vertical line) is about 8 MeV above the K^+K^- threshold [14].

The important K^-p FSI has been observed by comparing the invariant mass distributions of K^-p and K^+p [6, 15]. The ratio of K^-p to K^+p production changes by an order of magnitude within 50 MeV. The same holds true if one compares K^-pp and K^+pp. The K^-p FSI can be described by assuming an imaginary scattering length of 1.5 fm.

12th International Conference on Meson-Nucleon Physics and the Structure of the Nucleon (MENU 2010)
AIP Conf. Proc. 1374, 91-96 (2011); doi: 10.1063/1.3647104

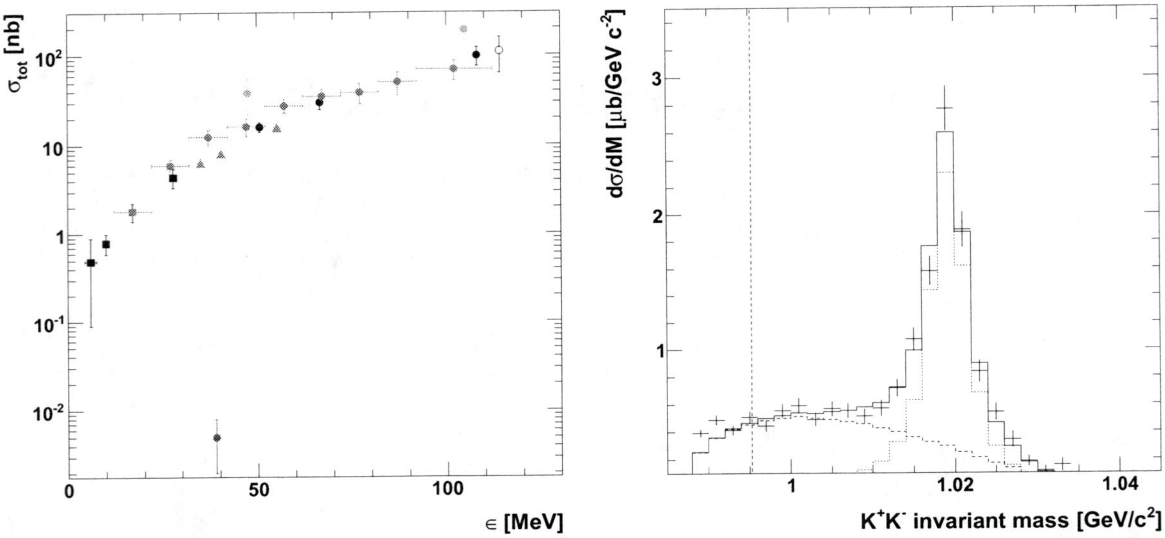

FIGURE 1. Left: Total cross sections for kaon-pair production as a function of the excess energy ε measured at COSY. Black: $pp \to ppK^+K^-$ [2, 3, 4, 5, 6]. Green: $pp \to dK^+\bar{K}^0$ [7, 8]. Red: $pn \to dK^+K^-$ [9, 10]. Pink: $pd \to {}^3HeK^+K^-$ [11]. Blue: $dd \to {}^4HeK^+K^-$ [12]. The high energy $pp \to ppK^+K^-$ result (open circle) has been measured at SATURNE [13]. Right: Invariant mass distribution of K^+K^- pairs measured by COSY-ANKE [6] at $\varepsilon = 51$ MeV ($T_p = 2.65$ GeV). Dotted and dashed histograms represent the ϕ and non-ϕ contributions. The solid histogram is the sum of both. The $K^0\bar{K}^0$ threshold is indicated by the dashed vertical line.

In Fig. 2, the left panel shows the total cross sections for $pp \to ppK^+K^-$ as a function of the excess energy ε. The data are from COSY-11 [2, 3, 4], COSY-ANKE [5, 6] and SATURNE-DISTO [13]. The dashed curve represents a pure 4-body phase-space calculation. The dot-dashed curve includes the pp FSI and the solid curve includes the K^-p FSI. The right panel shows the total cross section for ϕ production as a function of the corresponding excess energy ε. The data cannot be described by a pure three-body phase-space behavior. But the inclusion of the pp FSI is

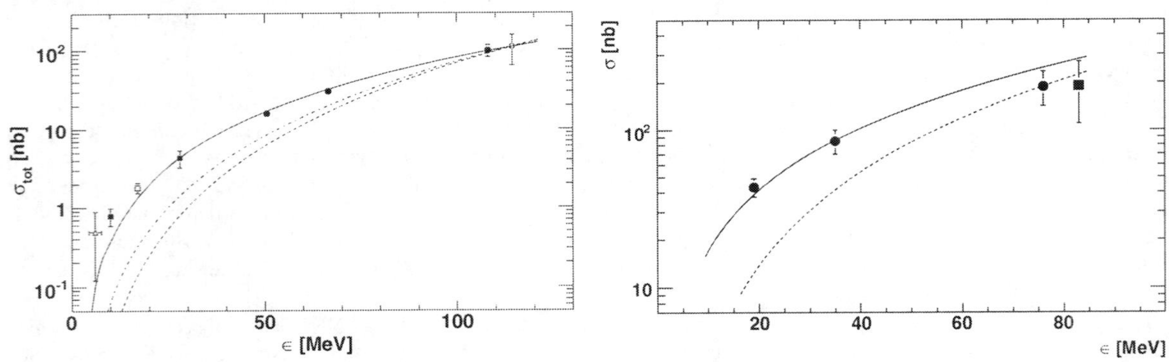

FIGURE 2. Left: Total cross section for $pp \to ppK^+K^-$ vs. excess energy ε [6]. Data from COSY-ANKE (closed circles) [6], COSY-11 (open triangle) [2], (open square) [3], (closed squares) [4] and DISTO (open circle) [13]. The dashed curve represents the energy dependence from four-body phase space. The dot-dashed includes the pp FSI and the solid curve includes the K^-p FSI. Right: Total cross sections for $pp \to pp\phi$ vs. excess energy ε [5]. Data from COSY-ANKE (closed circles) [5] and DISTO (closed square) [13]. The dashed curve represents a three-body phase-space calculation. The solid curve includes the pp FSI.

sufficient to reproduce the energy dependence [5]. Another important result is the angular distribution of the ϕ decay into K^+K^- at $\varepsilon = 18.5$ MeV which is a pure $\sin^2\Theta$ distribution in the Jackson frame. Thus, the ϕ meson is tensor polarized with $m = \pm 1$ along the beam axis as expected near threshold due to conservation laws [5]. Comparing the total cross sections for ϕ- and ω-production at corresponding excess energies [5, 16] yields a ratio $R_{\phi/\omega}$ which is about eight times larger than the prediction $R_{OZI} = 4.2 \cdot 10^{-3}$ [17] based on the Okubo-Zweig-Iizuka (OZI) rule.

The K^+K^-- and ϕ-production was also studied by the COSY-MOMO collaboration at the magnetic spectrograph BIG KARL using the $pd \to {}^3HeK^+K^-$ reaction [11]. The invariant mass distributions measured at three excess energies, 35, 40 and 55 MeV, can be described by pure phase-space distributions. Effects due to a possible $K^-{}^3He$ FSI are not visible. In comparison to the $pp \to pp\phi$ reaction [5] the ϕ peak is less pronounced. The surprising result of the $pd \to {}^3He\phi$ study is that the the ϕ meson is tensor polarized with $m = 0$ along the beam axis [11]. In contrast, the ω meson is unpolarized in the corresponding reaction $pd \to {}^3He\omega$ [18]. The ratio $R_{\phi/\omega}$ of total cross sections for $pd \to {}^3He\phi$ and $pd \to {}^3He\omega$ at corresponding excess energies is about a factor 20 larger [11] than predicted by the OZI-rule [17].

HYPERON-PRODUCTION EXPERIMENTS

Exclusive measurements of hyperon production have been performed using the associated strangeness reactions $pp \to K^+\Lambda p$, $pp \to K^+\Sigma^0 p$, $pp \to K^+\Sigma^+ n$ and $pp \to K^0\Sigma^+ p$ by the COSY-11, COSY-ANKE and COSY-TOF collaborations. The aim of those measurements is to understand (i) the reaction mechanism, (ii) to study the effect of N^* resonances and (iii) to study the FSI between the outgoing nucleon and hyperon.

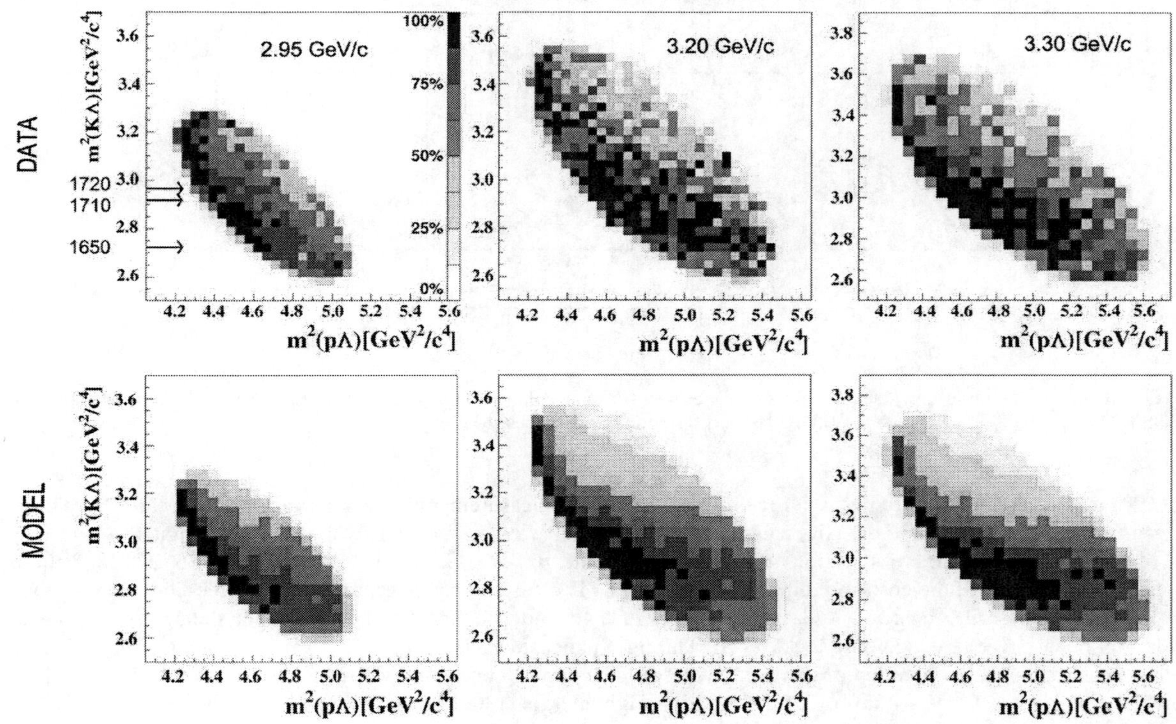

FIGURE 3. Dalitz plots of the reaction $pp \to K^+\Lambda p$ measured by COSY-TOF at 2.95, 3.20 and 3.30 GeV/c [19]. The model calculations are performed using the ISOBAR model of Sibirtsev [20].

The COSY-TOF detector enables exclusive measurements with a large solid angle in the laboratory system and almost 4π in the c.m. system. Recent measurements of the reaction $pp \to K^+\Lambda p$ at three bombarding energies [19] are shown in Fig. 3 in the form of Dalitz plots. The Dalitz plots show strong deviations from a homogeneous phase-space distribution. The data can be described using the ISOBAR model of Sibirtsev (concept outlined in [20]) which takes into account the Λp FSI and the contributions of three N^* resonances, $N(1650)$, $N(1710)$ and $N(1720)$. The exclusive measurements of $pp \to K^+\Lambda p$ and $pp \to K^+\Sigma^0 p$ by COSY-TOF provide also differential cross sections [21] in the

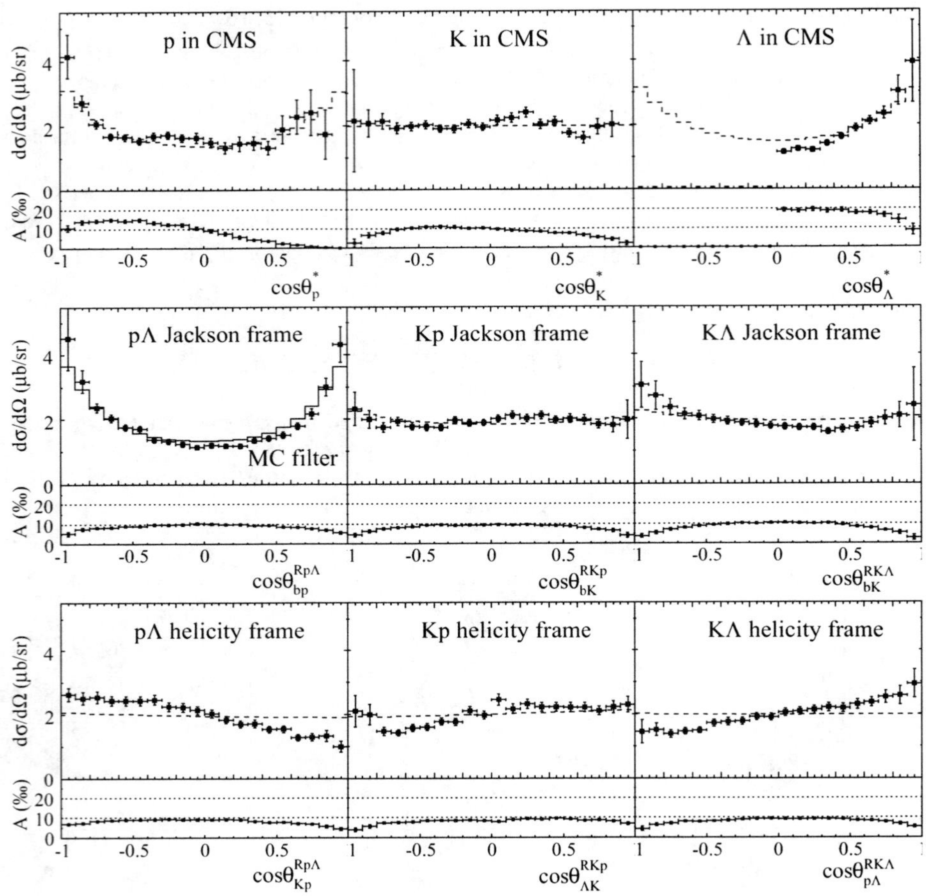

FIGURE 4. Differential cross sections in the CMS, Jackson and helicity frames for the reaction $pp \to K^+\Lambda p$ measured by COSY-TOF at $p_{beam} = 3.059$ GeV/c [21]. Note the differential acceptances A(‰).

CMS, Jackson and helicity frames, see Fig. 4. In addition, first measurements with polarized proton beams have been performed. The COSY-TOF data will be used for a detailed partial wave analysis by the Bonn-Gatchina PWA group.

In Fig. 5 [21] the total cross sections for $pp \to K^+\Lambda p$ and $pp \to K^+\Sigma^0 p$ [21, 22, 23, 24, 25, 26, 27, 28, 29] are shown as a function of the corresponding excess energy ε. The dashed curves representing a three-body phase-space ε^2 dependence describe the $pp \to K^+\Sigma^0 p$ data and there is no evidence for a $\Sigma^0 p$ FSI. However, the $pp \to K^+\Lambda p$ data can only be described by including the Λp FSI. The right panel shows the ratio of the $pp \to K^+\Lambda p$ to $pp \to K^+\Sigma^0 p$ total cross sections as a function of the excess energy. The new experimental values from COSY-TOF [21] confirm the general trend towards a ratio of 2.2 measured at high energies [30]. The large increase of the ratio towards low energies is well described by the Λp FSI.

The first measurement of the reaction $pp \to K^+\Sigma^+ n$ close to threshold has been performed by COSY-11 [31], by detecting the K^+ in coincidence with the neutron. The total cross sections at $\varepsilon = 13$ and 60 MeV were astonishingly high compared to those for $pp \to K^+\Sigma^0 p$. The energy dependence of the reaction $pp \to K^+\Sigma^+ n$ close to threshold has been studied recently by COSY-ANKE [28, 32] by detecting K^+ in coincidence with π^+ from the decay of the Σ^+ and by detecting $K^+ p$ coincidences. In contrast to the COSY-11 results [31] the total cross sections for $pp \to K^+\Sigma^+ n$ are very much smaller (factor 50 at $\varepsilon = 60$ MeV) and slightly smaller than those for $pp \to K^+\Sigma^0 p$ ($R(\Sigma^+/\Sigma^0) = 0.7 \pm 0.1$). The energy dependence can be described by a three-body phase-space calculation [32], with no evidence for a $\Sigma^+ n$ FSI. Details can be found in the contribution of Yu. Valdau to these proceedings.

A recent determination of the total cross section for $pp \to K^+\Sigma^+ n$ at $\varepsilon = 102.6$ MeV from inclusive K^+ data by

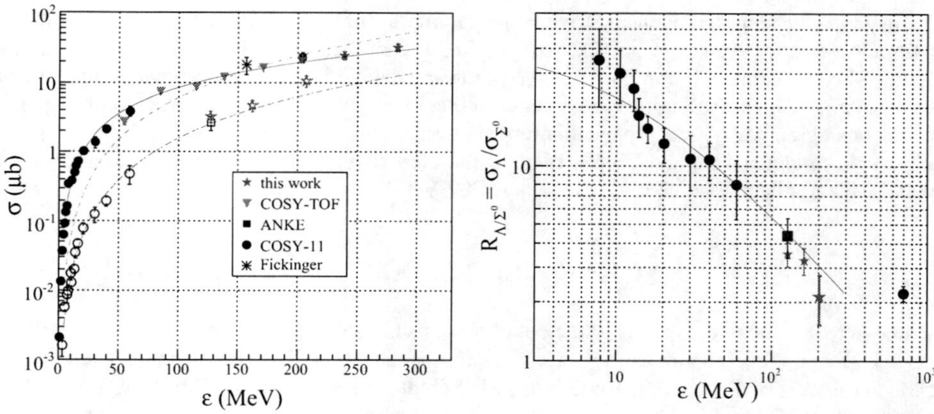

FIGURE 5. Left: Total cross sections for $pp \to K^+\Lambda p$ (solid symbols) and $pp \to K^+\Sigma^0 p$ (open symbols) [21, 22, 23, 24, 25, 26, 27, 28, 29] vs. corresponding excess energy ε. The new data (open and solid crosses, denoted 'this work' in the insert) are from COSY-TOF [21]. The dashed curves represent a three-body phase-space ε^2 dependence. The solid curve includes the Λp FSI. Right: Ratio of the $pp \to K^+\Lambda p$ to $pp \to K^+\Sigma^0 p$ total cross sections vs. ε. The ratio at $\varepsilon = 700$ MeV is an average value calculated from data given in [30]. The solid curve takes the Λp FSI into account.

[redacted]
with the exclusive measurement by COSY-ANKE [28, 32]. Possible reasons for this discrepancy are discussed in [34].

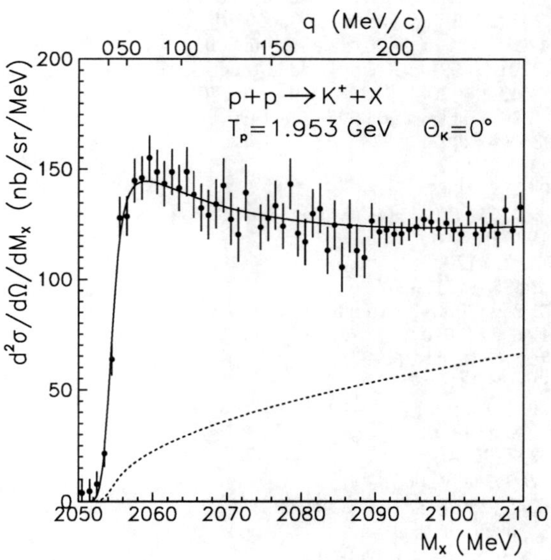

FIGURE 6. Missing mass spectrum of the reaction $pp \to K^+\Lambda p$ measured at $T_p = 1.953$ GeV and $\Theta_K = 0°$ [35]. Solid curve: fit with spin-averaged effective range parameters. Dashed curve: phase-space distribution.

The Λp FSI in the reaction $pp \to K^+\Lambda p$ has been studied with a high invariant mass resolution by the COSY-HIRES collaboration [35]. The kaons were detected at 0° using the magnetic spectrograph BIG KARL [36]. The double differential cross section is shown in Fig. 6 as a function of the missing mass, i.e. the invariant mass of the Λp system. The spectrum is characterized by a huge FSI enhancement near the Λp threshold. A narrow $S = -1$ dibaryon resonance predicted near 2100 MeV [37] is not visible. The data can be described by factorizing the reaction amplitude in terms of a production amplitude and a FSI enhancement factor (solid curve in Fig. 6). Taking for the enhancement the inverse Jost function yields the spin-averaged effective range parameters of the Λp interaction, i.e. the scattering length $\bar{a} = -2.43^{+0.16}_{-0.17}$ fm and the effective range $\bar{r} = 2.21 \pm 0.16$ fm. The dashed curve is the corresponding phase-space distribution without FSI enhancement. Taking the free Λp scattering data [38, 39] in a combined fit into account yields the possibility to disentangle the spin singlet and triplet effective range parameters of the Λp system [35]. However, this result should be considered with respect to the theoretical uncertainties of the Jost-function approach [40]. A direct determination of the spin singlet and triplet effective range parameters requires polarization measurements as proposed by Gasparyan et al. [41] and planned by COSY-ANKE and COSY-TOF.

ACKNOWLEDGMENTS

The results presented here are due to the efforts of the collaborations COSY-ANKE, COSY-11, COSY-HIRES, COSY-MOMO and COSY-TOF and the COSY machine crew. The authors thank K.-Th. Brinkmann, H. Clement, W. Eyrich, H. Freiesleben, M. Hartmann, J. Ritman, E. Roderburg, M. Schulte-Wissermann, A. Sibirtsev, H. Ströher and Yu. Valdau for their help and clarifying discussions.

REFERENCES

1. R. Maier et al., *Nucl. Instr. and Meth. Phys. Res. Sect. A* **390**, 1 (1997).
2. M. Wolke, "Production of associated strangeness in the reaction $pp \to ppK^+K^-$ close to threshold" *Jül-3532* (1998).
3. C. Quentmeier et al., *Phys. Lett. B* **515**, 276 (2001) [arXiv:nucl-ex/0103001].
4. P. Winter et al., *Phys. Lett. B* **635**, 23 (2006) [arXiv:hep-ex/0602030].
5. M. Hartmann et al., *Phys. Rev. Lett.* **96**, 242301 (2006) ibid. **97**, 029901 (2006) [arXiv:hep-ex/0604010].
6. Y. Maeda et al., *Phys. Rev. C* **77**, 015204 (2008) [arXiv:0710.1755 (nucl-ex)].
7. V. Kleber et al., *Phys. Rev. Lett.* **91**, 172304 (2003) [arXiv:nucl-ex/0304020].
8. A. Dzyuba et al., *Eur. Phys. J. A* **29**, 245 (2006) [arXiv:nucl-ex/0605030].
9. Y. Maeda et al., *Phys. Rev. Lett.* **97**, 142301 (2006) [arXiv:nucl-ex/0607001].
10. Y. Maeda et al., *Phys. Rev. C* **79**, 018201 (2009) [arXiv:0811.4303 (nucl-ex)].
11. F. Bellemann et al., *Phys. Rev. C* **75**, 015204 (2007) [arXiv:nucl-ex/0608047].
12. X. Yuan et al., *Eur. Phys. J. A* **42**, 1 (2009) [arXiv:0905.0979 (nucl-ex)].
13. F. Balestra et al., *Phys. Rev. C* **63**, 024004 (2001).
14. A. Dzyuba et al., *Phys. Lett. B* **668**, 315 (2008).
15. M. Silarski et al., *Phys. Rev. C* **80**, 045202 (2009).
16. M. Abdel-Bary et al., *Phys. Lett. B* **647**, 351 (2007) [arXiv:nucl-ex/0702059].
17. H.J. Lipkin, *Phys. Lett. B* **60**, 371 (1976).
18. K. Schönning et al., *Phys. Lett. B* **668**, 258 (2008).
19. S. Abd El-Samad et al., *Phys. Lett. B* **688**, 142 (2010).
20. A. Sibirtsev et al., *Eur. Phys. J. A* **27**, 269 (2006) [arXiv:nucl-th/0512059].
21. S. Abdel-Bary et al., *Eur. Phys. J. A* **46**, 27 (2010).
22. D. Grzonka and K. Kilian, *Nucl. Phys. A* **626**, 41C (1997).
23. J.T. Balewski et al., *Nucl. Phys. A* **626**, 85C (1997).
24. J.T. Balewski et al., *Phys. Lett. B* **420**, 211 (1998).
25. S. Severin et al., *Phys. Rev. Lett.* **83**, 682 (1999).
26. P. Kowina et al., *Eur. Phys. J. A* **22**, 293 (2004).
27. M. Abd El-Samad et al., *Phys. Lett. B* **632**, 27 (2006).
28. Yu. Valdau et al., *Phys. Lett. B* **652**, 245 (2007) [arXiv:nucl-ex/0703044].
29. W.J. Fickinger et al., *Phys. Rev.* **125**, 2082 (1962).
30. A. Baldini et al., *Landolt-Börnstein*, New Series, I/12b (1988).
31. T. Rożek et al., *Phys. Lett. B* **643**, 251 (2006).
32. Yu. Valdau et al., *Phys. Rev. C* **81**, 045208 (2010).
33. A. Budzanowski et al., *Phys. Lett. B* **692**, 10 (2010).
34. Yu. Valdau and C. Wilkin, submitted to *Phys. Lett. B* (2010) [arXiv:1008.3465v1 (nucl-ex)].
35. A. Budzanowski et al., *Phys. Lett. B* **687**, 31 (2010).
36. M. Drochner et al., *Nucl. Phys. A* **643**, 55 (1998).
37. A.T.M. Aerts and C.B. Dover *Nucl. Phys. B* **253**, 116 (1985).
38. G. Alexander et al., *Phys. Rev.* **173**, 1452 (1968).
39. B. Sechi-Zorn et al., *Phys. Rev.* **175**, 1753 (1968).
40. A. Gasparyan et al., *Phys. Rev. C* **72**, 034006 (2005).
41. A. Gasparyan et al., *Phys. Rev. C* **69**, 034006 (2004).

Overview of coupled-channels analyses of meson production reactions: The way to the N*s

Bruno Juliá-Díaz

Departament d'Estructura i Constituents de la Matèria, Diagonal 647, Barcelona, Spain.

Abstract. We provide a brief review of the main existing theoretical efforts to extract and interpret the properties of baryon resonances from the experimental data for hadro-, photo- and electro-production of mesons on the nucleon. The focus is set on the dynamical coupled-channels models. An effort is made to highlight the relevant aspects of the different approaches, within the space limitations of these proceedings.

Keywords: Baryon resonances, Dynamical coupled-channels, meson production reactions
PACS: 14.20.Gk, 13.75.Gx, 13.60.Le

INTRODUCTION

The physics of nucleon resonances is already more than 50 years old. Since the pioneering pion scattering experiments of Fermi and collaborators [1], which exposed the $\Delta(1232)$ resonance, plenty of experimental information has been gathered both on hadron and electromagnetic production of mesons in the so-called N^* region, for a review see [2, 3]. These experimental observables have been analyzed by several collaborations over the years extracting properties of light baryon resonances, which are collected periodically by the Particle Data Group, [4].

In the quark model picture, each of the excited states of the confined three constituent quark system should correspond to a baryon resonance. Most quark models predict the existence of many more resonances than the ones extracted from analyses of the π induced reactions [5]. This 'missing resonances' problem partially justified the experimental studies using electromagnetic probes, with the idea that the unseen resonances would have a larger coupling to the γN system than they do to the πN one.

The advent of high precision photo- and electro-production data from MIT-Bates, Graal, Jefferson Lab, and ELSA, has changed the experimental situation switching the largest amount of information on meson production reactions from the relatively older hadron induced reactions to electromagnetic induced ones. The latter are being constantly extended, with recent measurements of involved single and double-polarization observables which are especially sensitive to the partial wave content, e.g. [6], and with measurements of the Q^2 evolution of the observables [7].

As time has shown, however, the strong nature of the interaction and the appearance of several resonances in the same energy region has proven to be a difficult obstacle for disentangling the resonance content present in the experimental data. The signals from photo- and electro-production reactions unavoidably include strong backgrounds arising from multi-step strong processes which make the identification of resonances a very involved task. This is one of the reasons why very few low lying N^* or Δ^* have been uncontroversially discovered solely from the photo- or electro-production data. Despite the plethora of high precision data collected, still basic properties, e.g. the existence, of several low-lying nucleon resonances is not settled, as explicitly emphasized in Ref. [8].

Resonant states should provide clear (mathematical) correlations among the different meson production observables. Thus, it is nowadays acknowledged that a complete understanding of the low-lying baryon spectrum requires the simultaneous analysis of all the extant data on meson production reactions within an ambitious theoretical framework. The minimal requirements for the latter are: (1) the inclusion of relevant two- and multi-body channels, (2) the complete resolution of the scattering equations, (3) the analytic continuation of the transition amplitudes into the complex energy plane to extract the properties of baryon resonances.

Although several theoretical approaches are being used to extract the properties of low-lying baryon resonances, only few of them comply with the above requirements. The most advanced ones are as of today the dynamical coupled-channels models.

First, we present the basics of one such dynamical coupled-channels model, outlining the different terms. Then we provide a succinct description of some approaches which are being used today to learn about baryon resonances.

12th International Conference on Meson-Nucleon Physics and the Structure of the Nucleon (MENU 2010)
AIP Conf. Proc. 1374, 97-102 (2011); doi: 10.1063/1.3647105

Finally, we give some concluding remarks.

A BRIEF DESCRIPTION OF A DYNAMICAL COUPLED-CHANNELS FORMALISM

To illustrate the main elements entering in a dynamical coupled-channels description of meson production experimental data we consider some aspects of the model defined in Ref. [9].

The meson-baryon (MB) scattering equations are defined by,

$$T_{\alpha,\beta}(E) = t_{\alpha,\beta}(E) + t^R_{\alpha,\beta}(E), \tag{1}$$

where $\alpha, \beta = \gamma N, \pi N, \eta N, \pi\pi N$ are two- and three-body states defining the Hilbert space. The full amplitudes, e.g. $T_{\pi N,\pi N}(E)$, $T_{\eta N,\pi N}(E)$, $T_{\pi N,\gamma N}(E)$ are then used to calculate $\pi N \to \pi N$, $\pi N \to \eta N$ and $\gamma N \to \pi N$ scattering observables. The amplitude $t_{\alpha,\beta}(E)$ in Eq. (1) only contains meson-baryon dynamics, and is defined by the coupled-channels Lippmann-Schwinger equations,

$$t_{\alpha,\beta}(E) = V_{\alpha,\beta}(E) + \sum_{\delta} V_{\alpha,\delta}(E)\, G_{\delta}(E)\, t_{\delta,\beta}(E) \tag{2}$$

with $V_{\alpha,\beta}(E) = v_{\alpha,\beta} + Z^{(E)}_{\alpha,\beta}(E)$. The second term in the r.h.s of Eq. (1) is the so-called resonant term. It is defined by

$$t^R_{\alpha,\beta}(E) = \sum_{N^*_i, N^*_j} \bar{\Gamma}_{\alpha \to N^*_i}(E)[D(E)]_{i,j} \bar{\Gamma}_{N^*_j \to \beta}(E), \tag{3}$$

with $[D^{-1}(E)]_{i,j} = (E - M^0_{N^*_i})\delta_{i,j} - \bar{\Sigma}_{i,j}(E)$, where $M^0_{N^*}$ is a bare mass which can be included in the partial wave to allow for resonant states which are not only due to the meson-baryon dynamics. The properties of this bare particle are dressed by the meson-baryon dynamics. The self-energies are

$$\bar{\Sigma}_{i,j}(E) = \sum_{\delta} \Gamma_{N^*_i \to \delta} G_{\delta}(E) \bar{\Gamma}_{\delta \to N^*_j}(E). \tag{4}$$

The dressed vertex interactions in Eq. (3) and Eq. (4) are (defining $\Gamma_{\alpha \to N^*} = \Gamma^{\dagger}_{N^* \to \alpha}$)

$$\bar{\Gamma}_{\alpha \to N^*}(E) = \Gamma_{\alpha \to N^*} + \sum_{\delta} t_{\alpha,\delta}(E) G_{\delta}(E) \Gamma_{\delta \to N^*}, \tag{5}$$

$$\bar{\Gamma}_{N^* \to \alpha}(E) = \Gamma_{N^* \to \alpha} + \sum_{\delta} \Gamma_{N^* \to \delta} G_{\delta}(E) t_{\delta,\alpha}(E). \tag{6}$$

The previous system of equations is essentially what most approaches aim at employing to describe meson production reactions. The differences between the different groups strive usually on the prescriptions or approximations taken to solve the equations and on the way to define the transition vertexes [10]. In the following section we briefly describe the main current efforts providing a number of references. More detailed bibliography can be tracked back inside those.

DIFFERENT APPROACHES TO MESON PRODUCTION REACTIONS

Dynamical-Coupled-Channels of mesons and baryons

Our most elaborate tool to extract the properties of baryon resonances from meson production data are the dynamical coupled-channels models based on Lagrangian formulations of the interactions between the mesons and baryons.

The physics entering in these models can be summarized in three important pieces. First, the degrees of freedom considered are baryons and mesons, and their interactions are derived from phenomenological Lagrangian's. These are in some cases constrained by symmetries of the underlying microscopic theory, Quantum Chromodynamics (QCD). Secondly, the heavy multi-process character of the meson production reactions, which involves the intermediate

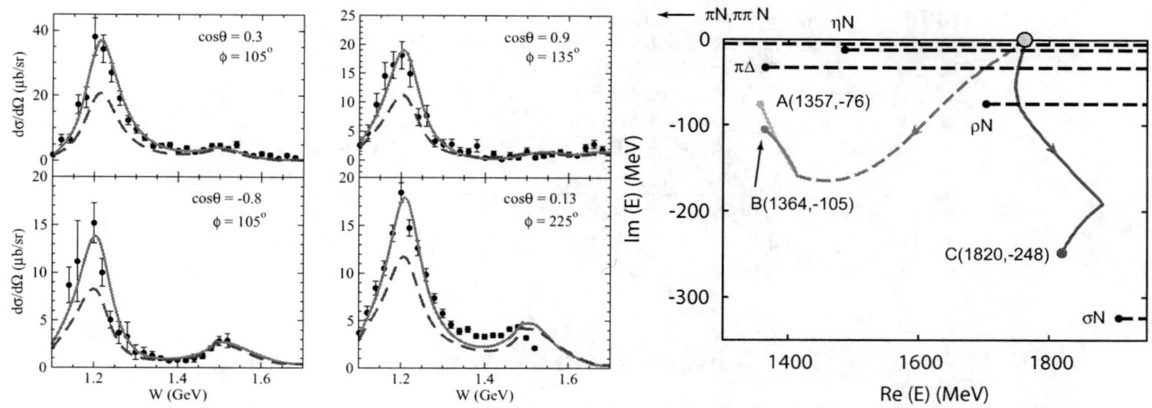

FIGURE 1. (left) Example of the coupled-channels effects on the five-fold differential cross sections $\Gamma_\gamma^{-1}[d\sigma^5/(dE_{e'}d\Omega_{e'}d\Omega^*_\pi)]$ of $p(e,e'\pi^0)p$ (upper panels) and $p(e,e'\pi^+)n$ (lower panels) at $Q^2 = 0.4$ (GeV/c)^2. Here $\theta \equiv \theta^*_\pi$ and $\phi \equiv \phi^*_\pi$. The solid curves are the full results, the dashed curves are the same as the solid curves but only the πN loop is taken into account, see Ref. [24] for details. The data are taken from Ref. [27]. (right) Trajectories of the evolution of P_{11} resonance poles A (1357,76), B (1364,105), and C (1820,248) from a bare N^* with 1763 MeV, as the couplings of the bare N^* with the meson-baryon reaction channels are varied from zero to the full strengths of the EBAC model, see [18] for details.

formation of multiple particle states, is taken into account employing state-of-the-art techniques in few-body quantum mechanics. This implies the use of multi-channels reaction theories.

Finally, these approaches do not perform any further simplification on the reaction theory equations (2, 3), thus allowing a correct mapping of the internal structures present in the problem. Possible N* states which nature is not solely due to meson baryon dynamics are introduced as bare N* states which properties are dressed by the meson-baryon dynamics, see Eq. (3). The N* vertex production functions are extracted from fits to the extant experimental data on different meson-baryon production reactions.

Importantly, in all these models special care is taken to preserve the correct analytical structure of the scattering amplitudes $T_{\pi N,MB}(E)$, Eq. (1). This, together with extrapolations of the latter into the complex energy plane, is needed to extract the properties of baryon resonances [11].

The main on going collaborations working along this direction include the Jülich group [12, 13, 14], the Dubna-Mainz-Taipei (DMT) group [15] and the EBAC@JLAB group [9, 16, 17, 18].

The DMT has studied πN reactions [15]. Their aim, as explicitly stated by the authors, is to extend the framework in the near future to the electromagnetic production of pions. The Jülich group has mostly focused on π induced reactions, and has very recently been extended to include kaon-hyperon channels [19, 14]. As occurred in the DMT model, the Jülich model is being extended to study single meson photo- and electro-production reactions [20].

The EBAC@JLAB group has played a prominent role since its foundation in 2006. The group has followed the reaction theory described in [21, 9], and briefly sketched in the previous section, and has obtained an accurate description of πN elastic data [16]. Then, the model has been extended to η production [22] (in collaboration with the Saclay group), π photo-production [23], π electro-production [24], $\pi\pi$ electro-[25] and hadron-production [26]. Currently, it is being extended to include KY channels and to perform a combined study of most production reactions [17]. A brief sample of the results obtained by the EBAC is given in Fig. 1.

It is important to note, however, than none of these models has been able to identify any resonance state by analyzing the photo- or electro-production data which was not already seen in the πN scattering data. Thus, up to now, despite the efforts by these collaborations, the precise and sizeable amount of experimental information on single and double meson production reactions measured at Graal (France), MAMI (Germany), ELSA (Germany), Spring-8 (Japan), and Jefferson Lab (USA) has not been profited to discover any new resonances, within these approaches. This situation is going to change very soon, with the current advances from the three groups reported at this conference [20, 17], specially dealing with electromagnetic production data, which should settle the properties, including even the existence of some states which are still controversial [8], of low-lying baryon resonances.

TABLE 1. Comparison of χ^2/data values (Norm/Unnorm) for normalized (Norm) and unnormalized (Unnorm) data used in the GWU/SAID model [8], EBAC[16], Giessen[48], and DMT[15]. Values for GWU/SAID correspond to a 2.46 GeV energy limit for W in c.m. EBAC up to 1.91 GeV, Giessen up to 2 GeV, and DMT up to 2.2 GeV. Table taken from Ref. [56].

Reaction	GWU/SAID χ^2/Data	EBAC χ^2/Data	Giessen χ^2/Data	Dubna-Mainz-Taipei χ^2/Data
$\pi^+ p \to \pi^+ p$	2.0/6.1	13.1/23.7	10.5/17.7	15.4/37.4
$\pi^- p \to \pi^- p$	1.9/6.2	4.9/16.0	12.1/34.1	9.0/23.0
$\pi^- p \to \pi^0 n$	2.0/4.0	3.5/6.3	6.3/15.2	6.5/16.7

Constituent Quark Models

Quark model degrees of freedom have been employed to study resonance properties since the early 1960s. In recent years, a number of collaborations are performing systematic studies of baryon resonances' static properties, such as masses, and also of excitation properties, e.g. electromagnetic structure of the resonance states. The main physical motivation is to employ degrees of freedom which are close to the ones present in QCD together with interactions among them which capture important aspects of the underlying theory, e.g. chiral symmetry, one gluon exchange, etc.

The largest current efforts are being carried out by the groups in Bonn/Gent [28, 29], Genoa [30], Graz [31], Helsinki [32, 33], Ljubljana/Coimbra [35], Saclay-IHEP [34, 36, 37], and Salamanca/Valencia [38]. These groups have research lines devoted to predicting resonance properties: masses, decay widths, etc., which are later compared to the information extracted by other groups from the meson production database. A number of them have also direct computation of the observables from hadro- or electro-production. Among the latter, a number of studies on π- and γ-induced η production have been performed by the Saclay group [34, 39].

The aim is twofold, first to explore the existence of new resonant states, by looking at their influence on the experimental data on meson production reactions, and, secondly, to investigate the sizeable presence of multi-quark components in the baryons [40]. The quark model predicts more resonance states than the ones which have been reported experimentally [5], it also permits consistent predictions for the direct reaction mechanisms involving the resonant states. A large group of quark model calculations only consider the direct production mechanisms. An example of such calculations is the recent work of the IHEP-Saclay group [37]. In a few cases, the direct production mechanisms have also been employed in a coupled-channels study of $K\Lambda$ production on the proton [41].

Chiral approaches

A significant amount of theorists have been concentrating their efforts on predicting properties of baryon resonances from one key feature of QCD, which is the almost fulfilled chiral symmetry of the theory in the non-strange sector. A very active community is the 'chiral dynamics' one.

The key theoretical input is the use of the lowest order chiral Lagrangian to generate the interactions between the ground state baryons and mesons, see for instance [42]. Then, the reaction theory is constructed by means of a Lippmann-Schwinger-like coupled-channels set of equations, which generates in some partial waves resonant states solely from the dynamics.

Important insights into the dynamical content of a well-known hyperon resonance, the Λ (1405) were obtained within this framework [43, 44]. In most cases only $S-$wave scattering is considered, which restricts the range of applicability to the first resonance region, e.g. N(1535) [45, 46]. These studies do not attempt to build a consistent framework to encompass the existing experimental data, but rather to analyze specific reaction mechanisms to isolate the effect of single resonances. Thus, they are important in general to seize which mechanisms are most relevant close to the different thresholds for particle production or in cases where the coupled-channels effects are minimized.

Approximate reaction theory methods

These category corresponds to a broad set of current collaborations which include: phenomenological K-matrix coupled-channels models: GWU/SAID [8], and Mainz/MAID [47], K-matrix models with meson-exchange interactions: Giessen [48], and KVI [49], K-matrix models with parameterizations: Bonn-Gatchina [50] and tree-level analyses: JLAB/MSU [51], Madrid [52] and, Valencia [53].

All of them share a number of simplifications in the reaction theory used to extract resonance properties. Some groups, GWU/SAID, Mainz/MAID, Giessen, KVI employ simplified versions of the scattering equations which can nowadays be largely improved. These simplifications were partly motivated by the tedious computations involved in solving exactly the reaction theory equations, such as the ones sketched in the previous section, which sometimes was almost impossible with the computer capabilities present at the time when they were performed.

Other approaches are interested in exploring specific production mechanisms in certain narrow ranges of energies, e.g. the groups at JLAB/MSU and Valencia. Their explorations help to understand aspects of the excitation of resonances and should be profited by more involved approaches such as dynamical coupled-channels approaches.

The GWU/SAID deserves a special mention as it has long been playing a prominent role in this field. Concerning meson-baryon analyses, their K-matrix parameterization of the πN scattering data has provided the international community with energy independent transition matrices for the different partial waves [8]. These PWA have been further employed by other models to constraint their free parameters. In this way, direct comparison to the full world database for πN scattering could partially be avoided. Other reactions have been analyzed with similar tools but not in a consistent way, e.g. single eta production. Only very recently, the SAID group has presented their first step towards a common description of strong- and photo-production of pions in some partial waves [54]. Their amplitudes are passed on to the MAID/Mainz group (and other groups), which aims at describing the photo- and electro-production data in the resonance region [47]. The MAID/Mainz group focuses on the evolution of the structure functions as the momentum transferred is increased. These are obtained from fits to the photo and electro-production database [55].

The Giessen and KVI models have made use of a vast amount of production data from many reactions, e.g. $\pi N \to \pi N$, $\pi N \to \eta N$, $\pi N \to KY$, etc. In their analyses, they derive transition potentials from phenomenological Lagrangians, similarly to the dynamical coupled-channels models The main drawback of both approaches arises from the simplifications introduced in the scattering equations. Briefly, an on-shell prescription is used for the integral Lippmann-Schwinger-like equations, avoiding in this way the exact calculation of their principle value parts. This, as acknowledged by the authors [48], will be improved.

The Bonn-Gatchina and JLAB/MSU analyses are performed by researches of CB-Elsa (Bonn) and Jefferson Lab (USA), respectively. In both cases, an important effort has been put on extracting new information from the latest experimentally measured data, as exemplified by the tree level model for double pion electro-production of the JLAB/MSU group [51]. Their theoretical inputs into the analyses are aimed at singling out the effects of new resonances on the observables for double pion electro-production. Bonn-Gatchina analyses have recently been improved by incorporating some coupled-channels effects and building analytical extrapolations of their amplitudes to extract information about baryon resonances. [50].

The GWU/SAID collaboration has collected amplitudes from several of the collaborations listed above. In this way, they can compare the different predictions by the collaborations. In table 1 we show a comparison of the quality of the description of the $\pi N \to \pi N$ world database attained by some of the groups listed above.

CONCLUDING REMARKS

A brief description of the different models used nowadays to analyze and learn from the extant and future meson production data has been presented. The existing and future amount of precise single and double meson production data, especially the involved polarization observables being measured by several experimental collaborations, will in the near future translate into equally precise information on the light baryon spectrum. This important step, going from the precise data to precise properties of baryon resonances, requires equally important efforts on developing theoretical analysis tools. The main aim of this presentation has been to provide a description of the state-of-the-art on the field emphasizing both the strong points, and the not so strong ones, which need to be improved in the near future in order to settle the low-lying baryon spectrum.

ACKNOWLEDGMENTS

In the last years I have benefited greatly from my colleagues, H. Kamano, T.-S.H. Lee, A. Matsuyama, T. Sato, B. Saghai and N. Suzuki. Also discussions with M. Döring, C. Fernández, J. Haidenbauer, K. Nakayama, S. Krewald and L. Tiator are aknowledged. Finally, I want to thank the organizers of the MENU conference for their kind invitation to review the different approaches. This work is supported by a CPAN CSD 2007-0042 contract and partially by Grants No. FIS2008-01661, and No. 2009SGR1289 from Generalitat de Catalunya.

REFERENCES

1. H. L. Anderson, E. Fermi, E. A. Long, and D. E. Nagle, Phys. Rev. **85**, 936 (1952).
2. F. Close, S. Donnachie, G. Shaw, 'Electromagnetic Interactions and Hadronic Structure', Cambridge U. Press (2007).
3. E. Klempt and J. M. Richard, Rev. Mod. Phys. **82**, 1095 (2010).
4. K. Nakamura *et al.* (Particle Data Group), J. Phys. G **37** 075021 (2010).
5. S. Capstick, W. Roberts, Prog. Part. Nucl. Phys. **45**, S241(2000).
6. N. Sparks, *et al.* (CBELSA/TAPS), Phys. Rev. C **81**, 065210 (2010).
7. I. G. Aznauryan, V. D. Burkert, CLAS Collaboration, Phys. Rev. C **80**, 055203, (2009).
8. R.A. Arndt, W.J. Briscoe, I.I. Strakovsky, R.L. Workman, Phys. Rev. C **74**, 045205 (2006).
9. A. Matsuyama, T. Sato and T. S. Lee, Phys. Rept. **439**, 193 (2007).
10. V. D. Burkert and T. S. H. Lee, Int. J. Mod. Phys. E **13**, 1035 (2004).
11. D. Morgan, M.R. Pennington, Phys. Rev. Lett.59, 2818, (1987).
12. A.M. Gasparyan, J. Haidenbauer, C. Hanhart, J. Speth, Phys. Rev. C **68**, 045207 (2003).
13. M. Doring, C. Hanhart, F. Huang, S. Krewald and U. G. Meißner, Nucl. Phys. A **829**, 170 (2009).
14. M. Doring, these proceedings, arXiv: 1009.5960.
15. G. Y. Chen, S.S. Kamalov, S. N. Yang, D. Drechsel, L. Tiator, Phys. Rev. C **76**, 035206 (2007).
16. B. Juliá-Díaz, T. S. Lee, A. Matsuyama and T. Sato, Phys. Rev. C **76**, 065201 (2007).
17. H. Kamano, these proceedings, arXiv:1010.0275.
18. N. Suzuki, B. Juliá-Díaz, H. Kamano, T. S. Lee, A. Matsuyama and T. Sato, Phys. Rev. Lett. **104**, 042302 (2010).
19. M. Doring, C. Hanhart, F. Huang, S. Krewald, U.-G. Meißner, D. Ronchen, arXiv:1009.3781.
20. F. Huang, M. Döring, H. Haberzettl, S. Krewald, K. Nakayama, arXiv:1010.0714.
21. H. Feshbach, 'Theoretical Nuclear Physics. Nuclear Reactions', Wiley Classics Library (1993).
22. J. Durand, B. Juliá-Díaz, T. S. Lee, B. Saghai and T. Sato, Phys. Rev. C **78**, 025204 (2008).
23. B. Juliá-Díaz, T. S. Lee, A. Matsuyama, T. Sato and L. C. Smith, Phys. Rev. C **77**, 045205 (2008).
24. B. Juliá-Díaz, H. Kamano, T. S. Lee, A. Matsuyama, T. Sato and N. Suzuki, Phys. Rev. C **80**, 025207 (2009).
25. H. Kamano, B. Juliá-Díaz, T. S. Lee, A. Matsuyama and T. Sato, Phys. Rev. C **80**, 065203 (2009).
26. H. Kamano, B. Juliá-Díaz, T. S. Lee, A. Matsuyama and T. Sato, Phys. Rev. C **79**, 025206 (2009).
27. K. Joo et al. (CLAS), Phys. Rev. Lett. **88**, 122001; H. Egiyan et al. (CLAS), Phys. Rev. C **73**, 025204, 2006.
28. U. Loering, B. C. Metsch, H. R. Petry, Eur. Phys. J. A **10**, 395 (2001).
29. T. Van Cauteren et al., Eur. Phys. J. A **20**, 283 (2004).
30. M. De Sanctis, M.M. Giannini, E. Santopinto, A. Vassallo, Phys. Rev. C **76**, 062201 (2007).
31. T. Melde, L. Canton, W. Plessas, Phys. Rev. Lett. **102**, 132002 (2009).
32. B. S. Zou, D. O. Riska, Phys. Rev. Lett. **95**, 072001, (2005).
33. B. Juliá-Díaz, D.O. Riska, Nucl. Phys. A **780**, 175, (2006).
34. J. He, B. Saghai, Z. Li, Phys. Rev. C **78**, 035204 (2008).
35. B. Golli, S. Sirca, M. Fiolhais, Eur. Phys. J. A **42**, 185, (2009).
36. C. S. An, B. Saghai, S. G. Yuan, J. He, Phys. Rev. C **81**, 045203 (2010).
37. X. Zhong, Q. Zhao, J. He, B. Saghai, Phys. Rev. C **76**, 065205 (2007).
38. P. González, J. Vijande, A. Valcarce, Phys. Rev. C **77**, 065213, (2008).
39. B. Saghai, Z. Li, Eur. Phys. J.A **11**, 217 (2001); J. He, B.Saghai, Phys. Rev.C **82**, 035206 (2010).
40. C. S. An, B.S. Zou, Eur. Phys. J.A **39**, 195, (2009).
41. B. Juliá-Díaz, B. Saghai, T.-S.H. Lee, F. Tabakin, Phys. Rev. C **73**, 055204 (2006).
42. J.A. Oller, E. Oset, A. Ramos, Prog. Part. Nucl. Phys. **45**, 157 (2000).
43. E. Oset, A. Ramos, Nucl. Phys. A **635**, 99 (1998).
44. V.K. Magas, E. Oset, A. Ramos, Phys. Rev. Lett. **95**, 052301 (2005).
45. T. Inoue, E. Oset, M.J. Vicente Vacas, Phys. Rev. C **65**, 035204, (2002).
46. M. Doring, K. Nakayama, Eur. Phys.J.A **43**, 83, (2010).
47. D. Drechsel, S.S. Kamalov, L. Tiator, Eur. Phys. J. A **34**, 69 (2007).
48. V. Shklyar, H. Lenske, U. Mosel, G. Penner, Phys. Rev. C **71**, 055206 (2005).
49. A. Usov, O. Scholten, (KVI), Phys. Rev. C **72**, 025205, (2005).
50. A.V. Anisovich, E. Klempt, V.A. Nikonov, M.A. Matveev, A.V. Sarantsev, and U. Thoma, Eur. Phys. J. A **44**, 203, (2010).
51. V. Mokeev, V. Burkert, T.-S, H. Lee, L. Elouadrhiri, G. V. Fedotov, and B. S. Ishkhanov, Phys. Rev. C **80**, 045212 (2009).
52. C. Fernández-Ramirez, E. Moya de Guerra, A. Udias, J.M. Udias, Phys. Rev. C **77**, 065212, (2008).
53. J.C. Nacher, E. Oset, M.J. Vicente, L. Roca, Nucl. Phys. A **695**, 295 (2001).
54. M. W. Paris and R. L. Workman, (GWU) Phys. Rev. C **82**, 035202 (2010).
55. L. Tiator, M. Vanderhaeghen, Phys. Lett. B **672**, 344, (2009).
56. R. A. Arndt, W.J. Briscoe, M.W. Paris, I.I. Strakovsky, R.L. Workman, arXiv:0906.3709.

What did we learn about GPDs from hard exclusive electroproduction of mesons

Peter Kroll

Universität Wuppertal, Fachbereich Physik
D-42097 Wuppertal, Germany

Abstract. It is reported on an analysis of electroproduction of light mesons at small Bjorken-x (x_{Bj}) within the handbag approach. The partonic subprocesses, meson electroproduction off quarks or gluons, are calculated within the modified perturbative approach (m.p.a.) in which quark transverse momenta are retained. The soft hadronic matrix elements, generalized parton distributions (GPDs), are constructed by means of double distributions. The constraints from parton distributions and sum rules are taken into account. Various moments of these GPDs are compared to recent results from lattice gauge theories.

Keywords: Handbag factorization, generalized parton distributions, electroproduction
PACS: 13.60Le, 12.38Bx, 12.39St, 12.38Qk

It has been shown [1] that, at large photon virtuality Q^2, meson electroproduction factorizes in partonic subprocesses, electroproduction off gluons or quarks, $\gamma^* g(q) \to M g(q)$, and in GPDs, representing soft proton matrix elements which encode the soft, non-perturbative physics. The calculation of the subprocess amplitudes requires the meson's wave function, i.e. a second soft, non-perturbative function. It has also been shown that in this so-called handbag approach which offers a partonic description of meson electroproduction, the dominant contribution is generated by transitions from longitudinally polarized virtual photons (γ_L^*) to like-wise polarized vector mesons (V_L). Other transitions are suppressed by inverse powers of the large scale, Q^2, although, as we had to learn, they are not small at experimentally accessible values of Q^2. Thus, for instance, the ratio of the longitudinal over the transverse cross sections for the production of ρ^0 or ϕ mesons is about 2 for $Q^2 \simeq 4$ GeV^2. Other clear signals for contributions from transverse photons come from asymmetries measured in π^+ electroproduction with a transversely polarized target [2] and from the transverse cross section for this process measured by the $F_\pi - 2$ collaboration [3].

In the following it is reported on an analysis [4, 5] of exclusive meson electroproduction within the handbag factorization scheme carried through in the kinematical regime of large energy in the photon-proton center of mass frame ($W \geq 4\,\mathrm{GeV}$), low skewness ($\xi \simeq x_{\mathrm{Bj}}/(2-x_{\mathrm{Bj}}) \leq 0.1$) and small momentum transfer ($-t \leq 0.6\,\mathrm{GeV}^2$). In this kinematical region the dominant helicity amplitudes for the process $\gamma^* p \to V p$ where V denotes a vector meson, read

$$\begin{aligned}
\mathcal{M}^V_{\mu+,\mu+} &= \frac{e_0}{2}\sum_a e_a \mathcal{C}^a_V \left\{ \langle H\rangle^g_{V\mu} + \langle H\rangle^a_{V\mu} + \langle \widetilde{H}\rangle^g_{V\mu} + \langle \widetilde{H}\rangle^a_{V\mu} \right\}, \\
\mathcal{M}^V_{\mu-,\mu+} &= -\frac{e_0}{2}\frac{\sqrt{-t}}{2m}\sum_a e_a \mathcal{C}^a_V \left\{ \langle E\rangle^g_{V\mu} + \langle E\rangle^a_{V\mu} \right\}. \qquad (1)
\end{aligned}$$

Explicit helicity labels refer to the proton while $\mu(=0,\pm1)$ denotes the helicity of the photon and the vector meson. The quark flavors are denoted by a and e_a is the corresponding charge. The non-zero flavor weight factors read for the vector mesons of interest

$$\mathcal{C}^u_{\rho^0} = -\mathcal{C}^d_{\rho^0} = \mathcal{C}^u_\omega = \mathcal{C}^d_\omega = 1/\sqrt{2}, \qquad \mathcal{C}^s_\phi = 1. \qquad (2)$$

The terms $\langle F\rangle$ denote convolutions of subprocess amplitudes and GPDs ($F = H, E, \widetilde{H}$) for the two relevant subprocesses, $\gamma^* g \to V g$ and $\gamma^* q \to V q$. Explicitly the convolutions read ($i = g, a, x_g = 0, x_a = -1$)

$$\langle F\rangle^i_{V\mu} = \sum_\lambda \int_{x_i}^1 dx\, \mathcal{H}^{Vi}_{\mu\lambda,\mu\lambda}(x,\xi,Q^2,t=0)\, F^i(x,\xi,t). \qquad (3)$$

The helicity of the parton is labeled by λ. Note that $\langle \widetilde{H}\rangle^i_V = 0$ for longitudinal photons. The GPD $\widetilde{E}$ contributes to the transverse amplitudes only to order ξ and is consequently neglected as is the contribution from E to the proton helicity non-flip amplitudes because it is proportional to ξ^2.

12th International Conference on Meson-Nucleon Physics and the Structure of the Nucleon (MENU 2010)
AIP Conf. Proc. 1374, 103-108 (2011); doi: 10.1063/1.3647106

The subprocess amplitudes $\mathcal{H}$ are calculated within the modified perturbative approach [6] in which quark transverse degrees of freedom as well as Sudakov suppressions are taken into account. This factorization scheme is based on work by [7, 8]. It is assumed in [4, 5] that the quarks and gluons are emitted and re-absorbed by the proton collinearly. The quark transverse momenta, $k_\perp$, are only taken into account in the subprocesses. This approximation is justified by the fact that the r.m.s. $k_\perp$ of the partons inside the proton is much smaller (the GPDs describe the full proton) than that in the meson for which only the rather compact valence Fock state is considered. Since the Sudakov factor is known only in the impact parameter space [6], canonically conjugated to the transverse momentum space, the subprocess amplitudes are calculated in the b space

$$\mathcal{H}^{Vi}_{\mu\lambda,\mu\lambda} = \int d\tau d^2b \hat{\Psi}_{V\mu}(\tau,-\mathbf{b})\, \hat{\mathcal{F}}^i_{\mu\lambda,\mu\lambda}(x,\xi,\tau,Q^2,\mathbf{b})\, \alpha_S(\mu_R) \exp[-S(\tau,\mathbf{b},Q^2)]\,. \tag{4}$$

Their t dependences are neglected for consistency since they provide corrections of order t/Q^2 which are generally neglected. On the other hand, the t dependence of the GPDs is taken into account since this t is scaled by a soft parameter. The hard scattering kernels $\mathcal{F}$, or their respective Fourier transform $\hat{\mathcal{F}}$, are calculated to leading-order of perturbative QCD including quark transverse momenta. The explicit expressions can be found in [4]. Also for the Sudakov factor S in (4) and the choice of the renormalization (μ_R) and factorization (μ_F) scales it is referred to these articles. In collinear approximation the amplitudes for transversely polarized photons and vector mesons suffer from infrared singularities [9, 10]. The quark transverse momenta, $\mathbf{k}_\perp$, provide an admittedly model-dependent regularization scheme of these singularities by replacements of the type

$$\frac{1}{dQ^2} \longrightarrow \frac{1}{dQ^2+\mathbf{k}_\perp^2} \tag{5}$$

in the parton propagators. Here, d is a momentum fraction or a product of two. As can be readily shown the transverse amplitudes are suppressed by $\langle k_\perp^2\rangle^{1/2}/Q$ as compared to the longitudinal ones in this regularization scheme.

The (longitudinal) amplitudes for electroproduction of pions are analogous to (1) with the replacement of H and E by $\widetilde{H}$ and $\widetilde{E}$, respectively. Of course, the gluonic subprocess is not allowed in this case. For the case of π^+ production pion exchange is to be taken into account as well.

The GPDs are constructed with the help of double distributions [11]. The main advantage of this construction is the guaranteed polynomiality of the GPDs. The double distribution is written as a product of a zero-skewness GPD and a weight function that generates the skewness dependence of the full GPD ($n_g = n_{\rm sea} = 2$, $n_{\rm val} = 1$)

$$f_i(\beta,\eta,t) = F_i(\beta,\xi=0,t)\,\frac{\Gamma(2n_i+2)}{2^{2n_i+1}\Gamma^2(n_i+1)}\,\frac{[(1-|\beta|)^2-\eta^2]^{n_i}}{(1-|\beta|)^{2n_i+1}}\,. \tag{6}$$

The zero-skewness GPD is parameterized as

$$F_i(\beta,\xi=0,t) = e^{b_i t}\,|\beta|^{-\alpha_i' t}\,h_i(\beta)\,. \tag{7}$$

The function h_i represents the forward limit, $\xi = t = 0$, of the GPD. For H and $\widetilde{H}$ the forward limits are the phenomenologically known unpolarized and polarized PDFs, respectively. They have to be suitably continued to negative values of β. For the other GPDs the forward limits are parameterized as

$$h_i(\beta) = N_i \beta^{\alpha(0)}(1-\beta)^{\gamma}\,. \tag{8}$$

As is well-known, at low β the PDFs behave power-like where the powers are given by the intercepts of appropriate Regge trajectories. It seems plausible to generate also the t-dependence of the GPDs by Regge ideas and to assume that such a Regge-like behavior holds for the other GPDs as well. Assuming linear Regge trajectories $\alpha_i(t) = \alpha_i(0) + \alpha_i' t$ ($i =$ g, sea, valence) and exponential t-dependencies of the Regge residues, one arrives at the parameterization (7). For large $-t$ one likely needs a more complicated t dependence [12].

The full GPDs are obtained by an integral over f_i

$$F^i(x,\xi,t) = \int_{-1}^{1} d\beta \int_{-1+|\beta|}^{1-|\beta|} d\eta\, \delta(\beta+\xi\eta-x)\, f_i(\beta,\eta,t)\,. \tag{9}$$

There are other methods to generate the skewness dependence, namely the Shuvaev transform [13] and the dual parameterization [14]. Both these methods lead to very similar results for the GPDs at small skewness.

In Ref. [4] the Regge parameters are fixed in the following way: In agreement with the HERA data [15] on the integrated cross section σ_L and with the CTEQ6 PDFs [16] the gluon ('Pomeron') trajectory [1] is taken as $\alpha_g = 1.10 + 0.06\ln(Q^2/4\,\text{GeV}^2) + 0.15\,\text{GeV}^{-1}t$. The increase of its intercept with Q^2 is a consequence of evolution. Since the sea quarks mix with the gluons under evolution, $\alpha_{\text{sea}}(t) = \alpha_g(t)$ is assumed. For the valence quark GPDs, H and E, on the other hand, a standard Regge trajectory is taken - $\alpha_{\text{val}} = 0.48 + 0.90\,\text{GeV}^{-2}t$. For the other GPDs there is no prominent Regge exchange, probably Regge cuts also play an important role. Therefore, effective trajectories are used to parameterize the low x behavior of these GPDs [5] whose parameters are fitted to experiment as it is done for the slope parameters b_i.

It has been checked in [4, 5] that the GPDs respect positivity bounds as well as the sum rules, i.e. their first moments are in agreement with the nucleon form factor data at small $-t$. The forward limit of E for the valence quarks is chosen in agreement with the form factor analysis performed in [12].

Given that E is not much larger than H and $\widetilde{H}$ much smaller (see below) the cross sections for vector meson electroproduction are dominated by contributions from the GPD H at small skewness and small $-t$. An exceptions is ρ^+ production for which the relevant flavor combination $E_v^u - E_v^d$ is indeed substantially larger than $H_v^u - H_v^d$ (v denotes valence quarks). The GPD E is for instance probed by transverse target asymmetries which are related to interference terms $\text{Im}[\langle E\rangle^*\langle H\rangle]$, and $\widetilde{H}$ by double spin asymmetries like A_{LL} and by electroproduction of pions.

In [4] a detailed analysis of cross sections and spin density matrix elements for ρ^0 and ϕ electroproduction has been performed in the kinematical range of $W \simeq 5 - 180\,\text{GeV}$ and $Q^2 \simeq 3 - 100\,\text{GeV}^2$. Generally very good results have been obtained. Two example of the results obtained in [4] are shown in Fig. 1. The ratio of the longitudinal and transverse cross sections increases $\propto Q^2$ due to the power suppression of the transverse amplitude. Despite this behavior the ratio is not large for Q^2 less than about $10\,\text{GeV}^2$ indicating a substantial contribution from the transverse amplitude. The longitudinal cross section for ρ^0 production shows a mild increase at large W since $\sigma_L \propto W^{4(\alpha_g(0)-1)}$. This signals the dominance of the gluonic subprocess (with a certain admixture from sea quarks). The valence quarks are only perceptible for W smaller than about $10\,\text{GeV}$. For HERMES kinematics ($W \simeq 5\,\text{GeV}$) the gluon (plus sea) contribution still amounts to about 50% of the cross section. Inspection of Fig. 1 reveals on the other hand that the handbag approach as proposed in [4] fails for low W: The sharp increase of σ_L between $W = 5$ and $2\,\text{GeV}$ [17, 18] is not reproduced if the cross section is evaluated from the above described low-skewness GPDs simply extrapolated to larger ξ and E being still neglected. The reason of this discrepancy is still unclear. In contrast to this result an analogous extrapolation to low W for ϕ production leads to fair agreement with experiment [19]. This may be regarded as a hint at a small gluonic (and sea quark) E.

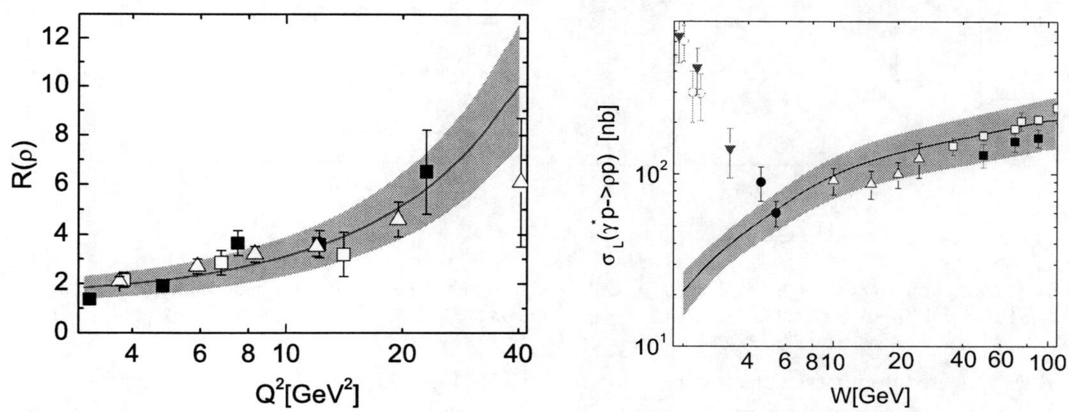

FIGURE 1. Left: Handbag predictions [4] of the ratio of longitudinal and transverse cross sections for ρ^0 production versus Q^2 at $W = 90$ GeV shown as a solid line. Data taken from H1 (solid squares) and ZEUS (open squares and triangles). Right: Predictions of the longitudinal cross section of ρ^0 production versus W at $Q^2 = 4\,\text{GeV}^2$. The open circles represent the recent CLAS data [17], the solid triangles the CORNELL data [18]. The shaded bands indicate the uncertainties of the theoretical analysis. For detailed references it is referred to [4].

[1] Note that the forward limit of the gluonic GPD H is defined as $xg(x)$ and analogously for the other GPDs.

In Fig. 2 the transverse target asymmetry for ρ^0 production is shown. The main contribution to it comes from an interference of E for valence quarks and H for gluons (plus sea). The zero-skewness GPD E, see (7), is taken from [12] (with $\alpha_{\rm val}$, $\gamma_e^u = 4.0$ and $\gamma_e^d = 5.6$). Given the errors of the HERMES data [20] a reasonable fit to experiment (solid line) is obtained if E^g and $E^{\rm sea}$ are ignored [2]. The other theoretical curves in Fig. 2 represent results for various variants of E where E^g and $E^{\rm sea}$ are estimated from positivity bounds and a combination of Ji's sum rule and the momentum sum rule of deep inelastic lepton-nucleon scattering [12, 22]. At present only extreme variants seem to be excluded. It is to be emphasized that with E and H at disposal one can evaluate Ji's sum rule. It turns out [23] that the total angular momenta of u quarks and gluons are large while those of d and strange quarks are very small. The results for u and d quarks are in very good agreement with lattice gauge theory [24].

The HERMES data on the cross section and the transverse target asymmetries for π^+ production have been analyzed in [5]. The relevant GPDs are $\widetilde{H}$ and $\widetilde{E}$ as well as pion exchange. However, this is not all. The $\sin\phi_s$ moment measured with a transversely polarized target [25] is very large and does not seem to vanish for forward scattering, see Fig. 2. Such a behavior can only be generated by the interference of the two helicity non-flip amplitudes

$$A_{UT}^{\sin\phi_s} \sim {\rm Im}\left[\mathcal{M}^*_{0-,++}\mathcal{M}_{0+,0+}\right]. \tag{10}$$

As has been advocated in [5] the amplitude $\mathcal{M}_{0-,++}$, describing a $\gamma_T \to \pi$ transition, can be modeled within the handbag approach as a twist-3 effect combining the leading-twist helicity-flip GPDs [26] with the twist-3 pion wave function [27]. Taking into account only the most important helicity-flip GPD, namely H_T, one has

$$\mathcal{M}^{\rm twist-3}_{0-,++} = e_0\langle H_T^u - H_T^d\rangle. \tag{11}$$

The convolution is to be calculated with the subprocess amplitude $\mathcal{H}_{0-,++}$ which is parametrically suppressed by μ_π/Q as compared to $\mathcal{H}_{0+,0+}$. The parameter μ_π takes on a value of about 2 GeV at a scale of 2 GeV. Hence, the twist-3 effect is sizeable for Q of the order of a few GeV. With this twist-3 effect a good description of all HERMES data has been achieved in [5].

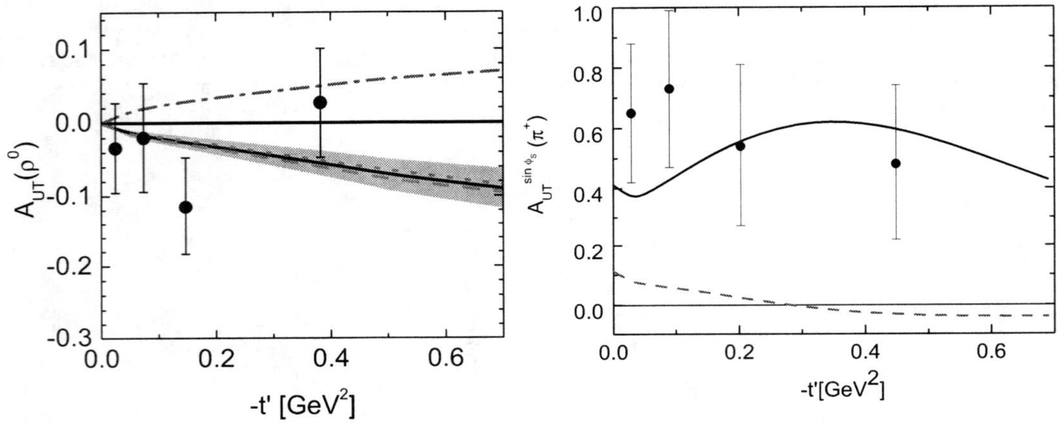

FIGURE 2. Left: The asymmetry A_{UT} for ρ^0 at $Q^2 = 3\,{\rm GeV}^2$ and $W = 5\,{\rm GeV}$. Right: The $\sin\phi_s$ moment for π^+ electroproduction at $Q^2 = 2.45\,{\rm GeV}^2$ and $W = 3.99\,{\rm GeV}$. The solid lines represent the predictions from the handbag approach [4, 5, 23]. Data taken from [20, 25].

The GPDs extracted from meson electroproduction data in Refs. [4, 5, 23] are valid for $\xi \lesssim 0.1$ and are probed by experiment for $x \lesssim 0.6$. The present status of these GPDs is summarized in Tab. 1 and the valence quark GPDs are displayed in Fig. 3. Since their parameterizations have no nodes except at the end-points and since they have similar t dependences, their well-known lowest moments at $t=0$ fix the relative signs and strength of these GPDs.

Comparison with recent lattice QCD studies [24, 28] where the lowest moments of the GPDs $H, \widetilde{H}, E, \widetilde{E}$ and H_T for u and d quarks have been calculated, reveals that in general there is good agreement with the relative strength of moments and their relative t dependences. At small t even the absolute values of the moments agree quite well but the

[2] Note that there are also preliminary data on this observable from COMPASS [21]. Both the sets of data together favor negative values of A_{UT}.

TABLE 1. Status of small-skewness GPDs as extracted from meson electroproduction data. The upper part is for gluons and sea quarks, the lower part for valence quarks. Except of H for gluons and sea quarks all GPDs are probed for scales of about $4\,\mathrm{GeV}^2$.

GPD	probed by	constraints	status
H	ρ^0, ϕ cross sections	PDFs	known
$\widetilde{H}$	-	polarized PDFs	probably small
E	$A_{UT}(\rho^0, \phi)$	sum rule for 2nd moment	probably small
others	-	-	unknown
H	ρ^0, ϕ cross sections	PDFs, Dirac ff	known
$\widetilde{H}$	π^+ data	pol. PDFs, axial ff	known
E	$A_{UT}(\rho^0, \phi)$	Pauli ff	known
$\widetilde{E}^{n.p.}$	π^+ data	-	uncertain
H_T	π^+ data	transversity PDFs [29]	known
others	-	-	unknown

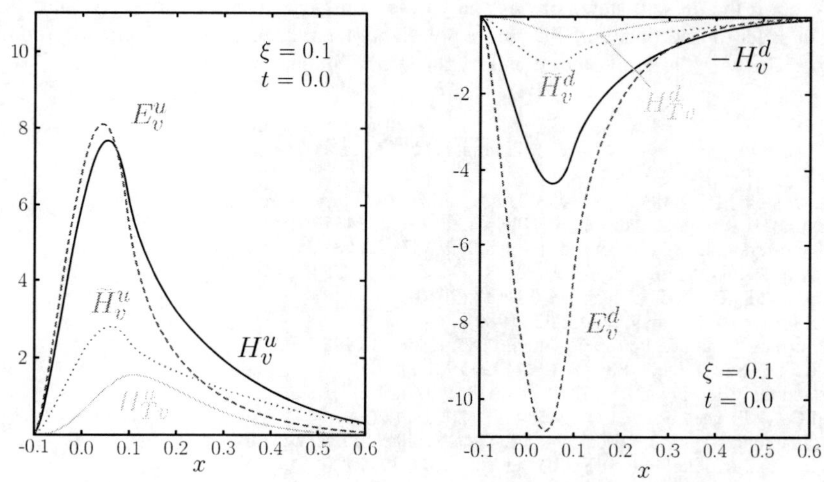

FIGURE 3. The valence-quark GPDs versus x at $\xi = 0.1$ and $t = 0$. The scale is $4\,\mathrm{GeV}^2$.

t dependence of the moments obtained from lattice QCD are usually flatter than the form factor data and the moments evaluated from the GPDs. An exception is the lowest moment of H_T for u quarks for which a value that is about 25% smaller than the lattice result has been found in [5]. In Fig. 4 the axial-vector and the pseudoscalar form factors are shown as examples. The form factors evaluated from the GPDs are compared to experimental data [30, 31] and to results from lattice QCD. For the pseudoscalar form factor only the pion-pole contribution

$$F_P^{\text{pole}}(t) = 4m^2 g_A\,[1 - t/\Lambda_N^2]^{-1}[m_\pi^2 - t]^{-1} \tag{12}$$

is shown for two values of the parameter Λ_N. Since one may also expect a flat behavior for this form factor from lattice QCD there is some room left for non-pole contributions from $\widetilde{E}$ at large $-t$.

In summary the handbag approach proposed in [4, 5, 23] which consists of GPDs constructed from double distributions and power corrections generated from quark transverse momenta in the subprocess describes quite well the data on meson electroproduction measured by HERMES, COMPASS, FNAL and HERA over a wide range of

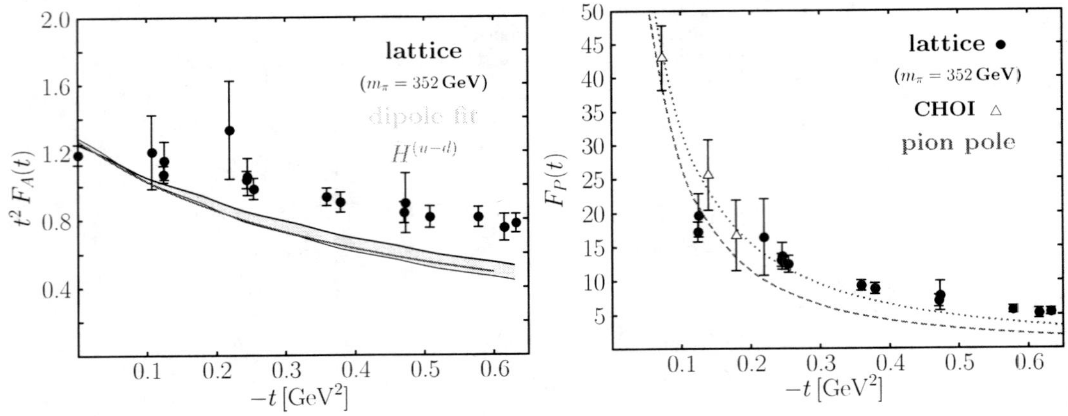

FIGURE 4. Left: The axial form factor of the nucleon scaled by t^2. The green band represents the dipole fit to the data [30], the solid circles the lattice results [24] for $m_\pi = 352\,\mathrm{MeV}$. The thick solid line is the form factor evaluated from $\widetilde{H}$. Right: The pseudoscalar form factor of the nucleon. Experimental data from [31], lattice results from [24]. The dashed (dotted) line represents the pion-pole contribution with $\Lambda_N = 0.51(0.8)\,\mathrm{GeV}$.

kinematics. In this report the present status of this analysis is summarized and described what we have learned about the GPDs from it. In order to improve the GPDs more polarization data and data on π^0 electroproduction are required.

Acknowledgments This work is supported in part by the BMBF under contract 06RY258.

REFERENCES

1. A.V. Radyushkin, Phys. Lett. **B385**, 333 (1996); J.C. Collins *et al.*, Phys. Rev. D56, 2982 (1997).
2. A. Airapetian *et al.* [HERMES Collaboration], Phys. Lett. **B682**, 345 (2010).
3. H. P. Blok *et al.* [Jefferson Lab Collaboration], Phys. Rev. **C**78 (2008) 045202.
4. S. V. Goloskokov and P. Kroll, Eur. Phys. J. **C**42, 281 (2005); *ibid.* **C**53, 367 (2008);
5. S. V. Goloskokov and P. Kroll, Eur. Phys. J. **C**65, 137 (2010).
6. J. Botts and G. Sterman, Nucl. Phys. B **325**, 62 (1989).
7. R. K. Ellis, W. Furmanski and R. Petronzio, Nucl. Phys. B **207**, 1 (1982).
8. J. C. Collins and D. E. Soper, Nucl. Phys. B**194**, 445 (1982).
9. L. Mankiewicz and G. Piller, Phys. Rev. D **61**, 074013 (2000).
10. I. V. Anikin and O. V. Teryaev, Phys. Lett. B **554**, 51 (2003).
11. D. Müller *et al*, Fortschr. Phys. **42**, 101 (1994); A.V. Radyushkin, Phys. Lett. **B449**, 81 (1999).
12. M. Diehl, T. Feldmann, R. Jakob and P. Kroll, Eur. Phys. J. **C39**, 1 (2005).
13. A. D. Martin, C. Nockles, M. G. Ryskin, A. G. Shuvaev and T. Teubner, Eur. Phys. J. C **63** (2009) 57.
14. M. V. Polyakov and K. M. Semenov-Tian-Shansky, Eur. Phys. J. A **40**, 181 (2009).
15. F. D. Aaron *et al.* [H1 Collaboration], JHEP **1005**, 032 (2010).
16. J. Pumplin *et al*, JHEP **0207**, 012 (2002).
17. S. A. Morrow *et al.* [CLAS Collaboration], Eur. Phys. J. A **39**, 5 (2009).
18. D. G. Cassel *et al*, Phys. Rev. D**24**, 2787 (1981).
19. J. P. Santoro *et al.* [CLAS Collaboration], Phys. Rev. C **78**, 025210 (2008).
20. A. Airapetian *et al.* [HERMES Collaboration], Phys. Lett. **B**679 (2009) 100.
21. A. Sandacz, talk presented at the workshop on 'Electromagnetic Interactions with Nucleons and Nuclei', Milos (2007).
22. M. Diehl and W. Kugler, Eur. Phys. J. C**39**, 933 (2007).
23. S. V. Goloskokov and P. Kroll, Eur. Phys. J. **C**59 (2009) 809.
24. Ph. Hagler *et al.* [LHPC Collaborations], Phys. Rev. **D**77, 094502 (2008).
25. A. Airapetian *et al.* [HERMES Collaboration], Phys. Lett. **B682**, 345 (2010).
26. M. Diehl, Eur. Phys. J. C**19**, 485 (2001); P. Hoodbhoy and X. Ji, Phys. Rev. D**58**, 054006 (1998).
27. V. M. Braun and I. E. Halperin, Z. Phys. C**48**, 239 (1990) [Sov. J. Nucl. Phys. **52**, 126 (1990)].
28. M. Gockeler *et al.* [QCDSF Collaboration and UKQCD Collaboration], Phys. Lett. B **627**, 113 (2005).
29. M. Anselmino *et al.*, Nucl. Phys. Proc. Suppl. **191**, 98 (2009).
30. T. Kitagaki *et al.*, Phys. Rev. **D**28, 436 (1983).
31. S. Choi *et al.*, Phys. Rev. Lett. **71**, 3927 (1993).

Recent Results from the WASA-at-COSY Experiment

Andrzej Kupsc for the WASA-at-COSY Collaboration

Department of Physics and Astronomy, Uppsala University,
Box 516, 751 20 Uppsala, Sweden

Abstract.
Studies of light meson decays are the key experiments for the WASA detector at COSY-Jülich. One of the world largest data samples of the η meson decays have been recently collected in the $pd \rightarrow {}^3\mathrm{He}\,\eta$ and in the $pp \rightarrow pp\eta$ reactions. The status of the analysis of various decay channels and the further plans for the light meson decay program are presented.

Keywords: Light meson decays.
PACS: 14.40.Be,13.40.Gp,13.20.Jf,13.25.Jx

INTRODUCTION

WASA is a 4π multidetector system for studies of light meson production and decays in light ion collisions. It allows detection of both photons and charged particles and it uses an internal hydrogen pellet target. The detector was built at TSL Uppsala where experiments were carried out until 2005 at the CELSIUS storage ring [1]. The facility is now located at COSY and has been in regular operation for the last three years (Fig. 1).

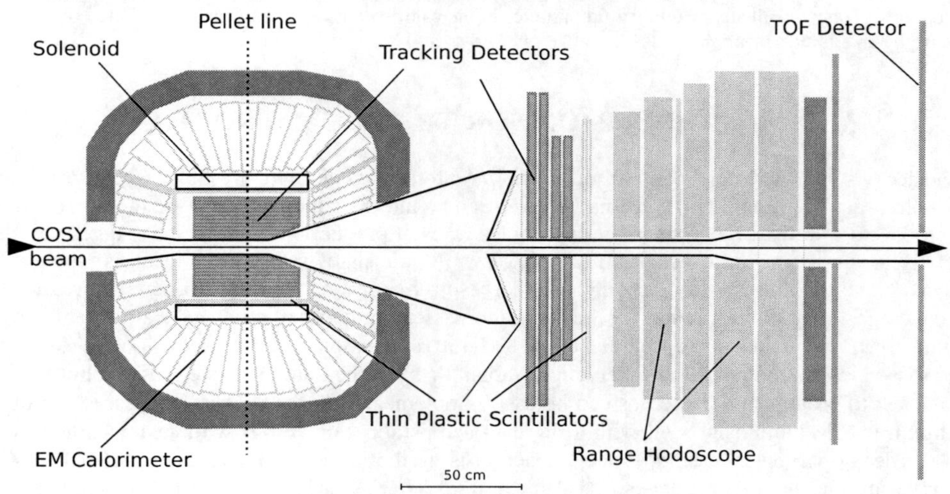

FIGURE 1. Schematic representation of the WASA detector setup.

The primary objectives of the WASA-at-COSY collaboration experimental programme are tests of symmetries and studies of symmetry breaking patterns in low–energy hadronic interactions and light meson decays. In addition, searches and further investigations of specific hadronic bound systems are carried out. The key experiment for the WASA detector at COSY is the study of (rare) η meson decays [2]. The η meson plays a special role in understanding low–energy Quantum Chromo Dynamics (QCD). Chiral symmetry, its realization in hadron physics at low energies, the role of quantum anomalies and explicit chiral symmetry breaking due to light quark masses can be investigated using η decays. In this paper, an overview of the η meson decay project is presented together with extensions to decays of other light neutral mesons: π^0, ω, and η'.

Status and preliminary results of the studies of isospin violation in the $dd \rightarrow \alpha\pi^0$ reaction were reported on this conference by A. Wrońska [3].

12th International Conference on Meson-Nucleon Physics and the Structure of the Nucleon (MENU 2010)
AIP Conf. Proc. 1374, 109-114 (2011); doi: 10.1063/1.3647107

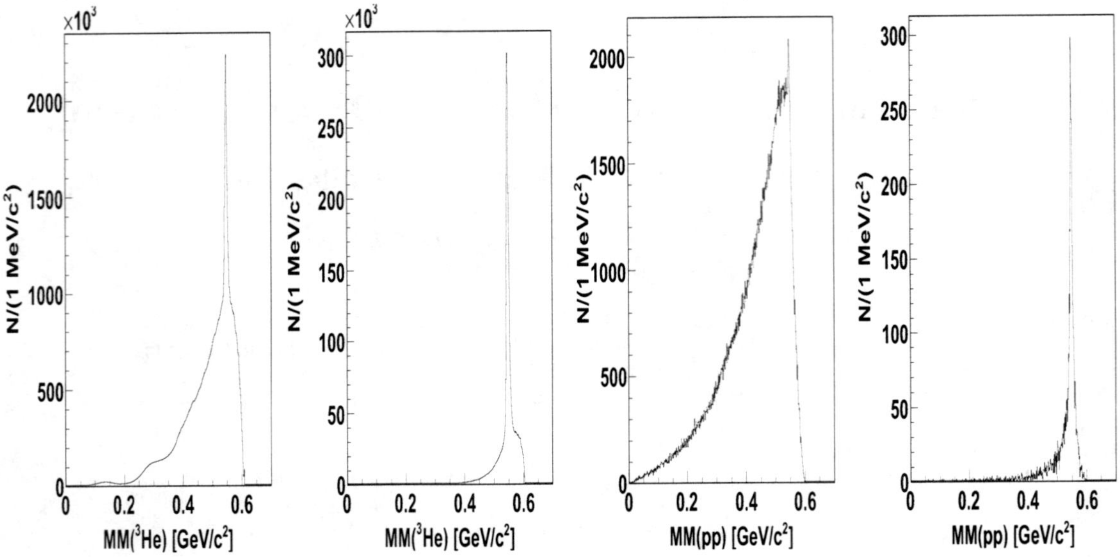

FIGURE 2. (Two panels on the left): Distribution of ^{3}He missing mass for the $pd \to ^3\mathrm{He}X$ reaction at 1.0 GeV. The trigger was based entirely on the ^{3}He signal in the forward detector without bias on the decay system. The plot shows all data collected in 2008 run period. (left) data analysis bases only on the forward detector track – there are about 11×10^6 events in the peak at the η meson mass (right) in addition invariant mass of two photons ≥ 300 MeV/c^2 is requested. (Two panels on the right): Proton–proton missing mass distributions in the $pp \to ppX$ reaction at 1.4 GeV with the trigger requiring two tracks from charged particles in the forward detector and at least two clusters from neutral particles in the calorimeter. (left) data analysis is based only on the forward detector tracks (right) in addition invariant mass of the two photons ≥ 300 MeV/c^2 is required.

MESON PRODUCTION FOR THE DECAY PROGRAM

Mesons for the decay studies at COSY come from *pp* and *pd* interactions close to threshold. A feature of the threshold experiments is that recoils are emitted in a small forward cone and their momenta can be measured in a dedicated detector which covers a limited range of scattering angles. Meson production is identified (tagged) by the missing mass. The COSY beam momentum resolution of 0.1% FWHM translates into a missing mass resolution of a few MeV/c^2 FWHM. The light decay particles are emitted more isotropically and their registration requires a detector with nearly 4π sr coverage. A typical resolution for the invariant masses of the decay products is a few tens MeV/c^2.

A clean source of η and ω mesons is provided by $pd \to ^3\mathrm{He}\eta, \omega$ reaction. The requirement of a ^{3}He ion in the final state selects only a small fraction of all *pd* interactions (about 0.1% of the total inclusive cross section). Identification of ^{3}He with respect to protons and deuterons can be easily implemented in the trigger. Doing this one obtains event rates at maximum usable luminosity well within the data acquisition capabilities without imposing any additional condition on the decay particles. The $pd \to ^3\mathrm{He}\,\eta$ reaction is used at a beam kinetic energy of 1.0 GeV where the ^{3}He ions are measured in the forward detector (FD) covering scattering angles from 3° to 18° and the cross section is about 400 nb [4, 5]. Similar kinematics is obtained for the $pd \to ^3\mathrm{He}\omega$ reaction at 1.45 GeV at a cross section of about 85nb [6]. Due to particle flux limitations of the calorimeter and the mini–drift chamber the maximal rate of η and ω mesons from the $pd \to ^3\mathrm{He}X$ reactions is about 10 η events/s and 2 ω events/s, respectively. In a test run WASA-at-COSY has collected a sample of the $pd \to ^3\mathrm{He}\eta'$ reaction at 1.8 GeV which corresponds to the predicted cross section maximum in the two step model [7]. The estimated value of the cross section from the data is below one nb and therefore the reaction is not competitive for η' meson decay studies.

Meson production cross sections in $pp \to ppX$ reactions are much larger [8, 9] (see tab. 1). In addition, the inclusive *pp* cross section is two times lower than for *pd* interactions. Therefore, the yield of the mesons could be higher for a given luminosity and one can use higher luminosities not exceeding the limits on the particle flux. A disadvantage of the $pp \to ppX$ reaction is the more difficult online identification and tagging. This is illustrated in Fig. 2 showing missing mass distributions for two reconstructed protons in the FD in a run for η meson decays at 1.4 GeV. The trigger condition in this case included requirement of at least two tracks in FD and at least two clusters from neutral

TABLE 1. List of reactions selected for production of light mesons for decay studies at COSY.

Reaction	Beam energy [GeV]	Excess energy [MeV]	cross section
$pd \to ^3\text{He}\eta$	1.00	59	400 nb
$pd \to ^3\text{He}\omega$	1.45	63	85 nb
$pp \to pp\pi^0$	0.55	122	1.3 mb
$pp \to pp\eta$	1.40	56	10 μb
$pp \to pp\omega$	2.14	86	10 μb
$pp \to pp\eta'$	2.54	45	300 nb

particles in the calorimeter. The missing mass plot on the left side was obtained based on all events selected by the trigger, where the right figure shows the pp missing mass after selection of the $\eta \to \gamma\gamma$ decay channel in the analysis. Dedicated trigger conditions for specific decay modes are needed. Due to limited trigger selectivity a production rate of about 100 η events/s has been achieved. A new more selective trigger system including digitized energy deposits is now being commissioned.

During the last three years data sets for η meson decay studies have been collected using both pp and pd interactions at COSY. A very first production run in April 2007 yielded 10^6 events for the $pp \to pp(\eta \to 6\gamma)$ reaction. In the next four week run period in October 2008, η decays were studied using the $pd \to ^3\text{He}\,\eta$ reaction. An unbiased data sample of $1.1\cdot 10^7$ η meson decays was collected. The data sample from this reaction was increased to $3\cdot 10^7$ η meson decays in an eight week continuation in August-September 2009. Recently, in spring 2010, an eight week pp production run was carried out yielding a data sample corresponding to about 10^8 η events and focused on the η decay modes with charged particles.

STATUS OF THE ANALYSIS

The decay width of the isospin-violating η meson decays into three pions is proportional to the light quark mass difference squared – $(m_d - m_u)^2$. The decays have the potential to provide one of the best constraints for the light quark mass ratios and are a sensitive test of convergence of Chiral Perturbation Theory (CHPT) and its extensions. The test can be accomplished by measurements of Dalitz plot densities of the two decay modes $\eta \to \pi^0\pi^0\pi^0$ and $\eta \to \pi^+\pi^-\pi^0$ (with BR $(32.57 \pm 0.23)\times 10^{-2}$ and $(22.74 \pm 0.28)\times 10^{-2}$ respectively [10]) decays as well as the $\Gamma(\eta \to \pi^0\pi^0\pi^0)/\Gamma(\eta \to \pi^+\pi^-\pi^0)$ ratio. The decay amplitude of η into three pions can be described using the following two normalized Dalitz variables:

$$x \equiv \frac{1}{\sqrt{3}} \frac{T_1 - T_2}{\langle T \rangle}; \; y \equiv \frac{T_3}{\langle T \rangle} - 1. \tag{1}$$

Here T_i are the kinetic energies of the pions in the rest frame of the η meson, $\langle T \rangle \equiv (T_1 + T_2 + T_3)/3$.

The $\eta \to \pi^0\pi^0\pi^0$ decay amplitude is constant in the leading order of CHPT. Experimentally, a small deviation from a uniform Dalitz Plot density distribution is observed with a maximum at the center point $(x = 0, y = 0)$. This variation of the density of about three percent was first observed by Crystal Ball at AGS [11] in 2001 and later confirmed by several independent measurements [12, 13, 14, 15, 16]. The CELSIUS/WASA [12] and the WASA-at-COSY collaborations [15] have contributed to the clarification of the experimental situation. The explanation of this effect poses a challenge for CHPT which in next–to–leading order predicts the opposite behavior of the density which decreases towards the Dalitz plot center [17]. The large uncertainty of the recently carried out next–to–next–to–leading order calculations [18] does not allow to decide the direction of the effect.

For the $\eta \to \pi^+\pi^-\pi^0$ decay channel the Dalitz plot density depends both on x and y variables. The recently obtained precise result from KLOE with $1.3 \cdot 10^6$ events [19] in the Dalitz plot represents an improvement of more than two orders of magnitude in the event statistics compared to all previous measurements. The KLOE Dalitz plot density cannot be reproduced by any of the CHPT calculations. On the experimental side it is important to bring the status of $\eta \to \pi^+\pi^-\pi^0$ decay to similar accuracy and consistency as for $\eta \to \pi^0\pi^0\pi^0$ decay. Therefore, new independent measurements with a precision similar to that of the KLOE experiment are needed. The WASA-at-COSY collaboration has collected data samples in both pp and pd interactions [20].

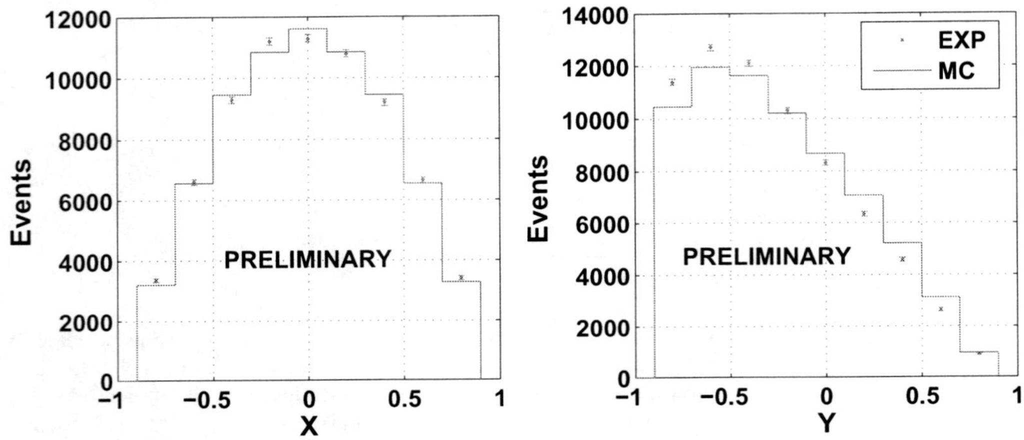

FIGURE 3. Dalitz plot projections: (left) x-projection (right) x-projection. The data are not corrected for the acceptance. Solid line indicates MC data with leading order CHPT amplitude and the points the experimental values. Only statistical uncertainties are shown.

The preliminary analysis of 2008 $pd \to {}^3\text{He}\,\eta$ data yields 149 000 $\eta \to \pi^+\pi^-\pi^0$ event candidates. The kinematic fit for the $pd \to {}^3\text{He}\pi^+\pi^-\pi^0$ reaction hypothesis was carried out. The number of $\eta \to \pi^+\pi^-\pi^0$ decays in each Dalitz plot bin was extracted by constructing the ^{3}He missing mass distribution and subtracting the direct $\pi^+\pi^-\pi^0$ production background estimated by a polynomial fit. The preliminary results for the x,y projections of the Dalitz plot are compared in Fig. 3 to the simulations assuming tree-level CHPT prediction.

The decays $\eta \to \pi^+\pi^-\gamma$ and $\eta \to \pi^0\gamma\gamma$ (with BR $(4.60 \pm 0.16)\times 10^{-2}$ and $(2.7 \pm 0.5)\times 10^{-4}$ respectively [10]) are driven by the chiral anomaly responsible for the $\eta \to \gamma\gamma$ decay [21] and are dominated by the intermediate states with the lightest vector mesons [22]. The measurements of the $\eta \to \pi^0\gamma\gamma$ decay is reported by K. Lalvani [23]. The analysis of the $\eta \to \pi^+\pi^-\gamma$ decay from 2008 $pd \to {}^3\text{He}\,\eta$ run [24] is nearly finished. The two and tree pion direct production and $\eta \to \pi^+\pi^-\pi^0$ decay are the main background reactions. After selecting the final state with two pions and at least one photon candidate, a kinematic fit has been performed to test the $pd \to {}^3\text{He}\pi^+\pi^-\gamma$ reaction hypothesis. Remaining background is subtracted bin-wise from the final distributions by determining the number of η mesons from the corresponding invariant mass distribution of the $\pi^+\pi^-\gamma$ system. The resulting data sample of 13740 ± 140 $\eta \to \pi^+\pi^-\gamma$ events is so far the largest collected in an exclusive measurement. The kinematic variables to represent the phase space of the decay are π^+ photon opening angle, θ, in the pion-pion rest frame and the photon energy in the η meson rest frame (E_γ). The preliminary results in Fig. 4 show the background subtracted and efficiency corrected $\cos\theta$ and E_γ distributions. The gray area indicates the systematic error, while the statistical error is indicated with the bars. The solid lines represent the simplest gauge invariant matrix element for the process corresponding to p – wave dipion phase space [25]. The photon energy distribution is also compared to the ρ meson dominated matrix element (dashed curve).

The WASA detector design was optimized for measurements of decays involving electron-positron pairs and photons such as Dalitz decays of π^0 and η mesons. The knowledge of the π^0 and η meson form factor is important for the calculations of the Standard Model contributions to $g-2$ of the muon and to the rare π^0 and η decays into a lepton-antilepton pair. The muon $g-2$ and the branching ratio for the $\pi^0 \to e^+e$ decay are currently the only observables where hints of a deviation from the Standard Model predictions are reported. A status of analysis of the $\eta \to e^+e^-\gamma$ decay ($BR = 7.0 \pm 0.7)\times 10^{-3}$ [10]) from the 2007 $pp \to pp\eta$ run is reported by H. Bhatt [26]. The analysis of the decays is carried out in parallel using the pd data. The decay $\eta \to \pi^+\pi^-e^+e^-$, related to $\eta \to \pi^+\pi^-\gamma$, allows for more precise tests of the chiral anomaly by comparison with existing CHPT calculations. A high statistics measurement of $\eta \to \pi^+\pi^-e^+e^-$ will provide constraints for a new kind of flavor conserving CP violation by measuring the asymmetry of the dihedral angle between the pion- and electron planes. The present value for the branching ratio for the $\eta \to \pi^+\pi^-e^+e^-$ decay is $(2.68 \pm 0.09 \pm 0.07)\times 10^{-4}$ [27] from the KLOE experiment. The measurement of double Dalitz η decays, $\eta \to e^+e^-e^+e^-$ allows the study the η meson form factor for two time–like virtual photons. Until recently only an upper limit of 6.9×10^{-5} (90% CL) [28] for the BR of the decay was available. Observation of the

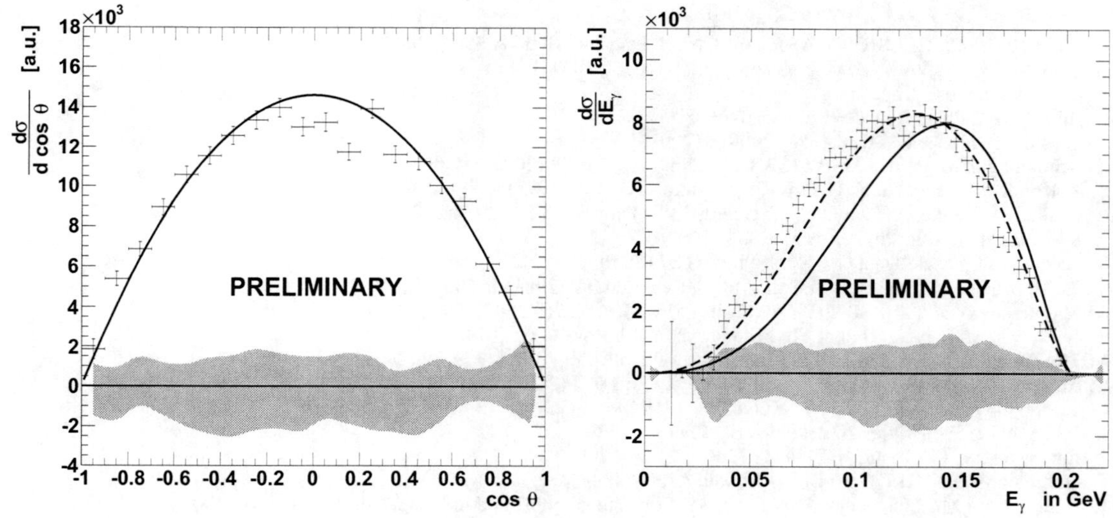

FIGURE 4. The background subtracted and corrected for the acceptance $\cos\theta$ distribution of the pions (left) and the centre-of-mass photon energy distribution (right). The shaded areas shows the systematic uncertainty. The contribution of the statistical uncertainty is indicated by the error bars. The solid lines are predictions of simplest gauge invariant matrix element and the dashed line VMD prediction.

signal from the $\eta \to e^+e^-e^+e^-$ decay in the KLOE experiment is reported by S. Giovannella at this conference [29]. WASA-at-COSY has carried out exploratory studies of the $\eta \to \pi^+\pi^-e^+e^-$ and $\eta \to e^+e^-e^+e^-$ decays using the 2008 $pd \to ^3$He η data yielding 180 (S:B=2.5) and 15 (S:B=1) event candidates, respectively [30]. The analysis of the recent pp data has just started.

The η decay into $\pi^0e^+e^-$ is forbidden to proceed via an intermediate $\pi^0\gamma^*$ state due to C-conservation. The present BR upper limit is 4×10^{-5} (90% CL) from 1975 experiment [31]. The feasibility of the measurement with the WASA detector was studied with the 2008 pd run period data. The expected sensitivity is close to the present upper limit.

The dominant mechanism for $\mathscr{P} \to \ell^+\ell^-$ decays within the Standard Model is a process involving two virtual photons and is additionally suppressed by helicity conservation. Recently the interest in the decays was revived due to observed excess for the $\pi^0 \to e^+e^-$ decay branching ratio [32, 33]. For the $\eta \to e^+e^-$ decay the best experimental limit (2.7×10^{-5} 90% CL) comes from $pd \to ^3$Heη CELSIUS/WASA experiment [34]. In WASA-at-COSY the feasibility of the measurement of the $\eta \to e^+e^-$ and $\pi^0 \to e^+e^-$ decays in pp interactions is being investigated. The sensitivity for the $\eta \to e^+e^-$ decay is close to the present upper limit. A test run of the $pp \to pp\pi^0$ reaction at a beam kinetic energy of 550 MeV (below the two–pion–production threshold) was carried out in 2010 for the $\pi^0 \to e^+e^-$ decay measurement.

ACKNOWLEDGMENTS

Travel grant by Swedish Research Council is acknowledged. This work was in part supported by the European Commission under the 7th Framework Programme through the 'Research Infrastructures' action of the Capacities'. Programme: FP7-INFRASTRUCTURES-2008-1, Grant Agreement N. 227431 and the Göran Gustafsson Foundation.

REFERENCES

1. C. Bargholtz *et al.* [CELSIUS/WASA Coll.], Nucl. Instrum. Meth. A **594** (2008) 339.
2. H. H. Adam *et al.* [WASA-at-COSY Coll.], arXiv:nucl-ex/0411038.
3. A. Wronska in these proceedings.
4. R. Bilger *et al.*, Phys. Rev. C **65** (2002) 044608.
5. T. Rausmann *et al.*, Phys. Rev. C **80** (2009) 017001.
6. K. Schonning *et al.* [CELSIUS/WASA Coll.], Phys. Rev. C **79** (2009) 044002.
7. L. A. Kondratyuk and Yu. N. Uzikov, Phys. Atom. Nucl. **60** (1997) 468.
8. P. Moskal, M. Wolke, A. Khoukaz, W. Oelert, Prog. Part. Nucl. Phys. 49 (2002) 1.
9. A. Khoukaz et al., Eur. Phys. J. A20 (2004) 345.
10. K. Nakamura [Particle Data Group], J. Phys. G **37** (2010) 075021.
11. W. B. Tippens *et al.* [Crystal Ball Coll.], Phys. Rev. Lett. **87** (2001) 192001.
12. M. Bashkanov *et al.* [CELSIUS/WASA Coll.], Phys. Rev. C **76** (2007) 048201.
13. S. Prakhov *et al.* [Crystal Ball at MAMI and A2 Coll.], Phys. Rev. C **79** (2009) 035204.
14. M. Unverzagt *et al.* [Crystal Ball at MAMI and A2 Coll.], Eur. Phys. J. A **39** (2009) 169.
15. C. Adolph *et al.* [WASA-at-COSY Coll.], Phys. Lett. B **677** (2009) 24.
16. F. Ambrosino *et al.* [KLOE Coll.], arXiv:1004.1319 [hep-ex].
17. J. Gasser and H. Leutwyler, Nucl. Phys. B **250** (1985) 539.
18. J. Bijnens and K. Ghorbani, JHEP **0711** (2007) 030.
19. F. Ambrosino *et al.* [KLOE Coll.], JHEP **0805** (2008) 006.
20. P. Adlarson and M. Zielinski [WASA-at-COSY Coll.], to be published in Int.J.Mod.Phys.A arXiv:1009.5508 [hep-ex].
21. S. L. Adler, Phys. Rev. **177**, (1969) 2426.
22. M. Gell-Mann, D. Sharp, and W. G. Wagner, Phys. Rev. Lett. **8** (1962) 261.
23. K. Lalvani in these proceedings.
24. C. F. Redmer [WASA-at-COSY Coll.], to be published in Int.J.Mod.Phys.A.
25. J. G. Layter, Phys. Rev. D **7** (1973) 2565.
26. H. Bhatt in these proceedings.
27. F. Ambrosino [KLOE Coll.], Phys. Lett. B **675** (2009) 283.
28. R. R. Akhmetshin *et al.* [CMD-2 Collaboration], Phys. Lett. B **501** (2001) 191.
29. S. Giovannella in these proceedings.
30. D. Coderre, M. Janusz and L. Yurev PhD theses in preparation.
31. M. R. Jane *et al.*, Phys. Lett. B **59** (1975) 99.
32. E. Abouzaid *et al.* [KTeV Coll.], Phys. Rev. D **75** (2007) 012004.
33. A. E. Dorokhov and M. A. Ivanov, Phys. Rev. D **75** (2007) 114007.
34. M. Berlowski *et al.* [CELSIUS/WASA Coll.], Phys. Rev. D **77** (2008) 032004.

Results on Charmonium(-like) and Bottomonium(-like) States from Belle and BaBar

Jens Sören Lange, for the Belle Collaboration

Justus-Liebig-Universität Gießen, II. Physikalisches Institut, Heinrich-Buff-Ring 16, 35392 Gießen, Germany

Abstract. Spectroscopy results for Belle and BaBar are reported. A particular focus is put on new results of the X(3872) state with its radiative decays to $J/\psi\gamma$ and $\psi'\gamma$, its decay into $J/\psi 3\pi$ and the search for production in radiative Upsilon decays. Another focus is L=2 mesons, in particlar a possible D-wave assignment to the X(3872) and the confirmation of an Upsilon D-wave state.

PACS: 14.40.Pq, 14.40.Rt

Conference Talk presented at MENU10, 12th International Conference on Meson-Nucleon Physics and the Structure of the Nucleon, 05/31-06/04, 2010, College of William and Mary, Williamsburg, Virginia.

INTRODUCTION

In this paper, spectroscopy results from the two B factories BaBar [1] at PEP-II and from Belle [2] at KEKB are reported. The data samples of the two experiments are summarized in Tab. 1.
There are quite a number of unresolved interesting questions in charmonium spectroscopy, such as:

- The charmonium potential is usually regarded as a static quark-antiquark potential (Cornell ansatz) [3]

$$V(r) = -\frac{4}{3}\frac{\alpha_S}{r} + k \cdot r \qquad (1)$$

  with a Coulomb-like term and a confinement term. As can be seen, α_S is assumed fixed over the total range $0<r<\simeq 1$ fm, which is an approximation. Also, as the mechanism of confinement is still one of the unanswered questions in QCD, it is still unproven if the linear approximation (corresponding to a constant string force) in the confinement term is *(a)* valid to all orders of α_S and *(b)* valid even in the far long range $r>\simeq 1.3$ fm (i.e. in the string breaking regime). In fact, studies of potential constructed with two-gluon exchanges [4] [5] lead to a number of additional terms with different r dependances. In addition, the high lying states (e.g. $L\leq 2$, $n\geq 3$) are sensitive to the string constant $k\simeq 1$ GeV/fm, which is the slope of the confinement term, and mass measurements can provide a precise measurement of k.
- The strong coupling constant in the charmonium system has a quite high value α_S=0.54 [6], so there might be non-perturbative effects becoming visible e.g. in the hyperfine splittings.
- What is the nature of the newly observed narrow states near thresholds, which do not fit into potential model calculations? Are they molecular states? Or tetraquarks? Or threshold effects? As an example, molecular potentails might contain r^{-2} and r^{-3} [7] terms, not present in the Cornell potential, and leading to eigenstates with masses different from the quark-antiquark potential.
- There are yet unobserved states, some of them expected narrow, e.g. the $^3D_2$(J^{PC}=2^{--}) state.

Similar questions apply to the bottomonium system. At B factories, studies of bottomonium require the change of the beam energies. A few results are reported down below.

12th International Conference on Meson-Nucleon Physics and the Structure of the Nucleon (MENU 2010)
AIP Conf. Proc. 1374, 115-120 (2011); doi: 10.1063/1.3647108

TABLE 1. Integrated luminosities for data sets at different beam energies for BaBar and Belle.

	Belle	**BaBar**
On-resonance		
$\Upsilon(4S)$	711 fb^{-1}	433 fb^{-1}
$\Upsilon(5S)$	121 fb^{-1}	– fb^{-1}
$\Upsilon(3S)$	3.0 fb^{-1}	30 fb^{-1}
$\Upsilon(2S)$	24 fb^{-1}	14 fb^{-1}
$\Upsilon(1S)$	5.7 fb^{-1}	– fb^{-1}
Off-resonance		
	87 fb^{-1}	54 fb^{-1}
Total	952 fb^{-1}	553 fb^{-1}

WIDTH OF η_c

Although the η_c is the ground state of the charmonium system (J^{PC}=0^{-+}, $^1S_0$) and already discovered in 1980 [11] [12], its width has been of particular interest recently. Previous width measurements [17] showed values around $\Gamma \simeq 15$ MeV from radiative J/ψ and ψ' decays, and values around $\Gamma \simeq 30$ MeV from B meson decays. However, this is probably not surprising, as in the radiative decays the cross section varies according to E_γ^a with an exponent a=3−7. This energy dependance modifies the lineshape and the width determination becomes non-trivial. On the other hand, in the reaction $\gamma\gamma \rightarrow \eta_c$ the Breit-Wigner line shape is an appropriate approximation. In a new high statistics measurement by BaBar with a data set of 469 fb^{-1} and 14090 η_c signal events, a high precision measurement of the mass m=2982.2±0.4±1.6 MeV and the width Γ=31.7±1.2±0.8 MeV of the η_c could be obtained. This measurement represents a factor $\simeq 3$ improvement in both statistical and systematic errors compared to the BaBar measurement in 2008 in B meson decays [9] and the Belle measurement in 2008 in $\gamma\gamma$ collisions [10].

Decays of X(3872) to $D\overline{D}^*$

The X(3872) state has been discovered in B meson decays in the decay X(3872)$\rightarrow J/\psi\pi^+\pi^-$ by Belle [13]. It was confirmed in the same process by BaBar [14] and confirmed in inclusive production in $p\overline{p}$ production at $\sqrt{s}$=1.8 TeV at CDF-II [15] and D0 [16]. Among the newly observed charmonium-like states (sometimes referred to as XYZ states) the X(3872) is the only one with several decay channels having been observed. It has a surprisingly very narrow width Γ<2.3 MeV [13], although its mass is above the open charm threshold. As its mass with 3871.56±0.22 MeV [17] is very close (within 1 MeV) to the D^0D^{0*} threshold, it was discussed as a possible S-wave $[D^0D^{0*}]$ molecule [18] [19]. The decay into $D^{\pm}D^{\mp *}$ is kinematically forbidden, but the decay into D^0D^{0*} is a strong decay and among the so far observed decays it represents the dominant one, i.e. the branching fraction is a factor $\simeq 9$ higher than for the decay into $J/\psi\pi^+\pi^-$. In this decay channel, BaBar measured surprisingly a high mass of the X(3872) as m=3875.1$^{+0.7}_{-0.5}$(stat.)±0.5(syst.) MeV [20]. This high value initiated discussion, that there might be two different X states, namely X(3872) and X(3875), which would fit to a tetraquark hypothesis [21] and the two different states $[c\overline{c}u\overline{u}]$ and $[c\overline{c}d\overline{d}]$. On the other hand Belle measured in the same decay channel the mass as m=3872.9$^{+0.6}_{-0.4}$(stat.)$^{+0.4}_{-0.5}$(syst.) MeV [22] and thus consistent with the world average [17]. A possible explanation of the discrepancy is the difficulty of performing fits to signals close to threshold. In fact, the two fits used two very different approaches:

- Babar used a 1-dimensional binned maximum likelihood fit [20] with the D^*D invariant mass as the only variable, where the signal probability density function was extracted from MC simulations and an exponential function was used for the background parametrization.
- Belle used an 2-dimensional unbinned maximum likelihood fit [22], i.e. on the one hand the beam constraint mass with a Gaussian signal and an Argus function for the background, and on the other hand a Breit-Wigner signal for the D^*D invariant mass with a square root function for the background.

Radiative Decays of X(3872)

The branching fraction of the rare decay X(3872)$\to J/\psi\gamma$ is a factor $\simeq$6 smaller than the one for X(3872)$\to J/\psi\pi^+\pi^-$. The first evidence by Belle [23] was based upon a data set of 256 fb^{-1} with 13.6$\pm$4.4 signal events. The signal was confirmed by BaBar [24] with a data set of 424 fb^{-1} and 23.0$\pm$6.4 signal events. Although rare, this decay channel is very important, as their observation clearly establishes a C=+1 charge parity assignment to the X(3872).
Recently BaBar found evidence for the decay X(3872)$\to\psi'\gamma$ [25] with 424 fb^{-1} and 25.4$\pm$7.4 signal events. The signal yield of this observation was surprising, as it implied a large ratio BR(X(3872)$\to\psi'\gamma$)/ BR(X(3872)$\to J/\psi\gamma$)=3.4$\pm$1.4. There was a priori no understanding of the fact, why the transition of the X(3872) to a n=2 charmonium state should be stronger than to n=1. In fact, quite the opposite behaviour was expected.
In the case of X(3872)$\to J/\psi\gamma$the photon energy is E_γ=775 MeV, and thus due to vector meson dominance ρ and ω can contribute to the amplitudes. However, in the case of X(3872)$\to\psi'\gamma$ with the smaller E_γ=186 MeV the transition can only proceed through light quark annihilation with an expected small amplitude. A new measurement by Belle of both radiative channels was based upon a data set of 711 fb^{-1} [26]. The background was studied in MC simulations and revealed peaking behaviour in some background components close to the signal region. The signal X(3872)$\to J/\psi\gamma$ was clearly reestablished with $30.0^{+8.2}_{-7.4}$ signal events (4.9σ significance) for $B^+\to K^+$X(3872) and $5.7^{+3.5}_{-2.8}$ signal events (2.4σ significance) for $B^0\to K^0$X(3872).

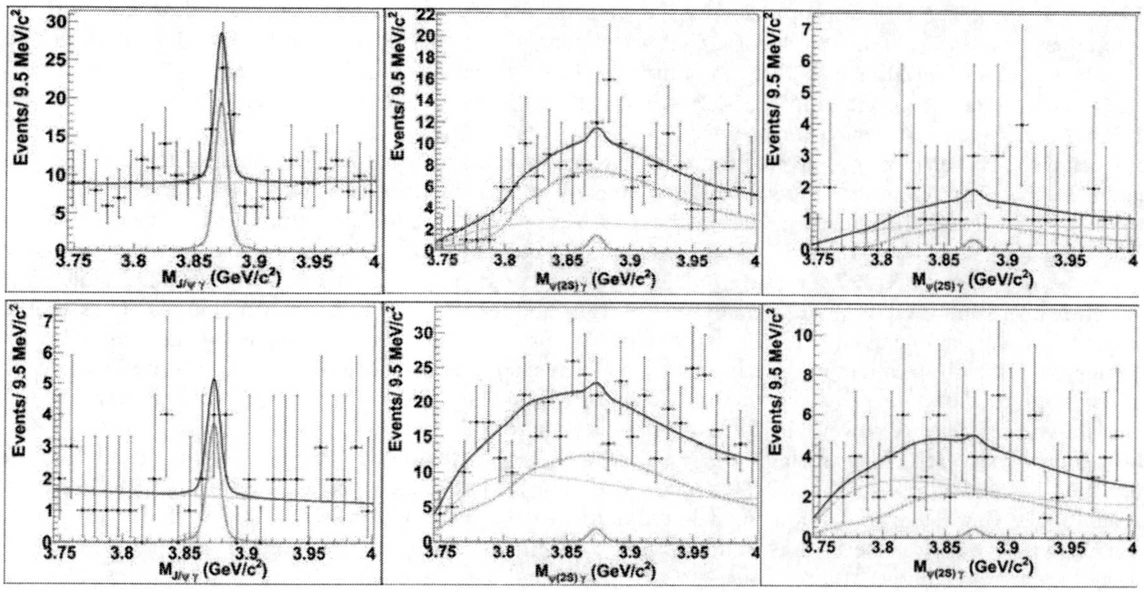

FIGURE 1. Preliminary Belle results on radiative decays of the X(3872). $J/\psi\gamma$ invariant mass in the X(3872) mass region for $B^+\to K^+$X(3872)($\to J/\psi\gamma$) (top left) and $B^0\to K^0_S$X(3872)($\to J/\psi\gamma$) (bottom left), and $\psi'\gamma$ invariant mass in the X(3872) mass region for $B^+\to K^+$X(3872)($\to\psi'\gamma$) (top center) $B^0\to K^0_S$X(3872)($\to\psi'\gamma$) (bottom center) with $\psi'\to e^+e^-,\mu^+\mu^-$ and $B^+\to K^+$X(3872)($\to\psi'\gamma$) (top right) $B^0\to K^0_S$X(3872)($\to\psi'\gamma$) (bottom right) with $\psi'\to J/\psi\pi^+\pi^-$. The solid, dotted and dot-dashed curves are the total, combinatorial background and combined $\psi'K^*$ and $\psi'K$ background, respectively. The background subtracted signal is also shown.

For X(3872)$\to\psi'\gamma$, the four decay channels $B^+\to K^+$X(3872) and $B^0\to K^0$X(3872) with X(3872)$\to\psi'\gamma$ with $\psi'\to l^+l^-$ and $\psi'\to J/\psi\pi^+\pi^-$. Charged and neutral B modes were treated separately, but the two ψ' subdecay modes were fitted simultaneously because of their different background shapes. The signal was treated as a double Gaussian, the combinatorial background was parameterized as a threshold function. The shape of the $\psi'K^*$ and $\psi'K$ background, and in particular the peaking structures, was modeled as a sum of bifurcated gaussians using a large MC sample. The signal yields were determined as $5.0^{+11.9}_{-11.0}$ signal events (0.4σ significance) for $B^+\to K^+$X(3872) and $1.5^{+4.8}_{-3.9}$ signal events (0.2σ significance) for $B^0\to K^0$X(3872). Thus, contrary to BaBar, Belle observed no signal, which would imply that there is no indication that the radiative transition from X(3872) to n=2 charmonium is stronger than to n=1

charmonium.
In the same analysis, the decay $\chi_{c1,2}\to J/\psi\gamma$ was used as a reference channel with signal yields of $32.8^{+10.9}_{-10.2}$ (3.6σ significance) for $B^+\to K^+\chi_{c2}$ and $2.8^{+4.7}_{-3.9}$ (0.7σ significance) for $B^0\to K^0\chi_{c2}$. In the charged mode, this represents the first observation of a $J^P=2^+$ state in a rare exclusive final state in a B meson decay, and thus a transition $0^-\to 0^-2^+$.

Decays of X(3872) and Y(3940) into $J/\psi 3\pi$

As the ρ meson carries isospin I=1, the X(3872) seems to violate isospin conservation in the decay X(3872)$\to J/\psi\rho(\to\pi^+\pi^-)$. One of the proposed explanations [27] is ρ/ω mixing, and therefore the investigation of the decay X(3872)$\to J/\psi\omega(\to\pi^+\pi^-\pi^0)$ is of importance. The difficulty hereby is the nearby Y(3940) state, which is also known to decay into the same final state. The latter decay was investigated be Belle with a data set of 275×10^6 B meson pairs. The mass of the Y(3940) was determined as 3943±11(stat.)±13(syst.) MeV with a width of 87±22(stat.)±26(syst.) MeV. Belle also observed a signal for X(3872)$\to J/\psi\omega(\to\pi^+\pi^-\pi^0)$, based upon a data set of 256 fb^{-1} [29]. The measured efficiency corrected ratio of X(3872)$\to J/\psi\pi^+\pi^-\pi^0$/ X(3872)$\to J/\psi\pi^+\pi^-$= 1.0±0.4(stat.)±0.3(syst.) indicated another case of isospin violation due to the additional π^0. BaBar published slightly different values for the Y(3940) with a data set of 383×10^6 B meson pairs mass [30], namely a mass of $3914.6^{+3.8}_{-3.4}$(stat.)±2.0(syst.) MeV and a width of 34^{+12}_{-8}(stat.)±5(syst.) MeV. However, a signal for X(3872)$\to J/\psi\omega(\to\pi^+\pi^-\pi^0)$ was not observed. In a recent re-analysis with 433 fb^{-1} by BaBar, a requirement on the 3-pion mass was adjusted, i.e. the lower offset was extended from 0.7695 GeV to 0.7400 GeV. With this change in the analysis technique BaBar was able to confirm the Belle signal for X(3872)$\to J/\psi\omega(\to\pi^+\pi^-\pi^0)$ and confirm the large isospin violation for the ratio X(3872)$\to J/\psi\pi^+\pi^-\pi^0$/ X(3872)$\to J/\psi\pi^+\pi^-$ measured as 0.7±0.3(stat.) and 1.7±1.3(stat.) for B^+ and B^0 decays, respectively.
In the re-analysis, BaBar also investigated the shape of the 3π mass distribution in order to determine the quantum number of the X(3872). A similar analysis for the 2π mass distribution in case of X(3872)$\to J/\psi\pi^+\pi^-$ was performed before by Belle [32] and CDF-II [33]. The result in both cases was that S-wave is preferred. However, in the new BaBar analysis the shape of the 3π mass distribution seems to indicate that P-wave is preferred. For the 2π case and S-wave, a parity of +1 for the X(3872) is preferred, leading to a tentative assignment of $J^{PC}=1^{++}$. This quantum number assignment is also supported by angular analyses [29] [34] and leads to a possible charmonium state assignment of χ'_{c1} ($^3P_1$), which is an n=2 state with a mass of 3953 MeV as predicted by potential models [6] and thus ≃70 MeV too high compared to the observation. For the 3π case and P-wave, a parity of −1 for the X(3872) is preferred, leading to a possible assignment of $J^{PC}=2^{-+}$. Then a possible charmonium assignment is η_{c2} ($^1D_2$), which is an n=1 state. The predicted mass is ≃100 MeV lower than for the χ'_{c1}. This state would be an L=2 meson.
In a different analysis, Belle investigated the $J/\psi\omega$ final state in $\gamma\gamma$ collisions based upon 694 fb^{-1} [35]. This analysis not only uses ϒ(4S), but also ϒ(3S) and ϒ(5S) data (see Tab. 1). The event selection uses a $p_T<0.1$ GeV/c balance requirement. The final state in $\gamma\gamma$ collisions is required to have isospin I=0. Fig. 2 shows the W distribution of the final candidate events, where W is defined as $W=m_5-m(l^+l^-)+m_{J/\psi}$. m_5 is the invariant mass of the system constructed from four charged tracks and a neutral pion candidate. A clear enhancement seen just above $J/\psi\omega$ threshold with 49±14(stat.)±4(syst.) events (7.7σ stat. significance). The fitted mass is 3915±3(stat.)±2(syst.) MeV, thus it might be the observation the Y(3940) state in a second production mode. However, the fitted width is Γ=17±10(stat.)±3(syst.) MeV, which is narrower than the width of the Y(3940) as measured in B decays. The production mode allows to establish the charge parity C=+1 for this state, same as the X(3872), but the determination of the other quantum numbers would require more statistics.

ϒ(1S) Radiative Decays to X(3872)

As shown in Tab. 1, Belle recorded an extensive data set with the beam energies adjusted to the ϒ(1S) resonance, the n=1 $^3S_1$ $b\bar{b}$ state with $J^P=1^-$ and a mass of 9.46 GeV. With this data set, radiative transistions $b\bar{b}\to c\bar{c}\gamma$ can be investigated. These rare events with an expected branching fraction of $\leq 10^{-5}$ [36] with interfering QED and QCD amplitudes. The transition may be from a 1^{--} state, such as the ϒ(1S), to a 1^{++} state, which is one of the most probable quantum number assignments for the X(3872).
Belle searched for the process ϒ(1S)$\to\gamma$X(3872)$(\to J/\psi\pi^+\pi^-)$ with a data set of 5.712 fb^{-1} [37], corresponding to 88×10^6 ϒ(1S) decays The photon detection required $E_\gamma^{lab}>3.5$ GeV and the recoil mass on four charged

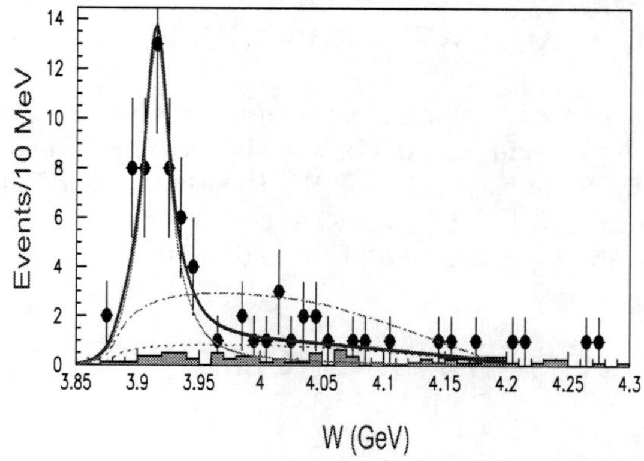

FIGURE 2. The W distribution of the final candidate events (dots with error bars) for $\gamma\gamma \to J/\psi\,\omega$ at Belle [35]. The shaded histogram is the distribution of non-J/ψ background estimated from the sideband distribution. The bold solid, thinner solid and dashed curves are the total, resonance and background contributions, respectively. The dot-dashed curve is the fit without a resonance.

tracks being consistent with zero, i.e. $-2<m_{recoil}<2$ GeV^2. Initial state radiation (ISR) was treated in two different ways. On the one hand, ISR events were rejected by a criterium on the cms polar angle of the photon, i.e. $|\cos\vartheta^*_\gamma|<0.9$. On the other hand, ISR events for ψ' production, with the same $J/\psi\pi^+\pi^-$ final state as the X(3872), were used as a crosscheck, and the cross section for this ISR process was determined as 20.2 ± 1.1(stat.) pb. For the X(3872) one event in the signal region was observed, resulting in an upper limit for the product branching fraction BR(Y(1S)$\to\gamma$X(3872)$\times$BR(X(3872)$\to J/\psi\pi^+\pi^-$)$<2.2\times10^{-6}$ at 90% CL.

THE ϒ(1D) STATE

D-wave mesons with orbital angular momentum L=2 are quite different, if we compare the charmonium and the bottonium system. For the charmonium system, there are two observed D-wave states *above* the $D\overline{D}$ threshold, namely the ψ(3770) (n=1) and the ψ(4153) (n=2). Both are $J^{PC}=1^{--}$ and broad, i.e. their experimentally measured width is >20 MeV. For the bottomonium system, there are two D-wave states predicted *below* threshold and thus narrow, i.e. their hadronic width ≤30 keV [38]. Consequently, in case of an observation it might even be possible to resolve the triplet of $J^{PC}=0^{--}$, 1^{--} and 2^{--}. An $\Upsilon(1^3D_2)$ candidate was observed by CLEO [39] based upon 5.8×10^6 ϒ(3S) decays with a mass of $10161.1\pm0.6\pm1.6$ MeV, consistent with a potential model prediction of 10158 MeV [40].
BaBar searched for the $\Upsilon(1^3D_2)$ in a data set of 122×10^6 ϒ(3S) decays [41]. The signal path has two radiative transitions ϒ(3S)→ϒ(2P)γ and ϒ(2P)→ϒ(1D)γ, used for tagging, with subsequent decay by emission of a $\pi^+\pi^-$ pair to ϒ(1S), which finally decays to e^+e^- or $\mu^+\mu^-$. The complete signal path was required for the identification. BaBar was able to clearly confirm the $J^{PC}=1^{--}$ state with $33.9^{+8.2}_{-7.5}$ signal events (5.8σ stat. significance). The fitted mass is $m=10164.5\pm0.8\pm0.5$ MeV, consistent with the CLEO measurement.

SUMMARY

The B factories continue to provide exciting results. Charmonium spectroscopy is studied in B meson decays, $\gamma\gamma$ collisions and ϒ(nS) decays. Bottomonium spectroscopy is studied in ϒ(nS) decays. Highly excited states such as L=2 states are clearly identified and provide accurate tests for potential models. States which are not consistent with any potential model, such as the X(3872), are studied in new ways, such as radiative decays or production in radiative decays. Surprising properties such as a large isospin violation are confirmed.

ACKNOWLEDGMENTS

We thank the KEKB group for excellent operation of the accelerator, the KEK cryogenics group for efficient solenoid operations, and the KEK computer group and the NII for valuable computing and SINET3 network support. We acknowledge support from MEXT, JSPS and Nagoya's TLPRC (Japan); ARC and DIISR (Australia); NSFC (China); MSMT (Czechia); DST (India); MEST, NRF, NSDC of KISTI, and WCU (Korea); MNiSW (Poland); MES and RFAAE (Russia); ARRS (Slovenia); SNSF (Switzerland); NSC and MOE (Taiwan); and DOE (USA).

REFERENCES

1. BaBar Collaboration, NIM A479(2002)1
2. Belle Collaboration, NIM A479(2002)117
3. E. Eichten, K. Gottfried, T. Kinoshita, K. D. Lane, T.-M. Yan, Phys. Rev. D17(1978)3090
4. S. N. Gupta, S. F. Radford, W. W. Repko, Phys. Rev. D26(1982)3305
5. S. Titard, F. J. Yndurain, Phys. Rev. D49(1994)6007
6. Barnes, Godfrey, Swanson, Phys. Rev. D72(2005)054026
7. C. Amsler, N. A. Törnqvist, Phys. Rep. 389(2004)61
8. BaBar Collaboration, Phys. Rev. D81(2010)052010
9. BaBar Collaboration, Phys. Rev. D78(2008)012006
10. Belle Collaboration, EPJ C53(2008)1
11. Mark-II Collaboration, Phys. Rev. Lett. 45(1980)1146
12. Crystal Ball Collaboration, Phys. Rev. Lett. 45(1980)1150
13. Belle Collaboration, Phys. Rev. Lett. 91(2003)262001
14. BaBar Collaboration, Phys. Rev. D71(2005)071103
15. CDF-II Collaboration, Phys. Rev. Lett. 93(2004)072001
16. D0 Collaboration, Phys. Rev. Lett. 93(2004)162002
17. K. Nakamura et al. (Particle Data Group), J. Phys. G37(2010)075021
18. N. A. Tornqvist, Phys. Lett. B590(2004)209
19. N. A. Tornqvist, Phys. Rev. Lett. 67(1991)556
20. BaBar Collaboration, Phys. Rev. D77(2008)011102
21. L. Maiani, A. D. Polosa, V. Riquer, Phys. Rev. Lett. 99(2007)182003
22. Belle Collaboration, Phys. Rev. D81(2010)031103
23. Belle Collaboration, arXiv:hep-ex/0505037
24. BaBar Collaboration, Phys. Rev. D74(2006)071101
25. BaBar Collaboration, Phys. Rev. Lett. 102(2009)132001
26. V. Bhardwaj, for the Belle Collaboration, QWG10, International Workshop on Heavy Quarkonium 2010, Fermilab, Batavia, May 18-21, 2010
27. K. Terasaki, Prog. Theor. Phys. 122(2010)1285
28. Belle Collaboration, Phys. Rev. Lett. 94(2005)182002
29. Belle Collaboration, hep-ex/0505037
30. BaBar Collaboration, Phys. Rev. Lett. 101(2008)082001
31. BaBar Collaboration, arXiv:1005.5190
32. Belle Collaboration, arXiv:hep-ex/0505038
33. CDF-II Collaboration, PRL 96(2006)102002
34. CDF-II Collaboration, PRL98(2007)132002
35. Belle Collaboration, Phys. Rev. Lett. 104(2010)092001
36. Y.-J. Gao, Y.-J. Zhang, K.-T. Chao, hep-ph/0701009
37. Belle Collaboration, Phys. Rev. D82(2010)051504
38. W. Kwong, Phys. Rev. D38(1988)279
39. CLEO Collaboration, Phys. Rev. D70(2004)032001
40. D. Ebert, R. N. Faustov, V. O. Galkin, Phys. Rev. D67(2003)014027
41. BaBar Collaboration, arXiv:1004.0175 [hep-ex]

The Jefferson Lab 12 GeV Upgrade

R. D. McKeown

Thomas Jefferson National Accelerator Facility, Newport News, Virginia 23606, USA
Department of Physics, College of William and Mary, Williamsburg, Virginia 23187, USA

Abstract. A major upgrade of the Continuous Electron Beam Accelerator Facility (CEBAF) at the Thomas Jefferson National Accelerator Facility is in progress. Construction began in 2008 and the project should be completed in 2015. The upgrade includes doubling the energy of the electron beam to 12 GeV, the addition of a new fourth experimental hall, and new experimental equipment in three of the experimental halls. A brief overview of this upgrade project is presented along with some highlights of the anticipated experimental program.

Keywords: Nucleon Structure, Exotic Mesons, Electron Beam Accelerator Facility
PACS: 14.20.-c, 14.40.Rt, 29.20.-c

INTRODUCTION

Since 1995, the CEBAF facility at Jefferson Laboratory has provided high-duty factor (continuous) beams of electrons to three experimental halls, each with a complementary set of experimental equipment. The accelerator was built using novel superconducting radiofrequency (SRF) accelerator technology, which has enabled unprecedented performance in beam quality. Due to advances in SRF technology the electron beam has exceeded the original 4 GeV energy specification, and beams with energies up to 6 GeV with currents up to 100 μA have been delivered for the experimental program. The timely development of advanced GaAs photoemission sources has enabled high quality polarized beam with polarizations up to 85%. CEBAF presently serves an international scientific user community of over 1200 scientists, and to date over 160 experiments have been completed.

Early in the history of this facility, plans were developed to upgrade the capability of the accelerator to enable beams up to 12 GeV in energy. The US nuclear physics community subsequently endorsed this concept, and the 2002 Long Range Plan of the US Nuclear Science Advisory Committee (NSAC) [1] contained a recommendation: "We strongly recommend the construction of CEBAF at Jefferson Laboratory to 12 GeV as quickly as possible". In March 2004 the US Department of Energy (DOE) granted CD-0 approval to develop a conceptual design for such a facility. Following further development of the design of the upgraded facility, the 2007 NSAC Long Range Plan [2] reaffirmed the community commitment to this project: "We recommend completion of the 12 GeV Upgrade at Jefferson Lab. The Upgrade will enable new insights into the structure of the nucleon, the transition between the hadronic and quark/gluon descriptions of nuclei, and the nature of confinement." The Jefferson Lab user community and the Laboratory staff completed the design, including detailed cost and schedule, in 2008 and DOE approved the start of construction in September 2008.

GENERAL DESCRIPTION

CEBAF is currently a recirculating linac, with 2 linac sections, each consisting of 20 cryomodules. A cryomodule contains 8 superconducting RF cavities and is capable of an average 25 MeV of acceleration. Thus each linac section is nominally capable of producing 0.5 GeV of energy gain. The recirculating arcs contain quadrupole and dipole magnets in separate beamlines that facilitate beam acceleration up to 5 times through both linacs, producing a nominal energy of 5 GeV with actual performance up to 6 GeV. Individual beam pulses can be "kicked" into an extraction line after the second (South) linac section and delivered into one of the 3 experimental halls. Thus the beam can be split into three simultaneous 499 MHz beams with energies in multiples of $\frac{1}{5}$ of the full 5-pass energy. The high duty factor associated with the 2 ns beam structure enables high luminosity experiments with coincident detection of multiple particles per event.

12th International Conference on Meson-Nucleon Physics and the Structure of the Nucleon (MENU 2010)
AIP Conf. Proc. 1374, 121-126 (2011); doi: 10.1063/1.3647109

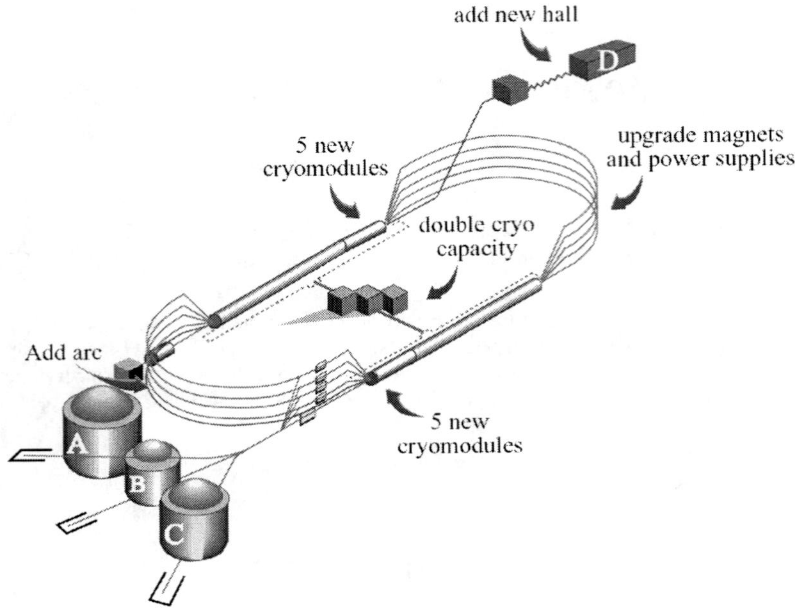

FIGURE 1. Jefferson Lab 12 GeV upgrade concept.

Fig. 1 illustrates the basic concept of the 12 GeV upgrade project. In addition to the upgrade of the accelerator system to enable delivery of 12 GeV beam, the experimental equipment will be enhanced to facilitate full exploitation of the higher energy beam. This includes substantial new equipment in Hall B and Hall C, and a completely new Hall D with a new detector/spectrometer system. The plan for Hall A also includes upgraded and new equipment that is outside the present construction project.

Accelerator system

The accelerator upgrade is based on the addition of 5 additional cryomodules in each linac section. The original CEBAF civil construction included extra drift space at the end of each linac section, thus enabling a future energy upgrade. The new cryomodules will contain 7-cell cavities (instead of the 5-cell cavities in the original CEBAF) with improved surface treatments to enable higher RF acceleration fields. As a result, each cryomodule will be capable of 100 MeV of acceleration (compared to 20 MeV in the original design specification), each linac section will now produce an energy gain of 1.1 GeV, and each recirculated beam can receive an energy increase of 2.2 GeV. Upgrades to the arc magnets and power supplies are also required. The upgraded CEBAF will be capable of beam delivery to all 3 existing experimental halls at energies up to 11 GeV. In principle, beam will be available in multiples of $\frac{1}{5}$ of the full 5-pass energy. For example at full energy, beam would be available at three of the values corresponding to 2.2 GeV, 4.4 GeV, 6.6 GeV, 8.8 GeV, and 11 GeV. The intensity of these beams may total up to 85 μA with high polarization (up to 85%) available.

To facilitate delivery of 12 GeV beam to Hall D, a fifth arc of magnets will be added to the 2nd recirculating arc, so that one beam can be accelerated through the north linac one more time before transmission to the new experimental Hall D. Operation of the additional cryomodules will also require an increase in the central helium liquifier capacity.

Hall A

The basic capability in Hall A will be retained, with 2 High Resolution Spectrometers (HRS), and new experimental equipment will be constructed for additional experiments in the future (funded outside the present scope of the 12 GeV

FIGURE 2. 3D view of CLAS12. The beam is incident from the left. The target is located inside the solenoid magnet.

upgrade project). In the nearer term, the proposed Super Big-Bite Spectrometer (SBS) will be constructed and sited in Hall A. This is a larger version of the successful Big-Bite Spectrometer that has been insrumental in extending the nucleon form factor measurements to higher momentum transfer. The SBS will utilize a large 48D48 dipole magnet followed by large area tracking (GEM detectors), time-of-flight scintillators, and particle ID detectors (Cerenkov and calorimeter). The solid angle acceptance is typically 50-100 msr (depending on the deployment details), and the spectrometer will be capable of operating at high luminosities, exceeding $10^{38}/cm^2$-s. The future experimental program in Hall A also includes high-precision parity violation experiments to provide stringent tests of the standard electroweak model.

Hall B

The CEBAF Large Acceptance Spectrometer (CLAS) has had a long and distinguished career of measurements with tagged photons and low intensity electron beams. This apparatus will be replaced in Hall B by the new spectrometer, CLAS12, shown in Fig. 2. This new spectrometer system will enable higher luminosity operation with electron beams, up to $10^{35}/cm^2$-s, and improved particle identification at forward angles. The magnetic system includes a 5 T superconducting solenoid for central tracking, and a superconducting toroid at forward angles. The solenoid also shields the detector system from the high flux of Møller electrons generated in the target.

The forward detector system includes a high threshold cerenkov counter (HTCC) with pion threshold of 4.9 GeV mounted in front of the toroid. The toroid and forward drift chamber system are designed to track charged particles in the range $5-40°$, and are followed by a low threshold cerenkov counter (LTCC). The existing electromagnetic calorimeter (EC) is being upgraded with a pre-shower counter. The central detector system includes a silicon vertex tracker and a scintillator time-of-flight array.

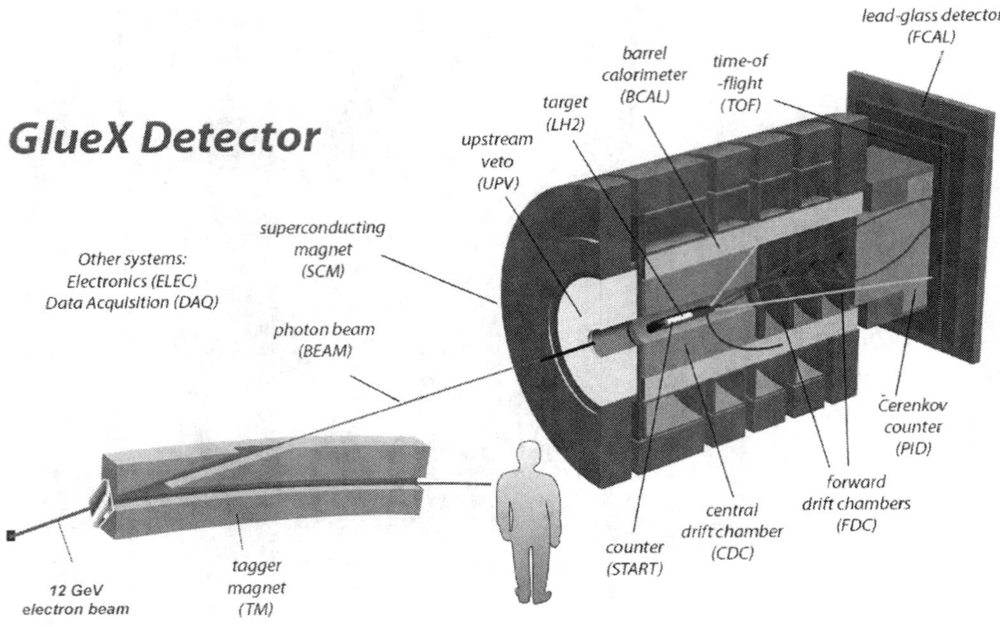

FIGURE 3. Schematic of GlueX apparatus.

Hall C

The existing high momentum spectrometer (HMS) in Hall C will continue operation in the 12 GeV program. A new super high momentum spectrometer (SHMS) will be contructed that will be capable of a maximum central momentum of 11 GeV. The momentum acceptance will cover -11 to +22% with a solid angle acceptance of 5 msr. The SHMS will achieve a momentum resolution of better than 10^{-3}. The SHMS detector package includes a drift chamber system for tracking, quartz and scintillator hodoscopes, a noble gas Cerenkov counter, a heavy gas Cerenkov counter, and a lead glass calorimeter.

Hall D

A major new experiment, known as GlueX (see Fig. 3), will be constructed and sited in Hall D. The main goal of the GlueX experiment is to search for exotic mesons produced via photoproduction on the nucleon. Linearly polarized photons will be produced upstream of Hall D by a coherent bremmstrahlung process using a thin ($\sim$ 20 micron) diamond wafer. The scattered electrons from 8.5-9 GeV bremmstrahlung photons will be tagged with scintillator detectors following a bending magnet, yielding a tagged photon resolution of 0.2% with fluxes expected to reach 10^8/s.

The GlueX spectrometer system is designed to provide hermetic detection of particles in the range 2-120° with excellent particle identification. Momentum measurement for charged particles is facilitated by a superconducting solenoid with 2.2 T field over a length of 495 cm and with an inner bore of 185 cm. A 30 cm-long liquid hydrogen target is surrounded by scintillation counters (START), a cylindrical drift chamber array (CDC) and an electromagnetic lead/scintillating fiber calorimeter with a barrel geometry (BCAL). Downstream of the target are an array of planar drift chambers (FDC). Outside and downstream of the inner bore of the magnet are a Cerenkov counter, a wall of scintillation counters (TOF) to measure time-of-flight, and an electromagnetic calorimeter (FCAL) consisting of lead-glass blocks.

PHYSICS PROGRAM

The physics program to be addressed with the Jefferson Lab 12 GeV upgrade has been developed in collaboration with the user community and with the guidance of the Program Advisory Committee. There are presently 34 approved experiments, and 13 additional proposals have conditional approval. These proposals fall into four general categories:

- the physical origins of quark confinement (including meson and baryon spectroscopy),
- the spin and flavor structure of the proton and neutron,
- the quark structure of nuclei,
- discovery of new physics through high precision tests of the Standard Model.

While it is not possible to provide a comprehensive review of this broad program here, a few selected highlights will be presented.

Meson Spectroscopy with GlueX

The primary goal of the GlueX experiment [3] is to test the concept of colored flux tubes between quarks (and antiquarks) as a model of confinement [4]. Excitation of the gluonic flux tube between a quark-antiquark pair can lead to mesonic states with exotic quantum numbers ($J^{PC} = 0^{+-}, 1^{-+}, 2^{+-}$) that cannot be described by only quark-antiquark states. These states and their properties have recently been studied in detail using lattice QCD methods [5], and these lattice calculations support the interpretation of excitation of gluonic degrees of freedom.

Nucleon Spin Structure

For the last 25 years, studies of the spin-dependent parton distributions have indicated that the quark helicities contribute a small fraction of the nucleon spin [6]. Recent data from RHIC and semi-inclusive Deep Inelastic Scattering (SIDIS) also indicate that the gluon spin contribution is small [7]. Although there has been considerable progress in measuring the spin-dependent parton distribution functions, it is quite unfortunate that a satisfactory accounting of the angular momentum of the nucleon in terms of its fundamental constituents is still lacking. There is now much speculation that the orbital angular momentum of the quarks may play a substantial role, and new experimental efforts are being proposed to address this issue. Previous treatments of inclusive deep inelastic scattering integrate over the transverse momenta, thus eliminating direct sensitivity to orbital angular momentum. Recently there has been a great deal of theoretical progress in the development of new experimental observables that could reveal aspects of the quark orbital angular momentum. These include generalized parton distributions (GPD) that can be accessed through deeply virtual Compton scattering (DVCS) [8, 9], and transverse momentum distributions [10] which can be studied in semi-inclusive deep inelastic scattering (SIDIS). The large acceptance available with the CLAS12 detector system makes it well-suited to the study of DVCS [11]. Measurements of TMD with SIDIS will be an important topic for the proposed SOLID spectrometer in Hall A.

Testing the Standard Model

Over the last decade the discoveries of dark energy (along with the improvement in our knowledge of dark matter), and the flavor oscillations of neutrinos (associated with their small but finite masses) are indications that the standard model of the strong and electroweak interactions requires modification. Moreover, protecting the Higgs mass from uncontrolled loop corrections also seems to require extending the standard model. Thus the idea that the standard model is part of a larger theoretical framework is now widely accepted, and such an extension of the theory should lead to observational consequences. With this motivation, there are substantial experimental activities associated with two general methods for probing for new physics beyond the standard model. One is to advance the high energy frontier with higher energy particle accelerators in an attempt to obtain evidence for new particles that can be produced at higher energy. Another method is to perform high precision measurements at lower energies that can provide information about the properties of a more complete theory. Indeed, such low energy precision measurements yield complementary information to that which can be obtained at the high energy frontier. A recent review of low energy tests of the standard model, including discussion of parity-violating electron scattering, provides a rather comprehensive discussion of the potential of these techniques [12].

The CEBAF 12 GeV upgrade will offer many new opportunities for precision measurements, particularly in studies of parity-violating electron scattering. The neutral weak interaction violates parity symmetry, and measurement of parity violation in electron scattering provides access to interference of neutral weak and electromagnetic amplitudes. Some extensions of the standard model involve the existence of massive new ($M > M_Z$) neutral bosons that couple to electrons and/or quarks. These and other similar phenomena would lead to changes in the parity-violating interaction with electrons and thus manifest themselves in parity violation experiments.

There are presently 2 proposals to perform new precision measurements of parity-violating electron scattering at the upgraded CEBAF. One would use the SOLID solenoidal magnetic spectrometer system to study parity-violating deep inelastic scattering [13]. The other proposal is to study parity-violating Møller scattering [14] utilizing a novel dedicated toroidal spectrometer. Both of these experiments are proposed to be sited in experimental Hall A.

SUMMARY

The 12 GeV upgrade project at Jefferson Lab will enable a powerful new experimental program that will advance our understanding of the quark/gluon structure of hadronic matter, the nature of Quantum Chromodynamics, and the properties of a new extended standard model of particle interactions. Construction is well underway, and commissioning of the upgraded beam will be begin in 2013. The upgraded experimental equipment associated with the construction project will be completed and ready for beam in 2015. Additional experimental equipment beyond the base equipment in the construction project is required for several new experiments; efforts are underway to develop these proposals further and to identify resources. New collaborators interested in participating in this exciting new physics program at Jefferson Lab are certainly welcome.

ACKNOWLEDGMENTS

I am grateful to Allison Lung, Kees de Jager, Volker Burkert, Steve Wood, and Eugene Chudakov for their assistance in preparing this presentation. This work was supported by DOE contract DE-AC05-06OR23177, under which Jefferson Science Associates, LLC, operates the Thomas Jefferson National Accelerator Facility.

REFERENCES

1. 2002 Long Range Plan, Nuclear Science Advisory Committee, http://dnp.nscl.msu.edu/nplinks/lrps/2002_lrp.pdf .
2. 2007 Long Range Plan, Nuclear Science Advisory Committee, http://dnp.nscl.msu.edu/nplinks/lrps/2007-Nuclear-Science.High-Res.pdf .
3. Jefferson Lab proposal E12-06-102.
4. N. Isgur and J. Paton, Phys. Rev. **D31**, 2910 (1985).
5. J. J. Dudek, *et al.,* Phys.Rev.**D82** 034508 (2010).
6. S. D. Bass, Rev. Mod. Phys.**77**, 1257-1302 (2005).
7. D. de Florian, *et al.*, Phys. Rev. Lett. **101** 072001, (2008).
8. X. Ji, Phys. Rev. Lett. **78**, 610 (1997); Phys. Rev. **D55**, 7114 (1997).
9. A.V. Radyushkin, Phys. Rev. **D56**, 5524 (1997).
10. J. C. Collins, Acta Phys. Pol. **B34**, 3103 (2003), and references therein.
11. Jefferson Lab proposal E12-06-119.
12. M. J. Ramsey-Musolf and S. Su, Phys. Rept. **456** 1 (2008).
13. Jefferson Lab proposal E12-10-007.
14. Jefferson Lab proposal E12-09-005.

Recent Results from CLEO

R. E. Mitchell

Indiana University, Bloomington, IN USA 47405

Abstract.
The CLEO Experiment, utilizing symmetric e^+e^- collisions provided by the Cornell Electron Storage Ring (CESR) in the charmonium and bottomonium regions, has a wide-reaching program in the study of hadrons and the strong force. This review briefly summarizes recent contributions the CLEO Collaboration has made in the study of: (1) light quark systems (rare decays of the η'); (2) charmonium radiative decays to light quarks (J/ψ and $\psi(2S)$ radiative decays to the η and η'); (3) mesons with open charm (the D and D_s decay constants); (4) charmonium transitions (a multipole analysis of $\psi(2S) \to \gamma\chi_{c1,2}$ and $\chi_{c1,2} \to \gamma J/\psi$); and (5) bottomonium spectroscopy (confirming evidence for the discovery of the $\eta_b(1S)$).

Keywords: QCD
PACS: 12.38.Qk,13.20.Gd,13.20.Fc,13.25.Gv,14.40.Pq

INTRODUCTION

Despite the recent conclusion of nearly thirty years of data-taking (1980–2008), the CLEO Collaboration continues to investigate a wide range of topics with e^+e^- collision data collected in the charmonium and bottomonium regions. This review highlights a few of these studies with a special emphasis on hadrons and the strong force. The topics range from light quarks (rare decays of the η') to heavy quarks (confirming evidence for the discovery of the $\eta_b(1S)$). All of the results presented here use the full CLEO datasets. Charmonium data is mostly from CLEO-c (2003–2008), which primarily includes: 27 million $\psi(2S)$ decays[1]; 818 pb^{-1} at the $\psi(3770)$; and 586 pb^{-1} at 4170 MeV. Bottomonium data is from CLEO III (2000–2003), which includes 15.5 fb^{-1} of data at the $\Upsilon(4S)$ as well as 6, 9, and 22 million decays of the $\Upsilon(3S)$, $\Upsilon(2S)$, and $\Upsilon(1S)$, respectively.

OBSERVATION OF $\eta' \to \pi^+\pi^-\pi^0$ AND $\eta' \to \pi^+\pi^-e^+e^-$

By fully reconstructing the process $\psi(2S) \to \pi^+\pi^- J/\psi; J/\psi \to \gamma\eta'; \eta' \to X$, where X is a decay mode of the η', CLEO has been able to make the first observations of $\eta' \to \pi^+\pi^-\pi^0$ and $\eta' \to \pi^+\pi^-e^+e^-$ and has set upper limits on several other rare or forbidden processes [1]. The branching ratio of $\eta' \to \pi^+\pi^-\pi^0$ was the last measurement needed to form the phenomenologically interesting ratio $r_\pm/r_0$, where:

$$r_\pm \equiv \frac{\mathscr{B}(\eta' \to \pi^+\pi^-\pi^0)}{\mathscr{B}(\eta' \to \pi^+\pi^-\eta)}$$

and

$$r_0 \equiv \frac{\mathscr{B}(\eta' \to \pi^0\pi^0\pi^0)}{\mathscr{B}(\eta' \to \pi^0\pi^0\eta)}.$$

The ratio is now experimentally known to be 1.11 ± 0.35, which can be compared to theoretical predictions ranging from 0.37 (based on η-π^0 mixing) to 5 (based on a model of $\pi^+\pi^-\pi^0$ rescattering through a $\rho(770)$ resonance) [2].

To reconstruct events of the form $\psi(2S) \to \pi^+\pi^- J/\psi; J/\psi \to \gamma\eta'; \eta' \to X$, CLEO performed a kinematic fit of the total 4-momentum of all of the decay products (the $\pi^+\pi^-$, the γ, and the particles of the η' decay, X) to the initial 4-momentum of the $\psi(2S)$ and used the quality of the fit as a filter to select events. The recoil mass of the $\pi^+\pi^-$ system was then used to select the J/ψ. This resulted in a clean sample of events, shown in Figure 1. Both the $\eta' \to \pi^+\pi^-\pi^0$

[1] Of the 27 million $\psi(2S)$ decays, 1.5 million were collected by CLEO III.

12th International Conference on Meson-Nucleon Physics and the Structure of the Nucleon (MENU 2010)
AIP Conf. Proc. 1374, 127-132 (2011); doi: 10.1063/1.3647110

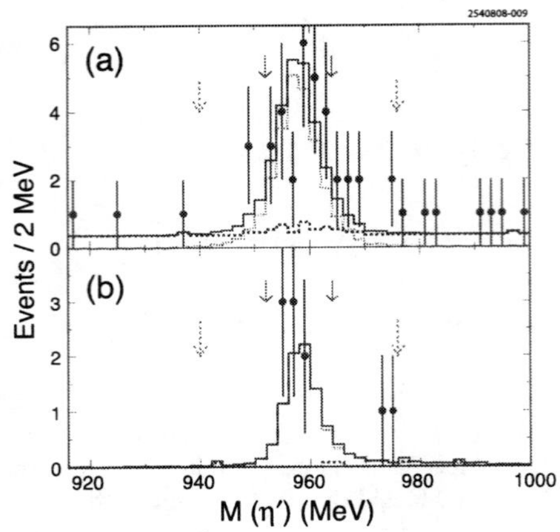

FIGURE 1. The observations of (a) $\eta' \to \pi^+\pi^-\pi^0$ and (b) $\eta' \to \pi^+\pi^-e^+e^-$. Each plot shows the invariant mass of the candidate η' decay products obtained from the process $\psi(2S) \to \pi^+\pi^- J/\psi; J/\psi \to \gamma\eta'; \eta' \to X$, where X is $\pi^+\pi^-\pi^0$ or $\pi^+\pi^-e^+e^-$. The lighter dashed lines are signal MC; the darker dotted lines are estimated cross-feeds from other η' decay channels. The solid arrows indicate the signal regions; the dotted arrows indicate the sideband regions, which extend to the limits of the plots.

and $\eta' \to \pi^+\pi^-e^+e^-$ decays were observed with more than 6σ significance. For normalization, the same procedure was used to reconstruct the more dominant process $\eta' \to \eta\pi^+\pi^-$ with $\eta \to \gamma\gamma$.

The sizes of the signals were obtained by counting events in the signal regions (shown by the solid arrows in Figure 1) and then subtracting estimated backgrounds. Two types of backgrounds were subtracted: linear backgrounds were estimated by using sidebands in the mass of X (indicated by dashed arrows), and peaking backgrounds from cross-feed from other η' decay channels were estimated with Monte Carlo (the dotted histograms). This resulted in $20.2^{+6.1}_{-4.8}$ and $7.9^{+3.9}_{-2.7}$ signal events for $\eta' \to \pi^+\pi^-\pi^0$ and $\eta' \to \pi^+\pi^-e^+e^-$, respectively, with corresponding branching fractions of $(37^{+11}_{-9} \pm 4) \times 10^{-4}$ and $(25^{+12}_{-9} \pm 5) \times 10^{-4}$.

J/ψ AND $\psi(2S)$ RADIATIVE DECAYS TO η AND η'

Using similar methods to those in the previous section, CLEO has also measured the branching fractions of radiative J/ψ and $\psi(2S)$ decays to the η and η' [3]. Again, a kinematic fit was performed matching the total 4-momentum of all decay particles to the $\psi(2S)$ 4-momentum and the fit quality was used to select events. The J/ψ decays were selected through the process $\psi(2S) \to \pi^+\pi^- J/\psi$ using the recoil mass of the $\pi^+\pi^-$ system. The resulting distributions of candidate η and η' masses are shown in Figure 2 for many different decay modes of the η and η'. The samples are clean. As before, signal events were counted in a signal region and backgrounds were subtracted using sidebands and a Monte Carlo estimate of cross-feed from other η and η' decay modes.

Defining a ratio of branching fractions, R_n, as:

$$R_n \equiv \frac{\mathscr{B}(\psi(nS) \to \gamma\eta)}{\mathscr{B}(\psi(nS) \to \gamma\eta')}$$

one finds the surprising result that $R_1 = (21.1 \pm 0.9)\%$, but $R_2 < 1.8\%$ at a 90% confidence level. The strong suppression of the decay $\psi(2S) \to \gamma\eta$ shows up as a depletion of events in the right-most column of Figure 2. Reminiscent of the so-called "rho-pi" puzzle (where there is an unexpected suppression of $\psi(2S) \to \rho\pi$), it is unclear why the $\psi(2S)$ and J/ψ behave so differently in their decays to $\gamma\eta$ and $\gamma\eta'$.

D AND D_s DECAY CONSTANTS: f_D AND f_{D_s}

The D and D_s decay constants provide a unique opportunity to directly compare precision experimental measurements with precision theoretical calculations in QCD. Experimentally, the decay constants, f_D and f_{D_s}, can be accessed

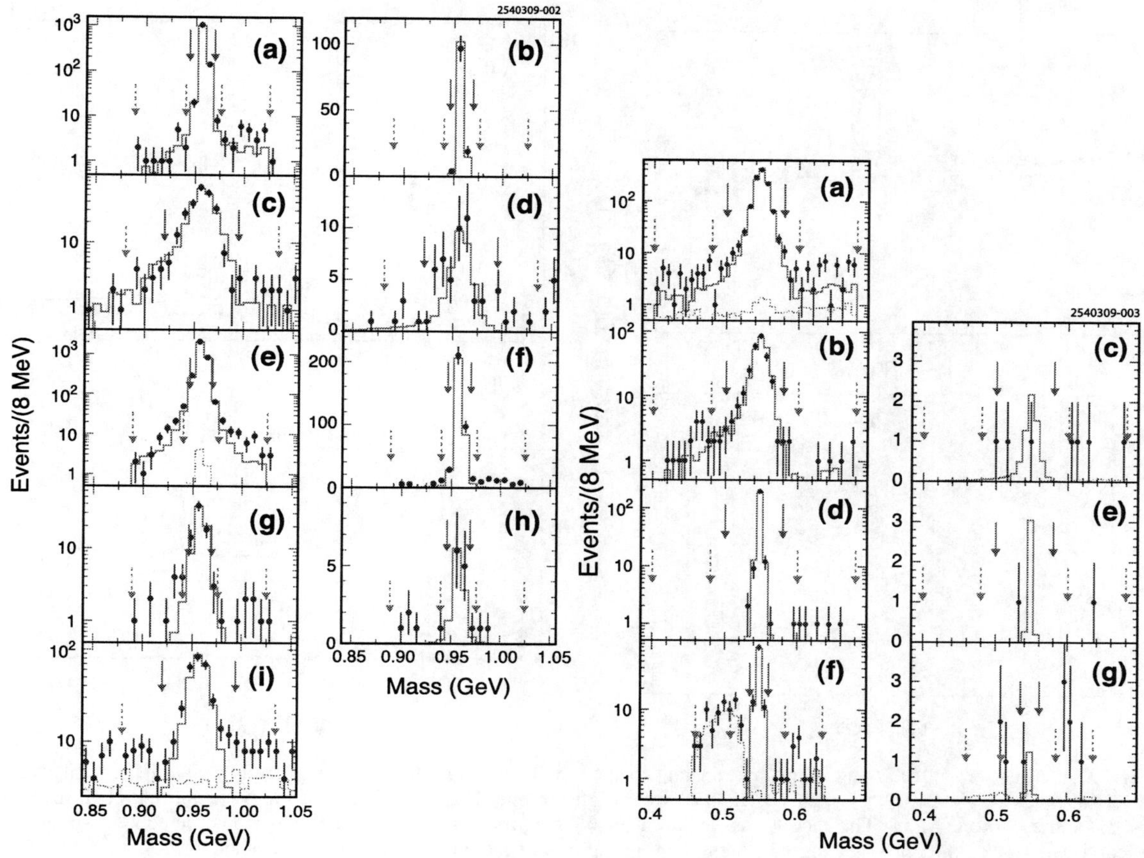

FIGURE 2. Signals for (first column) $J/\psi \to \gamma\eta'$, (second column) $\psi(2S) \to \gamma\eta'$, (third column) $J/\psi \to \gamma\eta$, and (fourth column) $\psi(2S) \to \gamma\eta$ in various decay modes of the η and η'. In the left plots, the η' is reconstructed in the following decay modes: (a), (b) $\pi^+\pi^-\eta(\gamma\gamma)$; (c), (d) $\pi^0\pi^0\eta(\gamma\gamma)$; (e), (f) $\gamma\pi^+\pi^-$; (g), (h) $\gamma\omega(\pi^+\pi^-\pi^0)$; and (i) $\gamma\gamma$. In the right plots, the η is reconstructed as: (a) $\gamma\gamma$; (b), (c) $3\pi^0$; (d), (e) $\pi^+\pi^-\pi^0$; and (f), (g) $\gamma\pi^+\pi^-$. In all cases, points are data, solid histograms are signal Monte Carlo, dashed histograms are estimated cross-feed from other channels, solid arrows indicate the signal regions, and dashed arrows indicate sidebands.

through leptonic decays of the D and D_s, respectively. For the D_s, the partial width can be written as:

$$\Gamma(D_s^+ \to l^+\nu) = \frac{G_F^2}{8\pi} f_{D_s^+}^2 m_l^2 M_{D_s^+} \left(1 - \frac{m_l^2}{M_{D_s^+}^2}\right)^2 |V_{cs}|^2,$$

where G_F^2 is the weak coupling, m_l is the mass of the lepton, M_{D_s} is the mass of the D_s meson, and $|Vcs|$ is a CKM matrix element. Since $|Vcs|$ can be taken from independent measurements, and all other constants are known, the decay rate is directly proportional to the decay constant squared. The same applies for the measurement of f_D. Theoretically, the decay constants can be calculated using lattice QCD. With partial samples of data, CLEO had found a more than 2σ deviation between experiment and theory for f_{D_s}, but agreement for f_D. With the full datasets [4, 5, 6, 7] and new lattice calculations, however, the discrepancy appears to have narrowed.

For illustration, Figure 3 shows the method used to measure $\mathscr{B}(D_s \to \mu\nu)$ and $\mathscr{B}(D_s \to \tau\nu)$ with $\tau \to \pi\nu$ using the process $e^+e^- \to D_s^{\pm *}D_s^{\mp} \to \gamma D_s^+ D_s^-$ at 4170 MeV [5]. First, one D_s is fully reconstructed using nine different "tag modes" (top left plot of Figure 3). Then, a second D_s is inferred using the recoil mass of the first D_s and the photon from the D_s^* decay (bottom left plot of Figure 3). Finally, one can look for the second D_s decaying to $\mu\nu$ or $\tau\nu$ (with $\tau \to \pi\nu$) by finding a μ or π candidate track and looking at the recoil mass (or the missing mass) of the first D_s, the photon, and the track (right plots of Figure 3). For $D_s \to \mu\nu$, the missing mass squared peaks at zero (the neutrino mass). For $D_s \to \tau\nu$ with $\tau \to \pi\nu$, the missing mass squared peaks above zero. Some discrimination between the μ

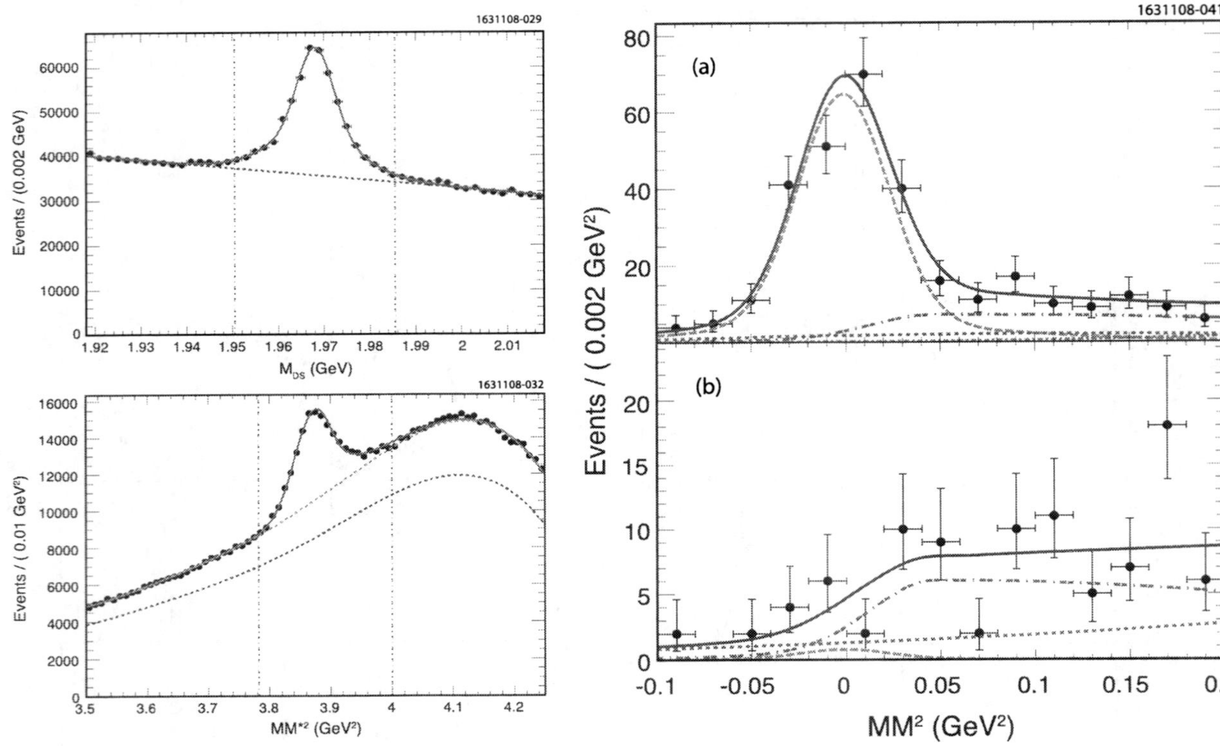

FIGURE 3. Measurement of D and D_s decay constants. (left top) The first D_s fully reconstructed in nine different decay modes from the process $e^+e^- \to D_s^{\pm *}D_s^{\mp} \to \gamma D_s^+ D_s^-$ at 4170 MeV. (left bottom) The second D_s inferred by using the recoil mass against the first D_s and a photon. (right) The recoil mass squared (or missing mass squared) of a fully reconstructed D_s, a photon, and a μ or π candidate track. The $D_s \to \mu\nu$ signal peaks at a missing mass squared of zero (the grey dashed curve), while the $D_s \to \tau\nu$ with $\tau \to \pi\nu$ signal peaks at a higher mass (the dot-dashed curve). The top plot uses events with calorimeter energy < 300 MeV (favorable for μ tracks), while the bottom plot uses calorimeter energy > 300 MeV (favorable for π tracks).

(from $D_s \to \mu\nu$) and the π (from $D_s \to \tau\nu$ with $\tau \to \pi\nu$) can be gained by looking at calorimeter energy: a calorimeter energy cut of < 300 MeV favors the μ and > 300 MeV favors the π, which is shown in the top and bottom right plots of Figure 3, respectively. The results from this method are $\mathscr{B}(D_s \to \mu\nu) = (0.565 \pm 0.045 \pm 0.017)\%$ (giving $f_{D_s} = 257.6 \pm 10.3 \pm 4.3$ MeV) and $\mathscr{B}(D_s \to \tau\nu) = (6.42 \pm 0.81 \pm 0.18)\%$ (giving $f_{D_s} = 278.0 \pm 17.5 \pm 4.4$ MeV).

The final average of all CLEO measurements of f_{D_s} is $259.0 \pm 6.2 \pm 3.0$ MeV. This can be compared to the lattice QCD predictions of 241 ± 3 MeV (with a 2010 update of 247 ± 2 MeV) from [8], and $249 \pm 3 \pm 16$ MeV (with a 2010 update of $261.4 \pm 7.7 \pm 5.0$ MeV) from [9]. Theory and experiment now agree within 2σ. For f_D, the agreement is better than 1σ.

HIGHER-ORDER MULTIPOLES IN $\psi(2S) \to \gamma\chi_{c1,2}$ AND $\chi_{c1,2} \to \gamma J/\psi$

The radiative decays $\psi(2S) \to \gamma\chi_{c1,2}$ and $\chi_{c1,2} \to \gamma J/\psi$ are dominated by electric dipole (E1) amplitudes, but are expected to have small additional contributions from the higher-order magnetic quadrupole (M2) amplitudes. The relative contributions of the M2 amplitudes are, at least in principle, sensitive to the anomalous magnetic moment of the charm quark. CLEO performed a high-statistics analysis of this process using the decay chains $\psi(2S) \to \gamma\chi_{cJ}; \chi_{cJ} \to \gamma J/\psi; J/\psi \to l^+l^- (l = e, \mu)$ for $J = 1, 2$, using clean samples of approximately 40 000 events for $J = 1$ and approximately 20 000 events for $J = 2$ [10], significantly larger event samples than previous measurements. Using an unbinned maximum likelihood fit to angular distributions, CLEO found the following normalized M2 admixtures: for $\psi(2S) \to \gamma\chi_{c1}$ ($\equiv b_2^{J=1}$), $(2.76 \pm 0.73 \pm 0.23) \times 10^{-2}$; for $\psi(2S) \to \gamma\chi_{c2}$ ($\equiv b_2^{J=2}$), $(1.0 \pm 1.3 \pm 0.3) \times 10^{-2}$; for $\chi_{c1} \to \gamma J/\psi$ ($\equiv a_2^{J=1}$), $(-6.26 \pm 0.63 \pm 0.24) \times 10^{-2}$; and for $\chi_{c2} \to \gamma J/\psi$ ($\equiv a_2^{J=2}$), $(-9.3 \pm 1.6 \pm 0.3) \times 10^{-2}$. For the quoted $J = 2$ measurements, the electric octupole (E3) moments were fixed to zero. There is clear evidence for

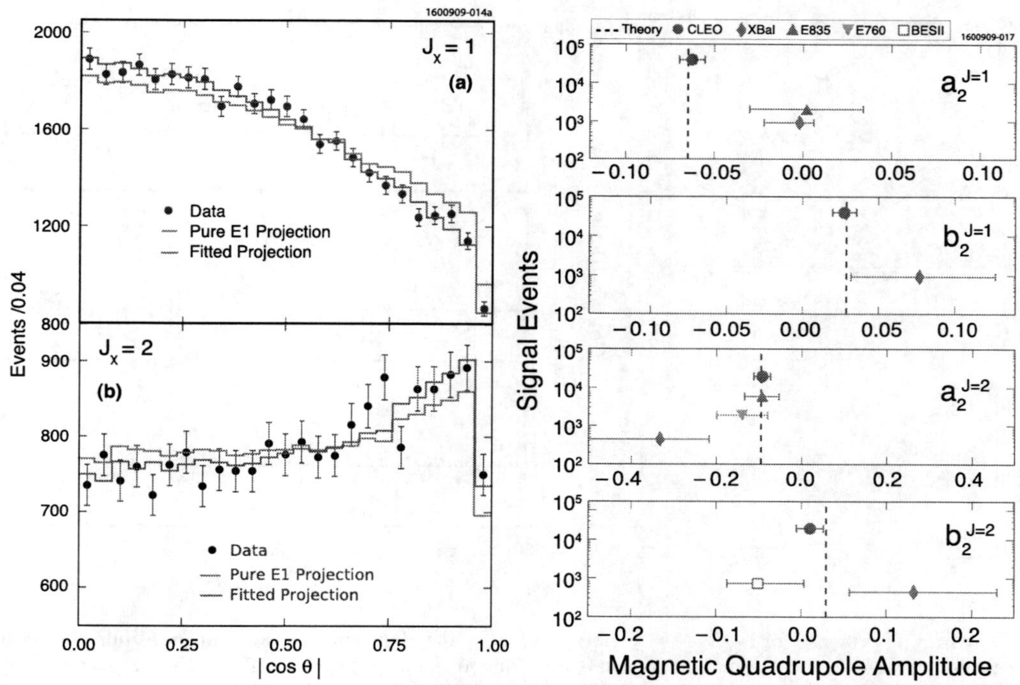

FIGURE 4. Higher-Order Multipoles in $\psi(2S) \to \gamma\chi_{c1,2}$ and $\chi_{c1,2} \to \gamma J/\psi$. (left) The cosine of the angle between the l^+ and the γ from the χ_{cJ} decay in the J/ψ rest frame from the process $\psi(2S) \to \gamma\chi_{cJ}; \chi_{cJ} \to \gamma J/\psi; J/\psi \to l^+l^- (l = e, \mu)$ for (top) $J = 1$ and (bottom) $J = 2$. This angle is sensitive to the M2 multipole. Points are data, the lighter (red) line is the projection from a fit using only E1 amplitudes, and the darker (blue) line is the projection when M2 amplitudes are included. (right) The CLEO results (circles) for the four normalized M2 amplitudes (defined in the text) compared to previous measurements (other points) and theoretical calculations assuming a zero anomalous magnetic moment of the charm quark and a charm quark mass of $1.5\,\mathrm{GeV}/c^2$ (dashed lines).

non-zero M2 amplitudes (left plot of Figure 4). These new measurements agree well with theoretical expectations when the anomalous magnetic moment of the charm quark is assumed to be zero and the mass of the charm quark is assumed to be $1.5\,\mathrm{GeV}/c^2$, as shown in the right plot of Figure 4.

CONFIRMING EVIDENCE FOR $\Upsilon(3S) \to \gamma\eta_b$

Finally, in addition to light quark mesons and charmonium, CLEO has also made recent contributions to our understanding of bottomonium. While the bottomonium system was first discovered more than thirty years ago, the ground state of the system, the $\eta_b(1S)$ was only recently discovered by the BaBar Collaboration in the transition $\Upsilon(3S) \to \gamma\eta_b$ [11]. Using 6 million $\Upsilon(3S)$ decays collected by the CLEO III detector (compared to BaBar's 109 million), CLEO confirmed the transition $\Upsilon(3S) \to \gamma\eta_b$ with a significance of approximately 4σ [12].

Both BaBar and CLEO measure this transition by inclusively searching for the monoenergetic photon from $\Upsilon(3S) \to \gamma\eta_b$ after suppressing electromagnetic backgrounds. The CLEO measurement divides the inclusive photon energy spectrum into three bins according to the angle the transition photon makes with the "thrust axis," the axis that maximizes the total momentum of all particles in an event when their momenta are projected onto it (left plot of Figure 5), and then fits all three bins simultaneously. The fit contains four components: a non-peaking background (polynomial), the $\chi_{bJ}(2P) \to \gamma\Upsilon(1S)$ lines (with shapes fixed from independent studies), initial state radiation to the $\Upsilon(1S)$ (with shape fixed using $\Upsilon(4S)$ data), and the signal shape (right plot of Figure 5). Details of the fitting procedure can be found in [12].

The branching ratio found is then $\mathscr{B}(\Upsilon(3S) \to \gamma\eta_b) = (7.1 \pm 1.8 \pm 1.3) \times 10^{-4}$. The η_b mass is also measured to be $9391.8 \pm 6.6 \pm 2.0\,\mathrm{MeV}/c^2$, which corresponds to a hyperfine splitting of $M(\Upsilon(1S)) - M(\eta_b) = 68.5 \pm 6.6 \pm 2.0\,\mathrm{MeV}/c^2$.

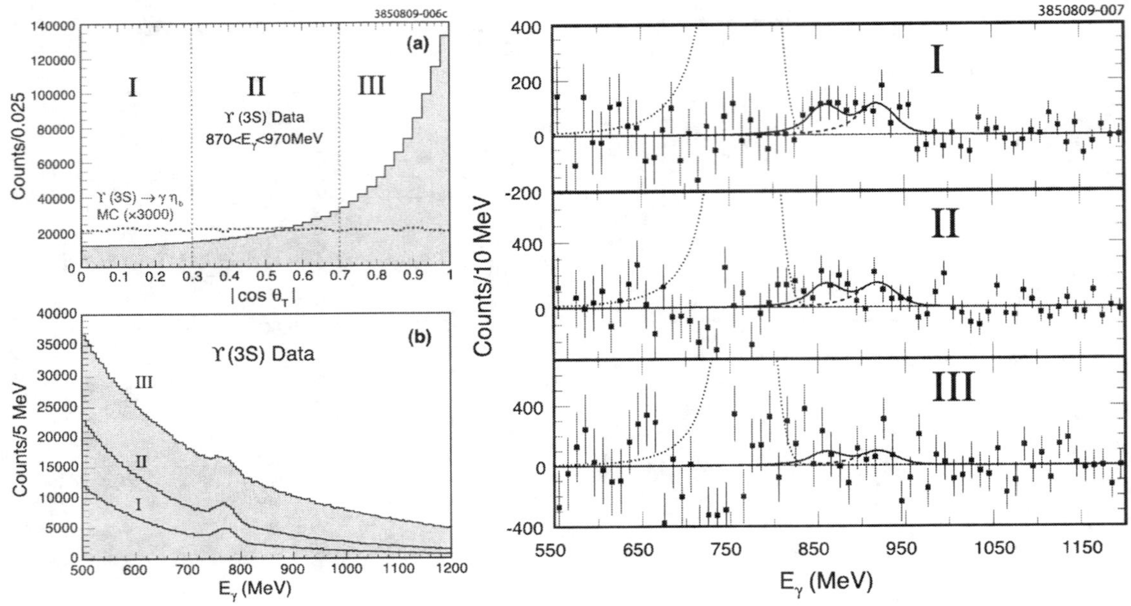

FIGURE 5. Confirming evidence for $\Upsilon(3S) \to \gamma\eta_b$. (left top) The cosine of the angle between the transition photon and the thrust axis of the event for data (histogram) and signal Monte Carlo (dotted line). (left bottom) The inclusive photon energy spectrum for the three different regions of thrust axis angle shown in the top left figure. (right) The background-subtracted inclusive photon energy spectrum. The dotted lines indicate the location of the subtracted $\chi_{bJ}(2P) \to \gamma\Upsilon(1S)$ lines, and the solid lines show the initial state radiation (lower energy) and signal (higher energy) components of the fit.

SUMMARY

The CLEO Collaboration continues to study a wide range of topics that span the energy spectrum from light quarks to bottomonium. The topics briefly reviewed here include: (1) observation of rare decays of the η', in particular $\eta' \to \pi^+\pi^-\pi^0$, whose measurement now allows the construction of important ratios of branching fractions; (2) measurements of radiative decays of the J/ψ and $\psi(2S)$ to the η and η', where there is a mysterious mismatch between J/ψ and $\psi(2S)$ branching fractions, reminiscent of the rho-pi puzzle; (3) precision measurements of D and D_s decay constants, which provide direct comparisons between experiment and lattice QCD calculations; (4) a multipole analysis of the radiative transitions $\psi(2S) \to \gamma\chi_{cJ}$ and $\chi_{cJ} \to \gamma J/\psi$, which finds no evidence for an anomalous magnetic moment of the charm quark; and (5) evidence for the $\eta_b(1S)$ through the transition $\Upsilon(3S) \to \gamma\eta_b$, providing confirmation of the BaBar discovery.

REFERENCES

1. P. Naik *et al.* [CLEO Collaboration], Phys. Rev. Lett. **102**, 061801 (2009) [arXiv:0809.2587 [hep-ex]].
2. B. Borasoy, U. G. Meissner and R. Nissler, Phys. Lett. B **643**, 41 (2006) [arXiv:hep-ph/0609010].
3. T. K. Pedlar *et al.* [CLEO Collaboration], Phys. Rev. D **79**, 111101 (2009) [arXiv:0904.1394 [hep-ex]].
4. B. I. Eisenstein *et al.* [CLEO Collaboration], Phys. Rev. D **78**, 052003 (2008) [arXiv:0806.2112 [hep-ex]].
5. J. P. Alexander *et al.* [CLEO Collaboration], Phys. Rev. D **79**, 052001 (2009) [arXiv:0901.1216 [hep-ex]].
6. P. U. E. Onyisi *et al.* [CLEO Collaboration], Phys. Rev. D **79**, 052002 (2009) [arXiv:0901.1147 [hep-ex]].
7. P. Naik *et al.* [CLEO Collaboration], Phys. Rev. D **80**, 112004 (2009) [arXiv:0910.3602 [hep-ex]].
8. E. Follana, C. T. H. Davies, G. P. Lepage and J. Shigemitsu [HPQCD Collaboration and UKQCD Collaboration], Phys. Rev. Lett. **100**, 062002 (2008) [arXiv:0706.1726 [hep-lat]].
9. C. Aubin *et al.*, Phys. Rev. Lett. **95**, 122002 (2005) [arXiv:hep-lat/0506030].
10. M. Artuso *et al.* [CLEO Collaboration], Phys. Rev. D **80**, 112003 (2009) [arXiv:0910.0046 [hep-ex]].
11. B. Aubert *et al.* [BABAR Collaboration], Phys. Rev. Lett. **101**, 071801 (2008) [Erratum-ibid. **102**, 029901 (2009)] [arXiv:0807.1086 [hep-ex]].
12. G. Bonvicini *et al.* [CLEO Collaboration], Phys. Rev. D **81**, 031104 (2010) [arXiv:0909.5474 [hep-ex]].

Extraction of the Compton Form Factor $\mathcal{H}$ from DVCS Measurements in the Quark Sector

H. Moutarde, on behalf of the CLAS group at Saclay

CEA, Centre de Saclay, IRFU/Service de Physique Nucléaire, F-91191 Gif-sur-Yvette, France

Abstract. Working at twist 2 accuracy and assuming the dominance of the Generalized Parton Distribution H we study the helicity-dependent and independent cross sections measured in Hall A, the beam spin asymmetries measured in Hall B at Jefferson Laboratory and beam charge, beam spin and target spin asymmetries measured by Hermes. We extract the real and imaginary parts of the Compton Form Factor $\mathcal{H}$, the latter being obtained with a 20–50 % uncertainty. We pay extra attention to the estimation of systematic errors on the extraction of $\mathcal{H}$. We discuss our results and compare to other extractions as well as to the popular VGG model.

Keywords: Generalized Parton Distributions, Compton Form Factors, Deeply Virtual Compton Scattering.
PACS: 13.60.Fz, 14.20.Dh

INTRODUCTION

The Deeply Virtual Compton Scattering (DVCS) process [1, 2] was early recognized as the cleanest way to access Generalised Parton Distributions (GPD) and has been so far a very active field of research. Theoretical developments are reviewed in Refs. [3, 4, 5, 6] and relevant experimental results are published in Refs. [7, 8, 9, 10, 11, 12, 13, 14, 15, 16, 17, 18, 19].

What we would like is to eventually produce an experimental 3d picture of the nucleon thanks to the knowledge of GPDs. Even if getting a precise experimental knowledge of GPDs is a long-term program, we can already perform the first fits to get a flavour of the actual sensitivity of observables to GPDs and elaborate robust and efficient extracting methods. We address these issues by studying recent JLab data, namely beam spin asymmetries [14] (BSA) and helicity-dependent and independent cross sections [12]. These data offer the interesting features of a large kinematic coverage and a fine kinematic binning. We also consider Hermes data : beam charge, beam spin and target spin asymmetries [15, 20].

The first section of this paper describes the evaluation of the $ep \to ep\gamma$ cross sections and outlines our hypothesis. In the second part we explain our fitting strategy. In the third section we discuss the extracted values of $\mathcal{H}$ and compare them to other similar studies.

PRELIMINARY ANALYSIS

DVCS at leading twist

Four GPDs H, E, $\tilde{H}$ and $\tilde{E}$ appear at twist 2, but the cross sections depend on the Compton Form Factors (CFF) $\mathcal{H}$, $\mathcal{E}$, $\tilde{\mathcal{H}}$ and $\tilde{\mathcal{E}}$. The convention of Ref. [21] is used to define the CFFs. The complex integration kernel yields a real and an imaginary part to a CFF. For example the CFF $\mathcal{H}$ satisfies :

$$Re\mathcal{H} = \mathcal{P}\int_{-1}^{+1} dx H(x,\xi,t)\left(\frac{1}{\xi-x} - \frac{1}{\xi+x}\right) \quad (1)$$

$$Im\mathcal{H} = \pi\Big(H(\xi,\xi,t) - H(-\xi,\xi,t)\Big) \quad (2)$$

where $\xi = x_B(1+t/(2Q^2))/(2-x_B+x_Bt/Q^2))$ is the generalised Bjorken variable [21, 24], x_B the standard Bjorken variable, Q^2 the virtuality of the initial photon and t the square momentum transfer between initial and final protons. $\mathcal{P}$ denotes the principal value of the integral.

12th International Conference on Meson-Nucleon Physics and the Structure of the Nucleon (MENU 2010)
AIP Conf. Proc. 1374, 133-138 (2011); doi: 10.1063/1.3647111

The use of dispersion relations relating real and imaginary parts of a CFF has been discussed by Anikin and Teryaev [25, 26, 27], Diehl and Yvanov [28] and Goldstein and Liuti [29, 30, 31]. Dispersions relations are used in the model-dependent global fit of Kumerički and Müller [32]. However the unknown subtraction (the D-term [33]) and the limited kinematic range of available data make this constraint rather weak in model-independent fitting. In this work we try to minimize the model-dependence of the extraction. Then we consider real and imaginary parts of CFFs as independent except when we use an explicit parametrisation of GPDs to compute CFFs.

$ep \to ep\gamma$ observables at twist 2

The harmonic analysis of $ep \to ep\gamma$ cross-sections has so far relied on the 2002 work of Belitsky, Müller and Kirchner [21] (BMK). In this formalism, the interference between the Bethe-Heitler (BH) and DVCS processes was treated with a leading order approximation of the BH part. This assumption was removed by Belitsky and Müller in the case of a spinless target [22] and for a (longitudinally polarised or unpolarised) spin 1/2 target [23] and by Guichon and Vanderhaegen in the case of a proton target [24]. In all the following, we will use the expressions from the latter. At last, the twist 2 approximation is motivated by the claim of early Q^2-scaling observed in Hall A [12]. Moreover the A_{DVCS}^{LU} asymmetry measured by Hermes [20] are small if not zero while they are supposed to vanish exactly at twist 2.

About current extractions methods

The methods to extract GPDs from measurements fall today into 4 groups :

local fits The real and imaginary parts of CFFs are the free parameters. A given kinematic bin (x_B, t, Q^2) is considered independantly of the others. The model dependence is low but some assumptions on the kinematics may be necessary to avoid underconstrained problems [38, 39, 40, 41]. It does not allow any extrapolations and in particular not those necessary to obtain spatial information in the transverse plane [42] or the quark orbital angular momentum in the nucleon [1].

global fits CFFs or GPDs have model-dependent expressions, the free parameters of which are fitted. See for example [32] for recent fit of DVCS data on unpolarized targets or [43, 44, 45] for recent studies of Deeply Virtual Meson Production data.

hybrid fits This is a combination of the two previous methods, used in [46] and detailed in the present text.

neural networks fits Work is in progress but preliminary results were recently presented [47].

H-dominance hypothesis

When we consider experiments on unpolarised proton targets we can neglect E, $\tilde{H}$ and $\tilde{E}$ (H-dominance). The contribution of these GPDs to helicity-dependent cross-sections is indeed kinematically suppressed [46] and we can check (for instance thanks to the VGG model [34, 35, 36, 37]) that $Im\mathcal{E}$ and $Im\tilde{\mathcal{H}}$ are usually smaller than $Im\mathcal{H}$. At last for small t and ξ we expect $\tilde{H}/H$ to be close to $\Delta q/q$ *i.e.* $\frac{1}{4}$.

Thus assuming H-dominance we may hope to extract information on H from BSAs or helicity-dependent cross sections with a systematic error of 20 % to 50 %[1], this approximation being better at small t. The advantage of this approach is the dramatic decrease of the number of degrees of freedom involved in fits. Guidal indeed showed [38] in the same kinematic region that it is not possible to extract sensible information about the real and imaginary parts of all CFFs by direct fits of helicity-dependent and independent cross sections.

---

[1] A direct test of the H-dominance assumption with the VGG model gives an upper bound of 25 %. This is comparable to the typical statistical uncertainty on BSAs.

FITTING STRATEGIES

The possibility to study GPDs in DVCS rests on factorisation theorems [48] which require a small value of $|t|/Q^2$. In the following, we restrict ourselves to kinematic configurations for which $|t|/Q^2 < 1/2$ and justify this choice *a posteriori*. As stated in the first section, we perform two kinds of fits :

Local fits $Re\mathcal{H}$ and $Im\mathcal{H}$ are the free parameters of the fits.

Global fit We use a "dual model"-like parametrization ; it consists basically in a simultaneous expansion in Gegenbauer and Legendre polynomials. See Ref. [46] for details on the fitting Ansatz.

We estimate the systematic errors associated to our H-dominance hypothesis by first fitting data setting the subdominant GPDs to 0, then fitting the same data setting the subdominant GPDs to their VGG value, and computing the difference. We estimate the systematic uncertainty related to the global fitting Ansatz by varying the number of polynomials involved in the expansion.

A careful fitting strategy is elaborated to handle the high number of free parameters involved in the fits and make sure that the numeric implementation is under control. We adopt an iterative (5 steps) fitting procedure. See Ref. [46] for full details.

RESULTS

Extraction of $Im\mathcal{H}$ and $Re\mathcal{H}$

Fig. 1 displays our results for $Im\mathcal{H}$ and $Re\mathcal{H}$. Both local fits and a global fit give results with comparable accuracy for $Im\mathcal{H}$, but as expected the results of the global fit are smoother. This is especially true concerning $Re\mathcal{H}$: in this case the local fits suffer from large fluctuations of $Re\mathcal{H}$ with values which fall outside the plot range. However, we could not reliably extract values of the CFFs for the larger values of ξ with the global fit.

The results for local and global fits are almost always compatible, which is a strong consistency check. Both rely on the assumptions of twist 2 accuracy and of H-dominance. On one hand, local fits suffer from numerical fluctuations (the 2-parameter local fits are not constrained enough on some bins) but are almost model-independent. On the other hand, global fits are smoother, but suffer from oscillations. That both methods give the same results indicates that fluctuations and oscillations are reasonably controlled in the bins for which results are displayed. Since, in both cases, the total error bars have the same size, we conclude that our estimation of systematic uncertainties on the global fit Ansatz is realistic even if not perfect.

All fits keep data satisfying $|t|/Q^2 < 1/2$. For the local fits, changing the maximal value of $|t|/Q^2$ amounts to dropping points. For the global fit, the whole results may be changed, but the good agreement between the results of both types of fits, and the slow Q^2-evolution of the extracted CFFs, indicate that this restricted kinematic region is suitable for an analysis in the GPD framework.

Our results are explicitly given in Ref. [46]. Error bars are dominated by systematic effects. Typically we obtain a relative accuracy of 20 to 50 % on $Im\mathcal{H}$, which is quite satisfactory under the assumption of H-dominance and given the statistical accuracy of JLab data. On the contrary, $Re\mathcal{H}$ is still largely undetermined, and is never extracted with a precision better than 50 %.

Discussion

Fig. 2 compares our results on Hall A kinematics to twist 2 model-independent extractions [38, 40] of Guidal and two extractions with the BMK formalism [12, 32]. Firstly, the use of the GV expressions creates important deviations to the extraction of Ref. [12]. Since the extracted combinations of GPDs are not the same, we will not make the argument more quantitative. Another discrepancy occur when comparing to the results of the global fit of the GPD H of Kumerički and Müller [32]. Secondly, we obtained results in good agreement with Ref. [38]. Note however the surprisingly big shift between the results of Ref. [38] and Ref. [40] induced by the inclusion of target spin asymmetries. These points need to be elucidated but we can already mention the following points :

- Large contributions of $\tilde{H}$ and $\tilde{E}$ are necessary in order to fit JLab Hall A data with the parametrization of Kumerički and Müller. Other DVCS data on unpolarized targets can be included in this global fit under the

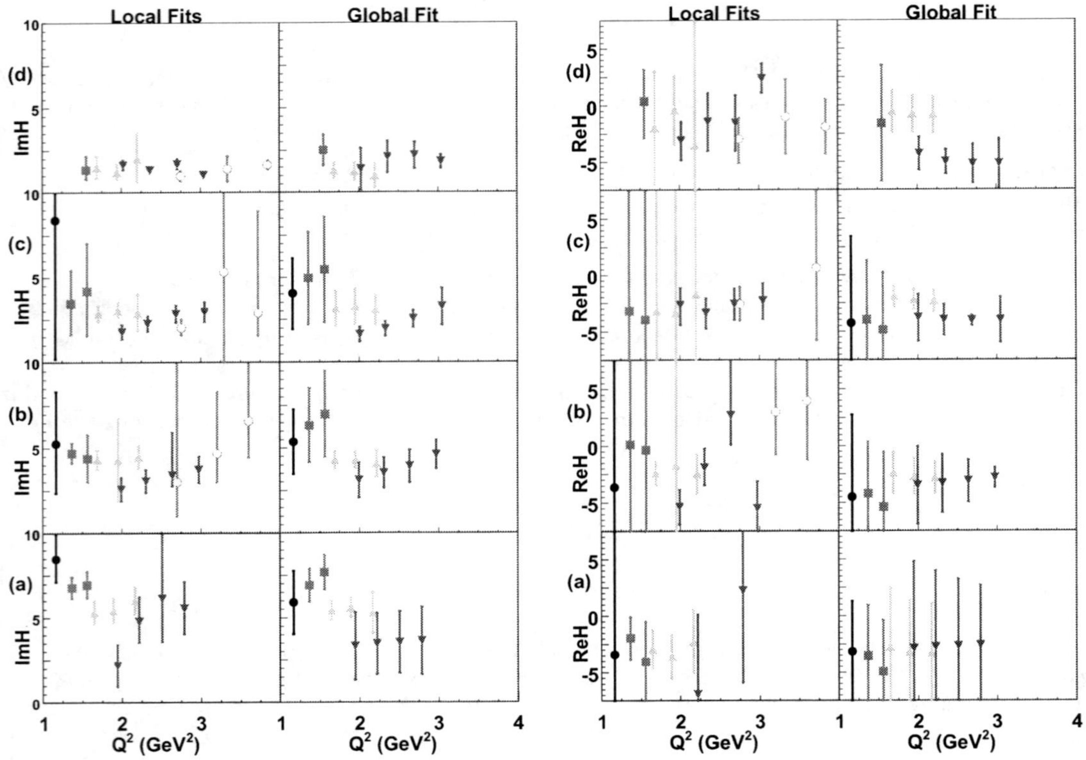

FIGURE 1. Q^2-behaviour ($1 < Q^2 < 4$ GeV^2) of the extracted values of $Im\mathscr{H}$ (left) and $Re\mathscr{H}$ (right) of local fits and global fit on Hall B kinematics : $0.09 < -t < 0.2$ GeV^2 (a), $0.2 < -t < 0.4$ GeV^2 (b), $0.4 < -t < 0.6$ GeV^2 (c), and $0.6 < -t < 1.$ GeV^2 (d). The error bars include both statistics and systematics. $Im\mathscr{H}$ ranges between 0 and 10 and $Re\mathscr{H}$ ranges between -7.5 and +7.5. The black full circles correspond to x_B=0.125, red squares to x_B=0.175, green up triangles to x_B=0.250, blue down triangles to x_B=0.360 and magenta open circles to x_B=0.491.

H-dominance assumption.

- The shift between the two extractions of Guidal is attributed to an $\tilde{H}$ contribution.
- Using VGG and the MIT Bag GPD model [49] we obtained a refined estimate of the systematic uncertainty caused by H-dominance. We obtained larger error bars for the extracted CFF on Hall A kinematics, and error bars compatible with previous estimates on Hall B kinematics.

These facts indicate that JLab Hall A data can certainly not be understood under the restricted set of assumptions : twist 2 + H-dominance. Confirming an explanation involving GPDs other than H will require the inclusion of observables more sensitive to these GPDs. Moreover it was already stressed in Ref. [46] that the twist 2 assumption need to be checked further. Interestingly the three different methods described in Refs. [32, 38, 40, 46] give results in good agreement on Hall B data. The reason of the distinction (discrepancy for Hall A data, agreement for Hall B data) need to be elucidated to make progress on GPD extraction in the valence region. Let us remind that JLab Hall A and Hall B data are compatible.

CONCLUSIONS

Working at the leading twist, and assuming H-dominance, we extracted $Im\mathscr{H}$ and $Re\mathscr{H}$ with 20–50 % accuracy on $Im\mathscr{H}$. The good agreement between the results of local and global fits is a strong consistency check. Moreover values of $Im\mathscr{H}$ on Hall B kinematics extracted by three different groups with three different methods are in good

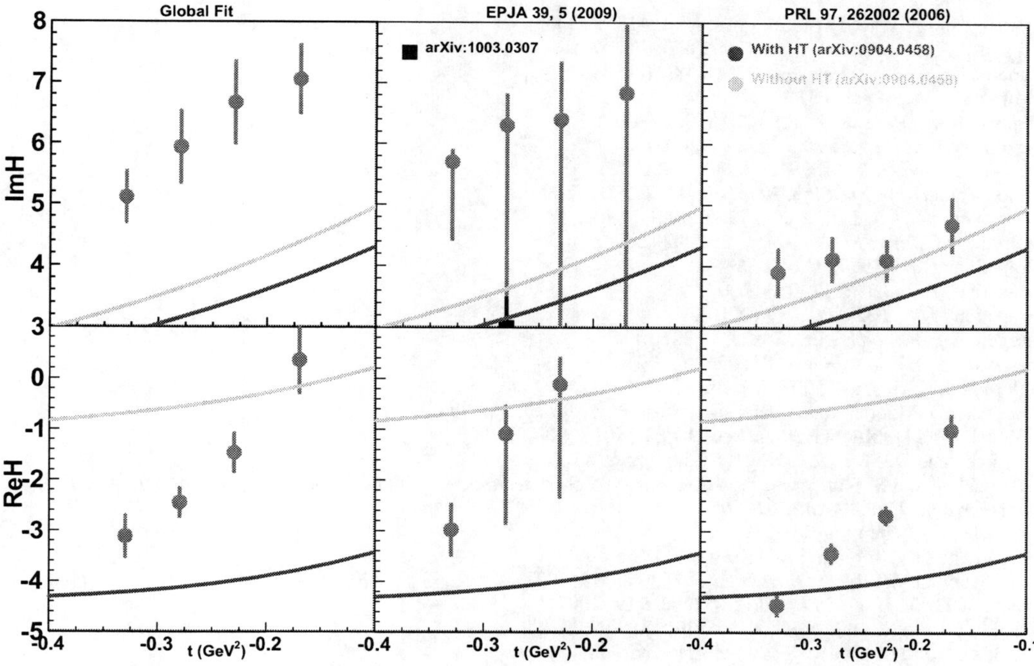

FIGURE 2. $Im\mathcal{H}$ (up) and $Re\mathcal{H}$ (down) vs t ($0.1 < -t < 0.4$ GeV^2) on Hall A kinematics (x_B=0.36 and $Q^2 = 2.3$ GeV^2). $Im\mathcal{H}$ ranges between 3 and 8, and $Re\mathcal{H}$ between -5 and 1. We compare our results (left column) to those of Guidal [38] (middle column) and Muñoz-Camacho *et al.* [12] (right column). In the latter column $\mathcal{H} + \frac{x_B}{2-x_B}\left(1+\frac{F_2}{F_1}\tilde{\mathcal{H}}\right) - \frac{t}{4M^2}\frac{F_2}{F_1}\mathcal{E}$ is plotted, and not $\mathcal{H}$. The error bars include both statistical and systematic uncertainties. The green (resp. blue) curve is the result of the model-dependent global fit of Kumerički and Müller [32] with (resp. without) $\tilde{\mathcal{H}}$. The full black square is the result of Ref. [40].

agreement. The comparison of local and global fits and the weak Q^2-dependency of the results also validates *a posteriori* the restriction to kinematic configurations with $|t|/Q^2 < 1/2$. Note however that the interpretation of JLab Hall A measurements is difficult and probably requires to go beyond the common H-dominance approximation.

Our results are dominated by systematic uncertainties. The systematic uncertainties related to H-dominance will decrease in the future using additional BSA measurements [50] on unpolarised and longitudinally polarised proton target which will put stronger constraints on the global fits. These future measurements will also help clarifying the interpretation of Hall A helicity-independent cross sections.

ACKNOWLEDGMENTS

The author would like to thank the CLAS group at Saclay, P. Guichon, M. Guidal, K. Kumerički, D. Müller and K. Passek-Kumerički for many fruitful and stimulating discussions. The author also thanks the organizers of the 12th International Conference on Meson-Nucleon Physics and the Structure of the Nucleon held in College of William and Mary (May 31-June 4, 2010), Williamsburg, Virginia. This work was supported in part by the Commissariat à l'Energie Atomique and the GDR n° 3034 Physique du Nucleon.

REFERENCES

1. X.-D. Ji., *Phys. Rev. Lett.* **78**, 610 (1997).
2. M. Diehl, Th. Gousset, B. Pire and J.P. Ralston, *Phys. Lett. B* **411**, 193 (1997).
3. K. Goeke, M.V. Polyakov and M. Vanderhaeghen, *Prog. Part. Nucl. Phys.* **47**, 401 (2001).
4. M. Diehl, *Phys. Rept.* **388**, 41 (2003).

5. A.V. Belitsky and A.V. Radyushkin, *Phys. Rept.* **418**, 1 (2005).
6. S. Boffi and B. Pasquini, *Riv. Nuovo Cim.* **30**, 387 (2007).
7. A. Airapetian *et al.*, *Phys. Rev. Lett.* **87**, 182001 (2001).
8. C. Adloff *et al.*, *Phys. Lett. B* **517**, 47 (2001).
9. S. Stepanyan *et al.*, *Phys. Rev. Lett.* **87**, 182002 (2001).
10. S. Chekanov *et al.*, *Phys. Lett. B* **573**, 46 (2003).
11. S. Chen *et al.*, *Phys. Rev. Lett.* **97**, 072002 (2006).
12. C. Muñoz Camacho *et al.*, *Phys. Rev. Lett.* **97**, 262002 (2006).
13. F.D. Aaron *et al.*, *Phys. Lett. B* **659**, 796 (2008).
14. F.X. Girod *et al.*, *Phys. Rev. Lett.* **100**, 162002 (2008).
15. A. Airapetian *et al.*, *JHEP* **06**, 066 (2008).
16. A. Airapetian *et al.*, *JHEP* **11**, 083 (2009).
17. A. Airapetian *et al.*, *JHEP* **06**, 019 (2010).
18. A. Airapetian *et al.*, *Nucl. Phys. B* **829**, 1 (2010).
19. A. Airapetian *et al.*, *Phys. Rev. C* **81**, 035202 (2010).
20. D. Zeiler *et al.*, arXiv:0810.5007 [hep-ex].
21. A.V. Belitsky, D. Mueller and A. Kirchner, *Nucl. Phys. B* **629**, 323 (2002).
22. A.V. Belitsky and D. Mueller, *Phys. Rev. D* **79**, 014017 (2009).
23. A.V. Belitsky and D. Mueller, arXiv:1005.5209 [hep-ph].
24. P.A.M. Guichon and M. Vanderhaeghen, *Analytic ee'γ cross section*, in *Atelier DVCS, Laboratoire de Physique Corpusculaire, Clermont-Ferrand, June 30 - July 01, 2008.*
25. O.V. Teryaev, arXiv:hep-ph/0510031.
26. I.V. Anikin and O.V. Teryaev, arXiv:0710.4211 [hep-ph].
27. I.V. Anikin and O.V. Teryaev, *Phys. Rev. D* **76**, 056007 (2007).
28. M. Diehl and D.Yu. Ivanov, *Eur. Phys. J. C* **52**, 919 (2007).
29. G.R. Goldstein and S. Liuti, *Phys. Rev. D* **80**, 071501 (2009).
30. G.R. Goldstein and S. Liuti, arXiv:0908.2215 [hep-ph].
31. G.R. Goldstein and S. Liuti, arXiv:1006.0213 [hep-ph].
32. K. Kumerički and D. Mueller, *Nucl. Phys. B* **841**, 1 (2010).
33. M.V. Polyakov and C. Weiss, *Phys. Rev. D* **60**, 114017 (1999).
34. P.A.M. Guichon and M. Vanderhaeghen, *Prog. Part. Nucl. Phys.* **41**, 125 (1998).
35. M. Vanderhaeghen, P.A.M. Guichon and M. Guidal, *Phys. Rev. Lett.* **80**, 5064 (1998).
36. M. Vanderhaeghen, P.A.M. Guichon and M. Guidal, *Phys. Rev. D* **60**, 094017 (1999).
37. M. Guidal, M.V. Polyakov, A.V. Radyushkin and M. Vanderhaeghen, *Phys. Rev. D* **72**, 054013 (2005).
38. M. Guidal, *Eur. Phys. J. A* **37**, 319 (2008).
39. M. Guidal and H. Moutarde, *Eur. Phys. J. A* **42**, 71 (2009).
40. M. Guidal, *Phys. Lett. B* **689**, 156 (2010).
41. M. Guidal, arXiv:1005.4922 [hep-ph].
42. M. Burkardt, *Phys. Rev. D* **62**, 071503 (2000), Erratum-ibid. D66, 119903 (2002).
43. S. Goloskokov and P. Kroll, *Eur. Phys. J. C.* **42**, 281 (2005).
44. S. Goloskokov and P. Kroll, *Eur. Phys. J. C.* **53**, 367 (2008).
45. S. Goloskokov and P. Kroll, *Eur. Phys. J. C.* **65**, 137 (2010).
46. H. Moutarde, *Phys. Rev D* **79**, 094021 (2009).
47. K. Kumerički and D. Müller, arXiv:1008.2762 [hep-ph].
48. J.C. Collins and A. Freund, *Phys. Rev. D* **59**, 074009 (1999).
49. X.-D. Ji, W. Melnitchouk and X. Song, *Phys. Rev. D* **56**, 5511 (1997).
50. V. Kubarovsky, *Deeply Virtual Exclusive Reactions with CLAS*, in *ICHEP Paris 2010.*

Modeling QCD for Hadron Physics

P. C. Tandy

Center for Nuclear Research, Department of Physics, Kent State University, Kent OH 44242, USA

Abstract. We review the approach to modeling soft hadron physics observables based on the Dyson-Schwinger equations of QCD. The focus is on light quark mesons and in particular the pseudoscalar and vector ground states, their decays and electromagnetic couplings. We detail the wide variety of observables that can be correlated by a ladder-rainbow kernel with one infrared parameter fixed to the chiral quark condensate. A recently proposed novel perspective in which the quark condensate is contained within hadrons and not the vacuum is mentioned. The valence quark parton distributions, in the pion and kaon, as measured in the Drell Yan process, are investigated with the same ladder-rainbow truncation of the Dyson-Schwinger and Bethe-Salpeter equations.

Keywords: Non-perturbative QCD, hadron physics, Dyson Schwinger equations, light quark mesons, pseudoscalar decay constants, quark condensate as a property of the chiral pion wavefunction, deep inelastic scattering, valence quark parton distributions of pion and kaon.
PACS: 24.85.+p, 12.38.Lg, 11.10.St, 14.40.-n

DYSON–SCHWINGER EQUATIONS OF QCD

A great deal of progress in the QCD modeling of hadron physics has been achieved through the use of the ladder-rainbow truncation of the Dyson-Schwinger equations (DSEs). The DSEs are the equations of motion of a quantum field theory. They form an infinite hierarchy of coupled integral equations for the Green's functions (n-point functions) of the theory. Bound states (mesons, baryons) appear as poles in the appropriate Green's functions, and, e.g., the Bethe-Salpeter bound state equation appears after taking residues in the DSE for the appropriate color singlet vertex. The Faddeev equation for the 3-quark structure of baryons follows in a similar way from the residues of the appropriate vertex DSE. For recent reviews on the DSEs and their use in hadron physics, see Refs. [1, 2, 3, 4].

In the Euclidean metric that we use throughout, the DSE for the dressed quark propagator is

$$S(p)^{-1} \quad = \quad Z_2\, i\not{p} + Z_4\, m(\mu) + Z_1 \int_q^\Lambda g^2 D_{\mu\nu}(p-q)\, \frac{\lambda^i}{2}\gamma_\mu\, S(q)\, \Gamma^i_\nu(q,p)\,, \tag{1}$$

where $D_{\mu\nu}(k)$ is the renormalized dressed-gluon propagator, $\Gamma^i_\nu(q,p)$ is the renormalized dressed quark-gluon vertex. We use $\int_q^\Lambda$ to denote $\int^\Lambda d^4q/(2\pi)^4$ with Λ being the mass scale for translationally invariant regularization. The renormalization condition is $S(p)^{-1} = i\gamma\cdot p + m(\mu)$ at a sufficiently large spacelike μ^2, with $m(\mu)$ the renormalized mass at renormalization scale μ. We use $\mu = 19\,\text{GeV}$. The $Z_i(\mu,\Lambda)$ are renormalization constants.

Bound state pole residues of the inhomogeneous Bethe-Salpeter equation (BSE) for the relevant vertex, yield the homogeneous BSE bound state equation

$$\Gamma^{a\bar{b}}(p_+,p_-) \quad = \quad \int_q^\Lambda K(p,q;P)\, S^a(q_+)\, \Gamma^{a\bar{b}}(q_+,q_-)\, S^b(q_-)\;, \tag{2}$$

where K is the renormalized $q\bar{q}$ scattering kernel that is irreducible with respect to a pair of $q\bar{q}$ lines. Quark momenta are $q_+ = q + \eta P$ and $q_- = q - (1-\eta)P$ where the choice of η is equivalent to a definition of relative momentum q; observables should not depend on η. The meson momentum satisfies $P^2 = -M^2$.

A viable truncation of the infinite set of DSEs should respect relevant (global) symmetries of QCD such as chiral symmetry, Lorentz invariance, and renormalization group invariance. For electromagnetic interactions and Goldstone bosons we also need to respect color singlet vector and axial vector current conservation. The rainbow-ladder (LR) truncation achieves these ends by the replacement $K(p,q;P) \to -4\pi\,\alpha_{\rm eff}(k^2)\, D^{\rm free}_{\mu\nu}(k)\frac{\lambda^i}{2}\gamma_\mu \otimes \frac{\lambda^i}{2}\gamma_\nu$ along with the replacement of the DSE kernel by $Z_1 g^2 D_{\mu\nu}(k)\Gamma^i_\nu(q,p) \to 4\pi\,\alpha_{\rm eff}(k^2)\, D^{\rm free}_{\mu\nu}(k)\,\gamma_\nu \frac{\lambda^i}{2}$ where $k = p-q$, and $\alpha_{\rm eff}(k^2)$ is an effective running coupling. This truncation is the first term in a systematic expansion [5, 6] of K; asymptotically, it reduces to leading-order perturbation theory. These two truncations are mutually consistent: together

12th International Conference on Meson-Nucleon Physics and the Structure of the Nucleon (MENU 2010)
AIP Conf. Proc. 1374, 139-144 (2011); doi: 10.1063/1.3647112

TABLE 1. DSE results [10, 11, 12, 13, 14, 15] for pseudoscalar and vector meson masses and electroweak decay constants, together with experimental data [16]. Units are GeV except where indicated. Quantities marked by † are fitted with the indicated current quark masses and the infrared strength parameter of the ladder-rainbow kernel.

	$m^{u=d}_{\mu=1\text{GeV}}$	$m^{s}_{\mu=1\text{GeV}}$	- $\langle\bar{q}q\rangle^0_{\mu=1\text{GeV}}$
expt	3 - 6 MeV	80 - 130 MeV	$(0.24\ \text{GeV})^3$
calc	5.5 MeV	125 MeV	$(0.241\ \text{GeV})^{3\dagger}$

	m_π	f_π	m_K	f_K	m_ρ	f_ρ	$m_K^\star$	$f_K^\star$	m_ϕ	f_ϕ
expt	0.138	0.131	0.496	0.160	0.770	0.216	0.892	0.225	1.020	0.236
calc	$0.138^\dagger$	$0.131^\dagger$	$0.497^\dagger$	0.155	0.742	0.207	0.936	0.241	1.072	0.259

	r_π^2	$r_K^{+\,2}$	$r_K^{0\,2}$	$g^{\rho\pi\pi}$	$g_{\phi KK}$	$g_{K^\star K\pi}$	$g_{\pi\gamma\gamma}$	$r^2_{\pi\gamma\gamma}$
expt	$0.44\ \text{fm}^2$	0.34	-0.054	6.02	4.64	4.60	0.50	$0.42\ \text{fm}^2$
calc	0.45	0.38	-0.086	5.4	4.3	4.1	0.50	0.41
	$\frac{g_{\rho\pi\gamma}}{m_\rho}$	$\frac{g_{\omega\pi\gamma}}{m_\omega}$	$\frac{g_{K^\star K\gamma}}{m_{K^+}}$	$\frac{g_{K^\star K\gamma}}{m_{K^-}}$	$\lambda_+(e3)$	$\Gamma(K_{e3})$	$\Gamma(K_{\mu3})$	
expt	0.74	2.31	0.83	1.28	0.028	$7.6\ 10^6\ s^{-1}$	5.2	
calc	0.69	2.07	0.99	1.19	0.027	7.38	4.90	

they produce color singlet vector and axial-vector vertices satisfying their respective Ward identities. This ensures that the chiral limit ground state pseudoscalar bound states are the massless Goldstone bosons from dynamical chiral symmetry breaking [7, 8]; and ensures electromagnetic current conservation [9]. The ladder-rainbow kernel found to be successful for light quark hadrons [8, 10] can be written $\alpha_{\text{eff}}(k^2) = \alpha^{\text{IR}}(k^2) + \alpha^{\text{UV}}(k^2)$. The IR term implements the strong infrared enhancement in the region $0 < k^2 < 1\,\text{GeV}^2$ required for sufficient dynamical chiral symmetry breaking. The UV term preserves the one-loop renormalization group behavior of QCD: $\alpha_{\text{eff}}(k^2) \to \alpha_s(k^2)^{1\text{loop}}(k^2)$ in the ultraviolet with $N_f = 4$ and $\Lambda_{\text{QCD}} = 0.234\,\text{GeV}$. The strength of α^{IR} along with two quark masses are fitted to $\langle\bar{q}q\rangle$, $m_{\pi/K}$.

Selected light quark meson results are displayed in Table 1. The charge radii are obtained from the calculated charge form factors of π and K [11, 12] in the triangle diagram approximation. This approximation is the dynamical partner of LR truncation: charge is automatically conserved. Vector meson dominance is a natural outcome here since it is generated by the inhomogeneous BSE for the dressed photon-quark vertex. Related to this, the decay constants for a vector meson to a pair of pseudoscalars are obtained by direct computation of the triangle diagram [13], or from analysis of the timelike electromagnetic form factors [14]. The Abelian axial anomaly coupling $g_{\pi\gamma\gamma}$ and associated interaction radius are obtained from electromagnetic transition form factors [14], which also yield the radiative decay constants. The three-body leptonic decays of the kaon [15] are also shown.

It has remained true for many years that efforts [17] to solve truncated versions of the DSEs for the QCD gauge sector n-point functions that determine the ladder-rainbow kernel have been unable to produce sufficient kernel strength. That is, an empirically acceptable amount of dynamical chiral symmetry breaking (and associated ground state hadron mass scales) has long been underestimated that way. Thus our infrared kernel component has remained semi-phenomenological while we seek to characterize it's deficiencies through application to new classes of observable. Some efforts to go beyond the semi-phenomenological ladder-rainbow truncation so as to progressively reduce the amount of phenomenology have employed an explicit 1-loop dressing of the gluon vertex: a gluon loop [18], or a pion loop [19]. However, an amount of phenomenology is still needed in these approaches to reproduce light quark physics. Furthermore, a study [20, 21] of this type of a diagram-by-diagram approach to the vertex and BSE kernel, within a schematic model of simplified momentum dependence, shows that the produced color factors run rapidly out of control.

It is only recently that large volume lattice-QCD calculations [22] and analysis have established that the gluon 2-point function is non-zero in the IR limit, corresponding to a gluon effective mass of ~ 0.5 GeV, and that the ghost 2-point function is no more singular than its perturbative behavior. These modern lattice-QCD results have recently [23] been combined to produce a kernel for the quark DSE that yields an empirically acceptable amount of dynamical chiral symmetry breaking. The gluon-quark vertex in that approach might be used in the recently developed method for generalized BSE kernel construction [24] to produce a physical and capable pathway to a wider variety of hadron bound state calculations.

NOVEL PERSPECTIVE ON THE QUARK CONDENSATE

Recently it has been advocated that the quark condensate, rather than being a spacetime independent constant filling all space, is really a property of the Bethe–Salpeter wavefunction of hadrons [25]. Clear evidence for such a perspective is provided by the exact QCD expression for the residue ρ_π at the pseudoscalar meson pole in the 3-point function describing the coupling of a pseudoscalar field with quarks [7]

$$i\rho_\pi = -\langle 0|\bar{q}i\gamma_5 q|\pi\rangle = Z_4(\zeta,\Lambda)\,\mathrm{tr}_{\rm cd}\int_{\Lambda_{\rm IR}}^{\Lambda}\frac{d^4q}{(2\pi)^4}\,\gamma_5\,S(q_+)\,\Gamma_\pi(q;P)\,S(q_-)\,. \tag{3}$$

Due to Goldsone's Theorem, and the resulting Goldberger-Treiman relation in which the chiral meson's invariant BSE amplitude associated with γ_5 is the chiral quark scalar self-energy divided by the chiral f_π, one can show rigorously that [7]

$$\lim_{\hat{m}\to 0} f_\pi\,\langle 0|\bar{q}\gamma_5 q|\pi\rangle = -Z_4(\zeta,\Lambda)\,\mathrm{tr}_{\rm cd}\int_{\Lambda_{\rm IR}}^{\Lambda}\frac{d^4q}{(2\pi)^4}\,S_0(q;\zeta) = \langle\bar{q}q\rangle_\zeta^0\,. \tag{4}$$

Thus the chiral condensate is a property of the $q\bar{q}$ projection of the chiral pion Bethe-Salpeter wavefunction, and thus from this perspective is not defined away from this finite-sized bound state. Although it is true that the (point) pion cloud effect on the pion leads to an infinite charge radius, the $q\bar{q}$ dressed quark core remains of finite size and that is where Eq. (4) places the condensate. Furthermore this is in accord with quark and gluon confinement; these virtual fields only exist inside their hadronic containers.

The quark condensate represents an enormous energy density, comparable to the interior of compact stellar objects. The above perspective would eliminate the 10^{45} overestimate of the cosmological constant from the QCD vacuum energy density. If the electroweak quanta turn out to be composites of higher mass entities, like technicolor quarks, the vacuum field we presently call the Higgs scalar is needed only inside those containers. This could eliminate the 10^{56} discrepancy from that sector of the Standard Model.

VALENCE QUARK DISTRIBUTIONS IN THE PION AND KAON

Data for the momentum-fraction probability distributions of quarks and gluons in the pion have primarily been inferred from Drell-Yan [26, 27, 28] and direct photon production [29] in pion-nucleon and pion-nucleus collisions, and semi-inclusive $ep \to eNX$ reactions [30]. For a recent review of nucleon and pion parton distributions see Ref. [31].

Lattice-QCD is restricted to low moments of the distributions, not the distributions themselves [37]. Model calculations of deep inelastic scattering (DIS) parton distribution functions are challenging because it is necessary to have perturbative QCD features (including the evolution of scale) coexisting with a covariant nonperturbative model made necessary by the bound state nature of the target. Aspects of chiral symmetry have led to DIS calculations within the Nambu–Jona-Lasinio model [38, 39, 40] but there are a number of difficulties with this approach [39, 40], among them a point structure for the pion BS amplitude at low model scale and a marked sensitivity to the regularization procedure due to the lack of renormalizability. Constituent quark models have also been employed [41, 42], with the difficulties encountered in such studies considered in Ref. [43]. Instanton-liquid models [44] have also been used. In these approaches, it is difficult to have pQCD elements join smoothly with nonperturbative aspects. All of these issues can in principle be addressed if parton distribution functions can be obtained from a model based on the DSEs, and in particular styled after the approach we have previously made for the related pion electromagnetic form factor [11, 12].

In the Bjorken kinematic limit, DIS selects the most singular behavior of a correlator of quark fields of the target with light-like and causal distance separation $z^2 \sim 0^+$. With incident photon momentum along the negative 3-axis, the kinematics selects $z^+ \sim z_\perp \sim 0$ leaving z^- as the finite distance conjugate to quark momentum component xP^+, where $x = Q^2/2P\cdot q$ is the Bjorken variable, $q^2 = -Q^2$ is the spacelike virtuality of the photon, and P is the target momentum. To leading order in the operator product expansion, the target structure functions are proportional to the probability distribution function $q_f(x)$. This in turn is given, in manifestly invariant form, by the correlator [45, 46, 47]

$$q_f(x) = \frac{1}{4\pi}\int d\lambda\, e^{-ixP\cdot n\lambda}\,\langle\pi(P)|\bar{\psi}_f(\lambda n)\,\gamma\cdot n\,\psi_f(0)|\pi(P)\rangle_c\,. \tag{5}$$

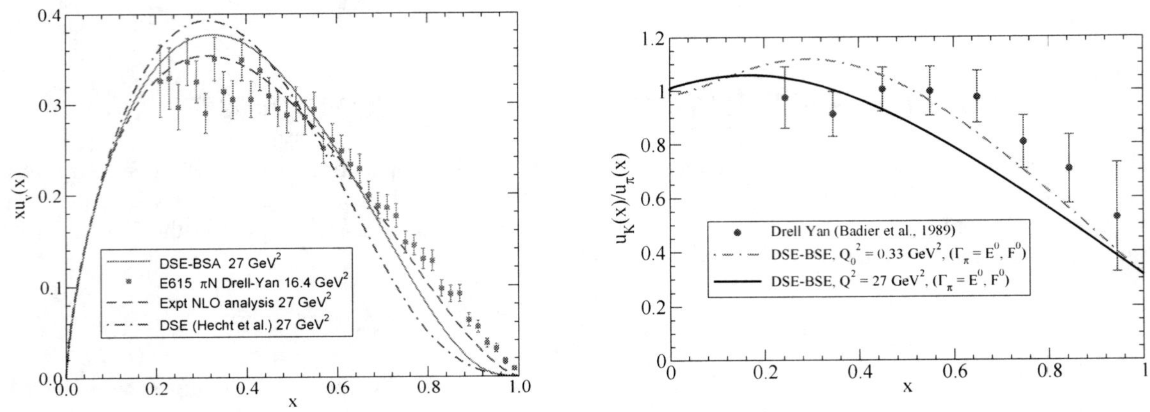

FIGURE 1. *Left Panel*: Pion valence quark distribution evolved to $(5.2\text{ GeV})^2$. Solid line is the full DSE-BSA calculation [32]; dot-dashed line is the semi-phenomenological DSE-based calculation of Hecht et al. [33]; experimental data points are from [28] at scale $(4.05\text{ GeV})^2$; the dashed line is the recent NLO re-analysis of the experimental data [34]. *Right Panel*: The ratio of u-quark distributions in the kaon and pion. The solid line is our preliminary result from DSE-BSE calculations [35, 32, 31]; the experimental data is from [36, 26].

where f is a flavor label. In the frame where the target has infinite momentum, $q_f(x)$ is the quantum mechanical probability that a single f-parton has momentum fraction $x = k\cdot n/P\cdot n$ [47]. In the above n^μ, and partner p^μ, are light-like vectors satisfying $n^2 = p^2 = 0$ and $n\cdot p = 2$ and they form a convenient basis for the longitudinal sector of 4-vectors. The dominant component of q is parallel to n. One can always choose a frame such that $k\cdot n = k^+$ and $k\cdot p = k^-$. Note that $q_f(x) = -q_{\bar f}(-x)$, and that the valence quark amplitude is $q_f^v(x) = q_f(x) - q_{\bar f}(x)$. It follows from Eq. (5) that $\int_0^1 dx\, q_f^v(x) = \langle\pi(P)|J_f(0)\cdot n|\pi(P)\rangle/2P\cdot n = F_\pi(0) = 1$. Approximate treatments should at least preserve vector current conservation to automatically obtain the correct normalization for valence quark number.

In ladder-rainbow truncation, which treats only the valence $q\bar q$ structure of the pion, Eq. (5) yields the explicit form

$$q_f^v(x) = \frac{i}{2}\int_p^\Lambda \mathrm{tr}_{\mathrm{cd}}[\Gamma_\pi(p,P)\,S(p)\,\Gamma^{(n)}(p;x)\,S(p)\,\Gamma_\pi(p,P)\,S(p-P)] \quad , \tag{6}$$

where $\Gamma^{(n)}(p;x)$ is a generalization of the dressed vertex that a zero momentum photon has with a quark. It satisfies the usual inhomogeneous BSE integral equation (here with a LR kernel) except that the inhomogeneous term is $\gamma\cdot n\,\delta(p\cdot n - xP\cdot n)$. This selection of LR dynamics exactly parallels the symmetry-preserving dynamics of the corresponding treatment of the pion charge form factor at $q^2 = 0$ wherein the vector current is conserved by use of ladder dynamics at all three vertices and rainbow dynamics for all 3 quark propagators [11, 12]. Here the number of valence u-quarks (or $\bar d$) in the pion is automatically unity since the structure of Eq. (6), along with the canonical normalization of the $q\bar q$ BS amplitude $\Gamma_\pi(p,P)$, ensures $\int_0^1 dx\, q_u^v(x) = 1$ because $\int_0^1 dx\,\Gamma^{(n)}(p;x)$ gives the Ward Identity vertex.

Eq. (6) is in Minkowski metric so as to satisfy the constraint on $p\cdot n$, but LR dynamical information on the various non-perturbative elements such as $S(p)$ and $\Gamma_\pi(p,P)$ is available only in Euclidean metric [10]. Since $q_f(x)$ is obtained from the hadron tensor $W^{\mu\nu}$ which in turn can be formulated from the discontinuity $T^{\mu\nu}(\varepsilon) - T^{\mu\nu}(-\varepsilon)$, we observe that all enclosed singularities from the difference of Wick rotations cancel except for the cut that defines the object of interest. With use of numerical solutions for dressed propagators and BS amplitudes that give an accurate account of light quark hadrons, our DIS calculations significantly extend the exploratory study made in Ref. [33]. That work employed phenomenological parameterizations of these elements.

In Fig. 1 we display our DSE result for the valence u-quark distribution evolved to $Q^2 = (5.2\text{ GeV})^2$ in comparison with πN Drell-Yan data [28] with a scale quoted as $Q^2 = (4.05\text{ GeV})^2$. We also compare with a recent NLO reanalysis of the data at scale $Q^2 = (5.2\text{ GeV})^2$. The distribution at the model scale Q_0^2 is evolved higher by leading order DGLAP. The model scale is found to be $Q_0 = 0.57$ GeV by matching the x^n moments for $n = 1,2,3$ to the experimental values given independently at $(2\text{ GeV})^2$ [48]. Our momentum sum rule result $\int_0^1 dx\, x\,(u_\alpha + \bar d_\alpha) = 0.74\ (\alpha = \pi), 0.76\ (\alpha = K)$

at Q_0 clearly shows the implicit inclusion of gluons as a dynamical entity in a covariant approach. The large x behavior in the form $u(x) \sim (1-x)^{\alpha}$ has long provided a contrast between previous parameterizations of experiment ($\alpha \sim 1.5$) and QCD-based approaches ($\alpha = 2+\gamma$) where γ represents anomalous dimensional behavior that increases with scale. At our model scale, $\alpha = 2.08$. Recently, the experimental data [28] has been reanalyzed [49] by an accounting for soft gluon processes, and the resulting experimental $u(x)$ at $Q^2 = (4\,\text{GeV})^2$, for $x > 0.4$ is essentially identical to the Hecht et al. DSE calculation shown in Fig. 1. This eliminates the discrepancy between QCD and experiment here.

The ratio u_K/u_π measures the dynamical effect of the local environment. In the kaon, the u-quark is partnered with a significantly heavier partner than in the pion and this shifts the probability to relatively lower x in the kaon. Our preliminary DSE model calculation [35, 32, 31] is shown in Fig. 1 along with available Drell Yan data [36, 26]. Here we include only the leading two invariants of the pion BS amplitude, $E(k;P)$ and $F(k;P)$, where k is $q\bar{q}$ relative momentum. For both amplitudes only the zeroth Chebychev moment in $k \cdot P$ is employed . This variable does not occur in static quantum mechanics, nor in the Nambu–Jona-Lasinio point-coupling field theory model [38] which also cannot accommodate the k^2 dependence. We do not make such a point meson approximation here; the k^2 dependence comes from the BSE solutions. Nevertheless, the qualitative features of the ratio u_K/u_π in Fig. 1 are adequately reproduced by a generalized Nambu–Jona-Lasinio model [31].

An explicit analysis of the exponent α in the large-x behavior $q(x) \sim (1-x)^{\alpha}$ can be made in the limit where the propagators in Eq. (6) have the dynamical quark self-energy replaced by a constant (constituent) mass and the vertex is taken to be $\gamma \cdot n\, \delta(p \cdot n - xP \cdot n)$. These limits maintain consistency with the Ward Identity mentioned above. We also truncate the pion Bethe-Salpeter vertex to just $\gamma_5 E_\pi(k^2)$ and take $E_\pi(k^2) = N_\pi/(k^2 - \Lambda_\pi^2)$ where the relative $q\bar{q}$ momentum is $k = p - P/2$. For $x > 1/2$ the pole in the spectator quark propagator is the only one in the upper half plane and the p^- integral may be easily evaluated by the residue theorem to yield

$$q^v(x) = \frac{4N_c\pi^2}{(2\pi)^4} \int_0^\infty d^2p_\perp \frac{p_\perp^2 + m_q^2}{[x(1-x)P^2 - p_\perp^2 - m_q^2]^2} E_\pi^2(k^2) \ , \tag{7}$$

where now k^2 is evaluated at the p^- pole. It is convenient to change the integration variable to $\mu = -p^2$ where the latter is the value at the p^- pole. Then

$$q^v(x) = \frac{4N_c\pi^2}{(2\pi)^4} \int_{\mu_m(x)}^\infty d\mu \, \frac{xP^2 + \mu + m_q^2}{[\mu + m_q^2]^2} E_\pi^2(-P^2/4 - (\mu + m_q^2)/2) \ , \tag{8}$$

where $\mu_m(x) = a/(1-x) - xm_\pi^2$, with $a = xm_q^2$. This divergence of the lower limit for large x renders the result totally dependent upon the ultraviolet behavior of the propagators and bound state amplitudes.

The leading large x behavior of $q^v(x) \sim (1-x)^{\alpha}$ comes from the leading divergent term of each factor of the integrand. The integral can be expressed

$$q^v(x) = N \int_0^\infty d\hat{\mu} \, \frac{\frac{a}{1-x} + b + \hat{\mu}}{[\frac{a}{1-x} + c + \hat{\mu}]^2} \left(\frac{a}{1-x} + d + \hat{\mu}\right)^{-n} \ , \tag{9}$$

where the physical case of Bethe-Salpeter bound state amplitudes E_π^2 determined by one gluon exchange has $n = 2$.

The quantities a, b, c, d depend on the mass-dimensioned quantities and scales in the system and have only a nonsingular dependence upon x. Their details are not important. A change of variable to $\bar{\mu} = (1-x)\hat{\mu}/a$ shows that the integral scales as $[(1-x)/a]^n$ when $a/(1-x)$ is greater than any physical mass scale in the system. Note that it is the pair of bound state vertices that totally determine the large x exponent: if the argument of the bound state amplitude E_π was not singular at large x, the combined scaling effect of the propagators would vanish giving $\alpha = 0$. The presence of true bound state vertices provides a natural regularization of the integral and allows us to ignore explicit mathematical regularization at a higher scale. In a renormalizable field theory such as QCD, there is an overall regularization mass scale that can be taken to be much larger than any scale at which physical observable questions are to be asked of the formalism. In the above case the renormalization scale is to be much larger than $a/(1-x)$ for values of x where an asymptotic exponent is to be quoted.

If one were to consider the limit of a Nambu–Jona-Lasinio model, the bound state vertices become constants and the model integral for $q^v(x)$ must be regularized, and the result, including the high x exponent becomes dependent on regularization scheme [31]. However, independent of scheme, any thus obtained exponent $\alpha \neq 0$ would be unconstrained by the physical properties of the system.

ACKNOWLEDGMENTS

The authors would like to thank C. D. Roberts, S. J. Brodsky and I. Cloet for helpful conversations and suggestions. This work has been partially supported by the U.S. National Science Foundation under grant no. PHY-0903991, part of which constitutes USA-Mexico collaboration funding in partnership with the Mexican agency CONACyT.

REFERENCES

1. C. D. Roberts, and A. G. Williams, *Prog. Part. Nucl. Phys.* **33**, 477–575 (1994), hep-ph/9403224.
2. P. C. Tandy, *Prog. Part. Nucl. Phys.* **39**, 117–199 (1997), nucl-th/9705018.
3. R. Alkofer, and L. von Smekal, *Phys. Rept.* **353**, 281 (2001), hep-ph/0007355.
4. P. Maris, and C. D. Roberts, *Int. J. Mod. Phys.* **E12**, 297–365 (2003), nucl-th/0301049.
5. A. Bender, C. D. Roberts, and L. Von Smekal, *Phys. Lett.* **B380**, 7–12 (1996), nucl-th/9602012.
6. M. S. Bhagwat, A. Holl, A. Krassnigg, C. D. Roberts, and P. C. Tandy, *Phys. Rev.* **C70**, 035205 (2004), nucl-th/0403012.
7. P. Maris, C. D. Roberts, and P. C. Tandy, *Phys. Lett.* **B420**, 267–273 (1998), nucl-th/9707003.
8. P. Maris, and C. D. Roberts, *Phys. Rev.* **C56**, 3369–3383 (1997), nucl-th/9708029.
9. C. D. Roberts, *Nucl. Phys.* **A605**, 475–495 (1996), hep-ph/9408233.
10. P. Maris, and P. C. Tandy, *Phys. Rev.* **C60**, 055214 (1999), nucl-th/9905056.
11. P. Maris, and P. C. Tandy, *Phys. Rev.* **C62**, 055204 (2000), nucl-th/0005015.
12. P. Maris, and P. C. Tandy, *Phys. Rev.* **C61**, 045202 (2000), nucl-th/9910033.
13. D. Jarecke, P. Maris, and P. C. Tandy, *Phys. Rev.* **C67**, 035202 (2003), nucl-th/0208019.
14. P. Maris, and P. C. Tandy, *Phys. Rev.* **C65**, 045211 (2002), nucl-th/0201017.
15. C.-R. Ji, and P. Maris, *Phys. Rev.* **D64**, 014032 (2001), nucl-th/0102057.
16. S. Eidelman, et al., *Phys. Lett.* **B592**, 1 (2004).
17. R. Alkofer, C. S. Fischer, and L. von Smekal, *Acta Phys. Slov.* **52**, 191 (2002), hep-ph/0205125.
18. P. Watson, W. Cassing, and P. C. Tandy, *Few Body Syst.* **35**, 129–153 (2004), hep-ph/0406340.
19. C. S. Fischer, and R. Williams, *Phys. Rev.* **D78**, 074006 (2008), 0808.3372.
20. H. H. Matevosyan, A. W. Thomas, and P. C. Tandy, *Phys. Rev.* **C75**, 045201 (2007), nucl-th/0605057.
21. H. H. Matevosyan, A. W. Thomas, and P. C. Tandy, *J. Phys.* **G34**, 2153–2164 (2007), 0706.2393.
22. A. Cucchieri, and T. Mendes, *Phys. Rev.* **D81**, 016005 (2010), 0904.4033.
23. A. C. Aguilar, and J. Papavassiliou (2010), 1010.5815.
24. L. Chang, and C. D. Roberts, *Phys. Rev. Lett.* **103**, 081601 (2009), 0903.5461.
25. S. J. Brodsky, C. D. Roberts, R. Shrock, and P. C. Tandy, *Phys. Rev.* **C82**, 022201 (2010), 1005.4610.
26. J. Badier, et al., *Z. Phys.* **C18**, 281 (1983).
27. B. Betev, et al., *Z. Phys.* **C28**, 15 (1985).
28. J. S. Conway, et al., *Phys. Rev.* **D39**, 92–122 (1989).
29. P. Aurenche, R. Baier, M. Fontannaz, M. N. Kienzle-Focacci, and M. Werlen, *Phys. Lett.* **B233**, 517 (1989).
30. C. Adloff, et al., *Eur. Phys. J.* **C6**, 587–602 (1999), hep-ex/9811013.
31. R. J. Holt, and C. D. Roberts, *Rev. Mod. Phys.* **82**, 2991–3044 (2010), 1002.4666.
32. T. T. Nguyen (2010), phD dissertation, Kent State University.
33. M. B. Hecht, C. D. Roberts, and S. M. Schmidt, *Phys. Rev.* **C63**, 025213 (2001), nucl-th/0008049.
34. K. Wijesooriya, P. E. Reimer, and R. J. Holt, *Phys. Rev.* **C72**, 065203 (2005), nucl-ex/0509012.
35. A. Bashir, T. Nguyen, C. D. Roberts, and P. C. Tandy (2010), in preparation.
36. J. Badier, et al., *Phys. Lett.* **B93**, 354 (1980).
37. C. Best, et al., *Phys. Rev.* **D56**, 2743–2754 (1997), hep-lat/9703014.
38. T. Shigetani, K. Suzuki, and H. Toki, *Phys. Lett.* **B308**, 383–388 (1993), hep-ph/9402286.
39. H. Weigel, E. Ruiz Arriola, and L. P. Gamberg, *Nucl. Phys.* **B560**, 383–427 (1999), hep-ph/9905329.
40. W. Bentz, T. Hama, T. Matsuki, and K. Yazaki, *Nucl. Phys.* **A651**, 143–173 (1999), hep-ph/9901377.
41. A. Szczepaniak, C.-R. Ji, and S. R. Cotanch, *Phys. Rev.* **D49**, 3466–3473 (1994), hep-ph/9309284.
42. T. Frederico, and G. A. Miller, *Phys. Rev.* **D50**, 210–216 (1994).
43. C. M. Shakin, and W.-D. Sun, *Phys. Rev.* **C51**, 2171–2176 (1995).
44. A. E. Dorokhov, and L. Tomio, *Phys. Rev.* **D62**, 014016 (2000).
45. R. L. Jaffe, *Nucl. Phys.* **B229**, 205 (1983).
46. R. L. Jaffe (1985), proceedings of the Los Alamos School on Quark Nuclear Physics, Los Alamos, N.Mex., Jun 10-14, 1985; Johnson, M. B. , (Ed.) and Picklesimer, A. , (Ed.), NEW YORK, USA: WILEY (1986) 861p.
47. R. K. Ellis, W. J. Stirling, and B. R. Webber, *Camb. Monogr. Part. Phys. Nucl. Phys. Cosmol.* **8**, 1–435 (1996).
48. P. J. Sutton, A. D. Martin, R. G. Roberts, and W. J. Stirling, *Phys. Rev.* **D45**, 2349–2359 (1992).
49. M. Aicher, A. Schafer, and W. Vogelsang (2010), 1009.2481.

An Overview of Meson-Nuclear Physics

Anthony W. Thomas

CSSM, School of Chemistry and Physics, University of Adelaide
Adelaide SA 5005, Australia

Abstract. In this opening talk at MENU2010 we outline some of the key achievements in the field over the past few years as well as some of its major challenges and opportunities.

Keywords: QCD, strange quarks, meson and baryon spectroscopy, hadron structure in medium, Standard Model tests, dark matter, new facilities
PACS: 12.38.-t, 11.30.-j, 12.39.-x, 13.25.-k, 13.30.Eg, 13.40.-f, 13.75.-n, 14.20.-c, 21.30.-x, 21.65.-f, 24.80.+y, 24.85.+p, 25.30.-c, 26.60.-c

INTRODUCTION

This conference covers an extremely broad range of topics and in just a few pages it is impossible to even touch on all the areas which will be discussed. We have chosen to summarise just one area where there has recently been impressive progress, namely our quantitative understanding of strangeness in the nucleon. We also discuss a couple of examples where theoretical progress is of direct importance for future experiments as well as for the interpretation of hitherto anomalous results. Finally we make some remarks on meson and baryon spectroscopy and the exciting array of new facilities that are coming on-line in the near future.

STRANGENESS

The absence of strange valence quarks in the nucleon means that they can only contribute to hadron properties through the creation of virtual quark-anti-quark pairs and thus the accurate test of their role plays a role for QCD analogous to that of the Lamb shift in QED. Some 5 years ago the use of indirect techniques [1], involving lattice QCD, chiral extrapolation and measured baryon properties led to extremely accurate determinations of the strange magnetic moment and charge radius of the proton [2, 3]. Just last year the magnetic moment calculation was beautifully confirmed by the first direct lattice calculation [4]. Contrary to much speculation over the preceding decade, these contributions turn out to be quantitatively quite small - below 1% for both the magnetic moment and the mean-square radius. The experimental results for the strange form factors of the proton [5, 6, 7, 8] are in remarkable agreement with the theoretical calculations [9], although for now the errors are perhaps an order of magnitude larger.

Studies of hadron structure as a function of quark mass have provided a very natural explanation of these results [10], namely that the finite size of the source of the meson clouds associated with chiral symmetry naturally suppress all Goldstone boson loops for meson masses above around 0.4 GeV [11]. The physical kaon mass falls above this limit and is naturally suppressed. The exploitation of this remarkable feature of nucleon structure through finite range regulated effective field theory has also led to a much more precise and again astonishingly small result for another property of strange quarks in the proton, namely the strange sigma commutator. Young and Thomas recently reported a value around 30MeV [12], which is an order of magnitude smaller than commonly used and believed. This result is also in excellent agreement with direct calculations that appeared around the same time [13, 14].

This new value for σ_s not only has consequences for speculation about possible kaon condensation but surprisingly it is also extremely important for searches for neutralinos – the favourite "beyond-the-Standard-Model" candidate for dark matter. Within the minimal supersymmetric extension it turns out that the biggest contribution to the neutralino-nucleon spin independent cross section came through σ_s, *until this new result*. As discussed by Giedt *et al.* [15], the new value solves a long-standing problem of increasing the accuracy of predictions of these cross sections but it also leads to values at least an order of magnitude smaller than hitherto expected. This has major implications for dark matter searches as well as for the interpetation of existing observations.

12th International Conference on Meson-Nucleon Physics and the Structure of the Nucleon (MENU 2010)
AIP Conf. Proc. 1374, 145-150 (2011); doi: 10.1063/1.3647113

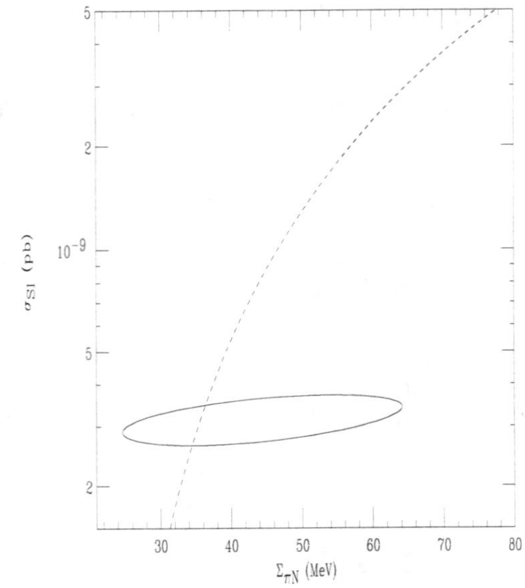

FIGURE 1. Illustration of the improvement in the precision with which the neutralino-nucleon cross section can be calculated within a particular minimal supersymmetric extension of the Standard Model when the strange quark sigma commutator is taken from the recent analysis of lattice data by Giedt *et al.* [15], rather than through the πN sigma commutator. Note that although the precision is much improved the absolute cross section is very much at the lower end of the range.

A final issue on which there has also been substantial recent progress concerns the parton distribution functions (pdfs) for strange quarks [17]. This is extremely important in connection with the NuTeV anomaly [16], where an asymmetry between s and $\bar{s}$ [18] quarks can provide a potentially large correction. Global analyses of pdfs seem to indicate a positive value for $\langle xs^- \rangle$, but the level of precision needs substantial improvement. For this particular issue, lattice QCD seems unlikely to make a significant contribution in the near future. However, for the combination $\langle x(s+\bar{s})\rangle$, the Kentucky group has recently been able to obtain quite an accurate value, namely 2.7 ± 0.3 % [19]. This may increase somewhat when the light quark masses take more physical values but the calculation nevertheless represents substantial progress.

SPIN DEPENDENDENT PHENOMENA

There continues to be a great deal of effort devoted to clarifying the relative importance of quark and gluon spin and orbital angular momentum in defining the spin of the proton. Data from RHIC has established that the total spin carried by gluons is less than about one half at a scale of a few GeV^2 [20, 21], and no doubt this will be pinned down even more precisely over the next few years. We already know that it is nowhere near large enough to play a significant role in resolving the famous spin crisis, where almost certainly it is the relativistic quark motion combined with gluon exchange and chiral symmetry that leads to the replacement of spin by quark orbital angular momentum [22, 23]. The debate over how best to define quark and gluon orbital angular momentum continues with vigour [24, 25] and this will be a topic for discussion at this meeting.

Within the experimental community there is tremendous excitement over the host of new observables which promise to bring new light to the study of hadron structure [26]. This includes a number of measurements involving semi-inclusive hadron production which will enable the determination of GPDs and TMDs [27]. On the theoretical side, we note the tremendous progress made in lattice QCD to measure moments of the GPDs relevant to the proton spin problem [28]. Critical issues for the near future include the need to develop a deep understanding of the systematic errors associated with finite volume corrections and extrapolations in both pion mass and momentum transfer needed to accurately pin down the quark orbital angular momentum [29, 30].

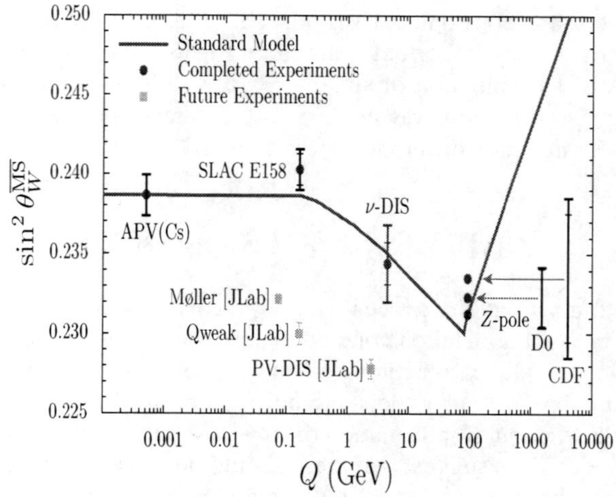

FIGURE 2. The curve represents the running of $\sin^2\theta_W$ in the $\overline{MS}$ renormalization scheme. The Z-pole point represents the combined results of six LEP and SLC experiments, while the CDF and D0 collaboration results (at the Z-pole) and the SLAC E158 result, are labelled accordingly. The atomic parity violating (APV) result has been shifted from $Q^2 = 0$ for clarity. The inner error bars represent the statistical uncertainty and the outer error bars the total uncertainty – see Ref. [42] for details and associated references.

Hadrons in-medium

The nuclear community shows conflicting attitudes to the role of QCD in providing the fundamental explanation of the properties of atomic nuclei. For some QCD exists simply to give us chiral symmetry and nuclear structure can then be computed in terms of a number of low energy constants fixed by data. This is a more formal and systematic version of the older approach, still very much in favour, where one calculates nuclear properties using two-, three- and more-body forces, the former fitted to NN data while the latter are fit to chosen sets of nuclear data. The most modern expression of this approach involves the use of the techniques of lattice QCD to compute the interaction of two and three hadrons directly from QCD [31, 32]. This is an ambitious long-term program, with current calculations limited to rather small lattice volumes. Nevertheless, as we shall see at the conference some remarkable results have already been obtained.

An alternate approach begins with the realization that at some density (perhaps 3 to 10 times nuclear matter density) nuclear matter will make a transition to quark matter – a phase transition which may have dramatic effects on the observable properties of neutron stars. This suggests that one might begin to build a theory of the nuclear many-body system starting with a description of hadron structure at the quark level and then considering the self-consistent modification of that structure in a nuclear medium. This is the approach taken within the QMC (quark-meson coupling) model [33, 34]. A remarkable advantage of this approach is that no new parameters are needed to calculate the effective density dependent forces [35] between any hadrons whose quark structure is known. Indeed, it has been possible to develop a remarkably successful derivation of realistic Skyrme forces [35, 36] for comparison with low energy nuclear phenomenology – while the fully relativistic underlying theory successfully predicts key features of hypernuclear physics and allows the study of the appearance of hyperons in dense matter [37]. Most recently, this model has also been used to study the possibility of binding of charm-anti-charm mesons in matter [38], with results very similar to those coming from direct lattice QCD simulations [39].

The QMC model has the additional advantage that one can address not only those low energy properties such as binding energies and charge densities but it can also be used to calculate the nuclear modification of the deep inelastic structure functions. Extensions of the QMC approach based upon a covariant, confining version of the NJL model have produced a satisfactory description of the EMC effect in finite nuclei [40]. It has also produced the remarkable prediction that should motivate quite a bit of experimental effort, namely that the EMC effect for the spin structure function should be roughly twice as large as the unpolarised EMC effect [40]. Another recent realisation of critical importance if one intends to use deep inelastic scattering data on nuclei in any way that assumes some sort of flavour symmetry, is that there will be a component of the EMC effect which is isovector in nature if one has a target with $N \neq Z$ [41]. Most importantly, because the EMC effect involves a change in the structure of the bound nucleon, that isovector EMC correction will persist even if one derives data for an effectively isoscalar nucleus by subtracting the contribution of the excess neutrons. A very important example of this is the corresponding correction to the Paschos-Wolfenstein relation in the NuTeV determination of $\sin^2\theta_W$. Figure 2 shows the result of a recent reanalysis of the NuTeV anomaly [42] in which this correction was applied to the data along with a correction for genuine charge symmetry violation associated with the mass difference between u and d quarks.

HADRON SPECTROSCOPY

Many presentations at this meeting will address recent developments in hadron spectroscopy. The Excited Baryon Analysis Center at JLab is now producing detailed coupled channels calculations for two pion production as well as kaon production in pion and photon induced reactions [43, 44]. In close collaboration with other groups at Jülich, Bonn and Mainz this has already clarified our understanding of some well established resonances [45]. It has also established a sound basis for the extraction of information on any new states that may appear in the mass region above 1.8–2.0 GeV, where old quark model ideas suggest that there should be more states than have hitherto been seen. This coupled channel analysis is being challenged by a considerable quantity of new, high quality data that will help to answer these questions.

On the theoretical side, lattice QCD has again proven to be a critical player. In particular, the JLab group has remarkable new results for both charm quark [46] and light quark [47] systems, with a careful blend of group theory and lattice techniques now allowing the clear identification of a large number of excited states for each choice of spin and parity. As the masses of the light quarks in these simulations approach the physical values more closely it will be necessary to deal with effects of channel coupling here too.

NEW FACILITIES

Across the world this field either has or soon will have the benefit of a large number of major new experimental facilities and there will be considerable discussion of their capabilities at this conference. At this time the most noteworthy is, of course, JPARC in Japan, which only recently delivered its first beam. While the beam intensity is currently low and the energy currently 30 GeV rather than the ultimate 50 GeV the accelerator is already delivering a neutrino beam to the Kamiokande detector and the high momentum kaon line is ready for experiments. The first of those experiments, using pions rather than kaons, will take one more look at the infamous pentaquark. This object, which would add remarkably to our zoo of fundamental particles if confirmed, currently looks rather sick, with only one of the initial groups that reported a sighting still claiming to have evidence for its existence. Hopefully we will have news of that test at the next meeting.

Apart from its role in neutrino oscillations, the facilities at JPARC are ideally suited for extending our knowledge of hypernuclei dramatically. Of particular importance is the discovery of Ξ-hypernuclei, which are expected, within the QMC model discussed earlier, to be bound by bound by up to 20 MeV [48]. Establishing their existence and eventually the systematics of their binding and structure is not only important because it will establish a totally new area of study in nuclear structure but because of its importance for the properties of dense matter, as found for example in neutron stars. Certainly, if Ξ-hypernuclei are confirmed it will be difficult to avoid the conclusion that the Ξ^- should appear at densities not far above $2\rho_0$ in matter that is in β-equilibrium. Their appearance, in turn, has the effect of dramatically softening the equation of state and hence the observable properties of neutron stars. Finally, we mention the promise of Drell-Yan for studies of the flavour structure of hadrons and nuclei as well as offering new types of spin-dependent measurements in which one can more easily detect processes which violate chiral symmetry.

Two future facilities, due to operate at a similar date, which will also undoubtedly be the source of intense interest are the 12 GeV upgrade at JLab (just a few kilometers away from this meeting) and the new facilities at GSI-FAIR. Hadron spectroscopy, especially the study of the existence and properties of the exotic mesons predicted within QCD, should be transformed by the Gluex experiment in Hall D. (Though we should also mention the important studies currently underway at COMPASS at CERN.) Critical to the success of this experiment, which should be able to access states up to as much as 2.6–2.8 GeV, is the dedicated 9 GeV photon beam with high linear polarisation for which Hall D was built, as well as the extremely high acceptance of the purpose built detector. GSI-FAIR also aims to produce and study Ξ-hypernuclei and both new facilities, as well as JPARC, should be able to develop the spectroscopy of strange baryons to a new level.

The development of a circulating anti-proton beam at GSI-FAIR will probably come a little after the opening of this new facility but the set of experiments already proposed for it, including precision charmonium spectroscopy, is very exciting. We leave the many other possibilities opened by these facilities for discussion at the meeting, except to emphasise the kinematic range opened for the study of the spin and flavour dependence of the parton distribution functions, as well as semi-inclusive deep inelastic processes by the 12 GeV upgrade.

We also expect that there will be lively discussion over the long term plans to build one or more electron-ion colliders in Europe and the United States. The very ambitious plans at BNL and JLab will come to a focus in the next long range plan for the United States, while the somewhat more modest plans at GSI-FAIR are still under development. Certainly the plans for an electron-ion collider at BNL or JLab offer a tremendous opportunity to look for non-perturbative gluonic effects as well as allowing one to probe the spin and flavour structure of matter at a deeper level than possible before. One should also mention the possibility of installing an electron accelerator at CERN so that one could have electron-ion collisions at the LHC, the LHeC project. This is extremely ambitious in terms of cost and energy reach (although it has no polarisation) but clearly would be motivated by discoveries at the LHC itself and therefore a decision must await the first discoveries there.

SUMMARY

In this brief welcome I have been able to do little more than touch on a few of the exciting physics challenges that will be discussed in detail at this meeting, as well as very briefly highlighting the wonderful array of new facilities that will be crucial to the resolution of those challenges. I wish all participants the best for a productive meeting!

ACKNOWLEDGMENTS

This work was supported by the Australian Research Council through an Australian Laureate Fellowship and by the University of Adelaide.

REFERENCES

1. D. B. Leinweber, A. W. Thomas, Phys. Rev. **D62**, 074505 (2000). [hep-lat/9912052].
2. D. B. Leinweber, S. Boinepalli, I. C. Cloet *et al.*, Phys. Rev. Lett. **94**, 212001 (2005). [hep-lat/0406002].
3. D. B. Leinweber, S. Boinepalli, A. W. Thomas *et al.*, Phys. Rev. Lett. **97**, 022001 (2006). [hep-lat/0601025].
4. T. Doi, M. Deka, S. -J. Dong *et al.*, [arXiv:0910.2687 [hep-lat]].
5. D. Androic *et al.* [G0 Collaboration], Phys. Rev. Lett. **104**, 012001 (2010). [arXiv:0909.5107 [nucl-ex]].
6. A. Acha *et al.* [HAPPEX Collaboration], Phys. Rev. Lett. **98**, 032301 (2007). [nucl-ex/0609002].
7. K. A. Aniol *et al.* [HAPPEX Collaboration], Phys. Rev. Lett. **96**, 022003 (2006). [nucl-ex/0506010].
8. F. E. Maas *et al.* [A4 Collaboration], Phys. Rev. Lett. **93**, 022002 (2004). [nucl-ex/0401019].
9. R. D. Young, J. Roche, R. D. Carlini *et al.*, Phys. Rev. Lett. **97**, 102002 (2006). [nucl-ex/0604010].
10. A. W. Thomas, R. D. Young, D. B. Leinweber, [nucl-th/0509082].
11. R. D. Young, D. B. Leinweber, A. W. Thomas, Prog. Part. Nucl. Phys. **50**, 399-417 (2003). [hep-lat/0212031
12. R. D. Young, A. W. Thomas, Phys. Rev. **D81**, 014503 (2010). [arXiv:0901.3310 [hep-lat]].
13. D. Toussaint *et al.* [MILC Collaboration], Phys. Rev. Lett. **103**, 122002 (2009). [arXiv:0905.2432 [hep-lat]].
14. H. Ohki, H. Fukaya, S. Hashimoto *et al.*, Phys. Rev. **D78**, 054502 (2008). [arXiv:0806.4744 [hep-lat]].
15. J. Giedt, A. W. Thomas, R. D. Young, Phys. Rev. Lett. **103**, 201802 (2009). [arXiv:0907.4177 [hep-ph]].
16. G. P. Zeller *et al.* [NuTeV Collaboration], Phys. Rev. Lett. **88**, 091802 (2002). [hep-ex/0110059].
17. S. Alekhin, S. A. Kulagin, R. Petti, Phys. Lett. **B675**, 433-440 (2009). [arXiv:0812.4448 [hep-ph]].
18. A. I. Signal, A. W. Thomas, Phys. Lett. **B191**, 205 (1987).
19. M. Deka, T. Streuer, T. Doi *et al.*, Phys. Rev. **D79**, 094502 (2009). [arXiv:0811.1779 [hep-ph]].
20. A. Adare *et al.* [PHENIX Collaboration], Phys. Rev. Lett. **103**, 012003 (2009). [arXiv:0810.0694 [hep-ex]].
21. B. I. Abelev *et al.* [STAR Collaboration], Phys. Rev. Lett. **100**, 232003 (2008). [arXiv:0710.2048 [hep-ex]].
22. F. Myhrer, A. W. Thomas, Phys. Lett. **B663**, 302-305 (2008). [arXiv:0709.4067 [hep-ph]].
23. A. W. Thomas, Phys. Rev. Lett. **101**, 102003 (2008). [arXiv:0803.2775 [hep-ph]].
24. M. Wakamatsu, Phys. Rev. **D81**, 114010 (2010). [arXiv:1004.0268 [hep-ph]].
25. X. -S. Chen, W. -M. Sun, X. -F. Lu *et al.*, Phys. Rev. Lett. **103**, 062001 (2009). [arXiv:0904.0321 [hep-ph]].
26. A. V. Belitsky, A. V. Radyushkin, Phys. Rept. **418**, 1-387 (2005). [hep-ph/0504030].
27. A. Afanasev, M. Anselmino, H. Avakian *et al.*, [hep-ph/0703288 [HEP-PH]].

28. J. D. Bratt *et al.* [LHPC Collaboration], [arXiv:1001.3620 [hep-lat]].
29. A. W. Thomas, ACasey, H. H. Matevosyan, Int. J. Mod. Phys. **A25**, 4149-4162 (2010).
30. P. Wang, A. W. Thomas, Phys. Rev. **D81**, 114015 (2010). [arXiv:1003.0957 [hep-ph]].
31. S. R. Beane, W. Detmold, K. Orginos *et al.*, [arXiv:1004.2935 [hep-lat]].
32. T. Hatsuda [HAL QCD Collaboration], PoS **CD09**, 068 (2009). [arXiv:0909.5637 [hep-lat]].
33. P. A. M. Guichon, K. Saito, E. N. Rodionov *et al.*, Nucl. Phys. **A601**, 349-379 (1996). [nucl-th/9509034].
34. K. Saito, K. Tsushima, A. W. Thomas, Prog. Part. Nucl. Phys. **58** (2007) 1-167. [hep-ph/0506314].
35. P. A. M. Guichon, H. H. Matevosyan, N. Sandulescu *et al.*, Nucl. Phys. **A772**, 1-19 (2006). [nucl-th/0603044].
36. P. A. M. Guichon, A. W. Thomas, Phys. Rev. Lett. **93**, 132502 (2004). [nucl-th/0402064].
37. P. A. M. Guichon, A. W. Thomas, K. Tsushima, Nucl. Phys. **A814**, 66-73 (2008). [arXiv:0712.1925 [nucl-th]].
38. G. Krein, A. W. Thomas, K. Tsushima, "J/Ψ mass shift in nuclear matter," [arXiv:1007.2220 [nucl-th]].
39. K. Yokokawa, S. Sasaki, T. Hatsuda *et al.*, Phys. Rev. **D74**, 034504 (2006). [hep-lat/0605009].
40. I. C. Cloet, W. Bentz, A. W. Thomas, Phys. Lett. **B642**, 210-217 (2006). [nucl-th/0605061].
41. I. C. Cloet, W. Bentz, A. W. Thomas, Phys. Rev. Lett. **102**, 252301 (2009). [arXiv:0901.3559 [nucl-th]].
42. W. Bentz, I. C. Cloet, J. T. Londergan *et al.*, Phys. Lett. **B693**, 462-466 (2010). [arXiv:0908.3198 [nucl-th]].
43. A. M. Sandorfi, S. Hoblit, H. Kamano *et al.*, [arXiv:0912.3505 [nucl-th]].
44. T. S. H. Lee, Int. J. Mod. Phys. **E18**, 1215-1225 (2009).
45. S. X. Nakamura, H. Kamano, T. -S. H. Lee *et al.*, [arXiv:1008.0214 [nucl-th]].
46. J. J. Dudek, R. Edwards, C. E. Thomas, Phys. Rev. **D79**, 094504 (2009). [arXiv:0902.2241 [hep-ph]].
47. J. J. Dudek, R. G. Edwards, M. J. Peardon *et al.*, Phys. Rev. Lett. **103**, 262001 (2009). [arXiv:0909.0200 [hep-ph]].
48. K. Tsushima, P. A. M. Guichon, R. Shyam *et al.*, [arXiv:0903.5478 [nucl-th]].

Hadron Physics at FAIR

Ulrich Wiedner[a]

[a]*Institut f. Experimentalphysik I, Ruhr-University Bochum*
D-44780 Bochum, Germany

Abstract. The new FAIR facility in Darmstadt has a broad program in the field of hadron and nuclear physics utilizing ion beams with unprecedented intensity and accuracy. The hadron physics program centers around the the high-energy storage ring HESR for antiprotons and the PANDA experiment that is integrated in it. The physics program includes among others topics like hadron spectroscopy in the charmonium mass region and below, hyperon physics, electromagnetic processes and charm in nuclei.

Keywords: *hadron spectroscopy, antiprotons, glueballs, hybrids, charmonium, formfactors*

PACS: 14.40.Lb, 14.20.Jn, 14.40.Rt, 14.20.Dh, 14.40.Be

THE NEW FAIR FACILITY

FAIR (Facility for Antiproton and Ion Research) was established as an international company on October 4th, 2010. It is set-up next to the existing GSI facility in Darmstadt, Germany. FAIR has a broad research program in hadron physics, nuclear structure physics, heavy-ion physics and atomic and plasma physics. This will be achieved by utilizing high-precision and high-intensity primary and secondary beams of stable and unstable ions with unprecedented quality. Particles from antiprotons up to uranium will be accelerated in an arrangement of synchrotrons using rapidly cycling superconducting magnets. FAIR will be built in several modules and the first experiments will be installed at FAIR in the year 2016. More than 3000 users from all around the world will finally be users of FAIR.

Hadron physics will mainly be done with antiprotons at FAIR. 30 GeV protons from the SIS100 are used to produce antiprotons that are subsequently collected, stored and cooled in two smaller storage rings. Those pre-cooled antiprotons are then transferred via the SIS100 into a dedicated antiproton storage ring called HESR. The HESR is equipped with one internal target station. The HESR will provide antiprotons with momenta between 1.5 and 15 GeV/c, giving a center-of-mass energy of up to 5.5 GeV. The beam will be stochastically cooled over the whole momentum range, giving a beam momentum spread of $\Delta p/p \sim 10^{-4}$. For the high-precision charmonium spectroscopy, additional high-energy electron cooling for momenta up to 8 GeV/c is foreseen. The momentum spread with this additional electron cooling is $\Delta p/p \sim 10^{-5}$ in a high-resolution mode of the storage ring operation. The antiproton program is only part of the whole proposed FAIR complex. However, the accelerator complex is optimized for a maximum parallel operation between the antiproton physics program and the other physics programs, to make best use of the facility.

12th International Conference on Meson-Nucleon Physics and the Structure of the Nucleon (MENU 2010)
AIP Conf. Proc. 1374, 151-156 (2011); doi: 10.1063/1.3647114

HADRON PHYSICS AT FAIR

Introduction

In particle physics, the so-called Standard Model has been a tremendous success. Quarks and leptons are considered to be the elementary building blocks of matter. Matter thus manifests itself in a hierarchy of composite systems. The stability of each system is determined by the interaction among the respective constituents. Hadron Physics is the physics of strongly interacting systems. In the past thirty years quantum chromodynamics (QCD) has been established as the theory of strong interactions that describes the interaction among the fundamental quarks through the exchange of gluons. QCD is believed to be well understood at short distance scales, but this is no longer the case as soon as the basic quark-gluon coupling is no longer weak. This region of strong QCD is governed by non-perturbative phenomena leading, e.g., to the formations of hadrons. The underlying processes like confinement and chiral symmetry breaking are however not very well understood. This presents a profound intellectual challenge for both experimentalists and theorists. Experiments with antiprotons have proven in the past to be a rich source of information in hadron physics. With the new proposed high-energy storage ring HESR for antiprotons at the future FAIR facility in Darmstadt, the physics of strange and charm quarks studied by antiproton annihilations will be accessible for hadron physics. This region is exactly the transition region between the perturbative QCD at short scales and strong QCD.

The PANDA experiment is designed to explore a broad physics program utilizing the cooled antiproton beams of the HESR with momenta up to 15 GeV/c. Some of the highlight research topics are presented in this talk:

- the study of gluonic degrees of freedom, e.g., hybrids and glueballs.
- spectroscopy of states containing charm quarks
- nucleon spectroscopy with electromagnetic processes
- spectroscopy of hypernuclei and double hypernuclei

Gluonic degrees of freedom

Hadron spectroscopy is the basis that inspired the SU(3) quark model and QCD as the generally accepted theory of strong interactions. QCD is a non-Abelian gauge theory and as a consequence, the gauge bosons, the gluons, can interact with each other. Therefore QCD predicts the existence of bound states of gluons, called glueballs (gg, ggg). Other types of hadronic matter in which gluons contribute to the overall quantum numbers, called hybrids ($\bar{q}qg$), could also exist. The gluonic excitation in a hybrid leads to new J^{PC} quantum numbers for this state, where J denotes the total angular momentum of the resonance, P its parity and C its charge conjugation. Some J^{PC} combinations cannot be formed by the fermion-antifermion system $\bar{q}q$, so their observation would be the cleanest experimental evidence for a non- $\bar{q}qg$ state. In any case, the precise measurement of the properties of several glueball or hybrid states compared to $\bar{q}qg$ mesons would help us understand QCD in the non-perturbative regime. Even though, glueballs have been mentioned in the literature for a long time, their nature is completely unknown. It is e.g. not clear how much perturbative gluons play a role or if the structure of a glueball is a closed flux of color with twists and knots in it [1].

Why is it so important to find the gluonic degrees of freedom? The elementary particles of the Standard Model gain their mass through the Higgs mechanism. However, only a few percent of the mass of the proton is due to the Higgs mechanism. The rest is created in an unknown way by the strong interaction. Glueballs would be massless without the strong interaction and their predicted masses arise solely from the strong interaction. The possibility to study a whole spectrum of glueballs might therefore be the key to understanding the mass creation by the strong interaction. What are the results in this field so far?

Several reactions are considered to be gluon-rich. The most prominent are radiative J/ψ decays, central production processes, and antiproton-proton annihilation. Because of the existence of LEAR at CERN, antinucleon-nucleon ($N\bar{N}$) annihilation data now dominate.

The study of $\bar{p}p$ annihilation has been underway for the past thirty years. Several bubble chamber experiments at CERN and BNL first investigated this topic. In 1983, with the low energy antiproton ring (LEAR) at CERN, a unique facility for antiproton physics came into operation. Until its closure at the end of 1996, LEAR provided pure and high-intensity antiproton beams ($2 \cdot 10^6$ $\bar{p}$/s) in the momentum range between 60 and 1940 MeV/c with a small momentum spread of $\Delta p/p \sim 10^{-3}$. The second generation of LEAR experiments were comprised of high-statistics 4π experiments for charged and neutral particles. This turned out to be the key to finding gluonic degrees of freedom. Not only was the most promising glueball groundstate candidate ($J^{PC}=0^{++}$), the $f_0(1500)$, discovered in $\bar{p}p$ annihilations [2], but also two states with exotic quantum numbers ($J^{PC}=1^{-+}$) are unambiguously seen [3,4,5]. Because their quantum numbers are exotic, those two particles cannot be ordinary mesons. What is especially striking in $\bar{p}p$ annihilations is the fact that these exotic particles are produced with the same strength as ordinary mesons. It therefore seems that the gluon richness of the annihilation process makes $\bar{p}p$ annihilations the prime search ground for gluonic excitations. Unfortunately the energy range of the LEAR machine was limited and therefore, e.g., higher-mass glueballs were out of reach. The majority of them are predicted by lattice calculations to have a mass between 3 and 5 GeV/c^2 [6]. Because of this high mass, most of them are not accessible for radiative J/ψ decays and central-production processes, and a new antiproton machine seems to be the only chance to find them and study their properties.

A few years ago the region above the open-charm threshold was mainly "terra incognita". Very few resonances were known from e^+e^- colliders and resonances were expected to be mainly broad since their decay into open charm is feasible. Consequently it came as a big surprise when mainly the BELLE and BaBar experiments at the B-factories discovered several states not fitting into the pattern. Nowadays they are referred to as X, Y, Z states. In most cases these states are associated with charmonium since the decay products contain charm quarks but their classification is far from being obvious. It is unexplained why, despite being above the open charm threshold, the strong decay mechanism into open charm is suppressed and the states decay rather into a charmonium ground state and light mesons. If some of them were hybrids this could be an indication of the long lifetime and small width of gluonic excitations. The Z^+ particles [7] must be multiquark states containing two lighter quarks together with the charm-anticharm quarks, so undoubtedly new degrees of freedom in the strong interaction have started to show. Current experiments cannot add much more information because the observed particles are produced in a decay chain and the knowledge of their properties is largely limited by detector resolution. Where tested, these states couple strongly to antiproton-proton annihilations as results from LEAR experiments but also from high-energy Fermilab experiments show. In most cases a limited detector resolution will not be an issue at FAIR due to the possibility to directly scan the resonances with a high-precision antiproton beam allowing PANDA to clarify the nature of the X, Y and Z states.

The appearance of additional narrow states in the charmonium mass region confirms the assumption that the identification of additional states is easier than in the light-quark sector due to reduced mixing. Mixing with conventional mesons is not at all an issue for hybrids with exotic quantum numbers, i.e. quantum numbers that could not be accommodated by a quark-antiquark pair. Several of the lowest-mass charmonium hybrids are predicted to have exotic quantum numbers. While usually the process of a complicated partial wave analysis leaves some ambiguities concerning the identification of an exotic nature, it can be easily established in antiproton-proton annihilations. In a production experiment, where the particle is usually produced together with a light meson, exotic quantum numbers are allowed for the unknown state. A subsequent formation experiment, with the center-of-mass energy of the experiment being exactly the particle mass should then allow to scan its properties. The exceptions are particles with exotic quantum numbers which cannot be reached in a formation experiment. Therefore the appearance of a resonance in a production experiment and its absence in a subsequent scan shows immediately its exotic nature.

The performance of the PANDA detector to detect these has been carefully studied, e.g. in the case of the Y(3940), which has been seen only in the unusual decay mode into J/ψ+ω [8]. Figure 1a shows the Monte Carlo simulation of the signal in PANDA, which is very clean. Background suppression on the level of better than 10^8 works superb by using the signal from the high-energetic electrons in the decay of the J/ψ(Fig. 1b).

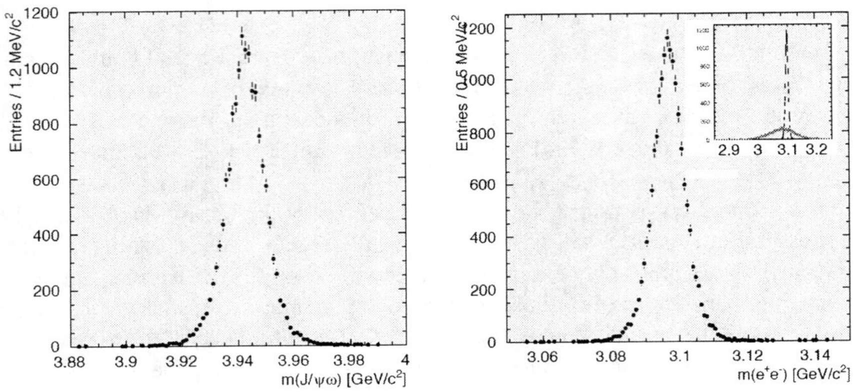

Fig. 1: a) left: the invariant J/ψω mass distribution obtained in PANDA MC studies b) right: the invariant e^+e^- mass spectrum before (inset) and after kinematic fitting.

An excellent electromagnetic calorimeter, as the one foreseen for PANDA, is mandatory for that.

But back to the question of glueballs. Glueballs are clearly among the most interesting particles in any interaction and at the very center of QCD. The vast majority of glueballs appear at masses overlapping with the charmonium spectrum. The above argument for a lesser mixing with conventional mesons in the charmonium energy region applies here also and glueballs may indeed be rather narrow states. The structure of a glueball is completely unknown and spectroscopy experiments may indeed be the only tool to reveal it. The knowledge then gained in this field in strong interactions might spill over to other fields and could e.g. influence developments of models of gravitation.

Spectroscopy of states containing charm quarks

The best understanding of charmonium has been achieved for the ψ states. These can be formed directly at electron-positron colliders. With an antiproton beam, however, charmonium states of all quantum numbers can be formed directly. A scan with varying beam energy through the charmonium resonance allows its precise width determination. The precision of the mass and width measurement depends solely on the beam quality, putting less demands on the detector resolution. This has been demonstrated by the excellent results achieved by the Fermilab experiments E760/835.

Potential models for $q\bar{q}$ interactions are tuned to the results of charmonium spectroscopy. The Fermilab experiments E760 and E835 showed that cooled antiproton beams were extremely well suited to do precision charmonium spectroscopy. In fact, the vast majority of high-precision charmonium data were measured in $\bar{p}p$ reactions [9]. Since the E835 experiment at Fermilab was shut down for good many important questions remain unanswered. Among them are:

- the precise mass and the width of the ground state η_c.
- the precise mass and properties of the first radial excitation of the η_c.
- the precise mass and the properties of the h_c resonance of the charmonium system
- radiative transitions of the χ_J charmonium states
- missing charmonium states above the $D\bar{D}$ threshold.

Why are antiproton annihilations so particularly well suited to study charmonium compared to other production modes? Antiproton annihilations allow formation experiments for all charmonium states. In particular charmonium states with quantum numbers different from $J^{PC}=1^{--}$ are best studied with antiproton annihilations, while those with $J^{PC}=1^{--}$ states will be studied intensively with the new BES III experiment.

Open charm channels are accessible via the formation of a charmonium state above the relevant open-charm threshold and the subsequent decay of this resonance into open charm. A recent calculation of the open-charm production cross section in antiproton annihilations [10] predicts rates for PANDA, which would put weak processes that serve as a test for physics beyond the Standard Model into range. This is currently under investigation by the PANDA collaboration.

Exploring the nucleon structure with electromagnetic processes.

An interesting theoretical framework for hard exclusive processes in lepton scattering experiments was developed with so-called Generalized Parton Distributions (GPDs). In $\bar{p}p$ annihilation processes channels with crossed kinematics can be accessed and should give good insights to the universal applicability of this approach.

Drell-Yan pairs, clearly detected in PANDA, allow determining all those structure functions precisely, that can be studied in the absence of polarization.

Finally PANDA will measure the electromagnetic form factors in the time-like region with unprecedented accuracy over the biggest Q^2 range for a single experiment. For the first time, a separate determination of the electric and magnetic form factors are within reach. Again, only the combination of the high-intensity antiproton beam with the envisaged outstanding detector will allow this.

Hypernuclear Spectroscopy

A hypernucleus contains one or more hyperons implanted into the nuclear medium and therefore adds a third dimension (strangeness) to the nuclear chart. Previous experiments in this field suffered from very low statistics and the limited detector resolution of the magnetic spectrometers used. But precisely one of the goals of hypernuclear physics is to determine as accurately as possible the level spectra and decay properties of (multi)strange hypernuclei. The combination of a high-luminosity antiproton storage ring with a modern detector system, consisting of novel solid-state micro trackers in connection with a high-resolution γ spectrometer, will remove the present obstacles to hypernuclear spectroscopy.

The experimental concept foresees the production of baryon-antibaryon pairs inside nuclear matter. The antibaryon (e.g., a $\bar{\Xi}$) could serve as a trigger for the reaction while the baryon (in this case a Ξ) is slowed down inside the nucleus and subsequently absorbed in a secondary active target. This capture reaction will then lead to the production of $\Lambda\Lambda$-hypernuclei. A modern highly segmented Ge-array detector will surround this target and allow high-precision spectroscopy of these double hypernuclei. The expected rates for the HESR experiments are approximately 300000 stopped Ξ per day, and with the typical γ branching ratios and detection efficiencies, one will observe several hundred γ transitions per day.

The PANDA Detector

Most of the experimental program will be done with a general-purpose detector, which is currently being designed by the PANDA collaboration consisting of 54 institutions worldwide.

To achieve the physics aims, the detector needs to cover the full solid angle. Good particle identification and excellent energy and angular resolution for charged particles and photons are mandatory. Charm particles decay often lead to di-lepton pairs. Thus, muon detection capabilities and a highly-segmented low-threshold electromagnetic calorimeter are important to tag and precisely reconstruct hidden-charm and to reduce background. Good vertex recognition and particle identification for charged kaons from very low energies up to a few GeV are required to reconstruct light hadronic and open-charm final states. At the same time, the detector must withstand large radiation dosage from hadrons emitted by the spallation process when using nuclear targets. These spallation products include neutrons down to thermal energies, which contribute most.

The planned PANDA detector is subdivided into two parts: i) a target spectrometer with a solenoid magnet surrounding the interaction region and ii) a forward spectrometer with a large-acceptance dipole magnet. The dipole magnet in forward direction bends as well the HESR beam, allowing for an electromagnetic calorimeter to be placed in 0° direction. Only this combination of two spectrometers therefore allows a full angular coverage and takes into account the wide range of energies. At the same time it still has sufficient flexibility, so that individual components can be exchanged or added for specific experiments, e.g. for the experiments with hypernuclei or for the special needs of CP violation studies.

The internal target, which could be a pellet target of frozen hydrogen droplets, a gas jet target or a wire used as nuclear target, is surrounded by Si-pixel detectors in the vertex region. The main vertex tracking further out is done with straw chambers or a high-rate TPC and mini drift chambers. Ring imaging Cherenkov counters will provide the particle identification. The proposed electromagnetic calorimeter is an arrangement of $PbWO_4$ crystals of the newest generation of this material, read out by avalanche photodiodes and vacuum phototriodes or tetrodes. The superconducting solenoid provides a field of 2 T. Particles emitted with polar angles below 10° in the horizontal and 5° in the vertical direction are measured with the help of a 1 m gap dipole. Mini drift chambers will be located before and behind the dipole for tracking. Particle identification will be obtained by a TOF-Stop detector and a dual-radiator RICH detector. Behind this there is a shashlyk electromagnetic calorimeter and a hadronic calorimeter followed by a muon detection system.

ACKNOWLEDGMENTS

This work is supported by grants from the German BMBF and the GSI and Jülich Helmholtz Centers.

REFERENCES

1. L. Faddeev, A. Niemi and U. Wiedner; Phys.Rev.D70:114033, 2004
2. Crystal Barrel Collaboration, C. Amsler et al., Phys. Lett. B 342, 433 (1995).
3. Crystal Barrel Collaboration, A. Abele et al., Phys. Lett. B446, 349 (1999).
4. Crystal Barrel Collaboration, A. Abele et al., Phys. Lett. B423, 175 (1998).
5. Crystal Barrel Collaboration, J. Reinnarth et al., Nucl. Phys. A692, 268c (2001).
6. C.J. Morningstar, M. Peardon, Phys. Rev. D60, 034509 (1999).
7. Belle Collaboration; PRL 100, 142001 (2008) arXiv:0708.1790 [hep-ex]
8. Belle Collaboration; PRL 94 (2005) 182002.
9. Review of Particle Properties, J. Phys. G 37, 075021 (2010).
10. A.Titov, B. Kaempfer; Phys. Rev. C 78, 025201 (2008).

PARALLEL SESSION CONTRIBUTIONS

Session A: Few Body

Elastic electron-deuteron scattering in the light-front Hamiltonian dynamics

Giovanni Salmè*, Tobias Frederico†, Adnei Marinho** and Emanuele Pace‡

*Istituto Nazionale di Fisica Nucleare, Sezione di Roma, P.le A. Moro 2, I-00185 Rome, Italy
†Instituto Tecnológico de Aeronáutica, DCTA, 12.228-900, São José dos Campos, SP, Brazil
**Istituto Nazionale di Fisica Nucleare, Sezione di Roma, P.le A. Moro 2, I-00185 Rome, Italy
‡Phys. Dep. "Tor Vergata" University and INFN Sezione di Tor Vergata, Via della Ricerca Scientifica 1, I-00133 Rome, Italy

Abstract. Within the Light-Front Hamiltonian Dynamics, we are extending our investigation of the deuteron electromagnetic properties by adding new, dynamical two-body terms to the one-body current operator, introduced in previous works. These two-body contributions are generated in a four-dimensional Yukawa model (two fermions interacting through a pion) from an analysis of the Light-Front-projected current that acts on the three-dimensional valence component and fulfills the Ward-Takahashi identity. Preliminary results for the magnetic moment and the non valence probability are shown.

Keywords: Deuteron, Poincaré invariance, Relativistic Hamiltonian Dynamics, Elastic electron scattering
PACS: 27.10.+h, 11.20.Cp,25.30.Bf

INTRODUCTION

Our aim is to describe the deuteron electromagnetic (em) observables with nucleon and meson degrees of freedom in a Light-Front Hamiltonian Dynamics (LFHD) framework (see, e.g., [1, 2] for a detailed presentation of LFHD) and, at the same time, to satisfy the extended Poincaré covariance, Hermiticity and current conservation [3]. Achieving such an accurate description could help us to robustly single out signatures of non standard effects (like the 6-quark bag [4, 5, 6]) representing possible targets of planned experimental efforts [7]. Moreover, the deuteron em observables can represent an excellent benchmark for different relativistic approaches (see, e.g., [6]). This work extends the investigation of Refs. [8, 9], where an em current operator was considered built up from the one-body current, by including two-body terms with a dynamical nature, generated by a one-pion exchange interaction [10].

An appealing feature of LFHD is the clean separation of the center of mass motion from the intrinsic one, as in the non relativistic case. Furthermore, combining LFHD with the Bakamjian-Thomas construction [11] of the Poincaré generators allows one to straightforwardly adopt the Schrödinger equation of a two-body system as the relativistic square mass operator of that system [12, 1], i.e

$$\frac{k^2}{m} + V(k^2) \to 4(m^2 + k^2) + 4mV(k^2)$$

This observation, together with the equivalence between non relativistic and LFHD analyses of the nucleon-nucleon phase shifts, makes it possible to embed the standard *non relativistic* deuteron wave function into a fully Poincaré covariant description, but properly treating the LF spin with the Melosh rotations [9]. In our approach, the choice of the reference frame is an important ingredient, since this allows our investigation to address new physical processes. Clearly, if the complete and calculable theory were at our disposal, the choice of the frame should not represent an issue, given the full covariance. Actually one is dealing with an approximate scheme, and therefore the choice of the reference frame, where the constrains for preserving the general properties can be more easily implemented, becomes strategic. In a Breit frame where the momentum transfer is longitudinal, i.e. $\vec{q}_\perp = 0$, the extended Poincaré covariance imposes to the current operator only a rotational covariance around the $\hat{z} \equiv \hat{q}$ -axis [3]. Then, any operator fulfilling the rotational covariance around the $z-axis$, as the one-body Dirac and Pauli currents, can be adopted for constructing a current operator. In particular the operator i) depends parametrically upon the CM momentum and ii) acts on the intrinsic variables. It is worth noting that in order to implement Hermiticity, a term that contains the dynamical generator of the transverse rotations is needed. Then, it is generated a two-body contribution to the current [3]. but such a contribution is distinct from the two-body contributions that we are going to discuss in what follows,

12th International Conference on Meson-Nucleon Physics and the Structure of the Nucleon (MENU 2010)
AIP Conf. Proc. 1374, 159-162 (2011); doi: 10.1063/1.3647115

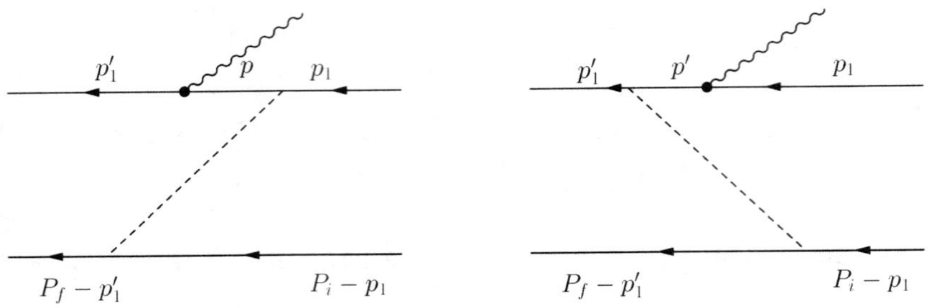

FIGURE 1. Diagrams for the first-order "interaction terms" of the 3D LF current, obtained within a Yukawa model in ladder approximation. The LF time flows from right to left . (Adapted from Ref.[13]).

where explicit dynamical effects are present. Our new, dynamical two-body currents are inspired by an exact analysis of the four-dimensional (4D) current corresponding to a Yukawa model in ladder approximation, where two fermions interact through a pseudoscalar, massive boson [13] (within our model there is no photon-boson coupling).

Let us sketch the framework within which the present 3D two-body currents have been devised. In Refs. [14, 15, 16, 13], by using two ingredients: the projection of the 4D quantities onto the 3D LF hyperplane, and the Quasi-Potential approach for the Transition Matrix [17], it has been established an exact, one-to-one correspondence between the 4D Bethe-Salpeter amplitude and the 3D LF *valence* wave function. From this relation, one can demonstrate the correspondence between the matrix elements of the 4D em current and the ones of the LF current, both fulfilling the proper Ward-Takahashi identity (WTI). This procedure allows one to define a 3D LF current operator that acts on the LF valence wave function. It is worth noting that the LF valence wave function is the eigensolution of a 3D dynamical equation, with an effective interaction determined by the 4D kernel of the Bethe-Salpeter equation. The formal correspondence between 4D and 3D quantities involves the 4D interaction and in order to have a workable approach, one has to resort to some approximation scheme. Adopting the Fock expansion of quantities like the effective interaction and the current operator onto the LF hyperplane, one is able to develop approximations, that can be improved order by order. An interesting feature of the approach of Ref. [13] is that the current operator fulfills the proper WTI, at any given order. Clearly, in a fully field-theoretical approach, namely with an infinite number of degrees of freedom, only the kinematical symmetries can be satisfied if a truncated Fock space is adopted. But, if one uses a finite number of degrees of freedom, and the Bakamjian-Thomas construction of the Poincaré generators, this approximation scheme can be made fully Poincaré covariant. In particular, if the truncated current operator satisfies the covariance around the z-axis in the Breit frame where $\mathbf{q}_\perp = 0$, then, following [3], the Poincaré covariance can be implemented.

In our present study, we are considering only the first-order LF current operator, that contains the contributions shown in Figs. 1, 2 and 3.

PRELIMINARY RESULTS FOR THE DEUTERON MAGNETIC MOMENT

In a Breit frame with $\mathbf{q}_\perp = 0$ ($\Rightarrow q^+ \neq 0$), the current operator for an interacting system can be defined in terms of the LF current, $\mathscr{J}^\mu(q\hat{e}_z)$, as follows

$$j^\mu(q\hat{e}_z) = \{\mathscr{J}^\mu(q\hat{e}_z) + L^\mu_\nu[r_x(-\pi)]\, e^{\imath\pi S_x}\, \mathscr{J}^\nu(q\hat{e}_z)^* e^{-\imath\pi S_x}\}/2 \tag{1}$$

where $L^\mu_\nu[r_x(-\pi)]$ is the 4D representation of the π-rotation around the x-axis and $\widehat{\mathbf{S}}_\perp$ the dynamical rotation generator (hat labels operator). The second term in Eq. (1) is necessary to implement Hermiticity [9]. In Eq. (1), the operator $\mathscr{J}^\mu(q\hat{e}_z)$ is given by

$$\mathscr{J}^\mu(q\hat{e}_z) = j^\mu_{one} + j^\mu_{two} = \overline{\Pi}_0 \mathscr{J}^\mu_0 \Pi_0 + \overline{\Pi}_0 \left[V\Delta_0 \mathscr{J}^\mu_0 + \mathscr{J}^\mu_0 \Delta_0 V\right] \Pi_0 \tag{2}$$

where Π_0 is the *free reverse LF projector* [13] and $\Delta_0 \equiv G_{free} - G_{glob}$, with G_{free} the free two-body propagator and G_{glob} an auxiliary Green's function (that takes into account the global-time propagation). The operator V is the pion mediated interaction [10]

$$V = i\, g^2\, \vec{\tau}_1 \cdot \vec{\tau}_2 \left[\gamma_1^5 \otimes \gamma_2^5\, \mathscr{F}^2[(\hat{p}_2 - \hat{p}_1)^2]\right] / \left[(\hat{p}_2 - \hat{p}_1)^2 - m_\pi^2 + i\varepsilon\right] \tag{3}$$

with $\mathscr{F}[(\hat{p}_2 - \hat{p}_1)^2]$ the 4D vertex counterpart of the nonrelativistic vertex function of Ref. [18]. The one-body contribution, j^{μ}_{one}, (see [8, 9] for details) contains Dirac, $F_{1N}[(\hat{p}' - \hat{p})^2]$, and Pauli, $F_{2N}[(\hat{p}' - \hat{p})^2]$, form factors (ff) $(N = p, n)$.

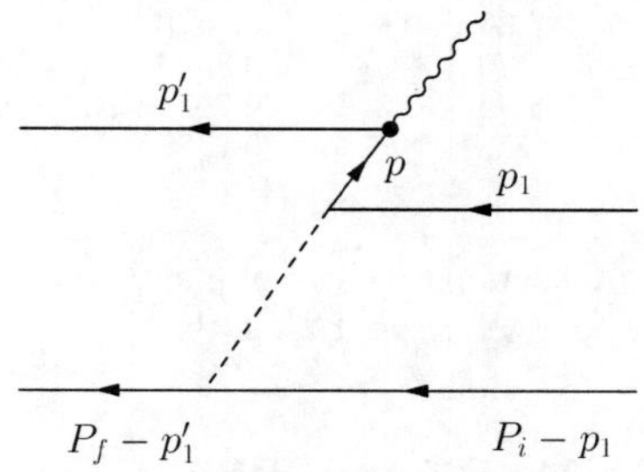

FIGURE 2. The same as in Fig. 1, but for the "pair term"

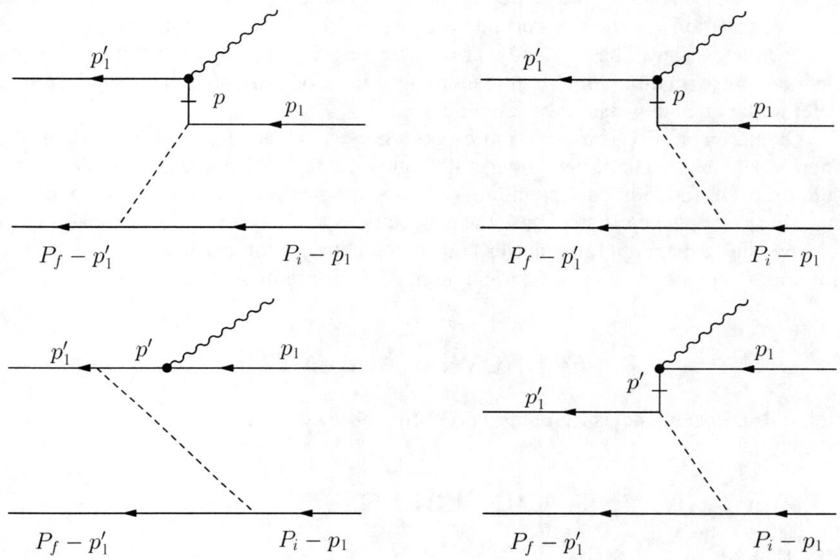

FIGURE 3. The same as in Fig. 1, but for the "instantaneous terms".

If WTI is fulfilled, the current is conserved, once matrix elements are taken between eigensolutions of the mass equation constructed from the proper Green's function. Although in our phenomenological calculations we are adopting deuteron wave functions corresponding to realistic interactions [18, 19, 20], current conservation is satisfied, since in elastic processes it follows from extended Poincaré covariance and Hermiticity [9]. The charge normalization can be fulfilled if one defines $\langle P_i, d|\mathscr{J}^-(0)|d; P_i\rangle = \langle P_i, d|\mathscr{J}^+(0)|d; P_i\rangle$. This leads to assume that such an equality holds for any momentum transfer. Remarkably, for evaluating the em observables only $j^+(q\hat{e}_z)$ and $j^{1(2)}(q\hat{e}_z)$ are relevant.

Preliminary results for the magnetic moment of the deuteron have been obtained by retaining only the two-body interaction terms corresponding to the diagrams depicted in Fig. 1. In the actual calculations, the nucleon ff of Ref. [21] are adopted. These ff are in nice agreement with the most recent measurements [22] and give a microscopical interpretation of the possible zero in the ratio $\mu_p G^p_E/G^p_M$ for $Q^2 > 8\ (GeV/c)^2$ in terms of interference between valence and nonvalence components of the proton state. It is worth noting that the behavior of the nucleon ff, in a wide kinematical region, enters in the evaluation of the static $(Q^2 \to 0)$ em properties in our approach [8, 9]. In Table 1, the preliminary values, for different deuteron wave functions corresponding to the CD-Bonn [18], RSC93 [19] and AV18 [20] interactions, are shown. The results corresponding to the one-body contribution with the Hermiticity term [8, 9] are also shown. The results appear encouraging. An accurate study of the numerical convergence of the multifold integrals for a vanishing momentum transfer is still necessary (see the analogous discussion for the one-body

TABLE 1. Deuteron magnetic moment for realistic NN interactions [Exp. 0.857406(1)]. Preliminary results.

Interaction	P_D	μ_D^{NR}	μ_{one}^{LFD}	μ_{1+2}^{LFD}
CD-Bonn	4.83	0.8523	0.8670	0.863 ± 0.004
RSC93	5.70	0.8473	0.8637	0.861 ± 0.004
Av18	5.76	0.8470	0.8635	0.860 ± 0.004

case in Refs. [8, 9]). From the charge normalization, one can obtain the probabilities of the valence and non valence components. At the present stage we have obtained $Prob_{NV} \sim 0.01$.

SUMMARY & PERSPECTIVES

We have presented preliminary results for the deuteron magnetic moment and non valence probability, including new, dynamical two-body contributions in the em current, within the LFHD approach proposed in Refs. [3, 8, 9]. The approach is fully Poincaré covariant. The two-body terms have been inspired by the analysis, performed in [13], of a Yukawa model in ladder approximation, for two interacting fermions. In particular, in Ref. [13], a LF current fulfilling the WTI at each order in the Fock expansion was obtained.

The analysis of the deuteron em ff is started. It represents a non trivial task from the numerical point of view, given the many, multidimensional integrals to be performed with high accuracy, in particular at low Q^2. First results, obtained by using i) the deuteron wave function corresponding to realistic potentials and ii) the two-body current interaction contribution shown in Fig. 1, appear consistent with the expectations, namely a very low probability for the component beyond the valence one, and a magnetic moment in fair agreement with the experimental value. Calculations of the quadrupole moment, more delicate from the numerical point of view, are in progress.

ACKNOWLEDGMENTS

Work supported by the Brazilian agencies CNPq and FAPESP and by the Italian MIUR.

REFERENCES

1. F. Lev, Riv. N. Cim. **16**, 1 (1993).
2. B. D. Keister and W. N. Polyzou, Adv. Nucl. Phys. **20**, 225 (1991).
3. F.M. Lev, E. Pace and G. Salmè , Nucl. Phys. **A 641**, 229 (1998).
4. C. E. Carlson, J. R. Hiller, R. J. Holt, Annu. Rev. Nucl. Part. Sci. **47**, 395 (1997).
5. S.J. Brodsky, PoS(LHC07)002, and references quoted therein.
6. R. Gilman and F. Gross, J. Phys. G **28**, R37 (2002), and references quoted therein.
7. www.jlab.org/exp_prog/PACpage/pac.html.
8. F. Lev, E. Pace and G. Salmè, Phys. Rev. Lett. **83**, 5250 (1999).
9. F. Lev, E. Pace and G. Salmè, Phys. Rev. C **62**, 064004 (2000).
10. T. Frederico and R.W. Schulze, Phys. Rev. C **54**, 2201 (1996).
11. B. Bakamjian and L. H. Thomas, Phys. Rev. **92**, 1300 (1953).
12. F. Coester, S. C. Pieper, and F. J. D. Serduke, Phys. Rev. C **11**, 1 (1975).
13. T. Frederico, J.A. Marinho, E. Pace, G. Salmè and P. Sauer, Phys. Rev. D **77**, 116010 (2008).
14. J. H. O. Sales, T. Frederico, B. V. Carlson, P. U. Sauer, Phys. Rev. C **61**, 044003 (2000).
15. J. H. O. Sales, T. Frederico, B. V. Carlson, P. U. Sauer, Phys. Rev. C **63**, 064003 (2001).
16. J. A. O. Marinho, T. Frederico, P. U. Sauer, Phys. Rev. D **76**, 096001 (2007).
17. R. M. Woloshyn and A. D.Jackson, Nucl. Phys. B **64**, 269 (1973).
18. R. Machleidt, Phys. Rev. C **63**, 024001 (2001).
19. V.G.J. Stoks, R.A.M. Klomp, C.P.F. Terheggen, J.J. de Swart, Phys. Rev. C **49**, 2950 (1994).
20. R.B. Wiringa,V.G.J. Stoks, R. Schiavilla, Phys. Rev. C **51**, 38 (1995).
21. J.P.B.C. De Melo, T. Frederico, E. Pace, S. Pisano and G. Salmè, Phys. Lett. B **671**, 153 (2009).
22. A.J.R. Puckett et al, Phys. Rev. Lett. **104**, 242301 (2010).

Polarization observables in ^{2}H(e,e'p) at GeV energies

Sabine Jeschonnek* and J. W. Van Orden†

*The Ohio State University, Physics Department, Lima, OH 45804
†Department of Physics, Old Dominion University, Norfolk, VA 23529 and
Jefferson Lab, 12000 Jefferson Avenue, Newport News, VA 23606

Abstract. We discuss results for polarization observables in exclusive electron scattering from deuteron targets. In our relativistic calculation, we use the Gross equation to describe the nuclear ground state, and we use the SAID parametrization of the full nucleon-nucleon scattering amplitude for our description of the final state interactions. We show results for polarized deuteron targets and for polarized ejectile nucleons, and discuss the relevance of the different parts of the nucleon-nucleon scattering amplitude for the final state interactions.

Keywords: electron scattering, final state interactions, polarization observables
PACS: 25.30.Fj, 21.45.Bc, 24.10.Jv

INTRODUCTION

Electron scattering from light nuclei is a great tool to investigate the short-range structure of nuclei, and to study the reaction mechanism. Indeed, when using the nucleus as a laboratory to measure e.g. neutron form factors or search for color transparency, a firm grip on all the details of the reaction mechanism is very important. At the GeV energies relevant to experiments at Jefferson Lab, the final state interactions are expected to be the dominant part of the reaction mechanism. In the calculations presented here, we use a relativistic ground state wave function, a relativistic positive-energy current operator, and all terms - spin-dependent and spin-independent - of the nucleon-nucleon scattering amplitudes to describe the final state interactions (FSIs). For a description of the formalism, see [1]. Our treatment of final state interactions works for any value of the four-momentum transfer Q^2, not just large Q^2. However, at lower momentum transfers, one expects meson exchange currents and isobar contributions to play a significant, though not dominant, role. These effects are not yet included in the calculations presented here. This is why we restrict ourselves to kinematics with larger values of Q^2.

Our calculation differs from many other high quality calculations using the generalized eikonal approximation [2, 3, 4] or a diagrammatic approach [5] by including all the spin-dependent pieces in the nucleon-nucleon amplitude. Full FSIs have recently been included in [6].

In [1], we have shown our calculation and data [7] for the unpolarized cross section measured in Jefferson Lab's Hall A. One main feature of our calculation is that it improves the agreement between theory and data at low missing momentum. In [1], we also have presented calculations of the asymmetry $A_{TL'}$, closely related to the fifth response function, at kinematics similar to the measurement at Jefferson Lab's Hall B [9]. Here, we will focus on results obtained for polarized deuteron targets and polarized ejectile protons.

POLARIZED DEUTERON TARGETS

A number of asymmetries are accessible with polarized deuteron targets and polarized electron beams. Our formalism, see [10] for details, handles arbitrary polarizations and polarization axes for the deuteron with a density matrix formalism coupled with a rotation. The single and double asymmetries for vector and tensor polarizations are defined as

$$A_d^V = \frac{v_L R_L(\widetilde{T}_{10}) + v_T R_T(\widetilde{T}_{10}) + v_{TT} R_{TT}(\widetilde{T}_{10}) + v_{LT} R_{LT}(\widetilde{T}_{10})}{\widetilde{T}_{10}\Sigma}$$

$$A_d^T = \frac{v_L R_L(\widetilde{T}_{20}) + v_T R_T(\widetilde{T}_{20}) + v_{TT} R_{TT}(\widetilde{T}_{20}) + v_{LT} R_{LT}(\widetilde{T}_{20})}{\widetilde{T}_{20}\Sigma}$$

12th International Conference on Meson-Nucleon Physics and the Structure of the Nucleon (MENU 2010)
AIP Conf. Proc. 1374, 163-166 (2011); doi: 10.1063/1.3647116

$$
\begin{aligned}
A_{ed}^{V} &= \frac{v_{LT'}R_{LT'}(\widetilde{T}_{10}) + v_{T'}R_{T'}(\widetilde{T}_{10})}{\widetilde{T}_{10}\Sigma} \\
A_{ed}^{T} &= \frac{v_{LT'}R_{LT'}(\widetilde{T}_{20}) + v_{T'}R_{T'}(\widetilde{T}_{20})}{\widetilde{T}_{20}\Sigma}
\end{aligned} \tag{1}
$$

where $\Sigma = v_L R_L(U) + v_T R_T(U) + v_{TT} R_{TT}(U) + v_{LT} R_{LT}(U)$. With time reversal and parity arguments, we have shown [10] that A_d^V and A_{ed}^T must both vanish in PWIA. For the other two asymmetries, FSIs still play a large role. In Fig. 1, we show our results for all four asymmetries for $Q^2 = 2$ GeV^2 in quasi-free kinematics, $x = 1$, as a function of the missing momentum p_m. In the figure, we show the full on-shell FSI result, the result without double spin-flip contributions in the final state interactions, and the result with the central part of the NN amplitude only. The two asymmetries that vanish in PWIA are clearly more sensitive to the details of the final state interactions. Including spin-dependent pieces is hugely important for these observables, especially at medium and large missing momentum. The other two asymmetries also show some sensitivity to the different parts of the NN amplitude in their peak and dip regions, and at large p_m, but not to the degree of A_d^V and A_{ed}^T. Our investigations show that one may glean very similar information from either pair of asymmetries - A_d^V and A_{ed}^T, or A_{ed}^V and A_d^T. This is of practical importance as measurements on vector polarized targets are easier to perform at Jefferson Lab.

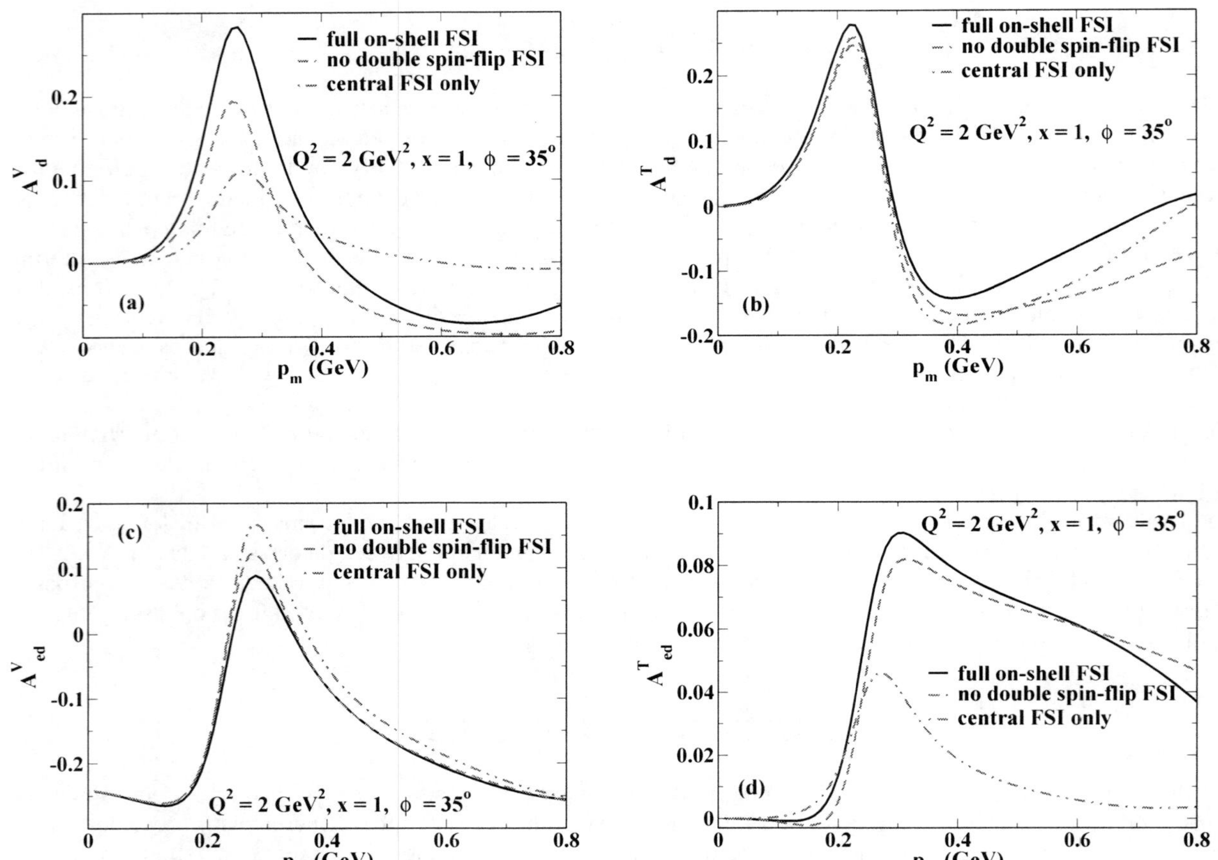

FIGURE 1. The asymmetries A_d^V (panel (a)), A_d^T (panel (b)), A_{ed}^V (panel (c)) and A_{ed}^T (panel (d)) for a beam energy of 5.5 GeV, $Q^2 = 2$ GeV^2, $x = 1$, and $\phi_p = 35°$ are shown calculated with the full on-shell FSI (solid line), without any double spin-flip terms in the on-shell FSI (dashed line), and with central on-shell FSI only (dash-double-dotted line) as a function of the missing momentum.

POLARIZED EJECTED PROTON

Another set of polarization observables is accessible by measuring the polarization of the outgoing proton. For a detailed description of the formalism, consult [11]. There, we defined single and double asymmetries according to

$$A_p^{\xi} = \frac{\sigma(\xi)}{\sigma(0)} \text{ and } A_{ep}^{\xi} = \frac{\sigma_h(\xi)}{\sigma(0)} \tag{2}$$

where $\xi = n', l', s'$ refer to the normal, longitudinal, and sideways directions, and for the A_{ep} asymmetries, a polarized electron beam is required. In order to avoid ambiguities for out-of-plane measurements if $\theta_p = 0, \pi$, we introduce a spectrometer-based coordinate system with the new unit vectors n', l', and s'.

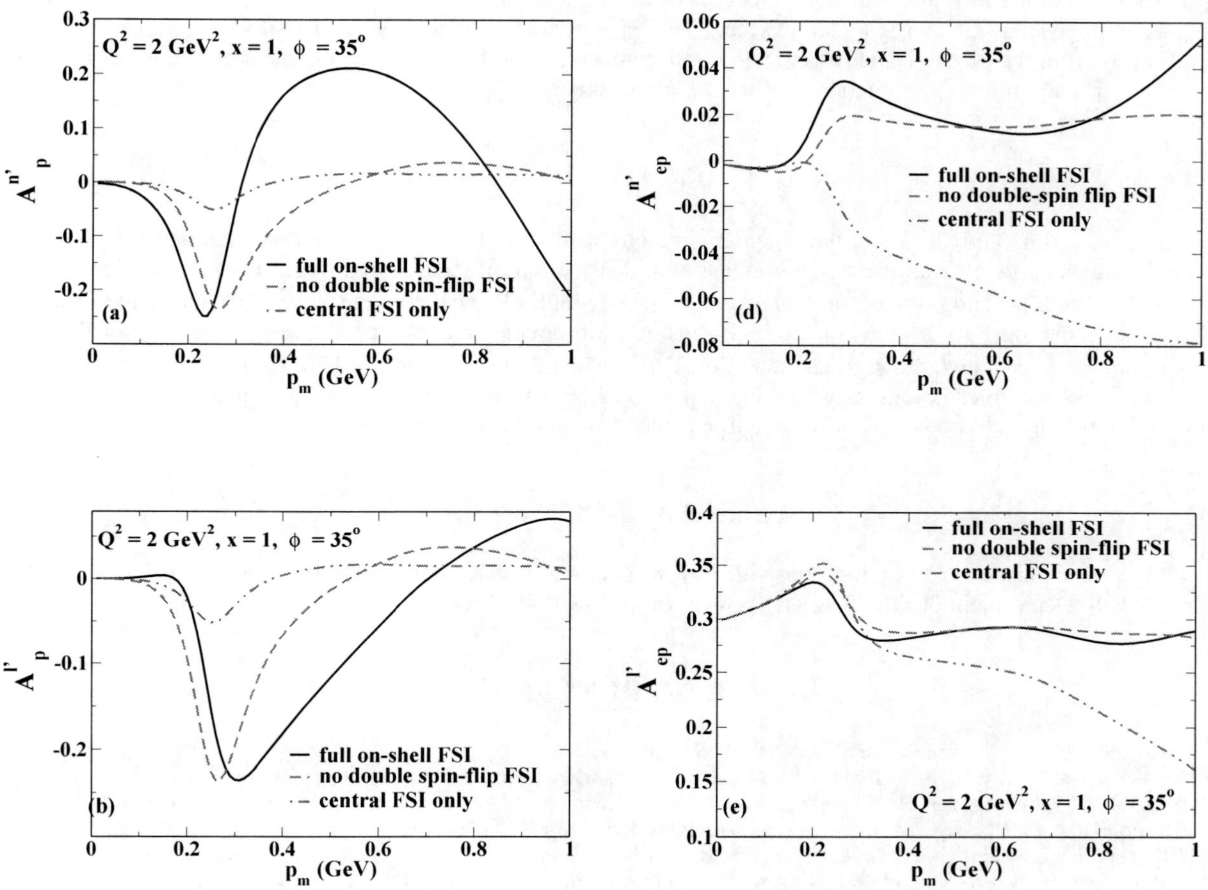

FIGURE 2. (Color online) The four panels show the four of the six asymmetries plotted versus the missing momentum p_m for a beam energy of 5.5 GeV, a transferred four-momentum of $Q^2 = 2$ GeV^2, $\phi_p = 35^o$, and $x = 1$. We show $A_p^{n'}$ (panel a), $A_p^{l'}$ (panel b), $A_{ep}^{n'}$ (panel d),and $A_{ep}^{l'}$ (panel e). The curves shown have been calculated with the full on-shell FSIs (solid), with central and single spin-flip FSIs (dashed), and with central FSI only (dash-dotted).

Here, due to space limitations, we show our results only for the normal and longitudinal asymmetries. Note that the asymmetries accessible with an unpolarized beam are zero in PWIA. FSI effects tend to be very large for the asymmetries accessible with an unpolarized beam, and are somewhat less dramatic, but still very important, for the asymmetries that are non-zero in PWIA already. From a theoretical point of view, the good news is that the off-shell FSI contributions are small even away from the quasi-free peak, at $x = 1.3$. There are some remaining theoretical uncertainties in the description of off-shell FSIs. Therefore, small off-shell FSI contributions will render the uncertainties less relevant, and allow for a cleaner comparison between theory and data.

Just like for the target polarization asymmetries, we show results for central FSI only, for FSI without double spin-flip contributions, and for the full on-shell FSI in Fig. 2, for the same kinematics as in Fig. 1. The clear message from this figure is that a calculation including only central FSIs will fail completely for missing momenta beyond $p_m = 0.2$ GeV. This is not surprising, as we are dealing with spin observables here. For the asymmetries accessible with an unpolarized beam, the central FSI on its own leads to a mostly zero asymmetry, whereas spin dependent FSIs lead to large structures in these observables. For the asymmetries accessible with a polarized electron beam only, the purely central FSI leads to non-zero results for the asymmetries, but the inclusion of spin-dependent FSIs leads to huge changes, both of the shape and size. The differences between the full FSIs and the FSIs without the three double spin-flip terms is largest for $A_p^{n'}$, where the double spin-flip contribution leads to a broad bump for missing momenta from $p_m \approx 0.3$ GeV to $p_m \approx 0.8$ GeV.

For the asymmetries that require a polarized beam, the influence of the double spin-flip terms is smaller. Fig. 2 gives the impression that an additional polarization, i.e. the beam polarization, may play a role similar to an additional spin-dependence in the FSI, and that at least one of them - either a spin-dependent FSI or a polarized beam - needs to be present to generate an approximately correct ejectile polarization asymmetry. A very similar picture emerges for $x = 1.3$, away from the quasi-elastic peak. The only difference is that the role of the double spin-flip terms becomes even more important for the asymmetries with an unpolarized beam.

OUTLOOK

If one is interested in studying FSIs, the kinematic regions with large FSI effects - both on-shell and off-shell - are attractive candidates. A measurement of the momentum distribution of A_d^V or A_{ed}^T at larger x, e.g. $x = 1.3$, would be highly interesting. Currently, there are no $D(e,e'\vec{p})n$ data at high Q^2 available, but an experiment could easily be performed at Jefferson Lab. In view of the high sensitivity to double spin-flip terms, we feel that one of the most interesting measurements would be to take data for $A_p^{n'}$ for $x = 1.3$, or even for $x = 1$. In addition, any measurement of polarization observables in regions where the unpolarized observables have already been measured might be very helpful in improving our knowledge of the reaction mechanism and the deuteron itself.

ACKNOWLEDGMENTS

This work was supported in part by funds provided by the National Science Foundation under grant No. PHY-1002478 and by the U.S. Department of Energy (DOE) under No. DE-AC05-84ER40150.

REFERENCES

1. S. Jeschonnek and J. W. Van Orden, Phys. Rev. C **78**, 014007 (2008).
2. M. M. Sargsian, arXiv:1010.1541 [nucl-th]; M. M. Sargsian, arXiv:0910.2016; M. M. Sargsian, Int. J. Mod. Phys. E **10**, 405 (2001); M. M. Sargsian, T. V. Abrahamyan, M. I. Strikman and L. L. Frankfurt, Phys. Rev. C **71**, 044614 (2005); L. L. Frankfurt, M. M. Sargsian and M. I. Strikman, Phys. Rev. C **56**, 1124 (1997).
3. J. Ryckebusch, D. Debruyne, P. Lava, S. Janssen, B. Van Overmeire and T. Van Cauteren, Nucl. Phys. A **728**, 226 (2003); D. Debruyne, J. Ryckebusch, W. Van Nespen and S. Janssen, Phys. Rev. C **62**, 024611 (2000); B. Van Overmeire and J. Ryckebusch, Phys. Lett. B **650**, 337 (2007); W. Cosyn and J. Ryckebusch, arXiv:0904.0914.
4. C. Ciofi degli Atti and L. P. Kaptari, Phys. Rev. C **71**, 024005 (2005).
5. J. M. Laget, Phys. Lett. B **609**, 49 (2005).
6. R. Schiavilla, O. Benhar, A. Kievsky, L. E. Marcucci and M. Viviani, Phys. Rev. C**72**, 064003 (2005).
7. P. E. Ulmer *et al.*, Phys. Rev. Lett. **89**, 062301 (2002).
8. W. Boeglin, "Probing microscopic structure of the lightest nuclei in electron scattering at JLab energies and beyond", Trento, Italy July 25-30, 2005, http://www.fiu.edu/ sargsian/ect05
9. G. Gilfoyle, spokesperson, Jefferson Lab Hall B, E5 run period; G.P. Gilfoyle, (the CLAS Collaboration), Bull. Am. Phys. Soc., Fall DNP Meeting, DF.00010(2006).
10. S. Jeschonnek and J. W. Van Orden, Phys. Rev. C **80**, 054001 (2009).
11. S. Jeschonnek and J. W. Van Orden, Phys. Rev. C **81**, 014008 (2010).

Configuration-space Faddeev calculation for proton-deuteron observables at energy E_{lab}=3 MeV

V.M. Suslov[a], M.A. Braun[a,b], I. Filikhin[a], B. Vlahovic[a], and I. Slaus[a]

[a] *Department of Physics, North Carolina Central University, 1801 Fayetteville Street, Durham, NC 27707, USA*
[b]*Department of High Energy Physics and Elementary Particles, Saint-Petersburg State University, 198504, Petrodvorets, St. Petersburg, Russia*

Abstract.. A new computational method for solving the nucleon-deuteron breakup scattering problem has been applied to study the elastic neutron- and proton-deuteron scattering on the basis of the configuration-space Faddeev-Noyes-Noble-Merkuriev equations. This method is based on the spline-decomposition in the angular variable and on a generalization of the Numerov method for the hyperradius. The Merkuriev-Gignoux-Laverne approach has been generalized for arbitrary nucleon-nucleon potentials and with an arbitrary number of partial waves taken into account. The nucleon-deuteron observables at the incident nucleon energy 3 MeV have been calculated using the charge-independent AV14 nucleon-nucleon potential including the Coulomb force for the proton-deuteron scattering. Results have been compared with those of other authors and with experimental proton-deuteron scattering data.

Keywords: Faddeev equations, pd elastic scattering, observables
PACS: 10.80. Jy, 21.45.+v,25.45.De

INTRODUCTION

There is an impressive amount of nucleon-deuteron scattering data: proton-deuteron and neutron-deuteron elastic nd breakup data: total, partial and differential cross sections and spin observables involving nucleon and deuteron beams. The data are compared with the rigorous three-body theory: Faddeev-equations-based theory using as input realistic high-precision nucleon-nucleon potentials, and including model three-nucleon forces [1]. In some calculations Coulomb force has been included [2].

The pd scattering has been studied by using hyperspherical harmonic method and Kohn Variational Principle [3] and by using the screening and renormalization procedure [4]. At energies above the threshold for the deuteron breakup only calculations using screening and renormalization have been done.

In this article we present the development of an alternative method for studying the proton-deuteron (pd) system based on the direct numerical solution of the Faddeev-Noyes-Noble-Merkuriev (FNNM) equations in configuration space. This approach was initiated by Merkuriev et al. (MGL) [5] who derived general formulae for nd breakup scattering. This method has been originally applied to study nd and pd elastic and breakup scattering but limited only to nuclear S-waves interaction and to simple NN potentials [6]. In the present work we generalize the MGL approach to any high precision realistic potential for both nd and pd for elastic processes.

THREE-NUCLEON FADDEEV CALCULATION IN CONFIGURATION SPACE

For proton-deuteron scattering the FNNM equations for partial components can be written in the following form):

$$[E+\frac{\hbar^2}{m}(\partial_x^2+\partial_y^2) - v_\alpha^{\lambda\ell}(x,y)]\ \Phi_\alpha^{\lambda_0 s_0 M_0}(x,y) = \Sigma_\beta [v_{1,\alpha\beta} + \Sigma_\tau (v_\tau^+ c_{\tau,\alpha\beta}^{M_0+} + v_\tau^- c_{\tau,\alpha\beta}^{M_0-})]\ \Phi_\beta^{\lambda_0 s_0 M_0}(x,y)$$

12th International Conference on Meson-Nucleon Physics and the Structure of the Nucleon (MENU 2010)
AIP Conf. Proc. 1374, 167-170 (2011); doi: 10.1063/1.3647117

$$+\Sigma_{\beta} v_{\alpha\beta} [\Phi_{\beta}^{\lambda_0 s_0 M_0}(x,y) + \int_{-1}^{1} du\, \Sigma_{\gamma}\, g_{\beta\gamma}(y/x,u)\, \Phi_{\gamma}^{\lambda_0 s_0 M_0}(x',y')]. \qquad (1)$$

Here Greek subindexes denote state quantum numbers: $\alpha=\{l,\sigma,j,s,\lambda,t,T\}$,where l, ,σ, j and t are the orbital, spin, total angular momenta and isospin of a pair of nucleons, λ is the orbital momentum of the third nucleon relative to the c.m.s. of a pair nucleons, and s is the total "spin" (**s =1/2+ j**). $\mathbf{M} = \boldsymbol{\lambda} + \mathbf{s}$ is the total three-particle angular momentum, and the value of total isospin is T.

In Equations (1) $v_{\alpha}^{\lambda\ell}$ is the centrifugal potential, $v_{\alpha\alpha'}(x)=\langle\alpha|v(\mathbf{x})|\alpha'\rangle= \delta_{\lambda\lambda'}\, \delta_{ss'}\, \delta_{\sigma\sigma'}\, \delta_{JJ}\, \upsilon_{\ell\ell'}^{\sigma J}$ are nucleon-nucleon potentials, where $\upsilon_{\ell\ell'}^{\sigma J}$ are the potential representatives in the two-body basis $\mathcal{Y}_{\ell\sigma}^{JJ_z}(\hat{x})$ (most often abbreviated as ${}^{2\sigma+1}\ell_J$), $v_{1,\alpha\beta}$ and coefficients $c_{\tau,\alpha\beta}^{M_0\pm}$ depending on quantum state numbers of channel combined with $v_{\tau}^{\pm}$ are matrix elements of the Coulomb potential projected onto the MGL basis. For given α and β summation over τ in Equations (1) is finite. If $\alpha=\{l,\sigma,j,s,\lambda,t,T\}$ and $\beta=\{l',\sigma',j',s',\lambda',t',T'\}$ then values of τ are restricted by the following inequality: $\max(|l-l'|, |\lambda-\lambda'|) \le \tau \le \min(l+l', \lambda+\lambda')$.This means that for a chosen set of basic states, Equations (1) take into account the Coulomb interaction exactly(although the latter has been expanded in partial waves).

The set of partial differential equation (1) must be solved for functions satisfying the regularity conditions

$$\Phi_{\alpha}^{\lambda_0 s_0 M_0}(x,0) = \Phi_{\alpha}^{\lambda_0 s_0 M_0}(x_{\max},0) = \Phi_{\alpha}^{\lambda_0 s_0 M_0}(0,y) = 0 \qquad (2)$$

The asymptotic conditions for pd elastic scattering has the following form [6]:

$$\Phi_{1,\tilde{\alpha}}^{\lambda_0 s_0 M_0}(x,y) \sim \{ \delta_{\lambda\lambda_0}\, \delta_{ss_0}\, \delta_{\sigma 1}\, \delta_{j1}\, e^{i\Delta_{\lambda}^{c}} F_{\lambda}^{c}(qy) + e^{-i\Delta_{\lambda}^{c}} (G_{\lambda}^{c}(qy) + i F_{\lambda}^{c}(qy))\, a_{\lambda s,\lambda_0 s_0}^{M_0} \}\, \varphi_{\ell}(x),$$

$$x\ \textit{finite},\quad y\to\infty, \qquad (3)$$

where $\Delta_{\lambda}^{c} = \arg\Gamma(\lambda+1+i\nu)$ is the Coulomb phase and ν is equal $n/\sqrt{3q}$, $\varphi_{\ell}(x)$ is l-th component of deuteron wave function (l=0,2), and F_{λ}^{c} and G_{λ}^{c} are the regular and irregular Coulomb functions, respectively.

The matrix of partial elastic amplitudes a has the structure

$$a = \frac{(e^{2i\Delta} - e^{2i\Delta^{c}})}{2i} = \frac{e^{2i\Delta^{c}}(e^{2i\delta} - 1)}{2i}, \qquad (4)$$

where Δ^{c} is a diagonal matrix of Coulomb phases and Δ is the Hermitean matrix of scattering phases. The phases $\delta = \Delta - \Delta^{c}$ are the contribution to the scattering phase due to the nuclear interaction.

In the case of neutron-deuteron elastic scattering one has to set the "charge" n equal to zero. This leads to equality to zero of the Coulomb phases Δ_{λ}^{c}, and the Coulomb functions F_{λ}^{c} and G_{λ}^{c} are reduced to the regularized spherical Bessel functions $\hat{j}_{\lambda}$ and $-\hat{y}_{\lambda}$, respectively.

RESULTS

To optimize numerical calculations, we have written down the Equations (1) in the polar coordinate system ($\rho^2 = x^2 + y^2$ and $\tan(\theta)$= y/x). The MGL approach [5] and the Numerov method [7] have been generalized for AV14 NN potential and for arbitrary number of partial waves taken into account. To calculate differential cross-section, proton and deuteron vector and tensor analyzing powers for the pd elastic scattering at 3 MeV, we have taken into account all values of the total three-nucleon angular momentum M up to 15/2 (1/2,3/2,...,15/2) with both parity values (±1) and values of the total two-nucleon angular momentum $j_{23} \le 4$.

Results of our calculations for the differential cross section and proton analyzing power A_y are shown in Fig. 1 together with those from the benchmark calculations of Deltuva et al. [8].

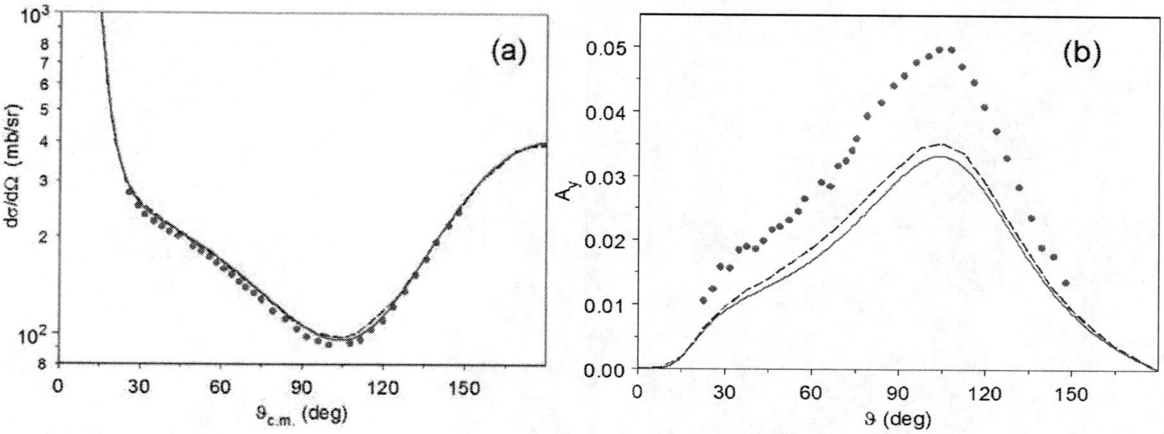

FIGURE 1 Differential cross section (a) and proton analyzing power A_y (b) for pd elastic scattering at 3 MeV lab energy as function of the c.m. scattering angle. The solid lines correspond to our results. The dashed lines correspond to results from [8]. The experimental data are from [9].

Our calculations have been performed using the AV14 NN potential and involving the correct asymptotic condition to take into account the Coulomb interaction while those of Ref. [8] used the AV18 NN potential and the screening and normalization procedure for the Coulomb force. All calculations are compared with the experimental data of Ref. [9]. All values of the total three-body angular momentum up to M = 15/2 are used in our calculation, while in [8] value of M is much larger. We chose values of j_{23} up to 4 whereas in [8] these values up to 5 have been used for the strong interaction. This truncation and the difference between AV14 and A V18 result in a small disagreement between our predictions for polarization observables and those from [8]. The results of calculations for the deuteron vector iT_{11} and tensor T_{20} analyzing powers as well as the experimental data [9] are shown in Fig. 2.

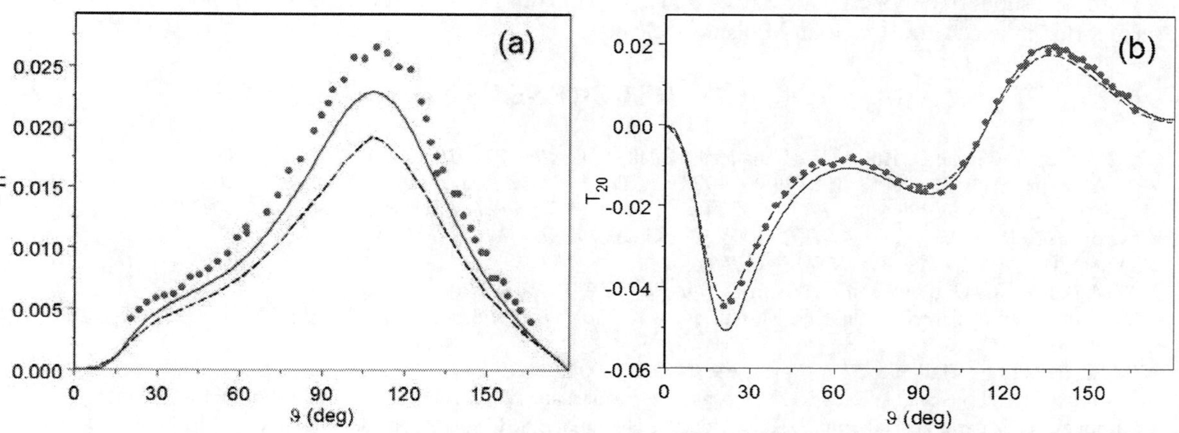

FIGURE 2 Deuteron vector iT_{11} (a) and tensor T_{20} (b) analyzing power for pd elastic scattering at 3 MeV lab energy as function of the c.m. scattering angle. The notations are the same as in **FIGURE 1**.

The results of calculations for the deuteron tensor T_{21} and T_{22} analyzing powers as well as the experimental data [9] are shown in Fig. 3. Predictions of our calculations and those of Ref. [8] are in reasonable agreement.

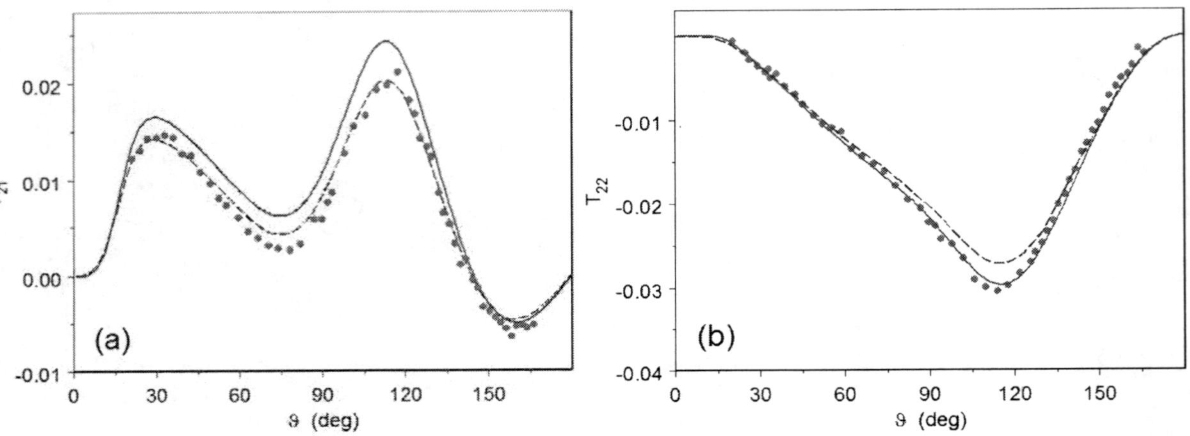

FIGURE 3 Deuteron tensor T_{21} (a) and T_{22} (b) analyzing power for pd elastic scattering at 3 MeV lab energy as function of the c.m. scattering angle. The notations are the same as in **FIGURE 1**.

Results using the AV18 NN potential give better agreement with experimental data for T_{20} and T_{21}. However, our calculations using the AV14 are in better agreement with the analyzing power iT_{11} and T_{22}, possibly indicating differences between AV14 and AV18 potentials. These facts are an evidence of inadequacy of modern potential models for pair nucleon-nucleon interaction. In this regard a new research is required.

ACKNOWLEDGMENTS

We are very grateful to Prof. H. Witala for courteously giving the computer code to calculate Nd observables. This work was supported by NSF CREST award HRD-0833184 and NASA award NNX09AV07A. The work of I.S. was supported in part by the Croatian Ministry of Science

REFERENCES

1. W. Glöckle, H. Witala, D. Hüber, H. Kamada, J. Golak, *Phys. Rep.* 274, 107-285 (1996).
2. E.O. Alt *et al.*, *Phys. Rev. C* 17, 1981-2005 (1978); A. Deltuva *et al.*, *Phys. Rev. C* **72**, 054004 (2005).
3. A. Kievsky *et al.*, *Nucl. Phys. A* 577, 511-527 (1994); A. Kievsky *et al.*, *Phys. Rev. C* **58,** 3085-3092 (1998); A. Kievsky *et al.*, *Phys. Rev. C* **64,** 024002 (2001); A. Kievsky *et al.*, *Phys. Rev. C* **69**, 014002 (2004).
4. E.O. Alt *et al.*, *Nucl. Phys. B* **2**, 167-180 (1967).
5. S.P. Merkuriev, C. Gignoux and A. Laverne, *Ann. Phys.* **99**, 30-71 (1976).
6. A.A. Kvitsinsky, Yu.A. Kuperin, S.P. Merkuriev, A.K. Motovilov and S.L. Yakovlev, *Fiz. Elem. Chastis At. Yadra* **17**, 267-317 (1986).
7. V.M Suslov and B. Vlahovic, *Phys. Rev. C* **69**, 044003 (2004).
8. A. Deltuva, A.C. Fonseca, A. Kievsky, S. Rosati, P.U. Sauer, and M. Viviani, *Phys. Rev. C* **71**, 064003 (2005).
9. S. Shimizu, K.Sagara, H. Nakamura, K. Maeda, N. Nishimori, S. Ueno, T. Nakashima, and S. Morinobu, *Phys. Rev. C* **52,** 1193-1202 (1995).

Time Delay in Neutron-Alpha Resonant Scattering

Bernard Hoop[a] and Gerald M. Hale[b]

[a]*Riley College of Education, Walden University, Minneapolis, Minnesota 55401-2511*
[b]*Theoretical Division, Los Alamos National Laboratory, Los Alamos, New Mexico 87545*

Abstract. Time delay analysis of neutron-alpha resonant scattering cross sections supports characterization of the lowest $3/2^+$ level in ^{5}He as fundamentally an n-α resonance on the second Riemann energy sheet of both n-α and deuteron-^{3}H channels, with an associated shadow pole on a different unphysical sheet that, through its associated zero on the physical sheet, contributes to the large ^{4}He(n,d)^{3}H cross section.

Keywords: time delay; neutron scattering; A=5; resonance; *S*-matrix poles; Riemann energy sheet
PACS: 21.10.Tg, 24.30.-v, 25.10.+s, 25.40.Dn, 25.40.Ny, 27.10.+h

INTRODUCTION

The lowest $3/2^+$ level in ^{5}He at excitation energy 16.84 MeV, about 48 keV above the deuteron-^{3}H threshold, is of importance in fusion energy applications. Barker [1] suggests that the observed properties of this level *"...can be understood as easily in terms of the R-matrix parameters themselves as from the positions and residues of the complex-energy poles of the S matrix.*" Hale and coworkers [2] derive an *S*-matrix pole structure from two-and four-level *R*-matrix parameters, which characterize this level as a main or conventional pole on the unphysical Riemann energy sheet closest to the physical sheet, where both n-α and d-^{3}H wave numbers have negative imaginary parts, and a shadow pole where the imaginary part of n-α wave number is negative, but d-^{3}H wave number is positive.

We apply a time delay method [3] to determine complex energies of *S*-matrix poles of this level. We further use the method to address issues raised by Barker on the smallness of the width of the shadow pole, specifically, the relationship between the partial widths for the conventional pole and total width for the shadow pole [1]. Three examples are used for illustration. 1) Resonant n-α elastic scattering cross section in $D_{3/2}$ partial wave is represented by an isolated Breit-Wigner resonance with deuteron and neutron partial widths taken from a 4-level *R*-matrix analysis, from which scattering amplitude trajectory is determined and used to calculate energy dependence of time delay. 2) Resonant n-α elastic scattering cross section is represented by a Breit-Wigner resonance using widths from conventional pole time delay vs energy derived in example 1), from which energy dependence of time delay is again determined. 3) Resonant n-α elastic cross section in example 2) is convoluted with an incident neutron energy spread of 40 keV, and time delay is again determined. In all three examples, ^{4}He(n,d)^{3}H reaction cross section calculated via reciprocity from all available d+^{3}H –> n+^{4}He data [4] is represented by an empirical function parameterized by ^{4}He(n,d)^{3}H threshold and n-α elastic resonance energies, by peak reaction cross section, and by reaction cross section ca. 1 MeV above threshold.

In each example, calculated elastic and reaction cross sections are added to yield total cross section and compared with all available neutron-α total and elastic scattering cross section data [5] via an overall squared error of fit [6]. Elastic cross sections derived from integrated angular distributions are multiplied by 1.1 to account for scale factor normalization uncertainty [5]. Experimental neutron energy spread of 40 keV is accounted for by convoluting a Gaussian energy distribution with calculated elastic cross section. Energy dependence of non-resonant $D_{3/2}$ phase shift in potential and interference terms, total cross section in other partial waves, and channel interaction radii are taken from Barker [1].

12th International Conference on Meson-Nucleon Physics and the Structure of the Nucleon (MENU 2010)
AIP Conf. Proc. 1374, 171-174 (2011); doi: 10.1063/1.3647118

TIME DELAY ANALYSIS

Real part of complex time delay is taken as sum of energy derivatives of resonant and non-resonant $D_{3/2}$ phase shifts, and transit time without n-α interaction [3], as determined from Argand diagrams of $D_{3/2}$ scattering amplitude derived from calculated total and elastic cross sections. Real part of time delay Re(Δt) vs n-α c.m. energy E is compared with best fits to the sum of two Lorentzians representing conventional (C) and shadow (S) pole time delays Δt_C and Δt_S, plus n-α transit time,

$$\mathrm{Re}(\Delta t) = \Delta t_C + \Delta t_S + \frac{2a_0}{v_{rel}} = \frac{2\hbar\Gamma_{C1}}{(E-E_{C+})(E-E_{C-})} + \frac{2\hbar\Gamma_{S1}}{(E-E_{S+})(E-E_{S-})} + \frac{2a_0}{v_{rel}}, \tag{1}$$

where S-matrix conventional and shadow pole complex energies are $E_{C\pm} = E_C \pm i\Gamma_C/2$, $E_{S\pm} = E_S \pm i\Gamma_S/2$, and Γ_{C1}, Γ_C, Γ_{S1}, Γ_S are, respectively, partial and total widths of time delay distributions, E_C and E_S are real parts of pole energies, a_0 (=5 fm) is n-α channel interaction radius, and v_{rel} is n-α c.m. relativistic velocity.

Example 1: Resonant Elastic Cross Section from 4-level *R*-Matrix Analysis

Resonant elastic cross section is calculated with deuteron and neutron partial widths $\Gamma_d = 0.0251$ MeV and $\Gamma_n = 0.0398$ MeV, respectively, and 22.125 MeV elastic resonance (neutron laboratory) energy [2]. Real part of resonant phase shift vs n-α c.m. energy derived from resonant scattering amplitude is shown in Figure 1(a). Corresponding resonant scattering amplitude is shown in Figure 1(b). Scattering amplitude trajectory encircles the center of the unitary circle, as the resonant phase shift in Figure 1(a) indicates. Real part of time delay is shown in Figure 1(c), compared with best fits to the sum of two Lorentzians representing conventional and shadow pole time delays, plus n-α transit time (Equation 1).

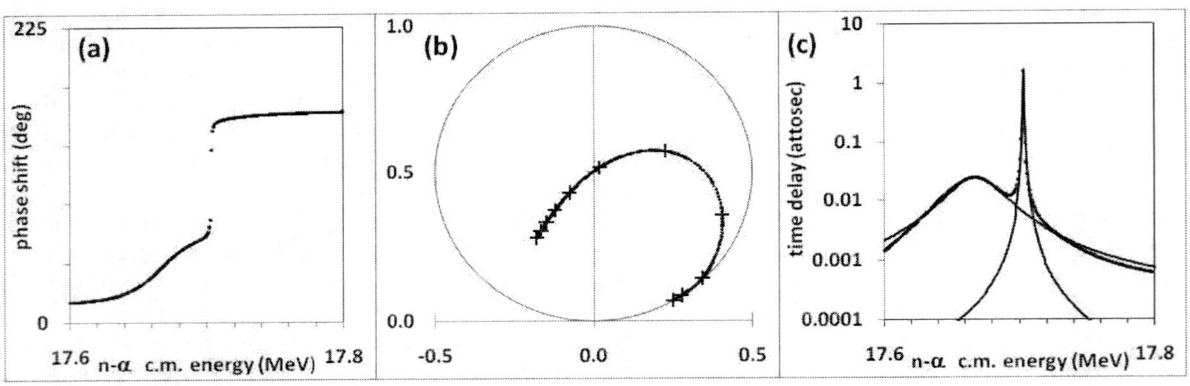

FIGURE 1. (a) Resonant phase shift in degrees vs n-α c.m. energy in MeV determined from resonant scattering amplitude. (b) Corresponding Argand diagram of resonant scattering amplitude. Vertical and horizontal axes are, respectively, imaginary and real parts of scattering amplitude. Large + symbols indicate values of n-α c.m. energy increasing (counterclockwise) from 17.60 to 17.80 MeV in 0.020 MeV steps. (c) Time delay in attoseconds vs n-α c.m. energy in MeV; small symbols: determined from Figure 1(b); solid lines: calculated from Δt_C and Δt_S terms in Equation 1.

Example 2: Resonant Elastic Cross Section from Conventional Pole Time Delay

Resonant n-α elastic cross section is calculated with deuteron and neutron partial widths $\Gamma_{C1} \equiv \Gamma_d = 0.0073$ MeV and $\Gamma_C \equiv \Gamma_n = 0.0351$ MeV taken from conventional pole time delay of Figure 1(c). Elastic resonance energy is 22.125 MeV. Values of Γ_{C1} and Γ_C characterize the conventional pole time delay [broad peak in Figure 1(c)]. Real part of resonant phase shift vs n-α c.m. energy is shown in Figure 2(a). Corresponding resonant scattering amplitude is shown in Figure 2(b). Scattering amplitude trajectory encircles the center of the unitary circle, crossing the imaginary axis twice, as the resonant phase shift plot indicates. Corresponding real part of time delay Re(Δt) is shown in Figure 2(c), again compared with best fits to the sum of two Lorentzians, plus n-α transit time calculated from Equation 1.

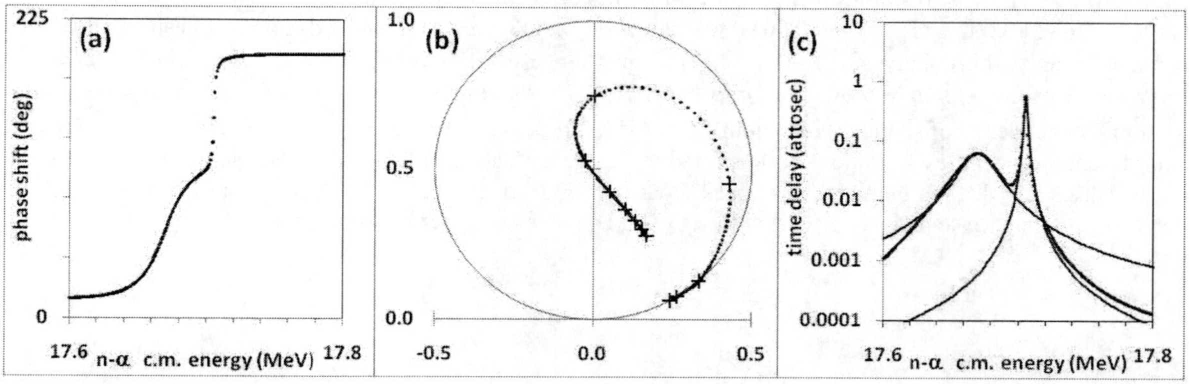

FIGURE 2. (a) Resonant phase shift in degrees vs n-α c.m. energy in MeV determined from resonant scattering amplitude. (b) Corresponding Argand diagram of resonant scattering amplitude. (c) Time delay in attoseconds vs n-α c.m. energy in MeV. Symbols and lines are as in Figure 1.

Example 3: Resonant Elastic Cross Section of Example 2 with 40-keV Energy Spread

Resonant n-α elastic cross section is calculated with deuteron and neutron partial widths $\Gamma_d = 0.0435$ MeV and $\Gamma_n = 0.0567$ MeV, respectively, derived from resonant elastic cross section of Example 2 convoluted with a 40-keV Gaussian experimental energy spread [5]. Resonance energy is 22.125 MeV. Scattering amplitude trajectory [Figure 3(b)] encircles the center of the unitary circle, as the resonant phase shift in Figure 3(a) indicates. As in Figures 1(c) and 2(c), real part of time delay is shown in Figure 3(c), compared with best fits to Equation 1.

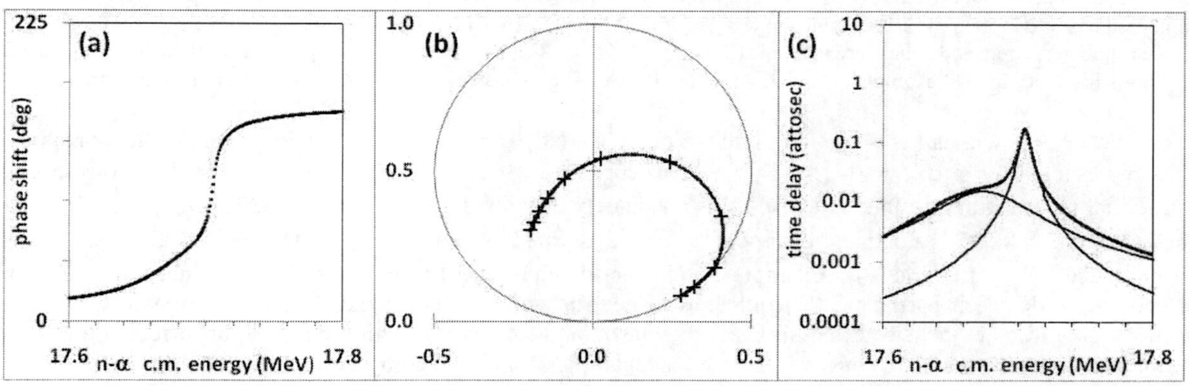

FIGURE 3. (a) Resonant phase shift in degrees vs n-α c.m. energy in MeV determined from resonant scattering amplitude. (b) Corresponding Argand diagram of resonant scattering amplitude. (c) Time delay in attoseconds vs n-α c.m. energy in MeV. Symbols and lines are as in Figure 1.

Maximum time delay $\Delta t_{C\max}$ and $\Delta t_{S\max}$ of conventional and shadow poles in the three examples are given in Table 1.

TABLE (1). Maximum time delay for conventional and shadow poles in three examples

Example no., Γ_d, Γ_n (MeV)	$\Delta t_{C\max}$ (attosec)	$\Delta t_{S\max}$ (attosec)
1) 0.0251, 0.0398	0.024	1.755
2) 0.0073, 0.0351	0.064	0.656
3) 0.0435, 0.0567	0.014	0.163

SUMMARY

In the three examples, resulting real parts of S-matrix conventional and shadow pole energies at maximum time delay are 17.671±0.003 and 17.705±0.001 MeV n-α c.m. (0.048 and 0.082 MeV d-^{3}H c.m.), respectively, in good agreement with previous analysis [2]. Barker [1] further suggests "*...that Γ_S for the shadow pole is approximately the difference of the R-matrix (observed) partial widths...*" Assuming $\Gamma_S = 4\Gamma_{S1}$ in the Δt_S term of Equation 1, logarithmic dependence of shadow pole width $\Gamma_{S1} = 0.5\hbar/\Delta t_{S\max}$ on Γ_d/Γ_n for five pairs of partial widths for each of the conditions of examples 1, 2 and 3 is illustrated in Figure 4. The proximity of the elastic resonance between the ^{4}He(n,d)^{3}H reaction threshold and its large peak cross section leads to loss of flux from the n-α to the d-^{3}H channel, which results in negative shadow pole delay times [3], i.e., values in upper half of Figure 4.

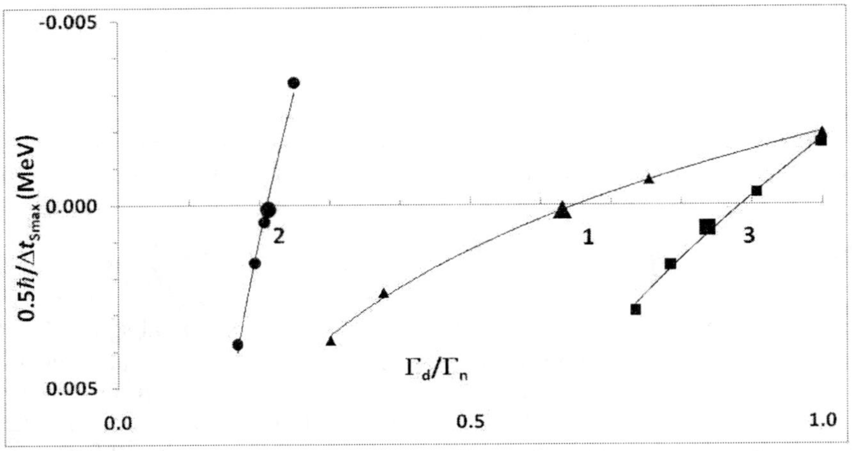

FIGURE 4. Shadow pole width, $\Gamma_{S1} = 0.5\hbar/\Delta t_{S\max}$ in MeV vs. partial width ratio Γ_d/Γ_n for five pairs of partial widths. Large numbered triangle is example 1 [cf. Table (1)]. Small triangles are Γ_d = 0.012, 0.015, 0.030, 0.0398 with Γ_n = 0.0398 MeV. Large numbered circle and square are examples 2 and 3 [cf. Table (1)]. Small circles and squares are calculated with partial width ratios for conditions of examples 2 and 3, respectively. Solid lines are $\Gamma_{S1} = c_1\ln(\Gamma_d/\Gamma_n)+c_2$, where c_1 and c_2 are constants.

For shadow pole widths Γ_{S1} plotted in Figure 4, peak ^{4}He(n,d)^{3}H reaction cross section σ relative to its maximum $\sigma_{\max}$ is represented by $\sigma/\sigma_{\max} = 1/[1+(\Gamma_{S1}/\Delta)^2]$, where $\sigma_{\max}$ = 91.703 mb and Δ = 0.034 MeV. Corresponding conventional pole maximum time delay $\Delta t_{C\max}$ is adequately described by $\Delta t_{C\max} = 0.028\Sigma/[(1-\Sigma)+\Gamma_d/\Gamma_n]-0.010$ attosec, where $\Sigma = 0.875$ is a conventional pole "strength" defined by Hale et al. [2]. The reciprocal of $\Delta t_{C\max}$ is also linearly related to Γ_d [cf. Table (1)]. Example 2 and the curve in Figure 4 shown as solid circles are characterized by relatively small deuteron partial width, compared to neutron width, which yields a narrow resonant elastic cross section. It should be emphasized that all available elastic and total cross section data [5], uncorrected for 40-keV experimental energy spread, are better represented by examples 1 and 3, than by example 2.

REFERENCES

1. F. C. Barker, *Phys. Rev. C* **56**, 2646-2653 (1997).
2. G. M. Hale, R. E. Brown and N. Jarmie, *Phys. Rev. Lett.* **59**, 763-766 (1987); L. N. Bogdanova, G. M. Hale and V. E. Markushin, *Phys Rev C* **44**, 1289-1295 (1991).
3. H. Haberzettl and R. Workman, *Phys. Rev. C* **76**, 059201 (2007) arXiv:0708.3989v1; N. G. Kelkar and M Nowakowski, *Phys. Rev. A* **78**, 012709 (2008) arXiv:0805.0608v1; N. Suzuki, T. Sato and T.-S. H. Lee. *Phys. Rev. C* **79**, 025205 (2009) arXiv:0806.2043v2.
4. Tabulated at www.nndc.bnl.gov/exfor/exfor11.htm
5. B. Haesner et al., *Phys. Rev. C* **28**, 995-999 (1983); R.E. Shamu and J.G. Jenkin, *Phys. Rev.* **135**, B99-B109 (1964); B. Hoop and H.H. Barschall, *Nucl. Phys.* **83**, 65-79 (1966); B. Hoop. "Scattering of Neutrons by Alpha Particles", Ph.D. Thesis, University of Wisconsin, Madison, 1966.
6. B. Hoop and P. Huber, *Helv. Phys. Acta* **40**, 710-712 (1967); R. A. Arndt et al., *Nucl. Sci. Engineering* **162**, 312-318 (2009) arXiv:0806.1198v2.

Session B: Strange Hadrons

Electromagnetic production of hyperon resonances

K. Hicks, D. Keller and W. Tang

Dept. of Physics & Astronomy, Ohio University, Athens OH 45701

Abstract. The study of hyperon resonances has entered a new era of precision with advent of high-statistics photoproduction data from the CLAS detector at Jefferson Lab. These data have multi-particle final states, allowing clean identification of exclusive reactions associated with strange mesons and baryons. Examples of physics results are: evidence for isospin interference in the decay of the $\Lambda(1405)$ resonance; a strong suggestion of meson cloud effects in the structure of the $\Sigma(1385)$ resonance; data from K^* photoproduction that will test the existence of the purported $K_0(800)$ meson. Properties of other hyperon resonances will also be studied in the near future.

Keywords: photoproduction, kaon, hyperon, resonance
PACS: 13.30.Ce, 13.60.-r 14.20.Gk, 14.20.Jn

INTRODUCTION

The mass of the strange quark has a peculiar value, being close to the scale parameter of QCD, Λ_{QCD} [1]. The light quarks (u and d) have a mass small enough so that theoretical tools, such as chiral perturbation theory, can be used to describe the dynamics of these quarks. On the other hand, the heavy quarks (c, b and t) have a mass large enough so that non-relativistic potential theories, such as heavy quark effective theory (HQET), can be used for the dynamic of these quarks. The strange quark is too heavy for chiral dynamics and too light for HQET, which makes it difficult to describe theoretically [1].

Because of the theoretical difficulty to describe the dynamics of the strange quark, it is important to measure the properties of particles with non-zero strangeness in order to provide guidance for theoretical models. For example, it has been debated for some time [2] whether the structure of the first excited state of the Λ baryon, the $\Lambda(1405)$ with $J^P = \frac{1}{2}^-$, is a $\bar{K}N$ bound state or a orbital excitation of the uds valence quarks in a conventional baryon quark model. The former would have strong 5-quark components in its wave-function, whereas the latter would have a predominantly 3-quark component. The fact that this issue is still not resolved is a testiment to the theoretical ambiguities in the dynamics of the strange quark.

Recently, the CLAS Collaboration has collected high-quality data for photoproduction of baryons with non-zero strangeness. The large statistics for these measurements forces constraints on theoretical models of strangeness production that were not possible previously. In particular, photoproduction of the ground state hyperons Λ and Σ shows evidence for a new nucleon resonance (N^*) that couples strongly to strangeness decay, which may explain why it was not seen in partial wave analysis of pion scattering experiments. The search for N^* resonances is important because significantly fewer N^*'s are observed than are predicted by quark models [3], known as the “missing" resonances problem.

In addition to data on photoproduction of the ground state hyperons, data on hyperon resonances (with an associated kaon) and on kaon resonances (with a ground state hyperon) are now being released. These new data allow further investigations into the dynamics of the strange quark in mesons and baryons.

CLAS DETECTOR AND RESULTS

The CLAS detector, located at the Thomas Jefferson National Accelerator Facility, is a large acceptance spectrometer with a toroidal magnetic field [4]. Beams of electrons or photons (produced via bremsstrahlung) with energies up to 6 GeV are incident on target of hydrogen, deuterium, or nuclear targets at the center of CLAS. For photoproduction, the energy of the photon is given by a tagging detector which records the momentum of scattered electrons. Multi-particle final states are detected at CLAS with high rates, providing billions of triggered events.

12th International Conference on Meson-Nucleon Physics and the Structure of the Nucleon (MENU 2010)
AIP Conf. Proc. 1374, 177-180 (2011); doi: 10.1063/1.3647119

Results for the reaction $\gamma p \rightarrow K^+\Lambda$ measured at CLAS have been published [5] and are described elsewhere [6]. Of special interest, as mentioned above, is the observation of new N^* resonances using both cross sections and polarization data. Curiously, the Λ is polarized nearly 100% in the direction of circularly polarized photons [6] which was not predicted by any theoretical reaction model before the CLAS measurements. The precision of the CLAS data has sparked new theoretical interest in strangeness production.

The primary focus of the current report is on hyperon resonances. Perhaps one of the most interesting results in the past decade from CLAS measurements is for the reaction $\gamma p \rightarrow K^+\Lambda(1405)$, where the decays $\Lambda^* \rightarrow \Sigma^+\pi^-$ and $\Lambda^* \rightarrow \Sigma^-\pi^+$ were measured simultaneously. In a constituent quark model, where the $\Lambda(1405)$ is described as an orbital excitation of the ground-state $\Lambda(1115)$, these two decays should be isospin symmetric. However, in models where the $\Lambda(1405)$ is dynamically generated [7], interference between the $\bar{K}N$ and $\Sigma\pi$ poles can produce asymmetric decays. The Carnegie Melon University group at CLAS [8] measured dramatically different decays for the $\Lambda(1405) \rightarrow \Sigma\pi$, showing that this hyperon resonance has a more complex structure than predicted by simple quark models. This is the first time that clear isospin interference has been observed in the strong decay of a well established baryon resonance.

$\Sigma^*(1385)$ Photoproduction

Few data exist for photoproduction of the Σ^* resonance, which is a partner to the well-known Δ resonance in the baryon decuplet. This is due in part to the difficulty of separating the Σ^{*0} at a mass of 1.385 GeV from the $\Lambda(1405)$ resonance, which overlap due to their natural decay widths. Hence, the first differential cross sections with reasonable statistics were measured for Σ^{*-} photoproduction from the neutron [9], where the neutral Λ resonances are not seen. More precise measurements are being analyzed for Σ^* photoproduction at CLAS, but are still very preliminary.

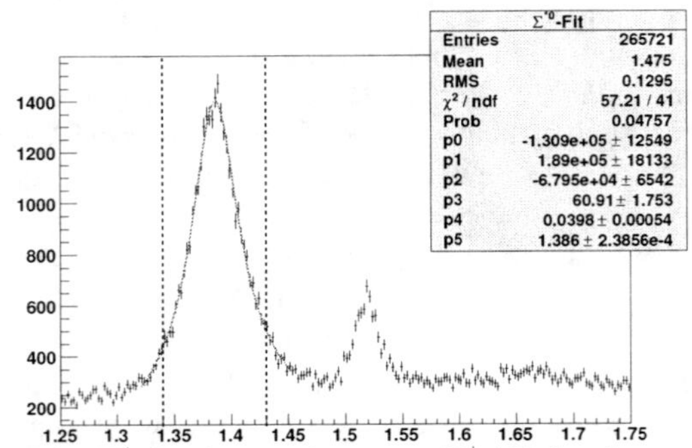

FIGURE 1. Missing mass (in GeV) of the K^+ for the reaction $\gamma p \rightarrow K^+\Lambda X$ showing peaks at the masses of the $\Sigma^0(1385)$ and $\Lambda(1520)$ resonances. The vertical scale shows counts.

One result of particular interest is the electromagnetic (EM) decay of the Σ^{*0} to the $\Lambda\gamma$ final state, which was published by CLAS [10]. The EM decay can be calculated exactly for a given theoretical model, and gives direct information on the wave-function of the baryon. Precise measurements have been made for the $\Delta \rightarrow N\gamma$ transition, yet no useful data were available for the $\Sigma^* \rightarrow Y\gamma$ transitions before the CLAS measurement.

Using a recent high-statistics data set, the $\Sigma^{*0} \rightarrow \Lambda\gamma$ EM decay has been remeasured at CLAS, reducing the statistical uncertainty from about 30% to about 10%. The smaller uncertainty is important, because effects from the pion cloud give a ~30% increase in the theoretical prediction [11] for $\Delta \rightarrow N\gamma$ at low Q^2, which is necessary for agreement with data.

The published CLAS decay width for $\Sigma^{*0} \rightarrow \Lambda\gamma$ is about a factor of two larger than theoretical predictions (for a range of theoretical models) [10], suggesting that meson cloud effects are significant. The new (preliminary) CLAS results [12] for this EM decay width are consistent with the earlier results, but have much better precision. To show the quality of the new data, the missing mass off the K^+ for the events of interest are shown in Fig. 1. Some background is seen under the Σ^{*0} peak at 1.385 GeV, which is mostly eliminated by a cut on the confidence level of a kinematic fit to the final state. The key step in the analysis is to separate the strong decay $\Sigma^{*0} \rightarrow \Lambda\pi^0$ from the EM decay, which can be done reliably using kinematic fits.

Knowledge of the quark structure of the Σ^* resonance is important, as it is the lowest-lying excited state of the strange baryons. Being a member of the baryon decuplet, along with the Δ resonance, it has spin $J = 3/2$. In the simple quark model, the decuplet baryons are visualized as a purely symmetric spin state of three valence quarks. Relationships between the decuplet baryons can be calculated using isospin (I-spin) and the analogue U-spin $SU(3)$

flavor symmetries. However, it is known that $SU(3)$ symmetries are broken, and the degree of symmetry breaking is something that experiments can measure. A broader program of measuring EM decays from a variety of the decuplet baryons with strangeness is in progress at CLAS, and will provide a window into the quark substructure of these baryons.

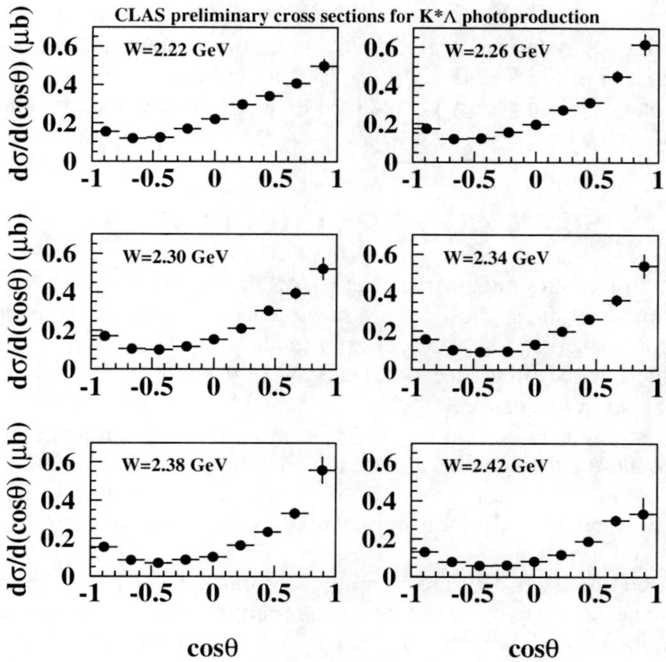

FIGURE 2. Preliminary angular distributions for photoproduction of the reaction $\gamma p \rightarrow K^{*+}\Lambda$ measured at CLAS.

K^*Y Photoproduction

Before a recent CLAS publication [13], few data existed for photoproduction of the $K^*(892)$ vector meson, the spin-1 partner to the scalar kaon. The interest in K^* photoproduction is, in part, to understand how spin is transferred from the photon to the strange quark. In the spin-0 kaon, the polarization of the $\bar{s}$-quark can only be inferred through measurements of Λ decay in the reaction $\gamma p \rightarrow K^+\Lambda$. As mentioned above, there appears to be nearly complete transfer of spin from a circularly polarized photon to the Λ. When the kaon is replaced by a K^*, the K^* now carries some of the spin, providing new reaction dynamics.

Preliminary differential cross sections [14] for the $\gamma p \rightarrow K^{*+}\Lambda$ reaction measured at CLAS are shown in Fig. 2. These angular distributions are for increasing photon energies near threshold, and have small statistical uncertainties (error bars are smaller than the symbols for most bins). These are the world's first high-statistics data for this reaction, and theorists are now developing models [15] for comparision with these data.

In addition to the K^{*+} cross sections, preliminary results for the Λ polarization from unpolarized beam show that there is less induced polarization compared with $K^+\Lambda$ photoproduction [5]. Results for $K^{*+}\Lambda$ production from polarized photons will be measured in the near future.

Because the K^* is a vector meson, it is possible to construct the spin-density matrix elements for $K^{*+}\Lambda$ photoproduction. Preliminary results show substantial spin effects near threshold, similar to those seen for ϕ-meson photoproduction. This suggests that significant spin transfer effects are seen to the K^*, as predicted by theory [15].

Within the context of a theoretical model, diagrams corresponding to t-channel exchange of virtual mesons show that the scalar meson, $K_0(800)$ with $J^P = 0^+$, can contribute to the forward-angle cross sections and spin observables. The existence of the $K_0(800)$ (also called the kappa-meson) has been under discussion for many years. If it exists, it is probably the strange partner to the well established $a_0(980)$ and $f_0(980)$ mesons in a scalar nonet. The $K_0(800)$ is not seen directly as a peak in the $K\pi$ invariant mass, presumably due to its wide width (since it can easily "fall apart" into $K\pi$ because of its $J^P = 0^+$ spin-parity). A similar argument is made for the purported σ-meson, which would be the singlet member of this scalar nonet.

The $K_0(800)$ does not contribute to kaon photoproduction t-channel, so only data for K^* photoproduction can support or weaken the case for the existence of the $K_0(800)$. So far, the case for the $K_0(800)$ is supported by the decay distributions of heavy mesons (such as $J/\psi \rightarrow \bar{K}^{*0}K^+\pi^-$ and $D^+ \rightarrow K^-\pi^+\pi^+$, see [16]) but the evidence remains ambiguous. Data for linearly polarized beam for K^{*0} photoproduction, which has been measured at LEPS, will provide an unambiguous signal of the $K_0(800)$, if it exists. Similarly, the ratio of K^{*+} and K^{*0} photoproduction total cross sections as a function of photon energy is predicted to be sensitive to the existence of the $K_0(800)$ meson. Analysis of these data are in progress.

SUMMARY AND CONCLUSIONS

The multi-particle data from high-statistics photon beams at CLAS have provided a rich source of results on hyperon resonance production. One example is the suggestion of a new nucleon resonance that couples strongly to the $K^+\Lambda$ final state, which was further identified by the study of spin-transfer to the Λ from circularly polarized beam. Of particular note is the isospin interference effects seen in the decays of the $\Lambda(1405)$ hyperon resonance.

Models of the structure of hyperon resonances can be constrained by measurements of EM decays, which do not suffer from the ambiguities of strong decays. Although EM decays are much harder to measure, due to their small branching ratios, the results are more interesting because of the unambiguous predictions within a given quark model. Results for both $\Delta \rightarrow N\gamma$ and $\Sigma^{*0} \rightarrow \Lambda\gamma$ are larger than simple quark model predictions. In the case of the Δ, the pion cloud makes a significant enhancement to the EM transition amplitude [11]. Similar meson cloud effects are now being investigated for the Σ^* EM decay.

For K^* photoproduction, high-precision data are just now becoming available. Current theoretical models are phenomenological, and data are necessary to constrain the various coupling constants in the model. One unconstrained parameter is the contribution of the $K_0(800)$ scalar meson, which is not firmly established. The existence of the $K_0(800)$ would fill in the missing places in the meson nonet that includes the well established $a_0(980)$ and $f_0(980)$ mesons, and hence it is important to measure K^* photoproduction reactions to see whether the data are described with diagrams that require $K_0(800)$ exchange in the t-channel.

ACKNOWLEDGMENTS

We gratefully acknowledge the CLAS Collaboration and the CEBAF accelerator staff, without which this work would not be possible. This work is supported in part by NSF grant PHY-0653454.

REFERENCES

1. R.L. Jaffe, *International Workshop on Nuclear and Particle Physics at J-PARC*, 2008.
2. J.W. Darewych, R. Koniuk and N. Isgur, Phys. Rev. D **32** (1985) 1765.
3. S. Capstick and W. Roberts, Phys. Rev. D **58** (1998) 074011.
4. B. Mecking *et al.*, Nucl. Instr. Meth. A **503** (2003) 513.
5. M. McCracken *et al.*, Phys. Rev. C **81** (2010) 025201.
6. R. Schumacher, Nucl. Phys. A **835** (2010) 231.
7. J.C. Nacher, E. Oset, H. Toki and A. Ramos, Phys. Lett. B **455** (1999) 55.
8. R. Schumacher and K. Moriya, these proceedings.
9. K. Hicks *et al.*, Phys. Rev. Lett. 102 (2009) 012501.
10. S. Taylor *et al.*, Phys. Rev. C **71** (2005) 054609.
11. B. Julià-Días, T.-S. H. Lee, T. Sato and L.C. Smith, ' Phys. Rev. C **75** (2007) 015205.
12. D. Keller *et al.* CLAS Analayis Note, 2010.
13. I. Hleiqawi *et al.*, Phys. Rev. C **75** (2007) 042201; **76** (2007) 039905.
14. W. Tang *et al.*, CLAS Analysis Note, 2010.
15. Y. Oh and H. Kim, Phys. Rev. C **74** (2006) 015208.
16. T. Komada (BES Collaboration), AIP Conf. Proc. **717** (2004) 337.

Properties of the Resonance $\Lambda(1520)$ as seen in the Forward Electroproduction at JLab Hall A

I.I. Strakovsky*, Y. Qiang†, Ya.I. Azimov**, W.J. Briscoe*, H. Gao†, D.W. Higinbotham‡ and V.V. Nelyubin§

*The George Washington University, Washington, DC 20052, USA
†Duke University Triangle Universities Nuclear Laboratory, Durham, NC 27708, USA
**Petersburg Nuclear Physics Institute, Gatchina, 188300, Russia
‡Thomas Jefferson National Acceleration Facility, Newport News, VA 23606, USA
§University of Virginia, Charlottesville, VA 22901, USA

Abstract. High-resolution spectrometer measurements of the reaction $H(e,e'K^+)X$ at small Q^2 are used to extract the mass and width of the $\Lambda(1520)$. We investigate dependence of the resonance parameters on different parametrizations of the background and the resonance peak itself. Our final values for the Breit-Wigner parameters are $M = 1520.4 \pm 0.6(\text{stat.}) \pm 1.5(\text{syst.})$ MeV and $\Gamma = 18.6 \pm 1.9(\text{stat.}) \pm 1(\text{syst.})$ MeV. The width appears to be more sensitive to the assumptions than the mass. We also estimate, for the first time, the pole position for this resonance and find that both the pole mass and width seem to be smaller than their Breit-Wigner values.

Keywords: Kaon electroproduction, hyperons, resonances, Breit-Wigner parameters, pole position
PACS: 13.30.Eg, 13.60.Rj, 14.20.Jn

INTRODUCTION

The $\Lambda(1520)$ is considered to be one of the best known baryon resonances. It is an excited hyperon state of the highest rank (4 stars) in the Review of Particle Physics (RPP) [1]. But its properties are still not exceedingly well understood. For example, its rather intensive decay to the $\pi\pi\Lambda$ channel [1] gives a hint at an unexpectedly strong coupling to the kinematically suppressed subchannel $\pi\Sigma(1385)$, stronger than to the main decay channels $\pi\Sigma$ and $\overline{K}N$. This could mean that the $\Lambda(1520)$ has a molecular nature (see, e.g., Ref. [2] and references therein). On the other hand, evidence has been suggested [3] that a new, previously unobserved, resonance $\Sigma(1380)$ with $J^P = 1/2^-$ and decay to $\Lambda\pi$ might influence the mode $\Lambda(1520) \to \pi\pi\Lambda$. These examples show that more detailed studies of the $\Lambda(1520)$, which may clarify both its nature and the hyperon spectroscopy, are needed.

As a first step, we investigate degree of precision available for the basic resonance parameters of the $\Lambda(1520)$ [4]. Surprisingly, all the experimental inputs shown in RPP [1] for its mass and width are dated before 1980. Many new experiments performed since then have not been accounted for. Meanwhile, they provided much more data with the $\Lambda(1520)$ observed either as the main goal or as a byproduct (even more data are expected to appear in near future).

Most of the earlier high-precision mass and width values, in particular those used for averaging in RPP [1], had come either directly, from production measurements with bubble chambers, or indirectly, from partial wave analyses of scattering data. High-energy spectrometers did not have good enough resolution for such measurements, and their results could not compete with bubble chamber data, even despite much higher statistics. Now, the High Resolution Spectrometers (HRS's) [5] in the JLab Hall A experiment [6] provided an exciting accuracy ($\sigma = 1.5$ MeV), comparable to the best previous resolutions. That is why their measurements are competitive with any previous ones.

Note also that experiments used for averaging in RPP, applied very different assumptions to extract the resonance parameters. However, influence of those assumptions on the parameter values has never been analyzed. We investigate this problem [4] on the base of the Hall A data set [6].

JLAB HALL A EXPERIMENT

High-resolution measurements of the missing mass (MM) spectra were developed at near-forward production angles in the reactions $ep \to e'K^{\pm}X$ and $e'\pi^+X$. The experiment [6] took place in Hall A at the Jefferson Lab using a 5.09 GeV

12th International Conference on Meson-Nucleon Physics and the Structure of the Nucleon (MENU 2010)
AIP Conf. Proc. 1374, 181-184 (2011); doi: 10.1063/1.3647120

electron beam incident on a 15 cm liquid hydrogen target. Scattered electrons were detected in one of the HRS's in coincidence with electroproduced hadrons in the second HRS. Each spectrometer was positioned at 12.5° relative to the beamline, but the use of additional septum magnets [7] allowed to reach smaller production angles, down to $\sim 6^{\circ}$. In this configuration, the spectrometers have an effective acceptance of approximately 4 msr in solid angle and $\pm 4.5\%$ in momentum, while still maintaining their nominal 10^{-4} full-width half-max momentum resolution [5]. To obtain the desired MM coverage, the central momentum of the electron HRS was varied between 1.85 and 2.00 GeV, while the central momentum of the hadron HRS was changed between 1.89 and 2.10 GeV. In such configuration, the average momentum transfer of the virtual photon was $< Q^2 > \approx 0.1$ (GeV/c)^2, and the average cm photon energy was $< E_{\gamma^*}^{cm} > = 1.1$ GeV which means that $< W > = 2.53$ GeV. For the kaon kinematics, the cm scattering angle was $5.6^{\circ} \leq \theta_{\gamma^* K}^{cm} \leq 11.4^{\circ}$, and the angular acceptance was $\Delta\Omega_{\gamma^* K}^{cm} \approx 38$ msr.

Calibration of HRS's was based on precise measurements of the known MM peaks for the neutron (in π^+X) and for the hyperons $\Lambda(1116)$, $\Sigma^0(1193)$ (both in K^+X). These three baryons decay through weak or electromagnetic interactions, and the proper widths of their peaks are negligible. Therefore, the observed widths of the peaks directly determine the MM resolutions at the corresponding momenta of the registered meson (π^+ or K^+). Then, the resolution of HRS for the $\Lambda(1520)$ may be determined by extrapolation of those resolution values to the MM region of 1520 MeV. Such a procedute gave the resolution $\sigma = 1.5$ MeV [6]. We apply this resolution when fitting the data. Further experimental details can be found in Refs. [6, 8, 4].

PRESENT ANALYSIS OF THE $\Lambda(1520)$

The main initial goal of the Hall A experiment [6] was to search for possible very narrow resonances with $\Gamma \sim 1$ MeV. Therefore, the MM spectra were scanned with 1 MeV steps. This is not adequate for the much broader resonances. To avoid strong fluctuations and excessively large statistical errors for every experimental point, we use now the 4 MeV bin size to analyze the data for $\Lambda(1520)$. Such a size is adequate for the resolution with $\sigma = 1.5$ MeV, it provides smaller statistical errors and fluctuations.

Generally, we describe the MM spectra in the form

$$\mathrm{Fit} = BW + BG\,, \tag{1}$$

where BW is the Breit-Wigner (BW) contribution for the resonance $\Lambda(1520)$, and the term BG combines all other contributions (including other possible resonances), which provide a background for the $\Lambda(1520)$.

The BW contribution may be written as

$$BW = A_{BW}\, \Gamma(M_X)\, D(M_X)\,. \tag{2}$$

In the non-relativistic form, we have

$$D_{nrel}^{-1}(M_X) = (M_X - M_0)^2 + \Gamma^2(M_X)/4\,. \tag{3}$$

In the relativistic form, this denominator should look like $|M_X^2 - [M_0 - i\Gamma(M_X)/2]^2|^2$, but usually $\Gamma^2 \ll M_0^2$, and D^{-1} may be approximately written in the form

$$D_{rel}^{-1}(M_X) = (M_X^2 - M_0^2)^2 + M_0^2\,\Gamma^2(M_X)\,. \tag{4}$$

With the energy-independent width, we just define $\Gamma(M_X) \equiv \Gamma_0$. Parameters M_0 and Γ_0 are the conventional BW mass and width of the $\Lambda(1520)$.

The energy-dependent width has a more complicated structure. Let us recall that the total width is the sum of partial ones for all decay modes, $\Gamma(M_X) = \Sigma_i \Gamma_i(M_X)$. Here we emphasize that every partial width should have its own energy dependence, corresponding to the threshold and kinematical properties of the particular decay channel.

Decays of the $\Lambda(1520)$ are dominated by only three channels, $\overline{K}N$, $\pi\Sigma$, and $\pi\pi\Lambda$. Branching ratios for two other modes, $\pi\pi\Sigma$ and $\Lambda\gamma$, are not more than 1% each [1], and we neglect them here. Also neglected is the channel $\Sigma^0\gamma$, with an even lower branching ratio [1]. Further, we define $\Gamma_0 = \Gamma(M_0)$. Here we also consider M_0 and Γ_0 as the BW mass and width of the $\Lambda(1520)$. In all cases M_0 and Γ_0 are fitting parameters.

Now, for extracting the mass and width of the $\Lambda(1520)$, we restrict ourselves to the mass interval from 1.45 to 1.65 GeV (this corresponds to 13,070 detected events). To the peak, we apply the BW term (2), while the background is described by various combinations. Further, for the best-fit procedure, we use two not quite equivalent methods, least-squares (min-χ^2) and log-likelihood (LL) ones. In such a way, we obtain several sets of numerical values for pairs (M_0, Γ_0) and can trace their dependence on the assumptions used.

RESULTS AND THEIR DISCUSSIONS

We begin with considering changes of the BW mass M_0. They appear to be rather small, though not always negligible. The non-relativistic expression (3) for the BW term provides M_0 values lower than the relativistic Eq.(4), at the level of ~ 0.05 MeV. Similarly, the LL procedure gives lower M_0 than min-χ^2; the difference may be as small as 0.01 MeV, but may reach ~ 0.15 MeV. Structure of the background can also shift M_0; the difference for various variants which we use is, again, not more than 0.15 MeV.

A more essential effect comes from energy dependence of the width. Description with the energy-dependent width results in M_0 about 0.6 MeV lower than for the energy-independent width. Interestingly, it is at the same level as our statistical uncertainty for M_0, which is ~ 0.6 MeV in all the studied cases.

Changes of the BW width Γ_0 also show some regularities. Shifts of results for using the (non-)relativistic Eqs.(3) or (4) are not more than 0.06 MeV. The LL fitting gives a lower width than min-χ^2; the difference may be 0.2 MeV, but may reach ~ 0.9 MeV (for comparison, our statistical uncertainty for the width is ~ 2 MeV).

More complicated is the influence of the background description. Most influential in fitting our data is contribution of the resonance $\Lambda(1405)$. Its exclusion diminishes Γ_0; the difference may be up to ~ 1 MeV (the corresponding shift of M_0 is much smaller, not more than 0.1 MeV).

Now we are able to formulate some conclusions, which may have a more general character.

- The width of $\Lambda(1520)$ is sufficiently small, so the relativistic and non-relativistic forms give practically the same values of M_0 and Γ_0 (at the present level of accuracy).
- By definition, the LL fitting always provides a larger value of χ^2, than the min-χ^2 fitting. However, formally they should be equivalent at asymptotically high statistics. In this sense, the present statistics is not asymptotical yet. Both M_0 and Γ_0 are different in the two fittings, the differences are comparable to the statistical uncertainties of those BW parameters. In terms of χ^2 per degree of freedom, χ^2/dof, which is typically ~ 1.5 in our studies here, the LL fitting is up to 0.05 higher than min-χ^2.
- The M_0 has a smaller statistical uncertainty and is less affected by any change of fitting procedure than the Γ_0. The width is especially sensitive to the background form.

For extracting our resulting BW parameters, we use the BW relativistic expression (2), (4) with energy-dependent width (both points are theoretically motivated to be more reasonable). The corresponding least-squares (min-χ^2) fit in the mass interval 1.45 - 1.65 GeV is shown in Fig. 1, together with all separate contributions. This fit corresponds to χ^2/dof=1.46 and gives

$$M_0 = 1520.4 \pm 0.6\,\text{MeV}\,, \quad \Gamma_0 = 18.6 \pm 1.9\,\text{MeV}\,, \tag{5}$$

with pure statistical uncertainties. Each of the two BW parameters has also a systematic uncertainty. It is about 1.5 MeV, mainly due to the mass scale uncertainty, for M_0, and about 1 MeV, mainly due to the fitting procedure ambiguity, for Γ_0.

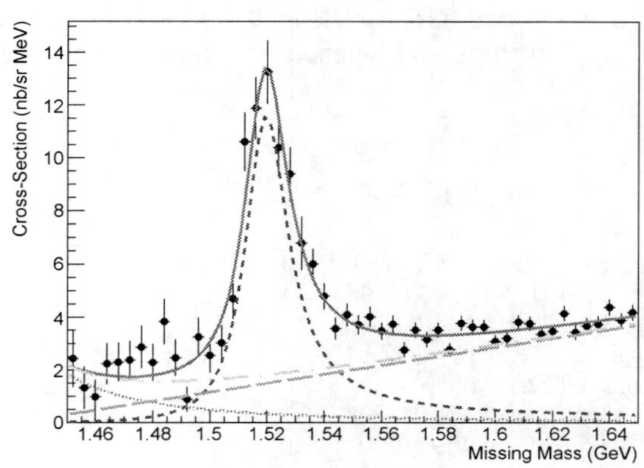

FIGURE 1. (Color on line) Fit to the experimental MM distribution is shown by the solid line. The short-dashed line is the BW contribution for the $\Lambda(1520)$ (note an asymmetric form, due to the energy-dependent width). The total background (dot-dashed line) is the sum of the linear binomial (long-dashed line) and the BW tail of the $\Lambda(1405)$ (dotted line).

For comparison, the same description with the LL fit gives slightly lower values

$$M_0 = 1520.3 \pm 0.6\,\text{MeV}\,, \quad \Gamma_0 = 17.8 \pm 1.9\,\text{MeV}\,, \tag{6}$$

and the higher value χ^2/dof=1.50 .

Even without accounting for the systematic uncertainties, both M_0 and Γ_0 are in reasonable agreement with previous experimental values and with their RPP average [1]. It is worth to note, however, that the uncertainties in the later works used in RPP are larger than those in the earlier works. This may hint that the uncertainties stated in the earlier works (and, therefore, in the average) are too optimistic.

Now we can discuss position of the S-matrix pole corresponding to the $\Lambda(1520)$, that has never been discussed for hyperons. It is determined by vanishing of the denominator (4), *i.e.*, by a solution of the equation

$$(W_p^2 - M_0^2)^2 + M_0^2\Gamma^2(W_p) = 0\,. \tag{7}$$

This non-linear equation may have non-unique solutions. Physically reasonable is only one of them, close to $(M_0 - i\Gamma_0/2)$. It is convenient, therefore, to rewrite Eq.(7) in the form

$$W_p = M_0\,[1 - i\Gamma(W_p)/M_0]^{1/2}\,. \tag{8}$$

Its complex solution gives the pole mass $M_p = \mathrm{Re}\,W_p$ and the pole width $\Gamma_p = -2\,\mathrm{Im}\,W_p$.

Numerical solution with values (5) for M_0 and Γ_0 gives

$$M_p = 1518.8\,\text{MeV}\,, \quad \Gamma_p = 17.2\,\text{MeV}\,. \tag{9}$$

Note that $M_p < M_{BW}$, $\Gamma_p < \Gamma_{BW}$, with the mass difference exceeding the statistical uncertainty. Such relation for masses may be rather general, as suggested by comparison with the mass pairs (BW and pole) shown for πN resonances in Listings of RPP [1]. This assumption is supported by the analysis of Eq.(8) presented in Ref. [4].

In summary, we have found the mass and width of the $\Lambda(1520)$ in near-forward electroproduction. The extracted BW parameters of the resonance are shown to depend not only on experimental data, but also on the way of their treatment. The extracted width is more sensitive to the various treatments than the mass. For the $\Lambda(1520)$, the non-resonance background should be accurately studied and understood if one intends to extract the mass and, especially, width with uncertainty of order 1 - 2 MeV. Having the BW mass and width (5), we also give the first estimate (9) of the pole parameters for the $\Lambda(1520)$. The pole values for both mass and width tend to be lower than the BW values.

ACKNOWLEDGMENTS

The authors express their gratitude to R. Arndt and to E. Pasyuk for useful discussions. This work was supported in part by the U. S. Department of Energy under Grants DE-FG02-99ER41110 and DE-FG02-03ER41231, by the Russian State grant SS-3628.2008.2, and by Jefferson Science Associates which operates the Jefferson Lab under DOE contract DE-AC05-06OR23177.

REFERENCES

1. K. Nakamura *et al.* (Particle Data Group), J. Phys. G **37**, 075021 (2010).
2. L.S. Geng, E. Oset, and B.S. Zou, Eur. Phys. J. **A 38**, 239 (2008).
3. J.J. Wu, S. Dulat, and B.S. Zou, Phys. Rev. C **81**, 045210 (2010).
4. Y. Qiang *et al.*, to be published in Phys. Lett. B (2010); arXiv:1003.5612 [hep-ph].
5. J. Alcorn *et al.*, Nucl. Instrum. Methods **A 522**, 294 (2004).
6. Y. Qiang *et al.* (JLab Hall A Collab.), Phys. Rev. **C 75**, 055208 (2007).
7. P. Brindza *et al.*, IEEE Trans Appl. Supercond. **11**, 1594 (2001).
8. Y. Qiang, PhD Thesis, MIT, 2007.

Photoproduction of $K\Sigma(1385)$ from the nucleon

Yongseok Oh

School of Physics and Energy Sciences, Kyungpook National University, Daegu 702-701, Korea

Abstract. The reactions of $K\Sigma(1385)$ photoproduction, i.e., $\gamma p \to K^+\Sigma^0(1385)$ and $\gamma n \to K^+\Sigma^-(1385)$, are investigated in the resonance energy region for studying the role of the nucleon and Δ resonances of masses around 2 GeV. The Lagrangians for describing the decays of these resonances into the $K\Sigma(1385)$ channel are constructed and the decay amplitudes are obtained, which allows us to determine the coupling constants using the predictions of quark models or the data listed by the Particle Data Group. The resulting cross sections are compared to the data from the Thomas Jefferson National Accelerator Facility and the SPring-8, which indicates nontrivial contributions from the two-star-rated resonances in the Particle Data Group as well as from some missing resonances predicted by a quark model.

Keywords: hyperon photoproduction, nucleon resonances
PACS: 25.20.Lj, 13.60.Le, 13.60.Rj, 14.20.Gk

Production of strange particles from photon-nucleon scattering has been extensively studied in recent experiments at electron/photon accelerator facilities [1, 2, 3, 4]. However, most of the data accumulated by these experiments are for the reactions of $K\Lambda(1116)$ and $K\Sigma(1193)$ production, and the photoproduction of the spin-$3/2$ hyperons such as the reaction of $\gamma p \to K\Sigma(1385)$ is virtually unknown until the recent experiments at JLab [5] and at SPring-8 [6]. The Σ^* or $\Sigma(1385)$ is the lowest mass hyperon in the baryon decuplet, so the analysis of its production mechanism will be valuable for testing the flavor SU(3) symmetry when combined with the analyzed production mechanism of $\Delta(1232)$ resonance.

Furthermore, this reaction is expected to have a nontrivial role in searching for the so-called missing resonances. This is based on the observation that, although the cross section of $\Sigma(1385)$ photoproduction is smaller than that of $K\Lambda(1116)$ and $K\Sigma(1193)$ photoproduction, the suppression factor is not large, which means that this reaction may have nontrivial role in a full coupled channel calculation for identifying new resonances and/or extracting the properties of baryon resonances.[1] In fact, the quark model of Ref. [10] predicts that several missing and not-well-established non-strange baryon resonances have large partial decay widths into the $K\Sigma(1385)$ channel, while the couplings of most well-established resonances to this channel are small. Then we may have a chance to see the effects of such missing or yet-to-be-established resonances through the analysis of $K\Sigma(1385)$ photoproduction process.

The recent experiments for $K\Sigma(1385)$ photoproduction were reported in Refs. [5, 6], which present the cross sections and the beam asymmetry of this reaction.[2] Theoretical investigation of $K\Sigma(1385)$ photoproduction, however, is rare. (See, for example, Refs. [13, 14].) Here we present the theoretical model of Ref. [15] for the reaction mechanisms of $\Sigma(1385)$ production in the reaction of $\gamma p \to K^+\Sigma^0(1385)$, which is based on the preliminary data of the JLab [5]. We then discuss the application of this model to the physical quantities in the reaction of $\gamma n \to K^+\Sigma^-(1385)$, of which data were reported in Ref. [6].

Our model for $K\Sigma^*$ photoproduction is depicted in Fig. 1. This includes the t-channel K and K^* meson exchanges [Fig. 1(a)], s-channel non-strange baryon exchanges [Fig. 1(b)], and u-channel hyperon exchanges [Fig. 1(c)]. It also contains the generalized contact term [Fig. 1(d)] that is introduced to restore gauge invariance. For the effective Lagrangians to calculate the production amplitudes from the diagrams for the t-channel K exchange, s-channel nucleon term, u-channel $\Sigma(1385)$ term, and the contact term, we refer the reader to Ref. [15], where the details of the effective Lagrangians and the coupling constants are discussed. Instead, we here focus on the Lagrangians which require more rigorous and detailed studies.

The t-channel K^* exchange involves the γKK^* and $K^*N\Sigma^*$ interactions. The Lagrangian for the former interaction is well known and its coupling constant can be determined from the measured width of the $K^* \to K\gamma$ decay. However, the Lagrangian for the latter interaction, which describes the vertex of $\frac{3}{2} \to 1 + \frac{1}{2}$ with the numbers representing the

---

[1] Note that similar conclusions were drawn also for $K^*\Lambda$ and $K^*\Sigma$ photoproduction [5, 7, 8, 9].

[2] The earlier experimental data for $K\Sigma(1385)$ photoproduction obtained in 1960's can be found, e.g., in Refs. [11, 12].

12th International Conference on Meson-Nucleon Physics and the Structure of the Nucleon (MENU 2010)
AIP Conf. Proc. 1374, 185-188 (2011); doi: 10.1063/1.3647121

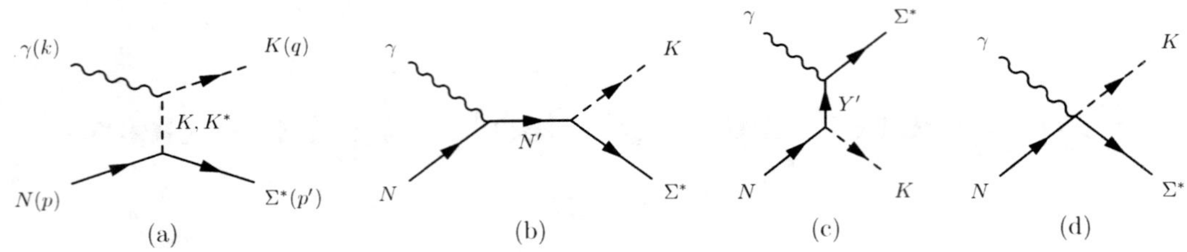

FIGURE 1. Feynman diagrams for $K\Sigma^*$ photoproduction. N' stands for the non-strange baryons and their resonances, while Y' the hyperons with strangeness -1 and their resonances. Diagram (d) includes the contact term of which form can be found in Ref. [15].

spin of the participating particles, has not been studied in detail. This interaction contains three terms in general, which can be written as

$$\mathscr{L}_{K^*N\Sigma^*} = \frac{ig_1}{2M_N}\overline{K}^{*\mu\nu}\overline{\mathbf{\Sigma}}^*_\mu \cdot \boldsymbol{\tau}\gamma_\nu\gamma_5 N + \frac{g_2}{(2M_N)^2}\overline{K}^{*\mu\nu}\overline{\mathbf{\Sigma}}^*_\mu \cdot \boldsymbol{\tau}\gamma_5\partial_\nu N - \frac{g_3}{(2M_N)^2}\partial_\nu\overline{K}^{*\mu\nu}\overline{\mathbf{\Sigma}}^*_\mu \cdot \boldsymbol{\tau}\gamma_5 N + \text{ H.c.}, \tag{1}$$

where $K^*_{\mu\nu} = \partial_\mu K^*_\nu - \partial_\nu K^*_\mu$. The coupling constants of this interaction is unknown and may be determined from the corresponding couplings of the $\rho N\Delta$ interaction if we invoke the SU(3) symmetry relation. However, even the couplings of the $\rho N\Delta$ interaction are poorly known and, in most analysis, only the g_1 term has been used [16, 17]. In this work, we fix the coupling g_1 of the $K^*N\Sigma^*$ interaction by using the SU(3) symmetry with that of the $\rho N\Delta$ interaction and treat g_2 and g_3 as free parameters. Fortunately, it turns out that in our case, the contribution from the K^* exchange is suppressed and the uncertainties of those couplings could be ignored. However, it would be worth while to analyze the data for estimating all three couplings of the $\rho N\Delta$ interaction.

For the u-channel diagrams, we consider the case with $Y' = \Lambda(1116)$ and $\Sigma(1385)$. The former is the lowest mass hyperon and the latter is required to fulfill the gauge invariance condition. Since one of the motivation of this study is to investigate the non-strange baryon resonances in the s-channel diagrams, we start with the most general expressions for the interactions of $RN\gamma$ and $RK\Sigma^*$, which read

$$\begin{aligned}
\mathscr{L}_{RN\gamma}(\tfrac{1}{2}^\pm) &= \frac{ef_1}{2M_N}\bar{N}\Gamma^{(\mp)}\sigma_{\mu\nu}\partial^\nu A^\mu R + \text{H.c.},\\
\mathscr{L}_{RN\gamma}(\tfrac{3}{2}^\pm) &= -\frac{ief_1}{2M_N}\bar{N}\Gamma^{(\pm)}_\nu F^{\mu\nu}R_\mu - \frac{ef_2}{(2M_N)^2}\partial_\nu\bar{N}\Gamma^{(\pm)}F^{\mu\nu}R_\mu + \text{H.c.},\\
\mathscr{L}_{RN\gamma}(\tfrac{5}{2}^\pm) &= \frac{ef_1}{(2M_N)^2}\bar{N}\Gamma^{(\mp)}_\nu\partial^\alpha F^{\mu\nu}R_{\mu\alpha} - \frac{ief_2}{(2M_N)^3}\partial_\nu\bar{N}\Gamma^{(\mp)}\partial^\alpha F^{\mu\nu}R_{\mu\alpha} + \text{H.c.},
\end{aligned} \tag{2}$$

and

$$\begin{aligned}
\mathscr{L}_{RK\Sigma^*}(\tfrac{1}{2}^\pm) &= \frac{h_1}{M_K}\partial_\mu K\bar{\Sigma}^{*\mu}\Gamma^{(\mp)}R + \text{H.c.},\\
\mathscr{L}_{RK\Sigma^*}(\tfrac{3}{2}^\pm) &= \frac{h_1}{M_K}\partial^\alpha K\bar{\Sigma}^{*\mu}\Gamma^{(\pm)}_\alpha R_\mu + \frac{ih_2}{M_K^2}\partial^\mu\partial^\alpha K\bar{\Sigma}^*_\alpha\Gamma^{(\pm)}R_\mu + \text{H.c.},\\
\mathscr{L}_{RK\Sigma^*}(\tfrac{5}{2}^\pm) &= \frac{ih_1}{M_K^2}\partial^\mu\partial^\beta K\bar{\Sigma}^{*\alpha}\Gamma^{(\mp)}_\mu R_{\alpha\beta} - \frac{h_2}{M_K^3}\partial^\mu\partial^\alpha\partial^\beta K\bar{\Sigma}^*_\mu\Gamma^{(\mp)}R_{\alpha\beta} + \text{H.c.}.
\end{aligned} \tag{3}$$

where A_μ is the photon field ($F_{\mu\nu} = \partial_\mu A_\nu - \partial_\nu A_\mu$), and R, R_μ, and $R_{\mu\nu}$ are the fields for the spin-1/2, 3/2, and 5/2 resonances, respectively, with

$$\Gamma^{(\pm)}_\mu = \begin{pmatrix} \gamma_\mu\gamma_5 \\ \gamma_\mu \end{pmatrix}, \qquad \Gamma^{(\pm)} = \begin{pmatrix} \gamma_5 \\ \mathbf{1} \end{pmatrix}. \tag{4}$$

The coupling constants f_i and h_i can be related to the theoretical predictions on the partial decay amplitudes of the resonances. This has a great advantage compared with the use of partial decay widths as the relative phase of the

TABLE 1. Resonances and their coupling constants h_i and f_i based on the predictions of Ref. [10]. The coupling constants are calculated using the resonance masses of PDG.

Resonance	**PDG** [18]	h_1	h_2	f_1	f_2
$N\frac{1}{2}^-(1945)$	$S_{11}^*(2090)$	−0.98	—	0.055	—
$N\frac{3}{2}^-(1960)$	$D_{13}^{**}(2080)$	0.24	−0.54	−1.25	1.21
$N\frac{5}{2}^-(2095)$	$D_{15}^{**}(2200)$	0.29	−0.08	0.37	0.28
$\Delta\frac{3}{2}^-(2080)$	$D_{33}^*(1940)$	−0.68	1.00	0.39	−0.57
$\Delta\frac{5}{2}^+(1990)$	$F_{35}^{**}(2000)$	−0.87	0.11	−0.68	−0.062
$N\frac{3}{2}^-(2095)$		0.99	0.27	0.49	−0.83
$N\frac{5}{2}^+(1980)$		0.59	0.24	0.019	−0.13
$\Delta\frac{3}{2}^-(2145)$		0.25	0.46	0.11	−0.059

coupling constants can be fixed. The explicit formulas for this relation were derived in Ref. [15], and the couplings fixed by the quark model prediction of Refs. [10] are listed in Table 1.[3] In this work, only the resonances which are predicted to have large couplings to the $K\Sigma^*$ channel in the model of Ref. [10] are considered.

Presented in Fig. 2 is the result of the total cross section for the $\gamma p \to K^+\Sigma^0(1385)$ reaction, which indicates nontrivial contribution from the missing and/or not-well-established resonances in production mechanism of this reaction. In particular, we find that nontrivial contributions from the $\Delta(2000)F_{35}$, $\Delta(1940)D_{33}$, and $N(2080)D_{13}$, as well as from the missing resonance $N\frac{3}{2}^-(2095)$ are required to describe the observed data of Ref. [5].

The same model is then applied to the reaction of $\gamma n \to K^+\Sigma^-(1385)$ with the same resonances and keeping the isospin symmetry. The predictions of this model for differential cross sections and beam asymmetry are compared with the data obtained by the LEPS Collaboration [6] in Fig. 3. Considering that only a few resonances are included and no additional parameters are allowed, the description of this model for the differential cross sections is reasonable. However, the difficulty in explaining the beam asymmetry indicates that more sophisticated and detailed analyses of the data are highly required to understand the production mechanism and to pin down the role of each baryon resonance.

In summary, we have explored the reactions of $K\Sigma(1385)$ photoproduction from the proton and the neutron targets.

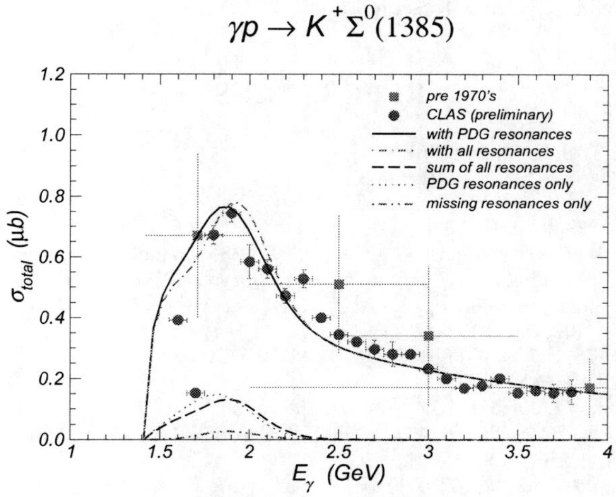

FIGURE 2. Total cross sections for the $\gamma p \to K^+\Sigma^0(1385)$ reaction with the resonance parameters listed in Table 1.

[3] The typographical errors committed in Ref. [15] are corrected here.

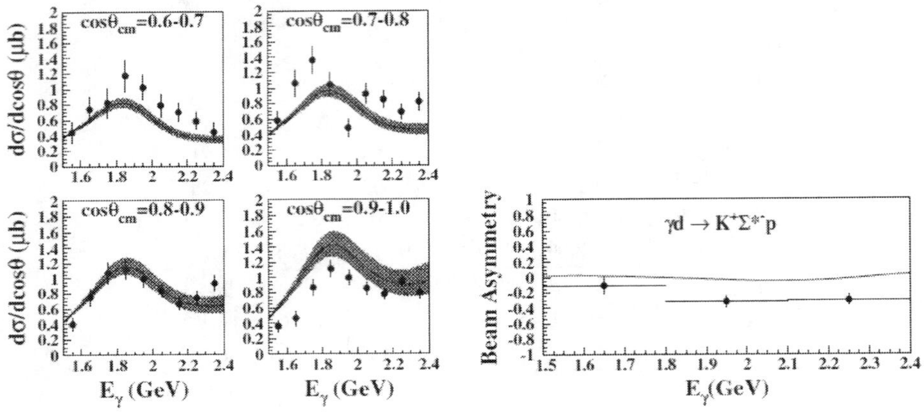

FIGURE 3. (Left panel) Differential cross sections for the $\gamma n \to K^+\Sigma^-(1385)$ reaction with the resonances listed in Table 1. (Right panel) Beam asymmetry of the same reaction.

We have constructed the Lagrangians for the interactions including baryon resonances up to spin-5/2, and the explicit formulas relating the coupling constants and the partial decay amplitudes were obtained. With the predictions of the quark model, we then calculate the cross sections and beam asymmetry of the reactions of $\gamma p \to K^+\Sigma^0(1385)$ and $\gamma n \to K^+\Sigma^-(1385)$, and the results are compared to the recent data measured by the CLAS and LEPS Collaborations. Our results for the cross sections manifestly indicate a nontrivial role of baryon resonances which are not-well-established or missing in the list of the Particle Data Group. However, the difficulty in explaining the measured beam asymmetry data for the $\gamma n \to K^+\Sigma^-(1385)$ reaction shows that much more detailed analyses are required to understand the production mechanisms of $\Sigma(1385)$.

ACKNOWLEDGMENTS

We are grateful to L. Guo, K. Hicks, C. M. Ko, K. Nakayama, and T.-S. H. Lee for fruitful discussions. This work was supported by Kyungpook National University Research Fund, 2010.

REFERENCES

1. R. Bradford, et al., *Phys. Rev. C* **73**, 035202 (2006).
2. R. G. T. Zegers, et al., *Phys. Rev. Lett.* **91**, 092001 (2003).
3. M. Q. Tran, et al., *Phys. Lett. B* **445**, 20 (1998).
4. K. Tsukada, et al., *Phys. Rev. C* **78**, 014001 (2008).
5. L. Guo, and D. P. Weygand, "Photoproduction of $K^{*+}\Lambda$ and $K^+\Sigma(1385)$ in the reaction $\gamma p \to K^+\Lambda\pi^0$ at Jefferson Lab," in *Proceedings of International Workshop on the Physics of Excited Baryons (NSTAR 05)*, edited by S. Capstick, V. Crede, and P. Eugenio, World Scientific, Singapore, 2006, pp. 306–309, `arXiv:hep-ex/0601010`.
6. K. Hicks, et al., *Phys. Rev. Lett.* **102**, 012501 (2009).
7. I. Hleiqawi et al., *Phys. Rev. C* **75**, 042201(R) (2007), **76**, 039905(E) (2007).
8. M. Nanova et al., *Eur. Phys. J. A* **35**, 333 (2008).
9. Y. Oh, and H. Kim, *Phys. Rev. C* **73**, 065202 (2006); **74**, 015208 (2006).
10. S. Capstick, and W. Roberts, *Phys. Rev. D* **58**, 074011 (1998); S. Capstick, *Phys. Rev. D* **46**, 2864 (1992).
11. J. H. R. Crouch, et al., *Phys. Rev.* **156**, 1426 (1967).
12. R. Erbe, et al., *Nuovo Cimento* **49A**, 504 (1967); R. Erbe, et al., *Phys. Rev.* **188**, 2060 (1969).
13. M. F. M. Lutz, and M. Soyeur, *Nucl. Phys. A* **748**, 499 (2005).
14. M. Döring, E. Oset, and D. Strottman, *Phys. Lett. B* **639**, 59 (2006).
15. Y. Oh, C. M. Ko, and K. Nakayama, *Phys. Rev. C* **77**, 045204 (2008).
16. H. Kamano, and M. Arima, *Phys. Rev. C* **69**, 025206 (2004).
17. Y. Oh, K. Nakayama, and T.-S. H. Lee, *Phys. Rep.* **423**, 49 (2006).
18. C. Amsler, et al., *Phys. Lett. B* **667**, 1 (2008), http://pdg.lbl.gov.

The energy dependence of the $pp \to K^+ n\Sigma^+$ reaction

Yu. Valdau[1] for the ANKE collaboration

Institut für Kernphysik, Forschungszentrum Jülich, D-52425 Jülich, Germany
High Energy Physics Department, Petersburg Nuclear Physics Institute, RU-188350 Gatchina, Russia

Abstract. The energy dependence of the total cross section for the $pp \to K^+ n\Sigma^+$ reaction has been investigated at the magnetic spectrometer COSY-ANKE. Signals from the production of the Σ^+ hyperon were searched for in three simultaneously measured spectra. The values obtained for the total production cross section $\sigma(\Sigma^+)$ are slightly below those of $\sigma(\Sigma^0)$ at the same excess energies. They follow a phase space dependence and do not show any evidence for strong threshold effects or a significant $n\Sigma^+$ final state interaction.

Keywords: Kaon production, Sigma production
PACS: 13.75.-n, 14.20.Jn, 14.40.Aq, 25.40.Ve

Light hyperon production in proton-proton collisions is an important part of the experimental program carried out at the COoler-SYnchrotron Jülich. The energy dependence of the total cross sections for Λ and Σ^0 production has been well investigated in the COSY range of excess energies ($\varepsilon < 1$ GeV) [1]. The ratio of the total cross sections for these two productions, $R(\Lambda/\Sigma^0) = \sigma_\Lambda/\sigma_{\Sigma^0}$, falls from a factor of 27 close to threshold down to 3 at high ε [2, 3]. At the moment, there is no agreed theoretical explanation for such a behavior. Calculations within the Jülich meson exchange model [4] suggest that the suppression of near-threshold Σ^0 production with respect to the Λ could be associated with differences between the production mechanisms and a destructive interference between K and π exchanges.

A recently published analysis [5] of the measured angular distributions for $pp \to K^+ p\Lambda$ and $pp \to K^+ p\Sigma^0$ reactions suggests that N^* resonances play a major role in Λ production whereas in the case of the Σ^0 no indication for such an effect has been observed. It is also concluded that kaon exchange is significant for Σ^0 production while it is less important for Λ [5]. However, these data were collected at relatively high excess energies, where the many partial waves and final-state interactions make the interpretation of the experimental data difficult.

The third light hyperon production channel, $pp \to K^+ n\Sigma^+$, has not been so well investigated. The primary reason for this is the presence of the neutron in the final state, which make any identification of the reaction complicated. On the other hand, this reaction channel is an isospin-3/2 filter for nucleon isobars [6]. Thus, the N^* resonances, which are believed to be important for Λ production [7], cannot contribute in the Σ^+ case.

Experimental data on $pp \to K^+ n\Sigma^+$ reaction close to threshold are scarce. Pioneering measurements, performed close to the reaction threshold at $\varepsilon = 13$ and 60 MeV using a neutron detector, reported surprisingly high values of the total cross sections [8]. These data could only be understood if the production mechanism and the final state interaction for the Σ^+ hyperon were completely different from those of other light hyperons.

The cross sections for the different Σ hyperon production channels can be linked together using a triangle inequality [9]. At high ε there is evidence that $\sigma(pp \to K^+ p\Sigma^0) \approx \sigma(pp \to K^0 p\Sigma^0)$ [10] and, assuming this, the inequality can be written as $1/6 < \sigma(pp \to K^+ p\Sigma^+)/\sigma(pp \to K^+ p\Sigma^0) < 6$ [11]. The ratio of total cross sections $R(\Sigma^+/\Sigma^0)$ obtained using the data from Ref. [8] is 127 and 60 at $\varepsilon = 13$ and 60 MeV, respectively. Thus, the values of Σ^+ total cross section would violate isospin invariance by a large factor unless there were a very striking threshold anomaly.

This unexpected result led to a series of investigations of the $pp \to K^+ n\Sigma^+$ reaction at COSY [11, 12]. These were based on a three-prong approach where the signal from the Σ^+ hyperon production was searched for in three simultaneously measured spectra. The measurements were done at the magnetic spectrometer ANKE, which is placed at an internal target station of the COSY-Jülich accelerator. The ANKE spectrometer consists of three dipole magnets. Two of them, $D1$ and $D3$, are used to deflect the circulating proton beam onto and away from the target, while the third one, $D2$, is the analyzing magnet. Positively charged particles from a reaction are deflected into the positive (Pd) and forward (Fd) detectors, while negatively charged particles can be registered in the so-called negative detector (not used in the analysis).

---

[1] e-mail adress:y.valdau@fz-juelich.de

12th International Conference on Meson-Nucleon Physics and the Structure of the Nucleon (MENU 2010)
AIP Conf. Proc. 1374, 189-192 (2011); doi: 10.1063/1.3647122

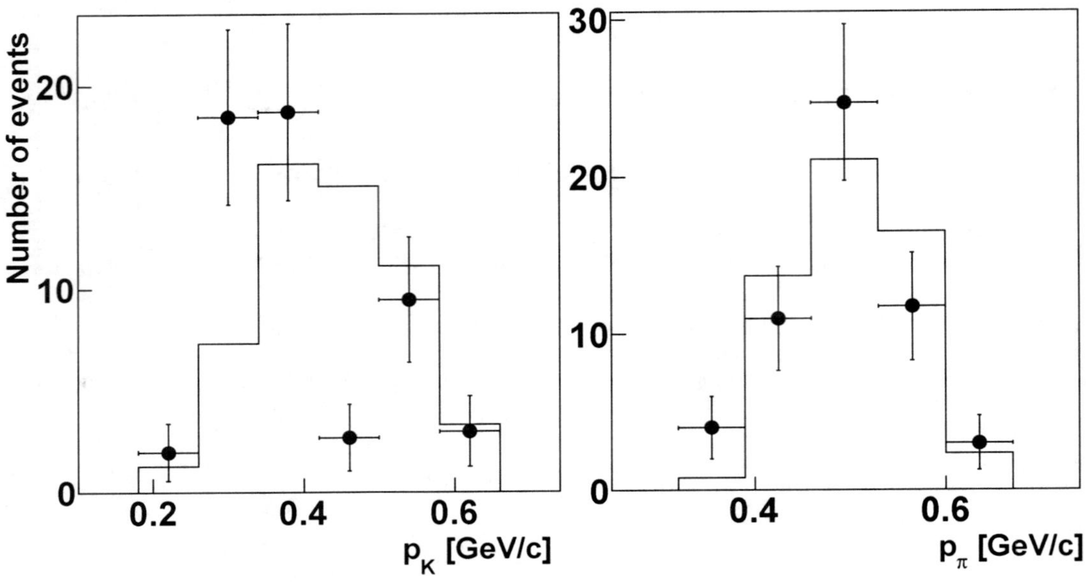

FIGURE 1. The K^+ and π^+ momentum spectra measured at $T_p = 1.958$ GeV in comparison to phase space simulations for the $pp \to K^+ n \Sigma^+$ reaction.

The Fd consists of a hodoscope of scintillator counters, two multi-wire proportional and one drift chamber. This was used for the identification of the high momentum protons from pp elastic scattering, used for normalization, and protons from the reactions of interest. In the analysis, the positive side detector allowed the π^+ and K^+ identification.

The Pd detector is made up of start counters, placed close to the exit window of the $D2$ vacuum chamber, two multi-wire proportional chambers, and stop counters. The first fifteen stop counters are part of the range telescopes used for K^+ identification on the basis of the delayed-veto technique. Each of the range telescopes consists of a stop counter, scintillator counter for measurements of the energy loss of the particles, a so-called veto counter for measurements of time, and two passive degraders. The thickness of the first degrader is chosen such that a K^+ deposits the maximum energy in the energy-loss counter and stops either at the edge of it or in the second degrader. The time signal from the K^+ decay products is then registered in the veto counter, allowing one to identify K^+ mesons against a 10^6 times higher background. The details of the ANKE detector analysis can be found in Ref. [13].

In contrast to the Λ and Σ^0 hyperons, the Σ^+ hyperon has a decay mode leading to a π^+ in the final state and this has a branching ratio of 50%. Thus, below the threshold of the $pp \to K^+ n \Lambda \pi^+$ reaction, the $pp \to K^+ n \Sigma^+$ reaction is the only source of $K^+\pi^+$ coincidences. Even at higher energies, the total cross section for $\Lambda\pi^+$ production is factor of seven smaller than the total cross section for Σ^+ production [9]. The identification of $K^+\pi^+$ coincidences therefore allows one to identify unambiguously Σ^+ production and to estimate the total cross section. The correlated K^+ and π^+ momentum spectra measured at 1.958 GeV are compared to phase-space simulations, corrected according to the ANKE acceptance, and presented in Fig. 1.

All three light hyperons, Λ, Σ^0 and Σ^+, have decay modes where there is a proton in the final state. The $pp \to K^+ p \Lambda/\Sigma^0$ reactions can be identified from the peaks in the experimental K^+p missing-mass spectra. These sit on physical backgrounds arising from hyperon decay. The total cross sections for Λ and Σ^0 production can be determined from the numbers of events in these peaks. However, since we have clean K^+ identification, the other events in the K^+p missing-mass spectra can also be analyzed and information about the total cross section for Σ^+ production extracted [12].

The K^+p missing-mass spectra measured at 2.062 and 2.157 GeV are compared to simulations in Fig. 2. Individual contributions from $\Lambda \to p\pi^-$, $\Sigma^0 \to \Lambda\gamma \to p\pi^-\gamma$ and $\Sigma^+ \to p\pi^0$ are shown by colored histograms. Despite the limited statistical precision in the spectra, it is already clear by eye that $R(\Sigma^+/\Sigma^0)$ is less than a factor of five. In fact, if one makes the drastic assumption that all the events above the kinematic limit for the Λ are to be associated with Σ^+ production, upper limits on $R(\Sigma^+/\Sigma^0)$ of 1 and 2 are obtained at 2.062 and 2.157 GeV, respectively. The data at 2.062 GeV were collected with a relatively low field in the spectrometer magnet $D2$, which means that the acceptance

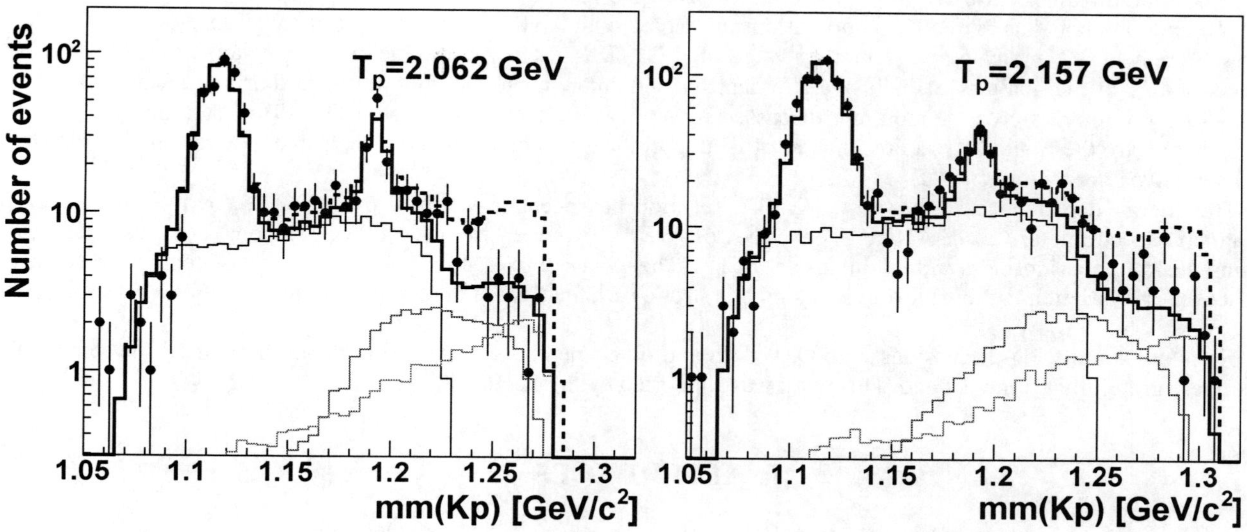

FIGURE 2. The K^+p missing-mass spectra measured at $T_p = 2.062$ and 2.157 GeV compared to simulations. Simulations for the Λ, Σ^0 and Σ^+ decay protons are shown by black, blue and red lines, respectively. Predictions for the complete spectra assuming $R(\Sigma^0/\Sigma^+) = 1$ and 5 are shown by solid and dashed lines, respectively. The numbers of events in the physical backgrounds are determined by the contents of the Λ and Σ^0 peaks.

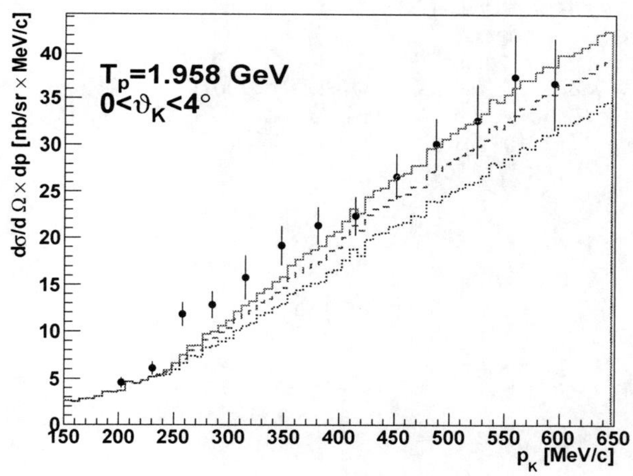

FIGURE 3. Inclusive K^+ momentum spectra for $\vartheta_{K^+} < 4°$ resulting from pp collisions at $T_p = 1.958$ GeV. The simulation of the $pp \to K^+p\Lambda$ reaction is shown by the black histogram. The addition of contributions from $pp \to K^+p\Sigma^0$ and $pp \to K^+n\Sigma^+$ are shown by blue and red histograms, respectively. The total cross sections for all reaction channels are taken from Ref. [12].

for the data accumulated at this energy significantly differs from the others.

Only very limited information about total cross section can be extracted from the K^+ inclusive spectra. Inclusive K^+ differential cross sections measured at $T_p = 1.958$ GeV and integrated over $\vartheta_{K^+} < 4°$ are presented in Fig. 3 in comparison to simulations. Individual contributions from different hyperon production channels, normalized to the corresponding total cross sections, are also shown.

A method to extract the Σ^+ total production cross section from a detailed analysis of K^+ inclusive spectra has, however, been proposed [14]. This was applied to new zero-degrees data in a recent paper by the HIRES collaboration and a value of $R(\Sigma^+/\Sigma^0) = 5$ at $T_p = 2.08$ GeV reported [15]. This high Σ^+ total production cross section follows the general trend of values extracted from measured K^+ inclusive spectra using this method [14].

The experimental data on the K^+p correlations, measured with ANKE spectrometer at very close energy 2.06 GeV, presented in Fig. 2, indicate a very modest value of $R(\Sigma^+/\Sigma^0)$, far below the factor of five claimed [15]. Possible uncertainties in the analysis of inclusive K^+ data, in particular in the assumptions regarding the behavior of the $pp \rightarrow K^+p\Lambda$ forward cross section, are discussed in detail in Ref. [16]. The new COSY-TOF data on differential observables for the Λ and Σ^0 reaction channels [5] support their conclusion that K^+ inclusive spectra can only lead to conservative upper limits on $\sigma(\Sigma^+)$ [16].

The energy dependence of the $pp \rightarrow K^+n\Sigma^+$ reaction has been investigated over a wide energy range at the magnetic spectrometer COSY-ANKE. The values of $\sigma(\Sigma^+)$, extracted from $K^+\pi^+$ correlations, are cross checked using the K^+p missing-mass and K^+ inclusive spectra. The values of $\sigma(\Sigma^+)$ thus obtained are consistent with a phase-space energy variation without any obvious sign of a strong $n\Sigma^+$ final state interaction or significant threshold anomaly.

Help and stimulating discussions with C. Wilkin and other members of ANKE collaboration on all the stages of work are gratefully acknowledged. This work was supported by the JCHP FFE.

REFERENCES

1. A. Sibirtsev, J. Haidenbauer, H. W. Hammer, U.-G. Meißner, *Eur. Phys. J. A* **29**, 363 (2006).
2. S. Sewerin et al., *Phys. Rev. Lett.* **83**, 682 (1999).
3. P. Kowina et al., *Eur. Phys. J. A* **22**, 293 (2004).
4. A. M. Gasparyan et al., *Phys. Lett. B* **480**, 273 (2000).
5. S. Abd El-Samad et al., *Eur. Phys. J. A* (2010) DOI 10.1140/epja/i2010-11023-0.
6. Ju-Jun Xie and Bing-Song Zou, *Phys. Lett. B* **649**, 405 (2007).
7. S. Abd El-Samad et al., *Phys. Lett. B* **632**, 27 (2006).
8. T. Rożek et al., *Phys. Lett. B* **643**, 251 (2006).
9. R.I. Loutitt et al., *Phys. Rev.* **123**, 1465 (1961).
10. M. Abdel-Bary et al., *Phys. Lett. B* **595**, 127 (2004).
11. Yu. Valdau et al., *Phys. Lett. B* **652**, 245 (2007).
12. Yu. Valdau et al., *Phys. Rev. C* **81**, 045208 (2010).
13. S. Barsov et al., *Nucl. Instrum. Methods Phys. Res. Sect. A* **462**, 354 (2001).
14. A. Sibirtsev, J. Haidenbauer, H. W. Hammer, U.-G. Meißner, *Eur. Phys. J. A* **32**, 229 (2007).
15. A. Budzanowski et al., *Phys. Lett. B* **692**, 10 (2010).
16. Yu. Valdau and C. Wilkin, arXiv:1008.3465v1 (2010).

Strangeness S=-3 and -4 baryon-baryon interactions in chiral EFT

Johann Haidenbauer

Institute for Advanced Simulation, Institut für Kernphysik, and Jülich Center for Hadron Physics, Forschungszentrum Jülich, D-52425 Jülich, Germany

Abstract. I report on recent progress in the description of baryon-baryon systems within chiral effective field theory. In particular, I discuss results for the strangeness $S = -3$ to -4 baryon-baryon systems, obtained to leading order.

Keywords: Hyperon-hyperon interaction, Effective field theory
PACS: 13.75.Ev, 12.39.Fe, 21.30.-x, 21.80.+a

INTRODUCTION

Chiral effective field theory (EFT) as proposed in the pioneering works of Weinberg [1] is a powerful tool for the derivation of nuclear forces. In this scheme there is an underlying power counting which allows to improve calculations systematically by going to higher orders in a perturbative expansion. In addition, it is possible to derive two- and corresponding three-nucleon forces as well as external current operators in a consistent way. For reviews we refer the reader to Refs. [2, 3, 4].

Over the last decade or so it has been demonstrated that the nucleon-nucleon (NN) interaction can be described to a high precision within the chiral EFT approach [5, 6]. Following the original suggestion of Steven Weinberg, in these works the power counting is applied to the NN potential rather than to the reaction amplitude. The latter is then obtained from solving a regularized Lippmann-Schwinger equation for the derived interaction potential. The NN potential contains pion-exchanges and a series of contact interactions with an increasing number of derivatives to parameterize the shorter ranged part of the NN force.

Recently, also hadronic systems involving the strange baryons Λ and Σ, and the $S = -2$ baryon Ξ were investigated within EFT by the group in Jülich [7, 8, 9, 10, 11]. Specifically, the interactions in the ΛN and ΣN channels [7] as well as those in the $S = -2$ sector ($\Lambda\Lambda$, $\Sigma\Sigma$, $\Lambda\Sigma$, ΞN) [8] were considered. In these works the same scheme as applied in Ref. [6] to the NN interaction is adopted. In the present contribution I focus on a recent extension of that study to systems with $S = -3$ and -4 [10].

FORMALISM

To leading order (LO) in the power counting, as considered in the aforementioned investigations [7, 8, 10], the baryon-baryon potentials involving strange baryons consist of four-baryon contact terms without derivatives and of one-pseudoscalar-meson exchanges, analogous to the NN potential of [6]. The potentials are derived using constraints from SU(3) flavor symmetry. Details on the derivation of the chiral potentials for the $S = -1$ to $S = -4$ sectors at LO using the Weinberg power counting can be found in Ref. [7]. The contributions of one-pseudoscalar-meson exchanges are identical to those already discussed extensively in the literature, see, e.g., [7]. The LO $\mathrm{SU}(3)_\mathrm{f}$ invariant contact terms for the octet baryon-baryon interactions that are Hermitian and invariant under Lorentz transformations follow from the Lagrangians

$$\begin{aligned}
\mathscr{L}^1 &= C_i^1 \langle \bar{B}_a \bar{B}_b (\Gamma_i B)_b (\Gamma_i B)_a \rangle \ , \quad \mathscr{L}^2 = C_i^2 \langle \bar{B}_a (\Gamma_i B)_a \bar{B}_b (\Gamma_i B)_b \rangle \ , \\
\mathscr{L}^3 &= C_i^3 \langle \bar{B}_a (\Gamma_i B)_a \rangle \langle \bar{B}_b (\Gamma_i B)_b \rangle \ . \qquad (1)
\end{aligned}$$

As described in [7], in LO the Lagrangians give rise to six independent low-energy coefficients (LECs) – C_S^1, C_T^1, C_S^2, C_T^2, C_S^3 and C_T^3 – that need to be determined by a fit to experimental data. Here S and T refer to the central and spin-spin

12th International Conference on Meson-Nucleon Physics and the Structure of the Nucleon (MENU 2010)
AIP Conf. Proc. 1374, 193-196 (2011); doi: 10.1063/1.3647123

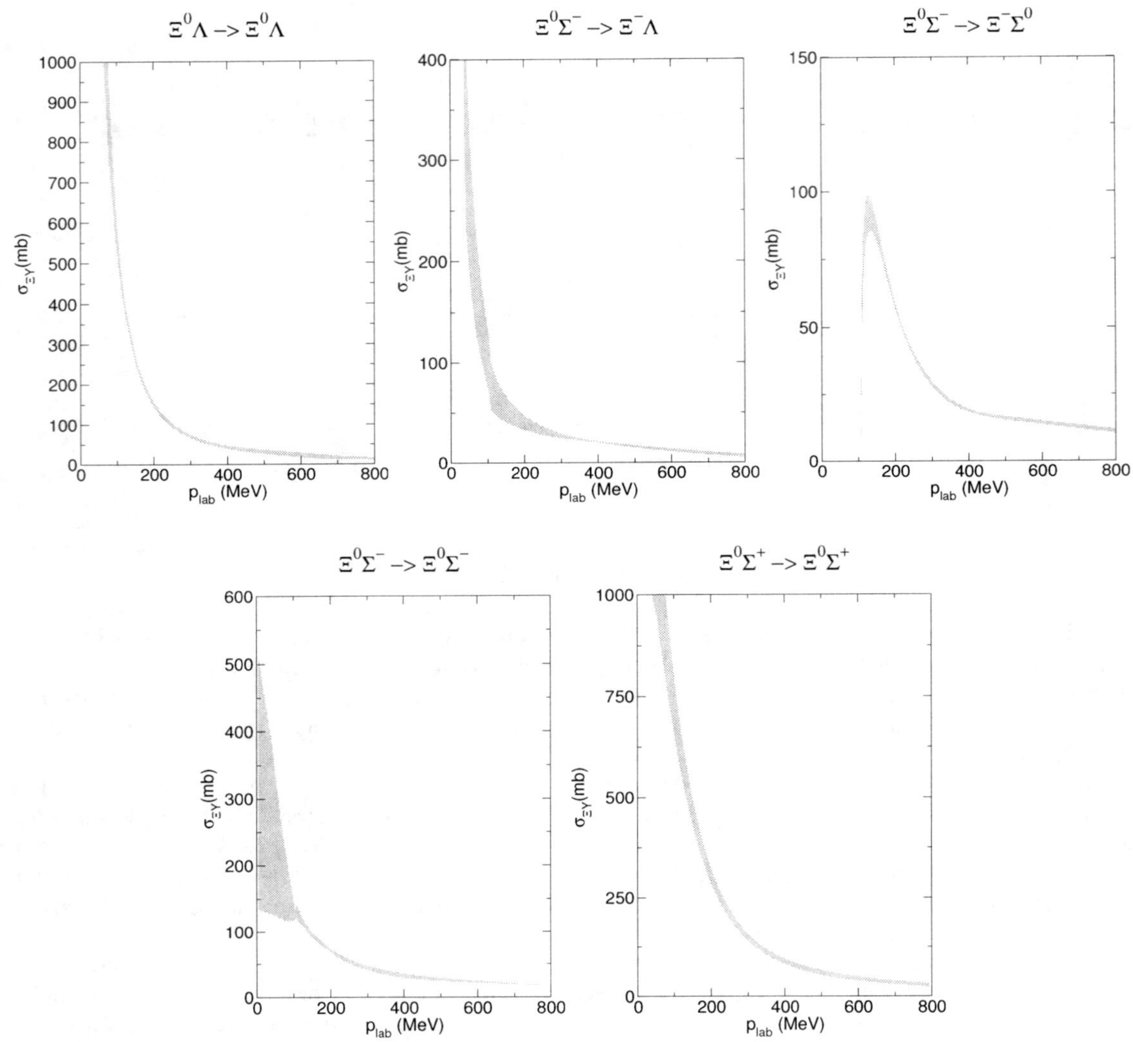

FIGURE 1. Total cross sections for various reactions in the strangeness $S = -3$ sector as a function of p_{lab}. The shaded band shows the chiral EFT results for variations of the cut-off in the range $\Lambda = 550 \ldots 700$ MeV.

parts of the potential, respectively. The scattering amplitude is obtained by solving a Lippmann-Schwinger equation for the LO potential. Thereby, the possible coupling between different baryon-baryon channels, $\Lambda N - \Sigma N$ or $\Xi\Lambda - \Xi\Sigma$, say, is taken into account. The potentials in the LS equation are cut off with a regulator function, $\exp\left[-\left(p'^4 + p^4\right)/\Lambda^4\right]$, in order to remove high-energy components of the baryon and pseudoscalar meson fields [6].

RESULTS AND DISCUSSION

The LO chiral EFT interaction for the $S = -3$ and -4 baryon-baryon sector depends only on those five contact terms that enter also in the YN interaction, cf. Table 1 in [10]. Thus, based on the values which were fixed in our study of the YN sector [7] we can make genuine predictions for the interaction in the $S = -3$ and -4 channels that follow from the imposed $SU(3)_f$ symmetry.

Corresponding results for the $\Xi^0\Lambda \to \Xi^0\Lambda$, $\Xi^0\Sigma^- \to \Xi^-\Lambda$, $\Xi^0\Sigma^- \to \Xi^-\Sigma^0$, $\Xi^0\Sigma^- \to \Xi^0\Sigma^-$, and $\Xi^0\Sigma^+ \to \Xi^0\Sigma^+$ scattering cross sections are presented in Fig. 1. Partial waves with total angular momentum up-to-and-including $J = 2$ are taken into account. The shaded bands show the cut-off dependence. From that figure one observes that the $\Xi^0\Lambda \to \Xi^0\Lambda$ and $\Xi^0\Sigma^+ \to \Xi^0\Sigma^+$ cross sections are rather large near threshold. Though the cross section for

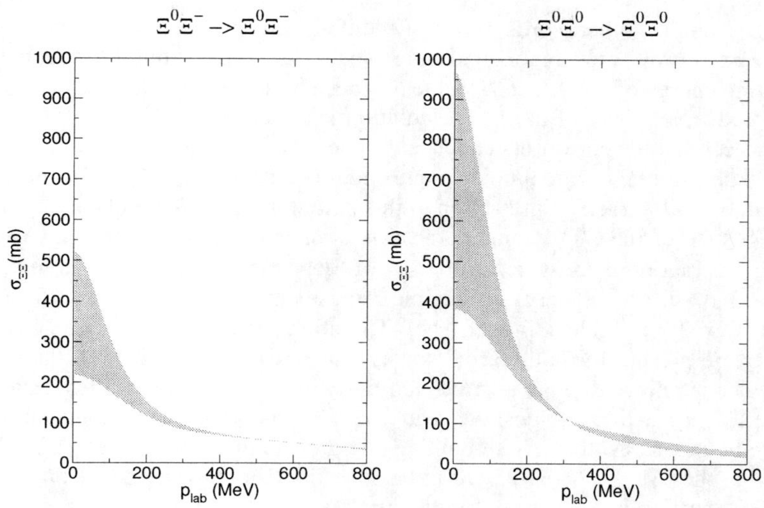

FIGURE 2. Total cross sections for the reactions $\Xi^0\Xi^- \to \Xi^0\Xi^-$ and $\Xi^0\Xi^0 \to \Xi^0\Xi^0$ as a function of p_{lab}. The shaded band shows the chiral EFT results for variations of the cut-off in the range $\Lambda = 550\ldots700$ MeV.

TABLE 1. Selected ΞY and $\Xi\Xi$ singlet and triplet scattering lengths a and effective ranges r (in fm) for various cut-off values Λ. The last columns show results for the Nijmegen potential (NSC97a, NSC97f) [15] and the model by Fujiwara et al. (fss2) [16].

	EFT				NSC97a	NSC97f	fss2
Λ (MeV)	550	600	650	700			
$a_s^{\Xi\Lambda}$	−33.5	35.4	12.7	9.07	−0.80	−2.11	−1.08
$r_s^{\Xi\Lambda}$	1.00	0.93	0.87	0.84	4.71	3.21	3.55
$a_t^{\Xi\Lambda}$	0.33	0.33	0.32	0.31	0.54	0.33	0.26
$r_t^{\Xi\Lambda}$	−0.36	−0.30	−0.29	−0.27	−0.47	2.79	2.15
$a_s^{\Xi^0\Sigma^+}$	4.28	3.45	2.97	2.74	4.13	2.32	−4.63
$r_s^{\Xi^0\Sigma^+}$	0.96	0.90	0.84	0.81	1.46	1.17	2.39
$a_t^{\Xi^0\Sigma^+}$	−2.45	−3.11	−3.57	−3.89	3.21	1.71	−3.48
$r_t^{\Xi^0\Sigma^+}$	1.84	1.72	1.70	1.70	1.28	0.96	2.52
$a_s^{\Xi\Xi}$	3.92	3.16	2.71	2.47	17.28	2.38	−1.43
$r_s^{\Xi\Xi}$	0.92	0.85	0.79	0.75	1.85	1.29	3.20
$a_t^{\Xi\Xi}$	0.63	0.59	0.55	0.52	0.40	0.48	3.20
$r_t^{\Xi\Xi}$	1.04	1.05	1.08	1.11	3.45	2.80	0.22

$\Xi^0\Sigma^- \to \Xi^-\Lambda$ rises too, in this case it is only due to the phase space factor $p_{\Xi^-\Lambda}/p_{\Xi^0\Sigma^-}$. There is a clear cusp effect visible in the $\Xi^0\Sigma^-$ cross section at $p_{lab} \approx 106$ MeV/c, i.e. at the opening of the $\Xi^-\Sigma^0$ channel. On the other hand, we do not observe any sizeable cusp effects in the $\Xi^0\Lambda$ cross section around $p_{lab} = 690$ MeV/c, i.e. at the opening of the $\Xi\Sigma$ channels. The latter is in line with the results reported by the Nijmegen group for their interactions [15], where a cusp effect in that channel is absent too. In this context I would like to remind the reader that the cusp seen in the corresponding strangeness $S = -1$ case, namely in the ΛN cross section at the ΣN threshold, is rather pronounced in our chiral EFT interaction [7] but also in conventional meson-exchange potential models [12, 13, 14].

Predicted cross sections for the $\Xi^0\Xi^0$ and $\Xi^0\Xi^-$ channels are shown in Fig. 2, again as a function of p_{lab} and with shaded bands that indicate the cut-off dependence.

Results for the $\Xi^0\Lambda$, $\Xi^0\Sigma^+$, and $\Xi\Xi$ scattering lengths and effective ranges are listed in Table 1. Here we also include predictions by other models [15, 16] for channels where pertinent results are available in the literature. This Table reveals the reason for the sizeable $\Xi^0\Lambda$ cross section predicted by the chiral EFT interactions, namely a rather large scattering length in the corresponding ${}^1S_0$ partial wave. It is obvious that its value is strongly sensitive to cut-off variations. It even changes sign (in other words, it becomes infinite) within the considered cut-off range. This means that a virtual bound state transforms into a real bound state, where the strongest binding occurs for the cut-off $\Lambda = 700$ MeV and leads to a binding energy of -0.43 MeV. While this behaviour is interesting per se, one certainly has to stress that in such a case the predictive power of our LO calculation is rather limited. One has to wait for at least an NLO calculation, where we expect that the cut-off dependence will become much weaker so that more reliable conclusions on the possible existence of a virtual or a real bound state should be possible. The $^1S_0$ scattering lengths of the other potentials suggest also an overall attractive interaction in this partial wave though only a very moderate one.

The results for the $^3S_1$ state of the $\Xi^0\Lambda$ channel are fairly similar for all considered interactions. Moreover, with regard to the chiral EFT interaction there is very little cut-off dependence. The S-waves in the $\Xi\Sigma$ $I = 3/2$ channel belong to the same (10^* and 27, respectively) irreducible representations where in the NN case real ($^3S_1 - {}^3D_1$) or virtual ($^1S_0$) bound states exist, cf. Table 1 in Ref. [10]. Therefore, one expects that such states can also occur for $\Xi\Sigma$. Indeed, bound states are present for both partial waves in the Nijmegen model, cf. the discussion in Sect. III.B in Ref. [15]. Their presence is reflected in the positive and fairly large singlet and triplet scattering lengths for $\Xi^0\Sigma^+$, cf. Table 1. The chiral EFT interaction has positive scattering lengths of comparable magnitude for $^1S_0$, for all cut-off values, and therefore bound states, too. These binding energies lie in the range of -2.23 MeV ($\Lambda = 550$ MeV) to -6.15 MeV (700 MeV). In the $^3S_1 - {}^3D_1$ partial wave the attraction is obviously not strong enough to form a bound state. The same is the case (but for both S waves) for the quark model fss2 of Fujiwara et al. [16].

The $^1S_0$ state of the $\Xi\Xi$ channel belongs also to the 27plet irreducible representation and also here the Nijmegen as well as the chiral EFT interactions produce bound states. In our case the binding energies lie in the range of -2.56 MeV ($\Lambda = 550$ MeV) to -7.28 MeV (700 MeV). The predictions of both approaches for the $^3S_1$ scattering length are comparable. The quark model of Fujiwara et al. exhibits a different behavior for the $\Xi\Xi$ channel, see the last column in Table 1. The small and negative $^1S_0$ scattering length signals an interaction that is only moderately attractive. The large and positive scattering length in the $^3S_1 - {}^3D_1$ partial wave, produced by that potential model, is usually a sign for the presence of a bound state, though according to the authors this is not the case for this specific interaction. Further results, and specifically $\Xi\Lambda$, $\Xi\Sigma$, and $\Xi\Xi$ phase shifts, can be found in [17].

SUMMARY

Our investigations show that the chiral EFT scheme, successfully applied in Ref. [6] to the NN interaction, also works well for the ΛN, ΣN [7, 11] and $\Lambda\Lambda$ [8] interactions. Moreover, as reported here, it can be used o make predictions for the $S = -3$ and -4 baryon-baryon interactions invoking constraints from SU(3) flavor symmetry. It will be interesting to see whether the new facilities J-PARC (Tokai, Japan) and FAIR (Darmstadt, Germany) allow access to empirical information about the interaction in the $S = -3$ and -4 sectors. Such information could come from formation experiments of corresponding hypernuclei or from proton-proton and antiproton-proton collisions at such high energies that pairs of baryons with strangeness $S = -3$ or $S = -4$ can be produced.

REFERENCES

1. S. Weinberg, Phys. Lett. B **251**, 288 (1990); S. Weinberg, Nucl. Phys. B **363**, 3 (1991).
2. P. F. Bedaque, U. van Kolck, Annu. Rev. Nucl. Part. Sci. **52**, 339 (2002).
3. E. Epelbaum, Prog. Part. Nucl. Phys. **57**, 654 (2006).
4. E. Epelbaum, H. W. Hammer and U.-G. Meißner, Rev. Mod. Phys. **81**, 1773 (2009).
5. D. R. Entem, R. Machleidt, Phys. Rev. C **68**, 041001 (2003).
6. E. Epelbaum, W. Glöckle, U.-G. Meißner, Nucl. Phys. A **747**, 362 (2005).
7. H. Polinder, J. Haidenbauer and U.-G. Meißner, Nucl. Phys. A **779**, 244 (2006).
8. H. Polinder, J. Haidenbauer and U.-G. Meißner, Phys. Lett. B **653**, 29 (2007).
9. J. Haidenbauer, Nucl. Phys. A **827**, 336c (2009).
10. J. Haidenbauer, U.-G. Meißner, Phys. Lett. B **684**, 275 (2010).
11. J. Haidenbauer, Nucl. Phys. A **835**, 168 (2010).
12. J. Haidenbauer, U.-G. Meißner, Phys. Rev. C **72**, 044005 (2005).
13. J. Haidenbauer, Eur. Phys. J. A **33**, 287 (2007).
14. T. A. Rijken, V. G. J. Stoks, Y. Yamamoto, Phys. Rev. C **59**, 21 (1999).
15. V. G. J. Stoks and T. A. Rijken, Phys. Rev. C **59**, 3009 (1999).
16. Y. Fujiwara, Y. Suzuki and C. Nakamoto, Prog. Part. Nucl. Phys. **58**, 439 (2007).
17. J. Haidenbauer, EPJ Web of Conferences **3**, 01009 (2010).

Effect of $\Lambda(1405)$ on structure of multi-antikaonic nuclei

Takumi Muto*, Toshiki Maruyama† and Toshitaka Tatsumi**

*Department of Physics, Chiba Institute of Technology, 2-1-1 Shibazono, Narashino, Chiba 275-0023, Japan
†Advanced Science Research Center, Japan Atomic Energy Agency, Tokai, Ibaraki 319-1195, Japan
**Department of Physics, Kyoto University, Kyoto 606-8502, Japan

Abstract. The second-order effects (SOE) including a pole contribution of the $\Lambda(1405)$ (Λ^*) and range terms on the structure of multi-antikaonic nuclei are investigated in a relativistic mean-field theory coupled with the nonlinear effective chiral Lagrangian. It is shown that, due to the attractive interaction from the Λ^*-pole contribution, the K^- mesons and the protons are more attracted each other than the case without the SOE. On the other hand, the density distribution for neutron is pushed outward due to the repulsive effect from the range terms. The SOE on the formation of the bound state of K^- mesons are also discussed associated with modification of relevant kaonic modes in a dense medium.

Keywords: multi-antikaonic nuclei, chiral symmetry, kaon condensation, subthreshold resonance $\Lambda(1405)$
PACS: 21.85.+d, 11.30.Rd, 21.65.Jk, 26.60.-c

INTRODUCTION

In a field of strangeness nuclear physics, in-medium kaon properties have been elucidated both theoretically basd on chiral symmetry and experimentally through K^- absorptions in nuclei and heavy-ion colisions etc[1]. Such studies of in-medium kaon properties have been motivated, on the one hand, by studies of kaon condensation in neutron stars from viewpoints of nuclear physics and astrophysics[2] and have, in turn, provided information on a possible existence of kaon condensation.

In a series of researches on kaon dynamics, deeply bound kaonic nuclei have been proposed based on the strong $\bar{K}-N$ attractive interaction[1, 3]. Stimulated by studies of kaon condensation and kaonic nuclei, multi-antikaonic nuclei (MKN), where several numbers of antikaons (K^-) are embedded in nuclei, have been investigated theoretically[4, 5, 6]. We have considered properties of the MKN in a relativistic mean-field theory (RMF) coupled with the nonlinear effective chiral Lagrangian[4]. Modification of kaon dynamics controlled by the number of the K^- mesons have been considered. Similarity and difference between the MKN and kaon condensation in neutron stars have also been discussed. It has been pointed out that repulsive $\bar{K}-\bar{K}$ interaction becomes sizable in comparison with the attractive $\bar{K}-N$ interaction as the number of embedded K^- mesons, $|S|$, increases. Thereby the lowest K^- energy, ω_{K^-}, increases with $|S|$, and it enters into the subthreshold resonance region of the $\Lambda(1405)$ (Λ^*), where $\omega_{K^-} \simeq \delta m_{\Lambda^* N}$ ($\equiv m_{\Lambda^*} - m_N$ = 467 MeV).

Much discussion has been made about the nature of the Λ^* including a two-pole structure for Λ^*[7, 8]. In this paper, we take into account the Λ^* as a pole contribution of a point particle as well as range terms and clarify these effects on the structure of the MKN and formation of bound states.

FORMULATION

The K^- mesons in the MKN are represented in the condensate approximation as a classical K^- field, $K^-(r) = f\theta(r)/\sqrt{2}$, where $\theta(r)$ is the chiral angle and f (=93 MeV) is the meson decay constant. On the assumption of spherical symmetry for the MKN, $\theta(r)$ is a function of the radial distance r from the center of the MKN. The mass number A, the number of protons Z, and the number of the embedded K^- mesons $|S|$ are kept conserved.

Within the RMF, the $N-N$ interactions are brought about through the exchange of σ, ω, and ρ mesons. On the other hand, we start with the nonlinear effective chiral Lagrangian for kaon dynamics including the $\bar{K}$-N and $\bar{K}$-$\bar{K}$ interactions. In this Lagrangian, the s-wave $\bar{K}-N$ scalar interaction is simulated by the KN sigma term, Σ_{KN} and the vector interaction (Tomozawa-Weinberg term) is incorporated model-independently. Then the contact interactions between the nonlinear K^- field and nucleons are replaced by the σ, ω, and ρ mesons-exchange through

12th International Conference on Meson-Nucleon Physics and the Structure of the Nucleon (MENU 2010)
AIP Conf. Proc. 1374, 197-200 (2011); doi: 10.1063/1.3647124

the relations, $g_{\sigma N}g_{\sigma K}/m_\sigma^2 = \Sigma_{KN}/(2m_K f^2)$, $g_{\omega N}g_{\omega K}/m_\omega^2 = 3/(8f^2)$, and $g_{\rho N}g_{\rho K}/m_\rho^2 = 1/(8f^2)$, where g_{iK}, g_{iN}, and m_i $(i = \sigma, \omega, \rho)$ are the coupling constants and meson masses, and m_K is the free kaon mass[4].

The thermodynamic potential Ω for the MKN is derived under the Thomas-Fermi approximation for the nucleons[4]. The Ω consists of contributions from the nucleons, K^- mesons, other mesons (σ, ω, ρ), and the Coulomb potential.

We further introduce a correction to the energy density obtained by the second-order perturbation with respect to the axial-vector hadronic current carrying the K^- quantum number, $\hat{A}_5^\mu = f\partial^\mu K^- + \cdots + (g_{\Lambda^*}/2)(\bar{\Lambda}^*\gamma^\mu p + \text{h.c.}) + \cdots$ with g_{Λ^*} being the coupling constant for $K^- p\Lambda^*$ vertex[2]. The effects of the energy density correction (we call the second-order effects, SOE) are separated into a pole term from the Λ^* and smooth parts as the range terms. These terms are absorbed into the effective nucleon masses. The classical K^- field equation is then given from $\delta\Omega/\delta\theta(r) = 0$ as

$$\begin{aligned} \nabla^2\theta(r) &= \sin\theta(r)\Bigg[m_K^{*2}(r) - 2\tilde{\omega}_{K^-}(r)X_0(r) - \tilde{\omega}_{K^-}^2(r)\cos\theta(r)\Big\{1 + d_n\rho_n^s(r)\Big\} \\ &- \tilde{\omega}_{K^-}^2(r)\cos\theta(r)\rho_p^s(r)\left\{d_p + \frac{g_{\Lambda^*}^2}{2f^2}\frac{\delta m_{\Lambda^*N} - \omega_{K^-}}{(\delta m_{\Lambda^*N} - \omega_{K^-})^2 + \gamma_{\Lambda^*}^2}\right\}\Bigg], \end{aligned} \quad (1)$$

where $m_K^{*2}(r) = m_K^2 - 2g_{\sigma K}m_K\sigma(r)$ with $\sigma(r)$ being the σ-mean field, and $\tilde{\omega}_K(r)$ $[\equiv \omega_K - V_{\text{Coul}}(\mathbf{r})]$ is the lowest energy of the K^- shifted by the Coulomb potential $V_{\text{Coul}}(\mathbf{r})$. The term including $X_0(r)$ stands for the KN vector interaction with $X_0(r) = g_{\omega K}\omega_0(r) + g_{\rho K}R_0(r)$, where $\omega_0(r)$ and $R_0(r)$ are the time components of the vector mean fields. The terms proportional to $d_p\rho_p^s, d_n\rho_n^s$ with $\rho_i^s(r)$ $(i = p, n)$ being the nucleon scalar density are the range terms, and the last term in the bracket of Eq. (1) is the pole term from the Λ^* with the width γ_{Λ^*}. The parameters, d_p, d_n, g_{Λ^*}, and γ_{Λ^*} are determined so as to reproduce the on-shell s-wave K-N scattering lengths[9]. One obtains $d_p = (0.35 - \Sigma_{KN}/m_K)/(f^2 m_K)$, $d_n = (0.23 - \Sigma_{KN}/m_K)/(f^2 m_K)$, g_{Λ^*}=0.58, and γ_{Λ^*}=12.4 MeV.

Together with Eq. (1) one obtains the coupled equations of motion (EOM) for the other mesons σ, ω, ρ, and the Poisson equation for the Coulomb potential $V_{\text{Coul}}(r)$ from $\delta\Omega/\delta\phi = 0$ ($\phi = \sigma$, ω_0, R_0, V_{Coul})[4, 5]. The values of g_{iN} $(i = \sigma, \omega, \rho)$, are determined so as to reproduce not only the properties of normal nuclear matter with saturation density ρ_0 = 0.153 fm^{-3}, but also the binding energy, proton-mixing ratio, and density distributions of proton and neutron for normal nuclei[4]. The coupling constants $g_{\omega K}$ and $g_{\rho K}$ are chosen from the quark and isospin coupling rule as $g_{\omega K} = g_{\omega N}/3$, $g_{\rho K} = g_{\rho N}$. The unknown parameter $g_{\sigma K}$ is related to the K^- optical potential U_K $[= -(g_{\sigma K}\sigma + g_{\omega K}\omega_0)]$ in symmetric nuclear matter. The value of U_K is set to be -80 and -120 MeV, which corresponds to Σ_{KN}=332 MeV and 754 MeV, respectively.

NUMERICAL RESULTS

Effect of the Λ^* pole on density distributions

Throughout this paper, a reference nucleus ($|S| = 0$) is taken to be the $^{15}_{8}$O (A=15, Z=8). First we summarize the results on the density distributions of protons, neutrons, and K^- mesons in the MKN. Without the SOE, even for $|S|$=1 and 2, the protons and neutrons are attracted to the K^- mesons in the central region due to the $\bar{K} - N$ attractive interaction, and the central baryon density $\rho_B^{(0)}$ (=$\rho_p(r = 0) + \rho_n(r = 0)$) slightly increases as compared with the case $|S|$=0[4]. The protons are attracted more than neutrons since the $\bar{K} - N$ attractive interaction is stronger for the isospin $I = 0$ than for $I = 1$. With the introduction of the SOE, due to the additional attraction from the Λ^*-pole contribution, the K^- mesons and protons are more attracted each other than the case without the SOE. On the other hand, the density distribution for neutron is pushed outward due to the repulsive effect from the range terms. These features become remarkable for large values of $|S|$. See Ref. [5] for details.

The gross structure of the MKN is characterized as follows: (I) In the central part of the MKN, densities of baryons and K^- mesons become as large as 3.5 ρ_0 for $|S| \sim 8$. (II) In the outer part of the MKN, neutron skin structure with a thickness δ_{np}, which we define as $\delta_{np} \equiv \sqrt{\langle r^2\rangle_n} - \sqrt{\langle r^2\rangle_p}$ with $\sqrt{\langle r^2\rangle_i}$ $(i = p, n)$ being the root-mean-square radii, becomes remarkable due to the SOE. [δ_{np}=(0.23−0.90) fm for $|S|$=2−8]. Observation of δ_{np} in the MKN may have a link to understanding properties of neutron-rich matter such as symmetry energy at subnuclear densities [10]. The gross structure of the MKN is little dependent upon the choice of the KN σ term, since the attraction from the $\bar{K} - N$ scalar interaction simulated by the Σ_{KN} and the repulsive effect from the range terms, both of which are of order $O(m_K^2)$, are compensated with each other [Eq. 1)].

Interplay of kaonic modes

To see the effects of the SOE on the ground state properties of the embedded K^- mesons in the MKN, we show, in Fig. 1, the lowest K^- energy ω_{K^-} as a function of $|S|$. The dashed-dotted lines are the results with the SOE, and the solid lines are our previous results without the SOE[4]. In the case of $U_K = -80$ MeV, the ω_{K^-} lies well below

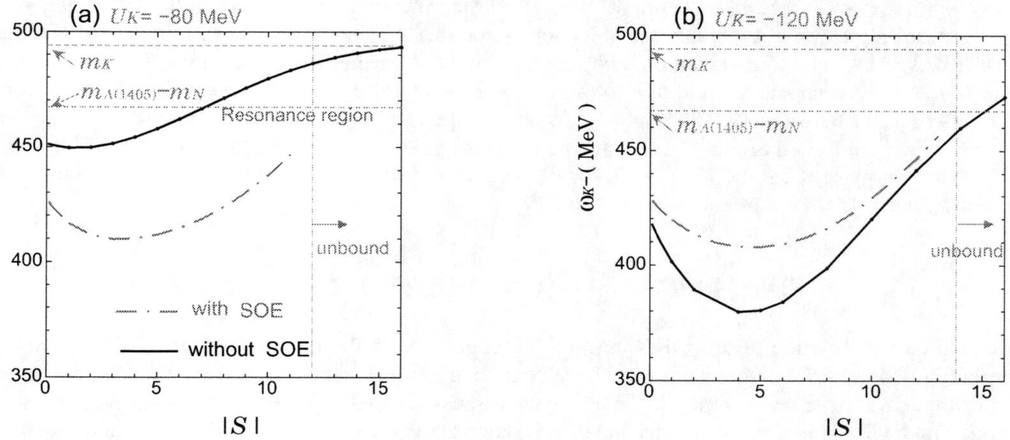

FIGURE 1. (a) The lowest K^- energy, ω_{K^-}, for the MKN with A=15, Z=8, as functions of $|S|$ in the case of U_K=−80 MeV. (b) The same as (a) but for U_K=−120 MeV.

the Λ^* resonance region as a result of the additional attraction from the Λ^* pole. The range terms ($\propto d_p\rho_p^s, d_n\rho_n^s$ where $d_p, d_n < 0$ in Eq. (1)) have repulsive contributions to ω_{K^-}, athough they have a minor effect. On the other hand, in the case of $U_K = -120$ MeV (Fig. 1 (b)), the repulsive effect from the range terms is large because of the large value of Σ_{KN} and overcomes the attraction from the Λ^* pole. As a result, the ω_{K^-} with the SOE is slightly pushed up as compared with that without the SOE. As seen from Figs. 1 (a) and (b), the dependence of ω_{K^-} on the value of Σ_{KN} or U_K is weak, as is the case with the result of the density distributions.

With increase in $|S|$, the repulsive $\bar{K}-\bar{K}$ interaction becomes as large as the attractive $\bar{K}-N$ interaction, and the lowest K^- energy ω_{K^-} increases with $|S|$ for $|S| \gtrsim 5$. For $|S| \geq 12$ in the case of $U_K = -$ 80 MeV [for $|S| \geq 14$ in the case of $U_K = -$ 120 MeV], the K^- mesons become unbound, where $\omega_{K^-} \gtrsim m_{\Lambda^*} - m_N$ above the Λ^*-resonance region.

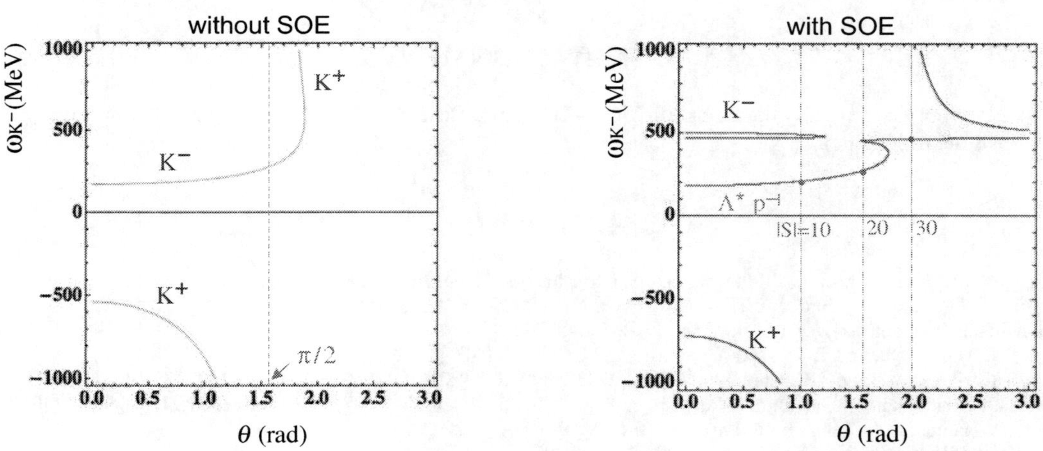

FIGURE 2. The lowest energy of the K^- for kaonic branches as functions of θ in matter with $\rho_p = Z/A \cdot \rho_B$ and $\rho_n = (A-Z)/A \cdot \rho_B$ for $\rho_B = 3.5\rho_0$ in the case of U_K=−80 MeV. The result without (with) the SOE is shown in the left (right) panel.

Here we consider interplay of the K^- and $\Lambda^* p^{-1}$ branches (the superscript "−1" denotes the hole state) and qualitatively explain a sudden disappearance of a bound state of the K^- mesons at a large value of $|S|$ by the help of a simple model, where we neglect finite-size effects of the MKN within a uniform matter approximation and omit the Coulomb potential $V_{Coul}(r)$. From the classical field equation for θ [Eq. (1) with neglect of r-dependence for all the relevant quantities], one obtains the lowest energy of K^-, $\omega_{K^-}(\rho_B, \theta)$ as functions of ρ_B and θ. In the right panel of Fig. 2, the lowest energies of the K^- for kaonic branches with the SOE are shown as functions of θ in the case of U_K=-80 MeV for $\rho_B = 3.5\rho_0$, which corresponds to the baryon density in the central region of the MKN with many K^- mesons. For comparison, the result without the SOE at the same density is shown in the left panel. Without the SOE, there appear only K^- and K^+ branches, which tend to be free K^- and K^+ mesons as $\rho_B \to 0$. Beyond $\theta = \pi/2$, K^+ branch makes a discontinuous transition, and the K^- and K^+ branches merge at a large θ. With the SOE (right panel), the $\Lambda^* p^{-1}$ branch appears in addition to the $K^\pm$ branches. These branches reveal complicated structure as a result of level-crossings between several branches. As seen in the right panel of Fig. 2, the lowest energy state of the bound K^- mesons is composed of the $\Lambda^* p^{-1}$ branch up to $|S| \lesssim 25$, while it transits to the K^- branch discontinuously for $|S| > 25$. The energy ω_{K^-} for the K^- branch for $|S| > 25$ is $\sim$ 500 MeV (the free kaon mass), so that one cannot obtain a bound state of the K^- mesons for such a large $|S|$.

SUMMARY AND CONCLUDING REMARKS

We have studied the structure of the MKN in the RMF coupled with the nonlinear effective chiral Lagrangian. A pole contribution from the Λ^* and range effects have been taken into account, and these effects on the structure and formation of the MKN have been clarified. We also have discussed modification of in-medium properties of the K^- mesons bound in the MKN in a simple model and its relevance to the gross structure of the MKN and possible binding of the K^- mesons in the MKN. The interplay of K^- and $\Lambda^* p^{-1}$ branches should be clarified more realistically based on our present framework for the MKN. The work will be reported elsewhere.

The framework should be improved to include hyperons toward a more realistic consideration of the MKN. For instance, hyperons may be relevant to the formation, structure, and decay of the MKN through inelastic channel-coupling processes, $\bar{K}N \to \pi\Lambda, \pi\Sigma$ etc. It has also been suggested in a liquid-drop picture that hyperon-mixing should lead to deeper binding of the MKN[11]. Possible existence of such deeply-bound objects composed of antikaons, nucleons, and hyperons is controversial[12]. More realistic consideration including finite-size effects based on the present framework is necessary. An extension of the RMF to include hyperons, taking into account the p-wave coupling between K^-, nucleon, and hyperons, may be important.

Formation mechanisms of the MKN in terrestrial experiments should also be explored in detail. Search for fragments including K^- mesons in heavy-ion collisions, multi-step fusions of single K^- nuclei, and so on, are simple scenarios for creating the MKN.

ACKNOWLEDGMENTS

This work is supported in part by the Grant-in-Aid for Scientific Research (No. 20028009) and by the funds provided by Chiba Institute of Technology.

REFERENCES

1. B. F. Gibson, K. Imai, T. Motoba, T. Nagae, and A. Ohnishi (Eds.) *Nucl.Phys.* **A 835**, 1(2010).
2. H. Fujii, T. Maruyama, T. Muto, and T. Tatsumi, *Nucl. Phys.* **A 597**, 645–671 (1996).
3. Y. Akaishi and T. Yamazaki, *Phys. Rev.* **C 65**, 044005-1–9 (2002).
4. T. Muto, T. Maruyama, and T. Tatsumi, *Phys. Rev.* **C 79**, 035207-1–16 (2009).
5. T. Muto, T. Maruyama, and T. Tatsumi, *Nucl Phys.* **A 835**, 370–373 (2010); *AIP Conf. Proc.* **1269**, 233–237 (2010).
6. D. Gazda, E. Friedman, A. Gal, and J. Mareš, *Phys.Rev.* **C 76**, 055204-1–11 (2007); *Phys.Rev.* **C 77**, 045206-1–8 (2008).
7. E. Oset, A. Ramos, and C. Bennhold, *Phys.Lett.* **B 527**, 99–105 (2002).
8. T. Hyodo and W. Weise, *Phys.Rev.* **C 77**, 035204-1–14 (2008).
9. A. D. Martin, *Nucl. Phys.* **B 179**, 33–48 (1981).
10. B. A. Brown,*Phys. Rev. Lett.* **85**, 5296–5299 (2000).
11. T. Muto, *Nucl.Phys.* **A 804**, 322–348 (2008).
12. D. Gazda, E. Friedman, A. Gal, and J. Mareš, *Phys.Rev.* **C 80**, 035205-1–8 (2009).

Low-lying spectra of $^{9}_{\Lambda}$Be and ^{9}Be within three-cluster model

I. Filikhin, V. M. Suslov and B. Vlahovic

North Carolina Central University, 1801 Fayetteville Street, Durham, NC 27707, USA

Abstract. An α-cluster model is applied to study the low-lying spectra of the $^{9}_{\Lambda}$Be and ^{9}Be nuclei. The $\alpha\alpha\Lambda$ and $\alpha\alpha n$ three-body problems are numerically solved by the Faddeev equations in configuration space using phenomenological pair potentials with spin-orbital $\alpha\Lambda$ and αn interactions taken into account. For the $^{9}_{\Lambda}$Be hypernucleus we found a set of the potentials that reproduces the experimental data for the ground state ($\frac{1}{2}^{+}$) binding energy and excitation energy of the $\frac{5}{2}^{+}$ and $\frac{3}{2}^{+}$ states, simultaneously. The *LS* coupling scheme is used for partial wave analysis. The total orbital momentum is fixed for each energy level. Under this assumption we calculated the ^{9}Be spectrum within $\alpha\alpha n$ model. The experiential data are well reproduced by the model, when a new classification for low-lying levels of ^{9}Be as members of spin-flip doublets is applied.

Keywords: Few-body systems,Cluster models,Hypernuclei
PACS: 21.45.-v, 21.60.Gx, 21.80.+a

INTRODUCTION

In the present work we are studying the $^{9}_{\Lambda}$Be and ^{9}Be nuclei within the phenomenological cluster models $\alpha+\alpha+\Lambda$ and $\alpha+\alpha+n$. The choice of pair potentials is the main problem for the cluster calculations [1, 2, 3, 4, 5]. In this work we are using the $\alpha\alpha$ and αn potentials, which simulate the Pauli principle by the repulsive core of the *s*-wave components. The $\alpha\alpha$ nuclear interaction is described by the Ali - Bodmer potential [6]. For the $\alpha\Lambda$ interaction the potential is rather uncertain, since the $\alpha\Lambda$ experimental data are limited by the value of the binding energy of the $^{5}_{\Lambda}$He hypernucleus, which is considered as the bound *s*-wave state of the $\alpha\Lambda$ system. In our calculations several phenomenological models are applied for the central part of $\alpha\Lambda$ interaction and we are proposing a new model, motivated by the Tang-Herndon potential [7]. Also we are focused on the spin-orbit component of $\alpha\Lambda$ interaction and introduce a new model, by adjusting the parameters of the $\alpha\Lambda$ spin-orbit potential from [4], which reproduces experimental data for the ground state ($\frac{1}{2}^{+}$) binding energy and excitation energy of the $\frac{5}{2}^{+}$ and $\frac{3}{2}^{+}$ states, simultaneously. The three-body model similar to $^{9}_{\Lambda}$Be is constructed for ^{9}Be nucleus using the *LS*-coupling scheme for partial wave and spin analysis. Within this approximation, each ^{9}Be energy level is defined as a member of spin-flip doublet, corresponding to the total orbital momentum of the system. A phenomenological αn potential adjusted to reproduce the $\alpha - n$ scattering data from [8] is used. Since the proposed model reproduces well the ^{9}Be spectral data, we are introducing a new classification for the ^{9}Be low-lying spectrum, as a set of spin-flip doublets. Mathematical basis of our calculations is the method of the Faddeev equations in the configuration space [9]. To estimate the energies and widths of low-lying resonance states applied is the method of analytical continuation in coupling constant. A variant of this method [10, 11] with an additional non-physical three-body potential is used. The strength parameter of this potential is considered as a variational parameter for analytical continuation of the bound state energy into the complex plane.

CALCULATIONS

Formalism

For the $\alpha\alpha\Lambda$ system including two identical particles the coupled set of the Faddeev equations is written as:

$$
\begin{aligned}
(H_0+V^{s}_{\alpha\Lambda}+V^{Coul.}-E)W &= -V^{s}_{\alpha\Lambda}(U+P_{12}W),\\
(H_0+V^{s}_{\alpha\alpha}+V^{Coul.}-E)U &= -V^{s}_{\alpha\alpha}(W+P_{12}W),
\end{aligned}
\qquad (1)
$$

where U is the Faddeev component corresponding to the rearrangement channel $(\alpha\alpha)-\Lambda$ and W corresponds to the rearrangement channel $(\alpha\Lambda)-\alpha$. The total wave function is expressed by the components U and W through relation

12th International Conference on Meson-Nucleon Physics and the Structure of the Nucleon (MENU 2010)
AIP Conf. Proc. 1374, 201-204 (2011); doi: 10.1063/1.3647125

$\Psi = U + (1+P_{12})W$. In above equations P_{12} is the permutation operator for the α particles (particles 1,2), $V^{Coul.}$ is the potential of the Coulomb interaction between the α particles, and $V^s_{\alpha\alpha}$ and $V^s_{\alpha\Lambda}$ are nuclear potentials of $\alpha\alpha$ and $\alpha\Lambda$ interactions, respectively. The total orbital angular momentum is given by $\vec{L} = \vec{\ell}_{\alpha\alpha} + \vec{\lambda}_{(\alpha\alpha)-\Lambda} = \vec{\ell}_{\alpha\Lambda} + \vec{\lambda}_{(\alpha\Lambda)-\alpha}$, where $\ell_{\alpha\alpha}$ ($\ell_{\alpha\Lambda}$) is the orbital angular momentum of $\alpha\alpha$ ($\alpha\Lambda$ pair) and $\lambda_{(\alpha\alpha)-\Lambda}$ ($\lambda_{(\alpha\Lambda)-\alpha}$) is the orbital angular momentum of a Λ hyperon (α particle) relatively to the center of mass of the $\alpha\alpha$ pair and $\alpha\Lambda$ pair, respectively. The LS coupling scheme for spin and partial wave analysis of Eq. (1) can be written as $\{(\lambda l)L, (\sigma s)S\}$, where σ is spin of the pair (1,2) and s is the spin of third particle [3].

The general functional form for the potentials, used in this work, is one or two range Gaussian:

$$V^l(r) = V^l_{rep}\exp(-(\beta^l_{rep}r)^2) - V^l_{att}\exp(-(\beta^l_{att}r)^2) \ . \tag{2}$$

Parameters of the $\alpha\alpha$, $\alpha\Lambda$ and αn potentials are given in Tabl. 1.

${}^9_\Lambda$Be three-body model: selection of the parameters for potentials

The ${}^9_\Lambda$Be hypernucleus has two bound states: ground state with total angular momentum $L=0$ (0^+ state) and excited state with $L=2$ (2^+ state). The latter is splitted into two states (spin-flip doublet) due to the $\alpha\Lambda$ spin-orbit interaction. Total orbital momenta of these states are $\frac{1}{2}^+$, $\frac{3}{2}^+$ and $\frac{5}{2}^+$, respectively.

In Table 2 presented are calculated energies for the bound states of the $\alpha\alpha\Lambda$ system for various potential models, three variants of the $\alpha\alpha$ Ali-Bodmer potential are used (ABa, ABd, ABe) [6]. Note that the model "a*" of the ABa potential was modified in [12]. For the $\alpha\Lambda$ interaction applied are the TH, Gibson, and Isle potentials [3].

We modified the parameters of the TH potential [7], in the Table 1 this model is denoted as TH(M). One can see that the TH(M) potential together with the ABe potential leads to good agreement with the experimental data for ${}^9_\Lambda$Be bound states. Orbital momentum configuration with $l=0$ of the core ^{8}Be nucleus contributes significantly to the ${}^9_\Lambda$Be ground state energy. The contributions of the $l=2,4$ configurations are found to be relatively small, whereas the opposite situation is observed for the second ${}^9_\Lambda$Be bound state. The $l=0,4$ configurations give relative small contribution to the binding energy, since the $l=2$ configuration forms this bound state.

The energy of spin-flip doublet splitting ΔE is calculated for the Scheerbaum (S) potential from [4] with (β^{so}_{att}=0.569 fm^{-1}) and for its modified version S(M). In the first case we obtained very large values for the energy splitting, as compared with the experimental data. This disagreement may be decreased by modifying the spin-orbit potential, motivated by the assumption that the range parameter of this potential has to be similar to that of the central part (see also Table 1). After this modification the energy of the spin-orbit splitting (TH(M) +ABe +S(M) model) decreases to the value that is comparable with the results of previous calculations: 160 keV [2] and 147 keV [4].

Our next modification of the spin-orbit potential, marked as S(MM), is further modification of S(M) obtained by an adjustment of the strength parameter V^{so}_{att}. The experimental splitting value of 43(5) keV [16] is reproduced by our calculations with V^{so}_{att}=2.25 MeV (this value of V^{so}_{att} is three times smaller than the initial value (the S potential)). The excitation energies for the doublet members are 3.016 MeV ($\frac{3}{2}^+$) and 2.972 MeV ($\frac{5}{2}^+$), the state $\frac{3}{2}^+$ is the upper member of this doublet, which agrees with previous calculation [18].

^{9}Be low-lying spectrum

The bound state of the $\alpha\alpha n$ system, having negative parity with $J^\pi = (\frac{3}{2})^-$ and L^π=1$^-$, is associated with ground state of the ^{9}Be nucleus. This state is lower member of the spin-flip doublet with orbital momentum L=1, the upper members of this doublet is state resonance $J^\pi = (\frac{1}{2})^-$ with energy 1.21±0.12 MeV. Calculated value of the energy spacing (0.9 MeV) for this doublet is close to the experimental data. The results of our calculations for these and other low-lying levels are listed in Tabl. 2. Classifying the calculated levels by the values of the total orbital momentum allows us to well reproduce experimental data. One can conclude that the low-lying ^{9}Be spectrum can be described in terms of spin-flip doublets. It means that each energy level of ^{9}Be is a member of spin-flip doublet with definite orbital momentum. Note that our calculations demonstrate relatively large discrepancy with experimental data [19] for the resonance with $J^\pi = (\frac{3}{2})^-$ and L^π=2$^-$. The same problem exists for other cluster calculations [20, 21] (see Tabl. 2). There is an experimental evidence [23] for a wide $J^\pi = (\frac{3}{2})^-$ resonance, with energy of 3.4±0.5 MeV, which is closer to the calculated value.

TABLE 1. Parameters of the $\alpha\alpha$ (ABe [6]), $\alpha\Lambda$, and αn potentials. The pair orbital momentum is denoted by l, V^l_{att} (V^l_{rep}) is given in MeV, and β^l_{att} (β^l_{rep}) is in fm^{-1}.

Interaction	Potential	l	V^l_{rep}	β^l_{rep}	V^l_{att}	β^l_{att}
$\alpha\alpha$	central	0	1050	0.8	150	0.5
		2	640	0.8	150	0.5
		4	–	–	150	0.5
$\alpha\Lambda$	central (TH(M))	0,1,2,...	–	–	54.36	0.733
	spin-orbit (S(MM))	–	–	–	2.25	0.733
αn	central	0	50.0	1/2.3 [13]	–	–
		1	40.0	1/1.67	63.0	1/2.3
		2	–	–	21.93	1/2.03 [14]
	spin-orbit	–	–	–	38.0	1/1.67

TABLE 2. Binding energy E_B (in MeV) and excited energy E_x (in MeV) of degenerate $(3/2^+, 5/2^+)$ spin doublet state of the $\alpha\alpha\Lambda$ system for different models of the $\alpha\alpha$ and $\alpha\Lambda$ interactions. The energy of spin-flip doublet splitting ΔE (in keV) is given for the S potential from [4] (V^{so}_{att}=-9 MeV, β^{so}_{att}=0.569 fm^{-1}). Modified spin-orbit potential S(M) is obtained by changing the α range parameter to correspond to that in $\alpha\Lambda$ potential. The energy is measured from the $\alpha+\alpha+\Lambda$ threshold. The experimental data are taken from [15, 16, 17].

(ls)		AB	TH	TH(M)	Gibson	Isle	Exp.
–	E_B	a*	-5.990	–	-6.709	-8.119	-6.62
		d	-5.994	–	-6.751	-8.348	
		e	-6.280	-6.602	-7.073	-8.714	
	E_x	a*	2.562	–	2.621	2.902	3.04
		d	2.738	–	2.756	2.927	
		e	2.990	2.991	3.032	3.193	
S	ΔE	a*	450	–	509	511	43
		d	346	–	447	512	
		e	352	378	457	525	
S(M)		e	126	172	–	–	
S(MM)		e	–	43	–	–	

CONCLUSION

The configuration space Faddeev equations are applied to calculate the energy spectrum of the ${}^9_\Lambda$Be hypernucleus within the $\alpha+\alpha+\Lambda$ cluster model. We found the set of phenomenological potentials that reproduces the ground state $\frac{1}{2}^+$ binding energy and excitation energy of the $\frac{5}{2}^+$ and $\frac{3}{2}^+$ states, simultaneously: the Ali-Bodmer potential of the model "e" for $\alpha\alpha$, modified Tang-Herndon potential (HT(M)) for $\alpha\Lambda$, and modified Scheerbaum potential (S(MM)) for spin-orbit $\alpha\Lambda$ interaction. In our model the total orbital momentum is fixed for each energy level. We showed that the ^{9}Be low-lying energy levels can be classified as members of spin-flip doublet corresponding to total orbital momentum ($L^\pi = 0^+, 2^+, 4^+, 1^-, 2^-, 3^-, 4^-$) of the system. Our calculations are in a good agreement with previous cluster calculations for ^{9}Be. In particular we confirmed that the $J^\pi = 9/2^-$ and $J^\pi = 7/2^+$ resonances have to be about 11 MeV and 7.5 MeV, respectively. Energy of the $J^\pi = 7/2^-$ resonance as upper member of the $L^\pi = 4^-$ level was predicted.

ACKNOWLEDGMENTS

This work is supported by the NSF award HRD-0833184 and NASA award NNX09AV07A. The numerical calculations were performed by HPC Center at the North Carolina State University.

TABLE 3. Energy levels in the $\alpha\alpha n$ system and low-lying ^{9}Be spectrum. Results of our calculations are presented in third column. The energy (in MeV) is measured from the $\alpha+\alpha+n$ threshold. Experimental data for ^{9}Be ($T=1/2$) are taken from [19].

L^π	J^π		[20] (CSM)	[21]	[22]*	Exp.
0^+	$\frac{1}{2}^+$	0.3(4)	–	–	–	0.111±0.007
1^-	$\frac{3}{2}^-$	-1.493	-2.16	-1.60	-1.48(4)	-1.5735
	$\frac{1}{2}^-$	0.9(0)	1.06	–	2.24(6)	1.21±0.12
2^-	$\frac{5}{2}^-$	0.7(2)	0.39	–	0.73(6)	0.8559±0.0013
	$\frac{3}{2}^-$	2.(7)	2.88	2.65	–	4.02±0.1 or 3.4±0.5[23]
2^+	$\frac{5}{2}^+$	1.(8)	1.75	–	–	1.476±0.009
	$\frac{3}{2}^+$	3.(0)	3.21	–	–	3.1312±0.025
3^-	$\frac{7}{2}^-$	4.(6)	5.02	–	5.03(6)	4.81±0.06
	$\frac{5}{2}^-$	6.(4)	6.57	–	–	6.36±0.08
4^+	$\frac{9}{2}^+$	5.(1)	5.04	–	–	5.19±0.06
	$\frac{7}{2}^+$	6.(6)	6.80	–	–	–
4^-	$\frac{9}{2}^-$	8.(6)	9.73	–	8.62(6)	–
	$\frac{7}{2}^-$	10.(7)	–	–	–	–

* Quantum Monte Carlo calculations for A=9

REFERENCES

1. T. Yamada, K. Ikeda, H. Bando, *Prog. Theor. Phys.* **73**, 397-413 (1985).
2. E. Hiyama, M. Kamimura, T. Motoba, T. Yamada, Y. Yamamoto, *Phys. Rev.* C **66**, 024007-024020 (2002).
3. I. N. Filikhin, A. Gal, V. M. Suslov, *Nucl.Phys.* A**743**, 194-207 (2004).
4. Y. Fujiwara, M. Kohno, K. Miyagawa, Y. Suzuki, *Phys. Rev.* C **70**, 047002-047006 (2004).
5. Y. Fujiwara, M. Kohno, Y. Suzuki, *Mod. Phys. Lett.* A **24**, 1031-1034(2009).
6. S. Ali and A. R. Bodmer, *Nucl. Phys.* **80**, 99-112 (1966).
7. Y. C. Tang, R. C. Herndon, *Phys. Rev.* **138**, B637-B643 (1965) ; R. H. Dalitz, B.W. Downs, *Phys. Rev.* **111**, 967-986 (1958).
8. K. M. Nollett, et al., *Phys. Rev. Lett.* **99**, 022502-022506 (2007).
9. L. D. Faddeev and S. P. Merkuriev, *Quantum Scattering Theory for Several Particle Systems*, Kluwer Academic, Dordrecht, 1993, pp. 398.
10. C. Kurokawa and K. Kato, *Phys. Rev.* C **71**, 021301-021306(R) (2005).
11. I. Filikhin, V. M. Suslov and B. Vlahovic, *Phys. Atom. Nucl.* **72**, 619-624 (2009).
12. D.V. Fedorov, A.S. Jensen, *Phys. Lett.* B **389**, 631-636 (1996).
13. M. V. Zhukov, B. V. Danilin, D. V. Fedorov, J. M. Bang, I.J. Thompson, and J. S. Vaagen, *Phys. Rep.* **231**, 151-199 (1993).
14. D. V. Fedorov, A. S. Jensen, and K. Riisager, *Phys. Rev.* C **49**, 201-212 (1994); A. Cobis, D. V. Fedorov and A. S. Jensen, *Nucl. Phys.* A **631**, 793-797 (1998).
15. D.H. Davis, in: *LAMPF Workshop on* (π,K) *Physics*, eds. B.F. Gibson, W.R. Gibbs, M.B. Johnson, *AIP Conf. Proc.*, **224**, AIP, New York, (1991), pp. 38-48.
16. H. Akikawa et al., *Phys. Rev. Lett.*, **88**, 082501-082505 (2002); H. Tamura, *Nucl. Phys.* A **754**, 58-69 (2005).
17. O. Hashimoto, H. Tamura, *Progress in Particle and Nuclear Physics* **57**, 564-653 (2006).
18. D.J. Millener, *Nucl. Phys.* A **754**, 48-57 (2005).
19. D.R. Tilley, et al., *Nucl. Phys.* A **745**, 155-362 (2004); http://www.tunl.duke.edu/nucldata/
20. K. Arai, P.Descouvemont, D. Baye, W. N. Catford, *Phys. Rev.* C 68, 014310-014323 (2003)
21. R. Álvarez-Rodríguez, A. S. Jensen, E. Garrido, and D. V. Fedorov, *Phys. Rev.* C 82, 034001-034012 (2010).
22. S. C. Pieper, K. Varga and R. B. Wiringa, *Phys.Rev.* C **66**, 044310-044324 (2002).
23. Y. Prezado, et al., *Phys. Lett.* B **618**, 43-50 (2005).

The Quark-Meson Coupling model as a description of dense matter

J. D. Carroll

Centre for the Subatomic Structure of Matter (CSSM), Department of Physics, University of Adelaide, SA 5005, Australia

Abstract. Quantum Hadrodynamics provides a useful framework for investigating dense matter, yet it breaks down easily when strangeness carrying baryons are introduced into the calculations, as the baryon effective masses become negative due to large meson field potentials. The Quark-Meson Coupling model overcomes this issue by incorporating the quark structure of the nucleon, thus allowing for a feedback between the the nuclei and the interaction with the meson fields. With the inclusion of this feature, QMC provides a successful description of finite nuclei and nuclear matter. We present the latest parameterization of QMC and discuss the predictions for dense nuclear matter and 'neutron' stars.

Keywords: QMC, EOS, dense matter, neutron stars
PACS: 26.60.Kp, 21.65.Qr, 12.39.-x

1. INTRODUCTION

The Quark-Meson Coupling model (QMC) [1, 2, 3, 4, 5, 6] has proven to be an extremely useful tool for calculating the properties of matter at various density scales—from finite nuclei [3] to neutron stars [4]—where the global properties of these objects can be accurately predicted and compared with experiment in an effort to refine the model parameters.

While other models for these systems have been investigated by many others [7, 8, 9], QMC uniquely distinguishes itself as a relativistic description at the quark level, constrained only via the quark-structure of baryons and a few baryon-meson coupling constants (for a thorough description, see [5]), fit to experimental data.

2. QMC

In order to calculate the properties of finite nuclei (including hypernuclei) and neutron stars, we require a parameterization of the baryon effective masses, M_B^* which is determined self-consistently using the QMC model, and which has a quadratic dependence on the scalar mean-field. This quadratic dependence is a distinguishing feature of QMC as compared to Quantum Hadrodynamics (QHD) which models a linear dependence of the effective mass on the scalar mean-field [7]; a feature which leads to negative effective masses at large baryon densities.

With a parameterization of M_B^*, we are able to calculate the properties of infinite nuclear matter, in which baryons are in beta-equilibrium with leptons, including attractive and repulsive potential contributions from scalar σ, and vector ω and ρ mesons. For example, using the parameterization found in Ref. [2] we calculate the properties of infinite matter under the constraints of global charge neutrality and conserved baryon density, as shown in Figure (1)(a). The couplings constants g_σ, $g_{\omega B}$, and g_ρ of the baryons B to the various mesons are determined by the standard procedure of fitting the properties of saturated nuclear matter.

We note that the fractional density of Σ^- baryons decreases above a certain density, while that of the remaining baryons tend toward a common value at large densities. We also note that the threshold density for Ξ hyperons to appear is well below that of the Σ^0 and Σ^+ hyperons.

The parameterization of the baryon effective masses has been refined over time to better reflect the properties of hypernuclei. Most recently in particular in 2008, by the self-consistent inclusion of one-gluon-exchange hyperfine terms that lead to the N-Δ and Σ-Λ mass splittings in free-space. A consequence of this is that the Σ hypernuclei are unbound in this model, which is consistent with the lack of evidence from the experimental searches for these. The parameterization is given in terms of the scalar mean-field $\langle\sigma\rangle$ by

12th International Conference on Meson-Nucleon Physics and the Structure of the Nucleon (MENU 2010)
AIP Conf. Proc. 1374, 205-207 (2011); doi: 10.1063/1.3647126

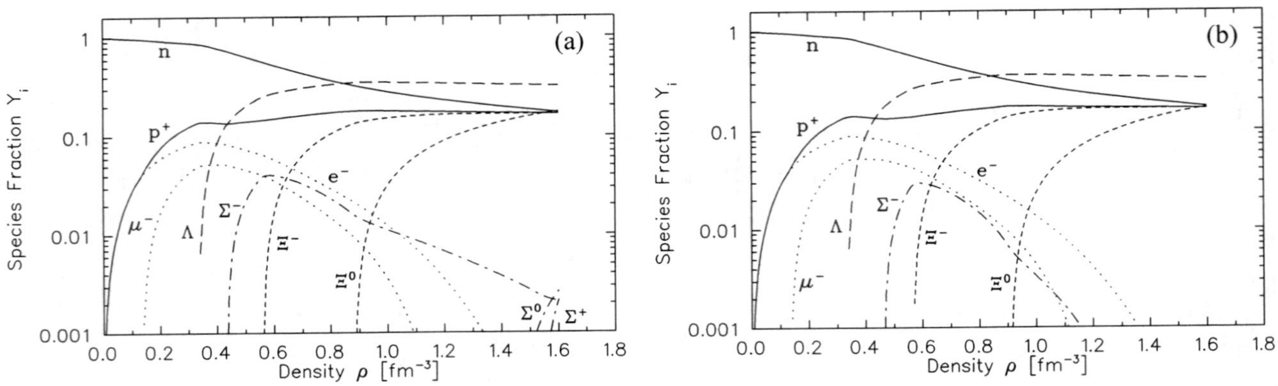

FIGURE 1. Species fractions Y_i for hyperonic QMC hadronic matter calculated using the 2007 ((a), from [2]), and 2008 ((b), from [3]) parameterizations of the effective masses. Note the suppression of the Σ baryon densities predicted using the 2008 parameterization.

$$\begin{aligned}
M_N^* &= M_N - g_{\sigma N}\langle\sigma\rangle \\
&\quad + \left[0.002143 + 0.10562 R_N^{\text{free}} - 0.01791\left(R_N^{\text{free}}\right)^2\right]\left(g_{\sigma N}\langle\sigma\rangle\right)^2, \\
M_\Lambda^* &= M_\Lambda - \left[0.6672 + 0.04638 R_N^{\text{free}} - 0.0022\left(R_N^{\text{free}}\right)^2\right] g_{\sigma N}\langle\sigma\rangle \\
&\quad + \left[0.00146 + 0.0691 R_N^{\text{free}} - 0.00862\left(R_N^{\text{free}}\right)^2\right]\left(g_{\sigma N}\langle\sigma\rangle\right)^2, \\
M_\Sigma^* &= M_\Sigma - \left[0.6653 - 0.08244 R_N^{\text{free}} + 0.00193\left(R_N^{\text{free}}\right)^2\right] g_{\sigma N}\langle\sigma\rangle \\
&\quad + \left[0.00064 + 0.07869 R_N^{\text{free}} - 0.0179\left(R_N^{\text{free}}\right)^2\right]\left(g_{\sigma N}\langle\sigma\rangle\right)^2, \\
M_\Xi^* &= M_\Xi - \left[0.3331 + 0.00985 R_N^{\text{free}} - 0.00287\left(R_N^{\text{free}}\right)^2\right] g_{\sigma N}\langle\sigma\rangle \\
&\quad + \left[-0.0032 + 0.0388 R_N^{\text{free}} - 0.0054\left(R_N^{\text{free}}\right)^2\right]\left(g_{\sigma N}\langle\sigma\rangle\right)^2,
\end{aligned} \tag{1}$$

where the value $R_N^{\text{free}} = 0.8$ fm is used (and the dependence on this value has been investigated and found to be small in Ref. [10]).

The relative densities of baryons and leptons in beta-equilibrium for the case of infinite nuclear matter for this new (2008) parameterization are shown in Figure (1)(b). We note that the most striking difference in comparison to the calculation using the 2007 parameterization—which does parameterize the additional hyperfine interactions—is that the contribution of the Σ hyperons has been reduced.

3. NEUTRON STARS

Calculating the mass and radius of a neutron star modelled with QMC involves solving the Tolman-Oppenheimer-Volkoff (TOV) equation

$$\frac{dP}{dR} = -\frac{G(P+\mathcal{E})\left(M(R)+4\pi R^3 P\right)}{R(R-2GM(R))}, \tag{2}$$

for a given Equation of State (EoS), in this case an EoS calculated using QMC (see Refs. [5, 6] for details). The TOV solutions for a range of central densities are shown in Figure (2) for the 2007 and 2008 parameterizations of the effective masses. We note that the choice of parameterization does not significantly affect the solutions to the TOV equation.

We further note that the maximum predicted stellar mass calculated using either parameterization of the effective baryon mass does not exceed 1.57 $M_\odot$, which can be attributed to the softness of the EoS due to the large number of

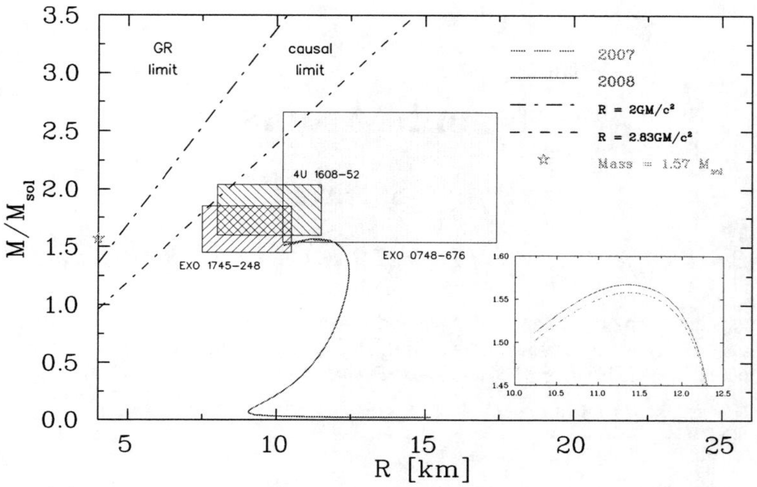

FIGURE 2. (Color online) Predicted total stellar mass and radius solutions of the TOV equation for hyperonic QMC using the 2007 and 2008 parameterizations of the effective mass. Also shown are the data points from Refs. [11] (EXO 0748-676), [12] (EXO 1745-248), and [13] (4U 1608-52), as well as the maximum mass predicted from either data set (star). Inset: Separation of the two data sets at the maximum stellar mass.

baryons present—at large densities, at least six of the octet baryons posess nontrivial fractional densities. This is in conflict with the observational evidence of larger mass neutron stars, though we caution using a direct comparison between these predictions (which correspond to static, spherically symmetric, non-rotating objects) and physical neutron stars which may not satisfy such approximations.

Future work will include investigating the effects of rotation on the stellar solutions, and inclusion of Fock (exchange) terms to the baryon self-energies, in an effort to increase the accuracy of our description of dense nuclear matter.

ACKNOWLEDGMENTS

This research was supported in part by DOE contract DE-AC05-06OR23177 (under which Jefferson Science Associates, LLC, operates Jefferson Lab), and in part by the Australian Research Council. The author would like to thank A. W. Thomas for his guidance and support, as well as D. B. Leinweber and A. G. Williams for their helpful discussions.

REFERENCES

1. P. A. M. Guichon, K. Saito, E. N. Rodionov *et al.*, Nucl. Phys. **A601**, 349-379 (1996). [nucl-th/9509034].
2. J. Rikovska-Stone, P. A. M. Guichon, H. H. Matevosyan, A. W. Thomas. Nucl. Phys. A792:341-369, 2007
3. P. A. M. Guichon, A. W. Thomas, K. Tsushima. Nucl. Phys. A814:66-73, 2008
4. J. D. Carroll, D. B. Leinweber, A. G. Williams, A. W. Thomas. Phys. Rev. C79:045810, 2009
5. J. D. Carroll. [arXiv:1001.4318 [hep-ph]].
6. J. D. Carroll, AIP Conf. Proc. **1261**, 226-231 (2010). [arXiv:1004.1217 [nucl-th]].
7. B. D. Serot, J. D. Walecka. Adv. Nucl. Phys.16:1-327, 1986.
8. S. C. Pieper, Riv. Nuovo Cim. **31**, 709 (2008) [arXiv:0711.1500 [nucl-th]].
9. P. Wang, S. Lawley, D. B. Leinweber, A. W. Thomas and A. G. Williams, Phys. Rev. C **72**, 045801 (2005) [arXiv:nucl-th/0506014].
10. P. A. M. Guichon, H. H. Matevosyan, N. Sandulescu and A. W. Thomas, Nucl. Phys. A **772**, 1 (2006) [arXiv:nucl-th/0603044].
11. F. Ozel. Nature 441, 1115 (2006)
12. T. Guver, F. Ozel, A. Cabrera-Lavers, P. Wroblewski. arXiv:0811.3979 [astro-ph].
13. F. Ozel, T. Guver, D. Psaltis. Astrophys. J. 693, 1775 (2009)

Kaonic Atoms

Ryugo S. Hayano

Department of Physics, The University of Tokyo, 7-3-1 Hongo, Bunkyo-ku, Tokyo 113-0033, Japan

Abstract. The precision spectroscopy of kaonic hydrogen and kaonic helium x-rays offers a unique possibility to precisely determine the K^-N strong interaction in the low energy limit. Results and status of modern experiments using silicon drift x-ray detectors (SDDs), namely, KEK E570, SIDDHARTA at DAΦNE and E17 at J-PARC, are discussed.

Keywords: kaonic x-rays, strangeness, $\bar{K}N$ interaction, DAΦNE, J-PARC
PACS: 13.75.Jz, 25.80.Nv, 36.10.Gv

INTRODUCTION

The measurement of the strong-interaction energy-level shift and width of kaonic atom x-rays offers a unique possibility to precisely determine the K^--nucleus strong interaction in the low energy limit. The most fundamental is the kaonic hydrogen, from which we deduce the K^-p scattering length at threshold. Until about 15 years ago, even the sign of the shift was uncertain, as experiments reported the $1s$-level shift to be "attractive"[1] [1–3], while theories (analyses of low-energy $\bar{K}N$ data) predicted "repulsive". However, new experiments, KpX [4, 5], DEAR [6] have now established that the $1s$-level shift is repulsive, and SIDDHARTA [11] will soon determine the $1s$ shift and width with higher precision.

The second simplest system, kaonic helium, has been also known to be puzzling. While most of the available kaonic-atom data can be fitted fairly well for $Z > 2$ by optical-potential models [7], there was a 5σ theory-vs-experiment discrepancy for the $2p$-level shift of kaonic ^{4}He. This "kaonic helium puzzle" has recently been solved by the KEK E570 experiment [19], and the J-PARC E17 experiment will soon measure the kaonic ^{3}He $2p$-level shift.

KAONIC HYDROGEN

KpX and DEAR

Before the KpX experiment [4, 5], the status of the kaonic-hydrogen study was quite puzzling due to the contradiction between the signs of the scattering lengths obtained by the previous x-ray measurements [1–3] and those extracted from the analyses of the low energy $\bar{K}N$ data, e.g., Refs. [8–10].

KpX was the first to observe a statistically-significant kaonic hydrogen $2p$–$1s$ peak with low background (Fig. 1 top panel). This was achieved by the use of (i) a gaseous hydrogen target (1.3% of liquid hydrogen density), (ii) sixty Si(Li) X-ray detectors each having an active area of 200 mm^2 and a FWHM resolution of some 400 eV, (iii) an elaborate vertex reconstruction and background suppression scheme, and (iv) an in-situ energy calibration scheme using the fluorescence from titanium (Ti) foils. The obtained shift and width were $\Delta E_{1s} = -323 \pm 63(\text{stat}) \pm 11(\text{syst})$ eV (repulsive) and $\Gamma_{1s} = 407 \pm 208(\text{stat}) \pm 100(\text{syst})$ eV, respectively.

The DEAR experiment [6] took advantage of the low-energy (16 MeV), nearly mono-energetic ($\Delta p/p = 0.1\%$) kaons from the decay of ϕ mesons produced by e^+e^- collisions at DAΦNE. The kaons were stopped in a gaseous hydrogen target (3% of liquid hydrogen density), and the x-rays were detected by sixteen CCD (charged-coupled device) x-ray detectors, each having a good FWHM resolution of 150 eV at 6 keV. However, due to the lack of timing information (the DEAR CCDs were read out every 90 s), the observed spectrum was dominated by a continuum background. After subtracting the background (about 10^5 counts per channel), the spectrum shown in

[1] Here, "attractive" means that the measured $2p$–$1s$ x-ray energy (at around 6.5 keV) is *larger* than that calculated by taking account of the Coulomb interaction only.

12th International Conference on Meson-Nucleon Physics and the Structure of the Nucleon (MENU 2010)
AIP Conf. Proc. 1374, 208-211 (2011); doi: 10.1063/1.3647127

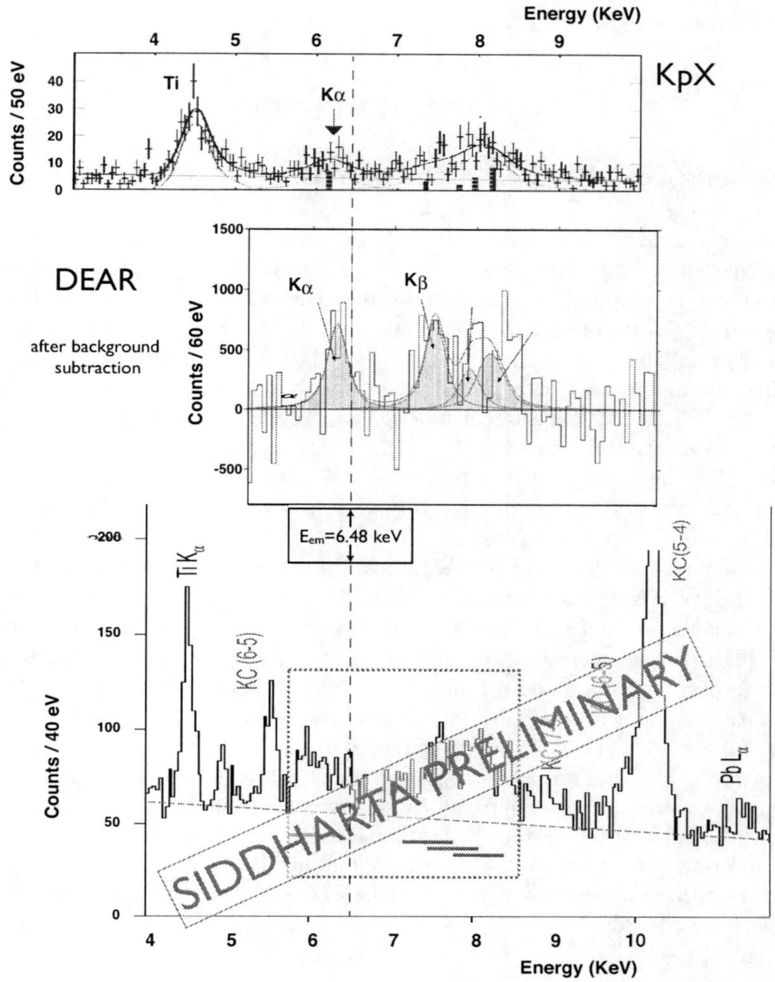

FIGURE 1. Comparison of the KpX (top), DEAR (middle) and SIDDHARTA (bottom: preliminary) kaonic hydrogen x-ray spectra.

the middle panel of Fig. 1 was obtained. The deduced shift and width were $\Delta E_{1s} = -193 \pm 37(\text{stat}) \pm 6(\text{syst})$ eV and $\Gamma_{1s} = 249 \pm 111(\text{stat}) \pm 30(\text{syst})$ eV, respectively.

SIDDHARTA

While both KpX and DEAR agreed on the sign of the shift to be repulsive, the shift values did not overlap with each other. A new measurement is clearly needed. The SIDDHARATA experiment at DAΦNE [11] therefore aims at determining the kaonic hydrogen $1s$ shift and width with a precision higher than in DEAR. To achieve this goal, 144 silicon drift detectors (SDDs, each having an area of 100 mm^2), each having a good energy ($\sim$ 150 eV) and time ($\sim 1\mu$s) resolution, were used. Two sets of thin plastic "kaon" counters were placed above and below the e^+e^- collision point to detect kaons, and data were collected with kaon–x-ray coincidence triggers. This, together with shield optimization, significantly reduced the background level.

A preliminary kaonic hydrogen spectrum is shown in the bottom panel of Fig. 1. Although the data still need to be carefully analyzed as there are background contributions from K^-C, K^-O, etc., produced by kaons stopped in the target vessel made of Kapton, the shift is definitely repulsive, and the $1s$ shift and width will soon be determined with higher precision than in previous experiments.

KAONIC HELIUM

Kaonic ^{4}He – KEK E570 has solved the kaonic helium puzzle

The kaonic-atom strong-interaction shift values for light elements, calculated using the best-fit optical potential [7], reproduce most of the measured x-ray shift values, except for kaonic ^{4}He (and ^{16}O). The (pre-E570) kaonic ^{4}He $2p$-level shift, $\Delta E_{2p} = -43 \pm 8$ eV is based on three past x-ray experiments [12–14], while theoretical calculations give $\Delta E_{2p} \sim 0$ eV [15, 16] . This 5-σ discrepancy has been known as the kaonic helium puzzle.

In 2005, Akaishi [17] predicted that a large ($\sim \pm 10$ eV) shift may appear near the resonance between atomic and nuclear poles. This calculation was made in conjunction with the Yamazaki-Akaishi prediction of deeply-bound kaonic nuclear states [18]. The E570 experiment at the KEK 12-GeV proton synchrotron, motivated by Akaishi's prediction, was designed to measure the kaonic ^{4}He $3d$–$2p$ x-rays to a precision of ~ 2 eV, so as to confirm or refute the large $2p$-level shift. This was done parasitically to the E549 experiment (a search for such deeply-bound states in the ^{4}He(stopped K^-, n/p)X reaction), by mounting x-ray detectors in the liquid helium target cryostat, and by mixing the x-ray triggers to the E549 $(K^-, p/n)$ triggers.

In order to achieve the precision goal, the E570 employed [19] (i) eight SDDs each having an area of 100 mm^2 and a resolution of $\sim$ 180 eV FWHM at 6 keV, (ii) a target fiducial cut on the reconstructed reaction vertices (from an incoming K^- track and an outgoing charged particle – pions from the (K^-, π^-) reaction on the target, or from hyperon decays), and (iii) an in-situ energy calibration using characteristic x-rays induced by charged particles (mostly π^- in the beam) on high-purity titanium and nickel foils placed just behind the target cell. The energy of the kaonic-helium $3d$–$2p$ x-ray, $\sim$ 6.4 keV, lies between the characteristic x-ray energies, 4.5 keV(Ti) and 7.5 keV(Ni). To obtain high-statistics energy calibration spectra, SDD self-triggered events (Fig. 2 top) were accumulated in addition to the stopped-K-triggered events (Fig. 2 bottom, which also includes the fiducial cut discussed above) were accumulated.

With these, the energy resolution was improved by a factor of 2, statistics by a factor of 3 and the signal-to-noise ratio by a factor of 6, as compared with Ref. [14]. The measured $2p$-level shift was 2 ± 2(stat) ± 2(syst) eV [19], thus agreeing with a majority of the theoretical calculations. This has solved the long-standing kaonic-helium puzzle. This has recently been confirmed by the SHIDDHARTA experiment, which obtained the $2p$-level shift to be $\Delta E_{2p} = 0 \pm 6$(stat) ± 2(syst) eV [20].

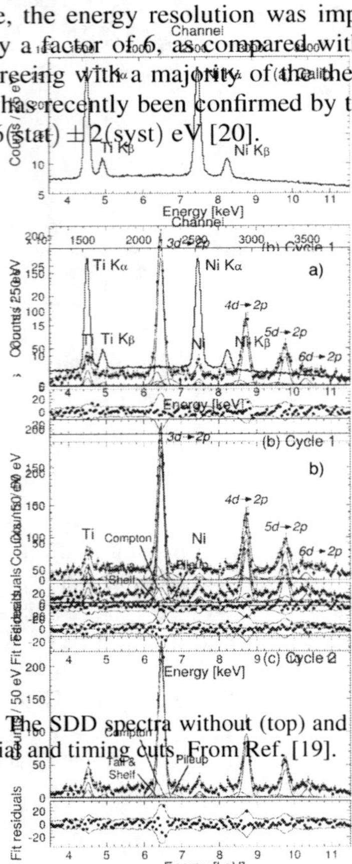

FIGURE 2. The SDD spectra without (top) and with (bottom) the fiducial and timing cuts. From Ref. [19].

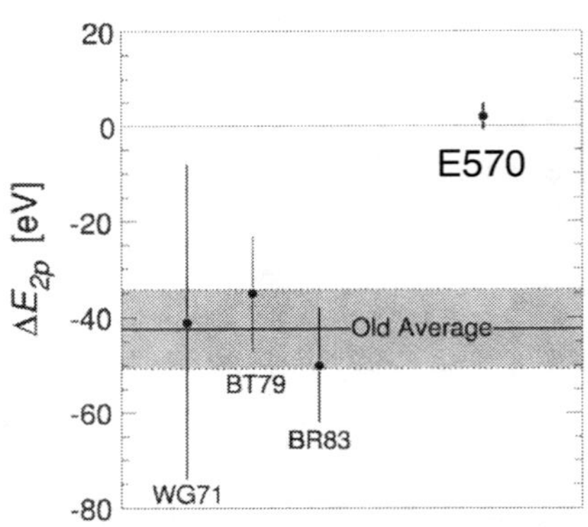

FIGURE 3. The $2p$-level shift of kaonic ^{4}He obtained by E570 [19] and the past three experiments (WG71 [12], BT79 [13], BR83 [14]).

Kaonic ^{3}He – J-PARC E17

Using the method established in the E570 experiment, the E17 group at J-PARC is now preparing for an experiment to measure the kaonic helium-3 x-rays. The experiment will be carried out at the K1.8BR beamline in the J-PARC hadron hall. The essential technical differences between the E570 and E17 experiments are:

1. The volume of the E570 liquid helium-4 target was close to 5000 cm^3, but it is about 500 cm^3 for the E17 liquid helium-3 target. This calls for a higher-intensity, better-quality kaon beam.
2. In order to minimize the cable length between the SDDs and the preamplifiers and also due to the space constraints, the preamplifiers are mounted inside the target cryostat. A careful energy and timing resolution optimization as a function of the SDD and amplifier operating temperatures has been done.

The E17 experiment was chosen as one of the "Day-1" experiments, and is waiting for the J-PARC main ring slow-extraction beam intensity reach $\gtrsim 5$ kW, which is needed to carry out the measurement.

SUMMARY

The precision spectroscopy of kaonic atom x-rays offers a unique possibility to precisely determine the K^--nucleus strong interaction in the low energy limit. The existing (slight) discrepancy between the KpX and DEAR results on the $1s$-level shift and width of kaonic hydrogen will soon be settled by a higher-precision data collected by the SIDDHARATA collaboration. For kaonic helium-4, the E570 experiment has established that the $2p$-level shift is small, ~ 0 eV, thereby solving the "puzzle". This has recently been confirmed by the SIDDHARATA collaboration. For these measurements, the use of silicon drift detectors (SDDs), having good energy and time resolutions, has played an essential role. At the newly inaugurated J-PARC, the E17 experiment will soon measure the kaonic helium-3 x-rays, also using SDDs.

ACKNOWLEDGMENTS

It is a pleasure to thank many colleagues participating in the experiments discussed herein, KpX, DEAR, E570, SIDDHARTA and E17, for illuminating discussions. This work was supported by MEXT Grant-in-Aid for Scientific Research No. 2002003.

REFERENCES

1. J. Davies, et al., *Phys. Lett. B* **83**, 55 (1979).
2. M. Izycki, et al., *Zeitschrift für Physik A* **297**, 11 (1980).
3. P. Bird, et al., *Nucl. Phys. A* **404**, 482 (1983).
4. M. Iwasaki, et al., *Phys. Rev. Lett.* **78**, 3067 (1997).
5. T. Ito, et al., *Phys. Rev. C* **58**, 2366 (1998).
6. G. Beer, et al., *Phys. Rev. Lett.* **94**, 212302 (2005).
7. C. J. Batty, E. Friedman, and A. Gal, *Phys. Reports* **287**, 385 (1997).
8. W. E. Humphrey, and R. R. Ross, *Phys. Rev.* **127**, 1305 (1962).
9. J. K. Kim, *Phys. Rev. Lett.* **19**, 1074 (1967).
10. A. D. Martin, *Nuclear Physics B* **179**, 33 (1981).
11. M. Cargnelli, et al., *Nucl. Phys. A* **835**, 27 (2010).
12. C. Wiegand, and R. Pehl, *Phys. Rev. Lett.* **27**, 1410 (1971).
13. C. Batty, et al., *Nucl. Phys. A* **326**, 455 (1979).
14. S. Baird, et al., *Nucl. Phys. A* **392**, 297 (1983).
15. C. J. Batty, *Nucl. Phys. A* **508**, 89 (1990).
16. S. Hirenzaki, et al., *Phys. Rev. C* **61**, 055205 (2000).
17. Y. Akaishi, "Kaonic Helium Atoms in relation to Kaonic Nuclear Bound States," in *Proceedings for EXA05*, Austrian Academy of Sciences Press, Vienna, 2005, p. 45.
18. Y. Akaishi, and T. Yamazaki, *Phys. Rev. C* **65**, 044005 (2002).
19. S. Okada, et al., *Phys. Lett. B* **653**, 387 (2007).
20. M. Bazzi, et al., *Phys. Lett. B* **681**, 310 (2009).

Precision spectroscopy of light kaonic atom X-rays in the SIDDHARTA experiment

M. Cargnelli*, M. Bazzi†, G. Beer**, L. Bombelli‡, A.M. Bragadireanu†, C. Curceanu (Petrascu)†, C. Fiorini‡, T. Frizzi‡, F. Ghio§, B. Girolami§, C. Guaraldo†, R. Hayano¶, M. Iliescu†, T. Ishiwatari*, M. Iwasaki‖, P. Kienle*,††, P. Lechner‡‡, P. Levi Sandri†, A. Longoni‡, V. Lucherini†, J. Marton*, S. Okada†, D. Pietreanu†, T. Ponta§§, A. Romero Vidal†, A. Scordo†, H.-X. Shi¶, D.L. Sirghi†, F. Sirghi†, H. Soltau‡‡, L. Strüder¶¶, H. Tatsuno¶, O. Vazquez Doce†, E. Widmann* and J. Zmeskal*

**Stefan Meyer Institut of the Austrian Academy of Sciences, Vienna, 1090, Boltzmanngasse 3, Austria*
†INFN, Laboratori Nazionali di Frascati, C.P. 13, Via E. Fermi 40, I-00044 Frascati (Roma), Italy
***Dep. of Physics and Astronomy, University of Victoria, P.O.Box 3055, Victoria B.C. Canada V8W3P6*
‡Politecnico di Milano, Dip. di Elettronica e Informazione, Via Ponzio, 34/5, I-20133 Milano, Italy
§INFN Sez. di Roma I and Instituto Superiore di Sanita I-00161, Roma, Italy
¶University of Tokyo,7-3-1, Hongo, Bunkyo-ku,Tokyo, Japan
‖RIKEN, Institute of Physical and Chemical Research, Saitama, Japan
††Physik Departement, Technische Universität München, Garching, Germany
‡‡PNSensors GmbH, Rönnerstr. 28, D-80803 München, German
§§IFIN-HH, P.O. box MG-6, R76900 Magurele, Bucharest, Romania
¶¶MPI for Extraterrestrial Physics, Giessenbachstr. , D-85748 Garching, Germany

Abstract. The $\overline{K}N$ system at rest makes a sensitive testing ground for low energy QCD. At the DAΦNE electron-positron collider of Laboratori Nazionali di Frascati we study kaonic atoms, taking advantage of the low-energy kaons from Φ-mesons decaying nearly at rest. The DEAR (DAΦNE Exotic Atom Research) experiment at LNF delivered the most precise data on kaonic hydrogen up to now. DEAR and its follow-up experiment SIDDHARTA (Silicon Drift Detector for Hadronic Atom Research by Timing Application) are using X-ray spectroscopy of kaonic atoms to measure the strong interaction induced shift and width of the ground state. SIDDHARTA is the first experiment on kaonic helium-3 and deuterium ever, and kaonic hydrogen was remeasured with improved precision.

Keywords: Kaonic atoms, low-energy QCD, Antikaon-nucleon physics, X-ray detection
PACS: 36.10.k 13.75.Jz 32.30.Rj 29.40.Wk

INTRODUCTION

Low energy phenomena in strong interaction can not (yet?) be described in terms of quarks and gluons. Instead, effective field theories are used which contain appropriate degrees of freedom to describe physical phenomena occurring at a chosen scale. Chiral perturbation theory was extremely successful in describing systems like pionic atoms, however it is not applicable for the kaonic hydrogen systems. This is due to the presence of resonances like the $\Lambda(1405)$, only slightly below the reaction threshold of 1432 MeV.

There are however non-perturbative coupled-channel techniques based on the driving terms of the chiral effective Lagrangian which proved to be useful. They generate the $\Lambda(1405)$ dynamically as a $\overline{K}N$ quasibound state and as a resonance in the $\pi\Sigma$ channel. A general feature of the theory in this field is that it relies heavily on input from experimental data. The data come from kaon scattering experiments, atomic x-ray measurements and the energy and width of known resonances. As compared to scattering data, the x-ray data have the advantage to measure at E=0, so no extrapolation is needed. The kaonic atom experiments measure the influence of the strong interaction on the x-ray transitions, namely the shift and width of the ground state. The vividness of the field was stimulated by recent and ongoing experimental and theoretical work: [1] [2] [3] [4] [5]

12th International Conference on Meson-Nucleon Physics and the Structure of the Nucleon (MENU 2010)
AIP Conf. Proc. 1374, 212-215 (2011); doi: 10.1063/1.3647128

EXPERIMENT

A kaonic atom is formed when a negative kaon enters a medium, looses its kinetic energy through ionization and excitation of the atoms and molecules and eventually is captured, replacing the electron, in an excited orbit ($n \simeq 25$). Via different cascade processes (Auger transitions, Coulomb deexcitation, elastic scattering) the kaonic atom deexcites to lower states. When the kaon reaches a low-n state with small angular momentum, strong interaction with the nucleus causes its absorption and modifies the binding and thus the energy level position.

For our experiments a new experimental technique was developed: large area soft-X-ray detectors with good energy and timing resolution, high counting rate capability and enough robustness to be operated in the radiation environment of an accelerator. This was done in the framework of the SIDDHARTA project (Silicon Drift Detectors for Hadronic Atom Research by Timing Application). The SDDs [6] [7] have an active area of 1 cm^2 and a thickness of 450 μm. We used 144 cm^2 of SDDs and 2 trigger scintillators to tag on the charged kaons.

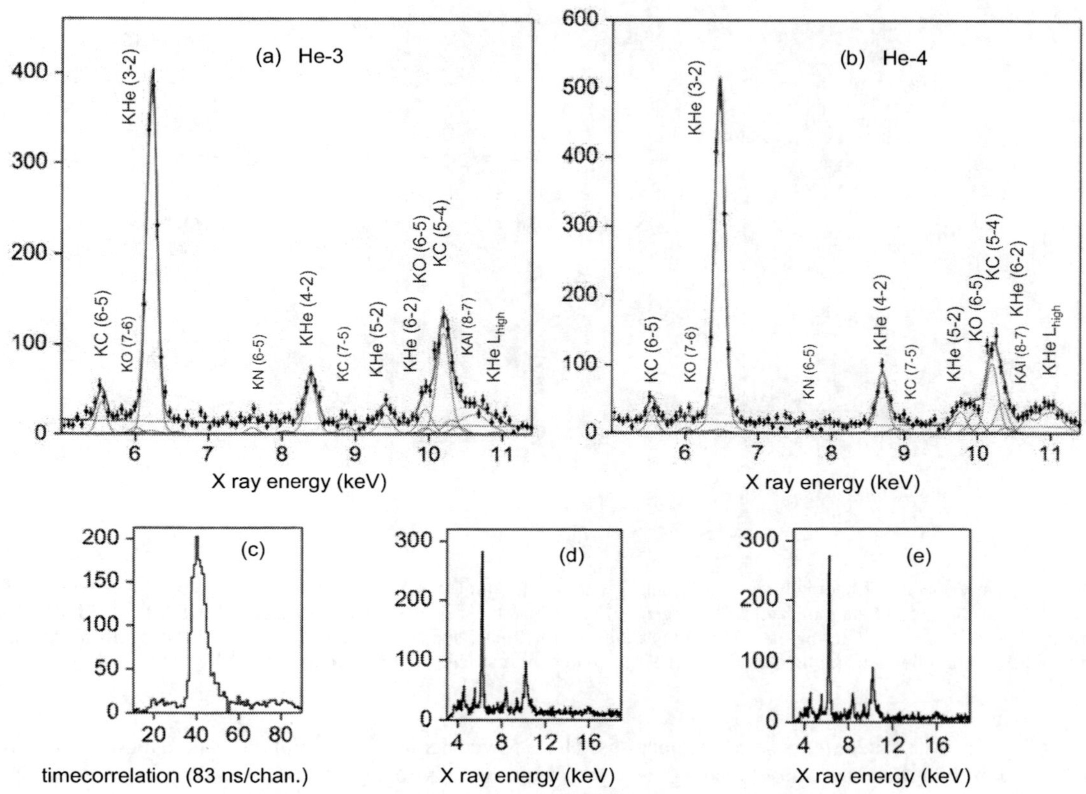

FIGURE 1. X-ray spectra of kaonic helium. (a) He-3, (b) He-4, (c) He-3 timecorrelation between kaon and X ray for E=6.2 keV, (d) He-3 x ray energies for a 1 μs timewindow, (e) ..after subtraction of the accidental background. The L-line pattern is easily seen, the other lines are from kaons stopped in the kapton target walls

The experiment was started at LNFrascati in 2008 and ended November 2009. At the end of our run period we got 8 pb^{-1} of SIDDHARTA-usable integrated luminosity per day (L(max) $\simeq 3\ 10^{32}\ cm^{-2}\ s^{-1}$). Future experimental efforts are under preparation.

We used a cryogenic gas target of 2.2 liter with about 0.01 of liquid hydrogen density (8.7 STP density), operated at 23 K. The entrance window was 20 cm above the interaction point (IP).

To test and optimize our experimental technique we measured the L-transitions of kaonic helium using different degraders, collimators and shieldings and we worked out the energy calibration procedure. Part of the data was taken

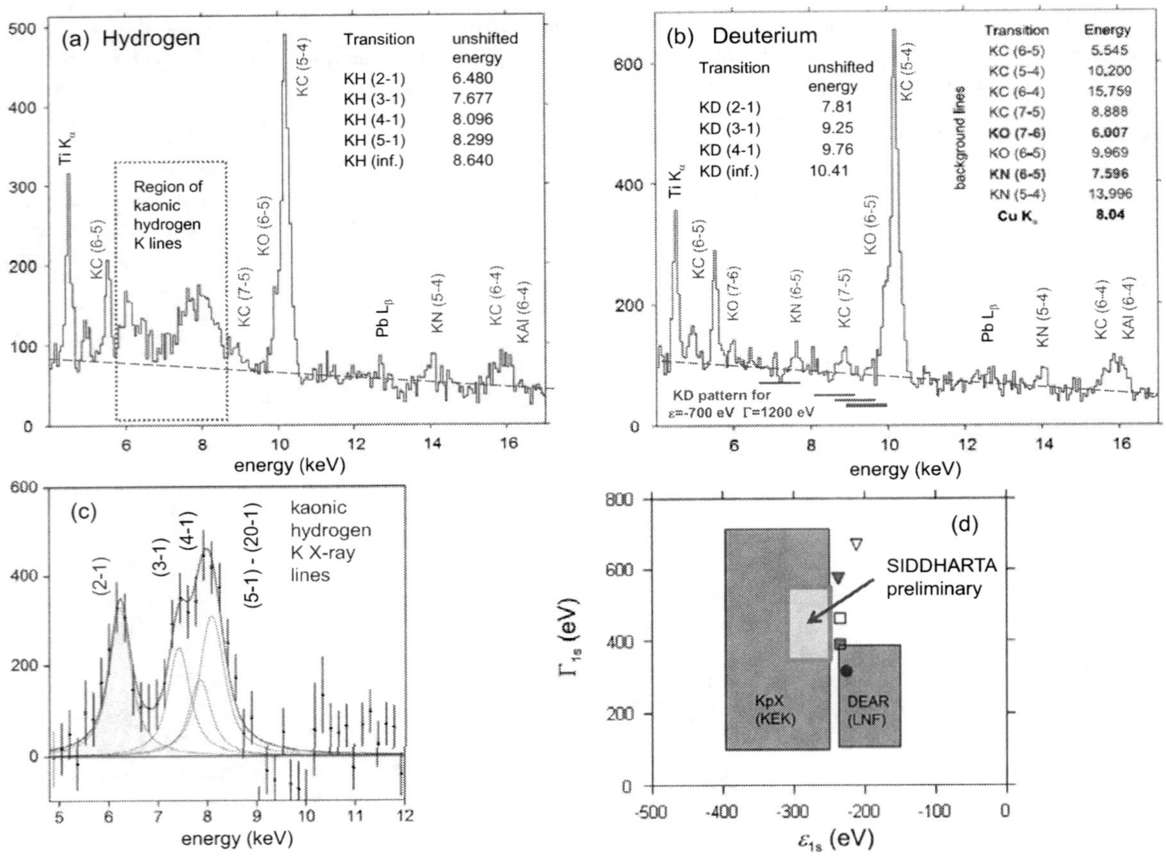

FIGURE 2. X-ray spectra of kaonic hydrogen (a) and kaonic deuterium (b). KH after subtraction of the KD spectrum (c). The shift and width of the ground state of kaonic hydrogen (d): gray boxes experimental results KpX [8], DEAR [9], SIDDHARTA preliminary, theory [1] using scattering data, branching ratios, the $\pi\,\Sigma$ mass spectrum and the kaonic atom data from DEAR shown as triangles, filled: including isospin breaking corrections, squares: fit restricted to DEAR data, dot: [10]

while a Fe-55 source was mounted inside the apparatus. This allowed detailed studies of the systematics of calibration. The high yield of the KHe L transitions enabled us to see meaningfull X-ray spectra within a day.

The kaonic helium has physics-interest in itself: Before the KEK E570 experiment [12] the experimental data were in contradiction with theory. The new value for the He-4 2p shift from E570 is now confirmed by our measurement [11]. Using our total He-4 data we will derive a yield pattern for the transitions. Since our target was gaseous helium, whereas E570 used liquid, the atomic cascades (and correspondingly the yields) are different and so complementary information will be gained. An additional advantage is, that in our case compton scattering of the X-rays is negligible and does not need to be corrected in the analysis.

During the last days of our run, we successfully measured kaonic He-3 [14], a system for which no experimental data existed up to now. A dedicated experiment for KHe-3 is planned at J-PARC [13]. In figure 1 the data from the He measurements are shown. The shift for the (3-2) line positions from their QED values is for both isotopes consistant with zero within the experimental errors (about 6 eV stat.+syst.).

We checked the performance of the experimental setup a number of times by filling helium. During the summer 2009 shutdown we modified and improved our setup. The data taken during September and October 2009 are the best from signal-to-background and also the DAFNE performance was very good in that period.

Comparing the hydrogen and deuterium spectra (fig. 2, a vs. b) the KH signal can be easily seen. The KD signal is expected to be much smaller than the KH one. Wether quantitative results for the KD X-rays can be obtained is still open, the access of events in the K-high region of about 9 keV (if a shift of 700 eV is assumed) is only about 2 sigma above background, in the KH (2-1) region there is no signal of KD seen nor expected. Therefore the kaonic deuterium spectrum can be used to determine the background lines in hydrogen, which originate from kaons stopping in the target windows made of kapton foils (C,O,N). We subtracted a deuterium spectrum normalized to equal KC line intensity, see fig.1c.

The signal lines were fitted using voigtians with given gaussian resolutions and free identical lorentz widths for (2-1),(3-1),(4-1) and a superposition of the higher transitions (5-1),... (20-1). The position pattern was fixed by the QED values. The intensities of (2-1),(3-1) and (4-1) were free, those of the higher transitions were coupled. An iterative procedure was applied to minimize the influence of the yield pattern, it is described elsewhere [15]. We also did a global fit of the unsubtracted spectra and obtained consistant results with the analysis of the KH minus KD spectum. The preliminary results for shift and width are shown in fig 1d.

RESULTS

The SIDDHARTA data taking is finished, results for kaonic helium-3 and helium-4 are published, confirming the zero shift. The analysis of the absolute KHe yields and the line pattern is under way. The kaonic hydrogen was clearly seen, the final precision is expected to be considerably better then the one of the existing experiments, a publication will soon follow. The kaonic deuterium measurement suffers from poor statistics, it will be published along with an outline of a future KD experiment at LNF using improved techniques.

ACKNOWLEDGMENTS

We thank C. Capoccia, B. Dulach, and D. Tagnani from LNF-INFN; and H. Schneider, L. Stohwasser, and D. Stückler from Stefan-Meyer-Institut, for their fundamental contribution in designing and building the SIDDHARTA setup. We thank as well the DAFNE staff for the excellent working conditions and permanent support. Part of this work was supported by Hadron- Physics I3 FP6 European Community program, Contract No. RII3-CT- 2004-506078; the European Community Research Infrastructure Integrating Activity -Study of Strongly Interacting Matter- (HadronPhysics2, Grant Agreement No. 227431) under the Seventh Framework Programme of EU; Austrian Federal Ministry of Science and Research BMBWK 650962/0001 VI/2/2009; Romanian National Authority for Scientific Research, Contract No. 2-CeX 06-11-11/2006; Grant-in-Aid for Specially Promoted Research (20002003), MEXT, Japan; and the Austrian Science Fund (FWF): [P20651- N20].

REFERENCES

1. B. Borasoy, R. Nißler, and W. Weise, Phys. Rev. Lett. 94, 213401 (2005).
2. U.-G. Meißner, U. Raha, and A. Rusetsky, Eur. Phys. J. C35, 349 (2004).
3. J.A. Oller, J. Prades, and M. Verbeni, Phys. Rev. Lett. 95 172502 (2005).
4. A.N. Ivanov et al., Eur. Phys. J. A23, 79-111 (2005).
5. U.-G. Meißner, U. Raha, and A. Rusetsky, Eur. Phys. J. C47, 473 (2006).
6. P. Lechner, et al., Nucl. Inst. and Meth. A 458 (2001) 281.
7. C. Fiorini, et al., Nucl. Inst. and Meth. A 568 (2006) 322.
8. M. Iwasaki et al. [KpX Collaboration], Phys. Rev. Lett. 78 (1997) 3067.
9. G. Beer et al. [DEAR Collaboration], Phys. Rev. Lett. 94, 212302 (2005).
10. A.N. Ivanov et al., Eur. Phys. J. A25 329 (2005), A.N. Ivanov et al., J. Phys. G31 (2005) 769.
11. M. Bazzi et al. [SIDDHARTA collaboration], Physics Letters B 681 (2009) 310-314.
12. S. Okada et at., Physics Letters B 653 (2007) 387-391.
13. R.S. Hayano et at., Proposal of J-Parc 50 GeV PS (J-PARC E17) 2006
14. M. Bazzi et al. [SIDDHARTA collaboration], submitted to Physics Letters B.
15. M. Bazzi et al. [SIDDHARTA collaboration], to be published in Physics Letters B.

Precision spectroscopy of kaonic ^{3}He X-rays at J-PARC

M .Sato*, H. Bhang†, M. Cargnelli**, Seonho Choi†, C. Curceanu‡, S. Enomoto§, H. Fujioka¶, Y. Fujiwara*, C. Guaraldo‡, T. Hashimoto*, R.S. Hayano*, T. Hiraiwa¶, M. Iio‖, K. Inoue§, T. Ishikawa*, S. Ishimoto††, T. Ishiwatari**, K. Itahashi‖, M. Iwasaki‖,‡‡, H. Kou‡‡, P. Kienle**,§§, J. Marton**, Y. Matsuda¶¶, H. Noumi§, H. Ohnishi‖, S. Okada‡, H. Outa‖, A. Romero Vidal‡, F. Sakuma‖, M. Sekimoto††, H. Shi*, D. Sirghi‡, F. Sirghi‡, T. Suzuki*, K. Tanida†, H. Tatsuno*, M. Tokuda‡‡, D. Tomono‖, A. Toyoda††, K. Tsukada‖, O. Vazquez Doce‡, E. Widmann**, B. Wünschek**, T. Yamazaki*,‖ and J. Zmeskal**

**Department of Physics, The University of Tokyo, Tokyo, Japan*
†Department of Physics, Seoul National University, Seoul, South Korea
***Stefan Meyer Institute für subatomare Physik, Vienna, Austria*
‡Laboratori Nazionali di Frascati, INFN, Fascati, Italy
§Research Center for Nuclear Physics (RCNP), Osaka University, Osaka, Japan
¶Department of Physics, Kyoto University, Kyoto, Japan
‖RIKEN Nishina Center, RIKEN, Saitama, Japan
††IPNS, High Energy Accelerator Research Organization (KEK), Ibaraki, Japan
‡‡Department of Physics, Tokyo Institute of Technology, Tokyo, Japan
§§Technische Universität München, Garching, Germany
¶¶Graduate School of Arts and Sciences, The University of Tokyo, Tokyo, Japan

Abstract. We will measure the X-rays from kaonic ^{3}He $3d \to 2p$ transition with a precision below 2 eV. It can provide crucial information on the kaon-nucleus strong interaction. The experiment (J-PARC E17) will be performed as Day-1, which is one of the first experiments in the J-PARC hadron facility in the year 2011. An overview and the present status of the J-PARC E17 experiment are described.

Keywords: Kaonic helium, Kaon-nucleus interaction, Silicon drift detector
PACS: 13.75.Jz, 25.80.Nv, 29.30.Kv, 36.10.Gv

INTRODUCTION

Kaonic atoms have been studied for a number of years to understand the low-energy kaon-nucleus interaction. When negative kaons stop in a matter they are captured by Coulomb field of the nucleus into an atomic state with a large principal quantum number to form a kaonic atom. They undergo de-excitation into lower atomic orbit by Auger effect and emission of X-rays. At the low-lying states where the kaon wave function significantly overlaps with that of the nucleus, kaons are absorbed by the nucleus. The energy levels of these states are appreciably shifted and broadened from the purely-electromagnetic value by the strong interaction between the kaon and the nucleus. The probability of kaon absorption increases drastically with decreasing angular momentum, and this strong interaction effect is pronounced in *last orbit*, where the strong-interaction absorption width becomes larger than that of the radiative transition. By measuring the X-rays emitted in radiative transitions to the last orbit, the strong interaction induced shift and width can be derived. Therefore, precision X-ray spectroscopies of kaonic atoms can provide direct information on the kaon-nucleus strong interaction.

There have been many experiments studied the kaonic X-rays of elements with a wide range of atomic number. In recent years there have been particular interests on light kaonic atoms, such as kaonic hydrogen and kaonic helium. In 2005, the KEK-PS E570 group measured the Balmar-series X-rays from kaonic ^{4}He atoms at KEK 12 GeV proton synchrotron [1]. The experiment was motivated by a long-standing discrepancy between experiments and theory: Prior to E570, three measurements on kaonic ^{4}He were performed around 1980's, and the average of the $2p$ level shift indicates a large repulsive-type shift of -43 ± 8 eV [2, 3, 4]. On the other hand, theoretical calculations based on

12th International Conference on Meson-Nucleon Physics and the Structure of the Nucleon (MENU 2010)
AIP Conf. Proc. 1374, 216-219 (2011); doi: 10.1063/1.3647129

optical models claim very small $2p$ level shift below 1 eV [5, 6]. This situation was called as *kaonic helium puzzle*. In E570, X-rays from kaonic ^{4}He atoms were precisely measured with a novel X-ray detector, silicon drift detector (SDD), and the strong-interaction induced shift was deduced to be,

$$\Delta E_{2p} = 2 \pm 2(\text{stat.}) \pm 2(\text{sys.})\ \text{eV}.$$

This result solves the kaonic helium puzzle and was confirmed very recently by SIDDHARTA group at LNF [7]. The large $2p$ level shift of kaonic ^{4}He was firmly excluded by these two measurements.

J-PARC E17 EXPERIMENT

Motivation

The aim of the J-PARC E17 experiment is to determine the shift and width of the $2p$ state of kaonic ^{3}He with a precision better than 2 eV. In a recent theoretical calculation, the $2p$ level shift of kaonic ^{3}He and ^{4}He are calculated by a coupled channel scheme with a phenomenological deep potential [9]. Contrary to the calculations with optical models, it claims possibilities of a large $2p$ level shift ($\sim$ 10 eV) for kaonic ^{3}He and/or ^{4}He atoms depending on the strength of the kaon-nucleus potential. From the result of the KEK-PS E570 experiment, the energy shift of the $2p$ level of kaonic ^{4}He is strictly constraint to be small, however, a large energy shift of the kaonic ^{3}He is still possible if the potential is sufficiently deep. In the framework of Ref. [9], such a deep potential accommodates with the existence of deeply-bound nuclear $\bar{K}$ states, which is one of the most controversial issues in the strangeness physics nowadays [10]. Therefore, a precision measurement of the energy shift of the $2p$ level of kaonic ^{3}He is a matter of great interest and can provide crucial information on the kaon-nucleus strong interaction.

Experimental apparatus

Experiment E17 will be performed at the K1.8BR beamline in the J-PARC hadron facility [8]. The experimental setup and technique are similar to those applied in the E570 experiment. Figure 1 shows a schematic side view of the experimental setup. It is consisted of 4 components; a beamline spectrometer, a liquid ^{3}He target, a cylindrical detector system (CDS) and silicon drift X-ray detectors (SDDs). Negative kaons are transfered through the beamline spectrometer in the K1.8BR beamline. In the experimental area, the timing and trajectories of the K^- beam are measured by a beam-defining counter (E0) and a beamline drift chamber (BLC), respectively. By Lucite cherenkov counters (LC), they are separated from a number of contaminating particles, such as π^-'s. With a carbon degrader, K^-'s are slowed down and are stopped inside the liquid ^{3}He target. After that K^-'s are captured with a ^{3}He nucleus, X-rays from radiative transitions of kaonic ^{3}He atoms are measured with SDDs, which are installed very close to the liquid ^{3}He target to have large geometrical acceptance ($\sim$ 1 % with 8 SDDs). For the background suppression, it is important to detect secondary charged particles after K^- nuclear absorption. That becomes possible with CDS, which is consisted with a cylindrical drift chamber (CDC) and scintillator hodoscope counters (CDH). Together with the trajectories of the K^- beam, the reaction vertex is reconstructed, and the fiducial target volume region is selected in the offline analysis to reject events K^-'s stop outside of the liquid ^{3}He target.

The most important issue to accomplish a precision measurement is accuracy of the absolute energy calibration. Since the gain of the X-ray detector is known to be fluctuated by its incident X-ray hit rate, an in-beam energy calibration is the most essential way to overcome this. The energy of the kaonic ^{3}He X-ray locates at 6.2 keV whose energy range is similar to those of K-series fluorescence X-rays of $3d$ transition metals. K_α lines for titanium (4.5 keV) and nickel (7.5 keV) are selected for the calibration sources. Since the K^- beam is contaminated with a number of other particles, calibration foils installed close to the target are activated by those particles, and fluorescence X-rays from the foils are obtained simultaneously with the kaonic ^{3}He X-rays.

In the following, the detail and the present status of each apparatus are described.

Beamline spectrometer. The K1.8BR beamline is a low-momentum secondary beamline with a single-stage electrostatic separator at the J-PARC 50 GeV proton synchrotron. We have installed a segmented beam hodscope counter and two-sets of drift chambers for the beam profile measurement and momentum analysis. The beam commissioning at K1.8BR has already started mainly at the momentum of 750 MeV/c. K^-'s are clearly identified with online hardware

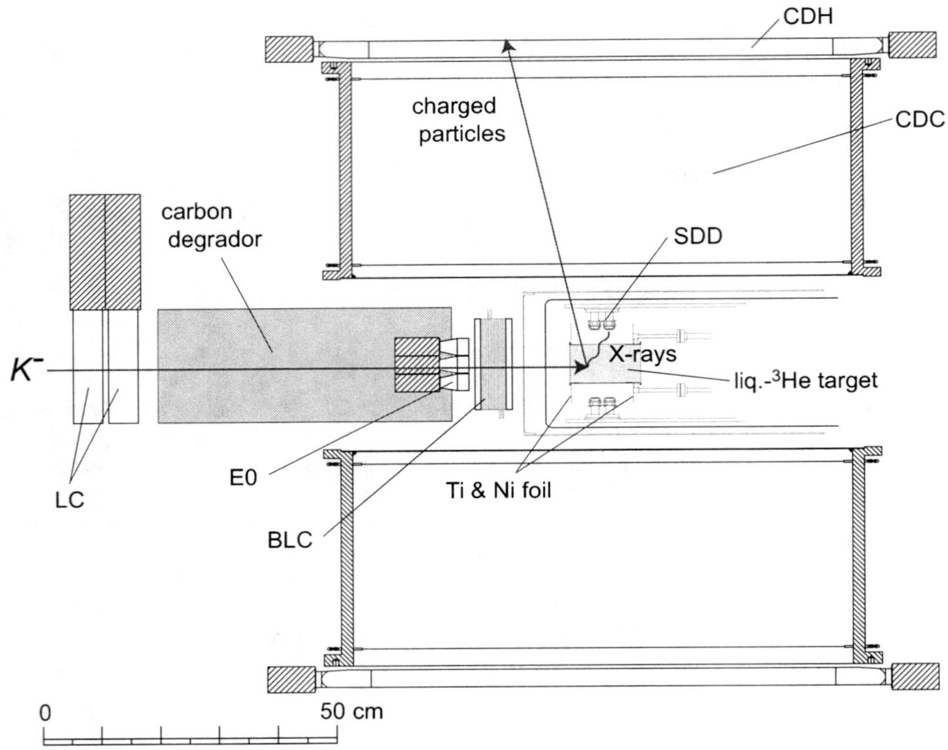

FIGURE 1. A schematic of the J-PARC E17 experimental setup

trigger. With the electrostatic separator, the yield of the K^-'s beam have been optimized by tuning the parameters, such as slit setting. We have studied the yield dependence on the beam momentum and the difference of the primary production target of nickel and platinum.

For the E17 experiment, the number of the stopped K^-'s will be optimized after the central momentum measurement and fine adjustment of the degrader thickness.

Liquid ^{3}He target. A liquid ^{3}He cryostat is based on a super-fluid ^{4}He cryostat used in KEK experiments [11]. It condenses gaseous ^{3}He by a heat contact with ^{4}He. Coolant ^{4}He is periodically supplied inside the cryostat from a 1000-litter Dewar placed nearby. Gaseous ^{3}He with the volume of 450 litter at STP is filled with buffer tanks which is hermetically connected to the target cell. During the cooling stage, a heat exchanger cooled down with 1.4 K ^{4}He condenses the ^{3}He gas and accumulates liquid ^{3}He inside the target cell. We have succeeded in an operation of liquid ^{3}He cryostat, and an achieved temperature is stable around 1.4 K. The pressure inside the target cell is consistent with well-known saturated vapor pressure curve of ^{3}He. This means that ^{3}He is definitively liquefied inside the target. Consumption rate of colorant is well suppressed to the operational level for the long-period beam time.

Cylindrical detector system. The cylindrical detector system is installed to detect secondary charged particles for the background suppression. It is consisted with a cylindrical drift chamber (CDC), hodscope counter (CDH) and a solenoid magnet. A magnetic field of the solenoid magnet is not applied in E17 because it is dispensable. The dimension of CDC is 95 cm length cylindrical shape with a diameter of 106 cm. The number of read-out wires of CDC is 1816. CDH provides a hardware hit timing of secondary charged particles. Performances of these detectors are evaluated with cosmic-ray events. A typical position resolution of CDC is about 200 μm, and the time resolution of CDH is around 70 ps. These resolutions are close to the design values.

Silicon drift detector. For an X-ray detection from kaonic ^{3}He atoms, we employed silicon drift detectors (SDDs) which were recently developed by KETEK for X-ray spectroscopes of kaonic atoms [12]. The effective area of an

SDD is 100 mm^2 with the thickness of 450 μm. For the better energy resolution, their preamplifiers are also installed inside the liquid ^{3}He cryostat to minimize the cable length between SDDs and preamplifiers. The operation of SDDs inside the cryostat is checked with an ^{55}Fe source. Typical energy resolution at the energy of 5.9 keV (Mn K_α) is obtained to be 140 eV in FWHM, which is much better than that of E570.

To apply a timing selection for the background suppression, the time resolution of the SDD is studied with electrons from a ^{90}Sr source. We obtained a large temperature dependence of the time resolution, which can be understood by an electron mobility inside silicon. Typical time resolution at the operational temperature region (120$\sim$130 K) is about 400$\sim$450 ns in FWHM. The detail of the temperature dependences of the energy and time resolution is described in a separate paper [13].

For the precise determination of the *2p* level shift of kaonic ^{3}He atom, basic studies of the line shape of detector response is extremely important, since the systematic errors of the shift of kaonic ^{4}He in E570 is dominated with uncertainty of the response functions. Incident angle and position dependences of the response function, energy/time resolutions and central peak position stability are under investigation taking into account those energy dependence. Results of these studies are discussed in dedicated papers to follow.

SUMMARY AND OUTLOOK

We will measure the X-rays of kaonic ^{3}He atoms at K1.8BR beamline in the J-PARC hadron facility. Most of the detector construction has been completed with the expected performances. The beam commissioning at the K1.8BR beamline has been started and we succeed in identifying the K^-'s clearly. The J-PARC E17 experiment will start its data taking in Day-1 of the J-PARC experimental program in 2011.

ACKNOWLEDGMENTS

This work is financially supported by the Ministry of Education, Science, Sports and Culture of Japan, Grand-in-Aid for Specially Promoted Research (200020003) and by Austrian Science Fund (FWF):[p20651].

REFERENCES

1. S. Okada, et al., *Phys. Lett.* **B653**, 387–391 (2007).
2. C. E. Wiegand and R. H. Pehl, *Phys. Rev. Lett.* **27**, 1410–1412 (1971).
3. C. J. Batty, et al., *Nucl. Phys.* **A287**, 455–462 (1979).
4. S. Baird, et al., *Nucl. Phys.* **A392**, 297–310 (1983).
5. C. J. Batty, *Nucl. Phys.* **A508**, 89–98 (1990).
6. S. Hirenzaki, Y. Okamura, H. Toki, E. Oset, and A. Ramos, *Phys. Rev.* **C61**, 055205 (2000).
7. M. Bazzi, et al., *Phys. Lett.* **B681**, 310–314 (2009).
8. R. S. Hayano, H. Outa, et al., *Proposal of J-PARC 50 GeV PS "Precision spectroscopy of Kaonic Helium-3 3d $\rightarrow$ 2p X-rays"*, (2006).
9. Y. Akaishi, "Kaonic helium atoms in relation to kaonic nuclear bound states", in *Proceeding for International Conference on Exotic Atoms and Related Topics (EXA05)*, edited by A. Hirtl, J. Marton, E. Widmann, and J. Zmeskal, Austrian Academy of Sciences Press, Vienna, pp.45–54 (2005).
10. Y. Akaishi, and T. Yamazaki, *Phys. Rev.* **C65**, 044005 (2002).
11. M. Sato, et al., *Nucl. Instrum. Meth.* **A606**, 233-237 (2009).
12. URL http://www.ketek.net/.
13. B. K. Wuenschek, et al., to be submitted to *Nucl. Instrum. Meth.*.

Study of the $\gamma d \to K^+K^-np$ reaction and an alternative explanation for the "Θ^+" peak

E. Oset and A. Martinez Torres

Departamento de Física Teórica and IFIC, Centro Mixto Universidad de Valencia-CSIC, Institutos de Investigación de Paterna, Aptd. 22085, 46071 Valencia, Spain.

Abstract. We look for an explanation of the peak in the K^+n invariant mass distribution of the $\gamma d \to K^+K^- n p$ reaction that was associated to the $\Theta^+(1540)$ pentaquark at LEPS. For this we develop a theoretical model adapted to the experiment. The study shows that the method used in the experiment to associate momenta to the undetected proton and neutron, together with the chosen cuts, necessarily creates an artificial broad peak in the assumed K^+n invariant mass in the region of the claimed $\Theta^+(1540)$, such that the remaining strength seen for the experimental peak is compatible with a fluctuation of 2σ significance.

Keywords: Alternative explanation for the "pentaquark" peak
PACS: 14.20.Pt

One of the issues in hadron physics that stimulated the largest excitement in the last few years was the work of [1], where the $\gamma\,^{12}C \to K^+K^-X$ reaction was studied and a peak was found in the K^+n invariant mass spectrum around 1540 MeV, which was identified as a signal for a pentaquark of positive strangeness, the "Θ^+". This finding stimulated a large number of theoretical papers, which in any case proved that the level of satisfaction reached with the use of quark models to describe hadron spectra was not fully justified. The unexpected finding lead also to a large number of poor statistics experiments where a positive signal was found, but gradually an equally big number of large statistics experiments showed no evidence for such a peak. A comprehensive review of these developments was done in [2], where one can see the relevant literature on the subject, as well as in the devoted section of the PDG [3].

More recently a new experiment was done at LEPS on a deuteron target, and with more statistics, and a clear peak was observed around 1526 MeV in the K^+n invariant mass distribution [4]. Yet, the experiment of [5] dealing with the same reaction as in LEPS but with ten times more statistics and with complete kinematics (but excluding small angles), failed to see any peak around the "Θ^+" region. The detail of complete kinematics should be stressed because in the LEPS experiment neither the proton nor the neutron were measured and some prescription had to be taken to guess their momenta.

In order to understand what is behind the peak seen in [4], we have constructed a theoretical model which contains the basic ingredients seen in the LEPS experiment, ϕ production on the proton and the neutron and $\Lambda(1520)$ production on the proton, together with rescattering of the kaons. The details of this model can be seen in [6]. To adapt the model to the set up of the LEPS experiment, we generate twenty random energies between 2 GeV to 2.4 GeV and implement the angular cuts and the mass cuts to eliminate the ϕ peak as in [4].

The LEPS detector is a forward magnetic spectrometer. Its geometry is implemented in our simulation by imposing that the angle of the kaons in the final state with respect the incident photon is not bigger than 20 degrees.

The nucleons are not detected at LEPS, therefore, some prescription is required in order to estimate the momentum of the p and n in the reaction $\gamma d \to K^+K^-np$ and determine the invariant mass of K^-p or K^+n. This is done using the minimum momentum spectator approximation (MMSA). For this purpose one defines the magnitude

$$p_{pn} = p_{miss} = p_\gamma + p_d - p_{K^+} - p_{K^-} \tag{1}$$

which corresponds to the four momentum of the outgoing pn pair. From there one evaluates the nucleon momentum in the frame of reference where the pn system is at rest, $\vec{p}_{CM}$. Boosting back this momentum to the laboratory frame, we will have a minimum modulus for the momentum of the spectator nucleon when the momentum $\vec{p}_{CM}$ for this nucleon goes in the direction opposite to $\vec{p}_{miss}$. Thus, the minimum momentum, p_{min}, is given by

$$p_{min} = -|\vec{p}_{CM}| \cdot \frac{E_{miss}}{M_{pn}} + E_{CM} \cdot \frac{|\vec{p}_{miss}|}{M_{pn}} \tag{2}$$

12th International Conference on Meson-Nucleon Physics and the Structure of the Nucleon (MENU 2010)
AIP Conf. Proc. 1374, 220-223 (2011); doi: 10.1063/1.3647130

where $E_{CM} = \sqrt{|\vec{p}_{CM}|^2 + M_N^2}$ is the energy of the nucleon in the CM frame. In this case, the momentum of the other nucleon will be in the direction of the missing momentum with a magnitude

$$p_{res} = |\vec{p}_{miss}| - p_{min} \tag{3}$$

In [4] the M_{K^+n} invariant mass for the reaction $\gamma d \to K^+K^-np$ is evaluated assuming the proton to have a momentum p_{min} (actually what one is evaluating is M_{K^+N}, with N the non spectator nucleon). Consequently, in this prescription, the momentum of the neutron in the final state will be

$$\vec{p}_n = p_{res} \cdot \frac{\vec{p}_{miss}}{|\vec{p}_{miss}|} \tag{4}$$

which is used to calculate the M_{K^+n} invariant mass for the reaction $\gamma d \to K^+K^-np$ in [4]. A cut is imposed at LEPS demanding that $|p_{min}| < 100$ MeV. This condition is also implemented in our simulation of the process.

In order to remove the contribution from the ϕ production at LEPS they consider events which satisfy that the invariant mass of the K^+K^- pair is bigger than 1030 MeV and bigger than the value obtained from the following expression

$$1020\,\text{MeV} + 0.09 \times (E_\gamma^{eff}(\text{MeV}) - 2000\,\text{MeV}) \tag{5}$$

where E_γ^{eff} is defined as the effective photon energy

$$E_\gamma^{eff} = \frac{s_{K^+K^-n} - M_n^2}{2M_n} \tag{6}$$

with $s_{K^+K^-n}$ the square of the total center of mass energy for the K^+K^-n system calculated using the MMSA approximation to determine the momentum of the neutron assuming the proton as spectator. In [4] only events for which 2000 MeV $< E_\gamma^{eff} <$ 2500 MeV are considered, a condition which is also incorporated in our simulation. The E_γ^{eff} of Eq. (6) with the MMSA prescription would correspond to the photon energy in the frame where the original non spectator (participant) nucleon is at rest.

The first result that we show is the distribution of K^+n invariant masses for the LEPS set up using the real momenta obtained from our Monte Carlo integral of the cross section versus the one obtained using the momenta determined with the MMSA prescription (Fig. 1 (left)). We see two blocks of points, one of them sticking around the diagonal and another one with points scattered around the plane. This is so, because actually the MMSA prescription reconstructs the K^+N invariant mass, where N is the participant nucleon, which about half of the times is the neutron and the other half the proton. The points around the diagonal correspond to the case where the participant is the neutron.

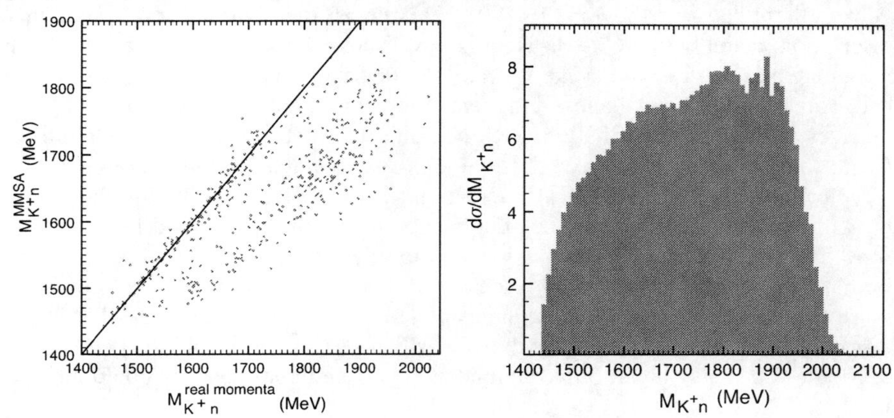

FIGURE 1. Left:M_{K^+n} calculated using the MMSA prescription versus M_{K^+n} obtained with the real momentum for the nucleons and the full model, i.e., ϕ production on the nucleons and $\Lambda(1520)$ production on the proton. Right: M_{K^+n} invariant mass distribution calculated using the real momenta and with a ϕ cut of $M_{K^+K^-} > 1050$ MeV.

The association of the K^+N spectrum to K^+n at LEPS has a repercussion in the assumed experimental K^+n distribution. The real distribution from the model is given in Fig. 1 (right). We have to take a cut for the K^+K^-

invariant mass slightly different than in [4] because the cut at LEPS involves E_γ^{eff} which relies for its definition on the MMSA prescription itself, so we must avoid that to use the real momenta. As we can see in Fig. 1 (right), there is no peak of the distribution around the "Θ^+" peak. Instead we show in Fig. 2 (left) the distribution obtained using the LEPS cuts and the MMSA prescription, normalized to the experimental data, which are also shown in the figure.

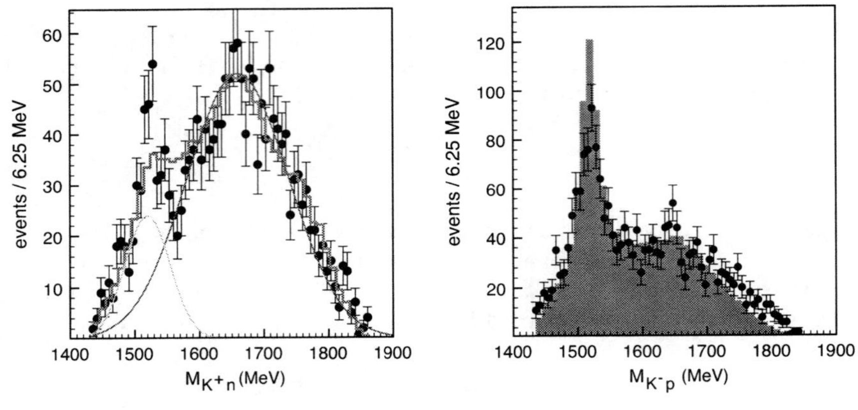

FIGURE 2. Left: M_{K^+n} invariant mass distribution obtained with the MMSA prescription and same cuts than LEPS normalized to the data of [4] (shown as dots), together with two gaussians functions: one peaking at 1520 MeV with a width of 80 MeV and another one peaking at 1660 MeV with a width of 185 MeV. Right: M_{K^-p} invariant mass distribution calculated with the MMSA prescription and same cuts than LEPS normalized to the data of [4] (shown as dots).

We use the full model of [6] including ϕ production on the proton and the neutron plus $\Lambda(1520)$ production on the proton. The MMSA prescription is used to compare to the data as explained before. As we can see in Fig. 2 (left), the combination of the LEPS cuts and the MMSA prescription, which also affects the cut, has produced an artificial peak below the region of the "Θ^+". We show in the figure how the shape of the distribution can be represented in terms of two gaussians, one of them peaking around the "Θ^+" peak. This means that in a large statistics experiment one would see this clear broad peak, which could be interpreted as a sign of a resonance. Yet, there is no resonance in that region in the model used. Coming back to the comparison of the "exact" distribution, generated from the theoretical model, with the data (taken from Fig. 12 a) from [4]), we see that the "Θ^+" peak has three points on top of the "exact" distribution. The accumulated strength of these three bins over the "exact" curve is about 35 events. The question now remains: could this peak, measured from the "exact" curve, be a statistical fluctuation due to the limited number of events of the experiment (around 2000 events)?. A hint to answer this question is provided by the LEPS experiment in fig. 10 a) of [4], where a peak followed by a dip is seen on top of the assumed background in the K^-p mass distribution. In order to compare with these data, we show in Fig. 2 (right) the mass distribution for K^-p obtained with our model and the MMSA prescription (actually the K^-N distribution, as discussed above), together with the data (Fig. 10 a) of [4]). The agreement with experiment is good, like for the K^+n distribution shown in Fig. 2 (left), indicating that we have indeed a realistic model. The discrepancies in the large momentum tail can be cured adding a small background which has minor changes in the K^+ n distribution. Yet, the point we want to make is that there is a peak in the data around 1650 MeV with four points over the "exact" curve with an accumulated strength of about 33 events, followed by a similar dip, a typical structure of a statistical fluctuation. Since for this distribution there are no discrepancies in the interpretation with respect to [4] we can take the background from there (solid line of Fig. 10 a) of [4]) and we also find about 40 events. This peak was not associated to any resonance in [4]. Instead it was dismissed as a statistical fluctuation. In fact, the peak disappeared when a slightly different cut was made (see Fig. 16 (Right) of [4]). The interpretation of this peak and dip as a fluctuation, as done in [4], is fair, as this peak has about 2σ significance over the background (σ is the statistical error of the data for which $\sqrt{N_{events}}$ is taken in [4]). But the peak of the "Θ^+" over the "exact" distribution has also about 35 accumulated events and consequently could also be considered as a statistical fluctuation.

Here we would like to make a more quantitative study of the statistical significance of the "Θ^+" peak. In [4] a best fit to the data was done assuming a background and a Gaussian peak in the "Θ^+" region. The best fit with these assumptions provided a background of about 22 events per bin below the "Θ^+" peak, as can be seen in Fig. 12 a) of [4]. With respect to this background, the "Θ^+" peak has a strength of about 5σ. According to this, the statistical significance of the peak would rule out the possibility of it being a statistical fluctuation.

Conversely, after the theoretical evaluation of the background, our argumentation goes as follows: The actual background below the "Θ^+" peak is bigger than the one provided by the LEPS best fit, around 36 events per bin instead of the 22 assumed in [4]. This makes the strength of the peak with respect to the background much smaller than in the LEPS best fit. It also makes σ larger and the statistical significance is now of about 2σ, something acceptable as a fluctuation, as in the case of the experimental $K^- p$ spectrum. This argumentation about the significance of the peaks is corroborated by further calculations which we have carried out. First, the best fit of LEPS is not the only good fit possible. We have seen that a fit to the data with a background and a fluctuation (a peak followed by a dip) gives the same reduced χ^2 than in [4], but returns a background nearly identical to the calculated one. Second, in order to know the actual errors of the limited LEPS statistical we have made 10 runs with the Von Neumann rejection method, producing about 2000 events each, like in [4]. This method proceeds like the experiment, generating events or not according to their probability to be produced, and we have checked that the statistical significance of the runs is equivalent to that of the experiment. From these runs we have evaluated the statistical errors of each run (see section VI of [6]). The errors found are of the order of 20% in the region of the "Θ^+". This means an error (σ) of about 7 events per bin, such that the difference of the peak to the background is indeed of the order of 2σ.

The other point we want to make is that it is possible to produce peaks by changing the cuts. This is shown explicitly in the experiment of [4] since the mentioned peak at 1650 MeV in the $K^- p$ mass distribution in Fig. 10 a) of that paper disappears when a slightly different cut is made in Fig. 16 (Right) of the same paper. This should be a warning for analyses in this kind of problems.

In summary, our theoretical study has shown that the background in the $\gamma\, d \longrightarrow K^+ K^- \, n\, p$ reaction, with the cuts and the use of the MMSA prescription of LEPS, generates artificially a broad peak below the "Θ^+" region. We also showed that this background is fairly larger than the one obtained at LEPS in their best fit to the data, assuming a background and a Gaussian peak in the region of the "Θ^+". We also mentioned that the fit of LEPS is not unique and other fits to the data, assuming a background and a fluctuation, are possible, producing the same reduced χ^2 and returning a background nearly identical to the calculated one. Based on the calculated background and the errors obtained from different Monte Carlo Von Neumann runs, we evaluated the statistical significance of the "Θ^+" peak and found it to be of about 2σ with respect to the background, compatible with a fluctuation. The larger statistical significance claimed in [4] was tied to the assumption of a significantly smaller background. Once this background has been accurately evaluated, the fit of LEPS, which grossly changes this background, has to be dismissed as unphysical. The fact that we find an alternative fit assuming a background and a fluctuation, and that this fit to the data returns the theoretical background, is certainly an extra test of consistency. The fact that the fluctuation obtained in that fit is basically identical to the one seen in the experiment in the $K^- p$ mass distribution, admitted as a fluctuation in the LEPS analysis, provides extra support to our conclusions that what one sees in the experiment is most likely a fluctuation on top of a large artificial background created by the cuts and the use of the MMSA at LEPS. On the other hand, a firm conclusion is that the claims of LEPS for a new state, which where based on a fit to the data that we have proved to be physically unacceptable, are not justified.

ACKNOWLEDGMENTS

This work is partly supported by the DGICYT contract FIS2006-03438, the Generalitat Valenciana in the program Prometeo and the EU Integrated Infrastructure Initiative Hadron Physics Project, n.227431. A. M. T thanks the support from a FPU grant of the MICINN.

REFERENCES

1. T. Nakano *et al.*, Phys. Rev. Lett. **91**, 012002 (2003).
2. K. H. Hicks, Prog. Part. Nucl. Phys. **55**, 647 (2005).
3. C. Amsler *et al.*, Phys. Lett. B **667**, 1 (2008).
4. T. Nakano *et al.*, Phys. Rev. C **79**, 025210 (2009).
5. B. McKinnon *et al.*, Phys. Rev. Lett. **96**, 212001 (2006).
6. A. Martínez Torres and E. Oset, Phys. Rev. C **81**, 055202 (2010).

Session C: Spin Structure

Nucleon spin structure at Jefferson Lab

Angela Biselli
for the CLAS collaboration

Fairfield University, Fairfield, CT 06828

Abstract. In the past decade an extensive experimental program to measure the spin structure of the nucleon has been carried out in the three halls at Jefferson Lab. Using a longitudinally polarized beam scattering off longitudinally or transversely polarized ^{3}He, NH_3 and ND_3 targets, the double spin asymmetries $A_{||}$ and $A_\perp$ were measured, providing data of impressively high precision that gives a better understanding of the structure of the nucleon in the deep inelastic scattering and the valence quarks regions. The virtual photon asymmetries $A_{1,2}$ and polarized structure functions $g_{1,2}$ were also extracted for the proton, neutron and deuteron over large kinematic ranges, allowing the extraction of first moments and the testing of sum rules and duality.

Keywords: Spin structure, Nucleon Spin
PACS: 13.60.Hb, 13.88.+e, 14.20.Dh

INTRODUCTION

After the "spin crisis" in 1980, large experimental programs have been devoted to unraveling the spin structure of the nucleon. Inclusive inelastic scattering of polarized electrons off polarized nucleons is a powerful tool to study the polarized structure functions and their moments. For a longitudinally polarized beam the measured asymmetries are

$$A_{||} = \frac{\sigma^{\uparrow\Uparrow} - \sigma^{\downarrow\Uparrow}}{\sigma^{\uparrow\Uparrow} + \sigma^{\downarrow\Uparrow}} \quad \text{and} \quad A_\perp = \frac{\sigma^{\uparrow\Rightarrow} - \sigma^{\downarrow\Rightarrow}}{\sigma^{\uparrow\Rightarrow} + \sigma^{\downarrow\Rightarrow}} \tag{1}$$

where $\sigma^{\uparrow\Uparrow}$ ($\sigma^{\uparrow\Downarrow}$) is the cross sections for beam and target polarization oriented parallel (antiparallel), and $\sigma^{\uparrow\Rightarrow}$ ($\sigma^{\downarrow\Rightarrow}$) is the cross section for a transversely polarized target with the electron spin aligned antiparallel (parallel) to the beam direction. The asymmetries $A_{||}$ and $A_\perp$ are related to the virtual photon asymmetries A_1 and A_2 by

$$\frac{A_{||}}{D} = A_1 + \eta A_2 \quad \text{and} \quad \frac{A_\perp}{d} = A_1 + \zeta A_2 \tag{2}$$

where $\eta = \varepsilon Q/(E - E'\varepsilon)$, $\zeta = \eta(1+\varepsilon)/(2\varepsilon)$, $D = (1 - E'\varepsilon/E)/(1+\varepsilon R)$, $d = D\sqrt{2\varepsilon/(1+\varepsilon)}$, $R = \sigma_L/\sigma_T$ is the ratio of longitudinal and transverse virtual photon-absorption cross sections, Q^2 is the squared 4-momentum transfer, and $\varepsilon = (1 + 2\frac{\vec{q}^2}{Q^2}\tan^2\frac{\theta_e}{2})^{-1}$ is the degree of transverse photon polarization. The asymmetry A_1 is related to the difference of virtual photo-absorption cross sections for total helicity between photon and nucleon of 1/2 and 3/2, while A_2 is related to the interference term between the transverse and longitudinal photon-nucleon amplitudes.

Given the virtual photon asymmetries, one can extract the spin structure function g_1 and g_2 using:

$$g_1(x,Q^2) = \frac{\tau}{1+\tau}(A_1 + \frac{1}{\sqrt{\tau}}A_2)F_1(x,Q^2) \quad \text{and} \quad g_2(x,Q^2) = \frac{\tau}{1+\tau}(-A_1 + \sqrt{\tau}A_2)F_1(x,Q^2) \tag{3}$$

where $\tau = \nu^2/Q^2$, ν is the energy transfer of the virtual photon, and F_1 is the unpolarized structure function. The spin structure function g_2 contains information about higher twists.

A quantity of particular interest is the first moment of g_1

$$\Gamma_1(Q^2) = \int_0^1 g_1(x,Q^2)dx. \tag{4}$$

At high Q^2, Γ_1^{p-n} is related to the spin carried by the quarks as stated by the Bjorken sum rule. For $Q^2 \to 0$ this quantity becomes $\frac{Q^2}{2M^2}I_{GDH}$, where I_{GDH} is the Gerasimov-Drell-Hearn sum rule and is equal to $-\kappa^2/4$, where κ is the nucleon anomalous magnetic moment.

12th International Conference on Meson-Nucleon Physics and the Structure of the Nucleon (MENU 2010)
AIP Conf. Proc. 1374, 227-230 (2011); doi: 10.1063/1.3647131

SELECTED RESULTS

The virtual photon asymmetry A_1 was measured in all three experimental Halls at Jefferson Lab in different kinematic ranges and for different nucleons. The Hall A experiment [1, 2] used the High Resolution Spectrometer (HRS) with a momentum resolution of 2×10^{-4} and a ^{3}He target, in which more than 87% of the spin is carried by the neutron, thereby providing an effective polarized neutron target. The target was oriented both longitudinally and transversely with respect to the direction of the beam, allowing the measurement of both $A_{||}$ and $A_{\perp}$. The result for A_1^n shows the first evidence of A_1^n becoming positive and large at high x, consistent with the relativistic constituent quark model (RCQM) predictions, the LSS 2001 perturbative quantum chromo-dynamics (PQCD) fit to previous data and a global NLO QCD analysis of deep inelastic scattering (DIS) data [1]. The Hall B experiment, known as the EG1 experiment [3, 4], used the CLAS detector, which gives a nearly 2π acceptance, achieving a kinematic coverage from 0.05 up to 5 GeV^2 in Q^2 and from 1.08 to 3 GeV in W. The targets were longitudinally polarized, solid, 1-cm-long, ammonia targets, $^{15}NH_3$ and $^{15}ND_3$, polarized up to 70% and 40% respectively via dynamic nuclear polarization [5]. Preliminary results are shown in Figure 1 for the virtual photon asymmetry A_1 for the proton (left) and deuterium (right) averaged in the deep inelastic region $Q^2 > 1$ GeV^2 and $W > 2$ GeV. These preliminary results confirm the trend of A_1 to exceed the SU(6) limit at x=1 already found in [4]. At higher x, where the valence quarks are expected to dominate, data are most consistent with the hyperfine perturbed quark model [6], but also very close to other models that consider different scenarios of SU(6) symmetry breaking [7].

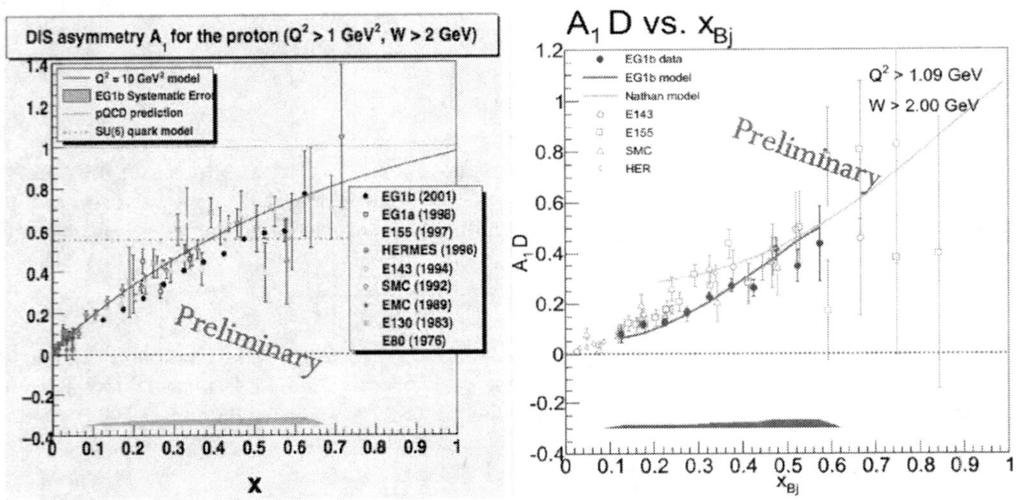

FIGURE 1. Virtual photon asymmetry A_1 vs. x for proton (left) and deuteron (right).

The structure function g_1 can be extracted from the virtual photon asymmetries using a parameterization of the world's data for the unpolarized structure function F_1. Figure 2 shows a sample of the results of the EG1 experiment for the deuterium polarized structure function g_1 for several Q^2 bins. Similar results were found for the proton. As can be seen at low Q^2 the $\Delta(1232)$ resonance drives the structure function towards negative values, due to dominance of the spin-3/2 amplitude. At higher Q^2, g_1 becomes positive everywhere.

The virtual photon asymmetry A_2 and the polarized structure function g_2 were measured both in Hall A [2] and Hall C [8] due to the ability of running there with a transversely polarized target. The RSS experiment in Hall C used the High Momentum Spectrometer and longitudinally and transversely polarized ammonia targets $^{15}NH_3$ and $^{15}ND_3$. The experiment focused on the resonance region $1.09 < W < 1.91$ GeV at an average Q^2 of 1.3 GeV^2. The structure function g_2 for the proton is clearly non-zero, providing strong evidence of higher-twist contributions as shown in Figure 3 [8].

The extensive kinematic range of these measurement allows the extraction of the first moments. Figure 4 shows the first moment of the structure function g_1 (see Eq. 4) for the proton using EG1 data. The world data parameterization was used to include the unmeasured part to the integral down to x =0.001. The low Q^2 data points are consistent with the I_{GDH} limit, and are in fairly good agreement with the phenomenological model by Burkert and Ioffe [9] that parameterizes the transition between the real photon point and the DIS regime using measured pion electro- and photo-production resonance amplitudes. The comparison of data at low Q^2 with the ChPT calculation of Ji (dashed

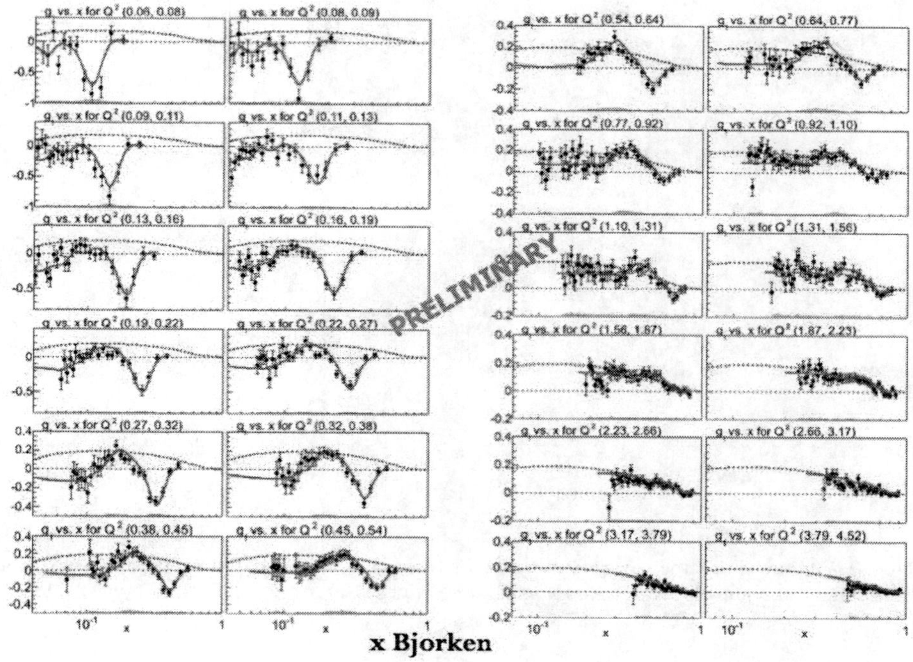

FIGURE 2. The structure function g_1 vs x for several Q^2 bins from 0.06 to 2.23 GeV^2.

curve) [10] shows a disagreement; however, this calculation is an expansion up to Q^4 and a fit of the data up to Q^8 shows agreement with Ji's Q^4 coefficient and a non negligible Q^6 term even at $Q^2 < 0.1$ GeV^2 [3]. The right panel of Figure 3 shows the first moment of the structure function g_1 for the neutron using Hall A data [11] including, the latest preliminary results at low Q^2, which are also in good agreement with the phenomenological model by Burkert and Ioffe.

The experimental program on the spin structure at Jefferson Lab has produced unprecedentedly precise measurements of the spin structure functions $A_{1,2}$, $g_{1,2}$, Γ_1 and related quantities for the proton, neutron and deuteron. New results are expected from more recent experiments such as the Hall B EG4 experiment and the Hall C SANE ex-

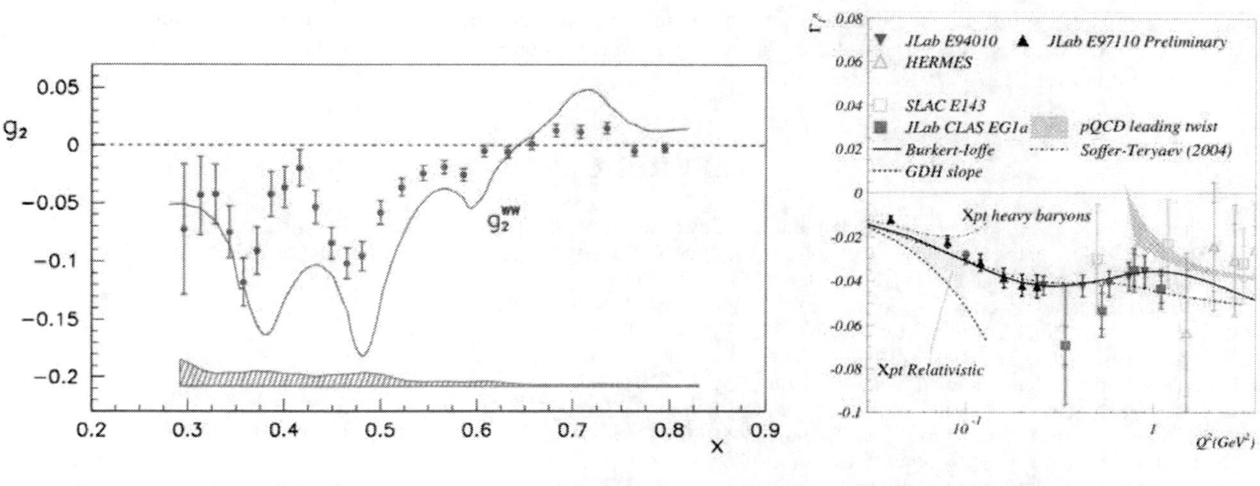

FIGURE 3. The structure function g_2 vs. x for proton (left) [8] and first moment of the structure function g_1 for the neutron using Hall A data [11] (right).

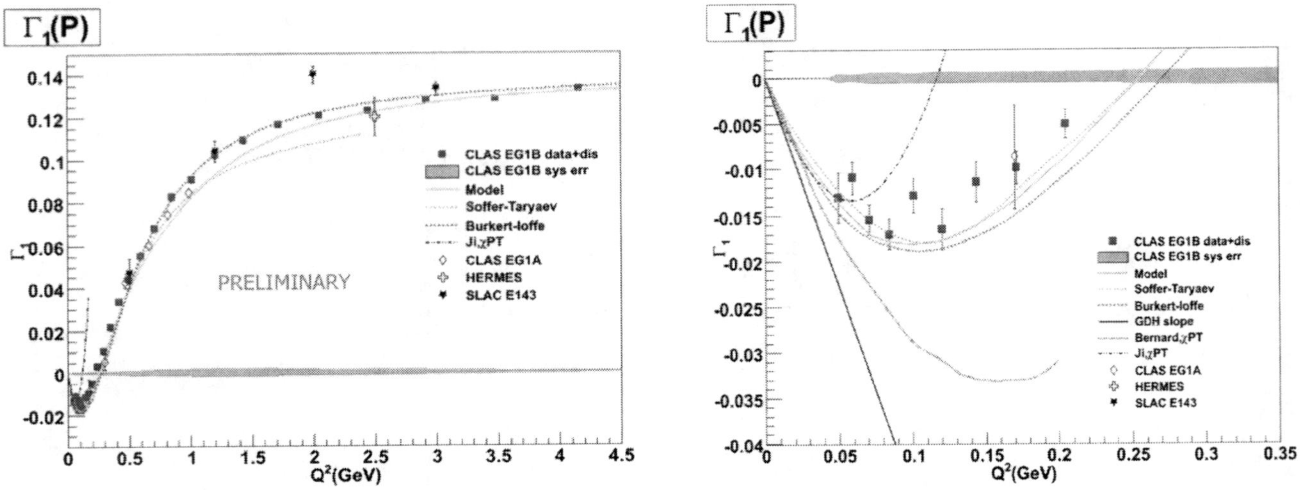

FIGURE 4. First moment of g_1 for the proton (Γ_1^p) vs Q^2. The left plot shows the full range in Q^2 while the right plot shows the lower Q^2 region. The figure shows the new preliminary data from EG1 (squares) and the previous EG1 (diamonds), HERMES (crosses) and SLAC E143 (stars) results. Phenomenological predictions (Burkert-Ioffe, Soffer-Teryaev, ChPT) and the GDH slope are also shown.

periment. The EG4 experiment, as a result of an additional Cherenkov counter placed at low angles, will extend the measurement of Γ_1 down to 0.015 GeV^2, allowing a better test of the ChPT calculations. The SANE experiment which uses the novel large solid angle electron telescope BETA detector will extract A_1 at high x, study the structure function g_1 and g_2 and twist 3 effects from their moments. In addition new experiments are planned for the 12 GeV upgrade of the accelerator. The higher energy and the improved detectors will increase the statistical precision of the current measurements, will extend the measurements to kinematic regions at the moment not accessible, such as high x, and will allow the extraction of fundamental quantities such as parton distribution functions.

ACKNOWLEDGMENTS

We would like to thank the staff of the Accelerator and Physics Divisions at the Jefferson Lab for their outstanding efforts to provide us with the high quality beam and the facilities for data analysis. This work was supported by the U.S. Department of Energy and the National Science Foundation, the French Commissariat a l'Energie Atomique, the Italian Istituto Nazionale di Fisica Nucleare, and the Korean Science and Engineering Foundation. Jefferson Science Associates (JSA) operates the Thomas Jefferson National Accelerator Facility for the United States Department of Energy under contract DE-AC05- 84ER40150.

REFERENCES

1. X. Zheng *et al.* [Jefferson Lab Hall A Collaboration], Phys. Rev. Lett. **92**, 012004 (2004).
2. X. Zheng *et al.* [Jefferson Lab Hall A Collaboration], Phys. Rev. C **70**, 065207 (2004) [arXiv:nucl-ex/0405006].
3. Y. Prok *et al.* [CLAS Collaboration], Phys. Lett. B **672**, 12 (2009).
4. K. V. Dharmawardane *et al.* [CLAS Collaboration], Phys. Lett. B **641**, 11 (2006).
5. C. D. Keith *et al.*, Nucl. Instrum. Meth. A **501**, 327 (2003).
6. N. Isgur, Phys. Rev. D **59**, 034013 (1999).
7. F. E. Close and W. Melnitchouk, Phys. Rev. C **68**, 035210 (2003).
8. F. Wesselmann *et al.* [RSS Collaboration], Phys. Rev. Lett., **98**, 132003, (2007).
9. V. D. Burkert and B. L. Ioffe, Phys. Lett. B **296**, 223 (1992).
10. X. D. Ji, C. W. Kao and J. Osborne, Phys. Lett. B **472**, 1 (2000).
11. M. Amarian *et al.* [Jefferson Lab E94-010 Collaboration], Phys. Rev. Lett. **92**, 022301 (2004).

New results on the neutron spin polarizabilities at low Q^2

Vincent A. Sulkosky[1]

12000 Jefferson Ave, Suite #4, Newport News, VA 23606, USA

Abstract. Thomas Jefferson National Accelerator Facility (Jefferson Lab) has become one of the premier facilities for the study of nucleon spin structure. Over the past two decades, several high precision measurements of the polarized structure functions $g_1(x,Q^2)$ and $g_2(x,Q^2)$ have been performed that have extensively covered the region between 0.02 and 5 GeV^2. Spin moments have been formed from the measured structure functions. These proceedings will concentrate on experiment E97-110, which was performed in Hall A to study the neutron and ^{3}He spin structure at four-momentum transfers squared (Q^2) = 0.02 - 0.26 GeV^2. This range allows us to make a benchmark-check of Chiral Perturbation Theory calculations. Previous results for the neutron spin polarizability, δ_{LT}, down to 0.1 GeV^2 have shown significant disagreement with the calculations. Preliminary results on the generalized spin polarizabilities, γ_0 and δ_{LT}, for the neutron were presented at the conference and will be discussed here.

Keywords: neutron,spin structure,chiral perturbation theory
PACS: 14.20.Dh

INTRODUCTION

Over the past thirty or so years, the study of nucleon spin structure has been an active field, both experimentally and theoretically [1, 2]. Spin-dependent sum rules, for example, the Gerasimov-Drell-Hearn (GDH) [3, 4, 5, 6] sum rule, are important tools available to study nucleon spin structure. The GDH sum rule was originally derived for real photon absorption, and the integral was later extended to virtual photon absorption [7]. An extension by Ji and Osborne [8] also forms a Q^2-dependent sum rule, which is constrained at the two ends of the Q^2 spectrum by well known sum rules: GDH at $Q^2 = 0$ and the Bjorken [9] sum rule at large Q^2. Ji's extension utilizes the relationship between the forward Compton scattering amplitudes and the spin-dependent structure functions $g_1(x,Q^2)$ and $g_2(x,Q^2)$ and serves as a natural extension of the sum rule to virtual-photon scattering. Here, x is the Bjorken scaling variable. This unique relation, valid at any Q^2, can be used to make comparisons between theoretical calculations and experimental data. The comparison allows us to study the transition between the non-perturbative and the perturbative regions of Quantum Chromodynamics (QCD).

Theoretical calculations from QCD are difficult to obtain due to the theory's non-perturbative nature in the low energy region. In this regime, low-energy effective field theories such as Chiral Perturbation Theory (ChPT) provide calculations for the sum rules. Experiment E97-110 [10] was specifically conducted to study the neutron spin structure functions at very low Q^2 and to make a benchmark check of ChPT calculations for these moments. In these proceedings, the data at low Q^2 are reviewed. Then the experimental details and preliminary results from experiment E97-110 are presented.

NUCLEON SPIN STRUCTURE RESULTS

Due to the recent technological advances in polarized beams and targets, experiments at Jefferson Lab have allowed precise measurements of the nucleon spin structure functions down to $Q^2 = 0.02$ GeV^2. Hall A measurements [11, 12, 13] have focused on polarized ^{3}He used as an effective polarized neutron target; whereas, Hall B and C measurements [14, 15, 16, 17, 18] have involved polarized proton and deuteron targets. In these proceedings, only the neutron results from the Hall A experiments will be discussed.

The generalized spin polarizabilities involve higher moments in Bjorken-x of the polarized structure functions. Due to the higher-x weighting, these integrals converge faster at small x than the first moments and therefore are

---

[1] On behalf of the Jefferson Lab Hall A and E97-110 Collaboration

12th International Conference on Meson-Nucleon Physics and the Structure of the Nucleon (MENU 2010)
AIP Conf. Proc. 1374, 231-234 (2011); doi: 10.1063/1.3647132

significantly less sensitive to the unmeasured low-x contributions. The γ_0 and δ_{LT} spin polarizabilites in terms of the structure functions can be expressed as

$$\gamma_0 = \frac{16\alpha M^2}{Q^6}\int_0^{x_0} x^2\left[g_1\left(x,Q^2\right) - \frac{4M^2}{Q^2}x^2 g_2\left(x,Q^2\right)\right]dx$$

$$\delta_{LT} = \frac{16\alpha M^2}{Q^6}\int_0^{x_0} x^2\left[g_1\left(x,Q^2\right) + g_2\left(x,Q^2\right)\right]dx,$$

where M is the mass of the target, $x_0 = \frac{Q^2}{2M\nu_0}$, ν is the energy transfer of the electron and α is the electromagnetic coupling constant. The inelastic threshold is represented by ν_0.

The generalized spin polarizabilities for δ_{LT} and γ_0 are shown in Fig. 1 and Fig. 2, respectively. The next-to-leading order ChPT calculations are shown as the dashed, heavy baryon ChPT (HBChPT), and the dot-dashed, relativistic baryon ChPT (RBChPT), lines in both plots. The light shaded band also includes the Δ resonance and vector meson contributions for the RBChPT [19, 20] calculation. The δ_{LT} spin polarizability is expected to serve as a firm testing

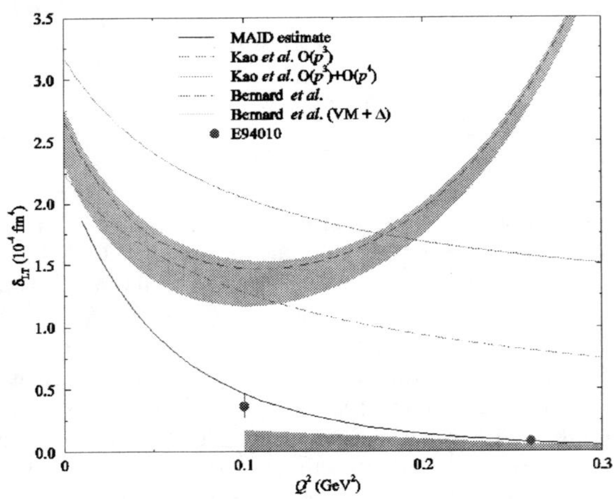

FIGURE 1. Spin polarizabilitie δ_{LT} of the neutron (see text for details).

ground for the ChPT calculations, since it is mostly insensitive to the Δ-resonance contribution compared to the γ_0 polarizability. In Fig 1, the data for δ_{LT} are compared with the theory calculations and the MAID model. In this case, we see the surprising result that the data down to 0.1 GeV^2 show a sizable discrepancy with both ChPT calculations. This disagreement has been referred to as the "δ_{LT} puzzle" and presents a real challenge to theorists. One of the major limitations of the available data was that the lowest Q^2 point might still be outside the region of ChPT's applicability. Additional experiments at Jefferson Lab have since obtained data for Q^2 down to approximately 0.02 GeV^2.

PRELIMINARY RESULTS FOR EXPERIMENT E97-110

Experiment E97-110 acquired data at Jefferson Lab in Hall A [21]. The specific details of the experiment and data analysis can be found in ref. [10]. The inclusive reaction $^3\vec{He}(\vec{e},e')$ was examined with a longitudinally polarized electron beam scattered from either a longitudinally or transversely (in-plane) polarized ^{3}He target [21]. Polarized cross-section differences were formed by combining the measured longitudinal and tranverse asymmetries $A_\parallel$ and $A_\perp$ with the unpolarized cross section σ_0: $\Delta\sigma_{\parallel,\perp} = 2\sigma_0 A_{\parallel,\perp}$. Then, the ^{3}He spin structure functions $g_1\left(x,Q^2\right)$ and $g_2\left(x,Q^2\right)$ were extracted from the polarized cross-section differences after radiative corrections were applied:

$$\Delta\sigma_\parallel\left(x,Q^2\right) = \frac{4\alpha^2}{M\nu Q^2}\frac{E'}{E}\left[\left(E+E'\cos\theta\right)g_1\left(x,Q^2\right) - \left(\frac{Q^2}{\nu}\right)g_2\left(x,Q^2\right)\right] \quad \text{and} \qquad (1)$$

$$\Delta\sigma_\perp(x,Q^2) = \frac{4\alpha^2}{M\nu Q^2}\frac{E'}{E}E'\sin\theta\left[g_1(x,Q^2) + \frac{2E}{\nu}g_2(x,Q^2)\right], \tag{2}$$

where E is the beam energy, E' the scattered electron energy and θ the electron scattering angle. Finally, after the structure functions have been extracted at constant Q^2, the generalized spin polarizabilities and other moments of the spin structure functions can be formed.

For experiment E97-110, the polarized ^{3}He target was used as an effective neutron target. The neutron moments were extracted from the measured ^{3}He quantities by using the method described in [22]. The systematic uncertainty due to this extraction was inferred to be 5% for $Q^2 > 0.1$ GeV^2 and 10% at lower Q^2. The deuteron data from the EG4 [23] experiment will also provide an important systematic check of the neutron extraction from light nuclei. The parameterization in ref. [24] was also utilized to account for the unmeasured Deep Inelastic Scattering (DIS) contribution.

In Fig. 2, the preliminary results for the γ_0 spin polarizability for the neutron are compared to theoretical calculations and earlier measurements. The solid circles show the preliminary data, and the squares indicate the results from a previous Hall A experiment [13] between 0.1 GeV^2 and 0.26 GeV^2. The error bars on the data indicate the size of the statistical uncertainties; whereas, the solid bands near zero show the size of the systematic uncertainties. The band on the negative side of the vertical axis is for the new data. We expect to significantly improve the (largely dominant) systematic uncertainties for the final results. As Q^2 approaches zero, the data indicate a smooth transition and agree reasonably well with the earlier data but shows a sizeable difference with the MAID model [25] (represented by the solid line) below 0.15 GeV^2 [25]. The two available calculations from Chiral Perturbation Theory [19, 20, 26] are

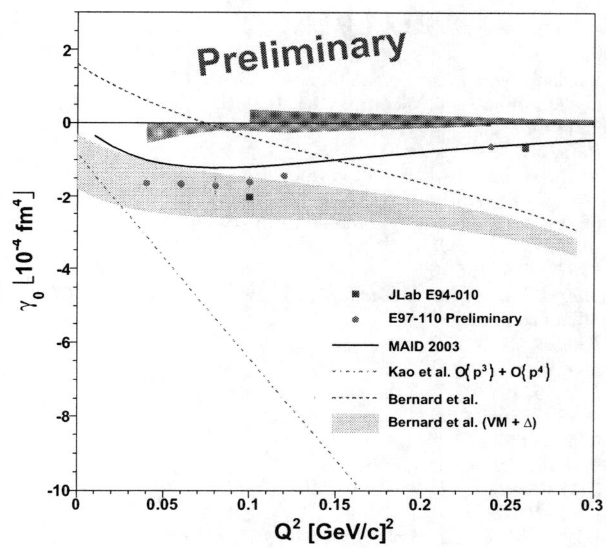

FIGURE 2. Preliminary neutron γ_0^n results (see text for details)

also shown on the figure. The calculation with an estimate of the Δ-resonance and vector meson contributions is represented by the light-shaded band. With the current uncertainties, the data show good agreement with the RBChPT calculation [19, 20] below 0.12 GeV^2. However, the HBChPT [27] calculation shows a large discrepancy, which might indicate the importance of the resonances. Preliminary results for the δ_{LT} polarizability were also shown in the conference. The results show good agreement with the E94-010 data [13] and a still significant disagreement with the ChPT calculations.

A few years ago, the EG4 experiment in Hall B also obtained high precision data in a similar Q^2 range as experiment E97-110 for the proton and deutron g_1 structure functions. On the other hand, the g_2 structure function for the proton is largely unknown, since traversely-polarized target data are not available at low Q^2. A new experiment [28] was recently approved to measure this structure function in the low-Q^2 regime. Since δ_{LT} requires knowledge of $g_2^p(x,Q^2)$, this measurement will also allow an isospin separation of the spin polarizabilities.

SUMMARY

Experiment E97-110 was performed at Jefferson Lab in Hall A to provide precise spin structure data for the neutron and ^{3}He at low Q^2 between 0.02 and 0.26 GeV^2. The first moments and higher moments of the spin structure functions have been formed from the measured polarized ^{3}He cross-section differences. At the measured low-Q^2 values, these quantities provide an important benchmark check of ChPT calculations. The preliminary results for the neutron generalized spin polarizabilities were shown and discussed. The γ_0 spin polarizability has very good agreement with the RBChPT calculation; on the other hand, there is a sizeable difference between the data and calculations without the resonance contributions. For the δ_{LT} polarizability, the preliminary results show a large disagreement compared with the ChPT calculations. Final neutron results are expected after the systematic checks have been completed by the end of 2010.

ACKNOWLEDGMENTS

We acknowledge all members of the Hall A collaboration for the excellent work of designing, running of the experiment and analyzing the data. We also acknowledge the outstanding support of the Hall A technical staff and Accelerator Division in accomplishing this experiment. This work was supported by the U.S. Department of Energy contract DE-AC05-84ER40150, Modification No. M175, under which the Southeastern Universities Research Association operates the Thomas Jefferson National Accelerator Facility.

REFERENCES

1. J. P. Chen, A. Deur, and Z.-E. Meziani, *Mod. Phys. Lett.* **A20**, 2745–2766 (2005).
2. S. E. Kuhn, J. P. Chen, and E. Leader, *Prog. Part. Nucl. Phys.* **63**, 1–50 (2009), 0812.3535.
3. S. B. Gerasimov, *Yad. Fiz.* **2**, 598 (1965).
4. S. B. Gerasimov, *Sov. J. Nucl. Phys.* **2**, 430–433 (1966).
5. S. D. Drell, and A. C. Hearn, *Phys. Rev. Lett.* **16**, 908–911 (1966).
6. M. Hosoda, and K. Yamamoto, *Prog. Theor. Phys. (Kyoto)* **30**, 425 (1966).
7. M. Anselmino, B. L. Ioffe, and E. Leader, *Sov. J. Nucl. Phys.* **49**, 136 (1989).
8. X.-D. Ji, and J. Osborne, *J. Phys.* **G27**, 127 (2001).
9. J. D. Bjorken, *Phys. Rev.* **148**, 1467–1478 (1966).
10. V. A. Sulkosky, *The Spin Structure of ^{3}He and the Neuton at Low Q^2: A Measurement of The Generalized GDH Integrand*, Ph.D. thesis, The College of William and Mary (2007).
11. M. Amarian, et al., *Phys. Rev. Lett.* **89**, 242301 (2002).
12. M. Amarian, et al., *Phys. Rev. Lett.* **92**, 022301 (2004).
13. M. Amarian, et al., *Phys. Rev. Lett.* **93**, 152301 (2004).
14. R. Fatmei, et al., *Phys. Rev. Lett.* **91**, 222002 (2003).
15. J. Yun, et al., *Phys. Rev. C* **67**, 055204 (2003).
16. K. V. Dharmawardane, et al., *Phys. Rev. B* **641**, 11 (2006).
17. Y. Prok, et al., *Phys. Rev. B* **672**, 12 (2009).
18. K. Slifer, et al., *Phys. Rev. Lett.* **105**, 101601 (2010).
19. V. Bernard, T. R. Hemmert, and U.-G. Meissner, *Phys. Lett.* **B545**, 105–111 (2002).
20. V. Bernard, N. Kaiser, and U. G. Meissner, *Phys. Rev.* **D48**, 3062–3069 (1993).
21. J. Alcorn, et al., *Nucl. Instrum. Meth.* **A522**, 294–346 (2004).
22. C. Ciofi degli Atti, and S. Scopetta, *Phys. Lett.* **B404**, 223–229 (1997).
23. M. Battaglieri, A. Deur, R. De Vita, and M. Ripani, JLab experiment E03-006 (2003).
24. N. Bianchi, and E. Thomas, *Nucl. Phys. Proc. Suppl.* **82**, 256 (2000).
25. D. Drechsel, O. Hanstein, S. S. Kamalov, and L. Tiator, *Nucl. Phys.* **A645**, 145–174 (1999).
26. X.-D. Ji, C.-W. Kao, and J. Osborne, *Phys. Lett.* **B472**, 1–4 (2000).
27. C. W. Kao, T. Spitzenberg, and M. Vanderhaeghen, *Phys. Rev.* **D67**, 016001 (2003).
28. A. Camsonne, J. P. Chen, and K. Slifer, JLab experiment E08-027 (2008).

Gauge invariance and canonical quantization in the nucleon internal structure[1]

Fan Wang*, Xiang-song Chen†, Xiao-fu Lü†, Wei-ming Sun* and T. Goldman**

*Department of Physics, Nanjing University and J-CPNPC, Nanjing, 210093, P.R. China
†Department of Physics, Sichuan University, Chendu, 610064, P.R. China
**Theoretical Division, Los Alamos National Laboratory, Los Alamos, NM 87545, USA

Abstract. It is unavoidable to deal with the quark and gluon momentum and angular momentum contributions to the nucleon momentum and spin in the study of nucleon internal structure. However, we never have the quark and gluon momentum, orbital angular momentum and gluon spin operators which satisfy both the gauge invariance and the canonical momentum and angular momentum commutation relations. A new set of quark and gluon momentum, orbital angular momentum and spin operators, which satisfy both the gauge invariance and canonical momentum and angular momentum commutation relations, are proposed. The key point to achieve such a proper decomposition is to separate the gauge potential into the pure gauge and the gauge covariant parts. The same conflicts also exist in QED and quantum mechanics and have been solved in the same manner. The impacts of this new decomposition to the nucleon internal structure study are discussed.

Keywords: gauge invariance, canonical quantization, nucleon internal structure, quark gluon momentum and angular momentum operator
PACS: 14.20.Dh, 11.15.-q, 12.38.-t, 11.10.Jj

INTRODUCTION

In quantum physics, any observable is expressed as a Hermitian operator in Hilbert space. The fundamental operators, such as momentum, orbital angular momentum, spin, satisfy the canonical momentum and angular momentum commutation relations. These commutation relations or Lie algebras define the properties of these operators.

Gauge invariance has been recognized as the first principle through the development of the standard model. In classical gauge field theory, gauge invariance principle requires any observable must be expressed in terms of gauge invariant variable. In quantum gauge field theory, in general one only requires the matrix elements of an operator in between physical states to be gauge invariant. However one usually requires the operators themselves to be gauge invariant. This is called the strict gauge invariance in[1]. We will restrict our discussion in strict gauge invariance in this paper and leave the other possibility to the future study[1, 2]

In the study of nucleon (atom) internal structure, it is unavoidable to deal with the quark, gluon (electron, photon) momentum, orbital angular momentum and spin contributions to the nucleon (atom) momentum and spin. However, we never have the quark, gluon (electron, photon) momentum, orbital angular momentum and spin operators which satisfy both the gauge invariance and canonical momentum and angular momentum commutation relations except the quark (electron) spin. Even it has been claimed in some textbooks that one can not define separately the photon spin and orbital angular momentum operators [3] and almost a proper gluon spin operator search has been given up in the nucleon spin structure study for the last ten years. This situation has left puzzles in quantum mechanics, quantum electrodynamics (QED) and quantum chromodynamics (QCD). For example, the expectation value of the Hamiltonian of hydrogen atom is gauge dependent under a time dependent gauge transformation[4]. The meaning of the multipole radiation analysis from atom to hadron spectroscopy would be obscure if the photon spin and orbital angular momentum operators were not well defined especially the parity of these microscopic states determined from the multipole radiation analysis would be obscure. There will be no way to compare the measured gluon spin contribution to nucleon spin with the theoretically calculated one if one has not a proper gluon spin operator, etc.

In section II the conflict between gauge invariance and canonical quantization of the usual quark gluon (electron photon) momentum, orbital angular momentum and spin operators are discussed from the simple quantum mechanics of a charged particle moving in an electromagnetic field to those of quark and gluon in QCD. In the third section a new

[1] Email address:fgwang@chenwang.nju.edu.cn

12th International Conference on Meson-Nucleon Physics and the Structure of the Nucleon (MENU 2010)
AIP Conf. Proc. 1374, 235-241 (2011); doi: 10.1063/1.3647133

set of momentum, orbital angular momentum and spin operators, which satisfy both the gauge invariance and canonical momentum and angular momentum commutation relations, are given. The key point to achieve this is to separate the gauge potential into pure gauge and gauge covariant (invariant) parts. The possible impacts of these modification to the nucleon internal structure study and prospect will be discussed in section IV.

GAUGE INVARIANCE AND CANONICAL QUANTIZATION OF MOMENTUM AND ANGULAR MOMENTUM OPERATORS OF FERMION AND GAUGE FIELD PARTS

The conflict between gauge invariance and canonical quantization of the momentum and orbital angular momentum operators of a charged particle moving in the electromagnetic field, a U(1) Abelian gauge field, has existed in quantum mechanics since the establishment of gauge invariance principle. Starting from the Lagrangian of a non-relativistic charged particle with mass m, velocity $\vec{v}$ and charge e moving in an electromagnetic field $A^{\mu} = (A^0, \vec{A})$,

$$\mathscr{L}(m, \vec{v}, e, A_{\mu}) = \frac{1}{2m}(m\vec{v})^2 - e(A^0 - \vec{v}\cdot\vec{A}), \tag{1}$$

one obtains the canonical momentum $\vec{p}$, the orbital angular momentum $\vec{L}$ and the Hamiltonian H

$$\vec{p} = m\vec{v} + e\vec{A}, \quad \vec{L} = \vec{r}\times\vec{p}, \quad H = \frac{1}{2m}(\vec{p} - e\vec{A})^2 + eA^0. \tag{2}$$

All of these three classical dynamical variables are gauge dependent. In the coordinate representation, the momentum operator $\vec{p}$ is quantized as $\vec{p} = \frac{\vec{\nabla}}{i}$, no matter what kind gauge is fixed on even though the classical canonical momentum operator is gauge dependent. The orbital angular momentum and Hamiltonian operators are quantized by replacing $\vec{p}$ by $\frac{\vec{\nabla}}{i}$. These quantized momentum and angular momentum operators satisfy the canonical commutation relation or the Lie algebra. In general, the $[p_i, H] \neq 0$, which is different from the Poincaré algebra of the total momentum $P_i, (i = 1, 2, 3)$ and total H of the whole system where $[P_i, H] = 0$.

However, after a gauge transformation, $\psi' = e^{-ie\omega(x)}\psi$, the matrix elements of the above operators transform as,

$$\langle\psi'|\vec{p}|\psi'\rangle = \langle\psi|\vec{p}|\psi\rangle - e\langle\psi|\vec{\nabla}\omega(x)|\psi\rangle, \; \langle\psi'|\vec{L}|\psi'\rangle = \langle\psi|\vec{L}|\psi\rangle - e\langle\psi|\vec{r}\times\vec{\nabla}\omega(x)|\psi\rangle, \; \langle\psi'|H'|\psi'\rangle = \langle\psi|H|\psi\rangle + e\langle\psi|\partial_t\omega(x)|\psi\rangle. \tag{3}$$

It is obvious that the matrix elements of these three operators are all gauge dependent. Therefore they are not measurable and so these operators do not correspond to observables. This problem has been left in quantum mechanics since the gauge principle was proposed.

The relativistic version of the quantum mechanics has the same problem. The gauge dependence of the expectation value of the Hamiltonian of a charged particle moving in electromagnetic field under a time dependent gauge transformation had been discussed by T. Goldman[4].

This conflict had been carried over to QED. Starting from a QED Lagrangian,

$$\mathscr{L} = \bar{\psi}[i\gamma^{\mu}(\partial_{\mu} + ieA_{\mu}) - m]\psi - \frac{1}{4}F_{\mu\nu}F^{\mu\nu}, \quad F_{\mu\nu} = \partial_{\mu}A_{\nu} - \partial_{\nu}A_{\mu}. \tag{4}$$

By means of the Noether theorem one obtains the momentum and angular momentum operators as follows:

$$\vec{P} = \vec{P}_e + \vec{P}_{ph} = \int d^3x\psi^{\dagger}\frac{\vec{\nabla}}{i}\psi + \int d^3xE^i\vec{\nabla}A^i, \tag{5}$$

$$\vec{J} = \vec{S}_e + \vec{L}_e + \vec{S}_{ph} + \vec{L}_{ph} = \int d^3x\psi^{\dagger}\frac{\vec{\Sigma}}{2}\psi + \int d^3x\vec{x}\times\psi^{\dagger}\frac{\vec{\nabla}}{i}\psi + \int d^3x\vec{E}\times\vec{A} + \int d^3x\vec{x}\times E^i\vec{\nabla}A^i. \tag{6}$$

Here $\Sigma^j = \frac{i}{2}\varepsilon_{jkl}\gamma^k\gamma^l$. These electron and photon momentums, orbital angular momentums and spin, after quantization, satisfy momentum and angular momentum Lie algebra. However, they are not gauge invariant except the electron spin.

The multipole radiation analysis is the basis of atomic, molecular, nuclear and hadron spectroscopy. The multipole field is based on the decomposition of the electromagnetic field into field with definite orbital angular momentum and spin quantum numbers. If the photon spin and orbital angular momentum operators were gauge dependent, then the physical meaning of the multipole field would be obscure especially the parity of these microscopic states determined by the measurement of the orbital angular momentum quantum number of the multipole radiation field would be obscure.

QCD has the same problem as QED. The quark gluon momentum, orbital angular momentum and spin operators derived from QCD Lagrangian by Noether theorem have the same form as those of electron and photon if one omits the color indices. They satisfy the momentum and angular momentum Lie algebra but they are not gauge invariant except the quark spin.

Because of the lack of gauge invariant quark, gluon momentum operators, the present operator product expansion (OPE) used the following two operators as quark gluon momentum operators,

$$\vec{P} = \vec{\mathscr{P}}_q + \vec{\mathscr{P}}_g = \int d^3x\psi^\dagger \frac{\vec{D}}{i}\psi + \int d^3x\vec{E}\times\vec{B}, \quad \vec{D} = \vec{\nabla} - ig\vec{A}. \tag{7}$$

Both the quark and gluon "momentum" operators $\vec{\mathscr{P}}_q$ and $\vec{\mathscr{P}}_g$ defined in Eq.(7) are gauge invariant but neither the quark "momentum" $\vec{\mathscr{P}}_q$ nor the gluon "momentum" $\vec{\mathscr{P}}_g$ satisfies the momentum algebra, for example,

$$[D_l, D_m] = -ig(\partial_l A_m - \partial_m A_l) - ig^2 C_{abc} A_l^a A_m^b T^c, \tag{8}$$

where C_{abc} is the $SU(3)$ group structure constant. The $\vec{\mathscr{P}}_g$ does not satisfy the momentum algebra either in the interacting quark-gluon field, i.e., QCD case. Therefore neither the $\vec{\mathscr{P}}_q$ nor the $\vec{\mathscr{P}}_g$ used in the OPE is the proper momentum operator.

The gluon spin contribution is under intensive study, PHENIX, STAR, COMPASS, HERMES, and others are measuring the gluon spin contribution to nucleon spin. However there is no gluon spin operator which satisfies both gauge invariance and angular momentum algebra. There is also no quark, gluon orbital angular momentum operator which satisfies the gauge invariance and orbital angular momentum algebra. These situations hindered the study of the nucleon spin structure.

A NEW SET OF MOMENTUM, ORBITAL ANGULAR MOMENTUM AND SPIN OPERATORS FOR THE FERMION AND GAUGE FIELD PARTS

III.1 Decomposing the gauge field A_μ into pure gauge part and gauge invariant (covariant) part

Let us start from the simpler QED case. It is well known that to use gauge potential A_μ to describe the electromagnetic field the A_μ is not unique, i.e., there is gauge freedom. Under a gauge transformation,

$$A'^\mu = A^\mu + \partial^\mu \omega(x), \tag{9}$$

one obtains a new gauge potential A'^μ from A^μ. A^μ and A'^μ describe the same electromagnetic field,

$$F_{\mu\nu} = \partial_\mu A_\nu - \partial_\nu A_\mu = \partial_\mu A'_\nu - \partial_\nu A'_\mu. \tag{10}$$

Such a gauge freedom is necessary because the gauge potential A^μ plays two role in gauge field theory: The first is to provide a pure gauge potential A_{pure} to compensate the induced potential due to the phase change in a local gauge transformation of the Fermion field $\psi'(x) = e^{-ie\omega(x)}\psi(x)$ which must be varied with the arbitrary changed phase parameter $\omega(x)$. The second is to provide a physical potential A_{phys} for the physical interaction between Fermion field and gauge field which should be gauge invariant under gauge transformation. The pure gauge potential A_{pure} should not contribute to electromagnetic field, $F_{pure}^{\mu\nu} = \partial^\mu A_{pure}^\nu - \partial^\nu A_{pure}^\mu = 0$. This equation can not fix the A_{pure}. One has to find additional condition to fix it. The spatial part of the above equation is $\nabla\times\vec{A}_{pure} = 0$, which means $\vec{A}_{pure}$ does not contribute to magnetic field. This equation can be expressed in another form,

$$\nabla\times\vec{A}_{phys} = \nabla\times\vec{A}. \tag{11}$$

A natural choice of the additional condition in QED case is

$$\nabla\cdot\vec{A}_{phys} = 0, \tag{12}$$

which is the transverse wave condition and we know that this part is the physical one from the Coulomb gauge quantization. Combining these two conditions, Eq.(11,12), under the natural boundary condition, $\vec{A}_{phys}(|x| \to \infty) = 0$, for any given set of gauge potential $\vec{A}$, one can decompose it uniquely as follows,

$$\vec{A}_{phys}(x) = \vec{\nabla} \times \frac{1}{4\pi} \int d^3x' \frac{\vec{\nabla}' \times \vec{A}(x')}{|\vec{x} - \vec{x}'|} = \frac{1}{4\pi} \int d^3x' \frac{\vec{B}(x')}{|\vec{x} - \ vec(x')|}, \quad \vec{A}_{pure} = \vec{A} - \vec{A}_{phy}. \tag{13}$$

We have to emphasize that for fixed $\vec{A}(x)$, the integration can be done and the obtained $\vec{A}_{phys}(x)$ is a local function of space-time x. It is easy to prove that these two parts transform as follows in a gauge transformation Eq.(9),

$$\vec{A}'_{phy} = \vec{A}_{phy}, \quad \vec{A}'_{pure} = \vec{A}_{pure} - \vec{\nabla}\omega(x). \tag{14}$$

The time component A^0 can be decomposed in the same manner. From the condition $F^{i0}_{pure} = 0$, one obtains

$$\partial_i A^0_{phys} = \partial_i A^0 + \partial_t (A^i - A^i_{phys}), \quad A^0_{phys} = \int_{-\infty}^{x} dx^i (\partial_i A^0 + \partial_t A^i - \partial_t A^i_{phys}). \tag{15}$$

To decompose the gauge potential $A_\mu = A^a_\mu T^a$ for the gluon field is more complicated than QED case. We first define the pure gauge potential A^μ_{pure} (hereafter we omit the color indices if not necessary) by the same condition, i.e., it does not contribute to color electromagnetic field,

$$F^{\mu\nu}_{pure} = \partial^\mu A^\nu_{pure} - \partial^\nu A^\mu_{pure} + ig[A^\mu_{pure}, A^\nu_{pure}] = 0. \tag{16}$$

In order to make this defining condition looks similar to Eq.(11), we introduce a notation,

$$\vec{D}_{pure} = \vec{\nabla} - ig\vec{A}_{pure}, \quad \vec{D}_{pure} \times \vec{A}_{pure} = \vec{\nabla} \times \vec{A}_{pure} - ig\vec{A}_{pure} \times \vec{A}_{pure} = 0. \tag{17}$$

The additional condition is even more complicated, i.e., one does not have a natural choice as Eq.(12) in QED. We make the following choice[5],

$$\vec{\mathscr{D}}_{pure} = \vec{\nabla} - ig[\vec{A}_{pure}, \ ], \quad \vec{\mathscr{D}}_{pure} \cdot \vec{A}_{phys} = \vec{\nabla} \cdot \vec{A}_{phys} - ig[A^i_{pure}, A^i_{phys}] = 0. \tag{18}$$

The summation over the vector component i has been assumed in the above equation and following ones. Please note that in the above adjoint representation of the new covariant derivative operator $\vec{\mathscr{D}}$, the bracket $[A^i_{pure}, A^i_{phys}]$ is not the quantum bracket but a color $SU^c(3)$ group one,

$$[A^i_{pure}, A^i_{phys}] = iC_{abc} A^{ib}_{pure} A^{ic}_{phys} T^a. \tag{19}$$

These equations can be rewritten as follows,

$$\begin{aligned} \vec{\nabla} \cdot \vec{A}_{phys} &= ig[A^i, A^i_{phys}], \\ \vec{\nabla} \times \vec{A}_{phys} &= \vec{\nabla} \times \vec{A} - ig(\vec{A} - \vec{A}_{phys}) \times (\vec{A} - \vec{A}_{phys}), \\ \partial_i A^0_{phys} &= \partial_i A^0 + \partial_t (A^i - A^i_{phys}) - ig[A^i - A^i_{phys}, A^0 - A^0_{phys}]. \end{aligned} \tag{20}$$

These equations can be solved perturbatively: in the zeroth order, i.e., assuming $g = 0$, these equations are the same as those of QED, one can obtain the zeroth order solution; then taking into account the nonlinear coupling through iteration one obtains a perturbative solution as a power expansion in g.

Under a gauge transformation,

$$\psi' = U\psi, \quad A'_\mu = UA_\mu U^\dagger - \frac{i}{g} U \partial_\mu U^\dagger, \tag{21}$$

where $U = e^{-ig\omega}$. The $\vec{A}_{pure}$ and $\vec{A}_{phys}$ will be transformed as

$$\vec{A'}_{phys} = U\vec{A}_{phys}U^\dagger, \quad \vec{A'}_{pure} = U\vec{A}_{pure}U^\dagger - \frac{i}{g} U \partial_\mu U^\dagger. \tag{22}$$

III.2 **Quantum mechanics**

We have mentioned in the introduction part that even in quantum mechanics, there are already puzzles related to the fundamental operators, the matrix elements of canonical momentum, orbital angular momentum and Hamiltonian of a charged particle moving in an electromagnetic field are all not gauge invariant. In order to get rid of these puzzles, gauge invariant operators have been introduced,

$$\vec{\mathscr{P}} = \vec{p} - e\vec{A}, \quad \vec{\mathscr{L}} = \vec{x} \times \vec{\mathscr{P}}. \tag{23}$$

However, the gauge invariant "momentum" $\vec{\mathscr{P}}$ does not satisfy the canonical momentum Lie algebra, so they are not the real momentum. The gauge invariant "orbital angular momentum" $\vec{\mathscr{L}}$ does not satisfy the angular momentum Lie algebra either.

Based on the gauge potential decomposition proposed in above section, we introduce a new set of momentum and orbital angular momentum operators which satisfy both gauge invariance and the corresponding commutation relations,

$$\vec{p}_{pure} = \vec{p} - e\vec{A}_{pure}, \quad \vec{L}_{pure} = \vec{x} \times \vec{p}_{pure}. \tag{24}$$

The long standing puzzle, the gauge non-invariance of the expectation value of the Hamiltonian[4] can be solved in the same manner. For the non-relativistic quantum mechanics, the new Hamiltonian is

$$H = \frac{(\vec{p} - e\vec{A})^2}{2m} + eA^0 - e\partial_t \frac{1}{\nabla^2} \nabla \cdot \vec{A}. \tag{25}$$

The last term is a pure gauge term, it cancels the unphysical energy appeared in eA^0 induced by the pure gauge term and then guarantees the expectation value of this Hamiltonian gauge invariant. It is a direct extension of Eq.(24) to the fourth momentum component.

The Dirac Hamiltonian has the same unphysical energy part and has to be canceled in the same manner as that for the Schr*ö*dinger Hamiltonian. A detailed discussion can be found in[6].

Our study shows that the canonical momentum, orbital angular momentum and the Hamiltonian used in quantum mechanics are not observables, one must subtract the pure gauge part, the unphysical one, from these operators as we did in Eq.(24,25) to obtain the observable ones.

III.3 **QED**

We have explained that the momentum and angular momentum operators of the Fermion and gauge field part, Eq.(5) and (6), derived from the QED Lagrangian by means of Noether theorem are not gauge invariant except the Fermion spin. One can obtain a gauge invariant decomposition by adding a surface term,

$$\vec{P} = \vec{\mathscr{P}}_e + \vec{\mathscr{P}}_{ph} = \int d^3x \psi^\dagger \frac{\vec{D}}{i} \psi + \int d^3x \vec{E} \times \vec{B}, \tag{26}$$

$$\vec{J} = \vec{S}_e + \vec{\mathscr{L}}_e + \vec{\mathscr{J}}_{ph} = \int d^3x \psi^\dagger \frac{\vec{\Sigma}}{2} \psi + \int d^3x \vec{x} \times \psi^\dagger \frac{\vec{D}}{i} \psi + \int d^3x \vec{x} \times (\vec{E} \times \vec{B}). \tag{27}$$

There are two problems with this decomposition: (1), $\vec{\mathscr{L}}_e$ and $\vec{\mathscr{J}}_{ph}$ do not satisfy the angular momentum commutation relation even though in free electromagnic field the photon total angular momentum $\vec{\mathscr{J}}_{ph}$ does; (2), there is no separate photon spin and orbital angular momentum operators and this feature will ruin the multipole radiation analysis as we discussed in the second section.

Based on the decomposition of the gauge field potential into pure gauge and the physical parts, Eq.(13), we obtain the following decomposition,

$$\vec{P} = \vec{P}_e + \vec{P}_{ph} = \int d^3x \psi^\dagger \frac{\vec{D}_{pure}}{i} \psi + \int d^3x E^i_{phys} \vec{\mathscr{D}}_{pure} A^i_{phys}. \tag{28}$$

$$\vec{J} = \vec{S}_e + \vec{L}_e + \vec{S}_{ph} + \vec{L}_{ph} = \int d^3x \psi^\dagger \frac{\vec{\Sigma}}{2} \psi + \int d^3x \vec{x} \times \psi^\dagger \frac{\vec{D}_{pure}}{i} \psi + \int d^3x \vec{E}_{phys} \times \vec{A}_{phys} + \int d^3x \vec{x} \times E^i_{phys} \vec{\mathscr{D}}_{pure} A^i_{phys}. \tag{29}$$

Here the operator $\vec{D}_{pure}$ and $\vec{\mathscr{D}}_{pure}$ are the same as giving in Eq.(17) and (18) but with g replaced by e. Because of the Abelian property of the $U(1)$ gauge field, the adjoint representation of the operator $\vec{\mathscr{D}}$ is simplified to be a simple $\vec{\nabla}$.

It is not hard to check that each operator in the above decomposition, Eq.(28) and (29) is gauge invariant and satisfies the momentum, angular momentum commutation relation.

The photon spin and orbital angular momentum operators are well defined as shown in Eq.(29). The multipole radiation analysis is theoretically sound now as it should be.

III.4 **QCD**

One can copy results for QED, the Eqs.(28,29), to QCD to obtain the quark, gluon momentum, orbital angular momentum and spin operators which satisfy both the gauge invariance and canonical momentum and angular momentum commutation relations.

A decomposition of the form of Eq.(26,27) has been used in the nucleon spin structure study for the last ten years[7]. Each operator in those decomposition is gauge invariant and so corresponding to observable. However, because they do not satisfy the momentum, angular momentum Lie algebra so the measured ones are not the quark, gluon momentum and orbital angular momentum and can not be compared to those used in hadron spectroscopy. But in fact such a comparison is doing and leading to puzzles in the nucleon spin structure study[8]

The gluon spin operator had been searched for more than ten years in the nucleon spin structure study and no satisfied one was obtained. Now one can calculate the matrix element of the gluon spin operator $\vec{S}_g = \int d^3x \vec{E} \times \vec{A}_{phys}$ between the polarized nucleon state $|N(p,s)\rangle$ to obtain the gluon spin contribution to nucleon spin and compare it with the measured ones.

SUMMARY AND PROSPECT

Since the establishment of gauge invariance principle, we enjoy that the total momentum, angular momentum and the Lorentz boosting operators of a gauge system satisfy both the gauge invariance and Poincaré algebra. However, we never have the separate momentum, orbital angular momentum operators of the Fermion (electron in QED, quark in QCD) and boson (photon in QED, gluon in QCD) part which satisfy both the gauge invariance and the canonical momentum, angular momentum Lie algebra. We have the electron, quark spin operator but we never have the photon, gluon spin operator which satisfy both the gauge invariance and spin Lie algebra. Even it had been claimed in some textbooks that it is impossible to have a well defined photon spin[3]. The nucleon spin structure study needs the gluon spin operator, but after about ten years effort in searching for a gluon spin operator since the so-called proton spin crisis such an effort has almost been given up for the last ten years. In this report we proposed a new set of quark (electron), gluon (photon) momentum, orbital angular momentum and spin operators, Eq.(28) and Eq.(29), which satisfy both the gauge invariance and the canonical momentum, angular momentum Lie algebra.

To achieve this a key point is to separate the gauge potential into pure gauge and physical parts: The former is unphysical and can be gauged away as in Coulomb gauge, it is used to compensate the induced unphysical gauge potential due to the local gauge transformation of the Fermion field to keep the gauge invariance. The physical part is responsible for the physical coupling between Fermion and boson field. It is physical and should be gauge invariant (covariant). We provide a method to do this separation both for the Abelian $U(1)$ and the non-Abelian $SU(3)$ gauge field.

Our proposed momentum operators for the Fermion part are different from the canonical ones, the latter ones are not gauge invariant and so do not represent observables because they include the unphysical pure gauge potential contribution. The new ones subtract the unphysical pure gauge potential contribution and so they are gauge invariant and represent the observables. The new momentum operator is also different from the kinematic one, Eq.(7), its effect on the quark parton distribution should be studied.

We achieved to obtain a gauge invariant orbital angular momentum and spin operators of the photon and gluon by means of the physical part of the gauge field, Eq.(29), which provides the theoretical basis of the widely used multipole radiation analysis, the gluon spin contribution in the nucleon spin structure study.

The Poincaré algebra can not be fully maintained for the momentum, angular momentum and Lorentz boosting operators of the individual Fermion and boson part of an interacting gauge field system. What is the meaning of these observables if they are not Lorentz covariant? The Lorentz covariance can be maintained to what extent are left for further study.

The gluon spin contribution to the nucleon spin is under measurement in different labs. A lattice QCD calculation with the gauge invariant gluon spin operator is called for.

In summary, the nucleon internal structure is better to be reexamined based on the new quark, gluon momentum, orbital angular momentum, spin operators and our picture of the nucleon internal structure might be modified.

REFERENCES

1. F. Strocchi and A.S. Wightman, J. Math. Phys. **15**, 2198 (1974).
2. X.S. Chen and F. Wang, hep-ph/9802346
3. V.B. Berestetskii, E.M. Lifshitz and L.P. Pitaevskii, Quantum Electrodynamics. (Oxford, Pergamon) 1982.
   C. Cohen-Tannoudji, J. Dupont-Roc and G. Grynberg, Photons and Atoms. (New York, Wiley) 1987.
4. T. Goldman, Phys. Rev. **D15**, 1063 (1977).
5. X.S. Chen, X.F. L*ü*, W.M. Sun, F. Wang and T. Goldman, Phys. Rev. Lett. **100**, 232002 (2008); **103**, 062001 (2009).
6. W.M. Sun, X.S. Chen, X.F. L*ü* and Fan Wang, Phys. Rev. **A82**, 012107 (2010).
7. X. Ji, Phys. Rev. Lett. **78**, 610 (1997); X.S. Chen, and F. Wang, Commun. Theor. Phys. **27**, 121 (1997).
8. M. Burkardt, C.A. Miller and W.D. Nowak, arXiv:0812.2208[hep-ph].

Recent advances of polarized ^{3}He target at Jefferson Lab

Yi Qiang

Jefferson Lab, Newport News, VA 23606

Abstract. Polarized ^{3}He targets have been widely used in nuclear and particle physics experiments to study neutron structure in the spin degree of freedom, as most of the ^{3}He spin is carried by the unpaired neutron. The Spin-Exchange Optical Pumping (SEOP) process is used in Jefferson Lab Hall A to polarize its ^{3}He target. In recent years, both the performance and corresponding polarimetry of such a target have been greatly improved. Several experiments recently carried out in Hall A have achieved record high figure of merit using this target.

Keywords: Polarized ^{3}He Target
PACS: 29.25.Pj

INTRODUCTION

Polarized targets have been widely used in nuclear and particles physics experiments to study nucleon structure in the spin degree of freedom. As a free neutron target cannot be made dense enough due to its short lifetime, 886 s, polarized nuclear targets such as ^{2}H or ^{3}He are commonly used to extract polarized neutron data.

Polarized ^{3}He targets are widely used in nuclear physics research and there exist two major methods of polarizing ^{3}He nuclei. The first one is based on the metastability-exchange optical pumping technique developed in the early 1960s at Rice university [1] and later used in HERMES, Mainz and MIT-Bates.

The second method is based on spin exchange optical pumping [2]. First an alkali metal vapor is polarized by optical pumping, then the polarization is transferred to ^{3}He nuclei through spin exchange collisions. Targets based on this principle are almost always designed to operate at quite high gas pressures, typically a few atmospheres. The polarized ^{3}He target using spin-exchange optical pumping was originally developed at SLAC [3] and has been successfully used in JLab Hall A since 1998.

SPIN EXCHANGE OPTICAL PUMPING (SEOP)

Rb Only System

The first step of SEOP is optical pumping of alkali-metal atoms using circularly polarized laser light in a magnetic field. Due to the availably of lasers matching the $D1$ transitions of various alkali atoms, Rb is commonly chosen for this kind of target and all experiments in JLab Hall A before 2005 [4, 5, 6, 7, 8, 9] used Rb-only ^{3}He cells. In reality, beside ^{3}He and alkali, a very small portion of N_2, $\sim$1% of the total density, is added into the cells to increase the optical pumping efficiency by absorbing the unpolarized D1 light emitted by de-excited Rb atoms.

The next step is spin-exchange, and the key process is the collisional transfer of polarization between optically pumped alkali-metal atoms and the nuclei of the noble-gas atoms. For ^{3}He, the transfer of angular momentum occurs mainly through binary collisions between the atoms [2]. The contribution from van der Waals forces is negligible. The spin-dependent interactions producing the spin transfer and relaxation in collisions are dominated by two processes. One process is the spin-rotation interaction between the electron spin and the rotational angular momentum of the ^{3}He-Rb system. The second is the isotropic hyperfine interaction between the electron spin and the noble-gas nuclear spin.

For a Rb only target cell, the spin exchange efficiency is defined as

$$\eta_{SE}^{\mathrm{Rb}} = k_{SE}^{\mathrm{Rb}}/\Gamma_{\mathrm{Rb}}, \tag{1}$$

where k_{SE}^{Rb} is the rate of spin exchange and Γ_{Rb} is the spin-destruction rate of the rubidium vapor, which is dominated by the spin-rotation interaction during collisions among Rb atoms or between Rb atoms and other gases. The spin

12th International Conference on Meson-Nucleon Physics and the Structure of the Nucleon (MENU 2010)
AIP Conf. Proc. 1374, 242-245 (2011); doi: 10.1063/1.3647134

destruction caused by collisions of Rb atoms with the cell wall and by randomly polarized photons emitted from the decay of excited electrons is small. In general, Γ_{Rb} can be expressed as

$$\Gamma_{\mathrm{Rb}} = k_{\mathrm{Rb-He}} + k_{\mathrm{Rb-Rb}} + k_{\mathrm{Rb-N_2}}. \tag{2}$$

Only a small part of the Rb relaxation is caused by the spin exchange between ^{3}He nuclei and Rb electrons: $k_{SE}^{\mathrm{Rb}} \ll k_{\mathrm{Rb-He}}$. Under the standard operating conditions of 170 °C and 10 amg, less than 4% of Rb atoms transfer their polarization to ^{3}He, while the rest is lost in other spin destruction processes.

Rb-K Hybrid System

Recent study suggests that the efficiency of SEOP could be increased if K or Na were used as the alkali-metal spin-exchange partner rather than Rb alone, due to their lower electronic spin-destruction rates.

For a Rb-K hybrid, the spin-exchange efficiency is now written as

$$\eta_{SE}^{\mathrm{Rb-K}} = \frac{k_{SE}^{\mathrm{Rb}} + k_{SE}^{\mathrm{K}}}{\Gamma_{\mathrm{Rb}} + \Omega\Gamma_{\mathrm{K}} + q_{\mathrm{Rb-K}}}, \tag{3}$$

where Ω is the vapor density ratio of K to Rb, and $q_{\mathrm{Rb-K}}$ is the spin-destruction rate constant for Rb-K collisions. SEOP with a Rb-K mixture is more efficient than SEOP with pure Rb because k_{SE}^{K} dominates Γ_K and is comparable to k_{SE}^{Rb}. In hybrid cells, rapid electronic spin-exchange between the K and Rb equalizes the polarization of the two alkali-metals, allowing for the benefits of the lower spin-destruction rate of K with no change in the optical pumping laser. The gain in efficiency relative to pure Rb mainly depends on the vapor density ratio Ω.

In practice, Rb-K hybrid cells run at higher temperature, 230 °C, in order to obtain enough K vapor pressure. The higher spin-exchange efficiency greatly shortens the spin-up time of such cells, 4 hours compared to 10 hours of Rb only cells. The first batch of hybrid cells used in Hall A during the G_E^n experiment [10] in 2006 achieved significantly higher in-beam polarization, 50% vs. 40%. Further study showed that Ω of 5~10 gives optimal polarization performance, and most of the cells used after 2008 in Hall A have Ω in that range.

Narrow Line-Width Lasers

The initial target system in Hall A used a high power Fiber Array Package (FAP) laser system from Coherent ®. The typical FWHM of such a laser is about 2 nm, whereas the Rb D1 absorption has a width of 0.3 nm at optical pumping temperature. Therefore, only part of the laser power will be absorbed by Rb atoms. The rest of the power depolarizes the target through thermal process. Simply adding more laser power will not necessarily increase the target polarization; on the contrary it will reduce the target performance once a certain limit exceeded.

Before the start of the Hall A experiments [11, 12, 13, 14, 15, 16] in 2008, a new type of high power narrow line-width diode laser system became available from Spectra-Physics ®. The laser from such a system has a typical width of 0.2 nm, which is even narrower than the Rb D1 line and provides a much higher optical pumping efficiency. By implementing this new laser, all of these experiments benefited greatly from the huge improvement of the target polarization, which generally exceeds 60%.

HARDWARE SETUP

The ^{3}He target system in Hall A was redesigned in 2008 to fit the requirements of the series of six experiments with minimum change-over and very tight space constraints. A number of modifications and upgrades were implemented in the new design. The target cells used in all of these experiments have a 3 inch pumping chamber and a 40 cm long target chamber with a density of about 10 amg as shown in Figure 1. The luminosity is about 3×10^{36} nuclei/s/cm^2 with 15 μA beam.

A third set of Helmholtz coils was added to the existing two pairs of coils to provide additional vertical magnetic field. Using all three sets of coils, target polarization in any direction in 3-D space is possible. A bigger oven that allows all three pumping directions (longitudinal, horizontal transverse and vertical) was built with CS85 structural

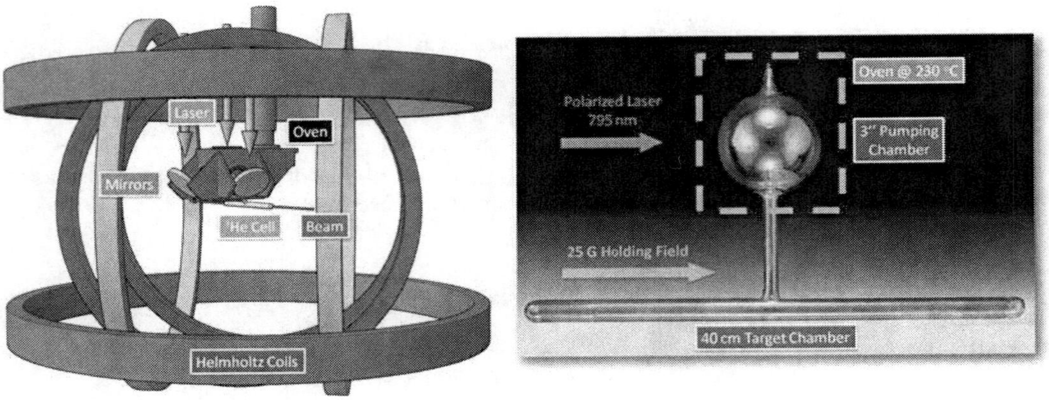

FIGURE 1. The polarized ^{3}He target setup (left) and a ^{3}He cell (right).

insulation material. This new material has just half of the density compared to the ceramics used in old ovens and much better insulation. The interior of the oven was painted black to reduce the internal laser reflection.

A total of fifteen long optical fibers connected the lasers from the remote laser room to the hall. They were then connected to three 5-to-1 optical fiber combiners so that up to five lasers could be used in each single polarizing direction. Three higher-power (30 W) narrow line-width lasers were used together to provide optimal pumping efficiency with the pumping chamber heated to 230 °C. The laser beam for each pumping direction was first polarized by passing through polarizing optics above the target and then reflected into the pumping chamber by two mirrors.

POLARIMETRY

Two kinds of polarimetry, NMR (Nuclear Magnetic Resonance) and EPR (Electron Paramagnetic Resonance), were used to measure the polarization of the target.

Two pairs of RF coils (vertical and longitudinal) and five pairs of NMR pickup coils (three on the pumping chamber and two on the target chamber) were mounted to perform NMR measurements for different spin directions. The magnetic flux of the polarized ^{3}He passing through the target chamber pickup coils was calibrated using water NMR measurements, for which the proton polarization is calculable. The statistical uncertainty of these measurements is much better than 1% and the total uncertainty of water NMR calibration is better than 3%.

The EPR measurement took place in the pumping chamber and both K and Rb EPR signals were used. Due to the much larger size of the oven (to have enough perturbative RF power) an EPR RF coil made of high temperature enameled wire was put inside the oven. A collection system with two lenses was installed to have better EPR D2 light detection. A typical EPR calibration has a statistical uncertainty better than 1% .

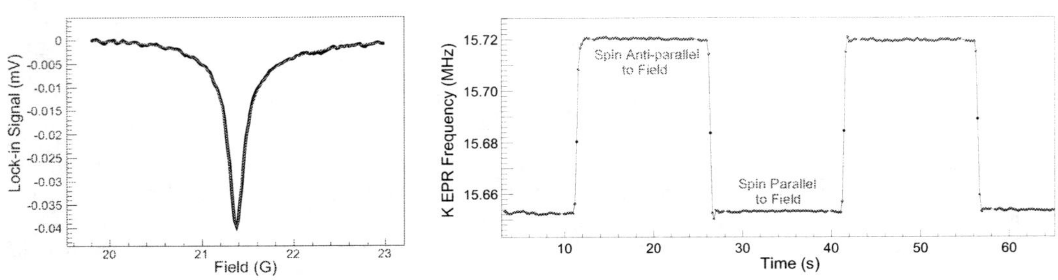

FIGURE 2. A water NMR signal (left) and an EPR measurement on K atom (right).

The target density and its distribution are crucial parameters for both NMR and EPR measurements. The densities of all production cells have been measured at JLab using pressure broadening with less than 2% statistical uncertainties.

SUMMARY

For a couple of decades of development and technology breakthroughs, the performance of polarized ^{3}He targets using SEOP technique has been improved steadily. Recent implementation of the hybrid Rb-K optical pumping and narrow line-width lasers further increases both the target polarization and luminosity and provides a greater figure-of-merit for nuclear physics research.

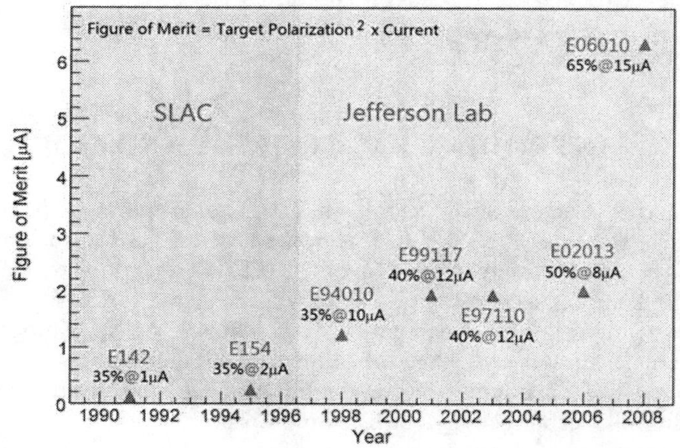

FIGURE 3. The evolution of the figure of merit for polarized ^{3}He target with SEOP technique.

ACKNOWLEDGMENTS

This work was supported by the US Department of Energy grant number DE-FG02-03ER41231 and the US Department of Energy contract DE-AC05-06OR23177, under which Jefferson Science Associates, LLC operates Thomas Jefferson National Accelerator Facility. I would like to thank my colleagues from the JLab polarized ^{3}He target group for their great contributions to the target development.

REFERENCES

1. F. D. Colegrove, L. D. Shearer, and G. K. Walters, *Phys. Rev.* **132** (1963).
2. T. E. Chupp, et al., *Phys. Rev.* **C36** (1987).
3. M. V. Romalis, Ph. D. thesis, Princeton University. (1998).
4. M. Amarian, et al., *Phys. Rev. Lett.* **89** (2002).
5. W. Xu, et al., *Phys. Rev. Lett.* **85** (2000).
6. X. Zheng, et al., *Phys. Rev. Lett.* **92** (2004).
7. K. Kramer, et al., *Phys. Rev. Lett.* **95** (2005).
8. P. Solvignon, et al., *Phys. Rev. Lett.* **101** (2008).
9. Jefferson Lab Experiment E97-110, Spokespersons: J. P. Chen, A. Deur, and F. Garibaldi. (1997).
10. Jefferson Lab Experiment E02-013, Spokespersons: G. Cates, N. Liyanaga, and B. Wojtsekhowski. (2002).
11. Jefferson Lab Experiment E06-010, Spokespersons: J. P.Chen, E. Cisbani, H. Gao, X. Jiang, and J. C. Peng. (2006).
12. Jefferson Lab Experiment E06-014, Spokespersons: S. Choi, X. Jiang, Z.-E. Meziani, and B. Sawatzky. (2006).
13. Jefferson Lab Experiment E05-015, Spokespersons: T. Averett, J. P. Chen, and X. Jiang. (2005).
14. Jefferson Lab Experiment E05-102, Spokespersons: S. Gilad, D. Higinbotham, W. Korsch, B. Norum, and S. Sirca. (2005).
15. Jefferson Lab Experiment E07-013, Spokespersons: T. Averett, T. Holmstrom, and X. Jiang. (2007).
16. Jefferson Lab Experiment E08-005, Spokespersons: T. Averett, D. Higinbotham, and V. Sulkosky. (2008).

Recent results on SIDIS from pion production at CLAS

P. Rossi

Laboratori Nazionali di Frascati
Via Enrico Fermi 40, 00044 Frascati, Italy
E-mail: rossi@lnf.infn.it

Abstract. Novel view on QCD dynamics inside hadrons is offered by the transverse momentum dependent parton distributions and fragmentation functions. They can be accessed in Semi-Inclusive Deep Inelastic Scattering (SIDIS) experiments. Recent results on SIDIS from pion production at CLAS is presented in this talk.

Keywords: Parton distributions, Transverse Momentum, Spin
PACS: 13.60.-r, 13.87.Fh, 13.88.+e, 14.20.Dh

INTRODUCTION TO TMDS AND SIDIS

The exploration of the internal structure of the hadrons has undergone enormous progress in the last decades. The investigation both theoretical and experimental of their partonic structure has been extended beyond the universal parton distribution functions (PDFs) and fragmentation functions (FFs) by exploring the parton's motion in the direction perpendicular to the parent hadron's momentum. This study allows the extraction of new types of parton distributions depending on the parton transverse momentum (TMDs). The role of correlations between partonic transverse momentum and spin are crucial for our understanding of the spin structure of the nucleon in terms of the quark and gluon degrees of freedom of QCD. Such spin-orbit correlations give rise to interesting single-spin phenomena in semi-inclusive lepton- and hadron-induced processes. TMDs are probability densities for finding a polarized/unpolarized parton with a longitudinal momentum fraction x and transverse momentum $\vec{k}_T$ in a polarized/unpolarized nucleon, see Table 1, where U, L, and T stand for transitions of unpolarized, longitudinally polarized, and transversely polarized nucleons (rows), and correspondingly for quarks (columns). The diagonal elements of the table (f_1, g_1, h_1) are the partonic momentum, longitudinal spin and transverse spin distribution functions. Similar quantities arise in the hadronization process. Two fundamental mechanisms have been identified leading to single spin asymmetries (SSAs) in hard processes: the Sivers mechanism [1, 2, 3, 4, 5], which generates an asymmetry in the distribution of quarks due to orbital motion of partons and the Collins mechanism [4, 6], which generates an asymmetry during the hadronization of quarks. Most of our information on TMDs is obtained from semi-inclusive deep inelastic scattering (SIDIS) experiments, $lN \to e'hX$ (with l = lepton, h = hadron). One of the main players in the game is CLAS at JLab.

TABLE 1. Leading-twist transverse momentum-dependent distribution functions

N/q	U	L	T
U	$\mathbf{f_1}$		$h_1^\perp$
L		$\mathbf{g_1}$	$h_{1L}^\perp$
T	$f_{1T}^\perp$	g_{1T}	$\mathbf{h_1}$ $h_{1T}^\perp$

PION TMDS WITH CLAS AT 6 GEV

Unpolarized target

Measurements of single spin asymmetries in semi-inclusive production of π^0 ($ep \to e\,\pi^0 X$) in hard scattering kinematics are presented using a 5.776 GeV polarized electron beam and the CLAS detector [7] in combination with

12th International Conference on Meson-Nucleon Physics and the Structure of the Nucleon (MENU 2010)
AIP Conf. Proc. 1374, 246-249 (2011); doi: 10.1063/1.3647135

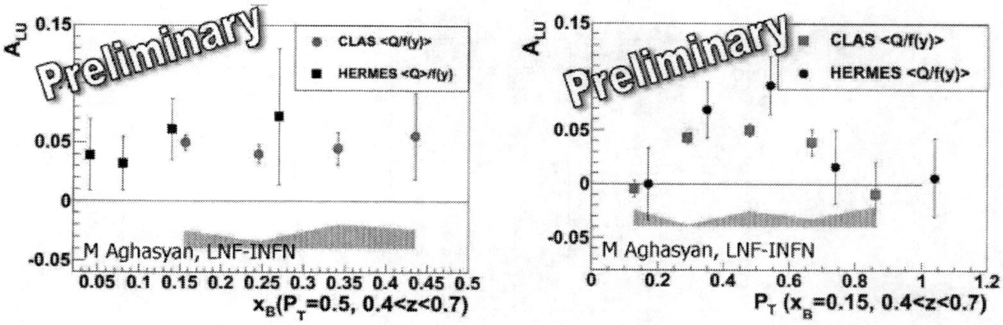

FIGURE 1. The measured π^0 beam-spin asymmetry amplitude $A_{UL}^{\sin\phi}$ for P_T=0.5 as function of x_B (left) and for x_B=0.15 as a function of P_T (right). Both amplitudes are integrated over $0.4 < z < 0.7$. Error bars present the statistical and the bands the systematic uncertainties. The CLAS data are compared with the HERMES ones [11].

the Inner Calorimeter (IC) (data of E01-113 experiment [8]). Polarized electrons with polarization between P =76% and 83% were scattered off a 2.5 cm long liquid unpolarized hydrogen target. CLAS and the inner calorimeter were used for the identification of photons from the π^0 decay and electrons. A solenoid magnet of 4.5 T field was used to shield the IC and CLAS drift chambers from Møller electrons, allowing to run at higher luminosity than in the standard configuration.

The beam-spin asymmetries in single-pion production off the unpolarized target are higher-twist by their nature [9, 10]. The higher twist observables are important for understanding the long-range quark-gluon dynamics. Different contributions to the beam SSAs, discussed so far, provide information on different leading and sub-leading parton distribution and fragmentation functions, related to both Collins and Sivers production mechanisms. At fixed and moderate values of the four momentum transfer Q^2 and at large values of x_B and z (where x_B is the fraction of the momentum of the nucleon carried by the struck quark and z is the fraction of the virtual photon energy carried by the detected π^0), the contribution of multiparton correlations or higher twist effects increases, eventually leading to a breakdown of the partonic description. Kinematical dependences of observables, will thus provide tests of the applicability of partonic descriptions.

In CLAS the kinematic dependences of the π^0 beam-spin asymmetry on x_B, z and P_T (where P_T is the transverse momentum of the π^0) have been measured. The $\sin\phi$ amplitude, integrating over the z-range $0.4 < z < 0.7$, has been extracted as a function of x_B and P_T. The asymmetry shows no significant x_B - dependence for fixed P_T range as shown in Fig. 1 (left). The $A_{UL}^{\sin\phi}$ moment as a function of P_T is shown in Fig. 1 (right). The $A_{UL}^{\sin\phi}$ moment increases with increasing P_T and reaches a maximum at P_T values of about 0.4 GeV. There is an indication, within available statistics, that the expected decrease of $A_{UL}^{\sin\phi}$ at larger P_T could start already at P_T values of about 0.7 GeV. In both plots the CLAS data are compared with the HERMES data [11] where, for the comparison a kinematic factor has been applied to take into account the different kinematical ranges covered by the two experiments.

Longitudinally Polarized Target

For a longitudinally polarized target the only azimuthal asymmetry arising at leading order is the $\sin 2\phi$ moment [12, 13, 14], involving the Collins fragmentation function $H_1^\perp$ and the Ralston-Soper-Mulders-Tangerman (RSMT) distribution function $h_{1L}^\perp$ [15, 12], describing the transverse polarization of quarks in a longitudinally polarized proton [12, 13, 14, 16].

Measurements of SSAs in SIDIS kinematics have been done at Jefferson Lab using a 5.7 GeV electron beam and the CLAS detector (EG1 experiment). Scattering of longitudinally polarized electrons off a polarized NH_3 and ND_3 targets was studied over a wide range of kinematics. The average beam polarization was $\sim 73\%$ and the average target polarization for NH_3 (ND_3) was $\sim 70\%$ ($\sim 30\%$). The scattered electrons and pions were detected in CLAS. In the DIS kinematics ($Q^2 > 1$ GeV^2, $W^2 > 4$ GeV^2), ~ 8 millions π^+ have been detected in the $y < 0.85, z > 0.4$ range.

The $\sin 2\phi$ moment $A_{UL}^{\sin 2\phi}$ as a function of x is plotted in Fig. 2 (full triangles) [17]. Calculations [18, 19] using $h_{1L}^\perp$ from the chiral quark soliton model [20] and the Collins function [21] extracted from HERMES [22] and Belle [23]

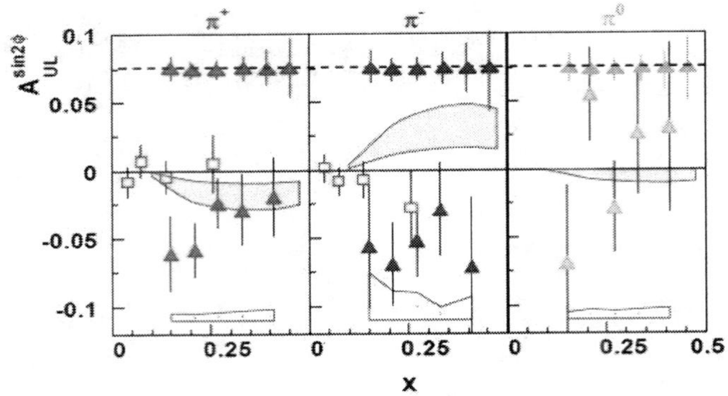

FIGURE 2. The measured x-dependence of the longitudinal target SSA $A_{UL}^{\sin 2\phi}$ (full triangles). The squares show the existing $A_{UL}^{\sin 2\phi}$ measurement from HERMES [22]. The lower band shows the systematic uncertainty. The upper band shows the existing theory predictions with uncertainties due to the Collins function [18, 19]. The full triangles in the upper part of the plots show the projected sensitivity of results for the experiment E05-133.

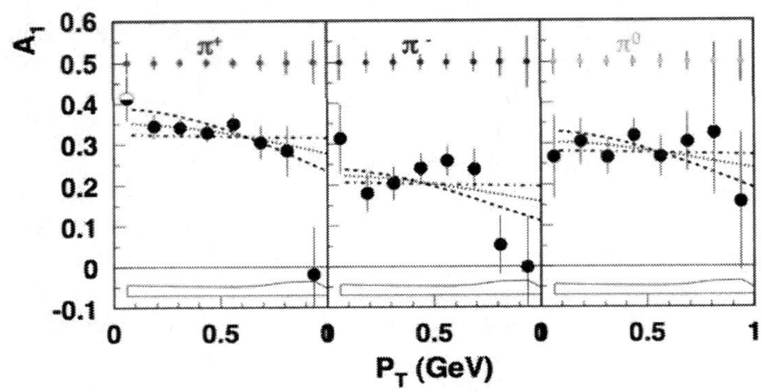

FIGURE 3. The double spin asymmetry A_1 as a function of transverse momentum P_T, integrated over all kinematical variables. The open band corresponds to systematic uncertainties. The dashed, dotted and dash-dotted curves are calculations for different values of the ratio of the transverse momentum widths for g_1 and f_1 (0.40, 0.68, 1.0) for a fixed width for f_1 (0.25 GeV^2) [24].

data, are plotted as filled bands. In the plot $A_{UL}^{\sin 2\phi}$ measurements from HERMES [22] (open squares) are also plotted. The kinematic dependence of the SSA for π^+ from the CLAS data is roughly consistent with these predictions. The interpretation of the π^- data, which tend to have SSAs with a sign opposite to expectations, may require accounting for additional contributions. By the way, current statistical errors for π^-, and in particular for π^0, are large and do not allow strong conclusions from the measured SSAs. Data at 6 GeV with ten times higher statistics have been collected in 2009 (experiment E05-133). They will allow statistically significant measurement of the $\sin 2\phi$ moment. The projected sensitivity of results for this experiment are shown as full triangles in the upper part of Fig. 2.

The data with longitudinally polarized beam and target provides important information input for the flavor decomposition through the measurement of double spin asymmetries, sensitive to the ratio of helicity and momentum distributions. The dependence of the double-spin asymmetry A_1 as a function of P_T and integrated over all x $(0.12 \div 0.48)$ for $Q^2 > 1$ GeV^2, $W^2 > 4$ GeV^2 and $y < 0.85$, is shown in Fig. 3. Although these plots are consistent with flat distributions, $A_1(P_T)$ may decrease with P_T at moderately small P_T for π^+. The slope for π^- could be positive for moderate P_T (ignoring the first data point). A possible interpretation of the P_T-dependence of the double-spin asymmetry may involve different widths of the transverse momentum distributions of quarks with different flavor and polarizations[24] resulting from a different orbital structure of quarks polarized in the direction of the proton spin and opposite to it [25, 26]. In Fig. 3 the measured A_1 is compared with calculations of the Torino group[24], which uses different values of the ratio of widths in k_T for partonic helicity, g_1, and momentum, f_1, distributions, assuming Gaussian k_T distributions with no flavor dependence.

Future data with transversely polarized target

In the last few years, first results on transverse SSAs have become available [27, 28]. HERMES measurements for the first time directly indicated significant azimuthal moments generated both by Collins and Sivers effects. An experiment using a trasversely polarized target in a frozen-spin state, such as the HD-Ice target has been approved at JLab (CLAS experiment E08-015) and is planned to run late 2011. The advantages of an HD-Ice target compared to the solid state dynamically polarized NH_3 and ND_3 targets are manifold: it requires a small holding field and this potentially allows the target to be located even in the center of the detector, thus dramatically increasing the acceptance; in addition the HD-Ice target has almost no diluition; the polarization are expected to be very high: 75% for hydrogen and 40% for deuterium.

CONCLUSIONS

Transverse Momentum Dependent distributions of partons contain novel information on the nucleon structure. They can be accessed in SIDIS with polarized leptons and nucleons by observing the azimuthal distributions of produced hadrons. Latest experimental data indicate that spin-orbit correlations are significant and lead to observable Single Spin Asymmetries. CLAS in Hall B at Jefferson Lab is playing a major role in these studies. Interesting results have been obtained from the π^0 Beam Spin Asimmetry measurement and new ones are expecting soon from the new experiment with longitudinally polarized NH3 and ND3 targets and from the future experiment with a transversely polarized target.

REFERENCES

1. D. W. Sivers, *Phys. Rev. D* **43**, 261 (1991).
2. M. Anselmino *and* F. Murgia, *Phys. Lett. B* **442**, 470 (1998), arXiv:hep-ph/9808426.
3. S.J. Brodsky, D.S. Hwang, *and* I. Schmidt, *Phys. Lett. B* **530**, 99 (2002), arXiv:hep-ph/0201296.
4. J.C. Collins, *Phys. Lett. B* **536**, 43 (2002), arXiv:hep-ph/0204004.
5. X.d. Ji *and* F. Yuan, *Phys. Lett. B* **543**, 66 (2002), arXiv:hep-ph/0206057.
6. P.J. Mulders *and* R.D. Tangerman, *Nucl. Phys. B* **461**, 197 (1996), arXiv:hep-ph/9510301.
7. B. Mecking *et al.*, Nucl. Instr. Meth. A **503**, 513 (2003).
8. CEBAF experiment E01-113 V. Burkert *et al.*.
9. A. Afanasev *and* C.E. Carlson, (2003), arXiv:hep-ph/0308163.
10. F. Yuan, Phys. Lett. B **589**, 28 (2004), arXiv:hep-ph/0310279.
11. A. Airapetian *et al.*, (HERMES) (2006), arXiv:hep-ex/0612059.
12. P.J. Mulders *and* R.D. Tangerman, Nucl. Phys. B **461**, 197 (1996).
13. A. Kotzinian, Nucl. Phys. B **441**, 234 (1995).
14. A. Kotzinian *and* P.J. Mulders, Phys. Rev. D **54**, 1229 (1996).
15. J.P. Ralston *and* D.E. Soper, Nucl. Phys. B **152**, 109 (1979).
16. E. Di Salvo, Int. J. Mod. Phys. A **22**, 2145 (2007), arXiv:hep-ph/0603005.
17. H. Avakian *et al.*, Phys. Rev. Lett. **105**, 262002 (2010), arXiv:1003.4549
18. H. Avakian *et al.*, Phys. Rev. D **77**, 014023 (2008), arXiv:hep-ph/0709.3253
19. A. V. Efremov, K. Goeke, and P. Schweitzer, Phys. Rev. D **67**, 114014 (2003), arXiv:hep-ph/0208124.
20. P. Schweitzer *et al.*, *Phys. Rev. D* **64**, 034013 (2001), arXiv:hep-ph/0101300.
21. A. V. Efremov, K. Goeke, *and* P. Schweitzer, Phys. Rev. D **73**, 094025 (2006), arXiv:hep-ph/0603054.
22. A. Airapetian *et al.*, (HERMES), Phys. Rev. Lett. **94**, 012002 (2005), arXiv:hep-ex/0408013.
23. K. Abe *et al.*, (Belle), Phys. Rev. Lett. **96**, 232002 (2006), arXiv:hep-ex/0507063.
24. M. Anselmino, A. Efremov, A. Kotzinian *and* B. Parsamyan, Phys.Rev. D **74**, 074015 (2006).
25. S.J. Brodsky *and* S.D. Drell, Phys. Rev. D **22**, 2236 (1980).
26. S.J. Brodsky, M. Burkardt *and* I. Schmidt, Nucl. Phys. B **441**, 197 (1995).
27. A. Airapetian *et al.*, Phys. Rev. Lett. **94**, 012002 (2005).
28. V.Y. Alexakhin *et al*, Phys. Rev. Lett. **94**, 202002 (2005).

A measurement of double spin asymmetry A_{LT} in semi-inclusive pion electroproduction on a transversely polarized ^{3}He target

Jin Huang[1]

Massachusetts Institute of Technology, Cambridge, Massachusetts, 02139, USA

Abstract. The double spin asymmetry A_{LT} for ^{3}He was measured for the first time at Jefferson Lab Hall A in the semi-inclusive $^3\mathrm{He}^{\uparrow}(\vec{e},e'\pi^{+/-})X$ reactions, using a 6 GeV polarized electron beam on a transversely polarized ^{3}He target. The produced charged pions were detected in coincidence with the scattered electrons by the high-resolution spectrometer and the BigBite spectrometer, respectively. The kinematic coverage was focused on the valence quark region, $x \sim 0.13$-0.41, $Q^2 \sim 1.31$-3.10 $(\mathrm{GeV}/c)^2$. When combined with the world data, this new data will provide constraints on the g_{1T}^q distribution functions, which describe the longitudinal polarization of quarks in a transversely polarized nucleon. The status of the current analysis and the preliminary results are presented.

Keywords: TMD DSA E06-010 Transversity A_{LT} "worm-gear" function g_{1T}
PACS: 13.60.-r; 13.87.Fh; 13.88.+e; 14.20.Dh; 24.85.+p

INTRODUCTION

High energy lepton-nucleon scattering is a powerful tool to study nucleon structure with an electromagnetic probe. Inclusive deep inelastic scattering (DIS) has been one of the primary methods employed to extract the parton distribution functions (PDFs). Semi-inclusive deep inelastic scattering (SIDIS), in which a hadron is detected in coincidence with the scattered lepton, provides additional and unique information about nucleon structure.

The transverse momentum dependent parton distributions (TMDs) [1, 2] describe nucleon structure in terms of quarks and gluons in all three directions of momentum. Of the eight leading-twist TMDs, the g_{1T}^q distribution functions (DFs) describe the longitudinal polarization of quarks in a transversely polarized nucleon. The g_{1T}^q DFs require an interference between quark wave function components that differ by one unit of quark orbit angular momentum, as explicitly shown in several models [3, 4, 5, 6]. Since the g_{1T}^q DFs link two perpendicular spin directions of the nucleon and quarks, they are also known as "worm-gear" functions [7].

All the eight leading twist TMDs can be accessed through SIDIS [1, 2]. With a longitudinally polarized lepton beam and a polarized target, whose spin is perpendicular to the momentum transfer, the leading-twist angular dependence of the beam helicity asymmetry can be expressed as:

$$A_{LT}(\phi_h,\phi_S) = \frac{1}{P_B P_T}\frac{Y^+_{\phi_h,\phi_S} - Y^-_{\phi_h,\phi_S}}{Y^+_{\phi_h,\phi_S} + Y^-_{\phi_h,\phi_S}} = A_{LT}^{\cos(\phi_h-\phi_S)}\cos(\phi_h-\phi_S), \tag{1}$$

where ϕ_h and ϕ_S are the azimuthal angles of the produced hadron and the target spin, defined according to the Trento conventions [8]. $P_{B(T)}$ is the polarization of the lepton beam (target). $Y^{+(-)} = \frac{d^2N/d\phi_h d\phi_S}{C\times L}$ is the yield density at the beam helicity of $+(-)1$, which also depends on ϕ_h and ϕ_S. $d^2N/d\phi_h d\phi_S$, C and L are event count density, beam charge and the DAQ livetime, respectively. At leading order, $A_{LT}^{\cos(\phi_h-\phi_S)}$ is proportional to the convolution of the g_{1T}^q DF and the unpolarized fragmentation function (FF) D_1 [2].

In recent years, active efforts have been made to study the theories and phenomenology of TMDs. Notably, g_{1T}^q DFs have been calculated in various models, as discussed in Refs. [9, 10, 5, 6, 11]. Common features of these model calculations suggest that g_{1T}^u is positive and g_{1T}^d is negative; the peak amplitude of g_{1T}^u is larger than that of g_{1T}^d; both of them reach maximum in the valence x-region at a level of a few percent relative to the unpolarized distribution f_1^q.

[1] Presented for the Jefferson Lab Hall A and E06-010 Collaboration

12th International Conference on Meson-Nucleon Physics and the Structure of the Nucleon (MENU 2010)
AIP Conf. Proc. 1374, 250-253 (2011); doi: 10.1063/1.3647136

In addition to the model studies, recent work [12, 13] explored g_{1T}^q DFs for the first time using lattice QCD, so far making use of a simplified definition of TMDs with straight gauge links.

The COMPASS collaboration carried out the first measurement of $A_{LT}^{\cos(\phi_h-\phi_S)}$ with a polarized muon beam on a polarized deuteron target [14]. The $A_{LT}^{\cos(\phi_h-\phi_S)}$ asymmetry was predicted at a level of a few to few tens of percent for pion production on a neutron target at Jefferson Lab kinematics [10, 5, 6, 15, 16, 17, 18].

EXPERIMENTAL SET-UP

In this talk, we presented a new experiment, E06-010 at Jefferson Lab (JLab) Hall A, measuring $A_{LT}^{\cos(\phi_h-\phi_S)}$ using a polarized electron beam and a transversely polarized ^{3}He target. The data for experiment E06-010 were successfully collected from October 2008 to February 2009. The analysis is ongoing.

The polarized electron beam was produced by illuminating a strained GaAs photocathode with a circularly polarized laser [19]. An electron beam with an energy of 5.9 GeV and an average current of 12 μA (ranging from 8 to 15 μA) was used. During E06-010, the polarization of the electron beam was around 80%. It was measured regularly by Möller polarimetry and partially monitored by Compton polarimetry. The beam helicity was flipped at a frequency of 30 Hz. The beam charge asymmetry between two helicity states was controlled below 150 ppm over a typical 20 minute data section. Therefore, by forming the beam helicity asymmetry in Eq. (1), the systematic uncertainties were significantly improved.

The polarized ^{3}He target is based on the principles of optical pumping and spin exchange. The in-beam part of the target cell is a 40 cm long glass vessel. The target density was about 10 amg [2]. To further reduce the systematic uncertainties, the spin state of ^{3}He nuclei was flipped every 20 minutes through the adiabatic fast passage [20]. The in-beam polarization, with the spin flips, was 60 to 65%, a world record for this type of target used in beam. To increase the $\phi_h - \phi_S$ coverage, the target spin was periodically switched between two directions: transverse in the scattering plane and perpendicular to the scattering plane (see Y. Qiang, these proceedings [21]).

To detect electrons, the BigBite Spectrometer consisted of a single dipole magnet as the analyzing magnet, three multi-wire drift chambers for tracking, a scintillator plane for the time-of-flight measurement and a lead-glass calorimeter system with a preshower/shower configuration for electron identification (ID) and triggering. Its angular acceptance was 64 msr and the momentum range was 700 MeV to 2.2 GeV. The left High Resolution Spectrometer (HRS) was used to detect the produced hadrons in coincidence with the BigBite Spectrometer. Its detector package included two drift chambers for tracking, two scintillator planes for timing and triggering, a gas Cerenkov detector and a lead glass system for the electron-pion separation. In addition, an aerogel Cerenkov detector and a ring imaging Cherenkov detector were used to further improve pion ID. The HRS central momentum was fixed at 2.35 GeV with a momentum acceptance of $\pm 4.5\%$ and an angular acceptance ~6 msr.

The SIDIS event sample was selected with particle identification and kinematic cuts, including four momentum transfer squared $Q^2 > 1\,(\mathrm{GeV}/c)^2$, invariant mass $W > 2.3$ GeV, missing mass $W' > 1.6$ GeV. The kinematic coverage focused on the valence quark region, $x \sim 0.13$-0.41, at $Q^2 \sim 1.31$-$3.10\,(\mathrm{GeV}/c)^2$ and $z \sim 0.5$. For the lower x-bins, the azimuthal angle distribution covered the regions with maximum and minimum A_{LT} asymmetries; it was closer to a 2π coverage for the higher x-bins.

RESULTS AND DISCUSSION

The detector calibration was completed within one year of data taking. Thereafter, the analysis efforts were separated into two teams (Red and Blue), each of which developed a method and a software package to extract the azimuthal asymmetries: a maximum likelihood method [22] was developed by the Red Team; the Blue Team developed a local-pair binning-fitting method [23]. An extensive comparison between results from both teams has been carried out and a good consistency was demonstrated. The preliminary $A_{LT}^{\cos(\phi_h-\phi_S)}$ results on ^{3}He are shown in Fig. 1. A non-zero asymmetry is favored for the π^- production channel.

The analysis teams are currently working on finalizing systematic uncertainties and extracting neutron asymmetries. Recently, a new JLab proposal [24] has been approved to extend this study during the Jefferson Lab 12 GeV era using

---

[2] amg is a density unit, defined as the number of ideal gas molecules per unit volume at 1 atm and 273.15 K.

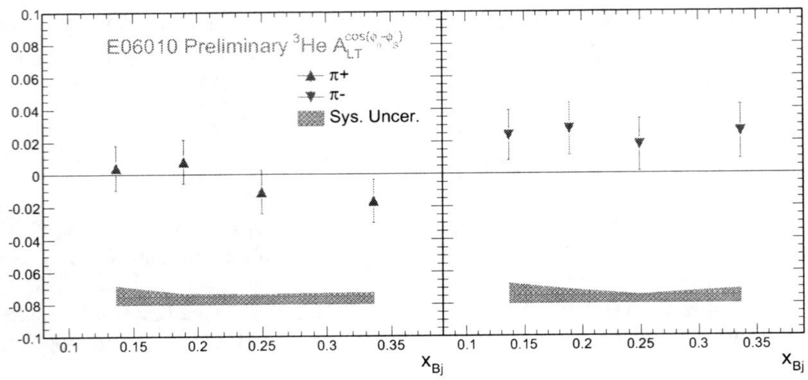

Figure 1. Preliminary results of ^{3}He double spin asymmetry $A_{LT}^{\cos(\phi_h-\phi_S)}$. The left panel shows the asymmetries of semi-inclusive π^+ electroproduction; the right panel shows that of π^-. The preliminary systematic uncertainties are drawn as the red bands below the corresponding data points.

a large acceptance spectrometer. The data are expected to cover a much larger kinematic range and the statistics will be improved by orders of magnitude.

ACKNOWLEDGMENTS

We acknowledge the outstanding support of JLab Hall A technical staff and Accelerator Division in accomplishing this experiment. This work was supported in part by the U. S. Department of Energy under contract number DE-FG-02-03ER41231, the U. S. National Science Foundation. This work was supported by DOE contract DEAC05-84ER40150 under which the Jefferson Science Association (JSA) operates the Thomas Jefferson National Accelerator Facility.

REFERENCES

1. P. J. Mulders, and R. D. Tangerman, *Nucl. Phys. B* **461**, 197 – 237 (1996).
2. A. Bacchetta, M. Diehl, K. Goeke, A. Metz, P. J. Mulders, and M. Schlegel, *JHEP* **2007**, 093 (2007).
3. X. Ji, J.-P. Ma, and F. Yuan, *Nucl. Phys. B* **652**, 383 – 404 (2003).
4. A. Bacchetta, F. Conti, and M. Radici, *Phys. Rev.* **D78**, 074010 (2008).
5. B. Pasquini, S. Cazzaniga, and S. Boffi, *Phys. Rev. D* **78**, 034025 (2008).
6. S. Boffi, A. V. Efremov, B. Pasquini, and P. Schweitzer, *Phys. Rev. D* **79**, 094012 (2009).
7. A. Bacchetta, "What can we learn from TMDs measurements," in *Spin 2008*, 2008.
8. A. Bacchetta, U. D'Alesio, M. Diehl, and C. A. Miller, *Phys. Rev. D* **70**, 117504 (2004).
9. R. Jakob, P. J. Mulders, and J. Rodrigues, *Nucl. Phys. A* **626**, 937 – 965 (1997).
10. A. Kotzinian, B. Parsamyan, and A. Prokudin, *Phys. Rev. D* **73**, 114017 (2006).
11. H. Avakian, A. V. Efremov, P. Schweitzer, and F. Yuan, *Phys. Rev. D* **81**, 074035 (2010).
12. J. W. N. Ph. Hägler, B. U. Musch, and A. Schäfer, *EPL (Europhysics Letters)* **88**, 61001 (2009).
13. B. U. Musch, P. Hagler, J. W. Negele, and A. Schafer (2010), arXiv:1011.1213v1.
14. B. Parsamyan, *Eur. Phys. J. - Special Topics* **162**, 89–96 (2008), 10.1140/epjst/e2008-00780-1.
15. A. Bacchetta, M. Radici, F. Conti, and M. Guagnelli, *Eur. Phys. J.* **A45**, 373–388 (2010).
16. H. Avakian, A. Kotzinian, B. U. Musch, A. Prokudin, and P. Schweitzer (2010), in Preparation.
17. J. She, J. Zhu, and B.-Q. Ma, *Phys. Rev.* **D79**, 054008 (2009).
18. J. Zhu, and B.-Q. Ma, *Phys. Lett. B* (2010), (Accepted).
19. C. Leemann, D. Douglas, and G. Krafft, *Annu. Rev. Nucl. Part. Sci.* **51**, 413–450 (2001).
20. A. Abragam, *Principles of Nuclear Magnetism*, Oxford University, New York, 1961.

21. Y. Qiang, "Recent progress in polarized ^{3}He targets," in *12th International Conference on Meson-Nucleon Physics and the Structure of the Nucleon*, 2010.
22. Y. Q. Jin Huang, Maximum likelihood estimation of asymmetry and angular modulation for transversity, Technote, Jefferson Lab E06010 Collaboration (2010), URL `http://www.jlab.org/~jinhuang/Transversity/MLE.pdf`.
23. X. Qian, *Measurement of Single Target-Spin Asymmetry in Semi-Inclusive* $n^{\uparrow}(e,e'\pi^{+/-})$ *Reaction on a Transversely Polarized* ^{3}*He Target*, Ph.D. thesis, Duke University (2010).
24. J. Chen, H. Gao, X. Jiang, J. Peng, X. Qian, and et al, JLab E12-10-006 : Target Single Spin Asymmetry in Semi-Inclusive Deep-Inelastic Electro Pion Production on a Trasversely Polarized 3He Target at 8.8 and 11 GeV, An update to PR12-09-014, Tech. rep., Jefferson Lab (2010), URL `http://www.jlab.org/exp_prog/proposals/10/PR12-10-006.pdf`.

Two Photon Exchange in Quasi-elastic and Deep-inelastic Scattering

Todd D. Averett, Joseph Katich, Bo Zhao

Department of Physics, College of William and Mary, Williamsburg, VA 23187
(on behalf of the Jefferson Lab Hall A and Polarized ^3He Collaborations)

Abstract. In this paper, I present an overview and preliminary results from three experiments at Jefferson Lab that were recently completed using a ^{3}He gas target with polarization oriented normal to the scattering plane of unpolarized incident electrons. A target single spin asymmetry was formed by periodically flipping the direction of the target spin. In the reaction $^{\uparrow 3}\mathrm{He}(e,e')$, the Born contribution is expected to be zero, giving direct sensitivity to two photon exchange. This asymmetry was measured in the quasi-elastic and deep-inelastic regimes with $0.1 < Q^2 < 1.0$ GeV^2. The asymmetry is predicted to decrease by two-orders of magnitude for deep-inelastic versus quasi-elastic scattering. Preliminary results from these experiments will be presented.

Keywords: two photon, polarized target, spin structure, Jefferson Lab
PACS: 25.30.Fj, 25.30.Bf, 29.25.Pj, 14.20.Dh

OVERVIEW

One of the most intriguing results from the 6 GeV program at Jefferson Lab was the unexpected disagreement between Rosenbluth separation and polarization transfer measurements of the proton elastic form factor ratio G_E^p/G_M^p with increasing Q^2 [1]. It was discovered that by applying corrections to the data sets to include contributions from two photon exchange, the discrepancy was largely resolved [2]. Based on this work, it was also realized that two photon exchange could be used as a new tool to study nucleon structure. A measurement of a non-zero asymmetry would provide confidence in the two-photon exchange formalism and models used to correct the proton form factor data. More importantly however, it provides a new mechanism to test models the nucleon. The only data that previously existed was taken several decades ago using a proton target. The new measurements presented here will provide precise data for the neutron, for which no data currently exists.

MOTIVATION AND EXPERIMENTS

In 2009-10, three experiments that are sensitive to two photon exchange were completed in Hall A at Jefferson Lab using polarized ^{3}He as an effective polarized neutron target, E05-015, E07-013 and E08-005 [3]. These experiments used a glass cell filled primarily with ^{3}He gas that was polarized using the technique of Spin-Exchange Optical-Pumping (SEOP) [4]. Unpolarized incident electrons were scattered from this target and detected using several magnetic spectrometers instrumented for electron detection. The target nuclei were polarized in a direction normal to the electron scattering plane (known as vertical polarization) and were flipped 180° every 20 minutes. This data was used to form a target single-spin asymmetry (SSA), A_y. The reaction $^{\uparrow 3}\mathrm{He}(e,e')$ was measured in the quasi-elastic and DIS regions.

The inclusive SSA is proportional to the imaginary part of the scattering amplitude squared. When only 1γ exchange is considered (Born approximation), the amplitude is real and $A_y = 0$ [5]. However, when two photon exchange is included, the asymmetry can be non-zero due to an imaginary contribution from the two photon exchange amplitude. The leading non-zero contribution to A_y comes from the $1\gamma + 2\gamma$ interference as shown below,

$$A_y \propto \frac{\mathscr{I}m\left(T_{1\gamma}^{\dagger}T_{2\gamma}\right)}{\left|T_{1\gamma}+T_{2\gamma}\right|^2} \tag{1}$$

12th International Conference on Meson-Nucleon Physics and the Structure of the Nucleon (MENU 2010)
AIP Conf. Proc. 1374, 254-257 (2011); doi: 10.1063/1.3647137

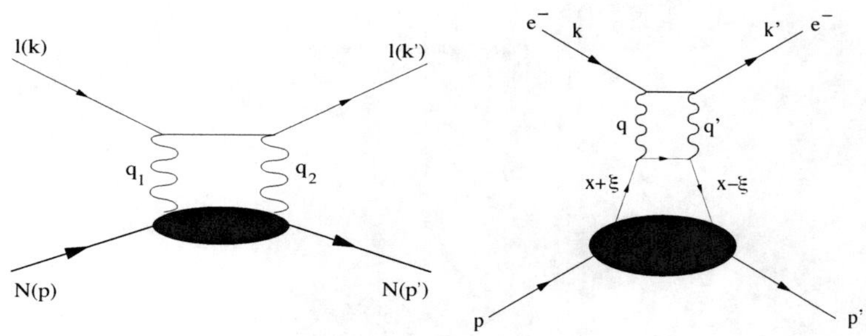

FIGURE 1. The left diagram shows two photon exchange with a nucleon. The black oval represents the response of the nucleon during scattering. The right diagram assumes that the two photons interact with only one asymptotically free quark.

where the $T_{1\gamma}$ and $T_{2\gamma}$ are the 1 and 2 photon exchange amplitudes. The interesting feature of 2γ exchange comes from the 2γ diagram which forms a closed loop with the nucleon intermediate state. Because it must be integrated over, the entire response of the nucleon is included. This provides us with a new tool to study models of the nucleon by inserting particular models for the nucleon response during scattering. See Figure 1. This nucleon intermediate state contains an elastic contribution, which is well-understood, and an inelastic contribution which must be estimated using model input.

For quasi-elastic scattering at large enough Q^2, where the 2γ handbag diagram is valid (Figure 1), the inelastic intermediate state contribution can be calculated using Generalized Parton Distributions (GPDs) [6, 7], and one can write [2],

$$A_y = \sqrt{\frac{2\varepsilon(1+\varepsilon)}{\tau}} \frac{1}{\sigma_R} \left[G_E \mathscr{I}(A_{GPD}) - \sqrt{\frac{1+\varepsilon}{2\varepsilon}} G_M \mathscr{I}(B_{GPD}) \right], \tag{2}$$

where $G_{E,M}$ are the nucleon elastic form factors, ε is a kinematic factor, σ_R is the unpolarized cross section and $\mathscr{I}$ denotes the imaginary part. The quantities A_{GPD} and B_{GPD} can be related to moments of the Generalized Parton Distributions. For the neutron, this asymmetry is particularly interesting because it is dominated by the term containing $\mathscr{I}(B_{GPD})$ since G_E^n is small. The quantity B_{GPD} is related to the GPD's H^q and E^q as follows,

$$B_{GPD} \equiv \int_{-1}^{1} \frac{dx}{x} K \sum_q e_q^2 (H^q - \tau E^q), \tag{3}$$

where K contains the hard scattering contribution and e_q is the quark charge. One can clearly see that a precise measurement of A_y provides a *unique new tool* to constrain GPD models through Equations 2 and 3. Though the GPD H^q is reasonably well-constrained by fits to unpolarized structure function data, relatively few constraints exist for E^q and in particular, for the sea quark contribution to these GPDs. Using a GPD model based on a parameterization of world data on parton distribution functions, modified by Regge behavior at low-x, Afanasev *et al.* [2] predict neutron asymmetries of $A_y^n \sim -2\%$ at $Q^2 = 1.0$ GeV^2. See Figure 2. Prior to these experiments, measurements on the neutron had not been made and measurements on the proton had large statistical uncertainties that did not allow one to test model predictions. Measurements of A_y^n in the quasi-elastic region were made during experiment E05-015 at $Q^2 = 0.13$, 0.46, 0.98 GeV^2 using the Hall A High Resolution Spectrometers (HRS). Preliminary data on ^{3}He at $Q^2 = 0.46$ and 0.98 GeV^2 are shown in Figure 3.

For the deep-inelastic measurement, the asymmetry probes helicity flip amplitudes at the quark level related to effects beyond the leading-twist picture of DIS. For scattering from a single, non-interacting massless quark, A_y is expected to be identically zero, even when 2γ exchange is included. In a recent calculation by Afanasev *et al.* [8], A_y was predicted to be $\sim 10^{-4}$ in the DIS region, compared to $\sim 10^{-2}$ in the quasi-elastic region. Concurrent with the transversity measurements discussed below, A_y was measured using singles electron triggers in the BigBite and left HRS spectrometers. First measurements of A_y^n in the DIS region were obtained at five points covering $Q^2 = 1-3$ GeV^2. Preliminary results for ^{3}He are shown in Figure 4 [9]. While these first measurements lack the precision to test predictions at the 10^{-4} level of Afanasev's prediction, they are consistently positive and non-zero at the 10^{-2} level.

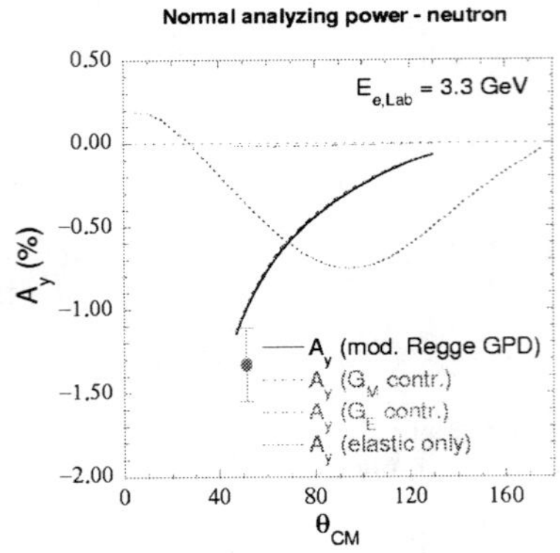

FIGURE 2. This plot shows the predicted target single-spin asymmetry A_y as a function of the center-of-mass electron scattering angle for experiment E-05-015. The blue curve is the contribution from the elastic intermediate state. The black curve is the inelastic contribution calculated using the GPD model of Afanasev *et al.* The predicted total asymmetry is the sum of the blue and black curves. The expected statistical uncertainty for this measurement at $Q^2 = 1.0$ GeV^2 is shown by the red circle at the predicted value for A_y.

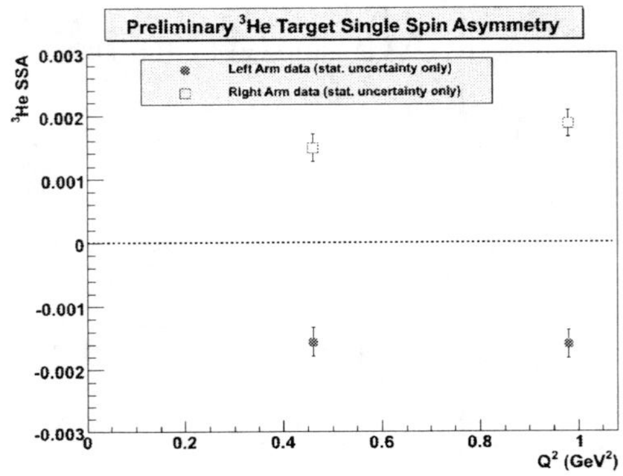

FIGURE 3. This plot shows preliminary results A_y for ^{3}He in the quasi-elastic region at $Q^2 \sim 0.5$ and 1.0 GeV^2. Note that as expected, the sign of the asymmetry changes when going from the beam-left to beam-right side. Uncertainties are statistical only. A correction to extract the neutron asymmetry from the data has not yet been applied.

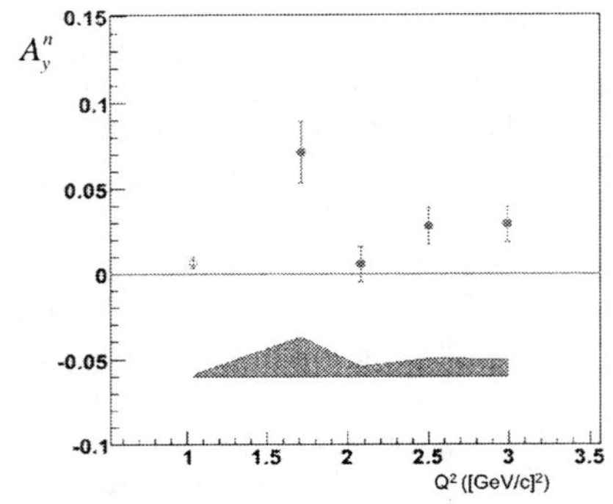

FIGURE 4. Preliminary results for A_y^n versus Q^2 in the reaction $^{\uparrow 3}\text{He}(e,e')$ in the DIS region. The open circle is data taken with the LHRS spectrometer. The solid circles are data taken with the BigBite spectrometer. Statistical uncertainties are shown on the data points. Systematic uncertainties are shown by the blue band.

QUASI-ELASTIC SCATTERING WITH RECOIL NEUTRON MEASUREMENT

Finally, there is a third measurement, E08-005, that ran concurrent with the quasi-elastic A_y experiment. The Hall A Neutron Detector (HAND) was used to detect the outgoing coincident neutron in the reaction $^{\uparrow 3}\text{He}(e,e'n)$. Because this reaction is not inclusive, the asymmetry is not required to be zero and is sensitive to the details of the ^{3}He nuclear wavefunction. In the PWIA however, this asymmetry is predicted to be zero. An unpublished measurement [10] from NIKHEF at $Q^2 = 0.2$ GeV^2, shows a large asymmetry of $\sim 50\%$. The data were best described using Fadeev calculations of ^{3}He by Golak [11] and included significantly larger Final State Interaction (FSI) effects than previous calculations. It is worth noting that this same model was successfully used to explain the discrepancy between measurements of G_E^n made using ^{3}He versus deuterium. This calculation predicts A_y will decrease rapidly with increasing Q^2 due to the decreasing contribution from FSI. This experiment remeasured near the NIKHEF point ($Q^2 = 0.14$ GeV^2) with significantly improved statistical precision, as well as two points at $Q^2 \sim 0.5$ and 1.0 GeV^2. Analysis of this data is underway. These precision measurements provide valuable constraints on the Q^2 dependence of the FSI and have generated significant interest in the few-body community.

POLARIZED TARGET

For these measurements, several improvements were made to the polarized target system in Hall A. First a third set of Helmholtz coils were added to provide a holding field in the vertical direction. Second, a system was developed to flip the direction of the nuclear spins every 20 minutes with minimal polarization loss. Hybrid target cells using a combination of potassium and rubidium were previously found to have significantly high efficiency for producing polarized ^{3}He nuclei for a given target geometry and laser intensity. Using this system, we were able to routinely achieve polarizations of around 65% with a beam current between 12-15 μA. Polarization was measured and calibrated using NMR and EPR systems, respectively. Online results for target polarization versus time are shown in Figure 5.

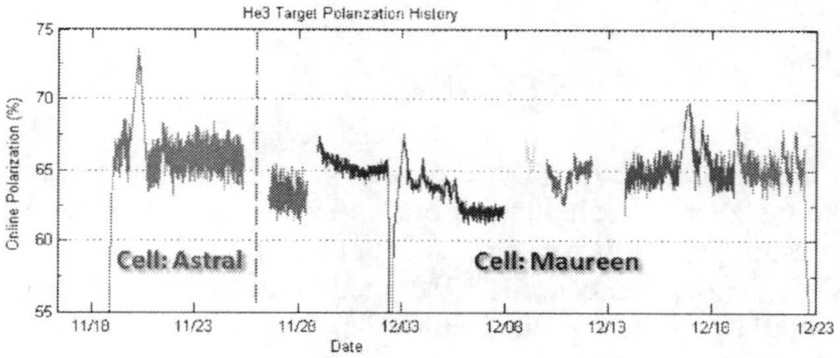

FIGURE 5. This plot shows the preliminary (online) results of the polarized ^{3}He target polarization over a one month period of time during the most recent set of experiments at Jefferson Lab. The large spikes occurred when the electron beam was not present. The relative flat regions near 65% show typical polarizations during the run periods with beam present.

REFERENCES

1. A.J.R. Puckett *et al.*, Phys. Rev. Lett. **104**, 242301 (2010).
2. A. Afanasev *et al.*, Phys. Rev. **D72**, 013008 (2005).
3. Jefferson Lab Experimental Proposals, http://www.jlab.org/div_dept/physics_division/experiments.
4. See for instance J. Alcorn *et al.*, Nucl. Instrum. Meth. A **522**, 294 (2004).
5. A. De Rujula, J.M. Kaplan and E. De Rafael, Nucl. Phys. **B**53 (1973) 545.
6. X. Ji, Phys. Rev. Lett. **78** (1997) 610; Phys. Rev. D **55** (1997) 7114.
7. A. Radyushkin, Phys. Lett. B **380** (1996) 417.
8. A. Afanasev, M. Strikman, C. Weiss, Phys. Rev. **D77**, 014028 (2008).
9. J. Katich, "Measurement of the Target-normal Single-spin Asymmetry, A_y^n in the Deep-Ineastic Region from the Reaction $^{\uparrow 3}\text{He}(e,e')$," Doctoral Thesis, College of William and Mary, 2010.
10. H. R. Poolman, "Quasifree spin-dependent electron scattering from a polarized he-3 internal target," NIKHEF Report, (1999).
11. W. Gloeckle, H. Witala, D. Huber, H. Kamada, and J. Golak, Phys. Rept. **274**, 107 (1996).

Highlights from RHIC Spin Physics Program

Ming X. Liu

P-25, MS H846, Los Alamos National Laboratory, Los Alamos, NM 87545

Abstract. The Relativistic Heavy Ion Collider (RHIC) at Brookhaven National Laboratory delivers the world's highest energy polarized proton-proton collisions at a center of mass energy up to 500 GeV and provides a unique opportunity to study the quark and gluon spin structure of the proton and QCD dynamics at high energy scale. RHIC has produced many exiting physics results in recent years. The latest data from RHIC significantly constrain the gluon spin contribution to the proton spin, and the parity violating single spin asymmetry are observed for the first time in W production by both the PHENIX and STAR collaborations. In this report, I present the latest results from the PHENIX and STAR experiments, followed by a brief discussion of the future prospects of transverse physics, particularly on the importance of the unique measurements of Drell-Yan single spin asymmetry.

Keywords: Spin, QCD, RHIC.
PACS: 24.70.+s, 24.85.+p

INTRODUCTION

Spin plays a key role in the determination of the properties of fundamental particles and their interactions. However, two outstanding spin puzzles have challenged our understanding of QCD over 30 years: (1) the challenge of "too small", i.e., the nucleon spin puzzle. It was believed that the spin of proton was carried mostly by the quarks that make up the proton. However, experiments in the 1980's led to the startling discovery that quark-spin contribute very little to the proton spin, setting off the "proton spin crisis". Where the rest of the proton spin coming from remains a major challenge to our understanding of the nucleon structure. Currently, there are several candidates that could make up the difference, the gluons' spin and quark and gluon's orbital angular momentum. Their contributions need to be determined by experimental measurements; (2) the challenge of "too big", i.e., the unexpectedly large left-right transverse single spin asymmetries (TSSA, A_N) in transversely polarized high-energy $p^\uparrow + p$ (and $l + p^\uparrow$) collisions. Prior to the early Fermilab E704 experimental measurements, it was believed based on collinear pQCD arguments that transverse single spin asymmetry should be very small in high-energy p+p collisions, $A_N \sim \frac{m_q}{p_T} \sim O(10^{-4})$, while the experimental data showed asymmetry order of 10%. Such large TSSAs are also observed at RHIC energies. Spin measurements have historically produced surprising results and are a stringent test to theories as spin is an intrinsically relativistic and quantum mechanical aspect of particle interactions.

Highlights from the Longitudinal Spin Physics Program

Measurements of the spin structure of the nucleon are a critical step in our understanding of strong interaction and nucleon structure. High energy polarized proton collisions at a center of mass energy up to $\sqrt{s}$ =500 GeV at RHIC provide a unique way to probe the proton structure. At RHIC energies, the NLO pQCD calculations have been applied successfully to describe the unpolarized cross section of hadrons and jet productions and also have been used to extract the polarized parton (quark and gluon) distribution functions.

12th International Conference on Meson-Nucleon Physics and the Structure of the Nucleon (MENU 2010)
AIP Conf. Proc. 1374, 258-261 (2011); doi: 10.1063/1.3647138

Since the discovery that quark's spin only contributes ~30% of the total proton spin, the gluon polarization inside the proton has been one of the major focus of experimental and theoretical investigations. Unlike the polarized DIS process, where gluons do not couple directly to the virtual phones, polarized p+p interactions provide direct access to the polarized quark and gluon distribution functions via the hard scattering processes at the leading order. The non-perturbative but universal polarized parton distribution functions $\Delta f_a(x,Q^2)$ can be accessed via the measurements of double spin asymmetries,

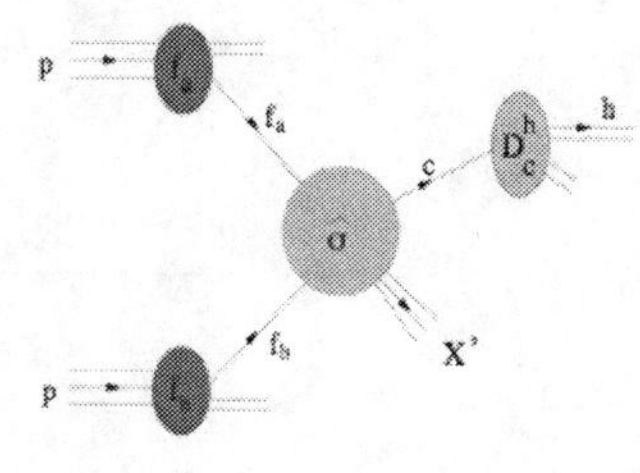

$$A_{LL} \equiv \frac{d\Delta\sigma}{d\sigma} \equiv \frac{d\sigma^{++} - d\sigma^{+-}}{d\sigma^{++} + d\sigma^{+-}}$$

with spin dependent particle *h* (or jet) production cross section given by,

$$d\Delta\sigma^h \sim \sum_{a,b} \int dx_a \int dx_b \Delta f_a(x_a, Q^2) \Delta f_b(x_b, Q^2) \times d\Delta\hat{\sigma}^{a+b\to c+X} \times D^{c\to h}$$

where the sum runs over all initial state partons *a* and *b*, with $d\Delta\hat{\sigma}^{a+b\to c+X}$ the corresponding partonic cross sections and $D^{c\to h}$ the fragmentation functions.

Recently, the PHENIX and STAR experiments have reported precision measurements of longitudinal double spin asymmetry A_{LL} in inclusive high p_T neutral pion production and inclusive high pT jet production in polarized p+p collisions at $\sqrt{s}$ = 200 GeV, respectively[1,2], see Figure 1.

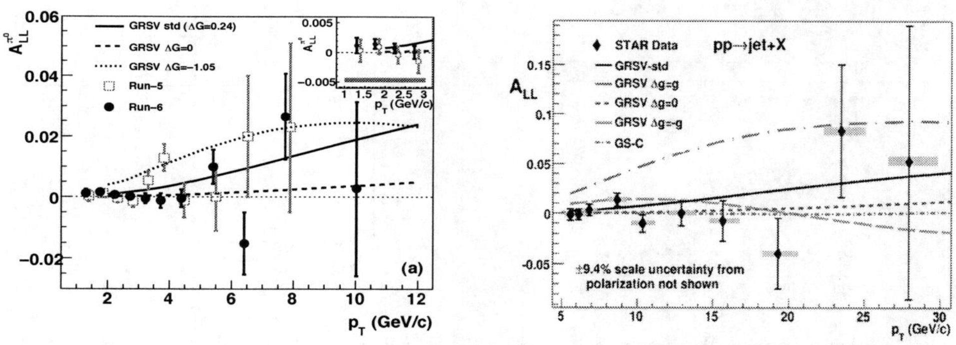

FIGURE 1. Left: Asymmetry A_{LL} in π^0 production as a function of p_T, from PHENIX. Right: Inclusive Jet A_{LL} as function of jet p_T, from STAR.

The results from PHENIX and STAR have put a strong constrain on the gluon spin contribution to the proton total spin. The details depend on the polarized parton distribution models used for the fit. In the GRSV framework[3], the PHENIX experiment has put a limit on gluon spin contribution, $-0.7 < \Delta G^{[0.02,0.3]} < 0.5$, at $\Delta\chi^2 = 9$. Much improved measurements are expected from future high luminosity runs in coming years.

Parity Violating Single Spin Asymmetry A_L in $W^{+/-}$ production

The parity violating longitudinal single spin asymmetry A_L in $W^{+/-}$ production in polarized p+p collisions is sensitive to the flavor dependent Δq and $\Delta\bar{q}$ distributions as W bosons only couple to the left-handed fermions and right-handed anti-fermions,

$$u_L + \bar{d}_R \rightarrow W^+$$
$$d_L + \bar{u}_R \rightarrow W^-$$

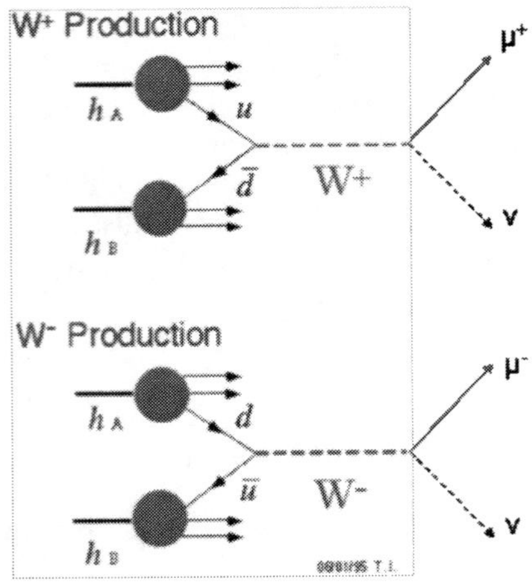

At leading order, the parity violating asymmetry is given by,

$$A_L(W^+) = \frac{\sigma^+ - \sigma^-}{\sigma^+ + \sigma^-} = \frac{\Delta\bar{d}(x_1)u(x_2) - \Delta u(x_1)\bar{d}(x_2)}{\bar{d}(x_1)u(x_2) + u(x_1)\bar{d}(x_2)}$$

$$A_L(W^-) = \frac{\sigma^+ - \sigma^-}{\sigma^+ + \sigma^-} = \frac{\Delta\bar{u}(x_1)d(x_2) - \Delta d(x_1)\bar{u}(x_2)}{\bar{u}(x_1)d(x_2) + d(x_1)\bar{u}(x_2)}$$

In 2010, the PHENIX and STAR collaborations reported the first observations of W boson production and significant non-zero parity violating single spin asymmetry from the 2009 run data[4,5]. Figure 2 shows the results from the STAR experiment, measurements are consistent with pQCD expectations. Future high statistic data will provide first clean measurements of flavor-identified sea-quark polarization that is almost impossible in DIS and further test pQCD predictions in a much wider kinematic range.

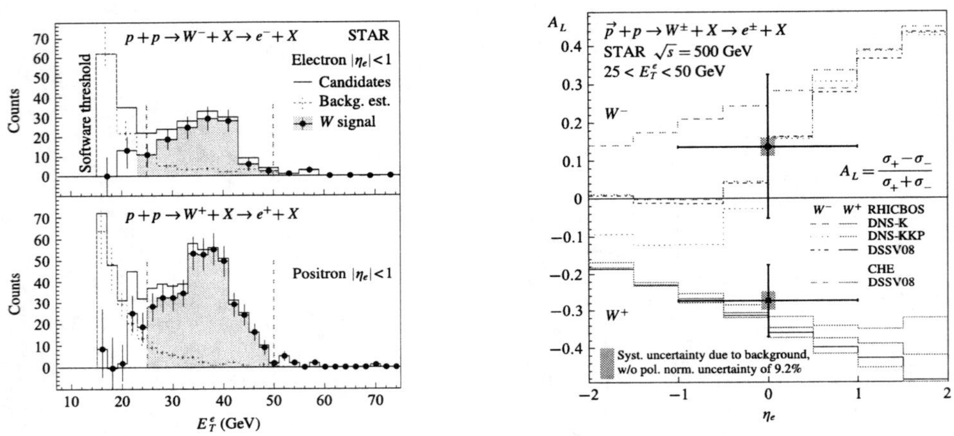

FIGURE 2. Left: high p_T electrons from W decays measured by the STAR experiment in the central rapidity; Right: high p_T electrons parity violationg single spin asymmetries for W events.

Highlights from the Transverse Spin Physics Program

Since the beginning of RHIC-Spin, interesting results on transverse single spin asymmetries and cross sections have been measured by the RHIC spin experiments (BRAHMS, PHENIX and STAR) in a very wide kinematic range in polarized p+p collisions. All three spin experiments have observed and confirmed significant non-zero TSSAs in inclusive single particle productions in the forward rapidities. In the middle and backward rapidities, TSSAs are consistent with zero within the experimental uncertainty.

The STAR collaboration has measured π^0 TSSA as a function of x_F and p_T, up to $p_T \sim 4.5$GeV from the latest transverse run in 2008. The PHENIX experiment has measured TSSAs in pion and inclusive charged hadron production as a function of x_F and p_T in the central and forward rapidities and confirmed previous results from other experiments. PHENIX also carried out the first measurements of TSSAs in open heavy quark and J/Psi production at $\sqrt{s}$=200 GeV, although statistically limited, the preliminary results indicate possible interesting new physics for transverse spin program in the future high luminosity runs.

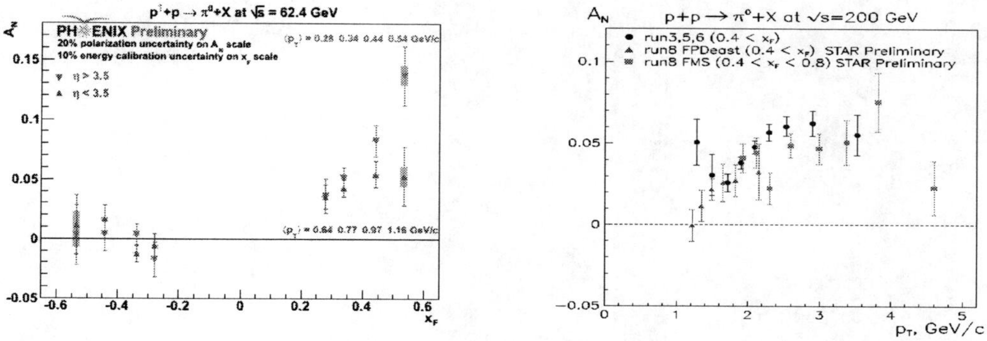

FIGURE 3. Left: TSSA A_N in π^0 production as a function of x_F at the center of mass energy $\sqrt{s}$=62 GeV, measured by the PHENIX experiment; Right: A_N in π^0 production as a function of p_T at the center of mass energy $\sqrt{s}$=200 GeV, from the STAR experiment.

With upcoming detector upgrades for both the PHENIX and STAR experiments and expected much improved accelerator performance, RHIC-Spin program will provide a great opportunity for new studies of polarized nucleon structure and the QCD dynamics. The future Drell-Yan A_N measurement is particularly important at RHIC[6]. There is a fundamental prediction of QCD that A_N of Drell-Yan should have an opposite sign compared to the $A_N(\pi^+)$ observed in the polarized DIS experiments, see Figure 4. Its verification or disproval will be an important milestone in our understanding of QCD.

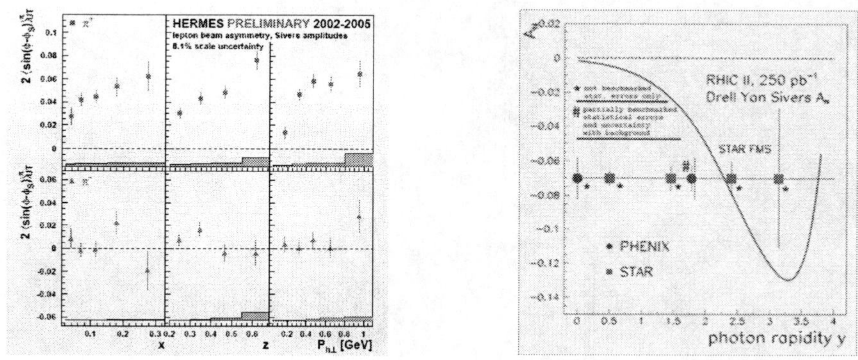

FIGURE 4. Left: Charged pion Sivers asymmetries measured by the HERMES experiment. Right: Expected Drell-Yan TSSA A_N in p+p at RHIC (based on HERMES fit), the sign of the asymmetry should be opposite to the one observed in π^+ production in the polarized DIS process.

ACKNOWLEDGMENTS

We thank the BNL Collider-Accelerator Department for developing the unique technologies enabling these measurements, and colleagues from PHENIX and STAR collaborations for helpful discussions. We acknowledge support from the Office of Nuclear Physics in DOE Office of Science.

REFERENCES

1. PHENIX collaboration, PRL 103, 012003 (2009).
2. STAR collaboration, PRL 100, 232003 (2008).
3. D. Florian *et. al.*, PRL 101, 072001 (2008).
4 PHENIX collaboration, PRL 106, 062001 (2011).
5. STAR collaboration, PRL 106, 062002 (2011).
6. RHIC Drell-Yan 2007 whitepaper, http://spin.riken.bnl.gov/rsc/write-up/dy_final.pdf

Session D: Meson Spectroscopy

The Measurement of χ_{cJ} Decays at BESIII

Guofa Xu

For the BESIII Collaboration
Institute of High Energy Physics, CAS

Abstract. Based on the 110M $\psi(3686)$ data sample which has been collected by BESIII at year 2009, many χ_{cJ} decay modes have been measured, such as $\chi_{cJ} \to \pi^0\pi^0$, $\chi_{cJ} \to \eta\eta$, $\chi_{cJ} \to VV$, $\chi_{cJ} \to \gamma V$, and so on. Results presented in this document are preliminary.

Keywords: $\psi(2S)$, double radiative decay, χ_{cJ}
PACS: 13.20.Gd, 14.40.Lb

INTRODUCTION

Charmonium decays provide a great opportunity for studying the gluon dynamics at relatively low energies. In the quark model, the χ_{cJ} ($J = 0,1,2$) mesons are $L = 1$ $c\bar{c}$ states. Since they cannot be produced directly in e^+e^- collisions, they are not as well studied as the ψ states. On the other hand, $\psi' \to \gamma\chi_{cJ}$ decays yield many χ_{cJ} mesons, providing a clean environment for χ_{cJ} investigations [1].

BEPCII is a two-ring e^+e^- collider designed for a peak luminosity of 10^{33} cm$^{-2}s^{-1}$ [2] at a beam current of 0.93 A. The BESIII detector is designed to fulfill the physics requirements discussed in Ref. [2], and the technical requirements for a high luminosity, multi-bunch collider. Details of each sub-detector and their performance can be found in Ref. [2].

Based on $(1.06 \pm 0.04) \times 10^8$ ψ' events collected with the BESIII detector, $\chi_{c0,c1} \to \pi^0\pi^0/\eta\eta$, $\chi_{cJ} \to \gamma V$, and $\chi_{cJ} \to VV$ decay modes have been measured in this paper.

MEASUREMENT OF $\chi_{c0,c1} \to \pi^0\pi^0, \eta\eta$

We do not consider χ_{c1} decays into these final states, as they are forbidden by spin-parity conservation. The optimization of the event selection and the estimation of physics backgrounds are performed through Monte Carlo simulations. A photon candidate is defined as a shower in the EMC with an energy deposit exceeding 50 MeV. The π^0 and η candidates are reconstructed from pairs of photon candidates, using the average event vertex of each run as the assumed origin of the photons. For $\pi^0 \to \gamma\gamma$, the $\gamma\gamma$ invariant mass is required to satisfy 0.075 GeV/$c^2 < M(\gamma\gamma) <$ 0.175 GeV/c^2. For $\eta \to \gamma\gamma$, the $\gamma\gamma$ invariant mass is required to satisfy 0.458 GeV/$c^2 < M(\gamma\gamma) <$ 0.608 GeV/c^2. The decay angle of a photon is the polar angle measured in the π^0 or η rest frame with respect to the π^0 or η direction in the ψ' rest frame. Real π^0 and η mesons decay isotropically, and their angular distributions are flat. However, the π^0 and η candidates that originate from a wrong photon combination do not have a flat distribution in this variable. To remove wrong photon combinations, the decay angle is required to satisfy $|\cos\theta_{decay}| < 0.95$. Candidate events for the final states of interest ($\gamma\pi^0\pi^0$ and $\gamma\eta\eta$) are selected using the following basic selection criteria. An event must have 5 or 6 photons and no charged tracks. All possible two photon pairings (the radiative photon from the ψ' decay which has $E < 0.4$ GeV is excluded) in the event are used to form π^0 and η candidates.

Backgrounds with missing final state particles are suppressed by requiring small transverse momentum squared $p_{t\gamma}^2$,

$$p_{t\gamma}^2 = 4p_{miss}^2 \sin^2(\theta_\gamma/2),$$

where p_{miss} is the missing momentum opposite to the $\pi^0\pi^0$ or $\eta\eta$ system and θ_γ is the angle between the radiative photon and the direction of the missing momentum p_{miss}. Figure1 shows the invariant mass of $\pi^0\pi^0$ (Figure1a) and $\eta\eta$ (Figure1b), and Table1 shows the comparison between our results[3] and that of CLEO-c and PDG, one can noticed that the precision of the branching fractions of $\chi_{c0,2} \to \pi^0\pi^0$ and $\chi_{c0,2} \to \eta\eta$ have been improved in our measurements.

12th International Conference on Meson-Nucleon Physics and the Structure of the Nucleon (MENU 2010)
AIP Conf. Proc. 1374, 265-268 (2011); doi: 10.1063/1.3647139

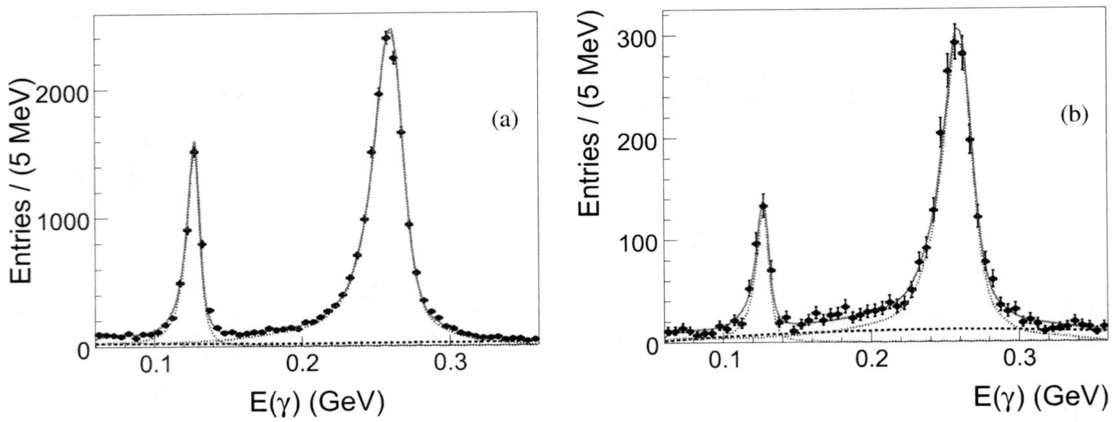

FIGURE 1. The radiative photon energy spectrum of selected $\chi_c \to \pi^0\pi^0$ (a) and $\chi_c \to \eta\eta$ (b) events. Dots with error bars are data. The solid curve is the result of the fit. The dotted curves are the χ_{cJ} signals. The dashed curve is the background polynomial.

TABLE 1. Branching fraction results (in units of 10^{-3}) for each decay mode. The uncertainties are statistical, systematic due to this measurement, and systematic due to the branching fractions of $\psi' \to \gamma\chi_{cJ}$, respectively. CLEOc results are determined using their own branching fractions for $\psi' \to \gamma\chi_{cJ}$, while ours are determined using branching fractions from the PDG. If we use the CLEOc branching fractions, we find $Br(\chi_{c0} \to \pi^0\pi^0) = 3.29 \times 10^{-3}$, $Br(\chi_{c0} \to \eta\eta) = 3.51 \times 10^{-3}$, $Br(\chi_{c2} \to \pi^0\pi^0) = 0.78 \times 10^{-3}$, and $Br(\chi_{c2} \to \eta\eta) = 0.58 \times 10^{-3}$.

Mode		χ_{c0}	χ_{c2}
$\pi^0\pi^0$	This Work	$3.23 \pm 0.03 \pm 0.23 \pm 0.14$	$0.88 \pm 0.02 \pm 0.06 \pm 0.04$
	CLEOc [5]	$2.94 \pm 0.07 \pm 0.32 \pm 0.15$	$0.68 \pm 0.03 \pm 0.07 \pm 0.04$
	PDG [4]	2.43 ± 0.20	0.71 ± 0.08
$\eta\eta$	This Work	$3.44 \pm 0.10 \pm 0.24 \pm 0.13$	$0.65 \pm 0.04 \pm 0.05 \pm 0.03$
	CLEOc [5]	$3.18 \pm 0.13 \pm 0.31 \pm 0.16$	$0.51 \pm 0.05 \pm 0.05 \pm 0.03$
	PDG [4]	2.4 ± 0.4	< 0.5

Study of χ_{cJ} in Charmonium Double Radiative Decays

In this paper, we mainly focus on the exclusive decay $\psi(3686) \to \gamma_l\chi_{cJ}$; $\chi_{cJ} \to \gamma_h V$, where $\gamma_l(\gamma_h)$ designates the characteristic lower (higher) energy photon in the signal topology and V is either a ϕ, ρ^0, or ω candidate. The ϕ, ρ^0, and ω candidates are reconstructed in the K^+K^-, $\pi^+\pi^-$, and $\pi^+\pi^-\pi^0$ decay modes, respectively. For the signal of vectors ϕ, ρ^0, and ω are selected as $|M_{K^+K^-} - M_\phi| \leq 0.01\text{GeV}/c^2$, $|M_{\pi\pi} - M_\rho| \leq 0.2$ GeV/c^2, and $|M_{\pi\pi\pi^0} - M_\omega| \leq 0.035$ GeV/c^2.

After all of the above criteria have been used, the invariant mass of $\gamma_h V$, where $V = \phi$, ρ^0, and ω are shown in Figure 2, one can see that there are exist an obvious structure in χ_{c1} region. In order to extract the event yield from the spectra, we first obtain a signal shape for each $\chi_{cJ} \to \gamma V$ channels (9 decay modes) by using the MC simulation of the signal where the mass, full width are taken from Ref. [4]. Each of the distributions in Figure 2 is fit to mass side-band(s) of vector plus a 2nd order polynomial as the background shape and a sum of three signal shapes, one for each of the χ_{cJ} states. Parameters of the polynomial function and the normalization factor for each of the χ_{cJ} lines are allowed to float in the fit. The fitted yields are summarized in Table 2. By examining the change in the fit likelihood when the signals yields are forced to zero, we estimate the significance of the $\chi_{c1} \to \gamma V$ signals to be much greater than 5σ. The upper limit branching ratios are estimated by Bayesian method [6] at 90% confidence level (CL). The effect of both the statistical and systematic uncertainties to the upper limit has been taken into account.

As we know, longitudinal (transverse) polarization exhibits a $cos^2\Theta(sin^2\Theta)$ dependence, where Θ is defined as the angle between the vector meson flight direction in the χ_{cJ} rest frame and either the π^+/K^+ direction in the ρ^0/ϕ rest

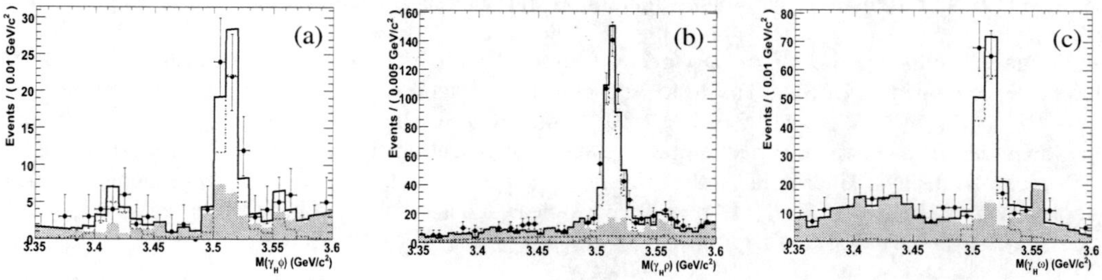

FIGURE 2. Invariant mass distributions of (a) $\gamma\phi$, (b) $\gamma\rho^0$, and $\gamma\omega$ and fit results (Dots with error bars for data; histogram for fitted curve; dashed line for signal shape; shadow for side-band(s) plus a 2nd order Polynomial background shape)

TABLE 2. Preliminary results of $\chi_{cJ} \to \gamma V$

Decay Modes	Events	Eff. (%)	Syst. Err.(%)	Br (10^{-6})	U.L.(90% C.L.) (10^{-6})	Stat. Significance
$\chi_{c0} \to \gamma\phi$	15.0 ± 6.6	32.4	8.1	$9.5 \pm 4.2 \pm 0.8$	<16.2	2.9σ
$\chi_{c1} \to \gamma\phi$	42.6 ± 8.6	34.6	7.8	$25.8 \pm 5.2 \pm 2.0$		6.4σ
$\chi_{c2} \to \gamma\phi$	4.6 ± 4.9	32.6	8.8		<8.0	
$\chi_{c0} \to \gamma\rho^0$	6 ± 12	22.6	7.4		<10.2	
$\chi_{c1} \to \gamma\rho^0$	432 ± 25	19.4	7.2	$228 \pm 13 \pm 16$		$\gg 10\sigma$
$\chi_{c2} \to \gamma\rho^0$	13 ± 11	15.7	7.9		<20.4	
$\chi_{c0} \to \gamma\omega$	5 ± 11	18.6	8.3		<12.7	
$\chi_{c1} \to \gamma\omega$	136 ± 14	22.7	8.0	$69.7 \pm 7.2 \pm 5.6$		$\gg 10\sigma$
$\chi_{c2} \to \gamma\omega$	1 ± 6	19.2	8.9		<6.0	

frame or the normal to the decay plane in the ω rest frame. The distributions of $cos\Theta$ (helicity) are shown in Figure 3, Finally, the fraction of the transverse component (f_T) can be obtained from the fit, and they are $0.29^{+0.13+0.10}_{-0.12-0.09}$ for $\chi_{c1} \to \gamma\phi$, $0.158 \pm 0.034^{+0.015}_{-0.014}$ for $\chi_{c1} \to \gamma\rho^0$, and $0.247^{+0.090+0.044}_{-0.087-0.026}$ for $\chi_{c1} \to \gamma\omega$, where the second is the systematic error. Our fit results of f_T indicates that the longitudinal component is dominance in the $\chi_{c1} \to \gamma V$ decay mode, and it would be expected for an axial-vector particle decaying into vector (ϕ, ρ^0, and ω) and γ in the framework of the vector dominance model taking into account the Landau-Yang theorem [7].

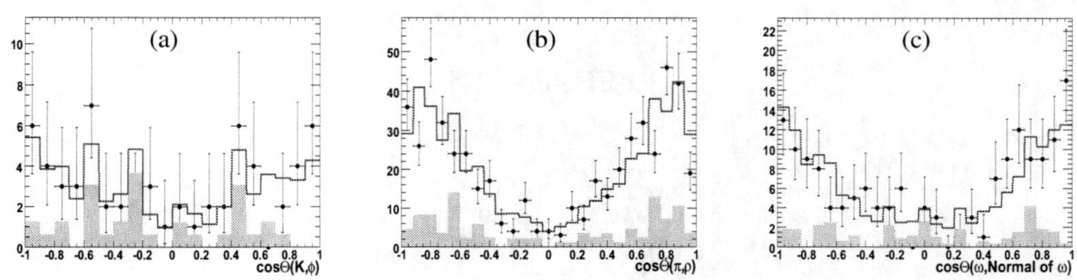

FIGURE 3. $cos\Theta$ (helicity) distributions and the fit curves of (a) $\chi_{c1} \to \gamma\phi$, (b) $\chi_{c1} \to \gamma\rho^0$, and (c) $\chi_{c1} \to \gamma\omega$. Dots with error bars for data; histogram for the fit function; and shadow for backgrounds

MEASUREMENT OF $\chi_{cJ} \to \phi\phi$, $\omega\omega$, AND $\omega\phi$

As for the decays of $\chi_{cJ} \to VV$ (V: vector meson, J=0,1,2), the measured branching fractions are still not well understood in QCD theory. For example, the decay of $\chi_{c1} \to \omega\omega, \phi\phi$ are suppressed due to the helicity selection rule [8]. However, the BESII measurements show that the χ_{cJ} has a larger branching fractions to decay into these final states [9].

First preliminary results for the $\chi^{cJ} \to \phi\phi$ decay (Figure 4)with both ϕ mesons reconstructed in the K^+K^- channel have been obtained at BESIII. The following branching fractions have been measured: BR($\chi_{c0} \to \phi\phi$) = $(8.0 \pm 0.4) \times 10^{-4}$, BR($\chi_{c1} \to \phi\phi$) = $(4.2 \pm 0.3) \times 10^{-4}$ and BR($\chi_{c2} \to \phi\phi$) = $(11.3 \pm 0.4) \times 10^{-4}$, only statistical uncertainties are given. The systematic uncertainties are under investigation. The $\phi\phi$ channel has also been analyzed in $\chi_{cJ} \to K^+K^-\pi^+\pi^-\pi^0$ decays. The results on the branching fractions from these studies are consistent with the results of the 4 kaons channel. In $\chi_{cJ} \to K^+K^-\pi^+\pi^-\pi^0$ decays clear signals have also been observed in the doubly-OZI suppressed channel $\chi_{c0} \to \phi\omega$ for the first time.

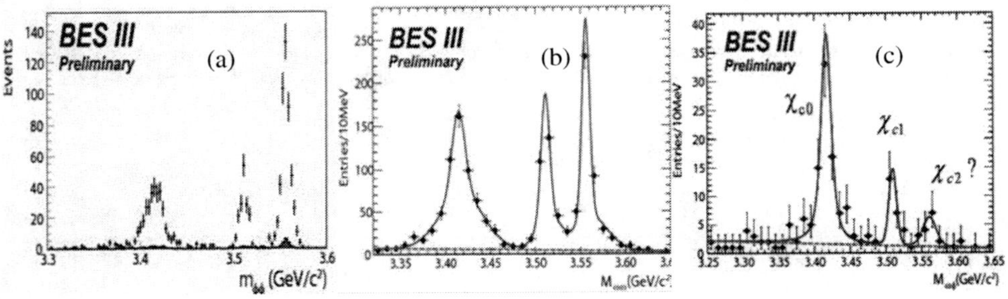

FIGURE 4. Invariant mass distributions of (a) $\phi\phi(,\phi \to K^+K^-)$, (b) $\omega\omega(,\omega \to \pi^+\pi^-\pi^0)$, and (c) $\omega\phi(,\omega \to \pi^+\pi^-\pi^0, \phi \to K^+K^-)$ (Dots with error bars for data; histogram for fitted curve; dashed line for background shape)

CONCLUSION

The measurement of the branching ratios of $\chi_{cJ} \to \pi^0\pi^0, \eta\eta$ have been published in Ref. [3]. Preliminary results on $\chi_{cJ} \to \gamma V, VV$ including first observations of $\chi_{c1} \to \gamma\phi$, $\chi_{c1} \to \phi\phi, \omega\omega$ and $\chi_{cJ} \to \phi\omega$ have been obtained, and more exciting results from BESIII will be coming soon.

ACKNOWLEDGMENTS

I would like to thank the staff of BEPC and the computing center for their hard efforts, and thank my colleagues of the BESIII Collaboration. I would also like to thank the organizer for warm hospitality at College of William and Mary.

REFERENCES

1. Q. Zhao, Phys. Rev. D **72**, 074001 (2005).
2. D. M. Asner *et al.*, Int. J. Mod. Phys. A **24**, Supp01 (2009).
3. M. Ablikim, *et. al.*, (BES Collaboration), Phys. Rev. D **81**, 052005 (2010).
4. K. Nakamura *et al.*, (Particle Data Group), Journal of Physics G **37**, 075021 (2010).
5. D. M. Asner *et al.*, (CLEO Collaboration), Phys. Rev. D **79**, 072007 (2009).
6. ZHU Yong-sheng, Chin. Phys. C (HEP & NP) **32**, 363-369 (2008).
7. D. V. Amelin, *et al.*, (VES Collaboration), Z. Phys. C **66**, 71 (1995).
8. S. J. Brodsky, G. P. Lepage, Phys. Rev. D **24**, 2848 (1981).
9. M. Ablikim, *et. al.*, (BES Collaboration), Phys. Lett. B **630**, 7 (2005).
   M. Ablikim, *et. al.*, (BES Collaboration), Phys. Lett. B **642**, 197 (2006).

Recent Results on Hadron Spectroscopy and Charmonium from BESIII

Fang Liu (for the BES Collaboration)

Institute of High Energy Physics, Chinese Academy of Sciences, Beijing 100049, China

Abstract. $p\bar{p}$ mass threshold enhancement is studied using the data sample of J/ψ and $\psi(2S)$ collected with BESIII detector in 2009. The enhancement is evident in J/ψ radiative decay, which is consistent with BESII results. No significance narrow enhancement is observed in $\psi(2S)$ radiative decay. The structure of X(1835) is also confirmed in $J/\psi \to \gamma\eta'\pi^+\pi^-$ at BESIII. The spin-singlet P-wave charmonium state $h_c(1P)$ is studied at BESIII, the mass spectra recoiling against the π^0 in the decay $\psi(2S) \to \pi^0 h_c$ both in h_c inclusive decay and in the final states with a tagged E1-transition photon from $h_c \to \gamma\eta_c$ are fitted to extract the number of h_c signals.

Keywords: Spectroscopy, Charmonium
PACS: 13.25Gv, 14.40Pq

INTRODUCTION

The upgraded Beijing Electron Positron Collider(BEPCII), is an electron-positron accelerator with separate storage rings for each beam. The BESIII [1] detector is designed to study the τ-charm physics [2]. Since July 19, 2008, the first collision happened, the BESIII has collected 106M $\psi(2S)$ events and 226M J/ψ events which are the largest $\psi(2S)$ and J/ψ data samples in e^+e^- collision experiment.

With such large data samples, the J/ψ and $\psi(2S)$ decay modes can be measured much more precisely than before. Using J/ψ decays, one can study light hadron spectroscopy and search for new hadronic states. With BESIII, the CKM matrix elements V_{cs}, V_{cd} and V_{us} will be extracted with an expected smaller systematic error. The $D-\bar{D}$ mixing measurement and searching for CP-violation will be possible.

CONFIRMATION OF $p\bar{p}$ MASS THRESHOLD ENHANCEMENT

An anomalously strong $p\bar{p}$ mass threshold enhancement was observed by the BESII experiment in the radiative decay process $J/\psi \to \gamma p\bar{p}$ [3]. An interesting feature of this enhancement is that corresponding structures are not observed in near-threshold $p\bar{p}$ cross section measurements [4]. These non-observations disfavor the attribution of the mass-threshold enhancement to the pure effects of $p\bar{p}$ final state interactions (FSI). This observation stimulated a number of theoretical speculations [5]. One of these is the intriguing suggestion that it is an example of a $p\bar{p}$ bound state [6], sometimes called baryonium [7].

After the selection [8], we get the similar characterize of the $p\bar{p}$ mass threshold enhancement in Fig.1 (a). The Dalitz plot in Fig.1 (b) for select events also shows that a band corresponding to the threshold enhancement evident in the upper right corner. Potential background processes are studied with an inclusive MC sample of 100M $\psi(2S)$ events and exclusive background decay $\psi(2S) \to \pi^+\pi^- J/\psi(J/\psi \to p\bar{p}\pi^0)$ in MC sample and real data. For these there is no enhancement at the threshold region of $p\bar{p}$ invariant mass spectrum.

We fit it with an acceptance weighted Breit-Wigner (BW) function of the form $BW(M) \propto \frac{q^{2L+1}k^3}{(M^2-M_0^2)^2+M_0^2\Gamma^2}$, where Γ is a constant (determined from fit), q is the proton momentum in the $p\bar{p}$ rest-frame, L is the $p\bar{p}$ orbital angular momentum, and k is the photon momentum, together with the background shape. Here, the background shape is described by the function of the form $f_{bkg}(\delta) = N(\delta^{1/2} + a_1\delta^{3/2} + a_2\delta^{5/2})$, where $\delta = M_{p\bar{p}} - 2m_p$ and the shape parameters a_1 and a_2 are determined from a fit to selected $\gamma p\bar{p}$ events for $\psi(2S) \to \pi^+\pi^- J/\psi(J/\psi \to \gamma p\bar{p})$ phase-space MC sample. The mass spectrum fitting with $S-$wave ($L=0$) in the threshold mass region of $M_{p\bar{p}} - 2m_p < 0.3$ GeV$/c^2$ is shown

12th International Conference on Meson-Nucleon Physics and the Structure of the Nucleon (MENU 2010)
AIP Conf. Proc. 1374, 269-272 (2011); doi: 10.1063/1.3647140

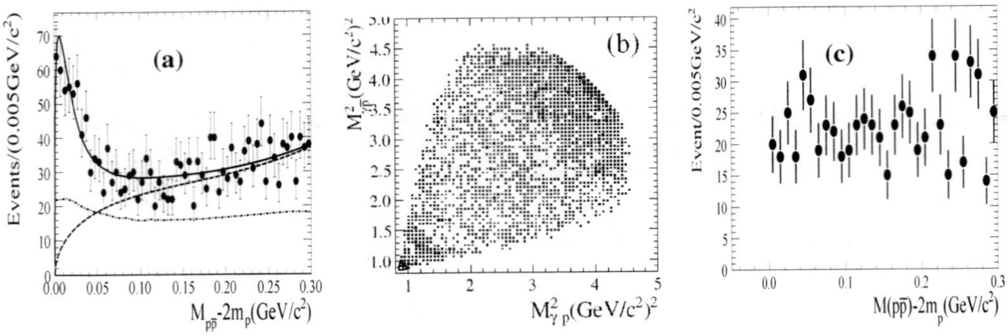

FIGURE 1. (a) The $p\bar{p}$ invariant mass spectrum and (b) Dalitz plot for the $\psi(2S) \to \pi^+\pi^- J/\psi(J/\psi \to \gamma p\bar{p})$ after final event selection. The solid curve is the fit result; the dashed curve shows the fitted background function, and the dash-dotted curve indicates how the acceptance varies with $p\bar{p}$ invariant mass. (c) The $p\bar{p}$ invariant mass spectrum in the threshold region for the selected $\psi(2S) \to \gamma p\bar{p}$ candidate events with $\psi(2S)$ data.

in Fig. 1(a), and it yields a peak mass of $M = 1861^{+6}_{-13}$ (stat)$^{+7}_{-26}$ (syst) MeV$/c^2$ and a width of $\Gamma < 38$ MeV$/c^2$ at the 90%$C.L.$

In the study of $\psi(2S) \to \gamma p\bar{p}$ with $\psi(2S)$ data, there is no significant narrow threshold enhancement as shown in Fig. 1 (c). It indicates that the strong $p\bar{p}$ threshold enhancement observed in J/ψ radiative decay is disfavored for the interpretation of pure final state interactions (FSI).

CONFIRMATION OF $X(1835)$ AT BESIII

A resonance, the X(1835), was observed in the $\pi^+\pi^-\eta'$ invariant-mass spectrum with a statistical significance of 7.7σ at BESII [9]. The mass and width of X(1835) are not compatible with any known meson resonance. With BESIII's larger data samples, it is important to confirm those measurements. The improved measurement will be useful to understand the truth of the threshold enhancement and X(1835).

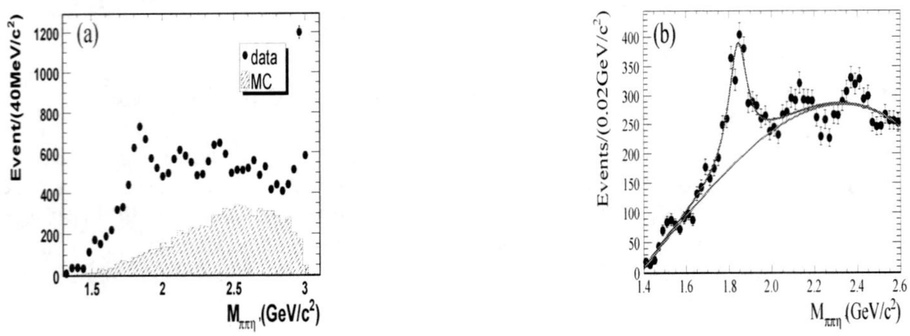

FIGURE 2. Candidate events after final event selection for $J/\psi \to \gamma\pi^+\pi^-\eta'(\eta' \to \gamma\rho)$. (a)Invariant mass spectrum of $\pi^+\pi^-\eta'$ after final selection, the solid circles are data and the shade histogram is from $J/\psi \to \gamma\pi^+\pi^-\eta'$ phase space MC events(with arbitrary normalization). (b) Mass spectrum fitting with one resonance and polynomial background shape.

For the $J/\psi \to \gamma\pi^+\pi^-\eta'(\eta' \to \gamma\rho)$ channel, the $\pi^+\pi^-\eta'$ invariant-mass spectrum for the selected events is shown in Fig.2.(a) and significant peak at $M \sim 1835$ MeV$/c^2$ is observed. If it is fitted with one resonance plus a polynomial background shape, the statistical significance of the resonance is about 9σ as shown in Fig.2.(b). In the $J/\psi \to \gamma\pi^+\pi^-\eta'(\eta' \to \pi^+\pi^-\eta, \eta \to \gamma\gamma)$ mode, the $\pi^+\pi^-\eta'$ invariant-mass spectrum for the selected events is shown in Fig.3(a) and significant peak at $M \sim 1835$ MeV$/c^2$ is also observed. If it is fitted with one resonance plus a polynomial background shape, the statistical significance of the resonance is about 18σ as shown in Fig.3(b).

Fig.3 shows the $\pi^+\pi^-\eta'$ invariant-mass spectrum for the combined $J/\psi \to \gamma\pi^+\pi^-\eta'(\eta' \to \gamma\rho)$ and $J/\psi \to \gamma\pi^+\pi^-\eta'(\eta' \to \pi^+\pi^-\eta)$. It is fitted with a Breit-Wigner function convolved with a Gaussian mass resolution function.

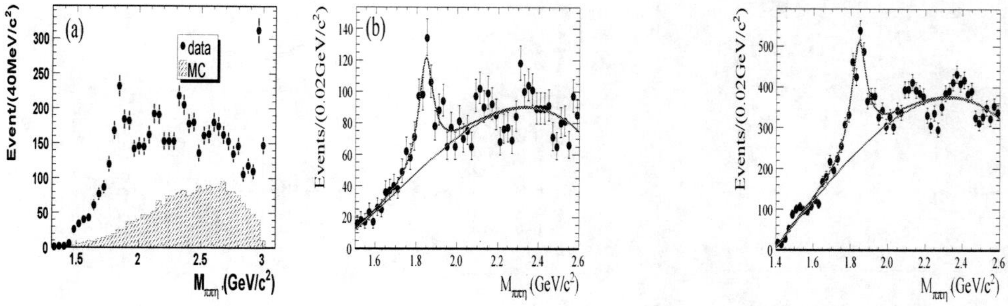

FIGURE 3. The $\pi^+\pi^-\eta'$ invariant-mass spectrum for the combined $J/\psi \to \gamma\pi^+\pi^-\eta'(\eta' \to \gamma\rho)$ and $J/\psi \to \gamma\pi^+\pi^-\eta'(\eta' \to \pi^+\pi^-\eta)$. It is fitted with a Breit-Wigner (BW) function convolved with a Gaussian mass resolution function.

The mass and width obtained from the fit are $M = 1842.4 \pm 2.8(stat)$ MeV$/c^2$ and $\Gamma = 99.2 \pm 9.2$ (stat) MeV/c^2 with a statistical significance of 21σ. These values are consistent with the published BESII results [9].

h_c MEASUREMENT

Although the charmonium family of mesons composed of a charmed quark and its own antiquark ($c\bar{c}$) has been studied for many years, knowledge is sparse on the singlet state $h_c(^1P_1)$. Early predictions for the properties of the h_c are found in Refs. [10, 11]. Many theoretical methods are used to study its branching ratios [12, 13, 14] and other physics [15]. The CLEO Collaboration observed the h_c and measured its mass [16, 17] using the decay chain $\psi(2S) \to \pi^0 h_c$, $h_c \to \gamma\eta_c$. The h_c are identified either by fullly reconstructing the event using many different hadronic decay channels of η_c, or by reconstructing the π^0 and γ in its decay chain. The E835 experiment [18] scanned antiproton energy and observed $p\bar{p} \to h_c \to \gamma\eta_c$. Recently, CLEO found evidence for decays of h_c to multi-pion final states [19]. The absolute branching ratios of $\psi(2S) \to \pi^0 h_c$ and $h_c \to \gamma\eta_c$ are very important to understand QCD theory and charmonium spectrum, but none of them have been measured in previous experiments.

In the E1-photon tagged analysis of $\psi(2S) \to \pi^0 h_c$, $h_c \to \gamma\eta_c$, the E1 photon emitted in $h_c \to \gamma\eta_c$ is tagged; no further constrains on the final states of η_c decay are needed.

Good photons should be selected. Candidates of signal π^0 are constructed using photons only in the barrel region, $|cos\theta| < 0.8$ with its invariant mass in the region of $120\text{MeV}/c^2 < M_{\gamma\gamma} < 145\text{MeV}/c^2$, and performed to $1-C$ kinematic fit to improve the resolution with no χ^2 cut. The daughter photons from the signal π^0 should not belong to other π^0s. In order to remove the following $\psi(2S) \to \pi^+\pi^- J/\psi$ and $\pi^0\pi^0 J/\psi$, events are rejected with $\pi^+\pi^-$ recoil mass in the region of 3097 ± 7MeV and $\pi^0\pi^0$ recoil mass in the region of 3097 ± 30MeV. E1-photon γ to tag $h_c \to \gamma\eta_c$ is selected with $450\text{MeV} < E_\gamma < 540\text{MeV}$ and not belonging to $\pi^0(0.10\text{GeV}/c^2 - 0.145\text{GeV}/c^2)$ and $\eta(0.53\text{GeV}/c^2 - 0.56\text{GeV}/c^2)$.

The E1-tagged and background subtracted spectra are shown in Fig. 4. When fitting the Fig. 4, the h_c signal is described by a Breit-Wigner function convoluted with the instrument resolution function obtained by fitting E1-tagged h_c MC simulation, the mass and width are float. The fit results are $N^{E1} = 3679 \pm 319$, $M(h_c) = 3525.40 \pm 0.13 \pm 0.18$ MeV$/c^2$, and $\Gamma(h_c) = 0.73 \pm 0.45 \pm 0.28$MeV with $\chi^2/d.o.f = 33.5/36.0$(p value 58.8%). The signal significance is 16.5 σ for the h_c signal.

In inclusive $\psi(2S) \to \pi^0 h_c$ analysis, the absolute branching ratio of $\psi(2S) \to \pi^0 h_c$ is determined by evaluating the inclusive π^0 recoiling mass spectrum.

Figure 4 shows the inclusive π^0 recoil mass spectra after the event selection while the background subtracted spectrum is shown in the inset. When fitting the Fig. 4, the mass and width of h_c are fixed to the values obtained from the E1-photon tagged analysis. The background is parameterized by a 4th-order Chebychev polynomial. The fit result is $N^{tot} = 9233 \pm 935$ with $\chi^2/d.o.f = 38.8/38.0$. The signal significance is 8.9σ.

Using those fit results and Monte Carlo efficiencies, we obtained $\mathscr{B}(\psi(2S) \to \pi^0 h_c) = (8.42 \pm 1.29 \pm 1.0) \times 10^{-4}$, $\mathscr{B}(\psi(2S) \to \pi^0 h_c) \times \mathscr{B}(h_c \to \gamma\eta_c) = (4.58 \pm 0.40 \pm 0.50) \times 10^{-4}$, and $\mathscr{B}(h_c \to \gamma\eta_c) = (54.3 \pm 6.7 \pm 5.2)\%$. $\mathscr{B}(\psi(2S) \to \pi^0 h_c)$ and $\mathscr{B}(h_c \to \gamma\eta_c)$ and the h_c width are measured for the first time. $M(h_c)$ and $\mathscr{B}(\psi(2S) \to$

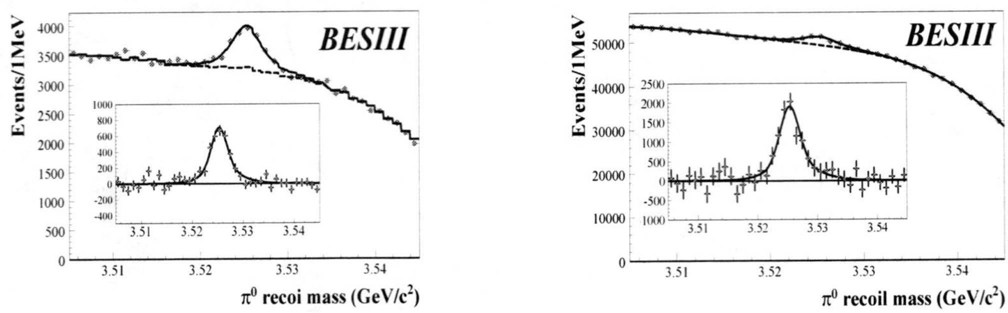

FIGURE 4. the π^0 recoil mass spectrum and fit for the inclusive analysis of $\psi(2S) \to \pi^0 h_c$. Fits are shown as solid lines, background as dashed lines. The inset shows the background-subtracted spectra.

$\pi^0 h_c) \times \mathscr{B}(h_c \to \gamma\eta_c)$ are consistent with those of CLEO results [17] and of comparable precision. The analysis results have been published [20].

SUMMARY

With the largest sample of J/ψ and $\psi(2S)$ collected with the BESIII detector, many measurements are reported. The $p\bar{p}$ threshold enhancement and $X(1835)$ are studied, which are consistent with the published BESII results. The $\Gamma(h_c)$, $\mathscr{B}(\psi(2S) \to \pi^0 h_c)$ and $\mathscr{B}(h_c \to \gamma\eta_c)$ are measured for the first time.

I would like to thank my colleagues of BESIII collaboration. I would also like to thank the organizer for warm hospitality at the College of William and Mary.

REFERENCES

1. BESIII Collaboration, M. Ablikim *et al.*, Nucl. Instrum. Meth. A. **614**, 3 (2010).
2. "Physics at BESIII", Edited by K. T. Chao, Y. F. Wang, Int. J. Mod. Phys. A **24**, No.1 supp (2009).
3. BES Collaboration, J.Z. Bai *et al.*, Phys. Rev. Lett. **91**, 022001 (2003).
4. M.Z. Wang *et al.*, Phys. Rev. Lett. **92**, 131801 (2004); S. Jin, Invited plenary talk at the XXXIIth International Conference on High Energy Physics(ICHEP04), Beijing (2004); BES Collaboration, M. Ablikim *et al.*, Phys. Rev. Lett. **99**, 011802 (2007); CLEO Collaboration, S.B. Athar *et al.*, Phys. Rev. D **73**, 032001 (2006); BES Collaboration, M. Ablikim *et al.*, Eur. Phys. J. C **53**, 15 (2008).
5. A. Datta *et al.*, Phys. Lett. B **567**, 273 (2003) ; M.L. Yan *et al.*, Phys. Rev. D **72** 034027 (2005); B. Loiseau *et al.*, Phys. Rev. C **72**, 011001 (2005); J. Ellis *et al.*, Phys. Lett. B **566**, 201 (2003); J.L. Rosner, Phys. Rev. D **68**, 014004 (2003); C.S. Gao *et al.*, Commun. Theor. Phys. **42**, 844 (2004); G.J. Ding *et al.*, Phys. Rev. C **72**, 015208 (2005).
6. E. Klempt *et al.*, Phys. Rep. **368**, 119 (2002); J.-M. Richard, Nucl. Phys. Proc. Suppl. **86**, 361 (2001) .
7. I.S. Shapiro, Phys. Rep. **35**, 129 (1978); C.B. Dover *et al.*, Phys. Rev. D **15**, 1997 (1977).
8. BESIII Collaboration, M. Ablikim *et al.*, Chinese Physics C **34**, 4 (2010).
9. BES Collaboration, M. Ablikim *et al.*, Phys. Rev. Lett. **95**, 262001 (2005).
10. Y. P. Kuang *et al.*, Phys. Rev. D **37**, 1210 (1988).
11. P. Ko, Phys. Rev. D **52**, 1710 (1995).
12. Y. P. Kuang, Phys. Rev. D **65**, 094024 (2002).
13. S. Godfrey *et al.*, Phys. Rev. D **66**, 014012 (2002).
14. J. J. Dudek *et al.*, Phys. Rev. D **73**, 074507 (2006).
15. E. S. Swanson, Phys. Rep. **429**, 243 (2006).
16. CLEO Collaboration, J. L. Rosner *et al.*, Phys. Rev. Lett. **95**, 102003 (2005); CLEO Collaboration, P. Rubin *et al.*, Phys. Rev. D **72**, 092004.
17. CLEO Collaboration, S. Dobbs *et al.*, Phys. Rev. Lett. **101**, 182003 (2008).
18. E-835 Collaboration, M. Andreotti *et al.*, Phys. Rev. D **72**, 032001 (2005).
19. CLEO Collaboration, G. S. Adams *et al.*, Phys. Rev. D **80**, 051106 (2009).
20. BESIII Collaboration, M. Ablikim *et al.*, Phys. Rev. Lett. **104**, 132002 (2010).

Diffractive Dissociation into $\pi^-\pi^-\pi^+$ Final States at COMPASS

Florian Haas for the COMPASS collaboration

Physik Department E18, Technische Universität München
James Franck Str., D-85748 Garching

Abstract. QCD predicts gluonic excitations like hybrids to contribute to the meson spectrum in addition to $q\bar{q}$ pair configurations. The most promising way to identify such states is the search for J^{PC} quantum number combinations which are forbidden in the constituent quark model. The fixed target COMPASS experiment at CERN offers the opportunity to search for such states in the light quark sector with an unprecedented statistics.
Diffractive reactions of 190 GeV/c pions on a lead target were studied by COMPASS during a pilot run in 2004. A Partial Wave Analysis (PWA) of the $\pi^-\pi^-\pi^+$ final state with 42 waves including acceptance corrections through a phase-space Monte Carlo simulation of the spectrometer was performed. The exotic $\pi_1(1600)$ meson with quantum numbers $J^{PC} = 1^{-+}$ has been clearly established in the rho-pi decay channel with a mass of 1660 ± 10(stat) MeV/c^2 and a width of 269 ± 21(stat) MeV/c^2. The improved detector performance in 2008 allows us to study this channel with significantly higher statistics. First results of the ongoing analysis of the 2008 data taking period, using a 190 GeV/c pion beam on a liquid hydrogen target are presented in this paper.

Keywords: hadron spectroscopy, light meson spectrum, diffraction, gluonic excitations, exotic mesons, hybrids
PACS: 13.25.-k, 13.85.-t, 14.40.Be, 29.30.-h

INTRODUCTION

A meson is described by a set of quantum numbers, which are isospin I and G-parity and the total spin J with parity P and charge conjugation parity C. In the quark model mesons can be described as bound states of quarks and antiquarks. Quantum Chromo Dynamics (QCD) predicts the existence of states which are not foreseen within the quark model. Possible states are hybrids, i.e. a system consisting of a color octet $q\bar{q}$ pair neutralized in color by a gluonic excitation, or glueballs, consisting only of glue. The experimental identification of such states, however, is difficult due to mixing with $q\bar{q}$ configurations with the same quantum numbers. The observation of exotic states, however, with quantum numbers not allowed in the simple quark model, e.g. $J^{PC} = 0^{--}, 0^{+-}, 1^{-+}, ...$, would give clear evidence that quark-gluon configurations beyond the quark model, as allowed by QCD, are realized in nature.
The lowest-lying hybrid is expected [1] to have exotic quantum numbers $J^{PC} = 1^{-+}$, and thus will not mix with ordinary mesons. Its mass is predicted in the region 1.3-2.2 GeV/c^2. There are three experimental candidates for a light 1^{-+} hybrid. The $\pi_1(1400)$ was observed by E852 [2], VES [3], and Crystal Barrel [4, 5]. Another 1^{-+} state, the $\pi_1(1600)$, decaying into $\rho\pi$ [6, 7, 8], $\eta'\pi$ [9, 10], $f_1(1285)\pi$ [11, 12] and $b_1(1235)\pi$ [12, 13] was observed in peripheral $\pi^- p$ interactions in E852 and VES. The resonant nature of both states, however, is still heavily disputed in the community [3, 12, 14]. A third exotic state, $\pi_1(2000)$, decaying to $f_1\pi$ and $b_1\pi$ was seen in only one experiment [11, 13]. Owing to its high statistics the COMPASS experiment can contribute significantly in the light meson sector to the search for exotic mesons and glueballs.

THE COMPASS EXPERIMENT

The **CO**mmon **M**uon and **P**roton **A**pparatus for **S**tructure and **S**pectroscopy [15] is located at the CERN SPS accelerator. It is a two stage magnetic spectrometer which provides a large angular acceptance over a wide momentum range. Besides calorimetry and particle identification COMPASS is equipped with a very precise charged particle tracking system. For the beam tracking, Silicon detectors around the target and Scintillating Fibres and PixelGEMs are used. Close to the beam large-size GEM detectors and Micromegas are the backbone of the tracking system. The periphery is covered by Drift Chambers and MWPCs.

12th International Conference on Meson-Nucleon Physics and the Structure of the Nucleon (MENU 2010)
AIP Conf. Proc. 1374, 273-277 (2011); doi: 10.1063/1.3647141

DIFFRACTIVE REACTIONS INTO 3 CHARGED PIONS ON PB IN 2004

The first years of COMPASS data taking, between 2002 and 2007, were dedicated to spin physics. In 2004 there was a short pilot run for the hadron physics program using a 190 GeV/c negative charged pion beam and a 3 mm lead target. During this short period a data sample with statistics, comparable to previous experiments, could be collected.

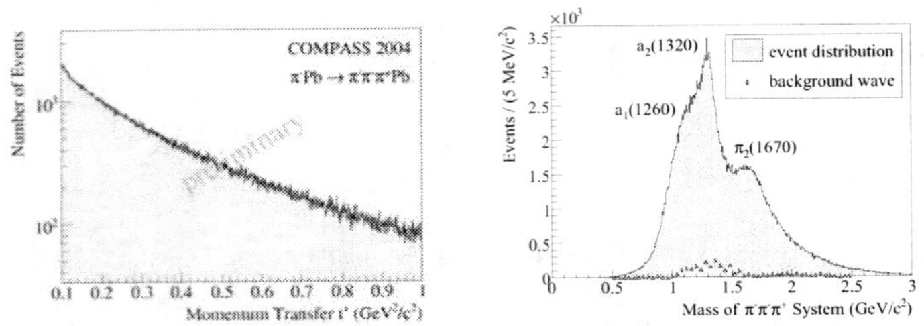

(a) Squared four-momentum transfer t' for diffraction of beam pions on single nucleons within the lead target, logarithmic scale.

(b) Invariant mass of the high t' data sample.

1: Kinematic Distributions.

Diffractive dissociation is a process where an incoming beam particle impinges on a target and is excited to a resonance X which finally decays into a final state of particles. Here we focus on the $\pi^-\pi^-\pi^+$ final state. The target particle stays intact, taking away the recoil momentum. In several days of data taking during the pilot run about $4 \cdot 10^6$ diffractive events were recorded. Fig. 1a shows the distribution of the kinematic variable t' which is given by $|t| - |t|_{min}$ where t is the square of the four-momentum transfer to the nucleus and $|t|_{min}$ the minimum value of $|t|$ allowed by kinematics for a given mass m_X. This analysis focuses on events in the "high t'" region between 0.1 GeV^2/c^2 and 1.0 GeV^2/c^2 (see Fig. 1a) where E852 [8] observed production of the exotic $\pi_1(1600)$ with $J^{PC} = 1^{-+}$. After this selection still $4.2 \cdot 10^5$ events remain. Fig. 1b shows the 3π invariant mass spectrum of this data sample and the intensity of the background wave with a flat distribution in 3-body phase space (triangles), obtained from a partial wave analysis, which will be described in the next section.

Isobar Model and Partial Wave Technique

The present analysis is based on the isobar model, assuming the resonance X to decay first into an intermediate isobar and a bachelor pion, wih the relative orbital angular momentum L between the two. The isobar subsequently decays into a $\pi^-\pi^+$ pair. A Partial Wave Analysis (PWA), performed in two steps, was applied to the 2004 data sample. In the first step, the mass-independent PWA, the data is divided into 40 MeV/c^2 wide bins in the 3π invariant mass m, and the cross section decomposed into:

$$\sigma_{\text{indep}}(\tau, m, t') = \sum_{\varepsilon=\pm1} \sum_{r=1}^{N_r} \left| \sum_i T_{ir}^{\varepsilon} f_i^{\varepsilon}(t') \psi_i^{\varepsilon}(\tau, m) / \sqrt{\int \left|\psi_i^{\varepsilon}(\tau', m)\right|^2 d\tau'} \right|^2 \qquad (1)$$

The production amplitudes T_{ir}^{ε} are the fitting parameters. They are obtained by a maximum likelihood fit. The real functions $f_i^{\varepsilon}(t')$ describes the t' dependence of the cross-section while τ represents the 5 phase-space variables of the 3-body decay. The decay amplitudes $\psi_i^{\varepsilon}(\tau', m)$, where the indices i and ε denote different partial waves, are characterized by a set of quantum numbers $J^{PC}M^{\varepsilon}[isobar]L$. M is the absolute value of the spin projection onto the beam direction; ε is the reflectivity [17], which describes the symmetry under a reflection through the production plane. The $\psi_i^{\varepsilon}(\tau, m)$ are described by Zemach tensors or D functions. In this analysis 41 partial waves were used. In addition an isotropic background wave was implemented.

In a second step a mass-dependent χ^2 fit is performed, parameterized by Breit-Wigner functions for the resonant part and a non-resonant background. In practice only a subset of waves is fitted. In this analysis the focus was on 6 waves, covering all important and clearly identified resonances and the exotic wave $1^{-+}1^{+}\rho\pi P$.

Fit Results

Figures 2(a)-(c) shows the intensity of the three most prominent waves $1^{++}0^{+}\rho\pi S$, $2^{++}1^{+}\rho\pi D$, and $2^{-+}0^{+}f_2\pi S$, described by Breit-Wigner functions and a coherent background, if necessary. Of peculiar interest are the fit results for the spin-exotic wave. The $1^{-+}1^{+}\rho\pi P$ intensity (Fig. 3a) has the shape of a broad bump, centered at 1.7 GeV/c^2, with a visible low-mass shoulder. One constant-width Breit-Wigner function (blue curve) and a non-resonant background (purple curve), possibly caused by a Deck-like effect, have been used to describe this wave. To clarify a possible resonant nature of this exotic amplitude, its interferences with well established states have been studied. Fig. 3b shows the phase difference to the $1^{++}0^{+}\rho\pi S$ wave (Fig. 2a), which clearly rises between 1.5 GeV/c^2 and 1.9 GeV/c^2. However, Fig. 3c, showing the phase difference to the $2^{-+}0^{+}f_2\pi S$ wave, exhibits a rather flat phase difference, which can be explained by the presence of two resonances, $\pi_1(1600)$ and $\pi_2(1670)$ with similar masses and widths. An overview of the fit results of all 6 partial waves used for the mass-dependent PWA and further details can be found in [16, 18].

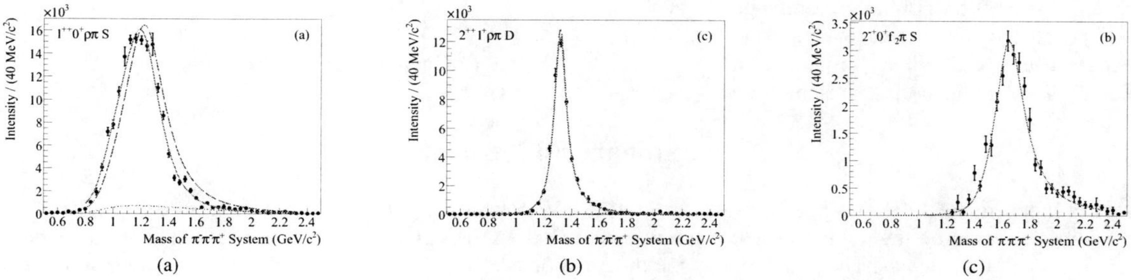

FIGURE 2. Intensities of major waves, $1^{++}0^{+}\rho\pi S$ (a), $2^{++}1^{+}\rho\pi D$ (b), and $2^{-+}0^{+}f_2\pi S$ (c), the red curve is the mass dependent fit.

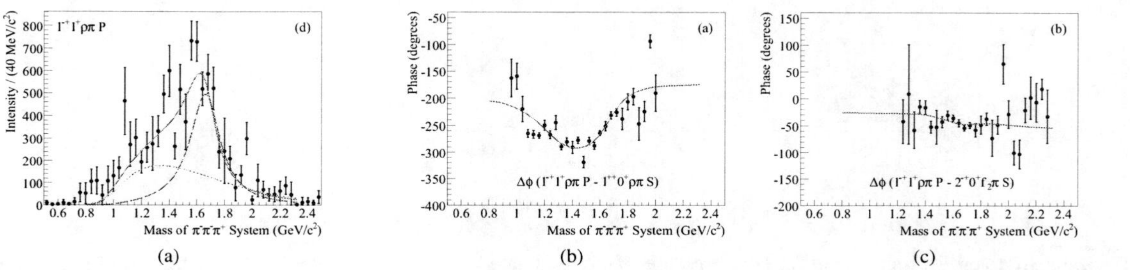

FIGURE 3. Intensity of the exotic wave $1^{-+}1^{+}\rho\pi P$ (a), and phase differences of this wave to the $1^{++}0^{+}\rho\pi S$ (b), and the $2^{-+}0^{+}f_2\pi S$ (c) waves.

SPECTROMETER UPGRADE

The 2008 and 2009 COMPASS data taking was mainly dedicated to meson spectroscopy. For that purpose the spectrometer was upgraded, also based on the experiences from the 2004 pilot run.

The negative hadron beam of COMPASS consists of 97% pions, 2.5% kaons and 0.5% anti-protons. In order to distinguish these incoming particle species, two CEDAR detectors, using the Cherenkov effect, were installed upstream of the target. Instead of the lead target, used in 2004, a completely new 40 cm long liquid hydrogen target was installed. Surrounding the target a new Recoil Proton Detector (RPD) was set up, triggering on recoil protons leaving the target. The RPD is one of the main components of the trigger system. Downstream of the target additional silicon microstrip detectors were placed, in order to have a good vertex resolution. Scintillating fibre detectors, used for beam tracking during the previous runs, were replaced by new PixelGEM detectors. These detectors are also able to track beam particles and particles in the vicinity of the beam, but they consist of less material. Also both electromagnetic calorimeters were upgraded with additional lead glass blocks and sampling ADCs.

DIFFRACTIVE REACTIONS INTO 3 CHARGED PIONS ON LH_2 IN 2008

In 2008 a negatively charged pion beam of 190 GeV/c momentum, the same as in 2004, impinged on the hydrogen target. Since the RPD was not yet implemented in the analysis the calculated beam energy E_a, as described in [16], was used to select exclusive events. Based on this calculation only events with E_a within $\pm$ 4GeV/c^2 were accepted (Fig. 4a). As in 2004 a cut on t' between 0.1 GeV^2/c^2 and 1.0 GeV^2/c^2 was applied. Fig. 4b shows the 3π invariant mass spectrum. During the whole 2008 campaign $96 \cdot 10^6$ exclusive diffractive events were collected. The next section shows first fit results of a PWA applied on a subsample of these events.

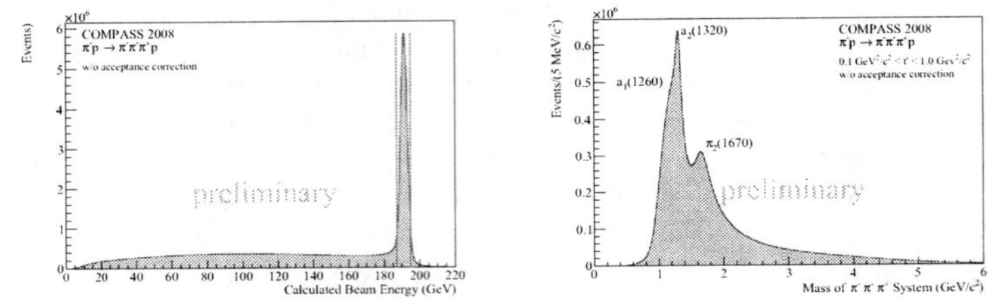

(a) Calculated beam energy, red lines correspond to the exclusivity cut.

(b) Invariant mass spectrum of the 3π system.

4: Kinematic Distributions.

Preliminary Fit Results

A mass-independent PWA has been carried out in 40 MeV/c^2 wide mass bins in a region from $0.5 - 2.5$GeV/c^2. This analysis is based on the same physical assumptions, including the same wave set, and the same PWA algorithm as used for the 2004 analysis. The intensities of the three major J^{PC} states, 1^{++}, 2^{++}, and 2^{-+}, are shown in Figures 5(a)-(c). At the present status of the analysis no acceptance correction was applied.

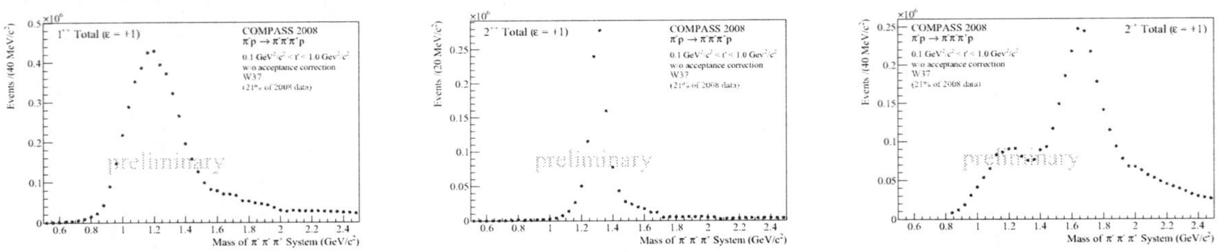

(a) (b) (c)

FIGURE 5. Total intensities for $J^{PC} = 1^{++}$ (a), 2^{++} (b), and 2^{-+} (c).

A striking first (preliminary) finding concerns the M-dependence of produced waves for the different target materials. The main difference between 2004 and 2008 data samples is the target material. While in 2004 two solid lead disks where used a liquid hydrogen target with nearly the same interaction length was installed in 2008. In order to get comparable results the fit of 2004 [18] is also not acceptance corrected. Due to different amounts of statistics (2004: $\approx 4 \cdot 10^5$ events, 2008/W37: $\approx 1.4 \cdot 10^7$ events) both data samples are normalized to the integral of the narrow $a_2(1320)$ resonance in the region between 1.1 GeV/c^2 and 1.6 GeV/c^2. A clear difference in population of M substates can be seen in comparison of the two data samples. As an example the total intensites for $J^{PC} = 1^{++}$ with different M projections are shown in Figures 6 and 7. The population of $M = 1$ seems to be significantly higher for the heavy nuclear target (Fig. 6b) than for hydrogen (Fig. 7b). At the same time, the population of $M = 0$ is reduced for lead (Fig. 6a) as compared to hydrogen (Fig. 7a). The total intensity for both projections, however, is rather similar (Fig. 6c and Fig. 7c).

Further studies were done in 2009 with the same spectrometer layout like in 2008 but with different targets. Data were taken with a new lead target composed of thin lead foils as well as with the Primakoff target consisting of Nickel and Tungsten. The analysis of these data could help to clarify the processes leading to this surprising M-dependence.

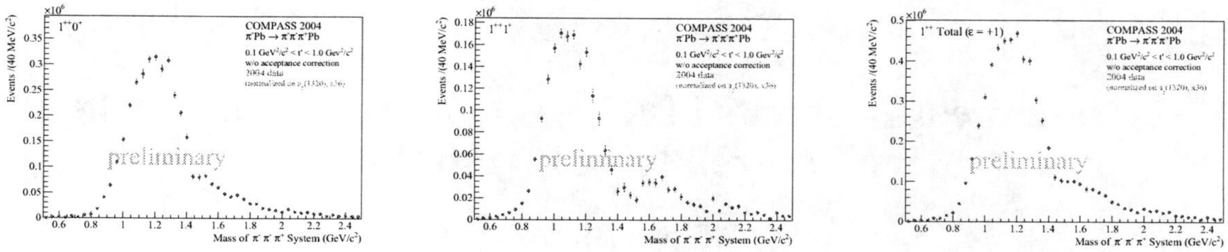

(a) (b) (c)

FIGURE 6. Total intensity for $J^{PC} = 1^{++}$ with $M = 0$ (a), $M = 1$ (b), and for both M projections (c), for the lead target.

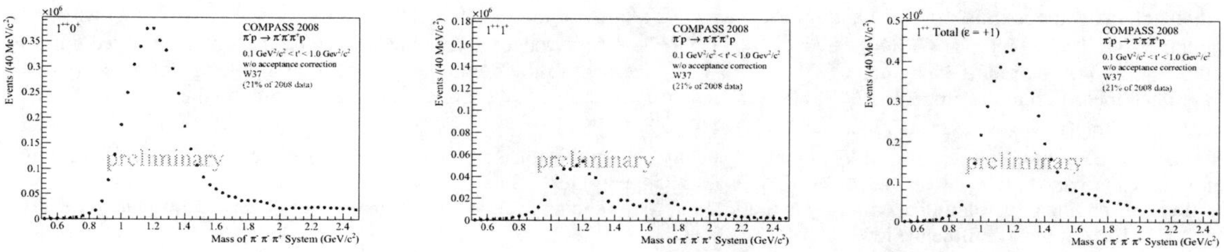

(a) (b) (c)

FIGURE 7. Total intensity for $J^{PC} = 1^{++}$ with $M = 0$ (a), $M = 1$ (b), and for both M projections (c), for the liquid hydrogen target, the same scale is used as in Fig. 6.

CONCLUSIONS

The COMPASS spectrometer is a powerful tool to investigate the light meson spectrum. Within a few days of data taking during a pilot run in 2004 enough statistics could be collected to perform a partial wave analysis of $\pi^-\pi^+\pi^-$ final states. A strong signal of the spin-exotic $\pi_1(1600)$ was observed in the $1^{-+}1^+\rho\pi P$ wave. A mass-dependent fit results in a mass of 1660 ± 10(stat) MeV/c^2 and a width of 269 ± 21(stat) MeV/c^2.
Two years of data taking, 2008 and 2009, dedicated to meson spectroscopy will improve the world statistics on diffractive by a factor of 10. First results of diffractive dissociation on hydrogen into $\pi^-\pi^+\pi^-$ final states were shown. A striking, but yet unexplained, dependence of the production strength for different M substates of the major waves on the target material was observed.

ACKNOWLEDGMENTS

This work is supported by the the German Bundesministerium für Bildung und Forschung, the Maier-Leibnitz-Labor der LMU und TU München, and the DFG cluster of excellence *Origin and Structure of the Universe*.

REFERENCES

1. K. J. Juge, J. Kuti and C. Morningstar, AIP Conf. Proc. **688**, 193 (2004)
2. D. R. Thompson *et al.*, Phys. Rev. Lett. **79**, 1630 (1997)
3. V. Dorofeev *et al.*, AIP Conf. Proc. **619**, 143 (2002)
4. A. Abele *et al.*, Phys. Lett. **B423**, 175 (1998)
5. A. Abele *et al.*, Phys. Lett. **B446**, 349 (1999)
6. G. S. Adams *et al.*, Phys. Rev. Lett. **81**, 5760 (1998)
7. S. U. Chung *et al.*, Phys. Rev. **D65**, 072001 (2002)
8. Y. Khokhlov, Nucl. Phys. **A663**, 596 (2000)
9. G. M. Beladidze *et al.*, Phys. Lett. **B313**, 276 (1993)
10. E. I. Ivanov *et al.*, Phys. Rev. Lett. **86**, 3977 (2001)
11. J. Kuhn *et al.*, Phys. Lett. **B595**, 109 (2004)
12. D. V. Amelin *et al.*, Phys. Atom. Nucl. **68**, 359 (2005)
13. M. Lu *et al.*, Phys. Rev. Lett. **94**, 032002 (2005)
14. A. R. Dzerbia *et al.*, Phys. Rev. **D73**, 072001 (2006)
15. P. Abbon *et al.*, Nucl. Instr. Meth. A **577**, 455 (2007)
16. A. Alekseev *et. al.*, COMPASS Collaboration, Phys. Rev. Lett. **104**, 241803 (2010)
17. S. U. Chung and T. L. Trueman, Phys. Rev. **D11**, 633 (1975)
18. Q. Weitzel, Dissertation Technische Universität München, `http://nbn-resolving.de/urn/resolver.pl?urn:nbn:de:bvb:91-diss-20080801-668430-1-0` (2008)

Precision measurements of $K^\pm \to \pi^\pm\pi^0\gamma$ decay at NA48/2 and search for CP violation

Andrea Bizzeti

Dipartimento di Fisica, Università di Modena e Reggio Emilia, via G. Campi 213/A, I-41100 Modena, Italy
Istituto Nazionale di Fisica Nucleare, Sezione di Firenze, via G. Sansone 1, I-50018 Sesto Fiorentino (FI), Italy

Abstract. The direct emission (DE) and interference (INT) terms of the radiative decay $K^\pm \to \pi^\pm\pi^0\gamma$ have been measured by the NA48/2 experiment at CERN SPS. About $6 \cdot 10^5$ $K^\pm \to \pi^\pm\pi^0\gamma$ decays have been selected from data collected during 2003 and 2004. The relative amounts of DE and INT with respect to the dominant Inner Bremsstrahlung (IB) contribution have been measured in the range $0 < T^*_\pi < 80$ MeV, where T^*_π is the kinetic energy of the charged pion in the kaon rest frame, obtaining: $\text{Frac(DE)} = (3.32 \pm 0.15_{\text{stat}} \pm 0.14_{\text{syst}}) \cdot 10^{-2}$, $\text{Frac(INT)} = (-2.35 \pm 0.35_{\text{stat}} \pm 0.39_{\text{syst}}) \cdot 10^{-2}$. This is the first observation of an interference term in $K^\pm \to \pi^\pm\pi^0\gamma$ decays. From this measurement the DE normalized electric and magnetic amplitudes are derived: $X_E = (-24 \pm 6)$ GeV^{-4}, $X_M = (254 \pm 9)$ GeV^{-4}.

In addition, the CP violating asymmetry in the K^+ and K^- branching ratios for this channel has been determined to be less than $1.5 \cdot 10^{-3}$ at 90% confidence level.

Keywords: Kaon decays; radiative decays; CP violation
PACS: 13.20Eb

INTRODUCTION

The radiative decay $K^\pm \to \pi^\pm\pi^0\gamma$ is a very interesting channel for studying the low energy structure of QCD. The total amplitude of this decay is the sum of two terms: the dominant one arises from the inner bremsstrahlung (IB) associated to the decay $K^\pm \to \pi^\pm\pi^0$ with the photon emitted from the outgoing charged pion, the other is due to direct emission (DE) in which the photon is radiated in the intermediate states of the decay.

The branching fraction of the IB component can be predicted from that of the $K^\pm \to \pi^\pm\pi^0$ decay by using the Low theorem[1] and QED corrections[2, 3]. As the $K^\pm \to \pi^\pm\pi^0$ decay is suppressed by the $\Delta I = 1/2$ rule, the IB component of the $K^\pm \to \pi^\pm\pi^0\gamma$ decay will be similarly suppressed, resulting in a relative enhancement of the DE contribution.

The DE term has been extensively studied in the framework of Chiral Perturbation Theory (ChPT)[4, 5, 6, 7, 8, 9]. Direct photon emission can occur through both electric and magnetic dipole transitions. The electric dipole transition interferes with the IB amplitude giving rise to an interference term (INT), which can have CP violating contributions. In ChPT, DE arises only at order p^4 and cannot be evaluated in a model-independent way. The magnetic part is the sum of two anomalous amplitudes: one reducible, that can be calculated using the Wess-Zumino-Witted functional[10, 11], and one direct amplitude, whose size is not model independent but is expected to be small. The electric transition amplitude depends on unknown constants that cannot be determined by symmetry requirements alone. A measurement of both DE and INT terms allows the determination of both the electric and magnetic contributions.

The properties of the $K^\pm \to \pi^\pm\pi^0\gamma$ decay can be conveniently described using the T^*_π and W kinematic variables, where T^*_π is the kinetic energy of the charged pion in the kaon rest frame and W is a Lorentz invariant variable given by:

$$W^2 = \frac{(P^*_K \cdot P^*_\gamma)\,(P^*_\pi \cdot P^*_\gamma)}{m_K^2\, m_\pi^2} \tag{1}$$

where P^*_K, P_π and P_γ are the 4-momenta of the kaon, the charged pion and the photon, respectively. The differential decay width can be written as:

$$\frac{\partial^2\Gamma^\pm}{\partial T^*_\pi \partial W} = \frac{\partial^2\Gamma^\pm_{\text{IB}}}{\partial T^*_\pi \partial W}\left[1 + 2\cos(\pm\phi + \delta^1_1 - \delta^2_0)\, m_\pi^2 m_K^2\, X_E\, W^2 + m_\pi^4 m_K^4\, \left(|X_E|^2 + |X_M|^2\right) W^4\right] \tag{2}$$

12th International Conference on Meson-Nucleon Physics and the Structure of the Nucleon (MENU 2010)
AIP Conf. Proc. 1374, 278-281 (2011); doi: 10.1063/1.3647142

where $\partial^2\Gamma^{\pm}_{\text{IB}}/\partial T^*_\pi \partial W$ is the differential rate for the IB component, ϕ is the CP violating phase, δ^I_l are the strong rescattering phases for a pion-pion state of isospin I and orbital angular momentum l, X_E and X_M are the normalized electric and magnetic amplitudes of the DE; ϕ, X_E and X_M are constants, while δ^1_1 and δ^2_0 are known functions of the $\pi\pi$ invariant mass.

The three terms in the sum represent IB, INT and DE contributions, respectively. Recently, a presence of a form factor in the pole part of the magnetic amplitude has also been suggested[18].

The different W dependence of the IB, INT and DE terms in eq. 2 allows the extraction of the different decay components. In particular, the ratio of the data W distribution to the one simulated for the Inner Bremsstrahlung (MC(IB)) component can be parametrized by:

$$\text{Data/MC(IB)} = c[1 + (a \pm e)W^2 + bW^4] \quad . \tag{3}$$

The c parameter represents an overall normalization factor, the b coefficient is related to the fraction of the Direct Emission and the coefficients a and e parametrize the CP conserving and CP violating parts of the Interference term, respectively. As it is expected that $e \ll a$, to a good approximation the CP violating contribution can be neglected in the study of the interference term. A comparison of the K^+ and K^- decays is needed in order to measure the small e term.

Up to now the DE contribution has only been measured[13, 14, 15, 16] ignoring the INT term, in a limited kinematic range 55 MeV $< T^*_\pi <$ 90 MeV, which is less affected by $K^\pm \to \pi^\pm\pi^0(\pi^0)$ background[1].

THE NA48/2 EXPERIMENT

The NA48/2 experiment has been taking data in the years 2003 and 2004, detecting in-flight decays of charged kaons in two simultaneous oppositely charged beams of (60 ± 3) GeV/c momentum produced as secondary beams at CERN SPS. The charged particle reconstruction is provided by a magnetic spectrometer, consisting of a dipole magnet and four drift chambers, with a spatial resolution of 100 μm and a momentum resolution $\Delta p/p = (1.0 \oplus 0.044p[\text{GeV/c}])\%$.

The energy and position of photons and electrons are precisely measured by a Liquid Krypton electromagnetic calorimeter, consisting of a $27X_0$ almost homogeneous ionization chamber with high-granularity tower read-out: its energy resolution is $\Delta E/E = 3.2\%/\sqrt{E[GeV]} \oplus 9\%/E[GeV] \oplus 0.42\%$ and its spatial resolution about 1.5 mm.

A scintillator hodoscope for fast triggering and precise time measurement, muon and photon veto counters and an iron-scintillator hadron calorimeter complete the experimental apparatus, a detailed description of which can be found in Ref. [19].

MEASUREMENT OF THE $K^\pm \to \pi^\pm\pi^0\gamma$ DECAY

In order to improve the sensitivity to DE and INT components, events have been selected in a wide kinematic region (defined by $T^*_\pi <$ 80 MeV and $E_\gamma >$ 5 GeV), containing a large fraction of DE and INT spectra (see Fig. 1). An almost background-free data sample of $6.0 \cdot 10^5$ $K^\pm \to \pi^\pm\pi^0\gamma$ candidate events has been selected.

The measured W distribution, shown in the left part of Fig. 2, has been fitted to the sum of MonteCarlo simulated W spectra for IB, DE and INT. By performing a Poissonian Maximum Likelihood fit in the range $0.2 < W < 0.9$ we obtain the following values for the fractions of DE and INT, normalized to IB:

$$\text{Frac(DE)} \equiv \text{BR(DE)/BR(IB)} = (\ 3.32 \pm 0.15_{\text{stat}} \pm 0.14_{\text{syst}}) \cdot 10^{-2} \tag{4}$$

$$\text{Frac(INT)} \equiv \text{BR(INT)/BR(IB)} = (-2.35 \pm 0.35_{\text{stat}} \pm 0.39_{\text{syst}}) \cdot 10^{-2} \tag{5}$$

with a correlation coefficient $\rho = -0.93$. The residuals of this fit are shown in the right part of Fig. 2. Contour plots corresponding to probabilities of 68.3% , 95.5% and 99.7% in the DE-INT plane are shown in the left part of Fig. 3

From these fractions the normalized electric and magnetic amplitudes are then extracted, obtaining: $X_E = (-24 \pm 4_{\text{stat}} \pm 4_{\text{syst}})\text{GeV}^{-4}$, $X_M = (254 \pm 6_{\text{stat}} \pm 6_{\text{syst}})\text{GeV}^{-4}$, with correlation coefficient $\rho = -0.83$.

---

[1] This cut is effective because $T^*_\pi(K^\pm \to \pi^\pm\pi^0) \sim 110$ MeV and $T^*_\pi(K^\pm \to \pi^\pm\pi^0) < 53$ MeV.

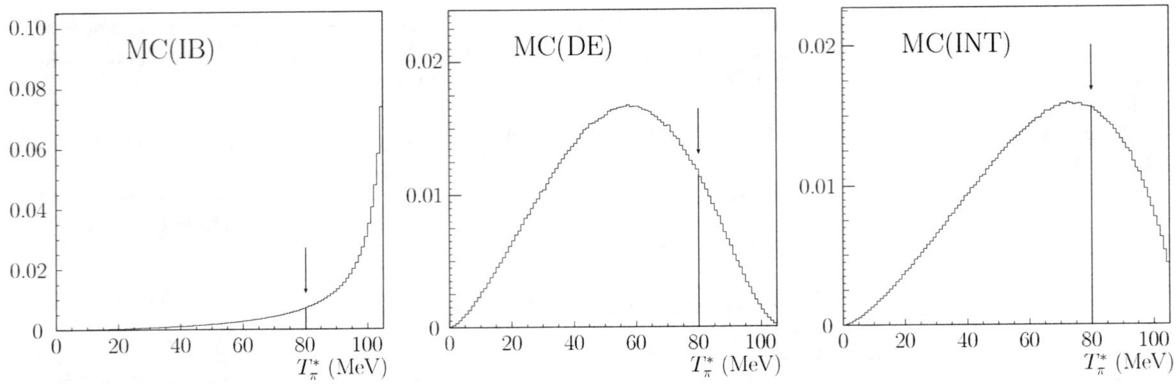

FIGURE 1. T^*_π distribution for Monte Carlo simulated IB, DE and INT components of the $K^\pm \to \pi^\pm\pi^0\gamma$ decay. The vertical line indicates the upper boundary of the region selected in this analysis ($0 < T^*_\pi < 80$ MeV).

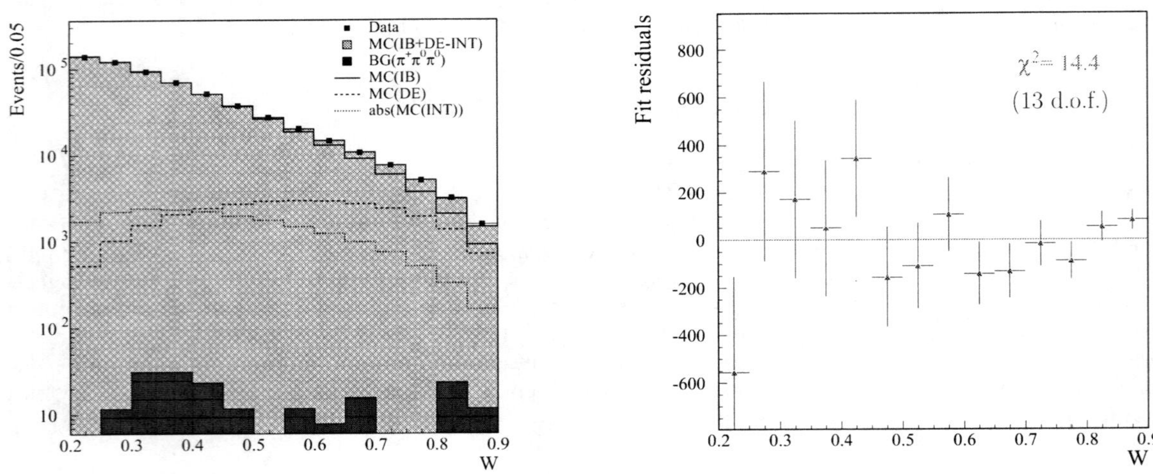

FIGURE 2. Left: W distributions for data (*squares*) and sum of simulated signal and background (*light grey* histogram). The *lines* show the signal contributions, the *dark grey* histogram the simulated $K^\pm \to \pi^\pm\pi^0\pi^0$ background. Right: Residuals of the Poissonian Maximum Likelihood fit.

As a crosscheck, the ratio of the W distributions of data to IB simulation, shown in the right part of Fig. 3, has been fitted to the function defined in (3). From the a and b parameters of the fit we obtain: Frac(DE) = $(3.19 \pm 0.16_{\text{stat}}) \cdot 10^{-2}$, Frac(INT) = $(-2.21 \pm 0.41_{\text{stat}}) \cdot 10^{-2}$, in good agreement with the result obtained with the Maximum Likelihood method.

SEARCH FOR CP VIOLATION

As the decay of $K^\pm \to \pi^\pm\pi^0\gamma$ with direct photon emission is not suppressed by the $\Delta I = 1/2$ rule, it as always been considered a good channel to search for CP violation[3, 21]. Present experimental knowledge of the asymmetry is $(0.9 \pm 3.3) \cdot 10^{-2}$[20].

In order to measure CP asymmetry in $K^\pm \to \pi^\pm\pi^0\gamma$ decays, the event selection has been modified by lowering the minimum photon energy to 3 GeV and removing the $W > 0.2$ cut. The data sample selected for this analysis contains $6.95 \cdot 10^5$ $K^+ \to \pi^+\pi^0\gamma$ and $3.86 \cdot 10^5$ $K^- \to \pi^-\pi^0\gamma$ events.

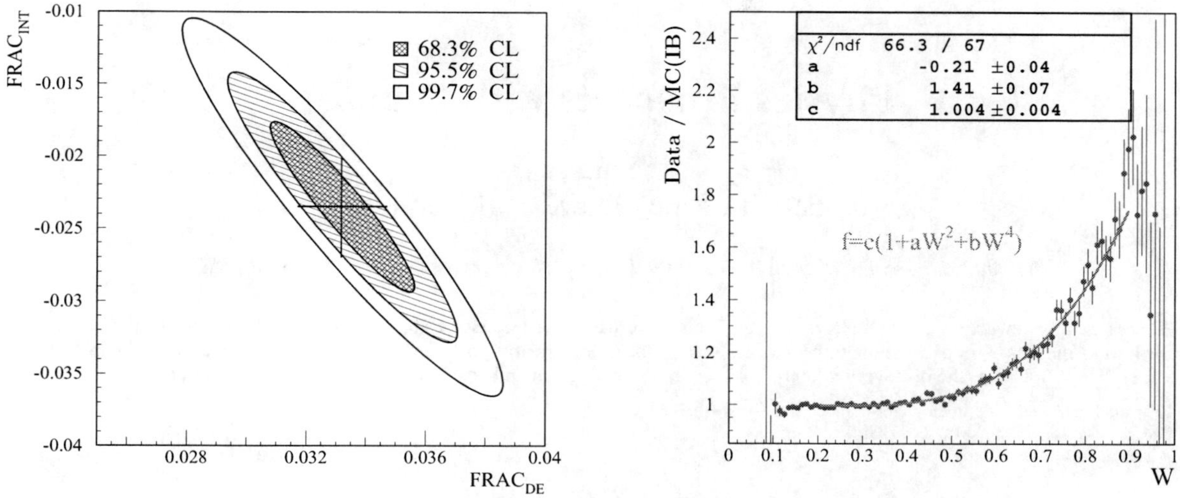

FIGURE 3. Left: The measured DE and INT components of the $K^{\pm} \to \pi^{\pm}\pi^{0}\gamma$ decay. Right: Ratio of W distributions for data events to IB simulated events. The *solid line* shows the result of the fit in the range $0.2 < W < 0.9$.

The asymmetry in the decay rates of K^{+} and K^{-} is defined as

$$A_N \equiv \frac{\Gamma^{+} - \Gamma^{-}}{\Gamma^{+} + \Gamma^{-}} = \frac{N_{+} - R \cdot N_{-}}{N_{+} + R \cdot N_{-}} \quad , \tag{6}$$

where Γ^{+} and Γ^{-} are the $K^{+} \to \pi^{+}\pi^{0}\gamma$ and $K^{-} \to \pi^{-}\pi^{0}\gamma$ partial decay widths, N_{+} and N_{-} are the numbers of these decays in the data sample, and R is the ratio of the number of K^{+} to that of K^{-} in the beam.

We obtain a rate asymmetry $A_N = (0.0 \pm 1.0_{\text{stat}} \pm 0.6_{\text{syst}}) \cdot 10^{-3}$, corresponding to a 90% CL limit $|A_N| < 1.5 \cdot 10^{-3}$. From this measurement the value of the CP violating phase ϕ is deduced: $\sin(\phi) = -0.01 \pm 0.43$, corresponding to a 90% CL limit $|\sin(\phi)| < 0.56$.

REFERENCES

1. F.E. Low, *Phys. Rev.* **110**, 974 (1958).
2. N. Christ, *Phys. Rev.* **159**, 1292 (1967).
3. G. D'Ambrosio, M. Miragliuolo and P. Santorelli, in *The DaΦne Physics Handbook*, ed. by L. Maiani, G. Pancheri and N. Paver, 1992.
4. H.Y. Cheng and S.C. Lee, *Z. Phys. C* **41**, 72 (1987).
5. H.Y. Cheng, *Phys. Rev. D* **42**, 72 (1990).
6. G. Ecker, A. Pich and E. De Rafael, *Nucl. Phys. B* **303**, 665 (1988).
7. G. Ecker, H. Neufeld and A. Pich, *Phys. Lett. B* **278**, 337 (1992).
8. G. Ecker, H. Neufeld and A. Pich, *Nucl. Phys. B* **413**, 321 (1994).
9. C. Bruno and J. Prades, *Z. Phys. C* **57**, 585 (1993).
10. J. Wess and B. Zumino, *Phys. Lett. B* **57**, 585 (1971).
11. E. Witten, *Nucl. Phys. B* **223**, 422 (1983).
12. R. J. Abrams *et al.*, *Phys. Rev. Lett.* **29**, 1118 (1972).
13. S. C. Adler *et al.*, *Phys. Rev. Lett.* **85**, 4856 (2000).
14. M. A. Aliev *et al.*, *Phys. Lett. B* **554**, 7 (2003).
15. M. A. Aliev *et al.*, *Eur. Phys. J. C* **46**, 61 (2006).
16. V. A. Uvarov *et al.*, *Phys. Atom. Nucl.* **69**, 26 (2006).
17. S.L. Glashow, J. Iliopoulos and L. Maiani, *Phys. Rev. D* **2**, 1285 (1970).
18. L. Cappiello and G. D'Ambrosio, *Phys. Rev. D* **75**, 094014 (2007).
19. V. Fanti *et al.*, *Nucl. Instr. Methods A* **574**, 433 (2007).
20. C. Amsler *et al.*, *Phys. Lett. B* **667**, 1 (2008).
21. G. Costa and P.K. Kabir, *Phys. Rev. Lett.* **18**, 429 (1967).

Physics Prospects with GlueX

A. Somov
(on behalf of the GlueX Collaboration)

Thomas Jefferson National Accelerator Facility, Newport News, Virginia 23606, USA

Abstract. The new experiment GlueX is currently being constructed at Jefferson Lab. The experiment was designed to search for hybrid mesons with exotic-quantum-numbers using a beam of linearly polarized photons incident on a liquid hydrogen target. We will discuss the discovery potential of the GlueX experiment and briefly overview its physics program.

Keywords: GlueX experiment, hybrid mesons
PACS: <12.38.Qk, 25.20.Lj, 29.27.Hj, 13.60.-r

INTRODUCTION

Understanding the role of gluons in the confinement of quarks in quantum chromodynamics (QCD) is one of the most tantalizing topics in modern particle physics to be explored. Perhaps the clearest way would be the experimental observations of the quark/gluon states whose quantum numbers (J^{PC}) are forbidden by Quark Model and comparison their properties with detailed QCD calculations. Such states could be glueballs, multi-quark molecules ($q\bar{q}q\bar{q}$), or so-called hybrid mesons ($q\bar{q}g$), where g represents an excitation of the gluonic field binding the $q\bar{q}$ system. Experimental observation of several hybrid meson candidates with exotic quantum numbers brought them into research focus of many theoreticians and stimulated further experimental search. The origin of exotic quantum numbers can be explained based on the so-called flux tube model [1], the model which has so far been successfully verified for heavy quarks using lattice QCD calculations [2]. In the flux tube model, the gluonic field, or the "flux tube" can have an angular momentum, m, which can be combined with the quantum numbers of a $q\bar{q}$ system. When the flux tube is in the ground state, m = 0, the fermion-antifermion system ($q\bar{q}$) can be characterized by specific quantum numbers, J^{PC}, corresponding to the conventional mesons, where J is the total angular momentum, P is parity and C is C-parity. If the flux tube is in the first excited state, m = 1, in addition to the quantum numbers of the normal mesons, the $q\bar{q}g$ system can have exotic quantum numbers, such as $J^{PC} = 0^{+-}$, 1^{-+}, 2^{+-}. Each of these J^{PC} states corresponds to a nonet of mesons. Recent lattice calculations predict masses of hybrid mesons with $J^{PC} = 0^{+-}$, 1^{-+}, 2^{+-} to be between 1.8 GeV and 2.8 GeV and $J^{PC} = 1^{-+}$ to be the lightest state. This predictions define the energy region for the experimental search of light hybrid mesons. Experimental observation of the nonets of the exotic-quantum-number mesons across several final states would certainly help in interpretation of the observed states and provide important consistency checks of theoretical models.

Exotic mesons have been extensively searched during the past few decades by several experiments using different beam particles and energies. Several experiments have reported on observing three exotic states $\pi_1(1400)$, $\pi_1(1600)$, and $\pi_1(2015)$ with the quantum number $J^{PC} = 1^{-+}$. The $\pi_1(1400)$ state has been seen by most experiments in $\eta\pi$ decay channel only. This does not agree with the expected decay modes of hybrid mesons. Furthermore, the $\pi_1(1400)$ mass is also smaller than the model predictions. The $\pi_1(1600)$ state is more consistent with the hybrid meson. It has been seen in $b_1\pi$, $f_1\pi$, and $\eta'\pi$ final states by VES and E852 collaboration. However, the $\rho\pi$ decay channel is controversial. This decay mode had originally been seen by the E852 experiment but was ruled out by the later analysis. Meantime, the COMPASS collaboration confirmed the $\rho\pi$ decay. No $\pi_1(1600) \rightarrow \pi^+\pi^-\pi^+$ signal has been observed by CLAS collaboration in photoproduction at about 6 GeV. The third hybrid meson candidate is $\pi_1(2015)$, whose observation was reported by the E852 experiment in $f_1\pi$ and $b_1\pi$ final states with limited statistics. Obviously, more data are needed to understand the nature of the discovered exotic states. More information on the current status of the meson spectroscopy can be found in [3]. Two new experiments dedicated to search for states with exotic quantum numbers are currently under construction: PANDA at GSI and GlueX at Jefferson Lab. Panda will perform a search for exotic states in $p\bar{p}$ annihilation while GlueX will use a beam of linearly polarized photons.

The production of exotic mesons is expected to be enhanced in γp interactions, where the experimental data is very limited. Naively, this can be explained by combining the quantum number of a beam photon $J^{PC} = 1^{--}$, which,

12th International Conference on Meson-Nucleon Physics and the Structure of the Nucleon (MENU 2010)
AIP Conf. Proc. 1374, 282-285 (2011); doi: 10.1063/1.3647143

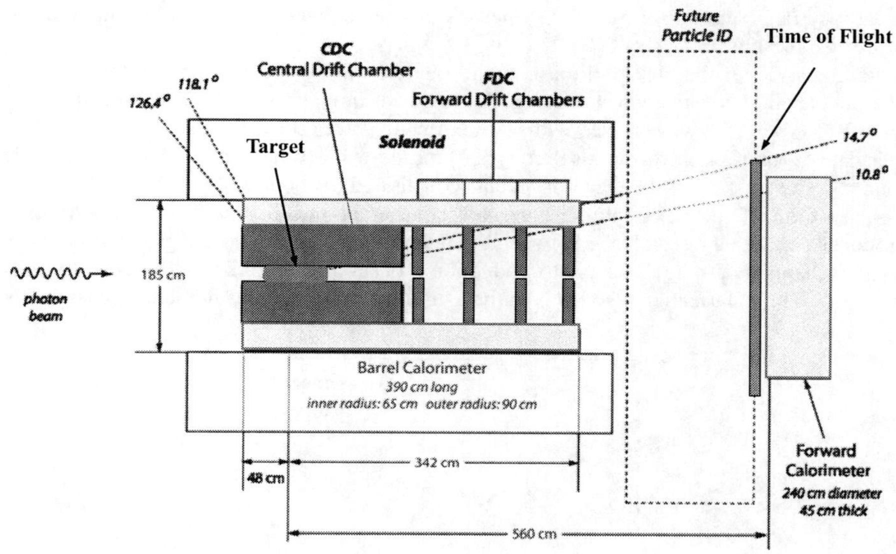

FIGURE 1. Layout of the GlueX detector.

according to the vector dominance model can be presented as a virtual $q\bar{q}$ pair with $L = 0$ and $S = 1$, with the quantum numbers of the excited flux tube, 1^{+-} or 1^{-+}. As a results, the states with exotic J^{PC} can be directly formed. However, in the case of the π beam, production of exotic mesons is expected to be suppressed because a spin flip of the quark is required. Large photoproduction cross-sections for exotic mesons are also predicted by recent lattice calculations of the radiative decay of charmonium states into hybrid mesons with exotic quantum numbers [4].

The upgrade of the CEBAF electron beam energy from 6 GeV to 12 GeV will provide a unique high-intensity beam of linearly polarized photons produced using a Bremsstrahlung technique. The GlueX experiment at Jefferson Lab has been designed to search for mesons with exotic-quantum-numbers using this beam. The GlueX experiment will collect a data sample a few order of magnitudes larger than all existing photoproduction data.

GLUEX EXPERIMENT

The photon beam used by the GlueX experiment will be produced via coherent Bremsstrahlung process by 12 GeV electrons incident on a $20\mu m$ thick diamond radiator. In order to increase the fraction of linearly polarized photons in the energy region of interest between 8.4 GeV and 9.1 GeV (the coherent peak region), the photon beam will be passed through a collimator situated about 75 m downstream from the radiator. The collimated photons will subsequently be sent to the GlueX liquid hydrogen target. The photon beam energy of ~ 9 GeV will provide enough statistics to study the spectrum of exotic meson candidates up to about 2.8 GeV/c^2. The photon energy will be determined with an accuracy of about 0.1% by measuring the energy of the electron after radiation. The GlueX detector will start operating using the flux of tagged photon of 10^7 γ/s (8.4 - 9 GeV) on the target. Later, after installing a computer farm that will be used in the Level-3 trigger system, the beam intensity will be increased to 10^8 γ/s.

GlueX is a magnetic spectrometer, which will be located in the newly constructed experimental Hall-D at Jefferson Lab. The detector design has been optimized to provide almost a 4π coverage for charged tracks and photons originating from the exotic meson candidates. The spectrometer layout is presented in Fig. 1. The central part of the detector will be situated inside the solenoid magnet operated at 2.2 T field. Charged tracks are reconstructed with the Central and Forward drift chambers, which cover the polar angle of the reconstructed tracks between $1^\circ \leq \theta \leq 155^\circ$ in the lab frame. Reconstruction of photons is performed using lead-scintillator (barrel) and lead glass (forward) calorimeters. The spectrometer will provide high detection efficiency for both charged tracks and photons and, therefore, will allow us to reconstruct exotic meson candidates in several final states. The angular acceptance of the detector is relatively large and uniform for most mesons of interest, which is particular important for the angular analysis of the observed mesons.

The quantum numbers J^{PC} of the observed mesons will be determined using a partial wave analysis (PWA). The GlueX PWA software is being intensively tested using Monte Carlo simulations. An example of the detector capability to reconstruct an exotic $J^{PC} = 1^{-+}$ wave mixed with six non-exotic waves corresponding to $\gamma p \to \pi^+\pi^-\pi^+ n$ events is presented in Fig.3. The exotic wave corresponds to $\pi_1(1600)$ meson, while other amplitudes were taken from the states such as a_1, a_2, and π_2. The fraction of the signal π_1 events constituted to about 2.5% of the total number of simulated events. Linear polarization of the beam photons provides important information for the PWA. In particular, the polarization introduces azimuthal dependence between the polarization vector and the production plane that depends on the parity of the exchanged particle in the photoproduction process and, therefore, allows one to constrain the final state quantum numbers. The polarization also leads to the azimuthal dependencies in the decay angle distributions.

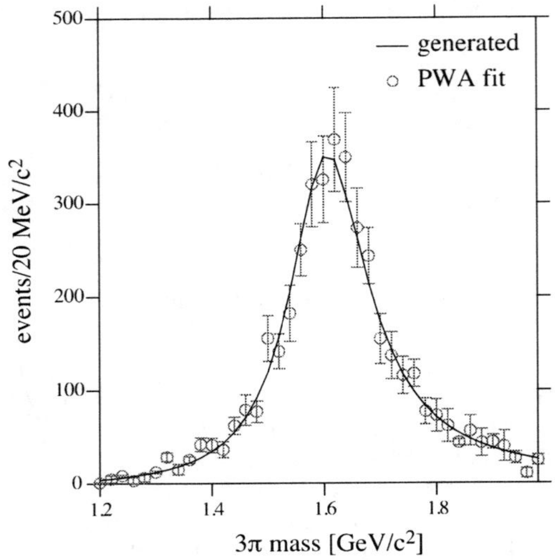

FIGURE 2. Results of the reconstruction of the $J^{PC} = 1^{-+}$ ($\pi_1(1600)$) exotic wave mixed with six non-exotic waves corresponding to the $\gamma p \to \pi^+\pi^-\pi^+ n$ events. The statistics correspond to a few days of running.

OTHER PHYSICS TOPICS WITH GLUEX

Being optimized for the meson spectroscopy, the GlueX detector is well-suited for studies of other physics topics such as photoproduction with nuclear targets and exclusive reactions at high-momentum transfer. The physics program of the GlueX is being continuously developing and is widely discussed [6]. Recently, a new experiment with the GlueX detector, PrimeX [7], has been approved by Jefferson Lab Program Advisory Committee. The experiment is aimed to measure the $\eta \to \gamma\gamma$ decay width via the Primakoff effect. The Primakoff effect represents production of mesons in interactions of the real beam photon with the virtual photon from the Coulomb field of the nucleus. The decay width can be extracted from the eta production angular distribution, which is presented in Fig.3. The measurements will allow to significantly improve the precision of the ratios of the light quark masses and the value of the $\eta - \eta'$ mixing angle.

OUTLOOK

GlueX is a new experiment at Jefferson Lab. whose physics program is intended to improve our knowledge of strong interactions. The main goal of the experiment is to search for gluonic excitations in photoproduction. The experiment is expected to collect a data sample a few order of magnitudes larger than all existing photoproduction data. The physics topics of the experiment spans from light meson spectroscopy to Primakoff production of pseudoscalar mesons. The construction of the experiment has started in 2009 and the commissioning stage is expected to be finished in 2015.

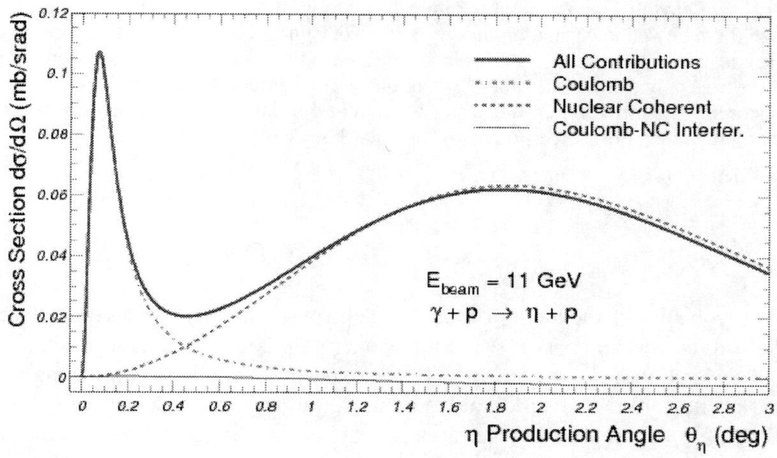

FIGURE 3. Differential cross sections for $\gamma p \rightarrow \eta p$ reaction. The curves show contribution from different processes: the dot-dashed curve corresponds to the Primakoff production, the dashed curve curve corresponds to the nuclear coherent process, the small solid curve is the interference between the Primakoff and nuclear coherent, and the solid curve is the total.

ACKNOWLEDGMENTS

This work was supported by the US Department of Energy contract DE-AC05-06OR23177, under which Jefferson Science Associates, LLC operates Thomas Jefferson National Accelerator Facility. I would like to thank my colleagues from the GlueX collaboration for their help preparing this proceedings.

REFERENCES

1. N. Isgur and J. Paton, Phys.Rev.D **31**, 2910, (1985),
   N. Isgur, R. Kokoski and J.Paton, Phys.Rev.Lett. **54**, 869, (1985).
2. G. Bali *et al.* [SESAM Collaboration], Nucl.Phys.Proc.Suppl. **63**, 209, (1998).
3. C. A. Meyer, and Y. Van Haarlem, *The Status of Exotic-quantum-number Mesons*, arXiv:1004.5516, (2010),
   E. Klempt and A. Zaitsev, Phys.Rep.**454**, 1, (2007).
4. J. J. Dudek, R. Edwards and C. E. Thomas, Phys.Rev.D **79**, 094504 (2009).
5. J. J. Dudek, R. Edwards, and C. E. Thomas, Phys.Rev.D bf 79,094504 (2009).
6. *Workshop on Photon-hadron physics with the GlueX detector*, March 8, (2008),
   `http://conferences.jlab.org/php2008/index.html`.
7. *A precision measurement of the η radiative decay width via Primakoff effect*, Proposal, Dec 2009.

Tetraquark states from lattice QCD

Nilmani Mathur

Department of Theoretical Physics,
Tata Institute of Fundamental Research,
Homi Bhabha Road, Mumbai 400005, India

Abstract. Recently there have been considerable interests in studying hadronic states beyond the usual two and three quark configurations. With the renewed experimental interests in $\sigma(600)$ and the inability of quark model to incorporate too many light scalar mesons, it is quite appropriate to study hadronic states with four quark configurations. Moreover, some of the newly observed charmed hadrons may well be described by four quark configurations. Lattice QCD is perhaps the most desirable tool to adjudicate the theoretical controversy of the scalar mesons and to interpret the structures of the newly observed charmed states. Here we briefly reviewed the lattice studies of four-quark hadrons.

Keywords: tetraquark, lattice QCD
PACS: 12.39.Mk, 14.40.Be, 14.40.Lb, 14.40.Rt, 12.38.Gc

INTRODUCTION

Besides meson and baryon spectra the theory of quantum chromodynamics allows the existence of other states, such as four-quark states, hybrid mesons and baryons (bound by excited gluon) along with glueballs. Indeed lattice calculations predict many such states, and various ongoing and future experimental investigations may shade light on the existence of those. A four-quark state could have either molecular or tetraquark in quark structure. By molecular state one means mesonic molecules of type $(\bar{q}q)(\bar{q}q)$, and a tetraquark state is interpreted with quark structure $(\bar{q}\bar{q})(qq)$ which may have a different decay and other properties than a molecular state [1]. A molecular or a tetraquark state has large four quark component in the Fock space in the same way a meson has large two quark or a baryon has large three quark component in the Fock space. A given observed state, in principle, could be a mixture of mesonic, four-quark or even glueball states of same quantum number in different proportion. Recently a lattice study has been performed in charm meson sector [2] to observe such mixing and that will be discussed later.

Phenomenologically the study of molecular/tetraquark states are relevant for two main reasons: old one–it is difficult to explain the light quark scalar resonances with only two quark mesons, and tetraquark phenomenology provide a possible solution to this problem [3, 1, 4, 5]. Beside that the new reason to study such states is the urgency to interpret the structures of many newly observed charmed states discovered recently at various charm factories [6].

Lattice QCD is perhaps the most desirable tool to adjudicate the theoretical controversy surrounding the existence of four-quark states and, if they exist, to study the structure of such states. However, a four-quark state will always be associated with the threshold scattering states and if these decay states lie closely to the resonance state one must separate them out convincingly before concluding for the existence of the bound four-quark states. This procedure is, in general, quite nontrivial and will be discussed in the next section.

IDENTIFYING A RESONANCE STATE ON LATTICE

On a lattice one calculate the mass of a particle from the large time behavior of the two point correlation function where one calculates the energy eigenvalues of the QCD Hamiltonian. However, resonance states are truly dynamical phenomena since they are poles of the S-matrix rather than the energy eigenstates of the Hamiltonian, and therefore, they cannot be identified with a single energy level of the Hamiltonian. Moreover, separation of threshold scattering states from a resonance state is essential to distinguish a resonance state, and on a lattice this procedure is, in general, nontrivial as it involves extraction of more than one states from two point correlation functions. Following methods are available to distinguish a resonance state from a scattering state which may be complimentary to each other.

• In a simple way, to identify a one particle state one can first study the spectrum in a few volumes and then compare those with its known multi-hadron decay channels. A one particle state will have no explicit volume dependence whereas scattering states will have inverse volume dependence and by studying mass and particularly, the spectral weight factor, one may be able to separate out a one particle state from the scattering states [7, 8]. However, such naive volume dependence holds only for non-interacting theory and not for interacting hadrons. Therefore, one has to be careful, unless effective mass plateau is very large, while identifying a one particle state by using such a method [9]. One can also use hybrid boundary condition on a finite-volume lattice to lift the threshold of the two-particle scattering state and hence can separate a bound state which will be unchanged due to such boundary effect.

• Luscher showed that the energies below the inelastic threshold and the scattering amplitude of a two particle system can be related to their energies in a finite volume [7, 10]. In this method, one calculates energy spectrum in several volumes and then relates the finite box energies to the infinite volume phase shifts by Luscher formula. One can show

12th International Conference on Meson-Nucleon Physics and the Structure of the Nucleon (MENU 2010)
AIP Conf. Proc. 1374, 286-289 (2011); doi: 10.1063/1.3647144

that an abrupt rearrangement of the finite volume energy levels happens when the elastic scattering phase shift in infinite volume quickly passes through $\pi/2$, which is a signature of resonance. On the lattice, one thus looks for an avoided level crossing to identify a resonance signal. Utilizing this method, it is possible to extract the phase shift and thereafter the position and the width of the resonance, and it has been demonstrated for various models [11] and recently for ρ resonance [12] in lattice QCD. Here also one may be able to observe change of sign of scattering length as a function of quark mass while extracting spectrum for various quark masses, and that could also be an indication of resonance state [13]. However, if the width of the resonance is large, which is often the case in reality, the above mentioned level crossing will not be very prominent and it will be difficult to get any resonance information.

• It was also shown recently that one can utilize Luscher's method to collect energies for several volumes into momentum bin in energy histograms leading to a probability distribution measuring the relative weight of energy levels for the resonance case to non resonance case [14]. This probability distribution will show peak at resonance positions. This method has shown to be effective to get resonance parameters in a one dimensional model and non-linear sigma model [15]. However, this method is yet to be shown effective for QCD data.

SCALAR RESONANCES OF LIGHT QUARKS

In the light scalar sector, there are too many experimental candidates, as shown in the fig. 1 of Ref. [17] (with 0^{++} column), than the SU(3) classification can produce in the $\bar{q}q$ nonet scheme. One viable solution is that low-lying scalars, such as the $\sigma(600), a_0(980)$, and $f_0(980)$, are tetraquark mesonia whose classification and spectroscopy have been studied in various models [3]. Another suggestion is that $a_0(980)$ and $f_0(980)$ are $K\bar{K}$ molecular states [16]. The inverted spectrum structure of Jaffe [3, 4] implies that the $I = 1$ state $(\bar{u}\bar{s}sd)$ is heavier than the $I = 1$ state $(\bar{u}\bar{d}ds)$ due to $m_s > m_d$, which is in agreement with experiment. Jaffe's tetraquark interpretation also naturally explains the large observed coupling of $a_0(980)$ and $f_0(980)$ to $K\bar{K}$, which is due to the additional valence pair of $\bar{s}s$. One naturally hopes that lattice QCD will be able to clarify the theoretical controversy surrounding the issue and to reveal the nature of the scalar mesons.

Lattice study of light scalar mesons

• First lattice calculation of light scalar mesons with $q^2\bar{q}^2$ configurations was performed by Alfred and Jaffe [4]. Using quenched lattices with Luscher-Weisz gluonic action and D234 fermionic action they looked for $0^{++} q^2\bar{q}^2$ bound states in I = 0 and 2 channels. They used a mass range above strange mass to take the advantage of lifting continuum threshold for the decay $q^2\bar{q}^2 \to (\bar{q}q)(\bar{q}q)$ so that the observation of $q^2\bar{q}^2$ bound state is possible. Their conclusion was that for I=2 channels there is only scattering states while for sufficiently heavy quarks there is a stable four-quark bound state with $J^{PC} = 0^{++}$ and non-exotic flavor quantum numbers.

• The first detailed study of scalar mesons both with $q^2\bar{q}^2$ and $q\bar{q}$ configurations was performed by Mathur *et al.* [17] by using the overlap fermion in the chiral regime with the pion mass as low as 182 MeV in the quenched approximation. In quenched approximation, $\eta'\pi$ state being the lightest state in the $q\bar{q}$ correlators, one needs to separate that to extract the physical scalar state. After that ghost states separation, they found that the chirally extrapolated results are consistent with $a_0(1450)$ being the $u\bar{d}$ meson and $K_0^*(1430)$ being the $u\bar{s}$ meson. With tetraquark interpolating field they found scattering states both for I = 0 and 2 and an additional state at 550 MeV for I = 0 channel. Through the study of volume dependence, they confirmed that this state is a one-particle state, in contrast to the two-pion scattering states and concluded that this extra observed state is a tetraquark mesonium which is quite possibly the $\sigma(600)$ meson.

• Suganuma *et al.* [18] used quenched anisotropic clover lattices in the quark mass range $m_s < m_q < 2m_s$ to study light tetraquark states. They employed the hybrid boundary condition (HBC) method, where anti-periodic and periodic boundary conditions are imposed on quarks and antiquarks, respectively. By applying HBC on a finite-volume lattice, they lifted the threshold of the two-meson scattering state while the mass of a compact four-quark resonance is almost unchanged due to the presence of two anti-quarks. They observed no tetraquark state in the above mass range.

• Loan *et al.* [19] studied the lowest-lying $q^2\bar{q}^2$ hadrons on a quenched improved anisotropic lattice by using the $\pi\pi$ diquark-antidiquark local and smeared operators, and calculated the energy shifts for $I = 0$ (without disconnected diagrams), $I = 1$ and $I = 2$, $\pi\pi$ states in the quark mass range corresponding to pion mass 480 MeV to 820 MeV. They did not observe any bound state in the above mass region.

• A more detailed calculation has recently been performed by Prelovsek *et al.* [20] by utilizing a set of various interpolating operators along with generalized eigenvalue method for extracting eigenstates. Both quenched overlap and dynamical chirally improved fermions were utilized for this calculation. They investigated whether the lightest scalar mesons $\sigma(600)$ and κ have a large tetraquark component $\bar{q}\bar{q}qq$, and searched for possible light tetraquark states with $J^{PC} = 0^{++}$ and $I = 0, 2, 1/2, 3/2$ on the lattice. Except for $I = 0$ and $I = 1/2$ channels only scattering states were observed. For $I = 0$ and $I = 1/2$ channels additional light states were observed, which may be interpreted as the observed resonances sigma and kappa with a sizable tetraquark component.

It is to be noted that in all lattice calculations for $I = 0$ channel, quark single- and double-annihilation (disconnected) diagrams were not included. To arrive at a clear conclusion one must take into account those diagrams in future calculations. Along with that one must separate out a resonance state and a scattering states as prescribed above. The feasibility for pursuing such comprehensive calculations in future is very encouraging [21].

NEWLY OBSERVED CHARM RESONANCES

Recently there have been tremendous activities in charmonium physics with the discovery of many resonances. While the charmonium model has been very successful in describing $c\bar{c}$ bound states, it is impossible to accommodate some of these new states in any conventional potential model, and naturally, these states are challenging our understanding of strong force [6]. However, a consensus is growing on that some of these states have exotic hadronic structure, for example, a meson-meson molecular state or a quark-gluon hybrid state or a tetraquark state.

Among these states most interesting are the following three : $X(3872)$, $Y(4260)$ and $Z^+(4430)$ (of course, a few other states also do not have clear understanding). Due to its given mass, width and J^{PC}, it is impossible to associate the $X(3872)$ with a $c\bar{c}$ state and naturally theorists speculated it as a more exotic state as mentioned above. However, some of its properties exclude it as a hybrid state and more convincing explanation comes from by speculating it as a D^0-$\bar{D}^{*0}$ molecular state or a $(\bar{c}\bar{q})_3(qc)_{\bar{3}}/(\bar{c}c)_8(\bar{q}q)_8$ tetraquark state [22, 1] or even possibly as a threshold cusp. It is also impossible to identify the $Y(4260)$ as a simple $c\bar{c}$ state due to the unavailability of a missing $c\bar{c}$ state with that quantum number at that mass region. Due to its large width, small leptonic coupling, small B decay branching fraction and non observation of identical meson decay modes, many theorists have identified it as a $\bar{c}gc$ hybrid state, though this identification is not conclusive yet. Lattice QCD also predicts a hybrid state above 4 GeV, though the exact mass of that state has not been settled yet.

Belle also reported the existence of a charged state with $c\bar{c}$ flavors and at mass 4433(4)(3) and width 45(18)(30) MeV [23] though it has not been reproduced in Babar [24]. Since it is a charged state it cannot be purely charmonium or $c\bar{g}c$ hybrid state, and a natural explanation for it will be a molecular or tetraquark state. Experimentally a detail study is very essential as that will provide the first proof for the existence of such a non trivial hadronic state of matter.

Here again, lattice QCD will be a very appropriate tool to study heavy four-quark and hybrid states of QCD and to discern the structure of such states. However, study of heavy four-quark states will be more difficult than the light quark sector as the threshold states for heavy quarks could lie very close to the resonance states: for example, the lowest DD^* scattering state has almost the same mass as X(3872). It will thus be highly non-trivial to separate a resonance state from a nearly degenerate scattering state. Below are the lattice attempts to study heavy four-quark states :

- With quenched overlap fermions TWQCD group [25] studied states with quark structures $c\bar{q}_1c\bar{q}_2$, with $q_{1,2}$ as u, d, s or c and with J^{PC} : 1^{++} and 1^{--}. They calculated the ground state energy in two spatial lattices and compared the ratio of the spectral weights to check the volume dependence of the extracted states. Their extracted masses for $c\bar{q}c\bar{q}$ 1^{++}, $c\bar{s}c\bar{s}$ 1^{++} and $c\bar{q}c\bar{q}$ 1^{--} states to be 3890 ± 30, 4100 ± 50, and 4238 ± 31 MeV respectively, with spectral weight ratio to be one for each case. They concluded these states to be X(3872) as $c\bar{q}c\bar{q}\,1^{++}$ and Y(4260) to be $c\bar{q}c\bar{q}\,1^{--}$. However, as mentioned above, one needs to be very careful to conclude these states as tetraquark state due to the close proximity of threshold states. One must find these scattering states along with bound resonance, if there is anything. Also it is to be noted that, as mentioned above, one should take special care, for example, to avoid shorter time effective mass plateau, to conclude the nature of a state from the volume dependence of spectral weights.
- Meng *et al* [26] studied tetraquark states with $J^P = 0^-, 1^-$, and 2^- channels on quenched anisotropic clover lattices. They studied D_1D^* threshold and tried to find a resonance-like state to identify $Z^+(4430)$. It is argued that, albeit the interaction between the two charmed mesons being attractive (scattering length parameter $a_0 \sim 2.52(47) fm$), it is unlikely that they can form a shallow bound state in the $J^P = 0^-$ channel.
- It is important to calculate the mixing between $c\bar{c}$ states near threshold with $D\bar{D}$ molecular (or tetraquark) states to get information about higher Fock state contributions to charmonium wavefunctions and on mass shifts. Bali *et al.* [2] studied such mixing effects and in the channels ($0^{-+}, 1^{--}$, and 1^{++}) they observed significant mixing effects between $c\bar{c}$ and four-quark states. The procedure adopted there will be very useful for future study regarding the Fock space components of various observed states and thus for their identification, particularly, in the light scalar meson sector where one should also include the glueballs.
- Liu *et al.* [13] studied scattering of charm mesons with various other hadrons including nucleon. They observed the sign change of the scattering length as a function of quark mass in the $DD^*(I = 1)$ channel, suggesting a loosely bound state in this channel. They also observed relatively strong attractive interaction in $DK(I = 1)$ channel. Furthermore,

scattering lengths of the charmed mesons ($\eta_c, J/\Psi, D$ and D_s) with the light hadrons (N, π and K) have also been calculated and in the charmonia-N channels they observed attractive interactions.

• T. Kawanai *et al.* [27] studied low energy charmonia-nucleon interaction in both quenched and full QCD simulations and they calculated potentials between the $\eta_c N$ states from the equal-time BS amplitude through the effective Schroedinger equation as well as the scattering lengths. They observed that the central potential in the $\eta_c N$ system is weakly attractive at short distances and exponentially screened at large distances. They also found $a^{J/\Psi N} \sim 0.35fm > a^{\eta_c N} > 0.25fm$, and concluded that the $J/\Psi N$ system is slightly more attractive than the $\eta_c N$ system at low energy.

• Recently Gong *et al.* [28], by using overlap valence quark on domain wall sea, studied $D^*_{s0}(2317)$ and concluded that it is just a $\bar{c}s$ scalar meson, and not a tetraquark state as speculated by various models.

SUMMARY AND OUTLOOK

The study of four-quark states by using lattice QCD method is very crucial for understanding the light scalar puzzle and to identifying the nature of various newly discovered charm resonances. Several lattice studies have been reported on the four-quark states both involving light and charm quarks. Various quenched lattice calculation suggested that the σ resonance possibly will have large four quark component though it needs to be confirmed with dynamical fermion and with disconnected diagrams. Establishing sigma as a four-quark state will have substantial implication on the chiral perturbation theory. Further study is needed to search for the four-quark states in charm sector. One should be careful in identifying a bound four-quark state from the threshold two meson states. In a lattice calculation one should also include disconnected diagram if it contributes to four-quark state. Mixing effects between mesons, four-quark and even with glueball of same quantum number should also be included. One hopes that in future all these will be incorporated and such efforts have already been started. To distinguish a molecular state from a tetraquark state, in a lattice calculation, one can use different spatial separation between sink positions of various quark fields and thereby obtain the information about the wavefunction to identify the structure of the four-quark state.

ACKNOWLEDGMENTS

I like to thank Keh-Fei Liu and Sasa Prelovsek for discussion and reading this manuscript. This work was done on a grant from DAE under grant No. DST-SR/S2/RJN-19/2007 of Ramanujan Fellowship.

REFERENCES

1. L. Maiani *et al.* Phys. Rev. Lett. 93, 212002 (2004); Phys. Rev. D71, 014028 (2005).
2. C. Ehmann and G. Bali, PoS(LAT2009)113, arXiv:0911.1238.
3. R.L. Jaffe, Phys. Rev. D15 (1977) 267 and 281; G. 't Hooft *et al.*, Phys. Lett. B662, 424 (2008).
4. M. Alford and R. Jaffe, Nucl. Phys. B578, 367 (2000).
5. K.F. Liu, Prog.Theor.Phys.Suppl.168, 160 (2007); Phys.Rev.D74, 094005 (2006).
6. M. Voloshin, Prog.Part.Nucl.Phys.61, 455 (2008); E.S. Swanson, Phys. Rept. 429, 243 (2006).
7. M. Luscher, Commun. Math. Phys. 105, 153 (1986).
8. N. Mathur *et al.* Phys. Rev. D70, 074508 (2004).
9. C. Alexandrou and A. Tsapalis, PoS(LAT2005)023, hep-lat/0509139; Z. Niu *et al.*, Phys. Rev. D82, 054501(2010).
10. M. Luscher, Nucl. Phys. B 354 (1991) 531; M. Luscher, Nucl. Phys. B 364, 237 (1991).
11. I. Montvay and P. Weisz, Nucl. Phys. B 290, 327 (1987); M. Luscher and U. Wolff, Nucl. Phys. B 339, 222 (1990).
12. S. Aoki *et al.*, Phys. Rev. D76, 094506 (2007); M. Gockeler *et al.*, arXiv:0810.5337.
13. L. Liu, PoS(LAT2009)099.
14. V. Bernard *et al.*, Nucl. Phys. B788, 1 (2008).
15. P. Giudice *et al.*, AIP Conf.Proc. 1257, 784 (2010), PoS LATTICE2010, 105 (2010).
16. K.F. Liu and C.W. Wong, Phys. Lett. B107, 391 (1981).
17. N. Mathur *et al.*, Phys. Rev. D76, 114505 (2007).
18. H. Suganuma *et al.*, Prog. Theor. Phys. Suppl. 168 (2007) 168.
19. M. Loan *et al.*, Eur. Phys. J. C57, 579 (2008).
20. S. Prelovsek *et al.*, Phys.Rev. D82, 094507 (2010).
21. J. Dudek *et al.*, Phys. Rev. Lett. 103, 262001 (2009); Phys. Rev. D82, 034508(2010); arXiv:1102.4299 [hep-lat], J. Bulava *et al.*, PoS LATTICE2010, 110 (2010); AIP Conf.Proc. 1257, 789 (2010).
22. L. Maiani *et al.* Phys. Rev. Lett. 99, 182003 (2007).
23. S.-K. Choi *et al.*, Phys. Rev. Lett. 100, 142001 (2008).
24. B. Aubert *et al.*, Phys. Rev. D79, 112001 (2009).
25. T. Chiu *et al.*, Phys. Lett. B 646 (2006) 95; Phys. Rev. D73, 111503 (2006), ibid 73, 094510 (2006).
26. C.G. Meng *et al.*, Phys. Rev. D80, 345030 (2009).
27. T. Kawanai *et al.*, Phys.Rev. D82, 091501 (2010); PoS LATTICE2010, 156 (2010).
28. M. Gong, A. Li, A. Alexandru, Y. Chen, T. Draper, PoS(Lattice 2010)106.

Highly excited and exotic meson spectroscopy from lattice QCD

Christopher E. Thomas
(for the Hadron Spectrum Collaboration)

Jefferson Laboratory, 12000 Jefferson Avenue, Suite #1, Newport News, VA 23606, USA

Abstract. We discuss recent progress in extracting highly excited and exotic meson spectra using lattice QCD. New results in the light meson sector are reviewed, where a combination of techniques have enabled us to confidently identify the spin of extracted states. Highlights of this, the most extensive excited meson spectrum calculated using lattice QCD, include many states with exotic quantum numbers and, for the first time in such calculation, spin-four mesons. We conclude with comments on future prospects.

Keywords: Lattice QCD, spectroscopy, light mesons, exotics
PACS: 12.38.Gc, 14.40.Be, 14.40.Rt

1. INTRODUCTION

Over the last few years there has been a resurgence of experimental interest in meson spectroscopy, particularly in the charmonium sector where a wealth of data has emerged from the B-factories, CLEO-c, BES and other experiments, in turn fuelling much theoretical work. This interest will continue in the light meson sector with the GlueX experiment, part of the Jefferson Lab's 12 GeV upgrade. GlueX will perform a systematic investigation of the spectrum of light mesons in photoproduction and, in particular, aims to produce mesons with exotic quantum numbers, such as $J^{PC} = 0^{+-}$, 1^{-+} and 2^{+-}. Such an exotic state signals physics beyond that of just a quark-antiquark pair, one possibility being a hybrid meson where the gluonic degrees of freedom play a nontrivial role.

A vital ingredient in understanding the strong interaction and testing QCD is the calculation of the spectrum of QCD and comparison against high quality experimental data. Lattice QCD provides one first-principles method for computing the spectrum and the calculation of the masses of the lowest-lying states has long been an important benchmark of such calculations. Recently, the Hadron Spectrum Collaboration has made significant progress in studying excited states using lattice QCD calculations, first testing techniques for extracting excited spectra [1] and radiative transition amplitudes [2, 3] using a quenched calculation in the charmonium sector, where there is a profusion of experimental data. These methods are now starting to be applied to the light meson sector [4, 5] with an eventual aim of calculating excited and exotic meson photocouplings, relevant, amongst other things, for the GlueX experiment.

In Section 2 we briefly outline the method used to extract excited meson spectra and in Section 3 we highlight some results, referring to Refs. [4, 5] for more details. We conclude in Section 4 with some comments on future prospects.

2. EXTRACTING EXCITED MESON SPECTRA FROM LATTICE QCD

The lattice provides a direct non-perturbative regularisation of QCD. In lattice QCD calculations, quantum fields in four-dimensional space-time are replaced by variables given values only on site and links of a lattice. In this way, calculating using the path integral formulation becomes an ordinary (but very large) integration problem. If an Euclidean space-time metric is used, the resulting integrals can be evaluated effectively using importance-sampling Monte Carlo methods. The spectrum can be extracted from calculations of two-point correlation functions of interpolating operators in Euclidean time, $C(t) =< 0|\mathcal{O}_i(t)\mathcal{O}_j(0)|0 >$. The interpolating operators, $\mathcal{O}$, are constructed from a fermion bilinear with gamma matrices and lattice-discretised gauge-covariant derivatives, overall projected onto definite momentum $\vec{p}$, $\mathcal{O} = \sum_{\vec{x}} e^{i\vec{p}\cdot\vec{x}}\bar{\psi}(x)\Gamma_i \overleftrightarrow{D}_j \overleftrightarrow{D}_k \ldots \psi(x)$, and are chosen to have the quantum numbers of the

12th International Conference on Meson-Nucleon Physics and the Structure of the Nucleon (MENU 2010)
AIP Conf. Proc. 1374, 290-293 (2011); doi: 10.1063/1.3647145

states under investigation[1]. Inserting a complete set of states, the correlator can be rewritten as

$$C(t) = \sum_n \frac{e^{-E_n t}}{2E_n} < 0|\mathcal{O}_i(0)|n >< n|\mathcal{O}_j(0)|0 > \ , \quad (1)$$

where the sum is over all states which overlap onto the operators, E_n are the energies of the states and $Z_i^{(n)} \equiv < n|O_i|0 >$ are the matrix elements (the overlaps of operators onto states). The ground state dominates at large t and so its energy and Z can be extracted by fitting to a single exponential for large t.

In order to extract excited state information, a multiple exponential fit to a single correlator could be attempted, but with noisy data and nearly degenerate energy levels this is a rather unstable procedure. A better alternative is to use a large basis of interpolating operators to form a matrix of correlators and then use a variational method, as implemented in Ref. [1]. The variational method boils down to solving a generalised eigenvector problem where the eigenvalues give E_n and the eigenvectors are related to the overlaps, $Z_i^{(n)}$. This method also gives the linear combination of operators, $\Omega^{(n)}$, which has the 'best' overlap onto a particular state n and the Z values can give information about the nature of states[1, 6, 5]. The orthogonality of the eigenvectors allows the variational method to separate states even if they are nearly degenerate.

A complication arises in lattice QCD where the theory is discretised on a finite four-dimensional Euclidean hypercubic grid. The three-dimensional rotational symmetry of the continuum is reduced to that of the octahedral group, equivalent to the symmetry group of a cube. Therefore, states are no longer labelled by continuum spin, J, but rather by lattice irreducible representations. An important advance, described in detail in Refs. [4, 5], has been the development of a method which allows the continuum spin of a state to be determined by using careful operator constructions. In the following section we show some results of the first applications of this method.

3. HIGHLIGHTS OF RESULTS

These calculations used dynamical anisotropic lattices[7, 8] with spatial lattice spacing, $a_s \sim 0.12$ fm and a finer temporal lattice spacing, $a_t = a_s/3.5$. The distillation method [9] was used to calculate the correlation functions. In the first results only connected Wick contractions were computed, giving access to isovector states and kaons. Four different light quark masses were used corresponding to $m_\pi \approx 700$, 520, 440 and 400 MeV. Two volumes were used, 16^3 and 20^3 in lattice units, corresponding to spatial extents of ≈ 2.0 fm and 2.4 fm respectively; in both cases the temporal extent is 128 in temporal lattice units. More details on the datasets are given in Ref. [5].

As an example of the results, in Fig. 1 we show the spin-identified isovector (octet) spectrum with $m_\pi \approx 700$ MeV on both volumes. Note that this m_π corresponds to degenerate light and strange quarks, i.e. $SU(3)$ symmetry, and so all mesons in an octet are degenerate ($m_\pi = m_K = m_{\eta_8}$). Boxes represent the extracted mass scaled by the Ω baryon mass and one sigma statistical uncertainties. It can be seen that this method has enabled the extraction and spin-identification of many states across all PC combinations. There are many states with exotic quantum numbers, extracted with a precision comparable to that of nonexotic states. For the first time in a lattice QCD calculation, spin-four mesons have been identified. There is really no statistically significant variation between the two volumes.

The spectrum of states shows many features of the $n^{2S+1}L_J$ pattern predicted by quark potential models. In the left pane of Fig. 1 there are candidates for a ground-state S-wave pair (0^{-+},1^{--}), a radial excitation ($m/m_\Omega \sim 1.1$) and a complete D-wave set (1^{--},2^{--},3^{--},2^{-+}) at ~ 1.2. The region ~ 1.6 contains parts of a D-wave radial excitation and G-wave set (3^{--},4^{--},5^{--},4^{-+})[2]. The middle pane contains a candidate P-wave set (0^{++},1^{++},2^{++},1^{+-}) at ~ 0.9, a radial excitation (~ 1.3) and a complete F-wave set (2^{++},3^{++},4^{++},3^{+-}) at ~ 1.4.

The right pane of Fig. 1 shows a set of states with exotic J^{PC}, not possible with just a quark-antiquark pair. Such states could be hybrid mesons, where the gluonic field is excited, or multiquark or molecular mesons where there is at least one extra quark-antiquark pair. These states have a large overlap onto operators containing the commutator of two covariant derivatives, i.e. the field strength tensor, suggesting that the gluonic field plays a non-trivial role in those states. This, along with there being no clear evidence for multi-meson states in the extracted spectrum, as argued in Ref. [5], suggests a hybrid nature for these states.

---

[1] Here we only consider $\vec{p} = \vec{0}$.

[2] We have no operators capable of producing a spin-5 state in the continuum

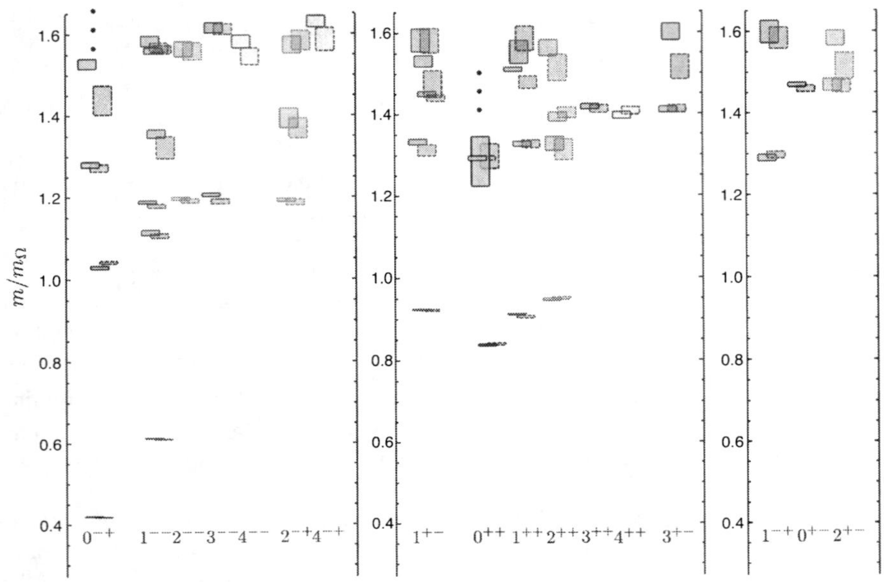

FIGURE 1. From Ref. [5]. Spin-identified isovector (octet) spectrum with $m_\pi \approx 700$ MeV on 16^3 (solid) and 20^3 (dashed) volumes. Ellipses indicate that there are heavier states with a given J^{PC} but that they are not well determined in this calculation.

There are also a number of states with nonexotic J^{PC} which do not appear to fit in to the quark model pattern. For example, the 0^{-+}, 1^{--} pair at ~ 1.3 is a bit light to be the second radial excitation of the S-wave and it is close in energy to a completely unexpected 2^{-+} state. These states again all have significant overlap onto operators containing the commutator of two covariant derivatives and are of comparable mass to the exotics, suggesting that they have a hybrid nature. Such nonexotic hybrids are predicted in models of hybrids and can mix with conventional mesons having the same J^{PC}.

Using the operator overlaps it is possible to perform a more detailed model-dependent interpretation of the spectrum. For example, we can determine the mixing between the conventional vector mesons and the vector hybrid and also the mixing between the 1^3D_1 and 2^3S_1 vectors. This work is ongoing.

In Ref. [5] we show how the spectrum varies as we move away from the $SU(3)$ flavour point, lowering the mass of the two degenerate light quarks and keeping the remaining strange quark heavy. Therein we also discuss kaons and "strangeonium". Here we will only highlight some results for the exotic states, shown in Fig. 2 with a comparison to some previous results for the 1^{-+}. It can be seen how using anisotropic lattices, a large basis of carefully constructed operators and high statistics have enabled us to extract the 1^{-+} more precisely, with precision comparable to that of the nonexotic states, and also reliably extract other states with exotic quantum numbers.

4. CONCLUSIONS

In summary, we have highlighted some recent progress in extracting excited meson spectra from dynamical lattice QCD calculations, referring to Refs. [4, 5] for more results and details. On all mass sets and volumes we are able to reliably extract and identify a large number of excited states across all *PC* combinations, including states with exotics quantum numbers (0^{+-}, 1^{-+} and 2^{+-}), hinting at explicit gluonic degrees of freedom. For the first time in such a calculation, we have extracted spin-four states (4^{++}, 4^{-+} and 4^{--}). We have argued that the pattern of extracted states shows features predicted by quark-potential models, along with other states, both exotic and nonexotic, that do not seem to lie in that classification.

In Ref. [5] we argue that there is no clear evidence for multi-particle states in our extracted spectrum and that to study such states we need to construct operators with a larger number of quark fields, e.g. operators which are the product of two fermion bilinears. Such constructions should give us access to multi-particle energy levels and a denser spectrum of states which can be interpreted in terms of resonances using techniques like Lüscher's and its extensions.

Further avenues of investigation include the computation of disconnected two-point correlation functions giving

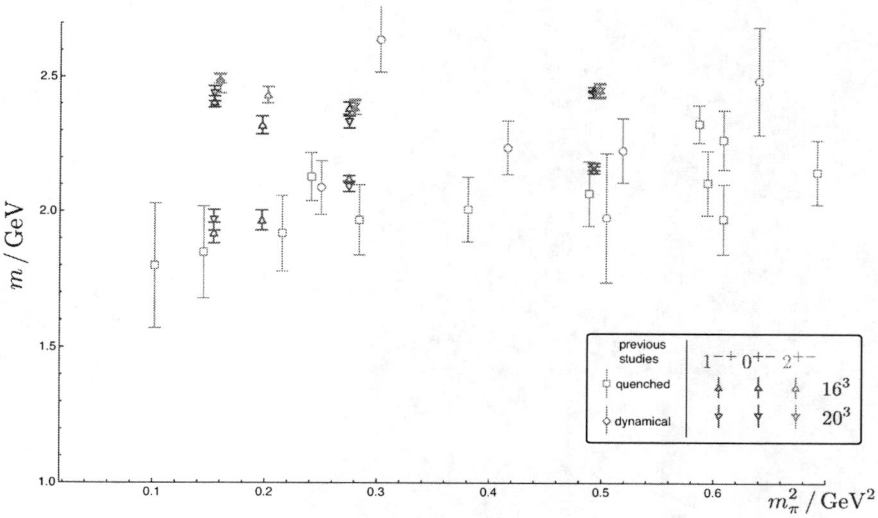

FIGURE 2. From Ref. [5]. Summary of extracted isovector exotic states with $J^{PC} = 1^{-+}, 0^{+-}, 2^{+-}$ as a function of m_π^2. For comparison we show some results for the 1^{-+} from Refs. [10, 11, 12, 13, 14, 15].

access to isoscalars and application of these methods to the baryon sector [16]. An eventual aim is to calculate the light meson photocouplings which will be relevant for, amongst other things, the GlueX experiment at Jefferson Lab's 12 GeV upgrade.

ACKNOWLEDGMENTS

I thank Jo Dudek, Robert Edwards, Mike Peardon, David Richards and other colleagues in the Hadron Spectrum Collaboration. The Chroma software suite [17] was used to perform this work on clusters at Jefferson Laboratory and Fermilab using time awarded under the USQCD Initiative. Part of this work used the CUDA GPU implementation of a mixed-precision iterative linear system solver for the Dirac equation by Michael Clark and Ronald Babich [18]. Authored by Jefferson Science Associates, LLC under U.S. DOE Contract No. DE-AC05-06OR23177. The U.S. Government retains a non-exclusive, paid-up, irrevocable, world-wide license to publish or reproduce this manuscript for U.S. Government purposes.

REFERENCES

1. J. J. Dudek, R. G. Edwards, N. Mathur, and D. G. Richards, *Phys. Rev.* **D77**, 034501 (2008), 0707.4162.
2. J. J. Dudek, R. G. Edwards, and D. G. Richards, *Phys. Rev.* **D73**, 074507 (2006), hep-ph/0601137.
3. J. J. Dudek, R. Edwards, and C. E. Thomas, *Phys. Rev.* **D79**, 094504 (2009), 0902.2241.
4. J. J. Dudek, R. G. Edwards, M. J. Peardon, D. G. Richards, and C. E. Thomas, *Phys. Rev. Lett.* **103**, 262001 (2009), 0909.0200.
5. J. J. Dudek, R. G. Edwards, M. J. Peardon, D. G. Richards, and C. E. Thomas, *Phys. Rev.* **D82**, 034508 (2010), 1004.4930.
6. J. J. Dudek, and E. Rrapaj, *Phys. Rev.* **D78**, 094504 (2008), 0809.2582.
7. H.-W. Lin, et al., *Phys. Rev.* **D79**, 034502 (2009), 0810.3588.
8. R. G. Edwards, B. Joo, and H.-W. Lin, *Phys. Rev.* **D78**, 054501 (2008), 0803.3960.
9. M. Peardon, et al., *Phys. Rev.* **D80**, 054506 (2009), 0905.2160.
10. C. Bernard, et al., *Phys. Rev.* **D68**, 074505 (2003), hep-lat/0301024.
11. C. W. Bernard, et al., *Phys. Rev.* **D56**, 7039–7051 (1997), hep-lat/9707008.
12. P. Lacock, and K. Schilling, *Nucl. Phys. Proc. Suppl.* **73**, 261–263 (1999), hep-lat/9809022.
13. P. Lacock, C. Michael, P. Boyle, and P. Rowland, *Phys. Lett.* **B401**, 308–312 (1997), hep-lat/9611011.
14. C. McNeile, and C. Michael, *Phys. Rev.* **D73**, 074506 (2006), hep-lat/0603007.
15. J. N. Hedditch, et al., *Phys. Rev.* **D72**, 114507 (2005), hep-lat/0509106.
16. J. Bulava, et al., *Phys. Rev.* **D82**, 014507 (2010), 1004.5072.
17. R. G. Edwards, and B. Joo, *Nucl. Phys. B. Proc. Suppl.* **140**, 832 (2005), hep-lat/0409003.
18. M. A. Clark, R. Babich, K. Barros, R. C. Brower, and C. Rebbi, *Comput. Phys. Commun.* **181**, 1517–1528 (2010), 0911.3191.

Session E: Transverse Momentum-Dependent Distributions

Transverse momentum dependent quark densities from Lattice QCD

B. U. Musch*, Ph. Hägler†, J. W. Negele** and A. Schäfer‡

*Theory Center, Jefferson Lab, 12000 Jefferson Avenue, Newport News, VA 23606, USA
†Theoretische Physik T39, TU München, James-Franck-Straße 1, 85747 Garching, Germany
**Massachusetts Institute of Technology, 77 Massachusetts Avenue, Bldg. 6-315, Cambridge, MA 02139, USA
‡Institut für Theoretische Physik, Universität Regensburg, 93040 Regensburg, Germany

Abstract. We study transverse momentum dependent parton distribution functions (TMDs) with non-local operators in lattice QCD, using MILC/LHPC lattices. We discuss the basic concepts of the method, including renormalization of the gauge link. Results obtained with a simplified operator geometry show visible dipole deformations of spin-dependent quark momentum densities.

Keywords: transverse momentum; parton distribution functions; lattice; QCD
PACS: 12.38.Gc, 13.88.+e, 13.85.Ni

INTRODUCTION

Generalized parton distribution functions (GPDs) and transverse momentum dependent parton distribution functions (TMDs) provide us with a picture of the internal quark distributions in a nucleon at the instant of an interaction, see illustration Fig. 1 a). GPDs and TMDs have their natural interpretation at large nucleon momentum $\mathbf{P} = (0,0,\mathbf{P}_z)$. The quark momentum k in terms of light cone coordinates $k^\pm \equiv (k^0 \pm k^3)/\sqrt{2}$, $\mathbf{k}_\perp = (\mathbf{k}_x, \mathbf{k}_y)$ scales like $k^+ : \mathbf{k}_\perp : k^- \sim P^+ : 1 : (P^+)^{-1}$ with the large momentum component P^+ of the nucleon. TMDs resolve the dependence on $x \equiv k^+/P^+$ and transverse momentum $\mathbf{k}_\perp$, but not on the suppressed component k^-. In spin-polarized channels at leading twist, TMDs encode dipole- or quadrupole-shaped deformations of the nucleon in the $\mathbf{k}_\perp$-plane. We have studied such deformations in first explorative lattice QCD calculations [1, 2, 3], see Fig. 1 and our discussion below. These studies have been motivated by a history of successful lattice computations of x-moments of GPDs, providing images of the nucleon in the impact parameter, $\mathbf{b}_\perp$-, plane, see [4] for a review. A remaining theoretical problem concerns the precise form of the correlator defining TMDs in the continuum, see [5, 6] and references therein. In its basic form, it is given by [7]

$$\begin{aligned}\Phi_q^{[\Gamma]}(x,\mathbf{k}_\perp;P,S;\mathscr{C}) &\equiv \int dk^- \int \frac{d^4 l}{(2\pi)^4}\, e^{-ik\cdot l} \underbrace{\frac{1}{2} \langle P,S|\; \bar{q}(l)\, \Gamma\, \mathscr{U}[\mathscr{C}_l]\; q(0)\; |P,S\rangle}_{\widetilde{\Phi}_q^{[\Gamma]}(l,P,S;\mathscr{C})} \Big|_{k^+=xP^+} \\ &= \frac{1}{P^+} \underbrace{\int \frac{d(l\cdot P)}{2\pi}\, e^{-i(l\cdot P)x}}_{\displaystyle \int_x} \underbrace{\int \frac{d^2\mathbf{l}_\perp}{(2\pi)^2}\, e^{i\mathbf{l}_\perp\cdot\mathbf{k}_\perp}}_{\displaystyle \int_{\mathbf{l}}} \widetilde{\Phi}_q^{[\Gamma]}(l,P,S;\mathscr{C})\Big|_{l^+=0}\end{aligned} \tag{1}$$

where Γ is a Dirac matrix. The Wilson line $\mathscr{U}[\mathscr{C}_l]$ running along a continuous path $\mathscr{C}_l$ from l to 0 ensures gauge invariance of the expression. For the SIDIS and Drell-Yan scattering process, the Wilson line extends to infinity along a direction v that needs to be chosen (almost) lightlike, such that the cross section factorizes into hard, perturbative parts and soft contributions, see, e.g., Ref. [8]. Based on its symmetry transformation properties, the above correlator can be parametrized in terms of TMDs [9, 10, 11], for example

$$2\rho_{TL}^{(q)} \equiv \Phi_q^{[\gamma^+ + \lambda\gamma^+\gamma^5]} = f_{1,q} + \lambda \frac{\mathbf{k}_\perp \cdot \mathbf{S}_\perp}{m_N} g_{1T,q} + \left[\frac{\mathbf{S}_j \varepsilon_{ji} \mathbf{k}_i}{m_N} f_{1T,q}^\perp \right]_{\text{odd}}, \tag{2}$$

12th International Conference on Meson-Nucleon Physics and the Structure of the Nucleon (MENU 2010)
AIP Conf. Proc. 1374, 297-300 (2011); doi: 10.1063/1.3647146

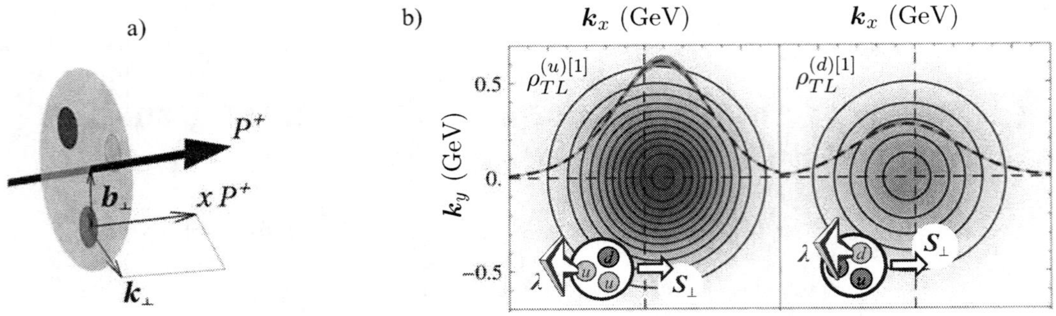

FIGURE 1. a) Illustration of quark degrees of freedom in the nucleon at large momentum. b) Dipole-deformed x-integrated densities obtained with straight gauge links at a pion mass $m_\pi \approx 500\,\text{MeV}$. The insets display the spin polarization of the quarks (red arrow) and of the nucleon (blue arrow).

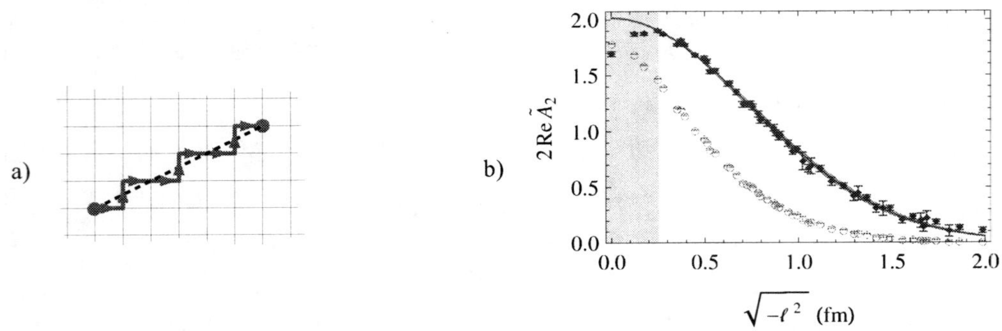

FIGURE 2. a) Representation of a straight Wilson line (dashed line) as a step-like product of link variables. b) Amplitude $\widetilde{A}_2(l^2,0)$ for up quarks at a pion mass $m_\pi \approx 500\,\text{MeV}$, using straight gauge links.

Here λ is the longitudinal quark polarization, and Λ and $\mathbf{S}_\perp$ are longitudinal and transverse nucleon polarization, respecively. The leading-twist TMDs $f_{1,q}$, $g_{1T,q}$, $f_{1T,q}^\perp$ are real-valued functions of x and $\mathbf{k}_\perp^2$. The "naively time-reversal odd" function $f_{1T,q}^\perp$ switches sign when comparing the SIDIS- with the Drell-Yan process, because the direction v of the Wilson line changes from future- to past-pointing [12].

STRAIGHT LINK TMDS FROM THE LATTICE

In light of the uncertainties about the precise form of the continuum correlator, and to develop our methods, our first lattice studies employ a simple operator geometry that does not relate to a specific scattering process: We connect the quark fields with a direct, straight Wilson line. For the resulting "process-independent" TMDs, the T-odd functions such as the Sivers function $f_{1T,q}^\perp$ vanish exactly.

In our approach, we calculate matrix elements $\langle P,S|\,O\,|P,S\rangle$ from ratios of three- and two-point functions using the same techniques as GPD calculations by the LHP collaboration in Ref. [13]. We also use the same sequential propagators and quark propagators, calculated by LHPC with domain-wall valence fermions on top of asqtad-improved staggered MILC gauge configurations [14, 15, 16] with 2+1 quark flavors at a lattice spacing $a \approx 0.12\,\text{fm}$. The difference with respect to GPD calculations is that we directly insert the non-local operator $O \equiv \bar{q}(l)\Gamma\mathscr{U}[\mathscr{C}_l]q(0)$ in our three-point function. The Wilson line $\mathscr{U}[\mathscr{C}_l]$ is approximated as a step-like product of HYP-smeared link-variables as illustrated in Fig. 2 a). See also Ref. [2, 3].

The connection between the matrix elements $\tilde{\Phi}^{[\Gamma]}$ and TMDs is established through a parametrization in terms of Lorentz-invariant amplitudes $\widetilde{A}_i(l^2, l\cdot P)$. For straight Wilson lines, we obtain in analogy to the parametrization in terms

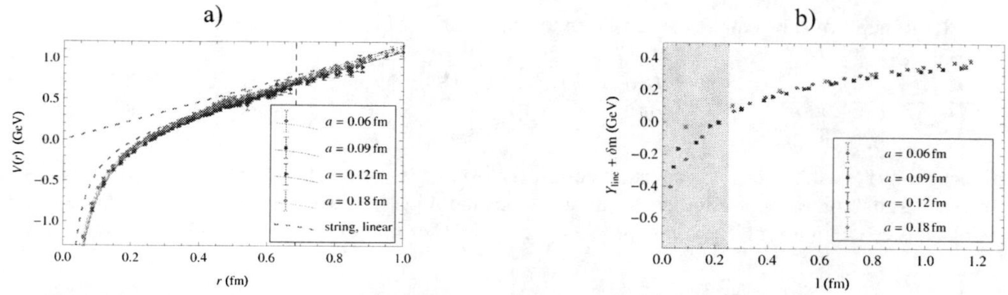

FIGURE 3. a) Static quark potential from MILC lattices at several lattice spacings a, matched to the string potential at $r \approx 0.7\,\text{fm}$. b) Test of the renormalization procedure with straight Wilson lines on a gauge fixed ensemble.

of amplitudes $A_i(k^2, k\cdot P)$ in Ref. [9] (here our sign conventions follow Ref. [11] with the substitution rule $k \longrightarrow i m_N^2 l$):

$$\tilde{\Phi}^{[\gamma^\mu]} = 2P^\mu \widetilde{A}_2 + 2i m_N{}^2 l^\mu \widetilde{A}_3\,, \qquad \tilde{\Phi}^{[\gamma^\mu\gamma^5]} = -2m_N S^\mu \widetilde{A}_6 - 2i m_N P^\mu (l\cdot S)\widetilde{A}_7 + 2m_N{}^3 l^\mu (l\cdot S)\widetilde{A}_8\,.$$

The TMDs are then obtained by

$$f_1(x,\mathbf{k}_\perp^2) = 2 \not{\mathcal{X}} \not{\mathcal{M}}\, \widetilde{A}_2(l^2, l\cdot P)\,, \qquad g_{1T}(x,\mathbf{k}_\perp^2) = 4m_N^2 \partial_{\mathbf{k}_\perp^2} \not{\mathcal{X}} \not{\mathcal{M}}\, \widetilde{A}_7(l^2, l\cdot P)\,.$$

In the equations above, $\not{\mathcal{X}}$ only acts on $l\cdot P$, while $\not{\mathcal{M}}$ only acts on l^2. Thus $x \leftrightarrow l\cdot P$ and $\mathbf{k}_\perp^2 \leftrightarrow l^2$ are pairs of conjugate variables. Our Euclidean lattice approach is restricted to the determination of amplitudes $\widetilde{A}_i$ for $l^0 = -il_4 = 0$, i.e., to the region $l^2 < 0$, $|l\cdot P| \leq \sqrt{-l^2}|\mathbf{P}|$, where $\mathbf{P}$ is the selected three-momentum of the nucleon on the lattice. The limited range in $|l\cdot P|$ prohibits us from a direct evaluation of $\not{\mathcal{X}}$. However, first studies of x- and $\mathbf{k}_\perp$- correlations are possible [17, 3]. Moreover, x-integrated TMDs and densities are directly accessible: Integrating Eq. (1) with respect to x removes $\not{\mathcal{X}}$ and sets $l\cdot P$ to zero. Correspondingly, the x-integral of, e.g., f_1 becomes $\int_{-1}^{1} dx\, f_1(x,\mathbf{k}_\perp^2) \equiv f_1^{[1]}(\mathbf{k}_\perp^2) = 2\not{\mathcal{M}}\, \widetilde{A}_2(l^2,0)$. In Fig. 2 b), open symbols correspond to unrenormalized lattice data for $\widetilde{A}_2(l^2,0)$.

To obtain results independent of our lattice spacing a and our lattice action, we must renormalize our data. The Wilson line $\mathscr{U}[\mathscr{C}_l]$ introduces a length dependent renormalization factor $\exp(-\delta m \sqrt{-l^2})$ [18, 19, 20]. To fix δm, we follow the strategy of Refs. [21, 22], and match the renormalized static quark potential $V^{\text{ren}}(r) = V(r) + 2\delta m$ to the string potential $V_{\text{string}} = \sigma r - \pi/(12r)$ [23] at a matching point $r = 1.5 r_0 \approx 0.7\,\text{fm}$. In Fig. 3 a), we test the method for several lattice spacings a on four MILC lattices with similar pion masses $m_\pi \approx 500\,\text{MeV}$. The renormalized lattice data agree very well with each other and are approximated well by the string potential (red dashed curve) near the matching point, indicated by a vertical dashed line. The procedure implements a gauge-invariant renormalization condition that we can formulate as the demand that the static quark potential asymptotically approach a straight line σr through the origin (shown as a red dashed line). In connection with TMDs, we lack at present an interpretation of this renormalization condition as a physical renormalization or factorization scale. In Figure 3 b), we check the applicability of the approach to Wilson lines by plotting $Y_{\text{line}}^{\text{ren}}(l) = \ln(U_{l-a/2}/U_{l+a/2})/a + \delta m$, where U_l is the expectation value of the color trace of a straight Wilson line of length l evaluated on a Landau gauge fixed ensemble, and where the length dependent renormalization has been carried out with the values δm obtained from the static quark potential. Only at short lengths, $l \lesssim 0.25\,\text{fm}$, we find significant differences between lattice data from different lattice spacings, a sign of lattice cutoff effects. For our TMD calculations discussed below we exclude data obtained in this region from our fits. For $l \gtrsim 0.25\,\text{fm}$, we assume that renormalization of the lattice operator can be carried out as in the continuum, $O^{\text{ren}} = Z_{\psi,z}^{-1} \exp(-\delta m \sqrt{-l^2})\, O$, where the renormalization constants $Z_{\psi,z}^{-1}$ and δm are independent of the Dirac structure Γ [19].

Figure 2 b) shows the renormalized lattice data for $\widetilde{A}_2(l^2,0)$ as solid data points. The curve and statistical error band correspond to a Gaussian fit to this data in the range $\sqrt{-l^2} \geq 0.25\,\text{fm}$. Note that the renormalization constant $Z_{\psi,z}^{-1}$ has been fixed (in the isovector, $u-d$-channel) such that the x-$\mathbf{k}_\perp$-integrated Gaussian density of unpolarized quarks yields the correct total number of valence quarks, $\int d^2\mathbf{k}_\perp f_{1,u-d}^{[1]} = 1$. Similar fits for $\widetilde{A}_7$ enable us to calculate the "worm-gear" function $g_{1T}^{[1]}$, and correspondingly, the dipole deformed x-integrated density $\rho_{TL}^{(q)[1]}$ defined in Eq. (2) and shown in Fig. 1 b). While the widths of our distributions depend strongly on our renormalization condition for δm, the average transverse quark momentum shift can be expressed in terms of ratios of the Gaussian amplitudes at $l^2{=}0$:

$$\langle \mathbf{k}_x \rangle_{TL} \equiv \left. \frac{\int d^2\mathbf{k}_\perp \, \mathbf{k}_x \, \rho_{TL}^{[1]}}{\int d^2\mathbf{k}_\perp \rho_{TL}^{[1]}} \right|_{\substack{\lambda=1, \\ \mathbf{S}_\perp=(1,0)}} = m_N \frac{\int d^2\mathbf{k}_\perp \, \mathbf{k}_\perp^2/(2m_N^2)\, g_{1T}^{[1]}(\mathbf{k}_\perp)}{\int d^2\mathbf{k}_\perp \, f_1^{[1]}(\mathbf{k}_\perp)} = -m_N \frac{\widetilde{A}_7(0,0)}{\widetilde{A}_2(0,0)} = \begin{cases} 67(5)\,\mathrm{MeV} & \text{(up)} \\ -30(5)\,\mathrm{MeV} & \text{(down)} \end{cases}$$

(errors statistical only). In these ratios, renormalization factors largely cancel. Reference [24] reveals a remarkable similarity of our results with a light-cone constituent quark model [25], despite the unphysically large quark masses employed in our lattice calculation: They find $\langle \mathbf{k}_x \rangle_{TL} = 55.8\,\mathrm{MeV}$ for up-, and $\langle \mathbf{k}_x \rangle_{TL} = -27.9\,\mathrm{MeV}$ for down-quarks.

CONCLUSIONS AND OUTLOOK

We have performed first lattice studies of TMDs using non-local operators with a simplified, straight gauge link. Resulting average momentum shifts $\langle \mathbf{k}_x \rangle_{TL}$ corroborate model results. An ongoing project with staple-shaped gauge links can potentially address TMDs specific to SIDIS or the Drell-Yan process, including T-odd functions responsible for single-spin asymmetries.

ACKNOWLEDGMENTS

We are grateful to the LHP and MILC collaborations, for providing us gauge configurations and propagators. We thank Vladimir Braun, Meinulf Göckeler, Gunnar Bali, Markus Diehl, Alexei Bazavov, and Dru Renner for helpful discussions. Our software uses the `Chroma`-library [26], and we use USQCD computing resources at Jefferson Lab. We acknowledge support by the Emmy-Noether program and the cluster of excellence "Origin and Structure of the Universe" of the DFG (Ph.H. and B.M.), SFB/TRR-55 (A.S.) and the US Department of Energy grant DE-FG02-94ER40818 (J.N.). Authored by Jefferson Science Associates, LLC under U.S. DOE Contract No. DE-AC05-06OR23177. The U.S. Government retains a non-exclusive, paid-up, irrevocable, world-wide license to publish or reproduce this manuscript for U.S. Government purposes.

REFERENCES

1. P. Hägler, et al., *Europhys. Lett.* **88**, 61001 (2009), 0908.1283.
2. B. U. Musch, Phd thesis, TU München (2010), 0907.2381.
3. B. U. Musch, P. Hägler, A. Schäfer, and J. W. Negele (tbp).
4. P. Hägler, *Phys. Rept.* **490**, 49–175 (2010), 0912.5483.
5. J. Collins, *PoS* **LC2008**, 028 (2008), 0808.2665.
6. I. Cherednikov, A. Karanikas, and N. Stefanis (2010), 1004.3697.
7. J. C. Collins, and D. E. Soper, *Nucl. Phys.* **B194**, 445 (1982).
8. X.-d. Ji, J.-p. Ma, and F. Yuan, *Phys. Rev.* **D71**, 034005 (2005), hep-ph/0404183.
9. P. J. Mulders, and R. D. Tangerman, *Nucl. Phys.* **B461**, 197–237 (1996), hep-ph/9510301.
10. D. Boer, and P. J. Mulders, *Phys. Rev.* **D57**, 5780–5786 (1998), hep-ph/9711485.
11. K. Goeke, A. Metz, and M. Schlegel, *Phys. Lett.* **B618**, 90–96 (2005), hep-ph/0504130.
12. J. C. Collins, *Phys. Lett.* **B536**, 43–48 (2002), hep-ph/0204004.
13. P. Hägler, et al., *Phys. Rev.* **D77**, 094502 (2008), 0705.4295.
14. C. W. Bernard, et al., *Phys. Rev.* **D64**, 054506 (2001), hep-lat/0104002.
15. C. Aubin, et al., *Phys. Rev.* **D70**, 094505 (2004), hep-lat/0402030.
16. A. Bazavov, et al. (2009), 0903.3598.
17. B. U. Musch, et al., *PoS* **LC2008**, 053 (2008), 0811.1536.
18. N. S. Craigie, and H. Dorn, *Nucl. Phys.* **B185**, 204 (1981).
19. H. Dorn, *Fortsch. Phys.* **34**, 11–56 (1986).
20. P. Boucaud, C. L. Lin, and O. Pene, *Phys. Rev.* **D40**, 1529 (1989).
21. M. Cheng, et al., *Phys. Rev.* **D77**, 014511 (2008), 0710.0354.
22. A. Bazavov, et al., *Phys. Rev.* **D80**, 014504 (2009), 0903.4379.
23. M. Lüscher, K. Symanzik, and P. Weisz, *Nucl. Phys.* **B173**, 365 (1980).
24. B. Pasquini, S. Boffi, A. Efremov, and P. Schweitzer (2009), 0912.1761.
25. B. Pasquini, S. Cazzaniga, and S. Boffi, *Phys. Rev.* **D78**, 034025 (2008), 0806.2298.
26. R. G. Edwards, and B. Joo, *Nucl. Phys. Proc. Suppl.* **140**, 832 (2005), hep-lat/0409003.

Phenomenological extraction of Transverse Momentum Dependent distributions

Alexei Prokudin*

Jefferson Lab, 12000 Jefferson Avenue, Newport News, VA 23606
**E-mail: prokudin@jlab.org*

Abstract. We discuss phenomenological extraction of Transverse Momentum Dependent Distributions (TMDs) from experimental data. At leading twist spin structure of spin-1/2 hadron can be described by 8 TMDs. TMDs reveal three-dimensional distribution of partons inside polarised nucleon. Experimentally these functions can be studied in polarised experiments using Spin Asymmetries in particular Single Spin Asymmetries (SSAs). We discuss transversity that measures distribution of transversely polarised quarks in a transversely polarised nucleon and Sivers distribution function that describes distribution of unpolarised quarks in a transversely polarised nucleon.

Keywords: Transverse Momentum Dependent distributions, Spin Asymmetries, Semi Inclusive Deep Inelastic Scattering
PACS: 13.88.+e, 13.60.-r, 13.15.+g, 13.85.Ni

INTRODUCTION

Transverse Momentum Dependent distribution functions (TMDs) depend on two independent variables: fraction of hadron momentum carried by parton, x, and intrinsic transverse momentum of the parton, $\mathbf{p}_T$.

The quark-quark distribution correlation function is defined as [1, 2]

$$\Phi_{ij}(x,\mathbf{p}_T) = \int \frac{d\xi^- d^2\xi_T}{(2\pi)^3}\, e^{ip\cdot\xi}\, \langle P,S_P|\bar{\psi}_j(0)\,\mathscr{U}^{n-}_{(0,+\infty)}\,\mathscr{U}^{n-}_{(+\infty,\xi)}\,\psi_i(\xi)|P,S_P\rangle\Big|_{\xi^+=0} \tag{1}$$

with $p^+ = xP^+$, $P^+ = (P^0+P^3)/\sqrt{2}$ is the big component of proton's momentum. $\mathscr{U}$ is the gauge link (Wilson line) that assures color gauge invariance of the correlator Φ_{ij}.

Corresponding distribution functions can be obtained by projecting the correlator onto the full basis of γ matrices $\Phi^{[\Gamma]} = \frac{1}{2}\mathrm{Tr}[\Phi\Gamma]$. At leading twist (expansion in P^+) the spin structure of the proton can be described by 8 TMDs (see [1, 2] and references therein):

$$\Phi^{[\gamma^+]}(x,p_T^2) = f_1(x,p_T^2) - \frac{\varepsilon_T^{\rho\sigma} p_{T\rho} S_{T\sigma}}{M} f_{1T}^{\perp}(x,p_T^2), \tag{2}$$

$$\Phi^{[\gamma^+\gamma_5]}(x,p_T^2) = S_L g_{1L}(x,p_T^2) - \frac{p_T\cdot S_T}{M} g_{1T}(x,p_T^2), \tag{3}$$

$$\begin{aligned}\Phi^{[i\sigma^{\alpha+}\gamma_5]}(x,p_T^2) &= S_T^{\alpha} h_1(x,p_T^2) + S_L \frac{p_T^{\alpha}}{M} h_{1L}^{\perp}(x,p_T^2)\\ &- \frac{p_T^{\alpha}p_T^{\rho} - \frac{1}{2}p_T^2 g_T^{\alpha\rho}}{M^2} S_{T\rho} h_{1T}^{\perp}(x,p_T^2) - \frac{\varepsilon_T^{\alpha\rho} p_{T\rho}}{M} h_1^{\perp}(x,p_T^2)\end{aligned} \tag{4}$$

here the projections are done with respect to the quark's polarisation, $\Phi^{[\gamma^+]}$ is distribution of unpolarised quarks, $\Phi^{[\gamma^+\gamma_5]}$ is distribution of longitudinally polarised quarks and $\Phi^{[i\sigma^{\alpha+}\gamma_5]}(x,p_T^2)$ is the distribution of transversely polarised quarks.

Experimental data are available from fixed target SIDIS $\ell P(S_P) \to \ell' hX$ experiments HERMES and COMPASS. BELLE experiment provides data on $e^+e^- \to h_1h_2X$ annihilation. RHIC data on SSA in hadron production $P(S_P)P \to hX$ is usually not used in phenomenological extractions, but potentially is a rich source of information on TMDs. Future experimental data will come from JLab, COMPASS and Electron Ion Collider.

12th International Conference on Meson-Nucleon Physics and the Structure of the Nucleon (MENU 2010)
AIP Conf. Proc. 1374, 301-304 (2011); doi: 10.1063/1.3647147

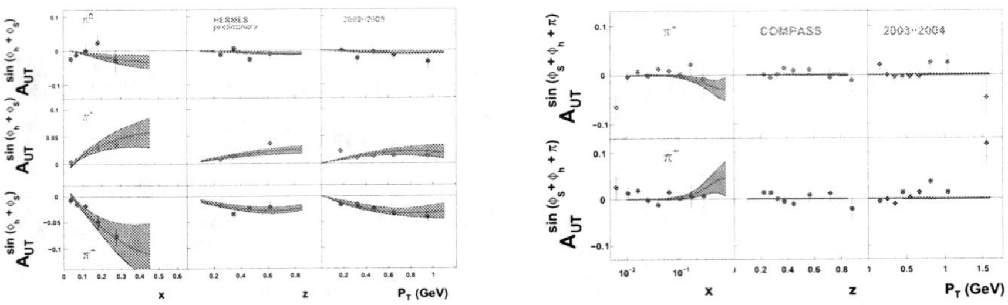

FIGURE 1. Fit[9] of HERMES [5] (left panel) and COMPASS [6] data (right panel).

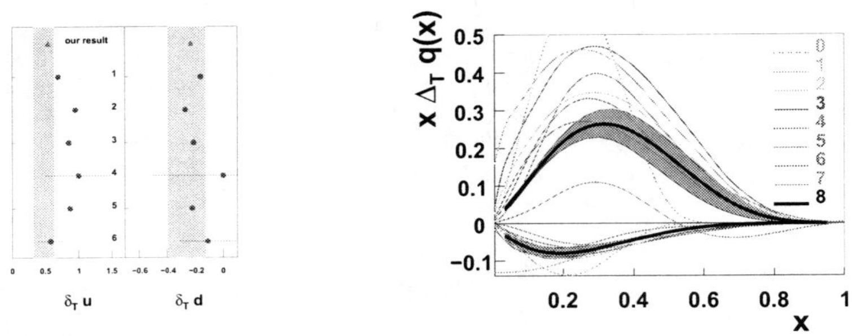

FIGURE 2. Left panel: extraction of tensor charge [9] compared to models [10, 11, 12, 13, 14, 15]. Right panel: transversity for u and d quarks [9] compared to models [16, 17, 18, 12, 19, 10, 11]

TRANSVERSITY

Transversity distribution $h_1(x, p_T^2)$ [3] describes distribution of transversely polarised quarks inside a transversely polarised nucleon.

In SIDIS transversity can be measured together with chiral-odd fragmentation function $H_{1q}^{\perp}(z, k_T^2)$ so-called Collins fragmentation function [4]

$$A_{UT}^{\sin(\Phi_h+\Phi_S)} \propto \frac{\sum_q e_q^2 h_{1q} \otimes H_{1q}^{\perp}}{\sum_q e_q^2 f_{1q} \otimes D_{1q}}, \tag{5}$$

where Φ_h and Φ_S are azimuthal angles of produced hadron and target polarisation vector with respect to lepton scattering plane.

Experimental data on SIDIS asymmetries $A_{UT}^{\sin(\phi_h+\phi_S)}$ from HERMES, COMPASS collaborations [5, 6] and $e^+e^- \to h_1h_2X$ data from BELLE collaboration [7] allow extraction[8, 9] of transversity and Collins fragmentation function.

In Fig. 1 we show the best fit to the HERMES [5] and COMPASS [6] data, respectively.

We include one of the two sets of Belle[7] data either A_{12} or A_0 data, see for details Refs.[8, 9]. In this analysis we use A_{12} data, the $\cos(\phi_1 + \phi_2)$ method.

$$A_{12}(z_1, z_2, \theta, \varphi_1 + \varphi_2) \propto \cos(\phi_1 + \phi_2) \frac{\sum_q e_q^2 H_{1q}^{\perp}(z_1) H_{1\bar{q}}^{\perp}(z_2)}{\sum_q e_q^2 D_{1q}(z_1) D_{1\bar{q}}(z_2)}, \tag{6}$$

here ϕ_1 and ϕ_2 are azimuthal angles of produced hadrons h_1 and h_2 with respect to the plane formed by the quark thrust axis and electron-positron pair, in a so-called Collins-Soper frame.

The extracted values[9] of tensor charge are $\delta_T u = 0.54^{+0.09}_{-0.22}$, $\delta_T d = -(0.23^{+0.09}_{-0.16})$ at $Q^2 = 0.8$ GeV^2.

In Fig. 2 we plot tensor charge and transversity distributions compared to models. As can be seen from Fig. 2 the experimental precision is still not good enough to discriminate among models. Anti-quark transversity distributions and Q^2 dependence of asymmetry is not yet known experimentally. Electron Ion Collider will be able to shed light on both anti-quark distributions and Q^2 dependence. JLab data will allow to expand x range of transversity extraction.

FIGURE 3. Left panel: extraction of Sivers function [22]. Right panel: three dimensional parton distribution.

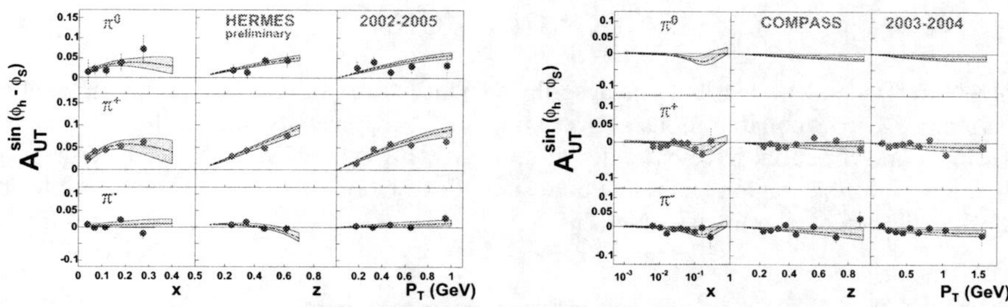

FIGURE 4. Fit[22] of pion HERMES [23] (left panel) and COMPASS [6] data (right panel).

SIVERS DISTRIBUTION FUNCTION

$f_{1T}^{\perp}(x, p_T^2)$ is so-called Sivers function [20], it describes correlation between orbital angular motion of quarks and the spin of the proton $\varepsilon_T^{\rho\sigma} p_{T\rho} S_{T\sigma}$. This function exists due to the presence of Final State Interactions of the struck quark and the remnant of the nucleon after the interaction.

Distribution of unpolarised quarks inside a transversely polarised nucleon can be written as

$$f_{q/p^\uparrow}(x, \mathbf{p_T}) = f_1(x, p_T^2) - f_{1T}^{\perp}(x, p_{\perp}^2) \frac{\mathbf{S} \cdot (\hat{\mathbf{P}} \times \mathbf{p_T})}{M}, \tag{7}$$

here $\mathbf{S} \cdot (\hat{\mathbf{P}} \times \mathbf{p_T})$ is correlation between quark motion and the spin of the nucleon. The function that measures this correlation is Sivers function $f_{1T}^{\perp}$.

This function obeys a modified universality, it changes sign from SIDIS to Drell-Yan [21] process

$$f_{1T}^{\perp}(x, p_T^2)^{SIDIS} = -f_{1T}^{\perp}(x, p_T^2)^{DY}, \tag{8}$$

the prediction of change of sign based on color gauge symmetry and parity and time reversal invariance $\mathscr{P}, \mathscr{T}$ of strong interactions. Experimental test of this relation is very important for our understanding of QCD.

The asymmetry associated with Sivers effect is

$$A_{UT}^{\sin(\Phi_h - \Phi_S)} \propto \frac{\sum_q e_q^2 f_{1T}^{\perp} \otimes D_{1q}}{\sum_q e_q^2 f_{1q} \otimes D_{1q}}. \tag{9}$$

In Fig. 3 we plot Sivers function extracted from the experimental data from HERMES[23] and COMPASS [6] collaborations, we show a three dimensional parton distribution at $x = 0.01$, as can be seen from Fig. 3 the distribution of partons in a transversely polarised hadron is not rotational symmetric, the distributions have dipole deformation with respect to the "center" of the hadron.

Fit[22] of HERMES [23] and COMPASS [6] data for $\pi^{\pm}$ and $K^{\pm}$ production is presented in Fig. 4 and Fig. 5.

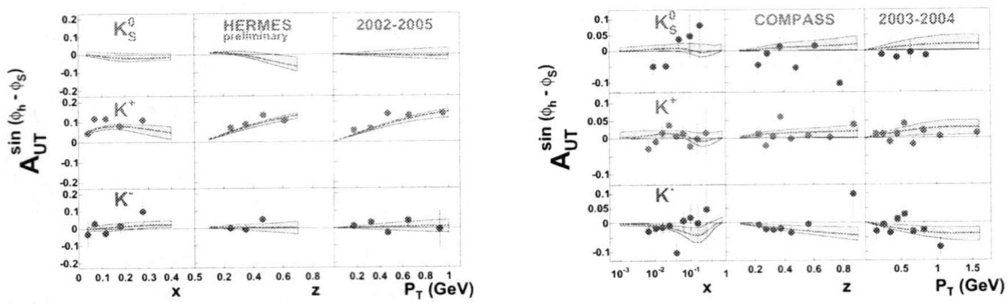

FIGURE 5. Fit[22] of kaon HERMES [23] (left panel) and COMPASS [6] data (right panel).

CONCLUSIONS

Transverse Momentum Dependent distributions describe spin structure of the proton. Experimental data from HERMES, COMPASS and BELLE collaborations allow extraction of transversity and Collins fragmentation functions. Sivers functions for u and d quarks were extracted from HERMES and COMPASS data. Future experimental data from JLab will widen explored x region and refine precision of extraction of TMDs. Q^2 range at Electron Ion Collider will allow to check evolution of asymmetries and TMDs.

ACKNOWLEDGMENTS

Authored by a Jefferson Science Associate, LLC under U.S. DOE Contract No. DE-AC05-06OR23177. The U.S. Government retains a non-exclusive, paid-up, irrevocable, world-wide license to publish or reproduce this manuscript for U.S. Government purposes.

REFERENCES

1. P. J. Mulders, and R. D. Tangerman, *Nucl. Phys.* **B461**, 197–237 (1996).
2. A. Bacchetta, et al., *JHEP* **02**, 093 (2007), hep-ph/0611265.
3. J. P. Ralston, and D. E. Soper, *Nucl. Phys.* **B152**, 109 (1979).
4. J. C. Collins, *Nucl. Phys.* **B396**, 161–182 (1993).
5. M. Diefenthaler (2007), arXiv:0706.2242 [hep-ex].
6. M. Alekseev, et al. (2008), arXiv:0802.2160 [hep-ex].
7. R. Seidl, et al., *Phys. Rev.* **D78**, 032011 (2008), 0805.2975.
8. M. Anselmino, et al., *Phys. Rev.* **D75**, 054032 (2007), hep-ph/0701006.
9. M. Anselmino, et al., *Nucl. Phys. Proc. Suppl.* **191**, 98–107 (2009), 0812.4366.
10. I. C. Cloet, W. Bentz, and A. W. Thomas, *Phys. Lett.* **B659**, 214–220 (2008), 0708.3246.
11. M. Wakamatsu, *Phys. Lett.* **B653**, 398–403 (2007), 0705.2917.
12. B. Pasquini, M. Pincetti, and S. Boffi, *Phys. Rev.* **D72**, 094029 (2005), hep-ph/0510376.
13. M. Gockeler, et al., *Phys. Lett.* **B627**, 113–123 (2005), hep-lat/0507001.
14. H.-x. He, and X.-D. Ji, *Phys. Rev.* **D52**, 2960–2963 (1995), hep-ph/9412235.
15. L. P. Gamberg, and G. R. Goldstein, *Phys. Rev. Lett.* **87**, 242001 (2001), hep-ph/0107176.
16. V. Barone, T. Calarco, and A. Drago, *Phys. Lett.* **B390**, 287–292 (1997), hep-ph/9605434.
17. B.-Q. Ma, I. Schmidt, J. Soffer, and J.-J. Yang, *Phys. Rev.* **D64**, 014017 (2001), hep-ph/0103136.
18. P. Schweitzer, et al., *Phys. Rev.* **D64**, 034013 (2001), hep-ph/0101300.
19. A. Bacchetta, F. Conti, and M. Radici, *Phys. Rev.* **D78**, 074010 (2008), 0807.0323.
20. D. W. Sivers, *Phys. Rev.* **D41**, 83 (1990).
21. J. C. Collins, *Phys. Lett.* **B536**, 43–48 (2002), hep-ph/0204004.
22. M. Anselmino, et al., *Eur. Phys. J.* **A39**, 89–100 (2009), 0805.2677.
23. A. Airapetian, et al., *Phys. Rev. Lett.* **94**, 012002 (2005), hep-ex/0408013.

Lattice Calculations of Nucleon Form Factors

S. N. Syritsyn, for the LHP Collaboration

Massachusetts Institute of Technology, 77 Massachusetts Ave, Cambridge, MA, 02139, USA

Abstract. We present recent results of calculation of the isovector electromagnetic and axial form factors of the nucleon using lattice QCD with three different lattice actions and pion masses down to $m_\pi \gtrsim 300$ MeV. Because of the precision of our high-statistics calculations, we can test predictions of baryon chiral perturbation theory for the charge and axial radii of the nucleon. We find that currently available baryon ChPT calculations disagree with our data, indicating that the corresponding effective theory approximations are not valid above $m_\pi \approx 300$ MeV.

Keywords: nucleon structure, lattice QCD
PACS: 12.38.Gc, 13.40.Gp

INTRODUCTION

In this paper, we present and discuss our calculations of the nucleon vector and axial form factors using three different types of quark action. In *Domain Wall* calculation [1], both sea [2] and valence quarks are chiral and pion masses are $m_\pi = 297$, 355 and 403 MeV. Since simulation of chiral quarks is expensive, previous (*Hybrid*) study [3] used non-chiral sea quarks [4, 5] and chiral valence quarks with the lightest pion mass $m_\pi = 293$ MeV. In addition, we show preliminary results from a recent calculation with non-chiral, Wilson-Clover, quarks and pion masses down to ≈ 200 MeV using gauge configurations provided by the *BMW* collaboration[6]. Because the isoscalar form factors contain disconnected contributions which are difficult to compute, we limit the discussion to the isovector quantities.

VECTOR FORM FACTORS

Electromagnetic form factors characterize fundamental aspects of the structure of protons and neutrons. In particular, they specify the spatial distribution of charge and magnetization. The nucleon Dirac $F_1(Q^2)$ and Pauli $F_2(Q^2)$ form factors are defined as follows:

$$\langle P',S'|\bar{q}\gamma^\mu\tau^a q|P,S\rangle = \bar{U}(P',S')\Big[\gamma^\mu F_1(Q^2) + \frac{i\sigma^{\mu\nu}(P'-P)_\nu}{2M_N}F_2(Q^2)\Big]\tau^a U(P,S), \tag{1}$$

where $U(P,S)$, $\bar{U}(P',S')$ are the initial and final nucleon spinors, and $Q^2 = -(P'-P)^2$ is the momentum transfer. To compare our results with experiment, in Fig. 1(left) we show the isovector electric Sachs form factor $G_E(Q^2) = F_1(Q^2) - \frac{Q^2}{(2M_N)^2}F_2(Q^2)$ from *Domain Wall* lattices together with the phenomenological fit [7]. Note that we *do not include* quark electric charges in the definition of the form factors (1). As expected, the lattice results are systematically higher, indicating that the charge distribution is narrower because the pion masses in our simulations are heavier than physical.

Momentum transfer dependence of the nucleon form factors at $Q^2 \to 0$ on the pion mass can be studied using effective low-energy theories, e.g. heavy baryon [8] chiral perturbation theory (ChPT) including $\Delta(123)$ [8, 9] and covariant baryon ChPT [10, 11]. The low-energy behavior of the vector form factors is characterized by the so-called mean-squared radii $\langle r_{1,2}^2\rangle$ defined as

$$F_{1,2}(Q^2) = F_{1,2}(0)(1 - \frac{1}{6}\langle r_{1,2}^2\rangle Q^2 + \mathscr{O}(Q^4)),\quad \langle r_{1,2}^2\rangle = -\frac{6}{F_{1,2}(0)}\frac{dF_{1,2}(Q^2)}{dQ^2}\Big|_{Q^2=0} \tag{2}$$

We extract the Dirac m. s. radii $\langle r_{1,2}^2\rangle^v$ by fitting a dipole form

$$F(Q^2) \approx \frac{F(0)}{(1+Q^2/M_D^2)^2},\quad \langle r^2\rangle \approx \frac{12}{M_D^2} \tag{3}$$

12th International Conference on Meson-Nucleon Physics and the Structure of the Nucleon (MENU 2010)
AIP Conf. Proc. 1374, 305-308 (2011); doi: 10.1063/1.3647148

to our data with remarkable agreement [1]. We show the results from *Domain Wall*, *Hybrid*, and *BMW* lattices in Fig. 1(right). These results show good agreement with each other indicating that discretization errors are small. However, the chiral fits shown in Fig. 2 fail to describe both our data and the experimental value simultaneously. Additionally, because the chiral formulas for $\langle r_{1,2}^2 \rangle$ radii share some low-energy constants (e.g., the $g_{\pi N\Delta}$ coupling in the heavy-baryon ChPT including $\Delta(1232)$), lattice calculations provide a unique test of whether a particular order of effective theory approximation is sufficient. For example, the solid line in Fig. 2 shows the best fit of the prediction in Ref. [9] to our data, giving $\chi^2/\text{dof} \approx 17$ and indicating that higher order terms should be included in the effective theory [12]. Alternatively, this disagreement proves that the effective theory at the given order should not be applied above $m_\pi \geq 300$ MeV, the lightest mass in our chiral fits.

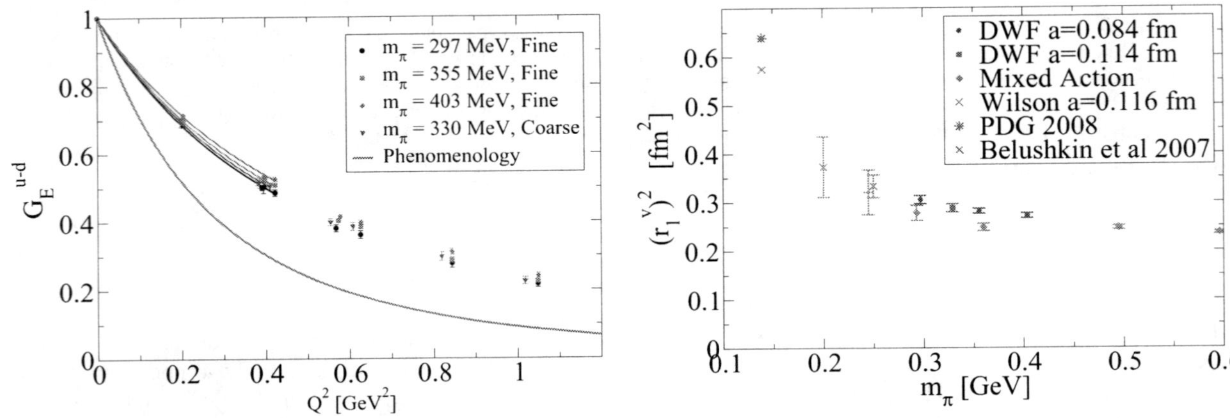

FIGURE 1. The isovector electric Sachs form factor $G_E^v(Q^2)$ (left) and Dirac radius $\langle r_1^2 \rangle^v$ (right) of the nucleon.

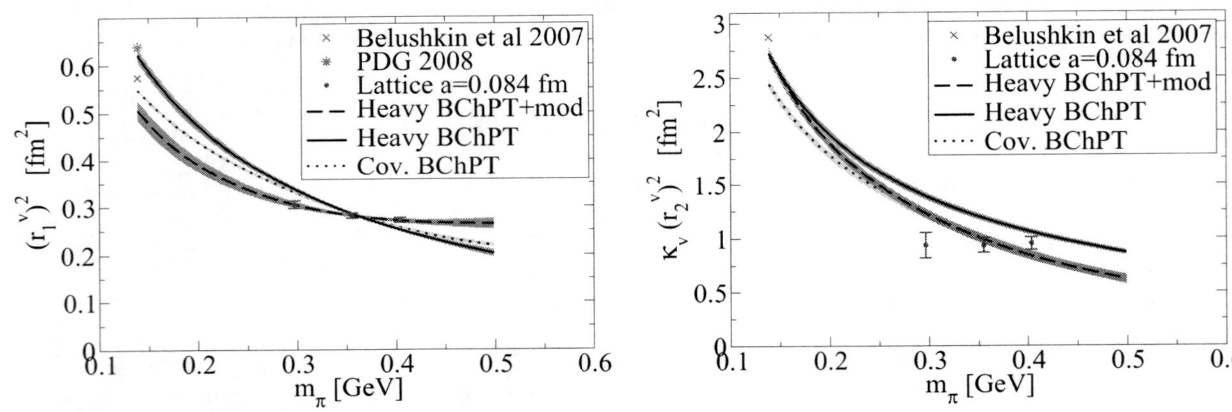

FIGURE 2. Simultaneous fits of the *Domain Wall* lattice results for the isovector Dirac (left) and Pauli (right) radii to the ChPT formulas in [12], [9], and [13], respectively.

AXIAL FORM FACTORS

Electroweak probes such as (anti)neutrinos interacting with quarks via chiral currents give access to the nucleon axial structure [14]. The axial structure of the nucleon is characterized by axial form factors,

$$\langle P', S' | A^{a\mu} | P, S \rangle = \bar{U}(P', S') \left[\gamma^\mu G_A(Q^2) + \frac{(P'-P)^\mu}{2M_N} G_P(Q^2) \right] \gamma^5 \tau^a U(P, S), \tag{4}$$

where $A^{a\mu} = \bar{q}\gamma^\mu \gamma^5 \tau^a q$ is the non-singlet axial current, which measures the correlation between the spin and isospin distributions inside a nucleon. $G_A(Q^2)$ is called the nucleon axial form factor and $G_P(Q^2)$ the induced pseudoscalar form factor at the momentum transfer $Q^2 = -(P'-P)^2$.

In phenomenology, it is common to parameterize the axial form factor $G_A(Q^2)$ for $Q^2 \lesssim 1$ GeV^2 with the dipole formula (3), where the parameter $M_D \equiv M_A$ is called axial mass. Similarly to the vector form factors in Eq. (2), the $Q^2 \to 0$ behavior of the axial form factor is characterized by so-called mean-squared axial radius $\langle r_A^2 \rangle \approx 12/M_A^2$.

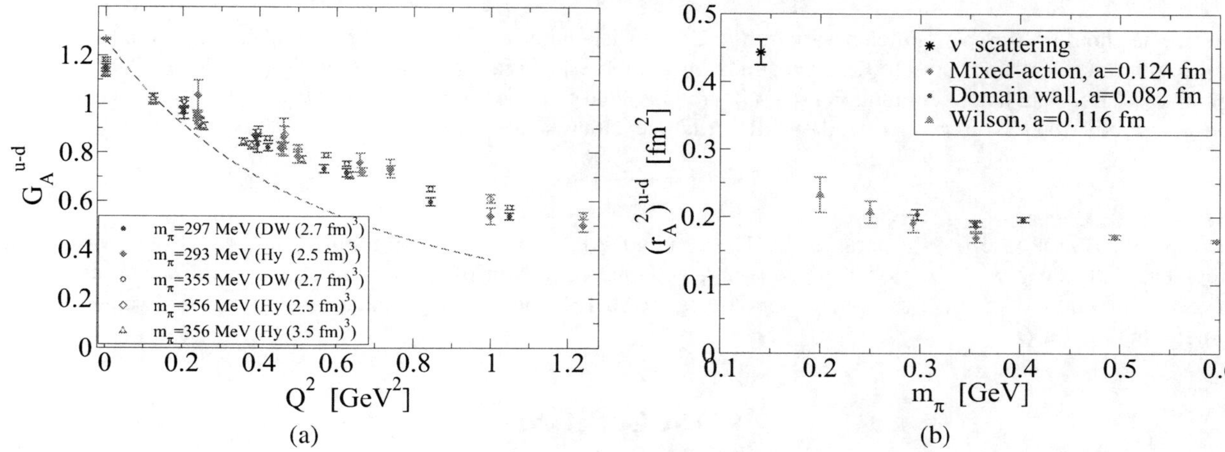

FIGURE 3. The axial form factor $G_A(Q^2)$ (a) and the axial radius $\langle r_A^2 \rangle$ (b) of the nucleon from lattice calculations using *Domain Wall* and *Hybrid* lattices. The dashed line on Panel (a) shows the dipole fit to experimental data. The star on Panel (b) shows the value for the nucleon axial radius from neutrino scattering experiments [15] as well as our preliminary results on *BMW* lattices.

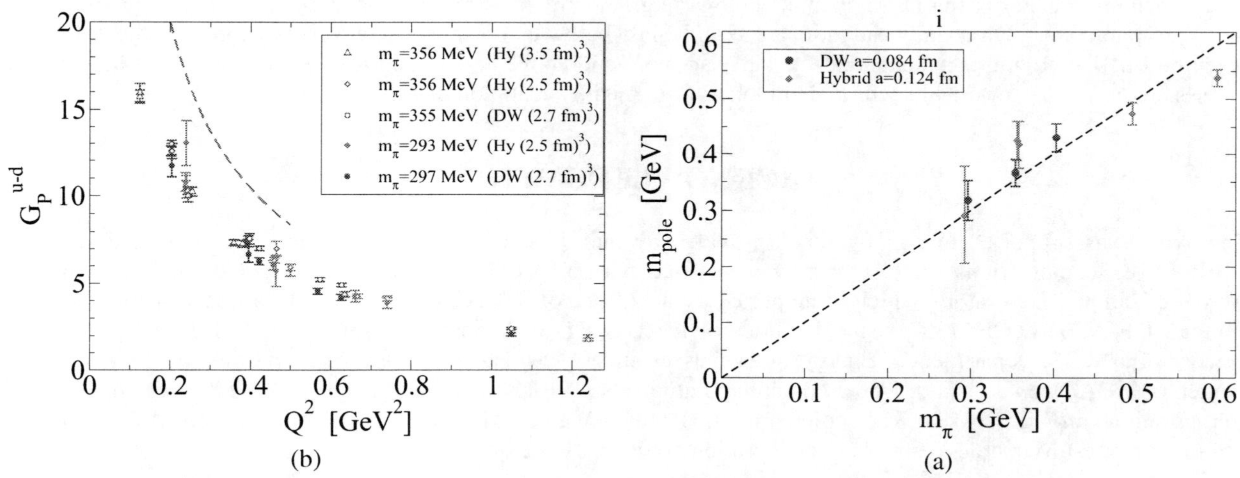

FIGURE 4. The induced pseudoscalar form factor $G_P(Q^2)$ of the nucleon (a) and the pole mass m_{pole} vs. m_π from fits (b) of our calculation results using *Domain Wall* and *Hybrid* lattices. The dashed curve on Panel (a) is the phenomenological prediction that includes only a pion pole term. The dashed line on Panel (b) is the $m_{\text{pole}} = m_\pi$ prediction.

In Figure 3(a) we show our results for the G_A form factor for both the *Domain Wall* and *Hybrid* lattices with the two lightest pion masses, $m_\pi \approx 300$ and 350 MeV. For $m_\pi \approx 350$ MeV, the results with two different spatial volumes ((2.5 fm)^3 and (3.5 fm)^3) show no noticeable difference indicating that finite volume effects are small. For comparison, we also show the dipole form (3), $G_A = g_A/(1+Q^2/M_A^2)^2$, with the axial mass $M_A \approx 1.026(21)$ obtained from neutrino scattering experiments [15].

In Figure 3(b) we show the axial radius for $300 \lesssim m_\pi \lesssim 600$ MeV. Although the results from different calculations agree, they significantly underestimate the phenomenological value (also shown in Fig. 3(b)). This is clearly a systematic trend, which can be ascribed to the heavy pion masses used in our simulations. Further, the $\langle r_A^2 \rangle$ values show very mild dependence on the pion mass. We expect this behavior to change close to the physical pion mass to accommodate both our results and the phenomenology. However, it will be in a striking disagreement with heavy-baryon ChPT at the next-to-leading order [9], which predicts no such dependence at all. Simulations at lighter pion masses are necessary to provide a basis to extend chiral perturbation theory accordingly.

The induced pseudoscalar form factor $G_P(Q^2)$ is expected to receive its dominant contribution from a pion pole,

$$G_P(Q^2) = \frac{4M_N F_\pi g_{\pi N}}{m_\pi^2 + Q^2} + \mathcal{O}(Q^0)\,. \tag{5}$$

Hence, one should observe significant variation of the Q^2-dependence of this form factor with the pion mass, which is an important check for correct QCD dynamics on a lattice. In Figure 4(a) we show $G_P(Q^2)$ for the two lightest pion masses. Because lattice computations of G_P are restricted to the off-forward case $Q^2 \neq 0$, it is hard to check the pion-pole prediction (5) directly. Instead, we fit our lattice data to a pole form

$$G_P(Q^2) = \frac{a}{m_{\text{pole}}^2 + Q^2} + b, \tag{6}$$

where a, b and m_{pole} are free fit parameters. The results for m_{pole} vs. m_π are presented in Fig. 4(b) showing good agreement with the pion pole model, $m_{\text{pole}} = m_\pi$, with the exception of the heaviest pion mass $m_\pi \approx 600$ MeV. However, in order to confirm the phenomenological form (6) one has to calculate G_P at smaller values of the momentum transfer Q^2.

CONCLUSIONS

We have presented the results of high-statistics, high-precision calculations of the isovector nucleon form factors on a lattice with $m_\pi \gtrsim 300$ MeV and preliminary results with the pion masses down to $m_\pi \approx 200$ MeV using cheaper non-chiral quarks. These results agree well with each other. Although the general dependence of the form factors on the momentum transfer is similar to phenomenology (e.g., the dipole form for G_E and G_A and the pion pole for G_P), the "size" parameters, Dirac and axial radii, are significantly below their experimental values. Moreover, the deviations from the ChPT predictions indicate that the latter are not valid above $m_\pi \gtrsim 300$ MeV. Simulations with lighter pion masses are necessary for credible chiral extrapolations to the physical pion mass.

ACKNOWLEDGMENTS

This work was supported in part by funds provided by the U. S. Department of Energy under grant DE-FG02-94ER40818. Computations for this work were carried out using the Argonne Leadership Computing Facility at Argonne National Laboratory, which is supported by the Office of Science of the U. S. Department of Energy under contract DE-AC02-06CH11357; using facilities of the USQCD Collaboration, which are funded by the Office of Science of the U. S. Department of Energy; and using resources provided by the New Mexico Computing Applications Center (NMCAC) on Encanto. The LHP collaboration also wishes to acknowledge the use of dynamical quark configurations provided by the RBC collaboration (Domain Wall quarks), the MILC collaboration (Asqtad quarks) and the Budapest-Marseille-Wuppertal collaboration (Wilson-Clover quarks).

REFERENCES

1. S. N. Syritsyn, et al., *Phys. Rev. D* **81**, 034507 (2010), 0907.4194.
2. C. Allton, et al., *Phys. Rev.* **D78**, 114509 (2008), 0804.0473.
3. J. D. Bratt, et al. (2010), 1001.3620.
4. C. W. Bernard, et al., *Phys. Rev.* **D64**, 054506 (2001), hep-lat/0104002.
5. A. Bazavov, et al. (2009), 0903.3598.
6. S. Durr, et al., *Science* **322**, 1224–1227 (2008), 0906.3599.
7. J. J. Kelly, *Phys. Rev.* **C70**, 068202 (2004).
8. T. R. Hemmert, B. R. Holstein, and J. Kambor, *J. Phys.* **G24**, 1831–1859 (1998), hep-ph/9712496.
9. V. Bernard, H. W. Fearing, T. R. Hemmert, and U. G. Meißner, *Nucl. Phys.* **A635**, 121–145 (1998), hep-ph/9801297.
10. J. Gasser, M. E. Sainio, and A. Svarc, *Nucl. Phys.* **B307**, 779 (1988).
11. T. Becher, and H. Leutwyler, *Eur. Phys. J.* **C9**, 643–671 (1999), hep-ph/9901384.
12. M. Gockeler, et al., *Phys. Rev.* **D72**, 054507 (2005), hep-lat/0506017.
13. T. Gail, *Chiral Analysis and Baryon Form Factors*, Ph.D. thesis, Technical University Munich (2007).
14. V. Bernard, L. Elouadrhiri, and U. G. Meissner, *J. Phys.* **G28**, R1–R35 (2002), hep-ph/0107088.
15. C. Amsler, et al., *Phys. Lett.* **B667**, 1 (2008).

Final state interactions & the Sivers function

Leonard Gamberg* and Marc Schlegel†

*Division of Science, Penn State University-Berks, Reading, Pennsylvania 19083, USA
†Institute for Theoretical Physics, Universität Tübingen, D-72076 Tübingen, Germany

Abstract. The non-vanishing of naive T-odd parton distributions function can be explained by the existence of the gauge link which emerges from the factorized description of the deep inelastic scattering cross section into perturbatively calculable and non-perturbative factors. This path ordered exponential describes initial / final-state interactions of the active parton due to soft gluon exchanges with the target remnants. Although these interactions are non-perturbative, studies of final state interactions have been approximated by perturbative one-gluon exchange in Abelian models. We include higher-order gluonic contributions from the gauge link by applying non-perturbative eikonal methods, incorporating color degrees of freedom in a calculation of the Sivers function. In this context we study the effects of color by considering the FSIs with with Abelian and non-Abelian gluon interactions. We confirm the large N_c QCD scaling behavior of Sivers functions and further uncover the deviations for finite N_c. Within this framework of FSIs we perform a quantitative check of approximate relations between T-odd TMDs and GPD which goes beyond the discussion of overall signs.

Keywords: Transverse momentum parton distributions, Final state interactions
PACS: 12.38.Lg, 12.38.-t, 13.88.+e, 13.85.Qk

Over the past two decades the transverse partonic structure of hadrons has been the subject of a great deal of theoretical and experimental study. Central to these investigations are the observations of large transverse single spin asymmetries (TSSAs) and azimuthal asymmetries in hadronic reactions– from inclusive hadron production [1, 2, 3] to Drell-Yan Scattering [4, 5], and in semi-inclusive deep inelastic lepton-nucleon scattering experiments [6, 7]. Two explanations to account for TSSAs in QCD have emerged which are based on the twist-three [8, 9, 10] and twist-two [11, 12, 13] approaches. We focus on the twist–two approach in the factorized picture of semi-inclusive deep inelastic lepton-hadron scattering (SIDIS) [13, 14] at small transverse momenta of the produced hadron, $P_T \sim k_T <<$ $\sqrt{Q^2}$, where $\sqrt{Q^2}$ is the hard scale. In this kinematic regime the Sivers effect describes a twist-two transverse target spin asymmetry through the "naive" time reversal odd (T-odd) structure, $\Delta f(x,\vec{k}_T) \sim S_T \cdot (P \times \vec{k}_T) f_{1T}^{\perp}(x,k_T^2)$ [11]. k_T is the quark intrinsic transverse momentum and P is the momentum of the target. The Sivers asymmetry has been the focus of much theoretical work on QCD factorization theorems. Among the most interesting results is that the Sivers function is not universal. It is predicted that there is a relative sign between the Sivers function from consideration of the gauge link dependence going from SIDIS to Drell-Yan scattering [15, 16]. The non-universality is further reflected in azimuthal asymmetries where one encounters the transverse moments of the quark correlator Φ [17]. Here $\Phi^i_\partial(x) = \int d^2k_T\, k^i_T\, \Phi(x,\vec{k}_T)$, where the non-trivial link dependence remains after integration over k_T. The correlator decomposes as $\Phi^\alpha_\partial(x) = \tilde{\Phi}^\alpha_\partial(x) + C_G \pi \Phi^\alpha_G(x,x)$ with calculable process-dependent gluonic pole factors C_G [18] ($e.g. C_G{}^{(SIDIS)} = -C_G{}^{(DY)}$ whereas $C_G{}^{(SIDIS)} = 1$). $\tilde{\Phi}_\partial$ contains the T-even operator combination, while Φ_G contains the T-odd operator combination. The latter is precisely the soft limit $x_1 \to 0$ of a quark-gluon correlator $\Phi_G(x,x_1)$. It was shown in Ref. [17] that the k_T-weighted Sivers function can be written in terms of the gluonic pole matrix element which we express

$$\langle k^i_T \rangle(x) = \int \frac{dz^-}{2\pi} e^{ixP^+z^-} \langle P,S_T | \bar{q}(-z^- n/2)\gamma^+ [-z^- n/2; z^- n/2] \hat{I}^i(z^- n/2) q(z^- n/2) | P,S_T \rangle, \tag{1}$$

where $\langle k^i_T \rangle(x) \equiv 2M\varepsilon^{ji}_T S^j_T f^{\perp(1)}_{1T}(x)$. The light-like vector $n = (n^- = 1, n^+ = 0, n_\perp = 0)$ represents a specific direction on the light-cone, and $[x;y]$ denotes a gauge link operator connecting the two locations x and y. The operator $\hat{I}$ originates from the time-reversal behavior of the FSI/ISI implemented by the gauge link operator in (1) and is expressed in terms of the gluonic field strength tensor

$$2\hat{I}^i(z^- n/2) = \int dy^- \,[z^- n/2, y^- n] g F^{+i}(y^- n)\, [y^- n, z^- n/2]. \tag{2}$$

12th International Conference on Meson-Nucleon Physics and the Structure of the Nucleon (MENU 2010)
AIP Conf. Proc. 1374, 309-313 (2011); doi: 10.1063/1.3647149

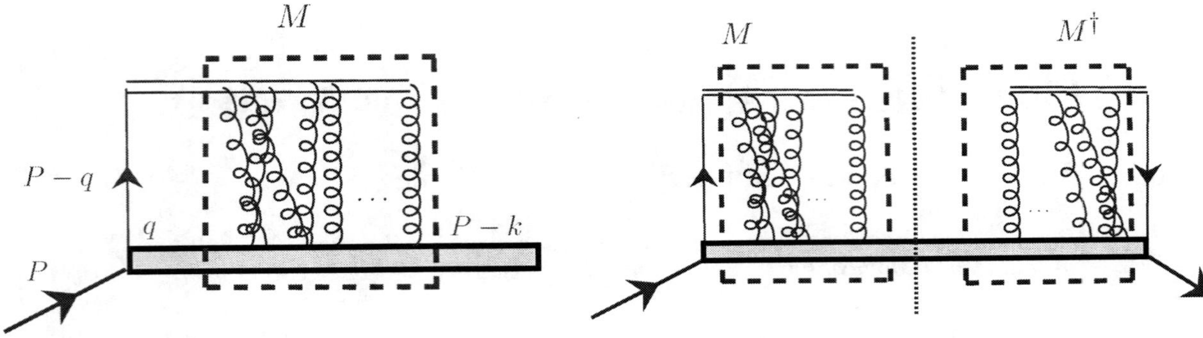

FIGURE 1. Left: The matrix element $W = \langle P-k|\,[\infty\ n;0]\,q(0)|P\rangle$ dressed with the FSIs. The FSIs are described by a non-perturbative scattering amplitude M that is calculated in a generalized ladder approximation [20]. Right: The quark-quark correlator with FSIs.

We exploit the model dependent relation of transverse distortion in momentum space with transverse distortion in impact parameter space [19] by inserting a complete set of momentum states and demand that the operator $\hat{I}$ is diagonal in momentum eigenstates, that is $\sum_{\lambda'_{\mathbf{P}}}\sum_{\lambda_{\mathbf{P}}}\cdots\langle\lambda'_{\mathbf{P}}|\hat{I}^i|\lambda_{\mathbf{P}}\rangle\delta^{(4)}(\lambda'_{\mathbf{P}}-\lambda_{\mathbf{P}})$. This leads us to model the target remnant of the quark-quark correlation function in terms of sum of spectators states $\{|\lambda_{\mathbf{P}}\rangle\}$ and to define the lensing function $I(\mathbf{P}) \equiv \sum_{\lambda'_{\mathbf{P}}}\langle\lambda'_{\mathbf{P}}|\hat{I}|\lambda_{\mathbf{P}}\rangle\delta^{(4)}(\lambda'_{\mathbf{P}}-\lambda_{\mathbf{P}})$. In this picture is it clear why spectator models can describe the gluonic pole matrix element as the factorization of FSIs convoluted with transverse impact parameter distortion (via impact parameter GPDs). While one cannot achieve a general factorization between TMDs and impact parameter GPDs; that is for the whole sum, the phenomenological relation may exist for each Fock space state separately. In this approximation the Dirac structure is carried by the distortion and the color dependence is determined by the chromodynamic lensing function [21]. This is most clearly displayed by transforming Eq. (1) into a mixed coordinate-momentum representation where the gluonic pole matrix element is expressed as

$$\langle k_T^i\rangle(x) = \int d^2b_T\int\frac{dz^-}{(2\pi)}\,e^{ixP^+z^-}\langle P^+,\vec{0}_T;S_T|\bar{q}(z_1)\gamma^+[z_1\,;z_2]\,\hat{I}^i(z_2)\,q(z_2)\,|P^+,\vec{0}_T;S_T\rangle. \tag{3}$$

The impact parameter b_T sits in the arguments of the quark fields, $z_{1/2} = (\mp\frac{z^-}{2\sqrt{2}},\vec{b}_T,\pm\frac{z^-}{2\sqrt{2}})$ and the lensing operator is $2\hat{I}^i(z_2) = \int dy^-\,[z_2\,;y]\,gF^{+i}(y)\,[y;z_2]$ where, $y^\mu = y^-n^\mu + b_T^\mu$. For each Fock space state we have the phenomenological relation,

$$\langle k_T^i\rangle(x) \simeq 2\int d^2b_T\,I^i(x,\vec{b}_T)\frac{\varepsilon_T^{ij}b_T^iS_T^j}{M}\frac{\partial}{\partial\vec{b}_T^2}\mathscr{E}(x,\vec{b}_T^2). \tag{4}$$

We utilize this picture to calculate the gluonic pole matrix element using a soft approximation for the lensing function [20, 22] which we then convolute with the parameterizations of impact parameter GPD $\mathscr{E}$ [23, 24].

First we model the target remnant in terms of the sum of spectators the quark correlation function

$$\Phi_{ij}(x,\vec{k}_T) = \frac{1}{2(2\pi)^3(1-x)P^+}\sum_{\delta,\beta}\bar{W}_j^{\beta\gamma}(P,k;\sigma)W_i^{\gamma\beta}(P,k;\sigma), \tag{5}$$

which is expressed in terms of the matrix element $W_i^{\alpha,\delta}(P,k;\sigma) = \langle P-k,\sigma,\delta|\,[\infty n\,;0]^{\alpha\beta}\,q_i^\beta(0)\,|P\rangle$ where σ and δ represent the helicity and color of the intermediate states (see Fig. 1). The FSIs– generated by the gauge link – are described by a non-perturbative amputated scattering amplitude $(M)^{\alpha\beta}_{\gamma\delta}$ with β, α (γ, δ) color indices of incoming eikonal quark and out going spectator remnant. In momentum space W is given in terms of M

$$\Delta W_i^{\alpha\beta}(P,k,S) = \int\frac{d^4q}{(2\pi)^4}\,ig_N\left((P-q)^2\right)\frac{\left[(\not{P}-\not{q}+m_q)u(P,S)\right]_i(M)^{\alpha\delta}_{\delta\beta}(q,P-k)}{[n\cdot(P-k-q)+i0]\left[(P-q)^2-m_q^2+i0\right][q^2-m_s^2+i0]}, \tag{6}$$

where $i(n\cdot(P-k-q)+i0)^{-1}$ represents the eikonal propagator and $\Delta W = W - W^0$, where W^0 denotes the contribution without final-state interactions. Tracing Eq. 5 with γ^+ and weighting, and integrating with respect to k_T yields the first

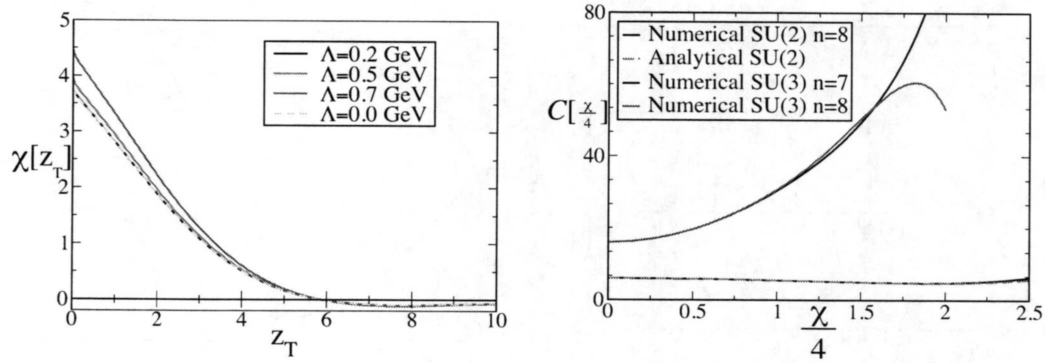

FIGURE 2. Left:The eikonal phase $\chi^{DS}(|\vec{z}_T|)$ vs. $|\vec{z}_T|$ with input from Dyson-Schwinger equations at scales $\Lambda_{QCD} = 0\,\text{GeV}, 0.2\,\text{GeV}, 0.5\,\text{GeV}, 0.7\,\text{GeV}$ [20]. Right: The function $C[\frac{\chi}{4}]$ of Eq. (13) as a function of the eikonal amplitude $\frac{\chi}{4}$. We compare the numerical result computed by means of Eq. (10) up to the order $n = 8$ with the analytical result for the $SU(2)$ color case. The numerical and analytical result agree up to $\frac{\chi}{4} \sim 2$. For $SU(3)$, we compare the numerical results for the orders $n = 7, 8$. The results are reliable for $\frac{\chi}{4} \sim 1.5$.

moment of the Sivers function [22]

$$\varepsilon_T^{ij} S_T^j f_{1T}^{\perp,(1)\text{sc}}(x) = -\frac{1}{2(1-x)M^2} \int \frac{d^2k}{(2\pi)^2}\, \varepsilon_T^{rl} k^r S_T^l I^i(x,\vec{k}) E(x,0,-\left(\frac{\vec{k}}{1-x}\right)^2), \tag{7}$$

where the lensing function I^i can be expressed in terms of the real and imaginary part of the scattering amplitude M [20].

We use functional methods to incorporate the color degrees of freedom for soft gauge boson coupling to highly energetic particles on the light-cone in the calculation of M, which is given by the expression

$$\frac{\left(M^{\text{eik}}\right)^{\alpha\delta}_{\delta\beta}(x,|\vec{q}_T+\vec{k}_T|)}{2(1-x)P^+} = \int d^2 z_T\, e^{-i\vec{z}_T\cdot(\vec{q}_T+\vec{k}_T)} \left[\int d^{N_c^2-1}\alpha \int \frac{d^{N_c^2-1}u}{(2\pi)^{N_c^2-1}}\, e^{-i\alpha\cdot u} \left(e^{i\chi(|\vec{z}_T|)t\cdot\alpha}\right)_{\alpha\delta} \left(e^{it\cdot u}\right)_{\delta\beta} - \delta_{\alpha\beta}\right]. \tag{8}$$

In Eq. (8) the N_c^2-1 dimensional integrals over the color parameters results from auxiliary fields $\alpha^a(s)$ and $u^a(s)$ that were introduced in the functional formalism of Ref. [25] in order to decouple the gluon fields from the color matrices. The eikonal phase $\chi(|\vec{z}_T|)$ in Eq. (8) represents the amount of soft gluon exchanges that are summed up into an exponential form, and is given in terms of the gluon propagator

$$\chi(|\vec{z}_T|) = g^2 \int_{-\infty}^{\infty} d\alpha \int_{-\infty}^{\infty} d\beta\, n^\mu \bar{n}^\nu \mathscr{D}_{\mu\nu}(z+\alpha n-\beta\bar{n}). \tag{9}$$

$\mathscr{D}$ denotes the gluon propagator, and g the strong coupling. In this form the four-vector v is related to the complementary light cone vector $\bar{n}$, $v = -((1-x)P^+/m_s)\bar{n}$, with $n\cdot\bar{n} = 1$ and $\bar{n}^2 = 0$. We evaluate the color integrals by deriving a power series representation for the expression in brackets in Eq. (8), the color function $f_{\alpha\beta}(\chi)$

$$f_{\alpha\beta}(\chi) = \sum_{n=1}^{\infty} \frac{(i\chi)^n}{(n!)^2} \sum_{a_1=1}^{N_c^2-1} \cdots \sum_{a_n=1}^{N_c^2-1} \sum_{P_n} (t^{a_1}...t^{a_n} t^{a_{P_n(1)}}...t^{a_{P_n(n)}})_{\alpha\beta}\,, \tag{10}$$

where P_n represents the sum over all permutations of the set $\{1,...,n\}$. If we had a direct ladder where gluons were not allowed to cross we would have only factors $(t^{a_1}...t^{a_n}t^{a_n}...t^{a_1})_{\alpha\beta} = C_F^n \delta_{\alpha\beta}$ with $C_F = N_c^2 - 1/2N_c$, and we could work in an Abelian theory with an effective replacement $\alpha \to C_F\alpha_s$ for the fine-structure constant. Since we allow generalized ladders with crossed gluons we have to sum over all permutations in (10), and the simple replacement is not possible. In a large N_c expansion the crossed gluons diagrams would be suppressed such that the direct ladder represents the leading order in $1/N_c$. In an Abelian theory, the generating matrices t reduce to identity and since we have $n!$ permutations of the set $\{1,...,n\}$, we recover the well-known Abelian result,

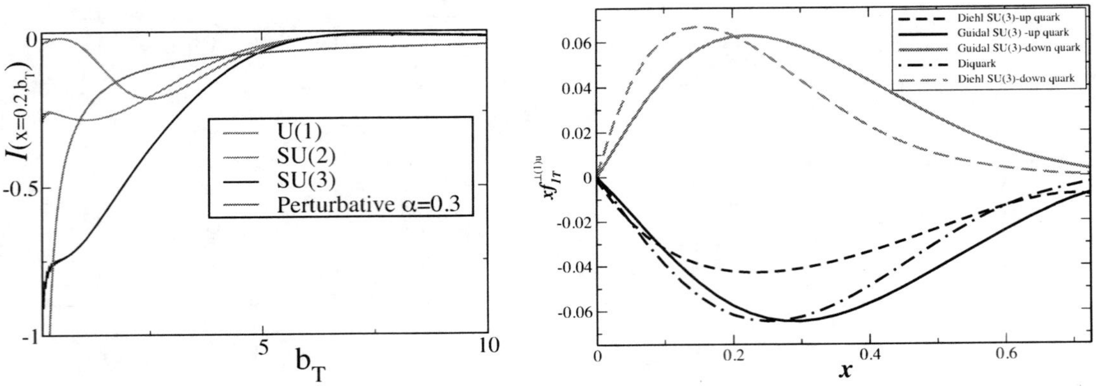

FIGURE 3. Left: The lensing function $I^i(x,\vec{b}_T)$ from Eq. (12) for U(1), SU(2) and SU(3) for $x = 0.2$ at a scale $\Lambda_{QCD} = 0.2\,\text{GeV}$. For comparison we also plot the perturbative result of Ref. [26] including the eikonalized antiquark spectator with an arbitrary value for the coupling, $\alpha = 0.3$. Right: The first moment of the Sivers function versus x using various models for the GPD E.

$f^{U(1)}(\chi) = \sum_{n=1}^{\infty}(i\chi)^n/n! = e^{i\chi} - 1$. For $N_c = 2$, $t^a = \sigma^a/2$ and we can calculate the integral analytically. We obtain, $f^{SU(2)}_{\alpha\beta}(\chi/4) = \delta_{\alpha\beta}(\cos\chi/4 - \chi/4\sin\chi/4 - 1) + i\delta_{\alpha\beta}(2\sin\chi/4 + \chi/4\cos\chi/4)$. We also calculate numerically the lowest coefficients in the power series (10), and they agree with the coefficients in an expansion in χ of the analytical result. This serves as a check of both numerical and analytical approaches. For $N_c = 3$, due to difficulty of integrating over the Haar measure we use the power series (10) to obtain the approximative color function which is valid when $a = \chi/4$ is small,

$$\Re[f^{SU(3)}_{\alpha\beta}](a) = \delta_{\alpha\beta}(-c_2a^2 + c_4a^4 - c_6a^6 - c_8a^8 + ...), \quad \Im[f^{SU(3)}_{\alpha\beta}](a) = \delta_{\alpha\beta}(c_1a - c_3a^3 + c_5a^5 - c_7a^7 + ...), \tag{11}$$

with the numerical values $c_1 = 5.333$, $c_2 = 6.222$, $c_3 = 3.951$, $c_4 = 1.934$, $c_5 = 0.680$, $c_6 = 0.198$, $c_7 = 0.047$, $c_8 = 0.00967$. Transforming to coordinate space we can express the lensing function directly in terms of the real and imaginary part of the color function f which is itself a function of the eikonal phase χ Eq. (9). This results in a lensing function of the form

$$I^i(x,\vec{b}_T) = \frac{(1-x)}{2N_c}\frac{b^i_T}{|\vec{b}_T|}\frac{\chi'}{4}C\left[\frac{\chi}{4}\right], \tag{12}$$

where we define the function,

$$C\left[\frac{\chi}{4}\right] \equiv \left[(\mathrm{Tr}\Im[f])'\left(\frac{\chi}{4}\right) + \frac{1}{2}\mathrm{Tr}\left[(\Im[f])'\left(\frac{\chi}{4}\right)(\Re[f])\left(\frac{\chi}{4}\right)\right] - \frac{1}{2}\mathrm{Tr}\left[(\Im[f])\left(\frac{\chi}{4}\right)(\Re[f])'\left(\frac{\chi}{4}\right)\right]\right], \tag{13}$$

and χ' denotes the first derivative with respect to $|\vec{z}_T|$, and $(\Im[f])'$ and $(\Re[f])'$ are the first derivatives of the real and imaginary parts of the color function f.

In Fig. 3 the function $C[\frac{\chi}{4}]$ is plotted versus $\frac{\chi}{4}$ for various approximations. While the convergence of the power series seems to be better for $SU(2)$ than in the $SU(3)$ case where the numerical result calculated with eight coefficients agrees with the analytical result up to $\frac{\chi}{4} \sim 2$, we can trust the numerical result computed with eight coefficients up to $\frac{\chi}{4} \sim 1.5$ for $SU(3)$.

In order to numerically estimate the lensing function and in turn the Sivers function we utilize the infrared behavior of the gluon and the running coupling in the non-perturbative regime where we infer that the soft gluon transverse momentum defines the scale at which the coupling is evaluated. These two quantities have been extensively studied in the infrared limit in the Dyson-Schwinger framework [27] and in lattice QCD [28]. We use calculations of these quantities from Dyson-Schwinger equations [27] where both α_s and $\mathscr{D}^{-1}$ are defined in the infrared limit (details can be found in a forthcoming publication). This determines the eikonal phase and thus the lensing functions (12) for a $U(1)$, $SU(2)$ and $SU(3)$ color function. We plot the results in Fig. 3 for a color function for $U(1), SU(2), SU(3)$. While we observe that all lensing functions are attractive and fall off at large transverse distances, they are very different in size at small distances.

Using the eikonal model for the lensing function together with different models for the GPD E we present the first moment of the Sivers function from Eq. (4). In Fig. (3) we compare the diquark model result [22], to two phenomenological models [24, 23]. We find that the lensing function grows with N_c which in turn predicts the growth of the Sivers function. This is consistent with the large N_c QCD scaling behavior of Sivers functions [29] and further we uncover the deviations for finite N_c. Further, the sign and size of the Sivers function is consistent with present extractions from data [30]. Within this framework of FSIs we have performed quantitative analysis of approximate relations between T-odd TMDs and GPD which goes beyond the discussion of overall signs.

ACKNOWLEDGMENTS

L.G. thanks the organizers of MENU for the opportunity to present this work. L.G. acknowledges support from the U.S. Department of Energy under contract DE-FG02-07ER41460.

REFERENCES

1. J. Adams, et al., *Phys. Rev. Lett.* **92**, 171801 (2004).
2. S. S. Adler, et al., *Phys. Rev. Lett.* **95**, 202001 (2005).
3. B. I. Abelev, et al., *Phys. Rev. Lett.* **101**, 222001 (2008).
4. M. Guanziroli, et al., *Z. Phys.* **C37**, 545 (1988).
5. J. S. Conway, et al., *Phys. Rev.* **D39**, 92–122 (1989).
6. A. Airapetian, et al., *Phys. Rev. Lett.* **103**, 152002 (2009).
7. M. Alekseev, et al., *Phys. Lett.* **B673**, 127–135 (2009).
8. A. V. Efremov, and O. V. Teryaev, *Sov. J. Nucl. Phys.* **39**, 962 (1984).
9. A. V. Efremov, and O. V. Teryaev, *Phys. Lett.* **B150**, 383 (1985).
10. J.-w. Qiu, and G. Sterman, *Phys. Rev. Lett.* **67**, 2264–2267 (1991).
11. D. W. Sivers, *Phys. Rev.* **D41**, 83 (1990).
12. J. C. Collins, *Nucl. Phys.* **B396**, 161–182 (1993).
13. D. Boer, and P. J. Mulders, *Phys. Rev.* **D57**, 5780–5786 (1998).
14. X.-d. Ji, J.-p. Ma, and F. Yuan, *Phys. Rev.* **D71**, 034005 (2005).
15. S. J. Brodsky, D. S. Hwang, and I. Schmidt, *Nucl. Phys.* **B642**, 344–356 (2002).
16. J. C. Collins, *Phys. Lett.* **B536**, 43–48 (2002).
17. D. Boer, P. J. Mulders, and F. Pijlman, *Nucl. Phys.* **B667**, 201–241 (2003).
18. C. J. Bomhof, and P. J. Mulders, *JHEP* **02**, 029 (2007).
19. S. Meissner, A. Metz, and K. Goeke, *Phys. Rev.* **D76**, 034002 (2007).
20. L. Gamberg, and M. Schlegel, *Phys. Lett.* **B685**, 95–103 (2010).
21. M. Burkardt, *Nucl. Phys.* **A735**, 185–199 (2004).
22. L. Gamberg, and M. Schlegel Final-state interactions and the Sivers function in a relativistic eikonal model, in preparation, 2010.
23. M. Guidal, M. V. Polyakov, A. V. Radyushkin, and M. Vanderhaeghen, *Phys. Rev.* **D72**, 054013 (2005).
24. M. Diehl, T. Feldmann, R. Jakob, and P. Kroll, *Eur. Phys. J.* **C39**, 1–39 (2005).
25. H. M. Fried, Y. Gabellini, and J. Avan, *Eur. Phys. J.* **C13**, 699–709 (2000).
26. S. Meissner, A. Metz, M. Schlegel, and K. Goeke, *JHEP* **08**, 038 (2008).
27. R. Alkofer, C. S. Fischer, F. J. Llanes-Estrada, and K. Schwenzer, *Annals Phys.* **324**, 106–172 (2009).
28. A. Sternbeck, and L. von Smekal, *Eur. Phys. J.* **C68**, 487–503 (2010).
29. P. V. Pobylitsa , Transverse-momentum dependent parton distributions in large-N(c) QCD, hep-ph/0301236.
30. M. Anselmino, et al., *Eur. Phys. J.* **A39**, 89–100 (2009).

Session F: NN Interactions

The nucleon-nucleon system in chiral effective theory

Daniel R. Phillips

Department of Physics and Astronomy, Ohio University, Athens, OH 45701

Abstract. I discuss the conditions under which the application of chiral perturbation theory to the NN potential gives reliable results for NN scattering phase shifts. χPT also yields a convergent expansion for the deuteron charge operator. For cutoffs < 1 GeV, this produces precise predictions for deuterium's quadrupole and charge form factors in the range $Q^2 < 0.25$ GeV^2.

Keywords: NN scattering, chiral symmetry, electron-deuteron scattering
PACS: 25.10.+s, 25.30.Bf, 12.39.Fe

INTRODUCTION

The use of chiral perturbation theory (χPT) to compute nuclear forces is a problem that has received much attention over the past twenty years. This effort began with the seminal papers of Weinberg [1] and is referred to in this contribution as "chiral effective theory" (χET). χET uses χPT to compute the NN potential up to some fixed order, n, in the chiral expansion in powers of $P \equiv (p, m_\pi)/\Lambda_0$. (Here the breakdown scale Λ_0 is nominally $m_\rho \sim 4\pi f_\pi$, but in reality is somewhat lower for reactions involving baryons.) This NN potential is then iterated, using the Schrödinger or Lippmann-Schwinger equation, to obtain the scattering amplitude. The resulting wave functions can be combined with, e.g. electromagnetic current operators, also derived from χPT, in order to derive results for few-nucleon-system observables that are grounded in QCD's pattern of chiral-symmetry breaking. In addition, χET results can be assessed for convergence as a function of the order, n, of the calculation. This enables specification of theoretical uncertainties based on the anticipated size of the $O(P^{n+1})$ corrections to the process of interest.

NN SCATTERING IN χET

To solve the Schrödinger equation with potentials derived from χPT we must introduce a cutoff, Λ, on the intermediate states, because χPT potentials grow with momenta. The low-energy constants (LECs) multiplying contact interactions in the nucleon-nucleon part of the chiral Lagrangian should then be adjusted to eliminate any Λ-dependence in the effective theory's predictions for low-energy observables. If this is not possible we conclude that χET is unable to give reliable predictions.

At leading order (LO) the χET potential contains a zero-derivative contact interaction that is operative only in NN partial waves with $L = 0$. One-pion exchange is also present in the LO potential, and is active in all partial waves. It couples $S = 1$ partial waves with $\Delta L = 2$ through its tensor part. Ref. [2] showed that the LO χET potential leads to a Λ-independent amplitude in the limit $\Lambda \to \infty$ in the $^3S_1$-$^3D_1$ channel. At fixed pion mass a Λ-independent result is also obtained in the $^1S_0$ channel [2]. Recently we showed how to "subtractively renormalize" the LO equations for NN scattering in these two channels [3]. This technique eliminates the contact interaction in favor of a low-energy observable (e.g. the relevant NN scattering length). This makes it straightforward to take the limit $\Lambda \to \infty$: no fine-tuning of the pertinent LEC is necessary. The resulting phase shifts (and one mixing parameter) do not provide anything like a precision description of NN data, but they are, at least, a well-defined, renormalized LO calculation.

In Refs. [4, 5] (Refs. [6, 7]) V was computed to $O(P^2)$ and $O(P^3)$ ($O(P^4)$), and the several NN LECs which appear in V were fitted to NN data for a range of cutoffs between 500 and 800 MeV. The $O(P^4)$ predictions contain very little residual cutoff dependence in this range of Λ's, and describe NN data with considerable accuracy.

However, several recent papers showed that a LO χET calculation does not yield stable predictions in partial waves with $L > 0$ once Λ is sufficiently large [8, 9, 10, 11]. This problem occurs because only one-pion exchange is present in these waves at LO. The resulting singular potential has no NN LEC that permits renormalization. Recently we examined this problem within the context of subtractive renormalization. We confirmed the conclusion of Ref. [9], i.e. any partial wave with $L > 0$ where one-pion exchange is attractive does not have stable LO results in χET. We also

12th International Conference on Meson-Nucleon Physics and the Structure of the Nucleon (MENU 2010)
AIP Conf. Proc. 1374, 317-320 (2011); doi: 10.1063/1.3647150

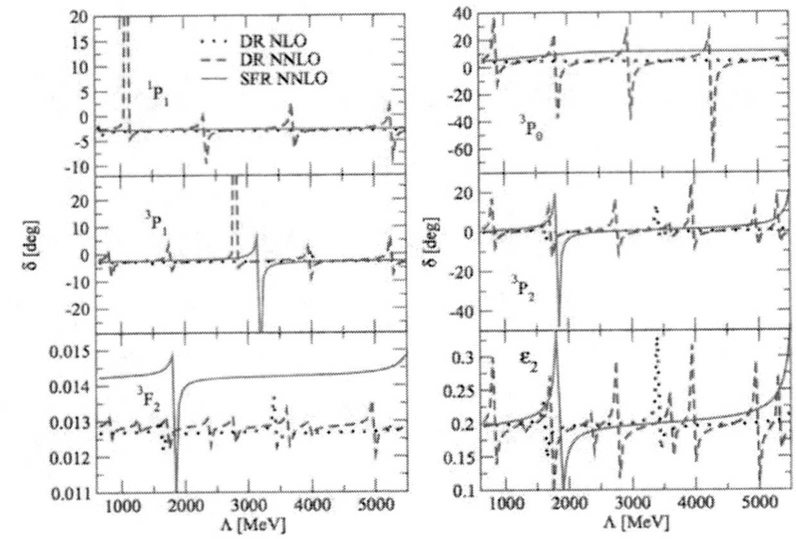

FIGURE 1. P-wave NN phase shifts at $T_{\rm lab} = 10$ MeV as a function of the Lippmann-Schwinger equation cutoff Λ. The black-dotted line is the $O(P^2)$ χET calculation with the potential obtained via dimensional regularization. The dashed red (solid green) line is for the $O(P^3)$ potential calculated using dimensional regularization (spectral-function regularization). From Ref. [12].

showed that this problem is not removed at $O(P^2)$ or $O(P^3)$. In particular, at $O(P^3)$ two-pion exchange produces a highly singular, attractive potential. The NN contact interactions needed to renormalize this potential are not present, e.g. in the $^3P_2$-$^3F_2$ channel. The resulting lack of stability with Λ of the NN phase shifts can be seen in Fig. 1. We identified the cutoff where such difficulties first occur as ≈ 1 GeV.

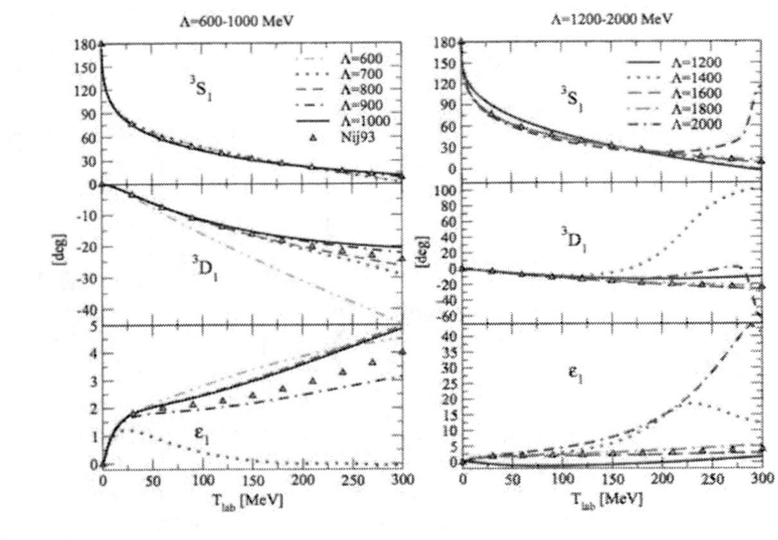

FIGURE 2. The phase shifts associated with NN scattering in the $J = 1$, $S = 1$ channel at various cutoffs Λ. The triangles are the 1993 Nijmegen PSA [14]. These results are for the $O(P^3)$ dimensionally regularized χET potential. From Ref. [13].

In Ref. [13] we used subtractive renormalization to calculate S-wave NN phase shifts from $O(P^2)$ and $O(P^3)$ χET potentials. Here too we found that the phase shifts are not stable once $\Lambda > 1$ GeV (see Fig. 2). In this case part of the problem is that the momentum-dependent contact interaction that appears at $O(P^2)$ has limited effect as $\Lambda \to \infty$ [15].

All of this indicates that χET as formulated above is not properly renormalized, i.e. the impact of short-distance physics on the results is not under control. Thus the NN potential obtained by straightforward application of χPT cannot be used over a wide range of cutoffs. It can perhaps be used if we employ Λ's in the vicinity of m_ρ [16, 17]. Since the short-distance physics of the effective theory for $p \gg m_\rho$ is different to the short-distance physics of QCD itself, it is not clear that considering $\Lambda \gg m_\rho$ yields any extra information about the real impact of short-distance physics on observables. Using low cutoffs also has the advantage that relevant momenta are demonstrably within the domain of validity of χPT. This justifies χET as a systematic theory, but at the cost of limiting us to $m_\pi \ll \Lambda < m_\rho$.

PREDICTIONS FOR ELASTIC ELECTRON-DEUTERON SCATTERING AT LOW Q^2

Chiral effective theory can also be used to organize electromagnetic current operators as an expansion in powers of P. The pertinent matrix elements are then constructed via:

$$\mathcal{M}_\mu = \langle \psi^{(f)} | \sum_{k=0}^{n} J_\mu^{(k)} | \psi^{(i)} \rangle, \tag{1}$$

where $|\psi^{(i)}\rangle$ and $|\psi^{(f)}\rangle$ are solutions of the Schrödinger equation for the χET potential V—in principle at order n. (For recent progress towards the completely consistent construction of V and J_μ see Refs. [18, 19].)

Here we focus on deuteron charge and quadrupole form factors, calculated as matrix elements of the NN charge operator J_0 using the standard (Breit-frame) formulae [20]. We compare χET predictions for G_C and G_Q with extractions from data for the deuteron structure function A and the tensor-polarization observable T_{20} [24]. (For χET calculations of the deuteron magnetic form factor, G_M, and deuteron photodisintegration see Refs. [20, 21, 22, 23].)

The leading piece of $J_0^{(s)}$ occurs at $O(e)$: $\langle \mathbf{p}' | J_0^{(s)}(\mathbf{q}) | \mathbf{p} \rangle = |e| \delta^{(3)}(p' - p - q/2) G_E^{(s)}(Q^2)$, with $G_E^{(s)}$ the nucleon's isoscalar electric form factor. Apart from tiny effects $\sim 1/M^2$ this is correct up to $O(eP^3)$ (counting $M \sim \Lambda_0$).

The LO calculation of G_C that uses this operator, together with the LO deuteron wave functions of Refs. [9, 26], provides a good description of the data out to $|\mathbf{q}| \approx 600$ MeV [27]. However, two-pion-exchange corrections affect the position of the form-factor minimum, and so must be considered for a precision description. If the potential at $O(P^3)$ is employed to calculate wave functions then the charge operator should be evaluated up to $O(eP^3)$. At that order J_0 contains a two-body contribution of one-pion range. The coefficient of this contribution is, however, determined by g_A, f_π, and the nucleon mass [20], and so results for G_C and G_Q are still predictions.

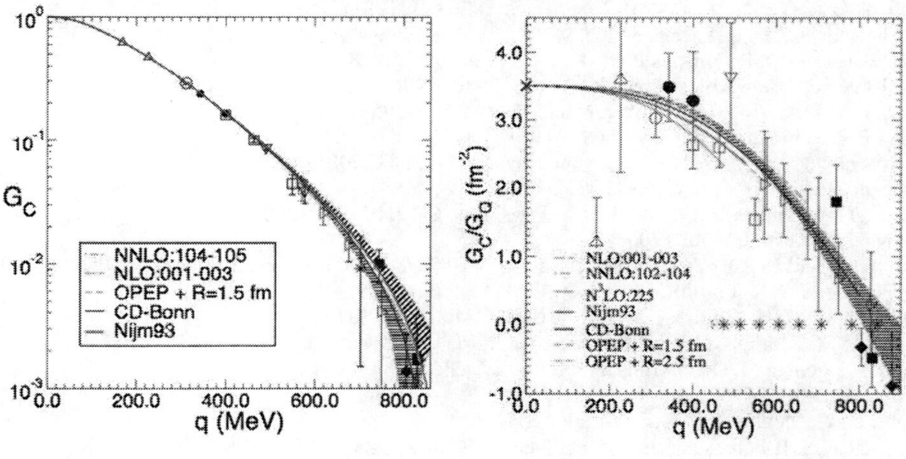

FIGURE 3. Predictions for G_C and G_C/G_Q with various (low-cutoff) wave functions, most of which include two-pion exchange. The bands represent the theoretical uncertainty and the stars are the values of $|\mathbf{q}|$ where BLAST has data. The data shown is from Ref. [24]. Figures from Ref. [25], which should be consulted for details.

The left panel of Fig. 3 shows that prediction for G_C for various wave functions. The consistent, $O(eP^3)$, χET result is obtained with the "NNLO" wave function, and is indicated by the diagonally shaded band. The variability comes from changes in the result as Λ is varied. In these calculations we keep $\Lambda < 1$ GeV, which ensures that the $O(P^2)$ and $O(P^3)$ corrections to the deuteron wave function are only small perturbations to the LO result. The $O(eP^3)$ χET result is, though, in markedly better agreement with the electron-deuteron scattering data than the LO $[O(e)]$ one.

In the $M \to \infty$ limit there are no two-pion-range two-body contributions to the isoscalar part of J_0 at $O(eP^4)$ [18, 28] or $O(eP^5)$ [28]. There are also no one-pion-range pieces of $J_0^{(s)}$ at $O(eP^4)$ [28]. At $O(eP^5)$ a short-distance operator

$\sim (N^\dagger N)^2 \nabla^2 A_0$ can affect the predictions for G_C. The coefficient of this operator is constrained by measurements of the deuteron charge radius. The NNLO wave functions of Ref. [7] with the $O(eP^4)$ J_0 yield $\langle r^2 \rangle_{\rm pt}^{1/2} = 1.975(1)$ fm. The uncertainty comes from short-distance differences between the wave functions. The deuteron isotope shift gives a value 1.9748(7) fm for the same quantity [29]. The very small difference between these two numbers constrains the LEC multiplying the $O(eP^5)$ short-distance contribution to $J_0^{(s)}$. This indicates that the prediction for G_C shown in Fig. 3 is accurate to $\pm 1.5\%$ at $Q^2 = 0.16$ GeV^2 [28]. This claim will be tested by forthcoming JLab data.

The situation is analogous, but plays out slightly differently, in the case of G_C/G_Q. At $O(eP^3)$ the quadrupole moment, Q_d, is quite sensitive to the short-distance physics included in the deuteron wave function. It varies by 2% when Λ is changed from 500 to 800 MeV. Intriguingly, this is roughly the size of the discrepancy between the Q_d predicted at $O(eP^3)$ and the experimental value $Q_d = 0.2859(3)$ fm^2. Ref. [25] considered the short-distance operator at $O(eP^5)$ that represents the contribution of modes above Λ_0 to G_Q. The LEC multiplying this short-distance $O(eP^5)$ operator was fixed by demanding that its coefficient be such that the experimental Q_d is reproduced. Thus, at $O(eP^5)$ χET cannot predict G_Q at $Q^2 = 0$, but it *can* predict the Q^2-dependence of G_Q. That prediction's theoretical uncertainty comes from the Q^2-dependence of short-distance NN physics, and is $\sim 3\%$ at $Q^2 = 0.16$ GeV^2. The inclusion of this short-distance, quadrupole operator leads to χET predictions for G_C/G_Q that have a different Q^2-dependence to that obtained in potential models. BLAST data will provide a significant test of this result.

ACKNOWLEDGMENTS

This research was supported by the U. S. Department of Energy, Office of Nuclear Physics, under contract No. DE-FG02-93ER40756 with Ohio University. I thank Charlotte Elster, Norbert Kaiser, Tae-Sun Park, Jerry Yang, and Evgeny Epelbaum, Jim Friar, and Stefan Kölling for fruitful collaboration and discussions on the topics presented here. I am also grateful to the organizers of MENU for putting together a broad and stimulating program.

REFERENCES

1. S. Weinberg, Phys. Lett. B **251**, 288 (1990); Nucl. Phys. B **363**, 3 (1991).
2. S. R. Beane, P. F. Bedaque, M. J. Savage and U. van Kolck, Nucl. Phys. A **700**, 377 (2002).
3. C.-J. Yang, Ch. Elster and D. R. Phillips Phys. Rev. C **77**, 014002 (2008).
4. C. Ordonez, L. Ray and U. van Kolck, Phys. Rev. C **53**, 2086 (1996).
5. E. Epelbaum, W. Glöckle and U.-G. Meißner, Nucl. Phys. A **671**, 295 (2000).
6. D. R. Entem and R. Machleidt, Phys. Rev. C **68**, 041001(R) (2003).
7. E. Epelbaum, W. Glöckle and U-G. Meißner, Nucl. Phys. A **747**, 362 (2005).
8. D. Eiras and J. Soto, Eur. Phys. J. A **17**, 89 (2003).
9. A. Nogga, R.G.E. Timmermans and U. van Kolck, Phys. Rev. C **72**, 054006 (2005).
10. M. C. Birse, Phys. Rev. C **74**, 014003 (2006).
11. M. Pavon Valderrama and E. Ruiz Arriola, Phys. Rev. C **74**, 064004 (2006) [Erratum-ibid. C **75**, 059905 (2007)].
12. C. J. Yang, C. Elster and D. R. Phillips, Phys. Rev. C **80**, 034002 (2009).
13. C. J. Yang, C. Elster and D. R. Phillips, Phys. Rev. C **80**, 044002 (2009).
14. V. G. J. Stoks, R. A. M. Klomp, M. C. M. Rentmeester and J. J. de Swart, Phys. Rev. C **48**, 792 (1993).
15. S. R. Beane, T. D. Cohen and D. R. Phillips, Nucl. Phys. A **632**, 445 (1998);
16. G. P. Lepage, arXiv:nucl-th/9706029.
17. E. Epelbaum and U.-G. Meißner, arXiv:nucl-th/0609037.
18. S. Kölling, E. Epelbaum, H. Krebs and U.-G. Meißner, Phys. Rev. C **80**, 045502 (2009).
19. S. Pastore, L. Girlanda, R. Schiavilla, M. Viviani and R. B. Wiringa, Phys. Rev. C **80**, 034004 (2009); S. Pastore, R. Schiavilla and J. L. Goity, Phys. Rev. C **78**, 064002 (2008).
20. D. R. Phillips, Phys. Lett. B **567**, 12 (2003).
21. M. Walzl and Ulf-G. Meißner, Phys. Lett. B **513**, 37 (2001).
22. T. S. Park, K. Kubodera, D. P. Min and M. Rho, Phys. Lett. B **472**, 232 (2000).
23. D. Rozpedzik, these proceedings.
24. D. Abbott *et al.*, Eur. Phys. J. A **7**, 421 (2000); D. M. Nikolenko *et al.*, Phys. Rev. Lett. **90**, 072501 (2003).
25. D. R. Phillips, J. Phys. G **34**, 365 (2007).
26. M. Pavon Valderrama and E. Ruiz Arriola, Phys. Rev. C **72**, 054002 (2005).
27. M. Pavon Valderrama, A. Nogga, E. Ruiz Arriola and D. R. Phillips, Eur. Phys. J. A **36**, 315 (2008).
28. T.-S. Park, N. Kaiser, D. R. Phillips, in preparation.
29. J. L. Friar, J. Martorell and D. W. L. Sprung, Phys. Rev. A **56**, 4579 (1997); J. L. Friar, private communication (2009).

Renormalization of one meson exchange potentials and their currents

A. Calle Cordón [1] and E. Ruiz Arriola

Departamento de Física Atómica, Molecular y Nuclear, Universidad de Granada, E-18071 Granada, Spain.

Abstract. The Nucleon-Nucleon One Meson Exchange Potential, its wave functions and related Meson Exchange Currents are analyzed for point-like nucleons. The leading N_c contributions generate a local and energy independent potential which presents $1/r^3$ singularities, requiring renormalization. We show how invoking suitable boundary conditions, neutron-proton phase shifts and deuteron properties become largely insensitive to the nucleon substructure and to the vector mesons. Actually, reasonable agreement with low energy data for realistic values of the coupling constants (e.g. SU(3) values) is found. The analysis along similar lines for the Meson Exchange Currents suggests that this renormalization scheme implies tremendous simplifications while complying with exact gauge invariance at any stage of the calculation.

Keywords: NN interaction, One Boson Exchange, Renormalization, Strong form factors, Large Nc , Chiral symmetry, Gauge invariance.
PACS: 03.65.Nk,11.10.Gh,13.75.Cs,21.30.Fe,21.45.+v

INTRODUCTION

The original idea of Yukawa that NN interaction at long distances is due to One Pion Exchange (OPE) was verified quantitatively by the Nijmegen benchmarking partial wave analysis for NN scattering in the elastic region with $\chi^2/\text{DOF} \sim 1$; a partial wave and energy dependent square well potential was considered for distances below 1.4-1.8 fm [1] while the neutral and charged pion masses could be determined from the fit to their currently accepted PDG values assuming OPE above such distances. The verification of other meson exchanges is less straightforward from NN elastic scattering since the shortest resolution distance probed at the pion production threshold is about $\lambda = \hbar/\sqrt{M_N m_\pi} \sim 0.5$fm and effectively short distance interactions admit a variety of parameterizations and forms [2, 3, 4] which show up quantitatively when nucleons are placed off-shell by the presence of a third particle. Already in the simplest case of a photon as the additional particle either in the initial or final state (see [5] and references therein) Gauge invariance relates meson exchange potentials and currents (MEC's) but also requires that Hamiltonian eigenstates are used to compute electroweak matrix elements. Clearly, one should not expect to know the longitudinal currents any better than potentials. Ambiguities in transverse, i.e. non-minimal coupling, terms of the current just reflect the composite character of Nucleons as well as their finite size. In the OBE potential, strong form factors are added to mimic this finite size [6], which is about 0.6fm. In the case of gauge invariance, the inclusion of a form-factor introduced by hand, i.e., not computed consistently within meson theory, implies a kind of non-locality in the interaction which could be made gauge invariant by introducing link operators between two points, thereby generating a path dependence, for which no obvious resolution has been found yet. Note that purely phenomenological potentials not based entirely on the Meson Exchange picture are inherently ambiguous. If the corresponding NN wave functions are combined with MEC's conflicting results with gauge invariance are eventually produced. Since the finite size of the nucleon is comparable to the minimal resolution probed in NN scattering, we do not expect to see the difference between a point-like nucleon and an extended one at sufficiently low energies. Recently [7] we have suggested replacing strong form factors in the NN potential by renormalization conditions on low energy scattering properties. In the lowest partial waves we have shown that *after* renormalization finite nucleon effects parameterized as strong form factors are indeed marginal. We suggest using renormalization ideas for potentials, wave functions and currents computed consistently. In the present contribution we review our findings [7] and apply them to analyze the radiative neutron capture, $n + p \to d + \gamma$, a nuclear reaction where the MEC are known to be essential [5].

[1] Speaker at 12th International Conference On Meson-Nucleon Physics And The Structure Of The Nucleon (MENU 2010) 31 May - 4 Jun 2010, Williamsburg, Virginia

12th International Conference on Meson-Nucleon Physics and the Structure of the Nucleon (MENU 2010)
AIP Conf. Proc. 1374, 321-324 (2011); doi: 10.1063/1.3647151

TABLE 1. Deuteron properties and low energy parameters in the ${}^3S_1 - {}^3D_1$ channel for OBE potentials including π,σ, ρ, ω mesons [7] as well as axial-vector meson a_1. We use the same numbers and notation as in Ref. [7]. Here AVMD means taking $m_{a_1} = \sqrt{2} m_\rho \simeq 1107$MeV and PDG taking $m_{a_1} = 1230$MeV. See also Ref. [14] and references therein.

	$\gamma(\mathrm{fm}^{-1})$	η	$A_S(\mathrm{fm}^{-1/2})$	r_m(fm)	$Q_d(\mathrm{fm}^2)$	P_D	$\langle r^{-1}\rangle$	α_0(fm)	$\alpha_{02}(\mathrm{fm}^3)$	$\alpha_2(\mathrm{fm}^5)$	r_0(fm)
$\pi\sigma\rho\omega a_1$ (AVMD)	Input	0.02557	0.8946	1.9866	0.2792	6.53%	0.639	5.467	1.723	6.621	1.722
$\pi\sigma\rho\omega a_1$ (PDG)	Input	0.02552	0.8937	1.9846	0.2780	6.58%	0.671	5.463	1.714	6.607	1.712
$\pi\sigma\rho\omega^* a_1$ (AVMD)	Input	0.02544	0.8966	1.9909	0.2788	5.90%	0.5087	5.477	1.720	6.604	1.734
$\pi\sigma\rho\omega^* a_1$ (PDG)	Input	0.02540	0.8951	1.9876	0.2773	6.01%	0.557	5.470	1.708	6.588	1.724
NijmII	Input	0.02521	0.8845(8)	1.9675	0.2707	5.635%	0.4502	5.418	1.647	6.505	1.753
Reid93	Input	0.02514	0.8845(8)	1.9686	0.2703	5.699%	0.4515	5.422	1.645	6.453	1.755
Exp.	0.231605	0.0256(4)	0.8846(9)	1.9754(9)	0.2859(3)	5.67(4)	–	5.419(7)	–	–	1.753(8)

MESON EXCHANGE POTENTIALS

A useful and simplifying assumption arises from our observation that the symmetry pattern of the sum rules for the old nuclear Wigner and Serber symmetries discussed in Refs. [8, 9] largely complies to the large N_c and QCD based contracted $SU(4)_C$ symmetry [10]. In the large N_c limit with $\alpha_s N_c$ fixed, nucleons are heavy, $M_N \sim N_c$, and the definition of the NN potential $\sim N_c$ makes sense. The tensorial spin-flavour structure was found to be [10]

$$V(r) = V_C(r) + \tau_1 \cdot \tau_2 \left[\sigma_1 \cdot \sigma_2 W_S(r) + S_{12} W_T(r) \right] \sim N_c \,. \tag{1}$$

Other operators such as spin-orbit or relativistic corrections are $\mathcal{O}(N_c^{-1})$ and hence suppressed by a relative $1/N_c^2$ factor. While these counting rules are directly obtained from quark-gluon dynamics, quark-hadron duality and confinement requires that above the confinement scale one can saturate Eq. (1) with multiple exchanges of mesons which have a finite mass for $N_c \gg 3$ [11]. We retain one boson exchange (OBE) with π,σ,ρ and ω and a_1 mesons [2]. The corresponding potential reads [3]

$$\begin{aligned}
V_C(r) &= -\frac{g_{\sigma NN}^2}{4\pi} \frac{e^{-m_\sigma r}}{r} + \frac{g_{\omega NN}^2}{4\pi} \frac{e^{-m_\omega r}}{r} \,, && (2) \\
W_S(r) &= \frac{g_{\pi NN}^2}{48\pi} \frac{m_\pi^2}{\Lambda_N^2} \frac{e^{-m_\pi r}}{r} + \frac{f_{\rho NN}^2}{24\pi} \frac{m_\rho^2}{\Lambda_N^2} \frac{e^{-m_\rho r}}{r} - \frac{g_{a_1 NN}^2}{6\pi} \frac{e^{-m_{a_1} r}}{r} \,, && (3) \\
W_T(r) &= \frac{g_{\pi NN}^2}{48\pi} \frac{m_\pi^2}{\Lambda_N^2} \frac{e^{-m_\pi r}}{r} \left[1 + \frac{3}{m_\pi r} + \frac{3}{(m_\pi r)^2} \right] - \frac{f_{\rho NN}^2}{48\pi} \frac{m_\rho^2}{\Lambda_N^2} \frac{e^{-m_\rho r}}{r} \left[1 + \frac{3}{m_\rho r} + \frac{3}{(m_\rho r)^2} \right] \\
&+ \frac{g_{a_1 NN}^2}{12\pi} \frac{e^{-m_{a_1} r}}{r} \left[1 + \frac{3}{m_{a_1} r} + \frac{3}{(m_{a_1} r)^2} \right] \,, && (4)
\end{aligned}$$

where $\Lambda_N = 3M_p/N_c$ and $g_{\sigma NN}, g_{\pi NN}, f_{\rho NN}, g_{\omega NN}, g_{a_1 NN} \sim \sqrt{N_c}$ and $m_\pi, m_\sigma, m_\rho, m_\omega, m_{a_1} \sim N_c^0$. To leading and subleading order in N_c one may neglect spin orbit, meson widths and relativity. The tensor force W_T is singular at short distances $\sim 1/r^3$ and requires renormalization. The renormalization is carried out in coordinate space using a boundary condition at a short distance cut-off r_c (see [7] for details) which makes the Hamiltonian self-adjoint for $r > r_c$. Besides being much simpler and efficient, this method allows to deal with cut-off independent potentials. In practice convergence is achieved for $r_c \sim 0.3$fm (see e.g. left panel Fig. 3 for the asymptotic D/S ratioη) [4].

Overall, the agreement is good for *realistic* couplings and masses as expected from other sources (see Ref. [7] for a short compilation) including a natural SU(3) value for $g_{\omega NN}$ coupling. The deuteron properties and low energy parameters are shown in table 1. The ${}^1S_0$ phase shift is reproduced for $m_\sigma \sim 500$MeV while the ${}^3S_1 - {}^3D_1$ phase shifts are plotted in Fig. 1. Space-like electromagnetic form factors in the impulse approximation [15] for $G_E^p(-\mathbf{q}^2) =$

---

[2] Other mesons such as η are sub-leading. Due to the $U_A(1)$ anomaly the η' meson appears to be heavy, but in the large N_c limit one it becomes degenerate with the pion $m_{\eta'} = m_\pi + \mathcal{O}(1/N_c)$. So, one might think that this meson would be as important as the pion itself. Actually this is not so since being an iso-scalar state it generates terms in the potential $V_S \sigma_1 \cdot \sigma_2$ and $V_T S_{12}$ which are $\mathcal{O}(1/N_c)$ and hence do not contribute to the leading potential in Eq. (1).

[3] In our previous work we left out the a_1 meson. We use the chiral Lagrangian [12] and take $g_{a_1 NN} = (m_{a_1}/m_\pi) f_{\pi NN} = 8.4$.

[4] Imposing a cut-off in momentum space generates an apparent delayed convergence due to the long distances distortion of the potential, so that unexpectedly large momentum cut-offs are needed. Renormalized results, however, agree in both $r-$ and $p-$spaces [13].

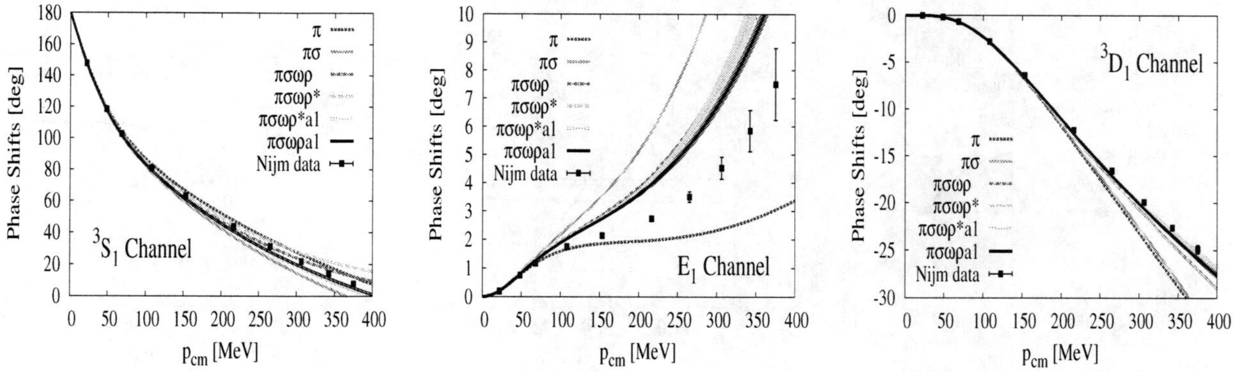

FIGURE 1. *np* spin triplet eigen phase shifts for the total angular momentum $j = 1$ as a function of the c.m. momentum compared to an average of the Nijmegen partial wave analysis and high quality potential models [2]. See table 1 for notation. The band in the case of the a_1 represents the error of changing m_{a_1} from the AVMD value to the PDG value.

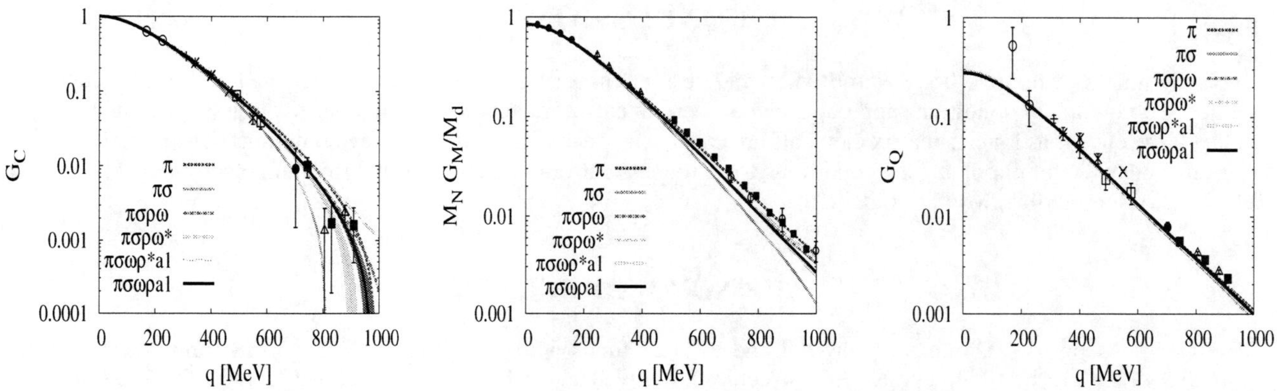

FIGURE 2. Deuteron charge (left), magnetic (middle) and quadrupole (right) form factors as a function of transfer q (in MeV). See Table 1 for notation. Data can be traced from Ref. [15] (see also references in [13]).

$1/(1+\mathbf{q}^2/m_\rho^2)^2$ and without MEC are plotted in Fig. 2 (see [13] for the π case). As we see, including shorter range mesons induces moderate changes, due to the expected short distance insensitivity embodied by renormalization, *despite* the short distance singularity and *without* introducing strong meson-nucleon-nucleon vertex functions [5].

MESON EXCHANGE CURRENTS

Gauge invariance is easily preserved within our coordinate space approach by keeping the *same* boundary condition as for the potential in the large N_c setup without need of new parameters (see also Ref. [18]). The simplest (purely transverse) MEC correction to the deuteron magnetic moment in the Impulse Approximation (IA) $\mu_d^{\rm IA} = (\mu_p + \mu_n) + \frac{3}{2}\left(\mu_p + \mu_n + \frac{1}{2}\right) P_D$ is shown in Fig. 3 (middle panel) as a function of r_c. Likewise, we show (right panel) the neutron capture cross section. The (longitudinal) MEC contribution yields a constant shift at relatively large distances. The different short distance behaviour between transverse and longitudinal MEC's will be elaborated elsewhere.

---

[5] We include only the OBE part of the leading N_c potential but multiple meson exchanges could also be added as well as Δ degrees of freedom to comply with large N_c counting rules [11]. Eq. (1) yields V_c at leading order in N_c. It is worth reminding that for chiral potentials $V_c = \mathcal{O}(g_A^4/(f_\pi^4 M_N))$ is Next-to-next-to-leading order (NNLO). Actually, as noted in Refs. [9, 16, 7] the expected large N_c behaviour [11] does not hold for the (Two Pion Exchange) chiral potentials *even* after inclusion of Δ [17]. Likewise, the Wigner symmetry pattern is not fulfilled for those chiral potentials [8, 9].

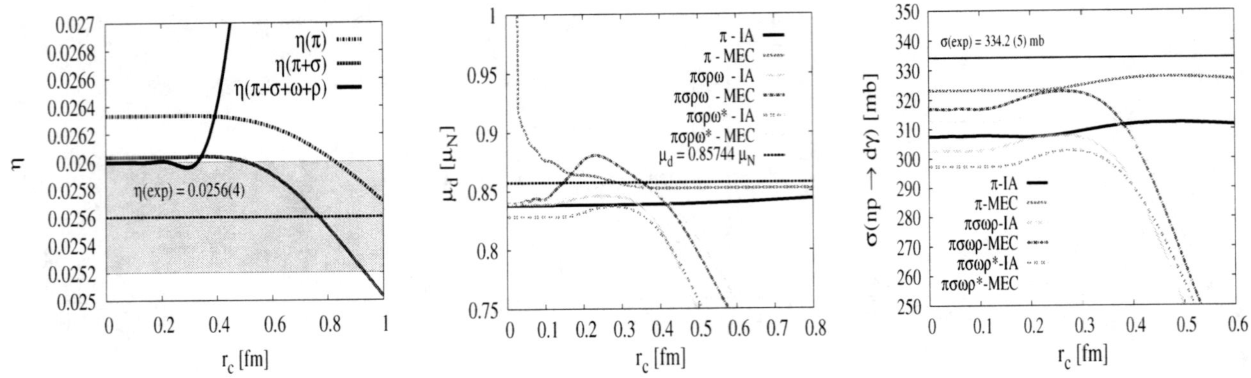

FIGURE 3. Short distance Cut-off, r_c (in fm), dependence of several observables including when necessary MEC in addition to Impulse Approximation (IA) compared with experimental bands. Asymptotic D/S ratio $\eta = 0.0256(4)$ (left). Deuteron magnetic moment $\mu_d = 0.85744\mu_N$ (middle). Neutron capture cross section and $\sigma(np \to d\gamma) = 334.2(5)$mb. (right)

CONCLUSIONS

Self-adjointness of the two-body Hamiltonian and current conservation are simply intertwined within the renormalization with boundary conditions approach above a certain cut-off distance. Current conservation is guaranteed at any value of the cut-off as long as matrix elements are consistently evaluated with NN wave functions constructed from the meson exchange Hamiltonian. The conditions for finiteness for both deuteron, scattering and electromagnetic matrix elements of longitudinal MEC's coincide.

ACKNOWLEDGMENTS

This work is supported by the Spanish DGI and FEDER funds with grant FIS2008-01143/FIS, Junta de Andalucía grant FQM225-05, and EU Integrated Infrastructure Initiative Hadron Physics Project contract RII3-CT-2004-506078.

REFERENCES

1. V. G. J. Stoks, R. A. M. Kompl, M. C. M. Rentmeester, and J. J. de Swart, *Phys. Rev.* **C48**, 792–815 (1993).
2. V. G. J. Stoks, R. A. M. Klomp, C. P. F. Terheggen, and J. J. de Swart, *Phys. Rev.* **C49**, 2950–2962 (1994).
3. R. B. Wiringa, V. G. J. Stoks, and R. Schiavilla, *Phys. Rev.* **C51**, 38–51 (1995), nucl-th/9408016.
4. R. Machleidt, *Phys. Rev.* **C63**, 024001 (2001), nucl-th/0006014.
5. D. O. Riska, *Phys. Rept.* **181**, 207 (1989).
6. R. Machleidt, K. Holinde, and C. Elster, *Phys. Rept.* **149**, 1–89 (1987).
7. A. Calle Cordon, and E. Ruiz Arriola, *Phys. Rev.* **C81**, 044002 (2010), 0905.4933.
8. A. Calle Cordon, and E. Ruiz Arriola, *Phys. Rev.* **C78**, 054002 (2008), 0807.2918.
9. A. Calle Cordon, and E. Ruiz Arriola, *Phys. Rev.* **C80**, 014002 (2009), 0904.0421.
10. D. B. Kaplan, and A. V. Manohar, *Phys. Rev.* **C56**, 76–83 (1997), nucl-th/9612021.
11. M. K. Banerjee, T. D. Cohen, and B. A. Gelman, *Phys. Rev.* **C65**, 034011 (2002), hep-ph/0109274.
12. V. G. J. Stoks, and T. A. Rijken, *Nucl. Phys.* **A613**, 311–341 (1997), nucl-th/9611002.
13. M. P. Valderrama, A. Nogga, E. Ruiz Arriola, and D. R. Phillips, *Eur. Phys. J.* **A36**, 315–328 (2008), 0711.4785.
14. J. J. de Swart, C. P. F. Terheggen, and V. G. J. Stoks (1995), nucl-th/9509032.
15. R. A. Gilman, and F. Gross, *J. Phys.* **G28**, R37–R116 (2002), nucl-th/0111015.
16. E. Ruiz Arriola, and A. Calle Cordon, *PoS* **EFT09**, 046 (2009), 0904.4132.
17. N. Kaiser, S. Gerstendorfer, and W. Weise, *Nucl. Phys.* **A637**, 395–420 (1998), nucl-th/9802071.
18. D. O. Riska, *Nucl. Phys.* **A710**, 55–82 (2002), nucl-th/0204016.

Large Q^2 electrodisintegration of the deuteron

Misak M. Sargsian

Department of Physics, Florida International University, Miami, FL 33199

Abstract. The break up of the deuteron is studied at high Q^2 kinematics, with main motivation to probe the deuteron at small internucleon distances. For this, two main issues are studied: electromagnetic interaction of the virtual photon with the bound nucleon and the strong interactions of the produced baryons in the final state of the reaction. Within virtual nucleon approximation we developed a new prescription to account for the bound nucleon effects in the electromagnetic interaction. The final state interaction at high Q^2 kinematics is calculated within generalized eikonal approximation (GEA). Comparison with the first experimental data confirm GEA's early prediction that the rescattering is maximal at $\sim 70^0$ of recoil nucleon production relative to the momentum of the virtual photon. Also the forward recoil nucleon angles are best suited for studies of the deuteron at small distances.

Keywords: Deuteron, High Energy Scattering
PACS: 25.10.+s, 11.80.Fv, 25.30.Fj, 25.30.Rw

Introduction

Deuteron represents an ideal micro-lab for studies of the structure of NN interaction ranging from the intermediate to very small distances (see e.g. [1, 2, 3].

The simplest reaction which could be used to investigate short-range NN interactions in nuclear targets is the exclusive electrodisintegration of the deuteron in which large magnitudes of the relative momentum of the pn system in the ground state are probed. Expectations that this can be achieved only at high-momentum transfer reactions[6, 4, 5, 7] was confirmed in a series of high energy experiments[8, 9, 10, 11, 12, 13] involving various nuclei.

Three experiments[14, 15, 16] have been performed recently using deuteron target at relatively high (up to 6 GeV) energy electron beam of the Jefferson Lab and more comprehensive experimental program will follow after the 12 GeV upgrade of the lab. This makes the development of theoretical approaches for the description of high Q^2 electro-nuclear processes involving deuteron a pressing issue. Since mid 90's there were an intensive efforts in developing such theoretical approaches [18, 19, 20, 21, 22, 23, 25, 26, 27, 24].

In this work[17] we study one particular process that is high Q^2 disintegration of the deuteron with probing large magnitudes of recoil nucleon momenta. The model is based on virtual nucleon approximation in which the struck nucleon is treated as an off-shell particle. The main theoretical framework is based on the generalized eikonal approximation (GEA)[20, 28, 7, 29, 30, 31] which allows us to represent the reaction through the set of covariant diagrams (Fig.1) for which effective Feynman diagram rules can be defined. Because of the covariance of the diagrams the virtualities involved in the scattering amplitudes are defined unambiguously. This allows us to develop a self consistent approach for accounting for the binding effects in the high Q^2 electromagnetic interaction off the bound nucleon. The second important feature of GEA is that final-state interaction of produced two nucleons is calculated without requiring stationary approximation for the recoil nucleons - this is important feature for calculating processes in which the recoil nucleon is produced with large momenta.

Main Assumptions of Virtual Nucleon Approximation

First, one considers only the pn component of the deuteron, neglecting inelastic initial state transitions. Since the deuteron is in a isosinglet state this will correspond to restricting the kinetic energy of recoil nucleon to

$$T_N < 2(m_\Delta - m_N) \sim (m_{N^*} - m) \sim 500 \text{ MeV} \tag{1}$$

where m, m_Δ and m_{N^*} are masses of the nucleon and low-lying non-strange baryonic resonances.

12th International Conference on Meson-Nucleon Physics and the Structure of the Nucleon (MENU 2010)
AIP Conf. Proc. 1374, 325-328 (2011); doi: 10.1063/1.3647152

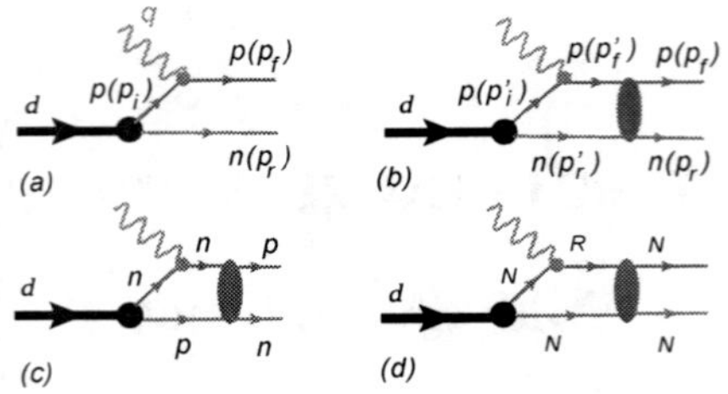

FIGURE 1. GEA diagrams

Then, one neglects by the negative energy projection of the virtual nucleon propagator. This can be justified if,

$$M_d - \sqrt{m^2 + p^2} > 0, \tag{2}$$

where M_d is the mass of the deuteron and p is the relative momentum of the bound pn system.

The third assumption which is made in the calculation is that at large Q^2 (> 1 GeV^2) the interaction of virtual photon with exchanged mesons are a small correction and can be neglected (see e.g. discussions in Ref.[7, 4]).

Generalized Eikonal Approximation

The assumptions discussed above allow us to restrict the consideration by the set of Feynman diagrams presented in Fig.(1), which can be calculated based on the effective Feynman diagram rules derived in GEA[7]. This work does not include calculation of the diagram of Fig.1d which is currently in progress.

Plane Wave Impulse Approximation Amplitude:
Starting with the covariant form of the amplitude of Fig1.a, we calculate explicitly the off shell part of the electromagnetic amplitude through the one-shell positive energy reduction of the propagator of the spectator nucleon[17]. This procedure results to the PWIA amplitude[17]:

$$\langle s_f, s_r \mid A_0^\mu \mid s_d \rangle = \sqrt{2}\sqrt{(2\pi)^3 2E_r}\sum_{s_i} J_N^\mu(s_f, p_f; s_i, p_i)\Psi_d^{s_d}(s_i, p_i, s_r, p_r), \tag{3}$$

with

$$J_N^\mu(s_f, p_f; s_i, p_i) = J_{N,on}^\mu(s_f, p_f; s_i, p_i) + J_{N,off}^\mu(s_f, p_f; s_i, p_i). \tag{4}$$

where $J_{N,on}^\mu$ is the covariant on-shell electromagnetic current of the nucleon and

$$J_{N,off}^\mu(s_f, p_f; s_i, p_i) = \bar{u}(p_f, s_f)\Gamma_{\gamma^* N}^\mu \gamma^0 u(p_i, s_i)\frac{E_i^{off} - E_i^{on}}{2m}, \tag{5}$$

in which $E_i^{off} = M_d - E_i^{on}$ and $E_i^{on} = \sqrt{m^2 + p_i^2}$, $\vec{p}_i = -\vec{p}_r$. Note that the total current in Eq.(4) is conserved since it is derived from the gauge invariant amplitude and no additional conditions are needed to restore the current conservation.

Forward Elastic and Charge Interchange Final State Interaction Amplitudes:
Applying the effective Feynman diagram rules to the diagrams of Fig.(1)b,c and projecting (similar to PWIA) the propagator of the spectator nucleon to its positive energy solution for the forward rescattering amplitude we obtain:

$$\langle s_f, s_r \mid A_1^\mu \mid s_d \rangle = \frac{i\sqrt{2}(2\pi)^{\frac{3}{2}}}{4} \sum_{s'_f, s'_r, s_i} \int \frac{d^2 p'_r}{(2\pi)^2} \frac{\sqrt{2\tilde{E}'_r}\sqrt{s(s-4m^2)}}{2\tilde{E}'_r |q|} \times$$
$$\langle p_f, s_f; p_r, s_r \mid f^{NN,on}(t,s) \mid \tilde{p}'_r, s'_r; \tilde{p}'_f, s'_f \rangle \cdot J_N^\mu(s'_f, p'_f; s_i, \tilde{p}'_i) \cdot \Psi_d^{s_d}(s_i, \tilde{p}'_i, s'_r, \tilde{p}'_r)$$

$$-\frac{\sqrt{2}(2\pi)^{\frac{3}{2}}}{2}\sum_{s'_f,s'_r,s_i}\mathscr{P}\int\frac{dp'_{r,z}}{2\pi}\int\frac{d^2p'_r}{(2\pi)^2}\frac{\sqrt{2E'_r}\sqrt{s(s-4m^2)}}{2E'_r|\mathbf{q}|}\times$$

$$\frac{\langle p_f,s_f;p_r,s_r\mid f^{NN,off}(t,s)\mid p'_r,s'_r;p'_f,s'_f\rangle}{p'_{r,z}-\tilde{p}'_{r,z}}J^{\mu}_N(s'_f,p'_f;s_i,p'_i)\cdot\Psi^{s_d}_d(s_i,p'_i,s'_r,p'_r), \qquad (6)$$

where $\tilde{p}'_r=(p_{r,z}-\Delta,p'_{r,\perp})$, $\tilde{E}'_r=\sqrt{m^2+\tilde{p}'^2_r}$, $\tilde{p}'_i=p_d-\tilde{p}'_r$ and $\tilde{p}'_f=\tilde{p}'_i+q$.

In high energy limit in which the helicity conservation of small angle NN scattering is rather well established the on-shell amplitude is predominantly imaginary and can be parameterized in the form

$$\langle p_f,s_f;p_r,s_r\mid f^{NN,on}(t,s)\mid \tilde{p}'_r,s'_r;\tilde{p}'_f,s'_f\rangle=\sigma^{pn}_{tot}(i+\alpha)e^{\frac{B}{2}t}\delta_{s_f,s'_f}\delta_{s_r,s'_r}, \qquad (7)$$

where $\sigma^{pn}_{tot}(s)$, $B(s)$ and $\alpha(s)$ are found from fitting of experimental data on elastic $pn\to pn$ scattering. For the half-off-shell part of the $f^{NN,off}$ amplitude we use the following parameterization:

$$f^{NN,off}=f^{NN,on}e^{B(m^2_{off}-m^2)}, \qquad (8)$$

where $m^2_{off}=(M_d-E'_r+q_0)^2-(p'_r+q)^2$.

For the charge-exchange final state interaction amplitude (Fig.1c) the derivation is similar and it is expressed through the charge-exchange $pn\to np$ scattering amplitude.

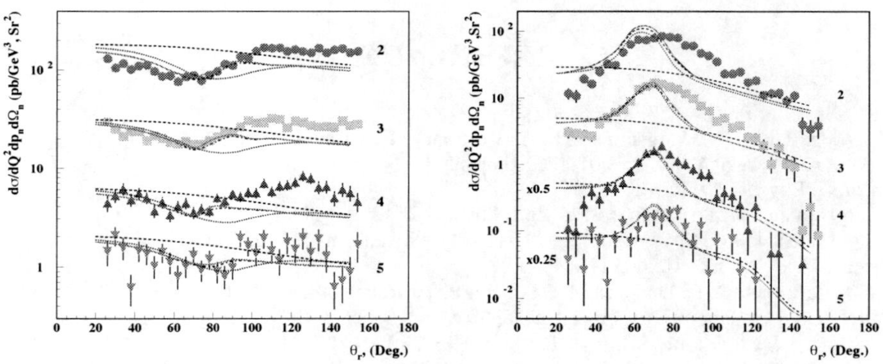

FIGURE 2. Dependence of the differential cross section on the direction of the recoil neutron momentum. The data are from Ref.[15]. Dashed line - PWIA calculation, dotted line - PWIA+ pole term of forward FSI, dash-dotted line - PWIA+forward FSI, solid line - PWIA + forward and charge exchange FSI. The momenta of the recoil neutron in (a) and (b) are restricted to $200<p_r<300$ MeV/c and $400<p_r<600$ MeV/c respectively. The labels 2, 3, 4 and 5 correspond to the following values of $Q^2=2\pm0.25;3\pm0.5;4\pm0.5;5\pm0.5$ GeV^2. The data sets and calculations for "4" and "5" in (b) are multiplied by 0.5 and 0.25 respectively

Deuteron Wave Function:

To fix the normalization of the deuteron wave function we use the fact that the deuteron elastic charge form-factor $G_C\to1$ at $Q^2\to0$. Using above condition in the elastic deuteron scattering amplitude one can relate the deuteron wave function in virtual nucleon approximation to the nonrelativistic deuteron wave function as follows[17]

$$\Psi_d(p)=\Psi^{NR}_d(p)\frac{M_d}{2(M_d-\sqrt{m^2+p^2})}. \qquad (9)$$

This relation provides a smooth transition to the nonrelativistic wave function Ψ^{NR} in the small momentum limit.

Comparison with Experimental Data and Some Conclusions

In the last few years three experiments[14, 15, 16] produced the first data at relatively large Q^2 kinematics. The experiment of Ref.[15]) covered the highest to date Q^2 range from $2-5$ GeV^2. Comparison with these experimental data are given in Fig.2 which allow us to make the following conclusions:

- the angular distribution clearly exhibits an eikonal feature, with the minimum (Fig.2(a)) or maximum (Fig.2(b)) at transverse kinematics due to the final state interaction. The most important result is that the maximum of FSI is at recoil angles of 70^0 in agreement with the GEA prediction of Ref.[28]. Note that the conventional Glauber theory predicted 90^0 for the FSI maximum.

- The disagreement of the calculation with the data at $\theta_r > 70^0$ appears to be due to the isobar contribution at the intermediate state of the reaction. This region corresponds to $x < 1$ and it is kinematically closer to the threshold of Δ-isobar electroproduction. The comparisons also indicate that the relative strength of the Δ-isobar contribution diminishes with an increase of Q^2 and at neutron production angles $\theta_r \rightarrow 180^0$.

- The forward direction of the recoil nucleon momentum, being far from the Δ-isobar threshold, exhibits a relatively small contribution due to FSI. This indicates that the forward recoil angle region is best suited for studies of PWIA properties of the reaction such as the off-shell electromagnetic current and deuteron wave function.

These comparisons clearly show that the forward angles of spectator nucleon production is best suited for isolating PWIA scattering off the virtual nucleon. Therefore this kinematic region provides the most optimal condition for probe and studying the NN interaction at short internucleon separations.

Acknowledgments:
This work is supported by the U.S. Department of Energy Grant under Contract DE-FG02-01ER41172.

REFERENCES

1. R. A. Gilman and F. Gross, J. Phys. G **28**, R37 (2002).
2. L. L. Frankfurt, G. A. Miller, M. M. Sargsian and M. I. Strikman, Phys. Rev. Lett. **84**, 3045 (2000).
3. C. G. Granados and M. M. Sargsian, arXiv:1007.4705 [hep-ph].
4. M. M. Sargsian *et al.*, J. Phys. G **29**, R1 (2003).
5. L. Frankfurt, M. Sargsian and M. Strikman, Int. J. Mod. Phys. A **23**, 2991 (2008).
6. L. L. Frankfurt and M. I. Strikman, Phys. Rept. **76**, 215 (1981); Phys. Rept. **160**, 235 (1988).
7. M. M. Sargsian, Int. J. Mod. Phys. E **10**, 405 (2001).
8. L. L. Frankfurt, M. I. Strikman, D. B. Day and M. Sargsian, Phys. Rev. C **48**, 2451 (1993).
9. K. Sh. Egiyan *et al.* [CLAS Collaboration], Phys. Rev. C **68**, 014313 (2003).
10. K. Sh. Egiyan *et al.* [CLAS Collaboration],Phys. Rev. Lett. **96**, 082501 (2006).
11. E. Piasetzky, M. Sargsian, L. Frankfurt, M. Strikman and J. W. Watson, Phys. Rev. Lett. **97**, 162504 (2006).
12. R. Shneor *et al.* [Jefferson Lab Hall A Collaboration], Phys. Rev. Lett. **99**, 072501 (2007).
13. R. Subedi *et al.*, Science **320**, 1476 (2008).
14. P. E. Ulmer *et al.*, Phys. Rev. Lett. **89**, 062301 (2002).
15. K. Sh. Egiyan *et al.* [the CLAS Collaboration], Phys. Rev. Lett. **98**, 262502 (2007).
16. W. U. Boeglin, Eur. Phys. J. A **24**, 77 (2005); L. Coman, " $^2H(e,e'p)$ Studies of the Deuteron at High Q^2", PhD. Thesis (*unpublished*), (2008).
17. M. M. Sargsian, Phys. Rev. C **82**, 014612 (2010).
18. L. Frankfurt, E. Piasetsky, M. Sargsian and M. Strikman, Phys. Rev. C **51**, 890 (1995).
19. A. Bianconi, S. Jeschonnek, N. N. Nikolaev and B. G. Zakharov, Phys. Lett. B **343**, 13 (1995).
20. L. L. Frankfurt, W. R. Greenberg, G. A. Miller, M. M. Sargsian and M. I. Strikman, Z. Phys. A **352**, 97 (1995).
21. L. L. Frankfurt, E. Piasetzky, M. M. Sargsian and M. I. Strikman,Phys. Rev. C **56**, 2752 (1997).
22. S. Jeschonnek, Phys. Rev. C **63**, 034609 (2001).
23. C. Ciofi degli Atti, L. P. Kaptari and D. Treleani, Phys. Rev. C **63**, 044601 (2001).
24. J. M. Laget, Phys. Lett. B **609**, 49 (2005).
25. S. Jeschonnek and J. W. Van Orden, Phys. Rev. C **78**, 014007 (2008); Phys. Rev. C **80**, 054001 (2009).
26. S. Jeschonnek and J. W. Van Orden, Phys. Rev. C **81**, 014008 (2010).
27. J.J. Adam, F. Gross, S. Jeschonnek, P. Ulmer and J. W. Van Orden, Phys. Rev. C **66**, 044003 (2002).
28. L. L. Frankfurt, M. M. Sargsian and M. I. Strikman, Phys. Rev. C **56**, 1124 (1997).
29. M. M. Sargsian, T. V. Abrahamyan, M. I. Strikman and L. L. Frankfurt, Phys. Rev. C **71**, 044614 (2005).
30. C. Ciofi delgi Atti and L. P. Kaptari,Phys. Rev. Lett. **100**, 122301 (2008).
31. M. Alvioli, C. Ciofi degli Atti and L. P. Kaptari, Phys. Rev. C **81**, 021001 (2010)

Two-pion exchange currents in photodisintegration of the deuteron

D. Rozpędzik*, J. Golak*, S. Kölling†,** and E. Epelbaum‡

*M. Smoluchowski Institute of Physics, Jagiellonian University, PL-30059 Kraków, Poland
†Forschungszentrum Jülich, Institut für Kernphysik (IKP-3), D-52425 Jülich, Germany
**Helmholtz-Institut für Strahlen- und Kernphysik (Theorie) and Bethe Center for Theoretical Physics, Universität Bonn, D-53115 Bonn, Germany
‡Institut für Theoretische Physik II, Ruhr-Universität Bochum, D-44780 Bochum, Germany

Abstract. Chiral effective field theory (ChEFT) is a modern framework to analyze the properties of few-nucleon systems at low energies [1]. It is based on the most general effective Lagrangian for pions and nucleons consistent with the chiral symmetry of QCD. For energies below the pion-production threshold it is possible to eliminate the pionic degrees of freedom and derive nuclear potentials and nuclear current operators solely in terms of the nucleonic degrees of freedom. This is very important because, despite a lot of experience gained in the past, the consistency between two-nucleon forces, many-nucleon forces and the corresponding current operators has not been achieved yet. In this presentation we consider the recently derived long-range two-pion exchange (TPE) contributions to the nuclear current operator which appear at next-to leading order of the chiral expansion [2]. These operators do not contain any free parameters. We study their role in the deuteron photodisintegration reaction and compare our predictions with experimental data from Refs [3]. The bound and scattering states are calculated using five different chiral N2LO nucleon-nucleon (NN) potentials which allows to estimate the theoretical uncertainty at a given order in the chiral expansion. For some observables the results are very close to the reference predictions based on the AV18 NN potential and the current operator (partly) consistent with this force.

Keywords: ChEFT, electromagnetic reactions, exchange currents
PACS: 21.30.-x, 21.45.Bc, 24.70.+s

INTRODUCTION

Chiral effective field theory (ChEFT) provides a systematic and model-independent framework to analyze hadron structure and dynamics in harmony with the spontaneously broken approximate chiral symmetry of QCD. This approach is a powerful tool for the derivation of the nuclear forces. Exchange vector and axial currents in nuclei have also been studied in the framework of ChEFT. Since the pioneering work of Park et al. [4], heavy-baryon chiral perturbation theory has been applied to derive exchange axial and vector currents for small values of the photon momentum. These calculations were carried out in time-ordered perturbation theory. The resulting exchange vector currents were, in particular, applied to analyze radiative neutron-proton capture within a hybrid approach.

ChEFT has been also used to study the electromagnetic properties of the deuteron [5]. One of the most fundamental processes on the deuteron is the photodisintegration reaction. It has been a subject of intensive experimental and theoretical research for several decades (see Refs [6]). However, very few applications to this reaction at small momentum transfer have been carried out so far. A recent review on the theoretical achievements in this field based on conventional framework can be found in Ref. [7]. A strong interest for this reaction within ChEFT, especially in view of planned experiments, resulted in the application of chiral framework. This requires a consistent derivation of the nuclear Hamiltonian and the electromagnetic current operator for a given few-nucleon system.

In the two-nucleon (2N) system, the leading contributions to the exchange current originate from one-pion exchange and are well known. The corrections to the 2N current operator at the leading one loop level in ChEFT were recently worked out by Pastore et al. in Ref. [8] based on time-ordered perturbation theory. Independently, the two-pion exchange 2N current operator was derived in Ref. [2] using the method of unitary transformation. The resulting current operator is consistent with the corresponding chiral two-nucleon potential [1] obtained within the same scheme. In the present work, we explore the effects of the leading two-pion exchange 2N operator [2] in the deuteron photodisintegration reaction. We emphasize, however, that the presented calculation is not yet complete. In particular, the corresponding expressions for the one-pion exchange and short-range contributions to the current operator within the method of unitary transformation are not yet available. The important outcome of our investigation is that various observables in the deuteron photodisintegration appear to be highly sensitive to the two-pion exchange current.

12th International Conference on Meson-Nucleon Physics and the Structure of the Nucleon (MENU 2010)
AIP Conf. Proc. 1374, 329-332 (2011); doi: 10.1063/1.3647153

FORMALISM

The formalism to describe 2N electromagnetic reactions requires the knowledge of the consistent potential and electromagnetic current. The NN potential based on ChEFT is well known up to next-to-next-to-next-to-leading order in the chiral expansion. As already pointed out, we restrict ourselves in this study to next-to-leading order (NLO) long-range contributions to the 2N current operator. To achieve a better accuracy, all calculations are made using the N2LO potential. At this order, it contains the one-pion-exchange (OPE) (1π) and two-pion exchanges (TPE) (2π) contributions as well as various contact interactions (*cont*) [1]

$$V_{2N} = V_{1\pi} + V_{2\pi} + V_{cont} \,. \tag{1}$$

The effective current operator for the 2N system is a sum of the single-nucleon operators $J^\mu(i), i = 1,2$ and two-nucleon operators of different type ($J^\mu(1,2)$)

$$J^\mu_{2N} = J^\mu(1) + J^\mu(2) + J^\mu_{1\pi}(1,2) + J^\mu_{2\pi}(1,2) + J^\mu_{cont}(1,2)\,, \tag{2}$$

where the expressions for the single-nucleon and OPE currents are well established.

In this contribution, we concentrate on a treatment of the long-range TPE contributions to the 2N current operator derived in Ref. [2]. The expressions for the TPE current operator in momentum space are rather involved and contain the standard loop functions and the three-point functions in a form suitable for numerical calculations.

The TPE four-current operator $J^\mu = (J^0, \vec{J})$ can be decomposed according to its isospin-spin-momentum structure and quite generally written in the form

$$J^0 = \sum_{\alpha=1}^{5} \sum_{\beta=1}^{8} f_\alpha^{\beta S}(\vec{q}_1, \vec{q}_2) T_\alpha O_\beta^S \,, \qquad \vec{J} = \sum_{\alpha=1}^{5} \sum_{\beta=1}^{24} f_\alpha^{\beta}(\vec{q}_1, \vec{q}_2) T_\alpha \vec{O}_\beta \,, \tag{3}$$

where $\vec{q}_i \equiv \vec{p}' - \vec{p}$ is the momentum transferred to nucleon *i*, T_α is the 2N isospin operator, O_β^S and $\vec{O}_\beta$ are the (momentum dependent) spin operators in the 2N space, $f_\alpha^{\beta S}$ and f_α^{β} are scalar functions. The explicit form of the scalar functions can be found in Ref. [2]. The expressions do not contain any free parameters. Here and in what follows, we use the operator basis for O_β^S and $\vec{O}_\beta$ given in Ref. [2]. Due to their isospin structure, not all combinations of (3) contribute to photodisintegration of the deuteron. The non-vanishing contributions emerge from

$$\vec{J} = \sum_{\beta=3}^{10} f_2^\beta(\vec{q}_1, \vec{q}_2) T_2 \vec{O}_\beta + f_3^2(\vec{q}_1, \vec{q}_2) T_3 \vec{O}_2 \,. \tag{4}$$

Further information about these operators can be found in Ref. [2].

We choose a chiral potential V_{2N} to generate the deuteron bound state, $|\Psi_{bound}\rangle$, and the proton-neutron (pn) scattering state $|\Psi^{pn}_{scatt}\rangle$. These are needed in order to obtain the nuclear matrix element N^μ

$$N^\mu \equiv \langle \Psi_{scatt} | J^\mu_{2N} | \Psi_{bound} \rangle \,, \tag{5}$$

from which all observables can be calculated. We use the solution of the Lippmann-Schwinger equation, $t = V_{2N} + tG_0V_{2N}$, in order to express N^μ as

$$N^\mu = \langle \vec{p}_0 \mid (1 + tG_0)\, J^\mu_{2N} \mid \Psi_{bound} \rangle, \tag{6}$$

where G_0 is the free 2N propagator and $\vec{p}_0$ is the relative pn momentum in the final state.

We work in momentum space and employ the standard partial wave decomposition of the chiral potential, see for example Ref. [9]. Thus we have to express the TPE current operator in the same partial wave basis. To this end, we first prepare all the spin and isospin matrix elements using *Mathematica*© and then calculate the resulting four-fold angular integrals (7) on the parallel supercomputer IBM Blue Gene/P of the Jülich Supercomputing Centre (JSC).

$$\begin{aligned} &\langle p'(l's')j'm'; t'm_{t'} | J^\mu_{\alpha\beta} | p(ls)jm; tm_t \rangle \\ = &\int d\hat{p}' \int d\hat{p} \sum_{m_l'} \sum_{m_l} C(l's'j'; m_{l'}, m' - m_{l'}, m') Y^*_{l'm_{l'}}(\hat{p}') C(lsj; m_l, m - m_l, m) Y_{lm_l}(\hat{p}) \\ \times \; & f_\alpha^\beta(\vec{q}_1, \vec{q}_2) \langle t'm_{t'} | T_\alpha | tm_t \rangle \langle s'm' - m_{l'} | O_\beta | sm - m_l \rangle \end{aligned} \tag{7}$$

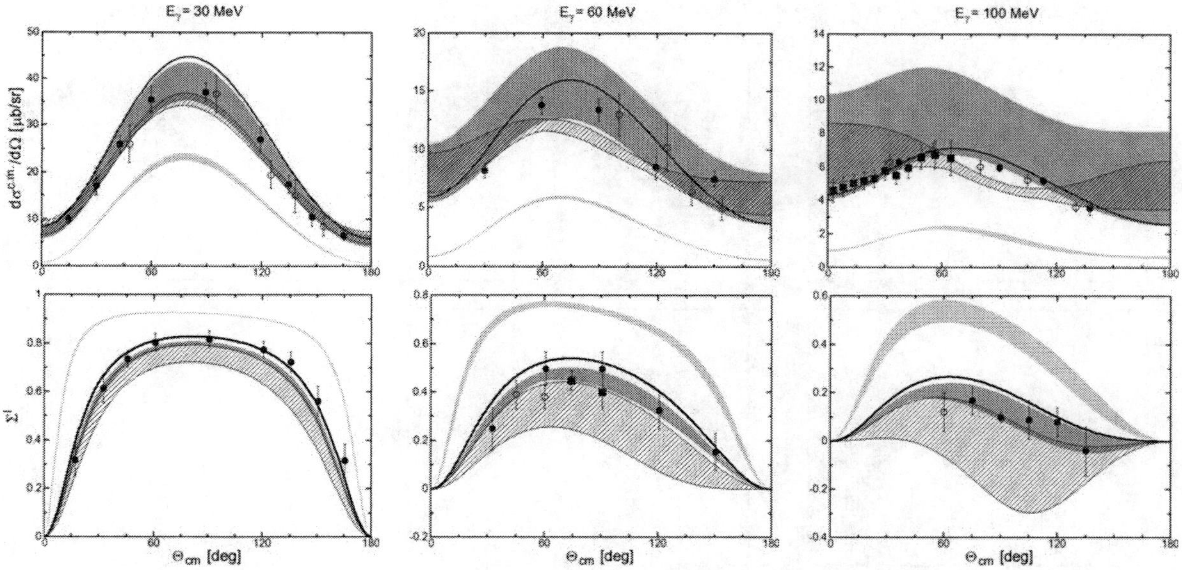

FIGURE 1. (color online) The results for the unpolarized cross section and the photon analyzing power in the deuteron photodisintegration process at the photon laboratory energies of E_γ= 30, 60, 100 MeV, displayed as functions of the proton emission angle. The solid black line refers to the standard calculation based on the AV18 potential,the blue (light) band covers results obtained with the single-nucleon current only, the dashed band represents the predictions based on the single-nucleon and OPE parts and the red (dark) band includes, in addition, the contributions of the TPE current. The experimental data are from Ying et al. [3].

RESULTS

We now discuss the results for the deuteron photodisintegration process for the unpolarized cross section and selected polarization observables. The results for the differential cross section and the photon analyzing power at the photon laboratory energies of $E_\gamma = 30, 60, 100$ MeV are shown in Figure 1. Note that the theoretical predictions depend on the two cut-off parameters that appear in the chiral potential. While the first cut-off parameter Λ appears in the regulator function for the Lippmann-Schwinger equation, the second parameter $\tilde{\Lambda}$ enters the spectral function regularization (SFR) and denotes the ultraviolet cut-off value in the mass spectrum of the two-pion-exchange potential. The bands reflect the dependence of the results on variations of Λ and $\tilde{\Lambda}$. Following Ref. [1], the cut-off values are varied between 450 and 600 MeV for Λ and between 500 and 700 MeV for $\tilde{\Lambda}$. The plots show the contributions from the different parts of the 2N current: single-nucleon current (blue band), one-pion exchange contribution (dashed band) and the long-range TPE contributions (red band). As a reference, we also show the results based on the phenomenological AV18 potential [10] and the corresponding exchange currents. We have also calculated the deuteron tensor analyzing powers as a function of the proton emission angle for three E_γ - energy bins. These results are visualized in Figure 2.

We see that in all observables the 2N current operator plays an important role and a restriction to the single nucleon current operator only leads to a strong disagreement with the data. One also observes a rather good agreement between the calculations based on the chiral potential/current operator and the ones based on the AV18 potential. In general, our results agree reasonably well with the experimental data. Definite conclusions can, however, only be drawn after carrying out a complete calculation including *all* ingredients present at this chiral order. This work is in progress. We also expect that such a complete calculation will yield a much narrower bands as the cut-off dependence should, to a large extent, be absorbed into the "running" of the corresponding short-range current operators.

CONCLUSIONS

We explored the effects of the TPE currents derived recently in the framework of ChEFT in the deuteron photodisintegration reaction. We also studied the effects of various other ingredients of the chiral 2N current operator in the unpolarized cross section and several polarization observables. As a main outcome of our study, we find that the new

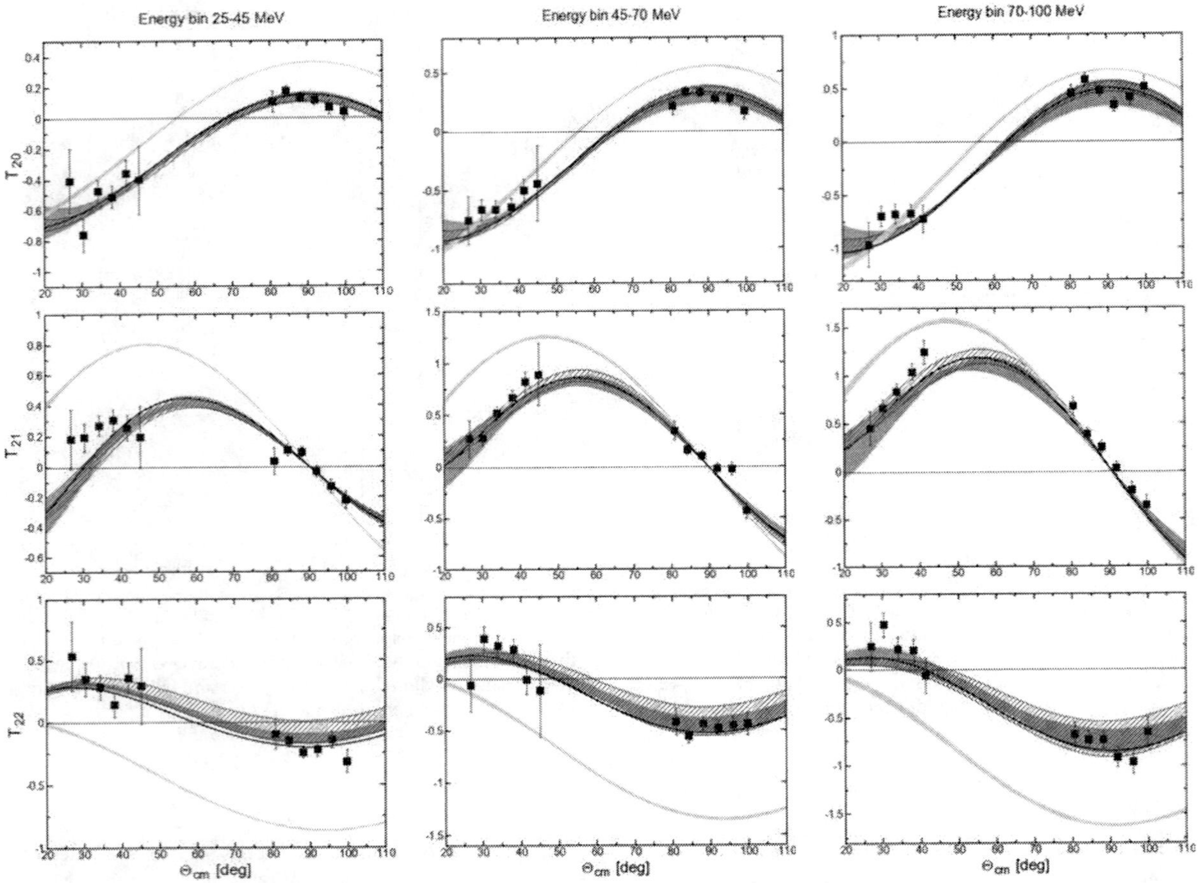

FIGURE 2. (color online) The deuteron tensor analyzing powers vs. proton emission angle for three E_γ- energy bins. The bands and lines have the same meaning as in Figure 1. The theoretical predictions are compared to the experimental data from Rachek et al. [3]. The pictures do not show systematic errors.

terms in the exchange current operator beyond the well-known one-pion exchange contribution play an important role in this reaction. Our dynamical framework is still in development. The inclusion of the consistent short-range contact and higher-order OPE contributions to the 2N current operator is in progress. We also plan to extend these calculations to the three-nucleon system. Finally, we emphasize that new, hight-quality data will be necessary to fix unknown parameters in the short-range part of the exchange current and to test the predictions of ChEFT.

This work was supported in part by the Polish Ministry of Science and Higher Education (grants N N202 104536, N N202 077435).

REFERENCES

1. E. Epelbaum, *Prog. Part. Nucl. Phys.* **57**, 654 (2006) and references therein.
2. S. Kölling et al., *Phys. Rev. C* **80**, 045502 (2009).
3. S. Ying, et al., *Phys. Rev. C* **38**, 1584–1600 (1988); I. A. Rachek et al., *Phys. Rev. Lett.* **98**, 182303 (2007).
4. T. S. Park et al. *Phys. Rept.* **233**, 341 (1993).
5. M. Walzl and U.-G. Meißner, *Phys. Lett. B* **513**, 37 (2001); D. Phillips, *PoS CD***09**, 066 (2009).
6. H. Arenhövel et al., Few-Body Sys. Suppl. **3**, 1 (1991); R. Gilman et al., *J. Phys. G, Nucl. Part. Phys.* **28**, R37 (2002).
7. J. Golak et al. *Phys. Rept.* **415**, 89 (2005) and references therein.
8. S. Pastore et al., *Phys. Rev. C*, **78**, 064002 (2008).
9. W. Glöckle et al. The QM Few-Body Problem (1983).
10. R. B. Wiringa et al., *Phys. Rev. C* **51**, 38 (1995).

Session G: Photoproduction

Regge-plus-resonance predictions for neutral-kaon photoproduction from the deuteron

P. Vancraeyveld[1], L. De Cruz and J. Ryckebusch

Department of Physics and Astronomy, Ghent University, Proeftuinstraat 86, B-9000 Gent, Belgium

Abstract. We present a Regge-inspired effective-Lagrangian framework for neutral-kaon photoproduction from the deuteron. Quasi-free kaon production is investigated using the Regge-plus-resonance elementary operator within the relativistic plane-wave impulse approximation. The Regge-plus-resonance model was developed to describe photoinduced and electroinduced charged-kaon production off protons. We show how this elementary operator can be transformed to account for the production of neutral kaons from both protons and neutrons. Our results compare favourably to the sole ${}^{2}\mathrm{H}(\gamma,K^{0})YN$ dataset published to date.

Keywords: Regge phenomenology, kaon photoproduction, nuclear reactions
PACS: 11.10.Ef, 11.55.Jy, 12.40.Nn, 13.60.Le, 25.20.Lj

Electromagnetic production of strangeness plays a prominent role in current efforts to chart the excitation spectrum of the nucleon. Since the production mechanism inevitably involves quark-antiquark components of the nucleon's sea, it offers the opportunity to improve on our understanding of the strong interaction in the confinement regime. Six different kaon photoproduction channels exist which can be treated within a single theoretical framework. The current kaon photoproduction database is heavily dominated by the $p(\gamma,K^{+})\Lambda$ and $p(\gamma,K^{+})\Sigma^{0}$ channels. A comprehensive survey of the four remaining reactions, namely $n(\gamma,K^{0})\Lambda$, $p(\gamma,K^{0})\Sigma^{+}$, $n(\gamma,K^{0})\Sigma^{0}$ and $n(\gamma,K^{+})\Sigma^{-}$, would provide useful complementary information to elucidate the strangeness-production reaction mechanism. In this respect a trustworthy description of kaon production from the deuteron is of chief importance, for the deuteron's weak binding energy makes it a prime neutron target. In this contribution, we focus on the semi-inclusive production of neutral kaons from a deuteron target as it is the only channel for which data is available [1].

Whilst strangeness production is an important tool in nucleon spectroscopy and is in part mediated by nucleon-resonance exchange, the smooth energy dependence of the measured observables hints at a dominant role for the background, i.e. non-resonant, processes. The Regge-plus-resonance (RPR) approach to kaon production seeks to decouple the determination of the coupling constants for the background and the resonant diagrams. This has resulted in a hybrid model which accounts for electromagnetic kaon production from threshold up to $E_{\gamma} = 16\,\mathrm{GeV}$ [2, 3, 4, 5]. The non-resonant contributions to the amplitude are efficiently modelled in terms of $K^{+}(494)$ and $K^{*+}(892)$ Regge-trajectory exchange in the t-channel [6]. The three coupling constants can be unambiguously determined given sufficient high-energy ($E_{\gamma} \gtrsim 4$ GeV) data [7]. The Regge am-

[1] Email address: Pieter.Vancraeyveld@UGent.be

12th International Conference on Meson-Nucleon Physics and the Structure of the Nucleon (MENU 2010)
AIP Conf. Proc. 1374, 335-338 (2011); doi: 10.1063/1.3647154

plitudes are supplemented with a selection of s-channel resonance-exchange diagrams, whose parameters are optimised to data in the resonance-region ($E_\gamma \lesssim 4$ GeV) while keeping the background amplitude anchored. For the $p(\gamma, K^+)\Lambda$ reaction, a set of established nucleon resonances turned out to be insufficient. The addition of a $N(1900)D_{13}$ resonance made it possible to accurately describe both photo- and electroproduction data [2, 4]. The $p(\gamma^{(*)}, K^+)\Sigma^0$ data, on the other hand, is properly described within the RPR framework considering only established nucleon and delta resonances [3, 4]. Accordingly, there is no direct hint for missing resonances in that channel.

In the $K^+\Lambda$ and $K^+\Sigma^0$ production channels, sufficient data is available to constrain the free parameters of the production operator. For the remaining channels ($K^0\Lambda$, $K^0\Sigma^+$, $K^0\Sigma^0$, $K^+\Sigma^-$) the small amount of data does not allow one to tune the model parameters. In the strong interaction vertex of the considered Feynman diagrams, the coupling constants can be transformed assuming SU(2) isospin symmetry. Converting the electromagnetic coupling constants requires experimental input. For the production of neutral kaons two alterations to the background amplitude are required. First, the $K(494)$-exchange t-channel diagram does no longer contribute and second, a value for the magnetic transition moment of the $K^{*0}(892)$ to the $K^0(494)$ is needed. Optimizing the RPR model to the $p(\gamma, K^0)\Sigma^+$ dataset [8, 9, 10], we obtain $\kappa_{K^{*0}(892)K^0(494)}/\kappa_{K^{*+}(892)K^+(494)} = 0.05 \pm 0.01$. For kaon photoproduction on a neutron, the nucleon-exchange diagrams need to be transformed using photocoupling helicity amplitudes, which are extracted from pion production reactions, amongst others. Confronting RPR predictions with $n(\gamma, K^+)\Sigma^-$ data has demonstrated that this approach gives fair results, albeit with considerable uncertainties that stem from the experimental errors on the employed helicity amplitudes [5]. For the Reggeized background, its three parameters can be transformed adopting only isospin symmetry arguments, yielding robust predictions.

In order to embed the RPR elementary operator in the nuclear medium, we invoke the impulse approximation. For now, we neglect final-state interactions and assume plane waves for the final-state particles. The fully Lorentz-invariant expression for the reaction amplitude is

$$\mathcal{M}^{\lambda_\gamma,\lambda_D}_{\lambda_Y,\lambda_N}(\gamma D \to KYN) =$$
$$\frac{1}{\sqrt{2}}\overline{u}(\vec{p}_Y,\lambda_Y)\varepsilon_{\lambda_\gamma}\cdot \hat{J}_{\text{elem}}\left(\gamma N' \to KY\right)\frac{m_N + \not{p}_D - \not{p}_N}{m_N^2 - (p_D - p_N)^2}\varepsilon_{\lambda_D}\cdot \Gamma_{\text{Dnp}}(p_N, p_D)\mathcal{C}\overline{u}^T(\vec{p}_N,\lambda_N), \tag{1}$$

where the factor $\sqrt{2}$ stems from isospin factors and the fact that the production operator acts on a single nucleon. The λ_i are the spin projections of the initial- and final-state particles and $\mathcal{C}$ is the charge conjugation matrix. Since the spectator nucleon (with momentum p_N) is on-mass shell, the covariant Dnp-vertex Γ_{Dnp} is defined by [11]

$$\Gamma^\mu_{\text{Dnp}}(p_N,p_D) = F(|\vec{p}|)\gamma^\mu - \frac{G(|\vec{p}|)}{m_N}p^\mu - \frac{m_N - (\not{p}_D - \not{p}_N)}{m_N}\left(H(|\vec{p}|)\gamma^\mu - \frac{I(|\vec{p}|)}{m_N}p^\mu\right), \tag{2}$$

with $p = \frac{1}{2}p_D - p_N$. The four scalar form factors F, G, H and I can be expressed in terms of the s-, p- and d-wave components of the deuteron wave function [12]. In our calculations we will adopt the relativistic wave functions obtained with the WJC-1 solution presented in Refs. [13, 14].

Within the relativistic plane-wave impulse approximation (RPWIA), the photoproduction of a neutral kaon from a deuteron target can proceed in three different elementary ways. For two of these, $n(\gamma,K^0)\Lambda$ and $n(\gamma,K^0)\Sigma^0$, the incoming photon interacts with a bound neutron, and no data is available to tune the RPR model. In figure 1, we compare the RPR predictions with the data obtained using the NKS at the Laboratory of Nuclear Science at Tohoku University [1]. They have measured the semi-inclusive neutral-kaon photoproduction cross section from deuterium in two photon-energy bins close to threshold. The results are presented as a function of the momentum of the outgoing kaon p_K and for a range of scattering angles θ_{K^0} in the forward direction:

$$\frac{d\sigma}{dp_{K^0}} = \sum_{Y=\Lambda,\Sigma^0,\Sigma^+} \int_{0.9}^{1} \cos\theta_{K^0} \int d\Omega_Y^* \frac{1}{32(2\pi)^4} \frac{|\vec{p}_Y^*||\vec{p}_{K^0}|^2}{M_D E_\gamma E_{K^0} W_{YN}} \frac{1}{6} \sum \left| \mathscr{M}_{\lambda_Y,\lambda_N}^{\lambda_\gamma,\lambda_D}(\gamma D \to K^0 YN) \right|^2 . \tag{3}$$

In this expression all quantities are expressed in the laboratory frame, except for those marked with an asterisk, and W_{YN} is the invariant mass of the final hyperon-nucleon system. In the lowest energy bin, where the Σ production channels have not opened yet, the RPR model underpredicts the data by a factor of two approximately. At the second photon energy, on the other hand, the description is far more satisfactory, since both the Λ and Σ quasi-elastic peaks are well reproduced.

In summary, we have presented results for semi-inclusive neutral-kaon photoproduction from the deuteron within the RPWIA. For the elementary process, we employ the Regge-plus-resonance model, which has been optimised against recent photoproduction data. We adopt symmetry considerations to transform this elementary operator to account for the production of neutral kaons from both protons and neutrons. With this technique, we obtain predictions for $^2\mathrm{H}(\gamma,K^0)YN$ that describe the data fairly well. Soon, things will improve on the experimental side with experiments at Jefferson Lab presently being analysed and others scheduled to run in the near future. At Tohoku University, new data has been taken with a polarised photon-beam using an improved spectrometer with a larger acceptance [15]. We intend to extend our formalism and investigate the role of hyperon-nucleon final-state interactions on exclusive and inclusive observables. Besides making our model more complete, this will give us access to the elusive hyperon-nucleon interaction by focussing on kinematic regions where one expects important contributions from hyperons rescattering off the spectator nucleon.

ACKNOWLEDGMENTS

This work was supported by the Research Foundation – Flanders (FWO) and the research council of Ghent University.

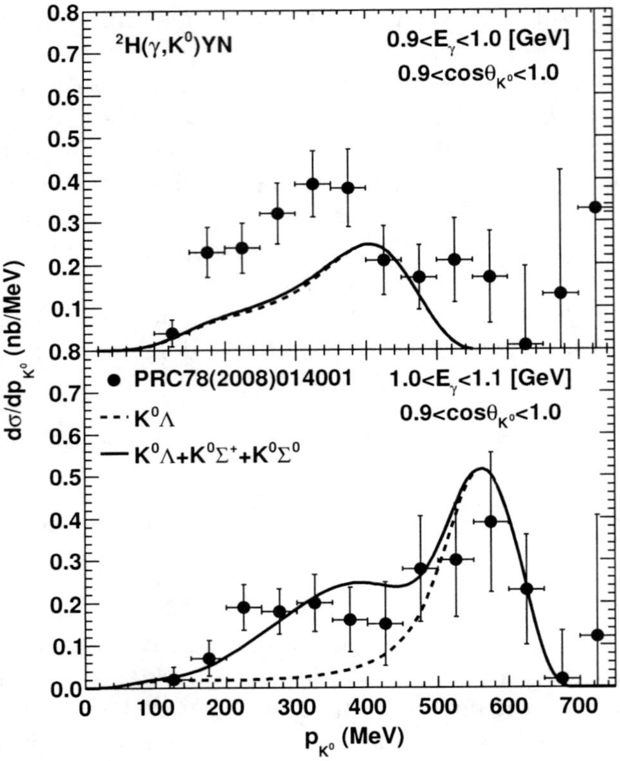

Figure 1. The semi-inclusive $^2H(\gamma,K^0)YN$ differential cross section as a function of the kaon momentum p_K after integrating over $0.9 \leq \cos\theta_{K^0} \leq 1$. The model results in the upper (lower) panel have been calculated at $E_\gamma = 950(1050)$ MeV. The solid line shows the full result, whereas the dashed line singles out the contribution of the $^2H(\gamma,K^0)\Lambda p$ reaction channel. The data are from Ref. [1].

REFERENCES

1. K. Tsukada, et al., *Phys. Rev.* **C78**, 014001 (2008).
2. T. Corthals, J. Ryckebusch, and T. Van Cauteren, *Phys. Rev.* **C73**, 045207 (2006).
3. T. Corthals, D. G. Ireland, T. Van Cauteren, and J. Ryckebusch, *Phys. Rev.* **C75**, 045204 (2007).
4. T. Corthals, T. Van Cauteren, P. Vancraeyveld, J. Ryckebusch, and D. G. Ireland, *Phys. Lett.* **B656**, 186 (2007).
5. P. Vancraeyveld, L. De Cruz, J. Ryckebusch, and T. Van Cauteren, *Phys. Lett.* **B681**, 428 (2009).
6. M. Guidal, J. M. Laget, and M. Vanderhaeghen, *Nucl. Phys.* **A627**, 645 (1997).
7. L. De Cruz, D. G. Ireland, P. Vancraeyveld, and J. Ryckebusch, *Phys. Lett.* **B** (2010), in press, doi:10.1016/j.physletb.2010.09.026.
8. R. Lawall, et al., *Eur. Phys. J.* **A24**, 275 (2005).
9. R. Castelijns, et al., *Eur. Phys. J.* **A35**, 39 (2008).
10. B. Carnahan, Ph.D. thesis, Catholic University of America, Washington, USA (2003).
11. R. Blankenbecler, and L. F. Cook, *Phys. Rev.* **119**, 1745 (1960).
12. W. W. Buck, and F. Gross, *Phys. Rev.* **D20**, 2361 (1979).
13. F. Gross, and A. Stadler, *Phys. Rev.* **C78**, 014005 (2008).
14. F. Gross, and A. Stadler, *Phys. Rev.* **C82**, 034004 (2010).
15. H. Kanda (2010), private communication.

Determination of the Linear Polarization of the Hall-B Tagged Photon Beam at Jefferson Lab

A. A. Sabintsev* and K. Livingston†

*The George Washington University, Washington, DC 20052, USA
†University of Glasgow, Glasgow G12 8QQ, Scotland, UK

Abstract. The JLab CLAS g9a/FROST experiments are double polarization measurements that have accumulated photoproduction data using linearly polarized, tagged photons incident on a longitudinally polarized, frozen spin butanol target. The analysis of the resulting coherent peaks was used to determine the photon polarization, which is in agreement with phenomenological calculations.

Keywords: Meson Photoproduction, Nuclear Instrumentation, Linear Polarization, Target Polarization, Baryon Spectroscopy
PACS: 14.20.Gk, 14.40.Be, 14.40.Df, 29.25.Pj, 29.27.Hj

BACKGROUND

The goal of all photoproduction experiments performed in Jefferson Lab's Experimental Hall-B is Baryon Spectroscopy, which is the study of excited states of nucleons. In these experiments, high-energy photons are absorbed by nucleons, creating these excited states, or Baryon Resonances. Many resonances are predicted by theory[1], but not all have been found experimentally. The goal of the Frozen Spin Target (FROST) experiments is to better determine the properties of known baryon resonances, to confirm the existence of those that are poorly known, and to look for those that are missing; resonances that have been predicted, but have yet to be observed.

THE FROST EXPERIMENTS

The FROST experiments are a set of five double polarization measurements performed between 0.7 GeV and 2.4 GeV on a super-cooled (<30 mK) butanol target. Both the photon beam and target are polarized. A complete measurement can be made with eight polarization observables[2], where the photon energy, E_γ and the meson production angle, θ are fixed. The experiments are split into two groups. The first set (g9a) used a longitudinally polarized target with linear and circularly polarized beams, and the second set (g9b) used a transversely polarized target with the same aforementioned beam polarizations. The first set of experiments were completed in the spring of 2008. The second set were completed in the summer of 2010.

The following five photoproduction channels were measured by FROST: **(1)** $\gamma p \to \pi^0 p$, **(2)** $\gamma p \to \pi^+ n$, **(3)** $\gamma p \to \eta p, \eta' p$, **(4)** $\gamma p \to \pi^+\pi^- p$, **(5)** $\gamma p \to K^+\Lambda, K^+\Sigma^0, K^0\Sigma^+$. From these experiments, we are able to obtain, via analysis, the unpolarized, single polarized, and beam-target double polarized observables as listed in Table 1. The recoil double polarization observables are obtained from the self-analyzing Λ and Σ hyperons.

DETERMINING THE LINEAR POLARIZATION

In the analysis of data from linear polarization experiments it is essential to know the orientation of the polarization plane, which is the alignment (parallel and perpendicular) of the electric field vector of the linearly polarized beam to the Hall-B floor. It is also essential to determine the degree of linear polarization as a function of photon energy for each of the two polarization plane settings. The polarization plane is known from the azimuthal orientation of the crystal radiator used to produce the photon beam from JLab's electron beam via coherent bremsstrahlung[3]. The degree of linear polarization as a function of photon energy has to be inferred by comparing the photon energy distribution with one obtained using a coherent bremsstrahlung calculation. More specifically, the comparison between data and calculation is made using enhancement plots. In the enhancement, the photon energy spectrum obtained

12th International Conference on Meson-Nucleon Physics and the Structure of the Nucleon (MENU 2010)
AIP Conf. Proc. 1374, 339-341 (2011); doi: 10.1063/1.3647155

TABLE 1. These sixteen polarization observables yield a complete set of measurements for pseudo-scalar meson photoproduction experiments. FROST is able to perform a complete measurement with the first eight observables.

#	Type	Observable	$\mathscr{P}_\gamma$	$\mathscr{P}_T$	$\mathscr{P}_R$	Transversity Representation
1.	Unpolarized	$\{d\sigma/d\Omega\}/\mathcal{N}$				$=\lvert b_1\rvert^2+\lvert b_2\rvert^2+\lvert b_3\rvert^2+\lvert b_4\rvert^2$
2.		$P\cdot\{d\sigma/d\Omega\}/\mathcal{N}$			$\hat{e}'_y$	$=\lvert b_1\rvert^2-\lvert b_2\rvert^2+\lvert b_3\rvert^2-\lvert b_4\rvert^2$
3.	Single Polarization	$\Sigma\cdot\{d\sigma/d\Omega\}/\mathcal{N}$	Linear			$=\lvert b_1\rvert^2+\lvert b_2\rvert^2-\lvert b_3\rvert^2-\lvert b_4\rvert^2$
4.		$T\cdot\{d\sigma/d\Omega\}/\mathcal{N}$		$\hat{e}_y$		$=\lvert b_1\rvert^2-\lvert b_2\rvert^2-\lvert b_3\rvert^2+\lvert b_4\rvert^2$
5.		$E\cdot\{d\sigma/d\Omega\}/\mathcal{N}$	Circular	$\hat{e}_z$		$=-2\mathrm{Re}\left(b_1b_3^*+b_2b_4^*\right)$
6.	Double Polarization	$F\cdot\{d\sigma/d\Omega\}/\mathcal{N}$	Circular	$\hat{e}_x$		$=2\mathrm{Im}\left(b_1b_3^*-b_2b_4^*\right)$
7.	(Beam & Target)	$G\cdot\{d\sigma/d\Omega\}/\mathcal{N}$	Linear	$\hat{e}_z$		$=2\mathrm{Im}\left(b_1b_3^*+b_2b_4^*\right)$
8.		$H\cdot\{d\sigma/d\Omega\}/\mathcal{N}$	Linear	$\hat{e}_x$		$=-2\mathrm{Re}\left(b_1b_3^*-b_2b_4^*\right)$
9.		$C_x\cdot\{d\sigma/d\Omega\}/\mathcal{N}$	Circular		$\hat{e}'_x$	$=2\mathrm{Im}\left(b_1b_4^*-b_2b_3^*\right)$
10.	Double Polarization	$C_z\cdot\{d\sigma/d\Omega\}/\mathcal{N}$	Circular		$\hat{e}'_z$	$=-2\mathrm{Re}\left(b_1b_4^*+b_2b_3^*\right)$
11.	(Beam & Recoil)	$O_x\cdot\{d\sigma/d\Omega\}/\mathcal{N}$	Linear		$\hat{e}'_x$	$=-2\mathrm{Re}\left(b_1b_4^*-b_2b_3^*\right)$
12.		$O_z\cdot\{d\sigma/d\Omega\}/\mathcal{N}$	Linear		$\hat{e}'_z$	$=-2\mathrm{Im}\left(b_1b_4^*+b_2b_3^*\right)$
13.		$T_x\cdot\{d\sigma/d\Omega\}/\mathcal{N}$		$\hat{e}_x$	$\hat{e}'_x$	$=2\mathrm{Re}\left(b_1b_2^*-b_3b_4^*\right)$
14.	Double Polarization	$T_z\cdot\{d\sigma/d\Omega\}/\mathcal{N}$		$\hat{e}_x$	$\hat{e}'_z$	$=2\mathrm{Im}\left(b_1b_2^*-b_3b_4^*\right)$
15.	(Target & Recoil)	$L_x\cdot\{d\sigma/d\Omega\}/\mathcal{N}$		$\hat{e}_z$	$\hat{e}'_x$	$=2\mathrm{Im}\left(b_1b_2^*+b_3b_4^*\right)$
16.		$L_z\cdot\{d\sigma/d\Omega\}/\mathcal{N}$		$\hat{e}_z$	$\hat{e}'_z$	$=2\mathrm{Re}\left(b_1b_2^*+b_3b_4^*\right)$

TABLE 2. Four polarization observables are obtained using a linearly polarized beam. The photon beam asymmetry, Σ, is acquired by determining the degree of linear polarization. The double polarization observables, G and H are obtained through analysis, and the the recoil baryon polarization, P, is determined by induction if Σ, G and H are known.

Beam Polarization ($\mathscr{P}_\gamma$)	Target Polarization($\mathscr{P}_T$)	Observable
Linear	Unpolarized	$-\Sigma$
Linear	$\hat{e}_x$	H
Linear	$\hat{e}_y$	$(-P)$
Linear	$\hat{e}_z$	G

from coherent bremsstrahlung is divided by a similar spectrum made with an amorphous radiator. This cancels out the strong channel to channel variations in the intensity due to the differing acceptances of the detectors on the tagger focal plane and also removes the $\sim 1/E$ slope which is characteristic in incoherent bremsstrahlung production. The coherent edge is defined to be the slope of the peak with the highest negative gradient. This peak is called the coherent peak [Fig. 1].

The first step in determining the linear polarization comes from the creation of coherent edge plots, which provide a visual history of the stability of the polarization over the entirety of all the runs performed at a specific energy and a specific polarization plane setting [Fig. 2]. After ascertaining that the beam was stable, the enhancement data was extracted and fit to a phenomenological bremsstrahlung calculation[4], which takes in 17 parameters that are all dependent on the experimental setup, such as the orientation of the polarization plane, the width of the collimator, the Z of the target and crystal radiator, etc... Once a phenomenological calculation was produced and seen to be in agreement with the experimental data, a set of tables that relate the degree of linear polarization as a function of the photon energy for every run period were produced[5], where a run period is determined by the photon energy level and the orientation of the polarization plane. These tables were then used to determine the beam asymmetry, Σ. Once the double polarization observables G and H are acquired from analysis, they can be used in conjunction with Σ to determine P, the recoil baryon polarization [Table 2] by inference.

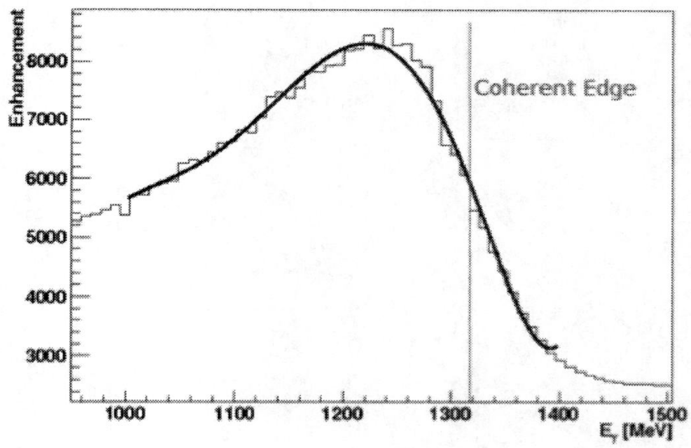

FIGURE 1. A sample enhance plot for a run performed at $E_\gamma = 1.3$ GeV. The peak is called the coherent peak, since it delineates the polarized peak. The coherent edge is the negative gradient of the coherent peak. The black line represents a phenomenological bremsstrahlung calculation that is seen to be in agreement with experimental data.

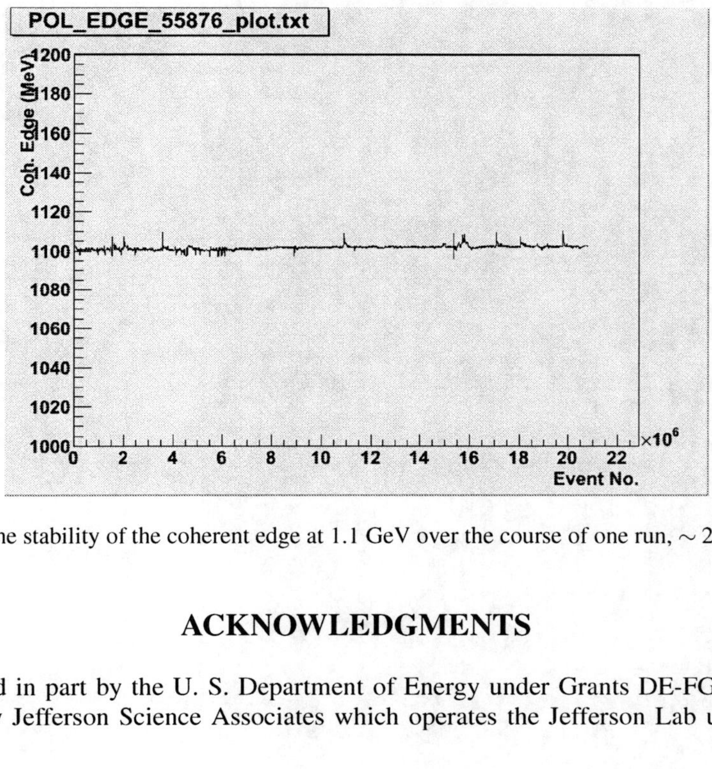

FIGURE 2. The stability of the coherent edge at 1.1 GeV over the course of one run, ~ 20 million events.

ACKNOWLEDGMENTS

This work was supported in part by the U. S. Department of Energy under Grants DE-FG02-99ER41110 and DE-FG02-03ER41231and by Jefferson Science Associates which operates the Jefferson Lab under DOE contract DE-AC05-06OR23177.

REFERENCES

1. S. Capstick and W. Roberts, Prog. Part. Nucl. Phys. **45**, S241 (2000).
2. I. S. Barker, A. Donnachie & J.K. Storrow, Nuclear Physics **B95** (1975).
3. U. Timm. Fortschritte der Physik, **765** (1969).
4. A. Natter *et al.*, Nuc. Inst. Meth. Research B **211** (2003).
5. A. A. Sabintsev and K. Livingston, JLAB CLAS-Note **9**, (2010).

Session H: Transition Form Factors

Transverse charge densities of N* resonances

L. Tiator and M. Vanderhaeghen

Institut für Kernphysik, Johannes Gutenberg-Universität, D-55099 Mainz, Germany

Abstract. The unitary isobar model MAID2007 has been used to analyze the recent data of pion electroproduction. The model contains all four-star resonances in the region below $W = 2$ GeV. Both single-Q^2 and Q^2 dependent transition form factors could be obtained for the Delta, Roper, $D_{13}(1520)$, $S_{11}(1535)$, $D_{15}(1675)$, and $F_{15}(1680)$ nucleon resonances. Furthermore, we show the empirical transverse charge densities for the Roper, the S11 and D13 resonances, which are evaluated from Fourier transforms of the transition form factors.

Keywords: pion photo- and electroproduction, non-strange baryons, transition form factors, transverse charge densities
PACS: 14.20.Gk, 13.30.Eg, 13.75.Gx

INTRODUCTION

With the unitary isobar model MAID we have analyzed pion electroproduction data and have obtained transition form factors for all four-star resonances below $W = 2$ GeV. For the proton target in most cases we could extract both single-Q^2 and Q^2-dependent transition form factors, for the neutron target we parameterized the Q^2 dependence in a simpler form as far as the existing data from the world data base allowed.

Furthermore, the precise e.m. FF data, extracted from experiment, allow us to map out the quark charge densities in a baryon. It was shown possible to define a proper density interpretation of the form factor data by viewing the baryon in a light-front frame. This yields information on the spatial distribution of the quark charge in the plane transverse to the line-of-sight. In this way, the quark transverse charge densities were mapped out in the nucleon [1, 2], and in the deuteron [3] based on empirical FF data. To understand the e.m. structure of a nucleon resonance, it is of interest to use the precise transition FF data to reveal the spatial distribution of the quark charges that induce such a transition. In this way, using the empirical information on the $N \to N^*$ transition form factors from the MAID analysis [4], the $N \to \Delta(1232)$ transition charge densities have been mapped out in Ref. [2], the $N \to N^*(1440)$ in Ref. [5] and the $N \to S_{11}(1535)$ and $N \to D_{13}(1520)$ in Ref. [6].

TRANSITION FORM FACTORS

In the unitary isobar model for pion photo- and electroproduction, the partial wave T-matrix amplitudes are constructed as

$$t_{\gamma\pi}(W) = t^{B}_{\gamma\pi}(W) + t^{R}_{\gamma\pi}(W), \tag{1}$$

a background and a resonance T-matrix element, where each of them is individually unitarized. The background is calculated from Born and vector meson exchange terms in the K-matrix unitarization technique. For the resonance contributions we follow Ref. [7] and assume Breit-Wigner forms for the resonance shape,

$$t^{R,\alpha}_{\gamma\pi}(W,Q^2) = \bar{\mathcal{A}}^{R}_{\alpha}(W,Q^2)\,\frac{f_{\gamma N}(W)\Gamma_{tot}\,M_R\,f_{\pi N}(W)}{M_R^2 - W^2 - iM_R\Gamma_{tot}}\,e^{i\phi_R}, \tag{2}$$

with total width $\Gamma_{tot}(W)$ and physical mass M_R.

In most cases the e.m. resonance couplings $\mathcal{A}^{R}_{\alpha}(W,Q^2)$ are independent of the total energy and depend only on Q^2. However, for the $\Delta(1232)$ resonance we observe the need of an energy dependence in terms of the virtual photon three-momentum $k(W,Q^2)$. For all other resonances, which are discussed here, we can assume a simple Q^2 dependence, $\mathcal{A}_{\alpha}(Q^2)$. These couplings can be taken as constants in a single-Q^2 analysis, e.g. in photoproduction, where $Q^2 = 0$ but also at any fixed Q^2, where enough data with W and θ variation are available. Alternatively they can also be

12th International Conference on Meson-Nucleon Physics and the Structure of the Nucleon (MENU 2010)
AIP Conf. Proc. 1374, 345-348 (2011); doi: 10.1063/1.3647156

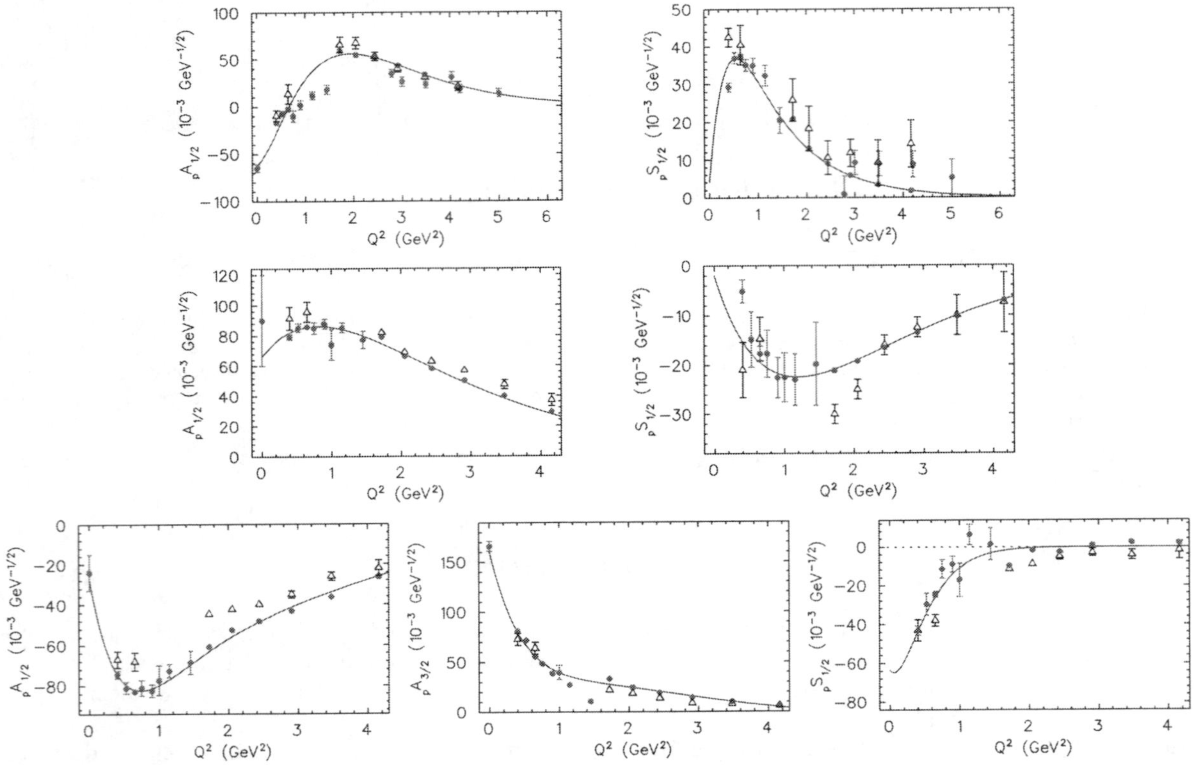

FIGURE 1. Transverse and longitudinal helicity transition form factors of the proton to the $P_{11}(1440)$ Roper resonance (top), $S_{11}(1535)$ resonance (middle) and $D_{13}(1520)$ resonance (bottom). The red circles are the MAID analysis of 2007 [4] and 2008, and the black triangles are the 2008 JLab analysis [8]. The data points of the transverse form factors at $Q^2 = 0$ are the PDG values [9].

parameterized as functions of Q^2 in a form like

$$\bar{\mathscr{A}}_\alpha(Q^2) = \bar{\mathscr{A}}_\alpha(0)(1 + a_1 Q^2 + a_2 Q^4 + a_3 Q^8)\, e^{-b_1 Q^2} . \tag{3}$$

With such an ansatz it is possible to determine the parameters $\bar{\mathscr{A}}_\alpha(0)$ from a fit to the world database of photoproduction, while the parameters a_i and b_1 can be obtained from a combined fitting of all electroproduction data at different Q^2. The latter procedure we call the 'superglobal fit'. In MAID the photon couplings $\bar{\mathscr{A}}_\alpha$ are direct input parameters. They are directly related to the helicity couplings $A_{1/2}, A_{3/2}$ and $S_{1/2}$ of nucleon resonance excitation. For further details see Ref. [4].

The best known nucleon resonance is naturally the $\Delta(1232)$. Due to the Watson theorem the photoproduction multipoles as well as the transition form factors could be determined in a near model-independent way. Our results of the form factors can be found in Ref. [4] and the transition densities are shown in Ref. [2].

The helicity amplitudes for the Roper resonance $P_{11}(1440)$ are shown on the left in Fig. 1. Our latest super-global solution (solid lines) is in reasonable agreement with the single-Q^2 analysis. The figure shows a zero crossing of the transverse helicity form factor at $Q^2 \approx 0.7$ GeV^2 and a maximum at the relatively large momentum transfer $Q^2 \approx 2.0$ GeV^2. The longitudinal Roper excitation rises to large values around $Q^2 \approx 0.5$ GeV^2 and in fact produces one of the strongest longitudinal transition form factor we can find in our analysis. This answers the question raised by Li and Burkert [10] whether the Roper resonance is a radially excited 3-quark state or a quark-gluon hybrid, because in the latter case the longitudinal coupling should vanish completely.

Fig. 1 also shows our results for the $S_{11}(1535)$ resonance. The red single-Q^2 data points show our 2007 analysis, while the black triangles are the 2008 analysis of Ref. [8]. Our Q^2 dependent analysis describes all data points quite well, except for the longitudinal form factor in the region around $Q^2 = 2$ GeV^2.

While the inclusion of more recent π^+ data [11] did not modify our 2007 solution for the S_{11} resonance, we find some significant deviations for the $D_{13}(1520)$ resonance, see bottom row in Fig. 1. Also for this resonance the JLab

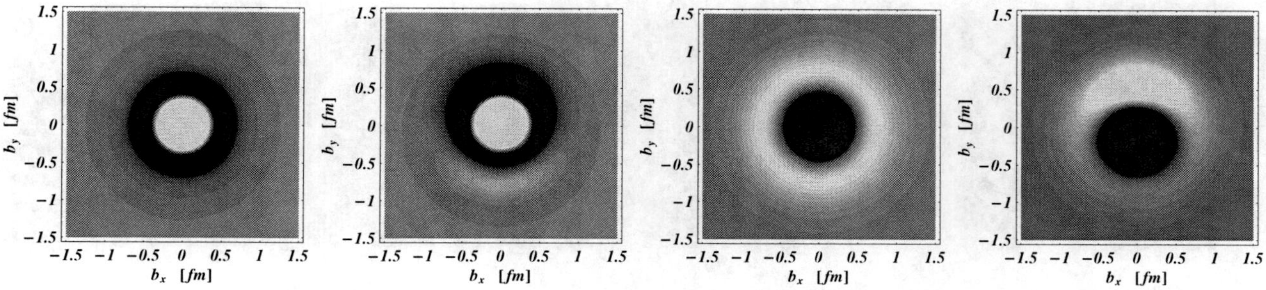

FIGURE 2. Quark transverse charge densities corresponding to the $N \to P_{11}(1440)$ e.m. transition. From left to right: unpolarized proton to Roper, $p \to R^+$, ($\rho_0^{pN^*}$), polarized $p \to R^+$, ($\rho_T^{pN^*}$), unpolarized $n \to R^0$, ($\rho_0^{nN^*}$) and polarized $n \to R^0$, ($\rho_T^{nN^*}$). The light (dark) regions correspond with positive (negative) densities. For the $p \to P_{11}(1440)$ e.m. transition FFs, we use the improved MAID2008 fit and for the neutron we use the MAID2007 fit.

partial wave analysis of 2008 agrees well with the MAID analysis for most cases, however a significant deviation remains for $A_{1/2}$ in the region of $Q^2 = 2$ GeV^2. The pion electroproduction data base used in this analysis are unpolarized cross sections from JLab/CLAS, available in the kinematical region of 1100 MeV $< W <$ 1680 MeV and 0.4 GeV^2 $< Q^2 <$ 1.8 GeV^2 [11–15].

EMPIRICAL TRANSVERSE CHARGE TRANSITION DENSITIES

In the following, we will consider the e.m. $N \to N^*$ transition when viewed from a light front moving towards the baryon. Equivalently, this corresponds with a frame where the baryons have a large momentum-component along the z-axis chosen along the direction of $P = (p+p')/2$, where p (p') are the initial (final) baryon four-momenta. Here we give only the relations between the form factors and the transverse densities. All the details can be found in Refs. [2, 3, 5].

First we will consider the case of $j = 1/2$ resonances, as P_{11} and S_{11}. These cases are very similar to the nucleon and can be worked out in an analogous way. The Fourier transform leads to

$$\rho_0^{NN^*}(\vec{b}) = \int_0^\infty \frac{dQ}{2\pi} Q J_0(bQ) F_1^{NN^*}(Q^2), \tag{4}$$

where J_0 denotes the cylindrical Bessel function of order zero. Note that $\rho_0^{NN^*}$ only depends on $b = |\vec{b}|$. It has the interpretation of the quark (transition) charge density in the transverse plane which induces the $N \to N^*$ excitation.

The above unpolarized transition charge density involves only one of the two independent $N \to N^*$ e.m. form factors. To extract the information encoded in $F_2^{NN^*}$, we consider the transition charge densities for a transversely polarized N and N^*. We denote this transverse polarization direction by $\vec{S}_\perp = \cos\phi_S \hat{e}_x + \sin\phi_S \hat{e}_y$. The Fourier transform of the transverse current can be worked out as

$$\rho_T^{NN^*}(\vec{b}) = \rho_0^{NN^*}(b) + \sin(\phi_b - \phi_S) \int_0^\infty \frac{dQ}{2\pi} \frac{Q^2}{(M^* + M_N)} J_1(bQ) F_2^{NN^*}(Q^2), \tag{5}$$

where the second term, which describes the deviation from the circular symmetric unpolarized charge density, depends on the orientation of $\vec{b} = b(\cos\phi_b \hat{e}_x + \sin\phi_b \hat{e}_y)$. In the following we choose the transverse spin along the x-axis ($\phi_S = 0$).

In Fig. 2 we show the results for the $N \to P_{11}(1440)$ transition charge densities both for the unpolarized case and for the case of transverse polarization for the proton (left) and for the neutron (right). We use the empirical information on the $N \to N^*(1440)$ transition FFs as given in our parametrization of this work.

It is seen that for the transition on a proton, which is well constrained by data, there is an inner region of positive quark charge concentrated within 0.5 fm, accompanied by a relatively broad band of negative charge extending out to about 1 fm. When polarizing the baryon in the transverse plane, the large value of the magnetic transition strength at the real photon point, yields a sizeable shift of the charge distribution, inducing an electric dipole moment. For the neutron, which is not very well constrained by data, the MAID2007 analysis yields charge distributions of opposite sign compared to the proton, with active quarks spreading out over even larger spatial distances.

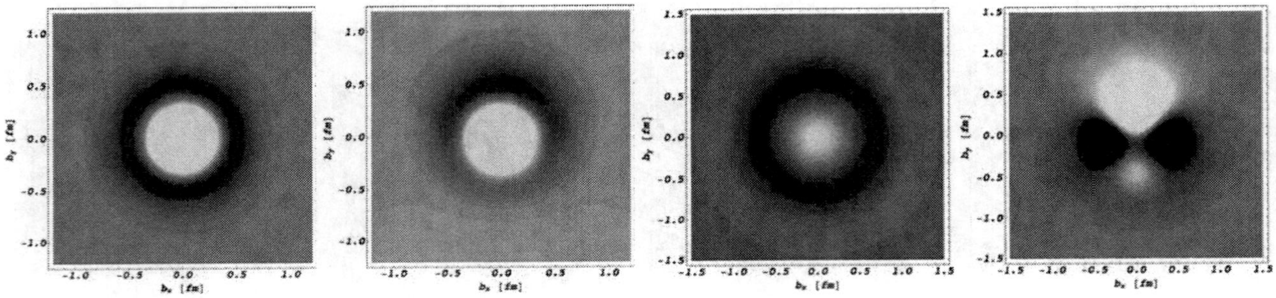

FIGURE 3. Quark transverse charge densities corresponding to the $p \rightarrow S_{11}(1535)$ and $p \rightarrow D_{13}(1520)$ e.m. transitions. From left to right: unpolarized $p \rightarrow S_{11}$, ($\rho_0^{pN^*}$), polarized $p \rightarrow S_{11}$, ($\rho_T^{pN^*}$), unpolarized $p \rightarrow D_{13}$, ($\rho_0^{pN^*}$) and polarized $p \rightarrow D_{13}$, ($\rho_T^{pN^*}$). The light (dark) regions correspond with positive (negative) densities. For the $p \rightarrow S_{11}(1535)$ e.m. transition FFs, we use the MAID2007 fit and for $p \rightarrow D_{13}(1520)$ we use the improved MAID2008 fit.

Fig. 3 shows on the left the unpolarized and polarized transition charge densities from the proton to the $S_{11}(1535)$ resonance. It can be compared to the corresponding Fig. 2 for the Roper. The up quarks are not so strongly localized and also the ring of down quarks is less pronounced, in particular for the density corresponding to $F_2^{NN^*}$ in the second panel.

Finally, Fig. 3 also shows on the right two transition densities from the proton to the $D_{13}(1520)$ resonance. Similar to the $N \rightarrow \Delta(1232)$ transition, also here we have 3 FFs leading to 3 densities. The unpolarized density is similar to the Roper with, however, more diffuse boundaries between up and down quarks. In addition to the dipole transition density, in this case we also get a quadrupole density which is shown in the last panel.

SUMMARY AND CONCLUSIONS

Using the world data base of pion photo- and electroproduction and recent data from Mainz, Bonn, Bates and JLab we have made a first attempt to extract all longitudinal and transverse helicity transition form factors of nucleon resonance excitation for four-star resonances below $W = 2$ GeV. For this purpose we have extended the unitary isobar model MAID and have parameterized the Q^2 dependence of the transition amplitudes. Comparisons between single-Q^2 fits and a Q^2 dependent superglobal fit give us confidence in the determination of the helicity couplings of the $P_{33}(1232), P_{11}(1440), D_{13}(1520), S_{11}(1535), D_{15}(1675)$, and the $F_{15}(1680)$ resonances, even though the model uncertainty of these amplitudes can be as large as 50% for the longitudinal amplitudes of the D_{13} and F_{15}.

These form factors were used to extract quark transverse charge densities inducing these transitions. The rings of up and down quarks in these two-dimensional representations show very different structures for the Roper, the S_{11} and the D_{13} resonances.

REFERENCES

1. G.A. Miller, *Phys. Rev. Lett.* **99**, 112001 (2007).
2. C.E. Carlson and M. Vanderhaeghen, *Phys. Rev. Lett.* **100**, 032004 (2008).
3. C.E. Carlson and M. Vanderhaeghen, *Eur. Phys. J. A* **41**, 1 (2009).
4. D. Drechsel, S.S. Kamalov, and L. Tiator, *Eur. Phys. J. A* **34**, 69 (2007); http://www.kph.uni-mainz.de/MAID/.
5. L. Tiator, M. Vanderhaeghen, *Phys. Lett. B* **642**, 344 (2009).
6. L. Tiator, D. Drechsel, S.S. Kamalov, and M. Vanderhaeghen, *CPC(HEP & NP)* **33**, 1051 (2009); [arXiv:0909.2335 [nucl-th]].
7. D. Drechsel, O. Hanstein, S.S. Kamalov, and L. Tiator, *Nucl. Phys. A* **645**, 145 (1999).
8. I.G. Aznauryan et al., *Phys. Rev. C* **78**, 045209 (2008).
9. K. Nakamura et al. (Particle Data Group), *J. Phys. G* **37**, 075021 (2010).
10. Z.P. Li, V. Burkert, and Z. Li, *J Phys. Rev. D* **46**, 70 (1992).
11. K. Park et al., *Phys. Rev. C* **77**, 015208 (2008).
12. K. Joo et al., *Phys. Rev. Lett.* **88**, 122001-1 (2002).
13. K. Joo et al., *Phys. Rev. C* **70**, 042201 (2004).
14. H. Egiyan et al., *Phys. Rev. C* **73**, 025204 (2006).
15. M. Ungaro et al., *Phys. Rev. Lett.* **97**, 112003 (2006).

Transition nucleon resonance electrocouplings from CLAS data on $\pi^+\pi^-p$ electroproduction off protons

Victor I. Mokeev

Thomas Jefferson National Accelerator Facility, Newport News, Virginia 23606, USA
Skobeltsyn Nuclear Physics Institute at Moscow State University, Moscow, Leninskie gory 119889, Russia

Abstract. Electrocouplings of excited proton states with masses less than 1.8 GeV were determined for the first time from the CLAS data on $\pi^+\pi^-p$ electroproduction at photon virtualities $Q^2 < 1.5$ GeV^2. Electrocouplings were obtained from a combined fit of all available observables within the framework of a phenomenological reaction model.

Keywords: meson electroproduction, nucleon resonance structure, transition form factors
PACS: PAC number(s): 11.55.Fv, 13.40.Gp, 13.60.Le, 14.20 Gk

INTRODUCTION

Studies of nucleon resonance electrocouplings $\gamma_{virt}NN^*$ from $ep \to e'\pi^+\pi^-p$ channel represent an important part of the N^* Program with the CLAS detector [1]. The $N\pi$ and $\pi^+\pi^-p$ exclusive electroproduction channels are major contributors in the resonance excitation region with entirely different non-resonant mechanisms. Therefore, they provide independent information on $\gamma_{virt}NN^*$ electrocouplings, that is vital for the reliable extraction of these fundamental quantities. Comprehensive information on $\gamma_{virt}NN^*$ electrocouplings of various excited proton states in a wide range of photon virtualities Q^2 gives access to active degrees of freedom in the N^* structure at various distances. This allow us to explore non-perturbative strong interaction mechanisms, that are responsible for baryon formation. Preliminary results on the Q^2-evolution of $\gamma_{virt}NN^*$ electrocouplings for excited proton states with masses less than 1.8 GeV are reported. They were obtained for the first time from $\pi^+\pi^-p$ electroproduction off protons [2, 3].

ANALYSIS APPROACH

The CLAS data on $\pi^+\pi^-p$ electroproduction [2, 3] for the first time provided information on nine independent one-fold differential and fully integrated cross sections in each bin of W and Q^2 in a mass range $1.31 < W < 2.1$ GeV and at photon virtualities from 0.25 to 1.5 GeV^2. Analysis of these data allowed us to establish all essential mechanisms contributing to $\pi^+\pi^-p$ electroproduction in this kinematical area. The presence and strengths of the contributing $\pi^+\pi^-p$ electroproduction mechanisms was established by studying the kinematical dependencies in differential cross sections and their correlations in a variety of available observables. This resulted in the development of meson-baryon reaction model JM for the description of $\pi^+\pi^-p$ electroproduction off protons [4, 5]. The primary objective is to determine $\gamma_{virt}NN^*$ electrocouplings and $\pi\Delta$ and ρp partial hadronic decay widths of N^*'s from a combined fit of all available observables. The model incorporates the full $\pi^+\pi^-p$ production amplitude of the $\pi^-\Delta^{++}$, $\pi^+\Delta^0$, ρp, $\pi^+D^0_{13}(1520)$, $\pi^-P^{++}_{33}(1600)$, $\pi^+F^0_{15}(1685)$ isobar channels and direct double pion production mechanisms, that account for all resonant and non-resonant partial waves combined. Direct double pion production mechanisms describe processes when the final $\pi^+\pi^-p$ state is created without formation of unstable hadrons in intermediate states. Direct 2π production mechanisms, required by general unitarity condition [8], were established for the first time in the analysis of CLAS $\pi^+\pi^-p$ electroproduction data. Implementation of direct 2π production processes into the JM model allowed us to move beyond the isobar approximation for the description of $\pi^+\pi^-p$ electroproduction off protons. The JM model incorporates all N^*'s with masses less than 2.0 GeV that couple to the two pion channel. Resonant amplitudes contribute to $\pi\Delta$ and ρp isobar channels, while the other isobar channels contain only non-resonant mechanisms. Breit-Wigner parametrization for resonance amplitudes, employed in the JM model (the regular Breit-Wigner ansatz), is described in [6, 7]. However, this ansatz doest not satisfy unitarity condition in a case, when various N^* states could be mixed in a dressed resonance propagator. Therefore, in the 2010 JM model version the resonant amplitudes were parametrized using a unitarized Breit-Wigner ansatz, that was initially proposed in [8]. This ansatz was modified to

12th International Conference on Meson-Nucleon Physics and the Structure of the Nucleon (MENU 2010)
AIP Conf. Proc. 1374, 349-352 (2011); doi: 10.1063/1.3647157

make it consistent with resonant amplitude parametrization adopted in the JM model. The unitarized Breit-Wigner ansatz allowed us to account for transitions between various N^* states in the dressed resonance s-channel propagator and to impose restrictions on resonant amplitudes required by general unitarity condition. Non-resonant amplitudes incorporated into the JM model were presented in the papers [4, 5].

A reasonable description of the $\pi^+\pi^-p$ electroproduction channel was achieved, allowing us to separate resonant and non-resonant contributions to the measured cross sections, which is needed for the evaluation of $\gamma_{virt}NN^*$ electrocouplings.

FITTING PROCEDURE

The N^* electrocouplings and their $\pi\Delta$ and ρp partial hadronic decay widths were obtained from fits to the $\pi^+\pi^-p$ electroproduction data [2, 3] within the framework of JM model [4, 5]. These resonance parameters, as well as parameters describing the non-resonant mechanisms in the JM model were varied simultaneously in the χ^2 minimization. In this way we accounted for correlations of resonant and non-resonant contributions. For each trial set of computed cross sections the χ^2/d.p. value was estimated in point by point comparison between measured and computed nine one-fold differential cross sections in all bins of W and Q^2 covered by measurements.

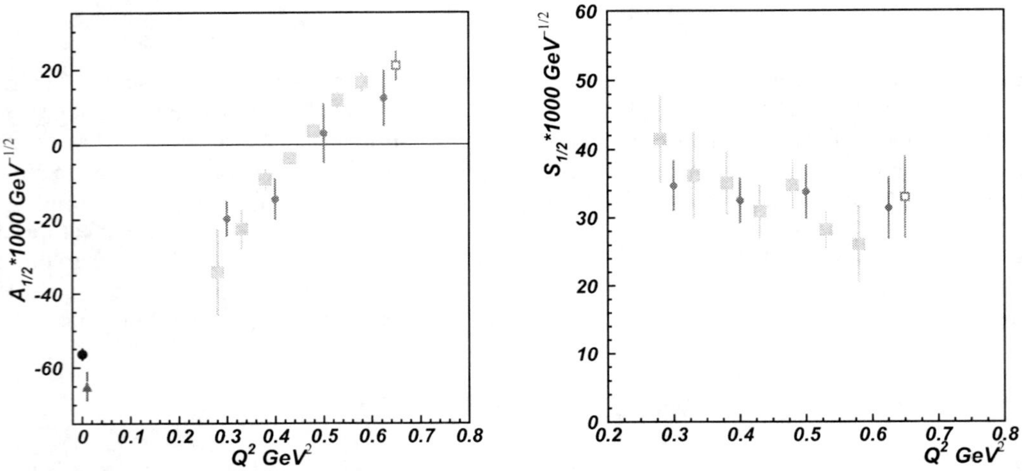

FIGURE 1. $A_{1/2}$ (left) and $S_{1/2}$ (right) electrocouplings of $P_{11}(1440)$ state determined from the CLAS data [3] on $\pi^+\pi^-p$ electroproduction (filled squares) and from $N\pi$ electroproduction [9] (filled circles). Results from a combined analysis of $N\pi$ and $\pi^+\pi^-p$ channels [10] are shown by open squares. Filled circle and triangle at the photon point correspond to CLAS [11] and PDG analyses of $N\pi$ channels, respectively.

$\gamma_{virt}NN^*$ electrocouplings, as well as resonance $\pi\Delta$ and ρp partial hadronic decay widths assigned to selected in the fit calculated cross sections were averaged, giving us the mean values of resonance parameters determined from the data fit. Dispersions in the sets determined in the fit resonance parameters were considered as their uncertainties.

RESULTS

Electrocouplings of the $P_{11}(1440)$ state determined from the analysis are shown in Fig. 1. The error bars include quadratic sums of statistical and systematic uncertainties. We also show electrocouplings of this state obtained from a combined analysis of the data on π^+n and π^0p electroproduction channels [9]. Electrocoupling values extracted from analyses of these exclusive channels are in good agreement. We also found consistent values for electrocouplings of $D_{13}(1520)$ and $F_{15}(1685)$ states determined independently from the two channels. As an example, the $A_{3/2}$ electrocouplings of these states are shown in Fig. 2.

$N\pi$ and $\pi^+\pi^-p$ exclusive electroproduction channels represent two major contributors to meson electroproduction in the N^* excitation region with completely different non-resonant mechanisms. The successful description of a large

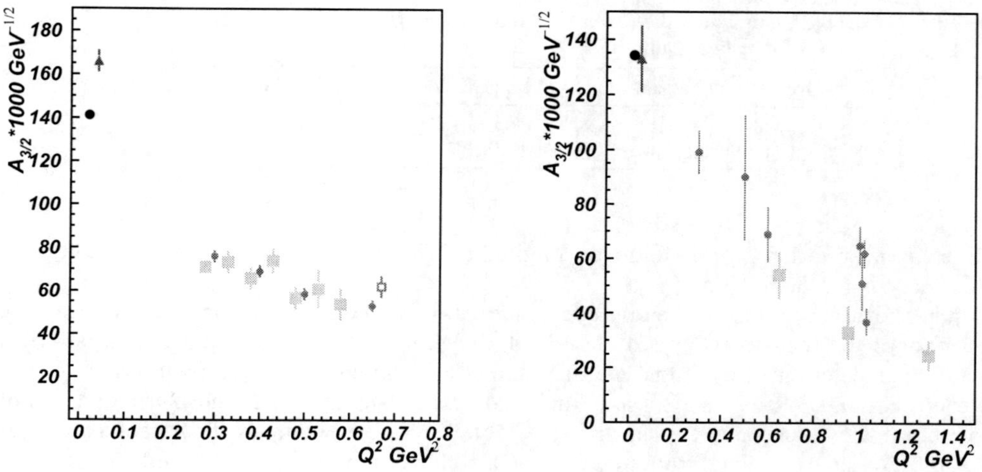

FIGURE 2. The $A_{3/2}$ electrocouplings of $D_{13}(1520)$ (left) and $F_{15}(1685)$ (right) resonances. The results from analysis of $\pi^+\pi^-p$ CLAS data [2, 3] are shown by filled squares. Electrocouplings of $D_{13}(1520)$ state determined from analysis of the CLAS data on $N\pi$ electroproduction [9] and electrocouplings of $F_{15}(1685)$ state obtained from analyses of previously available world data, taken from the compilation [12], are shown by filled circles. Filled circles and triangles at the photon point correspond to CLAS [11] and PDG analyses of $N\pi$ channels, respectively.

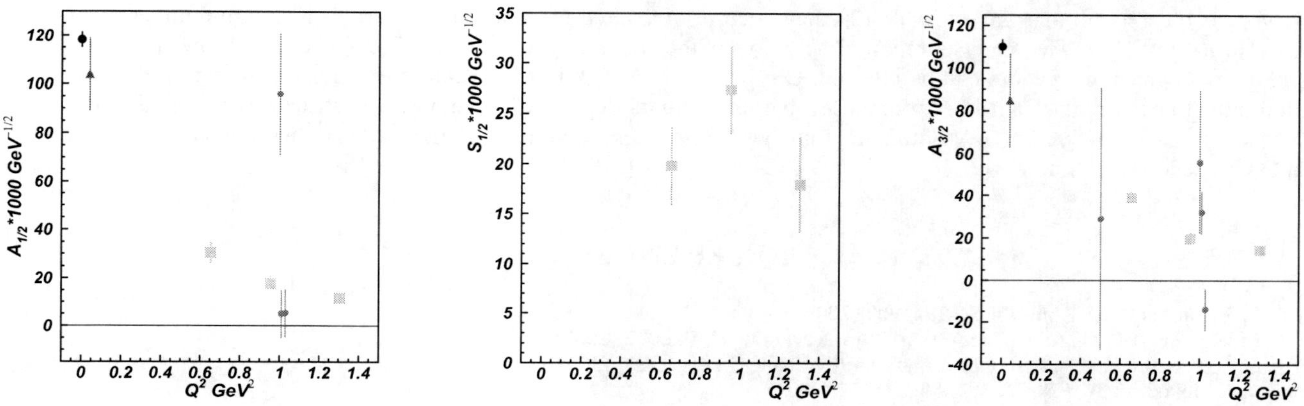

FIGURE 3. Electrocouplings of $D_{33}(1700)$ resonance. Results from analysis of $\pi^+\pi^-p$ CLAS data [2] are shown by filled squares, while filled circles represent the results obtained worldwide from analyses of $N\pi$ electroproduction data, taken from the compilation [12]. Filled circles and triangles at the photon point correspond to CLAS [11] and PDG analyses of $N\pi$ channels, respectively.

body of observables in these two exclusive channels using nearly the same electrocouplings of $P_{11}(1440)$, $D_{13}(1520)$ and $F_{15}(1685)$ states provides strong evidence for a reliable evaluation of these fundamental quantities.

Many excited states with masses above 1.6 GeV preferably couple to $N\pi\pi$ final states. Therefore, analysis of $\pi^+\pi^-p$ electroproduction is of particular importance to obtain reliable information on high lying N^* states.

Figure 3 shows $D_{33}(1700)$ electrocouplings determined in our analysis in comparison with previously available world data from analyses of $N\pi$ electroproduction. Large uncertainties of previous world data on $D_{33}(1700)$ electrocouplings are related to the fact that $N\pi$ is a minor decay channel (10-20%) of this state, that couples dominantly to $N\pi\pi$ final states (80-90%). Analysis of the $\pi^+\pi^-p$ data [2] within the framework of JM model allowed us to obtain accurate information on electrocouplings of this and also of the $S_{31}(1620)$, $S_{11}(1650)$, $F_{15}(1685)$, and $P_{13}(1720)$ states

TABLE 1. $\pi\Delta$ and ρp partial hadronic decay widths of $P_{13}(1720)$ state determined from analysis of the CLAS $\pi^+\pi^-p$ electroproduction data [2], employing the regular [6] and the unitarized [8] Breit-Wigner ansatzs for resonant amplitudes.

Breit-Wigner ansatz:	$\Gamma_{\pi\Delta}$, MeV	$\Gamma_{\rho p}$, MeV
Regular	1.5 ± 1.1	$114 \pm 12.$
Unitarized	10.9 ± 1.4	83 ± 3.3

at photon virtualities from 0.5 to 1.5 GeV^2. Moreover, longitudinal electrocouplings of these resonances have become available from this analysis for the first time.

In order to examine the effects of resonant amplitude unitarization, we compared resonance parameters determined within the frameworks of the regular [6, 7] and, described above, unitarized Breit-Wigner ansatzs. We found that $\gamma_{virt}NN^*$ electrocouplings determined by employing the regular and the unitarized formulations are consistent with each other. However, unitarization of resonant amplitudes affects substantially $\pi\Delta$ and ρp partial hadronic decay widths of excited proton states. An example is shown in the Table 1, where we compare the $\pi\Delta$ and ρp partial hadronic decay widths of $P_{13}(1720)$ state, obtained from the $\pi^+\pi^-p$ data fit by employing the JM model, that incorporate the regular and unitarized Briet-Wigner ansatzs, respectively.

CONCLUSION AND OUTLOOK

Accurate information on the Q^2-evolution of $\gamma_{virt}NN^*$ electrocouplings for many excited proton states with masses less than 1.8 GeV and at photon virtualities up to 1.5 GeV^2 have become available from CLAS data on $\pi^+\pi^-p$ electroproduction. These results open up new opportunities for theory to explore confinement mechanisms in the baryon sector through their manifestation in the structure of excited proton states of various quantum numbers, as it was outlined in [13]. The analysis reported here covers the range of photon virtualities, where both meson-baryon and quark degrees of freedom can be relevant. Our results on high lying N^* electrocouplings for the first time make it possible to explore the transition from meson-baryon to quark degrees of freedom in the structure of excited proton states with masses above 1.6 GeV within the framework of dynamical coupled channel approaches under development in EBAC at Jefferson Lab [14].

REFERENCES

1. V. D. Burkert, AIP Conf. Proc. **1182**, 900 (2009).
2. M. Ripani et al., CLAS Collaboration, Phys. Rev. Lett. **91**, 022002 (2003)
3. G. V. Fedotov et al., CLAS Collaboration, Phys. Rev. C **79**, 015204 (2009).
4. V. I. Mokeev et al., Phys. Rev. C **80**, 045212 (2009)
5. V. I. Mokeev and V. D. Burkert, J. Phys. Conf. Ser, **69**, 012019 (2007).
6. M. Ripani et al., Nucl. Phys. A **672**, 220 (2000).
7. V. I. Mokeev et al., Phys. Atom. Nucl. **64**, 1292 (2001).
8. I. J. R. Aitchison, Nucl. Phys. A **189**, 417 (1972).
9. I. G. Aznauryan et al., CLAS Collaboration, Phys. Rev. C **80**, 055203 (2009).
10. I. G. Aznauryan et al., Phys. Rev. C **72**, 045201 (2005).
11. M. Dugger et al., CLAS Collaboration, Phys. Rev. C **79**, 065206 (2009).
12. V. D. Burkert et al., Phys. Rev. C **67**, 035204 (2003).
13. I. G. Aznauryan et al., JLAB-PHY-09-0993, arXiv:0907.1901 [nucl-th].
14. N. Suzuki, T. Sato, T.-S.H. Lee, JLAB-THY-10-1162, arXiv:1006.2196 [nucl-th].

Valence quark contributions for the $\gamma N \to P_{11}(1440)$ transition

G. Ramalho* and K. Tsushima†

*Centro de Física Teórica de Partículas, Av. Rovisco Pais, 1049-001 Lisboa, Portugal
† Excited Baryon Analysis Center (EBAC) in Theory Center, Thomas Jefferson National Accelerator Facility, Newport News, Virginia 23606, USA
and
CSSM, School of Chemistry and Physics, University of Adelaide, Adelaide SA 5005, Australia

Abstract. A covariant spectator quark model is applied to estimate the valence quark contributions to the $F_1^*(Q^2)$ and $F_2^*(Q^2)$ transition form factors for the $\gamma N \to P_{11}(1440)$ reaction. The Roper resonance, $P_{11}(1440)$, is assumed to be the first radial excitation of the nucleon. The model requires no extra parameters except for those already fixed by the previous studies for the nucleon. The results are consistent with the experimental data in the high Q^2 region, and those from the lattice QCD. Finally the model is also applied to estimate the meson cloud contributions from the CLAS and MAID analysis.

Keywords: Covariant quark model, Roper electroproduction, Meson cloud
PACS: 12.39.Ki, 13.40.Gp, 14.20.Gk

INTRODUCTION

Study of the meson-nucleon reactions is one of the most important research topics associated with modern accelerators like CEBAF at Jefferson Lab, and defines new challenges for theoretical models. Although the electroproduction of nucleon resonances ($\gamma N \to N^*$) is expected to be governed by the short range interaction of quarks and gluons, i.e., perturbative QCD, at low and intermediate four-momentum transfer squared Q^2, one has to rely on some effective and phenomenological approaches such as constituent quark models.

The Roper [$P_{11}(1440)$] resonance is particularly interesting among the many experimentally observed nucleon resonances. Although usual quark models predict the Roper state as the first radial excitation of the nucleon, they usually result to give a much larger mass than the experimentally observed mass [1]. Experimentally the Roper has a large width which implies that it can be a πN or $\pi\pi N$ molecular-like system, or alternatively, a system of a confined three-quark core surrounded by a large amount of meson clouds. The system can be studied using dynamical coupled channel models for the meson-nucleon reactions [2].

To describe the $\gamma N \to P_{11}(1440)$ transition, we use a covariant spectator model [3, 4, 5, 6, 7]. The model has been successfully applied to the nucleon [4, 6, 8, 9] and Δ systems [7, 10, 11, 12]. A particular interest of this work is the model for the nucleon [4]. In the covariant spectator quark model a baryon is described as a three-valence quark system with an on-shell quark-pair or diquark with mass m_D, while the remaining quark is off-shell and free to interact with the electromagnetic fields. The quark-diquark vertex is represented by a baryon B wave function Ψ_B that effectively describes quark confinement [4]. To represent the nucleon system, we adopt the simplest structure given by a symmetric and antisymmetric combination of the diquark states combined to a relative S-state with the remaining quark [4]:

$$\Psi_N = \frac{1}{\sqrt{2}}\left[\Phi_I^0\Phi_S^0 + \Phi_I^1\Phi_S^1\right]\psi_N(P,k) \tag{1}$$

where $\Phi_S^{0,1}$ [$\Phi_I^{0,1}$] is the spin [isospin] state which corresponds to the diquark with the quantum number 0 or 1. The function ψ_N is a scalar wave function which depends exclusively on $(P-k)^2$, where P (k) is the baryon (diquark) momentum. As the Roper shares the same spin and isospin quantum numbers with the nucleon its wave function Ψ_R can also be represented by Eq. (1) except for the scalar wave function ψ_R.

The constituent quark electromagnetic current in the model is described by

$$j_I^\mu = \left(\tfrac{1}{6}f_{1+} + \tfrac{1}{2}f_{1-}\tau_3\right)\left(\gamma^\mu - \frac{\not q q^\mu}{q^2}\right) + \left(\tfrac{1}{6}f_{2+} + \tfrac{1}{2}f_{2-}\tau_3\right)j_2\frac{i\sigma^{\mu\nu}q_\nu}{2M}, \tag{2}$$

12th International Conference on Meson-Nucleon Physics and the Structure of the Nucleon (MENU 2010)
AIP Conf. Proc. 1374, 353-356 (2011); doi: 10.1063/1.3647158

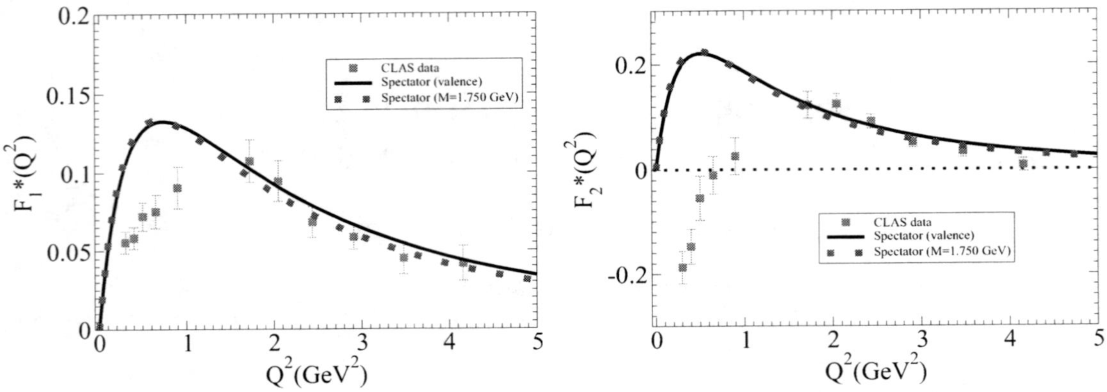

FIGURE 1. $\gamma N \to P_{11}(1440)$ transition form factors [14]. The solid [dotted] line represents the result for $M_R = 1.440$ GeV [1.750 GeV]. Data are taken from Ref. [15].

where M is the nucleon mass and τ_3 is the isospin operator. To parameterize the electromagnetic structure of the constituent quark in terms of the quark form factors $f_{1\pm}$ and $f_{2\pm}$, we adopt a vector meson dominance based parametrization given by two vector meson poles: a light one corresponds to the ρ or ω vector meson, and the heavier one corresponds to an effective heavy meson with mass $M_h = 2M$ to take accounts of the short range phenomenology. This parametrization allows us to extend the model for other regimes such as lattice QCD [5, 11, 13].

The $\gamma N \to P_{11}$ transition in the model is described by a relativistic impulse approximation in terms of the initial P_- and final P_+ baryon momenta, with the diquark (spectator) on-mass-shell [14]:

$$\begin{aligned} J^\mu &= 3\sum_\Lambda \int_k \bar{\Psi}_R(P_-,k) j_I^\mu \Psi_N(P_-,k), \\ &= \bar{u}_R(P_+)\left[\left(\gamma^\mu - \frac{\not{q}q^\mu}{q^2}\right)F_1^*(Q^2) + \frac{i\sigma^{\mu\nu}q_\nu}{M_R+M}F_2^*(Q^2)\right]u(P_-). \end{aligned} \tag{3}$$

In the first line the sum is over the diquark states $\Lambda = \{s,\lambda\}$, where s and $\lambda = 0,\pm 1$ stand for the scalar diquark and the vector diquark polarization, respectively, and the covariant integral is $\int_k \equiv \int d^3k/[2E_D(2\pi)^3]$, where E_D is the diquark energy. The factor 3 is due to the flavor symmetry. In the second line the transition form factors F_1^* and F_2^* are defined independent of the frame using the Dirac spinors of the Roper (u_R) and nucleon (u) with the respective masses M_R and M. For simplicity the spin projection indices are suppressed.

To represent the scalar wave functions as functions of $(P-k)^2$, we can use a dimensionless variable

$$\chi_B = \frac{(M_B - m_D)^2 - (P-k)^2}{M_B m_D}, \tag{4}$$

where M_B is the baryon mass. With this form, the nucleon and the Roper scalar wave functions are given by [14]:

$$\psi_N(P,k) = \frac{N_0}{m_D(\beta_1+\chi_N)(\beta_2+\chi_N)}, \tag{5}$$

$$\psi_R(P,k) = \frac{\beta_3-\chi_R}{\beta_1+\chi_R}\frac{N_1}{m_D(\beta_1+\chi_R)(\beta_2+\chi_R)}. \tag{6}$$

The nucleon scalar wave function is chosen to reproduce the asymptotic form predicted by pQCD for the nucleon form factors ($G_E, G_E \sim 1/Q^4$) and to describe the elastic nucleon form factor data [4]. The parameters β_1 and β_2 are associated with the momentum range. With $\beta_2 > \beta_1$, β_1 and β_2 are the long-range and short-range regulators, respectively. The expression for the Roper scalar wave function is inspired by the nonrelativistic wave functions in the spherical harmonic potentials [1, 18, 19], where the factor $\beta_3 - \chi_R$ (linear in the momentum variable χ_R) characterizes the radial excitation. The factors N_0 and N_1 are normalization constants given by the condition $\int_k |\psi_B|^2 = 1$ at $Q^2 = 0$ for $B = N,R$. The new parameter β_3, associated with the Roper, is fixed by the orthogonality with the nucleon wave function.

RESULTS

The results for the $\gamma N \to P_{11}(1440)$ transition form factors are presented in Fig. 1. Recall that all parameters associated with the Roper have been determined by the relation with the nucleon. Therefore, the results are genuine predictions. The agreement with the data is excellent for $Q^2 > 2$ GeV^2, in particular for F_1^*. Our results are also consistent with other work for the intermediate and high Q^2 region. See Ref. [14] for more details. In dynamical coupled channel models [20] the Roper appears as a heavy bare system (≈ 1.750 GeV) dressed by meson clouds that reduce the mass of the system to near the observed mass (≈ 1.440 GeV). To explore further this reduction of mass due to the meson clouds, we replace the physical mass by the 'bare' mass. The result is indicated by the dotted line. As one can observe from the figure the effect due to the Roper bare mass is small although it approaches very slightly to the data.

The disagreement in the low Q^2 region in Fig. 1 can be interpreted as a limitation of the approach, since only valence quark effects have been included, but not the quark-antiquark, or meson cloud effects. The meson cloud effects are expected to be important in the low Q^2 region [1, 2, 15]. This argument is supported by the fact that when the model extended to a heavy pion mass lattice QCD regime, where the meson cloud effect is suppressed, since the result agrees well the heavy pion mass lattice data [14, 17]. The importance of the meson cloud contributions in inelastic reactions was also observed in the $\gamma N \to \Delta$ reaction [7, 10, 11]. Furthermore, it was shown that the covariant spectator quark model can describe both the lattice and physical data [11]. Based on the successes in describing the heavy pion mass lattice QCD data and the high Q^2 region data, we have some confidence that the model can describe well the valence quark contributions. Thus, we can use the spectator quark model to estimate the meson cloud contributions.

To estimate the meson cloud contributions, we decompose the form factors as,

$$F_i^*(Q^2) = F_i^b(Q^2) + F_i^{mc}(Q^2) \quad (i = 1,2), \tag{7}$$

where F_i^b and F_i^{mc} represent the valence quark (bare) and meson cloud contributions, respectively. This decomposition is justified if the meson is created by the overall baryon but not by a single quark in the baryon core. Replacing F_i^b by the result of the spectator quark model, one can estimate the meson cloud contributions in the $\gamma N \to P_{11}(1440)$ transition. We estimate the meson cloud contributions in two different ways as presented in Fig. 2 (left panel). One is based on the MAID fit made for the whole Q^2 region, and estimates the bands associated with the meson cloud assuming the uncertainty of the CLAS data points (red region). The other estimate is from CLAS data for each data point by subtracting the valence quark contribution. The results are represented by circles.

Similarly, the meson cloud contributions can be determined for the helicity amplitudes, $A_{1/2}$ and $S_{1/2}$, associated with the photon polarizations of $+1$ and 0 respectively. Representing the corresponding amplitudes by $R = A_{1/2}$ and $S_{1/2}$, respectively, one can obtain the meson cloud contributions:

$$R^{mc}(Q^2) = R(Q^2) - R^b(Q^2). \tag{8}$$

The results are also shown in Fig. 2 (right panel). Both methods suggest the significant meson cloud effects for low Q^2 region ($Q^2 < 1$ GeV^2), and a fast falloff with increasing Q^2.

In conclusion, the spectator quark model can be effectively applied to study the nucleon resonances, particularly in the intermediate and high Q^2 region, where the valence quark effects are dominant, and both the relativity and covariance are essential. Another example of a successful application of the present approach can be found in Ref. [21].

ACKNOWLEDGMENTS

The authors would like to thank V. D. Burkert for the invitation. G. R. thanks the organizers for the financial support. G. R. was supported by the Portuguese Fundação para a Ciência e Tecnologia (FCT) under the grant SFRH/BPD/26886/2006. Notice: Authored by Jefferson Science Associates, LLC under U.S. DOE Contract No. DE-AC05-06OR23177. The U.S. Government retains a non-exclusive, paid-up, irrevocable, world-wide license to publish or reproduce this manuscript for U.S. Government purposes.

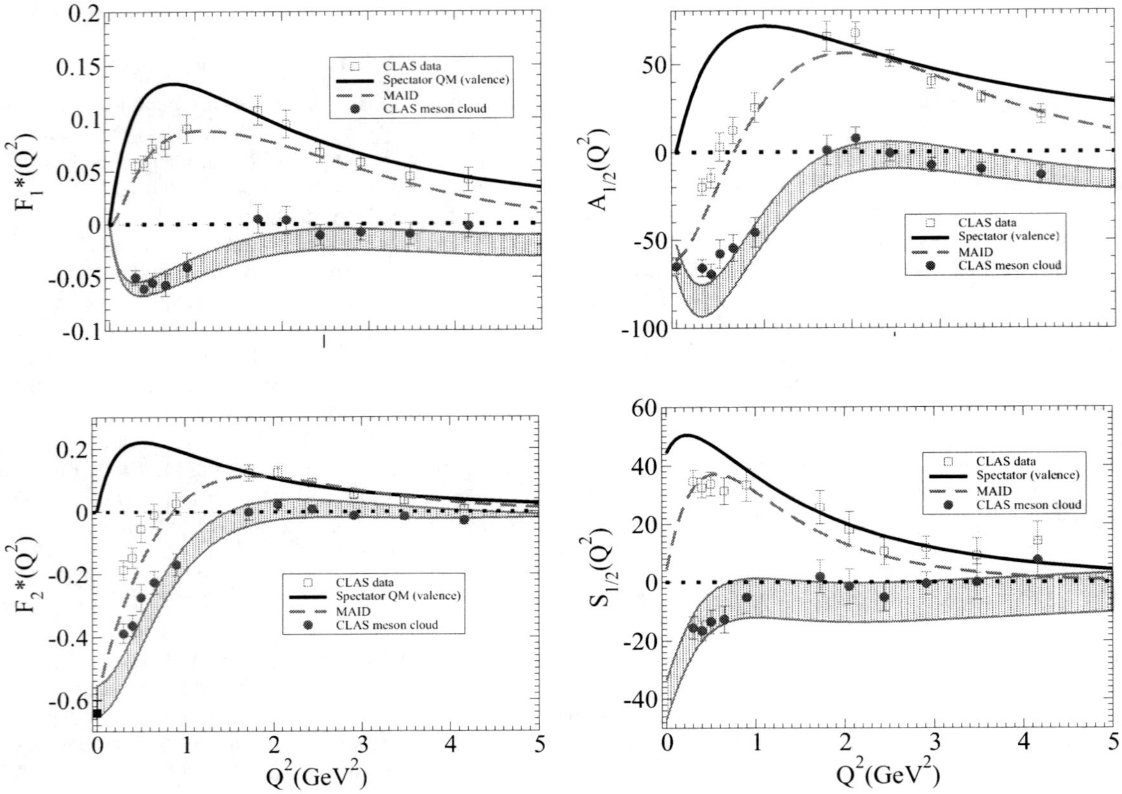

FIGURE 2. $\gamma N \to P_{11}(1440)$ transition form factors (left) and amplitudes (right) [14], compared with the CLAS data (squares) [15], and the meson cloud contributions estimated from CLAS data (circles). The dashed line is the MAID fit from Ref. [16]. The bands represent the limits of the meson cloud contributions estimated from the MAID fit [14].

REFERENCES

1. I. G. Aznauryan, Phys. Rev. C **76**, 025212 (2007).
2. V. D. Burkert and T. S. H. Lee, Int. J. Mod. Phys. E **13**, 1035 (2004).
3. F. Gross, Phys. Rev. **186**, 1448 (1969).
4. F. Gross, G. Ramalho and M. T. Peña, Phys. Rev. C **77**, 015202 (2008).
5. G. Ramalho, K. Tsushima and F. Gross, Phys. Rev. D **80**, 033004 (2009).
6. F. Gross, G. Ramalho and K. Tsushima, Phys. Lett. B **690**, 183 (2010).
7. G. Ramalho, M. T. Peña and F. Gross, Eur. Phys. J. A **36**, 329 (2008).
8. F. Gross, G. Ramalho and M. T. Peña, Phys. Rev. C **77**, 035203 (2008).
9. G. Ramalho, F. Gross, M. T. Peña and K. Tsushima, arXiv:1008.0371 [hep-ph].
10. G. Ramalho, M. T. Peña and F. Gross, Phys. Rev. D **78**, 114017 (2008).
11. G. Ramalho and M. T. Peña, Phys. Rev. D **80**, 013008 (2009).
12. G. Ramalho, M. T. Peña and F. Gross, Phys. Rev. D **81**, 113011 (2010).
13. G. Ramalho and M. T. Peña, J. Phys. G **36**, 115011 (2009).
14. G. Ramalho and K. Tsushima, Phys. Rev. D **81**, 074020 (2010).
15. I. G. Aznauryan *et al.* [CLAS Collaboration], Phys. Rev. C **80**, 055203 (2009).
16. L. Tiator and M. Vanderhaeghen, Phys. Lett. B **672**, 344 (2009).
17. H. W. Lin, S. D. Cohen, R. G. Edwards and D. G. Richards, Phys. Rev. D **78**, 114508 (2008).
18. S. Capstick and B. D. Keister, Phys. Rev. D **51**, 3598 (1995).
19. B. Julia-Diaz, D. O. Riska and F. Coester, Phys. Rev. C **69**, 035212 (2004) [Erratum-ibid. C **75**, 069902 (2007)].
20. N. Suzuki, B. Julia-Diaz, H. Kamano, T. S. Lee, A. Matsuyama and T. Sato, Phys. Rev. Lett. **104**, 042302 (2010).
21. G. Ramalho and K. Tsushima, Phys. Rev. D **82**, 073007 (2010).

Exclusive π^0, η electroproduction in the resonance region at high Q^2.

M. Ungaro and K. Joo

University of Connecticut, Storrs, CT, 06269, USA

Abstract. We present preliminary results from π^0 and η electroproduction data taken with CLAS at Jefferson Lab. These and other CLAS measurements, coupled with recent theoretical developments, will allow non-perturbative approaches to shed lights on the role of quarks and gluons in nuclei and address issues such as confinement and the non-zero quark mass in the chiral limit.

Keywords: CLAS; pi0; eta; barion resonances; transition form factors
PACS: 25.20.-x, 25.30.Rw

THE N* PROGRAM IN CLAS: BEYOND CQM AND PQCD

Constituent Quark Models (CQM) provide simple and effective interpretations of many phenomena [1, 2, 3, 4, 5, 6], especially near the photon point. On the other hand, an investigation of the fundamental structure of the nucleon and its excited states has to involve partonic degrees of freedom and Quantum Chromo-Dynamics (QCD). At very large Q^2, perturbative QCD (pQCD) prescribes scaling of the N* helicity amplitudes [7, 8]:

$$A_{1/2} \propto 1/Q^3,\ A_{3/2} \propto 1/Q^3,\ G^*_M \propto 1/Q^4,$$

However pQCD cannot answer fundamental questions such as:

- What is the nature of quark confinement?
- What is the dynamic of quark-gluon interactions at low energy?

Thanks to recent theoretical developments, understanding the microscopic nature of the nucleon using non-perturbative approaches is beginning to become a reality, and the crucial role played by high precision measurements of nucleon resonances is becoming increasingly clear. In particular these approaches are promising;

Lattice QCD: The Hadron Spectrum Collaboration involving the Lattice Group at Jefferson Lab, Carnegie Mellon University, Univ. of Maryland, and Trinity College (Dublin) has embarked on an ambitious program to compute the high lying excited state spectrum of baryons and mesons, as well as their electromagnetic transition form-factors up to Q^2 of 10 GeV^2 [9, 10]. New techniques on the lattice allow the calculation of baryon spectra not only in the ground state but also of the excited states [11, 12, 13] and first results have been obtained for the N$\rightarrow$P$_{11}$ transition form factor [14]. Although still exploratory, these results are very promising and underline the importance of experimental N* measurements.

Dynamical Chiral Symmetry Breaking and Dyson-Schwinger Equations: The complex of Dyson-Schwinger equations (DSEs) is a powerful tool that has been employed with marked success to study confinement and DCSB, and their impact on hadron observables [15, 16, 17, 18, 19]. International studies of nucleon electromagnetic form factors lead by Cloet, Bentz, Thomas are underway. High precision measurements of nucleon-resonance transition form factors, DSEs and DCSB can be used to chart the momentum evolution of the dressed-quark mass. In particular, the transition between current and constituent quark domains, believed to happen between 5-10 GeV^2.

Light Cone Sum Rule and Form Factors: One of the prediction of QCD is that the, at large momentum transferred, the form factor become dominated by the contribution of valence states with small transverse separation between them. Such distributions are dubbed Distribution Amplitudes (DA).

A non perturbative approach that allows quantitative descriptions of form factor using is the Light-Cone Sum Rule (LCSR): within this framework, the the DA are directly connected to the transition form factors.

12th International Conference on Meson-Nucleon Physics and the Structure of the Nucleon (MENU 2010)
AIP Conf. Proc. 1374, 357-361 (2011); doi: 10.1063/1.3647159

The DA can be calculated on the lattice. Recent effort from Warkentin, Braun, of Theoretical Groups in Regensburg and others from the QCDSF collaboration calculated the DA for both the proton and the S_{11} [20] in good qualitative agreement with experimental data, especially at $Q^2 > 2$. The group has plan in the next couple of years to a) decrease the pion mass for more precise measurement and b) to repeat this calculation for all the baryons in the octet and the decuplet.

The N* Program in CLAS is aimed at mapping the the $\gamma_v NN^*$ electro-couplings as a function of photon virtuality with the combined analysis of the major electro-production channels. Among the channels analyzed: $p\pi^0$, $p\eta$, $n\pi^+$, $p\eta$, $K\Lambda$ and double charged pion electro-production. The CLAS data for the first time provided information on many observables in these exclusive channels, including fully integrated cross sections and a variety of 1-fold differential cross sections complemented by single and double polarization asymmetries in a range of Q^2 from 0.2 to 5 GeV^2 .

In this article we present preliminary results for $p\pi^0$ electroproduction data and an outlook for $p\eta$ data. The experiment took place between October 2001 and January 2002 at Jeffeson Lab, using a 5.75 GeV electron beam impinging on a 5cm liquid Hydrogen target with an intensity of 10nA ($10^{34} cm^{-2} s^{-1}$). 3.7 billions triggers were collected.

ANALYSIS OVERVIEW

The selection analysis include electron and proton particle ID and fiducial cuts, and kinematic corrections. The absolute normalization was checked with elastic and inclusive events. The Bethe-Heitler events were removed using co-planarity and peaking approximation kinematics. This process has been also tested in the MonteCarlo simulation with π^0, elastic and radiative elastic events. $eP \to ePX$ events after particle ID, fiducial cuts and kinematic corrections are shown in Fig. 1 (left). The result of Bethe-Heitler events removal is show in Fig. 1 (center), while the missing mass spectrum after Bethe-Heitler removal for both $p\pi^0$ and $p\eta$ channels is show in Fig. 1 (right).

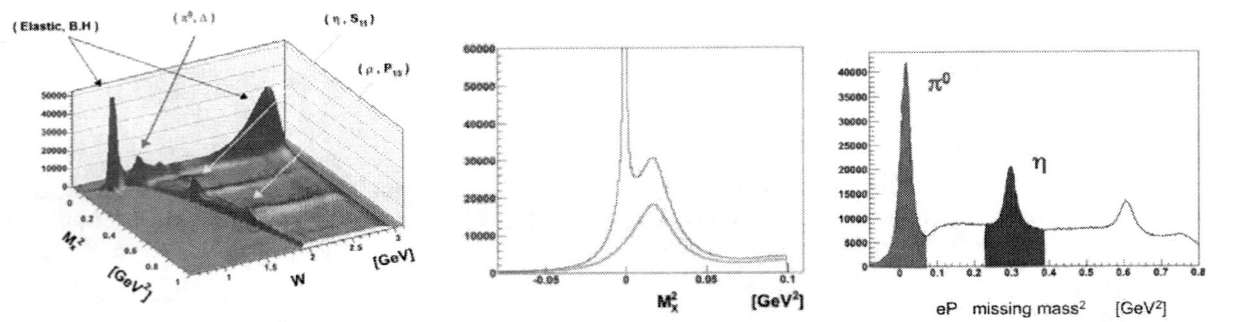

FIGURE 1. Left: $eP \to ePX$ events. Several resonances/decay channels are clearly visible. Center: Missing mass around π^0. Upper line: all events. Lower line: after Bethe-Heitler removal. Right: missing mass spectrum for both $p\pi^0$ and $p\eta$ after Bethe-Heitler removal.

Binning

The π^0 data was binned as follows:

W : 45 bins, $\Delta W = 20 MeV$ from 1.1 to 2.0 GeV
Q^2 : 7 bins, such that $\Delta Q^2/Q^2 \sim 0.18$ from 2.1 to 6 GeV^2
$\cos(\theta^*)$: 10 bins from -1 to 1
ϕ^* : 24 bins from 0 to 360 deg.

The η data was binned as follows:

W : 25 bins, $\Delta W = 20 MeV$ from 1.5 to 2.0 GeV
Q^2 : 7 bins, such that $\Delta Q^2/Q^2 \sim 0.18$ from 2.1 to 6 GeV^2
$\cos(\theta^*)$: 10 bins from -1 to 1
ϕ^* : 24 bins from 0 to 360 deg.

Several models were used to calculate the theoretical cross section. These include MAID2007 [21], MAID2003 [22], SAID [23]. The program exclurad [24] was used to calculate the radiative correction. Several theoretical models were used to calculate the systematic errors coming from the theoretical uncertainties.

To calculate the bin averaging correction each variables of each 4-dimensional bin was divided in 15 subdivisions. Several theoretical models were used to calculate the systematic errors coming from the theoretical uncertainties.

200 millions events were generated, simulated through a geant3 Monte Carlo program with the CLAS detectors and reconstructed in order to calculated the acceptance correction. The ϕ^* binning for the acceptance correction was quadrupled in order to account for steep variations due to the CLAS azimuthal segmentation.

PRELIMINARY RESULTS

In Fig. 2 the preliminary π^0 cross section is shown as a function of ϕ^* for 240 points at $W = 1.41$ *GeV* and $Q^2 = 3.0$ GeV^2 in 10 $\cos(\theta^*)$ bins. The cross section is compared with MAID2007 [21], MAID2003 [22] with and without the Roper resonance, and SAID [23]. While data and models agree very well on the Delta resonance (left), significant discrepancies exist at higher W (right). Each plot represent 240 out of 75000 data points. Each ϕ^* distribution is fitted with the function: $a + b\cos\phi + c\cos 2\phi$ in order to extract the response functions $\sigma_T + \varepsilon_L\sigma_L$, σ_{TT} and σ_{LT} from the general expression of the meson electroproduction *c.m.* cross section:

$$\frac{d^2\sigma}{d\Omega^*_\pi} = \sigma_T + \varepsilon_L\sigma_L + \varepsilon\sigma_{TT}\sin^2\theta^*_\pi\cos 2\phi^*_\pi + \sigma_{LT}\sqrt{2\varepsilon_L(\varepsilon+1)}\sin\theta^*_\pi\cos\phi^*_\pi. \tag{1}$$

where ε and ε_L are the transverse and longitudinal virtual photon polarizations respectively.

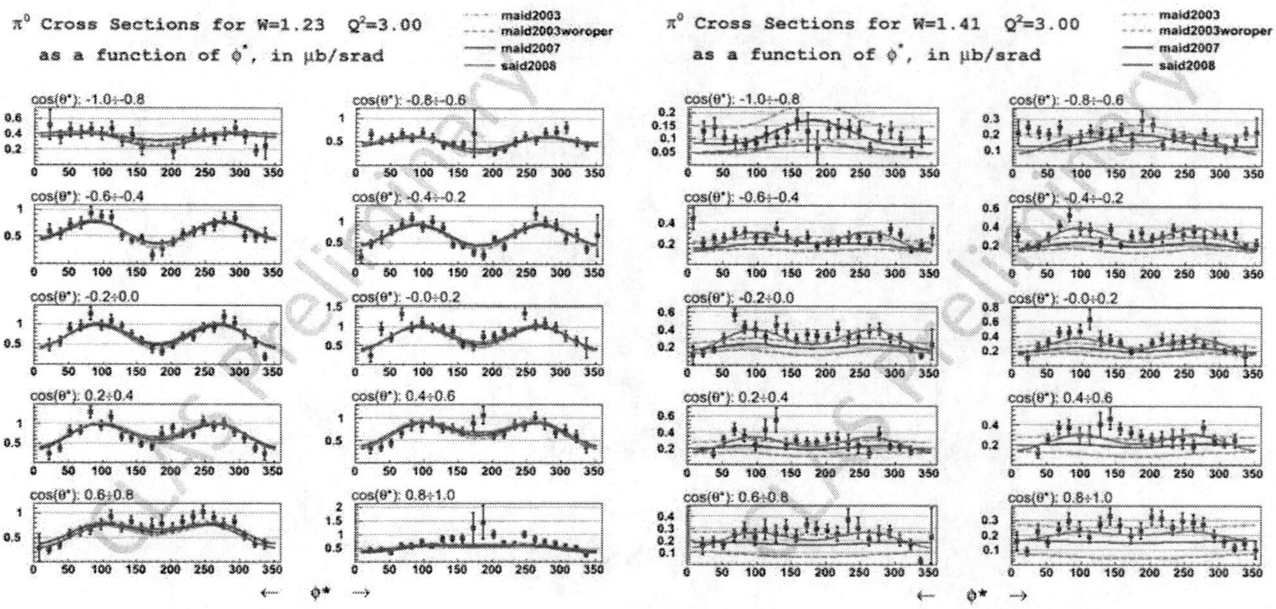

FIGURE 2. Preliminary π^0 cross section as a function of ϕ^* for $W = 1.23$ *GeV* (left) and $W = 1.41$ *GeV* (right), $Q^2 = 3.0$ GeV^2 in 10 $\cos(\theta^*)$ bins. The lines are theoretical calculations as indicated by the labels.

In Fig. 4 the response functions resulting from the ϕ^* fits are shown for $Q^2 = 2.4$ GeV^2. In order to extract all the resonance parameters however, the cross section will be analysed using the Lagrangian approaches developed at

JLAB. In Particular:

Unitary Isobar Model(UIM)[25] :

- non-resonant amplitudes incorporate nucleon and meson Born terms, vector meson exchanges, Regge terms at high energy.
- resonant terms incorporate relativistic Breit-Wigner amplitudes with energy-dependent widths.
- full amplitude unitarized in K-matrix approximation.

Fixed-t dispersion relations (DR)[25] :

- resonance amplitudes constructed in same way as in UIM
- non-resonant amplitudes from Born terms and dispersion relation.

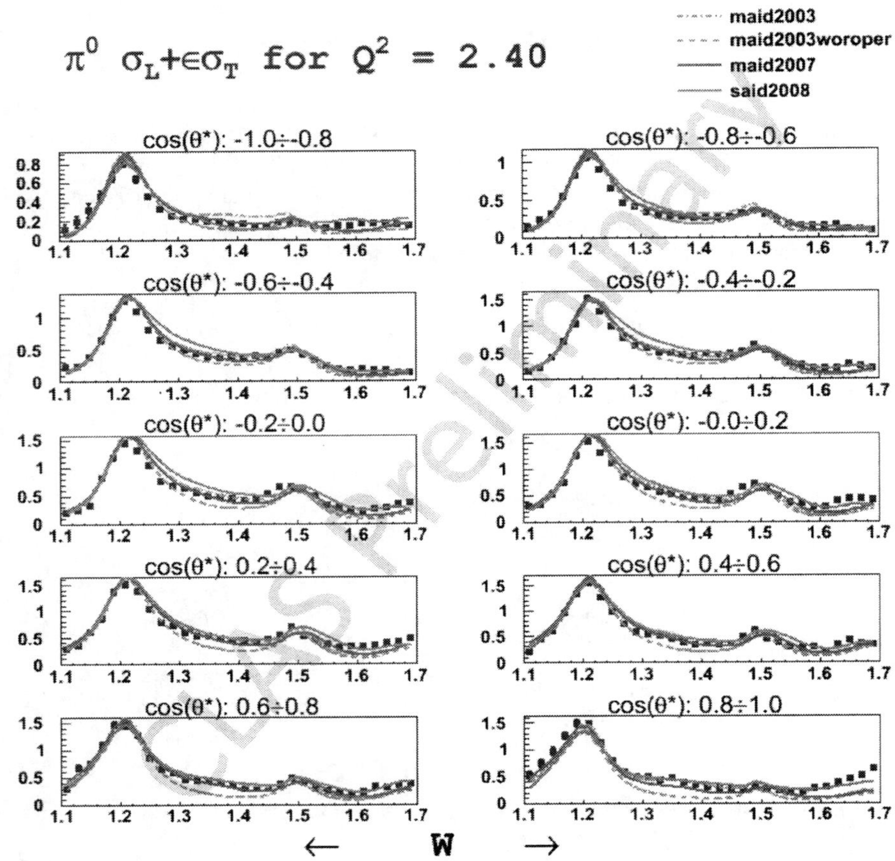

FIGURE 3. Results of the $a+b\cos\phi+c\cos2\phi$ to extract the $\sigma_T+\varepsilon_L\sigma_L$ response functions from Eq. 1.

CONCLUSIONS, OUTLOOK

Thanks to recent theoretical developments, one could argue that the N* resonances are "back in fashion": high precision measurements of N* amplitudes will be indispensable to non-perturbative approaches such as lattice-QCD, DSEs and LCSR as they are refined. Preliminary results for π^0 electroproduction and the outlook for η electroproduction in the resonance region looks very promising to reach these goals.

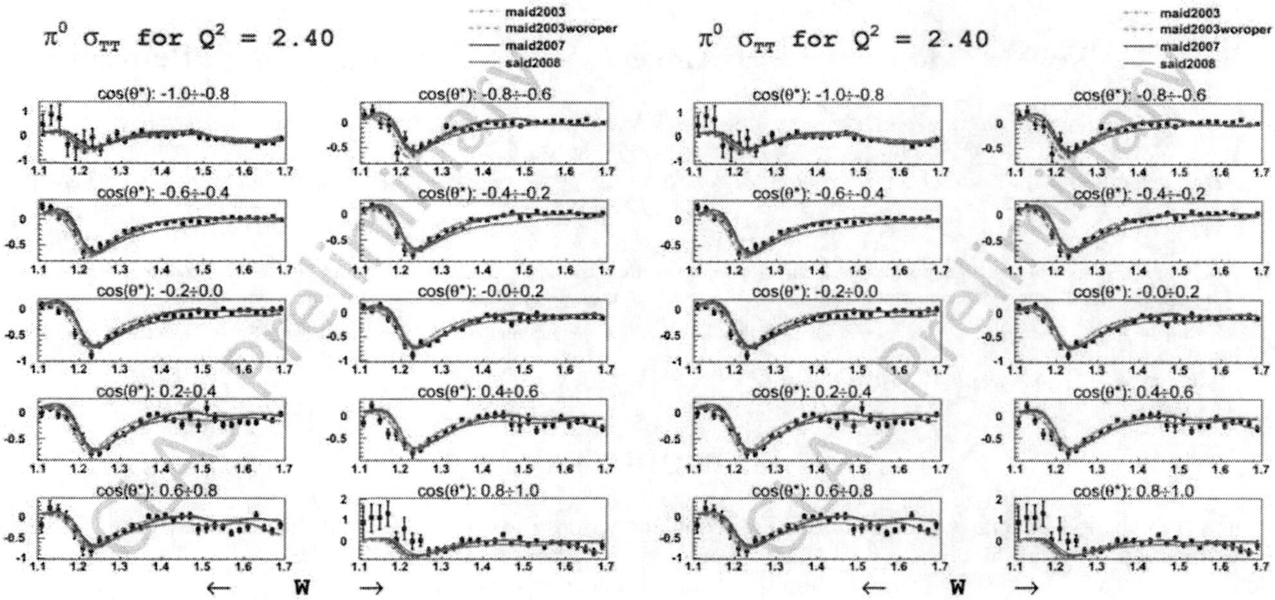

FIGURE 4. Results of the $a + b\cos\phi + c\cos 2\phi$ to extract the response functions from Eq. 1: Left: σ_{TT} . Right: σ_{LT}.

REFERENCES

1. M. D. Sanctis, M. M. Giannini, L. Repetto, and E. Santopinto, *Phys. Rev.* **C62**, 025208 (2000).
2. B. Julia-Diaz, D. O. Riska, and F. Coester, *Phys. Rev.* **C69**, 035212 (2004), hep-ph/0312169.
3. P. L. Chung, and F. Coester, *Phys. Rev.* **D44**, 229–241 (1991).
4. M. M. Giannini, *Rept. Prog. Phys.* **54**, 453–530 (1990).
5. N. Isgur, and G. Karl, *Phys. Rev.* **D18**, 4187 (1978).
6. F. Cardarelli, E. Pace, G. Salme, and S. Simula, *Phys. Lett.* **B357**, 267–272 (1995), nucl-th/9507037.
7. G. P. Lepage, and S. J. Brodsky, *Phys. Rev.* **D22**, 2157 (1980).
8. J. J. Kelly, *Phys. Rev.* **C70**, 068202 (2004).
9. I. Aznauryan, et al. **D79** (2009), 0907.1901.
10. H.-W. Lin, et al., *Phys. Rev.* **D79**, 034502 (2009), 0810.3588.
11. S. Basak, et al., *Phys. Rev.* **D72**, 074501 (2005), hep-lat/0508018.
12. J. M. Bulava, et al., *Phys. Rev.* **D79**, 034505 (2009), 0901.0027.
13. J. J. Dudek, R. Edwards, and C. E. Thomas, *Phys. Rev.* **D79**, 094504 (2009), 0902.2241.
14. H.-W. Lin, S. D. Cohen, R. G. Edwards, and D. G. Richards, *Phys. Rev.* **D78**, 114508 (2008), 0803.3020.
15. C. D. Roberts, and A. G. Williams, *Prog. Part. Nucl. Phys.* **33**, 477–575 (1994), hep-ph/9403224.
16. C. D. Roberts, and S. M. Schmidt, *Prog. Part. Nucl. Phys.* **45**, S1–S103 (2000), nucl-th/0005064.
17. R. Alkofer, and L. von Smekal, *Phys. Rept.* **353**, 281 (2001), hep-ph/0007355.
18. P. Maris, and C. D. Roberts, *Int. J. Mod. Phys.* **E12**, 297–365 (2003), nucl-th/0301049.
19. C. D. Roberts, M. S. Bhagwat, A. Holl, and S. V. Wright, *Eur. Phys. J. ST* **140**, 53–116 (2007), 0802.0217.
20. N. Warkentin, et al., *PoS* **LATTICE2008**, 154 (2008), 0811.2212.
21. D. Drechsel, S. S. Kamalov, and L. Tiator, *Eur. Phys. J.* **A34**, 69–97 (2007), 0710.0306.
22. D. Drechsel, O. Hanstein, S. S. Kamalov, and L. Tiator, *Nucl. Phys.* **A645**, 145–174 (1999), nucl-th/9807001.
23. R. A. Arndt, W. J. Briscoe, M. W. Paris, I. I. Strakovsky, and R. L. Workman, *Chin. Phys.* **C33**, 1063 (2009), 0906.3709.
24. A. Afanasev, I. Akushevich, V. Burkert, and K. Joo, *Phys. Rev.* **D66**, 074004 (2002), hep-ph/0208183.
25. I. G. Aznauryan, *Phys. Rev.* **C67**, 015209 (2003), nucl-th/0206033.

Photopion Physics at the HIγS Facility

Henry R. Weller [*], Mohammad W. Ahmed[*], Sean Stave[*] and Aron M. Bernstein [†]

[*] *Department of Physics and Triangle Universities Nuclear Laboratory, Duke University, Durham, NC, 27708-0308, USA*
[†] *Department of Physics and Laboratory for Nuclear Science, Massachusetts Institute of Technology, Cambridge, MA 02139, USA*

Abstract. Future photopion experiments at the HIγS facility are described. It is shown that they will provide accurate tests of ChPT and will be sensitive to isospin breaking dynamics due to the mass difference of the up and down quarks.

Keywords: Chiral Perturbation Theory, Photo-pion production, Isospin symmetry breaking, High Intensity γ-ray Source, Scattering lengths, ChPT
PACS: 29.50.+v,11.30.Rd,25.20.-x,14.80.Va,24.70.+s

INTRODUCTION

Chiral Pertubation Theory (ChPT) with one-pion-loop corrections (i.e., one-loop order) has been used to calculate the amplitudes for threshold photo- and electro-pion production from the nucleon [1, 2, 3, 4, 5, 6]. This theoretical development was driven by a series of threshold photoproduction experiments with high-duty-cycle accelerators carried out at Mainz [7, 8, 9] and Saskatoon [10]. Today there is impressive overall agreement between the ChPT calculations and the data. This agreement with a large body of data supports the concept of spontaneous chiral symmetry hiding in QCD. It is important to measure the near-threshold interactions because they are an explicit effect of chiral symmetry breaking and have been evaluated within ChPT.

PHOTOPION PRODUCTION NEAR THRESHOLD

There is a strong influence of the charged-pion channel through the two-step $\gamma p \to \pi^+ n \to \pi^0 p$ reaction in the region below the π^+n threshold. Since the ratio of the electric dipole amplitudes for the neutral and charged pion channels $E_{0+}(\gamma p \to \pi^+ n)/E_{0+}(\gamma p \to \pi^0 p) \simeq -20$, the two step reaction is as strong as the direct path. This leads to a unitary cusp in the $\gamma p \to \pi^0 p$ reaction.

The simplest dynamical way to understand the occurrence of the unitary cusp in the $\gamma^* p \to \pi^0 p$ reaction (where γ^* is a real or virtual photon) is to use the 3-channel S matrix for the open channels $\gamma^* p, \pi^0 p$ and $\pi^+ n$ [11]. Applying the constraints of unitarity and time-reversal invariance, one is then led to the coupled-channel result for the s-wave amplitude $E_{0+}(\gamma p \to \pi^0 p)$:

$$E_{0+}(\gamma p \to \pi^0 p : k) = e^{i\delta_0}[A(k) + i\beta q_+]$$
$$\text{with } \beta = E_{0+}(\gamma p \to \pi^+ n) \cdot a_{cex}(\pi^+ n \to \pi^0 p) \tag{1}$$

where δ_0 is the s-wave $\pi^0 p$ phase shift (predicted to be small), $A(k)$ is a smooth function of the photon energy k, β is the cusp parameter, $a_{cex}(\pi^+ n \to \pi^0 p)$ is the charge exchange s-wave scattering length for the $\pi^+ n \to \pi^0 p$ reaction and q_+ is the π^+ CM momentum , which is continued below the $\pi^+ n$ threshold as $i \mid q_+ \mid$. The cusp function βq_+ contributes to the real (imaginary) part of E_{0+} below (above) the $\pi^+ n$ threshold.

The expected value of β can be calculated [11] on the basis of unitarity using Eq. 1. The best experimental value of $a(\pi^- p \to \pi^0 n) = -(0.122 \pm 0.002)/m_\pi$, obtained from the observed width in the 1s state of pionic hydrogen [12], was used. This is in excellent agreement with ChPT predictions of $-(0.130 \pm 0.006)/m_\pi$ [13]. Assuming isospin is conserved, $a(\pi^+ n \leftrightarrow \pi^0 p) = -a(\pi^- p \leftrightarrow \pi^0 n)$. The latest measurement for $E_{0+}(\gamma p \to \pi^+ n) = 28.06 \pm 0.27 \pm 0.45$ [14] (where the first error is statistical and the second is systematic) is in good agreement with the ChPT prediction of 28.2 ± 0.6 [15]. Using these experimental values and the relationship given above leads to a value of $\beta = 3.43 \pm 0.08$. On the other hand, the ChPT prediction (which does not satisfy unitarity) is $\beta = 2.78$ [1, 2, 3].

12th International Conference on Meson-Nucleon Physics and the Structure of the Nucleon (MENU 2010)
AIP Conf. Proc. 1374, 362-365 (2011); doi: 10.1063/1.3647160

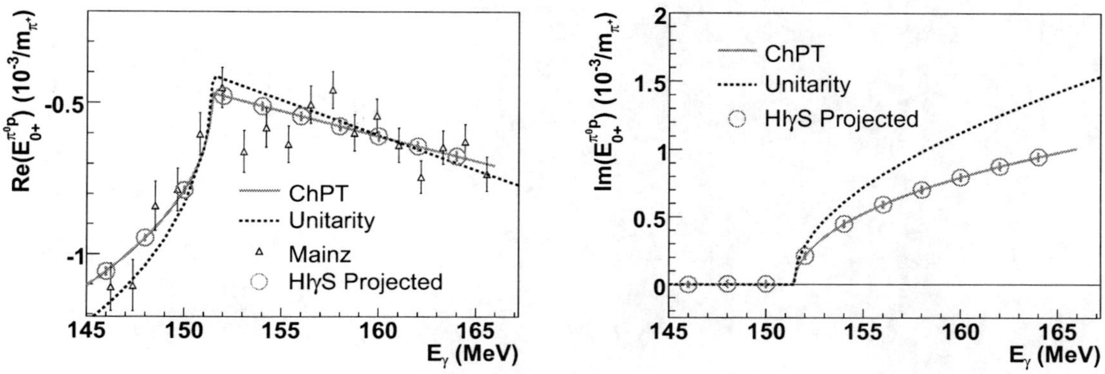

FIGURE 1. Left: Re (E_{0+}) for the $\gamma p \rightarrow \pi^0 p$ reaction [9]. The solid curves are ChPT calculation [1, 2, 3, 4, 15], and dashed curves are unitary fit [7, 8, 11] (see Eq. 1). Right: Im (E_{0+}) for the $\gamma p \rightarrow \pi^0 p$ reaction [11]. For both panels the projected data points show the estimated statistical errors for 100 hours of beam time at HIγS at each energy in four different beam-target polarization configurations.

The results for Re E_{0+} are presented in Fig. 1. The points are extracted from the latest data from Mainz [9]. The anticipated improvements in the determination of this multipole based on 100 hours of beam time per observable at HIγS are also shown.

There are as yet no measurements of Im $E_{0+}(\gamma p \rightarrow \pi^0 p)$. The theorectical predictions are presented in Fig. 1. Future experiments will provide a measurement at the 1-to-2% level and easily be able to distinguish between these predictions. From the results of the proposed measurement of Im E_{0+} one can extract $a_{cex}(\pi^+ n \rightarrow \pi^0 p)$ from the value of β using the experimental results for $E_{0+}(\gamma p \rightarrow \pi^+ n)$.

THE HIγS EXPERIMENT

The High Intensity Gamma-ray Source (HIγS) at Duke University is a nearly monochromatic Compton γ-ray source with a very high flux, a wide energy range, and switchable polarizations. The current HIγS electron accelerators include a 0.18 GeV linear accelerator pre-injector, a 0.18 – 1.2 GeV booster injector, a 0.24 – 1.24 GeV storage ring, and several storage ring based Free-Electron Lasers (FELs).The high flux performance is realized by colliding the electron beam in the storage ring with a temporally matched, high-power FEL beam built up inside a low-loss laser resonator. HIγS can be operated in a wide energy range from 1 MeV to about 100 MeV in the present configuration and with future upgrades above 100 MeV up to the pion threshold energy. Highly polarized gamma-ray beams can be produced with linear and circular polarizations using different FEL wiggler configurations.Given these outstanding features, HIγS is a world-class Compton gamma-ray source for a wide range of scientific research programs in nuclear physics, astrophysics, medicine, and industrial applications.

With the current accelerator configuration of HIγS, the maximum gamma-beam energy is limited to about 100 MeV due to the availability of commercial FEL mirrors at the shortest wavelength of about 190 nm. To extend HIγS operation toward and beyond the pion-threshold energy requires the development of highly reflective, radiation-resistive 150 nm FEL mirrors. By working closely with commercial optics companies, we project that useful 150 nm FEL mirrors will be available in the next few years. Once it has 150 nm FEL mirrors, by operating the storage ring and booster up to maximum energy of 1.2 GeV, HIγS will be able to produce gamma-beams at energies up to around 158 MeV. The projected initial performance is 1×10^8–1×10^9 γ/s after increasing the electron beam injection rate by upgrading the injectors.

The experimental setup for the proposed experiments at HIγS includes the following key resources : a) Crystal Box detector [16]; b) Neutral Meson Spectrometer (NMS) [17]; c) Liquid Scintillating Neutron Detectors; and d) HIγS Frozen-Spin Polarized Target (HIFROST). HIFROST is currently being installed at HIγS by a TUNL-University of Virginia collaboration under the direction of Donald Crabb. The HIγS target has a length of 7.3 cm, a diameter of 2.0 cm, and a target thickness of $\sim 3.5 \times 10^{23}$ polarized protons per cm^2. The polarization is expected to be 70-80 %.

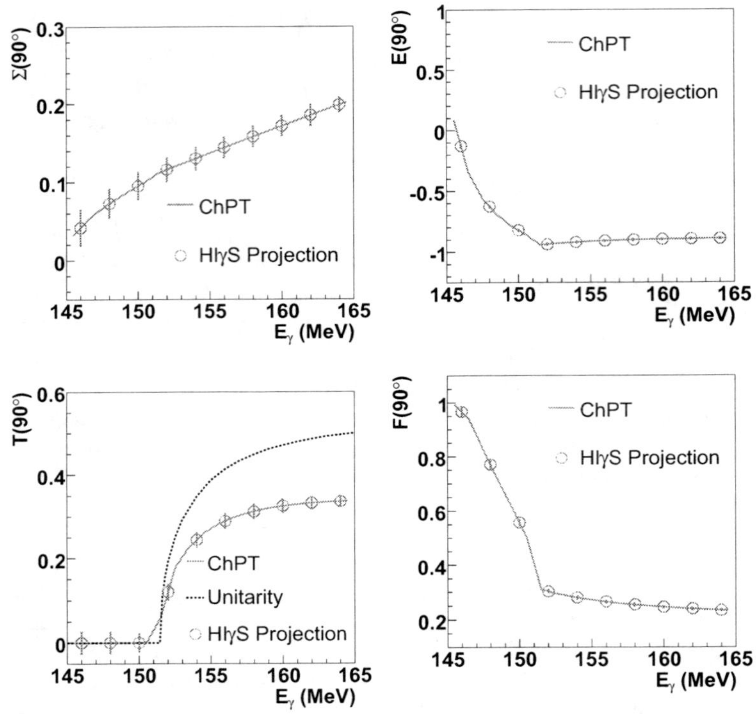

FIGURE 2. Predicted polarization asymmetries as a function of photon energy at a pion CM angle of 90^0 for the $\vec{\gamma}\vec{p} \rightarrow \pi^0 p$ reaction.The solid curves correspond to the ChPT predictions [4]. The dotted curve shown for the case of the $T(90^0)$ is the result of a fit based on unitarity. The points show projected statistical errors at various photon energies for 100 hours of running time each at HIγS.

The pions from the $^1\text{H}(\gamma,\pi^0)$ reaction will be detected primarily by the Crystal Box with possible additional coverage by the NMS. At E_γ of 158 MeV, this setup provides $\sim 3\pi$ solid angle coverage for π^0 detection. This could be enhanced to over $\sim$ 90% of 4π with the inclusion of the NMS in the setup. The energy of the neutral pion is inferred from its decay photons.

SIMULATED RESULTS

A full MonteCarlo simulation was performed for the anticipated experiments at HIγS. The predictions of ChPT were used for the $\pi^0 p$ channel. The experimental setup utilized the Crystal Box assembly for π^0 detection. The beam-on-target intensity was taken to be 10^7 γ/s, and the HIFROST target was assumed with a thickness of 3.5×10^{23} protons/cm^2. All observables were measured at all CM scattering angles. The observables for the case of the π^0p channel which were considered were $\sigma_T(\theta)$ (transverse differential cross section), $\Sigma(\theta)$(Linearly polarized γ-ray asymmetry), $T(\theta)$(Polarized target asymmetry), $E(\theta)$(Circular polarized beam - longitudinally polarized target asymmetry), and $F(\theta)$(same as E(θ) but transversely polarized target). The projected observables for the π^0p channel are shown in Fig. 2.

These curves show the sensitivity to β that will enable us to measure $a_{cex}(\pi^+ n \rightarrow \pi^0 p)$. The observables of Fig. 2 also lead to values for the three p-waves amplitudes. The results are shown in Fig. 3, along with to the predictions of ChPT [1, 2, 3, 4].

Last, but far from least, we discuss one case in which isospin breaking is $\simeq (m_d - m_u)/(m_d + m_u) \simeq 25\%$ rather then the usual $\simeq (m_d - m_u)/\Lambda_{QCD} \simeq 2\%$. The quantity of interest is $a(\pi^0 p)$, the s-wave π^0 scattering length on the proton [18, 19]. Clearly this quantity cannot be directly measured since π^0 beams cannot be constructed. Our present knowledge comes from data involving charged pions and constructing the isospin even s-wave scattering length

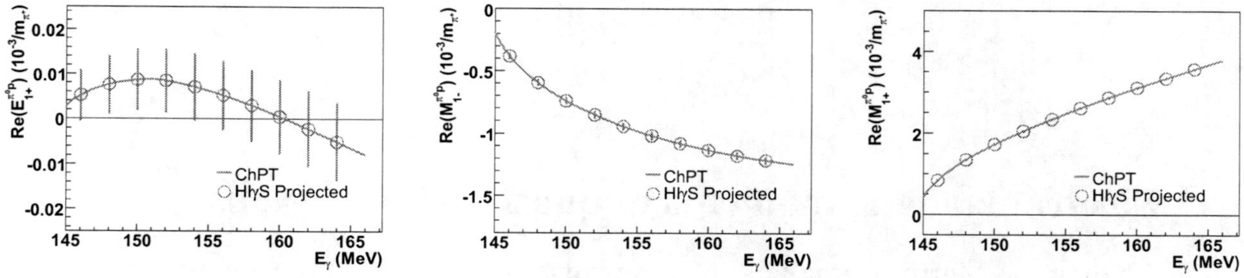

FIGURE 3. Left: Re (E_{1+}), Middle: Re (M_{1-}), Right: Re (M_{1+}) for the $\gamma p \to \pi^0 p$ reaction. The solid curves are a ChPT [1, 2, 3, 4] calculation. For all panels the projected data points show the estimated statistical errors for 100 hours of beam time at HIγS at each energy in four different beam-target polarization configurations (see Fig. 2).

$a^+ = (a_{\pi^- p} + a_{\pi^+ p})/2 = (a_{\pi^- p} + a_{\pi^- n})/2$. This can be obtained from scattering data or more accurately from pionic hydrogen and deuterium [20, 12]. If isospin symmetry holds, then $a^+ = a(\pi^0 p)$. In analyzing the pionic hydrogen and deuterium data it has been shown that the isospin breaking terms, which are primarily electromagnetic in origin, have to be taken into account [20]. When this is done, the latest reported data give the value $a^+ = (0.0069 \pm 0.0034) m_\pi^{-1}$ [12].

It has been suggested that $a(\pi^0 p)$ can be measured as a final state interaction in the $\gamma \vec{p} \to \pi^0 p$ reaction with transversely polarized protons in the energy region between the π^0 threshold of 144.7 MeV and the π^+ threshold of 151.4 MeV [11]. The HIγS facility may be capable of a direct measure of this provided that it can produce a beam $\simeq 10^9$ photons/s on target. A 1000 hour experiment will enable a measurement of $a(\pi^0 p)$ with a statistical accuracy $\simeq 10^{-3}/m_\pi$, which is comparable with the present determination of a^+. A comparison of $a(\pi^0 p)$ and a^+ will test the predicted violation of isospin symmetry by $\simeq 25\%$ [18, 19]. A full discussion of the content of this report can be found in Ref [21]

ACKNOWLEDGMENTS

The work described in this paper is supported in part by US Department of Energy, Office of Science grants DE-FG02-97ER41033 and DE-FG02-94ER40818.

REFERENCES

1. V. Bernard, N. Kaiser, and U. G. Meißner, *Int. J. Mod. Phys.* **E4**, 193 (1995).
2. V. Bernard, N. Kaiser, and U. G. Meißner, *Nucl. Phys.* **B383**, 442 (1992).
3. V. Bernard, N. Kaiser, and U. G. Meißner, *Phys. Rev. Lett.* **74**, 3752 (1995).
4. V.Bernard, N.Kaiser, and U.G.Meissner, *Eur. Phys. J.* **A11**, 209 (2001).
5. V. Bernard, et al., *Phys. Lett.* **B268**, 291 (1954).
6. V. Bernard, et al., *Z. Phys.* **C70**, 483 (1996).
7. A. M. Bernstein, et al., *Phys. Rev.* **C55**, 1509 (1997).
8. M. Fuchs, et al., *Phys. Lett.* **B368**, 20 (1996).
9. A. Schmidt, et al., *Phys. Rev. Lett.* **87**, 232501 (2001).
10. J. Bergstrom, et al., *Phys. Rev.* **C53**, R1052 (1996).
11. A. M. Bernstein, *Phys. Lett.* **B442**, 20 (1998).
12. D. Gotta, *AIP conf. Proc.* **1037**, 162 (2008).
13. V. Bernard, N. Kaiser, and U. Meissner, *Phys. Lett.* **B309**, 421 (1993).
14. E. Korkmaz, et al., *Phys. Rev. Lett.* **83**, 3609 (1999).
15. V. Bernard, N. Kaiser, and U.-G. Meissner, *Phys. Lett.* **B383**, 116 (1996).
16. R. D. Bolton, et al., *Phys. Rev.* **D38**, 2077 (1988).
17. M. W. Ahmed, et al., *Phys. Rev.* **C68**, 064004 (2003).
18. S. Weinberg, *Transactions of the New York Academy of Science Series II* **38**, 185 (1977).
19. N. Fettes, and U.G.Meißner, *Nucl. Phys.* **A693**, 693 (2001).
20. U.G.Meißner, et al., *Phys. Lett.* **B639**, 478 (2006).
21. A. Bernstein, et al., *Annu. Rev. Nucl. Part. Sci.* **59**, 115 (2009).

Electromagnetic moments of quasi-stable baryons

T. Ledwig*, J. Martin-Camalich†, V. Pascalutsa* and M. Vanderhaeghen*

*Institut für Kernphysik, Universität Mainz, D-55099 Mainz, Germany
†Departamento de Fisica Teorica and IFIC, Universidad de Valencia-CSIC, Spain

Abstract. We address electromagnetic properties of quasi-stable baryons in the context of chiral extrapolations of lattice QCD results. For particles near their decay threshold we show that an application of a small external magnetic field changes the particle's energy in a non-analytic way. The conventional electromagnetic moments are only well-defined when the background field B satisfies $|eB|/2M_* \ll |M_* - M - m|$ where M_* is the mass of the resonance and M, m the masses of the decay products. A direct application of this situation is the chiral extrapolation of the $\Delta(1232)$-isobar electromagnetic properties. We discuss such an extrapolation of the $\Delta(1232)$-isobar magnetic dipole, electric quadrupole and magnetic octupole moments by a covariant chiral effective field theory.

Keywords: electromagnetic moments, chiral extrapolation, resonances, one-photon approximation
PACS: 14.20.-c,13.40.Em

INTRODUCTION

Conventionally, the energy shift of a spin 1/2 particle in an external magnetic field $\vec{B}$ is given by:

$$\Delta E = -\vec{\mu} \cdot \vec{B} \ , \tag{1}$$

which defines the particle's static magnetic moment $\vec{\mu}$. However, for resonances near their decay threshold it was shown that the application of a small external magnetic field changes the resonance's energy in a non-analytic way [1]. As a consequence, electromagnetic moments of a resonance with mass M_* and decay product masses M and m are only well defined if the condition

$$|eB|/2M_* \ll |M_* - M - m| \tag{2}$$

is met. In order to use the approximation Eq. (1) for resonances the condition expressed by Eq. (2) has to be fulfilled. We sketch the derivation of this relation in the second section of this report.

Fields in which one encounters the above situation are lattice QCD calculations and chiral effective field theories. Lattice QCD calculations begin to obtain results for the $\Delta(1232)$-isobar electromagnetic moments [2, 3] where the lightest pion mass presently reached is around $m_\pi \simeq 300$ MeV and a chiral extrapolation to the physical point is needed. In doing so, one crosses the threshold value $m_\pi = M_\Delta - M_N$, with $M_\Delta - M_N$ the $\Delta(1232)$-nucleon mass gap, below which the $\Delta(1232)$-isobar is indeed an unstable particle. In the third section we investigate the chiral extrapolation of $\Delta(1232)$-isobar electromagnetic moments by a manifestly covariant chiral effective field theory [4]. Explicitly discussed are the $\Delta(1232)$-isobar magnetic dipole, electric quadrupole and magnetic octupole moments where cusps and singularities reflect the non-applicability of Eq. (2) at the point $m_\pi = M_\Delta - M_N$.

ANOMALOUS MAGNETIC MOMENT AND ENERGY SHIFT OF A RESONANCE

We consider a simple model of two spin 1/2 fields Ψ and ψ of masses M_* and M interacting with a scalar field ϕ of mass m by:

$$\mathscr{L}_{\text{int}} = g\left(\overline{\Psi}\psi\phi + \overline{\psi}\Psi\phi^*\right), \tag{3}$$

where g is a Yukawa coupling constant. The leading order electromagnetic vertex corrections to the anomalous magnetic moment (a.m.m.) of Ψ for charged ϕ and ψ are Feynman-diagrams of the type (D3) and (D4), depicted in the next section. For the above Lagrangian the double lines in these graphs denote the Ψ, the single lines the ψ and the dotted lines the ϕ. For the present report it is sufficient to consider only graph (D3), a discussion of graph (D4) is

12th International Conference on Meson-Nucleon Physics and the Structure of the Nucleon (MENU 2010)
AIP Conf. Proc. 1374, 366-369 (2011); doi: 10.1063/1.3647161

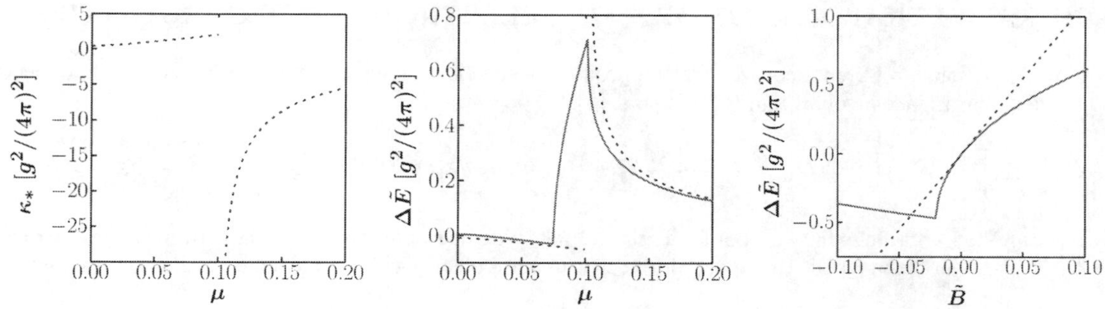

FIGURE 1. Left panel: The a.m.m. contribution κ_* as function of $\mu = m/M_*$ for fixed value $r = M/M_* = 0.9$. The dashed blue curve depicts the real part. The imaginary part exhibits a similar singularity but is not shown explicitly here. Middle panel: Energy shift of the particle Ψ in a constant field of $\tilde{B} = 0.05$ for $r = 0.9$ and as a function of μ. The dashed blue curve depicts the linear approximation Eq. (4) while the solid red line depicts the solution of Eq. (5). Right panel: Same quantities as in the middle panel, this time plotted for constant $\mu = 0.09$ and variable $\tilde{B}$.

analogous. Coupling the photon minimally to ϕ yields the following a.m.m. κ_* together with the conventional energy shift $\Delta\tilde{E}$:

$$\kappa_* = \frac{2g^2}{(4\pi)^2}\int_0^1 dx \frac{-(r+x)x(1-x)}{x\mu^2 - x(1-x) + (1-x)r^2} \quad , \quad \Delta\tilde{E} = -\frac{\kappa_*}{2}\tilde{B} \, , \tag{4}$$

with $r = M/M_*$, $\mu = m/M_*$ and the dimensionless variables: $\tilde{B} = \frac{eB_z}{M_*^2}$ and $\Delta\tilde{E} = \frac{\Delta E}{M_*} + \frac{1}{2}\tilde{B}$. In the left panel of Fig. 1 we plot κ_* as a function of the mass parameter μ. We obtain a singularity at the point $m = M_* - M$ ($\mu = 1 - r$), i.e. where m is equal to the mass gap of Ψ and ψ. The same singularity occurs for the graph (D4) with different overall factors, hence they do not cancel upon summation. Since an infinite a.m.m. would imply an infinite energy shift of the resonance when placed in an external magnetic field, this singularity can not be physical. The nature of this singularity can be seen by calculating the energy shift of the resonance Ψ with the background field technique of [5]. For this we consider a constant magnetic field in z-direction, $\vec{B} = B\vec{e}_z$, together with a Ψ spin projection of $s_z = +1/2$. The resulting energy-shift, to leading order in g, for the case Ψ and ϕ are positively charged is [1]:

$$\Delta\tilde{E} = \frac{g^2}{(4\pi)^2}\Big\{(r+\alpha)(\Omega+\mathscr{A}) - [(r+\alpha)(\Omega+\mathscr{A})]_{\tilde{B}=0}\Big\} \, , \tag{5}$$

where the $\mathscr{A}$ contribution is analytic in $\tilde{B}$ and the Ω is non-analytic :

$$\mathscr{A} = -2 + \alpha\ln r^2 + \beta\ln\mu^2 - \frac{\mu^2(1-\ln\mu^2) - r^2(1-\ln r^2)}{2(\alpha+r)(1-\tilde{B})} \quad , \quad \Omega = \lambda\,\ln\frac{(\alpha+\lambda)(\beta+\lambda)}{(\alpha-\lambda)(\beta-\lambda)} . \tag{6}$$

with

$$\alpha = \frac{1}{2(1-\tilde{B})}\left(1+r^2-\mu^2-\tilde{B}\right) \, , \; \beta = \frac{1}{2(1-\tilde{B})}\left(1-r^2+\mu^2-\tilde{B}\right) \, , \; \lambda = \left[\alpha^2 - r^2/(1-\tilde{B})\right]^{1/2} . \tag{7}$$

In the middle panel of Fig. 1 we plot Eq. (5) and Eq. (4). We see that the linear approximation Eq. (4) in the vicinity of the threshold $m = M_* - M$ does not accurately reproduce the energy shift of Ψ. Furthermore, plotting Eq. (5) as a function of $\tilde{B}$, right panel of Fig. 1, reveals a cusp which occurs at $\tilde{B} = 0$ for the case $m = M_* - M$. Hence, a definition of the magnetic moment as the derivative at $\tilde{B} = 0$ is not possible in such situation. From the λ term in Ω we can see that the energy shift $\Delta\tilde{E}$ can only be expanded, i.e. a magnetic moment defined by Eq. (1), when the condition Eq. (2) is fulfilled. The energy shift $\Delta\tilde{E}$ on the other hand is the physical observable, accessible e.g. in lattice QCD calculations.

CHIRAL EXTRAPOLATION OF $\Delta(1232)$ ELECTROMAGNETIC PROPERTIES

For the $\Delta(1232)$-isobar, a Lorentz-covariant decomposition of the electromagnetic $\gamma\Delta\Delta$ vertex with explicit electromagnetic gauge invariance involves four form factors [6]:

$$\langle\Delta'|V^\mu|\Delta\rangle = -\bar{u}_\alpha(p')\left\{\left[F_1^*(Q^2)\gamma^\mu + \frac{i\sigma^{\mu\nu}q_\nu}{2M_\Delta}F_2^*(Q^2)\right]g^{\alpha\beta} + \left[F_3^*(Q^2)\gamma^\mu + \frac{i\sigma^{\mu\nu}q_\nu}{2M_\Delta}F_4^*(Q^2)\right]\frac{q^\alpha q^\beta}{4M_\Delta^2}\right\}u_\beta(p) ,$$

where $u_\alpha(p)$ is the Rarita-Schwinger spinor for a spin-3/2 state with mass M_Δ. At $Q^2 = 0$, these form factors are related to the $\Delta(1232)$-isobar the magnetic dipole μ_Δ, the electric quadrupole $\mathscr{Q}_\Delta$ and the magnetic octupole $\mathscr{O}_\Delta$ moments:

$$\mu_\Delta = \frac{e}{2M_\Delta}\left[e_\Delta + F_2^*(0)\right] , \quad \mathscr{Q}_\Delta = \frac{e}{M_\Delta^2}\left[e_\Delta - \frac{1}{2}F_3^*(0)\right] , \quad \mathscr{O}_\Delta = \frac{e}{2M_\Delta^3}\left[e_\Delta + F_2^*(0) - \frac{1}{2}\left(F_3^*(0) + F_4^*(0)\right)\right] .$$

In order to study the chiral behavior of the $\Delta(1232)$-isobar electromagnetic moments, we use the chiral effective Lagrangian as given by the BχPT Lagrangian of [7] where the $\Delta(1232)$-isobar is included with the δ-power counting scheme of [8]. The explicit Lagrangian consisting of pion, nucleon, $\Delta(1232)$-isobar and photon fields can be found in [9, 10]. The power counting breaking terms as found in [7] are treated by the renormalization precription of [11], i.e. in addition to the divergent loop contributions also the finite power counting breaking terms are subtracted.

In Fig. 2 we show the Feynman diagrams that contribute to the one-loop level to the $\Delta(1232)$-isobar electromagnetic moments.

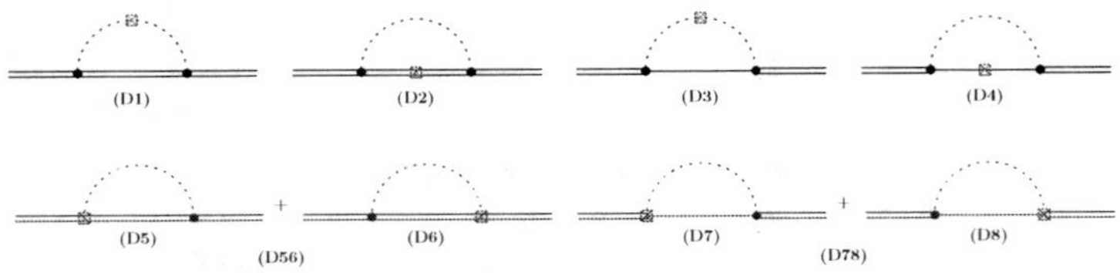

FIGURE 2. Diagrams contributing to $\Delta(1232)$ electromagnetic properties at the one-loop level. Solid-single lines represent nucleons, solid-double lines the $\Delta(1232)$-isobar and dashed lines the pions. The photon coupling is denoted by the blue square while the $N\pi$ and $\Delta\pi$ vertices by black dots.

In Fig. 3 we present some results of our investigation [4]. Depicted are the $\Delta(1232)^+$ magnetic dipole, electric quadrupole and magnetic octupole moments as functions of the pion mass squared m_π^2. The lattice QCD data points are taken from [2], green triangles, and [3] orange rectangles. The electromagnetic moments exhibit cusps and singularities at the threshold value $m_\pi = M_\Delta - M_N$, with M_N as the nucleon mass, which stem from the Feynman-graphs (D3) and (D4) of Fig. 2. The cusp in the magnetic moment was already seen in Ref. [9]. As discussed in the previous section, these pathologies are the consequence of the non-fulfillment of condition Eq. (2) for that m_π value. At this point one can not find any weak magnetic field which enables to define properly the electromagnetic properties of the $\Delta(1232)$-isobar. For lattice QCD calculations of electromagnetic moments there exist only the three point function or the background field method. In the case of the background field technique, the periodicity condition gives a lower bound for the applied magnetic field strength [2, 12]. Hence, there is a whole m_π region around the threshold value $m_\pi = M_\Delta - M_N$ in which Eq. (2) is not met. In order to give an explicit situation, we take two spatial lattices of $L = 24$ with spacing $a^{-1} = 2$ GeV and $L = 32$ with spacing $a^{-1} = 1$ GeV, and implement the magnetic field by $eBa^2 = 2\pi/L^2$ [2]. Further, we take Eq. (2) to be unity, i.e. a completely non-fulfillment of this relation, and solve for that pion mass region. In this region higher order terms in $\vec{B}$ of the resonance energy $E(\vec{B})$ cannot be neglected and a static electromagnetic moment is not well defined in the traditional way. We see that for the given lattices this region ranges from $m_\pi = 275.3$ MeV up to $m_\pi = 310.7$ MeV for the coarse lattice and from $m_\pi = 290.5 \sim 295.5$ MeV for the finer lattice. We represent these two regions as grey bands in Fig. 3.

A similar problem could also affect the lattice calculations of a.m.m. by the three point function method. There the results are limited by finite Q^2 and an extrapolation to $Q^2 = 0$ needs to be done. Qualitatively, a finite Q^2 would enter as an additional energy paramter and the singularities would be shifted away from the point $Q^2 = 0$ for pion masses other then $M_\Delta - M_N$. Hence, for a given m_π one would get a finite Q^2 value for which the form factor is singular. For practical lattice calculations this could mean that one extrapolates across this singularity to $Q^2 = 0$ when all data points are only on the right of the singularity. A quantitative discussion of this situation for the three point method is left as a topic for future work.

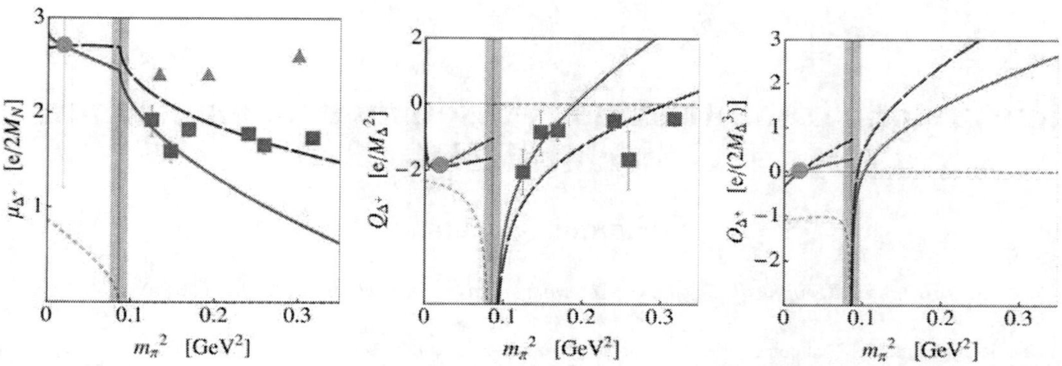

FIGURE 3. The magnetic dipole μ_{Δ^+}, electric quadrupole $\mathscr{Q}_{\Delta^+}$ and magnetic octupole $\mathscr{O}_{\Delta^+}$ moments of the $\Delta^+(1232)$. The curves correspond to: red solid and blue long-dashed curves are normalized to values at the physical pion mass and show our results with and without inclusion of the $\gamma\Delta\Delta$ non-minimal coupling, respectively; green short-dashed curve to the imaginary parts. The yellow circle correspond to the used values at the physical point: the experimental value $\mu_{\Delta^+} = (2.7 \pm 1.5)\ \mu_N$, a large-$N_c$ estimate $\mathscr{Q}_{\Delta^+} = -1.87\ e/M_\Delta^2$ and $\mathscr{O}_{\Delta^+} = 0$. The lQCD data of [2] are denoted by green triangles while those of [3] are depicted by orange rectangles. The grey bands are described in the text.

CONCLUSION

We addressed electromagnetic properties of quasi-stable baryons. For such particles we concluded that the application of a small external magnetic field changes the particle's energy in a non-analytic way. Static electromagnetic moments are only well defined when the condition Eq. (2) is fulfilled [1]. Explicit situations where Eq. (2) could be violated are lattice QCD calculations of electromagnetic properties and their chiral extrapolations by means of effective-field theories. Especially, we discussed such an extrapolation for the $\Delta(1232)$-isobar electromagnetic moments by means of a covariant chiral effective field theory [4] . The non-fulfillment of Eq. (2) is reflected by cusps and singularities at the point $m_\pi = M_\Delta - M_N$. Modern lattice QCD results are about to approach the pion mass region where this condition applies and the presented techniques can be used.

ACKNOWLEDGMENTS

The work of T.L. was partially supported by the Research Centre "Elementarkraefte und Mathematische Grundlagen" at the Johannes Gutenberg University Mainz.

REFERENCES

1. T. Ledwig, V. Pascalutsa, M. Vanderhaeghen, e-Print: arXiv:1004.5055.
2. C. Aubin, K. Orginos, V. Pascalutsa and M. Vanderhaeghen, Phys. Rev. D **79**, 051502 (2008).
3. C. Alexandrou *et al.*, Nucl. Phys. A **825**, 115 (2009); C. Alexandrou *et al.*, Phys. Rev. D **79**, 014507 (2009).
4. T. Ledwig, J. Martin-Camalich, V. Pascalutsa, M. Vanderhaeghen, to be submitted for publication.
5. C. M. Sommerfield, Annals of Physics **5**, 26 (1958).
6. S. Nozawa and D. B. Leinweber, Phys. Rev. D **42**, 3567 (1990).
7. J. Gasser, M. E. Sainio, A. Svarc, Nucl. Phys. B **307**, 779, (1988).
8. V. Pascalutsa and D. R. Phillips, Phys. Rev. C **67**, 055202 (2003).
9. V. Pascalutsa and M. Vanderhaeghen, Phys. Rev. Lett. **94**, 102003 (2005).
10. V. Pascalutsa, M. Vanderhaeghen, Phys. Rev. D **73**, 034003 (2006); V. Pascalutsa, M. Vanderhaeghen, Phys. Rev. D **77**, 014027 (2008); Pascalutsa, M..Vanderhaeghen, S. N. Yang, Phys. Rept. **437**, 125 (2007).
11. J. Gegelia, G. Japaridze, Phys. Rev. D **60**, 114038 (1999); J. Gegelia, G. Japaridze, X.Q. Wang, J. Phys. G **29**, 2303 (2003); T. Fuchs, J. Gegelia, G. Japaridze, S. Scherer, Phys. Rev. D **68**, 056005 (2003).
12. F. X. Lee, R. Kelly, L. Zhou, W. Wilcox, Phys. Lett. B **627**, 71 (2005).

Nucleon polarizabilities and Δ-resonance magnetic moment in chiral EFT

Vladimir Pascalutsa

Institut für Kernphysik, Johannes Gutenberg Universität, Mainz D-55099, Germany

Abstract. Recent chiral EFT calculations of nucleon polarizabilities reveal a problem in the current empirical determination of proton's electromagnetic polarizabilities. We also report on the progress in the empirical determination of the $\Delta(1232)$-resonance magnetic moment in the process of $\gamma p \to p\pi^0\gamma'$ measured at MAMI.

Keywords: chiral expansion, dispersion relations, electromagnetic polarizabilties, resonance magnetic moment
PACS: 14.20.Dh, 14.20.Gk, 13.60.Fz, 13.60.Le, 11.30.Rd, 11.55.Fv

1. INTRODUCTION

The chiral effective-field theory (χEFT) describes the physics of low-energy strong interaction in terms of hadronic degrees of freedom, rather than in terms of the underlying quarks and gluons. Originally χEFT dealt with only the Nambu-Goldstone bosons (pions, kaons) of the spontaneous chiral symmetry breaking [1, 2], but the lightest baryons have eventually been included too [3]. The inclusion of baryons has proved to be difficult and even until now there is no consensus on how to count the baryon contributions. However, my aim here is not to discuss the various counting schemes; this is done elsewhere [4]. I am concerned here with the question of what χEFT, combined with experiment, can tell us about the nucleon and its first excitation, the $\Delta(1232)$-isobar.

The two sets of quantities considered here (nucleon polarizabilities and Δ's magnetic dipole moment) bear at least one thing in common: they are not measured directly but rather are extracted from experimental data using theoretical modeling. In principle the χEFT should be a suitable framework for this task, and we are going to focus on it here. The results for polarizabilities have been obtained in collaboration with Vadim Lensky [5, 6], while the work on Δ's MDM is being done together with Marc Vanderhaeghen [7, 8]. In both cases I will present here some new, thusfar unpublished results.

2. COMPTON SCATTERING: NUCLEON POLARIZABILITIES

Proton's electric (α) and magnetic (β) polarizabilities are determined in low-energy Compton scattering ($\gamma p \to \gamma p$). The purely elastic process is described by Born contributions which give rise to the Powell cross section. Polarizabilities are characterized by the inelastic contribution. At the level of unpolarized differential cross section we have:

$$\begin{aligned}\frac{d\sigma}{d\Omega} - \frac{d\sigma^{(\mathrm{Born})}}{d\Omega} &= -\frac{e^2}{8\pi M_p}\left(\frac{\nu'}{\nu}\right)^2 \nu\nu' \\ &\times \left[(\alpha+\beta)(1+z)^2 + (\alpha-\beta)(1-z)^2\right] + O(\nu^3),\end{aligned} \tag{1}$$

where e and M_p are the proton charge and mass, ν (ν') is the incoming (outgoing) photon energy in the lab system, z is the cosine of the lab scattering angle, $d\Omega = 2\pi dz$. This is a model-independent result and the Born term is known, hence by measuring the angular distribution at very low energy one should be able to determine α and β.

However, the experiments at low energies are very tough. So far most of the data are obtained for energies above 100 MeV, where the low-energy expansion begins to diverge, due to proximity of the pion production cut which sets the convergence radius. The last and the best —in terms of statistics and coverage— experiment, performed by the A2/TAPS Collaboration at MAMI [9], employed a dispersion-relation (DR) model of L'vov [10] to extract α and β. The result of this extraction is displayed in Fig. 1 by a dashed ellipse labeled 'TAPS'. Applying the Baldin sum rule

12th International Conference on Meson-Nucleon Physics and the Structure of the Nucleon (MENU 2010)
AIP Conf. Proc. 1374, 370-375 (2011); doi: 10.1063/1.3647162

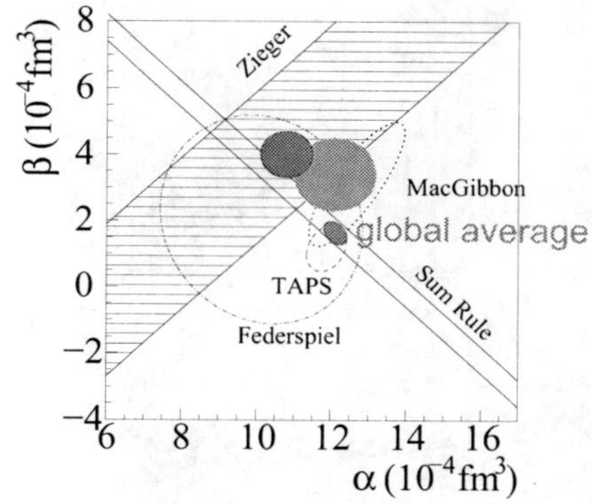

FIGURE 1. The NNLO BChPT result [6] is shown by the red blob and the Δ-less HBχPT result [12] is shown by the grey blob. 'Sum Rule' indicates the Baldin sum rule constraint on $\alpha+\beta$. Experimental results are from Federspiel et al. [13], Zieger et al. [14], MacGibbon et al. [15], and TAPS [9]. 'Global average' represents the PDG summary [11].

value for the sum of polarizabilities cuts out the 'global average' result which is quoted by the Particle Data Group [11], namely :

$$\text{PDG:} \quad \begin{aligned} \alpha &= (12.0 \pm 0.6) \times 10^{-4}\,\text{fm}^3, \\ \beta &= (1.9 \pm 0.5) \times 10^{-4}\,\text{fm}^3. \end{aligned} \tag{2}$$

What is remarkable is that the present χEFT calculations [6, 12] differ from PDG in the values for polarizabilities (see the red and the grey blobs) while agreeing with the cross-section data. For instance, the next-to-next-to-leading order (NNLO) calculation in baryon chiral perturbation theory (BχPT) yields [6]:

$$\text{B}\chi\text{PT}: \quad \begin{aligned} \alpha &= (10.8 \pm 0.7) \times 10^{-4}\,\text{fm}^3, \\ \beta &= (4.0 \pm 0.7) \times 10^{-4}\,\text{fm}^3, \end{aligned} \tag{3}$$

and at the same time gives the red solid curves for the cross section in Fig. 2. The cyan-colored band around these curves corresponds to uncertainty of $\pm 0.7 \times 10^{-4}$ fm^3 in both α and β, which is exactly the radius of the red blob in Fig. 1. The DR code of L'vov produces the dotted black curves for the same values of α and β, and the dashed blue curves for the PDG values given in Eq. (2). The TAPS data are shown by black dots. The other data are properly referenced in [6].

The figure shows that for the same values of α and β the DR and χEFT yield nearly the same results. Thus, the observed discrepancy between the χEFT and PDG results is not because of the difference in theoretical approaches—they agree where they should, but because the extraction from the TAPS data went bad apparently, at least in the error estimate. One can see with a bare eye that the cross-section data do not prefer the blue-dashed curves (PDG value) over the black-dotted curves (BχPT value) to any significance. In fact, at backward angles the BχPT curves are even in a better agreement with the data.

It is fair to conclude that the existing cross-section data do not corroborate the claimed accuracy of proton polarizabilities. This is quite unfortunate because we do need a precise knowledge of these quantities, whether it is to understand the atomic measurements or to provide a basis for determinations of proton's spin polarizabilities. A new accurate measurement of Compton scattering below pion-production threshold can of course remedy this situation. It would be particularly interesting to use polarized beam, as proposed at the HIγS facility [16]. The polarized beam allows one to separate α and β, which is very important since the smallness of β makes it difficult for observation in the α-dominated unpolarized cross section. We have checked that the cross sections for the photon beam linearly polarized in the scattering plane ($||$) or orthogonal to it ($\perp$) have the following expression at leading order in the low-energy expansion:

$$\begin{aligned} \frac{d\sigma_{||}}{d\Omega} - \frac{d\sigma_{||}}{d\Omega}^{(\text{Born})} &= -\frac{e^2}{2\pi M_p}\left(\frac{\nu'}{\nu}\right)^2 \nu\nu'(\alpha z^2 + \beta z), \\ \frac{d\sigma_{\perp}}{d\Omega} - \frac{d\sigma_{\perp}}{d\Omega}^{(\text{Born})} &= -\frac{e^2}{2\pi M_p}\left(\frac{\nu'}{\nu}\right)^2 \nu\nu'(\alpha + \beta z), \end{aligned} \tag{4}$$

and hence by measuring these cross sections one would be able to determine α and β independently.

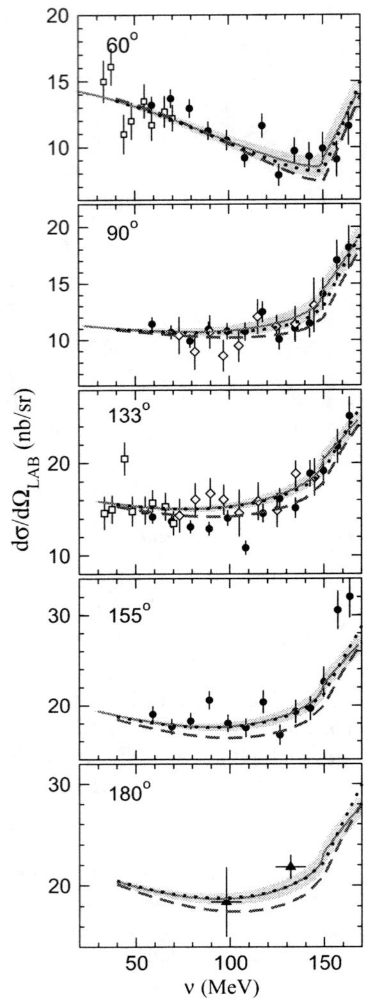

FIGURE 2. Energy dependence of $\gamma p \to \gamma p$ differential cross section, at fixed scattering angles. The curves are explained in the text.

3. RADIATIVE PION PHOTOPRODUCTION: Δ'S MDM

Recent measurement of the radiative pion photoproduction ($\gamma p \to p\pi^0\gamma'$) [17] aimed at a determination of the $\Delta(1232)$-resonance magnetic dipole moment (MDM). The obtained data, however, could not be well-described using the existing theoretical frameworks, including a χEFT calculation [7, 8]. Any extraction then becomes a moot point.

It is now better understood what the problem is. A large portion of the data lies outside the Δ-resonance region. This is best seen in by plotting the data as function of the incoming photon energy, see Fig. 3, rather than as function of outgoing photon energy as it's usually done. The red/black data points in this figure are from the dedicated experiment [17], while the green/brown points are from a preliminary analysis of newer data [18].

The χEFT calculation clearly shows the resonant structure but the data-binning is so wide that one cannot distinguish it. Of course, a comparison of theory with the data below or above the resonance should be applied with caution since at least the χEFT calculation cannot be applied at arbitrary energies. The present calculation [8] is done at a next-to-leading order in the resonance regime of the δ-counting scheme [19, 20], and as such works only in a narrow window around the resonance. This can be seen, e.g., from Fig. 4 where the corresponding description of pion-photoproduction observables is shown in the region of 50 MeV around the resonance.

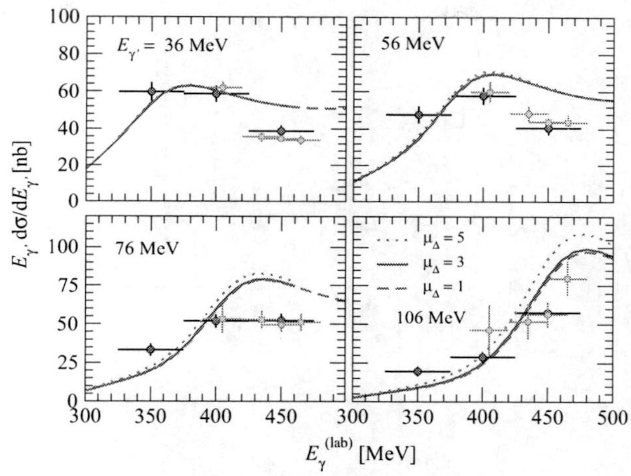

FIGURE 3. Differential cross section of $\gamma p \to p\pi^0\gamma'$ as function of photon beam energy. The curves are results of χEFT calculation [8] for different values of the Δ MDM (in units of nuclear magneton). The data points are from Crystal Ball @MAMI Collaboration: red—[17], green—[18].

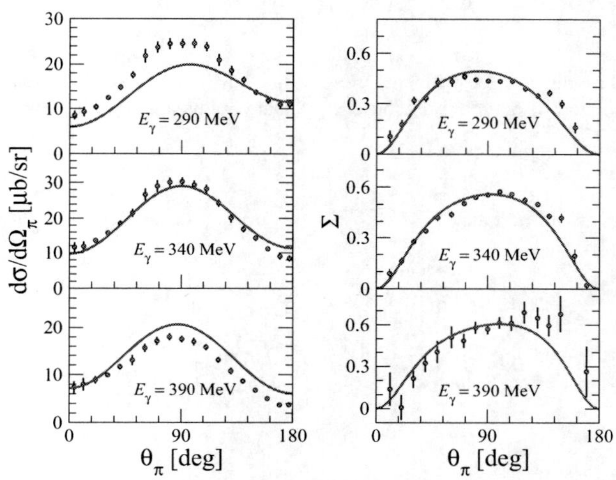

FIGURE 4. Differential cross-section and beam asymmetry of $\gamma p \to p\pi^0$ in NLO χEFT [8, 20].

The Fig. 3 also shows that the differential cross section is largely insensitive to the value of Δ^+ MDM. In fact, the dependence of this observable on the MDM is quadratic, i.e., an expansion of in powers of the outgoing photon energy (E'_γ) can be written as:

$$E'_\gamma\, d\sigma/dE'_\gamma = s_0 + s_1\, E'_\gamma + s_2\, \mu_\Delta^2\, E'^2_\gamma + O(E'^3_\gamma), \tag{5}$$

where real-valued s_i depend on the mass, width, charge and spin of the resonance, but are independent of its magnetic moment.

It is much easier and more feasible to extract the MDM from a linearly-dependent quantity. One example of such an observable is a circularly-polarized photon beam asymmetry defined in [8]. Unfortunately this asymmetry is very small and tough to measure. Unpolarized observables of this kind are therefore welcome.

We propose an asymmetry of the outgoing pion and photon landing into different forward/backward hemispheres versus landing the same forward/backward hemisphere. In terms of the 5-fold differential cross section in the center-

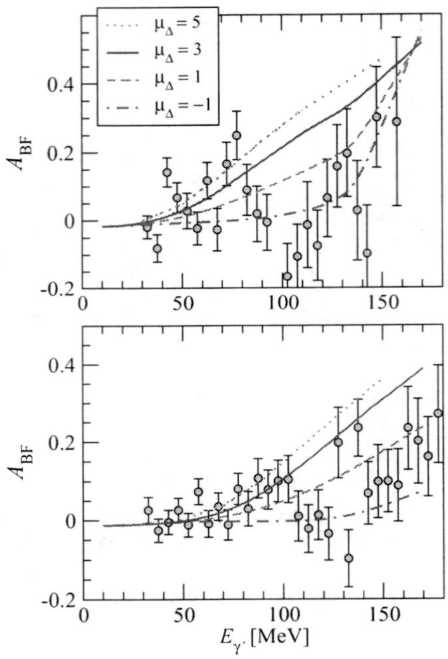

FIGURE 5. Backward-forward asymmetry for different values of the Δ MDM, at fixed beam energy: 350 MeV – upper panel, 400 MeV – lower panel, as a function of outgoing photon energy. Data points are from a preliminary analysis of CB@MAMI Collaboration [21].

of-mass system this asymmetry is defined as:

$$A_{BF} = \frac{1}{d\sigma/dE'_\gamma}\Big[\int_0^{\frac{\pi}{2}} d\Omega_{\gamma'} \int_{\frac{\pi}{2}}^{\pi} d\Omega_\pi + \int_{\frac{\pi}{2}}^{\pi} d\Omega_{\gamma'} \int_0^{\frac{\pi}{2}} d\Omega_\pi - \int_0^{\frac{\pi}{2}} d\Omega_{\gamma'} \int_0^{\frac{\pi}{2}} d\Omega_\pi - \int_{\frac{\pi}{2}}^{\pi} d\Omega_{\gamma'} \int_{\frac{\pi}{2}}^{\pi} d\Omega_\pi\Big] \frac{d^5\sigma}{dE'_\gamma d\Omega_{\gamma'} d\Omega_\pi}. \tag{6}$$

Its low-energy expansion goes as

$$A_{BF} = a_0 + a_1\,\mu_\Delta E'_\gamma + O(E'^2_\gamma), \tag{7}$$

and thus the MDM becomes seen at lower energy. Some χEFT results for this asymmetry are shown in Fig. 5, and can be compared to a preliminary analysis using the database of the dedicated CB@MAMI experiment [17]. One can see a significant sensitivity of this asymmetry on the MDM value. However, the present data are neither accurate nor consistent enough to determine the Δ^+ MDM with any good precision.

4. CONCLUSION

The existing data on proton Compton scattering cross-sections do not corroborate the claimed (e.g., in PDG tables) precision of our knowledge of proton's electric and magnetic polarizabilities. New experiments, using polarized photon beam, are needed to correct this situation.

A reliable extraction of the Δ-resonance magnetic dipole moment from the process of radiative pion production will require a substantial improvement in the experimental database, as well as further systematic refinements of the theoretical modeling.

ACKNOWLEDGMENTS

I am indebted to Jürgen Ahrens for providing me with the results of L'vov's dispersion relations shown in Fig. 2.

REFERENCES

1. S. Weinberg, Physica A **96**, 327 (1979).
2. J. Gasser and H. Leutwyler, Annals Phys. **158** (1984) 142.
3. J. Gasser, M. E. Sainio and A. Svarc, Nucl. Phys. B **307**, 779 (1988).
4. V. Pascalutsa, "Baryon Chiral Perturbation Theory: An Update," plenary talk at BARYONS'10, Osaka, Japan, Dec. 7-11, 2010.
5. V. Lensky and V. Pascalutsa, Pisma Zh. Eksp. Teor. Fiz. **89**, 127 (2009) [JETP Lett. **89**, 108 (2009)].
6. V. Lensky and V. Pascalutsa, Eur. Phys. J. C **65**, 195 (2010).
7. V. Pascalutsa and M. Vanderhaeghen, Phys. Rev. Lett. **94**, 102003 (2005).
8. V. Pascalutsa and M. Vanderhaeghen, Phys. Rev. D **77**, 014027 (2008).
9. V. Olmos de Leon *et al.*, Eur. Phys. J. A **10**, 207 (2001).
10. A. I. L'vov, Sov. J. Nucl. Phys. **34**, 597 (1981) [Yad. Fiz. **34**, 1075 (1981)].
11. K. Nakamura *et al.* [Particle Data Group], J. Phys. G **37**, 075021 (2010).
12. S. R. Beane, M. Malheiro, J. A. McGovern, D. R. Phillips and U. van Kolck, Phys. Lett. B **567**, 200 (2003) [Erratum-ibid. B **607**, 320 (2005)]; Nucl. Phys. A **747**, 311 (2005).
13. F. J. Federspiel *et al.*, Phys. Rev. Lett. **67**, 1511 (1991).
14. A. Zieger *et al.*, Phys. Lett. B **278**, 34 (1992).
15. B. E. MacGibbon *et al.*, Phys. Rev. C **52**, 2097 (1995).
16. M. W. Ahmed, talk at INT Workshop "Soft Photons and Light Nuclei," Seattle, June 16 - 20, 2008.
17. S. Schumann *et al.*, Eur. Phys. J. A **43**, 269 (2010).
18. Sergey Prakhov, private communication (May 2010).
19. V. Pascalutsa and D. R. Phillips, Phys. Rev. C **67**, 055202 (2003).
20. V. Pascalutsa, M. Vanderhaeghen and S. N. Yang, Phys. Rept. **437**, 125 (2007).
21. Sven Schumann, private communication (March 2010).

Measurement of the transition form factor of η meson with WASA-at-COSY

H. Bhatt for the WASA-at-COSY Collaboration

Indian Institute of Technology Bombay, India

Abstract. Reaction $\eta \to e^+e^-\gamma$ is used to investigate the transition form factor of η meson with WASA detector at COSY. Where the η meson is produced in pp collision at 1.4 GeV. We present the analysis techniques and preliminary results of η Dalitz decays.

Keywords: Transition Form Factor, Dalitz decay
PACS: 13.25.Jx

INTRODUCTION

The issue of confinement in Quantum Chromo Dynamics (QCD) has engaged the attention of physicists for a very long time. One of the ways to gain further insight into how colored quarks and gluons make colorless hadrons, is to study the phenomenological characteristics such as the the magnetic moments, form factors, polarizability of the meson. In this paper, we report the measurement of the transition form factor of the η meson through its Dalitz decay $\eta \to e^+e^-\gamma$. As the invariant mass squared of the leptons pairs is equal to the four momentum transfer squared ($q^2 = m^2_{l^+l^-}$), one can probe the spatial structure as a function of momentum transfer using the above decay mode of the eta meson.

One can investigate the agreement between theory and experiment by comparing the experimental results on q^2 dependency of the transition form factor with one of the theoretical models such as the Vector Meson Dominance (VMD)[1]. The model is based on the assumption that the interaction of photons with hadrons go through virtual vector mesons. The quantum numbers of the photon being same as that of vector mesons such as ρ, ω, ϕ, the photon can fluctuate for a fraction of time into such mesons. It is possible to experimentally determine the transition form factor F(q^2) by comparing the experimental measured mass spectrum of lepton pairs decay with the theoretical calculations assuming point like particles.

$$\frac{d\Gamma}{dq^2} = \mid \frac{d\Gamma}{dq^2} \mid_{pointlike} \mid F(q^2) \mid \tag{1}$$

Usually the form factor is fitted with a simple pole type formula:

$$F(q^2) \simeq 1 + b_\eta q^2 \tag{2}$$

Where b_η is the slope of the transition form factor.

TABLE 1. Experimental measured b_η values.

Experiment	$b_\eta\,[GeV^{-2}]$
Lepton-G	1.9±0.4 [2]
NA60	1.95 ± 0.17±0.05 [3]
SND	1.6 ± 2.0 [4]

The VMD model predicts $b_\eta = 1.8\ GeV^{-2}]$ [1]. TABLE 1 shows the b_η values from previous experiments.

EXPERIMENTAL SET UP

The experiment was done with WASA detector at COSY in Juelich, Germany [5]. The WASA (Wide Angle Shower Apparatus) provides nearly full solid angle coverage for both neutral and charged particles as shown in Figure 1. COSY

12th International Conference on Meson-Nucleon Physics and the Structure of the Nucleon (MENU 2010)
AIP Conf. Proc. 1374, 376-379 (2011); doi: 10.1063/1.3647163

is a cooler synchrotron and storage ring which delivers beams of polarized and unpolarized protons and deuterons in the momentum range between 0.3 GeV/c and 3.7 GeV/c. It consists of two parts: the forward part which is designed for detection and identification of the scattered particles like protons, deuterons and ^3He and the central part which detects and identifies the decay products of mesons, namely, photons, electrons and charged pions.

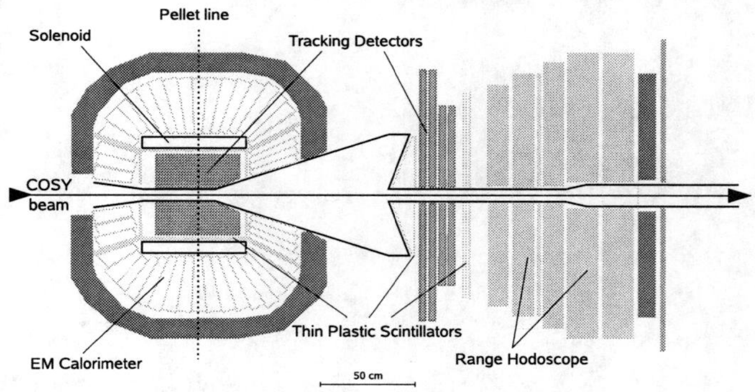

FIGURE 1. Schematic view of the WASA detector.

ANALYSIS OF $\eta \to e^+e^-\gamma$

$\eta \to e^+e^-\gamma$ has a very small branching ratio. Hence a number of decay channels could contribute to the background. We have performed extensive simulations to determined the kinematic conditions required to optimise the signal to background ratio. In Table 2 we present the cross-sections for signal and contributing background reactions obtained as a result of these studies.

TABLE 2. Cross-section for signal and each background reaction.

Channel	Cross-section (mb)
$\eta \to e^+e^-\gamma$ (Signal)	4.9×10^{-5}
$\eta \to \pi^+\pi^-\gamma$	4.68×10^{-4}
$\eta \to \pi^+\pi^-\pi^0$	2.26×10^{-3}
$\eta \to \gamma\gamma$	3.9×10^{-3}
$pp \to pp\pi^+\pi^-\pi^0$	0.02
$pp \to pp\pi^+\pi^-$	1

We have simulated 10 million events for each reaction and 22 Million events for $pp \to pp\pi^+\pi^-$ reaction. In the analysis, events with two oppositely charged tracks and one neutral track in the central detector (CD) and two charged tracks in forward detector (FD) have been selected.

The following criteria have been used to suppress the background :

1. E_γ, photon energy: The photon energy distribution is different for the signal and the background channels as can be seen from Figure 2(a).
2. MM_{PP}, Missing Mass of two scattered protons: The Missing Mass of two scattered protons, detected in the forward detector is defined in following way:

$$MM_{pp}^2 = (E_{beam} - E_{p_1} - E_{p_2})^2 - (\vec{p}_{beam} - \vec{p}_{p_1} - \vec{p}_{p_2})^2 \quad (3)$$

   where E_p and $\vec{p}_p$ are the energy and momentum of two scattered protons. As can be seen from Figure 2(b), a condition on this will be reduce the background substantially.
3. E_{dep}/P ratio: The ratio of the total energy deposited and momentum of the charged particles in the central detector is different for the leptons and the pion as shown in Figure 2(c).
4. ϕ_V, Angular orientation of electron pair: The reaction $\eta \to \gamma\gamma$ contributes as a background if one photon gets converted into an electron-positron pair in the beam tube wall. A method has been developed to separate the

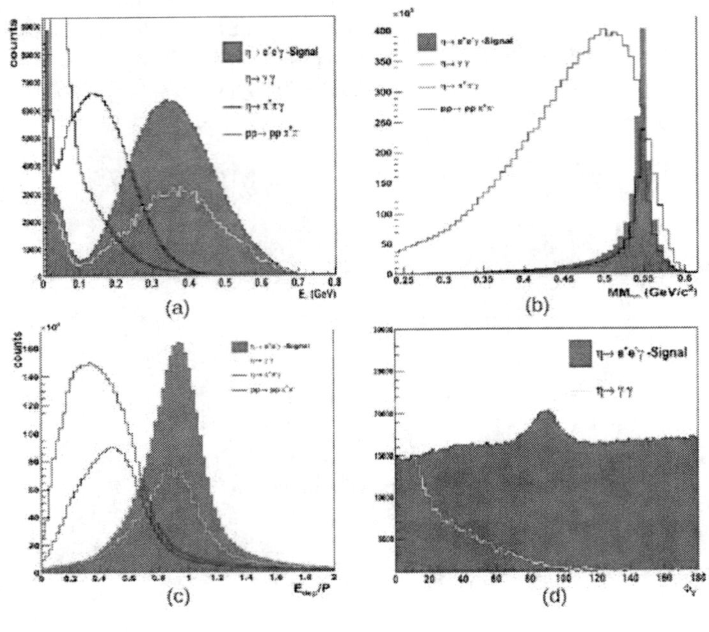

FIGURE 2. (a) Photon Energy (b) Missing Mass of two scattered proton (c) Ratio of deposited energy and momentum (d) Orientation angle ϕ_V, for signal and background reactions.

external conversion from the Dalitz decay by calculating the angular orientation ϕ_V of the e^+ e^- pair plane with respect to the magnetic field [6]. Figure 2 (d) shows the distribution of ϕ_V for both $\eta \to e^+e^-\gamma$ and $\eta \to \gamma\gamma$ reactions. Events due to external conversion on an average have a much lower value of ϕ_V.

To improve the signal to background ratio further, we have kinematically fitted each event to optimise experimental resolution. In the present work, for each event the hypothesis, $pp \to pp(\eta \to e^+e^-\gamma)$, has been ensured. The 4 constraints of energy and momentum have been applied. For each event the kinematic fitting routine estimates the χ^2_{kf} and its probability P(χ^2_{kf}). Figure 3 shows the probability distribution of the data sample with Monte Carlo distributions of the signal and each of the background reactions. As is obvious from the figure a condition on this distribution will reduce the background substantially.

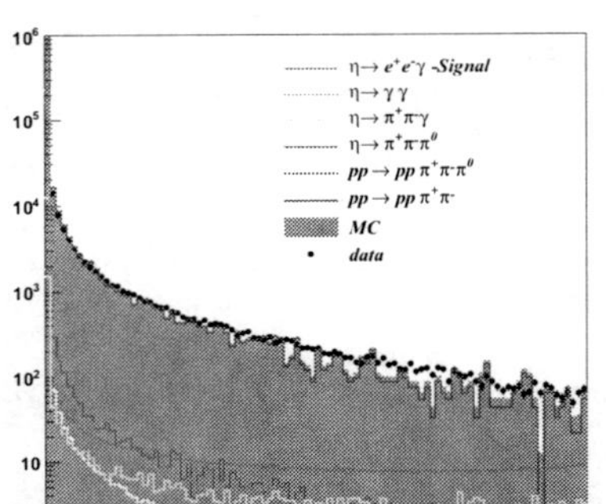

FIGURE 3. Comparison of the probability distribution of data and Monte Carlo simulations (for the signal and the background channels).

Based on the figures shown before, the following selecting criteria have been used while analysing the data:

1. Photon energy, $E_\gamma > 180$ MeV (Figure 2(a)).
2. Ratio of the Energy deposited and momentum, E_{dep}/P> 0.55 (Figure 2(b)).
3. Azimuthal angle, $\Phi_V > 1.0$ radians(Figure 2(c)).
4. Missing mass of two scattered protons, $0.530 < MM_{pp} < 0.570$ (Figure 2(d)).
5. Probability distribution, $P(\chi^2) > 0.2$ (Figure 3).

Figure 4 (a) shows the invariant mass of η meson and 4 (b) shows the invariant mass of lepton pair after applying the above conditions. In obtaining Figure 4(b), it is further required that the invariant mass of the η meson should be within 3σ to get a clean signal of the Dalitz events.

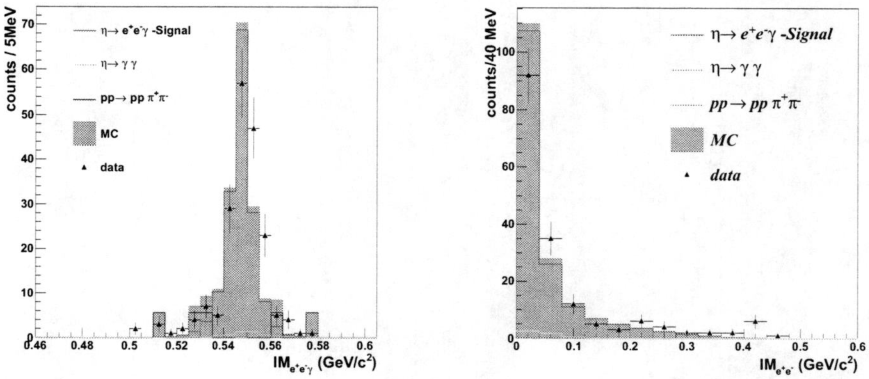

(a) (b)

FIGURE 4. (a) Invariant Mass of η meson (b) Invariant Mass of lepton pair (e^+e^-).

Preliminary Results

On the basis of simulations, we have reconstructed 160 $\pm$13 Dalitz events ($\eta \to e^+e^-$). It has been further estimated that the contribution from the reaction $\eta \to \gamma\gamma$ is $\sim$3 %, while $\sim$6 % of the events can be assigned to the reaction pp$\to$pp$\pi^+\pi^-$. Final analysis, to measure the transition form factor by fitting the invariant mass of lepton pair distribution is in progress.

ACKNOWLEDGMENTS

I sincerely thank the WASA-at-COSY collaboration for giving me the opportunity to take part in the experiment and for helping me in this analysis.

REFERENCES

1. Landsberg and L.G, *Phys. Rep.*, 128-301 (1985).
2. R. Djeliadin, et al., *Phys. Lett. B* **94**, 548 (1980).
3. R. Arnaldi et. al., , *Phys. Lett. B* **677**, 260-266 (2009).
4. M. N. Achsov et. al., , *Phys. Lett. B* **504**, 275-287 (2001).
5. Bo Hoistad, James Ritman,"Proposal for the Wide Angle Shower Apparatus (WASA) at COSY-Juelich " (2004).
6. T. Dahms, PHD Dissertation, Dilepton spectra in p + p and Au + Au collisions at RHIC, Stony Brook University (2008).

Session I: Structure Functions

The CTEQ-JLab global analysis of large-x PDFs

Alberto Accardi

Hampton University, Hampton, Virginia 23668, USA
and Jefferson Lab, Newport News, Virginia 23606, USA

Abstract. I discuss several corrections to leading twist calculations of nucleon structure functions which are needed to include experimental data at large parton fractional momentum x and at low scales Q^2 in global fits of parton distribution functions. In particular I discuss the results of the CTEQ-JLab global fit, and some work in progress. Topics covered include the interplay of target mass and higher-twist corrections, the importance of nuclear corrections for deuterium target data, and applications to the study of quark-hadron duality. Implications for collider physics are highlighted.

Keywords: Global fits, parton distribution functions, nuclear corrections
PACS: 12.38.Bx, 13.60.Hb, 15.65.-q, 14.70.Dj

Precise parton distribution functions (PDFs) at large parton fractional momentum x are vital for understanding the non perturbative structure of the nucleon and the effects of color confinement on its partonic constituents. For instance, the d/u quark distribution ratio near $x = 1$ is very sensitive to the nature of the quark-quark forces in the nucleon; the ratios of spin-polarized to spin-averaged PDFs $\Delta u/u$, and particularly $\Delta d/d$, in the limit $x \to 1$ reflect the non perturbative quark-gluon dynamics in the nucleon, and can shed light on the origin of the nucleon's spin. Precise PDFs at large x also have impact in other areas of nuclear and high-energy physics, e.g., by allowing precise computations of QCD background processes in searches of new physics signals at hadron colliders, and control over systematic uncertainties in neutrino oscillation experiments.

PDFs can be extracted from experimental data through global QCD fits which combine data from many different processes and observables, and analyze them by means of perturbative QCD calculations. Currently, however, the unpolarized PDFs are well determined only for $x \lesssim 0.5$ for valence quarks, $x \lesssim 0.3$ for gluons, and $x \lesssim 0.1$ for heavy quarks. To better constrain these at large x it is necessary to study hard scattering processes near kinematic thresholds, such as Deep Inelastic Scattering (DIS) at large Bjorken invariant x_B and low 4-momentum transfer squared Q^2, Drell-Yan (DY) lepton pair production and electroweak vector boson production at large rapidity. In these kinematic regimes several corrections to leading twist perturbative QCD calculations can become important because of the rapid fall-off of the cross section near the kinematic boundary. Examples are target (TMC) and jet mass corrections, threshold resummation, and higher-twists (HT) contributions. Moreover, data taken on nuclear targets must be corrected for nuclear effects such as shadowing, binding, Fermi motion and nucleon off-shellness, to access the partonic structure at the nucleon level. Accessing the highest values of x in DIS also requires understanding quark-hadron duality in order to utilize data in the resonance region.

All these effects need to be incorporated in a consistent framework, simultaneously computed for a wide range observables, and utilized in a global PDF fit. The CTEQ-JLab global fit published in References [1] and [2] took the first fundamental step in this program by considering the combined effect of TMC and HT corrections, alongside nuclear corrections for DIS data on deuterium targets needed for flavor separation of the up and down quark. In this talk, I discuss the results of this analysis and some recent work in progress. Details can be found in references [1, 2, 3].

TARGET MASS AND HIGHER-TWIST CORRECTIONS

It is the usual practice in global PDF fits to place kinematic constraints on the DIS data sets, typically $Q > 2$ GeV and $W > 3.5$ GeV, so that only leading twist massless QCD contributions need to be considered, thereby reducing the model dependent error on the extracted PDFs. As a byproduct of this procedure the PDFs are directly constrained by data only in the region $x < 0.7$. However, plentiful DIS data exist outside this region. In order to utilize them in global fits, one needs minimally to include TMCs, which scale as M_N^2/Q^2, with M_N the nucleon mass, and HT corrections, which scale as λ^2/Q^2, with λ a hadronic scale describing non perturbative parton-parton correlations inside the nucleon.

Several methods are available in the literature to perform TMCs. One is the well-known Georgi-Politzer formalism

12th International Conference on Meson-Nucleon Physics and the Structure of the Nucleon (MENU 2010)
AIP Conf. Proc. 1374, 383-386 (2011); doi: 10.1063/1.3647164

based on the Operator Product Expansion (OPE). However, this formalism leads to non-zero values of the structure functions in the unphysical region $x > 1$. Another prescription is a simple rescaling of the structure function, obtained by substituting x with the Nachtmann variable $\xi = 2x/(1+\sqrt{1+4x^2M_N^2/Q^2})$, which also leads to unphysical structure functions at $x > 1$. Lastly, working in Collinear Factorization (CF) the kinematic boundaries are naturally respected. One advantage of the CF formalism versus the OPE formalism is that the former can be also applied to semi-inclusive DIS, and indeed to any hard-scattering process.

However, TMCs do not exhaust all possible $O(1/Q^2)$ power corrections. These include dynamical higher-twist corrections (parton correlations) as well as all uncontrolled leading-twist power corrections, such as jet mass corrections. They also include higher-order perturbative terms, which are logarithmic in Q^2 but resemble a power law at low scales, and large-x resummation. Despite their disparate origin, these "residual" corrections are usually called "higher-twist". In the CTEQ-JLab fits, these are parameterized phenomenologically using a multiplicative factor:

$$F_2^{data} = F_2^{TMC}(1+C(x)/Q^2) \,, \tag{1}$$

where F_2^{TMC} denotes the structure function after the target mass corrections have been made. The function $C(x)$ is parametrized by $C(x) = ax^b(1+cx)$. After inclusion of TMCs, this functional form is sufficiently flexible to describe well the DIS data. To simplify the global fits, the HT corrections for protons and neutrons where taken equal.

NUCLEAR CORRECTIONS

In order to separate the d and u quark at large $x \gtrsim 0.6$ it is necessary to consider DIS data on deuteron targets, which are sensitive to a different linear combination of u and d quarks than the corresponding data on proton targets. However, at large x the deuteron deviates from a simple sum of a free proton and neutron due to significant effects of nuclear binding, Fermi motion and nucleon off-shellness. Since the deuteron is weakly bound one can approximate the bound nucleon structure function by its on-shell value, and write the deuteron (d) structure function as

$$F_2^d(x,Q^2) \approx \sum_{N=p,n} \int dy\, f_{N/d}(y,\gamma)\, F_{2N}^{TMC+HT}\left(\frac{x}{y},Q^2\right) . \tag{2}$$

Here F_{2N} is the proton p or neutron n structure function including TMC and HT corrections. The "smearing function" $f_{N/d}$ is computed from the deuteron wave function and implements nuclear binding and Fermi motion corrections; at $Q^2 \to \infty$ it can be interpreted as the light-cone momentum distribution of nucleons in the deuteron, and $\gamma^2 = 1+4x^2M_N^2/Q^2$ parametrizes the deviation from that limit. The variable $y = (M_d/M_N)(p_N \cdot q/p_d \cdot q)$ is the deuteron's momentum fraction carried by the struck nucleon, where q is the virtual photon four-momentum, $p_{N(d)}$ the nucleon (deuteron) four-momentum, and M_d the deuteron mass. Off-shell corrections to the bound nucleon F_{2N} can also be implemented under a few assumptions, which are outside the scope of this talk.

The rsulting nuclear corrections are strongly Q^2 dependent at $Q^2 \lesssim 20$ GeV^2, mostly due to the TMC and HT corrections to F_{2N}. It is also clear that nuclear smearing corrections do not disappear at large Q^2: in general, they are not a subleading effect, and cannot be avoided by kinematics cuts such as those commonly used in global PDF fits.

THE CTEQ-JLAB PARTON DISTRIBUTIONS AT LARGE *x*

The CTEQ-JLab global PDF fits [1, 2] were performed at NLO to a wide variety of data similar to that used in the determination of the CTEQ6M1 PDFs, except that neutrino data were exclued since their use would involve additional nuclear corrections beyond those for deuterium. In addition, the E-866 dimuon data were added as were data for the CDF $\gamma + jet$ production, the CDF W lepton asymmetry, and the DØ W asymmetry.

In the first paper [1] the effects of lowering the kinematic cuts and introducing TMC and HT were studied, and the effects of nuclear corrections were preliminarily studied. Initially , a reference fit was done using the standard $W > 3.5$ GeV and $Q > 2$ GeV, with TMC, HT and nuclear corrections turned off in order to compare to the CTEQ6M1 PDFs, and study the effect of the new data sets. Subsequently, several prescriptions for TMC, HT and nuclear corrections were considered, and the DIS kinematic cuts progressively relaxed to $W > 1.73$ GeV and $Q > 1.3$ GeV, in order to avoid most of the resonance region but be able to include a good number of Jefferson Lab data. As TMCs, the discussed

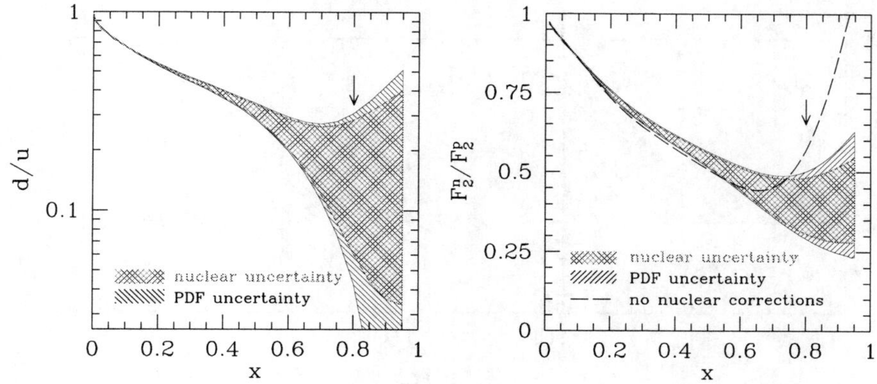

FIGURE 1. *Left:* Full range of variations of the d/u ratio due to deuteron wave function and off-shell corrections (cross-hashed band) and PDF fit uncertainties (diagonal hashed). The arrow indicates the region at large x where the fits are not directly constrained by data. *Right:* Same for the proton to neutron F_2 ratio. For reference, the dashed line shows the result using a fit with no nuclear corrections beside isospin symmetry. Plots taken from Ref. [2].

OPE, ξ-scaling and CF prescriptions were considered. Nuclear smearing was performed using the Paris wave function with on-shell nucleons or a simple model for off-shell corrections. The follow-up paper [2] investigated in detail the systematic nuclear uncertainty using a modern set of non-relativistic (AV18, CD-Bonn) and relativistic (WJC-1, WJC-2) wave functions, and a range of off-shell corrections. The main results of these analyses are the following:

- **Standard kinematic cuts.** When using the standard DIS kinematic cuts the PDFs are insensitive to TMC and HT corrections; however nuclear corrections are large and start at $x \gtrsim 0.45$, in a region well inside the cuts.
- **Enlarged kinematic cuts.** The PDFs are relatively stable against variations of the DIS cuts in the vicinity of the $W > 1.73$ GeV and $Q > 1.3$ cut. As a consequence of the enlarged data set, there is a substantial reduction in the uncertainty of these PDFs due to the increased data.
- **Stability with respect to TMCs.** The PDFs are nearly independent of the TMC prescription; this is very important for fitting leading-twist PDFs. Changes in TMCs are absorbed by the phenomenological HT term, for which TMC modeling induces a non-negligible systematic uncertainty .
- **Large sensitivity to nuclear corrections.** The d PDF is very sensitive to the nuclear correction model adopted (Fig. 1, left plot). This induces a large systematic uncertainty on the neutron structure functions, which can be calculated using the fitted PDFs (Fig. 1, right plot).

One very important result of [2] is that, surprisingly, the large-x *gluon* PDF is as sensitive to nuclear corrections as the d PDF; they are anticorrelated to each other due to an interplay of DIS, DY and jet data. This has potentially profound implications for future collider experiments, since the resulting variation of the gluon is significant even at values of x as low as 0.4. As an illustration, we can consider the parton luminosities,

$$L_{ij} = \frac{1}{(1+\delta_{ij})\hat{s}} \left[\int_{\hat{s}/s}^{1} \frac{dx}{x} f_i(x,\hat{s}) f_j\left(\frac{\hat{s}}{xs},\hat{s}\right) + i \leftrightarrow j \right], \tag{3}$$

where s ($\hat{s}$) is the hadronic (partonic) center of mass energy squared, and f_i is the PDF for a parton of flavor i at $Q^2 = \hat{s}$. As an example, the gg luminosity controls the total main channel for Higgs production, the gd luminosity controls the "standard candle" cross section for W^- production, and the $d\bar{u}$ luminosity is relevant to jet production. These are plotted in Fig. 2 for $\sqrt{s} = 7$ TeV, relevant to the current LHC runs. The nuclear uncertainty grows quickly above 5-10% as $\sqrt{\hat{s}}$ exceeds 1 TeV.

An interesting application of the CTEQ-JLab large-x PDFs to quark-hadron duality has been reported in [3, 4]. These PDFs have been fitted to $e+p$ and $e+d$ data as close as possible to the resonance region; therefore it is interesting to extrapolate them and check how well partonic degrees of freedom can describe (on average) the data in the resonance region where hadronic degres of freedom dominate. The preliminary results indicate that quark-hadron duality works within 5-10%. This opens the possibility of using resonance region data to extend the range of validity in x of the fits. A study is underway to further explore this issue [5].

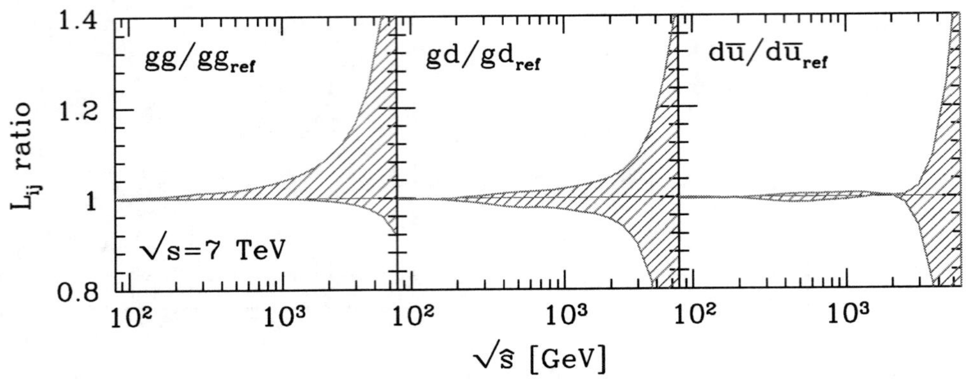

FIGURE 2. Relative nuclear uncertainties on parton luminosities at $\sqrt{s} = 7$ TeV normalized to an intermediate fit ("ref"). Shown is the full range of variations due to the use of different nuclear corrections in the global fits (diagonal hashed band). Plot taken from Ref. [3].

CONCLUSIONS

I have shown that a good control of global PDF fits can be achieved when including DIS data in the pre-asymptotic region of large x and small Q^2, if one considers TMC and HT corrections. The resulting PDFs are stable against available TMC prescriptions, with the modeling uncertainty absorbed in the phenomenologically extracted HT terms. This is very good for applications to collider and neutrino physics, and for comparing PDF moments to lattice calculations. However, the d-quark and, surprisingly, the gluon PDF turn out to be very sensitive to uncertainties in the modeling and calculation of nuclear effects. This in turn implies a large theoretical uncertainty, e.g., on neutron structure functions and parton luminosities at the LHC.

Progress in constraining the d quark and the gluon PDFs at large x requires a better theoretical understanding of nuclear corrections, combined with new data on free protons but sensitive to d, such as from parity violating DIS or from neutrino DIS on a hydrogen target. Alternatively one can use data minimizing nuclear corrections, such as from proton-tagged DIS on deuteron targets, for which the nuclear uncertainty is smaller than 2% [6]. Using quark-hadron duality to include resonance region data in the fits, thereby extending their x range, seems also feasible. Finally, data sensitive to large x gluons, such as from the longitudinal and charm structure functions, F_L and F_2^c, are required to constrain the gluons independently of the jet data.

ACKNOWLEDGMENTS

I am most grateful to M. E. Christy, C. E. Keppel, S. Malace, W. Melnitchouk, P. Monaghan, J. G. Morfín, J. F. Owens, and L. Zhu for their collaboration on the matter presented in this talk. This work has been supported by the DOE contract DE-AC05-06OR23177, under which Jefferson Science Associates, LLC operates Jefferson Lab, and NSF awards No. 0653508 and No. 1002644.

REFERENCES

1. A. Accardi, M. E. Christy, C. E. Keppel, P. Monaghan, W. Melnitchouk, J. G. Morfín and J. F. Owens, Phys. Rev. D **81**, 034016 (2010).
2. A. Accardi, W. Melnitchouk, J. F. Owens, M. E. Christy, C. E. Keppel, L. Zhu and J. G. Morfin, arXiv:1102.3686 [hep-ph].
3. A. Accardi, Proceedings of HiX2010, Jefferson Lab, USA, Oct 13-15, 2010, arXiv:1101.5148 [hep-ph].
4. S. P. Malace, "Neutron structure functions in the resonance region and tests of quark-hadron duality", these proceedings.
5. S .P. Malace and A. Accardi, *in preparation.*
6. A. Accardi, W. Melnitchouk, *in preparation.*

Analyzing Unfavored Fragmentation Functions Using NJL-Jet Model

Hrayr H. Matevosyan*, Anthony W. Thomas* and Wolfgang Bentz†

*CSSM, School of Chemistry and Physics,
University of Adelaide, Adelaide SA 5005, Australia
http://www.physics.adelaide.edu.au/cssm
†Department of Physics, School of Science,
Tokai University, Hiratsuka-shi, Kanagawa 259-1292, Japan
http://www.sp.u-tokai.ac.jp/

Abstract. The Nambu-Jona-Lasinio (NJL) - jet model provides a sound framework for calculating the fragmentation functions in an effective chiral quark theory, where the momentum and isospin sum rules are satisfied without the introduction of ad hoc parameters. The most recent version of the model includes the fragmentation of the light and strange quarks to pions, kaons, nucleons, and antinucleons; where the effects of the production of secondary pions and kaons from the vector mesons ρ, K^* and ϕ are also calculated. The results for the model fragmentation function exhibit a qualitative agreement with the empirical parameterizations. The results also allow to test, within the model assumptions, several assumption in parametrizations of the unfavored fragmentation functions used in empirical fits to the experimental data.

Keywords: Fragmentation functions, NJL-Jet
PACS: 13.60.Hb, 13.60.Le, 13.60.Rj, 12.39.Ki

INTRODUCTION

The recent studies on extraction of the quark fragmentation functions [1, 2] from a variety of experimental data on inclusive hadron production in hard scattering processes [3, 4, 5, 6, 7, 8, 9, 10, 11, 12, 13, 14], have generated a renewed interest in the subject [15, 16, 17, 18, 19, 20, 21]. It is reasonable to expect that in the near future an understanding of the fragmentation functions will be attained which is as precise as the present knowledge on quark distribution functions, although currently the empirical parametrizations of the fragmentation functions employ several assumptions to relate various fragmentation functions in order to constrain the number of independent parameters used in the fits to the experimental data. These assumptions are largely based on the light quark flavor and charge conjugation symmetries, while the relatively small unfavored fragmentation functions are often related to each other using the weaker $SU(3)$ flavor symmetry.

Recently developed NJL-jet model [22, 23, 24] allows to calculate the fragmentation functions of the light and strange quarks to pions, kaons, nucleons, and antinucleons; including the effects of the production of secondary pions and kaons from the vector mesons ρ, K^* and ϕ. Here the quark fragmentation process description is based on the quark-cascade model by Field and Feynman [15], schematically depicted in Fig. 1, modified to include the intermediate vector meson resonances. The advantages of the NJL-Jet model are the absence of ad-hoc parameters included, i.e. all the NJL parameters are fixed independent of any experimental data on fragmentation functions, where the momentum and isospin sum rules for the solutions of fragmentation functions are naturally satisfied at the model scale. The resulting fragmentation functions compare favorably to the empirical parametrizations of [1, 2]. Moreover, the unfavored fragmentation functions are naturally present in the model as the initial quark in the cascade can changes its flavor allowing for the emission of the unfavored hadrons.

UNFAVORED FRAGMENTATION FUNCTIONS FROM NJL-JET

Here we present some of our results for the unfavored fragmentation functions at the scale $Q^2 = 4\ \mathrm{GeV}^2$ evolved at next to leading order (NLO) from our model scale $Q_0^2 = 0.2\ \mathrm{GeV}^2$ using the software from Ref. [25] (the details of determination of the model scale are given in [23].)

12th International Conference on Meson-Nucleon Physics and the Structure of the Nucleon (MENU 2010)
AIP Conf. Proc. 1374, 387-390 (2011); doi: 10.1063/1.3647165

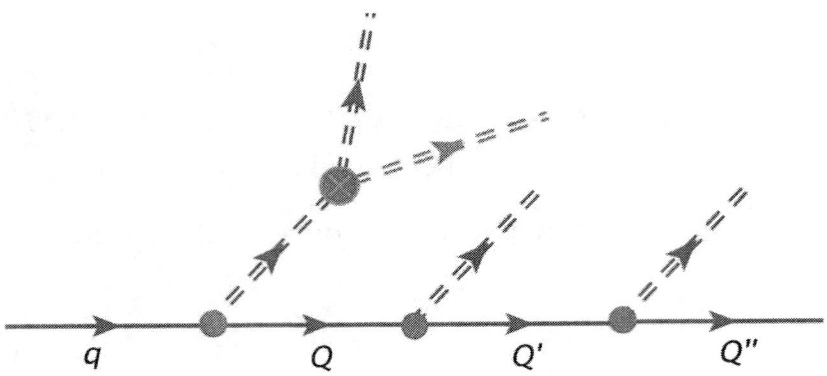

FIGURE 1. Quark cascade with vector meson decays.

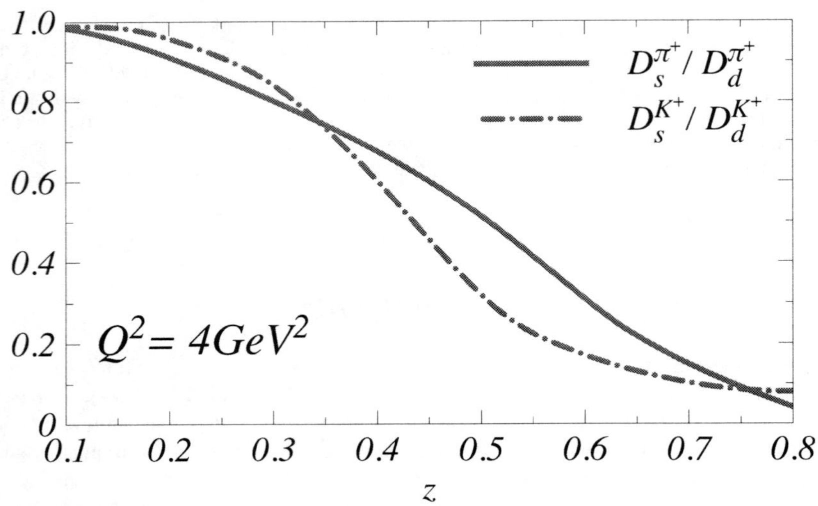

FIGURE 2. The ratios of unfavored fragmentation functions of d and s quarks to π^+ and K^+ respectively, calculated in NJL-jet model and evolved at NLO to $Q^2 = 4$ GeV^2.

First, we consider the widely used assumptions for unfavored fragmentation functions to π^+ and K^+, $D_d^{\pi^+} = D_s^{\pi^+}$ and $D_d^{K^+} = D_s^{K^+}$. The dependence of the ratios of the corresponding fragmentation functions on the initial quark's light-cone momentum fraction z is shown in Fig. 2. Here we see that the unfavored fragmentation functions of d and s quarks notably differ for large values of z, though the unfavored fragmentations themselves decrease as z increases and vanish as $z \to 1$. This behavior can be qualitatively explained in the quark-cascade picture. If we ignore the secondary pions and kaons coming from the decays of the vector mesons, then the leading contribution to the fragmentation functions at large values of z are generated from only the first or second emitted hadrons in the quark cascade, as demonstrated in Ref. [26] on the example of $D_u^{\pi^+}$. Thus, the leading contribution to the unfavored fragmentation functions come from the process where the initial quark emits a hadron, changing the quark's flavor in the cascade to one that can directly emit the hadron of interest. For example, the fragmentation function $D_d^{\pi^+}$ at large values of z gets its leading contribution from the process $d \to u\pi^-$ and a consequent $u \to d\pi^+$, where the leading contribution to $D_s^{\pi^+}$ arises from $s \to uK^-$ and a consequent $u \to d\pi^+$. In NJL-jet model, the elementary probability of a d quark directly emitting a π^- is larger than the similar probability of an s quark emitting a K^-, thus the resulting fragmentation function $D_d^{\pi^+}$ is larger than $D_s^{\pi^+}$. On the other hand, at values of small z both fragmentation functions get significant contributions from hadrons emitted further in the the quark cascade [26], which diminishes the effect of the initial quark's flavor.

In the recent studies of model-independent extraction of kaon fragmentation functions [27, 28] the authors proposed a test of a different assumption used in all the global fits of fragmentation functions, $D_d^{K^+} = D_d^{K^-}$, by examination of the dependence of cross-sections for several semi-inclusive processes on $D_u^{K^+} - D_u^{K^-}$. Here we present out model calculation of $D_d^{K^+} - D_d^{K^-}$ and compare it to $D_u^{K^+} - D_u^{K^-}$ in Fig. 3, showing that the $D_d^{K^+} - D_d^{K^-}$ is small for large values of z, but becomes notable at about 10% or more of $D_u^{K+} - D_u^{K-}$ for small values of z.

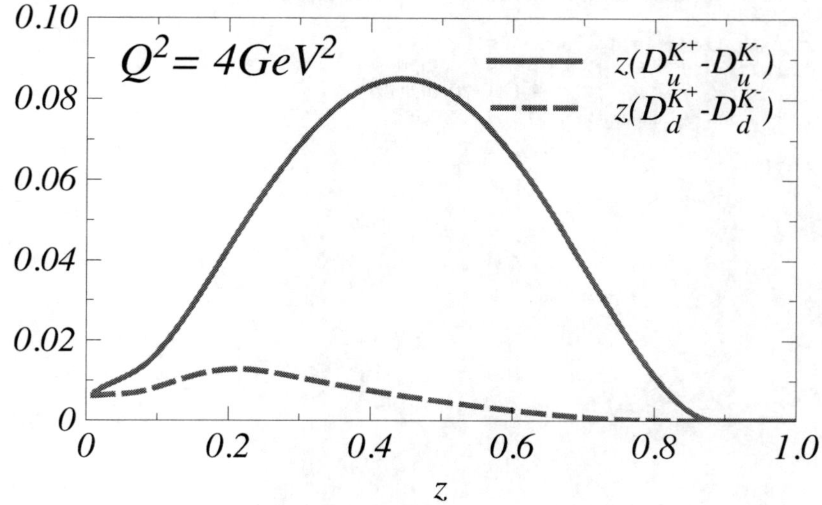

FIGURE 3. Comparison of $D_d^{K^+} - D_d^{K^-}$ to $D_u^{K^+} - D_u^{K^-}$, calculated in NJL-jet model and evolved at NLO to $Q^2 = 4$ GeV^2.

Thus the NJL-jet model proves to be a powerful tool for providing input, based on chiral quark models, to the properties of the quark fragmentation functions. This input might be valuable in constructing empirical parametrizations of fragmentation functions and estimating the magnitude of uncertainties introduced by employing assumptions in relating different fragmentation functions.

ACKNOWLEDGMENTS

This work was supported by the Australian Research Council through the grant of an Australian Laureate Fellowship (A.W. T.) and by a Subsidy for Activating Educational Institutions from the Department of Physics, Tokai University.

REFERENCES

1. M. Hirai, S. Kumano, T. H. Nagai, and K. Sudoh, *Phys. Rev.* **D75**, 094009 (2007), `hep-ph/0702250`.
2. D. de Florian, R. Sassot, and M. Stratmann, *Phys. Rev.* **D75**, 114010 (2007), `hep-ph/0703242`.
3. D. Buskulic, et al., *Z.Phys.* **C66**, 355–366 (1995).
4. P. Abreu, et al., *Eur.Phys.J.* **C5**, 585–620 (1998).
5. R. Akers, et al., *Z.Phys.* **C63**, 181–196 (1994).
6. G. Abbiendi, et al., *Eur.Phys.J.* **C16**, 407–421 (2000), `hep-ex/0001054`.
7. H. Aihara, et al., *Phys.Rev.Lett.* **61**, 1263 (1988).
8. H. Aihara, et al., *Phys.Lett.* **B184**, 299 (1987).
9. K. Abe, et al., *Phys.Rev.* **D59**, 052001 (1999), `hep-ex/9805029`.
10. S. S. Adler, et al., *Phys.Rev.Lett.* **91**, 241803 (2003), `hep-ex/0304038`.
11. J. Adams, et al., *Phys.Lett.* **B637**, 161–169 (2006), `nucl-ex/0601033`.
12. J. Adams, et al., *Phys.Rev.Lett.* **97**, 152302 (2006), `nucl-ex/0602011`.
13. I. Arsene, et al., *Phys.Rev.Lett.* **98**, 252001 (2007), `hep-ex/0701041`.
14. B. Abelev, et al., *Phys.Rev.* **C75**, 064901 (2007), `nucl-ex/0607033`.
15. R. D. Field, and R. P. Feynman, *Nucl. Phys.* **B136**, 1 (1978).
16. G. Altarelli, R. K. Ellis, G. Martinelli, and S.-Y. Pi, *Nucl. Phys.* **B160**, 301 (1979).
17. J. C. Collins, and D. E. Soper, *Nucl. Phys.* **B194**, 445 (1982).

18. R. L. Jaffe (1996), hep-ph/9602236.
19. R. K. Ellis, W. J. Stirling, and B. R. Webber, *Camb. Monogr. Part. Phys. Nucl. Phys. Cosmol.* **8**, 1–435 (1996).
20. V. Barone, A. Drago, and P. G. Ratcliffe, *Phys. Rept.* **359**, 1–168 (2002), hep-ph/0104283.
21. A. D. Martin, R. G. Roberts, W. J. Stirling, and R. S. Thorne, *Eur. Phys. J.* **C35**, 325–348 (2004), hep-ph/0308087.
22. T. Ito, W. Bentz, I. C. Cloet, A. W. Thomas, and K. Yazaki, *Phys. Rev.* **D80**, 074008 (2009), 0906.5362.
23. H. H. Matevosyan, A. W. Thomas, and W. Bentz (2010), 1011.1052.
24. H. H. Matevosyan, A. W. Thomas, and W. Bentz (To be published in 2011).
25. M. Botje (2010), 1005.1481.
26. H. H. Matevosyan, A. W. Thomas, and W. Bentz (2011), 1102.3322.
27. E. Christova, and E. Leader, *Phys. Rev.* **D79**, 014019 (2009), 0809.0191.
28. S. Albino, and E. Christova, *Phys. Rev.* **D81**, 094031 (2010), 1003.1084.
29. S. Albino, E. Christova, and E. Leader (2011), 1102.2305.

Polarized structure functions in nuclear matter

S. Atashbar Tehrani*, A. N. Khorramian*,†, S. Taheri Monfared*,† and F. Arbabifar†

*School of Particles and Accelerators, IPM (Institute for Studies in Theoretical Physics and Mathematics), P.O.Box 19395-5531, Tehran, Iran
†Physics Department, Semnan University, Semnan, Iran

Abstract. We present a comprehensive analysis of the structure functions for the polarized nuclei ^3He, and focus on the polarized distribution functions and light-cone momentum distributions of protons and neutrons. We demonstrate that the first moment of the ^3He structure function can provide an incisive test of the Bjorken sum rule.

Keywords: Nuclei polarized structure function, Bjorken sum rules
PACS: 43.35.Ei, 78.60.Mq

INTRODUCTION

A primary goal of Quantum Chromodynamics (QCD) has been the detailed investigation of the spin structure of nuclei, and the subsequent decomposition into partonic components. There are a variety of effects which will modify the properties of protons and neutrons embedded in nuclei as compared with those in a free nucleon. These effects include spin depolarization, nuclear binding and Fermi motion of nucleons, the virtuality of the nucleons, nuclear shadowing and anti-shadowing, and the presence of non-nucleonic degrees of freedom. All these effects contribute to the spin structure functions of ^3He and ^3H. While each of the above mentioned effects are considered in the literature [1], the goal of the present paper is to combine some of them in a single application.

We perform a QCD fit using the new experimental data [2] to extract the ^3H and ^3He structure functions from polarized nucleon data. In particular, we quantify the modification of the ^3He or ^3H spin structure functions due to the presence of nuclear effects.

NUCLEAR POLARIZED STRUCTURE FUNCTIONS

In this section we present the complete finite-Q^2 formulas for the nuclear structure functions, and then formulate the results in terms of the familiar light-cone convolutions. Using the polarized proton and neutron structure functions, we can then extract the polarized nuclear structure functions. If we treat $g_1^{^3He}$ and $g_1^{^3H}$ as three nucleon systems, we can relate the polarized nuclear and nucleon structure functions. The polarized structure functions $g_1^{^3He}$ and $g_1^{^3H}$ can be represented as the convolution of the polarized proton structure g_1^p and the polarized neutron structure g_1^n with the spin-dependent nucleon light-cone momentum distributions $\Delta f^N_{^3He}(y)$ and $\Delta f^N_{^3H}(y)$:

$$\begin{aligned} g_1^{^3He}(x,Q^2) &= \int_x^3 \frac{dy}{y}\Delta f^n_{^3He}(y) g_1^n(\frac{x}{y},Q^2) + 2\int_x^3 \frac{dy}{y}\Delta f^p_{^3He}(y) g_1^p(\frac{x}{y},Q^2) \\ &- 0.014[g_1^p(x,Q^2) - 4g_1^n(x,Q^2)] , \end{aligned} \quad (1)$$

$$\begin{aligned} g_1^{^3H}(x,Q^2) &= 2\int_x^3 \frac{dy}{y}\Delta f^n_{^3H}(y) g_1^n(\frac{x}{y},Q^2) + \int_x^3 \frac{dy}{y}\Delta f^p_{^3H}(y) g_1^p(\frac{x}{y},Q^2) \\ &+ 0.014[g_1^p(x,Q^2) - 4g_1^n(x,Q^2)] . \end{aligned} \quad (2)$$

Here, y is the ratio of the struck nucleon to nucleus light-cone plus-components of the momenta [1, 3].

The motion of the nucleons inside the nucleus (Fermi motion) and their bindings are parameterized through the light-cone momentum distributions $\Delta f^N_{^3He}(\Delta f^N_{^3H})$, which is readily calculated using the ground-state wave functions

12th International Conference on Meson-Nucleon Physics and the Structure of the Nucleon (MENU 2010)
AIP Conf. Proc. 1374, 391-394 (2011); doi: 10.1063/1.3647166

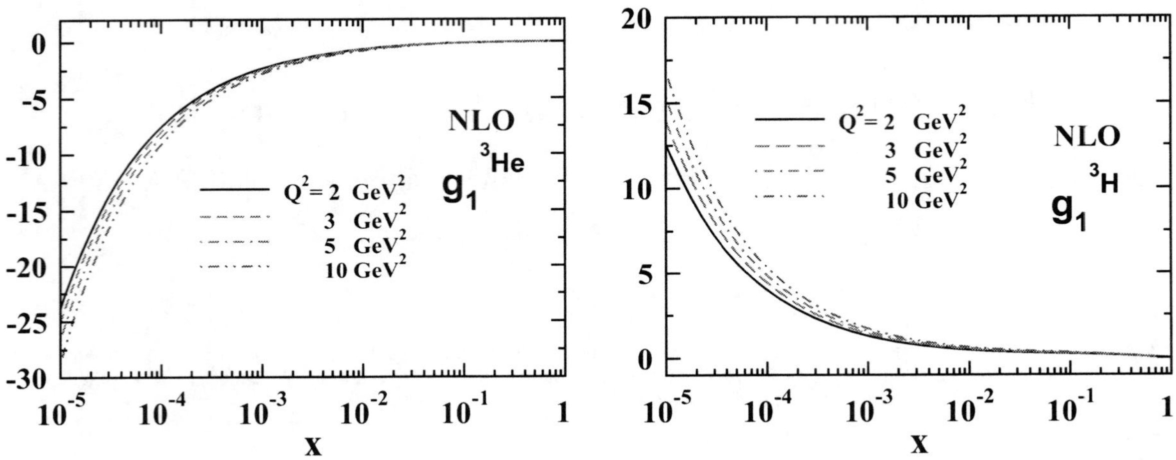

FIGURE 1. Analytical results for the polarized ^3He and ^3H structure functions as a function of x and for different values of Q^2.

of $^3He(^3H)$. Note that the last term in above equation is important only in the region $0.2 \leq x \leq 0.8$. Applying isospin symmetry, we find $\Delta f^p_{^3He}(y)$ is equal to $\Delta f^n_{^3H}(y)$, and $\Delta f^n_{^3He}(y)$ is equal to $\Delta f^p_{^3H}(y)$; therefore, we only need two independent functions $\Delta f^p_{^3He}(y)$ and $\Delta f^n_{^3He}(y)$. Using these results together with Eq. (1) we are able to extract the polarized structure functions $g_1^{^3He}(x,Q^2)$ and $g_1^{^3H}(x,Q^2)$. For the light-cone momentum distribution functions we use the explicit functional form:

$$\Delta f^n_{^3He}(y) = \frac{a^n e^{-\frac{0.5(1-d^n)(-b^n+y)^2}{c^{n2}}}}{1+\frac{d^n(-b^n+y)^2}{c^{n2}}} , \tag{3}$$

$$\Delta f^p_{^3He}(y) = \frac{\sum_{i=0}^4 a_i^p U_i(y)}{\sum_{i=0}^4 b_i^p U_i(y)} , \tag{4}$$

where U_n is the Chebyshev polynomials of the second kind. To get detailed information about our fitting procedure on nucleon structure function refer to Refs. [2, 6].

BJORKEN SUM RULE CHECKING

We will now make use of the Bjorken and Ellis-Jaffe sum rules to deduce information about the ^3He-^3H system in relation to that of the nucleon (p, n) system.

The Bjorken sum rule [7] relates the difference of first moments of the proton and neutron spin structure functions to the axial vector coupling constant $g_A = 1.2670 \pm 0.0035$: [8]

$$\int_0^1 [g_1^p(x,Q^2) - g_1^n(x,Q^2)]dx = \frac{1}{6}\, g_A[1+O(\frac{\alpha_s}{\pi})] . \tag{5}$$

Including the next-to-leading (NLO) QCD corrections, this sum rule has been confirmed experimentally within about 10 percent.

This sum rule can be straightforwardly generalized to the $^3He - ^3H$ system

$$\int_0^3 [g_1^{^3H}(x,Q^2) - g_1^{^3He}(x,Q^2)]dx = \frac{1}{6}\, \tilde{g}_A[1+O(\frac{\alpha_s}{\pi})] , \tag{6}$$

where $\tilde{g}_A = 1.211 \pm 0.002$ is the axial vector coupling constant of the triton β decay [9].

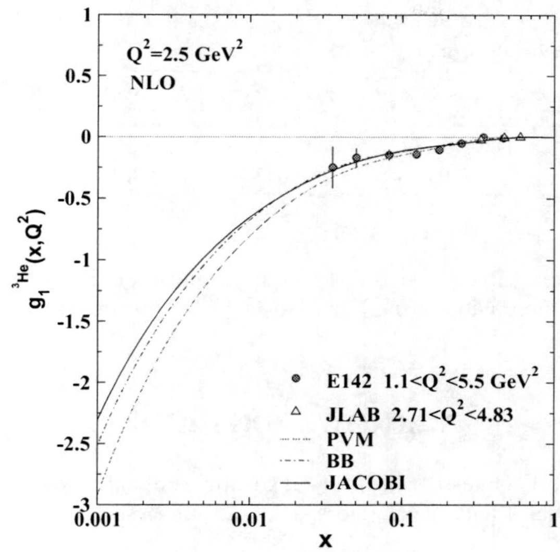

FIGURE 2. Analytical results for the polarized ^3He structure functions as a function of x and for fixed $Q^2 = 2.5$ GeV^2. The data are well described by our QCD NLO curve (solid curve). Also shown are the QCD NLO curve obtained by AK[13] (dashed) according to the polarized valon model (PVM)[3] and BB (dashed-dotted)[14] for comparison.

Taking the ratio of the Eqs. (5) and (6), we obtain

$$\frac{\int_0^3 [g_1^{^3H}(x,Q^2) - g_1^{^3He}(x,Q^2)]dx}{\int_0^1 [g_1^p(x,Q^2) - g_1^n(x,Q^2)]dx} = \frac{\tilde{g}_A}{g_A} \,. \tag{7}$$

Now by having the g_A and $\tilde{g}_A$ values, the above ratio is equal to 0.956[1]. Note that the QCD radiative corrections are expected to cancel exactly in above equation. This compares with a value of 0.924 obtained using the results of our QCD fit [2].

NUCLEAR STRUCTURE FUNCTION MOMENTS

Next we evaluate the first moment of the ^{3}He spin structure function. Using the Bjorken sum rule (7) and Eq. (1), one can show:

$$\Gamma_{^3He} = \int_0^1 g_1{}^{^3He}(x)dx = [p_n + 2p_p]\Gamma_p - \frac{1}{6}\frac{g_A}{g_V}p_n \tag{8}$$

where $\Gamma_p = \int_0^1 g_1^p(x)dx$. Using the effective nucleon polarizations of $p_p = -0.028 \pm 0.004$ and $p_n = 0.86 \pm 0.02$, one obtains $g_A/g_V = 1.259$ [12] and $\Gamma_{^3He} = -0.180 + 0.804\,\Gamma_p$. Using the results of our calculations, we can cross-check these relations; the results of the first moments are presented in Table 1 as a function of Q^2.

RESULTS

We have presented a comprehensive picture of the nuclear structure functions relevant for DIS on polarized ^3He for a wide kinematic range. Figure 1 displays our results for $g_1^{^3He}$ and $g_1^{^3H}$ for different values of Q^2. Our results are compared with the polarized valon model (PVM) and experimental data from $E142$ [15] and $JLAB$ [16] in Figure 2.

We also provide a `FORTRAN` package containing our polarized structure functions $xg_1^p(x,Q^2)$, $xg_1^n(x,Q^2)$, $xg_1^d(x,Q^2)$, $xg_1^{NS}(x,Q^2)$, $xg_2^p(x,Q^2)$, $g_1^{^3He}(x,Q^2)$ and $g_1^{^3H}(x,Q^2)$, as well as the parton densities $x\delta u_v(x,Q^2)$, $x\delta d_v(x,Q^2)$, $x\delta g(x,Q^2)$, and $x\delta\bar{q}(x,Q^2)$. These are calculated in fixed-order perturbation theory at NLO in the $\overline{\text{MS}}$–scheme. These packages are available on the web at: `http://particles.ipm.ir/links/QCD.htm`

TABLE 1. The first moments of g_1^p and $g_1^{^3He}$ in the NLO approximation for some different values of Q^2.

First Moment	Q^2=2 GeV^2	Q^2=3 GeV^2	Q^2=5 GeV^2	Q^2=10 GeV^2
Γ_1^p	0.129	0.132	0.135	0.138
$\Gamma_1^{^3He}$	-0.076	-0.073	-0.071	-0.069

The polarized distributions are obtained from a fit to the experimental polarized structure functions, as described above. The routines perform an interpolation using a grid in $\{x, Q^2\}$ with cubic splines in x and a linear interpolation in $\log(Q^2)$.

ACKNOWLEDGMENTS

The authors especially grateful to F. Olness for fruitful suggestions, critical remarks and for helpful discussions. We acknowledge the School of Particles and Accelerators, Institute for Research in Fundamental Sciences (IPM) for financially supporting this project.

REFERENCES

1. F. R. P. Bissey, V. A. Guzey, M. Strikman and A. W. Thomas, Phys. Rev. C **65**, 064317 (2002) [arXiv:hep-ph/0109069].
2. Ali N. Khorramian, S. Atashbar Tehrani, S. Taheri Monfared, F. Arbabifar, F. I. Olness, [arXiv:hep-ph/1011.4873].
3. M. M. Yazdanpanah, A. Mirjalili, S. Atashbar Tehrani and F. Taghavi-Shahri, Nucl. Phys. A **831**, 243 (2009).
4. F. R. P. Bissey, A. W. Thomas and I. R. Afnan, Phys. Rev. C **64**, 024004 (2001) [arXiv:nucl-th/0012081].
5. I. R. Afnan *et al.*, Phys. Rev. C **68**, 035201 (2003) [arXiv:nucl-th/0306054].
6. Ali N. Khorramian, S. Atashbar Tehrani, F. I. Olness,S. Taheri Monfared, F. Arbabifar, Nucl. Phys. B (Proc. Suppl.) **207-208**,65 (2010).
7. J. D. Bjorken, Phys. Rev. **148**, 1467 (1966).
8. C. Amsler et al. (Particle Data Group), Phys. Lett. B **667**,1 (2008).
9. B. Budick, J. S. Chen and H. Lin, Phys. Rev. Lett. **67**, 2630 (1991).
10. K. Ganesamurthy, R. Sambasivamr, Braz. J. Phys. [online] v.39, n.2, (2009).
11. W. Bentz, I. C. Cloet, T. Ito, A. W. Thomas and K. Yazaki, Prog. Part. Nucl. Phys. **61**, 238 (2008) [arXiv:0711.0392 [nucl-th]].
12. C. Ciofi degli Atti, E. Pace, G. Salme and S. Scopetta, arXiv:nucl-th/9310029.
13. S. Atashbar Tehrani and A. N. Khorramian, JHEP **0707**, 048 (2007) [arXiv:hep-ph/0705.2647].
14. J. Blumlein and H. Bottcher, arXiv:1007.2784 [hep-ph].
15. P. L. Anthony *et al.* [E142 Collaboration], Phys. Rev. D **54**, 6620 (1996) [arXiv:hep-ex/9610007].
16. X. Zheng *et al.* [Jefferson Lab Hall A Collaboration], Phys. Rev. C **70**, 065207 (2004) [arXiv:nucl-ex/0405006].

Session J: η/Nuclear Reactions

Search For η-Bound Nuclei

H. Machner[1]

Fakultät für Physik, Universität Duisburg-Essen, Lotharstr. 1, 47048 Duisburg, Germany

Abstract. The η meson can be bound to atomic nuclei. Experimental search is discussed in the form of final state interaction for the reactions $dp \rightarrow {}^3\mathrm{He}\eta$ and $dd \rightarrow {}^4\mathrm{He}\eta$. For the latter case tensor polarized deuterons were used in order to extract the s-wave strength. For both reactions complex scattering lengths are deduced: In a two-nucleon transfer reaction under quasi-free conditions, $p^{27}\mathrm{Al} \rightarrow {}^3\mathrm{He}X$, was investigated. The system X can be the bound ${}^{25}\mathrm{Mg}\otimes\eta$ at rest. When a possible decay of an intermediate $N^*(1535)$ is required, a highly significant bump shows up in the missing mass spectrum. The data give for a bound state a binding energy of 13.3±1.6 MeV and a width of σ=4.4±1.3 MeV.

Keywords: Meson Nucleon interaction
PACS: 21.85.+d, 13.75.-n

INTRODUCTION

In contrast to the pion-nucleon interaction, the η-nucleon interaction at small momenta is attractive and rather strong. This attraction can be seen from the fact that the η threshold (1488 MeV) is situated just below the $N^*(1535)$ resonance which couples strongly to the $\eta - N$ channel. Initial calculations by Bhalerao and Liu [1] obtained attractive s-wave $\eta - N$ scattering lengths. With these values, Haider and Liu [2] have shown that η can be bound in nuclei with A $\geq$ 10. Other groups have also found similar results [3, 4, 5], while [6, 7, 8] claimed that even light systems like ^{3,4}He might be bound.

On the experimental side results are scarce. The first experiments searching for η-mesic nuclei at BNL [9] and LAMPF [10] by using a missing-mass technique in the (π^+, p) reaction reached negative or inconclusive results. Later it became clear that the peaks are not necessarily narrow and that a better strategy of searching for η-nuclei is required. Furthermore, the BNL experiment was in a region far from the recoilless kinematics, so the cross section is substantially reduced [11]. More recently, the existence of η-mesic ^{3}He was claimed to have been observed in the reaction $\gamma^3\mathrm{He} \rightarrow \pi^0 pX$ using the photon beam at MAMI [12]. It has, however, been pointed out [13] on the basis of new high statistics data that the data of Ref. [12] does not permit an unambiguous determination of the existence of a ${}^3\mathrm{He}\eta$-bound state, because nucleon resonances produce opening angle dependent structures in excitation functions and subtraction of excitation functions for different opening angles can produce artificial structures almost anywhere.

Two different methods have been applied in the search for η bound states. One is the study of η production on nuclei and extraction of the η nucleus scattering length. This method can give the pole position but cannot distinguish between bound and unbound. The other is the direct production of the η meson in a bound state. This state is measured via missing mass technique. The GEM collaboration has used both methods, as will be shown in this paper.

SEARCHES VIA TWO BODY FINAL STATE INTERACTION

According to the Watson-Migdal theory [14, 15], when there is a weak transition to a system where there is a strong final state interaction (*FSI*), one can factorize the *s*-wave reaction amplitude, f_s, near threshold in the form

$$f_s = \frac{f_B}{1 + \frac{r_0}{2} a p^2 - i a p} \tag{1}$$

where p is the η c.m. momentum, a the complex scattering length and r_0 the effective range. The unperturbed production amplitude f_B is assumed to be slowly varying and is often taken to be constant in the near-threshold

[1] h.machner@fz-juelich.de

12th International Conference on Meson-Nucleon Physics and the Structure of the Nucleon (MENU 2010)
AIP Conf. Proc. 1374, 397-401 (2011); doi: 10.1063/1.3647167

region.

Unitarity demands that the imaginary part of the scattering length be positive, *i.e.*, $a_i > 0$. In addition, to have binding, there must be a pole in the negative energy half-plane, which requires that $|a_i|/|a_r| < 1$. Finally, in order that the pole lie on the bound- rather than the virtual-state plane, one needs also $a_r < 0$.

Recently, two different experiments at COSY Jülich measured η production in $pd \to \eta^3$He reactions very close to threshold with an extremely high precision of the data [16, 17]. The authors choose different methods in fitting the otherwise compatible data and therefore the values reported for the scattering length can not be compared with each other. We fitted Eq. 1 with $r_0 = 0$ and f_B=const. to the data of [17].

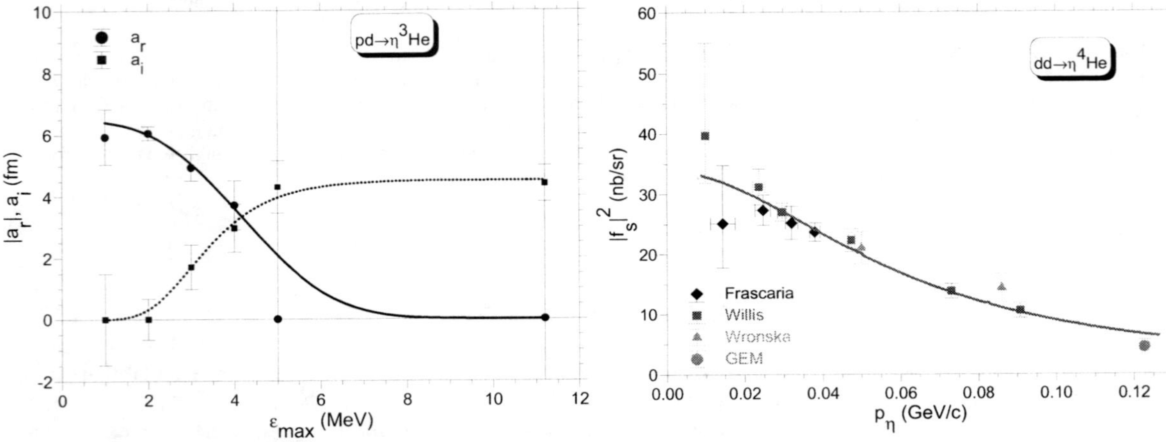

FIGURE 1. Left panel: The η^3He scattering length as function of energy bin (0-ε_{max}). The solid and dotted curves are to guide the eyes. Right panel: The spin averaged s-wave matrix element for $dd \to \eta\alpha$ reaction as function of the η momentum. The solid curve is a fit with Eq. 1 to the data [19], [23], [24], and [25].

In these fits we restricted the data to energies below ε_{max}. The results are shown in Fig. 1. For the two smallest energy settings one finds almost identical results. The imaginary part is almost zero. It then becomes larger while the real part becomes smaller. This may be a sign of an $s-p$ interference becoming important [18] for the effective range. Since the data are not sensitive to the sign of the real part of the scattering length, the quest for a bound state or an unbound pole can not be answered. From the above analysis of the data with $\varepsilon_{max} \leq 2$ MeV we get a pole position $|Q_{^3\mathrm{He}\eta}| \approx 1$ MeV. From the model calculations it is known that binding is more probable for heavier nuclei than for lighter nuclei. We therefore can expect the relation $|Q_{^3\mathrm{He}\eta}| < |Q_{^4He\eta}|$ to hold. In the following we study the FSI in the η ^4He system produced in the reaction $dd \to \eta\alpha$. The existing data before the GEM measurement were not sufficient to extract the s-wave contribution of the cross section. In order to do so GEM made use of a tensor polarized deuteron beam [19]. The beam momentum of 2385.5 MeV/c corresponds to an excess energy of $Q = 16.6$ MeV for this reaction when an η-meson mass of $m_\eta = 547.7$ MeV/c^2 is used [20]. The Big Karl magnetic spectrograph [21, 22] employed for this study is equipped with two sets of multi-wire drift chambers (MWDC) for position measurement and thus track reconstruction. Two layers of scintillating hodoscopes, 4.5 m apart, led to a more accurate time-of-flight measurement than previously achieved with Big Karl. They also provided the energy-loss information necessary for particle identification. More information on the experiment is given in Ref. [19].

In a first step the unpolarized cross section was measured. From Legendre polynomial fits to the angular distribution it became evident that up to d-waves contribute. In a second step tensor polarized deuteron beams were used and the angular distribution of the tensor analyzing power A_{xx} deduced. From these measurements the s-wave amplitude could be extracted and from this a spin-averaged matrix element. Wrońska et al. [23] could not distinguish whether the angular distribution is due to the $s-p$ interference or a d-wave. This puzzle is now solved, since the present data indicate the absence of a p-wave. Data for s-wave were extracted from the measurement of Willis et al. [25] by just dividing by 4π. This is, however, not justified in the region of their highest point. Assuming a dependence of the d-waves with p_η^2 we can extract a more correct value for the s-wave strength. All values for the spin averaged matrix element are shown in the right panel of Fig. 1 together with a fit of the complex scattering length $a_{^4He\eta} = [\pm(3.1 \pm 0.5) + i \cdot (0.0 \pm 0.5)]$ fm. This result can be converted into a pole position $|Q_{^4He\eta}| \approx 4$ MeV. Since the present results fulfill the conditions for a bound state, except for the sign of the real part of the scattering length, it may well be that $\eta - \alpha$ binding is observed.

PRODUCTION OF η BOUND NUCLEAR STATES

The following method was successfully employed in the production of Hypernuclei and pionic atoms. A nucleon is replaced by a hyperon or a pion. The maximum cross section is obtained when the momentum transfer from the projectile to the particle to be implanted is minimized, which means that the produced particle is almost at rest in the laboratory system. In pionic atom studies a proton transfer via the $(d,{}^3\mathrm{He})$ reaction was successfully applied [26]. However, although this reaction has reasonably large cross section it also has the disadvantage of a huge background due to the break-up of the deuterons. The resulting protons have the beam velocity and thus the same magnetic rigidity as the ^{3}He particles. This problem can be overcome by making use of a two-nucleon transfer reaction like $(p,{}^3\mathrm{He})$ but at the expense of a much smaller cross section. The choice of an odd-odd target nucleus would be ideal in order to avoid nuclear excitations as much as possible. Since there exist no solid material with these properties ^{27}Al was chosen as a compromise. The experiment was done by making use of the following reaction chain:

$$p+{}^{27}\mathrm{Al} \to {}^3\mathrm{He}+\mathrm{X} \tag{2}$$

where the ^{3}He carries the beam momentum away. The unobserved system X is therefore at rest. One of the possibilities is $X={}^{25}\mathrm{Mg}+\eta$. Since the η is also at rest it can undergo a second chain of reactions

$$\eta+N \leftrightarrow N^* \to N\pi. \tag{3}$$

The final fate of that chain is the decay to a pion and a nucleon. In the case of a neutral resonance the final state can be $p+\pi^-$, which have to be emitted almost back to back, if we ignore Fermi motion. The ^{3}He is detected with the magnetic spectrograph Big Karl [21, 22]. For the detection of the two decay particles a dedicated detector ENSTAR was built. It has cylindrical shape around the target and consists of three layers of scintillating material. The layers were subdivided into pieces making polar and azimuthal angle measurements possible. The read-out of the pieces was performed with scintillating fibres [27].

The experiment was performed at a beam momentum of 1745 MeV/c. At this momentum final states with small binding energies can be produced with small momentum transfer. States with binding energies from 0 MeV to -30 MeV will have momentum transfer below 30 MeV/c.

Two settings of the spectrograph were used with a large overlap. Excellent particle identification in Big Karl was achieved by measuring energy loss in the start detector of the TOF facility in the focal plane and the TOF. The identification of a back to back proton and pion with momenta corresponding to a $N^*(1535)$ resonance is mandatory to see structure in the missing mass spectra. More details can be found in [28].

The two spectra were then combined. Counts were transformed into differential cross sections by making use of the integrated luminosity and detector acceptance as well as its efficiency.

These cross sections are shown in Figure 2. Here we have converted the missing mass scale into binding energy BE. The data show an enhancement around $BE \approx -13$ MeV. At first, we test the hypothesis of peak structure being a fluctuation of background, i.e. the origin of the background is taken to be independent of the signal. The background outside the peak region, for simplicity approximated by a constant, was found to be 5.8±0.64. The significance is then given by $(N-BG)/\sqrt{BG+\sigma_{BG}}$ where N is the total counts in the region of interest, BG is the total background in this region as determined from the fit to the outside region and σ_{BG} is error in the estimation of background value as taken from the fit. This yields a value of significance which is 5.3σ. Here we have assumed Gaussian errors. For the assumption of Poisson errors the background is 6.2±1.0. This larger value is typical for Poisson distribution and hence the significance reduces to 4.9σ. Finally a Gaussian on top of the background was fitted to the whole data set. This yielded for the case of Poisson statistics 6.4±0.96 for the background, 8.3±3.6 for the amplitude, -12.0±2.2 MeV for the centroid and 4.7±1.7 MeV for the width. In the second method, the statistical significance is extracted by assuming the background events as well as the peak events on top of the background being Poisson distributed. Again a constant background and a Gaussian was assumed. In this way, we obtain a value of 6.20σ for the significance, assuming a simultaneous determination of amplitude, centroid and width of the signal. The fit gives for the centroid -13.13 ± 1.64 MeV and for the width 4.35 ± 1.27 MeV. These results compare favorably with those from the first method. We, therefore, consider the present experimental results to provide a strong hint of a nuclear η bound state.

In the following we take a different view on the data than in [28]. It may well be that the state observed is the $1p$ state and not the $1s$ state. This requires a rather large scattering length $a_{\eta N}$. The $1s$state is more strongly bound than

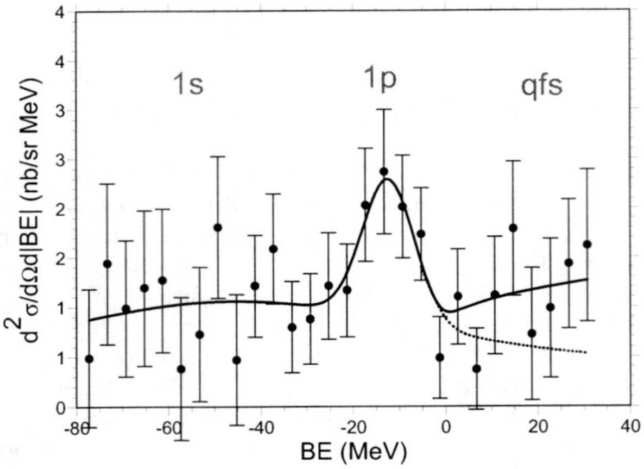

FIGURE 2. Missing mass spectrum converted to binding energy *BE* of a bound system $^{25}Mg \otimes \eta$ as measured with the magnetic spectrograph. All cuts and background subtraction have been applied. The solid curve is a fit with a constant background, two Gaussians and a phase space behavior for the unbound system. Dotted curve: same as solid curve but without phase space contribution.

the $1p$ and it is expected to have a larger width than the latter. We therefore fit two Gaussians on a constant background to the data. The quasi-free region is represented by phase space behavior. The resulting curve is shown in Fig. 2. The hypothetical $1s$ state has a very large width as expected and reaches into the unbound region. However, the evidence for this peak is small than the one for the peak at 12 MeV. Due to the low statistics of the data, which is a result of the small cross section, the uncertainties in the fitted parameters are large.

The author is grateful to the members of the GEM collaboration.

REFERENCES

1. R. S. Bhalerao, L. C. Liu, *Phys. Rev. Lett.*, **54**, 865 (1985).
2. Q. Haider, L. C. Liu, *Phys. Lett.*, **B 172**, 257 (1986).
3. C. Garcia-Recio, T. Inoue, J. Nieves, E. Oset, *Phys. Lett.*, **B 550**, 47 (2002).
4. R.S. Hayano, S. Hirenzaki, A. Gillitzer, *Eur. Phys. J.*, **A 6**, 99 (1999).
5. K. Tsushima, *Nucl. Phys.*, **A 670**, 198c (2000).
6. S. A. Rakityansky, S. A. Sofianos, M. Braun, V. B. Belyaev, and W. Sandhas, Phys. Rev. **53** R2043 (1996).
7. S. Wycech, A. M. Green and J. A. Niskanen, Phys. Rev. **C 52** 544 (1995).
8. N.N. Scoccola and D.O. Riska, Phys. Lett. **B 444** 21 (1998).
9. R. E. Chrien et al., *Phys. Rev. Lett.*, **60**, 2595 (1988).
10. J. B. Lieb, *Proc. International Conf. Nuclear Physics, Sao Paulo, Sao Paulo, Brazil*, (1988).
11. H. Nagahiro, D. Jido, S. Hirenzaki, *Phys. Rev.*, **C 80**, 025205 (2009).
12. M. Pfeiffer et al., *Phys. Rev. Lett.*, **92**, 252001 (2004).
13. B. Krusche, *priv. communication.*
14. K. M. Watson, *Phys. Rev.*, **88**, 1163 (1952).
15. A. B. Migdal, *JETP*, **1**, 2 (1955).
16. J. Smyrski, et al., *Phys. Lett.*, **B 649**, 258 (2007).
17. T. Mersmann et al., *Phys. Rev. Lett.*, **98**, 242301 (2007).
18. C. Wilkin et al., *Phys. Lett.*, **B 654** , 92 (2007).
19. A. Budzanowski et al. (The GEM Collaboration), *Nucl. Phys.*, **A 821**, 193 (2009).
20. K. Hagiwara et al. (PDG), *Phys. Rev.*, **D 66**, 010001 (2002).
21. M. Drochner, et al., *Nucl. Phys.*, **A 643**, 55 (1998).
22. H. Bojowald, et al., *Nucl. Instruments and Meth. in Phys. Res.*, **A 487**, 314 (2002).
23. A. Wrońska et al., *The European Physical Journal*, **A 26**, 421 (2005).
24. R. Frascaria et al., *Phys. Rev.*, **C 50**, R537 (1994).
25. N. Willis et al., *Phys. Lett.*, **B 406**, 14 (1997).
26. T. Yamazaki et al., *Z. Physik*, **A 355**, 219 (1996).
27. M. Betigeri et al., *Nucl. Instruments and Meth. in Phys. Res.*, **A 578**, 198 (2007).

28. A. Budzanowski et al. (COSY-GEM collaboration), *Phys. Rev.*, 2009 **C 79**, 012201R (2009).

Experimental study of ppη dynamics with WASA-at-COSY

Neha Shah for the WASA-at-COSY collaboration

Department of Physics, IIT Bombay, Powai, Mumbai - 76, INDIA.

Abstract. To investigate the interaction of η-meson with the nucleons, its production, near the kinematical threshold, in proton-proton collisions has been studied with the WASA detector at COSY storage ring in Juelich, Germany. The data has been taken at beam energy 1400 MeV (corresponding to excess energy (Q = 57 MeV). The η-meson was detected via its $3\pi^0$ decay in nearly 4π detector and two protons were measured in forward direction. The determination of four vectors of both protons and the η-meson in the final state allowed to derive complete kinematical information of the ppη-system. The analysis resulted in 9×10^6 events of $\eta\rightarrow3\pi^0$ giving total production cross-section $(8.87\pm0.03_{stat}\pm2.57_{sys})\mu$b. The angular distribution of η-meson in the center of mass frame is anisotropic and squared invariant mass distributions for proton-proton and proton-η shows deviation from pure phase space.

Keywords: η meson, WASA-at-COSY, Differential cross-section
PACS: PACS: 14.40.Aq, 13.60.Le

INTRODUCTION

Considerable theoretical and experimental efforts have been put in to achieve a better understanding of the η-meson properties and its interaction with other particles. This is primarily due to the special place that the η meson occupies in particle physics. The π^0 and the η meson have similar properties. Both are neutral, spin less, and have almost the same lifetime. They have the same principal decay mode, namely $\eta(\pi^0)\rightarrow2\gamma$. However, there is a major difference in the way they interact with the nucleons as the eta is much heavier than the pion. This gets manifested in the meson-nucleon dynamics, especially at low energies. The S_{11}-resonance, N(1535), can be formed in both π^0N and ηN collisions. Due to the large mass of the η-meson this resonance is very close to the ηN-threshold (excess energy only 49 MeV). Furthermore, as the width of the resonance is 150MeV, it covers the whole low energy region of the ηN interaction. This results in the interaction of nucleons with η mesons being much stronger than with pions. The N(1535)-resonance then decays with almost equal probabilities into the ηN and πN channels. The ηN and πN interactions couple to each other leading to considerable enhancement in the η production cross-section. Thus one of the ways to study the η N interaction would be to measure the production cross section at energies near the threshold. However, the most crucial issue in the study of production dynamics near the threshold is the determination of relevant degrees of freedom for the description of interactions. There are phenomenological models in which the quark-gluon degrees of freedom are considered for the calculation of production of η meson. However in general the production amplitude is calculated within the single meson exchange model. In this model the nucleons approach each other and undergo some initial state interactions, a virtual meson is exchanged between them exciting the nucleon to resonance, which then decays into η and nucleon. These outgoing particles are than subjected to the final state interactions. In table 1 we have reviewed previous measurements from PROMICE/WASA [1], COSY-TOF [2], COSY-11 [3, 4], DISTO [5] and CELSIUS/WASA [6] at various excess energies.

EXPERIMENT

The Wide Angle Shower Apparatus (WASA) is installed at the COoler SYnchrotron (COSY) in Juelich, Germany. The COSY facility provides phase space cooled beam of protons and deuteron with momenta up to 3.7 GeV/c. WASA is designed and optimized for the studies of production and decays of light mesons in hadronic interactions. It consists of the forward detector for the measurement of target-recoil particles and scattered projectiles over the angles 3°-18°, the central detector for measurement of decay products of the mesons over 20°-169° and the unique pellet target system as shown in figure 1. The pellet target system produces pellets of hydrogen or deuterium of thickness about 35 μm and density of the order of 10^{15} atoms/cm^2.

12th International Conference on Meson-Nucleon Physics and the Structure of the Nucleon (MENU 2010)
AIP Conf. Proc. 1374, 402-405 (2011); doi: 10.1063/1.3647168

TABLE 1. Review of previous measurements.

Excess Energy	Detector	Conclusions
16, 37 MeV	PROMICE/WASA [1]	Angular distributions shows anisotropy. They could explain the data by taking interference between Ss and Sd partial waves.
16, 41 MeV	COSY-TOF [2]	Angular distribution shows isotropy. S_{pp} and $S_{p\eta}$ distributions show deviation from phase space. The data could not be explained by taking pp FSI alone.
10, 15.5 MeV	COSY-11 [3, 4]	Angular distribution shows isotropy. S_{pp} and $S_{p\eta}$ distributions show deviation from phase space. The S_{pp} could be explained by taking pp FSI but not $S_{p\eta}$.
324, 412, 554 MeV	DISTO [5]	Angular distribution shows anisotropy at lowest energy and anisotropy disappears as energy increases. The data could be explain by taking N(1535) resonance.
40, 72 MeV	CELSIUS/WASA [6]	Angular distributions shows anisotropy and S_{pp} and $S_{p\eta}$ distributions show deviation from phase space. They could explain the data by taking contributions from higher partial waves.

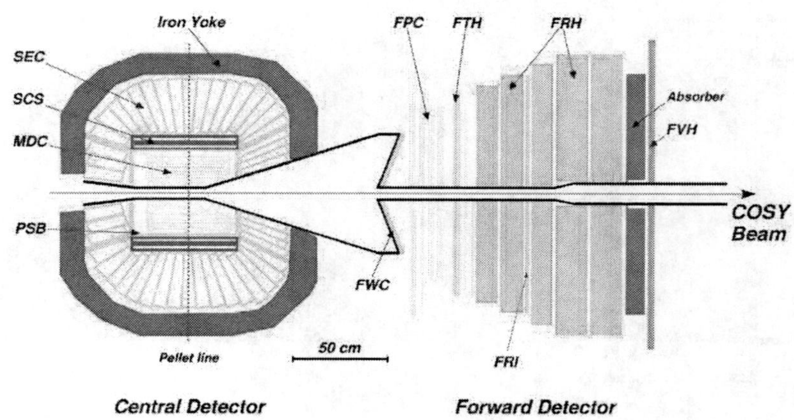

FIGURE 1. WASA Detector setup

ANALYSIS

We have analysed the data collected during the first production run for WASA-at-COSY which corresponds to 96 hours of data taking. The η-mesons have been produced in proton-proton interactions at a beam energy 1.4 GeV (corresponding to excess energy 57 MeV). The η-meson can be reconstructed in the central detector from its decay product by reconstructing their invariant mass or can be identified by using missing mass of two protons in the forward detector. For reconstruction in the central detector, we have selected $\eta \rightarrow 3\pi^0$ channel with branching ratio (32.51±0.29)%. Hence in the final state we have $pp \rightarrow pp\ 6\gamma$. The major background contributions are from

$pp \to pp\ 3\pi^0$ with cross section 1.09 μb as it has the same final state and $pp \to pp\ 2\pi^0$ with cross section 250 μb because of cluster split offs, pile-up or noise.

The events were selected by asking two charged tracks in the forward detector with energy deposit greater than 25 MeV for each track and six neutral clusters in the central detector with energy deposit more than 20 MeV for each cluster. In the central detector, to select $3\pi^0$ candidates from the six reconstructed photons, fifteen possible combinations of two photon pairs were made and the χ^2 method was used to choose the best three pions out of all fifteen possible combinations. To improve the accuracy of all measured quantities and to constrain the events to allowed phase space, a kinematic fit was applied. The full final state including two protons was considered in the fitting procedure. In addition we had a constraint over individual pion mass. The goodness of kinematic fit was tested by checking confidence level and pull distributions. In addition to basic selection criterion, the events were selected with the following conditions:

- $\chi^2 < 12.0$
- Confidence level > 0.1

Figure 2 shows the missing mass of two protons in the forward detector (left) and reconstructed invariant mass of six gammas in the central detector (right) for data as well as simulations after applying above mentioned conditions. The background contribution from $pp \to pp\ 2\pi^0$ was negligible after applying above mentioned criterion and relative contribution of background $pp \to pp\ 3\pi^0$ was 10 %. To extract the signal, we have fitted the invariant mass distribution by taking signal as Gaussian and background as a polynomial. Subtracting background contribution we have 97k $\eta \to 3\pi^0$ events, which when normalized to luminosity and with appropriate acceptance correction gives the total cross section of η-meson production $(8.87 \pm 0.03_{stat} \pm 2.57_{sys})$ μb, which is in good agreement with the previous measurement from CELSIUS/WASA [7], CELSIUS [8] and PINOT [9].

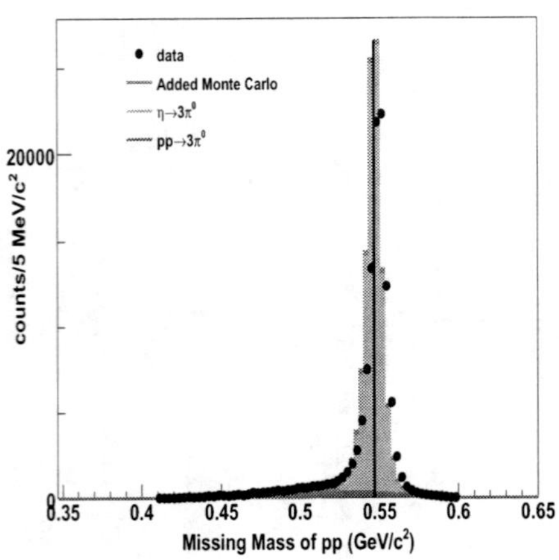

FIGURE 2. Missing mass of two protons in the forward detector (left) and reconstructed invariant mass of three π^0 in the central detector (right), for data with simulations, after applying above mentioned conditions.

RESULTS

The obtained differential cross-sections must be corrected for detector acceptance before they are compared to theoretical calculations. To derive the differential distributions we divide the range of the studied variable into bins with a width corresponding to the achieved experimental resolution. In each bin we have counted the number of ηs produced. This method was adopted to subtract the contribution of multi-pion background and to obtain the model-free distributions for differential cross-sections. The preliminary results for angular distribution of η-meson in the center of mass frame, squared invariant masses of pp and pη are presented here. Angular distribution of η-meson in the center of mass frame is found to be anisotropic as shown in the left panel of figure 3, where dots are data points and the line shows the polynomial fit of second order. The squared invariant mass distributions for pp and pη (middle and right panel in figure 3) shows deviation from phase space and to describe this deviation. The error bar shows the systematic error.

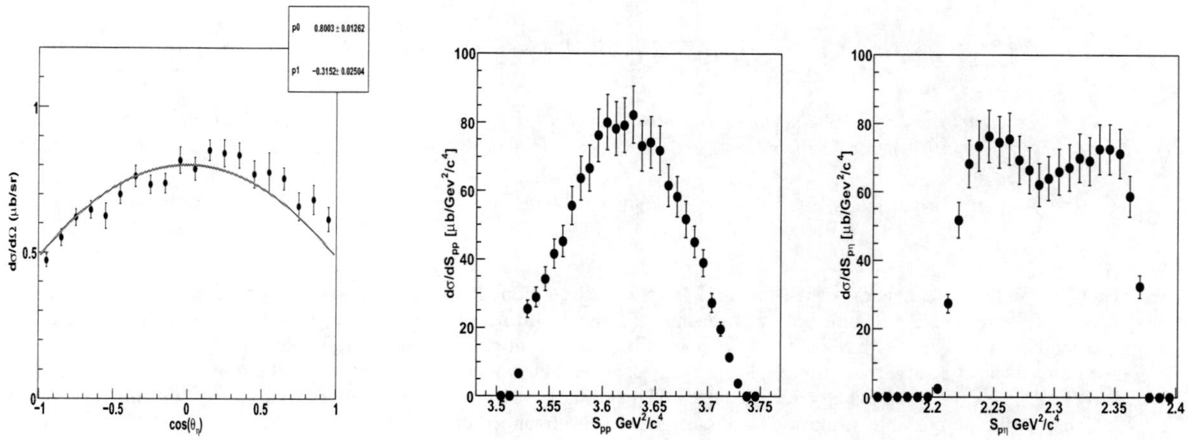

FIGURE 3. The left panel shows the angular distribution of η-meson in the center of mass frame. The dots are data point and the line shows the polynomial fit to the data. The middle and the right panel shows the squared invariant mass distributions for pp and pη.

ACKNOWLEDGMENTS

We acknowledge the financial support given by the DAAD-DST exchange program to participate in the experiment. We also want to thank the technical and administration staff at the Forschungszentrum Juelich for their help.

REFERENCES

1. H. Calen et al., Phys Lett. B 458 (1999) 190.
2. M. Abdel-Bary et al., Eur. Phys. J. A 16 (2003) 127.
3. P Moskal et al., Phys. Rev. C 69 (2004) 025203.
4. R. Czyzyjuewicz et al., Phys. Rev. Lett. 98 (2007) 122003.
5. F. Balestra et al., Phys Rev C 69 (2004) 064003.
6. H. Petern, η meson production in proton-proton collisions at Excess energies of 40 and 72 MeV, Uppsala Universitet, Sweden (2009).
7. C. Pauly, Light meson production in pp reactions at CELSIUS/WASA above the η threshold, Dissertation, Universität Hamburg, Germany (2006).
8. H. Calen et al., Phys. Lett. B 366 (1996) 39.
9. E. Chiavassa et al., Phys. Lett. B 322 (1994) 270.

η and η' photoproduction

F. Huang*, M. Döring†, H. Haberzettl** and K. Nakayama*,†

*Department of Physics and Astronomy, University of Georgia, Athens, GA 30602, USA
†Institut für Kernphysik and Jülich Center for Hadron Physics, Forschungszentrum Jülich, 52425, Jülich, Germany
**Center for Nuclear Studies, Department of Physics, The George Washington University, Washington, DC 20052, USA

Abstract. Recent high precision data of η and η' photoproduction in free and quasi-free processes are analyzed within an effective Lagrangian approach. Gauge invariance of the resulting amplitude is fully satisfied via the introduction of a generalized contact current that accounts for the effects of the final-state interaction. Moreover, the recently discovered enhancement of the η photoproduction yield on the quasi-free neutron compared to that on the quasi-free proton at energies around $\sqrt{s} \approx 1.67$ GeV is also analyzed within an SU(3) coupled-channel chiral unitary approach. Additional possibilities how to disentangle the possible scenarios for the origin of this enhancement are presented.

Keywords: Photoproduction of mesons, Baryon spectroscopy
PACS: 25.20.Lj, 13.60.Le, 25.24.-h, 25.24.Vc, 25.10.+s

INTRODUCTION

The current knowledge of most of the nucleon resonances comes primarily from the study of πN scattering and pion photoproduction reactions. In recent years, the study of other meson-production reactions has intensified in search of so-called missing resonances [1]. These are resonances that are predicted by quark models but not observed in the existing experimental data. Since it is conceivable that these resonances escape detection in the usual pion-production channels simply because they couple only weakly to the pion but might otherwise couple more strongly to other mesons, it is important to study such competing meson-production reactions in the hope of revealing at least some of the "missing" resonances. One of the most widely studied reactions in this respect concerns the photoproduction of the η meson. One of the appeals of η production is the fact that it serves as an isospin filter for the underlying resonances since the η is an isoscalar meson. One of the most prominent resonances that has been found to couple strongly to the ηN channel is the $S_{11}(1535)$ resonance. In recent years, η photoproduction has received renewed attention because of an observed bump structure around $\sqrt{s} \approx 1.67$ GeV in the ratio of the total cross sections, $\sigma_n/\sigma_p \equiv \sigma(\gamma n \to \eta n)/\sigma(\gamma p \to \eta p)$ [2–4] in the quasi-free processes. This bump structure was interpreted [5] as a potential signal of a non-strange member of an anti-decuplet pentaquarks.

Among the members of the pseudoscalar mesons in the flavor SU(3) classification, the η' meson with a mass of $m_{\eta'} \cong 958$ MeV is the heaviest. Of particular interest in investigating the η' production reactions is that it may be suited to extract information on nucleon resonances in the less-explored higher N^* mass region. Since the η' meson is much heavier than a pion or η, its production processes near threshold necessarily sample a much higher resonance-mass region than the corresponding lighter-meson production processes. Therefore, in addition to serving as an isospin filter (η' is also an isoscalar meson), the η' is well-suited for investigating high-mass resonances in low partial-wave states. Furthermore, studying η' production off the nucleon may perhaps offer the possibility of imposing a more stringent constraint on the strength of its coupling to the nucleon which at present is only poorly known. This has attracted much attention in connection with the so-called "nucleon-spin crisis" in polarized deep inelastic lepton scattering [6] since the knowledge of the $NN\eta'$ coupling constant in the zero-squared-momentum limit can be related to how much the quarks contribute to the "spin" of the proton through the flavor-singlet axial charge [7–9] which, in turn, constrains the gluonic contribution to the spin of the proton [8]. In contrast to lighter pseudoscalar meson photoproductions, not many data existed for the η' meson until recently. Indeed, only the total cross-section data from the ABBHHM Collaboration [10] were available until the late 1990s. More recently, the differential cross-section data from the SAPHIR [11] and CLAS [12] Collaborations — the latter of high-precision quality — became available. Quite recently, the CBELSA/TAPS Collaboration has measured the η' photoproduction cross sections on the free proton [13] as well as on the quasi-free proton and neutron [14]. The CLAS Collaboration [15] has also reported very-

12th International Conference on Meson-Nucleon Physics and the Structure of the Nucleon (MENU 2010)
AIP Conf. Proc. 1374, 406-409 (2011); doi: 10.1063/1.3647169

high-precision cross-section data. Currently, the beam asymmetry in this reaction is being measured by the CLAS Collaboration [16]. Theoretically, only few groups have analyzed the η' photoproduction reactions so far. In Ref. [17], a reggeized ρ and ω model was used to account for the dominant background contribution, while in Refs. [18, 19] the ρ and ω meson-exchange model was considered. In the two latter references the reggeized ρ and ω contributions were investigated as well.

In view of the considerable amount of new data being added to the world database for both the η' and η photoproduction processes, it is timely to investigate these reactions taking into account the new high-precision data. Here, we present some selected results for these photoproduction processes which are part of an ongoing effort to consistently describe not only the photon-induced but also the hadron-induced meson production reactions. Our fully gauge-invariant model for photoproduction based on an effective Lagrangian approach includes the nucleonic, mesonic and nucleon resonance currents. The details of the model are given in Refs. [18–20]. The issue of the observed bump structure in the cross section ratio in η photoproduction mentioned above is discussed here within an SU(3) coupled channel chiral unitary approach, where an alternative (to the pentaquark) explanation of this bump structure is given [21].

η' PHOTOPRODUCTION

Figure 1 (left panel) displays the results for the η' angular distribution in the center-of-momentum (c.m.) frame for the reaction $\gamma p \to \eta' p$ (free proton) in the energy range from threshold to $W = 2.35$ GeV. We note that a minimum of four resonances — S_{11}, P_{11}, P_{13}, and D_{13} — is necessary to reproduce the high-precision CLAS data [15]. As one can see, the fit quality is excellent with a $\chi^2/N \cong 0.5$. The new high-precision differential cross-section data impose a more stringent constraint on the resonance parameters compared to earlier cross-section data [12], however, they are not enough to determine these parameters uniquely. For that one requires spin observables. The predicted total cross section is shown in Fig. 1 (right panel). It is interesting to note that the shallow bump structures observed in the prediction based on the earlier data, especially, the one at around 2.1 GeV [19], seem to be nearly absent in the present result.

Recently, the CBELSA/TAPS Collaboration [14] has reported their (preliminary) differential cross-section data on the quasi-free proton and neutron. We have calculated these quasi-free processes within our present approach by folding the corresponding cross section in the free processes with the nucleon momentum distribution in the deuteron following the prescription of Ref. [22]. For the quasi-free proton case, no free parameters exist because all of them are fixed by the free proton case. For the quasi-free neutron, the only free parameters are the resonance-nucleon-gamma ($N^*N\gamma$) coupling constants which are adjusted to fit the data. Overall, we describe the data well with some minor deviations at energies close to threshold.

η PHOTOPRODUCTION

The η angular distributions in the c.m. frame of the system in the free proton process are shown in Fig. 2 (left panel). The model includes the known resonances $S_{11}(1535)$, $S_{11}(1650)$, $P_{11}(1710)$, $P_{13}(1720)$, $D_{13}((1520)$, $D_{15}(1675)$, and $F_{15}(1680)$. The results for the beam asymmetry are also shown in Fig. 2 (right panel). These illustrate the quality of fit of our model results ($\chi^2/N \cong 1.8$). Unlike in η' photoproduction, here the resonance parameters are much better constrained. The present model cannot describe the measured target asymmetry [24], in particular, at energies close to threshold. This failure is shared with all other existing models, and so this remains an open problem.

In the remainder of this section, we will address the issue of the bump structure observed in the cross-section ratio σ_n/σ_p in the quasi-free reactions $\gamma n \to \eta n$ and $\gamma p \to \eta p$ off the deuteron within an SU(3) coupled-channel chiral unitary approach. Within this approach, the bump structure arises naturally as a result of coupled-channel effects due to the strong coupling to the intermediate $K\Lambda$ and $K\Sigma$ channels. In the quasi-free proton case, the photon can couple to the charged Kaon in the intermediate $K^+\Lambda$ state, while in the quasi-free neutron case, this contribution is absent, for the corresponding intermediate state is $K^0\Lambda$ and the photon does not couple to a neutral Kaon. The resulting cross-section ratio is shown in Fig. 3 (left panel). It is a direct consequence of the dynamics driven by the Weinberg-Tomozawa interaction together with the strong coupling to $K\Lambda$ and $K\Sigma$ channels resulting from SU(3) symmetry. The present result does not rule out other explanations for the bump structure, such as the narrow resonances.

The prediction of our chiral unitary model for the quasi-free $\gamma N \to K\Lambda$ processes on the deuteron is shown in Fig. 3 (right panel). One can see, in particular, that, unlike for η photoproduction, the cross-section ratio σ_n/σ_p (insert)

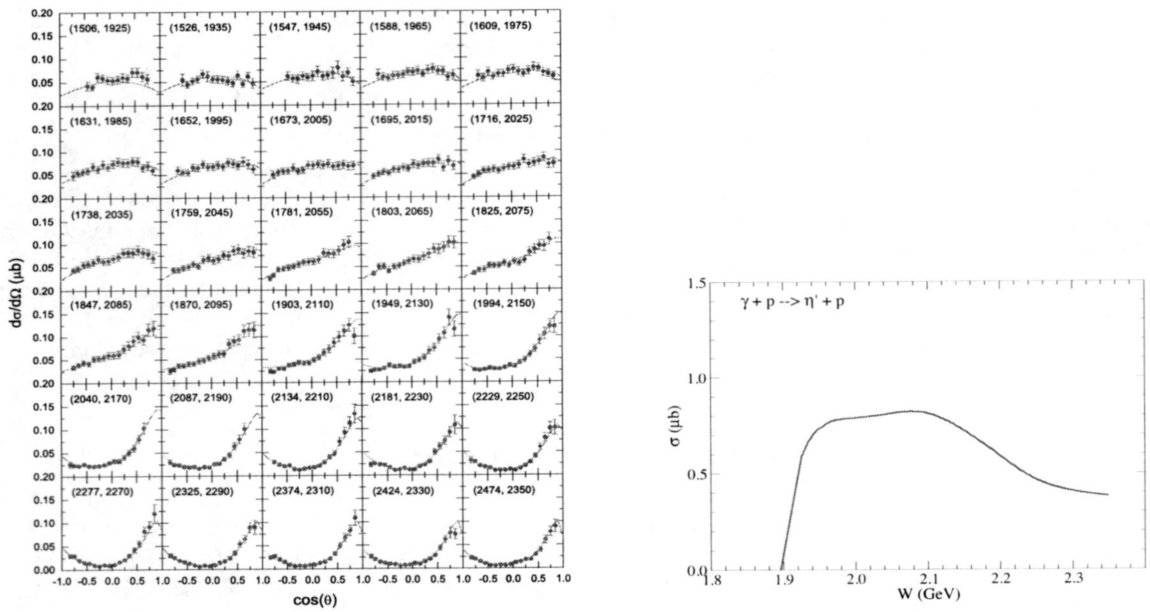

FIGURE 1. Results for $\gamma p \to \eta' p$ (free proton). Left panel: η' angular distribution in the c.m. frame of the system. The numbers in parenthesis are the photon incident energy (E_γ) and total energy (W) of the system (E_γ, W) in units of MeV. The following four resonances were included: $S_{11}(2120), P_{11}(2134), P_{13}(1924)$, and $D_{13}(2219)$. $\chi^2/N \simeq 0.5$. Cross-section data alone are unable to fix the resonance parameters uniquely. Data are from Ref. [15]. Right panel: Predicted total cross section as a function of total energy of the system W.

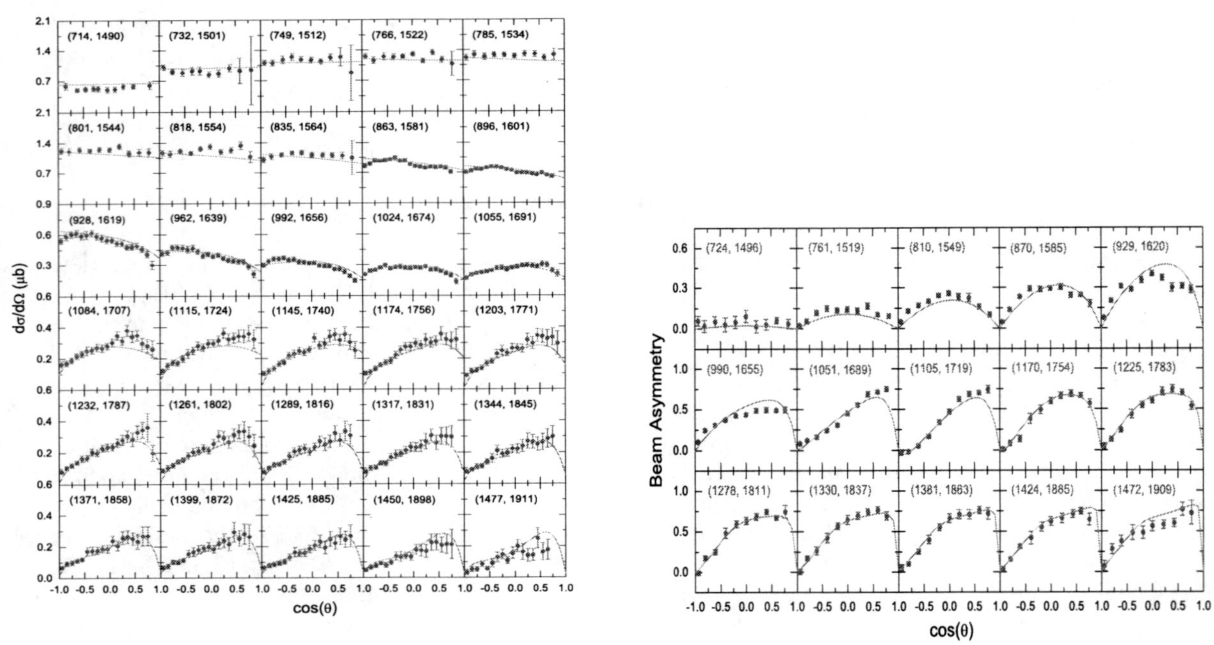

FIGURE 2. Results for $\gamma p \to \eta p$ (free proton). Left panel: η angular distribution in the c.m. frame of the system. The numbers in parenthesis are the photon incident energy (E_γ) and total energy (W) of the system (E_γ, W) in units of MeV. Data are from Ref. [23]. Right panel: Beam asymmetry as a function of η emission angle in the c.m. frame of the system.

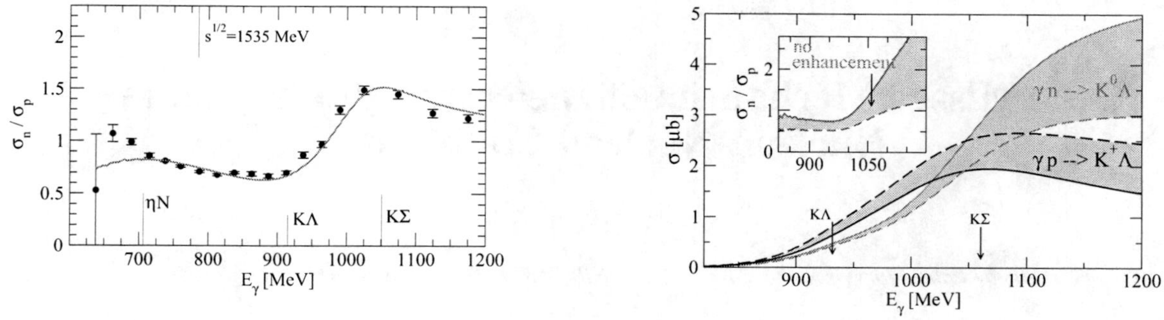

FIGURE 3. Left panel: Cross-section ratio σ_n/σ_p from the quasi-free processes $\gamma n \to \eta n$ and $\gamma p \to \eta p$ on the deuteron, as a function of photon incident energy E_γ. Data are from Ref. [4]. Right panel: Total cross section for the quasi-free $\gamma n \to K^0\Lambda$ (black curves) and $\gamma n \to K^+\Lambda$ (red curves). The shaded bands indicate the uncertainty associated with the theoretical predictions. Insert (right panel): Corresponding cross section ratio σ_n/σ_p. No bump structure is predicted here.

does not exhibit any bump structure. Therefore, measurements of the cross section ratio in $K\Lambda$ photoproduction can shed light on the accuracy of our chiral unitary model. Furthermore, it is interesting to note that a measurement of the longitudinal spin-transfer coefficient (L_z) in $\gamma N \to \eta N$, together with the corresponding unpolarized cross sections, can disentangle the coupled-channel scenario discussed above with the narrow-resonance scenario in a model-independent way. Of course, this poses an enormous experimental challenge.

ACKNOWLEDGMENTS

This work is supported by the FFE grant No. 41788390 (COSY-58) and the DFG grant No. DO 1302/1-2 and SFB/TR-16.

REFERENCES

1. S. Capstick and N. Isgur, *Phys. Rev. D34*, 2809 (1986); S. Capstick and W. Roberts, *ibid., 57*, 4301 (1998); ***58***, 074011 (1998).
2. V. Kuznetsov *et al.*, GRAAL Collaboration, *Phys. Lett. B**647***, 23 (2007).
3. F. Miyahara *et al.*, *Prog. Theor. B (Phys. Suppl.) 168* 90 (2007).
4. I. Jaegle *et al.*, CBELSA/TAPS Collaboration, *Phys. Rev. Lett.* **100**, 252002 (2008).
5. D. Diakonov, V. Petrov, and M.V. Polyakov, *Z. Phys. A***359**, 305 (1997). M.V. Polyakov and A. Rathke, *Eur. Phys. J. A* **18**, 691 (2003). A. Fix, L. Tiator, and M.V. Polyakov, *Eur. Phys. J. A***32**, 311 (2007).
6. J. Ashman et al., *Phys. Lett. B206*, 364 (1988).
7. G. M. Shore and G. Veneziano, *Nucl. Phys. B**381***, 23 (1992).
8. G. M. Shore and G. Veneziano, *Phys. Lett. B244*, 752 (1990).
9. T. Feldmann, *Int. J. Mod. Phys. A**15***, 159 (2000).
10. ABBHHM Collaboration, *Phys. Rev.* **175**, 1669 (1968); *Nucl. Phys. B***B108**, 45 (1976).
11. R Plötzke *et al.*, *Phys. Lett. B***444**, 555 (1998); J. Barth *et al.*, *Nucl. Phys. A***691**, 374c (2001).
12. M. Dugger *et al.*, *Phys. Rev. Lett.* **96**, 062001 (2006); *ibid.*, 169905(E).
13. V. Crede *et al.*, *Phys. Rev. C**80***, 055202 (2009).
14. I. Jaegle (for CBELSA/TAPS Collaboration). in *Workshop on the Physics of Excited Nucleons (NSTAR2009)*, Beijing, April 18-22, 2009; I. Jaegle *et al.*, CBELSA/TAPS Collaboration, in preparation.
15. M. Williams *et al.*, *Phys. Rev. C**80***, 045213 (2009).
16. P. Collins, private communication.
17. W. -T. Chiang *et al.*, *Phys. Rev. C**68***, 045202 (2003).
18. K. Nakayama and H. Haberzettl, *Phys. Rev. C***69**, 065212 (2004).
19. K. Nakayama and H. Haberzettl, *Phys. Rev. C***73**, 045211 (2006).
20. K. Nakayama, Yongseok Oh, and H. Haberzettl, arXiv:0803.3169.
21. M. Döring and K. Nakayama, *Phys. Lett. B**683***, 145 (2010).
22. A.V. Anisovich, *et al.*, *Eur. Phys. J. A**41***, 13 (2009).
23. O. Bartalini *et al.*, GRAAL Collaboration, *Eur. Phys. J. A33*, 169 (2007).
24. A. Bock *et al.*, PHOENICS Collaboration, *Phys. Rev. Lett. **81***, 534 (1998).

On a Missed Mechanism of Dielectron Production in Nucleon-Nucleon Collisions

Anatoliy S.Khrykin

Dzhelepov Laboratory of Nuclear Problems, JINR, 141980 Dubna, Moscow Region, Russia

Abstract. We examine a new mechanism of e^+e^- pair production in NN collisions associated with the NN-decoupled dibaryon $d_1^\star(1956)$ formation in the process $NN \to \gamma^\star d_1^\star$, where $\gamma^\star$ is the virtual photon which converts into a e^+e^- pair. It is shown that a substantial excess of dielectron yields from $Ca+Ca$ and $C+C$ collisions at 1 GeV/A in the dielectron mass spectra in the region from 0.2 to 0.5 GeV/c^2 measured by the *DLS Collaboration* as compared with calculated ones can be attributed to the contribution of this mechanism. A simple means for verification of the existence of such a mechanism is proposed.

Keywords: DLS puzzle; dilepton spectroscopy; mechanisms of dilepton production; dibaryon
PACS: 25.75-q; 25.75.Dw; 13.75.Cs; 14.20.Pt; 13.30Ce

INTRODUCTION

Experimental results on dielectron (e^+e^- pair) production in collisions of $Ca+Ca$, and $C+C$ [1] at 1.04 GeV/A and $p+d$, and $p+p$ at a number of energies from 1 to 4.88 GeV/A [2] reported by the *DLS Collaboration* more than 10 years ago have demonstrated that theoretical models at the time (see,e.g. Refs. [3, 4, 5] failed to give a satisfactory account of the measured e^+e^- pair invariant mass spectra. In particular, such spectra from $Ca+Ca$, and $C+C$ collisions substantially exceed those predicted by the model calculations in the dielectron mass M region $0.2 < M < 0.5$ GeV/c^2 and reference therein). This *DLS* finding called the "DLS puzzle" was recently confirmed by the *HADES Collaboration* [6] and now it is regarded as firmly established.

Since the shortage of dielectrons in calculated spectra became apparent, several theoretical groups have attempted to understand its origin. The emphasis mainly concentrated on the development of suitable models describing the conventional elementary processes of e^+e^- pair production [7]. These processes are believed to be associated with the production in NN collisions of neutral mesons (m), such as π^0, η, ω,..., and baryon resonances (R), such as $\Delta(1232)$, $N(1520)$,..., which can undergo the Dalitz decay $m(R) \to \gamma e^+e^-(Ne^+e^-)$, or the vector mesons ρ^0, and ω which can directly decay into into e^+e^- pairs, and the NN virtual (timelike) photon bremsstrahlung $NN \to NN\gamma^\star \to NNe^+e^-$. Nevertheless, despite considerable efforts the nature of the discrepancy between the calculated and measured spectra noted above is still obscure.

The shortage of dielectrons in calculated spectra could indicate the existance of an additional source of e^+e^- pairs that was not included in the model calculations. One of such possible sources is due to the NN-decoupled dibaryon resonance $d_1^\star(1956)$. This resonance [8] can be produced in the radiative process $NN \to \gamma d_1^*$ and decays in that $d_1^* \to NN\gamma$. It has been found that it manifests itself in a series of experimentally studied photon production processes induced by nucleon-nucleon and nucleon-nucleus collisions at intermediate energies [9, 10]. However, we know little about the physical nature of this resonance and do not yet know its quantum numbers. We only know that its spin J and parity P can be either from a set $J^P = 1^+, 3^+$, etc. if its isospin $I = 1$ or any if $I = 2$.

At the same time, if the resonance d_1^* really exists then together with its formation (decay) channel with a real photon the same channels with a virtual, massive photon should also take place. Conversion of such photons into e^+e^- pairs would give rise to a new source of dielectrons. However the virtual photons from the $d_1^\star$ decay should have relatively small masses which do not exceed the kinematical limit $M_{max} = M_R - M_{NN} \sim 0.08$ GeV$/c^2$, where M_R and M_{NN} are the mass of the resonance and the sum of masses of nucleons from this decay, respectively. Therefore this process cannot have to do with the shortage of dielectrons in question. In this paper we examine the e^+e^- pair production mechanism associated with the $d_1^\star$ formation in NN collisions due to virtual photon emission and show that its contribution to the invariant mass spectra of the dielectron from $p+p$, $Ca+Ca$, and $C+C$ collisions at 1.0 GeV/A can supply the observed shortages of e^+e^- pairs in these spectra in the region $0.2 < M < 0.5$ GeV/c^2.

12th International Conference on Meson-Nucleon Physics and the Structure of the Nucleon (MENU 2010)
AIP Conf. Proc. 1374, 410-413 (2011); doi: 10.1063/1.3647170

KINEMATICS AND MATRIX ELEMENT

The process to be considered is

$$N(p_1)+N(p_2)\rightarrow\gamma^{\star}(k)+d_1^{\star}\rightarrow e^{+}(p_3)+e^{-}(p_4)+d_1^{\star}(p_5), \tag{1}$$

with $p_1+p_2=p_3+p_4+p_5$, where p_1 and p_2 are the four-momenta of the colliding nucleons, p_3, p_4, p_5 are the four-momenta of the electron, the positron and the dibaryon, respectively, and $k=p_3+p_4$ is the four-momentum of the virtual photon. The kinematical region for its mass M is $2m_e\leq M\leq M_{max}=\sqrt{s}-M_R$, where m_e is the electron mass and $\sqrt{s}$ is the total CMS energy of colliding nucleons. This mass region overlaps with that where the shortage of dielectrons of interest was observed. For example, in the case of pp collisions at a kinetic energy of the incident protons $T_k=1.04$ GeV the upper limit for the dielectron mass is $M_{max}\simeq 0.38$ GeV.

The matrix element $\mathcal{M}$ for the process (1) can be written as

$$\mathcal{M}=\frac{e^2}{k^2}j^{\mu}J_{\mu}, \tag{2}$$

where $J_{\mu}=(J_0,\vec{J})=<d_1^{\star}|\hat{J}_{\mu}|NN>$ and $j_{\mu}=(j_0,\vec{j})=<e^+e^-|\hat{j}_{\mu}|0>$ are the hadronic and leptonic electromagnetic transition currents, respectively, and $\hat{J}_{\mu}$ and $\hat{j}_{\mu}$ are the operators of these currents. Note that the same current J_{μ} also describes the process $N+N\rightarrow\gamma+d_1^{\star}$, where γ is a real photon. In this case, however, the space part $\vec{J}$ of this current is always perpendicular to the momentum $\vec{k}$ of the photon. The space part $\vec{J}$ of the current J_{μ} for the process (1) can, in general, have both a transverse $\vec{J}_T$ and a longitudinal $\vec{J}_L$ ($\vec{J}_L||\vec{k}$) component, so that $\vec{J}=\vec{J}_L+\vec{J}_T$.

The square of the matrix element (2) can be represented in terms of a transverse and a longitudinal components of $\vec{J}$ as[11]:

$$|\mathcal{M}|^2=\frac{1}{4}\frac{e^4}{M^4}\frac{1}{2m_e{}^2}(M^2|\vec{J}_T|^2-|\vec{J}_T\vec{q}_T|^2+\frac{M^2}{k_0^2}(1-\frac{M^2}{k_0^2}|\vec{q}_L\vec{J}_L|^2-2\frac{M^2}{k_0^2}|\vec{q}_L\vec{J}_L||\vec{q}_T\vec{J}_T|, \tag{3}$$

where $\vec{q}_L$ and $\vec{q}_T$ are the longitudinal and transverse components of the momentum $\vec{q}=\vec{p}_3-\vec{p}_4$, respectively.

The quantities $\vec{J}_T$ and $\vec{J}_L$ in Eq. (3) depend on the quantum numbers of the $|d_1^{\star}>$ and its electromagnetic structure. We assume in this work that the d_1^{*} has spin zero and isospin two. The general structure of the electromagnetic transition from an initial $|pp>$ state to a final pointlike $|d_1^{\star}>$ state with such quantum numbers has been investigated in the work [12] and found to be of the magnetic type. The square of the matrix element for the magnetic dipole ($M1$) transition $p+p\rightarrow\gamma+d_1^{\star}$ derived in the [12] is given by

$$|\mathcal{M}(pp\rightarrow\gamma d_1^{\star})|^2=e^2|\vec{J}_T|^2=Ce^2\cdot(p_1\cdot k)(p_2\cdot k), \tag{4}$$

where p_1 and p_2 are the four-momenta of the colliding nucleons, and C is a certain normalization constant.

The fundamental property of an electromagnetic current for the magnetic type transition with a real or virtual photon is that its components satisfy the conditions $J_0=0$ and $\vec{J}\perp\vec{k}$. The square of the matrix element (3) for such a transition between an initial state $|NN>$ and a final state $|d_1^{\star}>$ takes the form

$$|\mathcal{M}|^2=\frac{1}{4}\frac{e^2}{M^2}\frac{1}{2m_e{}^2}\left(1-\frac{1}{2}\frac{|\vec{q}_T|^2}{M^2}\right)e^2|\vec{J}_T|^2 \tag{5}$$

SPECTRA CALCULATIONS AND COMPARISON WITH DLS DATA

The differential cross section $d\sigma/dM$ for the production of a dielectron of invariant mass M in the process (1) is given by

$$\frac{d\sigma}{dM}=\frac{2\pi^4}{4f}\int|\mathcal{M}|^2|F(M^2)|^2\prod_{i=3}^{5}\frac{d^3\vec{p}_i}{2E_i2\pi^3}\delta^4(p_1+p_2-\sum_{i=3}^{5}p_i)\delta(M-M(p_3,p_4)), \tag{6}$$

where $f=\sqrt{(p_1p_2)^2-m_1^2m_2^2}$, m_1 and m_2 are the masses of the colliding nucleons and $F(M^2)$ is the form factor for the $|NN>\rightarrow|d_1^{\star}>$ transition. In our calculation we used the form factor predicted by the vector dominance model:

$$F(M^2)=\frac{m_{\rho}{}^4+m_{\rho}{}^2\Gamma_{\rho}{}^2}{(m_{\rho}{}^2-M^2)^2+m_{\rho}{}^2\Gamma_{\rho}{}^2}, \tag{7}$$

where m_ρ and Γ_ρ are the mass and width of the ρ meson, respectively.

The integral (6) was calculated by the Monte Carlo method. The phase space region of the integration was limited by the *DLS* apparatus acceptance that was given by the corresponding filter provided by the *DLS collaboration*. In calculations the finite mass resolution of the *DLS* spectrometer $\sigma_M/M = 0.1$ was taken into account. In the case of $C+C$ collisions the calculation was done within the quasi-free NN collision approximation. To allow for the Fermi motion of nucleons in a ^{12}C nucleus the nucleon momentum distribution obtained in the work [13] was used.

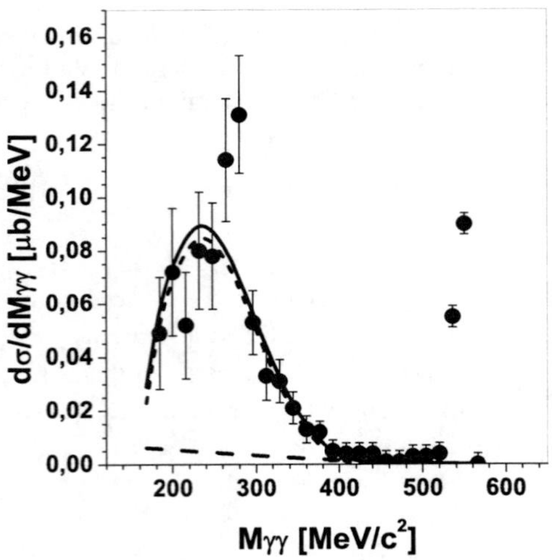

FIGURE 1. Experimentally observed two-photon invariant mass spectrum of the reaction $pp\gamma\gamma$ (full circles)[15] and calculated contributions to this spectrum at 1.36 GeV as a function of $\gamma\gamma$ mass $M_{\gamma\gamma}$. The short dashed line is the contribution from the dibaryon mechanism. The dashed line is the contribution from the $\gamma\gamma$ background. The solid line is the sum of the contributions from the dibaryon mechanism and the background.

The calculated spectrum was normalized in such a way that the ratio of the value of the calculated total cross section $\sigma_{ee}^{tot}(s)$ for the process (1) at the CMS energy $\sqrt{s}$ of the colliding nucleons to the total cross section $\sigma_\gamma^{tot}(s)$ for the process $NN \to \gamma d_1^\star$ is equal to the conversion coefficient R given by [14]

$$R = \frac{\sigma_{ee}^{tot}(s)}{\sigma_\gamma^{tot}(s)} = \frac{\alpha}{\pi}\left[\frac{2}{3}\ln\left(\frac{\sqrt{s}-M_R}{m_e}\right) - \frac{2}{9} + O\left(\left(\frac{m_e}{\sqrt{s}-M_R}\right)^2\right)\right]. \tag{8}$$

Using Eq. (4), for the total cross section $\sigma_\gamma^{tot}(s)$ we obtain

$$\sigma_\gamma^{tot}(s) = \frac{\omega}{64\pi s p}\frac{1}{4}\int|\mathcal{M}(pp\to\gamma d_1^\star)|^2 d\Omega = \frac{C\omega^3}{64\pi s p}[E^2 - p^2/3], \tag{9}$$

where $\omega = (s - M_R^2)/2\sqrt{s}$, p is the CMS momentum of the colliding protons, $E = \sqrt{p^2+m_p^2}$ and m_p is the proton mass. The constant C entering into Eq. (9) was found from the equality $\sigma_\gamma^{tot}(s) = \sigma_\gamma^{tot}(s_0)$ where s_0 is the CMS energy of the pp system for the kinetic energy of incident proton $T_k = 1.36$. GeV. The value $\sigma_\gamma^{tot}(s_0)$ was derived from data on the measurement of the two-photon ($\gamma\gamma$) invariant mass distribution of the $pp \to pp\gamma\gamma$ reaction at $T_k = 1.36$ GeV made by the *CELSIUS-WASA Collaboration* [15]. The surprising feature of this distribution is a pronounced resonant structure at a mass of about 280 MeV. It was shown in our work [10] that this structure is due to the dibaryon mechanism $pp \to \gamma d_1^\star \to pp\gamma\gamma$ of the studied reaction. The contribution of this mechanism to the $\gamma\gamma$ invariant mass distribution of the reaction $pp \to pp\gamma\gamma$ together with the experimentally measured distribution are shown in Fig.2. Here we also show the $\gamma\gamma$ background calculated under the assumption that its differential cross section $d\sigma_{back}/dM_{\gamma\gamma}$ is inversely proportional to the $\gamma\gamma$ invariant mass $M_{\gamma\gamma}$. A fit of the sum of the contributions from the dibaryon mechanism and the background to the experimental $\gamma\gamma$ invariant mass distribution gives for the cross section $\sigma_\gamma^{tot}(s_0) = 11\mu$b. The total cross section of the background in the mass region of the observed structure was estimated to be 1 μb.

The dielectron invariant mass spectra predicted by theoretical models that take into account only the conventional processes of dielectron production (solid lines) for the case of $p+p$ [5] (left panel) and $C+C$ [4] (right panel) collisions at 1.04 GeV compared to the corresponding *DLS* data (solid circles) [2] and [1], respectively are presented in Fig. 2. The short dashed lines show the dibaryon mechanism contributions calculated by Eq. 6 that includes the form factor (7), and dotted lines show such contributions calculated by Eq. 6 without it. The dashed and dot-dashed lines represent the sums of the corresponding spectra predicted by the theoretical models and the dibaryon mechanism contributions calculated with and without the form factor (7), respectively.

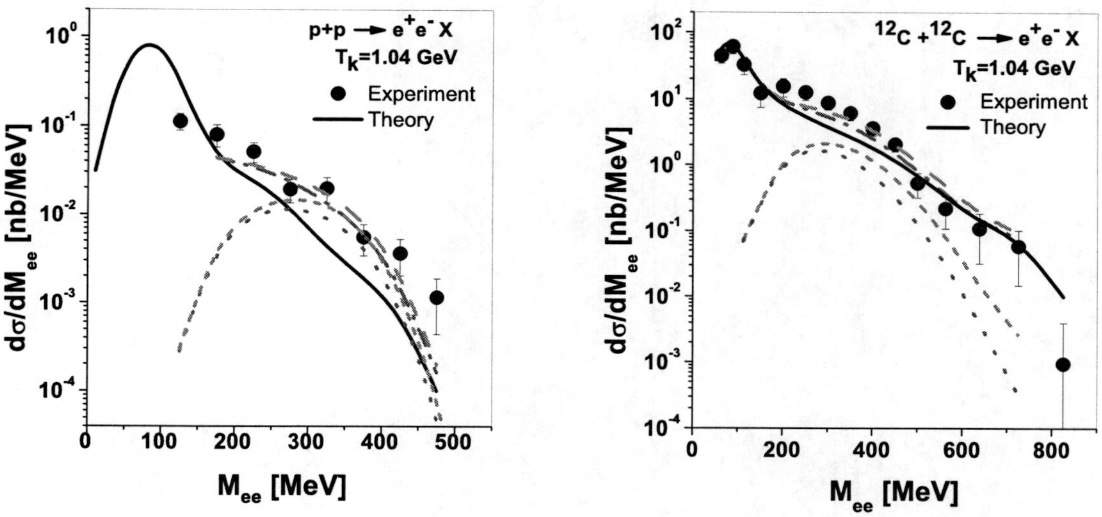

FIGURE 2. The dielectron spectra for $p+p$ (left panel)[2] and $C+C$ (right panel) [1] collisions at 1.04 GeV, as a function of the dielectron mass M_{ee} measured by the *DLS* compared to those predicted by the theoretical models that take into account only the conventional processes of dielectron production (solid lines) for the $p+p$ [5] and for $C+C$ [4] collisions, respectively. The short dashed and dotted lines correspond to the contributions of the dibaryon mechanism calculated by Eq. 6 that includes the form factor (7) and does not include it, respectively. The dashed and dot-dashed lines show the sums of the corresponding spectra predicted by theoretical models and the dibaryon mechanism contributions with the form factor (7) and without this form factor.

Fig. 2 shows that adding the dibaryon mechanism contributions to the dielectron invariant mass spectra predicted by the theoretical models that take into account only the conventional processes of dielectron production we get a very good agreement between resulting spectra and experimental data. Note that the reliability of the numerical values of the calculated contributions of the dibaryon mechanism of dielectron production $NN \to e^+e^-d_1^*$ is provided for by the fact that its total cross section was normalized to the experimentally measured total cross section of the process $pp \to \gamma d_1^*$. Similar results were obtained for the case of $Ca+Ca$ collisions at 1.04 GeV/c.

The presence or absence of the dibaryon mechanism of e^+e^- pair production can be established by measurement of the missing mass spectrum in the reaction $pp \to \gamma^* X \to e^+e^-X$. If this mechanism really exists it should manifest itself in such a spectrum as a narrow peak at a mass around 1956 MeV superimposed on a smooth background.

REFERENCES

1. R. J. Porter et al., DLS Collaboration, Phys. Rev. Lett. **79**, 1229 (1997).
2. W. K. Wilson et al.,Phys. Rev. **C57**, 1865(1998).
3. C. Ernst, S. A. Baas, M. Belkacem, H. Stöcker, and W. Greiner, Phys. Rev. C **58**,447 (1998).
4. E.L.Bratkovskaya, W. Cassing, R. Rapp and J. Wambach, Nucl. Phys. **A634**, 168 (1998).
5. A. Faessler, C. Fuchs, M. I. Krivoruchenko and B. V. Martemyanov, J. Phys. G **29**, 603 (2003).
6. G. Agakishev et al., HADES Collaboration, Phys. Lett. **B663**,43(2008), Phys. Rev. Lett. **98**,052302(2007).
7. L.P. Kaptari and B. Kampfer, Nucl. Phys. **A764**, 338 (2006); R. Shyam and U. Mosel, Phys. Rev. **C79**, 035203 (2009).
8. A. S. Khrykin et al., Phys. Rev. **C64**, 034002 (2001)
9. A. S. Khrykin, Nucl. Phys. **A721**, 625c (2003)
10. A.S.Khrykin and S.B.Gerasimov, in: *Proceeding of the 11th International Conference on Meson-Nucleon Physics and the Structure of the Nucleon,* IKP, Forschungentrum Juelich, Germany, September 10-14, 2007, edited by H. Machner and S. Krewald, eConf C070910(2008),250.
11. D. Van Neck, A.E.L. Dieperink, O. Scholton, Nucl. Phys. **A574**, 643 (1994).
12. O. Scholton and Ojwang J. G. O., Phys. Rev. **C71**, 034005 (2005).
13. O. Benhar, C. Ciofi Degli Atti and G. Salmè, Phys. Lett. B**177**, 135(1986).
14. B.E. Lautrup and J. Smith, Phys. Rev. **D3**, 1122 (1970).
15. M. Bashkanov, H. Clement, E. Doroshkevich, M. Kaskulov, R. Meier, T. Skorodko and G.J. Wagner, Int. Jour of Mod. Phys. **A20**, 554(2005);hep-ex/0406081.

Session K: Nonperturbative Models

m_π and N_c dependence of resonances form Unitarized Chiral Perturbation Theory

G. Ríos*, C. Hanhart† and J. R. Peláez*

*Departamento de Física Teórica II. Universidad Complutense de Madrid.
†Institut für Kernphysik and Jülich Center for Hadron Physics, Forschungzentrum Jülich GmbH.

Abstract. We review our studies on the ρ and σ resonances properties derived from the Inverse Amplitude Method. In particular, we study the leading $1/N_c$ behavior of the resonances masses and widths and their evolution with changing m_π. The $1/N_c$ expansion gives a clear definition of $\bar{q}q$ states, which is neatly satisfied by the ρ but not by the σ, showing that its *dominant component* is not $\bar{q}q$. The m_π dependence of the resonance properties is relevant to connect with lattice studies. We show that our predictions compare well with some lattice results and we find that the $\rho\pi\pi$ coupling constant is m_π independent, in contrast with the $\sigma\pi\pi$ coupling, that shows a strong m_π dependence.

Keywords: Scalar mesons, chiral lagrangians, $1/N_c$ expansion
PACS: 14.40.Cs, 12.39.Fe, 13.75.Lb,11.15.Pg

Light hadron spectroscopy lies beyond the realm of perturbative QCD. At low energies, however, one can use the QCD low energy effective theory, named Chiral Perturbation Theory (ChPT) [1], to describe the dynamics of the lightest mesons. ChPT describes the interactions of the pseudo Nambu-Goldstone bosons of the QCD chiral symmetry breaking, namely, the pions, by means of a effective lagrangian compatible with all QCD symmetries involving only the pion field. The infinite tower of terms in this lagrangian is organized as a low energy expansion in powers of p^2/Λ_χ^2, where p stands either for derivatives, momenta or masses, and $\Lambda_\chi \simeq 4\pi f_\pi$, where f_π denotes the pion decay constant. ChPT is renormalized order by order by absorbing loop divergences in the renormalization of higher order parameters, known as low energy constants (LECs), that parametrize the high energy QCD dynamics and *carry no energy or mass dependence*. They depend on a regularization scale μ but after renormalization the observables are independent of this scale. The value of the LECs depend on the underlying QCD dynamics and are determined from experiment. Up to the desired order, the ChPT expansion provides a *systematic and model independent* description of how observables depend on some QCD parameters like the light quark mass $\hat{m} = (m_u + m_d/2)$ or the number of colors, N_c [2].

The use of ChPT is limited to low energies and masses, nevertheless, combined with dispersion relations and elastic unitarity it leads to a successful description of meson dynamics up to energies around 1 GeV, generating resonant states not originally present in the lagrangian, without any a priori assumption on their existence or nature. In particular, we find the ρ and σ resonances as poles on the second Riemann sheet of $\pi\pi$ elastic scattering amplitudes. With this approach we can then study some of these resonances properties, like their spectroscopic nature through their mass and width dependence on N_c, or their dependence on the pion mass in order to connect with lattice studies. In the following sections we review this "unitarized ChPT" approach, named the Inverse Amplitude Method (IAM) [3, 4, 5], and then apply it to study the leading $1/N_C$ behavior and the chiral extrapolation of the ρ and σ mesons.

The ρ and σ resonances appear as poles on the second Riemann sheet of the $(I,J) = (1,1)$ and $(I,J) = (0,0)$ $\pi\pi$ scattering partial waves of definite isospin, I and angular momentum J, respectively. Elastic unitarity implies for these partial waves, $t(s)$, and physical values of s below inelastic thresholds, that

$$\mathrm{Im}\, t(s) = \sigma(s)|t(s)|^2 \;\Rightarrow\; \mathrm{Im}\,\frac{1}{t(s)} = -\sigma(s), \qquad \text{with} \quad \sigma(s) = 2p/\sqrt{s}, \tag{1}$$

where s is the Mandelstam variable and p is the center of mass momentum. Consequently, the imaginary part of the inverse amplitude is known exactly. However, ChPT amplitudes, being an expansion $t \simeq t_2 + t_4 + \cdots$, with $t_k = O(p^k)$, can only satisfy Eq. (1) perturbatively

$$\mathrm{Im}\, t_2(s) = 0, \quad \mathrm{Im}\, t_4(s) = \sigma(s) t_2^2(s) \;\ldots \tag{2}$$

and cannot generate poles. Therefore the resonance region lies beyond the reach of standard ChPT. This region however, can be reached combining ChPT with dispersion theory through the IAM [3, 4, 5].

12th International Conference on Meson-Nucleon Physics and the Structure of the Nucleon (MENU 2010)
AIP Conf. Proc. 1374, 417-420 (2011); doi: 10.1063/1.3647171

The analytic structure of the $\pi\pi$ scattering amplitude $t(s)$, consisting on a right cut extending from $s_{th} = 4m_\pi^2$ to ∞, and a left cut from $-\infty$ to 0, allows to write a dispersion relation for the auxiliary function $G(s) \equiv t_2^2(s)/t(s)$

$$G(s) = G(0) + G'(0)s + \tfrac{1}{2}G''(0)s^2 + \frac{s^3}{\pi}\int_{s_{th}}^{\infty} ds' \frac{\operatorname{Im} G(s')}{s'^3(s'-s-i\varepsilon)} + LC(G) + PC, \tag{3}$$

where the integral over the left cut has been abbreviated as $LC(G)$ and PC stands from possible pole contributions corresponding zeros of t. The different terms in Eq.(3) can be evaluated using unitarity and ChPT as follows: The right cut can be *exactly* evaluated taking into account the elastic unitarity conditions Eqs. (1), (2), $\operatorname{Im} G(s') = -\sigma(s')t_2^2(s') = -\operatorname{Im} t_4(s')$, for $s' \in (4m_\pi^2, \infty)$. The subtraction constants only involve the amplitude and its derivatives evaluated at $s = 0$, so they can be safely approximated with ChPT: $G(0) \simeq t_2(0) - t_4(0)$, $G'(0) \simeq t_2'(0) - t_4'(0)$, $G''(0) \simeq -t_4''(0)$. The left cut, which is suppresed by $1/s'^3(s'-s)$, is *weighted at low energies*, so it is appropriate to approximate it with ChPT: $LC(G) \simeq -LC(t_4)$. The pole contribution only appears in the scalar wave, which vanishes at the so called Adler zero. It counts $O(p^6)$ and and has been calculated explicitly [6] and is not just formally suppressed, but numerically negligible except near the Adler zero, away from the physical region.

Neglecting PC for the moment, and taking into account that $t_2(s)$ is just a first order polynomial in s, and that a dispersion relation can be also written for t_4, we can write Eq.(3) as

$$G(s) \equiv \frac{t_2^2(s)}{t(s)} \simeq t_2(0) - t_2'(0)s - t_4(0) - t_4'(0)s - \tfrac{1}{2}t_4''(0)s^2 - \frac{s^3}{\pi}\int_{s_{th}}^{\infty} ds' \frac{\operatorname{Im} t_4(s')}{s'^3(s'-s-i\varepsilon)} - LC(t_4) = t_2(s) - t_4(s), \tag{4}$$

which immediately leads to the IAM formula $t^{IAM}(s) = \frac{t_2^2(s)}{t_2(s)-t_4(s)}$. The IAM formula satisfies exact elastic unitarity and, when reexpanded at low energies, reproduces the ChPT expansion up to the order used to approximate the subtraction constants and the left cut. Here we have presented an $O(p^4)$ IAM but it can be generalized to higher chiral orders. Note that in the IAM derivation ChPT has been always used at low energies, to evaluate parts of a dispersion relation whose elastic unitarity cut has been taken into account exactly. Thus, there are no additional model dependencies in the approach, which is reliable up to energies where inelasticities become important. Taking the pole contribution into account leads to a modified IAM formula [6] which is almost indistinguishable from the ordinary one except in the Adler zero region, where it fixes some problems of the ordinary IAM with the Adler zero [6]. Actually, this modified IAM formula is the one used in this work since, as it will be shown below, one amplitude pole gets near the Adler zero region.

This simple IAM formula is able to reproduce $\pi\pi$ scattering phase shift data up to roughly 1 GeV and generates the poles associated to the ρ and σ resonances with values of the LECs compatible with standard ChPT [5]. The $1/N_c$ expansion is implemented in ChPT through the LECs, whose leading $1/N_c$ scaling is known from QCD. Also, the quark mass dependence implemented in the IAM agrees with that of ChPT up to the order used. Hence, it is straightforward to study the leading $1/N_C$ behavior and the $\hat{m}$ dependence of the resonances generated with the IAM, which we proceed to expose in the following sections.

NATURE OF RESONANCES FROM THEIR LEADING $1/N_c$ BEHAVIOR

The QCD $1/N_c$ expansion [2] provides a clear definition of $\bar{q}q$ bound states: their masses and widths scale as $O(1)$ and $O(1/N_c)$ respectively. The QCD leading $1/N_c$ behavior of the ChPT parameters (f_π, m_π and the LECs) is well known. Hence, by scaling with N_c the ChPT parameters in the IAM, the N_c dependence of the ρ and σ mesons mass and width has been determined [7, 8]. They are defined from the pole positions as $\sqrt{s_{pole}} = M - i\Gamma$. Note that we should not take too large N_C values, since the $N_c \to \infty$ is a weakly interacting limit, where the IAM approach is less reliable [9]. Also, for very large N_c values, even a tiny admixture of $\bar{q}q$ in the physical state would become dominant, but this does not give any information about the dominant component of the $N_c = 3$ physical state.

Fig. 1 (left) shows the ρ and σ mass and width N_c scaling. It can be clearly seen that the ρ follows remarkably well the expected behavior of a $\bar{q}q$ state, confirming that the method can obtain the correct N_c behavior of well known $\bar{q}q$ states. In contrast,the σ does not that $\bar{q}q$ pattern, allowing us to conclude that *the σ dominant component is not $\bar{q}q$*.

Loop contributions play an important role in determining the σ pole position. Since they are $1/N_c$ suppressed compared to tree level terms, it may happen that for larger N_c values they become comparable to tree level $O(p^6)$ contributions, which are subdominant in the chiral expansion, but not N_c suppressed. Thus we checked the $O(p^4)$ results with an $O(p^6)$ IAM calculation [8]. We defined a χ^2-like function to measure how close a resonance is from a

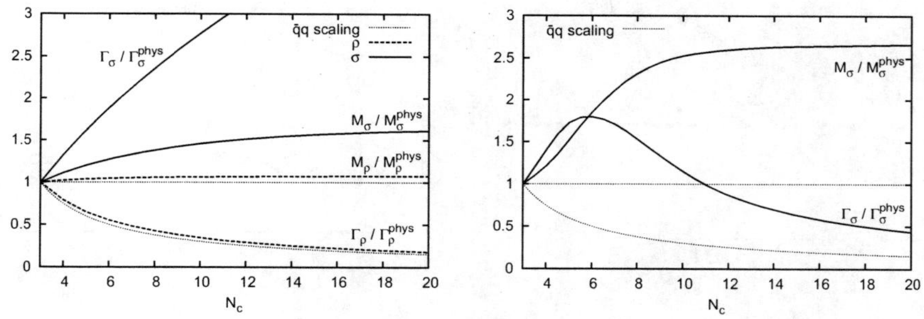

FIGURE 1. **Left:** ρ and σ $1/N_c$ scaling $O(p^4)$. **Right:** σ $1/N_c$ scaling $O(p^6)$.

$\bar{q}q$ behavior. First, we used it at $O(p^4)$ to show that it is not possible for the σ to behave predominantly as a $\bar{q}q$ state describing simultaneously the data and the ρ $\bar{q}q$ behavior. Next, we obtained an $O(p^6)$ data fit where the ρ $\bar{q}q$ behavior was imposed. Figure 1 (right) shows the M_σ and Γ_σ N_c scaling obtained from that fit. Note that both M_σ and Γ_σ grow *near* $N_c = 3$, confirming the $O(p^4)$ result of a non $\bar{q}q$ dominant component. However, for N_c between 8 and 15, where we still trust the IAM, M_σ becomes constant and Γ_σ starts decreasing. This may hint to a *subdominant* $\bar{q}q$ *component*, arising as loops become suppressed as N_c grows. Finally, by forcing the σ to behave as a $\bar{q}q$, we found that in the best case this subdominant component could become dominant around $N_c > 6-8$, but always with an $N_c \to \infty$ mass above 1 GeV instead of its physical ~ 450 MeV value. This supports the emerging picture of two low energy scalar nonets, one of exotic nature below 1 GeV and another of ordinary $\bar{q}q$ nature above 1 GeV.

CHIRAL EXTRAPOLATION OF THE ρ AND σ RESONANCES

ChPT also provides an expansion of m_π in terms of $\hat{m}$ (at leading order $m_\pi^2 \sim \hat{m}$). Thus, by changing m_π in the amplitudes we see how the IAM poles depend on $\hat{m}$. We report here our analysis of the ρ and σ properties dependence on m_π [10].

The values of m_π considered should fall within the ChPT applicability range and allow for some elastic regime below $K\bar{K}$, that would almost disappear if $m_\pi > 500$, which would be the most optimistic applicability range. We expect higher order corrections to be more relevant as m_π increases. Thus, our results become less reliable as m_π grows.

Fig. 2 (top left) shows the evolution of the σ and ρ pole positions as m_π is increased. In order to see the pole movements relative to the $\pi\pi$ threshold, which is also increasing, we use units of m_π, so the threshold is fixed at $\sqrt{s} = 2$. Both poles move closer to threshold and they approach the real axis. The ρ poles reach the real axis at the same time that they cross threshold. One of them jumps into the first sheet and becomes a bound state, while its conjugate partner remains on the second sheet practically at the very same position as that in the first. In contrast, the σ poles go below threshold with a finite imaginary part before they meet in the real axis, still on the second sheet, becoming virtual states. As m_π increases, one pole moves toward threshold and jumps through the branch point to the first sheet staying in the real axis below threshold, very close to it as m_π keeps growing. The other σ pole moves down in energies away from threshold and remains on the second sheet. Similar movements were found within quark models [12] and a finite density analysis [13].

Fig. 2 (top right) shows our results for the ρ mass dependence on m_π compared with some lattice results [14], and the PDG value for the ρ mass. Now M_ρ is defined as the point where the phase shift crosses $\pi/2$, except for those m_π values where the ρ becomes a bound state, where it is defined from the pole position. In view of the incompatibilities between different lattice collaborations, we find a qualitative good agreement with lattice results. The M_ρ dependence on m_π agrees also with estimations for the two first coefficients of its chiral expansion [15].

In Fig. 2 (bottom left) we compare the m_π dependence of M_ρ and M_σ, normalized to their physical values. The bands cover the LECs uncertainties. Both masses grow with m_π, but M_σ grows faster than M_ρ. Above $2.4\, m_\pi^{\text{phys}}$, we show two bands since the two σ poles lie on the real axis with two different masses.

In the bottom center panel of Fig. 2 we compare the m_π dependence of Γ_ρ and Γ_σ normalized to their physical values: note that both widths become smaller. We compare this decrease with the expected phase space reduction as

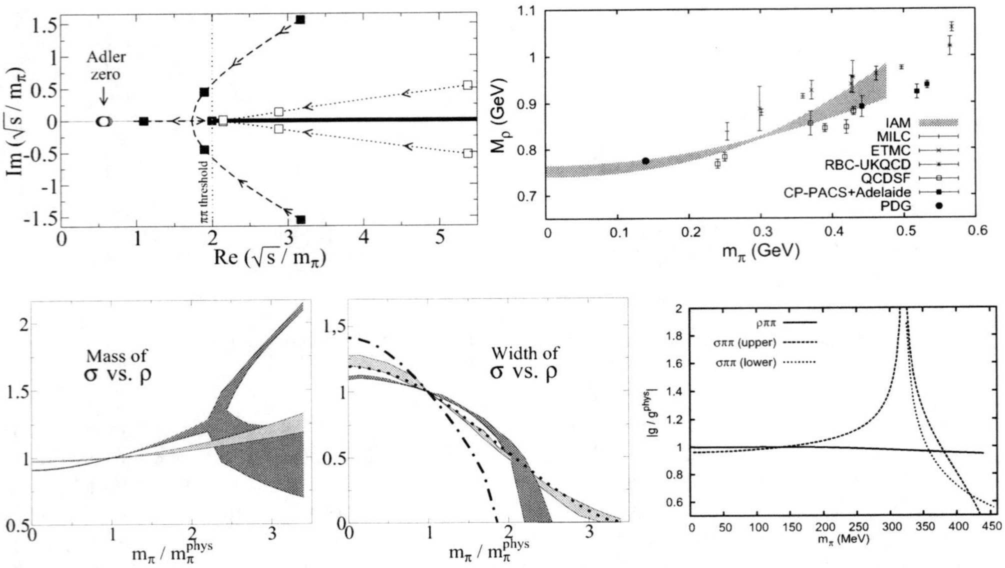

FIGURE 2. **Top Left:** Movement of the σ (dashed lines) and ρ (dotted lines) poles for increasing m_π (direction indicated by the arrows) on the second sheet. The filled (open) boxes denote the pole positions for the σ (ρ) at pion masses $m_\pi = 1$, 2, and $3 \times m_\pi^{phys}$, respectively. **Top Right:** Comparison the IAM M_ρ dependence on m_π with some recent lattice results[14]. **Bottom Left:** Comparison of the ρ (light) and σ (dark) mass dependence on m_π. **Bottom Center:** Comparison of the ρ (light) and σ (dark) width dependence on m_π. The dotted (ρ) and dot-dashed (σ) lines show the decrease due to only phase space assuming a constant coupling to $\pi\pi$. **Bottom Right:** ρ and σ couplings calculated from the pole residue. In all panels, the bands cover the LECs uncertainty.

resonances approach the $\pi\pi$ threshold. We find that Γ_ρ follows very well this expected behavior, which implies that the $\rho\pi\pi$ coupling is almost m_π independent. In contrast, Γ_σ deviates from the phase space reduction expectation. This suggests a strong m_π dependence of the σ coupling to two pions, which we confirm with a explicit calculation of the resonances couplings from the pole residues as shown in the bottom left panel.

REFERENCES

1. S. Weinberg, Physica **A96** (1979) 327. J. Gasser and H. Leutwyler, Annals Phys. **158** (1984) 142;
2. G. 't Hooft, Nucl. Phys. B **72**, 461 (1974). E. Witten, Annals Phys. **128**, 363 (1980).
3. T. N. Truong, Phys. Rev. Lett. **61** (1988) 2526. Phys. Rev. Lett. **67**, (1991) 2260; A. Dobado et al., Phys. Lett. **B235** (1990) 134.
4. A. Dobado and J. R. Peláez, Phys. Rev. D **47** (1993) 4883; Phys. Rev. D **56** (1997) 3057.
5. F. Guerrero and J. A. Oller, Nucl. Phys. B **537** (1999) 459 [Erratum-ibid. B **602** (2001) 641]. J. R. Peláez, Mod. Phys. Lett. A **19**, 2879 (2004) A. Gómez Nicola and J. R. Peláez, Phys. Rev. D **65** (2002) 054009 and AIP Conf. Proc. **660** (2003) 102.
6. A. Gómez Nicola, J.R. Peláez and G. Ríos, Phys. Rev. D **77** (2008) 056006.
7. J. R. Pelaez, Phys. Rev. Lett. **92**, 102001 (2004)
8. J. R. Pelaez and G. Rios, Phys. Rev. Lett. **97**, 242002 (2006)
9. J. R. Pelaez and G. Rios, arXiv:0905.4689 [hep-ph].
10. C. Hanhart, J. R. Pelaez and G. Rios, Phys. Rev. Lett. **100**, 152001 (2008)
11. D. Morgan, Nucl. Phys. A **543** (1992) 632; D. Morgan and M. R. Pennington, Phys. Rev. D **48** (1993) 1185.
12. E. van Beveren et al., AIP Conf. Proc. **660**, 353 (2003); Phys. Rev. D **74**, 037501 (2006).
13. D. Fernandez-Fraile, A. Gomez Nicola and E. T. Herruzo, Phys. Rev. D **76**, 085020 (2007)
14. Ph. Boucaud *et al.* [ETM Collaboration], Phys. Lett. B **650**, 304 (2007) C. Allton *et al.* [RBC and UKQCD Collaborations], Phys. Rev. D **76**, 014504 (2007) C. W. Bernard *et al.*,Phys. Rev. D **64**, 054506 (2001) C. R. Allton *et al.*Phys. Lett. B **628**, 125 (2005) M. Gockeler *et al.*[QCDSF Collaboration], [arXiv:hep-lat/0810.5337].
15. P. C. Bruns and U.-G. Meißner, Eur. Phys. J. C **40** (2005) 97.

Light and heavy at the end of the funnel

Eef van Beveren* and George Rupp†

*Centro de Física Computacional, Departamento de Física, Universidade de Coimbra, P-3004-516 Coimbra, Portugal
†Centro de Física das Interacções Fundamentais, Instituto Superior Técnico, Universidade Técnica de Lisboa, Edifício Ciência, P-1049-001 Lisboa, Portugal

Abstract. We show that, by taking a bare mass spectrum with constant spacings for the quark-antiquark propagators, which is subject to considerable mass shifts from meson loops, one adequately describes a large variety of mesonic resonances, from the light scalars to the $b\bar{b}$ states. All our results indicate that a harmonic-oscilator spectrum with universal frequency, in combination with coupled-channel effects, does a much better job than the $q\bar{q}$ spectrum of the funnel potential.

Keywords: Meson spectra, harmonic-oscillator potential, equal level splittings, unitarization through meson loops, threshold effects
PACS: 14.40.-n, 12.39.Pn, 11.80.Gw, 11.55.Ds

INTRODUCTION

For over three decades now, it has been widely assumed that the spectra of quarkonia can be described by a kind of *universal, funnel-type* potential [1], which basically stems from the naive picture of color-flux-tube formation at large interquark separations, while one-gluon exchange should represent the interactions at shorter distances. The Cornell potential [2] was the first and simplest version of such a potential. However, several mesonic resonances observed long ago, as well as some recently discovered ones, contradict such a description. In particular, the $\rho(1250\text{–}1300)$ (see e.g. Refs. [3–16]) does not at all fit in the $J^{PC} = 1^{--}$ isovector spectrum of, for instance, Ref. [17], which employed a semirelativistic version of a funnel-type confining $q\bar{q}$ potential. Moreover, evidence for the new vector charmonium resonances $\psi(5S)(4790)$ and $\psi(4D)(4870)$ [18–20], and also for an $\Upsilon(4S)(10735)$ [21], hints at masses that are again far from those predicted by the confining funnel potential. The same discrepancy between predictions of the funnel potential and experiment is observed e.g. for the firmly established $K^*(1410)$ [22], as well as the $f_2(1565)$ [22]. Note that the discrepancies here are not of a few tens of MeV, but rather of the order of 200 MeV!

Furthermore, the funnel potential does not accommodate the light scalar mesons $f_0(600)$, $K_0(800)$, $f_0(980)$, and $a_0(980)$, nor the heavy-light scalar $D_{s0}(2317)$, nor the $D_{s1}(2460)$ and the $D^*_{sJ}(2860)$, at least not without unitarization [23]. But instead of unitarizing one's favorite quark model, it has become fashion to fall back upon tetraquarks, meson molecules, hybrids, or glueballs, whenever a state does not fit the funnel potential [24, 25].

A better alternative to funnel confinement has been suggested three decades ago [26]. Based on the Weyl conformal invariance [27] property of the theory of quark dynamics (QCD), which leads, by a judicious choice of the time parameter, to anti-De Sitter confinement (AdS) [28], one obtains a harmonic-oscillator (HO) spectrum for $q\bar{q}$ systems [29]. The experimentally observed spectrum is reproduced via the inclusion of meson loops [26, 30]. Most remarkably, in the lowest-order approximation to AdS, one obtains an interquark potential consisting of a Coulomb-like term and a linear term [31], exactly as in the case of lattice QCD (LQCD). Nevertheless, full AdS yields an HO-like spectrum. So it seems the lowest-order term of LQCD does not dictate the spectrum of the $q\bar{q}$ propagator, as it does not provide a satisfactory description of how confinement should follow from full QCD.

HARMONIC OSCILLATOR AND MESON LOOPS

Unitarization of the quark model also reveals threshold enhancements in electron-positron annihilation [32], which may explain e.g. why the $\psi(3770)$ consists of two strictures [33], namely a $D\bar{D}$ threshold enhancement, interfering with a $c\bar{c}$ resonance [19]. Furthermore, this phenomenon explains the enhancement at 4.634 GeV observed by the Belle Collaboration in $e^+e^- \to \Lambda_c^+\Lambda_c^-$ [34], not as a resonance, but rather a threshold enhancement [19], or to be more precise, at least three interfering threshold enhancements, namely $\Lambda_c^+\Lambda_c^-$, $D_{s1}(2536)^{\pm}D_s^{*\mp}$, and $D_{s0}(2317)^+D_{s0}(2317)^-$.

12th International Conference on Meson-Nucleon Physics and the Structure of the Nucleon (MENU 2010)
AIP Conf. Proc. 1374, 421-424 (2011); doi: 10.1063/1.3647172

Further threshold enhancements are observed in data published by the BABAR Collaboration [35], though not as peaked structures, but rather as valleys due to depletion of the $e^+e^- \to J/\psi\pi^+\pi^-$ signal at the opening of open-charm thresholds. Moreover, at the positions of $c\bar{c}$ resonances, one also observes dips in the data where peaks should be expected [35], most noticeably at the $\psi(4S)$ resonance (see Fig. 1 of Ref. [36]). Assuming depletion as an explanation for the $e^+e^- \to J/\psi\pi\pi$, $e^+e^- \to J/\psi KK$, and $e^+e^- \to \psi(2S)\pi\pi$ signals, one is led to consider the corresponding new "resonances" $X(4260)$, $X(4660)$, ... leftovers from open-charm decay [36]. Hence, the well-established $J^{PC} = 1^{--}$ $c\bar{c}$ spectrum anno 2010 still only consists of J/ψ, $\psi(2S, 3S, 4S)$, and $\psi(1D, 2D)$.

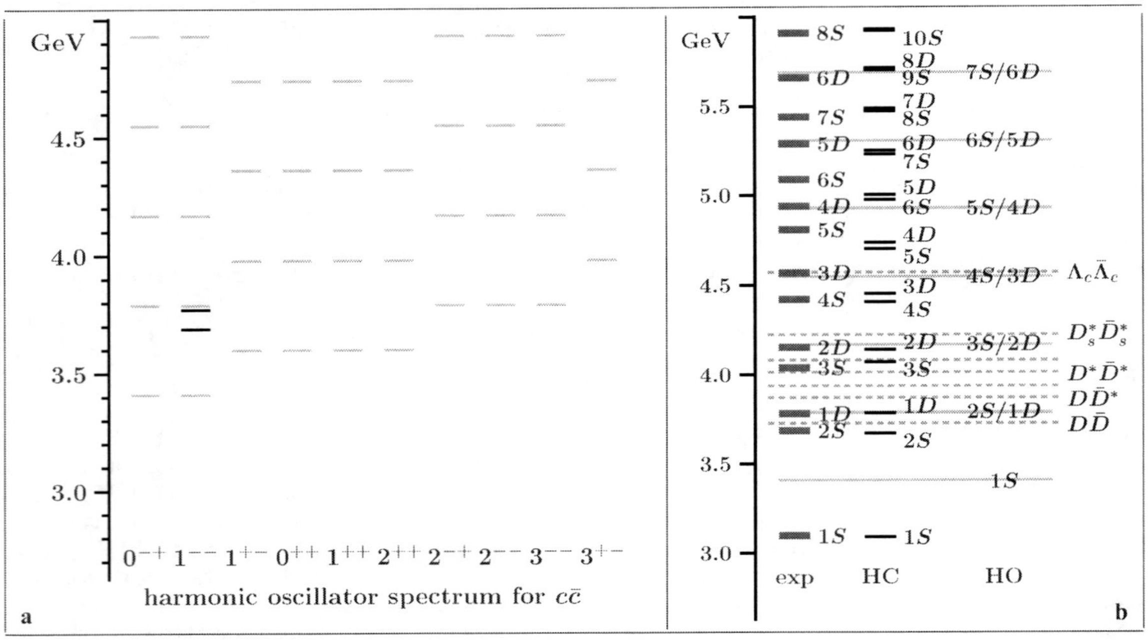

FIGURE 1. **a**: Charmonium level spectrum (▬) as obtained from the HO, for $m_c = 1.562$, $\omega = 0.190$ GeV. The two black dashes indicate where the central masses for the $\psi(2S)$ (lower dash) and the $\psi(1D)$ (upper dash) end up through the effect of meson loops. Similar mass shifts occur for other resonances and bound states. **b**: $J^{PC} = 1^{--}$ $c\bar{c}$ level spectrum observed by us in data from experiment (exp) (▬), as predicted by the funnel-type $c\bar{c}$ potential model of Ref. [37] (HC) (▬), and as predicted by pure HO confinement (HO) (▬). Meson and baryon loops shift the D states a few MeV down/up, whereas the S states shift 100–200 MeV downwards. For completeness, we also indicate the levels of the sharp, low-lying meson-meson and baryon-baryon thresholds (- - -) of the channels $D\bar{D}$, $D\bar{D}^*$, $D_s\bar{D}_s$, $D^*\bar{D}^*$, $D_s\bar{D}_s^*$, $D_s^*\bar{D}_s^*$, and $\Lambda_c^+\Lambda_c^-$.

Harmonic-oscillator confinement, with a radial level splitting of 380 MeV, leads to abundantly many states of the $q\bar{q}$ propagator. This is shown for $c\bar{c}$ in Fig. 1a. For other flavor combinations, the level scheme is exactly the same, just shifted up/down according to the sum of the effective quark masses. In Fig. 1a, we depict the effect of meson loops on the central mass positions of the $\psi(2S)$ and the $\psi(1D)$. In the HO spectrum these two states are degenerate. However, due to the interaction generated by the meson loops, the poles associated with the resonances repel each other in such a way that one of them is subject to a small mass shift, whereas the other shifts considerably and downwards. Higher up in the $c\bar{c}$ spectrum, the mass shift of the lower pole becomes of the order of 150–200 MeV [26]. As a consequence, the associated resonance acquires a central mass which is very similar to that of the P states. Observing and disentangling such states is not exactly an easy task, as the past three decades have shown.

A way out is to study a well-isolated system, with just one set of quantum numbers, like vector $c\bar{c}$, which can be produced in e^+e^- annihilation. However, with a radial level separation of about 380 MeV for the S and D states, combined with expected widths of the order of 50-100 MeV, one needs a lot of good data, binned in intervals of 5 MeV at most. Furthermore, the opening of numerous thresholds complicates such a task even more. As a consequence, studying the lower part of the spectrum is not the most adequate way to unravel the characteristics of quark confinement, since in particular the ground states suffer most from the coupling to open charm, while sharp open-charm thresholds give rise to noticeable enhancements. What one really needs are the higher radial excitations, very carefully distinguished from other effects, like, e.g., threshold enhancements. The latter occur less distinctly at higher energies, because they involve open-charm mesons have larger widths, so that thresholds are smeared out over several tens of MeV. Finally, the phenomenon of threshold enhancements also shows that specific open-charm channels decrease rapidly in amplitude for higher invariant masses [19]. Hence, studying the vector $c\bar{c}$ resonances in open charm leads to an extremely small set of data at higher energies. Nevertheless, although the quality of $e^+e^- \to c\bar{c}$ data is low for massive resonances, it is certainly justified to inspect such data carefully, in order to find signs of further resonances. Unfortunately, in practice lots of precious and, moreover, costly data are simply discarded in many analyses, by assuming an arbitrary background shape (see *e.g.* Fig. 3 of Ref. [38]). Three decades of very expensive accelerator physics without any further results in the $c\bar{c}$ vector spectrum is, to say the least, very embarrasing.

In Refs. [18–20, 36, 39–41] we have pointed at hints for the $\psi(3D)$, $\psi(5S)$, $\psi(4D)$, $\psi(6S)$, and $\psi(5D)$ $c\bar{c}$ resonances in several sets of reasonably good experimental data. Furthermore, we have found the first indications [18] of the $\psi(7S)$, $\psi(6D)$, and $\psi(8S)$ resonances in recent data on $e^+e^- \to D^*\bar{D}^*$, published by the BABAR Collaboration [42]. The resulting tentative spectrum of charmonium is shown in Fig. 1b. It confirms the prediction of the model of Ref. [30], which is based on HO confinement and meson loops, a model denoted by us as HORSE.

In the same figure we have indicated the prediction of the funnel-potential model of Ref. [37]. The data of Fig. 1b appear to contradict, in particular, the spin-orbit splittings as predicted in Ref. [37]. In the latter model, the *S-D* splittings for vector $c\bar{c}$ states become smaller for higher radial excitations, being only about 20 MeV for the *6D-7S* mass difference. From Fig. 1b we estimate this splitting to be roughly five to ten times larger. Now, in the HORSE, *S-D* mass differences are exactly zero at the quenched level, but get generated by meson loops. For the corresponding couplings, the three-meson vertices determined in Ref. [43] are employed, which involve the orbital and spin quantum numbers, not only of the $c\bar{c}$ pair, but also of the mesons in the loops. The resulting *S-D* splittings come out very different then, apart from the fact that the physical vector charmonium resonances naturally appear as mixtures of *S* and *D* states. We find that the combination dominated by the *D* wave shifts at most a few tens of MeVs away from the corresponding bare level. The dominantly *S*-wave combination shifts substantially more, viz. some 100–200 MeV, depending on the precise locations of nearby thresholds. This pattern is, to some extent, systematically repeated for higher radial excitations, which the present data seem to confirm.

For the light-quark spectrum the situation is even more complicated, as nonstrange and strange $q\bar{q}$ combinations have comparable spectra, which will come out on top of each other, besides possibly significant mixing of isoscalar $n\bar{n}$ ($n = u,d$) and $s\bar{s}$ states. To make things worse, decay channels involving kaons are common to both $n\bar{n}$ and $s\bar{s}$ resonances. The only system with a sufficient number of established states to find evidence (see Table 3 of Ref. [44]) for an HO level splitting of 380 MeV is given by the radially excited f_2 mesons. Moreover, the listing of experimental observations and/or published analyses of the data in the Particle Data Group (PDG) [22] tables is not very helpful in the light-meson sector. For example, whereas the HORSE clearly predicts the $\rho(2S)$ central mass at about 1.27 GeV, the $\rho(1250–1300)$ not even has its own entry. Observations of a ρ-type resonance at 1.25–1.3 GeV [8], which was moreover confirmed in a recent multichannel analysis [16], are listed under the $\rho(1450)$, with the "justification":

> Several observations on the $\omega\pi$ system in the 1200-MeV region may be interpreted in terms of either $J^P = 1^-$ $\rho(770) \to \omega\pi$ production, or $J^P = 1^+$ $b_1(1235)$ production. We argue that no special entry for a $\rho(1250)$ is needed. The LASS amplitude analysis showing evidence for $\rho(1270)$ [12] is preliminary and needs confirmation. For completeness, the relevant observations are listed under the $\rho(1450)$.

The resulting confusion is confirmed in Ref. [45], where the authors, using dispersion relations, fail to observe the $\rho(1270)$ resonance in data from Refs. [46–48]. Even more absurd is the lumping of $\phi(1500)$ and $\phi(1900)$ observations under the $\phi(1680)$ [22].

Finally, besides the resonance poles that follow directly from $q\bar{q}$ confinement, meson loops may give also rise to dynamically generated resonances, of which the low-lying scalar mesons $f_0(600)$, $K_0(800)$, $f_0(980)$, and $a_0(980)$ were the first to be discovered [23]. But also the heavy-light scalar $D_{s0}(2317)$ is an example of such a type of resonance. Their existence will further complicate the classification of experimentally observed resonances. However, they can be studied and predicted by using the coupling of HO confinement to the relevant meson-meson and possibly baryon-baryon channels [49].

CONCLUSION

We have argued that the $c\bar{c}$ spectrum provides the best source to infer the confinement mechanism, which then also allows to understand and classify the spectra of other flavor combinations. The available data clearly suggest that HO confinement, complemented with meson loops, is the best candidate.

ACKNOWLEDGMENTS

One of us (EvB) wishes to thank the organizers of *Menu 2010* for their warm hospitality, and for bringing together many different views on the strong interactions. This work was supported by the *Fundação para a Ciência e a Tecnologia* of the *Ministério da Ciência, Tecnologia e Ensino Superior* of Portugal, under contract CERN/FP/109307/2009.

REFERENCES

1. W. Lucha, F. F. Schoberl, and D. Gromes, Phys. Rept. **200**, 127 (1991).
2. E. Eichten, K. Gottfried, T. Kinoshita, J. B. Kogut, K. D. Lane and T. M. Yan, Phys. Rev. Lett. **34**, 369 (1975) [Erratum-ibid. **36**, 1276 (1976)].
3. R. E. Peierls, in Proceedings of the Glasgow conference on Nuclear and Meson physics (Pergamon Press, New York,1954), p 296.
4. J. Ballam *et al.*, Nucl. Phys. B **76**, 375 (1974).
5. S. Bartalucci *et al.*, Nuovo Cim. A **49**, 207 (1979).
6. V. K. Khenner, Yad. Fiz. **31**, 424 (1980).
7. N. M. Budnev and A. I. Orlov, Pisma Zh. Eksp. Teor. Fiz. **32**, 390 (1980).
8. D. Aston, Phys. Lett. B **92**, 211 (1980).
9. S. Dubnička, Lett. Nuovo Cim. **37**, 236 (1983).
10. A. E. Kaloshin, Yad. Fiz. **41**, 1002 (1985).
11. S. Dubnička, I. Furdik and V. A. Meshcheryakov, JINR-E2-88-521 (1988).
12. D. Aston *et al.* [LASS Collaboration], Nucl. Phys. Proc. Suppl. **21**, 105 (1991).
13. D. Aston *et al.* [LASS Collaboration], SLAC-PUB-5657 (1991).
14. D. Aston *et al.* [LASS Collaboration], SLAC-PUB-5606 (1994).
15. Yu. S. Surovtsev and P. Bydžovský, arXiv:0711.4748.
16. Yu. S. Surovtsev and P. Bydžovský, Nucl. Phys. A **807**, 145 (2008) [arXiv:hep-ph/0701274].
17. S. Godfrey and N. Isgur, Phys. Rev. D **32**, 189 (1985).
18. E. van Beveren and G. Rupp, arXiv:1005.3490.
19. E. van Beveren and G. Rupp, Phys. Rev. D **80**, 074001 (2009) [arXiv:0908.0242].
20. E. van Beveren, X. Liu, R. Coimbra, and G. Rupp, Europhys. Lett. **85**, 61002 (2009) [arXiv:0809.1151].
21. E. van Beveren and G. Rupp, arXiv:0910.0967.
22. C. Amsler *et al.* [Particle Data Group Collaboration], Phys. Lett. B **667**, 1 (2008).
23. E. van Beveren, T. A. Rijken, K. Metzger, C. Dullemond, G. Rupp and J. E. Ribeiro, Z. Phys. C **30**, 615 (1986) [arXiv:0710.4067].
24. D. Ebert, R. N. Faustov and V. O. Galkin, Phys. Rev. D **79**, 114029 (2009) [arXiv:0903.5183].
25. D. Ebert, R. N. Faustov and V. O. Galkin, Eur. Phys. J. C **66**, 197 (2010) [arXiv:0910.5612].
26. E. van Beveren, C. Dullemond, and G. Rupp, Phys. Rev. D **21**, 772 (1980) [Erratum-ibid. D **22**, 787 (1980)].
27. H. Weyl, Annalen der Physik **59**, 101 (1919).
28. C. Dullemond, T. A. Rijken and E. van Beveren, Nuovo Cim. A **80**, 401 (1984).
29. E. van Beveren, C. Dullemond and T. A. Rijken, Phys. Rev. D **30**, 1103 (1984).
30. E. van Beveren, G. Rupp, T. A. Rijken, and C. Dullemond, Phys. Rev. D **27**, 1527 (1983).
31. E. van Beveren, T. A. Rijken, C. Dullemond and G. Rupp, Lect. Notes Phys. **211**, 331 (1984).
32. E. van Beveren, and G. Rupp, Ann. Phys. **324**, 1620 (2009) [arXiv:0809.1149].
33. M. Ablikim *et al.* [BES Collaboration], arXiv:0807.0494.
34. G. Pakhlova *et al.* [Belle Collaboration], Phys. Rev. Lett. **101**, 172001 (2008) [arXiv:0807.4458].
35. B. Aubert *et al.* [BABAR Collaboration], Phys. Rev. Lett. **95**, 142001 (2005) [arXiv:hep-ex/0506081].
36. E. van Beveren, G. Rupp and J. Segovia, Phys. Rev. Lett. **105**, 102001 (2010) [arXiv:1005.1010].
37. T. Barnes, S. Godfrey and E. S. Swanson, Phys. Rev. D **72**, 054026 (2005) [arXiv:hep-ph/0505002].
38. G. Pakhlova *et al.* [Belle Collaboration], Phys. Rev. D **80**, 091101 (2009) [arXiv:0908.0231].
39. E. van Beveren and G. Rupp, arXiv:1004.4368.
40. E. van Beveren and G. Rupp, arXiv:0906.2278.
41. E. van Beveren and G. Rupp, arXiv:0904.4351.
42. B. Aubert [The BABAR Collaboration], Phys. Rev. D **79**, 092001 (2009) [arXiv:0903.1597].
43. E. van Beveren, Z. Phys. C **21**, 291 (1984) [arXiv:hep-ph/0602246].
44. E. van Beveren and G. Rupp, Eur. Phys. J. A **31**, 468 (2007) [arXiv:hep-ph/0610199].
45. A. P. Szczepaniak, P. Guo, M. Battaglieri and R. De Vita, Phys. Rev. D **82**, 036006 (2010) [arXiv:1005.5562].
46. S. D. Protopopescu *et al.*, Phys. Rev. D **7**, 1279 (1973).
47. B. Hyams *et al.*, Nucl. Phys. B **64**, 134 (1973) [AIP Conf. Proc. **13**, 206 (1973)].
48. P. Estabrooks and A. D. Martin, Nucl. Phys. B **79**, 301 (1974).
49. E. van Beveren and G. Rupp, Phys. Rev. Lett. **91**, 012003 (2003) [arXiv:hep-ph/0305035].

Matching the quark model to the $1/N_c$ expansion

Dan Pirjol* and Carlos Schat[1]†

* *National Institute for Physics and Nuclear Engineering, Department of Particle Physics, 077125 Bucharest, Romania*
†*Department of Physics and Astronomy, Ohio University, Athens, Ohio 45701, USA*
Departamento de Física, FCEyN, Universidad de Buenos Aires, Ciudad Universitaria, Pab.1, (1428) Buenos Aires, Argentina

Abstract. We compute the coefficients of the effective mass operator of the $1/N_c$ expansion for negative parity $L = 1$ excited baryons using the Isgur-Karl model in order to compare the general approach, where the coefficients are obtained by fitting to data, with a specific constituent quark model calculation. We discuss the physics behind the fitted coefficients for the scalar part of the most general two-body quark-quark interaction. We find that both pion exchange and gluon exchange lead to the dominance of the same operator at the level of the effective mass operator, which is also observed from data.

Keywords: Quark model, 1/N expansion of QCD, excited baryons
PACS: 12.39.-x, 12.38.-t, 14.20-c

INTRODUCTION

Quark models provide a simple picture of the physics of ground state baryons and their excitations [1, 2], but their connection with QCD is not so clear. An alternative approach to baryon phenomenology that is closer connected to QCD is provided by the $1/N_c$ expansion [3], which has been applied to the study of both the ground state and excited nucleons [4, 5, 6, 7, 8, 9, 10], leading to interesting insights that would be difficult to obtain in a particular model calculation.

In order to bridge the gap between the more intuitive quark model calculations and the effective operator approach, in a recent paper [11] we showed how to match an arbitrary quark model Hamiltonian onto the operators of the $1/N_c$ expansion. We made use of the transformation of the states and operators under the permutation group of N_c objects acting on the spin-flavor of the quarks. The approach is similar to the one discussed in Ref. [12] for the orbital part of the interaction and at $N_c = 3$. The main result of [11] can be summarized as follows: consider a two-body quark Hamiltonian $V_{qq} = \sum_{i<j} O_{ij} R_{ij}$, where O_{ij} acts on the spin-flavor quark degrees of freedom and R_{ij} acts on the orbital degrees of freedom. Then the hadronic matrix elements of the quark Hamiltonian on a baryon state $|B\rangle$ contains only the projections O_α onto irreducible representations of the spin-flavor permutations and can be factorized as $\langle B|V_{qq}|B\rangle = \sum_\alpha C_\alpha \langle O_\alpha \rangle$, where the coefficients C_α are reduced matrix elements of the orbital operators R_{ij}, given by overlap integrals of the quark model wave functions.

The matching procedure has been discussed in detail for the Isgur-Karl (IK) model in Ref. [13]. As an application of the general approach in Ref. [14, 15] we examine the spectrum of $L = 1$ negative parity baryons and discuss the implications for the spin-flavor structure of the most general two-body quark interaction. The present talk summarizes some aspects of this recent work that strive to understand the physics hidden in the numerical values of the fitted coefficients of the general operator expansion.

THE MASS OPERATOR OF THE ISGUR-KARL MODEL

The Isgur-Karl model is defined by the quark Hamiltonian

$$\mathscr{H}_{IK} = \mathscr{H}_0 + \mathscr{H}_{\rm hyp}\,, \tag{1}$$

[1] Speaker

12th International Conference on Meson-Nucleon Physics and the Structure of the Nucleon (MENU 2010)
AIP Conf. Proc. 1374, 425-428 (2011); doi: 10.1063/1.3647173

where $\mathscr{H}_0$ contains the confining potential and kinetic terms of the quark fields, and is symmetric under spin and isospin. The hyperfine interaction $\mathscr{H}_{\rm hyp}$ is given by

$$\mathscr{H}_{\rm hyp} = A \sum_{i<j} \left[\frac{8\pi}{3} \vec{s}_i \cdot \vec{s}_j \delta^{(3)}(\vec{r}_{ij}) + \frac{1}{r_{ij}^3} (3\vec{s}_i \cdot \hat{r}_{ij}\, \vec{s}_j \cdot \hat{r}_{ij} - \vec{s}_i \cdot \vec{s}_j) \right], \tag{2}$$

where A determines the strength of the interaction, and $\vec{r}_{ij} = \vec{r}_i - \vec{r}_j$ is the distance between quarks i, j. The first term is a local spin-spin interaction, and the second describes a tensor interaction between two dipoles. This interaction Hamiltonian is an approximation to the gluon-exchange interaction, neglecting the spin-orbit terms.

In the original formulation of the IK model [2] the confining forces are harmonic. We will derive in the following the form of the mass operator without making any assumption on the shape of the confining quark forces. We refer to this more general version of the model as IK-V(r).

The $L = 1$ quark model states include the following SU(3) multiplets: two spin-1/2 octets $8_{\frac{1}{2}}, 8'_{\frac{1}{2}}$, two spin-3/2 octets $8_{\frac{3}{2}}, 8'_{\frac{3}{2}}$, one spin-5/2 octet $8'_{\frac{5}{2}}$, two decuplets $10_{\frac{1}{2}}, 10_{\frac{3}{2}}$ and two singlets $1_{\frac{1}{2}}, 1_{\frac{3}{2}}$. States with the same quantum numbers mix. For the $J = 1/2$ states we define the relevant mixing angle θ_{N1} in the nonstrange sector as $N(1535) = \cos\theta_{N1} N_{1/2} + \sin\theta_{N1} N'_{1/2}$, $N(1650) = -\sin\theta_{N1} N_{1/2} + \cos\theta_{N1} N'_{1/2}$ and similar equations for the $J = 3/2$ states, which define a second mixing angle θ_{N3}.

We find [13] that the most general mass operator in the IK-V(r) model depends only on three unknown orbital overlap integrals, plus an additive constant c_0 related to the matrix element of $\mathscr{H}_0$, and can be written as

$$\hat{M} = c_0 \mathbf{1} + aS_c^2 + bL_2^{ab}\{S_c^a, S_c^b\} + cL_2^{ab}\{s_1^a, S_c^b\}, \tag{3}$$

where the spin-flavor operators are understood to act on the state $|\Phi(SI)\rangle$ constructed as a tensor product of the core of quarks 2,3 and the 'excited' quark 1, as given in [6, 11]. The coefficients are given by $a = \frac{1}{2}\langle R_S\rangle$, $b = \frac{1}{12}\langle Q_S\rangle - \frac{1}{6}\langle Q_{MS}\rangle$, $c = \frac{1}{6}\langle Q_S\rangle + \frac{1}{6}\langle Q_{MS}\rangle$. The reduced matrix elements R_S, Q_S, Q_{MS} for the orbital part of the interaction contain the unknown spatial dependence and are defined in Refs. [6, 13]. Evaluating the matrix elements using Tables II, III in Ref. [6] we find the following explicit result for the mass matrix

$$M_{1/2} = \begin{pmatrix} c_0 + a & -\frac{5}{3}b + \frac{5}{6}c \\ -\frac{5}{3}b + \frac{5}{6}c & c_0 + 2a + \frac{5}{3}(b+c) \end{pmatrix}, \quad M_{3/2} = \begin{pmatrix} c_0 + a & \frac{\sqrt{10}}{6}b - \frac{\sqrt{10}}{12}c \\ \frac{\sqrt{10}}{6}b - \frac{\sqrt{10}}{12}c & c_0 + 2a - \frac{4}{3}(b+c) \end{pmatrix}, \tag{4}$$

and $M_{5/2} = c_0 + 2a + \frac{1}{3}(b+c), \Delta_{1/2} = \Delta_{3/2} = c_0 + 2a$. Computing the reduced matrix elements with the interaction given by Eq. (2), one finds that the reduced matrix elements in the IK model with harmonic oscillator wave functions are all related and can be expressed in terms of the single parameter δ as $\langle Q_{MS}\rangle = \langle Q_S\rangle = -\frac{3}{5}\delta$; $\langle R_S\rangle = \delta$. This gives a relation among the coefficients a, b, c of the mass matrix Eq. (3): $a = \frac{1}{2}\delta, b = \frac{1}{20}\delta, c = -\frac{1}{5}\delta$. We recover the well known result that in the harmonic oscillator model, the entire spectroscopy of the $L = 1$ baryons is fixed by one single constant $\delta = M_\Delta - M_N \sim 300$ MeV, along with an overall additive constant c_0, and the model becomes very predictive. In Fig. 1 we show the result of a best fit of a, b, c in the IK-V(r) model together with the predictions of the IK model. The IK-V(r) spectrum is the best fit possible for a potential model with the spin-flavor interaction given in Eq. (2).

TESTING A MORE GENERAL SPIN-FLAVOR STRUCTURE

The most general quark Hamiltonian containing only two-body interactions has the form [14]

$$H_{qq} = H_0 + \sum_{i<j} (f_1(\vec{r}_{ij}) t_i^a t_j^a + f_2(\vec{r}_{ij}) \vec{s}_i \cdot \vec{s}_j + f_3(\vec{r}_{ij}) \vec{s}_i t_i^a \cdot \vec{s}_j t_j^a) + H_{s-o} + H_q, \tag{5}$$

where H_0 is the part of the quark Hamiltonian which does not depend on the quarks spin and flavor degrees of freedom. We show explicitly only the part of the Hamiltonian which transforms as a scalar ($\ell = 0$) under $SO(3)$, the group of orbital rotations - the scalar part of the quark Hamiltonian. The H_{s-o}, H_q denote the spin-orbit and the quadrupole interaction, which transform as a vector ($\ell = 1$) and a tensor ($\ell = 2$), respectively.

We will consider the case of contact scalar interactions

$$f_\nu(\vec{r}_{ij}) = A_\nu \delta^{(3)}(\vec{r}_{ij}), \qquad \nu = 1, 2, 3, \tag{6}$$

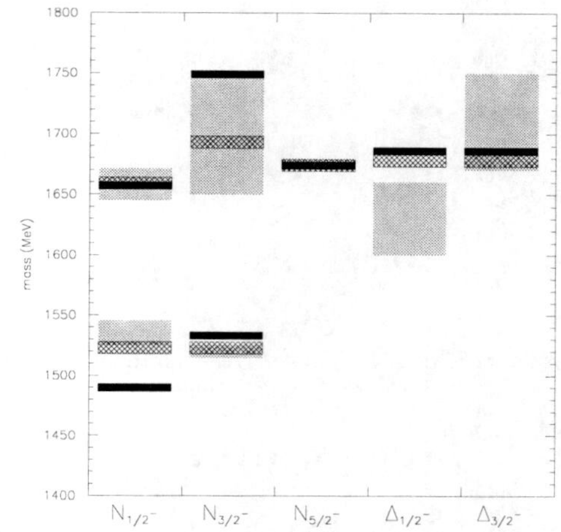

FIGURE 1. Masses predicted by the IK model (black bars), by the IK-V(r) model (hatched bars) and the experimental masses (green boxes) from Ref. [18].

and study the following question: what information can be obtained from the coefficients c_k of the $1/N_c$ studies of the spectrum of $L = 1$ negative parity baryons?

The quark Hamiltonian H_{qq} is matched onto the hadronic mass operator

$$M = c_0 \mathbf{1} + c_1 T^2 + c_2 S_c^2 + c_3 \vec{s}_1 \cdot \vec{S}_c + \cdots \tag{7}$$

where the ellipses denote terms arising from the tensor and spin-orbit interactions, which were considered in Ref. [14]. All three possible zero range two-body interactions $O_{ij}^{(v)} = t_i^a t_j^a, \vec{s}_i \cdot \vec{s}_j$ and $\vec{s}_i t_i^a \cdot \vec{s}_j t_j^a$ lead to $c_1 = c_3 = 0$ and are matched onto the same operator $O_2 = S_c^2$ in the effective theory. This means that there is no way to distinguish between these three types of scalar interactions as long as they are of very short range. Experimentally, at $N_c = 3$ one can determine only two linear combinations of the three coefficients $c_{1,2,3}$ (as functions of $\theta_{N1,N3}$) [14] from the mass spectrum and mixing angles of the negative parity $L = 1$ baryons. These linear combinations can be taken as $\tilde{c}_1 = c_1 - \frac{1}{2}c_3, \tilde{c}_2 = c_2 + c_3$ and their explicit expressions in terms of the nonstrange hadron masses and mixing angles can be found in Ref. [15].

A first estimate of $\tilde{c}_{1,2}$ can be made using the mixing angles $\theta_{N1,N3}$ determined from a fit to N^* strong decays and photoproduction data $(\theta_{N1}, \theta_{N3}) = (0.39 \pm 0.11, 2.82 \pm 0.11) = (22^\circ \pm 6^\circ, 162^\circ \pm 6^\circ)$ [16, 17]. Using these values and the hadron masses from the PDG [18] we obtain $\tilde{c}_1 = 3.9 \pm 11.0$ MeV, $\tilde{c}_2 = 129 \pm 18$ MeV. The corresponding ratio of the coefficients $\tilde{c}_i$ displays the relative suppression of $\tilde{c}_1$

$$\tilde{c}_1 / \tilde{c}_2 = 0.03 \pm 0.09 . \tag{8}$$

An alternative determination of this ratio can be made using only the excited baryon masses, as discussed in Ref. [14]. In this paper it was shown that in any quark model containing only two-body quark interactions there is a correlation between both mixing angles (up to a discrete ambiguity) that corresponds to the dashed and solid lines in Fig. 2. The scatter plot takes into account the experimental errors in the determination of the baryon masses. A second relation (Eq. (6) in Ref. [14]) expressing the spin-average of the $SU(3)$ singlet states $\bar{\Lambda} = \frac{1}{3}\Lambda_{1/2} + \frac{2}{3}\Lambda_{3/2}$ in terms of the nonstrange states further constrains the mixing angle and ratio to the dark (green) area in the plot and singles out the solid line as the preferred solution. In Fig. 2 the first determination of the ratio Eq. (8) is shown as the black point with error bars and falls within the dark (green) area.

Finally, it is interesting to notice [6, 15] that for the particular case ($v = 3$) of the spin-flavor structure corresponding to one-pion exchange $\tilde{c}_1$ vanishes regardless of the spatial range of the orbital part of the interaction (although for non-zero range $c_{1,3}$ do not vanish separately).

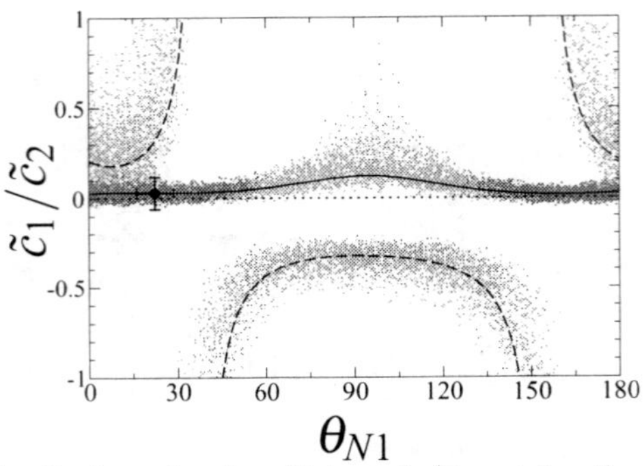

FIGURE 2. Scatter plot for the ratio of coefficients $\tilde{c}_1/\tilde{c}_2$ as a function of the mixing angle θ_{N1}.

CONCLUSIONS

We presented the effective mass operator of the Isgur-Karl model in the operator basis used in the $1/N_c$ expansion for excited baryons. This explicit model calculation of the operator coefficients (for details see Ref. [13]) connects two different approaches to baryon phenomenology and is an illustration of the general matching procedure discussed in Ref. [11] using the permutation group.

Considering the most general spin-flavor structure for the scalar part of the two-body quark interactions we find [15] that in the case of contact interactions all spin-flavor structures are matched onto the same effective operator in the $1/N_c$ expansion. For the particular case of a pion exchange interaction this holds independently of the range of the spatial part of the interaction. Comparing the values obtained for the operator coefficients from masses and independent determinations of mixing angles we find that a mixture of pion exchange and massless gluon exchange interactions is compatible with data, giving support to the physical picture considered in Refs. [19, 20].

ACKNOWLEDGMENTS

The work of C.S. was supported by CONICET and partially supported by the U. S. Department of Energy, Office of Nuclear Physics under contract No. DE-FG02-93ER40756 with Ohio University.

REFERENCES

1. A. De Rujula, H. Georgi and S. L. Glashow, Phys. Rev. D **12**, 147 (1975).
2. N. Isgur and G. Karl, Phys. Lett. B **72**, 109 (1977).
3. R. F. Dashen, E. Jenkins and A. V. Manohar, Phys. Rev. D **49**, 4713 (1994) [Erratum-ibid. D **51**, 2489 (1995)]; R. F. Dashen, E. Jenkins and A. V. Manohar, Phys. Rev. D **51**, 3697 (1995).
4. J. L. Goity, Phys. Lett. B **414**, 140 (1997).
5. D. Pirjol and T. M. Yan, Phys. Rev. D **57**, 1449 (1998); Phys. Rev. D **57**, 5434 (1998).
6. C. E. Carlson, C. D. Carone, J. L. Goity and R. F. Lebed, Phys. Lett. B **438**, 327 (1998); Phys. Rev. D **59**, 114008 (1999).
7. C. L. Schat, J. L. Goity and N. N. Scoccola, Phys. Rev. Lett. **88**, 102002 (2002); J. L. Goity, C. L. Schat and N. N. Scoccola, Phys. Rev. D **66**, 114014 (2002).
8. J. L. Goity, C. Schat and N. N. Scoccola, Phys. Lett. B **564**, 83 (2003).
9. D. Pirjol and C. Schat, Phys. Rev. D **67**, 096009 (2003); AIP Conf. Proc. **698**, 548 (2004).
10. N. Matagne and F. Stancu, Phys. Rev. D **71**, 014010 (2005); Phys. Lett. B **631**, 7 (2005); Phys. Rev. D **74**, 034014 (2006).
11. D. Pirjol and C. Schat, Phys. Rev. D **78**, 034026 (2008).
12. H. Collins and H. Georgi, Phys. Rev. D **59**, 094010 (1999).
13. L. Galeta, D. Pirjol and C. Schat, Phys. Rev. D **80**, 116004 (2009).
14. D. Pirjol and C. Schat, Phys. Rev. Lett. **102**, 152002 (2009).
15. D. Pirjol and C. Schat, arXiv:1007.0964 [hep-ph].
16. J. L. Goity, C. Schat and N. Scoccola, Phys. Rev. D **71**, 034016 (2005).
17. N. N. Scoccola, J. L. Goity and N. Matagne, Phys. Lett. B **663**, 222 (2008).
18. C. Amsler *et al.* [Particle Data Group], Phys. Lett. B **667**, 1 (2008).
19. A. Manohar and H. Georgi, Nucl. Phys. B **234**, 189 (1984).
20. S. Weinberg, arXiv:1009.1537 [hep-ph].

Session L: Nucleon Resonances

Pion photoproduction in a dynamical coupled-channel model

F. Huang*, M. Döring†, H. Haberzettl**,†, S. Krewald†,‡ and K. Nakayama*,†

*Department of Physics and Astronomy, University of Georgia, Athens, GA 30602, USA
†Institut für Kernphysik and Jülich Center for Hadron Physics, Forschungszentrum Jülich, 52425 Jülich, Germany
**Center for Nuclear Studies, Department of Physics, The George Washington University, Washington, DC 20052, USA
‡Institute for Advanced Simulations, Forschungszentrum Jülich, 52425 Jülich, Germany

Abstract. Pion photoproduction reactions are investigated in a dynamical coupled-channel approach based on the Jülich πN model, which presently includes the hadronic πN and ηN stable channels as well as the $\pi\Delta$, σN and ρN effective channels. This model has been quite successful in the description of $\pi N \to \pi N$ scattering for center-of-mass energies up to 1.9 GeV. The full pion photoproduction amplitude is constructed to satisfy the generalized Ward-Takahashi identity and hence, it is fully gauge invariant. The calculated differential cross sections and photon spin asymmetries up to 1.65 GeV center-of-mass energy for the reactions $\gamma p \to \pi^+ n$, $\gamma p \to \pi^0 p$ and $\gamma n \to \pi^- p$ are in good agreement with the experimental data.

Keywords: Meson photoproduction, gauge invariance, dynamical coupled-channel model
PACS: 25.20.Lj, 13.60.Le, 14.20.Gk, 13.75.Gx

INTRODUCTION

In the past few years, intensive experimental efforts have taken place at major laboratories around the world to study the baryon spectra using the electromagnetic probes. On the theoretical side, since the baryon resonances are unstable and couple strongly to baryon-meson continuum states, coupled-channel models are needed to analyze the experimental data and to interpret the extracted resonance parameters. For our present study, we employ the well-known Jülich πN model [1–3] that is based on time-ordered perturbation theory. It is a coupled-channel meson-exchange model which includes the πN and ηN stable channels as well as the $\pi\Delta$, σN and ρN effective channels accounting for the resonant part of the $\pi\pi N$ channel. The interaction kernel corresponding to the t- and u-channel diagrams is constructed based on the chiral Lagrangians of Wess and Zumino [4], supplemented by additional terms for the coupling of Δ, ω, η, a_0 and σ [2, 3]. For the s-channel diagrams, apart from the bare nucleon dressed by the coupling to the πN continuum state to reproduce the physical nucleon, the interaction kernel includes eight genuine bare resonances, namely $S_{11}(1535)$, $S_{11}(1650)$, $S_{31}(1620)$, $P_{31}(1910)$, $P_{13}(1720)$, $D_{13}(1520)$, $P_{33}(1232)$ and $D_{33}(1700)$. The bare genuine resonances get their dressed masses and widths from the re-scattering of the baryon-meson continuum states; the $P_{11}(1440)$ (Roper) resonance appears as a dynamically generated resonance due to the strong coupling of the πN and σN channels. This hadronic model has been quite successful in reproducing the πN partial-wave amplitudes up to the center-of-mass energy of $W = 1.9$ GeV and for the total angular momentum states with $J = 1/2$ and $3/2$.

In Ref. [5] a dynamical coupled-channel model for pseudoscalar meson photoproduction based on a field-theoretical approach of Haberzettl [6] has been introduced in conjunction with the Jülich hadronic coupled-channel model. This photoproduction model is distinguished from the majority of the existing dynamical models by the fact that, in addition to satisfying unitarity and analyticity as a matter of course, it also satisfies the full gauge-invariance condition dictated by the generalized Ward-Takahashi identity. By contrast, the vast majority of existing dynamical models satisfy only current conservation but not gauge invariance. In Ref. [5], a preliminary feasibility study of this field-theory approach was presented. In the present work, we carry out an extended and more quantitative calculation of the neutral and charged pion photoproduction reactions within this approach using the Jülich πN model [3]. We calculate cross sections as well as the beam asymmetries up to $W = 1.65$ GeV.

The full single-meson photoproduction amplitude, M^μ, is obtained by attaching a photon everywhere into the dressed $NN\pi$ vertex, which gives [5]

$$M^\mu = M_s^\mu + M_u^\mu + M_t^\mu + M_{\text{int}}^\mu, \tag{1}$$

12th International Conference on Meson-Nucleon Physics and the Structure of the Nucleon (MENU 2010)
AIP Conf. Proc. 1374, 431-434 (2011); doi: 10.1063/1.3647174

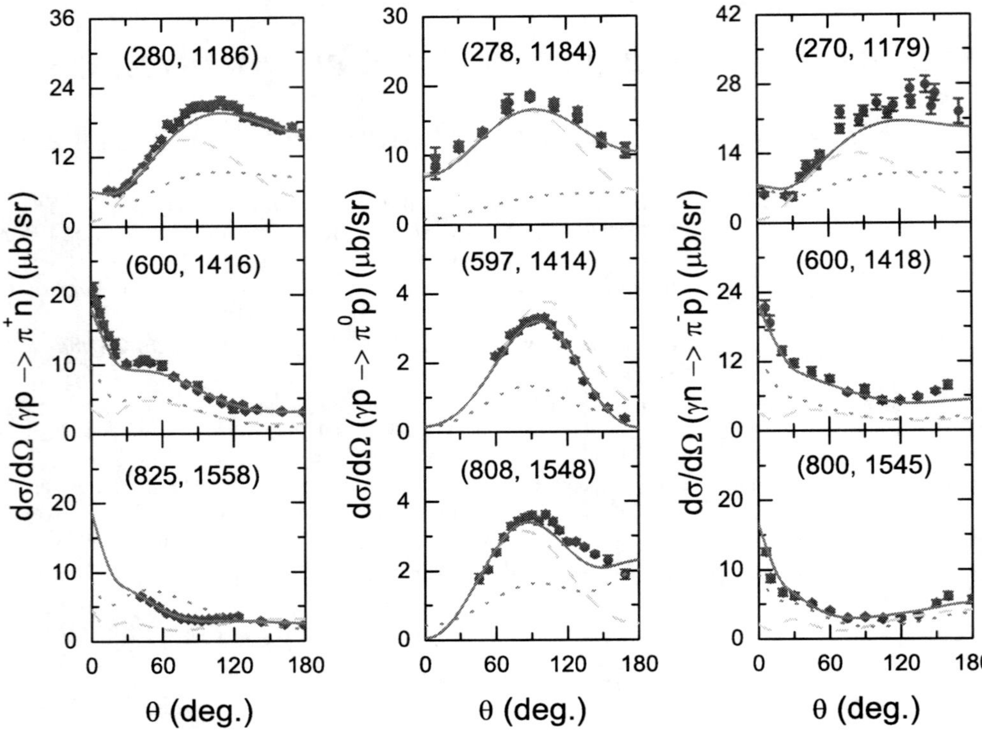

FIGURE 1. Differential cross sections for $\gamma p \to \pi^+ n$, $\gamma p \to \pi^0 p$ and $\gamma n \to \pi^- p$. The solid curves show the results from the full calculation. The dotted and the dashed curves are obtained by respectively switching off the loop integral or the contact current M_c^μ. Data are from Ref. [8].

where the first three terms describe the coupling of the photon to the external legs of the dressed $NN\pi$ vertex, with the subscripts s, u and t referring to the appropriate Mandelstam variables of their respective intermediate hadrons. The last term M_{int}^μ is interaction current, where the photon couples inside the $NN\pi$ vertex. In our work it is chosen as [5]

$$M_{\text{int}}^\mu = M_c^\mu + T^\mu + T^{NP} G_0 \left[\left(M_u^\mu + M_t^\mu \right)_T + T^\mu \right], \tag{2}$$

where $\left(M_u^\mu + M_t^\mu\right)_T$ denotes the transverse part of $\left(M_u^\mu + M_t^\mu\right)$; T^μ stands for an undetermined transverse contact current taken to be zero in the present work for simplicity; T^{NP} is the non-polar part of the hadronic scattering amplitude; G_0 describes the free propagation of the intermediate meson-baryon two-body system. M_c^μ is the generalized contact current as specified in Ref. [5]; it accounts for the complicated part of the interaction current which cannot be treated explicitly. The amplitude M_s^μ in Eq. (1) is given by

$$M_s^\mu = \Gamma_{RN\pi} S \Gamma_{RN\gamma}^\dagger, \tag{3}$$

where $\Gamma_{RN\pi}$ stands for the dressed $R \to N\pi$ vertex with R denoting either the nucleon or baryon resonance, S denotes the dressed baryon propagator. The dressed hadronic vertices and propagators are determined by the hadron dynamics of the Jülich πN coupled-channel model. $\Gamma_{RN\gamma}^\dagger$ is the full dressed $N\gamma \to R$ vertex as derived in Ref. [5].

RESULTS

The parameters of the present pion photoproduction model are determined by fitting the differential cross section and the beam asymmetry data of the charged and neutral pion photoproduction up to $W = 1.65$ GeV. They are listed in

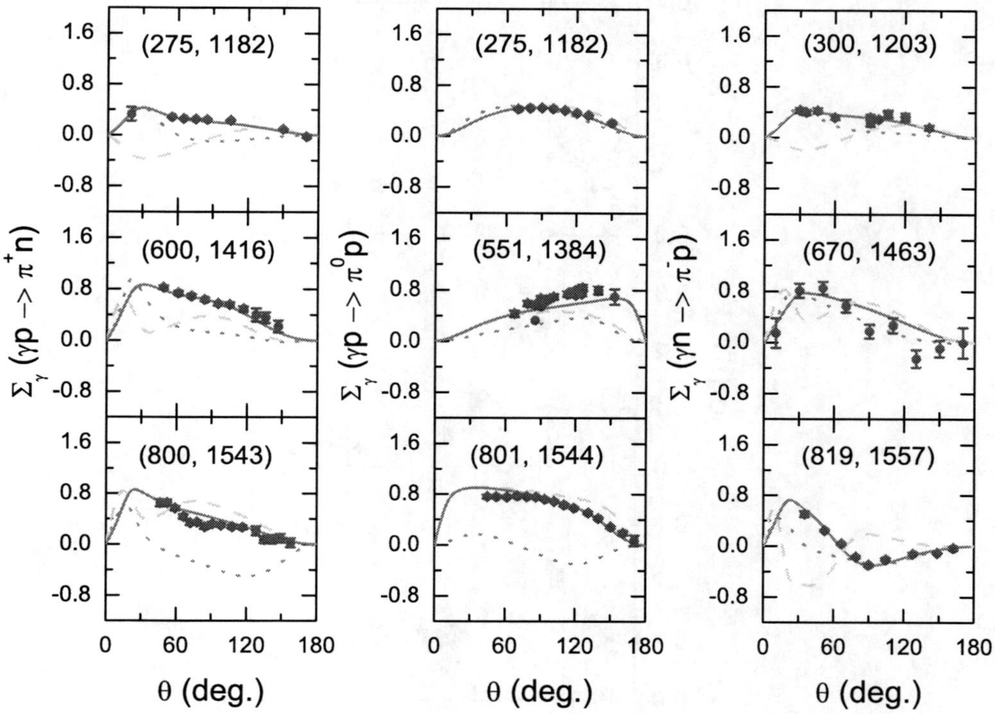

FIGURE 2. Photon spin asymmetries for $\gamma p \to \pi^+ n$, $\gamma p \to \pi^0 p$ and $\gamma n \to \pi^- p$. The solid curves show the results from the full calculation. The dotted and the dashed curves are obtained by respectively switching off the loop integral or the contact current M_c^μ. Data are from Ref. [8].

Ref. [7]. The calculated differential cross sections and beam asymmetries for $\gamma p \to \pi^+ n$, $\gamma p \to \pi^0 p$ and $\gamma n \to \pi^- p$ are shown in Figs. 1 and 2, respectively, together with the corresponding experimental data. More results for differential cross sections and beam asymmetries at 13 additional energies from the πN threshold up to $W = 1.65$ GeV can be found in Ref. [7]. Figure 3 shows the total cross sections for the three charged pion photoproduction channels. Note that the total cross-section data have not been included in our fit. In Figs. 1-3, the solid curves correspond to the results of our full calculation, while the dotted and dashed curves are obtained by respectively switching off the loop integral and the contact current M_c^μ in the full amplitude. One sees that our theoretical results for both the differential cross sections and beam asymmetries (Figs. 1 and 2) as well as the total cross sections (Fig. 3) are in good agreement with the corresponding experimental data. Juliá-Díaz *et al.* [9] have also performed an investigation of the single-pion photoproduction within the EBAC dynamical coupled-channel model. Compared to our model, they included two additional resonances, namely D_{15} and F_{15}; however, the $\gamma n \to \pi^- p$ channel was not analyzed in their work. Compared with the results of Ref. [9], the agreement of our results with the data is a little bit better for both the $\gamma p \to \pi^+ n$ and $\gamma p \to \pi^0 p$ channels. Figures 1-3 also show clearly that the loop integral and the contact current M_c^μ are important for reproducing the data, which means that the coupled-channel effects *and* gauge invariance are essential for understanding the single-pion photoproduction reactions. While the importance of coupled-channel effects was pointed out already in Ref. [9], we emphasize here that the majority of existing dynamical models of photoproduction reactions are actually not gauge invariant since their amplitudes obey current conservation only. The latter is shown in Ref. [6] to be necessary but not sufficient for maintaining full gauge invariance across all levels of a microscopic description of the photoreaction. The importance of fulfilling the gauge-invariance condition was also emphasized for the *NN* bremsstrahlung reaction [10].

Finally, we mention that the residues and pole positions of the full photoproduction amplitude are being evaluated in our model by making an analytic continuation of the amplitudes to complex energy plane. The results will be published

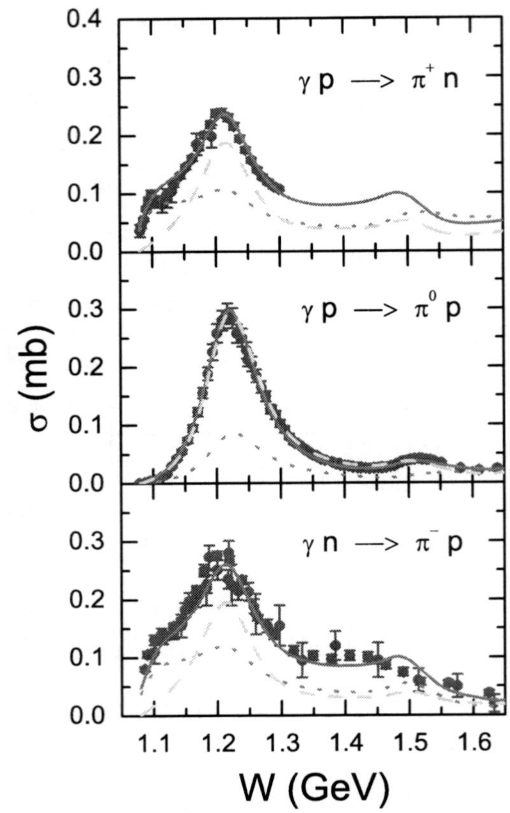

FIGURE 3. Total cross sections for $\gamma p \to \pi^+ n$, $\gamma p \to \pi^0 p$ and $\gamma n \to \pi^- p$. The solid curves show the results from the full calculation. The dotted and the dashed curves are obtained by respectively switching off the loop integral or the contact current M_c^μ. Data are taken from Ref. [8] but not included in the fit.

elsewhere.

ACKNOWLEDGMENTS

This work is supported by the FFE grant No. 41788390 (COSY-058). The authors acknowledge the Research Computing Center at the University of Georgia and the Jülich Supercomputing Center at Forschungszentrum Jülich for providing computing resources that have contributed to the research results reported within this proceeding.

REFERENCES

1. C. Schütz, J. Haidenbauer, J. Speth, and J.W. Durso, *Phys. Rev. C* **57**, 1464 (1998).
2. O. Krehl, C. Hanhart, S. Krewald, and J. Speth, *Phys. Rev. C* **62**, 025207 (2000).
3. A.M. Gasparyan, J. Haidenbauer, C. Hanhart, and J. Speth, *Phys. Rev. C* **68**, 045207 (2003).
4. J. Wess and B. Zumino, *Phys. Rev.* **163**, 1727 (1967).
5. H. Haberzettl, K. Nakayama, and S. Krewald, *Phys. Rev. C* **74**, 045202 (2006).
6. H. Haberzettl, *Phys. Rev. C* **56**, 2041 (1997).
7. F. Huang, M. Döring, H. Haberzettl, S. Krewald, and K. Nakayama, in preparation.
8. CNS Data Analysis Center, The George Washington University, `http://gwdac.phys.gwu.edu`.
9. B. Juliá-Díaz, T.-S.H. Lee, A. Matsuyama, T. Sato, and L.C. Smith, *Phys. Rev. C* **77**, 045205 (2008).
10. K. Nakayama and H. Haberzettl, *Phys. Rev. C* **80**, 051001 (2009).

Recent N* results from photoproduction experiments at CLAS

D. Sokhan, on behalf of the CLAS Collaboration

Institut de Physique Nucleaire, Bat. 100, Orsay 91406, France [1]

Abstract. The recent breakthroughs in the technology of polarised targets have enabled a new generation of meson photo-production experiments to be carried out. A measurement of a full set of polarisation observables off both polarised proton and neutron targets and in a large number of meson-production channels has come within sight. Such a measurement would very significantly reduce model-dependence in the analysis of the data and thus has the potential to resolve long-standing issues, such as the "missing resonance" problem, and shed new light on the nucleon excitation spectrum. This has formed the motivation for the recent N* experimental programme of CLAS.

Keywords: CLAS, photoproduction, meson, nucleon resonance
PACS: 25.20.Lj

1. INTRODUCTION

The nucleon excitation spectrum provides deep insight into the structure of the nucleon. Despite years of study, however, it is still relatively little known. Predictions are primarily from Lattice QCD (Quantum Chromodynamics) and from a variety of phenomenological models, such as the constituent quark model, the di-quark model and others. The plethora of different models predict spectra of different resonances, many more in fact than have so far been observed – a conundrum known as the "missing resonance" problem. Experimentally, the extraction of resonance parameters from the measurable observables is still heavily model-dependent. Thus the question remains – are the missing resonances not there in nature, or are we simply not seeing them?

A powerful technique for the study of the nucleon excitation spectrum is meson photoproduction. Since the first resonance region is above the combined invariant mass of a nucleon and the lowest-mass meson (pion), many resonances preferentially decay via the production of a meson. Sensitivity to different resonances can be observed in different meson-production reactions. Additionally, the electromagnetic (EM) interaction of real photons is well-understood and gives access to the EM properties of the nucleon.

Meson photoproduction can be described in terms of complex reaction amplitudes, which number four in the case of pseudo-scalar mesons. Experimentally, combinations of these yield 16 polarisation observables [1] – measurable quantities in experiments employing polarised photons, polarised targets and the ability to measure recoil polarisation. It has been shown [2] that a measurement of eight carefully chosen polarisation observables (which have to include cross-section and the three single polarisation observables, as well as four chosen from at least two of the groups of double-polarisation experiments: beam-target (BT), beam-recoil (BR) and target-recoil (TR)) is sufficient to remove ambiguities in the partial wave analysis (PWA), one of the main methods used to extract resonance parameters from the data. Additionally, measurements obtained from a variety of reaction channels are very important as input for coupled channel analyses where they enable a choice to be made between the large number of different models and can provide indication of strong quark correlations in the nucleon.

Moreover, since the electromagnetic interaction does not conserve isospin, the multipole amplitudes which feature in PWA contain both isoscalar and isovector contributions of the EM current. Measurements from both isospin partners, the proton and the neutron, are therefore required to study different isospin couplings. Since different meson production channels are sensitive to different resonances, data from a variety of channels and on both proton and neutron targets are required.

The recent experimental programme of the CLAS Collaboration at Jefferson Lab is contributing significantly, both to the quest for a "complete" measurement and in expanding and refining the dataset for coupled channel analyses.

[1] daria@ipno.in2p3.fr

12th International Conference on Meson-Nucleon Physics and the Structure of the Nucleon (MENU 2010)
AIP Conf. Proc. 1374, 435-438 (2011); doi: 10.1063/1.3647175

Due to enormous technological challenges in the construction and use of polarised targets, it is only recently that the beam-target and target-recoil groups of double-polarisation measurements have become accessible.

2. CLAS AT JEFFERSON LABORATORY

Jefferson Lab (JLab) is home to a 1.4 km long race-track accelerator, currently operating at energies up to 6 GeV and about to undergo an upgrade to 12 GeV within the next few years. 200 μA of continuous current is split and delivered simultaneously to three experimental halls, A, B and C. The experiments presented here were carried out in Hall B, where the electron beam is passed through a radiator, emitting, depending on the type of material used and the polarisation of the electron beam, either circularly or linearly polarised photons by the process of bremsstrahlung. Circular polarisations up to 80% and linear polarisations $>$ 90% are typical. The scattered electrons are momentum-analysed in a tagging spectrometer, which serves to "tag" the photons through a coincidence timing measurement between the deflected electrons and the event time in the detector, thus allowing the exact photon involved in an interaction to be identified. The target cell is positioned at the centre of CLAS, the CEBAF (Continuous Electron Beam Accelerator Facility) Large Acceptance Spectrometer for charged and neutral particles. It consists of near-spherical layers of scintillators, drift-chambers, and calorimeters, providing almost complete angular coverage in the azimuthal direction and from 8° to 140° in the polar angle. Coupled to a toroidal magnetic field from six superconducting coils, CLAS offers particularly excellent sensitivity to charged particles, while still allowing detection of neutral reaction products.

Although no recoil detector is currently installed, measurements of recoil polarisation are still possible for photoproduction channels involving the production of hyperons, such as $\gamma + p \rightarrow K^+ + \Lambda \rightarrow K^+ + p + \pi^-$. The procedure makes use of the "self-analysing" property of hyperons, whose polarisation can be determined from the angular distributions of their weak-decay products (in the case above, p and π^-). As a result, a complete measurement in kaon-photoproduction off the proton has already been possible in CLAS.

3. RECENT EXPERIMENTS

In this paper we present a selection of preliminary results from three recent experiments carried out at Jefferson Lab, which were aimed at expanding measurements of single and double polarisation observables in a number of meson photoproduction channels. The details of the experiments are discussed in the following sections.

3.1. The G8 experiment: Linearly polarised photons, LH_2 target

The G8 experiment, as it is labelled in JLab nomenclature, ran from June to Sept. 2005, using linearly polarised photons, in the range 0.9 - 2.1 GeV, incident on an unpolarised liquid H_2 target. This enabled measurements on the proton of the beam spin asymmetry from the pion (see M. Dugger *et al.* in this publication), η and η' photoproduction channels, and a number of observables from the kaon-production channels. The experiment complements a much earlier run (G1) in which similar measurements were carried out using a circularly polarised photon beam.

Fig. 1 presents a sample of preliminary results from measurement of three observables in the $\gamma(p,\Lambda)K^+$ channel [3]. The two measurements of the beam asymmetry (left) display very good agreement with previous data (GRAAL for photon energies in the range 1.15 - 1.5 GeV (Lleres *et al.*, 2007), and LEPS in the photon energy range 1.5 - 2.1 GeV (Zegers *et al.*, 2003)). The large statistics of the experiment allowed much finer angular binning, with reduced error bars, and at photon energies $>$ 1.5 GeV provided the first measurements at backward angles.

Very good agreement with past results from both CLAS and GRAAL is also observed in a measurement of the recoil polarisation (middle), where the much higher statistics allowed a finer binning and smaller error bars.

Significantly, a first measurement of the O_x beam-recoil observable has been possible, showing large polarisations. The data in the two right columns of Fig. 1 is displayed with a selection of Regge model predictions, calculated by the Gent group [4]. The Regge calculation including only the better-established "core" resonances, S11(1650), P11(1710), P13(1720) and P13(1900), does not appear to be sufficient. A test inclusion of additional "missing" resonances was made into the model and the prediction compared with the data. There is a tentative suggestion of an important role of the D13(1900) resonance, and no clear evidence for the P11(1900) state. Agreement at low energies, however, is poor.

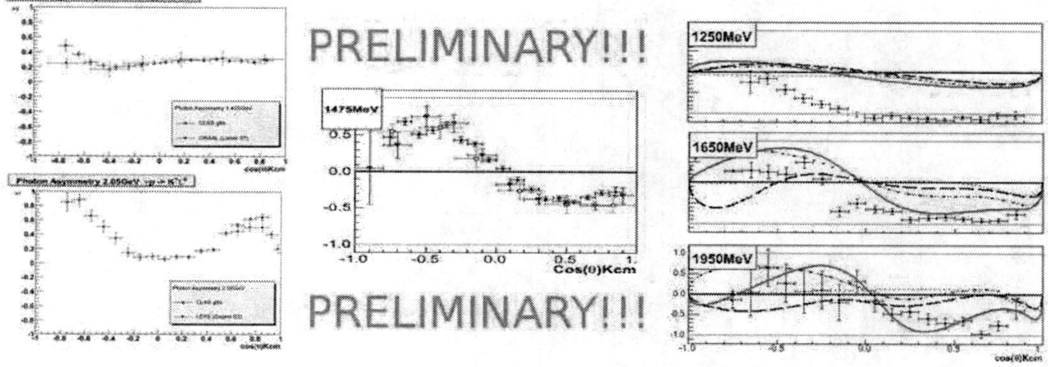

FIGURE 1. Single and double-polarisation measurements in $\gamma + p \rightarrow \Lambda + K^+$. Left: beam asymmetry for $1.40 < E_\gamma < 1.45$ GeV (top) and $2.0 < E_\gamma < 2.1$ GeV (bottom). Middle: recoil polarisation for $1.45 < E_\gamma < 1.5$ GeV, G8 data (blue diamonds) compared to past CLAS data (red down triangle, (McNabb *et al.*, 2004)) and GRAAL data (open green squares, (Lleres *et al.*, 2007)). Right: O_x double polarisation measurement, compared to Regge prediction (dotted green), Regge + core resonances (dot-dash blue), Regge + core + D13(1900) (solid red) and Regge + core + P11(1900) (dashed black) [3], [4].

3.2. The G13 experiment: Polarised photons, LD_2 target

A complementary experiment, G13, took data in autumn 2006 and March - June 2007, using photon beams of both circular (0.4 - 2.5 GeV) and linear polarisations (1.1 - 2.3 GeV) incident on a liquid deuterium target. This gave access to polarisation observable measurements off the neutron.

A comparison of the beam asymmetry measurement in $\gamma(p,\Lambda)K^+$ off the bound proton and a free proton (from the G8 experiment) showed good agreement, indicating that the quasi-free nucleons in deuterium can be assumed to provide a good approximation for free nucleons [5].

Some preliminary results of the first measurement of the beam asymmetry off the neutron in the channels $\gamma(n,\Lambda)K^0$ are shown in Fig. 2, where the data are compared to MAID PWA calculations [6] inclusive of different resonances. It is evident that the PWA fits the data rather badly, suggesting that the description is incomplete. A similar measurement performed for the channel $\gamma(n,\Sigma)K^0$ is also being prepared for publication.

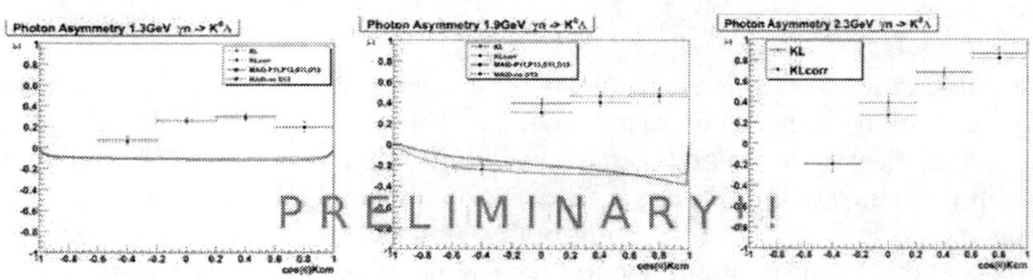

FIGURE 2. Beam asymmetry measurement from $\gamma(n,\Lambda)K^0$, in a sample of three photon energy bins [7]. Curves show MAID predictions [6]. The PWA is not valid for the energy of the third bin.

The preliminary results of a tremendously high statistics measurement of the beam asymmetry in the channel $\gamma(n,p)\pi^-$ are shown in Fig. 3 [8]. The world dataset has been increased from 52 previous data points to 1535. Comparisons to the MAID07 [6] and SAID09 [9] predictions indicate a slightly better agreement with MAID07 at backward angles and SAID09 in the forward direction. At invariant mass > 2 GeV/c^2, however, agreement is poor, indicating that physics previously unobserved in the PWA is at play.

3.3. The FROST experiment - polarised photons and a polarised target

The FROST (Frozen Spin Target) experiment employed a custom-designed dilution refrigerator to hold the spin of the butanol target (C_4H_9OH), aligned via Dynamic Nuclear Polarisation. Data was taken with both linear and circular

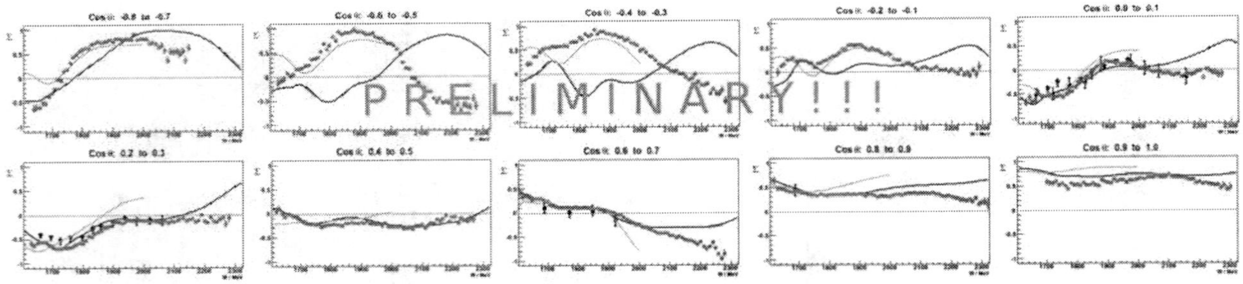

FIGURE 3. Beam asymmetry measurement in the $\gamma(n,p)\pi^-$ channel, overlaid with a MAID07 [6] prediction (short green) and SAID09 [9] (long blue). Previous data are shown in black [8].

photon beam polarisations in the energy range 0.5 - 2.4 GeV, with the target spin aligned parallel to the beam direction (longitudinally polarised, running Nov. '07 - Feb. '08) and perpendicular to it (transverse polarisation, from Mar. - July 2010). A selection of single and double polarisation observables are currently being extracted from the single and double pion, the η and the kaon photoproduction channels.

Very preliminary results from the measurement of the helicity asymmetry in $\gamma(p,n)\pi^-$ can be seen in Fig. 4, compared with predictions of two SAID models (SAID95 and SAID09) and MAID07. In the invariant mass range 1.35 - 1.5 GeV/c^2, SAID05 appears to describe the data best, while in the range 1.5 < W < 1.6 GeV/c^2 the SAID95 and MAID07 offer a better agreement. Most interestingly, above 1.85 GeV/c^2 all three PWAs appear to predict the opposite behaviour, indication that a fit inclusive of the current data may reveal signatures of different resonances.

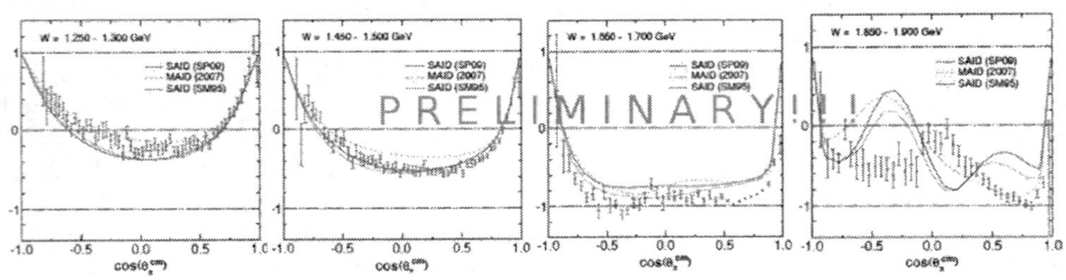

FIGURE 4. Helicity asymmetry measurement in $\gamma(p,n)\pi^-$ [10].

Preparations are currently underway for HDice, an experiment scheduled to run in 2011, which will employ an HD target with polarisation relaxation time on the order of months. This will not only enable the first ever measurements of target double-polarisation observables to be taken off the neutron, but will also allow simultaneous measurements on the proton, thus enabling comparison of proton and neutron data within the same setup.

With experiments on the proton and neutron, using both polarised and unpolarised targets and photon beams, a model-independent analysis of meson photoproduction data is fast approaching, promising greater insight into the nucleon resonance spectrum.

REFERENCES

1. I. Barker, A. Donnachie, and J. Storrow, *Nucl. Phys.* **B 95**, 347 (1975).
2. G. Keaton, and R. Workman, *Phys. Rev C* **53**, 1434 (1996); W.-T. Chiang, and F. Tabakin, *Phys. Rev. C* **55**, 2054 (1997).
3. C. Paterson, *PhD Thesis*, University of Glasgow, 2008.
4. P. Vancraeyveld, *Private communication*, 2010.
5. J. R. Johnstone, *PhD Thesis*, University of Glasgow, 2010.
6. D. Drechsel, S.S. Kamalov, and L. Tiator, *Eur. Phys. J. A* **34**, 69 (2007).
7. N. Hassall, *PhD Thesis*, University of Glasgow, 2010.
8. D. Sokhan, *PhD Thesis*, University of Edinburgh, 2009.
9. The SAID group at George Washington University, USA; M. Dugger *et al.*, *Phys. Rev. C* **79**, 065206 (2009).
10. S. Strauch, *Private communication*, 2010.

Measurement of the $\gamma p \to \eta p$ Reaction with the Crystal Ball Detector at the Mainz Microtron (MAMI-C)

I.I. Strakovsky*, S. Prakhov†, R.A. Arndt*, Ya.I. Azimov**, W.J. Briscoe*, B. Krusche‡, B.M.K. Nefkens†, M.W. Paris* and R.L. Workman
for the Crystal Ball Collaboration at MAMI*

*The George Washington University, Washington, DC 20052, USA
†University of California Los Angeles, Los Angeles, California 90095-1547, USA
**Petersburg Nuclear Physics Institute, Gatchina, 188300, Russia
‡Institut für Physik, University of Basel, CH-4056 Basel, Switzerland

Abstract. The $\gamma p \to \eta p$ reaction has been measured with the Crystal Ball and TAPS multiphoton spectrometers in the energy range from the production threshold to 1.4 GeV ($1.49 \leq W \leq 1.87$ GeV). Bremsstrahlung photons produced by the 1.5-GeV electron beam of the Mainz Microtron MAMI-C and momentum analyzed by the Glasgow Tagging Spectrometer were used for the η-meson production. Our accumulation of 3.8×10^6 $\gamma p \to \eta p \to 3\pi^0 p \to 6\gamma p$ events allows a detailed study of the reaction dynamics. The $\gamma p \to \eta p$ differential cross sections were determined for 120 energy bins and the full range of the production angles. Our data show a dip near $W = 1680$ MeV in the total cross section related to a substantial dip in η production at forward angles. The data are compared to predictions of previous SAID and MAID partial-wave analyses and to the latest SAID and MAID fits that included our data.

Keywords: Photoproduction, mesons, baryon resonances
PACS: 11.80.Et,13.60.Le,25.20.Lj

The N^* family of nucleon resonances has many well established members [1], several of which exhibit overlapping resonances, with very similar masses and widths, but different J^P spin-parity values. Apart from the $N(1535)1/2^-$ state, the known photo-decay amplitudes have been determined from analyses of single-pion photoproduction, while ηN branching ratios are in general poorly known. The data available for $\pi^- p \to \eta n$ are inadequate to study this question [2, 3]. Helpful, evidently, may be η-photoproduction data. The present work studies the region from threshold, where there are two closely spaced states: $N(1520)3/2^-$ and $N(1535)1/2^-$, up to CM energies of $W \approx 1800$ MeV, encompassing a sequence of six overlapping states: $N(1650)1/2^-$, $N(1675)5/2^-$, $N(1680)5/2^+$, $N(1700)3/2^-$, $N(1710)1/2^+$, and $N(1720)3/2^+$.

The MAMI–C energy range (E_γ up to 1.4 GeV) includes several well-established resonances as well as some more questionable ones. Indeed, the excellent photon-energy resolution offers the potential to illuminate any narrow states, possibly of exotic structure. Most of the states presently covered appear to have very small coupling to the ηN channel, and this in itself can be puzzling. For example, it is unclear why the ηN branching ratio is so small for the second S_{11}, $N(1650)1/2^-$, compared to the first $N(1535)1/2^-$. The reason for a small branching ratio of $N(1520)3/2^-$ to ηN [1, 4] has to be understood, too. Further, the PDG estimate for the $A_{3/2}$ decay amplitude of the $N(1720)3/2^+$ state is consistent with zero, while the recent SAID determination gives a small but non-vanishing value [5]. The reason for the disagreement between the PDG estimate for the $A_{1/2}$ decay amplitude and the recent SAID determination [5] is also unclear. Other unresolved issues are related to the second P_{11} and D_{13} resonances [$N(1710)1/2^+$ and $N(1700)3/2^-$] that are not seen in the recent πN partial-wave analysis (PWA) [2], contrary to other PWAs used by the PDG [1]. The ηN decay channel could be more favorable than πN for revealing these states. The present data have very good precision which may allow reliable extraction of the ηN partial waves related to those resonances. This would enhance our understanding of their internal dynamics. In addition, since the present data have good coverage of the ηp-threshold region, the S-wave dominance of the threshold behavior can also be checked.

The process $\gamma p \to \eta p$ was measured using the Crystal Ball (CB) [6] as the central spectrometer and TAPS [7] as a forward spectrometer. These detectors were installed in the tagged bremsstrahlung photon beam of the Mainz Microtron (MAMI) [8], with the photon energies determined by the Glasgow Tagging Spectrometer [9]. Further details on the experimental setup can be found in Ref. [10].

The photoproduction of η was observed by using the $3\pi^0$ decay mode. Process $\gamma p \to \eta p \to 3\pi^0 p \to 6\gamma p$ was

12th International Conference on Meson-Nucleon Physics and the Structure of the Nucleon (MENU 2010)
AIP Conf. Proc. 1374, 439-442 (2011); doi: 10.1063/1.3647176

investigated by analysis of events having six and seven "software" clusters reconstructed in both the CB and TAPS. The six-cluster sample was used to search for the events in which only six photons were detected, while for seven clusters the recoiling proton was also detected. The kinematic-fitting technique was used to test the reaction hypotheses needed in our analysis and to select good candidates for the events of interest. The MC-simulation of the $\gamma p \to \eta p \to 3\pi^0 p$ reaction that was used for the determination of the differential cross sections assumed an isotropic distribution of the production angle and independence of the reaction yield on the incident-photon energy. The simulation of the $\eta \to 3\pi^0$ decay was made according to phase space. The small deviation of the actual $\eta \to 3\pi^0$ decays from phase space was not significant in our analysis. More details on the data handling can be found in Ref. [10].

The large number of events accumulated allowed the division of the data into 120 bins in E_γ. The $\cos\theta$ spectrum at every energy was divided into 20 bins. The results presented for the total, $\sigma_t(\gamma p \to \eta p)$, and differential cross sections include only statistical uncertainties. The largest contributions to the systematic uncertainty come from the calculation of the experimental acceptance by the MC-simulation and from the determination of the photon-beam flux. The general systematic uncertainty in our $\gamma p \to \eta p$ cross sections was estimated to be 4%.

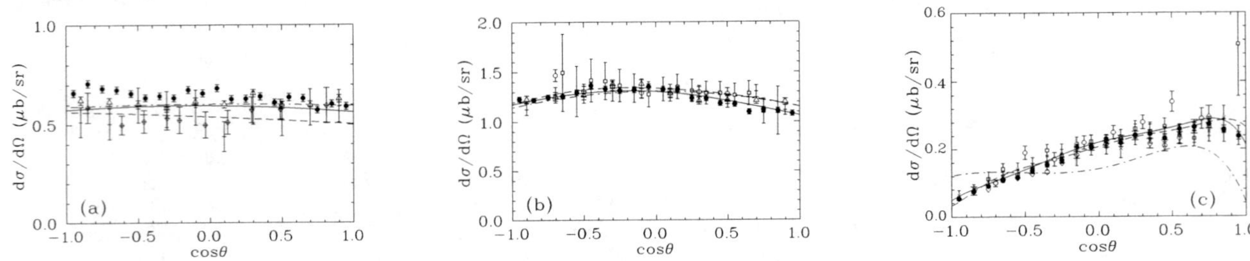

FIGURE 1. Differential cross sections for $\gamma p \to \eta p$ as a function of $\cos\theta$, where θ is the production angle of η in the c.m. frame. The present data (solid circles) are shown for three energy bins: (a) $E_\gamma = 714.5 \pm 2.1$ MeV, (b) 772.9 ± 2.1 MeV, (c) 1376.2 ± 9.7 MeV. Previous data are shown for experiments at MAMI-B [11] for 715.9 ± 5 MeV and 775.3 ± 5 MeV (open triangles); CLAS-g1c [12] for 775 ± 25 MeV, and 1375 ± 25 MeV (open circles); CLAS-g11a [13] for 1384 ± 10 MeV (open diamonds); GRAAL [14] for 714 ± 9 MeV (open diamonds with crosses); LNS [15] for 718.0 ± 10 MeV and 768.8 ± 10 MeV (horizontal bars); CB-ELSA [16] for 774 ± 25 MeV, and 1374 ± 25 MeV (open squares); and CB-ELSA/TAPS [17] for 1375 ± 25 MeV (crosses). Plotted uncertainties are statistical only. The curves denote SAID solution GE09 (solid line), SAID solution E429 (dashed line), and η-MAID solution (dotted line)

In Fig. 1, our differential cross sections for three incident-photon energies are compared to previous measurements made at similar energies [11, 12, 13, 14, 15, 16, 17]. The lowest energy shown, $E_\gamma = 714.5$ MeV ($W = 1490.3$ MeV), is close to the η-production threshold. The second energy, $E_\gamma = 772.9$ MeV ($W = 1526.7$ MeV), is at the maximum of the total cross section. The third energy, $E_\gamma = 1376.2$ MeV ($W = 1860.9$ MeV) is close to the upper end of our incident-photon energy range. As seen in Fig. 1, all our results are in reasonable agreement with the previous measurements, but our statistical uncertainties are much smaller and the energy binning much finer.

FIGURE 2. (a) Total cross section for $\gamma p \to \eta p$ as a function of the c.m. energy. Uncertainties plotted for our data are statistical only. Notation of different data is the same as in Fig. 1. (b) Dominant Legendre coefficients from the fits to our differential cross sections. The coefficients are plotted as a function of the c.m. energy; A_1 is shown by solid circles, A_2 by open triangles, and A_3 by open circles.

The present total cross sections for $\gamma p \to \eta p$ are obtained by integration of the differential cross sections. In Fig. 2(a), our total cross sections are compared to previous measurements [11, 12, 13, 14, 15, 16, 17] over the full energy range presently measured. Our results for the total cross sections are in general agreement with the major

TABLE 1. Individual contributions of different measurements of the differential cross sections to the total χ^2 value for the E429, GE09, η-MAID, and new Regge-MAID analyses of $\gamma p \to \eta p$ data (corresponding numbers are given in 7th column).

Experiment	E_γ (MeV)	W (MeV)	θ (deg)	Stat/Syst (%)	Data	χ^2/data	Ref
CB-MAMI-C	707−1402	1487−1875	0−180	2/4	2400	5.5−1.4−6.8−3.2	Present
GRAAL	714−1477	1491−1913	32−162	10/3	487	1.2−0.7−2.1−1.2	[14]
TAPS-MAMI-B	716− 790	1493−1539	26−154	4/4	100	1.4−1.4−1.2−4.9	[11]
LNS	718−1142	1494−1740	26−154	2/6	180	0.9−1.5−0.9−2.3	[15]
CB-ELSA	774−2887	1530−2511	18−139	2/15	631	1.3−1.3−3.1−2.6	[16]
CLAS-g1c	775−1925	1530−2121	46−134	3/5	190	2.3−2.3−5.6−5.4	[12]
CB-ELSA/TAPS	875−2522	1590−2372	18−162	4/10	680	2.6−2.6−11.2−3.3	[17]
CLAS-g11a	1044−2861	1690−2502	33−148	7/11	979	4.3−5.7−12.0−9.5	[13]

previous results. The energy range lying above a c.m. energy of 1640 MeV is especially important for untangling the six overlapping N^* states and investigation of a possible narrow N^* state in the mass range $\sim$ 1680 MeV. The $N^*(1680)$ was extracted in Ref. [19] from the πN PWA and suggested to be a member of the exotic anti-decuplet. Indeed a resonant bump at $\sim$ 1680 MeV is observed in quasi-free $\gamma n \to \eta n$ [20, 21, 22]. However, for this reaction, the measured width of the bump was dominated by the experimental energy resolution. Inspection of our $\sigma_t(\gamma p \to \eta p)$ energy dependence shows no evidence for a narrow bump related to a $N^*(1680)$ state in η photoproduction on a free proton. Rather our data show the existence of a shallow dip near $W = 1680$ MeV. However, such a situation may not contradict to the existence of a narrow bump in η photoproduction on a neutron, since γn and γp couplings of the $N^*(1680)$ can be essentially different (as for an anti-decuplet member).

The full angular coverage of our differential cross sections allied with the small statistical uncertainties allows a reliable determination of the Legendre coefficients A_i, which was difficult to do with the previous data. This unprecedented detail of the energy dependence of the Legendre coefficients will be indispensable in untangling the properties of the N^* states lying in the present energy range. In Fig. 2(b), we illustrate Legendre coefficients $A_1 - A_3$ (higher orders are relatively insignificant) as a function of the c.m. energy. The swing in A_1 from negative to positive values in the vicinity of $W = 1680$ MeV is intriguing.

To gauge the influence of our data and their compatibility with previous measurements, our differential cross sections have been included in a number of fits using the full SAID database for $\gamma p \to \eta p$ up to $E_\gamma = 2.9$ GeV. The impact of our data on the SAID PWA can be understood from the comparison of the new SAID fit GE09, which involves our data, with the previous SAID fit E429. The other data included in the GE09 fit involve all previously published data except recent CLAS-g11a [13], CB-ELSA/TAPS [17], and LEPS [18] differential cross sections. Our data were also included in the PWA under the Reggeized η-MAID model (Regge-MAID) [24] that was extended to a photon energy $E_\gamma = 3.7$ GeV (W=2.8 GeV) by adding new resonances in the s-channel. The η-MAID analysis involves only the data published up to 2002. The details of the new SAID and Regge-MAID PWAs will be the subject of future publications. The separate contributions of individual data sets to the total χ^2 value are listed for each of the four $\gamma p \to \eta p$ analyses in Table 1. As seen the more recent data sets display a greater degree of consistency.

For completeness, in Fig. 3 we compare the GE09, E429, η-MAID, and new Regge-MAID solutions for the $\gamma p \to \eta p$ excitation function at the extreme production angles: forward ($\theta = 0°$) and backward ($\theta = 180°$). The new data along with the new fits definitely indicate the existence of a dip structure around $W = 1670$ MeV, which has been already seen in our total cross section (Fig. 2(a)) and becomes even more pronounced at forward production angles of η (see Fig. 3). This feature was missed or questionable in the analysis of the previous data.

In summary, the $\gamma p \to \eta p$ differential cross sections have been measured at the tagged photon facility of the Mainz Microtron MAMI-C using the Crystal Ball and TAPS spectrometers. The data span the photon-energy range 707 to 1402 MeV and the full angular range in the c.m. frame. The accumulation of 3×10^6 $\gamma p \to \eta p \to 3\pi^0 p \to 6\gamma p$ events allows the fine binning of the data in energy and angle, which will enable the reaction dynamics to be studied in greater detail than previously possible. The present data agree well with previous equivalent measurements, but are markedly superior in terms of precision and energy resolution. The present cross sections for the free proton show no evidence of enhancement in the region $W \sim 1680$ MeV, contrary to recent measurements on the quasi-free neutron [20, 21, 22], though this does not exclude existence of the $N^*(1680)$ state as hypothesized in Ref. [19]. In the region around $W = 1680$ MeV, we rather observe a dip structure that becomes more pronounced at forward production angles of η. This feature was missed or questionable in the analysis of the previous data. The interpretation of this dip depends on

FIGURE 3. PWA predictions for the $\gamma p \to \eta p$ excitation function at the extreme [(a) forward ($\theta = 0°$) and (b) backward ($\theta = 180°$)] production angles of η, shown as a function of the c.m. energy W. Notation of the PWA predictions is the same as in Fig. 1.

dynamics. Our $\gamma p \to \eta p$ data points have been included in a new SAID (GE09) and Regge-MAID PWAs, to which they made substantial contributions, particularly for forward angles. Comparing to the previous SAID fit, E429, and to the η-MAID fit, the description of all existing data by the new solutions, GE09 and Regge-MAID, is more satisfactory in the entire energy range.

ACKNOWLEDGMENTS

This work was supported by the US DOE (Grants No. DE–FG02–99ER41110 and DE–FG02–01ER41194), NSF (Grant No. PHY 0652549), the Jefferson Laboratory, the Southeastern Universities Research Association (DOE Contract No. DE–AC05–84ER40150), SNF of Switzerland, and the Russian State grant SS-3628.2008.2.

REFERENCES

1. K. Nakamura *et al.* (Particle Data Group), J. Phys. G **37**, 075021 (2010).
2. R. A. Arndt, W. J. Briscoe, I. I. Strakovsky, and R. L. Workman, Phys. Rev. C **74**, 045205 (2006).
3. R. A. Arndt, W. J. Briscoe, M. W. Paris, I. I. Strakovsky, and R. L. Workman, Chinese Phys. C **33**, 1063 (2009). 045205 (2006).
4. R. A. Arndt, W. J. Briscoe, T. W. Morrison, I. I. Strakovsky, R. L. Workman, and A. B. Gridnev, Phys. Rev. C **72**, 045202 (2005).
5. M. Dugger *et al.* (CLAS Collab.), Phys. Rev. C **79**, 065206 (2009).
6. A. Starostin *et al.*, Phys. Rev. C **64**, 055205 (2001).
7. R. Novotny, IEEE Trans. Nucl. Sci. **38**, 379 (1991); A. R. Gabler *et al.*, Nucl. Instrum. Methods A **346**, 168 (1994).
8. H. Herminghaus *et al.*, IEEE Trans. Nucl. Sci. **30**, 3274 (1983); K.-H. Kaiser *et al.*, Nucl. Instrum. Methods A **593**, 159 (2008).
9. I. Anthony *et al.*, Nucl. Instrum. Methods A **301**, 230 (1991); S. J. Hall *et al.*, Nucl. Instrum. Methods A **368**, 698 (1996); J. C. McGeorge *et al.*, Eur. Phys. J. A **37**, 129 (2008).
10. Crystal Ball Collab. at MAMI (E.F. McNicoll *et al.*), Phys. Rev. C **82**, 035208 (2010).
11. B. Krusche *et al.*, Phys. Rev. Lett. **74**, 3736 (1995).
12. M. Dugger *et al.* (CLAS Collab.), Phys. Rev. Lett. **89**, 222002 (2002).
13. M. Williams *et al.* (CLAS Collab.), Phys. Rev. C **80**, 045213 (2009).
14. O. Bartalini *et al.* (GRAAL Collab.), Eur. Phys. J. A **33**, 169 (2007).
15. T. Nakabayashi *et al.*, Phys. Rev. C **74**, 035202 (2006).
16. V. Credé *et al.* (CB-ELSA Collab.), Phys. Rev. Lett. **94**, 012004 (2005).
17. V. Credé *et al.* (CB-ELSA/TAPS Collab.), Phys. Rev. C **80**, 055202 (2009).
18. M. Sumihama *et al.* (LEPS Collab.), Phys. Rev. C **80**, 052201 (2009).
19. R. A. Arndt, Ya. I. Azimov, M. V. Polyakov, I. I. Strakovsky, and R. L. Workman, Phys. Rev. C **69**, 035208 (2004).
20. V. Kuznetsov *et al.* (by GRAAL Collab.), Phys. Lett. B **647**, 23 (2007).
21. I. Jaegle *et al.* (CB-ELSA/TAPS Collab.), Phys. Rev. Lett. **100**, 252002 (2008).
22. F. Miyahara *et al.*, Prog. Theor. Phys. Suppl. **168**, 90 (2007).
23. W.-T. Chiang, S.-N. Yang, L. Tiator, and D. Drechsel, Nucl. Phys. **A700**, 429 (2002).
24. W.-T. Chiang, S.-N. Yang, L. Tiator, M. Vanderhaeghen and D. Drechsel, Phys. Rev. C **68**, 045202 (2003).
25. R. A. Arndt, W. J. Briscoe, I. I. Strakovsky, and R. L. Workman, Phys. Rev. C **66**, 055213 (2002).

Radiative decay into γ-baryon of dynamically generated resonances from the vector-baryon interaction

E. J. Garzon*, Bao-Xi Sun† and E. Oset*

*Departamento de Física Teórica and IFIC, Centro Mixto Universidad de Valencia-CSIC, Institutos de Investigación de Paterna, Aptdo. 22085, 46071 Valencia, Spain
†Institute of Theoretical Physics, College of Applied Sciences, Beijing University of Technology, Beijing 100124, China

Abstract. We study the radiative decay into γ and a baryon of the SU(3) octet and decuplet of nine and ten resonances that are dynamically generated from the interaction of vector mesons with baryons of the octet and the decuplet respectively. We obtain quite different partial decay widths for the various resonances, and for different charge states of the same resonance, suggesting that the experimental investigation of these radiative decays should bring much information on the nature of these resonances. For the case of baryons of the octet we determine the helicity amplitudes and compare them with experimental data when available.

Keywords: radiative decay, resonances, VV interaction
PACS: 14.20.Gk, 12.40.Vv

INTRODUCTION

Radiative decay of resonances into γN is one of the observables traditionally calculated in hadronic models. The underlying theory of the studies of [1, 2] is the local hidden gauge formalism for the interaction of vector mesons developed in [4, 5, 6, 7], which has the peculiar feature, inherent to vector meson dominance, that the photons couple to the hadrons through the conversion into a vector meson as ρ^0, ω, ϕ.

The extension of the work of [1, 2] to the interaction of vector mesons with baryons of the octet of the proton has also been successful [3] and nine resonances, degenerated in spin-parity $1/2^-$ and $3/2^-$, appear dynamically generated. We also extend the present work to study the radiative decay of these resonances into a photon and a baryon of the octet. In this case we can also evaluate helicity amplitudes and compare them with experimental results when available.

RADIATIVE DECAY

We study the radiative decay into $B\gamma$ of the resonances dynamically generated in Ref. [2, 3] with B a baryon of the decuplet and the octet. The radiative width will not depend on the spin of the resonance, but only on the coupling which is the same for all spin states due to the degeneracy. This would be an interesting experimental test of the nature of these resonances.

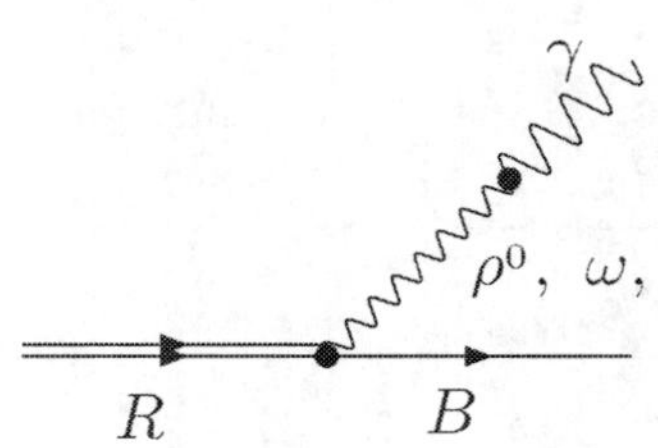

FIGURE 1. Diagram on the radiative decay of the resonance in a baryon and a photon.

12th International Conference on Meson-Nucleon Physics and the Structure of the Nucleon (MENU 2010)
AIP Conf. Proc. 1374, 443-446 (2011); doi: 10.1063/1.3647177

The peculiarity of this theory concerning photons is that they couple to hadrons by converting first into a vector meson, ρ^0, ω, ϕ. The amplitude of Fig. 1 requires the γV conversion Lagrangian, which comes from Refs. [4, 5, 6] and is given by

$$\mathscr{L}_{V\gamma} = -M_V^2 \frac{e}{\tilde{g}} A_\mu \langle V^\mu Q \rangle \tag{1}$$

with A_μ the photon field, V_μ the SU(3) matrix of vector fields and Q the charge matrix. Our amplitude for the $R \to B\gamma$ transition is given by

$$t_\gamma = -\frac{e}{\tilde{g}} \sum_{j=\rho^0,\ \omega,\ \phi} g_j F_j. \tag{2}$$

with $F_j = \frac{1}{\sqrt{2}}$ for ρ^0, $\frac{1}{3\sqrt{2}}$ for ω and $-\frac{1}{3}$ for ϕ. The radiative decay width of the resonance for decay into VB is given in the Mandl and Shaw normalization by

$$\Gamma_\gamma = \frac{1}{2\pi} \frac{2}{3} \frac{M_B}{M_R} q\, |t_\gamma|^2. \tag{3}$$

The couplings g_j for different resonance can be found in Ref. [2, 3] and we use them here for the evaluation of Γ_γ.

RESULTS FOR RADIATIVE DECAYS INTO γ AND BARYON

The couplings of the resonances to the different VB channels are given in Ref. [2, 3] in the isospin basis. For the case of ωB and ϕB, there is no change to be done, but for the case of ρB, one must project over the $\rho^0 B$ component. In Table 1 we summarize all the results obtained for the baryon decuplet making an association of our states to some resonances found in the PDG[8]. For the case of a baryon octet in the final state we compile the results in Table 2. As one can see, there is a large variation in the radiative width of the different states, which should constitute a good test for the model when these widths are measured.

TABLE 1. The predicted radiative decay widths of the 10 dynamically generated resonances for different isospin projection I_3. Their possible PDG counterparts are also listed.

S, I	Theory	PDG data		Predicted width (KeV) for I_3						
	pole position (MeV)	name	J^P	$-3/2$	-1	$-1/2$	0	$1/2$	1	$3/2$
$0, 1/2$	$1850+i5$	$N(2090)$	$1/2^-$			722		722		
		$N(2080)$	$3/2^-$							
$0, 3/2$	$1972+i49$	$\Delta(1900)$	$1/2^-$	1582		203		143		1402
		$\Delta(1940)$	$3/2^-$							
		$\Delta(1930)$	$5/2^-$							
$-1, 0$	$2052+i10$	$\Lambda(2000)$	$?^?$				583			
$-1, 1$	$1987+i1$	$\Sigma(1940)$	$3/2^-$		20		199		561	
		$\Sigma(2000)$	$1/2^-$							
	$2145+i58$	$\Sigma(2250)$	$?^?$		2029		206		399	
	$2383+i73$	$\Sigma(2455)$	$?^?$		537		277		182	
$-2, 1/2$	$2214+i4$	$\Xi(2250)$	$?^?$			54		815		
	$2305+i66$	$\Xi(2370)$	$?^?$			1902		320		
	$2522+i38$	$\Xi(2500)$	$?^?$			165		44		
$-3, 1$	$2449+i7$	$\Omega(2470)$	$?^?$				330			

TABLE 2. The predicted radiative decay widths of the nine dynamically generated resonances for different isospin projection I_3. The values in the bracket for $I_3 = 0$ denote widths for the radiative decay into $\Lambda\gamma$, while the values outside the bracket denote widths for $\Sigma\gamma$.

S, I	Theory	PDG data		Predicted width (KeV) for I_3				
	pole position (MeV)	name	J^P	-1	$-1/2$	0	$1/2$	1
$0, 1/2$	1696	$N(1650)$	$1/2^-$		334		253	
		$N(1700)$	$3/2^-$					
	$1977 + i53$	$N(2080)$	$3/2^-$		196		79	
		$N(2090)$	$1/2^-$					
$-1, 0$	$1784 + i4$	$\Lambda(1690)$	$3/2^-$			65 (166)		
		$\Lambda(1800)$	$1/2^-$					
	$1907 + i70$	$\Lambda(2000)$	$?^?$			321 (21)		
	$2158 + i13$					0 (17)		
$-1, 1$	$1830 + i40$	$\Sigma(1750)$	$1/2^-$	363		69 (240)		7
	$1987 + i240$	$\Sigma(1940)$	$3/2^-$	307		27 (90)		426
		$\Sigma(2000)$	$1/2^-$					
$-2, 1/2$	$2039 + i67$	$\Xi(1950)$	$?^?$		400		89	
	$2082 + i31$	$\Xi(2120)$	$?^?$		212		84	

HELICITY AMPLITUDES

For the case of the vector-baryon octet states it is customary to express the experimental information in terms of helicity amplitudes $A_{1/2}$ and $A_{3/2}$. It is interesting to note that the values of $A_{1/2}$ for $J = 1/2$, $3/2$ and $A_{3/2}$ for $J = 3/2$ are all related and the ratio of $1/\sqrt{3}$ between $A_{1/2}^{J=3/2}$ and $A_{3/2}^{J=3/2}$ could be contrasted with experiment. We compile in Table 3 all the results obtained for the resonances that are likely to be associated to states in the PDG for which there are data. About the signs of the helicity amplitudes, we should mention that the g_j couplings appearing there are all determined with their relative sign, but one of them has an arbitrary sign. Thus we have an arbitrary sign in the helicity amplitudes, but the relative sign between the $A_{1/2}^{J=3/2}$ and $A_{3/2}^{J=3/2}$ is well determined. Experimentally the phases of the helicity amplitudes are arbitrary.

For the case of $A_{1/2}^n$ the results obtained are larger than experiment but the sign is good. In the case of the $N^*(1700)$, $A_{1/2}^p$ can be considered qualitatively fine within theoretical and experimental errors. The case of the $N^*(2080)$ and $N^*(2090)$ is more unclear since there are apparently different states compiled, it would be possible that the averages of the PDG were done for measurements on different states. The experimental situation is hence unclear but the results obtained here should be a motivation for further reanalysis.

CONCLUSIONS

We have studied the radiative decay into γB, with B a baryon of the octet and decuplet of SU(3), of the dynamically generated resonances obtained within the framework of the local hidden gauge mechanism for vector interactions. The rates obtained are large and the radiative widths are of the order of 1 MeV. On the other hand, one of the appealing

TABLE 3. Comparison with experiments and other theories

PDG data		Helicity amplitudes $10^{-3}(GeV^{-1/2})$				
name	J^P	Decay	Theory	Exp.[8] PDG	Exp.[9] Barbour	Exp.[10] Devenish
$N(1650)$	$1/2^-$	$A^p_{1/2}$	64 ± 7	53 ± 16		
		$A^n_{1/2}$	-74 ± 7	-15 ± 4		
$N(1700)$	$3/2^-$	$A^p_{1/2}$	-46 ± 5	-18 ± 13		
		$A^p_{3/2}$	-79 ± 9	-2 ± 24	-14 ± 25	
		$A^n_{1/2}$	52 ± 5	0 ± 50	50 ± 42	
		$A^n_{3/2}$	91 ± 9	-3 ± 44	35 ± 30	
$N(2080)$	$3/2^-$	$A^p_{1/2}$	21 ± 5	-20 ± 8		26 ± 52
		$A^p_{3/2}$	36 ± 8	17 ± 11		128 ± 57
		$A^n_{1/2}$	29 ± 5	7 ± 13		53 ± 83
		$A^n_{3/2}$	50 ± 8	-53 ± 34		100 ± 141
$N(2090)$	$1/2^-$	$A^p_{1/2}$	-30 ± 6			
		$A^n_{1/2}$	-41 ± 6			

features of the results is the large difference, of about one order of magnitude, that one finds between the widths for different charge states of the same particle.

In the case of the octet one can define the helicity amplitude $A_{1/2}$ and $A_{3/2}$ for the n and p type states of the N^*. We have found good agreement with data in some cases and rough in others, but we have warned about the large experimental uncertainties and the possibility that the PDG averages are done over different states. Further experimental work is most desirable and we hope the present work stimulates work in this direction.

ACKNOWLEDGMENTS

This work is partly supported by DGICYT contract number FIS2006-03438. This research is part of the EU Integrated Infrastructure Initiative Hadron Physics Project under contract number RII3-CT-2004-506078. B. X. Sun acknowledges support from the National Natural Science Foundation of China under grant number 10775012.

REFERENCES

1. P. Gonzalez, E. Oset and J. Vijande, Phys. Rev. C **79**, 025209 (2009) [arXiv:0812.3368 [hep-ph]].
2. S. Sarkar, B. X. Sun, E. Oset and M. J. V. Vacas, Eur. Phys. J. A **44**, 431 (2010)
3. E. Oset and A. Ramos, Eur. Phys. J. A **44**, 445 (2010) [arXiv:0905.0973 [hep-ph]].
4. M. Bando, T. Kugo, S. Uehara, K. Yamawaki and T. Yanagida, Phys. Rev. Lett. **54**, 1215 (1985).
5. M. Bando, T. Kugo and K. Yamawaki, Phys. Rept. **164**, 217 (1988).
6. M. Harada and K. Yamawaki, Phys. Rept. **381**, 1 (2003) [arXiv:hep-ph/0302103].
7. U. G. Meissner, Phys. Rept. **161**, 213 (1988).
8. C. Amsler *et al.* [Particle Data Group], Phys. Lett. B **667**, 1 (2008).
9. I. M. Barbour, R. L. Crawford and N. H. Parsons, Nucl. Phys. B **141** (1978) 253.
10. R. C. E. Devenish, D. H. Lyth and W. A. Rankin, Phys. Lett. B **52** (1974) 227.

The Search for Missing Resonances: the Neutron Anomaly.

Igal Jaeglé for CB-ELSA/TAPS and A2 Collaborations

Department of Physics, University of Basel, Klingelbergstrasse 82, CH-4056 Basel, Switzerland

Abstract. We report the observation of a narrow structure in the invariant mass spectrum of the η−neutron pairs produced in the reaction $\gamma d \rightarrow \eta n(p)$ which is absent in the η−proton pairs produced in the reaction $\gamma d \rightarrow \eta p(n)$. These results are obtained from a liquid deuterium target, the data sample was collected with the Crystal Barrel and TAPS detectors at the electron accelerator ELSA. This structure is also seen in preliminary analyses of the data samples taken by the Crystal Ball and TAPS detectors at the electron accelerator MAMI with a liquid deuterium target and a liquid helium 3 target ruling out a nuclear effect. We report the measurements of the position and width of this narrow structure in the invariant mass spectrum at W $\approx$ 1665 MeV and with a FWHM of Γ = 25 MeV.

Keywords: photoproduction, η-neutron pairs, narrow structure, ELSA, MAMI, Crystal Ball, Crystal Barrel, TAPS
PACS: 13.60.Le, 14.20.Gk, 14.40.Aq

INTRODUCTION

At low energy, the nucleon can be considered to be a bound system made of three quarks, (uud) quarks for the proton and (udd) quarks for the neutron, and is an excellent laboratory to study QCD in the non-pertubative regime. One of the main question concerning this system and which various QCD-based models fail to answer clearly is: what are the degrees-of-freedom of this system ?

Experimentally, the degrees-of-freedom of the nucleon are studied by exciting the nucleon by an external probe. The excited states of the nucleon were first observed in πN scattering in which their contribution was clearly evident as bumps in the total cross section. These measurements allowed a first classification of the excitation spectrum of the nucleon, providing measurement of the masses, widths, quantum numbers, and branching ratios of many baryon resonances [1]. In spite of the large amount of information collected by these experiments, the number of states that were identified was less than that predicted by the standard quark model [2]. A possible explanation is that such "missing" states may decouple from the πN channel, making them undetectable in experiments with pion beams. Other explanations come from theoretical models that are able to predict a smaller number of states based on a reduced set of degrees-of-freedom [3].

The construction of high intensity and high duty cycle electron and photon facilities such as the Elektronen-Stretcher-Anlage (ELSA) in Bonn [4][5] and the Mainzer Mikrotron (MAMI) in Mainz [6][7], opened new possibilities for the study of baryon resonances using electromagnetic probes. These provide information about the resonance and nucleon wavefunctions through the measurement of the helicity amplitudes, i.e. the electromagnetic couplings between nucleon ground state and initial states. In addition electroproduction also allows us to explore baryon structure for different distance scales by varying the photon virtuality. Nowadays electroexcitation processes are a fundamental tool to pursue these studies. However due to the complexity of the baryon spectrum, the proximity and overlapping nature of the various excited states, the measurement of a single channel is not sufficient to complete this research program. To the contrary a thorough study of resonance properties requires the measurement of cross sections, angular distributions, Dalitz plots, as well as polarization observables for different final states. This program is carried since more than ten years by the CB-ELSA/TAPS and A2 Collaborations. In addition the interpretation of the observables extracted from the data requires sophisticated partial waves analyzes.

Most of the data available and taken with a photon probe comes from meson photoproduction off the proton, for the simple reason that there are no free neutron targets. However, the study of neutron excited states is of some interests, since the only way to access the isospin composition of the nucleon resonances is to measure both proton and neutron excited states. Furthermore some excited states might couple stronger to γn for example due to SU(3) symmetries like the Moorhouse selection rules [8].

12th International Conference on Meson-Nucleon Physics and the Structure of the Nucleon (MENU 2010)
AIP Conf. Proc. 1374, 447-451 (2011); doi: 10.1063/1.3647178

The non-existence of free neutron targets is solved by using a light nucleus (deuterium or/and helium). The difficulty with using a light nucleus as a target to learn about neutrons, however, is that the kinematics are "altered", when it is bound inside a nucleus, by Fermi motion and other nuclear effects. While the complexity of the baryon spectrum is simplified by using a specific meson photoproduction that can tag specific resonances as it turned out with the photoproduction of η-meson off the nucleons. The η works as an isospin filter, due to isospin conservation: only N^* resonances contribute to $N\eta$ while resonances in $\Delta\eta$ belong to the Δ^* series.

The η-meson photoproduction off the proton was widely studied [10, 11, 12, 13, 14, 15, 16, 17, 18, 19, 20]. It was found that the resonance $S_{11}(1535)$ is completely dominating the photoproduction of η-meson at threshold and shows up as bump in the total cross section. Above 1600 MeV in mass location the situation is less clear. The eta-MAID model [21] considers for example the contribution of the resonance $P_{11}(1710)$ while the Bonn-Gatchina model [17, 20, 22] considers the resonance $P_{13}(1730)$. Although the preliminary results of the CB-ELSA/TAPS Collaboration for the double polarization observable E seems to indicate contribution of partial waves with spin 1/2 (for more details see the contribution of R. Beck to this conference proceedings). In general, it is considered that the background terms are pretty low and that the photoproduction of η-mesons involves less than 12 resonances below 2.5 GeV (in invariant mass).

On the other hand the η-meson photoproduction off the neutron is less studied (for the reasons mentioned above) but since a few years it received more attention because four different experiments GRAAL in Grenoble [23], LNS in Sendai [24], CB-ELSA/TAPS in Bonn [9] and more recently A2 in Mainz [25] show a bump structure in the total cross section around 1700 MeV which is not seen in the total cross section of the η-meson photoproduction off the proton. This structure seems to get narrower if the Fermi motion is removed. In absence of double polarization observables, there are no explanations of this so-called "neutron-anomaly". Nevertheless there are several scenario proposed. The eta-MAID model [21] attributes this bump to the resonance $D_{15}(1675)$ but the $N\eta$ branching ratio of this state is 17 % in this model. This value is contradictory with other results which give an upper limit of $< 1\%$ [26]. An analysis done with the Bonn-Gatchina model [27] can reproduce the neutron data by taking into account the Fermi motion effect with three completely different scenarios, by either adding a 'conventionally' broad P_{11} resonance, a very narrow P_{11} state, or even by adjusting the interference pattern for the *S*-wave amplitudes. The Giessen coupled channel model [28] can also produce a bump in the neutron excitation around 1 GeV (in photon energy) due to the S_{11} - P_{11} sector, without introducing any additional resonance. Döring and Nakayama [29] can create a bump by using an *S*-wave coupled channel model in the neutron total cross section which is related to the opening of strangeness thresholds of $K\Lambda$ and $K\Sigma$ photoproduction around 900 MeV and 1050 MeV (for details see the proceeding of M. Döring in which these results were presented). Finally, the chiral soliton model [30, 31] predicted a state in this energy range, which has a much stronger photon coupling to the neutron than to the proton and a large decay branching ratio into $N\eta$. This state is the nucleon-like member of the anti-decuplet of pentaquarks, which would be a P_{11} state but the evidence for the existence of the Θ^+-pentaquark is strongly doubted.

The results reported here correspond to a new analysis of the data already published [9] where the Fermi motion was removed by reconstructing the η-neutron pairs invariant mass (for all cos(Θ_η^{cm}) unlike [9] where it was done for cos(Θ_η^{cm}) < -0.1). In addition very preliminary analysis of the A2 Collaboration will be discussed. But first the experimental setups and then the reaction identification of $\gamma + d \rightarrow \eta + n_{participant} + p_{spectator}$ will be shortly described.

CRYSTAL BARREL AND TAPS AT ELSA SETUP

The Bonn real photon beam was made by directing the electron beam (with can reach the maximum energy of 3.5 GeV) on a copper foil, where it produces photons by the Bremsstrahlung process. The photon energies were tagged via the momentum analysis of the scattered electron by a magnetic spectrometer. The Bremsstrahlung photons are nearly collinear with the incident electron beam and pass through a hole in the magnet yoke, after which they are collimated, and then impinge on the deuterium target (of 5.275 cm length) which sits in the center of an almost 4π detection system. It was composed of: the Crystal Barrel (CB, 1290 CsI crystals covering the full azimuthal angle for polar angles between $30^o$and 168° and the TAPS detectors (528 BaF2 crystals mounted as a hexagonal forward wall covering polar angles down to 4.5^o), and their respective Charge Particle Counters (CPC), the inner detector (three

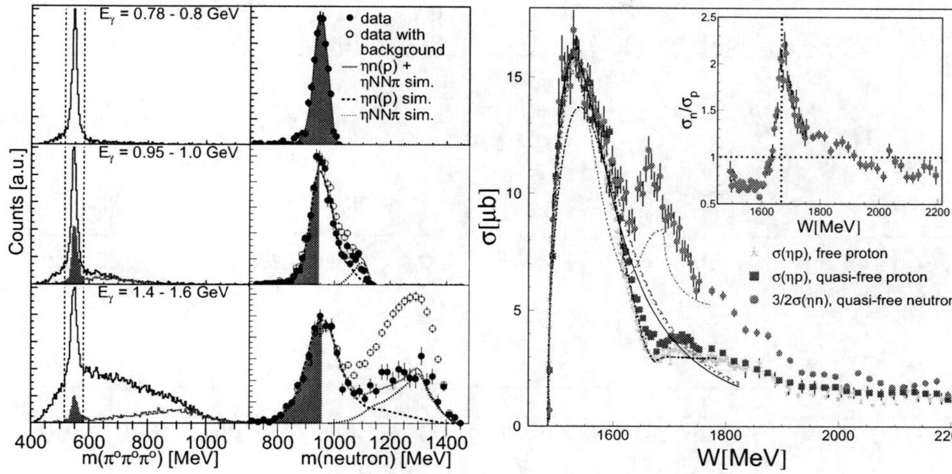

FIGURE 1. Left: Invariant (left hand column) and missing mass (right hand column) spectra for η-mesons in coincidence with recoil neutrons for three ranges of incident photon energy. Invariant masses: shaded (blue) signal after missing mass cut. Dashed lines: applied invariant mass cut. Missing mass data: open symbols represents data for indicated cut on invariant mass. Black dots: background subtracted by fitting invariant mass spectra for each bin of missing mass. Simulations: dashed curves (dotted) curves: simulation of η ($\eta\pi$) final states. Solid (red) curves: sum of simulations. Shaded (blue) areas: accepted events. Right: Total cross sections as function of final state invariant mass W without cut on spectator momentum. (Red) dots: quasi-free neutron, (blue) squares: quasi-free proton, (green) stars: free proton data. Curves: fitted (up to W = 1600 MeV) $S_{11}(1535)$ line shapes. (Black) solid: free proton, (blue) dashed: quasi-free proton, (red) dotted: quasi-free neutron. Insert: ratio of quasi-free neutron - proton data. Dash-dotted curves: model results from [29].

layers of plastic scintillating fibers) and the veto wall (528 plastic scintillators). More details can be found in [32].

CRYSTAL BALL AND TAPS AT MAMI SETUP

The Mainz real photon beam was produced in a similar way with a maximum electron energy of 1508 MeV but the beam intensity was an order of magnitude higher than in Bonn. The MAMI-Glasgow tagging spectrometer [33] was used to detect the electrons deflected by a magnetic field. The photon was incident on a 4.6 cm liquid deuterium target or a 5.3 cm liquid helium 3 target installed in the center of the Crystal Ball (CB) [34] made of 672 NaI(Tl) triangular-pyramidal crystals arranged in the form of two hemispheres that has two 21° openings. The downstream opening was covered by the TAPS [35, 36] wall. which consists of 360 BaF_2 crystals, up to 1°. Both detectors had their Charge Particle Identifier (CPI), the particle identification detector (PID) and the veto wall (320 plastic scintillators).

REACTION IDENTIFICATION

Only the reaction identification of $\gamma + d \to \eta + n_{participant} + p_{spectator}$ (with η decaying into $3\pi^0$) for the CB-ELSA/TAPS experiments will be discussed. We consider all events with seven neutral hits and no charged hits i.e. the CPC did not record a hit. The best combination of $3\pi^0$ is built by using a χ^2-test on the π^0 mass. The left over hit is considered as a neutron candidate. To check if the $3\pi^0$ and the neutron indeed are the η-neutron pair a missing mass analysis is performed where the detected neutron candidate is treated as a missing neutron and the neutron target is assuming to be at rest: $m(neutron) = \sqrt{(E_\gamma + m_{neutron} - E_\eta)^2 - (\vec{p}_\gamma - \vec{p}_\eta)^2}$. Figure 1 illustrates the missing mass analysis. In the left hand column of the left figure 1, invariant mass spectra of the $3\pi^0$ triplets are plotted for different incident photon energies. The η-resonance is clearly seen. A cut around the η-resonance is done (illustrated by the vertical dashed lines) and the corresponding missing neutron mass spectrum is plotted in the right hand column of the left figure on the left for the same incident photon beam. A peak at the neutron mass is present as expected. With increasing photon beam energy a tale on the missing neutron spectra start to appear. It corresponds to competitive

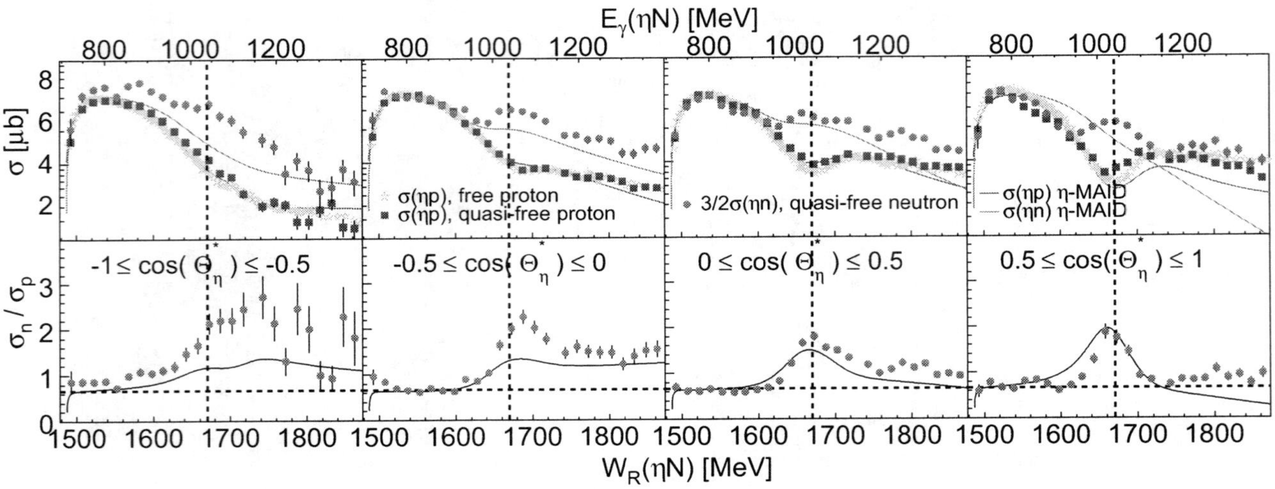

FIGURE 2. First raw: excitation functions for different bins of η cm polar angle. (Blue) open squares: quasi-free proton data, (black) stars: free proton data from [40] (also presented at this conference by Igor Strakovsky), (red) dots: quasi-free neutron data scaled up by 3/2. (Blue) solid lines: η-MAID [21] for proton, (red) dashed lines: η-MAID for neutron. Second raw: ratio of neutron and proton cross section for data and η-MAID.

background channels - $\eta\pi$-, where the pion (either charged or neutral) was not detected. The line shape of the missing neutron mass spectrum was reproduced by a GEANT-based Monte Carlo simulation [37] to model the response of the detector by taking into account the signal and the background channels. A conservative cut was chosen represented by the blue area on the right hand figure. The $3\pi^0$ signal remaining after the cut on the missing neutron spectra are also in blue (left hand figure). The remaining combinatorial background below the η-resonance was removed by a fit estimating the η signal.

RESULTS

A summary of the results is plotted in Figure 1 (right). The η-nucleon pairs are reconstructed using the fact that the reaction kinematics is exactly determined. The nucleon energy is calculated by the relation linking the following measured or reconstructed quantities: the η four vector, the incident photon beam energy and the nucleon angles. Details about the efficiency corrections and the absolute normalization are discussed in [38, 39]. The η-proton pairs invariant mass spectrum $\sigma(\eta p)$ (blue square) and the free proton data (green star) are in good agreement showing that the Fermi motion can be effectively removed and that the other nuclear effects are negligible. Furthermore a preliminary analysis of the liquid ^{3}He data taken by the A2 Collaboration [41] also shows a bump like structure which is ruling out any nucleus effect. The η-neutron pairs invariant mass spectrum $\sigma(\eta n)$ (red point) is scaled by a factor of 3/2. $\sigma(\eta n)$ and $\sigma(\eta p)$ have the same shape as expected in the $S_{11}(1535)$ energy region. Around 1.7 GeV a narrow structure is clearly seen in $\sigma(\eta n)$ and is absent from $\sigma(\eta p)$.

However, the excitation functions for different bins of η cm polar angle, see Figure 2, highlight an intriguing opposite behavior between $\sigma(\eta n)$ and $\sigma(\eta p)$. A bump-like structure is visible for all bins of η cm polar angle at $\approx$ 1670 MeV which is changing for the different bins, while no bump-like structure is seen for the proton but a dip starts to appear for forward η cm polar angles. At very forward angles the bump and the dip are symmetric. This behavior finally creates the bump - dip structures of the total cross sections. A possible interpretation is that two different processes occures: an interference between between amplitudes belonging to different partial waves (excitation fuctions - Figure 2) and a single amplitude or an interference between amplitudes belonging to the same partial waves (total cross section - Figure 1). Finally the η-neutron pairs invariant mass spectrum was fitted: in the $S_{11}(1535)$ energy region a parameterization of a Breit-Wigner curve with energy dependent width [42] and two further simple Breit-Wigner curves with constant width ($x \equiv 1$). The results of fit give W $\approx$ 1665 MeV and with a FWHM Γ = 25 MeV.

CONCLUSION

The Fermi motion can be effectively removed as shown by the good agreement between the free proton data and the quasi-free proton data. A narrow structure at W $\approx$ 1665 MeV and with a FWHM pf Γ = 25 MeV is seen in the invariant mass spectrum of the η-neutron pairs which is not due to a nucleus effect. This structure is not seen in the invariant mass spectrum of the η-proton pairs but an opposite behavior is seen in the excitation functions for different bins of η cm polar angle.

ACKNOWLEDGMENTS

We wish to acknowledge the outstanding support of the accelerator groups and operators of ELSA and of MAMI. This work was supported by Schweizerischer Nationalfonds and Deutsche Forschungsgemeinschaft (SFB/TR-16).

REFERENCES

1. D. E. Groom et al., *Eur. Phys. J.* C **15**, 1 (2000).
2. M. Kirchbach, *Mod. Phys. Lett.* A **12**, 3177 (1997).
3. B. Mecking et al., *Nucl. Instr. and Meth.* **94**, 262 (2003).
4. D. Husman, W.J. Schwille, *Phys. BL.* **44**, 40 (1988).
5. W. Hillert, *Eur. Phys. J.* A **28**, 139 (2006).
6. H. Herminghaus et al.,*IEEE Trans. Nucl. Sci.* **30**, 3274 (1983).
7. K.-H. Kaiser et al., *Nucl. Instrum. Methods* A **593**, 159 (2008).
8. Moorhouse, *Phys. Rev. Lett.* **16**, 771 (1966)
9. Jaegle et al., *Phys. Rev. Lett.* 100, 252002 (2008).
10. B. Krusche et al., *Phys. Rev. Lett.* 74, 3736 (1995).
11. J. Ajaka et al., *Phys. Rev. Lett.* **81**, 1797 (1998).
12. A. Bock et al., *Phys. Rev. Lett.* **81**, 534 (1998).
13. C.S. Armstrong et al., *Phys. Rev.* D **60**, 052004 (1999).
14. R. Thompson et al., *Phys. Rev. Lett.* **86**, 1702 (2001).
15. F. Renard et al., *Phys. Lett.* B **528**, 215 (2002).
16. M. Dugger et al., *Phys. Rev. Lett.* **89**, 222002 (2002).
17. V. Crede et al., *Phys. Rev. Lett.* **94**, 012004 (2005).
18. T. Nakabayashi et al., *Phys. Rev.* C **74**, 035202 (2006).
19. O. Bartholomy et al., *Eur. Phys. J.* A **33**, 133 (2007).
20. D. Elsner et al., *Eur. Phys. J.* A **33**, 147 (2007).
21. W.-T. Chiang et al., *Nucl. Phys.* A **700** ,429 (2002).
22. V.A. Anisovich et al., *Eur. Phys. J.* A **25**, 427 (2005).
23. V. Kuznetsov et al., *Phys. Lett.* B **647**, 23 (2007).
24. F. Miyahara et al., *Prog. Theor. Phys. Suppl.* **168**, 90 (2007).
25. D. Werthmüller for the Crystal Ball/TAPS collaborations, *Chinese Physics* C **33** (12): 1345-1348 (2009)
26. K. Nakamura et al.,, *Journal of Physics* G **37**, 075021 (2010).
27. V.A. Anisovich et al., *Eur. Phys. J.* A **41**, 13 (2009).
28. V. Shklyar, H. Lenske, U. Mosel, *Phys. Lett.* B **650**, 172 (2007).
29. M. Döring and K. Nakayama, *Phys. Lett.* B **683**, 145 2010.
30. M. Polyakov and A. Rathke, *Eur. Phys. J.* A **18**, 691 (2003).
31. R.A. Arndt et al., *Phys Rev* C **69**, 035208 (2004).
32. T. Mertens et al., *Eur. Phys. J.* A **38** 195 (2008).
33. I. Anthony et al.,*Nucl. Inst. Meth.* A **301**, 230 (1991).
34. M. Oreglia et al.,*Phys. Rev.* D **25**, 2259 (1982).
35. R. Novotny, *IEEE Trans. on Nucl. Science* **38**, 379 (1991).
36. A.R. Gabler et al.,*Nucl. Instr. and Meth.* A **346**, 168 (1994).
37. R. Brun et al., GEANT 3.21, CERN DD/EE/84-1, 1984
38. Jaegle et al., *Eur. Phys. J.* A **47**, 11 (2011).
39. Jaegle et al., Quasi-free photoproduction of η-mesons off the deuteron to be submitted in EPJA.
40. E.F. McNicoll et al., *Phys. Rev.* C **82**, 035208 2010.
41. L. Witthauer et al., under preparation,
42. B. Krusche and S. Schadmand, *Prog. Part. Nucl. Phys.* **51**, 399 (2003).

Session M: Nuclear Structure Functions

EMC effect for light nuclei: new results from Jefferson Lab

A. Daniel

Dept. of Physics and Astronomy, Ohio University, Athens OH 45701, USA
(for the JLab E03−103 collaboration).

Abstract.
High energy lepton scattering has been the primary tool for mapping out the quark distributions of nucleons and nuclei. Measurements of deep inelastic scattering in nuclei show that the quark distributions in heavy nuclei are not simply the sum of the quark distributions of the constituent proton and neutron, as one might expect for a weakly bound system. This modification of the quark distributions in nuclei is known as the EMC effect. I will discuss the results from Jefferson Lab (JLab) experiment E03-103, a precise measurement of the EMC effect in few-body nuclei with emphasis on the large x region. Data from the light nuclei suggests that the nuclear dependence of the high x quark distribution may depend on the nucleon's local environment, rather than being a purely bulk effect. In addition, I will also discuss about a future experiment at the upgraded 12 GeV Jefferson Lab facility which will further investigate the role of the local nuclear environment and the influence of detailed nuclear structure to the modification of quark distributions.

Keywords: EMC effect, structure functions, light nuclei
PACS: 13.60.Hb,25.30.Fj,24.85.+p

INTRODUCTION

Deep inelastic scattering (DIS) is one of the most effective tools to study the quark-gluon substructure of the nucleon. The question of the nuclear dependence of the quark structure of nuclei, as measured in DIS has been of great interest since the European Muon Collaboration (EMC) [1] found significant deviation between the structure functions of heavy (iron) and light (deuterium) nuclei. Since then, the nuclear dependence of structure functions has been extensively studied, both experimentally and theoretically. Even though it has been more than two decades since the discovery of the EMC effect and there are extensive data on the x and A dependence of the EMC effect, its origin still remains a mystery. These measurements show a universal shape for the EMC ratios in the valence region, and a weak dependence on A. However, the nuclear dependence can be well fit with either a simple A-dependent or density dependent form [2], thus making it difficult to test models of the EMC effect based on the A dependence. In addition, the EMC effect at very large x values is not well measured. The constraint that the data be taken in the DIS region ($W^2 > 4$ GeV^2) forced measurements to be taken at higher Q^2, led to measurements being limited by the cross section at large x, where the parton distribution functions fall relatively rapidly.

Because the nuclear dependence appears to have contributions from multiple effects, it has been difficult to determine exactly what causes the observed behavior. The formalisms used to explain the observed effect range from traditional nuclear descriptions in terms of pion exchange or binding energy shifts, to QCD inspired models such as dynamical rescaling, multi quark clustering, de-confinement in nuclei, etc. (see Refs [3, 4]). Conventional models that attempt to explain the observed effect use the convolution formalism to incorporate nuclear physics effects, while another approach tries to explain the EMC effect using the changes to the internal structure of the nucleon. As mentioned earlier, despite much theoretical progress, no unique and universally accepted explanation of the EMC effect has emerged. It should be noted that the conventional nuclear physics effects are not only significant at large x, but lead to a modification of the distributions at all values of x. These effects are the dominant contributions at large x, but the limited data at large x, coupled with the lack of precise data on light, easily calculable nuclei, has made it difficult to determine how well these 'traditional' nuclear physics effects are being included in models of the nuclear structure function.

12th International Conference on Meson-Nucleon Physics and the Structure of the Nucleon (MENU 2010)
AIP Conf. Proc. 1374, 455-458 (2011); doi: 10.1063/1.3647179

RESULTS FROM JEFFERSON LAB EXPERIMENT E03-103

The present data is from experiment E03103 [5], which ran in Hall C at Jefferson Lab. The main goals of the experiment were to improve the precision of the target ratios at large x and to provide data on light nuclei for few-body calculations. The majority of the data were taken with an electron beam energy of 5.766 GeV with beam currents ranging from 30 to 80 μA. The cryogenic targets ^{1}H, ^{2}H,^{3}He,^{4}He and solid targets Be, C, Cu and Au were studied. Data for all targets were taken at electron scattering angles of 40° and 50°, covering a Q^2 range 3 (GeV/c)^2 at $x = 0.3$ and 6 (GeV/c)^2 at $x = 0.9$. At high x, the kinematics were not in the conventional DIS region, so additional data at 4 other angles covering 18° to 32° were collected at a beam energy of 5.766 GeV. Data were also collected for a detailed Q^2 dependency study at 4 angles on C and ^{2}H, at a beam energy of 5.01 GeV. Scattered electrons were detected in the High Momentum Spectrometer. Details about the experiment and the EMC ratio extraction can be found in [6].

Kinematic coverage of the E03-103 experiment is shown in Fig. 3. The data are in the conventional DIS region up to $x \approx 0.6$. Detailed scaling studies of the structure function were needed since the data at large x for E03103 was at a lower W^2 than the previous measurements. Although there were indications [7] that the nuclear structure functions in the resonance region shows the same global behavior as in the DIS region, this was explicitly tested using E03-103 data. These studies shows that the cross section ratios scale to low values of Q^2 and W. Also, the C to ^{2}H ratio for the five highest Q^2 settings shows that there is negligible Q^2 dependence in the cross section ratios, even at the largest x values. This implies that the contribution from higher twist effects and additional scaling violation mechanisms to the ratios are small even in the high x region.

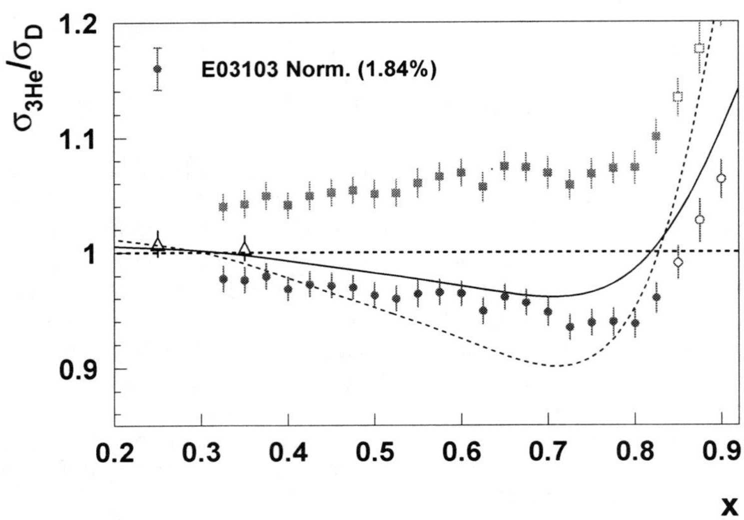

FIGURE 1. EMC ratio for ^{3}He. The upper squares are the raw ^{3}He/^{2}H ratios, while the bottom circles show the isoscalar corrected EMC ratio. The triangles are the HERMES results [8] which use a different isoscalar correction. The solid (dashed) curves are the SLAC A-dependent fits to ^{3}He (Carbon).

Figure 1 shows the EMC ratio for ^{3}He, with the low-x data from HERMES [8]. Since ^{3}He contains two protons and a neutron, the raw cross section ratios must be corrected for the proton excess to obtain the isoscalar corrected EMC ratios. In the figure the upper squares are the raw ^{3}He/^{2}H ratios while the bottom circles show the isoscalar corrected EMC ratios. There is a significant uncertainty in the neutron cross section in the large x region, hence, the extracted EMC ratios are very sensitive to these isoscalar corrections. Previous measurements (SLAC E139) used a correction based on the high Q^2 measurements of the free F_{2n}/F_{2p} ratio. E03-103 results are extracted using global fits of free proton and neutron cross sections and broadened using the convolution procedure mentioned in [9]. These distributions are then used to get the bound F_{2n}/F_{2p} ratios in nuclei.

The full results for the EMC ratios for ^{3}He, ^{4}He, Be and C are available in Ref. [6]. The results for all these nuclei are consistent with the SLAC measurements, with much better precision for ^{4}He and new measurements for ^{3}He. Figure 2 shows the size of the EMC effect for these light nuclei as a function of the scaled nuclear density. To avoid contributions from the normalization uncertainty in the measurements, especially important for nuclei with a small EMC effect, we quantify the size of the EMC effect based on the slope of the EMC ratio in the linear region between

$x = 0.35$ and $x = 0.7$. For the density in these light nuclei, the assumption of a uniform sphere density distribution, as used in the SLAC fits, is not a good approximation. We take the density distributions calculated from the *ab initio* Green's Function Monte Carlo calculations [10] to calculate the average density for each nucleus. We choose to scale down this average nuclear density by a factor of (A-1)/A, based on the idea that each nucleon is only influenced by the other (A-1) nucleons.

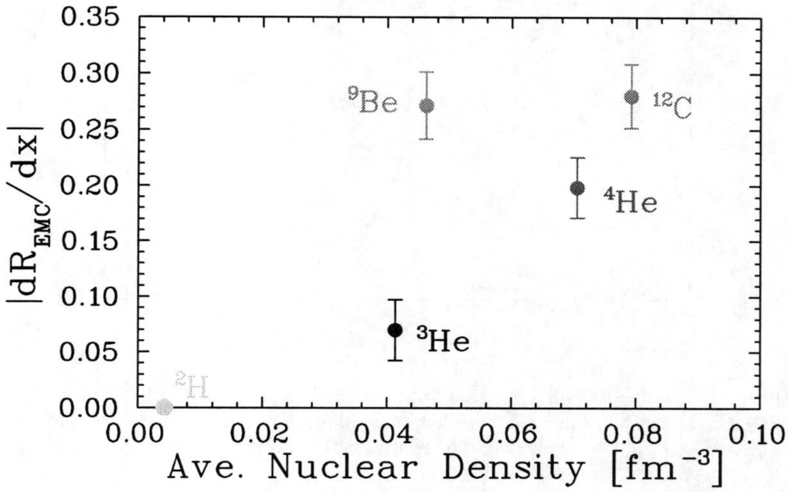

FIGURE 2. The figure shows the slope of the isoscalar EMC ratios for $0.35 < x < 0.7$ as a function of scaled nuclear density.

The observed difference in the EMC effect for ^{3}He and ^{4}He does not support the previous mass number dependent fits. However, the observed nuclear effects are similar for C and ^{4}He, and since the average nuclear density of C and ^{4}He are similar, it is possible that the modification is driven by the average nuclear density. But the effect in Be is also similar to what is seen in the C data, though the average density of Be is much lower than C. This contradicts the average density dependent picture of the magnitude of the EMC effect. These observations suggest that neither the mass-dependent or density-dependent parameterizations describe these light nuclei.

While neither of these simple models for the scaling of the EMC effect accurately represent the data in light nuclei, the data are consistent with the idea that the nearby nucleons are most important. The structure of ^{9}Be includes a significant component with two alpha clusters and one excess neutron. In this picture, most of the nucleons are bound in these tight clusters, and thus the local environment of the nucleons is similar to ^{4}He, even though the average density is quite low. This suggests that these clustering effects and the local environment may be important, and clearly shows that calculations of the EMC effect must take into account the detailed nuclear structure.

SUMMARY AND OUTLOOK

We have measured the nuclear dependence of the EMC effect for a series of light nuclei, and the data from E03-103 have provided important new information on the EMC effect. We have improved the precision of the data at high x by relying on the extended scaling that we observe from our data. Also, this experiment provides the first valence-region measurement of EMC effect in ^{3}He. E03-103 provides benchmark data for models of the EMC effect, allowing comparison to few-body nuclei with well understood structure, as well as providing high precision data at larger x values than were available from the SLAC measurements.

The nuclei included in E03103 were sufficient to examine the simple A-dependent and density-dependent scaling models, and show that detailed calculations including realistic nuclear structure will be important in explaining the EMC effect. A recently approved experiment [11] will provide further information on the detailed behavior of the observed nuclear dependence with an expanded set of light nuclei, including nuclei with significant clustering contributions. This experiment will provide data on hydrogen, deuterium, ^{3}He, ^{4}He, ^{6,7}Li, ^{9}Be, ^{10,11}B, ^{12}C,^{40,48}Ca, and ^{63}Cu. The kinematics of the proposed measurements are shown in Fig. 3. The higher beam energy available after the upgrade of the Jefferson Lab will increase the region of precise scaling to larger x values; going from x=0.6 to 0.8 for $W^2 > 4$ GeV^2, and up to $x = 0.92$ for $W^2 > 2$ GeV^2, where precise scaling was observed at the lower Q^2 values of E03-103.

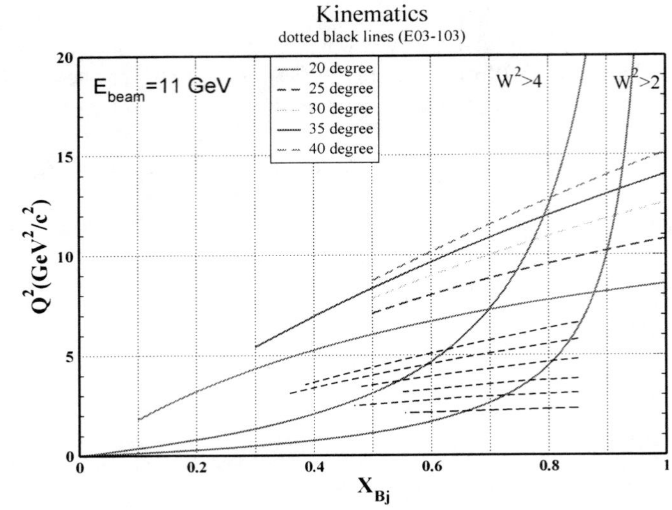

FIGURE 3. Overview of the kinematics for the approved experiment E12-10-008 at Jefferson Lab [11]. The black dotted lines are the kinematics from E03-103. Data on a small subset of targets at additional angles will be used to study the Q^2 dependence of the ratios (dashed lines). The solid purple lines correspond to contours of fixed W^2 values as mentioned in the figure.

In addition, measurements at higher beam energy will reduce the uncertainty related to the isoscalar correction for ^{3}He and allow for better comparisons of the x dependence of the EMC effect in light nuclei. One entirely new aspect of this measurement is the ability to look for a flavor dependence to the EMC effect. There have been recent suggestions of a significant flavor dependence in the EMC effect [12]. Measurements of ^{40}Ca and ^{48}Ca will provide a significant variation of the n/p ratio in the nucleus, while maintaining a comparison between nuclei of similar mass and density. Also, comparisons of nuclei which differ by just one nucleon (for example ^{11}B-^{10}B, ^{7}Li-^{6}Li, ^{12}C-^{11}B) will allow the extraction of the structure function of a single nucleon in the nucleus. This can be used to a check the isoscalar corrections applied in these nuclei.

ACKNOWLEDGMENTS

This work is supported in part by research grant PHY-0653454 from the National Science Foundation. The Southeastern Universities Research Association operates the Thomas Jefferson National Accelerator Facility under the DOE contract DE-AC05-84ER40150.

REFERENCES

1. J. J. Aubert, et al., *Phys. Lett. B* **123**, 123 (1983).
2. J. Gomez, et al., *Phys. Rev.* **D49**, 4348–4372 (1994).
3. D. F. Geesaman, K. Saito, and A. W. Thomas, *Ann. Rev. Nucl. Part. Sci.* **45**, 337–390 (1995).
4. P. R. Norton, *Rept. Prog. Phys.* **66**, 1253–1297 (2003).
5. J. Arrington, and D. Gaskell, spokespersons, Jefferson Lab experiment E03-103.
6. J. Seely, et al., *Phys. Rev. Lett.* **103**, 202301 (2009).
7. J. Arrington, R. Ent, C. E. Keppel, J. Mammei, and I. Niculescu, *Phys. Rev.* **C73**, 035205 (2006).
8. K. Ackerstaff, et al., *Phys. Lett.* **B475**, 386–394 (2000).
9. J. Arrington, F. Coester, R. J. Holt, and T. S. H. Lee, *J. Phys.* **G36**, 025005 (2009).
10. S. C. Pieper, and R. B. Wiringa, *Ann. Rev. Nucl. Part. Sci.* **51**, 53–90 (2001).
11. J. Arrington, A. Daniel, and D. Gaskell, spokespersons, Jefferson Lab experiment E12-10-008, http://www.jlab.org/exp_prog/proposals/10/PR12-10-008.pdf.
12. I. C. Cloet, W. Bentz, and A. W. Thomas, *Phys. Lett.* **B642**, 210–217 (2006).

Asymmetries in electron-induced breakup of polarized ^{3}He

S. Širca, for the E05-102 and Hall A Collaborations

Dept. of Physics, University of Ljubljana, 1000 Ljubljana, Slovenia

Abstract. The Jefferson Lab Experiment E05-102 "*Measurement of A'_x and A'_z asymmetries in the quasi-elastic* ^{3}He(e, e′d) *reaction*" was performed in Hall A in 2009. The main physics motivation of the experiment was to investigate the effects of small components of the ^{3}He ground-state wave-function by a simultaneous measurement of double-polarization asymmetries in quasi-elastic kinematics for three exclusive channels, (e, e′d), (e, e′p), and (e, e′n), at almost identical momentum transfers, as well as for (e, e′). This experiment will help map the spin structure of the ^{3}He nucleus onto the picture of the "free" polarized neutron. As such, it is of great relevance to the polarized-neutron programs at Jefferson Lab and beyond.

Keywords: Few-body systems,double-polarization asymmetries,electron-induced proton and deuteron knockout
PACS: 21.45.-v,25.30.Dh,25.30.Fj

INTRODUCTION

A variety of experiments based on ^{3}He targets seeking to extract neutron information (for example, the charge and magnetic form-factor) rely on an almost perfect theoretical knowledge of the spin structure of the ^{3}He nucleus in its ground state. The manifestations of S'- and D-state components, specifically the degrees of neutron and proton polarizations within the polarized ^{3}He nucleus, represent a most important part of this knowledge.

The statistical precision of double-polarization experiments on ^{3}He has reached a level that has to be accompanied by excellent control of systematic uncertainties and, among those, needs to be matched by the best theoretical models of the ^{3}He nucleus. The E05-102 experiment at Jefferson Lab [1] was devised as a comprehensive measurement of a set of double-polarization observables with enough kinematical lever arm to constrain the theories.

THE E05-102 EXPERIMENT AT JEFFERSON LAB

The physics case

The main physics case of the E05-102 experiment is the exploration of the ^{3}He system by studying the S'-state and D-state contributions to the ^{3}He ground-state wave-function. From state-of-the-art calculations of ^{3}He it became clear over the years that double-polarized asymmetries in the two-body breakup of ^{3}He are the most powerful tool to perform such studies. The proton and deuteron channels are particularly interesting because of the interference of the PWIA proton process and the deuteron process, together constituting PWIAS which stands for PWIA with the symmetrized final state. These processes in turn result in distinct proton vs. deuteron recoil-momentum distributions. To design the optimal setup of the experiment, we have used the calculations of three groups: the Bochum/Krakow group, the Hannover group, and the Pisa group.

The theories are expected to perform quite reliably in the proton knockout channel, at both low and high values of the momentum transfer q. At low q, the final-state interactions (FSI) certainly play a role; the nuclear forces used in all available calculations are the most up-to-date — in terms of pd-scattering, they are all phase-shift-equivalent; in addition, the meson-exchange current effects (MEC) are known to be very small. On the other hand, at high momentum transfers q, the PWIA mechanism is nothing but an overlap of the deuteron and the ^{3}He state at relatively low deuteron momenta. Yet, even for proton knockout, there is no comprehensive data set that would cover the entire angular region. The situation is even worse in the case of deuteron knockout, where even unpolarized cross-section data exhibit strong discrepancies with respect to the theory. For example, even the unpolarized five-fold cross-section at zero recoil momentum (at $q = 412\,\mathrm{MeV/c}$ momentum transfer and $50\,\mathrm{MeV}$ energy transfer) is over-predicted by the theory by approximately $20\,\%$ [5] !

12th International Conference on Meson-Nucleon Physics and the Structure of the Nucleon (MENU 2010)
AIP Conf. Proc. 1374, 459-462 (2011); doi: 10.1063/1.3647180

Details of calculations

The Bochum/Krakow calculations are full Faddeev calculations based on the AV18 nucleon-nucleon force [2, 3, 4]. They apply a complete treatment of FSI and MEC, as well as three-nucleon effects based on the Urbana IX three-nucleon force which is employed in both bound and scattering states. At present, it is not clear which ingredient of the Bochum/Krakow calculation may be responsible for the disagreement in the unpolarized cross-section mentioned above, but there are indications that the use of the Sachs form-factors instead of the Pauli form-factors, the three-nucleon force, as well as the inclusion of MEC in the charge-density operator could lead to a solution.

The Hannover group has also performed full Faddeev calculations of the two-body and three-body breakups of ^{3}He which include full FSI and MEC. An important feature of these calculations is the addition of the Δ isobar as an active degree of freedom providing a mechanism for an effective three-nucleon force and exchange currents [6, 7]. Three-particle scattering equations are solved exactly by using a Chebyshev expansion of the two-baryon transition matrix as the interpolation method. The underlying purely nucleonic reference potential is an improved charge-dependent CD Bonn potential. Point Coulomb interaction is added in the partial waves involving two charged baryons [8, 9, 10].

The third set of calculations was performed recently by the Pisa group. These computations are also based on the AV18 nucleon-nucleon interaction and the Urbana IX three-nucleon force, but the FSI are included by means of the variational pair-correlated hyper-spherical harmonic (PHH) expansion. In addition, MECs are included [11]. While these calculations are not of the Faddeev type, they certainly have completely equivalent precision and are expected to account for all relevant reaction mechanisms [12].

Figure 1 shows the predicted transverse and longitudinal double-polarization asymmetries for the ^{3}He(e,e′d) channel by all three groups. The Pisa calculation result is given only on the positive recoil momentum branch. The asymmetries are small (on the order of a few percent) in the kinematic region that has been covered by the E05-102 experiment. One of the most striking features is that even at small missing momenta, the theoretical predictions for the longitudinal asymmetry exhibit significant differences. One of the aims of E05-102 was to probe and try to distinguish between them with sufficient experimental precision.

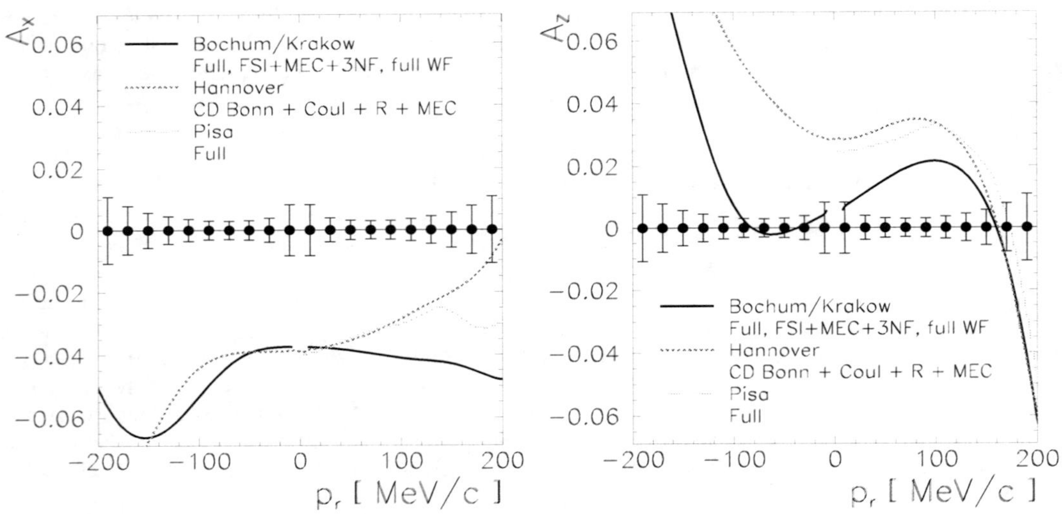

FIGURE 1. The double-polarization transverse and longitudinal asymmetries for the ^{3}He(e,e′d) channel as computed by the Bochum/Krakow, Hannover, and Pisa groups (full calculations). Quasi-elastic kinematics for the deuteron with 2400 MeV beam energy and 15° scattering angle. The error bars show the anticipated statistical uncertainties.

One of the main aims of the E05-102 experiment is to investigate the manifestations of small components of the ^{3}He ground-state wave-function. The double-polarization asymmetries can be computed with truncated wave-functions, for example, by excluding the D and S' components. However, the implication of this procedure may be that a deliberate omission of a part of the wave-function can introduce inconsistencies in the treatment of the three-nucleon scattering state. Figure 2 shows the dependence of A'_x and A'_z asymmetries in the two-body breakup of polarized ^{3}He upon various truncation approximations within the Bochum/Krakow framework. The main message of this plot is the distinct roles different truncations play in the proton channel as opposed to the deuteron channel.

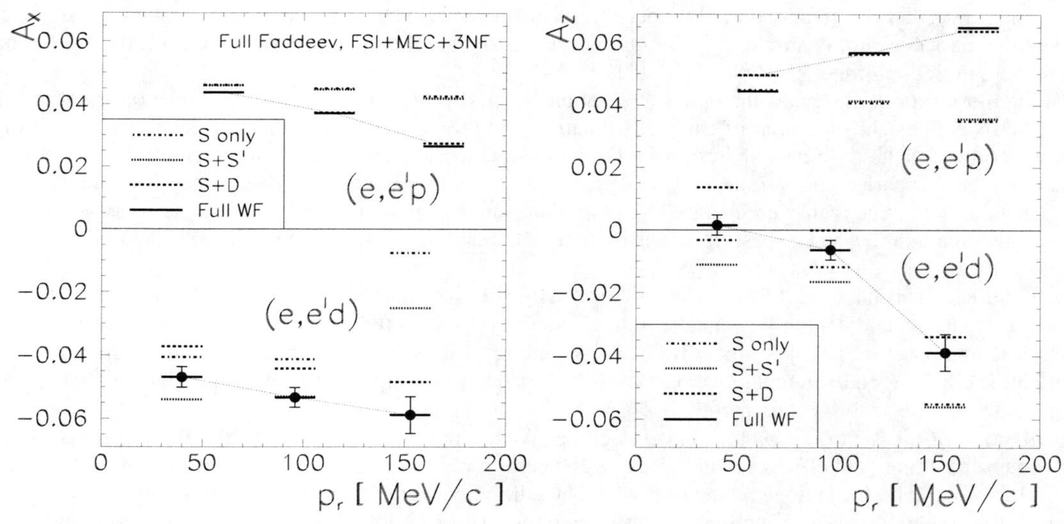

FIGURE 2. The transverse and longitudinal asymmetries for two-body breakup of polarized ^{3}He (proton and deuteron channel). Full Faddeev calculations of the Krakow/Bochum group are shown together with the anticipated uncertainties of our experiment. Different ground-state wave-function components have very distinct effects on the proton asymmetries versus those for the deuteron. For example, in the proton channel, the difference between the *S*-only and $S+S'$ approximations is negligible for both A_x and A_z, at all recoil momenta, in contrast to the deuteron channel where the effect is clearly pronounced.

Details of the experimental setup

In 2009, the E05-102 experiment has taken data for double-polarization asymmetries in quasi-elastic kinematics (and in some cases into the dip region) on all three exclusive channels $(e,e'd)$, $(e,e'p)$, and $(e,e'n)$ at almost identical momentum transfers of $\approx 0.37\,\mathrm{GeV}^2$. The proton and deuteron channels will be used to investigate the effects of small wave-function components as described above. The neutron channel will be used to determine the electric form-factor of the neutron. But the main strength of this simultaneous measurement of asymmetries in all exclusive channels is the tremendous constraint these highly sensitive and selective observables impose on the state-of-the-art theories. In addition to the coincidence measurements, we have accumulated ample single-arm (e,e') data that reaches well beyond the value of 2 in the Bjorken scaling variable. These data will be used to extract the neutron magnetic form-factor.

All data was taken with the $\approx 80\,\%$-polarized, $\approx 10\,\mu\mathrm{A}$ current beam of CEBAF, in conjunction with the 40-cm-long high-pressure polarized ^{3}He target routinely achieving polarizations in excess of $50\,\%$. The polarization of the target was monitored continuously by the NMR and EPR techniques. The in-plane target spin angle was controlled to a precision of $\approx 0.5^\circ$ by means of two pairs of Helmholtz coils.

Two High-Resolution Spectrometers (HRS) with standard detector packages were used to detect the scattered electrons, while protons and deuterons were detected by the BigBite spectrometer with the hadron package. The measurement of the elastically scattered electrons in the HRS has also allowed us to monitor independently the product of the beam and target polarizations. The neutrons were detected by the Hall A Neutron Detector (HAND).

WORK IN PROGRESS AND OUTLOOK

At present, the data analysis is in its calibration stages. The calibration of the target holding field angle in terms of the currents in the Helmholtz coils has been performed, and it was found that the uncertainty of the target in-plane spin angle typically does not exceed 0.5° for most angular settings that were used in the calibration runs. The remaining task is to map the currents used in the experiment to the true angles of the production settings. This is particularly important since hardware constraints prevented us from measuring the A_x' and A_z' asymmetries directly (orientations perpendicular and parallel to the momentum transfer vector were not allowed, only a rotated setup was possible). The calibration of the target densities and of the absolute target polarizations by means of the NMR and EPR methods is underway and shows that polarization is well under control, with typical values of about 55 %. Further work is needed to achieve the final values, for example, the calibration of the magnetic flux for the NMR pick-up coils, which comes as input to the calibration of the NMR on water [13].

One of the most important steps is the optics calibration of the BigBite spectrometer. While the optics of the HRS spectrometers is well established, that of BigBite (in particular at the drift distance and momentum coverage of the E05-102 experiment) is not. We have therefore started with a manual determination of the first few low-order matrix elements from the comparison of various BigBite detector plots with the target plots determined from the HRS data. We have used these baseline matrix elements to develop more sophisticated methods with automatic convergence and elimination of redundant terms. The best-fit reconstruction matrices now have only approximately 30 non-zero entries. For details of the optics calibration, see [14].

The calibration of particle identification capabilities of BigBite, which are quite important for a reliable and clean separation of the proton and deuteron channels, is also underway. Initial, crude gain matching of the photo-multiplier tubes was performed before the experiment by using cosmic rays. Presently, a more refined procedure which takes into account the effects of light attenuation along scintillator paddles is being performed by introducing corresponding correction factors in the analysis software.

The calibration of the neutron detector is also ongoing. We are presently studying walk effects in the scintillator bars, their gain-matching, and effects related to the presence or absence of the lead shielding in front of the neutron array [15]. Initial work has been done on the time-of-flight calibration. Simultaneously, we are trying to determine the dependence of the single-arm cross-sections for different targets (for example, ^{3}He and ^{12}C, as well as their ratio) on the Bjorken scaling variable x [15].

First attempts to determine single-arm asymmetries and extract the neutron magnetic form-factor have been made on the basis of helicity state scalers. We are now starting to look into radiative corrections. For updated information on this analysis, see [16].

ACKNOWLEDGMENTS

This work was supported in part by the U.S. Department of Energy and the U.S. National Science Foundation. It is supported by DOE contract DE-AC05-06OR23177, under which Jefferson Science Associates, LLC, operates the Thomas Jefferson National Accelerator Facility.

REFERENCES

1. S. Gilad, D. W. Higinbotham, W. Korsch, B. E. Norum, S. Širca (spokespersons), TJNAF Experiment E05-102, "*Measurement of A'_x and A'_z asymmetries in the quasi-elastic* ^{3}He(e,e′d) *reaction*.
2. J. Golak, R. Skibiński, H. Witała, W. Glöckle, A. Nogga, H. Kamada, Phys. Reports **415** (2005) 89.
3. W. Glöckle, H. Witała, D. Hüber, H. Kamada, J. Golak, Phys. Reports **274** (1996) 107.
4. W. Glöckle, J. Golak, R. Skibiński, H. Witała, H. Kamada, A. Nogga, Eur. Phys. J. A **21** (2004) 335.
5. C. M. Spaltro et al., Nucl. Phys. A **706** (2002) 403.
6. L. P. Yuan et al., Phys. Rev. C **66** (2002) 054004.
7. L. P. Yuan et al., Few-Body Systems **32** (2002) 83.
8. A. Deltuva, R. Machleidt, P. U. Sauer, Phys. Rev. C **68** (2003) 024005.
9. A. Deltuva et al., Phys. Rev. C **69** (2004) 034004.
10. A. Deltuva et al., Phys. Rev. C **70** (2004) 034004.
11. L. E. Marcucci, M. Viviani, R. Schiavilla, A. Kievsky, S. Rosati, Phys. Rev. C **72** (2005) 014001.
12. M. Viviani and L. Marcucci, private communication (2009).
13. Y. Zhang, private communication (2010).
14. M. Mihovilovič, contribution to these Proceedings.
15. E. Long, private communication (2010).
16. J. Ge, contribution to these Proceedings.

Configuration and calibration of the BigBite spectrometer

M. Mihovilovič, for the E05-102 and Hall A Collaborations

Jožef Stefan Institute, Jamova cesta 39, 1000 Ljubljana, Slovenia

Abstract. Optical calibration of the BigBite spectrometer for the E05-102 experiment at Jefferson Lab has been performed. In this paper properties of the BigBite spectrometer and its detector package will be described. Basic ideas and techniques for optical calibration will be explained. For the reconstruction of the BigBite target variables the transport-matrix formalism has been considered. The process of calibration is explained in detail, together with the latest results. σ-resolutions of the reconstruction are 1.2 cm for vertex position, 11 mrad for in-plane angle and 17 mrad for out-of-plane angle. Robustness of the calibration algorithm has also been investigated. Results of the scintillation plane gain matching will be shown. Precise calibration of this detector is crucial for particle-identification and energy determination.

Keywords: Physics; Magnetic spectrometer; Magneto-optics; Optical calibration
PACS: 28.41.Rc; 29.30.Aj; 41.85.-p

INTRODUCTION

The E05-102 experiment in Jefferson Lab's Hall A studied the $^3\vec{\mathrm{He}}(\vec{e},e'd)$ and $^3\vec{\mathrm{He}}(\vec{e},e'p)$ reactions in the quasi-elastic region. The purpose of the experiment was to use Faddeev calculations of the three-body system to better understand the effects of S'- and D-state contributions to the ^{3}He ground-state wave-functions [1]. The beam-target asymmetries A_x and A_z were measured in the range of the recoil momenta from 0 to approximately 200 MeV/c. In the experiment a 60% polarized ^{3}He target was used in conjunction with a polarized 2 GeV electron beam. The scattered electrons were detected with the High-Resolution Spectrometer [2] in coincidence with the deuterons and protons in the large-acceptance spectrometer BigBite [3]. For that a good calibration of the BigBite detector package is required. Precise understanding of its optical properties is also required.

TABLE 1. BigBite characteristics

Configuration	Single Dipole
Momentum range	$(200-900)\ \frac{\mathrm{MeV}}{\mathrm{c}}$
Momentum acceptance	$-0.6 \leq \frac{\delta \mathrm{p}}{\mathrm{p}} \leq 0.8$
Angular acceptance	96 msr
Flight path*	≈ 3 m
Maximum field	0.92 T
Maximum current	518 A

*during experiment E05-102

FIGURE 1. BigBite Spectrometer

BIGBITE SPECTROMETER

BigBite is a non-focusing spectrometer with large momentum and angular acceptance (see table 1). It consists of a singe normal-conducting clam-shell dipole magnet, followed by the detector package. In the E05-102 experiment it was used with the hadron detector package, which consists of two Multi-Wire-Drift-Chambers (MWDC) for tracking and a scintillation detector for particle identification.

OPTICAL CALIBRATION

The idea of optics calibration is to determine target variables that have physical meaning from the detector variables that can be directly measured. In BigBite two position coordinates (x_{det} and y_{det}) and two angles (θ_{det} and ϕ_{det}) are

12th International Conference on Meson-Nucleon Physics and the Structure of the Nucleon (MENU 2010)
AIP Conf. Proc. 1374, 463-466 (2011); doi: 10.1063/1.3647181

measured. Using this information the vertex position y_{Tg}, in-plane and out-of-plane scattering angles, ϕ_{Tg} and θ_{Tg}, and the relative particle momentum δ_{Tg} are reconstructed. This can be done in many different ways. For BigBite various analytical models have been implemented. The simplest one is the effective plane approach. The position of this plane and its inclination depend on the experimental setup and the magnitude of the magnetic field. A bit more sophisticated analytical model is the circular-arc approximation. In this approach an arc of a particle traveling through the magnet is calculated via extrapolation of a track through the detectors and the distance of the magnet relative to the target. From the radius of the arc and the density of the magnetic field, the momentum of the particle can be directly calculated, using the Lorentz equation. Unfortunately none of the considered analytical models consider fringe fields at the entrance and exit face of the magnet, which affect the resolution. In addition, the parameters used in these models need to be precisely known in order for the model to work.

Therefore a different approach has been considered, using the transport matrix formalism. Here a matrix is determined which transforms the detector variables directly to the target variables. Various parameterizations are possible. A polynomial expansion of the form

$$\Omega_{Tg} = \sum_{i,j,k} \theta_{det}^{i} y_{det}^{j} \phi_{det}^{k} \sum_{l=0}^{7} a_{ijkl}^{\Omega_{Tg}} x_{det}^{l}, \qquad \Omega_{Tg} \in \left\{\delta_{Tg}, \theta_{Tg}, \phi_{Tg}, y_{Tg}\right\}, \tag{1}$$

has been considered since the code for this has already been written and used for the parameterization of target variables in the High-Resolution Spectrometers. Knowing the optics of the spectrometer means knowing the parameters a_{ijkl} and being aware of the limits where such a parameterization works. The polynomial expansion is easy to handle, but one must precisely understand the contribution of the high-order terms. Uncontrolled behavior of these terms can cause wild oscillations of the reconstructed variables, especially on the edges of the acceptance. Our goal is therefore to find a well behaved low-order optical matrix that has as few high-order terms as possible.

For the precise determination of the matrix elements a_{ijkl}, various calibration measurements were made during the E05-102 experiment. y_{Tg} was calibrated using quasi-elastic carbon data taken with a 7-foil optics target. The position of the foils with respect to the spectrometer are well known and this enables precise determination of the coresponding matrix elements. For the calibration of θ_{Tg} and ϕ_{Tg} we considered carbon and deuterium measurements with a sieve-slit in front of the BigBite magnet and tried to reconstruct all visible sieve-holes. In addition hydrogen elastic data were used for the absolute positioning of the sieve-slit with respect to the optical axis of BigBite, which cannot be determined directly from the quasi-elastic carbon data. Finally the determination of the matrix elements for δ_{Tg} is made by using missing mass peak reconstruction. For that hydrogen and deuterium data are being considered.

The optical calibration began with a manual determination of the low-order matrix elements by comparing various BigBite detector plots to the target plots determined from the HRS-L data. This comparison was possible because only coincidence events in BigBite and HRS-L are being used. The resolution of those results was too poor for further analysis. However, since low-order terms are very robust, they were used to check the convergence of the following more sophisticated methods. After this initial step, an automated method was developed, which considers also higher-order terms and gives results that can be used in a physical analysis. This approach considers up to fifth order matrix elements. The relevant terms for each target variable were chosen using a combination of a Monte-Carlo simulation of BigBite optics and manual selection. The determined set of accepted matrix elements is not unique. Other sets could exist that would give the same or even better calibration results. However, it is impossible to know which set is the optimal one. To calculate matrix elements a χ^2-minimization written in Matlab was used. In this algorithm the calculated target variables (1) were compared with the directly measured values

$$\chi^2\left(a_i^{\Omega_{Tg}}\right) = \sqrt{\left(\Omega_{\mathrm{Tg}}^{\mathrm{True}} - \Omega_{\mathrm{Tg}}^{\mathrm{Optics}}\left(x_{det}, y_{det}, \theta_{det}, \phi_{det}; a_i^{\Omega_{Tg}}\right)\right)^2}\,. \tag{2}$$

The use of approximately 30 matrix elements for each target variable means that a global minimum in a thirtyone-dimensional space must be found. Numerically this is a very complex problem and it is not certain that the minimization method will not stop in one of the possible local minima instead of the global minimum. Therefore the robustness of the method needs to be examined. This has been done by checking the convergence of the minimization algorithm for a large number of a randomly chosen initial sets of parameters (see Figure 2 for the test results for y_{Tg}).

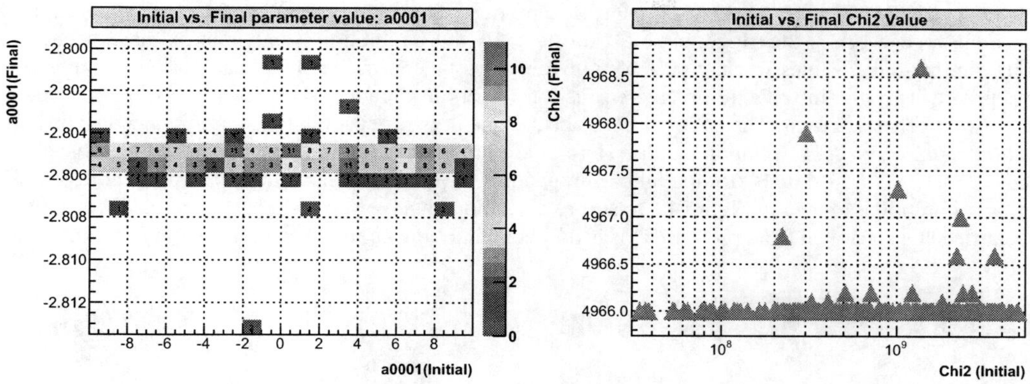

FIGURE 2. Transport matrix element a_{0001} and χ^2-function before and after minimization. The analysis was done for 250 initial randomly chosen points. The fact that the vast majority of the initial conditions converge to the same location is an indication of the robustness of the method.

It shows the value of the matrix element $a_{0001} = \langle \phi_{Fp} | \phi_{Tg} \rangle$ before and after the minimization. In the beginning the matrix element is uniformly distributed on $[-10, 10]$, while in the end it is normally distributed around the value of -2.81. The final value of the χ^2 function is more than four orders of magnitude smaller than before the minimization.

The results of the optical calibration for y_{Tg}, θ_{Tg} and ϕ_{Tg} are summarized in Table 2. A comparison with the NIKHEF calibration results, where BigBite was used with a different detector package as an electron spectrometer, before its arrival at Jefferson Lab, are also presented. It shows that the NIKHEF resolution was for a factor two better. This deterioration of the resolution was expected since BigBite is now used for hadron detection instead of electrons, with a different detector package. The determination of the matrix elements for δ_{Tg} is still in progress. For the precise missing mass peak reconstruction a well-working particle identification (PID) is required to distinguish quasi-elastic protons from elastic deuterons in ^{2}H calibration data. Therefore the behavior of scintillation detectors needs to be understood before δ_{Tg} determination could be preformed. Now that the scintillation detector calibration is done, the determination of δ_{Tg} matrix elements should be straightforward.

TABLE 2. Resolution of BigBite optical calibration and comparison with NIKHEF results.

Target variable	E05-102 calibration	NIKHEF*
$\sigma(y_{Tg})$	1.1 cm	$(0.3-0.8)$ cm†
$\sigma(\theta_{Tg})$	17 mrad	$(3-6)$ mrad
$\sigma(\phi_{Tg})$	11 mrad	$(3-6)$ mrad
$\sigma(\delta_{Tg})$	Not calibrated	0.8 %

* BigBite used as electron spectrometer with different detector package
† depending on the energy of e^-

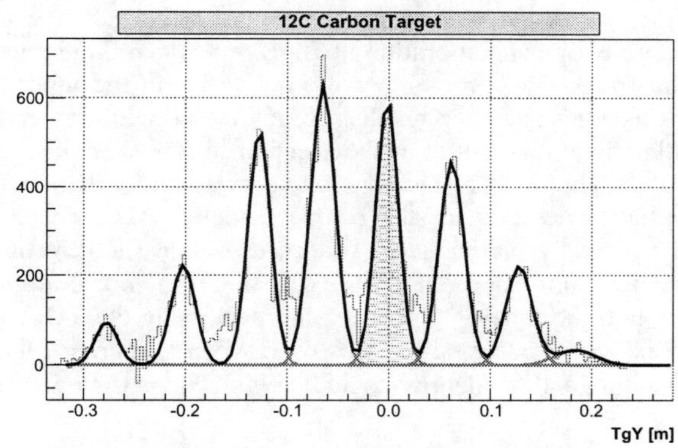

FIGURE 3. Reconstructed vertex position

SCINTILLATION PLANE CALIBRATION

The BigBite scintillation detector is made of two segmented scintillation planes, a thin 3 mm dE-plane and a thick 3 cm E-plane. Each plane is 2 m long and is made of 24 equal scintillation paddles. The signal from each paddle is read by two photo-multiplier-tubes mounted on each end of a scintillation bar. The calibration of this detector therefore means matching gains on each of the 96 PMTs. First gain matching was done before the experiment where actual high voltages on the PMTs were properly set using cosmic rays. This is important for the correct discrimination of the signals and proper work of the trigger circuit. Unfortunately this calibration is only approximate, because precise calibration with low-rate cosmics was not possible due to strict time constraints. Final calibration was therefore done after the experiment by the introduction of the correction factors in the analysis software. For this real production data on various targets were used. The calibration has been done in two steps. First we gain-matched signals from the two PMTs mounted on each scintillation paddle. Here light output attenuation effects along each bar have also been considered. In the final step signals from neighboring paddles have been compared and properly adjusted. The end effect of this calibration is shown in Figure 4. Deuterons with momenta between 340 MeV/c and 580 MeV/c can now be well identified and separated from protons. This is of crucial importance for the success of the E05-102 experiment.

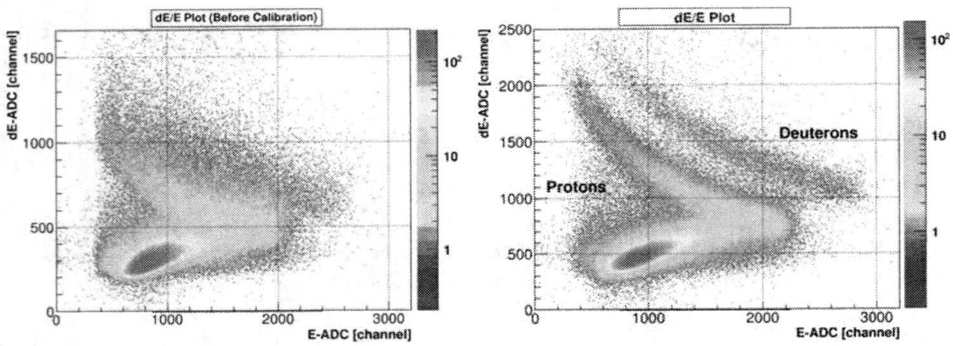

FIGURE 4. Energy deposition in dE and E scintillation planes before and after the calibration. After the process of gain matching deuterons can be well separated from protons.

CONCLUSION AND OUTLOOK

During the optical calibration of BigBite we learned that use of analytical models is not very promising, mostly due to the fringe-field problems and ambiguities in the input parameters required by the model. The transport matrix formalism gives better results. A matrix with low-order terms is preferred, since higher order terms can cause oscillations on the edges of the acceptance. The described calibration method already gives nice results for y_{Tg}, θ_{Tg} and ϕ_{Tg}. The analysis for δ_{Tg} is underway and will hopefully be completed soon. The calibration of the BigBite scintillation detector has also been considered. ADC signals from all PMTs have been properly gain-matched and can now be used for particle identification as well as for the estimation of the particle momentum via the energy deposition in the bars using the Bethe-Bloch equation [4]. Another option for the particle identification and energy determination is through the time-of-flight measurement using the TDC information. The analysis of these signals will be done in the next step of calibration. However, when all parts of the calibration are finished, precise information on particle momentum and its identity could be obtained via three independent techniques.

ACKNOWLEDGMENTS

This work was supported in part by the U.S. Department of Energy and the U.S. National Science Foundation. It is supported by DOE contract DE-AC05-06OR23177, under which Jefferson Science Associates, LLC, operates the Thomas Jefferson National Accelerator Facility.

REFERENCES

1. *Measurement of A_x and A_z asymmetries in the quasi-elastic* $^3\vec{\mathrm{He}}\,(\vec{e},e'd)$, S. Širca, S. Gilad, D. W. Higinbotham, W. Korsch, B. E. Norum (spokespersons), JLab Experiment E05-102, June 2005.
2. J. Alcorn *et. al.*, Nucl. Instr. and Meth. **A 522** (2004) 294.
3. D.J.J. Lange *et. al.*, Nucl. Instr. and Meth. **A 406** (1998) 182.
4. W.R. Leo, Techniques for Nuclear and Particle Physics Experiments, Springer-Verlag, Heidelberg, 1987.

Coherent photoproduction of π^+ from ^{3}He with CLAS at Jefferson Lab

Rakhsha Nasseripour and Barry Berman

The George Washington University, Washington, DC 20052, USA

Abstract. We have measured the differential cross section for the $\gamma^3\text{He}\rightarrow \pi^+ t$ reaction. This reaction was studied using the CEBAF Large Acceptance Spectrometer (CLAS) at Jefferson Lab. Real photons produced with the Hall-B bremsstrahlung tagging system in the energy range from 0.5 to 1.55 GeV were incident on a cryogenic liquid ^{3}He target. The differential cross sections for the $\gamma^3\text{He}\rightarrow \pi^+ t$ reaction were measured as a function of photon-beam energy and pion-scattering angle. Theoretical predictions to date cannot explain the large cross sections except at backward angles, showing that additional components must be added to the model.

Keywords: coherent pion photoproduction, few-body force, nuclear reactions
PACS: 13.40.-f, 13.60.Rj, 13.88.+e, 14.20.Jn, 14.40.Aq

INTRODUCTION

Reactions such as $\gamma^3\text{He}\rightarrow \pi^+ t$, $\gamma^3\text{He}\rightarrow \pi^0\ {}^3\text{He}$, $\gamma t \rightarrow \pi^-\ {}^3\text{He}$, and $\gamma t \rightarrow \pi^0 t$ have been studied by both experimental and theoretical groups over the last four decades. Studying these processes is useful in developing our understanding of nuclear structure and the long-range part of the nucleon-nucleon interaction described by the one-pion exchange model. However, all the previous measurements and calculations were done near threshold and in the Δ resonance region [1]-[9]. The goal of the present analysis is to measure the differential cross section for the $\gamma^3\text{He} \rightarrow \pi^+ t$ reaction for energies above the Δ resonance region. The ^{3}He target is the lightest nucleus on which one can observe coherent π^+ photoproduction with charge exchange and which also leads to a well defined final state that can be easily identified. This channel is one of the most important pion-production channels because it is an isoelastic nuclear transition within the isodoublet (^{3}H,^{3}He) with the same quantum numbers as the elementary reaction on the nucleon. The same nuclear wave functions can be used for the initial and final states (except for Coulomb effects). Studying this production process is ideal for understanding the interaction of pions with nuclei and to search for possible effects mediated by nucleon resonances in nuclear matter.

A model was developed by Tiator *et al.* [7] based on realistic three-body Faddeev functions in plane-wave impulse approximation (PWIA). Good agreement was found with low-momentum-transfer data (up to 3.1 fm^{-2}) from Ref. [3], however, the PWIA could not explain the data at higher momentum transfer.

In a later calculation preformed by Kamalov *et al.* [8], the intermediate pion scattering between two nucleons also has been taken into account. In this model, the coherent π^0 and π^+ photoproduction and elastic and charge-exchange pion scattering on ^{3}He have been calculated in a consistent way. In this model, realistic three-body Faddeev wave functions have been used and full nonlocal distorted-wave impulse-approximation (DWIA) results for pion photoproduction were obtained. Comparison with experimental data shows good agreement over a wide range of momentum transfer for the photon energy range between 230 and 450 MeV.

In 1995, the two-body mechanisms were explicitly included in the model [10] where the photon is absorbed by one nucleon and the pion is emitted from the other nucleon. The inclusion of these processes resulted in a better agreement between the calculations and the previous data at higher momentum transfers. However, after all of the considered effects and pion distortions, the model could not account for the large enhancement seen in the experimental data at large Q^2 ($Q^2 > 6$ fm^{-2})[3].

EXPERIMENT AND DATA ANALYSIS

The $\gamma^3\text{He} \rightarrow \pi^+ t$ reaction was measured during the g3a experiment in December 1999 with the CEBAF Large-Acceptance Spectrometer (CLAS) at Jefferson Lab [11], [12]. CLAS is a large acceptance multi-layer spectrometer

12th International Conference on Meson-Nucleon Physics and the Structure of the Nucleon (MENU 2010)
AIP Conf. Proc. 1374, 467-470 (2011); doi: 10.1063/1.3647182

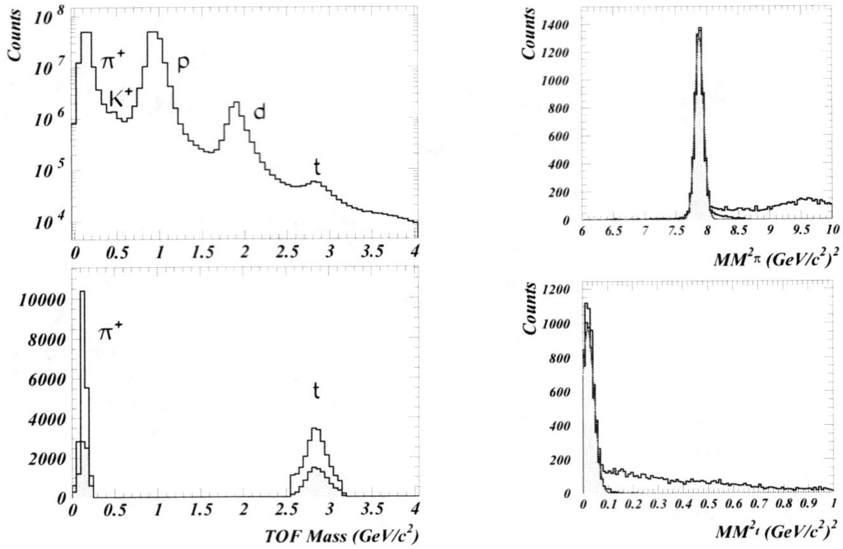

FIGURE 1. Left: Hadron mass calculated from the time-of-flight information. The histogram in the top panel shows the mass distribution for all the positively charged hadrons. The solid histogram in the lower panel is the selected sample of pions and tritons that are detected in coincidence. The shaded histogram shows the same distribution after applying all the kinematical cuts to remove the background (see text). Right: The calculated values for the missing mass squared for the detected pion (top) and the detected triton (bottom), before (solid histogram) and after (shaded histogram) applying the kinematical cuts. The background is completely removed by the kinematical cuts.

used to detect multiparticle final states. Six superconducting coils generate a toroidal magnetic field around the target with azimuthal symmetry about the beam axis. The coils divide CLAS into six sectors, each instrumented with three regions of drift chambers, time-of-flight detectors, and, in the forward region, Cherenkov counters, and electromagnetic calorimeters. Real photons produced with the Jefferson Lab Hall-B bremsstrahlung-tagging system [13] in the energy range from 0.35 to 1.55 GeV were incident on an 18-cm-thick liquid ^{3}He target. The primary electron beam operating at 1.645 GeV was incident on the thin radiator of the Hall-B photon Tagger. Tagged photons were produced with 20-95% of the energy of the primary electron beam. About 10^9 triggers were collected at the production current of 10 nA.

In order to associate the reaction of interest with the triggering tagged photon, the coincidence time between the Tagger and CLAS was required to be within 1 ns. A cut was applied to the time difference between the CLAS start time at the interaction point recorded by the CLAS start counter and the Tagger. The particles were identified by determining their charge, momentum, and velocity. Charge and momentum were obtained from the drift-chamber tracking information and the velocity from the time of flight and path length to the time-of-flight detectors. Figure 1(left) shows the reconstructed mass distribution of positively charged particles. The events of interest were those with two and only two positively charged particles detected in coincidence. A triton candidate was required to have a positive charge and a reconstructed mass squared between 6.5 and 10.0 (GeV/c^2)^2. A pion candidate was required to have a positive charge and a reconstructed mass squared between 0.05 and 0.3 (GeV/c^2)^2. In order to assure that the events of interest are produced within the ^{3}He target volume, a cut was applied to the z-component of the interaction vertex along the beam line.

Energy-loss corrections were applied to the selected particles because they lose a non-negligible part of their energy in the target material and start counter before they reach the drift chambers. Also, fiducial-volume cuts were applied to ensure that the particles are detected within those parts of the volume of CLAS where the detection efficiency is high and uniform. These cuts select regions of CLAS where simulations reproduce the detector response reasonably well.

In order to select cleanly the $\gamma^3\text{He} \to \pi^+ t$ channel, two-body kinematics were used. The two-body final-state kinematics for real events requires that the missing energy, missing momentum, and missing-mass squared for $t\pi^+$ events be zero. Also, the opening angle between the three-vectors of the detected pion and triton $\theta_{t\pi^+}$ should be close to 180° in the center-of-mass frame. The sample of events used after these kinematical cuts is therefore essentially background-free. This also can be confirmed by calculating the missing-mass squared of either the detected pion or

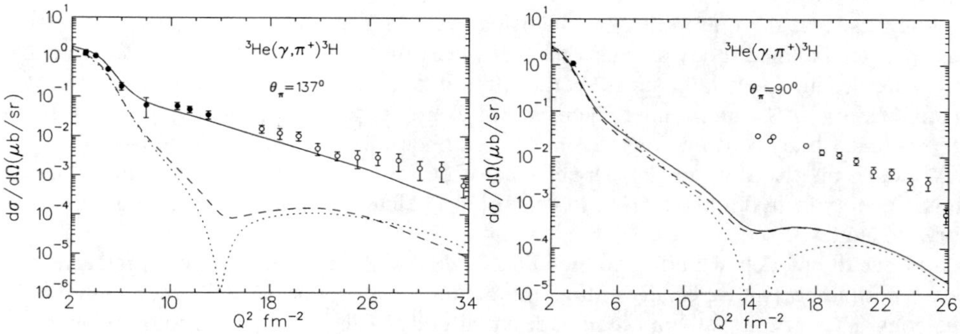

FIGURE 2. Momentum-transfer dependence of the differential cross section for fi xed pion angle of 137 (left) and 90 degrees (right) in the c.m. frame. The curves show the calculations by Tiator and Kamalov for three different assumptions: plane-wave impulse approximation PWIA (dotted lines); distorted-wave impulse approximation DWIA (dashed lines); and DWIA + 2-body mechanism [10](solid lines). Our data are from CLAS are shown as open circles and from Ref. [3] (left) and Ref. [9] (right) as fi lled circles.

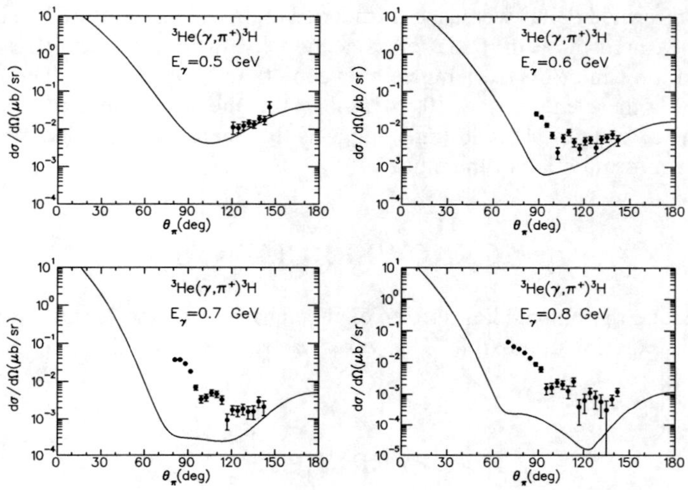

FIGURE 3. Comparison of the model calculations from Tiator and Kamalov with the differential cross section as a function of pion scattering angle in the c.m. frame for various photon-energy bins. The model includes DWIA + 2-body mechanism.

the detected triton. These distributions are shown before and after the above cuts in Fig. 1 (right), and show that the background has been removed completely. The clean sample of pions and tritons that are detected in coincidence is also shown within the shaded areas of Fig. 1 (left).

Results and Comparison with Model Calculations

Our results are compared with the model calculations by Tiator and Kamalov and with previous measurements. The calculations were originally suited only for the energies from threshold to the Δ resonance region. Recently this model has been extended (with the MAID) to higher energies [14] (see Fig. 2 and 3).

There is good agreement between the calculations and experimental data for small momentum transfers. For larger momentum transfers the calculations can describe the data only at backward angles. The old measurement of Bachelier *et al.* at 137 degrees [3] can be nicely extended with our data up to $Q^2 = 34$ fm^{-2} or 1.4 GeV^2 (Fig. 2). In fact with the new elementary production operator from MAID the agreement with data from Ref. [3] is much improved compared to the previous calculations in 1995. For other angles a huge discrepancy is found, *e.g.*, at 90 degrees (Fig.2 (right)) and at forward angles.

Our new data suggest that there are other mechanisms that produce much larger contributions than the 1-body (impulse approximation) and the 2-body mechanisms that were proposed in Ref. [10]. It is possible that two- or even three-body effects are driving those large cross sections, but it is not precisely known to what extent. In fact, strong evidence from analyzing CLAS data in other channels, for example, $\gamma^3\text{He} \to ppn$ [15], $\gamma^3\text{He} \to pd$ [16], and $\gamma^4\text{He} \to pt$ [17], suggests that 3-body contributions become more important, especially at E_γ=0.6-0.8 GeV.

Figure 3 show the comparison of angular dependence of our cross sections with the full model calculations for four bins of photon energy from 0.5 to 0.8 GeV. In general, the calculations fail to describe our data at higher photon energies and forward angles.

One way to improve the models would be to include 2-body and 3-body meson-exchange currents (MEC). These processes become more important especially at high momentum transfers because the momentum is shared between two or three nucleons. The very first attempt to include a two-body MEC in the pion-photoproduction model, where a pion is exchanged between the two nucleons, was considered by Raskin *et al.* [18]. In that model, a formalism for the pion-photoproduction amplitude with binding-induced contributions was given. However, no numerical calculation was ever performed. The two- and three-body MEC were also considered in the calculations for the ^{3}He and ^{3}H form factors [19, 20].

Another possible process to include in the model would be the photo-induced reaction $\Delta(\gamma,\pi N)$ on a free Δ that is created from $NN \to \Delta N$ reaction. The existence of these preformed deltas were investigated by studying reactions such as $A(\gamma,\pi^+ p)B$ [21]. Preformed deltas were also introduced in the calculations of ^{3}He and ^{3}H form factors [20].

In summary, we have measured the differential cross section for the $\gamma^3\text{He} \to \pi^+ t$ reaction in the energy range from 0.5 to 1.55 GeV, for pion center-of-mass angles between 40 and 140 degrees. The comparison with the only available model calculations at these energies [8, 10, 14]. shows that the calculations cannot describe our data at large momentum transfer and forward angles. This strongly suggests that there are additional production mechanisms that are not included in the current formulation of the model.

ACKNOWLEDGMENTS

We would like to acknowledge the outstanding efforts of the staff of the Accelerator and the Physics Divisions at Jefferson Lab that made this experiment possible. This work was supported by the U.S. Department of Energy under grant DE-FG02-95ER40901.

REFERENCES

1. J. R. O'Fallon *et al.*, *Phys. Rev.* **141**, 889 (1966).
2. P. Argan *et al.*, *Phys. Rev.* C **20**, 242 (1979).
3. D. Bachelier *et al.*, *Phys. Lett.* B**44**, 44 (1973), D. Bachelier *et al.*, *Nucl. Phys.* **A251**, 433 (1975).
4. C. Lazard *et al.*, *Nuovo Cimento*, A**16**, 605 (1973).
5. B. Bellinghausen *et al.*, *Nucl. Phys.* **A470**, 429 (1987).
6. J. L. Sanchez-Gomez and P. Pascual, *Nucl. Phys.* B**9**, 153 (1969).
7. L. Tiator *et al.*, *Nucl. Phys. A* **A333**, 343 (1980).
8. S.S. Kamalov, L. Tiator, and C. Bennhold, *Few-Body Systems* **10**, 143 (1991).
9. N. d'Hose *et al.*, *Nucl. Phys.* **A554**, 679 (1993).
10. S. Kamalov *et al. Phys. Rev. Lett.* **75**, 1288 (1995).
11. C.W. Leemann *et al.*, *Ann. Rev. Nucl. Part. Sci.* **51**, 413 (2001).
12. B.A. Mecking *et al.*, *Nucl. Instr. Meth.* **A503**, 513 (2003).
13. D.I. Sober *et al.*, *Nucl. Instr. Meth.* **A440**, 263 (2000).
14. L. Tiator and S. Kamalov, private communications (2010).
15. S. Niccolai *et al.*, *Phys. Rev.* C **70**, 064003 (2004).
16. Y. Ilieva, private communications (2010).
17. R. Nasseripour, *et al.*, *Phys. Rev.* C **80**, 044603(2009).
18. A. S. Raskin *et al.*, *Few-Body Systems* **17**, 71 (1994).
19. D. Drechsel *et al.*, *Phys. Lett.* **108B**, 3 (1982).
20. W. Strueve *et al.*, *Nucl. Phys.* **A465**, 651 (1987).
21. L. Wright and C. Chang, proceedings for the BLAST workshop (2000).

Session N: Meson-Baryon Interactions

Meson-baryon interaction in the meson exchange picture

M. Döring

Institut für Kernphysik and Jülich Center for Hadron Physics, Forschungszentrum Jülich, D-52425 Jülich, Germany

Abstract. Elastic πN scattering and the reaction $\pi^+ p \to K^+\Sigma^+$ are described simultaneously in a unitary coupled-channels approach which respects analyticity. SU(3) flavor symmetry is used to relate the t- and u- channel exchanges that drive the meson-baryon interaction in the different channels. Angular distributions, polarizations, and spin-rotation parameters are compared with available experimental data. The pole structure of the amplitudes is extracted from the analytic continuation.

Keywords: Dynamical coupled-channels models, Baryon spectroscopy
PACS: 11.30.Hv, 11.80.Gw, 13.75.Gx, 14.20.Gk, 24.10.Eq

INTRODUCTION

The excitation spectrum of baryons and mesons is expected to reveal important information on the mechanism of confinement as well as the instrinsic structure of hadrons. Properties of baryon resonances have been obtained by lattice calculations [1, 2, 3], mostly for the ground states but also for some excited states [2]. In quark models [4, 5, 6], a rich spectrum of excited states is predicted. Many of these resonances could be identified in elastic πN scattering, while at higher energies, usually more states are predicted than seen, a fact commonly referred to as the "missing resonance problem". Since resonances not seen in the πN channel might predominantly couple to other channels, there are intensive experimental efforts to measure, among others, multi-pion or KY final states, where $KY = K\Lambda$ or $K\Sigma$.

The reaction $\pi^+ p \to K^+\Sigma^+$ provides access to a pure isospin $I = 3/2$ two-body reaction channel in meson-nucleon dynamics. Moreover, the weak decay $\Sigma^+ \to p\pi^0$ allows to determine the polarization of the produced Σ^+. Thus, we have performed a first combined analysis of the reactions $\pi N \to \pi N$ and $\pi^+ p \to K^+\Sigma^+$ within the unitary dynamical coupled-channels framework [7], with the results summarized in these proceedings.

Dynamical coupled-channels models [8, 9, 10, 11, 12, 13, 14] are particularly suited for combined data analyses: E.g., in the work presented here [7], based on the "Jülich model" [8, 9, 10, 11], the SU(3) flavor symmetry for the exchange processes allows to relate different final states. The t- and u-channel diagrams connect also different partial waves and the respective backgrounds. Thus, the treatment of the interaction via meson and baryon exchange is expected to lead to a realistic background, with strong restrictions on the free parameters. In view of this, the strategy to perform baryon spectroscopy is to introduce only a minimum number of bare resonance states in order to obtain a good description of the data.

The coupled-channels scattering equation [8, 9, 10, 11] used in the present formalism fulfills two-body unitarity, as well as some requirements of three-body unitarity. Furthermore, it fulfills analyticity and takes into account the dispersive parts of the intermediate states as well as the off-shell behavior dictated by the interaction Lagrangians. This integral equation which is solved in the JLS-basis is given by

$$\langle L'S'k'|T^{IJ}_{\mu\nu}|LSk\rangle = \langle L'S'k'|V^{IJ}_{\mu\nu}|LSk\rangle + \sum_{\gamma L''S''} \int_0^\infty k''^2\, dk'' \langle L'S'k'|V^{IJ}_{\mu\gamma}|L''S''k''\rangle \frac{1}{z - E_\gamma(k'') + i\varepsilon} \langle L''S''k''|T^{IJ}_{\gamma\nu}|LSk\rangle \tag{1}$$

where z is the scattering energy, $J(L)$ is the total angular (orbital angular) momentum, $S(I)$ is the total spin (isospin), $k(k', k'')$ are the incoming (outgoing, intermediate) momenta, and μ, ν, γ are channel indices. In Eq. (1), E_γ is the on-mass shell energy in channel γ [11]. The pseudo-potential V iterated in Eq. (1) is constructed from an effective interaction based on the Lagrangians of Wess and Zumino [15, 16], supplemented by additional terms [9, 10] for including the Δ isobar, the ω, η, a_0 meson, and the σ. The channel space is given by $N\pi, N\eta, N\sigma, \Delta\pi$, $N\rho$ [8, 9, 10] and — as the novelty presented here — ΛK and ΣK [7].

12th International Conference on Meson-Nucleon Physics and the Structure of the Nucleon (MENU 2010)
AIP Conf. Proc. 1374, 473-476 (2011); doi: 10.1063/1.3647183

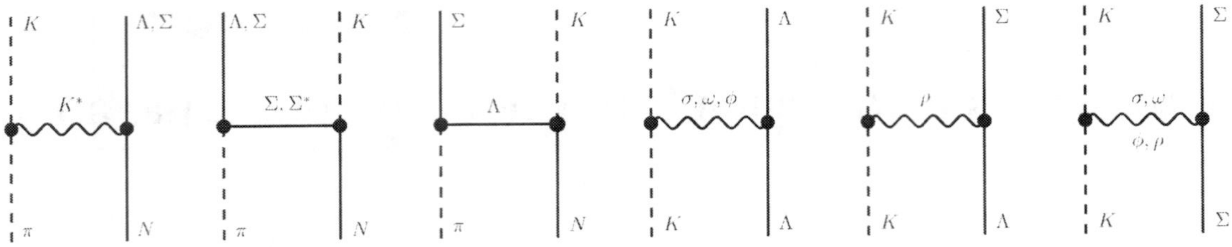

FIGURE 1. $\pi N \to KY$ and $KY \to KY$ transitions. For the other transitions used in this approach, see Refs. [9, 10].

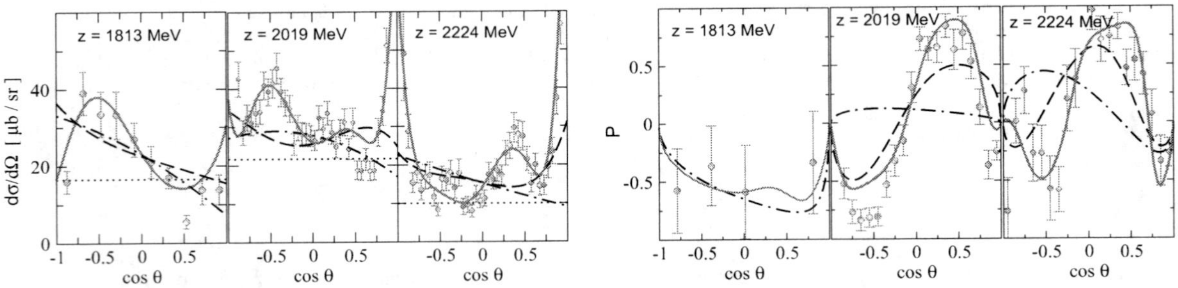

FIGURE 2. Selected results of the presented study (red lines) for three typical energies, for differential cross section and polarization. Data: Ref. [17] and references therein. Also, contributions from different partial waves are shown: Dotted lines: S wave. Dash-dotted lines: $S+P$ waves. Dashed lines: $S+P+D$ waves.

The t- and u-channel processes provide the non-resonant interaction in the meson exchange picture. The transition potentials without participation of KY have been derived in Refs. [9, 10] and explicit expressions can be found in these references. In Fig. 1, we show the extension to the KY channels. Most of the vertices present in these diagrams are related to the ones of Refs. [9, 10] using SU(3) symmetry; details can be found in Ref. [7].

Apart from this non-resonant interaction, s-channel diagrams have been introduced, which can be regarded as resonances (for details, see Ref. [7]. In the present study, we concentrate on the reaction $\pi^+ p \to K^+\Sigma^+$ and the isospin $I = 3/2$ elastic πN scattering. Thus, one bare s-channel state is included in each of the $I = 3/2$ partial waves S31, P31, D33, D35, F35, F37. Two are present in P33. These states were allowed to couple to all $I = 3/2$ channels πN, $K\Sigma$, $\pi\Delta$ and ρN. Together with these four bare couplings, the bare mass has to be left free as a fit parameter. Thus, there are altogether 40 parameters, apart from some free parameters in the $t-$ and $u-$channel diagrams; the latter parameters in the new diagrams from Fig. 1 have been chosen closely to their counterparts without KY channels from Refs. [10]. Thus, in the present approach there is explicit SU(3) breaking from the different hadron masses, but also a small SU(3) breaking from the t- and $u-$channel diagrams, mostly in terms of different form factors [7].

RESULTS

Results for three typical scattering energies z are shown in Fig. 2. The full results for differential cross section $d\sigma/d\Omega$ and polarization P, at 32 different energies from the $K\Sigma$ threshold up to $z = 2.3$ GeV, can be found in Ref. [7]. The red solid lines show the result of this study. Overall, the data are well described over the entire energy range. For energies above 2 GeV, we do not claim validity of the present model, because the analysis of Refs. [10, 11] has been limited to that energy. Consequently, at the highest energies, the $K^+\Sigma^+$ data have not been fitted, but up to $z \sim 2.25$ GeV the description of the data is still good, as Fig. 2 shows. In Fig. 2, also the influence of individual partial waves is illustrated. At lower energies, $S-P$ wave interference is enough to describe the polarization, but not entirely the differential cross section (dash-dotted lines at $z = 1813$ MeV). At higher energies, the figure shows that indeed all partial waves are needed to quantitatively describe the data; in particular, F waves are important.

Fig. 3 shows the predicted spin-rotation parameter β [7] (solid lines). The prediction from the isobar analysis of Ref. [19] is also shown. It is interesting to note that while $d\sigma/d\Omega$ and P are described to a similar precison in both

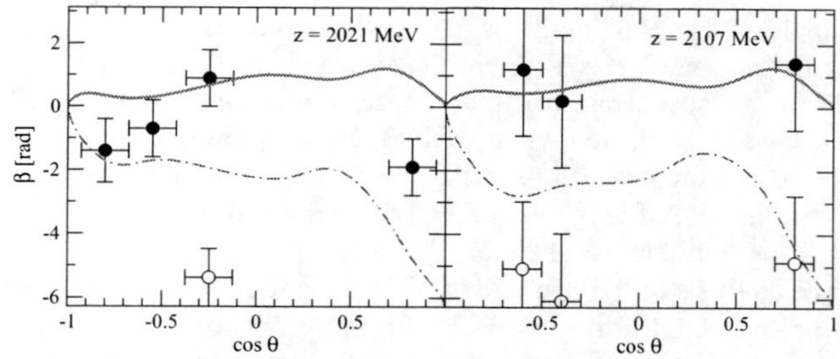

FIGURE 3. Spin rotation parameter β of $\pi^+ p \to K^+\Sigma^+$ at $z = 2021$ and $z = 2107$ MeV. Note that β is 2π cyclic which leads to additional data points at shifted values shown by the empty circles. Data: Ref. [18]. (Red) solid lines: Prediction from present solution. (Blue) dash-dotted lines: Prediction from Ref. [19].

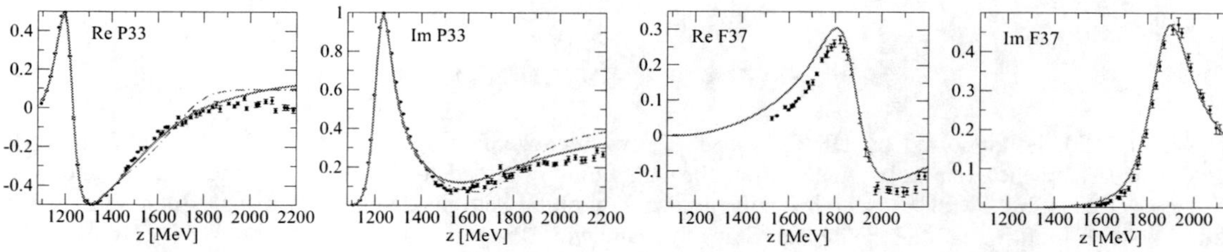

FIGURE 4. Elastic $\pi N \to \pi N$ partial waves P33 and F37. (Red) solid lines: Present solution. (Green) dash-dotted lines: Jülich model, solution 2002 from Ref. [10]. For the other partial waves see Ref. [7]. Data points: GWU/SAID [20].

studies, the outcome for β is so different. This demonstrates that higher precision data for β would help further pin down the partial wave content. In any case, what distinguishes the present analysis [7] from Ref. [19] is (apart from conceptual differences) the simultaneous consideration of elastic πN scattering. The results of the presented study [7] for the P33 and F37 partial waves are shown in Fig. 4, for the other partial waves see Ref. [7]. The partial waves for the reaction $\pi^+ p \to K^+\Sigma^+$ are shown in Ref. [7].

The resonance content of the amplitude is extracted using the analytic continuation following Refs. [11, 21] (see also Refs. [22, 23]). In Table 1, pole positions, residues, and branching ratios for the most prominent resonances in the reaction $\pi^+ p \to K^+\Sigma^+$ are shown and compared to results from Refs. [19, 24, 25]. The full list of resonances and their properties can be found in Ref. [7]. As for the well-established 4-star resonances, it is no surprise that most of the pole positions found in this study are in good agreement with the values from the GWU/SAID analysis [20], because

TABLE 1. Left: Resonance pole positions z_0 in the presented study [7]. Center: $\pi^+ p \to K^+\Sigma^+$ residues from Ref. [7]. Right: Transition branching ratios.

	Re z_0 [MeV] -2 Im z_0 [MeV]	$\lvert r\rvert$ [MeV] θ [^{0}]	$(\Gamma_{\pi N}^{1/2}\Gamma_{K\Sigma}^{1/2})/\Gamma_{\rm tot}$ [%] Ref. [7]	 Ref. [19]	 Refs. [24, 25]
$\Delta(1905)F_{35}$ $5/2^+$ ****	1764 218	1.4 -313	1.23	1.5(3)	<1
$\Delta(1910)P_{31}$ $1/2^+$ ****	1721 323	5.5 -6	2.98	<3	1.1
$\Delta(1920)P_{33}$ $3/2^+$ ***	1884 229	5.9 -38	5.07	5.2(2)	2.1(3)
$\Delta(1930)D_{35}$ $5/2^-$ ***	1865 147	1.6 -43	2.14	<1.5	
$\Delta(1950)F_{37}$ $7/2^+$ ****	1873 206	2.7 -255	2.54	5.3(5)	—

the partial waves from that analysis serve as input for the present study. However, the $\Delta(1920)P_{33}$ and $\Delta(1930)D_{35}$ resonances show no or very small resonance signals in the GWU/SAID analysis of elastic πN scattering. Their position is, thus, barely fixed from elastic πN scattering. It is then interesting to note that the constraints from the $K^+\Sigma^+$ data lead to resonance positions in vicinity to those quoted in the PDG [26], rated with 3 stars. Thus, we can accumulate further evidence for these states and their positions. Also, we find a $\Delta(1600)P_{33}$ resonance, dynamically generated mainly from the $\pi\Delta$ channel (see also Ref. [27]). Qualitatively, it shares properties with the corresponding state quoted in the PDG [26] like a strong decay into the $\pi\Delta$ channel.

In summary, a first combined analysis of the reactions $\pi N \to \pi N$ and $\pi^+ p \to K^+\Sigma^+$ within the dynamical coupled-channels framework has been presented. In this approach, for both πN and $K\Sigma$ a realistic and structured background can be provided. Consequently, only a minimal set of s-channel resonances is needed to obtain a good fit to the combined data sets. This is also tied to the fact that in this field-theoretical, Lagrangian based approach, the dispersive parts are fully included and thus, analyticity is ensured. Apart from the well-established 4-star resonances, there is a clear need for the three-star $\Delta(1920)P_{33}$ resonance. This state is found to couple only weakly to πN but stronger to $K\Sigma$, and thus evidence for a "missing resonance state" could be accumulated.

ACKNOWLEDGMENTS

The work of M.D. is supported by DFG (Deutsche Forschungsgemeinschaft, GZ: DO 1302/1-2). This work is supported in part by the Helmholtz Association through funds provided to the virtual institute "Spin and Strong QCD" (VH-VI-231), by the EU-Research Infrastructure Integrating Activity "Study of Strongly Interacting Matter" (HadronPhysics2, grant n. 227431) under the Seventh Framework Program of EU and by the DFG (TR 16). F.H. is grateful to the COSY FFE grant No. 41788390 (COSY-058).

REFERENCES

1. S. Dürr *et al.*, Science **322**, 1224 (2008).
2. J. Bulava *et al.*, Phys. Rev. D **82**, 014507 (2010).
3. G. P. Engel, C. B. Lang, M. Limmer, D. Mohler and A. Schäfer [BGR Collaboration], Phys. Rev. D **82**, 034505 (2010).
4. N. Isgur and G. Karl, Phys. Rev. D **18**, 4187 (1978).
5. S. Capstick and N. Isgur, Phys. Rev. D **34**, 2809 (1986).
6. U. Löring, B. C. Metsch and H. R. Petry, Eur. Phys. J. A **10**, 395 (2001).
7. M. Döring, C. Hanhart, F. Huang, S. Krewald, U.-G. Meißner and D. Rönchen, arXiv:1009.3781 [nucl-th].
8. C. Schütz, J. Haidenbauer, J. Speth and J. W. Durso, Phys. Rev. C **57**, 1464 (1998).
9. O. Krehl, C. Hanhart, S. Krewald and J. Speth, Phys. Rev. C **62**, 025207 (2000).
10. A. M. Gasparyan, J. Haidenbauer, C. Hanhart and J. Speth, Phys. Rev. C **68**, 045207 (2003).
11. M. Döring, C. Hanhart, F. Huang, S. Krewald and U.-G. Meißner, Nucl. Phys. A **829**, 170 (2009).
12. B. Juliá Díaz, T. S. Lee, A. Matsuyama and T. Sato, Phys. Rev. C **76**, 065201 (2007).
13. N. Suzuki, B. Juliá Díaz, H. Kamano, T. S. Lee, A. Matsuyama and T. Sato, Phys. Rev. Lett. **104**, 042302 (2010).
14. L. Tiator, S. S. Kamalov, S. Ceci, G. Y. Chen, D. Drechsel, A. Svarc and S. N. Yang, arXiv:1007.2126 [nucl-th].
15. J. Wess and B. Zumino, Phys. Rev. **163**, 1727 (1967).
16. U.-G. Meißner, Phys. Rept. **161**, 213 (1988).
17. D. J. Candlin *et al.*, Nucl. Phys. B **226**, 1 (1983).
18. D. J. Candlin *et al.*, Nucl. Phys. B **311**, 613 (1989).
19. D. J. Candlin *et al.* [Edinburgh-Rutherford-Westfield Collaboration], Nucl. Phys. B **238**, 477 (1984).
20. R. A. Arndt, W. J. Briscoe, I. I. Strakovsky and R. L. Workman, Phys. Rev. C **74**, 045205 (2006).
21. M. Döring, C. Hanhart, F. Huang, S. Krewald and U.-G. Meißner, Phys. Lett. B **681**, 26 (2009).
22. M. Döring and K. Nakayama, Eur. Phys. J. A **43**, 83 (2010).
23. M. Döring, D. Jido and E. Oset, Eur. Phys. J. A **45**, 319 (2010).
24. G. Penner and U. Mosel, Phys. Rev. C **66**, 055211 (2002).
25. V. Shklyar, G. Penner and U. Mosel, Eur. Phys. J. A **21**, 445 (2004).
26. K. Nakamura *et al.*, Journal of Physics G **37**, 075021 (2010).
27. E. Oset and A. Ramos, Eur. Phys. J. A **44**, 445 (2010).

Calculation of the $\pi^+\Sigma^+$ and $\pi^+\Xi^0$ Scattering Lengths in Lattice QCD

Aaron Torok

Indiana University Department of Physics
Swain Hall West
Bloomington, Indiana 47405

Abstract. The $\pi^+\Sigma^+$ and $\pi^+\Xi^0$ scattering lengths were calculated in mixed-action Lattice QCD with domain-wall valence quarks on the asqtad-improved coarse MILC configurations at four light-quark masses, and at two light-quark masses on the fine MILC configurations. Heavy Baryon Chiral Perturbation Theory with two and three flavors of light quarks was used to perform the chiral extrapolations. To NNLO in the three-flavor chiral expansion, the kaon-baryon processes that were investigated show no signs of convergence. Using the two-flavor chiral expansion for extrapolation, the pion-hyperon scattering lengths are found to be $a_{\pi^+\Sigma^+} = -0.197 \pm 0.017$ fm, and $a_{\pi^+\Xi^0} = -0.098 \pm 0.017$ fm, where the comprehensive error includes statistical and systematic uncertainties.

Keywords: Lattice QCD, Heavy Baryon Chiral Perturbation Theory, Pion-Hyperon Scattering, Meson-Baryon Scattering
PACS: 12.38.Gc,25.80.Dj,13.75.Gx,13.75.Jz,12.39.Fe

INTRODUCTION

Meson-baryon scattering has been extensively studied in the pion-nucleon channels, with other channels being more difficult to access experimentally. With regard to QCD, any hadron-hadron scattering process, or in fact, the mass of any hadron, is not well understood quantitatively, since the fundamental degrees of freedom are quarks and gluons, and the observables are mesons and baryons. Calculating the mass of a hadron, or the phase shift of hadron-hadron scattering (at low energies) from continuum QCD is intractable. The choices we have to calculate such quantities are: effective field theory and/or Lattice QCD (LQCD).

Almost all LQCD calculations use bare quark masses that are greater than the physical bare quark masses [1, 2, 3], which are themselves defined at a particular renormalization scale and subtraction scheme. Eventually, this will not be the case, and calculations will be done at the physical value of the light quark masses. The observables in LQCD are for instance, the masses of the mesons and baryons, and the energies that exist due to their interactions via the background gauge field.

Lattice QCD calculations of meson-meson interactions have yielded predictions for physical scattering lengths at the few percent level [4, 5, 6]. Lattice QCD calculations involving baryons have the additional feature, and challenge, of an exponentially decreasing signal-to-noise (s/n) ratio in both the single-particle, and multi-particle correlation functions. For the case of $\pi^+\pi^+$ scattering, the s/n ratio is a constant.

Meson-baryon scattering is interesting from a LQCD perspective partly because it is scattering with at least one baryon, and one important aspect of present and future LQCD calculations is to make quantitative predictions relevant to nuclear physics from LQCD. In this respect, LQCD calculations are still in the early stages, although there have been several LQCD calculations that are specifically focused toward nuclear physics by the NPLQCD collaboration [7, 8, 9], HAL QCD [10], and PACS-CS [11].

METHODOLOGY

In calculating the meson-baryon scattering lengths, the mixed-action lattice QCD scheme was used in which domain-wall quark [12, 13, 14, 15, 16] propagators are generated from a smeared source on $n_f = 2+1$ asqtad-improved [17, 18] rooted, staggered sea quarks [19]. The masses of the pions computed with the domain-wall propagators in this study are equal within a few percent to those of the lightest, staggered pions computed from staggered propagators that were generated with the same parameters as the given MILC gauge configuration [20]. This process has been done in

12th International Conference on Meson-Nucleon Physics and the Structure of the Nucleon (MENU 2010)
AIP Conf. Proc. 1374, 477-480 (2011); doi: 10.1063/1.3647184

several other mixed-action studies by NPLQCD and by LHP [21]. To improve the chiral symmetry properties of the domain-wall quarks, hypercubic-smearing (HYP-smearing) [22, 23, 24] was used in the gauge links of the valence-quark action. In the sea-quark sector, there has been significant debate regarding the validity of taking the fourth root of the staggered fermion determinant at finite lattice spacing [25, 26, 27, 28, 29, 30, 31, 32, 33, 30, 34, 35, 36, 37]. While there is no proof, there are arguments to suggest that taking the fourth root of the fermion determinant recovers the contribution from a single Dirac fermion. The results of this study assume that the fourth-root trick recovers the correct continuum limit of QCD.

The present calculations were performed predominantly with the coarse MILC lattices with a lattice spacing of $b \sim 0.125$ fm, and a spatial extent of $L \sim 2.5$ fm. On these configurations, the strange quark was held fixed near its physical value while the degenerate light quarks were varied over a range of masses. See Ref. [38] for further details. Results were also obtained on a coarse MILC ensemble with a spatial extent of $L \sim 3.5$ fm. However, this data is statistics limited. In addition, calculations were performed on two fine MILC ensembles at $L \sim 2.5$ fm with $b \sim 0.09$ fm. On the coarse MILC lattices, Dirichlet boundary conditions were implemented to reduce the original time extent of 64 down to 32, which saved a nominal factor of two in computational time. While this procedure leads to minimal degradation of a nucleon signal, it does limit the number of time slices available for fitting meson properties. By contrast, on the fine MILC ensembles, anti-periodic boundary conditions were implemented and the full time extent of the lattice is available. The s-wave scattering amplitude for two particles below inelastic thresholds can be determined using Lüscher's method [39], which entails a measurement of one or more energy levels of the two-particle system in a finite volume.

RESULTS

Since the calculations of meson and baryon masses, and the interaction energy of the two-particle meson-baryon system are all done at larger than physical pion masses, we need chiral perturbation theory in order to extrapolate to the physical value of the meson and baryon masses. In the case of $\pi^+\Sigma^+$and $\pi^+\Xi^0$scattering, there are expressions for the scattering lengths in both SU(2) [40] and SU(3) [41, 42, 43] heavy baryon chiral perturbation theory.

We use the continuum theory for extrapolation of the $\pi^+\Sigma^+$and $\pi^+\Xi^0$scattering lengths due to the level of precision of this calculation. Ultimately, lattice spacing effects, mixed action corrections, and finite volume effects need to be included, but at the 10-20% level of this calculation, we suspect that these effects are small as compared to the comprehensive error that we quote.

To $\mathcal{O}(m_\pi^3)$ in the two-flavor chiral expansion, $a_{\pi^+\Sigma^+}$ and $a_{\pi^+\Xi^0}$ are given by [40]

$$a_{\pi^+\Sigma^+} = \frac{1}{4\pi}\frac{m_\Sigma}{m_\pi+m_\Sigma}\left[-\frac{2m_\pi}{f_\pi^2}+\frac{2m_\pi^2}{f_\pi^2}C_{\pi^+\Sigma^+}+\frac{m_\pi^3}{\pi^2 f_\pi^4}\log\frac{m_\pi}{\mu}+\frac{2m_\pi^3}{f_\pi^2}h_{\pi^+\Sigma^+}(\mu)\right]; \quad (1)$$

$$a_{\pi^+\Xi^0} = \frac{1}{4\pi}\frac{m_\Xi}{m_\pi+m_\Xi}\left[-\frac{m_\pi}{f_\pi^2}+\frac{m_\pi^2}{f_\pi^2}C_{\pi^+\Xi^0}+\frac{m_\pi^3}{2\pi^2 f_\pi^4}\log\frac{m_\pi}{\mu}+\frac{m_\pi^3}{f_\pi^2}h_{\pi^+\Xi^0}(\mu)\right], \quad (2)$$

where the explicit forms —in terms of Lagrangian parameters— of the LECs $C_{\pi^+\Sigma^+}$, $h_{\pi^+\Sigma^+}$, $C_{\pi^+\Xi^0}$ and $h_{\pi^+\Xi^0}$ are given in Ref. [40]. The mixed-action modification to the SU(2) scattering length formula would begin with corrections to the $m_\pi^3\ln(m_\pi)$ terms, with the mixed valence-sea pions having the known additive mass shift [44]. The scale is set at: $\mu = \Lambda_\chi = 4\pi f_\pi$ and f_π is evaluated at its lattice physical value. Table 1 shows the values of the LECs.

TABLE 1. SU(2) LECs fit from each process at NLO and at NNLO. The first uncertainty in parentheses is statistical, and the second is the statistical and systematic uncertainty added in quadrature.

	NLO fit	NNLO fit
$C_{\pi^+\Sigma^+}$	0.66(04)(11) GeV^{-1}	1.98(17)(24) GeV^{-1}
$C_{\pi^+\Xi^0}$	0.69(06)(22) GeV^{-1}	2.01(24)(68) GeV^{-1}
$h_{\pi^+\Sigma^+}$	-	-0.65(36)(40) GeV^{-2}
$h_{\pi^+\Xi^0}$	-	-0.6(0.5)(1.1) GeV^{-2}

TABLE 2. SU(2) extrapolated scattering lengths using the LECs from Table 1. The first uncertainty in parentheses is statistical, and the second is the statistical and systematic uncertainty added in quadrature.

Quantity	LO (fm)	NLO (fm)	NLO (NNLO fit) (fm)	NNLO (fm)
$a_{\pi\Sigma}$	-0.2294	-0.208(01)(03)	-0.166(05)(08)	-0.197(06)(08)
$a_{\pi\Xi}$	-0.1158	-0.105(01)(04)	-0.083(04)(11)	-0.098(05)(12)

The scattering lengths are shown at each order in Table 2, and the final values for the $\pi^+\Sigma^+$and $\pi^+\Xi^0$scattering lengths, with the statistical and systematic errors added in quadrature, are:

$$a_{\pi^+\Sigma^+} = -0.197 \pm 0.017 \text{ fm} ; \quad (3)$$

$$a_{\pi^+\Xi^0} = -0.098 \pm 0.017 \text{ fm} . \quad (4)$$

CONCLUSION

An analysis of the scattering lengths of these two-body systems using HBχPT has leads to the conclusion that the three-flavor chiral expansion does not converge over the range of light quark masses that were investigated. While the kaon-baryon scattering lengths appear perturbative at NLO, a comparison of NNLO with NLO calls into question the convergence of the three-flavor chiral expansion. The $\pi^+\Sigma^+$ and $\pi^+\Xi^0$ scattering lengths appear to have a well-controlled chiral expansion in two-flavor HBχPT, however. The results, $a_{\pi^+\Sigma^+} = -0.197 \pm 0.017$ fm, and $a_{\pi^+\Xi^0} = -0.098 \pm 0.017$ fm, deviate from the LO (current algebra) predictions at the one- and two-sigma level, respectively.

The HBχPT analyses performed in this work support a general observation about convergence in the three-flavor chiral expansion, at least for the processes studied here. As the pion masses considered in this lattice calculation are comparable to the physical kaon mass, the distinct convergence patterns of the two- and three-flavor chiral expansions found in this work are suggestive that the breakdown in the three-flavor case is not due to the relative largeness of the strange-quark mass as compared to the light quark masses, but rather due to some other enhancement in the coefficients of the loop contributions, possibly related to a scaling with powers of n_f, the number of flavors.

In this study, the lowest-lying baryon decuplet is not considered, however, one interesting process for future work is the $\pi^-\Omega^-$ system. It does not involve disconnected diagrams since the pions have no valence quarks with the same flavor as the Ω^- constituents. It has been argued that there is a bound state [45] in this channel, and therefore, it would be of interest to determine whether this state appears bound on the lattice at the available quark masses.

ACKNOWLEDGMENTS

Thanks to U.G. Meißner for useful discussions, and R. Edwards and B. Joo for help with the QDP++/Chroma programming environment [46] with which the calculations discussed here were performed. I gratefully acknowledge the computational time provided by NERSC (Office of Science of the U.S. Department of Energy, No. DE-AC02-05CH11231), the Institute for Nuclear Theory, Centro Nacional de Supercomputación (Barcelona, Spain), Lawrence Livermore National Laboratory, and the National Science Foundation through Teragrid resources provided by the National Center for Supercomputing Applications and the Texas Advanced Computing Center. Computational support at Thomas Jefferson National Accelerator Facility and Fermi National Accelerator Laboratory was provided by the USQCD collaboration under *The Secret Life of a Quark*, a U.S. Department of Energy SciDAC project (`http://www.scidac.gov/physics/quarks.html`). The work of AT was supported in part by the National Science Foundation CAREER grant No. PHY-0645570. Part of this work was performed under the auspices of the US DOE by the University of California, Lawrence Livermore National Laboratory under Contract No. W-7405-Eng-48. The work of AP was partly supported by the EU contract FLAVIAnet MRTN-CT-2006-035482, by the contract FIS2008-01661 from MEC (Spain) and FEDER and by the Generalitat de Catalunya contract 2005SGR-00343. Thanks to Steven Gottlieb for his generous travel support, the MENU 2010 Local Organizing Committee, and their support.

REFERENCES

1. A. Bazavov, et al., *Rev. Mod. Phys.* **82**, 1349–1417 (2010), 0903.3598.
2. Z. Fodor, *J. Phys. Conf. Ser.* **230**, 012013 (2010).
3. S. Aoki, et al., *Phys. Rev.* **D81**, 074503 (2010), 0911.2561.
4. S. R. Beane, et al., *Phys. Rev.* **D77**, 014505 (2008), 0706.3026.
5. S. R. Beane, et al., *Phys. Rev.* **D74**, 114503 (2006), hep-lat/0607036.
6. S. R. Beane, et al., *Phys. Rev.* **D77**, 094507 (2008), 0709.1169.
7. S. R. Beane, et al., *Phys. Rev.* **D81**, 054505 (2010), 0912.4243.
8. S. R. Beane, et al., *Phys. Rev.* **D79**, 114502 (2009), 0903.2990.
9. S. R. Beane, et al., *Phys. Rev.* **D80**, 074501 (2009), 0905.0466.
10. T. Inoue, et al. (2010), 1007.3559.
11. T. Yamazaki, Y. Kuramashi, A. Ukawa, and f. t. P.-C. Collaboration, *Phys. Rev.* **D81**, 111504 (2010), 0912.1383.
12. D. B. Kaplan, *Phys. Lett.* **B288**, 342–347 (1992), hep-lat/9206013.
13. Y. Shamir, *Phys. Lett.* **B305**, 357–365 (1993), hep-lat/9212010.
14. Y. Shamir, *Nucl. Phys.* **B406**, 90–106 (1993), hep-lat/9303005.
15. Y. Shamir, *Phys. Rev.* **D59**, 054506 (1999), hep-lat/9807012.
16. V. Furman, and Y. Shamir, *Nucl. Phys.* **B439**, 54–78 (1995), hep-lat/9405004.
17. K. Orginos, D. Toussaint, and R. L. Sugar, *Phys. Rev.* **D60**, 054503 (1999), hep-lat/9903032.
18. K. Orginos, and D. Toussaint, *Phys. Rev.* **D59**, 014501 (1999), hep-lat/9805009.
19. C. W. Bernard, et al., *Phys. Rev.* **D64**, 054506 (2001), hep-lat/0104002.
20. S. R. Beane, P. F. Bedaque, K. Orginos, and M. J. Savage, *Phys. Rev.* **D73**, 054503 (2006), hep-lat/0506013.
21. W. Schroers, et al., *Nucl. Phys. Proc. Suppl.* **129**, 907–909 (2004), hep-lat/0309065.
22. A. Hasenfratz, and F. Knechtli, *Phys. Rev.* **D64**, 034504 (2001), hep-lat/0103029.
23. T. A. DeGrand, A. Hasenfratz, and T. G. Kovacs, *Phys. Rev.* **D67**, 054501 (2003), hep-lat/0211006.
24. T. A. DeGrand, *Phys. Rev.* **D69**, 014504 (2004), hep-lat/0309026.
25. S. Durr, C. Hoelbling, and U. Wenger, *Phys. Rev.* **D70**, 094502 (2004), hep-lat/0406027.
26. S. Durr, and C. Hoelbling, *Phys. Rev.* **D71**, 054501 (2005), hep-lat/0411022.
27. M. Creutz (2006), hep-lat/0603020.
28. C. Bernard, *Phys. Rev.* **D73**, 114503 (2006), hep-lat/0603011.
29. C. Bernard, M. Golterman, Y. Shamir, and S. R. Sharpe, *Phys. Lett.* **B649**, 235–240 (2007), hep-lat/0603027.
30. M. Creutz, *Phys. Lett.* **B649**, 241–242 (2007), 0704.2016.
31. C. Bernard, M. Golterman, and Y. Shamir, *Phys. Rev.* **D73**, 114511 (2006), hep-lat/0604017.
32. C. Bernard, M. Golterman, and Y. Shamir, *PoS* **LAT2006**, 205 (2006), hep-lat/0610003.
33. M. Creutz, *Phys. Lett.* **B649**, 230–234 (2007), hep-lat/0701018.
34. S. Durr, and C. Hoelbling, *Phys. Rev.* **D74**, 014513 (2006), hep-lat/0604005.
35. A. Hasenfratz, and R. Hoffmann, *Phys. Rev.* **D74**, 014511 (2006), hep-lat/0604010.
36. Y. Shamir, *Phys. Rev.* **D75**, 054503 (2007), hep-lat/0607007.
37. S. R. Sharpe, *PoS* **LAT2006**, 022 (2006), hep-lat/0610094.
38. S. R. Beane, K. Orginos, and M. J. Savage, *Int. J. Mod. Phys.* **E17**, 1157–1218 (2008), 0805.4629.
39. M. Luscher, *Commun. Math. Phys.* **105**, 153–188 (1986).
40. M. Mai, P. C. Bruns, B. Kubis, and U.-G. Meissner, *Phys. Rev.* **D80**, 094006 (2009), 0905.2810.
41. N. Kaiser, *Phys. Rev.* **C64**, 045204 (2001), nucl-th/0107006.
42. Y.-R. Liu, and S.-L. Zhu, *Phys. Rev.* **D75**, 034003 (2007), hep-ph/0607100.
43. Y.-R. Liu, and S.-L. Zhu, *Eur. Phys. J.* **C52**, 177–186 (2007), hep-ph/0702246.
44. K. Orginos, and A. Walker-Loud, *Phys. Rev.* **D77**, 094505 (2008), 0705.0572.
45. W. L. Wang, F. Huang, Z. Y. Zhang, Y. W. Yu, and F. Liu, *Eur. Phys. J.* **A32**, 293–297 (2007), nucl-th/0612007.
46. R. G. Edwards, and B. Joo, *Nucl. Phys. Proc. Suppl.* **140**, 832 (2005), hep-lat/0409003.

Meson–baryon scattering in manifestly Lorentz invariant chiral perturbation theory

Maxim Mai[1*], Peter C. Bruns[*], Bastian Kubis[*] and Ulf-G. Meißner[*,†]

[*] *Universität Bonn, Helmholtz-Institut für Strahlen- und Kernphysik (Theorie) and Bethe Center for Theoretical Physics, D-53115 Bonn, Germany*
[†] *Forschungszentrum Jülich, Institut für Kernphysik, Institute for Advanced Simulation, and Jülich Center for Hadron Physics, D-52425 Jülich, Germany*

Abstract. We analyze meson–baryon scattering lengths in the framework of covariant baryon chiral perturbation theory at leading one-loop order. We compute the complete set of matching relations between the dimension-two low-energy constants in the two- and three-flavor formulations of the theory. We derive new two-flavor low-energy theorems for pion–hyperon scattering that can be tested in lattice simulations.

Keywords: meson baryon, scattering length, perturbation theory, chiral perturbation theory, low-energy constant, low-energy theorem, effective Lagrangian
PACS: 12.39.Fe, 13.75.Gx, 13.75.Jz, 14.20.Jn

INTRODUCTION

At energies well below the chiral symmetry breaking scale, chiral perturbation theory (ChPT) is a useful tool to investigate various hadronic interactions, such as meson–baryon scattering (throughout this paper, mesons and baryons refer to the octet of Goldstone bosons and the lowest-lying baryon octet, respectively). The theory relies on an expansion of the QCD S-matrix and transition currents in powers of small external momenta and quark masses; the organization of the different orders of any calculation exploits the power counting in these parameters. For a recent review on baryon ChPT, see Ref. [1].

However, especially for meson–baryon systems, calculations in the three-flavor sector of up, down, and strange quarks are often hampered by a slow convergence due to the large kaon-loop contributions or even require a non-perturbative resummation due to the appearance of subthreshold resonances. In Ref. [2] an analysis of meson–baryon scattering lengths at third order in the framework of heavy-baryon chiral perturbation theory (HBChPT) was presented (see also Ref. [3] for an earlier calculation of kaon–nucleon scattering in that framework). This issue may be readdressed in the context of a covariant formulation of baryon ChPT, as the resummation of the kinetic energy terms might help improve the convergence. We will thus repeat the calculation of Ref. [2] in the framework of infrared regularization (IR) [12], which also allows to take the strict heavy-baryon limit. This gives us the opportunity to investigate the question how strongly the convergence of the chiral expansion is affected by the heavy-baryon limit in meson–baryon scattering at threshold.

On the other hand, integrating out the strange quarks from the effective field theory leaves one with pion–nucleon ChPT, which in general shows a better convergence due to the much smaller pion mass. Evidently, the two- and three-flavor theories are not independent – the pertinent low-energy constants (LECs) in the corresponding chiral effective Lagrangian are related by *matching*. As there are more LECs in SU(3) than in SU(2), matching allows to construct constraints for combinations of three-flavor LECs. Such constraints can e.g. be used in the unitarized coupled channel analysis of the existing and upcoming photo-kaon threshold production data from ELSA, JLab, and MAMI [4]. As meson–baryon scattering is an important ingredient for photoproduction calculations because of final-state interactions, a general matching analysis for the LECs related to meson–baryon and pion–nucleon scattering is called for. This indeed is one of the topics of this work. Furthermore, it was already stressed in the context of pion–kaon scattering that there exist certain observables in the three-flavor effective theory that are not afflicted by large kaon-mass corrections [5]. We therefore systematically analyze two-flavor versions of pion–hyperon scattering,

---

[1] mai@hiskp.uni-bonn.de

12th International Conference on Meson-Nucleon Physics and the Structure of the Nucleon (MENU 2010)
AIP Conf. Proc. 1374, 481-484 (2011); doi: 10.1063/1.3647185

extending earlier work on two-flavor ChPT for hyperons [6, 7, 8]. Such a framework allows to formulate low-energy theorems, which are useful in the analysis of lattice QCD data [9].

MESON–BARYON SCATTERING TO ONE LOOP

Framework

The meson–baryon scattering up to third chiral order is fully determined by chiral Lagrangian of the following form

$$\mathscr{L} = \mathscr{L}_{\phi}^{(2)} + \mathscr{L}_{\phi}^{(4)} + \mathscr{L}_{\phi B}^{(1)} + \mathscr{L}_{\phi B}^{(2)} + \mathscr{L}_{\phi B}^{(3)} , \quad (1)$$

where the power denotes the chiral order. For the minimal form of the chiral Lagrangian we refer to the original publication [10] and the references therein.

At first and second order in the chiral expansion, the T-matrix elements are determined by tree graphs with vertices from $\mathscr{L}_{\phi B}^{(1)}$ and $\mathscr{L}_{\phi B}^{(2)}$, and thus are independent of any regularization scheme. Born type graphs contribute to the first chiral order of the scattering amplitude, however at threshold their leading contribution is shifted to second order. We still count these terms as part of the first-order corrections. This is in contrast to the discussions in HBChPT [3, 2], where a strict $1/m_0$ expansion applies and Born term contributions are part of the second order. At second order only three symmetry breakers $\{b_0, b_D, b_F\}$ as well as four independent combinations $\{B_1, B_2, B_3, B_4\}$ of the dynamical LECs $\{b_1, \ldots, b_{11}\}$ of $\mathscr{L}_{\phi B}^{(2)}$ appear in meson–baryon threshold amplitudes, for which we will quote fit results below. To third chiral order both one-loop diagrams as well as contact terms from $\mathscr{L}_{\phi B}^{(3)}$ contribute. As there is no sufficient experimental information to fix the latter, we restrict ourselves to the one-loop corrections at this order.

Loop corrections are in general divergent and require regularization and renormalization. In the baryon sector, a straightforward application of dimensional regularization in the covariant theory is not possible as the large scale related to the baryon mass in the chiral limit invalidates the power counting [11]. This problem can be overcome utilizing the so-called infrared regularization as developed in Ref. [12]. For further technical details concerning the regularization procedure we refer to the original publication [10].

Results and discussion

We neglect isospin breaking and use the average octet baryon mass $m_0 = 1.150$ GeV for all baryons. Moreover we use following numerical input: $M_\pi = 139.57$ MeV, $M_K = 493.68$ MeV, $M_\eta = 547.75$ MeV, $F_\pi = 92.4$ MeV, $F_K = 113.0$ MeV, $F_\eta = 120.1$ MeV, $D = 0.8$, and $F = 0.5$. We vary the scale μ in a reasonable range around m_0, 0.938 GeV $\leq \mu \leq 1.314$ GeV. Moreover we use $a_{\pi N}^{+} := (2a_{\pi N}^{3/2} + a_{\pi N}^{1/2})/3 = -0.002 \pm 0.007$ fm [13] and $a_{KN}^{(0)} = +0.02$ fm, $a_{KN}^{(1)} = -0.33$ fm [14] to fix the $\{B_1, B_2, B_3, B_4\}$ using a particular combination of LECs $-2B_1 + 2B_3 + 2B_4$ restricted to a heavy-baryon LEC $d_0 = -0.996\ \text{GeV}^{-1}$, as well as $m_N = 0.938$ GeV, $m_\Lambda = 1.115$ GeV, $m_\Sigma = 1.192$ GeV, $m_\Xi = 1.314$ GeV, and $\sigma_{\pi N} = 45 \pm 8$ MeV [15] to fix the symmetry breakers.

Truncating the one–loop contributions at finite chiral order we obtain corresponding result of HBChPT, denoted by HB. The result for both approached read

$$b_0^{\text{IR[HB]}} = -0.45^{+0.01}_{-0.01}\,[-0.48^{+0.00}_{-0.00}]\ \text{GeV}^{-1}, \qquad b_D^{\text{IR[HB]}} = +0.05^{+0.00}_{-0.00}\,[+0.05^{+0.00}_{-0.00}]\ \text{GeV}^{-1},$$

$$b_F^{\text{IR[HB]}} = -0.45^{+0.01}_{-0.01}\,[-0.47^{+0.00}_{-0.00}]\ \text{GeV}^{-1},$$

$$B_1^{\text{IR[HB]}} = +0.43^{+0.08}_{-0.12}\,[+0.29^{+0.02}_{-0.04}]\ \text{GeV}^{-1}, \qquad B_2^{\text{IR[HB]}} = -0.34^{+0.30}_{-0.25}\,[-0.35^{+0.14}_{-0.13}]\ \text{GeV}^{-1},$$
$$B_3^{\text{IR[HB]}} = +0.46^{+0.34}_{-0.40}\,[+0.10^{+0.16}_{-0.18}]\ \text{GeV}^{-1}, \qquad B_4^{\text{IR[HB]}} = -0.53^{+0.22}_{-0.32}\,[-0.31^{+0.18}_{-0.22}]\ \text{GeV}^{-1}. \quad (2)$$

TABLE 1. *Full*[*HB*] result for the fit to $a^{+}_{\pi N}$, $a^{(1)}_{KN}$, $a^{(0)}_{KN}$ and $d_0 = -0.996\ \mathrm{GeV}^{-1}$. All scattering lengths are given in units of fm. Σ denotes the sum of all contributions from first, second, and third order.

Channel	=	$\mathcal{O}(q^1)$	$\mathcal{O}(q^2)_{IR[HB]}$		$\mathcal{O}(q^3)_{IR[HB]}$		$\Sigma_{IR[HB]}$	
$a^{(3/2)}_{\pi N}$	=	-0.12	$+0.05^{+0.06}_{-0.06}$	$[+0.05]$	$+0.04^{+0.01}_{-0.01}$	$[-0.06]$	$-0.04^{+0.07}_{-0.07}$	$[-0.13]$
$a^{(1/2)}_{\pi N}$	=	$+0.21$	$+0.05^{+0.06}_{-0.06}$	$[+0.05]$	$-0.19^{+0.01}_{-0.01}$	$[+0.00]$	$+0.07^{+0.07}_{-0.07}$	$[+0.26]$
$a^{(3/2)}_{\pi\Xi}$	=	-0.12	$+0.04^{+0.06}_{-0.07}$	$[+0.04]$	$+0.10^{+0.00}_{+0.00}$	$[-0.09]$	$+0.02^{+0.06}_{-0.07}$	$[-0.17]$
$a^{(1/2)}_{\pi\Xi}$	=	$+0.23$	$+0.04^{+0.06}_{-0.07}$	$[+0.04]$	$-0.24^{+0.02}_{-0.03}$	$[-0.03]$	$+0.02^{+0.08}_{-0.10}$	$[+0.23]$
$a^{(2)}_{\pi\Sigma}$	=	-0.24	$+0.10^{+0.02}_{-0.03}$	$[+0.07]$	$+0.15^{+0.02}_{-0.01}$	$[-0.07]$	$+0.01^{+0.04}_{-0.04}$	$[-0.24]$
$a^{(1)}_{\pi\Sigma}$	=	$+0.22$	$+0.09^{+0.15}_{-0.15}$	$[+0.11]$	$-0.21^{+0.01}_{-0.02}$	$[+0.00]$	$+0.10^{+0.16}_{-0.17}$	$[+0.33]$
$a^{(0)}_{\pi\Sigma}$	=	$+0.46$	$+0.11^{+0.15}_{-0.17}$	$[-0.01]$	$-0.47^{+0.02}_{-0.03}$	$[+0.04]$	$+0.10^{+0.17}_{-0.19}$	$[+0.49]$
$a^{(1/2)}_{\pi\Lambda}$	=	-0.01	$+0.03^{+0.03}_{-0.03}$	$[+0.03]$	$-0.03^{+0.01}_{-0.01}$	$[-0.11]$	$-0.01^{+0.04}_{-0.04}$	$[-0.09]$
$a^{(1)}_{KN}$	=	-0.45	$+0.60^{+0.14}_{-0.20}$	$[+0.40]$	$-0.48^{+0.18}_{-0.12}$	$[-0.28]$	$-0.33^{+0.32}_{-0.32}$	$[-0.33]$
$a^{(0)}_{KN}$	=	$+0.04$	$-0.15^{+0.59}_{-0.61}$	$[+0.04]$	$+0.13^{+0.05}_{-0.03}$	$[-0.06]$	$+0.02^{+0.64}_{-0.64}$	$[+0.02]$
$a^{(1)}_{\bar{K}N}$	=	$+0.20$	$+0.22^{+0.36}_{-0.40}$	$[+0.22]$	$-0.26^{+0.02}_{-0.03}+0.18i$	$[-0.02+0.36i]$	$+0.16^{+0.39}_{-0.44}+0.18i$	$[+0.40+0.36i]$
$a^{(0)}_{\bar{K}N}$	=	$+0.53$	$+0.97^{+0.42}_{-0.51}$	$[+0.58]$	$-0.40^{+0.05}_{-0.08}+0.22i$	$[+0.41+0.22i]$	$+1.11^{+0.47}_{-0.59}+0.22i$	$[+1.52+0.22i]$

Some selected results for the scattering lengths are collected in Table 1. The errors[2] quoted there are of twofold origin: they comprise the uncertainties of the experimental input, as well as the change due to the variation of the renormalization scale μ.

The conclusion to be drawn based on the numerical evaluation above is that the chiral series of the scattering lengths does not appear to converge in general. The best convergence rate is observed for the pion channels in the truncated approach. The reason for the slower convergence for SU(3) compared to SU(2) is clearly the magnitude of the expansion parameter, being $M_K/m_0 \sim M_\eta/m_0 \sim 1/2$ and $M_\pi/m_0 \sim 1/7$ for the SU(3) and SU(2) case, respectively. Moreover for the truncated (HBChPT) result, the convergence of the chiral series is improved at least for the pion channels in comparison to the full results. A similar observation was made in Ref. [17] in the analysis of pion–nucleon scattering in both schemes. Of course, for a definitive statement on the convergence of the chiral expansion one needs to include the so far neglected finite pieces of the local dimension-three operators. These certainly are expected to lead to cancellations at third order, reducing the overall size of these corrections.

MATCHING TO SU(2)

Pion–nucleon scattering can be analyzed in the framework of SU(2) as well as SU(3) ChPT, thus restrictions on the LECs in these approaches can be found. Let us denote c_i the SU(2) and b_i the SU(3) LECs respectively, then the matching procedure can be represented symbolically as

$$\sum c_i = \sum b_i + \mathscr{F}[\text{SU(3)/SU(2)-states}] + \mathcal{O}(M_K^2)\,, \tag{3}$$

where the loop functional $\mathscr{F}$ contains contributions from the loop diagrams as well as wave-function renormalization with SU(3) but not SU(2) intermediate states. For proper evaluation of the loop integrals we use the so called double-scale expansion [18], where we assume $m_0 \gg M_{K,\eta} \gg M_\pi$. As an example for the outcome the matching relation for c_1 reads

$$c_1 = b_0 + \frac{b_D}{2} + \frac{b_F}{2} + \frac{M_K}{256\pi F_\pi^2}\left[5D^2 - 6DF + 9F^2 + \frac{2}{3\sqrt{3}}(D-3F)^2\right] + \mathcal{O}(M_K^2). \tag{4}$$

This result is fully consistent with the analysis of the baryon masses [19]. Moreover we wish to remark here that the shifts Δc_i of $\mathcal{O}(M_K)$ in the matching relations are finite and in particular rather sizeable[3].

In the same sense we calculate the matching relations on the c_i^{Σ}, c_i^{Λ}, and c_i^{Ξ}, denoting the LECs of the corresponding pion–hyperon subgroup. For this we construct first the *complete* next-to-leading order Lagrangians for the $\pi\Sigma$, $\pi\Lambda$,

---

[2] For the sake of shortness we omit to give the error bars for the HB–result as well as hyperon channels

[3] For further details, techinques and complete result see [10]

and $\pi\Xi$ systems. The leading-order matching relations for these constants are already determined by the Lagrangians themselves. The next-to-leading-order is fully determined by the analysis of the one-loop corrections in the SU(3) and SU(2) theories in the sense of (3). For complete results and a more detailed discussion see [10].

THRESHOLD AMPLITUDES IN SU(2) AND LOW-ENERGY THEOREMS

An interesting application of the above results is the representation of the pion–hyperon threshold amplitudes in chiral SU(2). These can be used to prove low-energy theorems for these amplitudes, comparable to the well-known one for the isovector pion–nucleon scattering length.We identify isospin even and odd combinations of the scattering amplitudes in the pion–hyperon sector. For the isospin odd combination the result reads

$$T^{-}_{\pi\Sigma} = \frac{2M_\pi}{F_\pi^2}\left\{1+\frac{g_\Sigma^2 M_\pi^2}{16m_\Sigma^2}+\frac{g_{\Sigma\Lambda}^2 M_\pi^2}{4(m_\Lambda+m_\Sigma)^2}+M_\pi^2 d^r_{\pi\Sigma}(\mu)+\frac{M_\pi^2}{8\pi^2F_\pi^2}\left(1-2\log\frac{M_\pi}{\mu}\right)+\mathcal{O}(M_\pi^3)\right\}, \tag{5}$$

where we add a generic $\mathcal{O}(q^3)$ counterterm $\propto d_{\pi\Sigma}(\mu)$ to the isovector scattering length that cancels divergence and scale dependence of the loop contributions. Moreover if we take into account the pion-mass dependence of F_π,

$$F_\pi = F\left\{1+\frac{M_\pi^2}{F^2}\left(l_4^r-\frac{1}{8\pi^2}\log\frac{M_\pi}{\mu}\right)\right\}+\mathcal{O}(M_\pi^4), \tag{6}$$

where F is the pion decay constant in the SU(2) chiral limit, we can rewrite (5) as

$$T^{-}_{\pi\Sigma} = \frac{2M_\pi}{F^2}\left\{1+M_\pi^2 d^{\text{eff}}_{\pi\Sigma}+\mathcal{O}(M_\pi^3)\right\}. \tag{7}$$

Thus it may be expected to be similarly stable to variations of the pion mass as $T^{-}_{\pi N}$, which holds for $T^{-}_{\pi\Xi}$ as well, see [10]. This emphasizes the usefulness of low-energy theorems for lattice calculation, where typically pion masses larger than the physical ones are employed. For an interesting application see [9].

ACKNOWLEDGMENTS

The speaker thanks the organisers for the opportunity to participate on the conference, and also the Bonn Cologne Graduate School for the finanical support.

REFERENCES

1. V. Bernard, Prog. Part. Nucl. Phys. **60** (2008) 82 [arXiv:0706.0312 [hep-ph]].
2. Y. R. Liu and S. L. Zhu, Phys. Rev. D **75** (2007) 034003 [arXiv:hep-ph/0607100].
3. N. Kaiser, Phys. Rev. C **64** (2001) 045204 [Erratum-ibid. C **73** (2006) 069902] [arXiv:nucl-th/0107006].
4. B. Borasoy, P. C. Bruns, U.-G. Meißner and R. Nißler, Eur. Phys. J. A **34** (2007) 161 [arXiv:0709.3181 [nucl-th]].
5. A. Roessl, Nucl. Phys. B **555** (1999) 507 [arXiv:hep-ph/9904230].
6. S. R. Beane, P. F. Bedaque, A. Parreno and M. J. Savage, Nucl. Phys. A **747** (2005) 55 [arXiv:nucl-th/0311027].
7. B. C. Tiburzi and A. Walker-Loud, Phys. Lett. B **669** (2008) 246 [arXiv:0808.0482 [nucl-th]].
8. F. J. Jiang and B. C. Tiburzi, Phys. Rev. D **80** (2009) 077501 [arXiv:0905.0857 [nucl-th]].
9. A. Torok *et al.*, Phys. Rev. D **81** (2010) 074506 [arXiv:0907.1913 [hep-lat]].
10. M. Mai, P. C. Bruns, B. Kubis and U. G. Meissner, Phys. Rev. D **80** (2009) 094006 [arXiv:0905.2810 [hep-ph]].
11. J. Gasser, M. E. Sainio and A. Švarc, Nucl. Phys. B **307** (1988) 779.
12. T. Becher and H. Leutwyler, Eur. Phys. J. C **9**, 643 (1999) [arXiv:hep-ph/9901384].
13. H.-Ch. Schröder *et al.*, Eur. Phys. J. C **21** (2001) 473.
14. A. D. Martin, Nucl. Phys. B **179** (1981) 33.
15. J. Gasser, H. Leutwyler and M. E. Sainio, Phys. Lett. B **253** (1991) 252.
16. V. Bernard, N. Kaiser and U.-G. Meißner, Z. Phys. C **60** (1993) 111 [arXiv:hep-ph/9303311].
17. K. Torikoshi and P. J. Ellis, Phys. Rev. C **67** (2003) 015208 [arXiv:nucl-th/0208049].
18. J. Gasser, C. Haefeli, M. A. Ivanov and M. Schmid, Phys. Lett. B **652** (2007) 21 [arXiv:0706.0955 [hep-ph]].
19. M. Frink and U.-G. Meißner, JHEP **0407**, 028 (2004) [arXiv:hep-lat/0404018].

A unitary and causal effective field theory

A.M. Gasparyan[*,†] and M.F.M. Lutz[*]

[*]*GSI Helmholtzzentrum für Schwerionenforschung GmbH, Planckstrasse 1, 64291 Darmstadt, Germany*
[†]*SSC RF ITEP, Bolshaya Cheremushkinskaya 25, 117218 Moscow, Russia*

Abstract. We report on a novel scheme based on the chiral Lagrangian. It is used to analyze pion-nucleon scattering, pion photoproduction, and nucleon Compton scattering. Subthreshold partial-wave amplitudes are calculated in chiral perturbation theory and analytically extrapolated with constraints imposed by electromagnetic-gauge invariance, causality and unitarity. Experimental quantities are reproduced up to energies $\sqrt{s} \simeq 1300$ MeV in terms of the parameters relevant at order Q^3.

Keywords: <chiral symmetry, unitarity, causality, gauge invariance>
PACS: 11.55.Fv,13.60.Le,13.75.Gx,12.39.Fe

INTRODUCTION

Chiral perturbation theory has been extensively and successfully used as a systematic tool for studying the low energy QCD dynamics. Particular cases of pion-nucleon scattering, pion photoproduction and nucleon Compton scattering were considered in [1, 2, 3, 4, 5, 6, 7]. The application of χPT is however limited to the near threshold region.

In this talk we report on a recently proposed scheme based on the chiral Lagrangian [8]. We describe photon and pion scattering off the nucleon up to energies $\sqrt{s} \approx 1300$ MeV where the Lagrangian was truncated to order Q^3. An explicit isobar field is not considered. The physics of the isobar resonance enters our scheme by an infinite summation of higher order counter terms in the chiral Lagrangian. The particular summation is performed in accordance with unitarity and causality. The scheme is based on an analytic extrapolation of subthreshold scattering amplitudes that is controlled by constraints set by electromagnetic-gauge invariance, causality and unitarity. Unitarized scattering amplitudes are obtained which have left-hand cut structures in accordance with causality. The latter are solutions of non-linear integral equations that are solved by N/D techniques.

DESCRIPTION OF THE METHOD

The starting point of our method is the chiral Lagrangian involving pion, nucleon and photon fields [5, 2]. A strict chiral expansion of the amplitude to the order Q^3 includes tree-level graphs, loop diagrams, and counter terms. Counter terms depend on a few unknown parameters, which we fit to the empirical data on πN elastic scattering and pion photoproduction. The Compton scattering amplitude contains no counter terms to the order Q^3.

Our approach is based on partial-wave dispersion relations, for which the unitarity and causality constraints can be combined in an efficient manner. For a suitably chosen partial-wave amplitude with angular momentum J, parity P and channel quantum numbers a, b (in our case πN and γN channels) we separate the right-hand cuts from the left-hand cuts

$$T_{ab}(\sqrt{s}) = U_{ab}(\sqrt{s}) + \sum_{c,d} \int_{\mu_{\rm thrs}}^{\infty} \frac{dw}{\pi} \frac{\sqrt{s}-\mu_M}{w-\mu_M} \frac{T_{ac}(w)\,\rho_{cd}(w)\,T^*_{db}(w)}{w-\sqrt{s}-i\varepsilon}, \tag{1}$$

where the generalized potential, $U^{(JP)}_{ab}(\sqrt{s})$, contains left-hand cuts only, by definition. The sum in (1) runs over all possible intermediate states, and $\rho^{(JP)}_{cd}(\sqrt{s})$, is the phase-space matrix. It is emphasized that the separation (1) is gauge invariant. The non-linear integral equation (1) can be solved by means of N/D techniques [11]. The amplitude is represented as

$$T_{ab}(\sqrt{s}) = \sum_{c} D^{-1}_{ac}(\sqrt{s})\,N_{cb}(\sqrt{s}), \tag{2}$$

12th International Conference on Meson-Nucleon Physics and the Structure of the Nucleon (MENU 2010)
AIP Conf. Proc. 1374, 485-488 (2011); doi: 10.1063/1.3647186

where $D_{ab}(\sqrt{s})$ has no singularities but the right-hand s-channel unitarity cuts. The coupled-channel unitarity condition is a consequence of the ansatz

$$D_{ab}(\sqrt{s}) = \delta_{ab} - \sum_{c} \int_{\mu_{\rm thr}}^{\infty} \frac{dw}{\pi} \frac{\sqrt{s}-\mu_M}{w-\mu_M} \frac{N_{ac}(w)\,\rho_{cb}(w)}{w-\sqrt{s}}\,, \tag{3}$$

where $N_{ab}(\sqrt{s})$ obeys the linear integral equation

$$\begin{aligned} N_{ab}(\sqrt{s}) &= U_{ab}(\sqrt{s}) \\ &+ \sum_{c,d} \int_{\mu_{\rm thr}}^{\infty} \frac{dw}{\pi} \frac{\sqrt{s}-\mu_M}{w-\mu_M} \frac{N_{ac}(w)\,\rho_{cd}(w)\,[U_{db}(w)-U_{db}(\sqrt{s})]}{w-\sqrt{s}}\,. \end{aligned} \tag{4}$$

The key observation we exploit in this work is the fact that the solution of the nonlinear integral equation (1) requires the knowledge of the generalized potential for $\sqrt{s} > \mu_{\rm thr}$ only. Conformal mapping techniques may be used to approximate the generalized potential in that domain efficiently based on the knowledge of the generalized potential in a small subthreshold region. Given a suitable conformal mapping, $\xi(\sqrt{s})$ (transforming a domain Ω into a unit circle), we seek to establish a representation of the generalized potential of the form

$$\begin{aligned} &U(\sqrt{s}) = U_{\rm inside}(\sqrt{s}) + U_{\rm outside}(\sqrt{s})\,, \\ &U_{\rm outside}(\sqrt{s}) = \sum_{k=0}^{\infty} U_k\,[\xi(\sqrt{s})]^k\,, \qquad U_k = \frac{d^k U_{\rm outside}(\xi^{-1}(\xi))}{k!\,d\xi^k}\Big|_{\xi=0}\,, \end{aligned} \tag{5}$$

where we allow for an explicit treatment of cut structures that are inside the domain Ω. Then the expanded potential in (5) is regular enough for the integral equation (1) to be well defined and amenable to a solution via (4).

The decomposition (5) is faithful for energies $\sqrt{s} \in \Omega$ only. Nevertheless, the full scattering amplitude can be derived by a crossing transformation of the solution to (1). Such a construction is consistent with crossing symmetry if the solution to (1) and its crossing transformed form coincide in a matching region around $\sqrt{s} \simeq \mu_M$. With our choice

$$\mu_M = m_N\,, \tag{6}$$

this is clearly the case (see also [12]).

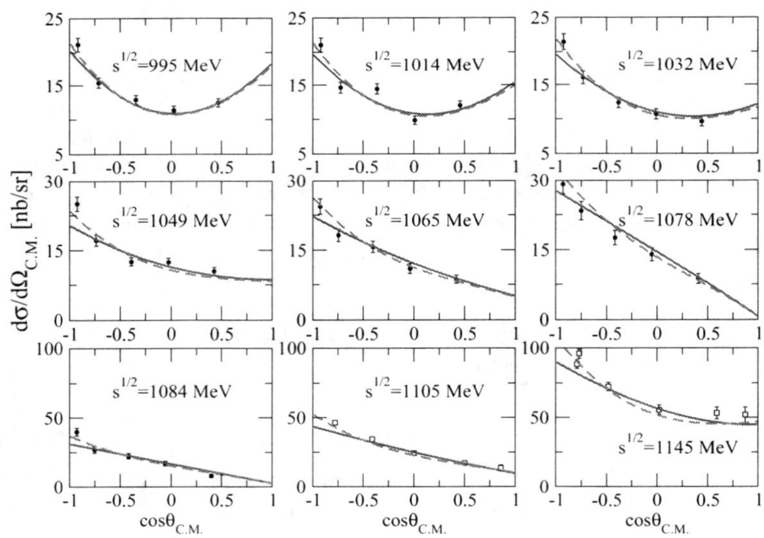

FIGURE 1. Differential cross section for Compton scattering off the proton as a function of the center-of-mass scattering angle. The data are from [9] and [10]. The solid lines follow from our partial-wave amplitudes with $J \leq 3/2$. The dashed lines show the effect of partial-wave contributions with $J > 3/2$.

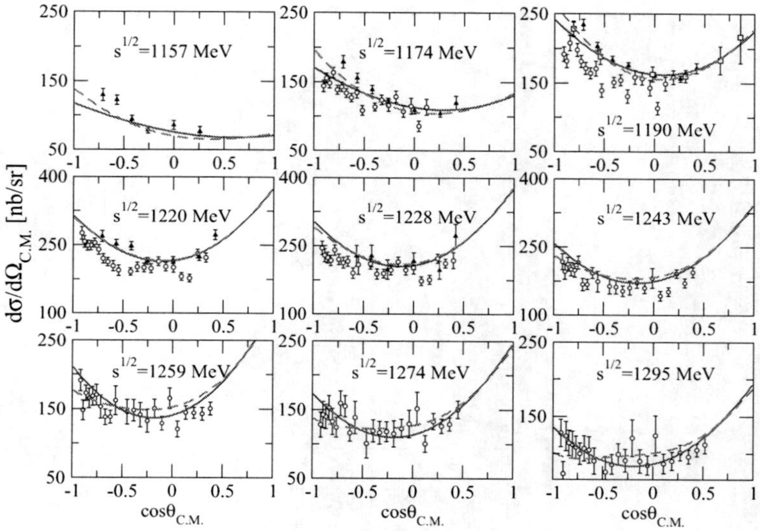

FIGURE 2. Differential cross section for Compton scattering off the proton as a function of the center-of-mass scattering angle. The data are from [10],[13], [14] and the lines are as in Fig. 1.

RESULTS

We determined the counter terms of the πN sector as to reproduce the empirical s- and p-wave πN elastic phase shifts. On this example we demonstrated reasonable convergence properties of the method with respect to the chiral expansion parameter Q up to total energies of about 1300 MeV. Additional parameters are relevant for pion photoproduction. Having adjusted those parameters we are able to reproduce a wide variety of pion photoproduction data, such as differential cross sections and spin observables for different reaction channels with good accuracy. Most predictive is proton Compton scattering, which involves no additional free parameter. Our results for the differential cross section and beam asymmetry are presented in Figs. 1-3 against empirical data. We find agreement with the data taking into account some discrepancy of the different data sets. The photon threshold region, the pion-production threshold

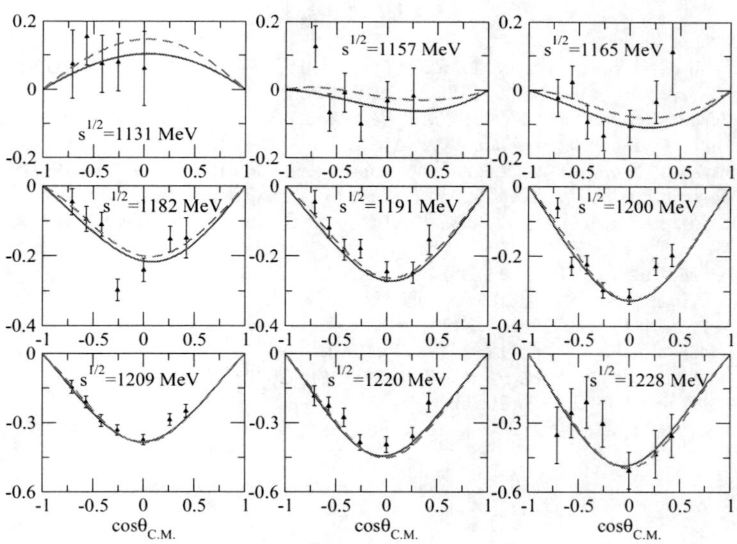

FIGURE 3. Beam asymmetry for Compton scattering off the proton as a function of the center-of-mass scattering angle. The data are from [13] and the lines are as in Fig. 1.

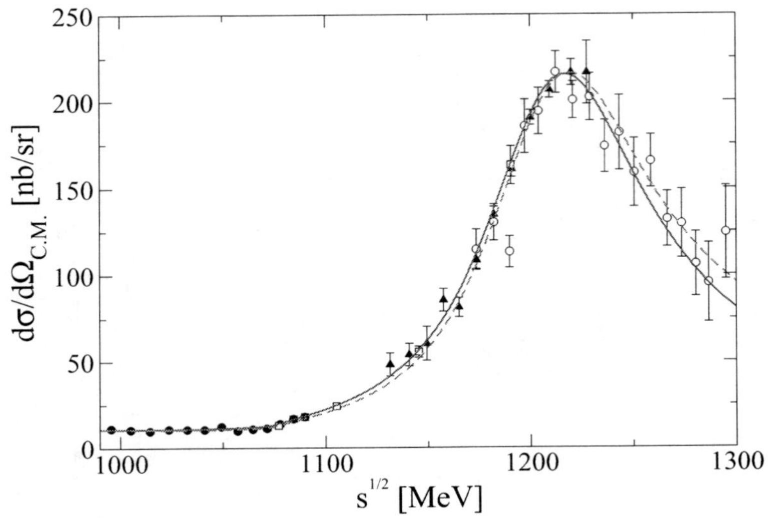

FIGURE 4. Compton scattering off the proton as a function of energy at scattering angle $\theta = 90^\circ$. The data are from [9], [10], [13], [14] and the lines are as in Fig. 1. The experimental values correspond to scattering angles closest to $\theta = 90^\circ$ taken in the interval $86^\circ < \theta < 94^\circ$.

region, and the isobar region are equally well reproduced. This is nicely illustrated by Fig. 4, which shows the energy dependence of the cross section at fixed scattering angle $\theta \simeq 90^\circ$.

SUMMARY

We reviewed recent progress on a unified description of pion and photon-nucleon scattering up to and beyond the isobar resonance region. A novel scheme that combines causality, unitarity and gauge invariance and that is based on the chiral Lagrangian was discussed. We focused on results obtained for proton Compton scattering.

REFERENCES

1. V. Bernard, N. Kaiser, and U.-G. Meissner, *Int. J. Mod. Phys.* **E4**, 193–346 (1995), hep-ph/9501384.
2. V. Bernard, *Prog. Part. Nucl. Phys.* **60**, 82–160 (2008), 0706.0312.
3. V. Pascalutsa, M. Vanderhaeghen, and S. N. Yang, *Phys. Rept.* **437**, 125–232 (2007), hep-ph/0609004.
4. V. Bernard, N. Kaiser, and U.-G. Meissner, *Nucl. Phys.* **A615**, 483–500 (1997), hep-ph/9611253.
5. N. Fettes, U.-G. Meissner, and S. Steininger, *Nucl. Phys.* **A640**, 199–234 (1998), hep-ph/9803266.
6. N. Fettes, and U.-G. Meissner, *Nucl. Phys.* **A676**, 311 (2000), hep-ph/0002162.
7. S. R. Beane, M. Malheiro, J. A. McGovern, D. R. Phillips, and U. van Kolck, *Nucl. Phys.* **A747**, 311–361 (2005), nucl-th/0403088.
8. A. Gasparyan, and M. F. M. Lutz (2010), 1003.3426.
9. V. Olmos de Leon, et al., *Eur. Phys. J.* **A10**, 207–215 (2001).
10. E. L. Hallin, et al., *Phys. Rev.* **C48**, 1497–1507 (1993).
11. G. F. Chew, and S. Mandelstam, *Phys. Rev.* **119**, 467–477 (1960).
12. M. F. M. Lutz, and E. E. Kolomeitsev, *Nucl. Phys.* **A700**, 193–308 (2002), nucl-th/0105042.
13. G. Blanpied, et al., *Phys. Rev.* **C64**, 025203 (2001).
14. S. Wolf, et al., *Eur. Phys. J.* **A12**, 231–252 (2001), nucl-ex/0109013.

Jülich-Georgia meson-baryon model

M. Döring*, C.Hanhart*, F. Huang†, S. Krewald*, K. Nakayama† and U.-G. Meißner**

*Institut für Kernphysik, Forschungszentrum,D-52425 Jülich, Germany
†Department of Physics and Astronomy, University of Athens, Georgia, USA
**Bethe Center for Theoretical Physics, University of Bonn, 53115 Bonn, Germany and Institut für Kernphysik, Forschungszentrum,D-52425 Jülich, Germany

Abstract. Meson-baryon scattering and photon-nucleon reactions are described in a coupled-channel approach for energies below 2 GeV. Poles and residues of the reaction amplitudes are obtained by analytical continuation. The dependence of the resonance poles on the pion mass is investigated.

Keywords: Meson-baryon scattering, Baryon resonances
PACS: 14.20.Gk; 13.75.Gx

In the energy region above 2 GeV, the extraction of resonance parameters via partial wave analysis is getting more difficult due to the increasing number of partial waves, the opening of new decay channels, and the overlapping of different resonances. Sophisticated theoretical analyses including the coupling of different reaction channels are presently developped by the Excited Baryon Analysis Center (EBAC) at JLAB and in Jülich. In the energy region above the first resonance, the Breit-Wigner parameterization of the resonances suffers from systematic uncertainties. The poles of the S-matrix, in contrast, are well-defined, though less intuitive. We therefore have extended the Jülich coupled-channel approach to the complex plane. This was possible because the Jülich meson-exchange model is based on field theoretical Lagrangians, respects two-body unitarity and the analytical structure of the underlying field theory. It couples the $N\pi$ and $N\eta$ reaction channels with effective $N\pi\pi$ channels summarized by the $N\sigma$, $N\rho$, and $\Delta\pi$ channels, which involve quasiparticles. While the analytical extension into the complex plane is straightforward for channels with stable particles ($N\pi$ and $N\eta$), the effective $N\pi\pi$ channels require special attention, as the two quasiparticle singularities induce new structures in the amplitudes, such as branch points in the complex plane in addition to the right hand $N\pi\pi$ cut along the physical axis. A full discussion of the sheet structure is given in Refs.[1, 2]. In Fig.1 we show both real and imaginary parts of partial wave amplitudes with angular momenta less than 5/2 obtained within the Jülich meson-exchange model for energies from threshold till approximately 2 GeV in comparison with the SAID amplitudes, Ref.[3]. The S_{11} shows two clear bumps near the real energies of 1500 MeV and 1650 MeV, which correspond to the $N^*(1535)$ and $N^*(1650)$. Note the cusp in the vicinity of 1500 MeV which is due to the opening of the $N\eta$ threshold. The extension to the Kaon-Hyperon channels will be discussed in the M.Döring's talk. Applications to photonuclear reactions employing the methods developed in Ref.[4] to ensure gauge invariance are presented by Fei Huang.

Fig.2 shows the analytical continuation to the second Riemann sheet. Two poles are

12th International Conference on Meson-Nucleon Physics and the Structure of the Nucleon (MENU 2010)
AIP Conf. Proc. 1374, 489-492 (2011); doi: 10.1063/1.3647187

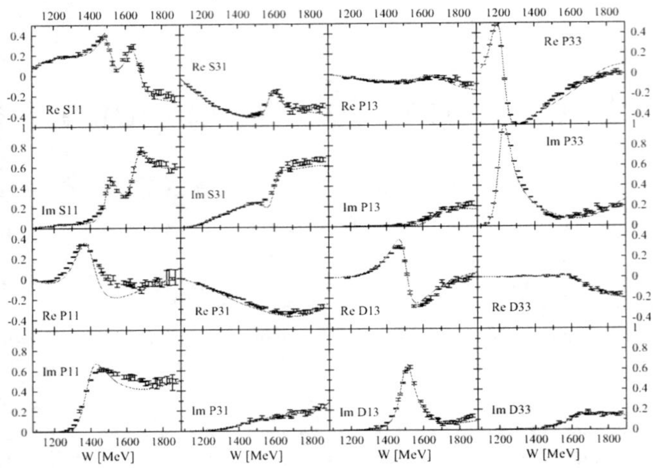

FIGURE 1. Real and imaginary parts of selected partial wave amplitudes.

visible which correspond to two overlapping resonances. The treatment of the underlying physical background is therefore non-trivial, see Ref.[2] for an analysis of the interplay between background and resonant amplitudes. After having achieved a quantitative fit to the experimental observables, the coupled channels approach allows to determine the precise pole positions and the residues, see Table 1. The established partial wave analyses agree concerning the mass of the $S_{11}(1535)$ within 30 MeV. There is rather strong disagreement concerning the widths and the residues, however. Likewise, there is reasonable agreement for the mass of the $S_{11}(1650)$, while the width is less well established. An detailed discussion of other resonances can be found in Refs.[1, 5].

Lattice gauge calculations are making great progress in baryon physics. In addition to the nucleon mass, lattice results of the first and even second excited state have been published. In the case of the Δ_{33} resonance, both mass and width are under investigation. Details can be found in Refs. [6, 7, 8, 9]. For numerical reasons, present lattice cal-

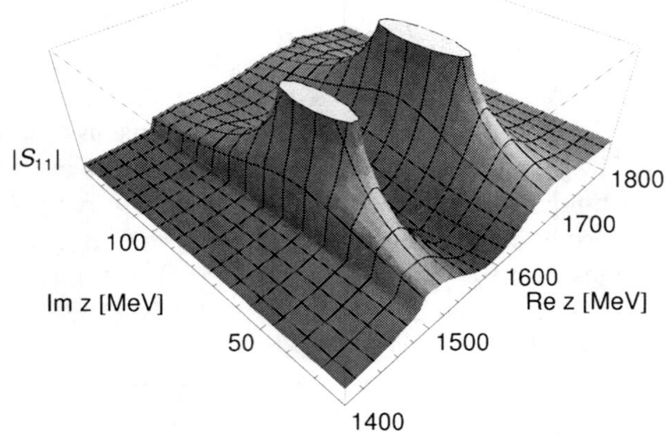

FIGURE 2. Modulus of the S_{11} partial wave as a function of the complex energy z.

TABLE 1. Poles and residues of the S_{11} resonances.

	Re z_0 (MeV)	Im z_0 (MeV)	Res (MeV)	Θ (deg)
$S_{11}(1535)$	1519	129	31	-3
ARN	1502	95	16	-16
HOE	1487			
CUT	1510	260	120	+15
$S_{11}(1650)$	1669	136	54	-44
ARN	1648	80	14	-69
HOE	1670	163	39	-37
CUT	1640	150	60	-75

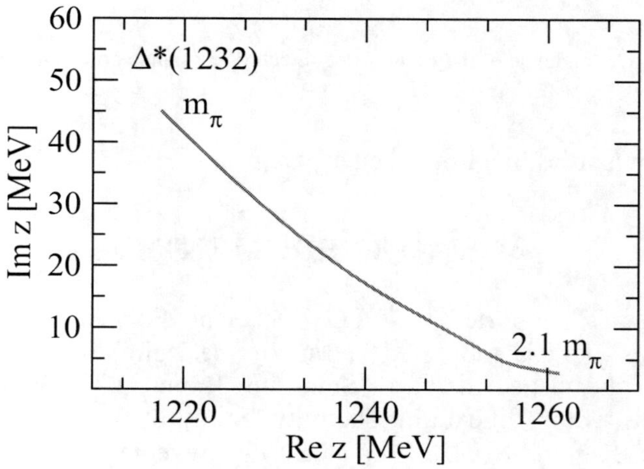

FIGURE 3. Dependence of the Δ_{33} resonance pole on the pion mass.

culations are performed with quark masses that correspond to unphysical pion masses. Therefore, lattice calculations have to be extrapolated to the physical pion mass. In the case of the nucleon mass, chiral perturbation theory has given valuable guidance to the extrapolation process. For higher masses, effects due to meson-baryon decay channels are expected to play a non-negligible role. Within the Jülich model, we can investigate the effect of increasing pion masses because the approach is based on effective Lagrangians. The Δ_{33} resonance decays into a pion and a nucleon. An increase of the pion mass modifies the pion-nucleon propagator so that one expects a reduction of the width of the resonance. This is indeed confirmed by the actual calculations shown in Fig.3 which shows the path of the resonance pole in the complex plane. A more detailed investigation of other resonances shows some non-trivial energy dependence of the resonance poles on the pion mass. The $S_{11}(1650)$ is nearly independent of the mass of the pion, in contrast to the $S_{11}(1535)$, as shown in Fig.4. Does it make sense to assume that the $S_{11}(1535)$ may have a strong admixture of a dynamically generated structure? We may not know, but at least we can hope for a reasonable phenomenological description

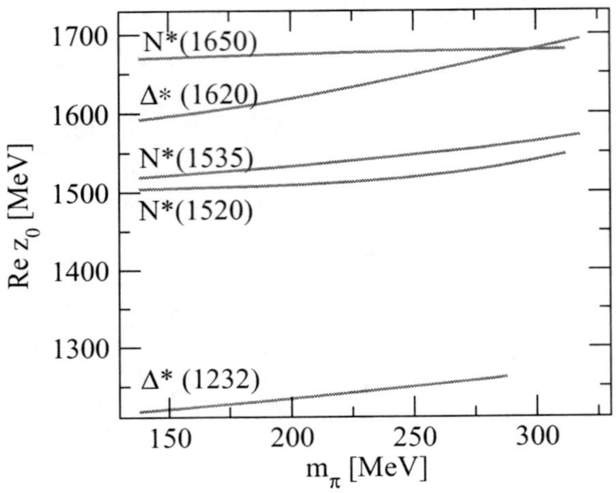

FIGURE 4. Dependence of the real part of selected resonance poles on the pion mass.

of meson-hadron dynamics in a limited energy region.

ACKNOWLEDGMENTS

The work of M.D. is supported by DFG (Deutsche Forschungsgemeinschaft, Gz: DO 1302/1-1). This work is supported in part by the Helmholtz Association through funds provided by the virtual institute "Spin and Strong QCD"(VH-VI-231), by the EU-Research Infrastructure Integrating Activity "Study of Strongly Interactiong Matter"(HadronPhysics2, grant No.227431) under the Seventh Program of EU and by the DFG(TR16). F.H. is grateful to the support from the Alexander von Humboldt Foundation and the COSY FFE grant No.41445282(COSY-58).

REFERENCES

1. M. Döring et al., *Nucl. Phys.* **A829**, 170(2009).
2. M. Döring et al., *Phys. Lett.* **B681**, 26(2009).
3. R. A. Arndt et al., *Phys. Rev.* **C74**, 045205(2006).
4. H. Haberzettl et al., *Phys. Rev.* **C74**, 045202(2006).
5. M. Döring et al., *arXiv:*1009.3781[nucl-th].
6. J.Bulava et al., *Phys. Rev.* **D82**, 014507(2010).
7. G. P. Engel et al., *Phys. Rev.* **D82**, 034505(2010).
8. M. Lage et al., *Phys. Lett.* **B681**, 439(2009).
9. V. Bernard et al., *JHEP* **0808**, 024(2008).

Recent results from Bonn-Gatchina partial wave analysis

A.V. Sarantsev

Helmholtz-Institut für Strahlen- und Kernphysik, Universität Bonn, Germany
Petersburg Nuclear Physics Institute, Gatchina, Russia

Abstract. The result of our recent analysis of the pion and photo induced reactions with $K\Lambda$, $K\Sigma$ and ηn final states is presented. The analysis provides evidence for $N(1710)P_{11}$, $N(1875)P_{11}$, $N(1900)P_{13}$, $\Delta(1600)P_{33}$, $\Delta(1910)P_{31}$, and $\Delta(1920)P_{33}$ which were not seen in the latest GWU analysis of elastic data [1].

Keywords: baryon, spectroscopy,resonances
PACS: 11.80.Et, 11.80.Gw, 13.30.-a, 13.30.Ce, 13.30.Eg, 13.60.Le, 14.20.Gk

INTRODUCTION

The spectrum of radially excited resonances is one of the most controversial issues in baryon spectroscopy. The Particle Data Group bases the evidence for the existence of states nearly entirely on four analyses of pion-induced reactions [1, 2, 3, 4]. However, these analyses do not include data on πN transition into the ηN, $K\Lambda$ and $K\Sigma$ final states which provide a vital information about baryon resonances in the mass region above 1600 MeV.

In this paper we study the consistency of the $\pi N \to \pi N$ scattering amplitudes with the πN and γN transition amplitudes into kaon-hyperon and ηN final states.

ANALYSIS OF THE DATA

The list of the fitted pion-induced reactions with two particle final states is given in Table 1. The full list of fitted reactions can be found in [5].

The contributions of the dominant waves to the reaction $\pi^- p \to K^0\Lambda$ are shown in Fig. 1a. At threshold, the dominant wave is S_{11} which decreases fast with increasing energy. Below 1800 MeV, the P_{11} wave is the second strongest wave. It shows a clear resonance behavior peaking at 1720 MeV. The P_{13} wave increases steadily from threshold reaching its maximum at 1900 MeV.

The description of the differential cross section, recoil asymmetry and rotation parameter is shown in Fig. 2. Indeed, near the threshold the angular distribution ($d\sigma/d\Omega$) rises nearly linearly in $\cos\Theta$ due to the interference of S_{11} and P waves. A quadratic part which appears at higher energies separates contributions from the P_{11} and P_{13} waves.

The two largest contributions to the reaction $\pi^+ p \to K^+\Sigma^+$ are assigned to the P_{33} and F_{37} partial waves (see Fig. 1b). When one of the waves is assumed to be non-resonant, angular and recoil distributions cannot be reproduced satisfactorily. The energy dependence of the F_{37} wave in this reaction is compatible with the well known $F_{37}(1950)$ state. The P_{33} wave has a maximum in the region 1900 MeV which is expected if the $\Delta(1920)P_{33}$ state provides a large contribution.

The contribution from Δ-states to $\pi^- p \to K^0\Sigma^0$ can be fixed, using the isotopic relations, from a fit of the $\pi^+ p \to K^+\Sigma^+$ reaction. These partial waves provide about 70% to the total cross section and contributions from nucleon partial waves are needed to describe the remaining part of the data (see Fig. 1c).

The total cross section for $\pi^- p \to \eta n$ calculated from the fitted differential cross section is shown in Fig. 1d. At low energies the cross section is dominated by the S_{11} contribution. In the 1700-1900 MeV region, the P-wave amplitudes describe about 50% of the total cross section. However the lack of polarization information does not allow us to determine them uniquely. There are two sets of solutions which lead to rather different amplitudes for the P_{11} and P_{13} partial waves.

Indeed, in spite of the large data base, the partial wave analysis does not converge to a unique solution. Instead, we found two classes of solutions which have rather different imaginary parts for the third P_{11} pole: the first set of solutions has a pole at $1860 \pm 20 - i55^{+15}_{-5}$ MeV (the best solution from this class is called BG2010-01) and second set

12th International Conference on Meson-Nucleon Physics and the Structure of the Nucleon (MENU 2010)
AIP Conf. Proc. 1374, 493-496 (2011); doi: 10.1063/1.3647188

TABLE 1. Fitted pion induced reactions with two particle final states and χ^2 contributions for the solution BG2010-02.

$\pi N \to \pi N$	**Wave**	N_{data}	χ_i^2/N_{data}	$\pi N \to \pi N$	**Wave**	N_{data}	χ_i^2/N_{data}
[1]	S_{11}	104	1.95	[1]	S_{31}	112	2.07
	P_{11}	112	2.12		P_{31}	104	3.86
	P_{13}	112	1.22		P_{33}	120	2.87
	D_{13}	96	2.87		D_{33}	108	2.68
	D_{15}	96	3.67		F_{35}	62	1.48
	F_{37}	72	2.76		F_{17}	82	1.98
	G_{17}	102	2.61				
$\pi^- p \to \eta n$	**Observ.**	N_{data}	χ_i^2/N_{data}	$\pi^- p \to K^0\Sigma^0$	**Observ.**	N_{data}	χ_i^2/N_{data}
[6]	$d\sigma/d\Omega$	68	1.68	[8]	$d\sigma/d\Omega$	259	0.77
[7]	$d\sigma/d\Omega$	84	2.50	[8]	P	90	1.36
$\pi^- p \to K^0\Lambda$	**Observ.**	N_{data}	χ_i^2/N_{data}	$\pi^+ p \to K^+\Sigma^+$	**Observ.**	N_{data}	χ_i^2/N_{data}
[9]	$d\sigma/d\Omega$	298	2.31	[13]	$d\sigma/d\Omega$	609	1.27
[10, 11]	$d\sigma/d\Omega$	299	0.90	[13]	P	304	1.58
[10, 11]	P	354	1.98	[14]	β	7	1.97
[12]	β	72	2.45				

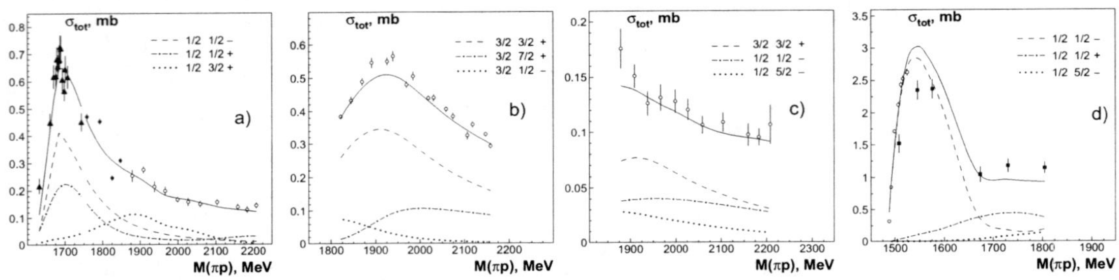

FIGURE 1. The total cross sections for the reactions a) $\pi^- p \to K^0\Lambda$ [9] (triangles),[10], [11] (open circles), b) $\pi^+ p \to K^+\Sigma^+$ [13], c) $\pi^- p \to K^0\Sigma^0$ and d) $\pi^- p \to \eta n$ and contributions from leading partial waves. The full curve shows the integrated fit of the differential cross section from the solution BG2010-02.

has a pole at $1850^{+20}_{-50} - i180 \pm 20$ (best solution: BG2010-02). For most resonances, the two classes of solutions give compatible results. Masses, widths, residues and their respective errors are listed in Table 2.

The P_{11} wave has a maximum in the $\pi^- p \to K^0\Lambda$ reaction at 1720 MeV which suggests a contribution from the $N(1710)P_{11}$ state, the long tail may indicate the need for a further state at even higher mass. We therefore used a 4-pole K-matrix amplitude to fit the data. The pole around 1700 MeV we identify with $N(1710)P_{11}$ and a pole around 1870 MeV as $N_{1/2^+}(1875)$, a resonance which was first suggested in [15]. In the first solution, $N_{1/2^+}(1875)$ is a rather narrow state with a small coupling to the πN channel and a rather small helicity coupling. In the second solution this state has appreciable πN and helicity couplings. In both solutions the hadron couplings for the $N(1710)P_{11}$ state are found to be very similar, however the helicity coupling differs in sign.

The $N(1900)P_{13}$ resonance was first introduced in our analysis [16]. It was required to achieve a good description of the double polarization observables C_x, C_z [17] in photoproduction of kaon-hyperon final states. $N(1900)P_{13}$ provides one of the dominant contributions to the $\gamma p \to K\Lambda$ and a notable contribution to the $\gamma p \to K^+\Sigma^0$ total cross sections. In the $\pi N \to K\Lambda$ reaction the P_{13} partial wave has a clear peak around 1900 MeV. We consider the existence of $N(1900)P_{13}$ to be solidly confirmed by the present analysis, moreover splitting it into two resonances is a likely possibility.

In a coupled-channel analysis of a large set of data [18], the mass of $\Delta(1920)P_{33}$ was found to be 1950 ± 40 MeV. The present analysis defines the pole position around 1925 MeV. The $\Delta(1600)P_{33}$ state contributed less to the data description and for a fit of separate reactions can be replaced by a background term. However it plays an important role for the combined description of all fitted data: if this state is excluded, systematic discrepancies are observed in

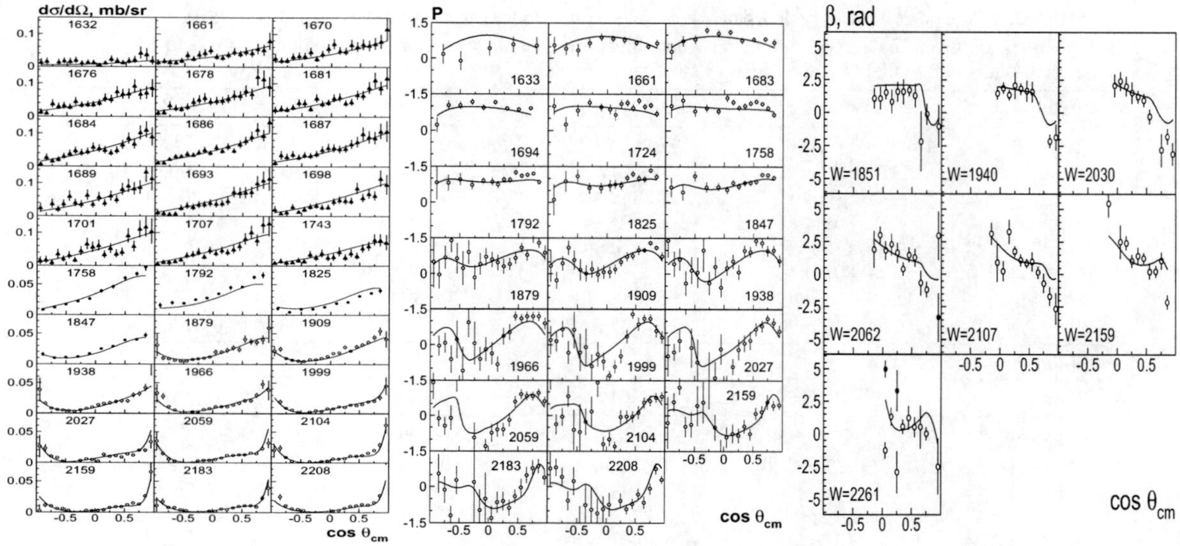

FIGURE 2. The differential cross section for the reaction $\pi^- p \to K^0 \Lambda$ (the triangles denote the data from [9] and open circles the data from [10, 11]), the recoil asymmetry [10, 11] and the rotation parameter [12]. The curves represent our fit BG2010-02.

the description of many observables.

The P_{31} partial wave provides only a rather broad contribution to the $\pi^+ p \to K^+ \Sigma^+$ reaction with a maximum around 2 GeV. Moreover, we found a very small photoproduction contribution from this wave. First we tried a fit with a two-pole K-matrix parameterization: one pole in the 1600 MeV region and second one in the 2000 MeV region. However if the first pole is taken out of the fit and replaced by non-resonant terms for direct two-pion production, we do not observe a notable change in the quality of the data description. Thus we conclude that the present data base does not provide a proof for the existence of a state with $(I)J^P = (3/2)1/2^+$ quantum numbers in the 1500-1800 MeV mass region.

DISCUSSION AND CONCLUSIONS

We have studied the mass spectrum of P-wave nucleon and Δ resonances in a coupled-channel analysis of a large body of data on pion- and photo-induced reactions. Individual partial waves are typically described by 6-channel K-matrices with up to five poles. The coupled-channel fit minimizes the number of parameters needed to represent resonant and background contributions. The πN elastic amplitudes from energy-independent partial wave analyses are included in the data base. Hence we fit the same data as [2, 1] but in our fits, the inelasticity is not a black box but constrained by data from pion- and photo-induced inelastic reactions in which the final states are fully reconstructed.

The surprising result of our analysis is that the P-wave resonances reported by [2, 3] and challenged by [1] are found with properties which are not very different from those reported by [2, 3]. In particular we confirm existence and properties of $N(1710)P_{11}$ and $N(1900)P_{13}$ and of $\Delta(1600)P_{33}$ and $\Delta(1920)$ P_{33}. We do not find evidence for $\Delta(1750)P_{31}$, a resonance not seen by [2, 3] (but in some other analyses) and we confirm with good evidence $N_{1/2^+}(1875)$ [15, 19].

ACKNOWLEDGMENTS

We would like to thank the members of SFB/TR16 for continuous encouragement. We acknowledge financial support from the Deutsche Forschungsgemeinschaft (DFG) within the SFB/ TR16 and from the Forschungszentrum Jülich within the FFE program.

TABLE 2. Pole positions and residues of the transition amplitudes (given in MeV as $|r|/\Theta$ with $ResA = |r|e^{i\Theta}$). The phases are given in degrees. The helicity couplings are given in $10^{-3}GeV^{-1/2}$. The errors are defined from the spread of results found in both classes of solutions. The PDG values are given in parentheses.

State	$N(1900)P_{13}$	$N_{3/2^+}(1975)$	$\Delta(1910)P_{31}$	$\Delta(1920)P_{33}$
Re(pole)	1890±50 (∼1900)	1970±25 (-)	1900±40 (1900±50)	1930±40 (1900±50)
-2Im(pole)	270^{+180}_{-100} (-)	250±60 (-)	500^{+50}_{-100} (300±100)	300±50 (300±100)
$A(\pi N \to \pi N)$	3±2 /-20±40°	2±1 /80±40°	38±8 /−125±15°	10±6 / 20±60°
$A(\pi N \to \eta N)$	6±3 /80±40°	4±2.5 /10±60°		
$A(\pi N \to K\Lambda)$	14±6 /145±25°	5±3 /-10±40°		
$A(\pi N \to K\Sigma)$	6±4 /80±50°	1.5±0.5 /-30±50°	19±4 /−80±20°	10±5 /150±30°
$A^{1/2}(\gamma p)$	70±25 / 130±40°	-8±8 /-20±50°	55±20 / 80±70°	120±35 /-120±35°
$A^{3/2}(\gamma p)$	110±60 / 170±35°	50±30 / 25±45°		110±30 /-170±30°
Solution	01	02	01	02
State	$N(1710)P_{11}$	$N(1710)P_{11}$	$N_{1/2^+}(1875)$	$N_{1/2^+}(1875)$
Re(pole)	1690^{+25}_{-10} (1720±50)	1695±15 (1720±50)	1860±20 ( )	1850^{+20}_{-50} ( )
-2Im(pole)	210±25 (230±150)	220±30 (230±150)	110^{+30}_{-10} ( )	360±40 ( )
$A(\pi N \to \pi N)$	5±4 /-80±40°	7±3 /170±25°	2±2 /30±30°	15±7 /75±30°
$A(\pi N \to \eta N)$	8±4 /20±20°	6±4 /40±15°	8±4 / −40±30°	11±4 / 0±20°
$A(\pi N \to K\Lambda)$	17±4 /-110±20°	16±5 /-110±15°	0.5±0.5 /10±30°	14±5 /90±30°
$A(\pi N \to K\Sigma)$	3±2 /20±40°	5±3 / 20±20°	3±2 / 160±25°	10±3 /-120±20°
$A^{1/2}(\gamma p)$	80±25 /100±30°	35±20 /-35±20°	12±6 /-80±40^0	55±15 /70±30^0
State	$N(1720)P_{13}$	$N(1720)P_{13}$	$\Delta(1600)P_{33}$	$\Delta(1600)P_{33}$
Re(pole)	1695±30 (1675±15)	1670±30 (1675±15)	1480±30 (1600±100)	1480±30 (1600±100)
-2Im(pole)	400±60 (190±85)	420±60 (190±85)	240±40 (300±100)	230±40 (300±100)
$A(\pi N \to \pi N)$	22±5 /-85±20°	28±6 /-95±20°	14±3 /-170±15°	14±3 /-170±15°
$A(\pi N \to \eta N)$	9±4 /-70±25°	8±3 /-65±25°		
$A(\pi N \to K\Lambda)$	15±8 /-125±30°	12±4 /-80±30°		
$A(\pi N \to K\Sigma)$	10±6 /-80±25°	10±4 /-30±40°	2±1 /-80±20°	3±1 /-110±20°
$A^{1/2}(\gamma p)$	110±40 /20±40°	95±30 /0±30°	26±6 /-80±30°	16±5 / 110±30°
$A^{3/2}(\gamma p)$	130±50 /70±40°	115±30 /70±25°	15±6 / 0±20°	10±5 /-150±30°

REFERENCES

1. R. A. Arndt, W. J. Briscoe, I. I. Strakovsky and R. L. Workman, *Phys. Rev.* C **74**, 045205 (2006)
2. G. Höhler, F. Kaiser, R. Koch and E. Pietarinen, "Handbook Of Pion Nucleon Scattering," Published by Fachinform. Zentr. Karlsruhe 1979, 440 P. (Physics Data, No.12-1 (1979)).
3. R. E. Cutkosky, C. P. Forsyth, J. B. Babcock, R. L. Kelly and R. E. Hendrick, "Pion - Nucleon Partial Wave Analysis," *4th Int. Conf. on Baryon Resonances*, Toronto, Canada, Jul 14-16, 1980. Published in Baryon 1980:19 (QCD161:C45:1980)
4. D. M. Manley and E. M. Saleski, *Phys. Rev.* D **45**, 4002 (1992).
5. A. V. Anisovich, E. Klempt, V. A. Nikonov, A. V. Sarantsev and U. Thoma, "P-wave excited baryons from pion- and photo-induced hyperon production," *arXiv*:1009.4803 [hep-ph].
6. W. B. Richards *et al.*, *Phys. Rev.* D **1**, 10 (1970).
7. S. Prakhov *et al.*, *Phys. Rev.* C **72**, 015203 (2005).
8. J. C. Hart *et al.*, *Nucl. Phys.* B **166**, 73 (1980).
9. T. M. Knasel *et al.*, *Phys. Rev.* D **11**, 1 (1975).
10. R. D. Baker *et al.*, *Nucl. Phys.* B **141**, 29 (1978).
11. D. H. Saxon *et al.*, *Nucl. Phys.* B **162**, 522 (1980).
12. K. W. Bell *et al.*, *Nucl. Phys.* B **222**, 389 (1983).
13. D. J. Candlin *et al.*, *Nucl. Phys.* B **226**, 1 (1983).
14. D. J. Candlin *et al.*, *Nucl. Phys.* B **311**, 613 (1989).
15. R. Castelijns *et al.*, *Eur. Phys. J.* A **35**, 39 (2008).
16. V. A. Nikonov, A. V. Anisovich, E. Klempt, A. V. Sarantsev and U. Thoma, *Phys. Lett.* B **662**, 245 (2008).
17. R. Bradford *et al.*, *Phys. Rev.* C **75**, 035205 (2007).
18. A. V. Anisovich, E. Klempt, V. A. Nikonov, M. A. Matveev, A. V. Sarantsev and U. Thoma, *Eur. Phys. J.* A **44**, 203 (2010).
19. A. V. Anisovich, V. Kleber, E. Klempt, V. A. Nikonov, A. V. Sarantsev and U. Thoma, *Eur. Phys. J.* A **34**, 243 (2007).

Toward a unified description of hadro- and photoproduction amplitudes

M. W. Paris, R. A. Arndt, R. L. Workman, W. J. Briscoe and I. I. Strakovsky

Department of Physics, Data Analysis Center at the Center for Nuclear Studies, The George Washington University, Washington, D.C. 20052, USA

Abstract. The near-term objectives of the research program at the Data Analysis Center are established within the context of the existing partial wave analyses available through the online suite of analysis and database codes accessible through SAID, the Scattering Analysis Interactive Database. This presentation reviews the efforts to determine a model independent method to obtain sets of partial wave amplitudes for strong and electromagnetic reactions, the interpretation of the amplitudes in terms of the excited states of the nucleon, the role of new precision unpolarized and polarized data, and new developments aimed at determining the photoproduction mulitpoles in a unitary, coupled-channel approach. The Chew-Mandelstam technique is discussed and applied to the problem of the S-wave pion- and eta-photoproduction amplitudes. The resulting eta production amplitudes exhibit the expected resonant behavior near the eta production threshold. Application of this method to a unified description of the hadro- and photoproduction amplitudes is discussed.

Keywords: Formal scattering theory, photoproduction multipoles, Chew-Mandelstam approach, nucleon resonance
PACS: 13.60.-r, 11.55.Bq, 11.80.Et, 11.80.Gw

INTRODUCTION

Non-perturbative features of quantum chromodynamics are crucial in determining the excitation spectrum of the nucleon. The model independent determination of scattering and reaction amplitudes is a necessary component to this end. A range of hadroproduction data, including $\pi N \to \pi N$, $\pi N \to \eta N$, $\pi N \to \omega N$, and other inelastic processes used to constrain theoretical models and phenomenological parametrization of the scattering and reaction amplitudes. Currently, however, a renaissance is underway in meson production and resonance physics with reaction data issuing from a number of precision electromagnetic facilities[1, 2, 3, 4, 5, 6, 7]. Phenomenological efforts to analyze these data consistent with some subset of constraints imposed by quantum field theory are under current study. The quality and quantity of data in electromagnetic induced reactions is becoming sufficient to rival and possibly surpass the hadroproduction data. Since the electromagnetic reactions proceed mainly through the hadronic channels, the new data offers the possibility of "back-constraining" the hadronic amplitudes, conventionally determined only in fits to the hadroproduction data.

This presentation describes an exploratory study of the S–wave π– and η–photoproduction multipoles in the "Chew-Mandelstam" approach, related to the N/D representation, to the electromagnetic reaction amplitude. The novel concept, which provokes and permits this exploratory study, is the generalization of the Chew-Mandelstam approach to the electromagnetic sector. We have developed a new form for the amplitude that incorporates multichannel hadronic rescattering effects in a complete manner consistent with unitarity[8]. The near-term objective is to develop a framework in which to analyze the hadro- and electroproduction reactions simultaneously in a global framework.

CHEW-MANDELSTAM PARAMETRIZATION

Previous work in the determination of the η photoproduction amplitudes[9, 10] has shown that an approach which includes the coupling of the electromagnetic channel to the πN and ηN channels in the region of energies near the center-of-mass energy, $W = 1535$ MeV gives a reasonably good description of the data and a plausible form for the amplitudes. However, as our ultimate objective is the simultaneous parametrization of hadro- and photoproduction scattering and reaction observables, we will go beyond the two-channel treatment for this study of the E_{0+}^{η} multipole amplitude.

12th International Conference on Meson-Nucleon Physics and the Structure of the Nucleon (MENU 2010)
AIP Conf. Proc. 1374, 497-500 (2011); doi: 10.1063/1.3647189

The form of the Chew-Mandelstam parametrization, which we employ in this study follows as a consequence of the analytic structure imposed by the unitarity[11] of the S matrix in the physical region, $W > m_i + m_t$, where W is the center-of-mass energy and m_i and m_t are the masses of the incident and target particles. Confining our attention to two-particle initial and final states, the S matrix is unitary $S^\dagger S = SS^\dagger = 1$. In the partial wave representation with $S = 1 + 2i\tilde{\rho}T = 1 + 2iT'$ we have

$$\text{Im}\, T'^{-1} = -\theta(W - M_+), \tag{1}$$

where M_+ is the diagonal matrix of threshold openings. Since this equation isolates the imaginary part of the inverse-T matrix, we may write

$$T'^{-1} = \text{Re}\, T'^{-1} + i\text{Im}\, T'^{-1}, \tag{2}$$

$$= K'^{-1} - i\theta(W - M_+), \tag{3}$$

where we've defined $\text{Re}\, T'^{-1} = K'^{-1}$. Multiplying from one side by T' and the other by K' gives the Heitler integral equation[12]

$$T' = K' + K' i\theta(W - M_+)T'. \tag{4}$$

This is the starting point for the Chew-Mandelstam parametrization of the reaction amplitude. Equation (1) implies discontinuities in the derivative of the imaginary part at each channel threshold $W = m_{\sigma+}$. The violation of the Cauchy-Riemann equations at threshold indicates the presence of a branch point.

The partial wave amplitude has the following singularities. There are branch points in the physical region at the channel-opening thresholds as in Eq.(1), branch points in the region $W < 0$, and possible poles consistent with causality[13]. An efficient parametrization following Ref.[14], which encodes these singularities, involves the factorization of the partial wave amplitude. This is referred to as the "N/D" approach. We will use the N/D language to clarify the nature of the singularities of the T matrix which are included and those neglected in our Chew-Mandelstam approach.

The preceding discussion of unitarity and the N/D approach provides the context for our present parametrization. The Chew-Mandelstam parametrization developed here is similar to those of Refs.[15]and [16]. We consider Eq.(3) and rewrite it, confining our attention to the S wave multipole as

$$T^{-1} = K^{-1} - i\tilde{\rho} \tag{5}$$

$$= (K^{-1} + \text{Re}C) - (\text{Re}C + i\tilde{\rho})$$

$$= \overline{K}^{-1} - C, \tag{6}$$

where $\tilde{\rho} = \rho\theta(W - M_+)$ and $\text{Im}C = \tilde{\rho} = \theta(W - M_+)\rho$. The transition matrix is given in terms of the "Chew-Mandelstam" (CM) K matrix, $\overline{K}$ by

$$T = \overline{K} + \overline{K}CT. \tag{7}$$

Equation (7) fixes our Chew-Mandelstam parametrization. In the language of the N/D approach, we have neglected the $W < 0$ branch points of N and made the approximation $N(W) = \overline{K}(W)$, an entire function. The "Chew-Mandelstam" function, C_α is determined solely by the unitarity constraint, Eq.(1) The Chew-Mandelstam function is given by a Cauchy integral over the discontinuity of C_α in the physical region with a single subtraction.

The relationship between the Heitler K matrix and the CM K matrix, $\overline{K}$ is given by

$$K = \overline{K} + \overline{K}[\text{Re}C]K. \tag{8}$$

This demonstrates a possible advantage of using the CM K matrix. If we consider a polynomial parametrization of a given CM K matrix element then we see, by solving Eq.(8) for K

$$K = \frac{1}{1 - \overline{K}[\text{Re}C]}\overline{K} \tag{9}$$

that poles may appear in the K matrix. Attempts to relate the K matrix poles to resonances have been made[17]. Here, we simply point out that, though K matrix poles are not simply related to T matrix poles[18], Eq.(8) shows that one need not explicitly include pole terms in $\overline{K}$ in order to have poles in K.

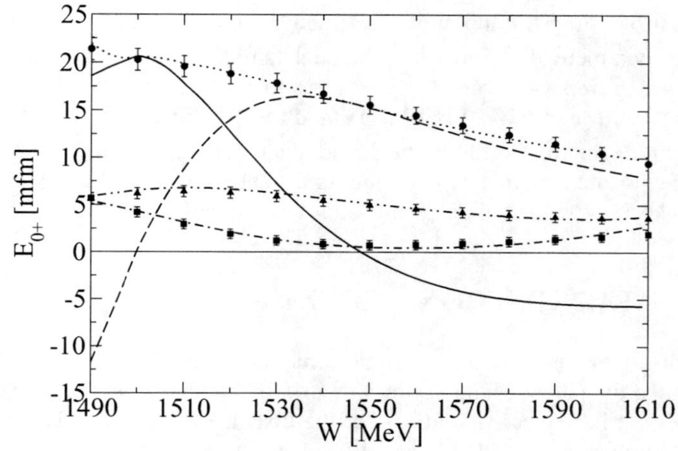

FIGURE 1. The predicted values for the real (solid curve) and imaginary (dashed curve) for E^{η}_{0+} versus the energy, W. The modulus $|E^{\eta}_{0+}|$ (dotted curve), the real (dot-dashed curve), and the imaginary (double dot-dashed curve) parts of the π–photoproduction, E^{π}_{0+} were fit to pseudodata generated from the SAID solution[20] with the parametrized form Eq.(11) using 8 parameters.

RESULTS

The Chew-Mandelstam parametrization for the T matrix, described in the preceding section, has been applied recently[19] to a coupled-channel fit for the πN elastic scattering and $\pi N \to \eta N$ reaction. It gives a realistic description of the data with χ^2 per datum better than any other parametrization or model, to our knowledge. The current SAID parametrization used in this fit is given as

$$T_{\alpha\beta} = \sum_{\sigma}[1-\overline{K}C]^{-1}_{\alpha\sigma}\overline{K}_{\sigma\beta} \tag{10}$$

where α, β and σ are channel indices for the considered channels, $\pi N, \pi\Delta, \rho N$ and ηN. Given the success of this approach in the hadronic two-body sector, the application to the study of meson photoproduction is warranted.

The central result of the current exploratory study is to show that this form can be extended to include the electromagnetic channel,

$$T_{\alpha\gamma} = \sum_{\sigma}[1-\overline{K}C]^{-1}_{\alpha\sigma}\overline{K}_{\sigma\gamma} \tag{11}$$

where γ denotes the electromagnetic channel, γN. Note that Eqs.(10) and (11) share the common factor, $[1-\overline{K}C]^{-1}_{\alpha\sigma}$ which encodes, at least qualitatively speaking, the hadronic channel coupling (or rescattering) effects.

The form Eq.(11) for photoproduction should be contrasted with that currently employed in the π–photoproduction studies of Refs.[20, 10]

$$T_{\pi\gamma} = A(W)(1+iT_{\pi\pi}(W)) + B(W)T_{\pi\pi}(W) \tag{12}$$

where the "structure functions" $A(W)$ and $B(W)$ are parametrized as polynomials in the energy, W, $T_{\pi\gamma} = T_{\pi N,\gamma N}$ and $T_{\pi\pi} = T_{\pi N,\pi N}$, and the factor $A(W)$ contains a contribution from tree-level Born diagrams. This satisfies Watson's theorem[21] (as does Eq.(11)), and is derived via the considerations discussed in Ref.[22]. While resulting in a realistic description of the data and being comparable, at least qualitatively, with other parametrization such as MAID[23] for π–photoproduction, it does not satisfy the full multichannel unitarity constraint imposed by Eq.(1). This deficiency led us to consider the form in Eq.(11), which manifestly satisfies the multichannel unitarity constraint, Eq.(1).

In the present study we perform a coupled-channel fit of the modulus $|E^{\eta}_{0+}(W)|$ and (the real and imaginary parts of) the existing SAID and MAID π–photoproduction amplitudes, E^{π}_{0+} in the $S_{11}(1535)$ resonance region. (See Ref.[8] for comparisons to MAID.) The fit was carried out by taking the factor $[1-\overline{K}C]^{-1}_{\alpha\sigma}$ in Eq.(11) as determined in the the hadronic study of Ref.[19] and adjusting the parameters of $\overline{K}_{\sigma\gamma}$ (discussed in detail in the subsections below). The phase of the E^{η}_{0+} multipole in this study gives a resonant wave and encourages us to continue with this approach.

Figure (1) shows the result of fitting the modulus $|E_{0+}^{\eta}|$ and the real and imaginary parts of the SAID SP09 E_{0+}^{π} multipole [6] using an eight-parameter fit. The Chew-Mandelstam $\overline{K}$ matrix was assumed to have four channels, $\pi N, \pi\Delta, \rho N$, and ηN. Eight parameters were varied in the fit to a total of 113 pseudodata points including the modulus $|E_{0+}^{\eta}|$ over the energy range 1490 MeV $\leq W \leq$ 1610 MeV and the amplitude E_{0+}^{π} over the energy range 1120 MeV $\leq W \leq$1610. The χ^2 per datum over for the fits to the pseudodata, generated with the SAID interactive code facility were less than one in all of the fits made in this work including those in the region 1120 MeV $\leq W \leq$ 1490 MeV which are not displayed in order to keep the figures manageable and focus attention on the $S_{11}(1535)$ resonance region.

CONCLUSION AND ONGOING WORK

We reviewed the implication of unitarity on the analytic structure of the single meson production scattering and reaction amplitudes. We related the Chew-Mandelstam K-matrix parametrization to the N/D approach, showing that the parametrization of the $\overline{K}$ matrix neglects the effects of the distant left-hand cut. Using the Chew-Mandelstam K matrix $\overline{K}$, we performed a simultaneous coupled-channel fit of the η–photoproduction S_{11} multipole modulus, $|E_{0+}^{\eta}|$ and the π–photoproduction amplitude, E_{0+}^{π}. The parametrization was restricted only to the CM K matrix elements $\overline{K}_{\sigma\gamma}$ in Eq.(11), while the $[1-\overline{K}C]^{-1}$ factors were taken from the existing SAID fits to the hadronic data. The anticipated resonant structure for the phase of the E_{0+}^{η} multipole was demonstrated in fits to SAID amplitudes.

The results of the exploratory study indicate that this is a reasonable approach toward the objective of determining a complete set of scattering and reaction amplitudes for $\pi N \to \pi N$, $\pi N \to \eta N$, $\gamma N \to \pi N$, and $\gamma N \to \eta N$ processes in a multichannel unitary formalism. The ultimate objective of the study will be a simultaneous fit to both the hadronic and electromagnetic scattering and reaction observables and will constitute, at least for two-body unitarity, a global description of the hadro- and photoproduction amplitudes.

ACKNOWLEDGMENTS

This work is dedicated to the memory of R. Arndt without whom it would not have been possible; it was supported by the US Department of Energy Grant No. DE-FG02-99ER41110.

REFERENCES

1. J. Ajaka, et al., *Phys. Rev. Lett.* **81**, 1797–1800 (1998).
2. V. Crede, et al., *Phys. Rev. Lett.* **94**, 012004 (2005), hep-ex/0311045.
3. T. Nakabayashi, et al., *Phys. Rev.* **C74**, 035202 (2006).
4. D. Elsner, et al., *Eur. Phys. J.* **A33**, 147–155 (2007), nucl-ex/0702032.
5. J. C. McGeorge, et al., *Eur. Phys. J.* **A37**, 129–137 (2008), 0711.3443.
6. M. Dugger, et al., *Phys. Rev.* **C79**, 065206 (2009), 0903.1110.
7. M. Williams, et al., *Phys. Rev.* **C80**, 045213 (2009), 0909.0616.
8. M. W. Paris, and R. L. Workman, *Phys. Rev.* **C82**, 035202 (2010), 1004.0455.
9. A. M. Green, and S. Wycech, *Phys. Rev.* **C55**, 2167–2170 (1997), nucl-th/9703009.
10. R. A. Arndt, A. M. Green, R. L. Workman, and S. Wycech, *Phys. Rev.* **C58**, 3636–3640 (1998), nucl-th/9807009.
11. W. Zimmerman, *Nuovo Cim.* **21**, 249–273 (1961).
12. W. Heitler, *Math. Proc. Camb. Phil. Soc.* **37**, 291–300 (1941).
13. R. J. Eden, P. V. Landshoff, D. I. Olive, and J. C. Polkinghorne, *The Analytic S-matrix*, Cambridge University Press, Cambridge, 1966, pp. 231–232.
14. G. F. Chew, and S. Mandelstam, *Phys. Rev.* **119**, 467–477 (1960).
15. J. L. Basdevant, and E. L. Berger, *Phys. Rev.* **D19**, 239 (1979).
16. R. A. Arndt, J. M. Ford, and L. D. Roper, *Phys. Rev.* **D32**, 1085 (1985).
17. S. Ceci, A. Svarc, B. Zauner, M. Manley, and S. Capstick, *Phys. Lett.* **B659**, 228–233 (2008), hep-ph/0611094.
18. R. L. Workman, R. A. Arndt, and M. W. Paris, *Phys. Rev.* **C79**, 038201 (2009), 0808.2176.
19. R. A. Arndt, W. J. Briscoe, I. I. Strakovsky, and R. L. Workman, *Phys. Rev.* **C74**, 045205 (2006), nucl-th/0605082.
20. R. A. Arndt, R. L. Workman, Z. Li, and L. D. Roper, *Phys. Rev.* **C42**, 1853–1863 (1990).
21. K. M. Watson, *Phys. Rev.* **88**, 1163–1171 (1952).
22. R. L. Workman, *Phys. Rev.* **C74**, 055207 (2006), nucl-th/0510025.
23. W.-T. Chiang, S.-N. Yang, L. Tiator, and D. Drechsel, *Nucl. Phys.* **A700**, 429–453 (2002), nucl-th/0110034.

Dynamical coupled-channels study of meson production reactions from EBAC@JLab

Hiroyuki Kamano

Excited Baryon Analysis Center (EBAC), Thomas Jefferson National Accelerator Facility, Newport News, VA 23606, USA

Abstract. We present the current status of a combined and simultaneous analysis of meson production reactions based on a dynamical coupled-channels (DCC) model, which is conducted at Excited Baryon Analysis Center (EBAC) of Jefferson Lab.

Keywords: Dynamical coupled-channels analysis, meson production reactions
PACS: 14.20.Gk, 13.75.Gx, 13.60.Le

INTRODUCTION

An understanding of the spectrum and structure of the excited nucleon (N^*) states is a fundamental challenge in the hadron physics. The N^* states, however, couple strongly to the meson-baryon continuum states and appear only as resonance states in πN and γN reactions. Such strong couplings to the meson-baryon continuum states influence significantly the N^* properties and cannot be neglected in extracting the N^* parameters from the data and giving physical interpretations. It should also be emphasized that at present even the existence is still uncertain for most of the N^* states higher than 1.6 GeV [1]. This will be because most of the N^* information were extracted only from the $\pi N \to \pi N$ and $\gamma N \to \pi N$ analysis. It is thus well recognized nowadays that the comprehensive study of *all* relevant meson production reactions with πN, ηN, $\pi\pi N$, KY, ωN, $\cdots$ final states based on a coupled-channels framework is inevitable for reliable extraction of such higher N^* states.

To make a progress to this direction, the Excited Baryon Analysis Center (EBAC) of Jefferson Lab is conducting a dynamical coupled-channels (DCC) analysis of πN, γN, and $N(e,e')$ reactions in the resonance region. The analysis is based on an unitarized coupled-channels model, the EBAC-DCC model (see Ref. [2] for the details), within which the couplings among relevant meson-baryon reaction channels are fully taken into account.

In this model, the reaction amplitudes $T_{\alpha,\beta}(p,p';E)$ are calculated from the following coupled-channels integral equations,

$$T_{\alpha,\beta}(p,p';E) = V_{\alpha,\beta}(p,p') + \sum_{\gamma} \int_0^\infty q^2 dq V_{\alpha,\gamma}(p,q) G_\gamma(q,E) T_{\gamma,\beta}(q,p',E), \tag{1}$$

$$V_{\alpha,\beta} = v_{\alpha,\beta} + \sum_{N^*} \frac{\Gamma^\dagger_{N^*,\alpha}\Gamma_{N^*,\beta}}{E - M^*}, \tag{2}$$

where $\alpha,\beta,\gamma = \gamma N, \pi N, \eta N, \pi\pi N, K\Lambda, K\Sigma, \omega N$ (the $\pi\pi N$ channel contains the quasi two-body $\pi\Delta, \rho N, \sigma N$ channels); $v_{\alpha,\beta}$ is a meson-exchange interaction including only ground state mesons and baryons, which is derived from phenomenological Lagrangians; $\Gamma_{N^*,\beta}$ describes the excitation of the nucleon to a bare N^* state with a mass M^*; $G_\gamma(q,E)$ is the meson-baryon Green function for the channel γ. The second term of Eq. (2) thus describes the s-channel exchange of bare N^* states. Through the reaction processes, the bare N^* states couple to the meson-baryon continuum states (reactions channels) and become resonance states. On the other hand, the meson-exchange potential $v_{\alpha,\beta}$ can also generate molecule-like resonances dynamically. The physical nucleon resonances will be a "superposition" of these two pictures in general.

During the developing stage of EBAC in 2006-2009, hadronic and electromagnetic parameters of the EBAC-DCC model were determined by analyzing $\pi N \to \pi N$ [3] and $\pi N \to \eta N$ [4] up to $W = 2$ GeV, and $\gamma N \to \pi N$ [5] and $N(e,e'\pi)N$ [6] up to $W = 1.6$ GeV. Then, the model was applied to $\pi N \to \pi\pi N$ [7] and $\gamma N \to \pi\pi N$ [8] to predict cross sections and examine consistency of the coupled-channels framework. Also, nucleon resonance poles were extracted from the model and a new interpretation for the dynamical origin of P_{11} nucleon resonances was proposed [9].

12th International Conference on Meson-Nucleon Physics and the Structure of the Nucleon (MENU 2010)
AIP Conf. Proc. 1374, 501-504 (2011); doi: 10.1063/1.3647190

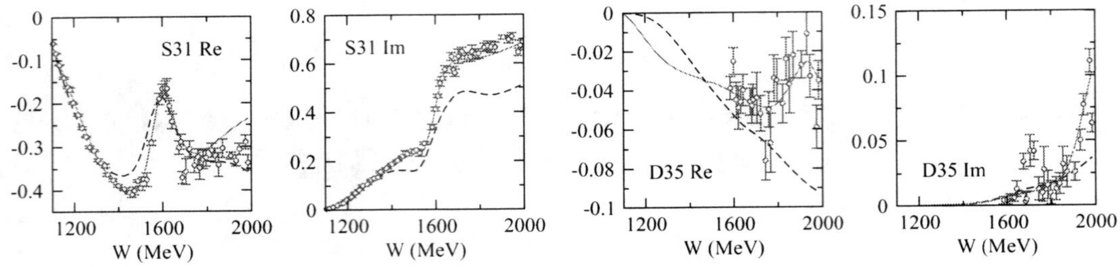

FIGURE 1. The partial wave amplitudes of the πN scattering. (Red solid curves) The current result (*preliminary*). (Black dashed curves) Our previous model [3]. The data points are the energy-independent solution of SAID, which are taken from Ref. [10].

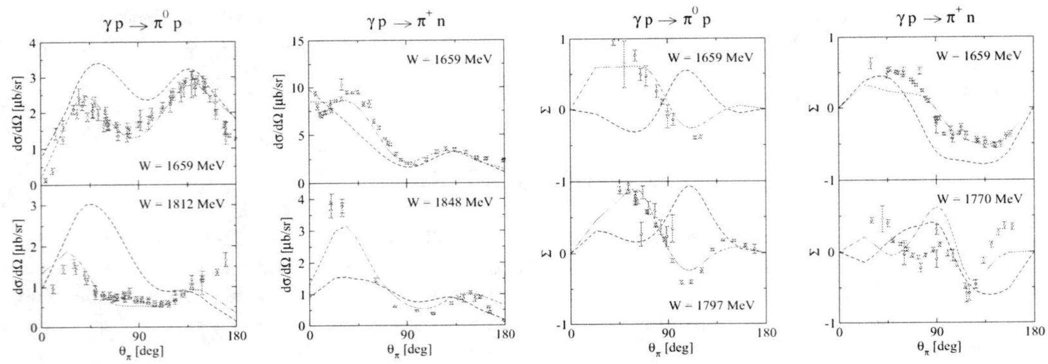

FIGURE 2. The differential cross sections (left two columns) and the photon asymmetry (right two columns) of the $\gamma p \rightarrow \pi^0 p, \pi^+ n$ reactions at high energies $W > 1.6$ GeV. (Red solid curves) The current result (*preliminary*). (Black dashed curves) Our previous model [5]. The data are taken from the database of Ref. [10].

Although the model constructed in the early analysis successfully describes the data in the wide energy region, it is still far from our goal because only πN and ηN channels are taken into account in the fit and each reaction is analyzed rather separately (e.g., in the analysis of $\gamma N \rightarrow \pi N$, the hadronic parameters fitted to the πN scattering data are used and, only the electromagnetic parameters are varied). To proceed further, we have started a full combined analysis of πN, ηN, $\pi\pi N$, KY, ωN channels recently, in which all parameters are varied simultaneously. We are currently performing a combined analysis of $\pi N, \gamma N \rightarrow \pi N, \eta N, K\Lambda, K\Sigma$ reactions as a first step. In this contribution, we present the status of our current effort for the analysis.

CURRENT STATUS OF THE EBAC-DCC ANALYSIS

The $\pi N \rightarrow \pi N$ and $\gamma N \rightarrow \pi N$ reactions: The partial wave amplitudes of the πN scattering are shown in Fig. 1. Here we present only the result of two partial waves, S_{31} and D_{35}, where a significant improvement has been achieved from our early model [3]. Other partial waves are the same quality as the early model. The improvement in the $D35$ partial wave is mainly due to an inclusion of one bare N^* state.

The current $\gamma N \rightarrow \pi N$ analysis has two major progresses over our previous analysis constructed in Ref. [5]: (a) the energy region for the analysis is extended from $W \leq 1.6$ GeV to $W \leq 2$ GeV, and (b) data of the unpolarized differential cross section and all measured polarization observables are taken into account in the fit, while only differential cross sections and photon asymmetry Σ are used in our previous analysis. In Fig. 2, our preliminary result (red solid curves) of $\gamma N \rightarrow \pi N$ at the energies $W > 1.6$ GeV is compared to our previous model (black dashed curves) as well as the experimental data. We can actually observe a visible improvement of the model in $W > 1.6$ GeV.

The $\pi N \rightarrow \eta N$ reaction: It is known that for the $\pi^- p \rightarrow \eta n$ reaction underlying inconsistencies exist among data sets from different experimental groups [4]. In our analysis, we follow the strategy described in Ref. [4] for how to

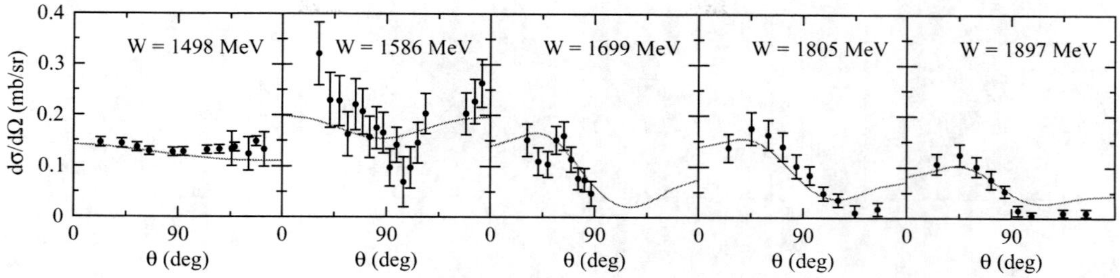

FIGURE 3. The differential cross sections of the $\pi^- p \to \eta n$ reactions. Red solid curves are the current result (*preliminary*). See Ref. [4] for the details of the data used.

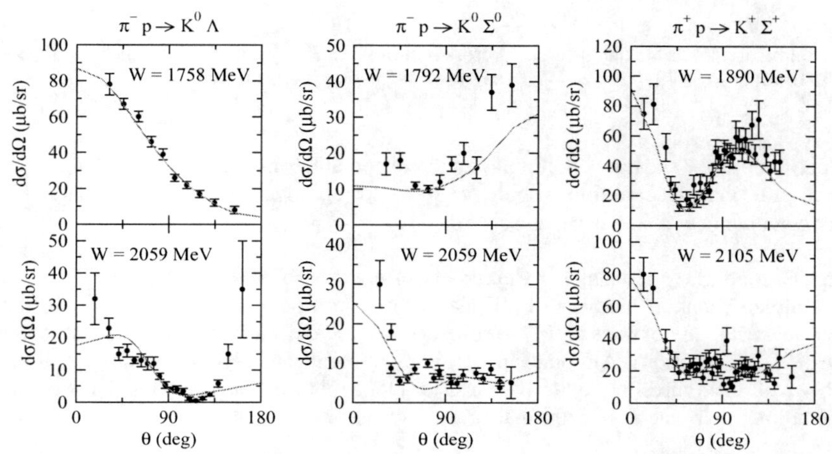

FIGURE 4. The preliminary result for the differential cross sections of $\pi^+ p \to K^+\Sigma^+$ (left panels), $\pi^- p \to K^0\Sigma^0$ (middle panels), and $\pi^- p \to K^0\Lambda^0$ (right panels). The data are from Refs. [11, 12].

select the data to be used for our analysis. In Fig. 3, we present the current status of the $\pi^- p \to \eta n$ reactions. We find that at present our result describes reasonably well the differential cross section data up to $W = 1.9$ GeV. The analysis of $\gamma p \to \eta p$ is also underway.

The $\pi N \to KY$ and $\gamma N \to KY$ reactions: Now we move to showing the current status of the analysis for the KY reactions. In Fig. 4, we present the differential cross sections of the $\pi N \to KY$ reactions for three different charge states. At present we have included the data up to $W = 2.1$ GeV for the analysis. The current result describes the data of the considered energy region reasonably well.

Finally, we present the current status of the $\gamma p \to K^+\Lambda$ analysis. This strangeness-production reaction is expected to be one of the most promising reactions to provide critical information for confirming/rejecting not well-established N^*s and/or discovering new N^*s. Because of this, measurement of the polarization observables, which are more exclusive than unpolarized cross section, for $\gamma p \to K^+\Lambda$ becomes very active at electron beam facilities. For example, first measurement of the $O_{x'}$, $O_{z'}$, and T asymmetries has been reported recently by GRAAL [16]. Furthermore, the so-called "(over-) complete experiments" is planned at CLAS, in which *all* 15 polarization observables are measured, and so extensive data will be available in near future.

In Fig. 5, the differential cross sections and the polarization observables $P, C_{x'}, C_{z'}$ are compared with those measured in the CLAS-g11a [14] and CLAS-g1c [15] experiments. Here we note that care must be taken in calculating the polarization observables because incompatibility exist in the expressions of the observables in the literature (see Ref. [13] for the detail). In our analysis, we follow the definitions explicitly described in Ref. [13]. Although it must be further improved, our model describes qualitatively the available differential cross sections and polarization

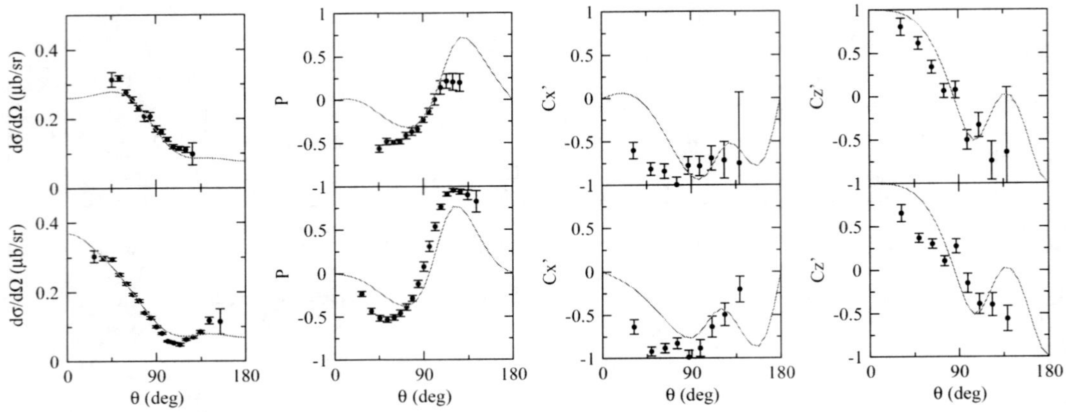

FIGURE 5. The preliminary result for the differential cross sections and P, $C_{x'}$, and $C_{z'}$ asymmetries of $\gamma p \to K^+\Lambda$. The upper (lower) panels are the result at $E_\gamma = 1250$ MeV ($E_\gamma = 1650$ MeV). The data are from Refs. [14, 15].

observables. In order to describe the data of all polarization observables measured in upcoming experiments at CLAS, however, we may need to introduce additional bare N^* states to reproduce such extensive data. Then the introduced bare N^* states may generate new resonance states.

In summary, the Excited Baryon Analysis Center of Jefferson Lab makes a continuous effort for a combined and simultaneous coupled-channels analysis of all relevant meson-production reactions toward the ultimate goal of establishing the N^* spectrum and extracting N^* parameters. We are currently performing a combined analysis of the $\pi N, \gamma N \to \pi N, \eta N, K\Lambda, K\Sigma$ reactions. Although further improvements are necessary, our current model describes the reactions reasonably well from threshold up to $W \sim 2$ GeV. Once this analysis is completed, we will gradually extend our analysis by including other reactions with final states such as $\pi\pi N$ and ωN.

ACKNOWLEDGMENTS

The author would like to thank B. Juliá-Díaz, T.-S. H. Lee, A. Matsuyama, S. X. Nakamura, T. Sato, and N. Suzuki for their collaborations at EBAC. This work was supported by the U.S. Department of Energy, Office of Nuclear Physics Division, under Contract No. DE-AC05-06OR23177 under which Jefferson Science Associates operates the Jefferson Lab.

REFERENCES

1. R. A. Arndt, W. J. Briscoe, I. I. Strakovsky, and R. L. Workman, *Phys. Rev. C* **74**, 045205 (2006).
2. A. Matsuyama, T. Sato, and T.-S. H. Lee, *Phys. Rep.* **439**, 193 (2007).
3. B. Juliá-Díaz, T.-S. H. Lee, A. Matsuyama, and T. Sato, *Phys. Rev. C* **76**, 065201 (2007).
4. J. Durand, B. Juliá-Díaz, T.-S. H. Lee, B. Saghai, and T. Sato, *Phys. Rev. C* **78**, 025204 (2008).
5. B. Juliá-Díaz, T.-S. H. Lee, A. Matsuyama, and T. Sato,and L. C. Smith, *Phys. Rev. C* **77**, 045205 (2008).
6. B. Juliá-Díaz, H. Kamano, T.-S. H. Lee, A. Matsuyama, T. Sato, N. Suzuki, Phys. Rev. C **80**, 025207 (2009).
7. H. Kamano, B. Juliá-Díaz, T.-S. H. Lee, A. Matsuyama, and T. Sato, Phys. Rev. C **79**, 025206 (2009).
8. H. Kamano, B. Juliá-Díaz, T.-S. H. Lee, A. Matsuyama, and T. Sato, Phys. Rev. C **80**, 065203 (2009).
9. N. Suzuki, B. Julia-Diaz, H. Kamano, T.-S. H. Lee, A. Matsuyama, T. Sato, Phys. Rev. Lett. **104**, 042302 (2010).
10. CNS Data Analysis Center, GWU, http://gwdac.phys.gwu.edu .
11. B. Juliá-Díaz, B. Saghai, T.-S. H. Lee, F. Tabakin, Phys. Rev. C **73**, 055204 (2006), references therein.
12. D. J. Candlin *et al.*, Nucl. Phys. **B226**, 1 (1983).
13. A. Sandorfi, S. Hoblit, H. Kamano, T.-S. H. Lee, arXiv:0912.3505 [nucl-th]; in preparation.
14. M. E. McCracken *et al.* (CLAS collaboration), Phys. Rev. C **81**, 025201 (2010).
15. R. K. Bradford *et al.* (CLAS collaboration), Phys. Rev. C **75**, 035205 (2007).
16. A. Lleres *et al.* (GRAAL collaboration), Eur. Phys. J. A **39**, 149 (2009).

Extraction of P_{11} Resonance from πN Data and Its Stability

S. X. Nakamura

Excited Baryon Analysis Center (EBAC), Jefferson Laboratory, Newport News, Virginia 23606, USA

Abstract. We address a question about how much resonance poles and residues extracted from data depend on a model used for the extraction, and on the precision of data. We focus on the P_{11} πN scattering and use the dynamical coupled-channel (DCC) model developed in Excited Baryon Analysis Center (EBAC) at JLab. We examine the model-dependence of the poles by varying parameters largely within the EBAC-DCC model. We find that two poles associated with the Roper resonance are fairly stable against the variation. We also study the stability of the Roper poles against different analytic structure of the P_{11} amplitude below πN threshold by using a bare nucleon model. We again find a good stability of the Roper poles.

Keywords: Dynamical coupled-channels analysis, meson production reactions, Roper resonances
PACS: 14.20.Gk, 13.75.Gx, 13.60.Le

INTRODUCTION

Extraction of N^* information, such as pole positions and vertex form factors, is an important task in hadron physics. This is because they are necessary information to address a question whether we can understand baryon resonances within QCD. In order to extract the N^* information, first, one needs to construct a reaction model through a comprehensive analysis of data. Then, pole positions and vertex form factors are extracted from the model with the use of the analytic continuation. Therefore, the N^* information extracted in this manner is inevitably model-dependent. There are several different approaches to extract the N^* information. Although almost all 4-stars nucleon resonances listed by Particle Data Group (PDG) are found in all approaches, existence of some N^* states, in particular those in the higher mass region, is controversial. Thus, commonly asked questions are how much model-dependent the extracted resonance parameters are, and how precise data have to be for a stable resonance extraction. These are the questions we address at Excited Baryon Analysis Center (EBAC) at JLab[1], within a dynamical coupled-channels model (EBAC-DCC) [2]. We focus on the πN P_{11} partial wave and the stability of its pole positions, particularly those corresponding to the Roper resonance. In the region near Roper $N(1440)$, two poles close to the $\pi\Delta$ threshold were found in our recent extraction [3] from the JLMS model [4] (JLMS is one of EBAC-DCC model), while only one pole in the similar energy region was reported in some other analyses. We examine the stability of this two-pole structure against the following variation, keeping a good reproduction of SAID single-energy (SAID-SES) solution [5] unless otherwise stated.

- Large variation of the parameters of the meson-baryon and bare N^* parameters of the EBAC-DCC model.
- Inclusion of a bare nucleon state: The analytic structure of this model is rather different from the original EBAC-DCC model, in particular in the region near the nucleon pole [6], .
- Fit to the solution based on the Carnegie-Mellon University-Berkeley model (CMB) [7] which has rather different behavior from SAID-SES for higher W.

DYNAMICAL COUPLED-CHANNELS MODELS

Here, we briefly describe dynamical coupled-channels models; the EBAC-DCC model and the bare nucleon model.

The EBAC-DCC model contains πN, ηN and $\pi\pi N$ channels and the $\pi\pi N$ channel has $\pi\Delta$, ρN and σN components. These meson-baryon (MB) channels are connected with each other by meson-baryon interactions ($v_{MB,M'B'}$), or excited to bare N^* states by vertex interactions ($\Gamma_{MB\leftrightarrow N^*}$). With these interactions, the partial-wave amplitude for the $M(\vec{k})+B(-\vec{k})\rightarrow M'(\vec{k}')+B'(-\vec{k}')$ reaction can be written by the following form:

$$T_{MB,M'B'}(k,k',E) \quad = \quad t_{MB,M'B'}(k,k',E)+t^R_{MB,M'B'}(k,k',E), \qquad (1)$$

12th International Conference on Meson-Nucleon Physics and the Structure of the Nucleon (MENU 2010)
AIP Conf. Proc. 1374, 505-508 (2011); doi: 10.1063/1.3647191

where the first term is obtained by solving the following coupled-channels Lippmann-Schwinger equation:

$$t_{MB,M'B'}(k,k',E) = v_{MB,M'B'}(k,k') + \sum_{M''B''} \int_{C_{M''B''}} q^2 dq v_{MB,M''B''}(k,q) G_{M''B''}(q,E) t_{M''B'',M'B'}(q,k',E). \quad (2)$$

Here C_{MB} is the integration contour in the complex-q plane used for the channel MB. The second term of Eq. (1) is associated with the bare N^* states, and given by

$$t^R_{MB,M'B'}(k,k',E) = \sum_{i,j} \bar{\Gamma}_{MB \to N^*_i}(k,E)[D(E)]_{i,j} \bar{\Gamma}_{N^*_j \to M'B'}(k',E), \quad (3)$$

where the dressed vertex function $\bar{\Gamma}_{N^*_j \to M'B'}(k,E)$ is calculated by convoluting the bare vertex $\Gamma_{N^*_j \to M'B'}(k)$ with the amplitudes $t_{MB,M'B'}(k,k',E)$. The inverse of the propagator of dressed N^* states in Eq. (3) is

$$[D^{-1}(E)]_{i,j} = (E - m^0_{N^*_i})\delta_{i,j} - \Sigma_{i,j}(E), \quad (4)$$

where $m^0_{N^*_i}$ is the bare mass of the i-th N^* state, and the N^* self-energy is defined by

$$\Sigma_{i,j}(E) = \sum_{MB} \int_{C_{MB}} q^2 dq \bar{\Gamma}_{N^*_j \to MB}(q,E) G_{MB}(q,E) \Gamma_{MB \to N^*_i}(q,E). \quad (5)$$

Defining $E_\alpha(k) = [m^2_\alpha + k^2]^{1/2}$ with m_α being the mass of particle α, the meson-baryon propagators in the above equations are: $G_{MB}(k,E) = 1/[E - E_M(k) - E_B(k) + i\varepsilon]$ for the stable πN and ηN channels, and $G_{MB}(k,E) = 1/[E - E_M(k) - E_B(k) - \Sigma_{MB}(k,E)]$ for the unstable $\pi\Delta$, ρN, and σN channels. The self energy $\Sigma_{MB}(k,E)$ is calculated from a vertex function defining the decay of the considered unstable particle in the presence of a spectator π or N with momentum k. For example, we have for the $\pi\Delta$ state,

$$\Sigma_{\pi\Delta}(k,E) = \frac{m_\Delta}{E_\Delta(k)} \int_{C_3} q^2 dq \frac{M_{\pi N}(q)}{[M^2_{\pi N}(q) + k^2]^{1/2}} \frac{|f_{\Delta \to \pi N}(q)|^2}{E - E_\pi(k) - [M^2_{\pi N}(q) + k^2]^{1/2} + i\varepsilon}, \quad (6)$$

where $M_{\pi N}(q) = E_\pi(q) + E_N(q)$ and $f_{\Delta \to \pi N}(q)$ defines the decay of the $\Delta \to \pi N$ in the rest frame of Δ, C_3 is the corresponding integration contour in the complex-q plane.

To examine further the model dependence of resonance extractions, it is useful to also perform analysis using models with a bare nucleon, as developed in Ref. [8]. Within the formulation of EBAC-DCC model, such a model can be obtained by adding a bare nucleon (N_0) state with mass m^0_N and $N_0 \to MB$ vertices and removing the direct $MB \to N \to M'B'$ in the meson-baryon interactions $v_{MB,M'B'}$. All numerical procedures for this model are identical to that used for the EBAC-DCC model, except that the resulting amplitude must satisfy the nucleon pole condition:

$$t^R_{\pi N,\pi N}(k \to k_{\rm on}, k \to k_{\rm on}, E \to m_N) = -\frac{[F_{\pi NN}(k_{\rm on})]^2}{E - m^0_N - \tilde{\Sigma}(m_N)}. \quad (7)$$

with

$$m_N = m^0_N + \tilde{\Sigma}(m_N) \qquad \text{and} \qquad F_{\pi NN}(k_{\rm on}) = F^{\rm phys.}_{\pi NN}(k_{\rm on}) \ . \quad (8)$$

Here we have used the on-shell momentum defined by $E = \sqrt{m^2_N + k^2_{\rm on}} + \sqrt{m^2_\pi + k^2_{\rm on}}$. Also, $\tilde{\Sigma}(m_N)$ is the self-energy for the nucleon. More details for the calculational procedure following Afnan and Pearce is found in Refs. [1, 8].

RESULTS

Now we show our numerical results to examine the stability of the P_{11} poles. In the following subsections, we present results from various fits by varying the dynamical content of the EBAC-DCC model, and by using a model with a bare nucleon. We show in figures the quality of fits of these models, and in Table 1 the pole positions from the models as well as χ^2 per data point (χ^2_{pd}). We find the poles with the method of analytic continuation discussed in detail in Refs. [3, 9]. In Table 1, we also present pole positions from JLMS[4] and SAID-EDS (energy-dependent)[5].

TABLE 1. The resonance pole positions M_R for P_{11} [listed as (ReM$_R$, $-$ImM$_R$) in the unit of MeV] extracted from various parameter sets. The location of the pole is specified by, e.g., $(s_{\pi N}, s_{\eta N}, s_{\pi\pi N}, s_{\pi\Delta}, s_{\rho N}, s_{\sigma N}) = (upuupp)$, where p and u denote the physical and unphysical sheets for a given reaction channel, respectively. χ^2_{pd} is χ^2 per data point.

Model	*upuupp*	*upuppp*	*uuuupp*	*uuuuup*	χ^2_{pd}
SAID-EDS	(1359, 81)	(1388, 83)	—	—	2.94
JLMS	(1357, 76)	(1364, 105)	—	(1820, 248)	3.55
$2N^*$-3p	(1368, 82)	(1375, 110)	—	(1810, 82)	3.28
$2N^*$-4p	(1372, 80)	(1385, 114)	(1636, 67)	(1960, 215)	3.36
$1N_0 1N^*$-3p	(1363, 81)	(1377, 128)	—	(1764, 137)	2.51

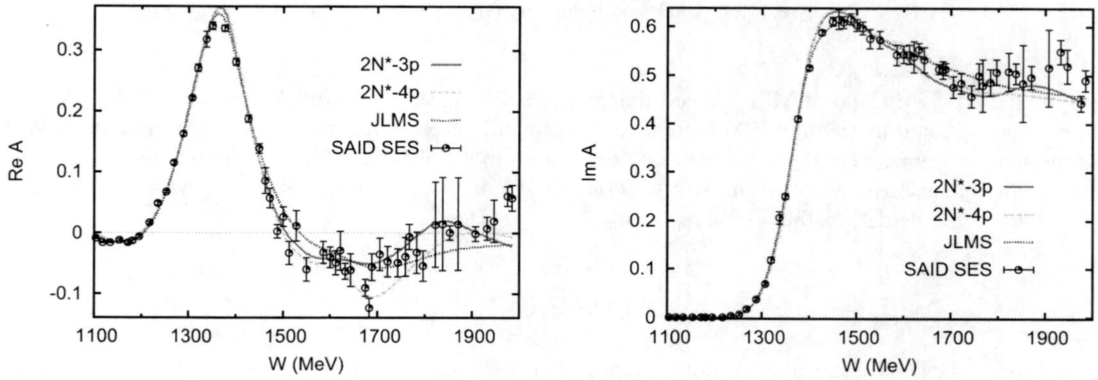

FIGURE 1. The real (left) and imaginary (right) parts of the on-shell P_{11} amplitudes as a function of the πN invariant mass W (MeV). A is unitless in the convention of Ref. [5].

$2N^*$-3p and $2N^*$-4p fits

We varied both the parameters for the meson-baryon interactions ($v_{MB,M'B'}$) and parameters associated with bare N^* states ($m^0_{N^*}$, $\Gamma_{N^* \leftrightarrow MB}$). The obtained meson-baryon interactions are quite different from those of JLMS. We obtained several fits which are different in how the oscillatory behavior of SAID-SES amplitude for higher W is fitted. The results from the $2N^*$-3p (dotted curves) and $2N^*$-4p (dashed curves) fits are compared with the JLMS fit (solid curves) in Fig. 1. The resulting resonance poles are listed in the 3th and 4th rows of Table 1. Here we see again the first two poles near the $\pi\Delta$ threshold from both fits agree well with the JLMS fit. This seems to further support the conjecture that these two poles are mainly sensitive to the data below $W \sim 1.5$ GeV where the SAID-SES has rather small errors. However, the $2N^*$-4p fit has one more pole at $M_R = 1630 - i45$ MeV. This is perhaps related to its oscillating structure near $W \sim 1.6$ GeV (dashed curves), as shown in the Figs. 1. On the other hand, this resonance pole could be fictitious since the fit $2N^*$-3p (dotted curve) with only three poles are equally acceptable within the fluctuating experimental errors. Our result suggests that it is important to have more accurate data in the high W region for a high precision resonance extraction.

$1N_0 1N^*$-3p

Here we show our results obtained with the bare nucleon model, and then address the question whether difference in the analytic structure of the πN amplitude below πN threshold strongly affects the resonance extractions. The bare nucleon model is fitted to SAID-SES, and at the same time, to the nucleon pole conditions Eqs. 8. Meanwhile, the original EBAC-DCC model has different singular structure below the πN threshold. The question is whether such differences can lead to very different resonance poles. Our fit of the bare nucleon model is shown in Fig. 2 and

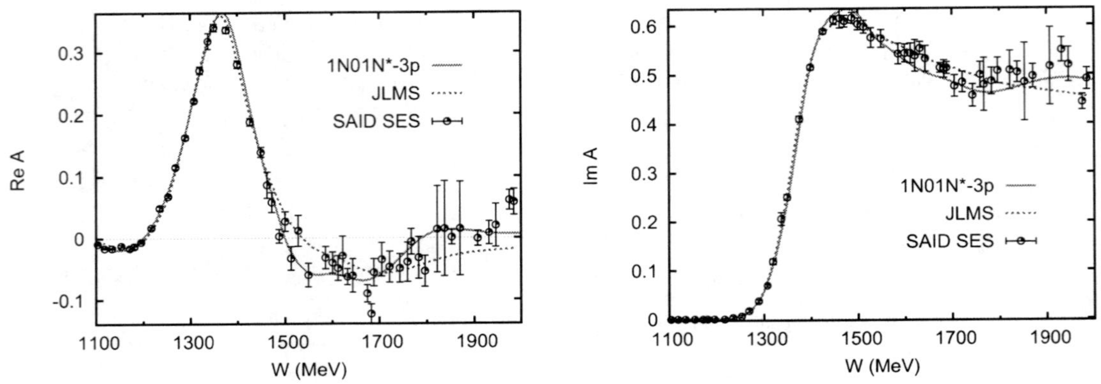

FIGURE 2. The real (left panel) and imaginary (right panel) parts of the P_{11} amplitudes.

compared with SAID-SES and JLMS. We see that the two fits agree very well below $W = 1.5$ GeV, while their differences are significant in the high W region. The corresponding resonance poles are given in Table 1. We also see here that the first two poles near the $\pi\Delta$ threshold are close to those of JLMS. Our results seem to indicate that these two poles are rather insensitive to the analytic structure of the amplitude in the region below πN threshold, and are mainly determined by the data in the region $m_N + m_\pi \leq W \leq 1.6$ GeV.

CONCLUSION

We have examined the stability of the two-pole structure of the Roper resonance. We showed that two resonance poles near the $\pi\Delta$ threshold are stable against large variations of parameters of meson-exchange mechanisms within EBAC-DCC model [2]. This two-pole structure is also obtained in an analysis based on a model with the bare nucleon state. Our results indicate that the extraction of P_{11} resonances is insensitive to the analytic structure of the amplitude in the region below πN threshold. Although we did not show a result, we have also fitted to the old CMB amplitude, which is rather different from SAID-SES for $W \geq 1.5$ GeV, and still found that the Roper two poles are stable.

ACKNOWLEDGMENTS

This work is supported by the U.S. Department of Energy, Office of Nuclear Physics Division, under Contract No. DE-AC02-06CH11357, and Contract No. DE-AC05-06OR23177 under which Jefferson Science Associates operates Jefferson Lab, and by the Japan Society for the Promotion of Science, Grant-in-Aid for Scientific Research(C) 20540270. This research used resources of the National Energy Research Scientific Computing Center, which is supported by the Office of Science of the U.S. Department of Energy under Contract No. DE-AC02-05CH11231.

REFERENCES

1. H. Kamano, S. X. Nakamura, T.-S. H. Lee, and T. Sato, *Phys. Rev. C* **81**, 065207 (2010).
2. A. Matsuyama, T. Sato, and T.-S. H. Lee, *Phys. Rep.* **439**, 193 (2007).
3. N. Suzuki, B. Juliá-Díaz, H. Kamano, T.-S. H. Lee, A. Matsuyama, and T. Sato, *Phys. Rev. Lett.* **104**, 042302 (2010).
4. B. Juliá-Díaz, T.-S. H. Lee, A. Matsuyama, and T. Sato, *Phys. Rev. C* **76**, 065201 (2007).
5. R. A. Arndt, W. J. Briscoe, I. I. Strakovsky, and R. L. Workman, *Phys. Rev. C* **74**, 045205 (2006).
6. B. Juliá-Díaz, H. Kamano, T.-S. H. Lee, A. Matsuyama, T. Sato, and N. Suzuki, *Chin. J. Phys.* **47**, 142 (2009).
7. R. E. Cutkosky, and S. Wang, *Phys. Rev. D* **42**, 235 (1990).
8. B. C. Pearce, and I. R. Afnan, *Phys. Rev. C* **34**, 991 (1986); *Phys. Rev. C* **40**, 220 (1989).
9. N. Suzuki, T. Sato, and T.-S. H. Lee, *Phys. Rev. C* **79**, 025205 (2009); arXiv:1006.2196[nucl-th].

Session O: Form Factors and Two-Boson Exchange

Measurement of the Proton Elastic Form Factor Ratio at Low Q^2

Xiaohui Zhan [1]

Argonne National Lab, Argonne, IL 60439, USA

Abstract. A new measurement of the proton elastic form factor ratio $\mu_p G_E/G_M$ in the range of $Q^2 = 0.3 - 0.7$ (GeV/c)^2 was made at Jefferson lab (JLab) Hall A using recoil polarimetry. With unprecedented ($\sim 1\%$) precision, the new data show that the proton form factor ratio strongly deviates from unity in this low momentum transfer region, indicating a significant difference between the proton charge and magnetic distribution at large length scale. The new result is a few percent lower than previous world polarization data. The updated global fit indicates a 2% decrease in G_E and 1% increase in G_M in this low Q^2 region.

Keywords: proton, form factor ratio, low Q^2
PACS: 14.20.Dh, 13.40.Gp, 24.70.+s, 25.30.Bf

INTRODUCTION

Elastic electron-proton scattering has been a powerful tool in probing the nucleon structure. The nucleon electromagnetic form factors are fundamental quantities which relate to the charge and magnetization distributions within the nucleon. For over 40 years, the form factors have been studied extensively by Rosenbluth separations [1] of the unpolarized electron-proton (ep) scattering cross section measurements. Recent advances in the technology of intense polarized beams, polarized targets and polarimetry have ushered in a new generation of experiments by measuring the double polarization observables [2, 3, 4]. Although the proton electric to magnetic ratio $\mu_p G_E/G_M$ determined by unpolarized measurements shows no dramatic Q^2 dependence up to $Q^2 \approx 6$ (GeV/c)^2, with much better precision, experiments at JLab with high-quality polarized electron beams measuring recoil polarization [5] revealed that ratio $\mu_p G_E/G_M$ drops almost linearly with momentum transfer at $Q^2 > 1$ (GeV/c)^2. The unexpected finding has prompt intense experimental and theoretical activities on the subject of the proton electromagnetic form factors [6, 7, 8]. The validity of analyzing data in the one-photon-exchange (OPE) approximation has been questioned, and the two-photon-exchange (TPE) processes are now considered as an significant correction to the unpolarized data and mostly account for the discrepancy at large momentum transfer [9, 10, 11].

While the high Q^2 behavior of ratio $\mu_p G_E/G_M$ reveals the quark dynamics, the low Q^2 behavior is more closely related to the proton size and large scale structure. High precision measurements at low Q^2 have been motivated by a recent semi-phenomenological fit [12], which suggests that structure might be present in all four nucleon form factors and the ratio $\mu_p G_E/G_M$ at $Q^2 \approx 0.3$ (GeV/c)^2 due to pion cloud effect. However, the world existing high precision data ($\sigma_{tot} < 2\%$) appear to be less conclusive about the form factor behavior at low momentum transfer. The measurement reported from Bates BLAST [13] via beam-target asymmetry are more consistent with 1 in the region of $0.15 < Q^2 < 0.5$ (GeV/c)^2, but a later JLab experiment LEDEX [14] using recoil polarimetry observed two high precision points substantially lower than 1 in the vicinity of $Q^2 \approx 0.4$ (GeV/c)^2. Interestingly, both the BLAST and LEDEX data are inconsistent with the structure suggested in [12].

THE EXPERIMENT

JLab experiment E08-007 measured the proton form factor ratio $\mu_p G_E/G_M$ in the range of $0.3 < Q^2 < 0.7$ (GeV/c)^2 via the polarization transfer reaction ${}^1\mathrm{H}(\vec{e}, e'\vec{p})$. In the one-photon-exchange (Born) formalism, the induced polarization is zero, and the ratio of the transferred transverse to longitudinal polarizations relates to the electromagnetic form factors

[1] For Jefferson Lab E08-007 collaborations.

12th International Conference on Meson-Nucleon Physics and the Structure of the Nucleon (MENU 2010)
AIP Conf. Proc. 1374, 511-514 (2011); doi: 10.1063/1.3647192

TABLE 1. Measured kinematic settings. θ_e (θ_p) is the scattered electron (proton) angle in the lab frame, P_p is the proton central momentum, ε is the virtual photon polarization.

Q^2 $(\mathrm{GeV}/c)^2$	θ_e (deg)	θ_p (deg)	P_p (GeV/c)	ε	Analyzer thickness (inches)
0.298	28.5	60.0	0.565	0.88	2.25
0.346	31.3	57.5	0.616	0.85	2.25
0.402	34.2	55.0	0.668	0.82	3.75
0.449	36.7	53.0	0.710	0.80	3.75
0.495	39.2	51.0	0.752	0.78	3.75
0.547	41.9	49.0	0.794	0.75	3.75
0.599	44.6	47.0	0.836	0.72	3.75
0.695	49.8	50.0	0.913	0.66	3.75

through the equation:

$$R \equiv \mu_p \frac{G_E}{G_M} = -\mu_p \frac{E_e + E'_e}{2M_p} \tan\left(\frac{\theta_e}{2}\right) \frac{P_t}{P_l}, \tag{1}$$

where μ_p is the proton magnetic moment, M_p is the proton mass, E_e (E'_e) is the incident (scattered) electron energy, θ_e is the electron scattering angle and P_t (P_l) is the transverse (longitudinal) component of the recoil proton polarization.

In this experiment, a 1.19 GeV electron beam with polarization 83% was incident on a 6 cm liquid hydrogen target. The recoil proton was detected by the left High-Resolution-Spectrometer (LHRS), in coincidence with the elastic scattered electron detected by the large acceptance spectrometer "BigBite". A focal plane polarimeter installed in the LHRS was used to measure the proton polarization. Since the elastic kinematics can be well determined by the proton, only the shower counter of the BigBite spectrometer was used to tag the electrons and form the coincidence trigger. The measured kinematic settings are given in Table 1.

For the recoil protons, the polarization precesses as the particle is transported through the spectrometer. At the FPP, the transverse polarization components lead to azimuthal asymmetries in the rescattering in the analyzing material due to spin-orbit coupling of the interaction. The alignment of the FPP chambers was determined with straight-through trajectories without the analyzer, and gave a calibration of the second scattering angles. The false asymmetry from instrumental inefficiency and the chamber misalignment is largely canceled in the helicity difference that determine the polarization-transfer observables. The transferred polarization was determined by a weighted-sum method [15] using the difference of the azimuthal angle distributions corresponding to the two beam helicity states, and the full spin precession matrix was calculated by COSY [16].

RESULTS

Figure 1 shows the final error budget of the measured kinematic points. The total uncertainty is $\sim 1-1.5\%$. As Q^2 increases, systemtic uncertainty starts to dominate. The ratio $\mu_p G_E/G_M$ falls off slowly from 0.95 to 0.90 with no indication of any strucutre in the measured region. The whole data set is in excellent agreement with another measurement which was run in a different time and analyzed indepedently [17]. Many existing theoretical form factor calculations employed the framework of Constituent Quark Model (CQM) or Vector Meson Dominance (VMD) picture [18, 19, 20, 21, 22, 23], and most of them can not reproduce the new data. Miller [19] did a calculation in the light front form based on a constituent quark model including the pion cloud effect. More recently, de Melo *et al.* [23] performed a VMD calculation by considering the non-valence contribution of the nucleon state in the light front frame. Both of them predicted the large deviation from 1, emphasizing the importance of the pion cloud effect in the nucleon structure at low Q^2.

We updated the global fit using the same approach in [11] by including the new data. The new fit suggested a 2% decrease in G_E and 1% increase in G_M in this low momentum transfer region.

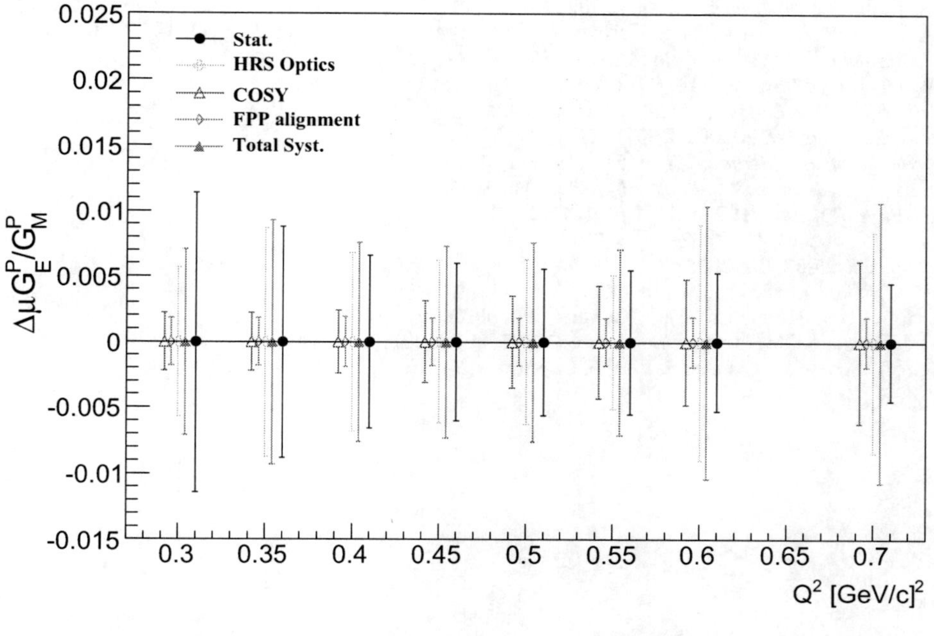

FIGURE 1. Comparison of the major contributions to the systematic uncertainties and the statistical uncertainty for each kinematics.

SUMMARY

In summary, we made a high precision proton form factor ratio measurement at low Q^2 via recoil polarimetry. For the measured region, the ratio $\mu_p G_E/G_M$ strongly deviates from unity, and is a few percent lower than previous world polarization measurements. The second phase of this experiment will access region down to $Q^2 = 0.015$ (GeV/c)^2 using beam-target asymmetry technique. It is a great opportunity to see the form factor behavior extending to even lower Q^2 region, and examine the systemtic uncertainty in different measurements. None of the current theoretical calculations adequately describe the new ratio result and the extracted individual form factors, necessitating a deeper theoretical understanding of the nucleon properties at low momentum transfer.

ACKNOWLEDGMENTS

We thank the Jefferson Lab physics and accelerator divisions. This work was supported by the U.S. Department of energy, the U.S. National Science Foundation, Argonne National Laboratory under Contract No. DE-AC02-06CH11357. Jefferson Science Associates operates the Thomas Jefferson National Accelerator Facility under DOE Contract No. DE-AC05-06OR23177. The polarimeter was funded by the U.S. National Science foundation, Grants No. PHY 9213864 and No. PHY 9213869.

REFERENCES

1. M. N. Rosenbluth, *Phys. Rev.* **79**, 615 (1950).
2. A. I. Akhiezer, L. N. Rozentsweig, I. M. Shmushkevich, *Sov. Phys. JETP* **6**, (1958).
3. N. Dombey, *Rev. Mod. Phys.* **41**, 236 (1969).
4. R. G. Arnold, C. E. Carlson, F. Gross, *Phys. Rev. C* **23**, 363 (1981).
5. M. K. Jones *et al. Phys. Rev. Lett.* **84**, 1398 (2000).
6. H. Gao, *Int. J. Mod. Phys.* **A20**, 1595 (2005).
7. J. Arrington, C. D. Roberts, J. M. Zanotti, *J. Phys.* **G34**, S23 (2007).

8. C. F. Perdrisat, V. Punjabi, M. Vanderhaeghen, *Prog. Part. Nucl. Phys.* **59**, 694 (2007).
9. P. G. Blunden, W. Melnitchouk, J. A. Tjon, *Phys. Rev. Lett.* **91**, 142304 (2003); *Phys. Rev. C* **72**, 034612 (2005).
10. Y. C. Chen, A. Afanasev, S. J. Brodsky, C. E. Carlson, M. Vanderhaeghen, *Phys. Rev. Lett.* **93**, 122301 (2004).
11. J. Arrington, W. Melnitchouk, J. A. Tjon, *Phys. Rev. C* **76**, 035205 (2007).
12. J. Friedrich, T. Walcher, *Eur. Phys. J. A* **17**, 607 (2003).
13. C. Crawford, *et al.*, *Phys. Rev. Lett.* **98**, 052301 (2007).
14. G. Ron *et al.*, *Phys. Rev. Lett.* **99**, 202002 (2007).
15. D. Besset *et al.*, *Nucl. Instrum. Methods* **A166**, 515 (1979).
16. K. Makino, B. Berz, *Nucl. Instrum. Methods. Phys. Res.* **A654**, 543 (1999).
17. M. Paolone *et al.*, *Phys. Rev. Lett.* **105**, 072001 (2010).
18. S. Boffi *et al. Eur. Phys. J. A* **14**, 17 (2002).
19. G. Miller, *Phys. Rev. C* **66**, 032201 (2002).
20. A. Faessler *et al. Phys. Rev. D* **58**, 114021 (2006).
21. E. Lomon, *Phys. Rev. C* **66**, 045501 (2002); arXiv:nucl-th/0609020v2 (2006).
22. M. Belushkin, H. Hammer, U. Meissner, *Phys. Rev. C* **75**, 035202 (2007).
23. J. P. B. C. de Melo, T. Frederico, E. Pace, S. Pisano, G. Salme, *Phys. Lett. B* **671**, 153 (2009).

γZ exchange corrections in parity-violating electron scattering

P. G. Blunden

Department of Physics and Astronomy, University of Manitoba, Winnipeg, MB, Canada R3T 2N2

Abstract. I report on recent work on two-boson exchange contributions ($\gamma\gamma$ and γZ) and their implications for parity-violating electron scattering measurements. These calculations are particularly relevant for high precision measurements of the strange electric and magnetic form factors, and the weak charge of the proton.

Keywords: electroweak corrections
PACS: 12.15.Lk, 13.60.Fz, 13.60.Hb, 14.20.Dh

TWO-BOSON EXCHANGE CORRECTIONS

Parity-violating (PV) electron–proton elastic scattering has become a standard tool with which to probe the strangeness content of the proton. Recent high-precision experiments at Jefferson Lab [1] and elsewhere [2] have provided important constraints on the strange electric and magnetic form factors [3].

By polarizing the incident electron and measuring the difference between right- and left-handed electrons scattering from unpolarized protons, the parity-violating asymmetry can be defined in terms of the differential cross sections as

$$A^0_{\rm PV} = \frac{\sigma_R - \sigma_L}{\sigma_R + \sigma_L} = \frac{2\Re e\{\mathcal{M}_\gamma^\dagger \mathcal{M}_Z\}}{|\mathcal{M}_\gamma|^2}\,, \tag{1}$$

where $\sigma_{R(L)}$ is the cross section for a right- (left-) hand polarized electron, and $\mathcal{M}_\gamma$ and $\mathcal{M}_Z$ are one-boson exchange (OBE) amplitudes for γ and Z, respectively. The purely electromagnetic contribution cancels in the numerator, so that the asymmetry is sensitive to the parity-violating part of $2\Re e\left(\mathcal{M}_\gamma^* \mathcal{M}_Z\right)$, involving the interference of $\mathcal{M}_\gamma$ with the product of vector and axial-vector currents in $\mathcal{M}_Z$. The denominator is dominated by the electromagnetic term, $|\mathcal{M}_\gamma|^2$. The correction to the PV cross section arising from the $\gamma\gamma$ and γZ two-boson exchange (TBE) contributions can be obtained from Eq. (1) by the replacements $\mathcal{M}_\gamma \to \mathcal{M}_\gamma + \mathcal{M}_{\gamma\gamma}$ and $\mathcal{M}_Z \to \mathcal{M}_Z + \mathcal{M}_{\gamma Z} + \mathcal{M}_{Z\gamma}$. The correction to the Born level PV asymmetry $A^0_{\rm PV}$ can then be represented as

$$A_{\rm PV} = (1+\delta)A^0_{\rm PV} \equiv \left(\frac{1+\delta_{Z(\gamma\gamma)}+\delta_{\gamma(Z\gamma)}}{1+\delta_{\gamma(\gamma\gamma)}}\right)A^0_{\rm PV}\,, \tag{2}$$

where $A_{\rm PV}$ is the full asymmetry, including TBE corrections, and the interference between one- and two-boson exchange is explicitly indicated. Since the purely electromagnetic correction $\delta_{\gamma(\gamma\gamma)}$ is typically only a few percent, the full correction δ can be written approximately as $\delta \approx \delta_{Z(\gamma\gamma)} + \delta_{\gamma(Z\gamma)} - \delta_{\gamma(\gamma\gamma)}$. In the model discussed here (see also Refs. [4, 5, 6]), the TBE amplitudes $\mathcal{M}_{\gamma\gamma}$ and $\mathcal{M}_{\gamma Z}$ contain contributions from both nucleon elastic and $\Delta(1232)$ isobar intermediate states. These depend on both the vector and axial-vector form factors. For the nucleon these are well-known. For the Δ, we use the vector form factors from Kondratyuk *et al.* [7], and the axial-vector form factors from Lalakulich and Paschos [8].

In Fig. 1 we show the various contributions to the TBE correction as a function of the virtual photon polarization ε at the low Q^2 value of 0.01 GeV^2, and also at $Q^2 = 1.0$ GeV^2. At small Q^2 the $\gamma(\gamma\gamma)$ and $Z(\gamma\gamma)$ contributions are very similar, and considerably smaller in magnitude than the $\gamma(Z\gamma)$ component. Hence the $\gamma(\gamma\gamma)$ and $Z(\gamma\gamma)$ will partially cancel in their effect on $A_{\rm PV}$, which will be determined mostly by the $\gamma(Z\gamma)$ component. The latter is related to the correction to the proton weak charge computed by Marciano and Sirlin [9], which will be discussed in more detail in the next section. At larger Q^2 the $\gamma(Z\gamma)$ component decreases in magnitude, while the $\gamma(\gamma\gamma)$ and $Z(\gamma\gamma)$ pieces become large and more negative. The ε-dependence, or equivalently the energy-dependence, is clearly important to consider in experimental analyses.

12th International Conference on Meson-Nucleon Physics and the Structure of the Nucleon (MENU 2010)
AIP Conf. Proc. 1374, 515-518 (2011); doi: 10.1063/1.3647193

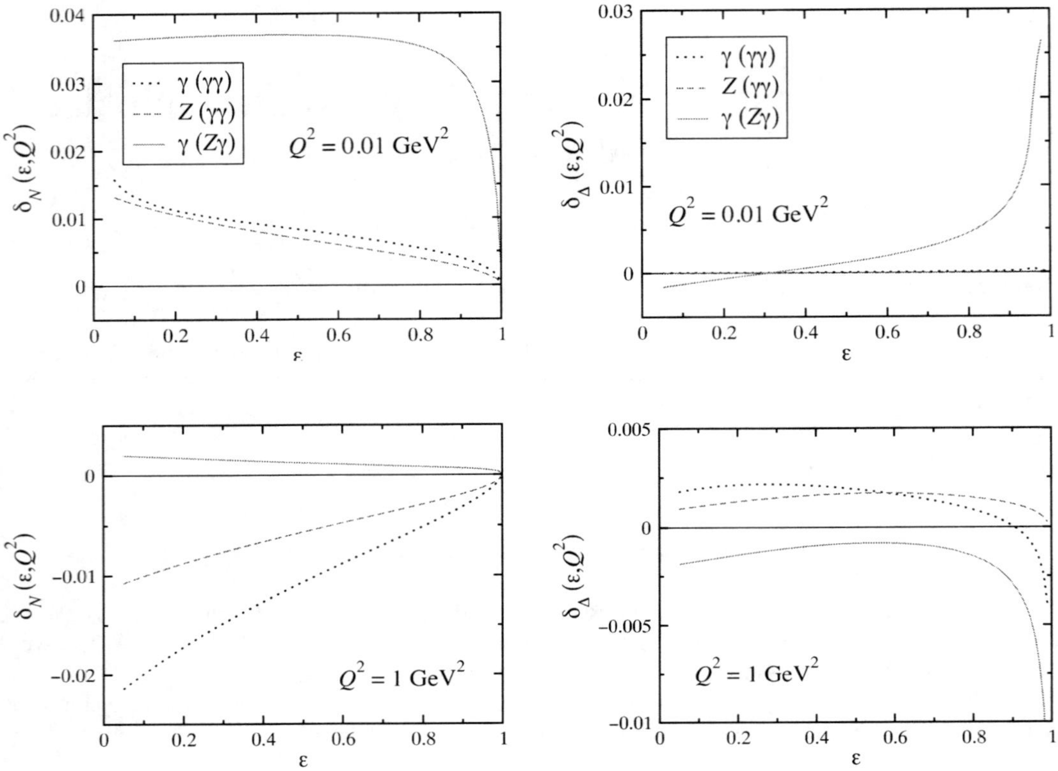

FIGURE 1. TBE corrections for nucleon (δ_N) and Δ (δ_Δ) intermediate states as a function of the virtual photon polarization ε at Q^2 of 0.01 and 1.0 GeV^2. Figures from Ref. [6].

Several interesting features can be noted for the Δ contribution. Firstly, the magnitude and shape of the Δ corrections are very different to the nucleon corrections. For the Δ contribution, at low Q^2 the $\gamma\gamma$ interference with either the γ or Z exchange is almost negligible, increasing somewhat at larger Q^2. The $\gamma(Z\gamma)$ contribution is also relatively small at low ε, and none of the corrections exceed $\sim 1\%$ in magnitude for $\varepsilon < 0.8$ and $Q^2 < 1\ \text{GeV}^2$. At larger ε, however, the $\gamma(Z\gamma)$ correction increases rapidly, becoming even bigger than the nucleon correction.

γZ CORRECTIONS AT FORWARD ANGLES

The Q_{weak} experiment at Jefferson Lab [10] is designed to measure the weak charge of the proton using PV to a higher level of precision than previously possible. This experiment will measure the PV asymmetry from protons at $E = 1.165$ GeV electron energy and very low momentum transfer squared, $t = -0.026\ \text{GeV}^2$.

In the $t \to 0$ and low energy limit, A_{PV} is related to the weak charge of the proton Q_W^p [11]:

$$A_{\text{PV}} \to \frac{G_F}{4\pi\alpha\sqrt{2}}\, t\, Q_W^p\,. \tag{3}$$

At tree level the proton's weak charge is given by $Q_W^p = 1 - 4\sin^2\theta_W$. Including electroweak radiative corrections, the proton weak charge is defined at zero energy and momentum transfer as [11]

$$Q_W^p = (1 + \Delta\rho + \Delta_e)(1 - 4\sin^2\theta_W(0) + \Delta_e') + \Box_{WW} + \Box_{ZZ} + \Box_{\gamma Z}(0)\,, \tag{4}$$

where $\sin^2\theta_W(0)$ is the weak mixing angle at zero momentum, and the corrections $\Delta\rho$, Δ_e and Δ_e' are given in Ref. [11]. The contributions $\Box_{WW}$ and $\Box_{ZZ}$ from the WW and ZZ box diagrams can be computed perturbatively, while the γZ

interference correction $\Box_{\gamma Z}(E)$ depends on physics at both low and high momentum scales [9, 11]. The current best theoretical estimate from Ref. [11] is $Q_W^p = 0.0713(8)$.

In Eq. (4) we have explicitly introduced a dependence of $\Box_{\gamma Z}(E)$ on the electron energy E. The E-dependence of the other radiative corrections in Eq. (4) is not expected to be important at the energy scales relevant for electron scattering. The Q_W^p extracted from $A^{\rm PV}$ in Eq. (3) will then differ from Q_W^p in Eq. (4) by an amount $\Box_{\gamma Z}(E) - \Box_{\gamma Z}(0)$, which we will refer to as a correction to Q_W^p at the particular kinematics of the electron scattering experiment.

As with the results of the previous section, the γZ term has contributions from both the axial-vector and the vector hadron coupling of the Z boson. The former, which we denote as $\Box_{\gamma Z}^A$, was computed by Marciano and Sirlin [9], and is the dominant correction at the low energies appropriate to parity-violation in atoms. The vector term $\Box_{\gamma Z}^V$ vanishes at zero energy, justifying its neglect in atomic parity-violation.

Although the contribution of the vector coupling is included in the calculations reported in the previous section, the importance of these terms at high energies was not explicitly recognized. Gorchtein and Horowitz [12] pointed out the implications of this term for $Q_{\rm weak}$, including the contributions of both the resonance and the deep inelastic regions. Ref. [12] used forward angle dispersion relations to evaluate the correction $\Box_{\gamma Z}^V(E)$, and found it to be large, with an uncertainty potentially capable of jeopardizing the interpretation of the $Q_{\rm weak}$ experiment. The analysis in the previous section of the γZ contribution, involving only nucleon and Δ intermediate states, finds smaller but similarly non-negligible effects. However, the advantage of the dispersion approach for forward scattering kinematics is that one can use cross section data instead of relying on models.

In the work reported here and in Ref. [13] we have revisited the $\Box_{\gamma Z}^V(E)$ correction using experimental constraints from recent JLab data on electroproduction in the resonance region, and HERA data in the deep-inelastic region. The imaginary part of the PV γZ exchange contribution can be written in terms of the cross section for all possible final hadronic states $ep \to eX$, which in turn can be expressed as a product of leptonic and hadronic tensors. The hadronic tensor depends on the interference electroweak structure functions $F_1^{\gamma Z}$ and $F_2^{\gamma Z}$. We find

$$\Im m \Box_{\gamma Z}^V(E) = \frac{\alpha}{(s-M^2)^2} \int_{W_\pi^2}^{s} dW^2 \int_0^{Q^2_{\max}} \frac{dQ^2}{1+Q^2/M_Z^2} \left[F_1^{\gamma Z} + F_2^{\gamma Z} \frac{s\,(Q^2_{\max}-Q^2)}{Q^2\,(W^2-M^2+Q^2)} \right], \tag{5}$$

where $s = M(M+2E)$ is the total c.m. energy squared. The structure functions $F_{1,2}^{\gamma Z}$ are functions of the exchanged boson virtuality, Q^2, and of the invariant mass W of the exchanged boson and proton (or alternatively of the Bjorken variable $x = Q^2/(W^2-M^2+Q^2)$). The lower limit of the W integration is given by the mass $W_\pi = M + m_\pi$ of the pion production threshold, and the upper limit of the Q^2 integration is given by $Q^2_{\max} = 2ME(1-W^2/s)$. At large Q^2 and W, the $F_2^{\gamma Z}$ structure function can be written (at leading order) in terms of parton distributions q and $\bar{q}$ as $F_2^{\gamma Z} = x\sum_q 2e_q g_V^q\,(q+\bar{q})$, with $e_{u,d}$ and $g_V^{u,d}$ the electromagnetic and weak vector charges for u and d quarks, respectively. In the limit $2g_V^q \to e_q$, the interference structure functions $F_2^{\gamma Z} \to F_2$, where $F_2 = x\sum_q e_q^2\,(q+\bar{q})$.

We note that the expression (5) is a factor of 2 larger than that quoted in Ref. [12] (a result that has now been independently verified). In the region of low intermediate state hadronic masses, $W < 2.5$ GeV, inclusive scattering is dominated by nucleon resonances. While there is an abundance of electroproduction data in the resonance region, there are no direct measurements of $F_{1,2}^{\gamma Z}$. We therefore use $F_{1,2}$, noting that for most prominent $I = \frac{3}{2}$ and $I = \frac{1}{2}$ states the magnitudes of the Z-boson transition couplings are equal to the respective photon couplings to within a few percent.

We fit all of the available data, including the latest from SLAC [14] and JLab [15], using the isobar model for each Q^2, taking into account the contributions from four resonances: $P_{33}(1232)$, $D_{13}(1520)$, $F_{15}(1680)$ and $F_{37}(1950)$. The fit allows the inclusive transition form factors for each resonance to be constrained accurately up to $Q^2 \approx 3$ GeV^2.

In the DIS region the interference structure functions can be expressed in terms of leading twist parton distribution functions (PDFs). However, the range of integration in Eq. (5) extends to low W and Q^2, beyond the region of validity of a PDF description. To proceed we follow Ref. [12] and approximate $F_{1,2}^{\gamma Z}$ by their electromagnetic analogs $F_{1,2}$ at very small x, where the light-quark PDFs are approximately flavor-independent. For F_2 we use the parametrization from Refs. [16, 17], which is motivated by Regge theory and valid at both high Q^2 and low Q^2. At larger x ($x > 0.4$) the flavor dependence of the PDFs makes the interference function $\sim$ 30–40% smaller than F_2. The electromagnetic structure functions therefore provide an upper limit on $F_2^{\gamma Z}$.

Performing the dispersion integral, the result for $\Re e \Box_{\gamma Z}^V(E)$ is shown in Fig. 2 as a function of the incident electron energy E. Although the integration in principle involves an infinite range of W and Q^2, in practice we find that around

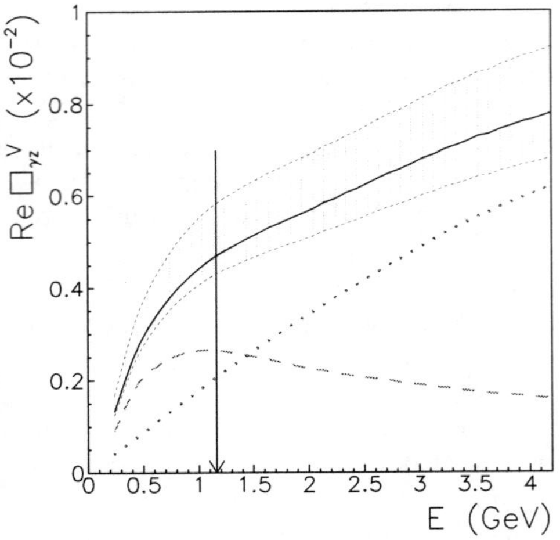

FIGURE 2. γZ box correction $\Box^V_{\gamma Z}(E)$ to Q^p_W as a function of electron energy E, showing the resonant (dashed) and nonresonant (dotted) contributions, as well as the sum (solid) and the overall (asymmetric) uncertainty (shaded). The vertical arrow at $E =$ 1.165 GeV indicates the energy of the $Q_{\rm weak}$ experiment. Figure from Ref. [13].

80% of the value of $\Re e\,\Box^V_{\gamma Z}(E)$ at the energy relevant to $Q_{\rm weak}$ comes from energies below 4 GeV, where the Q^2 range extends to ~ 6 GeV^2, and W to ~ 3 GeV.

The resonant and nonresonant contributions to $\Re e\,\Box^V_{\gamma Z}(E)$ are 0.0026 and 0.0021, respectively, at the energy relevant for the $Q_{\rm weak}$ experiment, $E = 1.165$ GeV. We should note, however, that this separation is somewhat arbitrary, as only the physically meaningful, total cross section enters into our fit. The combined correction to Q^p_W at the $Q_{\rm weak}$ energy is then $0.0047^{+0.0011}_{-0.0004}$, or $6.6^{+1.5}_{-0.6}$% of the Standard Model value 0.0713(8) for Q^p_W.

ACKNOWLEDGMENTS

The work reported here was done in collaboration with W. Melnitchouk, A. Sibirtsev, A.W. Thomas, and J.A. Tjon.

REFERENCES

1. K. Nakamura *el al.*, J. Phys. G **37**, 075021 (2010).
2. S. Baunack *et al.*, Phys. Rev. Lett. **102**, 151803 (2009).
3. R. D. Young, J. Roche, R. D. Carlini and A. W. Thomas, Phys. Rev. Lett. **97**, 102002 (2006).
4. H. Q. Zhou, C. W. Kao and S. N. Yang, Phys. Rev. Lett. **99**, 262001 (2007).
5. J. A. Tjon and W. Melnitchouk, Phys. Rev. Lett. **100**, 082003 (2008).
6. J. A. Tjon, P. G. Blunden and W. Melnitchouk, Phys. Rev. C **79**, 055201 (2009).
7. S. Kondratyuk, P. G. Blunden, W. Melnitchouk and J. A. Tjon, Phys. Rev. Lett. **95**, 172503 (2005).
8. O. Lalakulich and E. A. Paschos, Phys. Rev. D **71**, 074003 (2005).
9. W. J. Marciano and A. Sirlin, Phys. Rev. D **27**, 552 (1983); *ibid.* **29**, 75 (1984) [Erratum *ibid.* D **31**, 213 (1985)].
10. JLab experiment E-08-016 ($Q_{\rm weak}$), R. D. Carlini *et al.* spokespersons.
11. J. Erler, A. Kurylov and M. J. Ramsey-Musolf, Phys. Rev. D **68**, 016006 (2003); J. Erler and M. J. Ramsey-Musolf, Phys. Rev. D **72**, 073003 (2005).
12. M. Gorchtein and C. J. Horowitz, Phys. Rev. Lett. **102**, 091806 (2009).
13. A. Sibirtsev, P. G. Blunden, W. Melnitchouk and A. W. Thomas, Phys. Rev. D **82**, 055201 (2009).
14. L. W. Whitlow *et al.*, Phys. Lett. B **282**, 475 (1992).
15. M. Osipenko *et al.*, Phys. Rev. D **67**, 092001 (2003); I. Niculescu *et al.*, Phys. Rev. Lett. **85**, 1186 (2000); Y. Liang *et al.*, arXiv:nucl-ex/0410027; S. P. Malace *et al.*, Phys. Rev. C **80**, 035207 (2009).
16. A. Capella, A. Kaidalov, C. Merino and J. Tran Thanh Van, Phys. Lett. B **337**, 358 (1994).
17. A. B. Kaidalov and C. Merino, Eur. Phys. J. C **10**, 153 (1999).

Two-Boson Exchange Corrections in PVES

Chung Wen Kao

Department of Physics, Chung-Yuan Christian University,
Chung-Li 32023, Taiwan

Abstract. I summarize the recent progress on the understanding of the two-boson exchange effect in the parity-violating scattering. I report the results obtained in a hadronic model and a model using Nucleon General Parton Distributions(GPDS), respectively. Both of calculations show that the approximation adopted by Marciano and Sirlin brings significant deviations.

Keywords: strangeness, two-boson exchange.
PACS: PACS number:13.40 Ks

THE STRANGENESS OF THE PROTON AND PARITY-VIOLATING EP SCATTERING

Parity-violating ep scattering provides a unique probe to extract proton strange form factors by measuring the parity-violating asymmetry $A_{PV} = (\sigma_R - \sigma_L)/(\sigma_R + \sigma_L)$ with polarized electrons. At the tree level, the parity asymmetry A_{PV} arises from the interference of diagrams with one-photon-exchange and Z-boson exchange shown in Fig. 1(a) and (b), respectively. When combined with proton and neutron electromagnetic form factors and with the use of charge symmetry, one obtains the following relation,

$$A_{PV}(\rho,\kappa) = A_1 + A_2 + A_3,\ A_1 = -a\rho\left[(1-4\kappa\sin^2\theta_W) - \frac{\varepsilon G_E^{\gamma,p}G_E^{\gamma,n} + \tau G_M^{\gamma,p}G_M^{\gamma,n}}{\varepsilon(G_E^{\gamma,p})^2 + \tau(G_M^{\gamma,p})^2}\right],$$
$$A_2 = a\rho\frac{\varepsilon G_E^{\gamma,p}\tilde{G}_E^s + \tau G_M^{\gamma,p}\tilde{G}_M^s}{\varepsilon(G_E^{\gamma,p})^2 + \tau(G_M^{\gamma,p})^2},\ A_3 = a(1-4\sin^2\theta_W)\frac{\varepsilon' G_M^{\gamma,p}G_A^Z}{\varepsilon(G_E^{\gamma,p})^2 + \tau(G_M^{\gamma,p})^2}. \tag{1}$$

where $a = \frac{G_F Q^2}{4\pi\alpha\sqrt{2}}$, $\varepsilon \equiv [1+2(1+\tau)\tan^2\theta_{Lab}/2]^{-1}$ and $\tau = \frac{Q^2}{4M_N^2}$. ρ and κ represent the radiative corrections. The linear combination of the strange form factors, $G_E^s + \beta G_M^s$, has been extracted from A_2 in Eq. (1). Four experimental programs SAMPLE [1], HAPPEX [2], A4 [3], and G0 [4] have been designed to measure A_{PV}. Their results demonstrate that the sum of A_1 and A_3 is much larger than A_2 and the theoretical uncertainty is comparable in size to A_2. This calls for greater efforts to reduce the theoretical uncertainty to arrive at a more precise determination of the strange form factors. Leading order radiative corrections to A_{PV}, including the box diagrams Fig. 1(d) and other diagrams, have been extensively studied [5] and widely used in the global analyses. They evaluated the γZE diagram at the limit of $Q^2 = 0$ and $E_{lab} = m_e$. They then separated the loop integral into high-momentum and low momentum parts. The high-momentum part has been estimated with the handbag approximation. On the other hand, the low-momentum part has been calculated in a simple hadronic model including only the nucleon intermediate state with the form factors inserted. However our recent works [9, 10, 11, 12] have demonstrated that such an approximation actually causes significant deviations either in the high and low momentum parts. Consequently the extracted values of the strange form factors should be reevaluated in the refined framework.

THE TWO-BOSON EXCHANGE EFFECTS IN HADRONIC MODEL

The two-boson exchange (TBE) corrections to A_{PV} have been investigated in a hadronic model first with only the elastic nucleon intermediate states in [9]. Later the contributions of the $\Delta(1232)$ intermediate states have also been included in [10, 11]. It is found that both of the $2\gamma E$ and γZE corrections to A_{PV} depends strongly on Q^2 and ε, and can reach a few percent and are comparable in size with the current experimental measurements of strange quark effects in the proton weak neutral current. It has been shown that the interplay between the nucleon and Δ contributions depend strongly on the kinematics. The result is summarized in the Fig 2. Since we have explicitly calculated the effect

12th International Conference on Meson-Nucleon Physics and the Structure of the Nucleon (MENU 2010)
AIP Conf. Proc. 1374, 519-522 (2011); doi: 10.1063/1.3647194

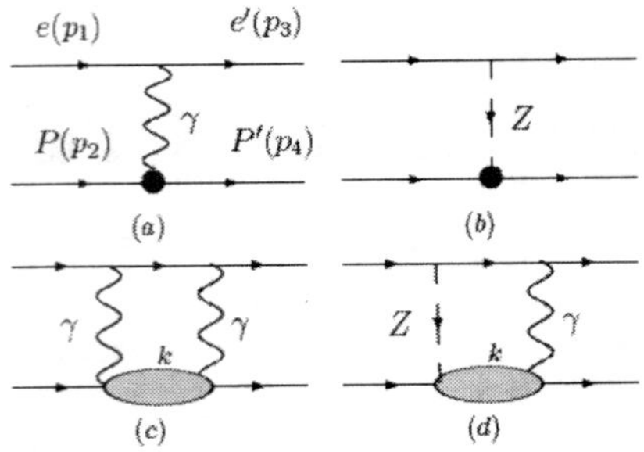

FIGURE 1. (a) One-photon exchange, (b) Z-boson exchange, (c) Two-photon exchange, and (d) γZ-exchange diagrams for elastic ep scattering. Corresponding cross-box diagrams are implied.

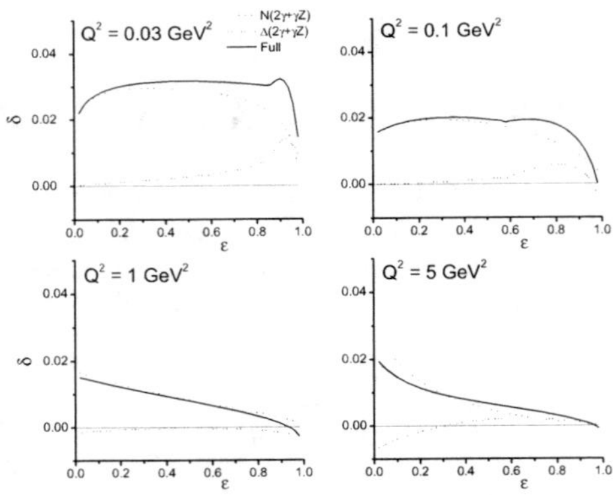

FIGURE 2. Two-boson exchange corrections with nucleon (dotted) and Δ (dashed) intermediate states to A_{PV}, as functions of ε from 0.1 to 0.9 at $Q^2 = 0.03$, 0.1, 1.0, and 5.0 GeV^2, respectively. The solid lines denote their sums $\delta = \delta_N + \delta_\Delta$

of TBE effects, we should then subtract $\Delta\rho = \rho_{\gamma Z}$ and $\Delta\kappa = \kappa_{\gamma Z}$ from ρ and κ and use $\rho' = \rho - \Delta\rho$ and $\kappa' = \kappa - \Delta\kappa$ in Eq. (1) instead. We set the experimental parity asymmetry $A_{PV}^{(Exp)}$ as follows:

$$A_{PV}^{(Exp)} \equiv A_{PV}(1\gamma + Z + 2\gamma + \gamma Z) = A_{PV}(\rho', \kappa')(1 + \delta_N + \delta_\Delta). \tag{2}$$

With the value we obtain for δ, we can determine $A_{PV}(\rho', \kappa')$ and extract the strange form factors. We also introduce $\overline{G}_E^s + \beta \overline{G}_M^s = (G_E^s + \beta G_M^s)(1 + \delta_G)$ to quantify the effects of the $2\gamma E$ and $\gamma Z E$ to the extracted values of $G_E^s + \beta G_M^s$, where $G_E^s + \beta G_M^s$ and $\overline{G}_E^s + \beta \overline{G}_M^s$ are extracted from $A_{PV}(\rho, \kappa)$ and $A_{PV}(\rho', \kappa')$, respectively. We also list the values of $G_s \equiv G_E^s + \beta G_M^s$. The resultant change in the values of G_s after TBE effects are properly taken into account, i.e., $\Delta G_s \equiv (\bar{G}_E^s + \beta \bar{G}_M^s) - (G_E^s + \beta G_M^s)$, are given in the last column of Table 1.

TWO-BOSON EXCHANGE EFFECT IN GPD CALCULATION

Next we report on the result of the partonic calculation of TBE corrections to A_{PV} in the same parton model developed in [7]. We first calculate the subprocess on a quark $e(k) + q(P_q) \to e(k') + q(P_{q'})$, denoted by the scattering amplitude H in Fig 3. Subsequently we embed the quarks in the proton as described through the nucleon's generalized parton distributions (GPDs). The quark-level parity-violating amplitudes of the γZ exchange diagrams are

$$\mathcal{M}_{\gamma Z}^{PV}(eq \to eq) = \frac{-iG_F}{2\sqrt{2}} \sum_{q=u,d} [t_q^1 (\bar{u}_e \gamma_\mu \gamma_5 u_e)(\bar{q}\gamma^\mu q) + t_q^2 (\bar{u}_e \gamma_\mu u_e)(\bar{q}\gamma^\mu \gamma_5 q)], \tag{3}$$

TABLE 1. The values of δ_N, δ_Δ, and their sum δ for the HAPPEX , G0, and A4 data. We also give the values of δ_G, G_s, and ΔG_s.

Exp	$Q^2(GeV^2)$	ε	$\delta_N(\%)$	$\delta_\Delta(\%)$	$\delta_G(\%)$	$G_s(10^{-2})$	$\Delta G_s(10^{-2})$
HAPPEX	0.477	0.974	0.18	-0.27	-2.54	1.4	-0.04
HAPPEX	0.109	0.994	0.21	-0.80	-20.63	0.7	-0.14
G0	0.144	0.9916	0.20	-0.64	14.14	-1.1	-0.16
G0	0.192	0.9886	0.19	-0.52	-16.05	0.8	-0.12
G0	0.210	0.9875	0.19	-0.48	48.25	-0.3	-0.14
G0	0.232	0.9860	0.19	-0.44	-20.25	0.6	-0.12
A4	0.108	0.83	1.07	0.53	2.00	7.1	0.14
A4	0.23	0.83	0.66	0.14	2.85	3.9	0.11

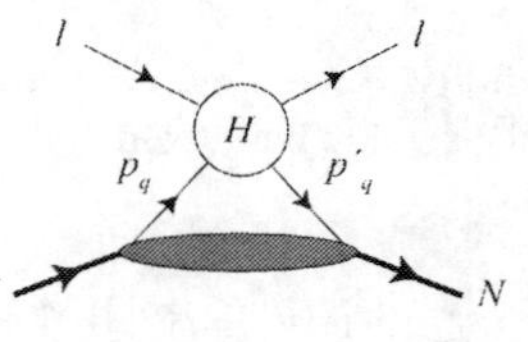

FIGURE 3. Handbag approximation for the elastic lepton-nucleon scattering. In the partonic process indicated by H, the lepton scatters from quarks within the nucleon, with momenta P_q and $P_{q'}$. The lower blob represents the GPDs of the nucleon.

where t_1^q=$Q_q[c_1 g_A^e g_V^q + c_2 g_A^q g_V^e]$ and t_2^q=$Q_q[c_1 g_V^e g_A^q + c_2 g_V^q g_A^e]$. Here g_V^e=$-1+4\sin^2\theta_W$, g_A^e=$+1$. c_1 and c_2 are defined as

$$c_1 = \frac{-e^2}{4\pi^2}\left[\ln\left(\frac{\lambda^2}{Q^2}\right)+\frac{\pi^2}{2}\right]+\frac{3e^2}{16\pi^2}\ln\left(\frac{\hat{u}}{\hat{s}}\right),\ c_2 = \frac{e^2}{16\pi^2}\left[-7+3\ln\left(\frac{\hat{s}}{M_Z^2}\right)+3\ln\left(\frac{\hat{u}}{M_Z^2}\right)\right], \tag{4}$$

where $\hat{s} = (P_q+k)^2, \hat{u} = (P_q-k')^2$ and λ is the infrared cut-off input by infinitesimal photon mass. The amplitudes are separated into the soft and hard parts, the soft part corresponds with the situation where the photon carries zero four momentum and one obtains $c_1^{soft} = \frac{-e^2}{4\pi^2}\left[\ln\left(\frac{\lambda^2}{Q^2}\right)+\frac{\pi^2}{2}\right]$, the hard part is $c_1^{hard} = \frac{3e^2}{16\pi^2}\ln\left(\frac{\hat{u}}{\hat{s}}\right)$. The ep parity-violating amplitudes are obtained as a convolution between an electron-quark hard scattering and a soft nucleon matrix element. The procedure is similar to the one in [7]. The result is presented in Fig 4. The γZE effects are insensitive to ε but sensitive to Q^2. The $2\gamma E$ effects are same with [6]. It also shows that the γZE effect is dominant.

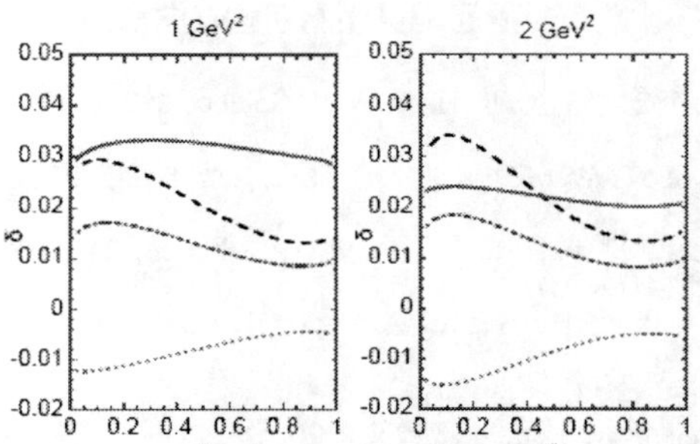

FIGURE 4. TPE and γZ exchange corrections to parity-violating asymmetry as functions of ε. The left panel for Q^2=1 GeV^2, the right for Q^2=2 GeV^2. The various contributions are $\gamma(\gamma\gamma)$(dashed line), $Z(\gamma\gamma)$(dotted lines), TPE total contribution $\delta_{2\gamma}$(dash-dotted lines) and $\delta_{\gamma Z}$ (solid lines).

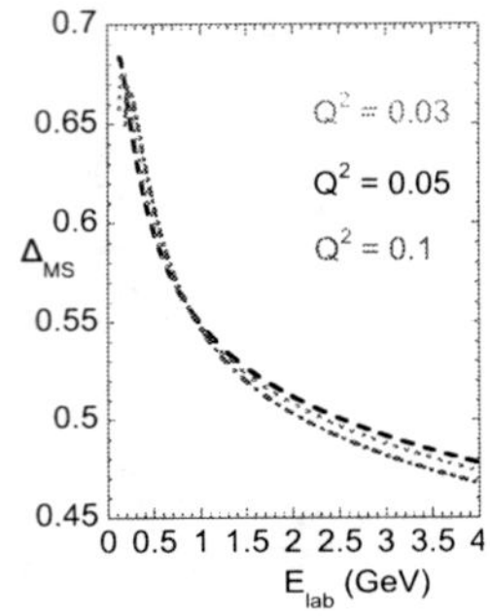

FIGURE 5. Δ_{MS} as the function of E_{lab}. The dotted line denotes Q^2=0.03 GeV^2. The dashed line represents Q^2=0.05 GeV^2 and the dotted-dashed line represents Q^2=0.1 GeV^2

Another important issue to address is the validity of the previous estimate of γZ exchange effects made by Marciano and Sirlin [5] because their estimate is widely used. They evaluated the γZ box diagrams in Fig. 3 in the limit of $k = k' = 0$ and $P_q = P_{q'}$. Their result shows that

$$\mathcal{M}_{\gamma Z}^{PV,MS}(eq \to eq) = \frac{-iG_F}{\sqrt{2}} \sum_{q=u,d} [C_{1q}(\bar{u}_e\gamma_\mu\gamma_5 u_e)(\bar{q}\gamma^\mu q) + C_{2q}(\bar{u}_e\gamma_\mu u_e)(\bar{q}\gamma^\mu\gamma_5 q)], \quad (5)$$

We therefore define the following quantity to characterize the difference between our partonic calculation and the previous estimate taken in the limit $k = k' = 0$ and $P_q = P_{q'}$:

$$\Delta_{MS} = \frac{A_{PV}^{Parton}(\gamma Z)}{A_{PV}^{MS}(\gamma Z)} = \frac{Re[\mathcal{M}_{1\gamma}^{\dagger}\mathcal{M}_{\gamma Z}^{PV,Parton}]}{Re[\mathcal{M}_{1\gamma}^{\dagger}\mathcal{M}_{\gamma Z}^{PV,MS}]}. \quad (6)$$

It shows that MS approximation always overestimates γZE exchange effect. One learns from Fig 5 that Δ_{MS} decreases fast as E_{lab} increases. We conclude that MS approximation indeed overestimates the TBE effects.

ACKNOWLEDGMENTS

This work is supported by the National Science Council of Taiwan under grants nos. NSC096-2112-M033-003-MY3. I would also like to express my gratitude to H.Q. Zhou, Y.C. Chen, K. Nagata, S.N. Yang and M. Vanderhaeghen for their collaboration.

REFERENCES

1. B. Mueller *et al.*, Phys. Rev. Lett. **78**, 3824 (1997); R. Hasty *et al.*, Science **290**, 2117 (2000); D. T. Spayde *et al.*, Phys. Lett. B **583**, 79 (2004).
2. K. A. Aniol *et al.* (HAPPEX), Phys. Rev. C **69**, 065501 (2004), Phys. Lett. B **635**, 275 (2006); A. Acha *et al.* (HAPPEX), Phys. Rev. Lett. **98**, 032301 (2007).
3. F. E. Maas *et al.* (A4), Phys. Rev. Lett. **93**, 022002 (2004); Phys. Rev. Lett. **94**, 152001 (2005); B. Glaser (for the A4 collaboration) Eur. Phys. J. A. **24, S2**, 141(2005).
4. D. S. Armstrong *et al.* (G0), Phys. Rev. Lett. **95**, 092001 (2005); C. Furget for the G0 collaboration, Nucl. Phys. Proc. Suppl. **159**, 121 (2006).
5. W. J. Marciano and A. Sirlin, Phys. Rev. D **29**, 75 (1984).
6. A. V. Afanasev and C. E. Carlson, Phys. Rev. Lett. **94**, 212301 (2005).
7. Y. C. Chen, A. V. Afanasev, S. J. Brodsky, C. E. Carlson, M. Vanderhaeghen, Phys. Rev. Lett. **93**, 122301 (2004).
8. M. K. Jones *et al.*, Phys. Rev. Lett. **84**, 1398 (2000); O. Gayou *et al.*, Phys. Rev. Lett. **88**, 092301 (2002).
9. H. Q. Zhou, C. W. Kao, and S. N. Yang, Phys. Rev. Lett. **99**, 262001 (2007); *ibid.* **100**, 059903(E) (2008).
10. K. Nagata, H. Q. Zhou, C. W. Kao, and S. N. Yang, Phys. Rev. C **79**, 062501(R) (2009).
11. H. Q. Zhou, C. W. Kao, S. N. Yang and K. Nagata, Phys. Rev. C **81** (2010) 035208
12. Y. C. Chen, C. W. Kao and M. Vanderhaeghen, arXiv:0903.1098 [nucl-th].

Measuring the 2γ Exchange Effect in Electron-Proton Scattering at CLAS

Robert Paul Bennett

Old Dominion University - Norfolk, VA USA

Abstract. Recent results from experiments at Jefferson Laboratory, Newport News VA, which measure the ratio of the electric to magnetic form factors of the proton, G_E/G_M, have forced us to reexamine the single photon exchange approximation in lepton-proton elastic scattering. Discrepancies between the ratio obtained via the time-tested Rosenbluth separation method and new polarization transfer measurements, which differ by as much as a factor of three, may be resolved by considering the effect of two photon exchange (TPE). Theoretical corrections for the TPE effect are difficult due to the large number of excited nucleon states that can contribute to the process. The TPE effect can, however, be determined directly by measuring the ratio of the positron-proton to electron-proton elastic scattering cross sections, $R = \sigma(e^+)/\sigma(e^-)$, since the TPE effect changes sign with the charge of the incident particle. The CLAS TPE experiment at Jefferson Laboratory, will determine the effect of two-photon exchange in elastic lepton-proton scattering by precisely measuring the ratio of positron-proton to electron-proton elastic cross sections over a large kinematic range ($0.1 < \varepsilon < 0.96$, $0.02 \leq Q^2 \leq 2.0$). We will accomplish this by directing the 5.5 GeV primary electron beam, provided by the Continuous Electron Beam Accelerator Facility (CEBAF), onto a set of radiators and converters to produce simultaneous and identical beams of electron and positrons. Acceptance and efficiency concerns are minimized by only considering the ratios of the elastic cross sections and switching polarity of magnets in the beam line and the spectrometer. Guided by the results of a short 2006 test run and extensive GEANT based modeling, new shielding and beam line components were designed to maximize luminosity. The unique experimental design and challenges of the TPE experiment and the projected luminosities will be presented.

Keywords: Form factor, two photon exchange, positron, electron
PACS: 13.40.-f, 13.40.Gp, 13.88.+e, 14.20.Dh, 25.30.Bf

INTRODUCTION

The two-photon exchange (TPE) effect is believed to be responsible for the observed discrepancy between the ratio of proton electric and magnetic form factors measured by the Rosenbluth [1, 2] and polarization transfer methods [3, 4, 5, 6]. This disagreement grows with Q^2 and reaches a factor of three at Q^2 of 5.6 GeV^2 as seen in Fig 1. Both the Rosenbluth separation and the polarization transfer formalism assume only the single photon is exchanged between the proton and lepton. Though recognized early on, TPE effects were generally assumed to be small and ignored in standard radiate corrections[7, 8, 9, 10]. However, recent theoretical calculations [11, 12, 13] suggest that the two-photon exchange correction, can in fact introduce an error in calculating the form factors from the cross sections. A direct model-independent measurement of the TPE correction can be achieved experimentally by finding the ratio of the positron-proton to electron-proton elastic cross sections. Keeping only terms up to order $O(\alpha_{em}^2)$, amplitude of elastic ep-scattering can be written as

$$A_{total} = e_e e_p A_{Born} + e_e^2 e_p A_{p,br} + e_e e_p^2 A_{p,br} + e_e^2 e_p^2 A_{2\gamma} \tag{1}$$

where the amplitudes A_{Born}, $A_{e,br}$, $A_{p,br}$, and $A_{2\gamma}$ represent the one-photon exchange, electron bremsstrahlung, proton bremsstrahlung and two-photon exchange, respectively. Squaring the above amplitude and keeping the only terms up to order α_{em} that have odd powers of electron charge, we have

$$|A_{ep\to ep}|^2_{odd} = e_e^2 e_p^2 \left[|A_{Born}|^2 + e_e e_p A_{Born} 2Re(A^*_{2\gamma}) + e_e e_p 2Re(A_{e,br} A^*_{p,br}\right] \tag{2}$$

where Re denotes the real part of the amplitude. After correcting for the interference between electron and proton bremsstrahlung, one can isolate the TPE term by taking the ratio of positron to electron-proton cross sections:

$$\sigma(e^\pm) = \sigma_{Born}(1 \mp \delta_{2\gamma}) \tag{3}$$

$$R = \frac{\sigma(e^+)}{\sigma(e^-)} \approx 1 - 2\delta_{2\gamma} \tag{4}$$

12th International Conference on Meson-Nucleon Physics and the Structure of the Nucleon (MENU 2010)
AIP Conf. Proc. 1374, 523-526 (2011); doi: 10.1063/1.3647195

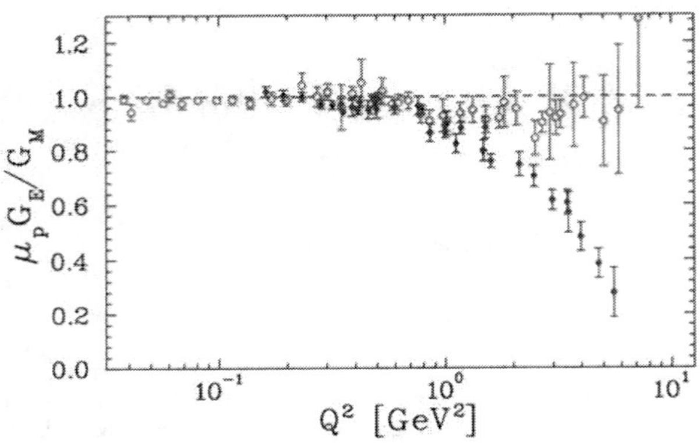

FIGURE 1. Comparison of G_E/G_M via the Rosenbluth separation method (hollow circles) and the polarization transfer method (solid diamonds) [14]

where $\delta_{2\gamma}$ is the two photon exchange correction factor.

ACCESSING TPE IN CLAS

The ratio, R, is measured at CLAS by producing simultaneous identical positron and electron beams in Hall B and and using the CEBAF Large Acceptance Spectrometer (CLAS) [15] to detect the scattered leptons and protons. The electron and positron beams are made by scattering the primary electron provided by CEBAF off of a a gold radiator foil to produce a photon beam. The remaining electrons are bent out of the beam line by the Hall B photon-tagger magnet. The photon beam then strikes a gold converter foil to produce e^+/e^- pairs. The leptons are subsequently bent out of the beam line and away from the residual photon beam by a three-dipole chicane. The photon beam is stopped by a tungsten block while the leptons are recombined into a single beam at the third dipole before proceeding on to the target. Fig.2 shows a schematic of the beam line layout.

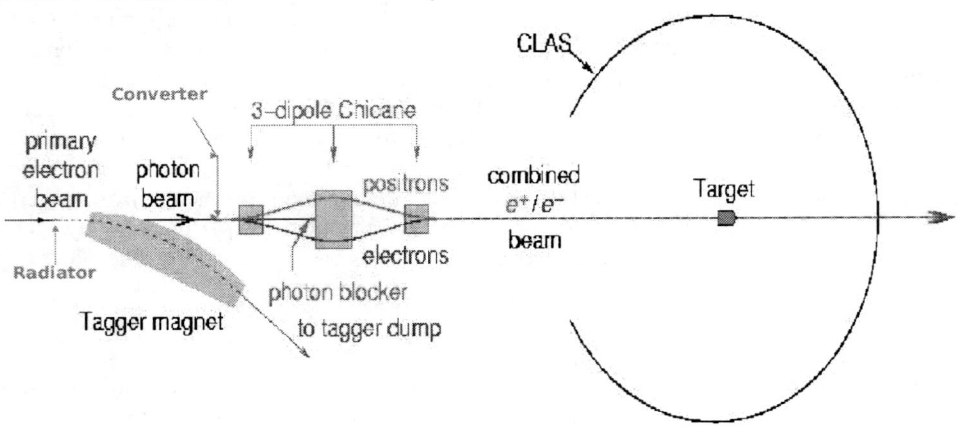

FIGURE 2. Beam line sketch of the TPE experiment in CLAS

MAXIMIZING LUMINOSITY

In 2006 a short test run was taken. By combining the results of this run with extensive GEANT simulations of the beam line and hall, we have designed new shielding and beam line components. Optimization of these new components will allow us to maximize the luminosity we can achieve in an upcoming TPE run. The main constraint in maximizing the

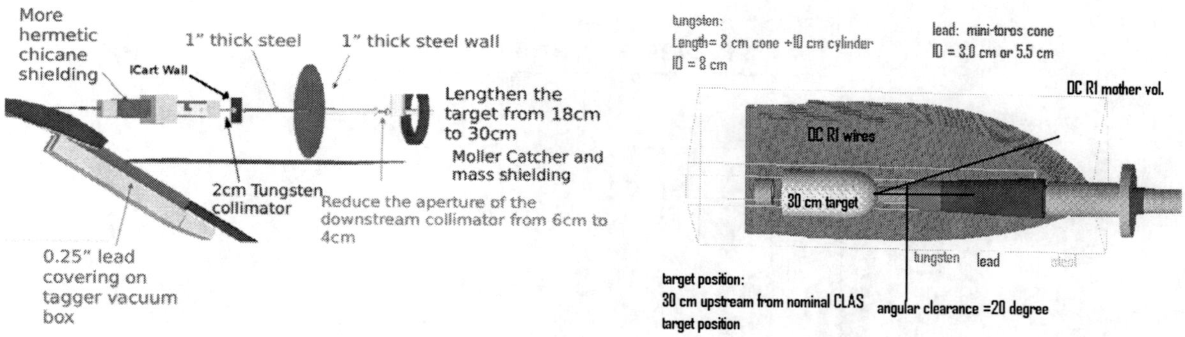

FIGURE 3. New shielding and beam line components

luminosity in the TPE experiment is the Hall B drift chamber occupancy. If the occupancy is above 3% the tracking efficiency of the CLAS detectors begins to suffer.

We simulate over 1000 detector and beam line components in Hall B in GEANT. The simulation was verified by comparing the change in occupancy in the innermost (Region 1) and the outermost (Region 3) regions of the CLAS drift chambers when the beam was blocked at different points of the beam line or new shielding was added to the hall. Table 1 shows a good agreement between data and simulation in Region 1 and the occupancies in the simulation scale with the data in Region 3.

TABLE 1. The effect of beam line modifications in test run data and simulations

	Type	Region 1 (% Occ)	Region 3 (% Occ)	R3 Nominal/Mod
Nominal	Data	2.05	0.97	1
	Simulation	2.54	3.57	1
Beam blocked at Rad-phi wall	Data	0.04	0.49	2.0
	Simulation	0.14	1.97	1.8
Insert ICart wall	Data	1.27	0.38	2.5
	Simulation	1.91	2.55	1.4
Beam blocked at DS Coll.	Data	0.14	0.34	2.8
	Simulation	0.10	1.84	1.9

New Beam Line Components and Shielding

Using the GEANT simulation we are able to identify several potential sources of backgrounds in the CLAS detector and design shielding to control these backgrounds. Modifications to the beam line will be made for the 2010-2011 TPE run in Hall-B. The major additions to the hall are shown in Fig. 3. They include not only extensive amounts of shielding, but also the lengthening of the target to 30cm, from the 18 cm used in the test run, and the inclusion several new beam colimators and the CLAS mini-torus to eliminate Moller electrons.

EXPECTED UNCERTAINTY AND KINEMATIC COVERAGE IN THE e^+/e^- RATIO

The new shielding additions will allow us to push the primary electron beam luminosity to over 100nA. We will also increase the primary beam energy from 3.5GeV, in the 2006 test run, to the maximum obtainable in Hall-B, 5.6 GeV. Taking the convolution of the estimate beam flux with the elastic electron-proton scattering cross sections we estimate that in two months of beam time a statistical uncertainty $< 1\%$ in the measurement of R can be achieved, over a large kinematic phase space ($0.1 < \varepsilon < 0.96$, $0.02 \leq Q^2 \leq 2.0$). Fig. 4 shows projected kinematics of the Hall-B TPE experiment along with several other experiments looking to measure R.

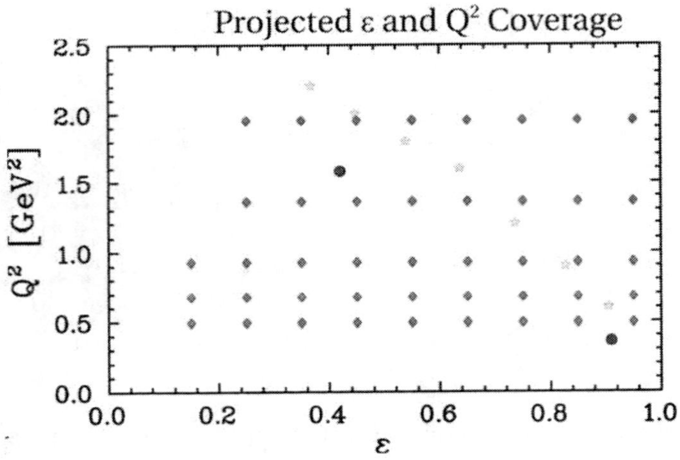

FIGURE 4. Projected kinematic coverage of the Hall-B experiment (red points) and two other proposed experiments Olympus at DESY (light blue) and VEPP-III at Novosibirsk (dark blue).

REFERENCES

1. L. Andivahis, and et al., *Phys. Rev. D* **50**, 5491–5517 (1994).
2. J. Arrington, *Phys. Rev. C* **68**, 034325 (2003).
3. V. Punjabi, and et al., *Phys. Rev. C* **71**, 055202 (2005).
4. M. K. Jones, and et al., *Phys. Rev. Lett.* **84**, 1398–1402 (2000).
5. O. Gayou, and et al., *Phys. Rev. Lett.* **88**, 092301 (2002).
6. A. J. R. Puckett, and et al., *Phys. Rev. Lett.* **104**, 242301 (2010).
7. S. D. Drell, and M. A. Ruderman, *Phys. Rev.* **106**, 561–563 (1957).
8. S. D. Drell, and S. Fubini, *Phys. Rev.* **113**, 741–744 (1959).
9. N. R. Werthamer, and M. A. Ruderman, *Phys. Rev.* **123**, 1005–1013 (1961).
10. F. Guerin, and C. A. Piketty, *Nuovo Cimento* **32** (1964).
11. A. V. Afanasev, and C. E. Carlson **94**, 212301 (2005).
12. P. G. Blunden, W. Melnitchouk, and J. A. Tjon **72**, 034612 (2005).
13. P. A. M. Guichon, and M. Vanderhaeghen **91**, 142303 (2003).
14. J. Arrington, W. Melnitchouk, and J. A. Tjon, *Phys. Rev. C* **76**, 035205 (2007).
15. B. Mecking, et al., *Nucl. Instr. Methods* **A 503/3**, 513 (2003).

Status of the OLYMPUS Experiment at DESY

M. Kohl for the OLYMPUS collaboration

Hampton University, Hampton, VA 23668 and Jefferson Lab, Newport News, VA 23606, USA

Abstract. The OLYMPUS experiment at DESY, Hamburg, Germany will definitively determine the effect of two-photon exchange in elastic lepton-proton scattering by precisely measuring the ratio of positron-proton to electron-proton elastic unpolarized cross sections. In this paper, the physics motivation is summarized, aspects of the experimental setup are discussed and the current status and timeline of the project are reviewed.

Keywords: Form factor, two photon exchange, positron, electron
PACS: 13.40.-f, 13.40.Gp, 13.88.+e, 14.20.Dh, 25.30.Bf

Recent determinations of the proton electric to magnetic elastic form factor ratio from polarization transfer measurements at Jefferson Lab [1, 2] have indicated an unexpected and dramatic discrepancy with the elastic form factor ratio obtained using the Rosenbluth separation technique in unpolarized cross section measurements [3]. The discrepancy has been explained as the effect of hard two-photon exchange beyond the usual one-photon exchange approximation in the calculation of the elastic electron-proton scattering cross section. [5, 6, 7, 8, 9, 10, 11]. The results from both methods are mostly based on the single photon exchange assumption including standard radiative corrections [4], which account for two-photon exchange only to the extent that one of the photons is soft.

Most of our understanding of the structure of the nucleon and of nuclei is based upon lepton scattering analyzed in terms of the single photon approximation, hence it is essential to precisely quantify the effect from the exchange of two and more photons. Contrary to standard radiative corrections of the lepton arm, which are calculable based on pure QED principles, calculations of the two-photon exchange amplitude are necessarily model-dependent in order to describe the intermediate off-shell hadronic state between the two photon vertices. Such calculations have been carried out e.g. in hadronic [7, 8], or partonic frameworks of generalized parton distributions and perturbative QCD [9, 10]. Also the importance of higher-order radiative effects, not necessarily through two or more photons, has been emphasized [12].

The general structure of two-photon exchange introduces three new complex amplitudes [11]. While the imaginary parts give rise to small single-spin asymmetries, which can be measured with transversely polarized electron beam, transversely polarized target, or induced transverse recoil polarization, the real part of the two-photon exchange amplitude can manifest itself indirectly in a modified, linear or non-linear dependence on the virtual photon polarization ε of the elastic cross section or polarization transfer. In particular, the real part of the two-photon amplitude matters for proton form factor extractions. While most calculations tend to remove the discrepancy in the form factor data, different model assumptions generally do not lead to a consistent determination of the individual amplitudes. For a satisfactory solution for the concurrent interpretation issue of lepton scattering experiments, it is essential to definitively verify the contribution of multiple photon exchange.

The only stringent observable that directly probes the size of the two-photon exchange amplitude is the difference of unpolarized elastic electron-proton and positron-proton cross sections. Experimentally, it is more advantageous to measure the ratio of such cross sections. To leading order, the interference of the single and two-photon amplitudes yield a contribution $\propto \alpha^3$ ($\alpha = 1/137$) to the cross section. This interference effect is odd in the number of lepton vertices and therefore changes its sign when switching between electron and positron probes. In the presence of two-photon exchange, the e^+/e^- cross section ratio will deviate from unity. This effect is expected to be angular dependent, to increase with the scattering angle, likewise to decrease with ε, and to increase with Q^2, in order to explain the form factor discrepancy. The effect disappears for $\varepsilon \to 1$. Existing measurements of the ratio were carried out in the 1960's without showing significant evidence for an effect [13], as displayed in Fig. 2. Most of these data were however measured either at low Q^2 or at large ε, where the effect is also expected to be small.

The OLYMPUS experiment has been proposed at the storage DORIS at DESY [14]. It aims to precisely measure the ratio of elastic e^+p and e^-p scattering cross sections to better than 1% total error in a kinematic region where the ratio

12th International Conference on Meson-Nucleon Physics and the Structure of the Nucleon (MENU 2010)
AIP Conf. Proc. 1374, 527-530 (2011); doi: 10.1063/1.3647196

is expected to deviate from 1 on the order 5-10%, if two-photon exchange is indeed responsible for the form factor discrepancy between Rosenbluth and recoil polarization measurements. OLYMPUS has been designed to minimize systematic and statistical uncertainties

- by running the experiment at the DORIS storage ring at DESY, Hamburg, Germany, which provides both electron and positron beams with rapid switching between both running modes, each with currents of > 100 mA,
- by scattering leptons from an internal, unpolarized, isotopically pure, essentially background-free hydrogen gas target of $3 \cdot 10^{15}$ atoms/s, resulting in a luminosity of $2 \cdot 10^{33}/(\text{cm}^2\text{s})$,
- by deploying the available symmetric, large acceptance detector BLAST from the MIT-Bates Linear Accelerator Center [15, 16, 17], which features a toroidal magnetic field allowing to rapidly switch between both polarities and excellent tracking capability over a wide range of scattering angles of $\approx 20^\circ - 80^\circ$ and $\pm 15^\circ$ out of plane,
- and by redundantly measuring the relative luminosity of OLYMPUS between electron and positron running and alternating toroidal magnet polarity based on small-angle elastic *ep* and symmetric Moller/Bhabha scattering.

OLYMPUS has been proposed at DESY in summer 2008 and was conditionally approved by the DESY Physics Research Committee (PRC) in October 2008. After a technical review in September 2009, main funding from the US agencies DOE and NSF has been secured in fall 2009, followed by the official approval of OLYMPUS by DESY in January 2010. Preparations of the experiment were started immediately, with the cleanup of the experimental area at DORIS and the transfer of the BLAST detector components to DESY completed in summer 2010. The detector has since been reassembled and will be finally mounted into the DORIS ring by mid 2011. Assembly and commissioning in park position can be carried out simultaneously with regular synchrotron operations of DORIS. The internal target has been newly designed and constructed at MIT using a customized target cell from INFN Ferrara. The new target will be installed early in 2011 and commissioned during DORIS synchrotron operations. Luminosity monitoring systems are under construction at Hampton University (GEM detector telescopes), Petersburg Nuclear Physics Institute (multi-wire proportional chambers, MWPC), and at Mainz University (symmetric Moller/Bhabha counter arrays). A total of three months of production running at a beam energy of 2.0 GeV and a luminosity of $L = 2 \cdot 10^{33}$ /(cm^{2}s) (500 h at 100 mA in each beam species) is required to provide the statistical accuracy of $< 1\%$ up to $Q^2 \leq 2.2$ (GeV/c)^2. It is noted that the region near $Q^2 = 2.5$ (GeV/c)^2 is of particular interest, where the two-photon amplitude has recently been experimentally overdetermined [2, 6]. Commissioning of the experiment at the standard operating energy of 4.5 GeV will provide an additional precise measurement of the cross section ratio at a large value of $\varepsilon \approx 0.9$ at a similar value of Q^2 in a fraction of the time, which will allow to infer a Rosenbluth-like separation of the e^+p cross section by assuming the known e^-p cross section. In the present scenario, production running of OLYMPUS will be completed

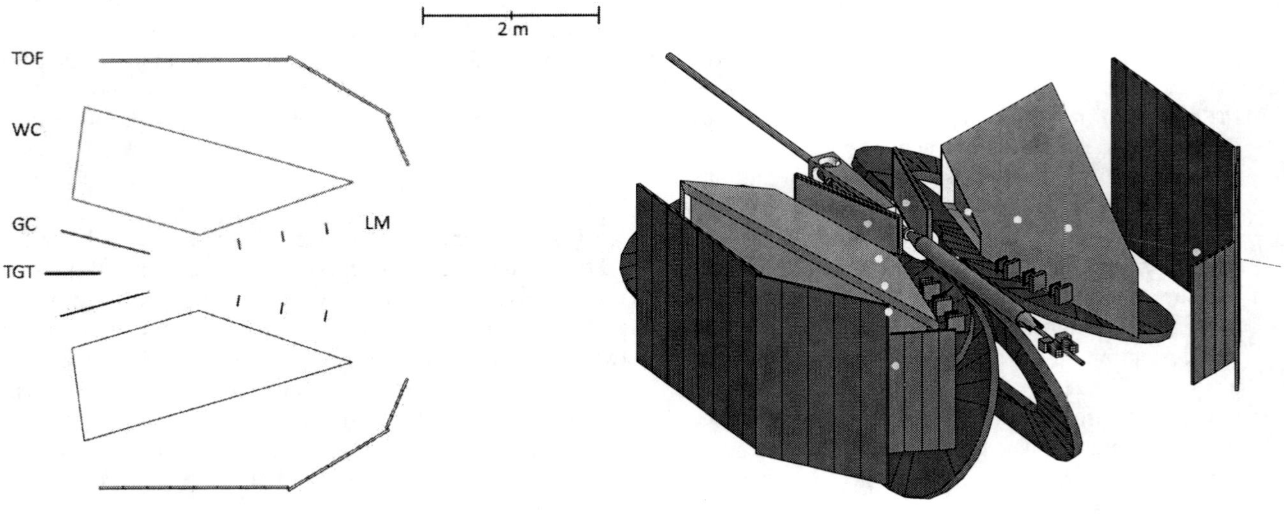

FIGURE 1. Left: Schematic layout of OLYMPUS in top view, with the beam coming from the left passing through the extended internal target. The displayed components are Time-of-Flight Scintillators (TOF), Wire Chambers (WC), Gas Electron Multiplier Chambers (GC), Luminosity Monitors (LM) and Target (TGT). In addition, there will be Moller/Bhabha monitors near the beamline for redundancy in luminosity monitoring. Right: Three-dimensional view of the OLYMPUS detector in Geant4.

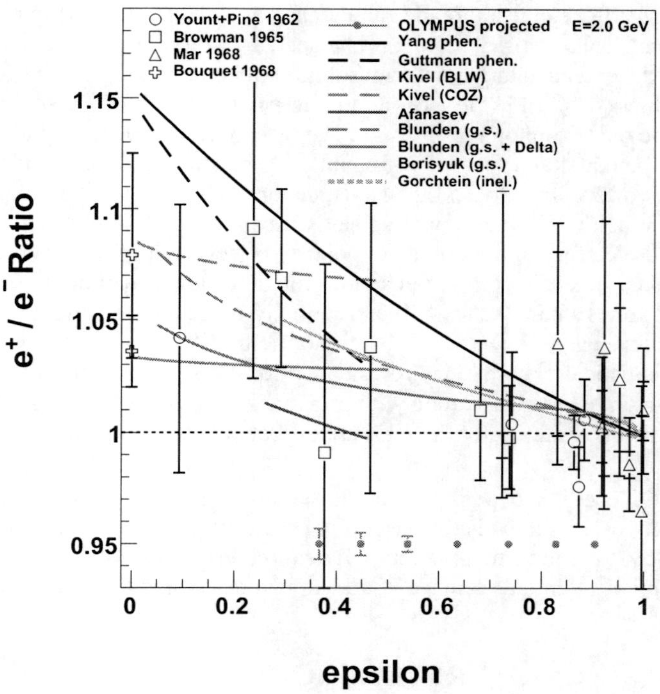

FIGURE 2. Projected precision of OLYMPUS for the e^+p to e^-p cross section ratio versus ε along with previous data [13] and various theoretical predictions [5, 6, 7, 8, 9, 10, 11], which have been evaluated for a constant beam energy of E = 2 GeV.

by end of 2012. It should be noted that besides OLYMPUS, there are two proposals for experiments to measure the same cross section ratio, yet at slightly different kinematics (generally lower in Q^2) with different techniques and systematics, at the VEPP-3 storage ring in Novosibirsk [18] (data taking started in fall 2009) and at Jefferson Lab [19] (scheduled to run in 2011).

The schematic layout of OLYMPUS is shown in Figure 1. Of the original BLAST setup [15], the toroidal magnet, the wire chambers (WC) and the time-of-flight scintillators (TOF) will be used. Both tracks of the scattered lepton and recoiling proton will be measured in coincidence by the BLAST detector. The left-right symmetry of the detector with respect to the beam axis effectively doubles the acceptance for coincident lepton-proton events, i.e. two redundant experiments are carried out simultaneously.

The BLAST wire chambers embedded in the horizontal octants of the toroid provide three tracking points of the curved track in the region of maximum magnetic field. The chambers have been restrung with new wires after arrival at DESY in summer 2010. The tracking capability will be enhanced with an additional tracking point close to the target where only little magnetic field is present. One GEM (Gas Electron Multiplier) chamber (GC) will be built in each sector for this purpose. The additional near-target tracking point will decouple the determination of scattering angle and vertex location from the momentum reconstruction and will result in improved track resolution, more powerful background suppression and a simplified analysis scheme. The internal target has been newly built, with a storage tube of ≈ 60 cm length and ellipical cross section, optimized for DORIS beam parameters. Also, a new vacuum scattering chamber has been designed and built allowing access to smaller scattering angles than in the previous geometry of BLAST. Much of the BLAST trigger and data acquisition system will be replaced as well.

In order to control the systematic uncertainties to the percent level, the relative luminosity of the experiment in a cycle of four combinations of beam species and magnet polarity will be measured precisely based on small-angle elastic scattering. The lepton is detected at a forward angle of $\approx 12^\circ$ in coincidence with the recoil proton accepted by the BLAST detector. For the lepton arm, a GEM tracking telescope based on three tracking planes is constructed for each sector, to be embedded into the forward cone between the coils of the horizontal octant and the downstream end of the

wire chamber (see Fig. 1). Each tracking plane is built as a 10×10 cm^2 triple-GEM detector with stereo-strip readout, following a design similar to that of a prototype GEM detector recently realized at MIT [20]. The GEM elements will be interleaved with additional MWPC elements to add redundancy, which can be used to cross-calibrate efficiencies or to be operated as an independent luminosity monitor. The spatial resolution of better than 100 μm will allow to reconstruct the lepton vertex at the extended target with a resolution of better than 1 cm, limited by multiple scattering. The kinematics correspond to low $Q^2 \approx 0.17$ (GeV/c)^2 and large $\varepsilon \approx 0.98$ at 2.0 GeV beam energy, where the cross section is well known and where the two-photon exchange effect is expected to be negligible. The e^+p and e^-p cross sections are hence expected to be equal and the ratio of count rates becomes a monitor of the respective luminosity ratio. With the given luminosity of $2 \cdot 10^{33}$ / (cm^{2}s) the statistical error of this monitor will be better than 1% in five hours at 4.5 GeV, with at least a factor 20 higher counting statistics at 2.0 GeV. It is anticipated that the beam will be switched between positrons and electrons and the magnet polarity be reversed of the order once per day. By cycling through four states, the detection efficiencies for lepton and proton detection cancel out to first order. This way, about 100 random cycles through the four running states will substantially reduce systematic uncertainties associated with slowly varying control parameters such as detector efficiencies or field drifts.
Figure 2 shows the projected statistical uncertainties for the e^+p to e^-p cross section ratio at a beam energy of 2.0 GeV as a function of virtual photon polarization along with previous data [13] and various theoretical expectations [5, 6, 7, 8, 9, 10, 11]. The systematic uncertainties of the ratio are expected to be less than 1%.

To summarize, the proposed OLYMPUS experiment at the lepton storage ring DORIS at DESY will provide a definitive determination of the two-photon exchange effect by precisely measuring the e^+p / e^-p unpolarized cross section ratio up to $Q^2 = 2.2$ (GeV/c)^2 and virtual photon polarization down to $\varepsilon = 0.37$.
This work has been supported by the National Science Foundation under grants PHY-0855473 and PHY-0959521.

REFERENCES

1. A.J.R. Puckett *et al.*, Phys. Rev. Lett. **104**, 242301 (2010); V. Punjabi *et al.*, Phys. Rev. C **71**, 055202 (2005); Erratum-ibid. Phys. Rev. C **71**, 069902(E) (2005) superseding M. Jones *et al.*, Phys. Rev. Lett. **84**, 1398 (2000); O. Gayou *et al.*, Phys. Rev. Lett. **88**, 092301 (2002); O. Gayou *et al.*, Phys. Rev. C **64**, 038202 (2001).
2. M. Meziane *et al.*, arXiv:1012.0339v1 [nucl-ex].
3. I.A. Qattan *et al.*, Phys. Rev. Lett. **94**, 142301 (2005); M.E. Christy *et al.*, Phys. Rev. C **70**, 015206 (2004); R.C. Walker *et al.*, Phys. Rev. D **49**, 5671 (1994); L. Andivahis *et al.*, Phys. Rev. D **50**, 5491 (1994); G.G. Simon *et al.*, Nucl. Phys. **A333**, 381 (1980); F. Borkowski *et al.*, Nucl. Phys. **A222**, 269 (1974); Nucl. Phys. **B93**, 461 (1975); J.J. Murphy, Y.M. Shin, and D.M. Skopik, Phys. Rev. C **9**, 2125 (1974); W. Bartel *et al.*, Nucl. Phys. **B58**, 429 (1973); C. Berger *et al.*, Phys. Lett. **B35**, 87 (1971); J. Litt *et al.*, Phys. Lett. **B31**, 40 (1970); T. Janssens *et al.*, Phys. Rev. **142**, 922 (1966).
4. Y.S. Tsai, Phys. Rev. **122**, 1898 (1961); L.W. Mo, Y.S. Tsai, Rev. Mod. Phys. **41**, 205 (1969); Y.S. Tsai, SLAC report, no. SLAC-PUB-848, 1971 (unpublished); L.C. Maximon, J.A. Tjon, Phys. Rev. C **62**, 054320 (2000).
5. Y.C. Chen, C.W. Kao, and S.N. Yang, Phys. Lett. **B652**, 269 (2007).
6. J. Guttmann, N. Kivel, M. Meziane, and M. Vanderhaeghen, arXiv:1012.0564v1 [hep-ph].
7. P.G. Blunden, W. Melnitchouk, and J.A. Tjon, Phys. Rev. Lett. **91**, 142304 (2003); Phys. Rev. C **72**, 034612 (2005); S. Kondratyuk, P. Blunden, W. Melnitchouk, and T.A. Tjon, Phys. Rev. Lett. **95**, 172503 (2005).
8. D. Borisyuk and A. Kobushkin, Phys. Rev. C **78**, 025208 (2008); M. Gorchtein, Phys. Lett. **B644**, 322 (2007).
9. Y.C. Chen, A.V. Afanasev, S.J. Brodsky, C.E. Carlson and M. Vanderhaeghen, Phys. Rev. Lett. **93**, 122301 (2004); A.V. Afanasev, S.J. Brodsky, C.E. Carlson, Y.C. Chen, and M. Vanderhaeghen, Phys. Rev. D **72**, 013008 (2005).
10. N. Kivel and M. Vanderhaeghen, Phys. Rev. Lett. **103**, 092004 (2009).
11. P.A.M. Guichon and M. Vanderhaeghen, Phys. Rev. Lett. **91**, 142303 (2003); M.P. Rekalo and E. Tomasi-Gustafsson, Eur. Phys. J. A **22**, 331 (2004).
12. Yu.M. Bystritskiy, E.A. Kuraev, E. Tomasi-Gustafsson, Phys. Rev. C **75**, 015207 (2007).
13. D. Yount and J. Pine, Phys. Rev. **128**, 1842 (1962); A. Browman, F. Liu, and C. Schaerf, Phys. Rev. **139**, B1079 (1965); J. Mar *et al.*, Phys. Rev. Lett. **21**, 482 (1968); B. Bouquet *et al.*, Phys. Lett. **B26**, 178 (1968).
14. *A Proposal to Definitively Determine the Contribution of Multiple Photon Exchange in Elastic Lepton-Nucleon Scattering* (2008); *Technical Design Report for the OLYMPUS Experiment* (2009); `http://web.mit.edu/OLYMPUS`.
15. D. Hasell *et al.*, Nucl. Instr. and Methods in Physics Research **A603**, 247 (2009).
16. C.B. Crawford *et al.*, Phys. Rev. Lett. **98**, 052301 (2007).
17. E. Geis *et al.*, Phys. Rev. Lett. **101**, 042501 (2008).
18. J. Arrington *et al.*, *Two photon exchange of electrons/positrons on the proton*, Proposal at VEPP-3, nucl-ex/0408020 (2004).
19. J. Arrington *et al.*, *Beyond the Born Approximation: A Precise Comparison of e^+p and e^-p Elastic Scattering in CLAS*, PR04-116, PAC26 (2004).
20. F. Simon *et al.*, Nucl. Instr. and Methods in Physics Research **A598**, 432 (2009).

Session P: Hard Exclusive Reactions

Hard exclusive neutral pion production at Jefferson Lab Hall A

E. Fuchey

Clermont Université, Université Blaise Pascal, CNRS/IN2P3, Laboratoire de Physique Corpusculaire, 63000 Clermont-Ferrand, France

Abstract. We present measurements of the $ep \to ep\pi^0$ cross section extracted at two values of four-momentum transfer $Q^2 = 1.9$ GeV^2 and $Q^2 = 2.3$ GeV^2 at Jefferson Lab Hall A. The kinematic range allows to study the evolution of the cross section as a function of Q^2 and W. Results will be confronted with Regge inspired calculations and GPD predictions. An intepretation of our data within the framework of semi-inclusive deep inelastic scattering has also been attempted.

Keywords: Deeply virtual pion production; Regge; Generalized parton distributions; Semi-inclusive DIS.
PACS: 13.60.Hb, 13.60.Le, 13.87.Fh, 14.20.Dh, 25.30.Rw

The past decade has shown a strong evolution of the study of hadron structure through exclusive processes (such as Deeply Virtual Compton Scattering (DVCS) and Deeply Virtual Meson Production (DVMP)), allowing access to the three-dimensional structure of hadrons. This document focuses on deeply virtual neutral pion production.

We present measurements of the differential cross section for the forward exclusive electroproduction reaction $ep \to ep\pi^0$. Results will be presented for four kinematics, two of them defined by the same value of $x_{Bj} = 0.36$ and two values of Q^2, 1.9 GeV^2 (Kin2) and 2.3 GeV^2 (Kin3), and the two other defined by the same value of $Q^2 = 2.1$ GeV^2 and two values of x_{Bj}, 0.33 (KinX3) and 0.4 (KinX2). The cross sections will be compared to different models that are available to describe π^0 electroproduction, including the Regge Model and the Generalized Parton Distribution (GPD) framework.

Forward photo-production at asymptotically high energies can be described by Regge theory, which exploits the analytic properties of the scattering amplitude in the limit $t/s \to 0$ [1]. Recent computations with Regge-inspired models exist for our kinematics. Among these, there is the t-channel meson exchange (TME) model by J.M. Laget *et al.* [2, 3]. This model is able to reproduce π^0 photoproduction data, low Q^2 π^0 electroproduction data and π^0 beam spin asymmetries for $Q^2 > 1$ GeV^2 [4], as well as the longitudinal charged pion electroproduction cross section for $Q^2 > 1$ GeV^2 [5].

In the Bjorken limit $Q^2 \to \infty$, and $t/Q^2 \ll 1$ at fixed x_{Bj}, the scattering amplitude is dominated by the leading twist amplitude of GPDs and the pion Distribution Amplitude (DA) [6, 7, 8]. The GPDs are light-cone matrix elements of non-local bilinear quark and gluon operators [9, 10, 11], unifying the elastic electro-weak form factors with the forward parton distributions of deep inelastic lepton scattering. With the definitions of [6, 7, 8], the cross sections are predicted to scale as $\sigma_L \sim Q^{-6}$ and $\sigma_T \sim Q^{-8}$ in the GPD framework.

An interpretation of exclusive data with semi-inclusive mechanisms also exists to explain transverse cross sections of hard exclusive charged pion electroproduction [12].

The present data were acquired as part of Jefferson Lab Hall A experiment E00-110 [13]. Additional details about the experimental configuration, calibrations, and analysis can be found in [14]. This document reports on the analysis of the triple coincidence $H(e,e'\gamma\gamma)X$ events. A 5.75 GeV electron beam was incident on a liquid hydrogen target. Electrons were detected in a High Resolution Spectrometer (HRS), and the photons in an 132 element PbF_2 calorimeter. The validation threshold for the data acquisition trigger was set to about 1 GeV for each photon cluster. Thanks to the resolution of the spectrometer and the calorimeter, one can use the missing-mass squared to ensure exclusivity. The pion selection relies on the invariant mass of the two photons.

In order to extract the differential cross section, it is advantageous to incorporate all model-independent kinematic dependences of the differential cross section into the experimental simulation. To this end, we express the differential cross section in terms of structure functions as described in the paper of Drechsel and Tiator [15]. We define the differential phase-space element $d^5\Phi = dQ^2 dx_{Bj} d\phi_e d[t_{min} - t] d\phi_\pi$ and the equivalent real photon energy in the CM frame $k_\gamma^{CM} = (W^2 - M_p^2)/2W$.

12th International Conference on Meson-Nucleon Physics and the Structure of the Nucleon (MENU 2010)
AIP Conf. Proc. 1374, 533-536 (2011); doi: 10.1063/1.3647197

To lowest order in the fine structure constant α, the differential cross section for an electron of helicity h is:

$$\frac{d^5\sigma(h)}{d^5\Phi} = \frac{\alpha}{2\pi^2}\frac{k'}{k}\frac{k_\gamma}{Q^2}\frac{1}{1-\varepsilon}\frac{d^2\sigma_v(h)}{dt d\phi_\pi}, \tag{1}$$

with $k_\gamma = (W^2 - M_p^2)/2M_p$, and $\varepsilon = 1/[1+2(q^2/Q^2)\tan^2\theta_e/2]$ the degree of linear polarization of the virtual photons. The virtual photo-absorption cross section is expanded, with $\varepsilon_L/\varepsilon = 4M_p^2 x_{Bj}^2/Q^2$, as

$$\begin{aligned}\frac{d^2\sigma_v(h)}{dt d\phi_\pi} &= \frac{1}{2\pi}\left\{\frac{d\sigma_T}{dt} + \varepsilon_L\frac{d\sigma_L}{dt} + \sqrt{2\varepsilon_L(1+\varepsilon)}\frac{d\sigma_{TL}}{dt}\cos\phi_\pi + \varepsilon\frac{d\sigma_{TT}}{dt}\cos 2\phi_\pi + h\sqrt{2\varepsilon_L(1-\varepsilon)}\frac{d\sigma_{TL'}}{dt}\sin\phi_\pi\right\};\\ &= \frac{1}{2\pi}\frac{1}{2q^{CM}k_\gamma^{CM}}\left\{\frac{W_{xx}+W_{yy}}{2} + \varepsilon_L W_{zz} - \sqrt{2\varepsilon_L(1+\varepsilon)}\Re\mathrm{e}W_{xz} + \varepsilon\frac{W_{xx}-W_{yy}}{2} - h\sqrt{2\varepsilon_L(1-\varepsilon)}\Im\mathrm{m}W_{yz}\right\}\end{aligned} \tag{2}$$

in terms of the usual hadronic tensor $W^{\mu\nu}$. The interference terms σ_{TL} and $\sigma_{TL'}$ have a leading $\sin\theta_\pi^{CM}$ dependence, and the linear polarization interference term σ_{TT} has a leading $\sin^2\theta_\pi^{CM}$ dependence. For this reason, we define reduced structure functions r_Λ which are equal to the hadronic tensors divided by their leading order angular dependence.

We divide the acceptance into 24 equal bins in $\phi_\pi \in [0, 2\pi]$ and 8 bins in $t_{min} - t \in [0, 0.3]$ GeV^2 for both the helicity dependent and independent parts of the cross section. We compare our experimental yields N_d in these bins, noted j_d, with yields obtained from a full simulation of the detectors acceptance and response, N_s which depend on the r_Λ values averaged over the bin. The values of the reduced structure functions r_Λ are those which minimize (σ_d the statistical error on the bin j_d)

$$\chi^2 = \sum_{j_d}\frac{(N_d - N_s(r_\Lambda))^2}{\sigma_d^2}. \tag{3}$$

The results will be presented as the four separated cross sections following the usual decomposition available in Eq. (2).

The external radiative effects on the incident electron, and internal real radiative effects at the vertex are treated in the equivalent radiator approximation [16, 17]. Pre-radiation is modeled by generating an event-by-event energy loss ΔE_{in} of the incident electron (E_0) following a distribution ($b \simeq 4/3$):

$$I_{in}(E_0, \Delta E_{in}, t_{in}) = \frac{bt_{in} + \delta_S/2}{\Delta E_{in}}\left[\frac{\Delta E_{in}}{E_0}\right]^{bt_{in}+\delta_S/2} \tag{4}$$

with

$$\delta_S = \frac{2\alpha}{\pi}\left[\ln\frac{Q^2}{m_e} - 1\right] \tag{5}$$

where t_{in} is the event-by-event target thickness (in radiation lengths) traversed by the electron before the scattering vertex. The Schwinger term δ_S models the internal pre-radiation. The scattered energy at the vertex is $E'_v = E_0 - \Delta E_{in} - Q^2/(2M_p x_{Bj})$. A similar distribution of the post radiated energy ΔE_{out} is used to model internal post-radiation.

External post-radiation by the scattering electron is modeled with the GEANT3 simulation. Internal virtual radiative correction factors, including vertex renormalization, vacuum polarization and the cut-off independent real radiation effects are calculated according to [18].

Two sources of systematic errors lie on the missing mass squared cut and in calorimeter threshold. The influence of the missing mass squared cut comes from the competition of the inclusive channels with the exclusive $\mathrm{H}(e, e'\pi^0)p$ reaction which cannot be reproduced in the simulation. The cross section stability relative to this cut indicates that these inclusive channels are not contributing. The systematic error from the missing mass cut is 1 % for ($x_{Bj} = 0.36$, $Q^2 = 2.3$ GeV^2) and ($x_{Bj} = 0.33$, $Q^2 = 2.1$ GeV^2), and 3 % for ($x_{Bj} = 0.36$, $Q^2 = 1.9$ GeV^2) and ($x_{Bj} = 0.4$, $Q^2 = 2.1$ GeV^2). The influence of the calorimeter threshold comes from the competition of the hardware threshold, which determines the event trigger, with the software analysis threshold. The cross section stability relative to the threshold indicates that the software threshold is set safely above the hardware threshold. The systematic error from the calorimeter threshold is 1 % for all kinematics.

The other sources of kinematic errors have experimental or instrumental origins. Among these sources of systematic errors, the most significative contributions come from the HRS acceptence (2.2 %), and the radiative corrections (1.5 %). For the helicity dependent cross section, the contribution from the beam polarization (2.0 %) has also to be

TABLE 1. Systematic errors for both helicity dependent and independent cross sections for all kinematics.

	Kin3/KinX3	Kin2/KinX2
helicity-independent	3.1 %	4.2 %
helicity-dependent	3.7 %	4.6 %

mentioned. All sources of systematic errors are independent and are added quadratically. The total quadratic systematic error is given for each kinematics in Tab. 1.

The exclusive π^0 electroproduction cross sections were extracted for Kin2 and Kin3, and for KinX2 and KinX3. Our statistics allowed us to achieve, for the ϕ_π-independent cross section, a statistical precision of 2.5% for $x_{Bj} = 0.36$ (Kin2, Kin3), and of 5% for $Q^2 = 2.1$ GeV^2 (KinX2, KinX3).

Fig. 1 shows $\sigma_T + \varepsilon_L \sigma_L$ plotted as a function of $t_{min} - t$, for $x_{Bj} = 0.36$ on the left and for $Q^2 = 2.1$ GeV^2 on the right. It also shows the Q^2 and the W-dependences for the total cross section $\sigma_T + \varepsilon_L \sigma_L$, which behaves as $1/Q^{4.5}$ and

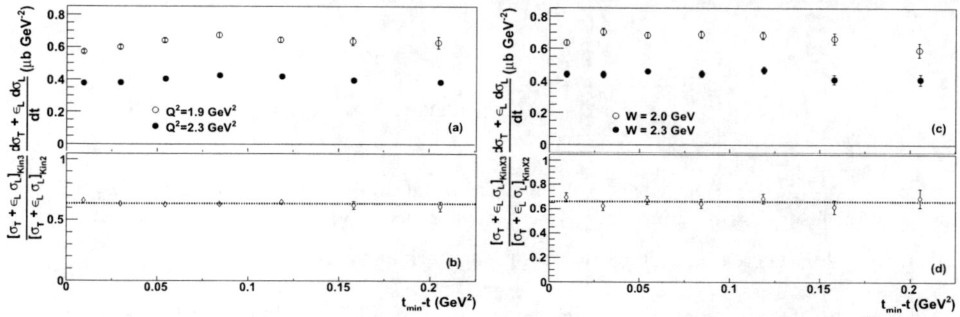

FIGURE 1. Separated H$(e,e'\pi^0)p$ cross section $\sigma_T + \varepsilon_L \sigma_L$ as a function of $t_{min} - t$: (a)for $x_{Bj} = 0.36$; (c) for $Q^2 = 2.1$ GeV^2. (b) and (d) show the respective ratios of $\sigma_T + \varepsilon_L \sigma_L$ for the considered kinematics as a function of $t_{min} - t$. The fits of these ratios (dashed line) indicate the Q^2-dependence (b) and the W-dependence (d) of the cross section.

$1/W^{3.5}$. The t-channel meson exchange model by J.M. Laget [3] gives a good agreement with $\sigma_T + \varepsilon_L \sigma_L$ and $\sigma_{TL'}$ for both Q^2 values at x_{Bj} fixed. However, it shows a disagreement in order of magnitude for σ_{TT} and disagreement in sign for σ_{TL} for both of those kinematics.

The cross section $\sigma_T + \varepsilon_L \sigma_L$ behaves roughly as $1/Q^{4.5}$. Assuming $\sigma_L \gg \sigma_T$, σ_L would behave roughly as $1/Q^3$, which is far from the QCD leading twist predicted by GPD formalism.

The data we extracted yield two conclusions with regard to the available models.

- The model by J.M. Laget is able to describe $\sigma_T + \varepsilon_L \sigma_L$ and $\sigma_{TL'}$, but neither σ_{TL} nor σ_{TT}.
- The Q^2-dependence of the cross sections (Fig. 1) demonstrates that we are far from the QCD leading twist prediction of $d\sigma_L/dt$, which behaves as $1/Q^6$. On the other hand, it is similar to the Q^2-dependence of the transverse cross section for charged pion electroproduction published by Hall C [5].

Moreover, the π^0 has no charge, no spin, so a direct coupling with a virtual photon is suppressed, which removes the pion-pole contribution to the longitudinal cross section. This suggests that the transverse $ep \to ep\pi^0$ cross section is likely to dominate, and transverse $ep \to en\pi^+$ cross sections have already been described by quark fragmentation mechanisms usually used to describe semi inclusive processes.

T. Horn *et al.* measured the exclusive π^+ electroproduction cross section at $Q^2 = 1.60$ and 2.45 GeV^2, with σ_T and σ_L separation [5]. The t-channel meson exchange model by J.M. Laget reproduces the σ_L component. However, the σ_T component does not follow the TME model prediction. M. Kaskulov *et al.* performed PYTHIA-JETSET calculations using the Lund model applied to π^+ transverse cross sections at Hall C kinematics [12]. These calculations applied to Hall C π^+ transverse cross sections are in excellent agreement with the data. This gives evidence that the π^+ transverse cross section at $Q^2 > 1$ GeV^2 above the resonance region is described by a partonic process. This suggests that the present π^0 data could similarly be described by incoherent scattering on the partonic structure of the nucleon target.

For these reasons, we consider our data within the context of Semi-Inclusive Deep Inelastic Scattering (SIDIS). We attempted to fit our data with a SIDIS formalism written by Anselmino *et al.* [19]. The equation (38) of [19] gives the cross section at $\mathscr{O}(k_\perp/\sqrt{Q^2})$ for semi-inclusive production for any hadron. It also provides the order of magnitude for the $\cos\phi_\pi$ contribution of a SIDIS process (Cahn effect). The idea is to adjust the ratio of the $\cos\phi_\pi$ contribution (identified to σ_{TL}) over the constant term (identified to $\sigma_T + \varepsilon_L\sigma_L$) by adjusting only the parameter $\langle p_\perp^2\rangle/\langle k_\perp^2\rangle$:

$$\frac{\sqrt{2\varepsilon_L(1+\varepsilon)}\sigma_{TL}}{\sigma_T+\varepsilon_L\sigma_L} = \frac{4(2-y)\sqrt{1-y}z_\pi p_{\pi\perp}}{\left(\frac{\langle p_\perp^2\rangle}{\langle k_\perp^2\rangle}+z_\pi^2\right)\sqrt{Q^2}(1+(1-y)^2)} \tag{6}$$

This adjustment is shown in Figure 2. Our exclusive data (where particle multiplicity is unity) is reproduced by this

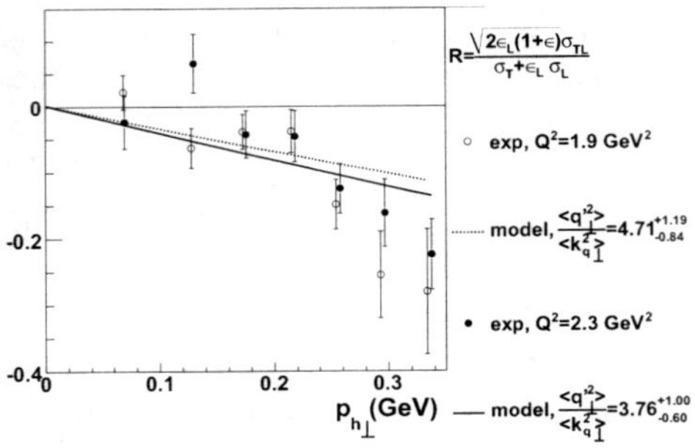

FIGURE 2. Ratio $\frac{\sqrt{2\varepsilon_L(1+\varepsilon)}\sigma_{TL}}{\sigma_T+\varepsilon_L\sigma_L}$ for Kin2 (open circles) and Kin3 (solid circles) plotted as a function of $p_{\pi\perp}$. Error bars represent statistical errors only. The fits to these quantities are represented by a dotted curve (solid curve) for Kin2 (Kin3).

model if $\langle p_\perp^2\rangle/\langle k_\perp^2\rangle \sim 4.0$, whereas at high particle multiplicity (inclusive data), $\langle p_\perp^2\rangle/\langle k_\perp^2\rangle \sim 0.8$.

We extracted the separated differential π^0 cross section at Jefferson Lab, Hall A, at four kinematic settings, with a 2.5 % statistical precision at x_{Bj} fixed, and with a 5 % statistical precision at Q^2 fixed. We studied the Q^2- and W-dependence of the hadronic tensor.

The model of J.M. Laget is able to describe the total cross section, but still needs improvement. There is a contradiction between our data and the leading twist QCD prediction for high Q^2 ($\sigma_T + \varepsilon_L\sigma_L \propto 1/Q^{4.5}$ instead of $1/Q^8$). The $\cos\phi_\pi$ cross section contribution is compatible in sign and shape with the Cahn effect of SIDIS.

To improve the understanding of our data, we plan to run another π^0 experiment in Fall 2010, at two beam energies, allowing us to disentangle σ_L from σ_T.

REFERENCES

1. G. R. Goldstein, and J. F. Owens, *Phys. Rev.* **D7**, 865–879 (1973).
2. J. M. Laget, *Phys. Rev.* **C76**, 052201 (2007).
3. J. M. Laget (2010), 1004.1949.
4. R. De Masi, et al., *Phys. Rev.* **C77**, 042201 (2008).
5. T. Horn, et al., *Phys. Rev.* **C78**, 058201 (2008).
6. J. C. Collins, L. Frankfurt, and M. Strikman, *Phys. Rev.* **D56**, 2982–3006 (1997).
7. K. Goeke, M. V. Polyakov, and M. Vanderhaeghen, *Prog. Part. Nucl. Phys.* **47**, 401–515 (2001).
8. M. Vanderhaeghen, P. A. M. Guichon, and M. Guidal, *Phys. Rev.* **D60**, 094017 (1999).
9. D. Mueller, D. Robaschik, B. Geyer, F. M. Dittes, and J. Horejsi, *Fortschr. Phys.* **42**, 101 (1994).
10. X.-D. Ji, *Phys. Rev. Lett.* **78**, 610–613 (1997).
11. A. V. Radyushkin, *Phys. Rev.* **D56**, 5524–5557 (1997).
12. M. M. Kaskulov, K. Gallmeister, and U. Mosel, *Phys. Rev.* **D78**, 114022 (2008).
13. P. Y. Bertin, C. E. Hyde, R. Ransome, and F. Sabatié (2000), jefferson Lab Proposal, E00-110.
14. C. Munoz Camacho, et al., *Phys. Rev. Lett.* **97**, 262002 (2006).
15. D. Drechsel, and L. Tiator, *J. Phys.* **G18**, 449–497 (1992).
16. L. W. Mo, and Y.-S. Tsai, *Rev. Mod. Phys.* **41**, 205–235 (1969).
17. Y.-S. Tsai, *Rev. Mod. Phys.* **46**, 815 (1974).
18. M. Vanderhaeghen, et al., *Phys. Rev.* **C62**, 025501 (2000).
19. M. Anselmino, et al., *Phys. Rev.* **D71**, 074006 (2005).

Exclusive electroproduction of the ρ^+ meson on the proton @ CLAS

A. Fradi

Univ Paris-Sud, Institut de Physique Nucléaire d'Orsay
91405 Orsay, France
*E-mail: fradi@ipno.in2p3.fr

Abstract. We present preliminary results of the exclusive electroproduction of ρ^+ on the proton at CLAS. We discuss the interpretation of the cross sections in terms of t-channel Reggeon exchanges and in terms of Generalized Parton Distributions (GPDs) formalism.

Keywords: Nucleon structure; Exclusive vector meson electroproduction; Generalized Parton Distribution
PACS: 25.30.Rw

INTRODUCTION

The exclusive electroproduction of mesons on the nucleon is an important tool to better understand the nucleon structure and, more generally, the transition between the low energy hadronic and high energy partonic domains of the Quantum Chromodynamics (QCD) theory. The CEBAF ("Continous Electron Beam Accelerator Facility") of the Jefferson Lab (JLab) at Newport News (USA) with its high intensity electron beam and its large acceptance spectrometer CLAS [1] (CEBAF Large Acceptance Spectrometer) offers a great opportunity to study the exclusive electroproduction of mesons, in particular those that have multi-particle decays.

This proceeding presents some first results for the exclusive electroproduction of the vector mesons ρ^+ on the proton at CLAS. These results come from the e1-dvcs experiment: the data were collected between March and May 2005 with a beam energy of 5.75 GeV and an integrated luminosity of ~ 40 fb^{-1}. It is the first ever measurement of cross section for the ρ^+ electroproduction channel. The presented results are still preliminary and more details can be found in ref. [2].

DATA ANALYSIS

The final state that is analyzed is $ep \to en\pi^+\pi^0$. The final state particles e, π^+ and π^0 were detected in CLAS and the exclusive process was identified by selecting a neutron with the missing mass technique. Figure 1 shows the acceptance-corrected invariant mass $M_{\pi^+\pi^0}$ distributions for each (Q^2, x_B) bin. One sees clearly the ρ^+ peaks around 775 MeV. They are, however, sitting on top of a non-negligeable non-resonant, two-pion background. We have used two contributions to fit these spectra: a skewed Breit-Wigner distribution to describe the resonant structure of the ρ^+ and a parametrization of the non-resonant two-pion continuum determined from simulations (mostly $ep \to en\pi^+\pi^0$ phase space). In order to extract the total cross section of the reaction $\gamma^* p \to n\rho^+$, we have calculated for each (Q^2, x_B) bin the area of the Breit-Wigner (green curve in Fig. 1), the distributions being already normalized and corrected for the acceptance. A separation of the longitudinal cross sections from the transverse ones was performed by analysing the pion decay angles of the ρ^+ and relying on the SCHC (s-channel helicity conservation), which was checked experimentally.

INTERPRETATION

We will compare in the following the extracted cross sections of the exclusive electroproduction of the ρ^+ with two theoretical approaches: on the one hand, the hadronic approach based on Regge theory and meson trajectory exchanges in the t-channel and on the other hand, the partonic approach based on the handbag diagram and GPDs. These two

12th International Conference on Meson-Nucleon Physics and the Structure of the Nucleon (MENU 2010)
AIP Conf. Proc. 1374, 537-541 (2011); doi: 10.1063/1.3647198

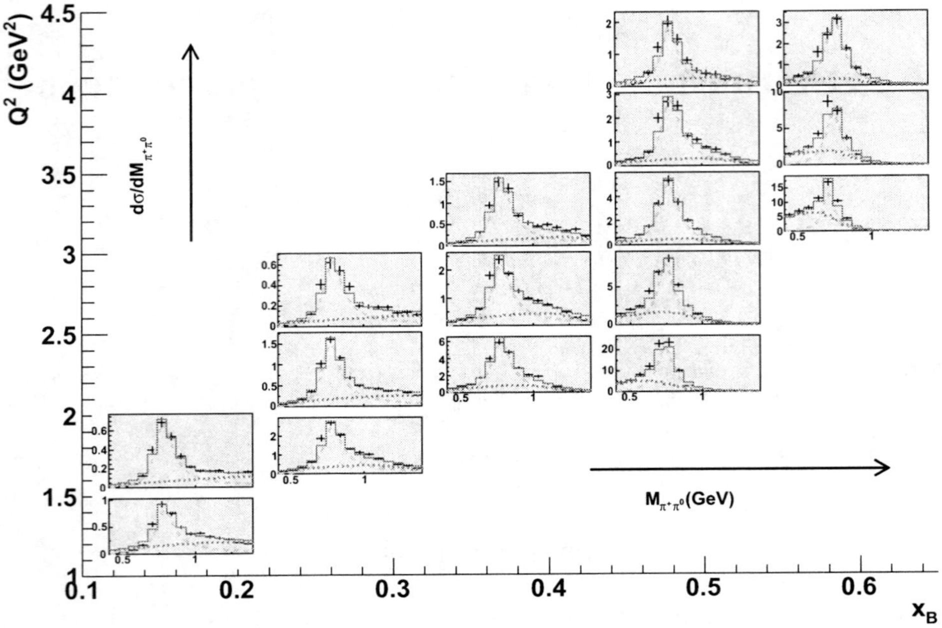

FIGURE 1. $M_{\pi^+\pi^0}$ acceptance-corrected distributions, showing fits for the background subtraction. In black: experimental data; in dashed green: Breit-Wigner of ρ^+; in dotted purple: $M_{\pi^+\pi^0}$ projection of the the non-resonant continuum $\gamma^* p \longrightarrow n\pi^+\pi^0$ reaction; in red: total fit result. In this proceeding, units are arbitrary for the vertical scale.

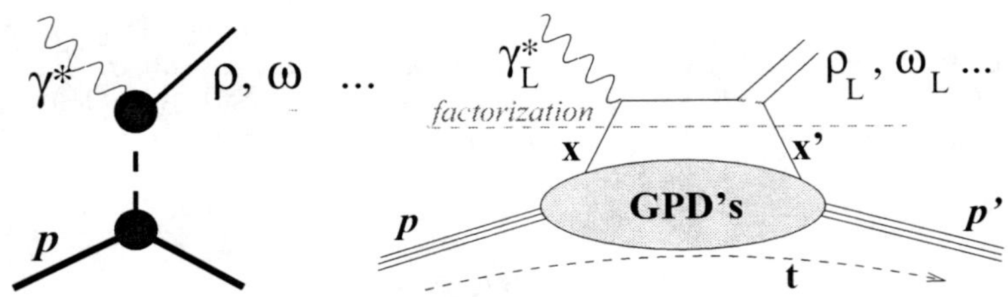

FIGURE 2. Shematic representations of the Reggeon t-channel exchange (left) and of the handbag diagram (right) for exclusive vector meson electroproduction.

approaches are illustrated in Fig. 2. This comparison will be useful in order to better understand the domain of validity of these two approaches and constrain their inputs.

The Regge "hadronic" approach

The Regge approach is most appropriate at low Q^2 above the resonance region and at forward angles (where the cross section is largest). It consists of exchanges of meson "trajectories" in the t-channel. In the following, we will use the "JML" acronym to refer to the specific model developed by J.-M. Laget [3].

For the ρ^+ channel, the exchange trajectories are those of the π^+ and of the ρ^+. The free parameters in this model are the coupling constants at the hadronic vertices (most of them being well constrained) and the mass scales of the electromagnetic form factors.

Figure 3 shows the total cross section $\sigma(\gamma^* p \longrightarrow n\rho^+)$ as a function of W for fixed Q^2 compared with the JML calculation shown with the red curve. The JML model can successfully reproduce the cross sections for almost all of our (Q^2, W) range. The decreasing cross section with W can be explained in the Regge theory by an intercept $(\alpha(0) = 0.5)$ of the ρ^+ trajectory which is dominant: $\sigma_{TOT} \propto s^{\alpha(0)-1}$.

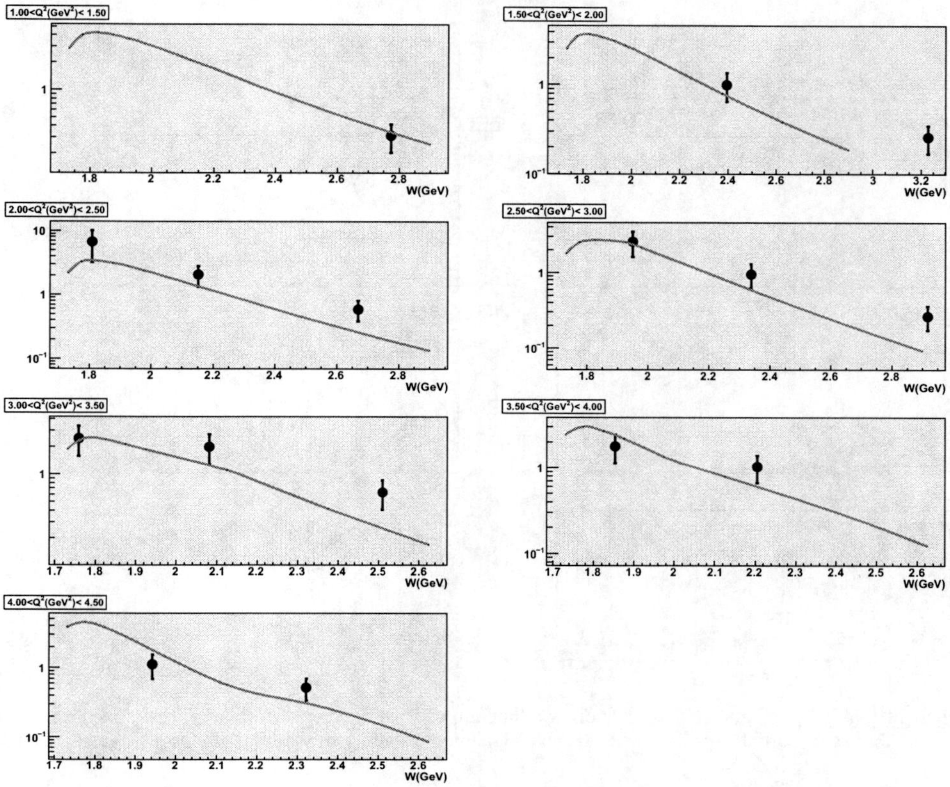

FIGURE 3. PRELIMINARY total cross section as a function of W at fixed Q^2, for the reaction $\gamma^* p \to n\rho^+$. The red line curves are the result of the JML model. Units are arbitrary on the y axis as the results are still preliminary.

The GPD "partonic" approach

The formalism of GPDs is valid in the so-called Bjorken regime, i.e. $Q^2, \nu \to \infty$ with $x_B = \frac{Q^2}{2M\nu}$ finite. It was proven [4] that the dominant process for exclusive meson electroproduction, in the Bjorken limit, is given by the so-called handbag diagram represented in Fig. 2. The handbag diagram is based on the notion of factorization in leading-order/leading-twist pQCD between a hard scattering process, exactly calculable in pQCD, and a nonperturbative nucleon structure part that is parametrized by the GPDs. For mesons, the factorization of the handbag diagram is only valid for the longitudinal part of the cross section. In the following we discuss the two particular GK [5] and VGG [6] GPD-based calculations that provide quantitative results for the longitudinal part of the exclusive meson cross sections.

Figure 4 shows longitudinal cross section of $\gamma^* p \to n\rho^+$ as a function of W and for fixed Q^2. The results of the calculations of the GK and VGG model are shown, respectively, with the red and the blue curves. We see that both models fail to reproduce the data. This discrepancy can reach an order of magnitude at the lowest W values. The trend of these particular GPD calculations is to decrease as W decreases, whereas the data increase. The same behavior was observed, in the low W region, for the exclusive electroproduction of the ρ^0 [7].

One can conclude that either the GPD formalism is not valid for our Q^2 coverage (due to important higher twist contributions for instance) or there is a fundamental contribution, which is missing in the way the GPDs are parametrized in the GK and VGG models.

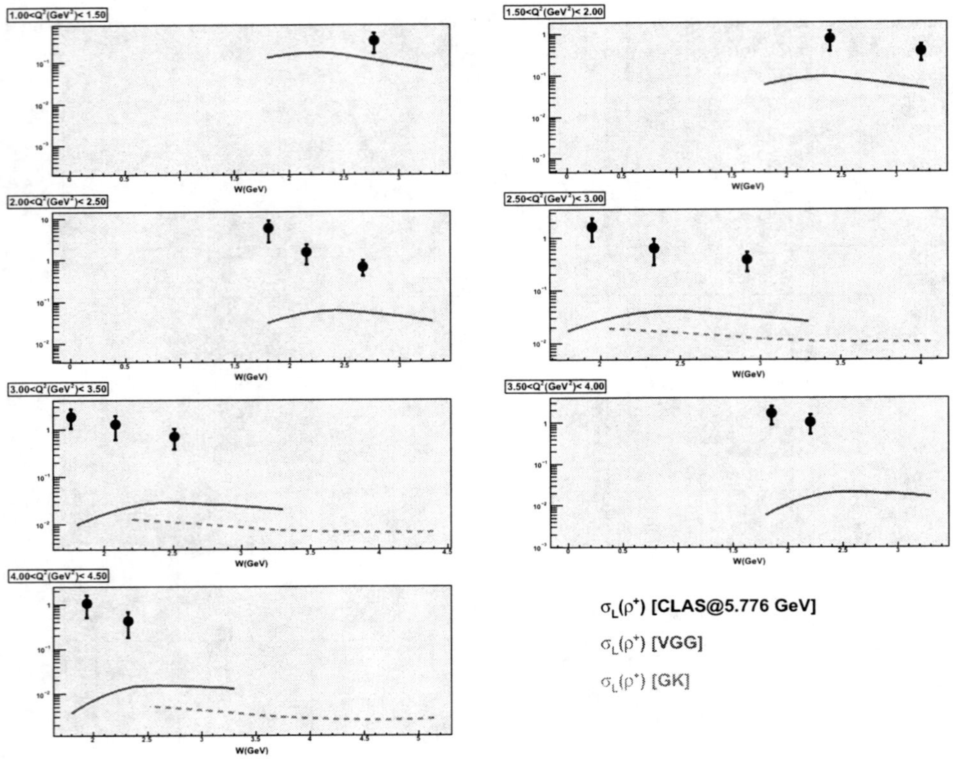

FIGURE 4. PRELIMINARY longitudinal cross section as a function of W at fixed Q^2, for the reaction $\gamma^* p \to n\rho^+$. The dashed red curve and the solid blue curve are the results of the GK and VGG models. Units are arbitrary on the y axis as the results are still preliminary.

Comparison of the t slope for the ρ^+, ρ^0, ω and ϕ channels

The CLAS collaboration has measured the largest ever set of data for the exclusive electroproduction on the proton of the vector mesons ρ^0 [7], ω [8] and ϕ [9].

Figure 5 shows the slope of the differential cross section $d\sigma/dt$ for the ρ^+, ρ^0, ω and ϕ channels as a function of W (in the top part) and as a function of Q^2 (in the bottom part). One can see the same trends of this slope, b, for all meson channels, which can be interpreted in simple and intuitive terms in the following way:

- b increases with W: the size of the nucleon increases as one probes the high W values (i.e. the sea quarks), which could mean that the sea quarks tend to extend to the periphery of the nucleon.
- b decreases with Q^2: as we go to large Q^2, the resolution of the probe increases and we tend to see smaller and smaller objects.

SUMMARY

We have presented (preliminary) first ever measurement of cross sections for the exclusive electroproduction of ρ^+ on the proton using the CLAS detector at JLab. These cross sections can be interpreted with two approaches:

- hadronic approach: the JML model describes well most of the features of the ρ^+ cross sections up to $Q^2 \sim 4.5$ GeV^2.
- partonic approach: GPD models fail to describe longitudinal ρ^+ cross sections especially for low W.

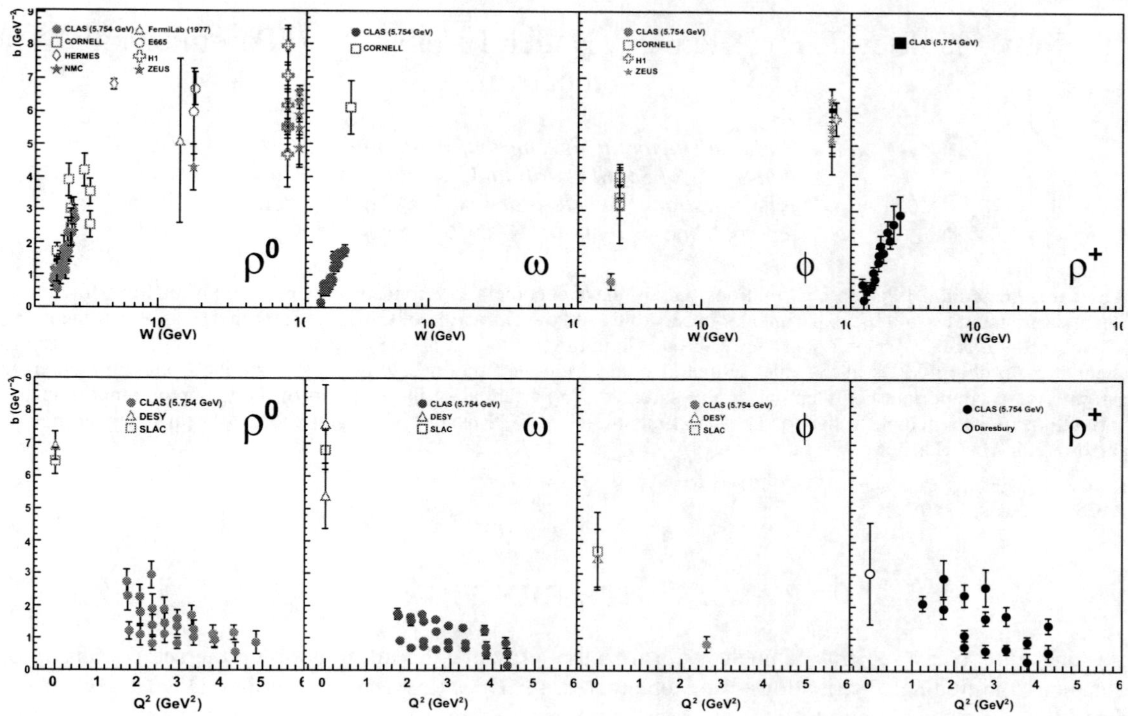

FIGURE 5. The slope b as a function of W (on the top) and as a function of Q^2 (on the bottom) for the ρ^0, ω, ϕ and ρ^+ channels.

We also found the same trends of the variation of the t slope as a function of W and as a function of Q^2 for all ρ^+, ρ^0, ω and ϕ channels.

REFERENCES

1. B.A. Mecking et al., Nucl. Instr. Meth. A **503**, (2003) 513 .
2. A. Fradi, thesis Univ. Paris-Sud at Orsay, 2009.
3. J.-M. Laget, Phys. Lett. B **489** (2000) 313.;
   F. Cano and J.-M. Laget, Phys. Rev. D **65** (2002) 074022. ;
   J. M. Laget, Phys. Rev. D **70** (2004) 054023.;
   F. Cano and J. M. Laget, Phys. Lett. B **551** (2003) 317.
4. J.C. Collins, L. Frankfurt and M. Strikman, Phys.Rev.D **56** (1997) 2982.
5. S.V. Goloskokov and P. Kroll, Eur. Phys. J.C **42** (2005) 281; Eur.Phys. J.C **50** (2007) 829.
6. M. Vanderhaeghen, P.A.M. Guichon, M. Guidal, Phys. Rev. D **60**, 094017 (1999).
7. S. Morrow et al., Eur. Phys. J. A **39**, (2009) 5.
8. L. Morand et al., Eur. Phys. J. A **24** (2005) 445.
9. J. P. Santoro et al., Phys. Rev. C **78** (2008) 025210.

Timelike Compton Scattering

T. Horn*, Y. Illieva†, F.J. Klein*, P. Nadel-Turonski**, R. Paremuzyan‡,** and S. Stepanyan**

*The Catholic University of America, Washington, DC 20064
†University of South Carolina, Columbia, SC 29208
**Thomas Jefferson National Accelerator Facility, Newport News, VA 23606
‡Yerevan Physics Institute, 375036 Yerevan, Armenia

Abstract. Generalized Parton Distributions (GPDs) have become a key concept in our studies of hadron structure in QCD. The measurement of suitable experimental observables and the extraction of GPDs from these data is one of the high priority 12 GeV programs at Jefferson Lab. Deeply Virtual Compton Scattering (DVCS) is generally thought of as the most promising channel for probing GPDs in the valence quark region. However, the inverse process, Timelike Compton Scattering (TCS) can provide an important complementary measurement, in particular of the real part of the Compton amplitude and power corrections at intermediate values of Q^2. The first studies of TCS using real tagged and quasi-real untagged photons were performed in Hall B at Jefferson Lab.

Keywords: Timelike Compton, DVCS, GPD
PACS: 12.38.Qk, 13.60Fz

INTRODUCTION

For a long time, measurements of nucleon form factors provided the information on the spatial density of partons, while their longitudinal distributions were obtained from Deep Inelastic Scattering (DIS). Only recently was it realized that these correspond to limiting cases of a more general way to characterize the structure of the nucleon. Indeed, Generalized Parton Distributions (GPDs) combine aspects of elastic nuclear form factors and the traditional Parton Distribution Functions (PDFs). Measuring the transverse spatial distributions in the nucleon has thus become an important goal. Detailed descriptions of the GPD formalism can be found, for instance, in Refs. [1, 2, 3]. Experimentally, this relies on QCD factorization, which allows the hard, short-distance interaction of the experimental probe with one parton to be unambiguously separated from the residual soft interaction of the struck parton with the rest of the hadron, which contains the long distance information about nucleon structure described by the GPDs.

TIMELIKE COMPTON SCATTERING AND NUCLEON STRUCTURE

The mapping of nucleon structure through the measurements of GPDs is undertaken by, for instance, Compton scattering. Efforts have largely been focused on Deeply Virtual Compton Scattering (DVCS), which is the simplest exclusive process giving access to GPDs. This process is also a key measurement in the 12 GeV Thomas Jefferson National Accelerator Facility (Jefferson Lab) program. It is, however, only a special case of the reaction $\gamma^* p \to \gamma^* p$, where the final state photon is real. Timelike Compton Scattering (TCS), which is the "inverse" reaction where the incoming photon is real and the final state has a timelike virtuality providing the hard scale in the process, offers a complementary way of probing GPDs. Indeed, it can be seen as a first step towards the most general Compton measurement of GPDs. Known as double DVCS (with virtual photons in both the initial and final states), this would allow to probe any kinematic point with a longitudinal momentum fraction x_B and the skewness, ξ, without the need for deconvolution, thus minimizing the model dependence in the determination of GPDs. However, due to the lower statistics and experimental challenges related to identical particles, such a measurement would only supplement DVCS and TCS results.

The main advantage of TCS is that the process provides a straightforward way to measure the real part of the Compton amplitude with small systematic uncertainties. The imaginary part of the Compton amplitude can relatively easily be studied through DVCS beam spin asymmetries, but access to the real part of the DVCS amplitude requires direct

12th International Conference on Meson-Nucleon Physics and the Structure of the Nucleon (MENU 2010)
AIP Conf. Proc. 1374, 542-545 (2011); doi: 10.1063/1.3647199

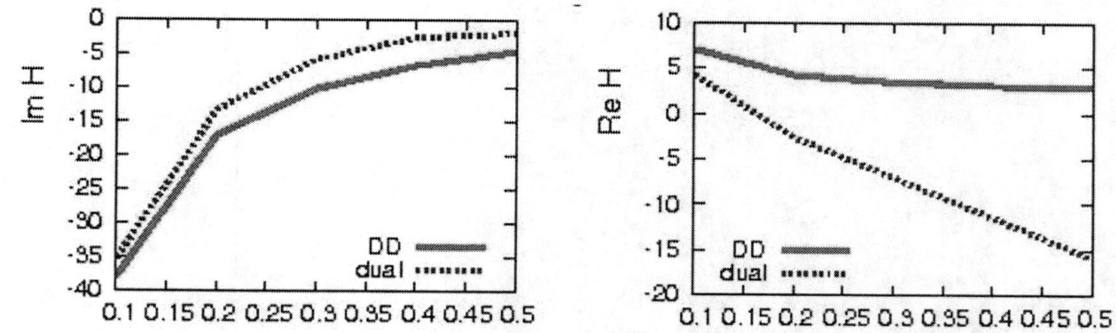

FIGURE 1. Two GPD model predictions for the imaginary (left) and the real (right) parts of the Compton amplitude as a function of the quantity τ, the equivalent of x_B in DVCS. Here, $\tau = Q'^2/s - M_p^2$, where Q'^2 is the virtuality of the final state photon measured as the invariant mass squared of the produced lepton pair, s is the center of mass energy, and M_p is the proton mass. The hard scale for this process is given by $Q'^2 = M^2_{e^+e^-}$. The blue arrow indicates the kinematic region accessible with the CLAS detector in Hall B at Jefferson Lab.

cross section measurements with a good understanding of, for instance, the Bethe-Heitler (BH) and π^0 backgrounds. Figure 1 shows recent GPD model calculations [5, 6] suggesting a considerable sensitivity to the real part, which could be probed through TCS asymmetries. TCS and DVCS processes are complementary and together could also make an important contribution to the ongoing theoretical debate on the contribution of power corrections at intermediate Q^2 and their effect on GPD extractions. For instance, since the TCS and DVCS amplitudes are equivalent only to leading order, a combination of these measurements could reduce model uncertainties in GPD extractions at finite values of Q^2. However, due to the timelike virtuality of the final state, TCS comes with some limitations. For example, the TCS cross section is smaller than for DVCS, and in particular smaller than BH over the full kinematic range. TCS measurements thus focus on the interference term, which is enhanced by the BH. Furthermore, resonances in the timelike final state limit the range in Q'^2, the virtuality of the final state photon measured as the invariant mass squared of the produced lepton pair. In kinematics accessible at Jefferson Lab, these contributions fall between the ϕ and the J/ψ.

The theoretical methods to extract GPDs from Compton scattering data are complex. The most recent additions include partial wave expansions [7] and s-channel dispersion relations [8]. The extraction procedure requires knowledge of both the imaginary and real Compton amplitudes. The imaginary part probes the GPD along the line where ξ is equal to x_B. The real part probes at a fixed skewness all values of x_B. Though advanced methods are being developed, current GPD models show considerable discrepancies in the real part of the Compton amplitude, in particular at large values of x_B. To minimize the uncertainty in the extraction of GPDs it is thus essential to have reliable measurements on both the real and imaginary parts of the amplitude.

The process of exclusive photoproduction of a lepton pair can proceed through either TCS or BH. As mentioned above, the TCS cross section is smaller than BH in all accessible kinematics and cannot be measured directly. However, the interference with BH can be large, and is easy to isolate. Indeed, while the TCS and BH amplitudes are both even under the exchange of lepton momenta, the interference term is odd. Observables that change sign will thus project out only the interference term. An example of such an observable is the azimuthal angular distribution of the lepton pair, where the ϕ is the angle between the reaction and the lepton decay planes.

The structure of the interference term allows for accessing both the real and the imaginary part of the TCS amplitude through the polarization of the incoming photon [9]. If the photon beam is unpolarized, only the real part of the amplitude contributes. Circular photon polarization also gives access to the imaginary part. The helicity amplitudes enter the TCS-BH interference term with different trigonometric dependencies on the azimuthal angle allowing to form asymmetries that project out only one specific contribution. These azimuthal e^+e^- asymmetries in TCS are also of interest for GPD model development. For example, the ratio of weighted cross sections, whose numerator is proportional to the cos ϕ part of the interference term, can compared directly with GPD models, testing the contribution of the Polyakov-Weiss D-term [4]. As illustrated in Figure 2, the sensitivity is quite large, making it an attractive measurement also before large amounts of statistics are collected.

TCS asymmetry measurements thus provide an attractive opportunity access the real part of the Compton amplitude, and in particular its real part. As such TCS has recently resulted in a considerable theoretical interest, also beyond $e-p$ scattering, for instance, in the context of ultraperipheral $p-p$ collisions [10], and e^+e^- radiative annihilations [11].

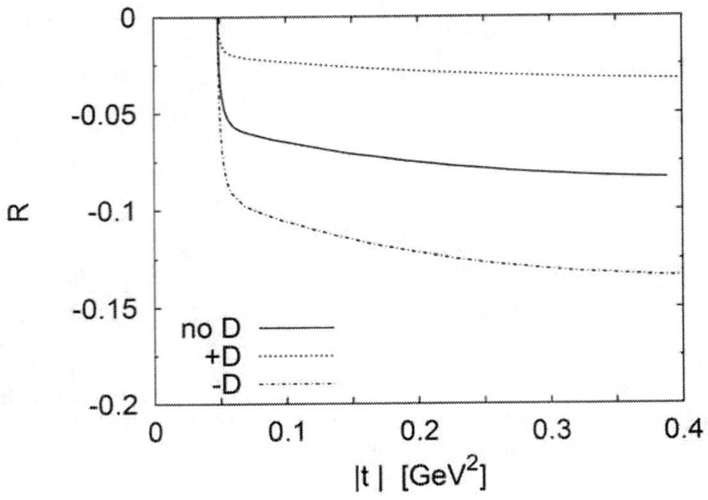

FIGURE 2. Asymmetry related to the helicity amplitude M^{--} calculated for photon energies E_γ of 13 GeV and Q'^2 of 5 GeV^2 with two model assumption for the Polyakov-Weiss D-term, and without it.

TIMELIKE COMPTON SCATTERING MEASUREMENTS

The first dedicated effort to study TCS with tagged real photons was made in Hall B at Jefferson Lab in the spring of 2008 during the g12 run period. The measurement used the CEBAF Large Acceptance Spectrometer (CLAS) [12], a 40 cm long LH_2 target and a photon tagger [13]. The latter tagged all photons with energies between 20% and 95% of the electron beam energy of 5.7 GeV, resulting in tagged real photon energies ranging between 3.6 and 5.4 GeV. Particle identification of the final state leptons was provided by the CLAS Cerenkov detectors and the electromagnetic calorimeter. Based on the earlier g7 data, this allows for a pion rejection of 10^{-4} with one detected lepton and 10^{-7} with both leptons detected. The photon beam was circularly polarized.

There are also several Hall B CLAS data sets using electron beams of energies between 5 and 6 GeV that can be used for TCS analyses. A recent analysis focuses on data from the e1-6 and e1f run periods, which also have circularly polarized photons [5]. TCS analysis with quasi-real photons requires reconstruction of the beam electron, which is scattered at a very small angle and cannot be detected. Detection of the other three final-state particles allows one to kinematically constrain the beam electron by requiring the missing momentum to be aligned with the beam direction, and the missing mass to be very small. In the case when $Q^2 \sim 0$ and Q'^2 is sufficiently large, the electron from the produced lepton pair will usually be scattered at a much larger angle. Figure 3 shows the e^+e^- invariant mass spectrum for data taken at electron beam energies of 6 GeV. The vector meson peaks are clearly visible, supporting the kinematic constraint method on the beam electron. Figure 3 also illustrates the available statistics for lepton pairs above the ϕ mass in the e1-6 data set. Further details on the analysis of these data sets can be found in Ref. [5].

The results from the TCS studies with tagged real photons and quasi-real photons can be used as a basis for the development of a program at 12 GeV Jefferson Lab. The new CLAS12 detector, which will be constructed in Hall B, will be ideal for TCS measurements with quasi-real photons with circular polarization. Such measurements could run in parallel with other parts of the Hall B 12 GeV physics program and accumulate sizable statistics. Furthermore, the CLAS12 will provide good particle identification and resolution. With the addition of a low-Q^2 tagger, it could also be possible to study the reaction with incoming photons having linear polarization and a small, but finite virtuality (double DVCS). It should, however, be noted that the acceptance of this spectrometer is rather complicated. It would thus be of great interest to compare with a hermetic detector such as the GlueX detector constructed in Hall D at 12 GeV Jefferson Lab. This natural extension of the TCS measurements in Hall B would also provide access to linearly polarized tagged real photons.

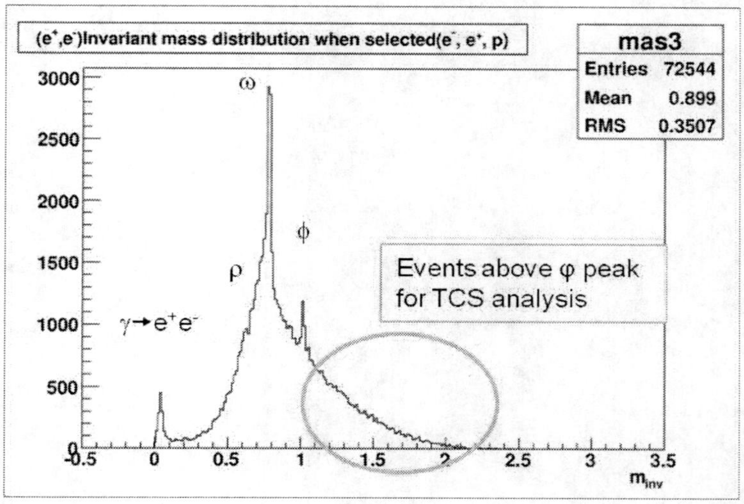

FIGURE 3. The e^+e^- invariant mass, Q' spectrum from Hall B at Jefferson Lab. Shown is the CLAS 6 GeV e1-6 data set showing the vector mesons, and the statistics for TCS candidates above the ϕ peak.

ACKNOWLEDGMENTS

This work is in part supported by the National Science Foundation through grant PHY-NSF-1019521.

REFERENCES

1. K. Goeke, V. Polyakov and M. Vanderhaeghen, Prog. Part. Nucl. Phys. **47**, 401 (2001).
2. M. Diehl, Phys. Rep. **388**, 41 (2003).
3. A. Belitsky and A. V. Radyushkin, Phys. Rep. **418**, 1 (2005).
4. M.V. Polyakov, C. Weiss, *Phys. Rev. D* **60**, 114017 (1999).
5. R. Paremuzyan, Ph.D. thesis, Yerevan Physics Institute, 2010.
6. V. Guzey, private communication (2010).
7. M.V. Polyakov, K.M. Semenov-Tian-Shansky, *Eur. Phys. J. A* **40**, 181 (2009).
8. K. Kumericki, D. Mueller, K. Passek-Kumericki, *Nucl. Phys. B* **794**, 244 (2008).
9. E. R. Berger, M. Diehl, and B. Pire, *Eur. Phys. J* **C23**, 675 (2002).
10. B. Pire, L. Szymanowski, and J. Wagner, *Phys. Rev. D* **79**, 014010 (2009).
11. A. Afanasev, S. Brodsky, C.E. Carlson, and A. Mukherjee, *Phys. Rev. D* **81**, 034014 (2010).
12. B. Mecking et. al, *Nucl. Instr. Meth. A* **503**, 513 (2004).
13. D. Sober et. al, *Nucl. Instr. Meth. A* **440**, 263 (2000).

Session Q: Heavy Hadrons

Prospects for X(3872) Detection at $\overline{\text{P}}$ANDA

Jens Sören Lange, Martin Galuska, Thomas Geßler, Wolfgang Kühn, Stephanie Künze, Yutie Liang, David Münchow, Björn Spruck, Matthias Ullrich, and Marcel Werner

Justus-Liebig-Universität Gießen, II. Physikalisches Institut, Heinrich-Buff-Ring 16, 35392 Gießen, Germany

Abstract. Monte-Carlo simulations for a resonance scan of the charmonium-like state X(3872) at $\overline{\text{P}}$ANDA are performed. Final state radiation hadronic background reactions are taken into account. The signal reconstruction uses a realistic pattern recognition (track finder and track fitter) and electron/pion discrimination.

PACS: 14.40.Pq, 14.40.Rt

Conference Talk presented at MENU10, 12th International Conference on Meson-Nucleon Physics and the Structure of the Nucleon, 05/31-06/04, 2010, College of William and Mary, Williamsburg, Virginia.

THE X(3872) STATE

The X(3872) is a charmonium(-like) state which was first observed in its decay X(3872)$\rightarrow$J/$\psi\pi^+\pi^-$ in B decays [1] [2] and inclusive production in $p\overline{p}$ collisions [3] [4]. Meanwhile more decays such as final states $D^0\overline{D}^{0*}$ or $J/\psi\omega$ have been found. In particular, the observation of its radiative decay into $J/\psi\gamma$ allows the assignment of positive charge parity C=+1. The interesting observation of isospin violation in the decay X(3872)$\rightarrow J/\psi\rho(\rightarrow\pi^+\pi^-)$ raised the question, if it might be a charmonium state at all. As its mass is within Δm<1 MeV of the $D^0\overline{D}^{0*}$ threshold, it might be interpreted as an S-wave molecular state [6] [7]. Many properties of the X(3872) are being studied at ongoing experiments. However, the width is unknown and probably can only be determined at $\overline{\text{P}}$ANDA. The current upper limit is Γ<2.3 MeV [1], whereas $\overline{\text{P}}$ANDA might be able to set an upper limit in the order of a few hundred keV. The quantitative investigation of this upper limit on the width by detailed Monte-Carlo (MC) simulations incl. background is topic of this paper.

CHARMONIUM PRODUCTION AT $\overline{\text{P}}$ANDA

The $\overline{\text{P}}$ANDA (Anti-Proton Annihilation at Darmstadt) experiment will be located at the future FAIR (Facility for Antiproton and Ion Research) facility [8] at GSI in Darmstadt, Germany. The primary synchrotron will have a circumference of 1.084 km, a proton beam with $30\leq E_{beam}\leq 90$ GeV and an intensity $\leq 2\times 10^{13}$/s. Antiprotons are produced by a secondary target and then stored and cooled in the HESR (High Energy Storage Ring). There are two HESR modes. In the high intensity mode (high resolution mode), there are 10^{11} (10^{10}) stored antiprotons, and stochastic cooling (electron cooling) will provide a beam momentum resolution of $\Delta p/p\geq 1\cdot 10^{-4}$ ($\Delta p/p\geq 4\cdot 10^{-5}$). $\overline{\text{P}}$ANDA will investigate $p\overline{p}$ collisions with an antiproton beam momentum $p_{beam}\leq 15$ GeV/c, corresponding to $\sqrt{s}\leq 5.5$ GeV. In e^+e^- collisions, only states with quantum numbers J^{PC}=1^{--} can be produced directly due to the intermediate virtual photon. As the X(3872) has positive C parity, thus, it can not be produced directly in e^+e^-. However, in $p\overline{p}$ collisions two or three gluons are produced in the annihilation, and states with any quantum number can be formed.

At the $\overline{\text{P}}$ANDA interaction region, three different types of targets are foreseen, namely *(a)* hydrogen pellets with a thickness of $\simeq$25 μm and a falling velocity of $\geq$60 m/s, *(b)* a cluster jet target and *(c)* fixed nuclear targets such as *Be*, *C*, *Si* or *Al*. For the pellet target, a peak luminosity of $\mathcal{L}=2\times 10^{32}$ cm^{-2} s^{-1} is expected. If adjusted to a single resonance, such as e.g. the J/ψ this would correspond to about 2×10^9 produced J/ψ per year.

A resonance scan with the cooled beam could provide a measurement of the X(3872) width. This technique was pioneered by the Fermilab experiments E760 and E835. With a beam momentum resolution of $\Delta p/p$=$2\cdot 10^{-4}$,

12th International Conference on Meson-Nucleon Physics and the Structure of the Nucleon (MENU 2010)
AIP Conf. Proc. 1374, 549-552 (2011); doi: 10.1063/1.3647200

corresponding to a $\sqrt{s}$ resolution of 0.5 MeV (FWHM), the measurement of the width of the J/ψ with $\Gamma(J/\psi)$=99±12(stat.)±6(syst.) keV and the width of the ψ' with $\Gamma(\psi')$=306±36(stat.)±16(syst.) keV could be performed [9].
The requirements for the $\overline{\text{P}}$ANDA experiment are quite high: capability for high rate detection and data acquisition for $\leq 2\times 10^7$ interactions/s, $\simeq 4\pi$ solid angle coverage (incl. a high resolution forward detector, as $\overline{\text{P}}$ANDA is a fixed target experiment), secondary vertex detection of e.g. D mesons with a resolution σ<100 μm, charged track momentum resolution Δp/p$\simeq$1%, charged particle identification for $e^\pm$, $\mu^\pm$, $\pi^\pm$, $K^\pm$, $p^\pm$ using e.g. Cherenkov detectors with internal reflection, and electromagnetic calorimetry in the range 10 MeV<E_γ<5 GeV. Details about the experiment can be found elsewhere [10]. For the down below results, MC simulations were performed with all $\overline{\text{P}}$ANDA subdetectors implemented, but the reconstruction used only the following $\overline{\text{P}}$ANDA subdetectors:

- a Time Projection Chamber (TPC) with 135 padrows (corresponding to $\leq$135 hits for a charged particle track) and 135,169 pads of 2×2 mm^2 area,
- a Micro Vertex Detector (MVD) with 120 silicon pixel modules (100×100 μm^2 pixel size and in total 10^7 readout channels), and 400 silicon strip modules ($\simeq$0.5 m^2 active area and 7×10^4 readout channels), and
- an Electromagnetic Calorimeter (EMC) with $\simeq$17,200 crystals of $PbWO_4$, a radiation hard scintillation material with a fast decay constant of $\simeq$6 ns. The thickness corresponds to $\simeq 28X_0$ radiation lengths.

The MC simulation was performed with the PandaRoot [11] simulation, digitization, reconstruction and analysis framework. It consists of $\simeq$43,000 geometry volumes (incl. details such as e.g. the beampipe for the pellet target) and $\geq$400,000 lines of C++ code. Transport engines are Geant3, Geant4, and Fluka. Event Generators are EvtGen, DPM, PYTHIA, and UrQMD. References can be found elsewhere [12]. PandaRoot is used on $\geq$15 Linux platforms.
Recent improvements, in particular since the $\overline{\text{P}}$ANDA Physics Report [13] are in particular: *(a)* for the first time X(3872) simulations, not investigated before as a physics channel for $\overline{\text{P}}$ANDA, *(b)* usage of detailed field maps of a homogenous B_z=2 T in the central region and dipole field of 2 Tm in the forward region (with flux effects in the iron of the $\overline{\text{P}}$ANDA muon subdetector and interference of the fields taken into accout), *(c)* usage of a realistic track finder and track fitter based upon a conformal map technique, and *(d)* simulation of final state radiation [14]. In about $\simeq$30% of all $J/\psi\to e^+e^-$ decays an additional photon is radiated.
The particle indentification uses the ratio E/p as a variable for discrimination of charged pions and leptons. E is the deposited shower energy in the EMC, p is the reconstructed track momentum by the track finder and track fitter using the MVD and the TPC. For leptons E/p is $\simeq$1. For charged pions, which generate hadronic showers, E/p can take any value $0\leq E/p\leq 1$. By applying an E/p cut of $0.8\leq E/p\leq 1.2$ for electrons, the combinatorial background is reduced by $\geq$90%.

ESTIMATED RATES FOR X(3872) FORMATION AT $\overline{\text{P}}$ANDA

For resonant X(3872) formation, the required beam momentum is p_{beam}=6.99100 GeV/c. Our baseline assumption is a peak cross section of $\sigma_{X(3872)}$=50 nb, in the same order of magnitude as e.g. the cross section of ψ' production [9]. Note, that, if the X(3872) is a molecular $D^0\overline{D}^{0*}$ state, there are estimates [15] that the cross section may increase up to $\leq$443 nb. The ratio of branching fractions of the X(3872) decays into the final states $D^0\overline{D}^{0*}:J/\psi\pi^+\pi^-:J/\psi\gamma$ is assumed to be 9:1:0. All other decays are assumed to have a zero branching fraction. The branching fraction for $J/\psi\to e^+e^-$ and $\mu^+\mu^-$ is $\simeq$6% each. The reconstruction efficiency of $\simeq$50% is dominated by the track reconstruction efficiency for the low momentum charged pions. All these factors lead to a reconstructable cross section of 250 pb. For resonance scans, the HESR will be operated in the high resolution mode with Δp/p=10^{-5}, which corresponds to a luminosity of $\mathscr{L}=2\times 10^{31}$ cm^{-2}s^{-1}. Assuming an accelerator duty factor of 50%, this leads to an integrated luminosity of $\mathscr{L}_{int}$=0.86 pb^{-1}/day. For a resonance scan with 20 energy points and 2 days/point, this would correspond to a yield of $\simeq$215 events of $p\overline{p}\to$X(3872)$\to J/\psi\pi^+\pi^-$ per day at peak.

BACKGROUND

The main background is meson production in processes such as $p\overline{p}\to\pi^+\pi^-\pi^+\pi^-$ with two misidentified charged pions (as leptons) in the EMC. The cross section for this process is 50 μb [16], compared to an estimated signal cross section of 50 nb. For comparison, the total $p\overline{p}$ cross section is $\simeq$70 mb. It is important to investigate the shape of the background e.g. in the 2-particle mass spectrum. For this purpose, a DPM (dual parton model) event generator [17] [18] [19] was used. In fact, a varying background shape in the region of the J/ψ was found and fitted with a first order Chebyshev polynome. The fit was performed differently for each beam momentum in the resonance scan.

DETERMINATION OF A RESONANCE WIDTH BY RESONANCE SCAN

The Breit-Wigner cross section for the formation and subsequent decay of a $c\overline{c}$ resonance R of spin J, mass M_R and total width Γ_R formed in the reaction $\overline{p}p \to R$ is

$$\sigma_{BW}(E_{cm}) = \frac{(2J+1)}{(2S+1)(2S+1)} \frac{4\pi(\hbar c)^2}{(E_{cm}^2 - 4(m_p c^2)^2)} \times \frac{\Gamma_R^2 BR(\overline{p}p \to R) \times BR(R \to f)}{(E_{cm} - M_R c^2)^2 + \Gamma_R^2/4} \tag{1}$$

where S is the spin of the (anti-)proton.

$$\sigma(E_{cm}) = \int_0^\infty \sigma_{BW}(E')G(E' - E_{cm})dE' \tag{2}$$

is a convolution of a Breit-Wigner term for the resonance and the function G for the beam resolution. The area under the resonance peak is given by

$$A = \int_0^\infty \sigma(E_{cm})dE_{cm} = \frac{\pi}{2}\sigma_{peak}\Gamma_R \tag{3}$$

which importantly is independent of the form of $G(E)$. σ_{peak} is the cross section at E_{cm}=$M_R c^2$ given by

$$\sigma_{peak} = \frac{(2J+1)}{(2S+1)(2S+1)} \frac{16\pi\hbar^2 BR(\overline{p}p \to R) \times BR(R \to f)}{(M_R - 4m_p^2)c^2} \quad . \tag{4}$$

By measuring A using a fit to the excitation function and inserting σ_{peak} into Eq. 3, the resonance width Γ_R can be determined.

PRELIMINARY RESULTS

MC simulations for $p\overline{p} \to$X(3872)$\to J/\psi\pi^+\pi^-$ with background were performed for 9 different beam momenta in the resonance region. Fig. 1 shows the e^+e^- invariant mass with PID applied and background subtracted. The final state is exclusive, i.e. no other particle is produced except for photons from final state radiation (visible as radiative tail at masses smaller than the J/ψ mass). Each data point corresponds to 2 days of data taking. A Gaussian fit was used to extract the signal yield. This J/ψ yield is equal to the tagged X(3872) yield. Then data points were fitted with an excitation function. The preliminary results are as follows:

- The fitted width of the excitation function is $\simeq$20% larger than the width which was used as input in the MC simulation, due to the analysis technique such as the background subtraction and the reconstruction technique (e.g. the track fitter). Note, that the beam resolution is not the limit for the width measurement. Even with a beam momentum resolution of $\Delta p_{beam} \simeq$0.5 MeV, the E760 experiment was able to determine the factor of $\simeq$5 smaller width of the J/ψ [9]. The difference between generated and reconstructed width would only be a systematic error on the width measurement.
- The width of the excitation function is still $\leq$100 keV, if the assumed resolution of the anti-proton beam is $\Delta p/p$=$2\cdot 10^{-5}$ (HESR high resolution mode). However, in this analysis all distributions were assumed to have a Gaussian shape, while the correct resonance shape would be given by a Breit-Wigner parametrisation and the shape of the function $G(E)$ in Eq. 2 might be non-Gaussian. In addition, for our analysis, we approximated the function $\sqrt{s}$=$\sqrt{2m_p^2 + 2m_p\sqrt{p_{beam}^2 + m_p^2}}$ by a linear function in the small range of the resonance region. As long as the beam resolution is accurately known, our results indicate that the width of the X(3872) could be determined at $\overline{\text{P}}$ANDA, if it is larger than $\Gamma_R \geq$100 keV and if the DPM generator describes the background shape correctly.

These results are still preliminary, as the unfolding (integral equation, Eq. 2) is ongoing work.

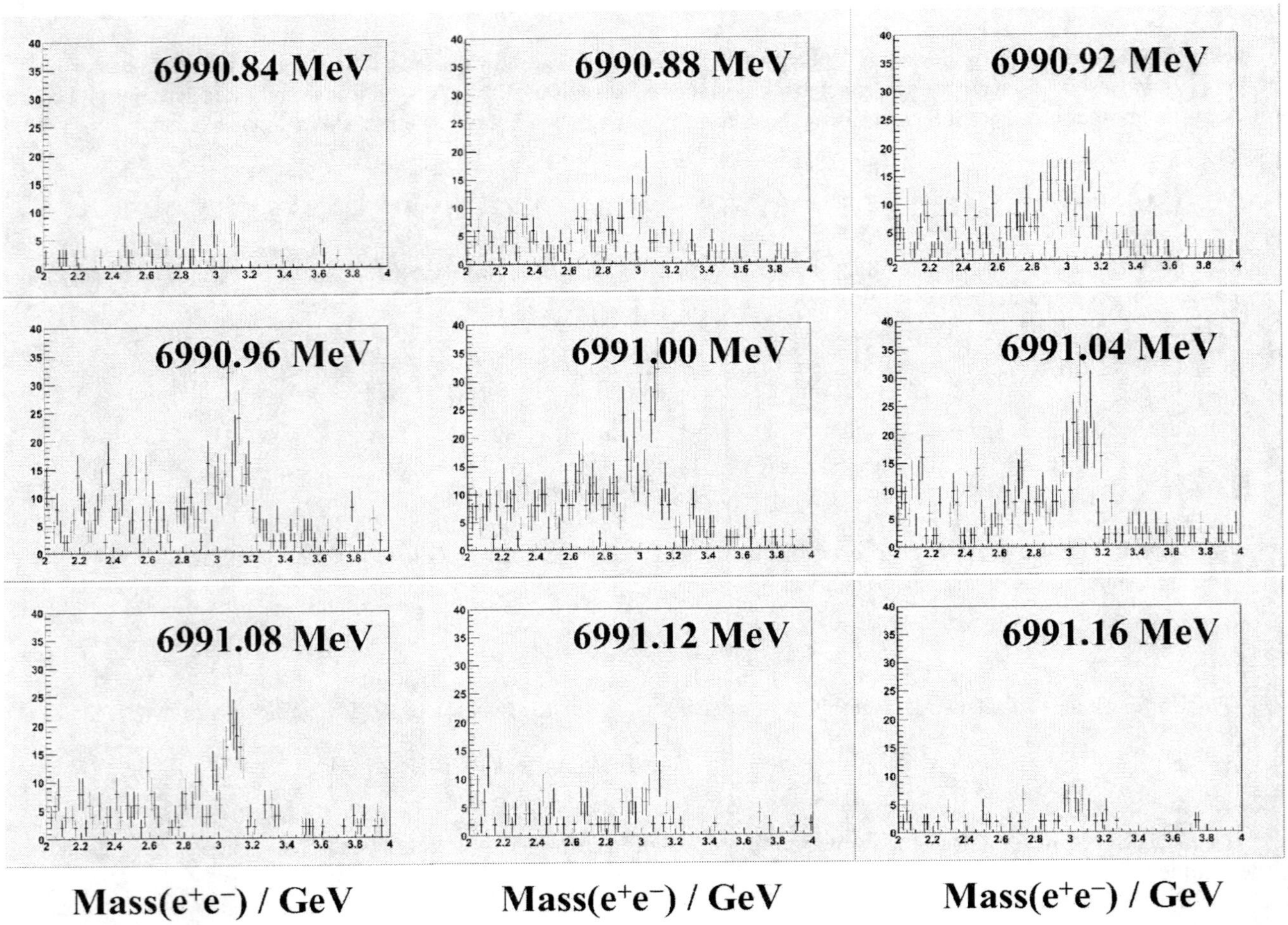

FIGURE 1. MC simulation of a resonance scan of the X(3872) at $\overline{\text{P}}$ANDA, with 9 scan points of different anti-proton beam momenta and data taking of 2 days for each point. The plots show the background subtracted, tagged J/ψ signal from the X(3872)$\rightarrow J/\psi\pi^+\pi^-$ decay. Only statistical errors are shown.

REFERENCES

1. Belle Collaboration, Phys. Rev. Lett. 91(2003)262001
2. BaBar Collaboration, Phys. Rev. D71(2005)071103
3. CDF-II Collaboration, Phys. Rev. Lett. 93(2004)072001
4. D0 Collaboration, Phys. Rev. Lett. 93(2004)162002
5. K. Nakamura et al. (Particle Data Group), J. Phys. G37(2010)075021
6. N. A. Tornqvist, Phys. Lett. B590(2004)209
7. N. A. Tornqvist, Phys. Rev. Lett. 67(1991)556
8. FAIR Baseline Technical Report, March 2006 download available at `http://www.fair-center.de`
9. E760 Collaboration, Phys. Rev. D47(1993)772
10. See for example `http://www-panda.gsi.de`;
    B. Ketzer, Int. J. Mod. Phys. A21(2006)5675; J. S. Lange, Int. J. Mod. Phys. A24(2009)369
11. `http://panda-wiki.gsi.de/cgi-bin/view/Computing/PandaRoot`
12. S. Spataro, Journal of Physics 119(2008)032035
13. $\overline{\text{P}}$ANDA Collaboration, arXiv:0903.3905 [hep-ex]
14. E. Barberio, B. van Eijk, Z. Was, Comp. Phys. Comm. 66(1991)115
15. G. Y. Chen, J. P. Ma, Phys. Rev. D77(2008)097501
16. V. Flaminio et al., Compilation of Cross-Sections, Part 3, p and anti-p Induced Reactions, High-Energy Reactions Analysis Group, CERN-HERA-79-03, 1979
17. A. Capella, U. Sukhatme, C.-I Tan, and J. Tran Thanh Van, Phys. Rept. 236(1994)225
18. A. B. Kaidalov, P. E. Volkovitsky, Z. Phys. C63(1994)517
19. V. Uzhinsky, A. Galoyan, hep-ph/0212369

Partial Wave Analysis at BES III, harnessing the power of GPUs

Niklaus Berger

IHEP, Beijing

Abstract. Partial wave analysis (PWA) is a core tool for determining resonance properties in hadron spectroscopy. The unbinned likelihood fits employed are however computationally very expensive especially in view of the very large data samples being collected at the Beijing Spectrometer (BES) III experiment and expected from future experiments such as Glue-X and PANDA. Using the power of massively parallel floating point calculations on graphics processing units (GPUs), we have managed to speed up PWA by more than two orders of magnitude compared to legacy implementations. The resulting software framework allows for the fast creation and running of PWA code. Modern hardware and software thus solve the PWA speed problems — many questions concerning fit methods and especially the reliability of fit results however remain open.

Keywords: Partial Wave Analysis, Hadron Spectroscopy, Parallel Computing, BES III
PACS: 14.40.Rt, 13.25.Gv, 13.20.Gd, 29.85.Fj

PARTIAL WAVE ANALYSIS AND BES III

Partial wave analysis (PWA) is a key technique in hadron spectroscopy. Especially in cases where multiple interfering broad resonances contribute to the multi-particle final state under study, PWA is required to make inroads on the properties of the intermediate resonances. PWA of large data samples is computationally very expensive. The advent of cheap massively parallel computing infrastructures such as Graphics Processing Units (GPUs) however now allows for very fast fits to be performed, exchanging the requirement of large computing resources with the need for additional programming efforts in order to access the massively parallel resources. In order to reduce these efforts to a minimum and to produce a shared code base for PWA at BES III, we created the GPUPWA software framework, which we continue do develop in order to make use of the ever changing hardware and software environment and to adapt to the users' needs. Whilst computational limitations are no longer a major burden on PWA, issues pertaining to fits with often dozens of free parameters as well as questions of what amplitude models are adequate remain open and a challenge to both the experimental and theoretical community in hadron spectroscopy.

The Beijing Spectrometer III experiment [1, 2] at the recently upgraded Beijing Electron-Positron collider (BEPC) II has started taking data in 2008 and already collected the world's largest data samples on the J/ψ and ψ' resonances (for an overview of results, see the *Recent Results from BES III* contribution in these proceedings). These samples and the harvest from the ongoing and future runs contain a wealth of spectroscopic information, a lot of which is only accessible through partial wave analysis, thus making efficient and reliable PWA of paramount importance to the BES III spectroscopy program.

PARTIAL WAVE ANALYSIS AS A COMPUTATIONAL PROBLEM

In a typical PWA (we use the radiative decay $J/\psi \longrightarrow \gamma X, X \longrightarrow K^+K^-$ [3] as an example), the decay is modelled by coherently summing the contributions from a variety of intermediate resonances X. The relative magnitudes and phases of these resonances are determined from a fit and the fit result is compared to the data. The set of resonances and their properties are changed until a sufficient agreement with the data is found.

The intensity I (relative number of events) at a particular point Ω in phase space can be expressed as

$$I(\Omega) = \left| \sum_{\alpha} V_\alpha A_\alpha(\Omega) \right|^2 , \tag{1}$$

12th International Conference on Meson-Nucleon Physics and the Structure of the Nucleon (MENU 2010)
AIP Conf. Proc. 1374, 553-556 (2011); doi: 10.1063/1.3647201

where the sum runs over all intermediate resonances α, V_α is the complex production amplitude for α and $A_\alpha(\Omega)$ the complex decay amplitude at a particular point in phase space. The likelihood for a particular model is

$$\mathscr{L} \propto \prod_{i=1}^{N_{Data}} \frac{I(\Omega_i)}{\int \varepsilon(\Omega) I(\Omega) d\Omega}, \tag{2}$$

where the product runs over all N_{Data} events in the sample and the integral is proportional to the total cross section, corrected for the detector efficiency $\varepsilon(\Omega)$. The integral is usually performed numerically by summing over a large number N_{MC}^{Gen} of simulated events (Monte Carlo, MC) generated evenly in phase space[1]. The limited acceptance and efficiency of the detector can be taken into account by summing only over simulated events that pass the reconstruction and analysis cuts. In a fit, a maximum of the logarithm of the likelihood, corresponding to the best set of parameters for the used model is searched for;

$$\ln \mathscr{L} \propto \sum_{i=1}^{N_{Data}} \ln \left(\sum_\alpha \sum_{\alpha'} V_\alpha V_{\alpha'}^* A_\alpha(\Omega_i) A_{\alpha'}^*(\Omega_i) \right) - \ln \left(\sum_\alpha \sum_{\alpha'} V_\alpha V_{\alpha'}^* \left(\frac{1}{N_{MC}^{Gen}} \sum_{j=1}^{N_{MC}^{Acc}} A_\alpha(\Omega_j) A_{\alpha'}^*(\Omega_j) \right) \right). \tag{3}$$

The first sum runs over all data events, the second over all MC events. If the widths and masses of resonances are kept constant in the fit (i.e. the V_α's are the only free parameters), the last (inner) bracket and the $A_\alpha(\Omega_i)A_{\alpha'}^*(\Omega_i)$ term for each data event can be pre-calculated.

The number of floating point operations required is dominated by the sum over the data events and scales with $N_{iterations} \times N_{data} \times N_{waves}^2$, whilst the lookup table takes up storage space scaling with $N_{data} \times N_{waves}^2$. The storage space problem can be addressed by increasing the memory of the relevant machine ($\approx$1.5 GB are required per million events for a model with 20 partial waves), or with appropriate caching mechanisms for data samples with several million events. If the required floating point operations are performed sequentially, the time required can however become very long.

Floating point precision can be an issue, the minimizing programs tend to fail if the delivered precision does not correspond to the expected one. For large amounts of data and simulated events however, the precision of the individual amplitude is not decisive, it can safely be computed in single precision. The final sum however has to be computed in double precision and the numerical result is strongly dependent on the summing algorithm; tree sums as usually implemented on GPUs perform much better here than accumulating loop summing employed in traditional CPU based programs.

There is no way of knowing whether the fit found just a local or the searched for global maximum of the likelihood. To gain confidence in the result, the fits are usually repeated with various sets of starting parameters. In addition, various models have to be tried out, especially in the study of possible new resonances and systematic effects. The thousands of fits needed in a typical partial wave analysis should thus be as fast as possible, especially as feedback from the physicist is required between the fits and they thus cannot easily be ran in parallel.

PARTIAL WAVE ANALYSIS ON GPUS

The mass market for computer games with realistic real-time rendering of simulated environments has driven the development of extremely powerful but cheap hardware in the form of graphics processing units (GPUs). The calculation of the colour of millions of independent pixels with three colours and one transparency value has a lot of parallels to high energy physics computing with many independent events and the use of four-vectors. It turns out that GPUs are indeed formidable devices for scientific computing, providing teraflops in a single workstation for a few hundred dollars extra. Whilst in the early days of general purpose computing on GPUs, computational problems had to be brought into a form very close to rendering an image, powerful computation frameworks have become available in the meantime. The two major vendors of GPUs, AMD/ATI and Nvidia have offered competing frameworks named Brook+ [4] and CUDA [5] respectively. Whilst Brook+ offers the cleaner programming model and first introduced double precision, CUDA has acquired a larger user base and development is more vibrant. In the meantime, the platform independent OpenCL standard [6] has been released and there are implementations from both vendors.

---

[1] In case of a very uneven phase-space population, e.g. due to the presence of narrow resonances, it is advisable to generate a non-equidistributed MC sample and weight events to a flat distribution in order to reduce the sampling error caused by the strongly varying amplitude

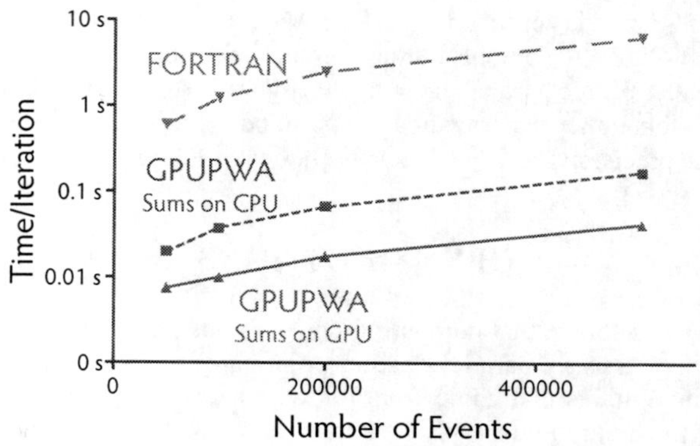

FIGURE 1. Execution times for one fit iteration of a $J/\psi \longrightarrow \gamma K^+K^-$ PWA with four waves and five free parameters for the legacy FORTRAN code (∇), an older GPUPWA version where the sums were performed on the CPU ($\Box$) and GPUPWA performing the sums on the GPU (Δ) versus the number of events used in the fit.

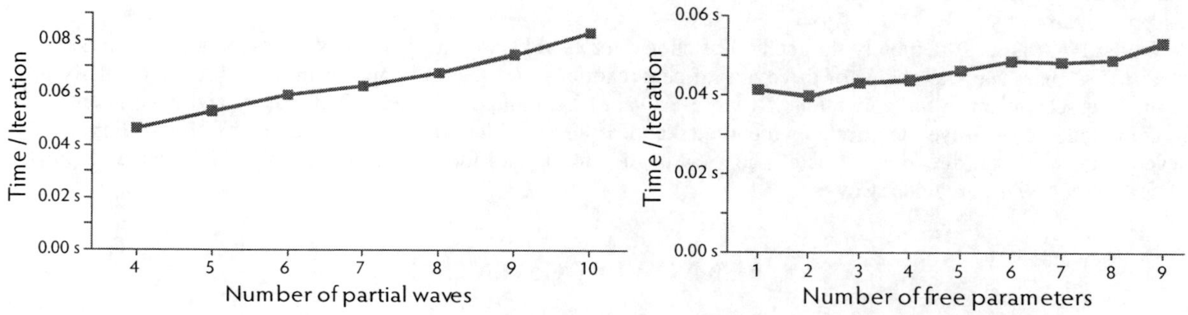

FIGURE 2. Execution times for one fit iteration of a $J/\psi \longrightarrow \gamma K^+K^-$ PWA with 500'000 events and five free parameters in GPUPWA versus the number of partial waves used in the fit (left) and for a fit with four waves versus the number of free parameters in the fit (right).

In order to abstract the intricacies of GPU based computing and to create a shared code base for PWA at BES III (and maybe beyond), we have created the GPUPWA framework (see also [7]). Originally based on the Brook+ framework by AMD, we are currently in the process of porting it to OpenCL; as the GPU interface is completely abstracted, this will only require minimal changes in the user code.

GPUPWA provides a framework where users can specify amplitudes in the covariant tensor formalism [8, 9] (in which case all the tensor manipulations are performed on the GPU) or arbitrary amplitudes to be read from files. Using these amplitudes, MC integration and the likelihood calculation are performed on the GPU. There is an interface to the `ROOT::Minuit2` fitters [10] for the likelihood minimization. We mostly use a modified version of the FUMILI fitter [11] because of its very fast convergence; Fumili error estimates are however notoriously inaccurate, so either the MINOS technique in Minuit [12] or parameter scans are employed. For plotting of results, GPUPWA interfaces to the ROOT histogramming package [13].

Performance

All the performance measurements quoted here were performed on an Intel Core 2 Quad 2.4 GHz workstation with 2 GB RAM running under Scientific Linux 5.2. The GPU is a ATI Radeon 4870 with 512 MB RAM. We compared a FORTRAN implementation used in BES II and the GPUPWA implementation of a $J/\psi \longrightarrow \gamma K^+K^-$ partial wave analysis, both minimizing using Fumili with analytical gradients and Hessian. The time for one fit iteration was measured with the system timer for various sizes of the data samples (see figure 1). For large enough samples, the

GPUPWA code is more than 150 times faster than the FORTRAN code.

The execution times of the GPUPWA code also show benign behaviour when increasing the number of partial waves (figure 2, left) or the number of free parameters in the fit (figure 2, right). This shows that the speed is limited by memory accesses and data transfers rather than raw floating point power – in fact we reach about 9 GFlop/s (counting also transcendentals and double precision operations as 1 Flop), below 1% of the theoretical power of the card.

OPEN QUESTIONS

The experience with GPUPWA has shown that computing power is not the limiting factor for PWA, even with very large statistics, especially if the massively parallel floating point capabilities of GPUs are brought into play. PWA is however saddled with the ambiguities that come from fits with many free parameters; often 10 and more waves are employed, each coming with an amplitude, a phase, a mass and a width. In addition, the choice of waves is also ambiguous; should we leave all waves in the fit or drop those that are not significant (and does significant start at three or at five sigma)?

In most fits, amplitudes are calculated in an isobar model. In many channels however, rescattering believed to be important; also effects of inelasticity should be included [14]. These additional features of the amplitudes will, more often than not, also add additional fit parameters; thus contributing both to a clearer physics message and more uncertainty from the fit.

Whilst PWA is most definitely the method of choice for establishing the properties of resonance that are seen as "bumps" in mass spectra; claims of discovery of new (exotic) states that only appear in the PWA are regularly met with great scepticism in the community. If we ever want to eliminate this scepticism, theory and experiment have to come together to solve the questions mentioned above[2] and establish rigorous procedures for fitting. Fortunately several efforts in that direction, especially in view of the present and future spectroscopy facilities such as BES III, GlueX and PANDA, are under way.

ACKNOWLEDGMENTS

We would like to thank the partial wave analysis group at BES III (the users of GPUPWA) for their efforts in testing the framework, their very helpful suggestions and bug reports.

The work presented was supported by the Chinese Academy of Sciences and the Swiss National Science Foundation.

REFERENCES

1. M. Ablikim, et al., *Nucl. Instr. Meth.* **A614**, 345–399 (2010), `0911.4960`.
2. D. M. Asner, et al. (2008), `0809.1869`.
3. J. Z. Bai, et al., *Phys. Rev.* **D68**, 052003 (2003), `hep-ex/0307058`.
4. Advanced Micro Devices, AMD stream computing, software stack, AMD white paper (2007), URL `http://ati.amd.com/technology/streamcomputing/firestream-sdk-whitepaper.pdf`.
5. T. R. Halfhill, *Microprocessor Report* **01/28/08**, 1–6 (2008).
6. M. Aaftab, The OpenCL specification, version 1.0, Tech. rep., The Khronos Group (2008), URL `http://www.khronos.org/registry/cl/specs/opencl-1.0.33.pdf`.
7. N. Berger, L. Beijiang, and W. Jike, *J. Phys. Conf. Ser.* **219**, 042031 (2010).
8. B. S. Zou, and D. V. Bugg, *Eur. Phys. J.* **A16**, 537–547 (2003), `hep-ph/0211457`.
9. S. Dulat, and B.-S. Zou, *Eur. Phys. J.* **A26**, 125–134 (2005), `hep-ph/0508087`.
10. F. James, and M. Winkler, MINUIT Users Guide, Tech. rep., CERN (2004).
11. FUMILI, Tech. rep., CERN Prorgram Library, D510 (1971).
12. F. James, and M. Roos, *Comput. Phys. Commun.* **10**, 343–367 (1975).
13. R. Brun, and F. Rademakers, *Nucl. Instrum. Meth.* **A389**, 81–86 (1997).
14. P. Guo, R. Mitchell, and A. P. Szczepaniak (2010), `1006.4371`.

---

[2] At least as far as this is possible — multi-parameter fits being *NP*-complete problems, we will never be sure whether the true minimum has been found.

Prediction of narrow N^* and Λ^* with hidden charm

Jia-Jun Wu[*,†], R. Molina[†,**], E. Oset[†,**] and B. S. Zou[*,**]

[*] *Institute of High Energy Physics, CAS, Beijing 100049, China*
[†] *Departamento de Física Teórica and IFIC, Centro Mixto Universidad de Valencia-CSIC, Institutos de Investigación de Paterna, Aptdo. 22085, 46071 Valencia, Spain*
[**] *Theoretical Physics Center for Science Facilities, CAS, Beijing 100049, China*

Abstract. The interaction between various charmed mesons and charmed baryons, such as $\bar{D}\Sigma_c$-$\bar{D}\Lambda_c$, $\bar{D}^*\Sigma_c$-$\bar{D}^*\Lambda_c$, and related strangeness channels, are studied within the framework of the coupled channel unitary approach with the local hidden gauge formalism. Six narrow N^* and Λ^* resonances are dynamically generated with mass above 4 GeV and width smaller than 100 MeV. These predicted new resonances definitely cannot be accommodated by quark models with three constituent quarks. We make estimates of production cross sections of these predicted resonances in $\bar{p}p$ collisions for PANDA at the forthcoming FAIR facility.

Keywords: hidden charm, resonance, coupled channel unitary approach
PACS: 14.20.Gk, 13.30.Eg, 13.75.Jz

INTRODUCTION

The use of chiral Lagrangians in combination with unitary techniques in coupled channels of mesons and baryons has been a very fruitful scheme to study the nature of many hadron resonances. The poles found in the analysis of meson baryon scattering amplitudes in the complex plane of $\sqrt{s}$ are identified with existing baryon resonances. Some dynamically generated baryons are considered as a bound state of meson baryon components, such as $N^*(1535)$, [1, 2, 3]. Another point of view is that this resonance can be considered as a hidden strangeness state, with large $(qqqq\bar{q})$ components[4, 5]. These two views are both very different from the usual 3-quark(qqq) model. But it is difficult to confirm that $N^*(1535)$ is beyond the 3-quark(qqq) model, because there are always some adjustable ingredients in various quenched quark models with 3 constituent quarks to fit the experimental data around this energy region. Therefore, it is important to find some resonances which definitely can't be described by the 3-quark(qqq) model.

In this paper, we want to generate dynamically baryon states in coupled channels of meson-baryon in the hidden charm sector. The study is done both with pseudoscalar mesons and vector mesons and we obtain six bound states with the hidden charm. They are predicted with mass above 4 GeV and width smaller than 100 MeV. If confirmed, of course they cannot be accommodated by quark models with three constituent quarks. We also make estimates of production cross sections with $\bar{p}$ collisions that could be carried out at the future FAIR facility within the PANDA project.

FORMALISM FOR MESON-BARYON INTERACTION

We follow the recent approach of Ref. [6] and extend it from three flavors to four. We consider the $PB \to PB$ and $VB \to VB$ interaction by exchanging a vector meson, as shown by the Feynman diagrams in Fig. 1.

The effective Lagrangians for the interactions involved are [6]:

$$\mathscr{L}_{VVV} = ig\langle V^\mu [V^\nu, \partial_\mu V_\nu]\rangle, \quad \mathscr{L}_{PPV} = -ig\langle V^\mu [P, \partial_\mu P]\rangle, \quad \mathscr{L}_{BBV} = g(\langle \bar{B}\gamma_\mu [V^\mu, B]\rangle + \langle \bar{B}\gamma_\mu B\rangle\langle V^\mu\rangle) \tag{1}$$

where $g = M_V/2f$. P and V stand for matrices containing pseudoscalar and vector mesons, respectively.

Under the low energy approximation, the three momentum versus the mass of the meson can be neglected. The transition potential are given by

$$V_{ab(P_1(V_1)B_1)\to P_2(V_2)B_2)} = \frac{C_{ab}}{4f^2}(E_{P_1(V_1)} + E_{P_2(V_2)}), \tag{2}$$

where the a, b stand for different channels of $P_1(V_1)B_1$ and $P_2(V_2)B_2$, respectively. The E is the energy of corresponding particle. The C_{ab} coefficients can be obtained from [7].

12th International Conference on Meson-Nucleon Physics and the Structure of the Nucleon (MENU 2010)
AIP Conf. Proc. 1374, 557-560 (2011); doi: 10.1063/1.3647202

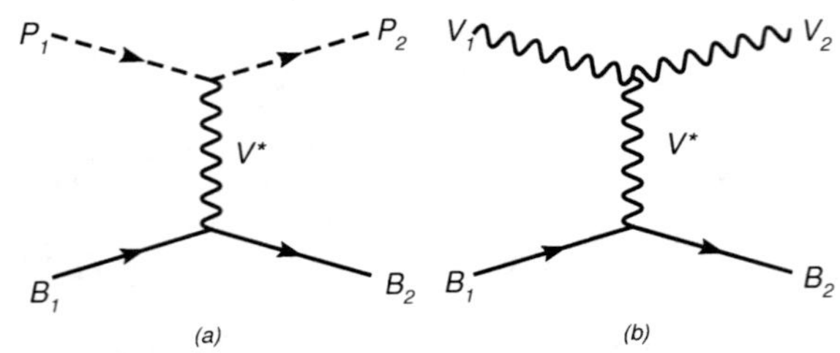

FIGURE 1. The Feynman diagrams of pseudoscalar-baryon (a) or vector- baryon (b) interaction via the exchange of a vector meson. P_1, P_2 is D^-, $\bar{D}^0$ or D_s^-, and V_1, V_2 is D^{*-}, $\bar{D}^{*0}$ or D_s^{*-}, and B_1, B_2 is Σ_c, Λ_c^+, Ξ_c, Ξ_c' or Ω_c, and V^* is ρ, K^*, ϕ or ω.

The G function is a loop function of a meson (P) and a baryon(B) which we regularize by putting a cutoff in the three-momentum. The formula is:

$$G_{(P,B)} = i2M_B \int \frac{d^4q}{(2\pi)^4} \frac{1}{(P-q)^2 - M_B^2 + i\varepsilon} \frac{1}{q^2 - M_P^2 + i\varepsilon} = \int_0^\Lambda \frac{q^2 dq}{4\pi^2} \frac{2M_B(\omega_P + \omega_B)}{\omega_P \omega_B ((P^0)^2 - (\omega_P + \omega_B)^2 + i\varepsilon)}, \tag{3}$$

where q is the four-momentum of the meson, P is the total momentum of the meson and the baryon. $\omega_i = \sqrt{\vec{q}^{\,2} + M_i^2}$ and Λ is the cutoff parameter in the three-momentum of the function loop.

We also can calculate it in dimensional regularization by means of the formula

$$\begin{aligned} G_{(P,B)} &= \frac{2M_B}{16\pi^2}\{a_\mu + \ln\frac{M_B^2}{\mu^2} + \frac{M_P^2 - M_B^2 + s}{2s}\ln\frac{M_P^2}{M_B^2} + \frac{\bar{q}}{\sqrt{s}}[\ln(s - (M_B^2 - M_P^2) + 2\bar{q}\sqrt{s}) + \ln(s + (M_B^2 - M_P^2) + 2\bar{q}\sqrt{s}) \\ &\quad - \ln(-s - (M_B^2 - M_P^2) + 2\bar{q}\sqrt{s}) - \ln(-s + (M_B^2 - M_P^2) + 2\bar{q}\sqrt{s})]\}, \end{aligned} \tag{4}$$

where $s = P^2$ and $\bar{q}$ denotes the three momentum of the meson or baryon in the center of mass frame. The μ is a regularization scale, for which we put 1000 MeV. Changes in the scale are reabsorbed in the subtraction constant a_μ to make results scale independent. When $\sqrt{s}$ is above the threshold in Eq. (4), we must change $\bar{q}$ by $-\bar{q}$ to find poles [8].

For these two types of G function, the free parameters are a_μ in Eq. (4) and Λ in Eq. (3). When we choose a_μ or Λ, the shapes of these two functions are almost the same close to threshold and they take the same value at threshold.

Then we can get the unitary T amplitudes by solving the coupled channels Bethe-Salpeter equation in the on shell factorization approach of [9, 10, 11] $T = [1 - VG]^{-1}V$. When we look for poles in the complex plane of $\sqrt{s}$, poles in the T matrix that appear in the first Riemann sheet below threshold are considered as bound states whereas those located in the second Riemann sheet and above the threshold of some channel are identified as resonances. These amplitudes behave close to the pole as $T_{ab} = g_a g_b/(\sqrt{s} - z_R)$. We can use the residue of T_{aa} to determine the value of g_a, except for a global phase. Then, the other couplings are derived from $g_b = \lim_{\sqrt{s}\to z_R}(g_a T_{ab}/T_{aa})$.

RESULT AND DISCUSSION

Here we show the results for the different sectors. By using the two G functions, the poles appear in both cases below threshold and therefore they are bound states. We show the pole positions for different values of $\alpha(\Lambda)$ in Table 1. Then we take intermediate value of $\alpha = -2.3$ to study them in detail. The coupling constants are listed in Table 2.

TABLE 1. Pole position from $PB \to PB/\ VB \to VB$ using the two different G functions of Eqs. (3) and (4). The units are in MeV.

(I,S)	$\alpha = -2.2(\Lambda = 0.7)$	$\alpha = -2.3(\Lambda = 0.8)$	$\alpha = -2.4(\Lambda = 0.9)$
	z_R	z_R	z_R
$(1/2, 0)$	4291(4273)/4438(4410)	4269(4236)/4418(4372)	4240(4187)/4391(4320)
$(0, -1)$	4247(4120)/4399(4256)	4213(4023)/4370(4155)	4170(3903)/4330(4030)
	4422(4394)/4568(4532)	4403(4357)/4550(4493)	4376(4308)/4526(4441)

We find six poles in our calculation. From Table 2, we see that both the $N^*(4269)$ and the $\Lambda^*(4403)$ depend on one channel, $\bar{D}\Sigma_c$ and $\bar{D}\Xi'_c$, respectively. As shown in Table 1, the uncertainties of these pole position are of the order of 100 MeV, which are typical in any hadron model. These two poles are rather stable. However, the $\Lambda^*(4213)$ depends on two channels, $\bar{D}_s\Lambda_c^+$ and $\bar{D}\Xi_c$, the uncertainties are much larger and the pole position is very unstable.

TABLE 2. Pole positions, z_R and coupling constants, g_a, for the states from $PB \to PB/\ VB \to VB$.

(I,S)	z_R (MeV)	g_a		
$(1/2,0)$		$\bar{D}\Sigma_c/\ \bar{D}^*\Sigma_c$	$\bar{D}\Lambda_c^+/\ \bar{D}^*\Lambda_c^+$	
	4269/ 4418	2.85/ 2.75	0/ 0	
$(0,-1)$		$\bar{D}_s\Lambda_c^+/\ \bar{D}_s^*\Lambda_c^+$	$\bar{D}\Xi_c\ /\ \bar{D}^*\Xi_c$	$\bar{D}\Xi'_c/\bar{D}^*\Xi'_c$
	4213/ 4370	1.37/ 1.23	3.25/ 3.14	0/0
	4403/ 4550	0/ 0	0/ 0	2.64/2.53

As all the states that we find have zero width, we consider the decay of the states to a light baryon plus either a light meson or a charmonium through heavy charmed meson exchanges by means of box diagrams as it was done in [12, 13]. The results are given in Table 3. Because an anomalous VVP vertex is very small [12], so we do not consider the transitions between VB and PB states. The transitions between VB and PB states through t-channel pseudoscalar meson exchanges are also found to be very small. As an example, we estimate the partial decay width of $N_{c\bar{c}}^{*+}(4265)$ decaying to the $J/\psi\, p$ is about 0.01 MeV, which is 3 orders of magnitude smaller than the decay to $\eta_c p$ of 23.4 MeV.

TABLE 3. Mass (M), total width (Γ), and the partial decay width (Γ_i) for the states from $PB\ /\ VB$ with units in MeV.

(I,S)	M	Γ	Γ_i					
$(1/2,0)$			$\pi N/\ \rho N$	$\eta N/\ \omega N$	$\eta' N/\ \phi N$	$K\Sigma/\ K^*\Sigma$	$\eta_c N/\ J/\psi N$	
	4261/ 4412	56.9/ 47.3	3.8/ 3.2	8.1/ 10.4	3.9 / 0	17.0/ 13.7	23.4/ 19.2	
$(0,-1)$			$\bar{K}N/\ \bar{K}^*N$	$\pi\Sigma/\ \rho\Sigma$	$\eta\Lambda/\ \omega\Lambda$	$\eta'\Lambda/\ \phi\Lambda$	$K\Xi/\ K^*\Xi$	$\eta_c\Lambda/\ J/\psi\Lambda$
	4209/ 4368	32.4/ 28.0	15.8/ 13.9	2.9/ 3.1	3.2/ 0.3	1.7/ 4.0	2.4/ 1.8	5.8/ 5.4
	4394/ 4544	43.3/ 36.6	0/ 0	10.6/ 8.8	7.1/ 9.1	3.3/ 0	5.8/ 5.0	16.3/ 13.8

It is very interesting that the six N^* and Λ^* states are all above 4200 MeV, but with small decay widths. Because of the large $c\bar{c}$ components in these states, all these decays necessarily involve the exchange of a heavy charmed vector meson and hence are suppressed. Of course, these predicted narrow resonances are beyond the 3-quark model.

PRODUCTION CROSS SECTION IN $\bar{p}p$ COLLISIONS

In order to look for these predicted new N^* and Λ^* states, we estimate the production cross section of these resonances at FAIR. A $\bar{p}$ beam of 15 GeV/c allows one to observe N^* resonances in $\bar{p}X$ production up to a mass $M_X \simeq 4538\ MeV$ or Y^* hyperon resonances in $\bar{\Lambda}Y$ production up to a mass $M_Y \simeq 4355\ MeV$. We estimate two reactions $p\bar{p} \to p\bar{p}\eta_c$ and $p\bar{p} \to p\bar{p}J/\psi$. The contribution of new resonances for these two reaction are shown in Fig. 2(a,b,e,f) and the conventional mechanism are shown in Fig. 2(c,d,g,h). It is found that the cross section of the conventional mechanism is much smaller than that of the new mechanism. The total cross sections of the reaction $p\bar{p} \to p\bar{p}\eta_c$ and $p\bar{p} \to p\bar{p}J/\psi$ are about $0.07-0.7\mu b$ and $0.002-0.037\mu b$ in the Fig.3, corresponding to the case with and without form factors. With the designed luminosity of about $10^{31}cm^{-2}s^{-1}$ for the $\bar{p}$ beam at FAIR [14], this corresponds to an event production rate of more than 60000 and 1700 events per day.

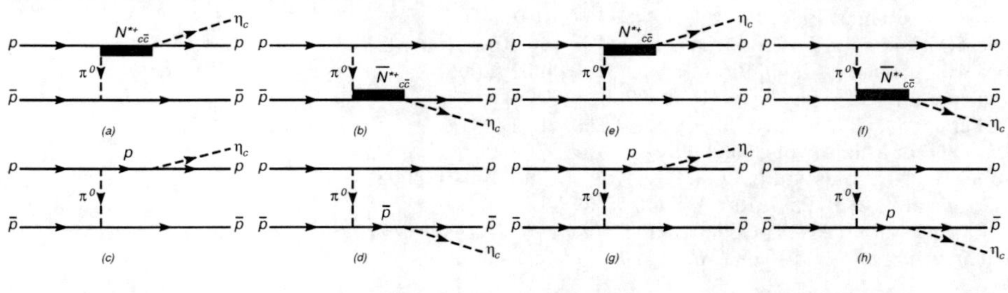

FIGURE 2. The different Feynman diagrams of the reaction $p\bar{p} \to p\bar{p}\eta_c$(right) and $p\bar{p} \to J/\psi p\bar{p}$(left)

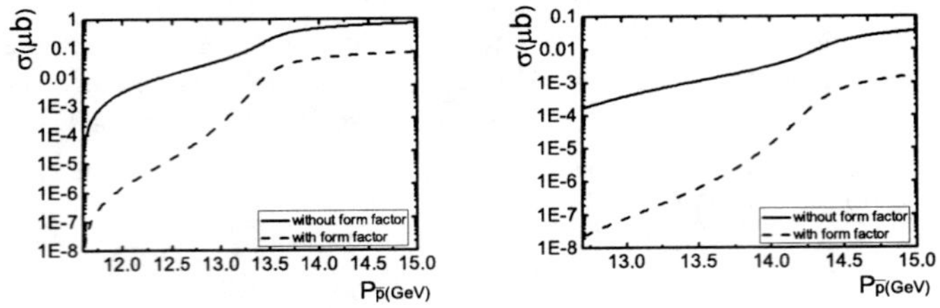

FIGURE 3. The total cross section vs the beam momentum of $\bar{p}$ for reaction $p\bar{p} \to p\bar{p}\eta_c$(right) and $p\bar{p} \to p\bar{p}J/\psi$(left). The solid lines are calculated with the form facts and the dashed lines are obtained without form facts.

SUMMARY

In summary, we find six states from PB and VB channels by using the local hidden gauge Lagrangian in combination with unitary techniques in coupled channels. All of these states have large $c\bar{c}$ components, so their masses are all larger than 4200 MeV. The width of these states decaying to light meson and baryon channels without $c\bar{c}$ components are all very small. On the other hand, the $c\bar{c}$ meson - light baryon channels are also considered and found to contribute to the width of these states. Then $\eta_c N$ and $\eta_c \Lambda$ are added to the PB channels, while $J/\psi N$ and $J/\psi\Lambda$ are added in the VB channels. The widths from decay to these channels are not negligible, in spite of the small phase space for the decay, because the exchange $D^*(orD_s^*)$ mesons were less off-shell than the corresponding one in the decay to light meson - light baryon channels. The total widths of these states are still very small. We made some estimates of cross sections for production of these resonances at the upcoming FAIR facility. The cross sections of the reaction $p\bar{p} \to p\bar{p}\eta_c$ and $p\bar{p} \to p\bar{p}J/\psi$ are about $0.07 - 0.7\mu b$ and $0.002 - 0.037\mu b$, in which the main contribution comes from the predicted $N^*_{c\bar{c}}(4265)$ and $N^*_{c\bar{c}}(4418)$ states, respectively. With this theoretical results, one can estimate about 60000 and 1700 events per day at the PANDA/FAIR facility.

ACKNOWLEDGMENTS

This work is partly supported by DGICYT Contract No. FIS2006-03438, the Generalitat Valenciana in the project PROMETEO and the EU Integrated Infrastructure Initiative Hadron Physics Project under contract RII3-CT-2004-506078. This work is also partly supported by the National Natural Science Foundation of China (NSFC) under grants Nos. 10875133, 10821063, and by the Chinese Academy of Sciences under project No. KJCX3-SYW-N2, and by the Ministry of Science and Technology of China (2009CB825200).

REFERENCES

1. N. Kaiser, P. B. Siegel and W. Weise, Phys. Lett. B **362**, 23 (1995) [arXiv:nucl-th/9507036].
2. T. Inoue, E. Oset and M. J. Vicente Vacas, Phys. Rev. C **65**, 035204 (2002) [arXiv:hep-ph/0110333].
3. J. Nieves and E. Ruiz Arriola, Phys. Rev. D **64**, 116008 (2001)
4. B. C. Liu and B. S. Zou, Phys. Rev. Lett. **96**, 042002 (2006) [arXiv:nucl-th/0503069].
5. L. S. Geng, E. Oset, B. S. Zou and M. Doring, Phys. Rev. C **79**, 025203 (2009) [arXiv:0807.2913 [hep-ph]].
6. E. Oset and A. Ramos, Euro. Phys. J. A **44**, 445 (2010).
7. J. J. Wu, R. Molina, E. Oset and B. S. Zou, arXiv:1007.0573 [nucl-th].
8. L. Roca, E. Oset and J. Singh, Phys. Rev. D **72**, 014002 (2005)
9. J. A. Oller and E. Oset, Phys. Rev. D **60**, 074023 (1999)
10. J. A. Oller and U. G. Meissner, Phys. Lett. B **500**, 263 (2001).
11. J. Nieves and E. Ruiz Arriola, Nucl. Phys. A **679**, 57 (2000)
12. R. Molina, D. Nicmorus and E. Oset, Phys. Rev. D **78**, 114018 (2008)
13. L. S. Geng and E. Oset, Phys. Rev. D **79**, 074009 (2009)
14. J. G. Messchendorp [PANDA Collaboration], *In the Proceedings of 11th International Conference on Meson-Nucleon Physics and the Structure of the Nucleon (MENU 2007), Julich, Germany, 10-14 Sep 2007, pp 123*

Heavy quark spin symmetry and heavy hadronic molecules

Feng-Kun Guo

Institut für Kernphysik and Jülich Center for Hadron Physics, Forschungszentrum Jülich, D–52425 Jülich, Germany

Abstract. In the heavy quark limit there is heavy quark spin symmetry. As a consequence, heavy hadrons or heavy quarkonia form spin multiplets. We argue that due to the symmetry, also heavy flavor hadronic molecules form spin multiplets, and the hyperfine splittings are the same as those between the heavy hadrons which are components of the molecules. Hence it provides a method to identify the nature of heavy hadronic molecules. As an example, we give predictions of the properties of an $\eta_c' f_0(980)$ bound state assuming the $Y(4660)$ to be an S-wave $\psi' f_0(980)$ shallow bound state. We also show that the enhancement observed in the $\Lambda_c^+\Lambda_c^-$ invariant mass distribution might be the $Y(4660)$.

Keywords: Heavy quark spin symmetry, hadronic molecules, $Y(4660)$
PACS: 12.39.Hg, 12.39.Mk, 14.40.Gx

There is a renaissance in hadron spectroscopy, and lots of new hadronic states were observed in recent years. Many of them do not fit into the expectations from conventional quark models, and hence were proposed to be exotics. Here we will focus on hadronic molecules which are bound states of hadrons bound together by strong interactions.

Hadronic molecules have special model-independent properties which can be used to identify them:

- For an S-wave loosely bound state, whose binding energy is small so that the scattering length of the constituents is large compared with the force range, the effective coupling, at the leading order (LO), to its constituents is fully determined by the binding energy [1, 2], i.e.

$$\frac{g^2}{4\pi} = 4(m_1+m_2)^2\sqrt{\frac{2\varepsilon}{\mu}}\left[1+\mathcal{O}\left(\frac{r}{a}\right)\right], \tag{1}$$

  where m_1 and m_2 denote the masses of the constituents, the binding energy is $\varepsilon = m_1+m_2-M$ with M the mass of the bound state, $\mu = m_1m_2/(m_1+m_2)$, $a = -1/\sqrt{2\mu\varepsilon}$ is the scattering length and r is the force range. Historically, Weinberg used this relation to identify the deuteron as a bound state of the proton and neutron [1]. Note that the above equation is valid for a pure S-wave loosely bound state. If there are another components, such as a genuine quark-model state, then the coupling is smaller.
- For a hadronic molecule made of a heavy hadron or a heavy quarkonium and a light hadron system, one can predict the properties of its spin partner with the help of the heavy quark spin symmetry [3]. Because the leading interaction of the heavy quark does not depend on its spin, and the chromo-magnetic interaction is suppressed by $1/m_Q$, heavy hadrons or heavy quarkonia form spin multiplets [4], such as the $\{D,D^*\}$, $\{\Sigma_c,\Sigma_c^*\}$, $\{\eta_c,J/\psi\}$ and so on. In the heavy quark limit they are degenerate and the hyperfine splittings start from the order $\mathcal{O}(1/m_Q)$ with m_Q being the heavy quark mass. For a hadronic molecule made of a heavy hadron and a light hadron system, because the LO interaction does not depend on the heavy quark spin, it should have its spin partner made of the same light hadron system and the spin partner of the heavy hadron. More importantly, these two hadronic molecules should have the same binding energy with corrections of the order $\mathcal{O}(\Lambda_{\rm QCD}/m_Q)$. There is a nice example, $M_{D_{s1}(2460)} - M_{D_{s0}^*(2317)} \approx M_{D^*} - M_D$, which can be easily explained if the $D_{s0}^*(2317)$ and the $D_{s0}^*(2460)$ are DK and D^*K bound states, respectively. In fact, it was shown in Ref. [5] that the spin symmetry breaking term at $\mathcal{O}(\Lambda_{\rm QCD}/m_Q)$ is very small, and hence makes the deviation from the above equality a $\mathcal{O}([\Lambda_{\rm QCD}/m_Q]^2)$ effect. For a hadronic molecule made of a heavy quarkonium and a light hadron system, the suppression of the spin-dependent interactions between a heavy quarkonium and light hadrons is at least $\mathcal{O}([\Lambda_{\rm QCD}/m_Q]^2)$. This is because the number of the exchanged gluons between the heavy quarkonium and the light hadron system should be at least two, and the two gluons should both be chromo-magnetic ones (a chromo-magnetic gluon and a chromo-electric one are CP-odd). As a result, a bound state of a heavy quarkonium and light hadrons will have partner(s) whose constituents are the same light hadrons and the spin-multiplet partner(s) of the heavy quarkonium. The mass

12th International Conference on Meson-Nucleon Physics and the Structure of the Nucleon (MENU 2010)
AIP Conf. Proc. 1374, 561-564 (2011); doi: 10.1063/1.3647203

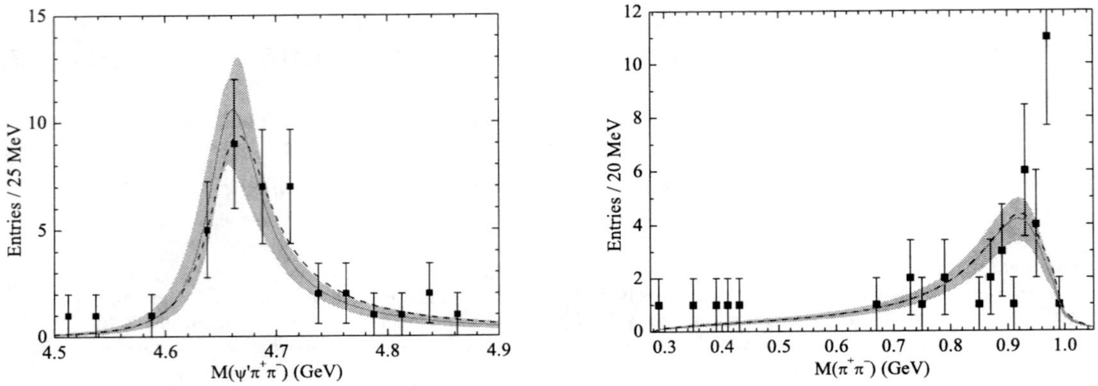

FIGURE 1. Left: Comparison of the line shape in the $\psi'\pi^+\pi^-$ invariant mass distribution derived from the molecular model with the data. Right: Comparison of the predicted $\pi\pi$ invariant mass distribution with the data.

splitting within the molecular spin-multiplet will be almost the same as the heavy quarkonia hyperfine splitting, and the binding energy of such a hadronic molecule will be the same as its spin partner(s) up to corrections of order $\mathscr{O}([\Lambda_{\rm QCD}/m_Q]^2)$.

The heavy quark spin symmetry enables us to predict new heavy meson molecules and provides us with a method to test heavy hadron molecule assumptions of some newly observed states. In the following, we will give predictions on the properties of an $\eta_c' f_0(980)$ bound state assuming the $Y(4660)$ to be an S-wave $\psi' f_0(980)$ bound state.

The $Y(4660)$ was observed by the Belle Collaboration in the $\psi'\pi^+\pi^-$ final state using the method of initial state radiation (ISR) [6]. Hence its quantum numbers are $J^{PC} = 1^{--}$. From fitting the line shape using a P-wave Breit-Wigner amplitude, the mass and width were reported as $4664 \pm 11 \pm 5$ MeV (i.e. well above $D\bar{D}$ threshold) and $48 \pm 15 \pm 3$ MeV, respectively. One can expect that a vector charmonium with such a large mass would decay mainly into the channels with charmed and anti-charmed mesons or baryons. However, the structure was neither observed in $e^+e^- \to \gamma_{ISR}\pi^+\pi^- J/\psi$ [7], nor in the exclusive $e^+e^- \to D\bar{D}, D\bar{D}^*, D^*\bar{D}^*, D\bar{D}\pi$ cross sections [8], nor in the process $e^+e^- \to J/\psi D^{(*)}\bar{D}^{(*)}$ [9]. The only possible evidence in open charm channels was observed in the $\Lambda_c^+\Lambda_c^-$ invariant mass distribution in $e^+e^- \to \gamma_{\rm ISR}\Lambda_c^+\Lambda_c^-$ [10]. The reported mass is 4634^{+8+5}_{-7-8} MeV, and the width is 92^{+40+10}_{-24-21} MeV. However, one does not know whether or not this structure is from the $Y(4660)$. Furthermore, the $\pi^+\pi^-$ invariant mass spectrum for the decay $Y(4660) \to \psi'\pi^+\pi^-$ shows a single peak at the high end, i.e. towards the mass region of the scalar meson $f_0(980)$. In Ref. [11], we notice that these experimental facts can be understood easily if the $Y(4660)$ is an S-wave $\psi' f_0(980)$ hadronic molecule. Since the ψ' is relatively stable, such a bound state would decay through the decays of the $f_0(980)$, and hence $Y(4660) \to \psi'\pi^+\pi^-$, which is just the observation channel, is a natural decay mode.

The nominal threshold of the ψ' and the $f_0(980)$ coincides with the mass of the $Y(4660)$. Assuming that the latter is a loosely bound state Eq. (1) implies that the coupling of the $Y(4660)$ to the ψ' and the $f_0(980)$ can be related to the small binding energy. By implementing this constraint, we fit the measured line shape of the $Y(4660)$ in the $\psi'\pi^+\pi^-$ mass distribution with only two parameters: the mass of the $Y(4660)$, M_Y, and an overall normalization constant N, for details see Ref. [11]. The best fit is shown as the solid line in the left panel of Fig. 1. The band reflects the uncertainty emerging from the fit. The resulting parameters are

$$M_Y = 4665^{+3}_{-5}\,\text{MeV}, \quad N = 10 \pm 2\,\text{GeV}^3. \tag{2}$$

Correspondingly, the coupling constant is in the range of $11 - 14$ GeV. The dashed line shows the best-fit result without the bound state constraint. In this case, we found $M_Y = (4672 \pm 9)$ MeV and $g = (13 \pm 2)$ GeV, i.e. values which are consistent with the bound-state fitting results. Furthermore, we checked that a larger width in a range of 30 ± 30 MeV is still compatible with the data with the bound state condition. The additional width may come from decays into open charmed mesons or baryons and so on.

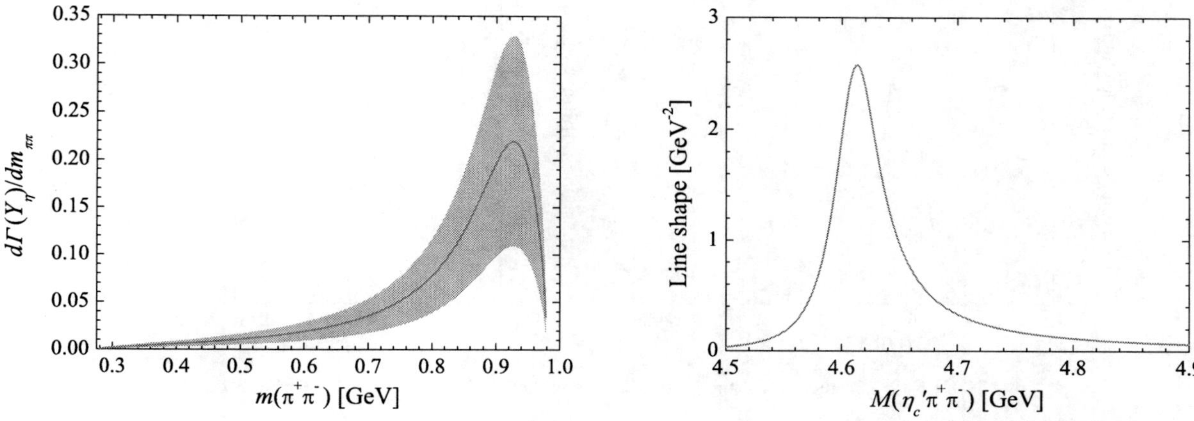

FIGURE 2. Left: The $\pi^+\pi^-$ invariant mass distribution for the $Y_\eta \to \eta_c'\pi^+\pi^-$ decay. Right: The predicted line shape of the Y_η in the $\eta_c'\pi^+\pi^-$ invariant mass distribution.

Using the parameters determined from fitting to the line shape, the $\pi^+\pi^-$ invariant mass distribution can be predicted as shown in the right panel of Fig. 1. The prediction agrees with the measured distribution very well, which supports the $\psi' f_0(980)$ bound state assumption.

With the help of the heavy quark spin symmetry as discussed before, we can predict the properties of the spin partner of $\psi' f_0(980)$ hadronic molecule [3], called Y_η in the following. It is an $\eta_c' f_0(980)$ bound state with quantum numbers $J^{PC} = 0^{-+}$ and a mass of

$$M_{Y_\eta} = M_{Y(4660)} - \left(M_{\psi'} - M_{\eta_c'}\right) = 4616^{+5}_{-6} \pm 1 \text{ MeV}. \quad (3)$$

The first uncertainty is from the mass of the $Y(4660)$ as given in Eq. (2) and the experimental uncertainties of the charmonia masses. The second uncertainty is from the $\mathcal{O}([\Lambda_{\rm QCD}/m_c]^2)$ spin-dependent effects. Using the Y_η mass given above and the coupling constant as given by the relation in Eq. (1), we predicted the width $\Gamma(Y_\eta \to \eta_c'\pi\pi) = 60 \pm 30$ MeV and the $\pi^+\pi^-$ invariant mass distribution for the $Y_\eta \to \eta_c'\pi^+\pi^-$ decay. The result is shown in the left panel of Fig. 2, where the solid line was obtained using the central values of all parameters, and the band reflects the 50% uncertainty from the bound state constraint in Eq. (1). Assuming that the width of the Y_η is saturated by the decays $Y_\eta \to \eta_c'\pi\pi(KK)$, the line shape of the Y_η in the $\eta_c'\pi^+\pi^-$ invariant mass distribution is predicted as the spectral function of the Y_η, shown in the right panel of Fig. 2. The Y_η can be searched for in B decays. For instance, one can measure the $\eta_c'\pi^+\pi^-$ invariant mass distribution of the $B^\pm \to \eta_c' K^\pm\pi^+\pi^-$, the branching fraction of which is estimated to be $\mathcal{B}(B^\pm \to \eta_c' K^\pm\pi^+\pi^-) \sim 1 \times 10^{-3}$ [3]. Such a large branching fraction offers an excellent opportunity for finding the Y_η in B decays.

In Ref. [12] we demonstrated that the $X(4630)$ observed in the $\Lambda_c^+\Lambda_c^-$ mass distribution using ISR could be consistent with the $\psi' f_0(980)$ molecular picture of the $Y(4660)$ taking into account the $\Lambda_c^+\Lambda_c^-$ final state interaction [13]. A combined fit to the cross section data of the reactions $e^+e^- \to \gamma_{\rm ISR}\psi'\pi^+\pi^-$ [6] and $e^+e^- \to \gamma_{\rm ISR}\Lambda_c^+\Lambda_c^-$ [10] was obtained with $\chi^2/\text{d.o.f.} = 1.4$ (for details, see Ref. [12]). If the binding energy is very small, the $\psi' f_0(980)$ bound state may even decay mainly into the $\Lambda_c^+\Lambda_c^-$. From the best fit, we got $\Gamma(Y \to \Lambda_c^+\Lambda_c^-)/\Gamma(Y \to \psi'\pi^+\pi^-) = 11.5$. The important point is that predictions for the Y_η can be made based on the spin symmetry. In Fig. 3, the predictions for the Y_η line shapes in the $\eta'\pi^+\pi^-$ and $\Lambda_c^+\Lambda_c^-$ channels are shown in arbitrary units, however, with the relative normalization fixed. The solid and dashed curves represent results with and without FSI, respectively.

In summary, the heavy quark spin symmetry is very useful in the context of heavy exotic hadrons. The heavy flavor hadronic molecules spin multiplet partners, and their hyperfine splittings should be approximately the same as those between the heavy hadrons they are composed of. In this way, the approximate equality of $M_{D^*} - M_D$ and $M_{D_{s1}(2460)} - M_{D_{s0}^*(2317)}$ can be understood without introducing additional mechanism. If the $Y(4660)$ observed in the $\psi'\pi\pi$ mass distribution can indeed be understood as an S-wave $\psi' f_0(980)$ bound state, we proposed the existence of an $\eta_c' f_0(980)$ bound state as the spin multiplet partner of the $Y(4660)$. Spin symmetry also allows us to predict the width from that of the $Y(4660)$. To search for it one should measure the $\eta_c'\pi^+\pi^-$ invariant mass distribution of the

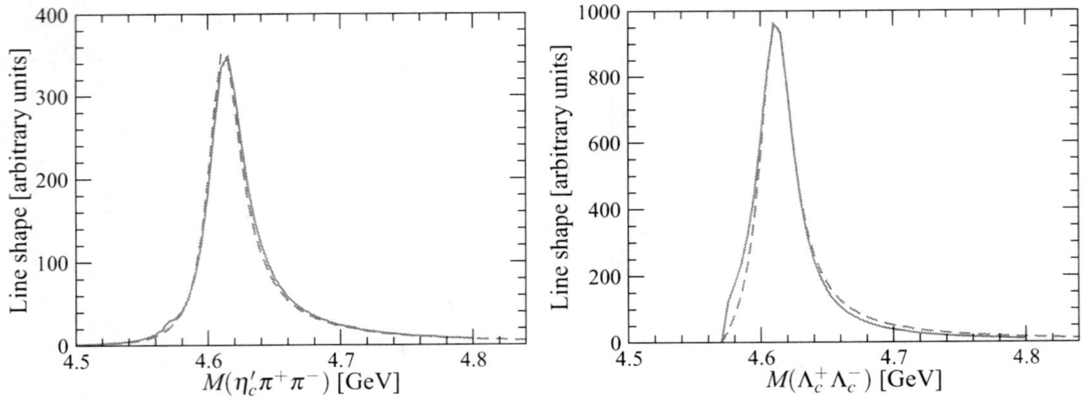

FIGURE 3. Predictions of the Y_η line shapes in the $\eta'\pi^+\pi^-$ and $\Lambda_c^+\Lambda_c^-$ in arbitrary units.

$B^\pm \to K^\pm \eta_c'\pi^+\pi^-$. Data with better quality for the $B^\pm \to K^\pm \Lambda_c^+\Lambda_c^-$ decays are necessary to study the structures in the $\Lambda_c^+\Lambda_c^-$ mass distribution in more detail, which might contain both the Y_η and $Y(4660)$. If the Y_η will be discovered in the future with properties predicted, it would be a direct support of the $\psi' f_0(980)$ bound state interpretation of the $Y(4660)$. We also demonstrated that the $X(4630)$ might be, but not necessarily, the same state as the $Y(4660)$, and can be likewise described in the $\psi' f_0(980)$ bound state scenario. Predictions of the Y_η were also given based on this conjecture. At last, besides the heavy flavor bound states, the heavy quark spin symmetry should also have similar implications for other dynamically generated systems, such as resonances, involving one heavy hadron.

ACKNOWLEDGMENTS

I am very grateful to J. Haidenbauer, C. Hanhart and U.-G. Meißner for the fruitful collaboration. I would like to thank the organizers for their invitation and the nice organization. This work is partially supported by the Helmholtz Association through funds provided to the virtual institute "Spin and strong QCD" (VH-VI-231), by the DFG (SFB/TR 16, "Subnuclear Structure of Matter"), by the European Community-Research Infrastructure Integrating Activity "Study of Strongly Interacting Matter" (acronym HadronPhysics2, Grant Agreement n. 227431) under the Seventh Framework Programme of the EU, and by BMBF (grant 06BN9006).

REFERENCES

1. S. Weinberg, *Phys. Rev.* **130**, 776–783 (1963); *Phys. Rev.* **131**, 440–460 (1963); *Phys. Rev.* **137**, B672–B678 (1965).
2. V. Baru et al., *Phys. Lett. B* **586**, 53–61 (2004).
3. F. K. Guo, C. Hanhart, U-G. Meißner, *Phys. Rev. Lett.* **102**, 242004 (2009).
4. M. Neubert, *Phys. Rept.* **245**, 259–396 (1994).
5. M. Cleven, F. K. Guo, C. Hanhart and U-G. Meißner, arXiv:1009.3804 [hep-ph].
6. X. L. Wang et al. [Belle Collaboration], *Phys. Rev. Lett.* **99**, 142002 (2007).
7. C. Z. Yuan et al. [Belle Collaboration], *Phys. Rev. Lett.* **99**, 182004 (2007).
8. G. Pakhlova et al. [Belle Collaboration], *Phys. Rev. Lett.* **98**, 092001 (2007); G. Pakhlova et al. [Belle Collaboration], *Phys. Rev. Lett.* **100**, 062001 (2008); G. Pakhlova et al. [Belle Collaboration], *Phys. Rev. D* **77**, 011103 (2008); B. Aubert et al. [BABAR Collaboration], arXiv:0710.1371 [hep-ex].
9. P. Pakhlov et al. [Belle Collaboration], *Phys. Rev. Lett.* **100**, 202001 (2008).
10. G. Pakhlova et al. [Belle Collaboration], *Phys. Rev. Lett.* **101**, 172001 (2008).
11. F. K. Guo, C. Hanhart, U.-G. Meißner, *Phys. Lett. B* **665**, 26–29 (2008).
12. F. K. Guo, J. Haidenbauer, C. Hanhart and U.-G. Meißner, arXiv:1005.2055 [hep-ph].
13. J. Haidenbauer and G. Krein, *Phys. Lett. B* **687**, 314–319 (2010).

Charmonium properties in a renormalization scheme

J. Segovia*, D. R. Entem*, F. Fernández* and E. Ruiz Arriola†

*Grupo de Física Nuclear, IUFFyM, Universidad de Salamanca, E-37008 Salamanca, Spain
†Departamento de Física Atómica, Molecular y Nuclear, Universidad de Granada, E-18071 Granada, Spain.

Abstract.
Constituent quark models, while successful, require a great deal of fine tunning of the short distance interactions by introducing phenomenological gluonic form factors which are ultimately designed to accurately reproduce the spectrum. We apply renormalization ideas to reduce the short distance sensitivity and show that, as naively expected, but not explicitly implemented in the models, the physics of binding is entirely linked to the string tension whereas leptonic decays depend more on the gluon exchange potential.

Keywords: Charmonium, renormalization, quark models
PACS: 14.40.Gx, 12.39.Pn, 11.10.Gh

INTRODUCTION

Since the early days of QCD, many models have been used to understand the physics of charmonium states as a non-relativistic $c\bar{c}$ system. Almost all models incorporate a confinement potential supplemented by a one gluon exchange contribution. The relativistic corrections of such potentials typically involve singular terms. These contributions have been studied in two ways, treating them as perturbative contributions [1] or non-perturbatively with the help of ad-hoc regulators [2]. For definiteness we will consider Ref. [3] as our model of reference since it has been applied to the study of the 1^{--} charmonium states [4] within a non-relativistic framework. There, regulators have indeed been incorporated to handle the singular part of the potential stemming from relativistic corrections.

Actually, an alternative way to deal with singular potentials uses the so-called renormalization with boundary conditions approach [5]. In this contribution we re-analyze the model of Ref. [4] within the new framework in order to obtain regularization independent results. The use of this scheme allows us to further study the dependence of the states on the model parameters, namely the strong coupling constant and the string tension, in a cleaner way since the regulator dependence has been replaced by a suitable renormalization condition.

UNREGULARIZED AND REGULARIZED POTENTIALS

The $q\bar{q}$ potentials are obtained from a non-relativistic reduction of one-gluon exchange (OGE) plus a confinement piece. The unregularized OGE potential reads

$$V_{OGE}^{C}(\vec{r}_{ij}) = \frac{1}{4}\alpha_s(\vec{\lambda}_i^c\cdot\vec{\lambda}_j^c)\frac{1}{r_{ij}} \qquad (1)$$

$$V_{OGE}^{T}(\vec{r}_{ij}) = -\frac{1}{16}\frac{\alpha_s}{m_i m_j}(\vec{\lambda}_i^c\cdot\vec{\lambda}_j^c)\frac{1}{r_{ij}^3}S_{ij} \qquad (2)$$

$$V_{OGE}^{SO}(\vec{r}_{ij}) = -\frac{1}{16}\frac{\alpha_s}{m_i^2 m_j^2}(\vec{\lambda}_i^c\cdot\vec{\lambda}_j^c)\frac{1}{r_{ij}^3}\left[((m_i+m_j)^2+2m_im_j)(\vec{S}_+\cdot\vec{L})+(m_j^2-m_i^2)(\vec{S}_-\cdot\vec{L})\right] \qquad (3)$$

Note that the tensor and spin-orbit terms are singular and when one treats literally the δ-function of the central part it leads to collapse of the system. All these OGE terms can be suitably regularized as follows [3]

12th International Conference on Meson-Nucleon Physics and the Structure of the Nucleon (MENU 2010)
AIP Conf. Proc. 1374, 565-568 (2011); doi: 10.1063/1.3647204

$$V_{OGE}^{C}(\vec{r}_{ij}) = \frac{1}{4}\alpha_s(\vec{\lambda}_i^c\cdot\vec{\lambda}_j^c)\left[\frac{1}{r_{ij}} - \frac{1}{6m_im_j}(\vec{\sigma}_i\cdot\vec{\sigma}_j)\frac{e^{-r_{ij}/r_0(\mu)}}{r_{ij}r_0^2(\mu)}\right] \tag{4}$$

$$V_{OGE}^{T}(\vec{r}_{ij}) = -\frac{1}{16}\frac{\alpha_s}{m_im_j}(\vec{\lambda}_i^c\cdot\vec{\lambda}_j^c)\left[\frac{1}{r_{ij}^3} - \frac{e^{-r_{ij}/r_g(\mu)}}{r_{ij}}\left(\frac{1}{r_{ij}^2}+\frac{1}{3r_g^2(\mu)}+\frac{1}{r_{ij}r_g(\mu)}\right)\right]S_{ij} \tag{5}$$

$$\begin{aligned} V_{OGE}^{SO}(\vec{r}_{ij}) &= -\frac{1}{16}\frac{\alpha_s}{m_i^2m_j^2}(\vec{\lambda}_i^c\cdot\vec{\lambda}_j^c)\left[\frac{1}{r_{ij}^3} - \frac{e^{-r_{ij}/r_g(\mu)}}{r_{ij}^3}\left(1+\frac{r_{ij}}{r_g(\mu)}\right)\right]\times \\ &\times\left[((m_i+m_j)^2+2m_im_j)(\vec{S}_+\cdot\vec{L})+(m_j^2-m_i^2)(\vec{S}_-\cdot\vec{L})\right] \end{aligned} \tag{6}$$

Moreover, the confinement interaction is taken to be [3]

$$V_{CON}^{C}(\vec{r}_{ij}) = \left[-a_c(1-e^{-\mu_c r_{ij}})+\Delta\right](\vec{\lambda}_i^c\cdot\vec{\lambda}_j^c) \tag{7}$$

$$\begin{aligned} V_{CON}^{SO}(\vec{r}_{ij}) &= -\left(\vec{\lambda}_i^c\cdot\vec{\lambda}_j^c\right)\frac{a_c\mu_c e^{-\mu_c r_{ij}}}{4m_i^2m_j^2r_{ij}}\left[((m_i^2+m_j^2)(1-2a_s)+\right. \\ &\left.+4m_im_j(1-a_s))(\vec{S}_+\cdot\vec{L})+(m_j^2-m_i^2)(1-2a_s)(\vec{S}_-\cdot\vec{L})\right] \end{aligned} \tag{8}$$

where Δ is a global constant fixing the origin of energies. At short distances this potential presents a linear behavior with an effective confinement strength $\sigma = -a_c\,\mu_c\,(\vec{\lambda}_i^c\cdot\vec{\lambda}_j^c)$ while it becomes constant at large distances.

RENORMALIZED CHARMONIUM MODEL

In the standard approach one takes the regularized potential and the regular solution at the origin. The short distance regulators of the OGE potential are then fine tuned as to reproduce the binding energy of the ground state. On the other hand, the regularization does not operate at long distances. Actually, in such a situation the confinement potential dominates. This suggests taking the ground state mass as an *input* of the calculation (the renormalization condition), integrate in the Schrödinger equation from long distances to a given short distance, r_c, and construct by orthogonality the excited states by matching logarithmic derivatives. We expect that taking r_c much smaller than any other length scale, results become insensitive to short distance details. This procedure defines the renormalized charmonium model. Note that the remarkable thing is that the potential and the ground state binding energy are *independent* variables.

Our first goal is to re-analyze the calculation of the charmonium spectrum. The charmonium masses in this scheme depend on the short distance cut-off, r_c until a small value is taken. Table 1 shows the masses predicted in the limit $r_c \to 0$ comparing the renormalization and the standard procedure. We find a perfect agreement between both schemes which provides confidence on the way the original model took into account the unknown short distance dynamics.

Next, we study the dependence of the mass with respect two important model parameters, the strong coupling constant (α_s) and our effective string tension of the confinement potential (σ). We study the dependence in the latter

TABLE 1. Masses in *MeV* of charmonium states

State	n	M_{RSC} (MeV)	M_{CQM} (MeV)	M_{EXP} (MeV)
$^3S_1$	1	$3096^\dagger$	3096	3096.916 ± 0.011
	2	3703	3703	3686.093 ± 0.034
	3	4097	4097	4039.6 ± 4.3
	4	4389	4389	-
	5	4614	4614	-
$^3D_1$	1	$3796^\dagger$	3796	3772.92 ± 0.35
	2	4153	4153	4153 ± 3
	3	4426	4426	4421 ± 4
	4	4641	4641	-

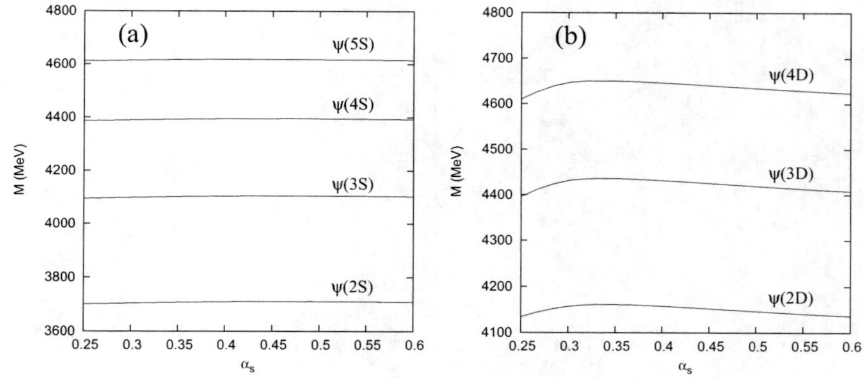

FIGURE 1. Dependence of the mass excited charmonium states on the strong coupling constant.

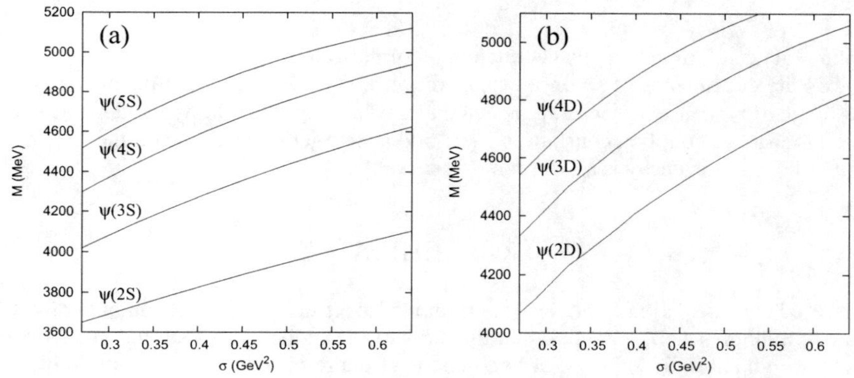

FIGURE 2. Dependence of the mass excited charmonium states on the effective string tension of our confinement potential, $\sigma = -a_c \mu_c (\vec{\lambda}_i^c \cdot \vec{\lambda}_j^c)$.

parameter without changing our threshold for the confinement. Fig. 1 shows the dependence of the mass with the strong coupling constant for S and D-wave states. For S-wave states the mass changes about 15 MeV along the range of the strong coupling constant, this change means a 0.4% in the mass. This is the reason why one can see a flattening behavior in the mass along the range of α_s. We find the same behavior for the D-wave states. Although in this case the change in mass is larger than in the case of S-wave states, the picture shows again a flattening behavior.

Fig. 2 shows the dependence effective on the string tension for the S and D-wave states, and we can see that the masses change hundreds of MeV. Since the range for σ is porcentually equal than the range of α_s we can conclude that at least the masses are almost given by the confinement potential. Unlike the standard approach, this feature reflects a sort of short distance insensitivity which does not allow to fix α_s from the spectrum.

A different scenario arises when leptonic decays are analyzed. A complete calculation of $V(\text{vector meson}) \to e^+e^-$ widths involves radiative and relativistic contributions. Thus for S-wave states one can write the leptonic decay as [6]

$$\Gamma\left(n^3S_1 \to e^+e^-\right) = \frac{4\alpha^2 e_c^2 |R_{nS}(0)|^2}{M_n^2}\left(1 - \frac{16\alpha_s}{3\pi}\right) \tag{9}$$

where $e_c = 2/3$ and M_n is the mass of the charmonium state. Decay widths depend on the value of the wave function at short distances. Of course while the decay may be triggered by a short distance operator, we may predict decay *ratios*,

$$\mathscr{R} = \frac{\Gamma\left(n^3S_1 \to e^+e^-\right)}{\Gamma\left(1^3S_1 \to e^+e^-\right)} = \frac{|R_{nS}(0)|^2}{|R_{1S}(0)|^2}\frac{M_1^2}{M_n^2} \tag{10}$$

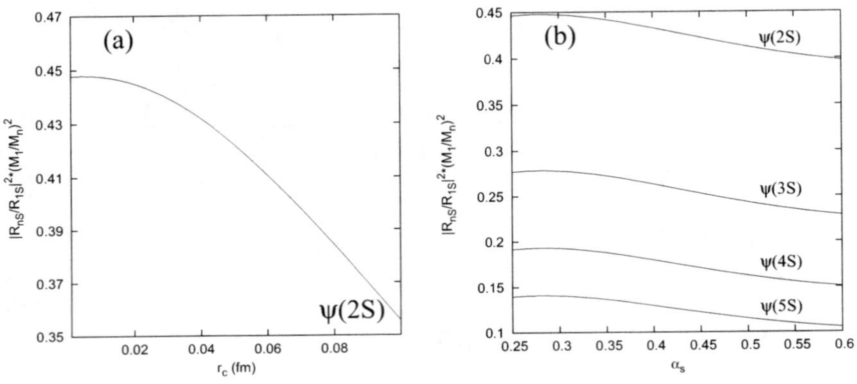

FIGURE 3. Left picture shows the dependence of the ratio $\mathscr{R}$ with respect to short distance cut-off. Right picture shows the dependence of the ratio $\mathscr{R}$ for the excited S-wave charmonium states on the strong coupling constant.

corresponding in practice to implement wave function renormalization.

Fig. 3(a) shows the dependence of $\mathscr{R}$ on the short distance cut-off for the first radial excitation of S-wave states. At some small range of r_c, the ratio does not depend on its value. Fig. 3(b) shows the dependence of this ratio for the S-wave states on the strong coupling constant, α_s. We find a stronger dependence on the strong coupling constant as expected because the leptonic decay is a short range observable.

CONCLUSIONS

We have re-analyzed the calculation of the 1^{--} charmonium spectrum in a constituent quark model using a renormalization boundary condition scheme. We find that the mass of the excited states strongly depend on the string tension parameter as one would naturally expect. Further, there is also a remarkable insensitivity to the strong coupling constant entering the one gluon exchange contribution of the potential. This avoids a great deal of unphysical fine tuning which suggested taking for this parameter unnaturally large values $\alpha_s = 0.3 - 0.4$. In contrast, the leptonic decay widths depend strongly on the strong coupling constant, also as expected because is a short range observable. In summary, the renormalization approach disentangles quite naturally and explicitly the long and short distance dependence of hadronic properties.

ACKNOWLEDGMENTS

Supported by Ministerio de Ciencia y Tecnología under Contracts No. FPA2007-65748, by Junta de Castilla y León under Contract GR12, by the European Community-Research Infrastructure Integrating Activity "Study of Strongly Interacting Matter" (HadronPhysics2 Grant no. 227431), the Spanish Ingenio-Consolider 2010 Program CPAN (CSD2007-00042), the Spanish DGI and FEDER funds with grant no. FIS2008-01143/FIS, Junta de Andalucía grants no. FQM225-05.

REFERENCES

1. E. Eichten, and F. Feinberg, *Phys. Rev D* **23**, 2724–2744 (1981).
2. S. Godfrey, and N. Isgur, *Phys. Rev. D* **32**, 189–231 (1985).
3. J. Vijande, F. Fernández, and A. Valcarce, *J. Phys. G: Nucl. Part. Phys.* **31**, 481–506 (2005).
4. J. Segovia, A. M. Yasser, D. R. Entem, and Fernández, *Phys. Rev. D* **78**, 114033 (2008).
5. M. Pavón Valderrama, and E. Ruiz Arriola, *Phys. Lett. B* **580**, 149–156 (2004).
6. R. Barbieri, R. Gatto, K. R., and K. Z., *Phys. Lett. B* **57**, 455–459 (1975).

Bottom Spectroscopy at CDF

C. Calancha, *on behalf of the CDF collaboration*

CIEMAT (SPAIN)
email: calancha@fnal.gov

Abstract. The Tevatron, with $p\bar{p}$ collisions at $\sqrt{s} = 1.96$ TeV, can produce all flavours of B hadrons and allows for unprocedented studies in the B physics sector. The CDFII detector [1] currently have more than 7fb^{-1} of data recorded.
This paper presents a selection of recent results on heavy hadron physics using data upto 4.2 fb^{-1}.

Keywords: Bottom spectroscopy, charmoniumlike, quarkonia polarization, heavy hadrons, heavy baryons.
PACS: 14.20.Mr, 14.20.Nd, 14.40.Rt, 14.40.Pq

HEAVY HADRON SPECTROSCOPY DURING CDF RUN II

The Tevatron accelerator is a suitable place to study heavy hadron spectroscopy. This is due to all heavy hadrons are copiously produced, including some states not accesible to B factories. In addition, the fact that these heavy hadrons are produced boosted, allows to measure the separation between produced and decay hadron vertex for those hadrons decaying via the weak interaction. CDF has a strong program on heavy hadron spectroscopy that have yielded many key results during the last years.

In 2007 CDF performed the first observation of the states $\Sigma_b^{(*)\pm}$ [2] in the decay channel $\Sigma_b^{(*)\pm} \to \Lambda_b^0\pi^\pm$, $\Lambda_b^0 \to \Lambda_c^+\pi^-$, $\Lambda_c^+ \to PK^-\pi^+$[1]. This analysis exploit the excellent vertex resolution of the CDFII detector to select high statistics sample of decays having two displaced tracks from the primary vertex.

In 2008 CDF reported a very precise measurement of the $B_c^\pm$ mass [3] in the decay channel $B_c^\pm \to J/\psi\pi^\pm$. The measured value $m = 6275.6 \pm 2.9(\text{stat}) \pm 2.5(\text{syst})$ MeV/c^2 agrees with the predictions from the non-relativistic potential models: $6247 - 6286$ MeV/c^2 [4] but it is significantly lower than the lattice QCD prediction: $6304 \pm 12^{+18}_{-0}$ MeV/c^2 [5].

CDF also has performed important contributions on the study of the $X(3872)$ in the decay channel $X(3872) \to J/\psi\pi^+\pi^-$. An angular analysis of this decay allowed the first determination of the quantum numbers of this exotic state [6]. Thereby, CDF concludes that the only assumptions compatible with data are $J^{PC} = 1^{++}$ or 2^{-+}. In addition, the last CDF mass measurement of the $X(3872)$ [7], $m = 3871.61 \pm 0.16(\text{stat}) \pm 0.19(\text{syst})$ MeV/c^2, is the most precise mass measurement to date. This analysis also examines the hypothesis that the $X(3872)$ stems from two different states. The conclusion is that the data support the one state hypothesis. A lower limit is set for the mass difference if $X(3872)$ is composed of two states: $\Delta m < 3.6$ MeV/c^2 at 95% C.L.

In 2008 CDF reported the first observation of the B_{s1} state [8] in the fully reconstructed decay $B_s^{**} \to B^+K^-$ or in the decay $B_s^{**} \to B^{+*}K^-$ $(B^{+*} \to B^+\gamma)$, where the photon is not detected. The B^+ mesons are fully reconstructed in two decay channels, $B^+ \to J/\psi K^+$ $(J/\psi \to \mu^+\mu^-)$, and $B^+ \to D^0\pi^+$ $(D^0 \to K^-\pi^+)$.

Figure 1 shows the $B_c^\pm$ and $X(3872)$ invariant masses together with the B_s^{**} Q-value distribution, $Q = m(B^+K^-) - m(B^+) - m(K^-)$. Two clear peaks are visible in the B_s^{**} Q-value distribution, both with a statistical significance higher than 5σ, which correspond to the states B_{s1} and B_{s2}^*. The measured masses, $m(B_{s1}) = 5829.41 \pm 0.21(\text{stat}) \pm 0.14(\text{syst}) \pm 0.6(PDG)$ MeV/c^2 and $m(B_{s2}^*) = 5839.64 \pm 0.39(\text{stat}) \pm 0.14(\text{syst}) \pm 0.5(PDG)$ MeV/c^2, are consistent with theoretical predictions [9].

[1] Unless otherwise stated all references to the specific charge combination imply the charge conjugate combination as well.

12th International Conference on Meson-Nucleon Physics and the Structure of the Nucleon (MENU 2010)
AIP Conf. Proc. 1374, 569-572 (2011); doi: 10.1063/1.3647205

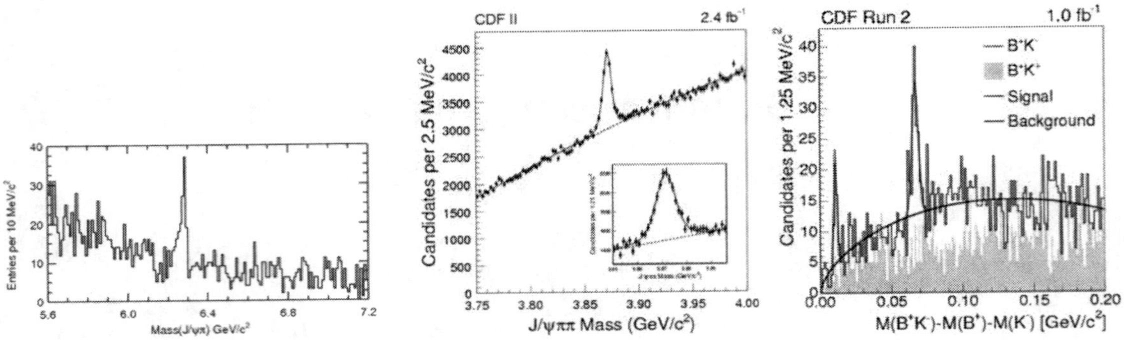

FIGURE 1. $B_c^{\pm}$ invariant mass distribution (left), $X(3872)$ invariant mass distribution (center), and B_s^{**} Q-value distribution (right) for two channels added together.

LATEST RESULTS

In this section we present the more recent results on heavy hadrons by CDF:

- Evidence for a new charmoniumlike state $Y(4140)$.
- The observation of the new heavy baryons Ξ_b^- and Ω_b^-.
- A measurement of the $\Upsilon(1S)$ polarization.

A New Charmoniumlike State

CDF has published recently evidence for a new charmoniumlike state [10], labeled by $Y(4140)$, decaying to $J/\psi\,\phi$. This is the first charmoniumlike structure decaying into two heavy quarkonium states ($c\bar{c}$ and $s\bar{s}$) which is accessible to exotic mesons such as a four-quark state, hybrid and glueball [11, 12, 13]. This analysis study the decay channel $B^+ \to Y(4140)K^+, Y(4140) \to J/\psi\,\phi$ ($J/\psi \to \mu^+\mu^-$, $\phi \to K^-K^+$). The analysis exploit the kaon particle identification and the long B lifetime and narrow B mass window to reduce the combinatorial background.

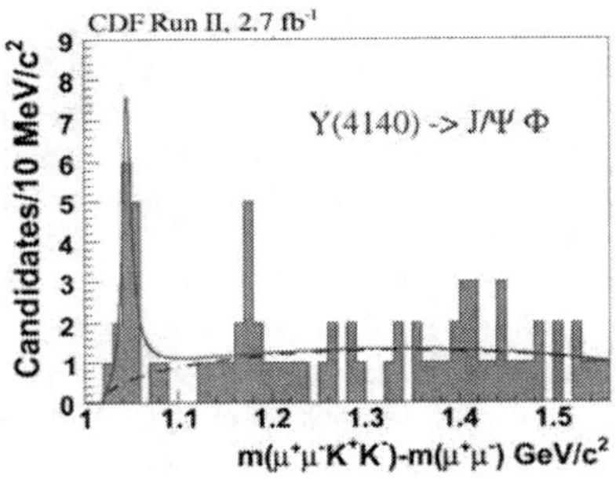

FIGURE 2. $Y(4140)$ mass diference distribution. The significance of the signal is estimated as 3.8σ.

Figure 2 shows the mass difference $m(\mu^+\mu^-K^+K^-) - m(\mu^+\mu^-)$. A clear enhancement is visible near the threshold. The significance of the signal is estimated as 3.8σ and its mass and width are measured to be: $m = 4143.0 \pm 2.9(\text{stat}) \pm 1.2(\text{syst})$ MeV/c^2, $\Gamma = 11.7^{+8.3}_{-5.0}(\text{stat}) \pm 3.7(\text{syst})$ MeV/c^2.

New Heavy Baryons

CDF has published the observation of the heavy baryons Ξ_b^- and Ω_b^- [14] in the decay channels $\Xi_b^- \to J/\psi\,\Xi^-$ ($J/\psi \to \mu^+\mu^-$, $\Xi^- \to \Lambda\pi^-$) and $\Omega_b^- \to J/\psi\,\Omega^-$ ($J/\psi \to \mu^+\mu^-$, $\Omega^- \to \Lambda K^-$). These states were observed at first time by DØ [15, 16] in the same decay channels.

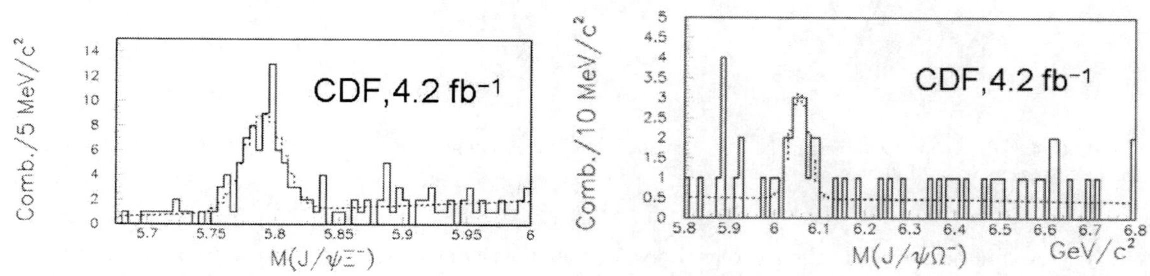

FIGURE 3. Ξ_b^- invariant mass distribution (left) and Ω_b^- invariant mass distribution (right). The statistical significance of the Ω_b^- signal is estimated as 5.5σ.

Due to the similar topology, CDF follows the same analysis strategy for both Ξ_b^- and Ω_b^-. The analysis exploits the long lifetime of the daughter hyperons by tracking these particles in the silicon vertex detector. This improve significantly the purity of the samples. Figure 3 shows the invariant mass distributions for both heavy baryons. The measured mass for the Ξ_b^-, $m = 5790.9 \pm 2.6(\text{stat}) \pm 0.8(\text{syst})$ MeV/c^2 is consistent with the theoretical predictions [17] and with the DØ discovery [15]. CDF measure the Ω_b^- mass to be $m = 6054.4 \pm 6.8(\text{stat}) \pm 0.9(\text{syst})$ MeV/c^2, which is in good agreement with the theory ($6010 - 6070$ MeV/c^2) [18], but disagrees with the DØ value ($m = 6165 \pm 10(\text{stat}) \pm 13(\text{syst})$ MeV/c^2) [16] by more than 6σ. Both experiments also disagree in the production rates for this particle,

$$\mathscr{R} = \frac{f(b \to \Omega_b^-)\mathscr{B}(\Omega_b^- \to J/\psi\,\Omega^-)}{f(b \to \Xi_b^-)\mathscr{B}(\Xi_b^- \to J/\psi\,\Xi^-)}$$

DØ reports a value, $\mathscr{R}_{D0} = 0.80 \pm 0.32^{+0.14}_{-0.22}$, that is significantly higher than CDF, $\mathscr{R}_{CDF} = 0.27 \pm 0.12 \pm 0.01$.

DØ is working on an update of this Ω_b^- analysis with much more data which hopefully will solve the question. In this analysis, CDF also perform the first exclusive Ξ_b^- lifetime measurement, $\tau = 1.56^{+0.27}_{-0.25} \pm 0.02$ ps, and the first Ω_b^- lifetime measurement ever, $\tau = 1.13^{+0.53}_{-0.40} \pm 0.8$ ps.

$\Upsilon(1S)$ Polarization

The production cross section and polarization of the heavy quarkonia is discussed within the non-relativistic QCD (NRQCD). This theory fit fairly well the experimental cross section spectra measured by CDF for all the quarkonia, but fails predicting the observed polarizations [19].

CDF has presented recently a measurement of the $\Upsilon(1S)$ polarization [20] in the decay channel $\Upsilon(1S) \to \mu^+\mu^-$ using the helicity frame. The analysis found a longitudinal polarization at high $\Upsilon(1S)$ p_T. Figure 4 shows the measured $\Upsilon(1S)$ polarization together with the expectations from NRQCD.

This measurement follows similar techniques as a DØ $\Upsilon(1S)$ polarization measurement [21], but totally disagrees with it. The DØ measurement is also inconsistent with NRQCD. CDF and DØ are working on update analysis with more data including an alternative frame (Collins-Soper) and explicitely comparing in the same pseudorapidity region.

CONCLUSION

CDF continues to make significant contributions to the field of flavour physics. The results presented here have shown the possibilities of hadron colliders for measurements in flavour physics despite the high combinatoric backgrounds.

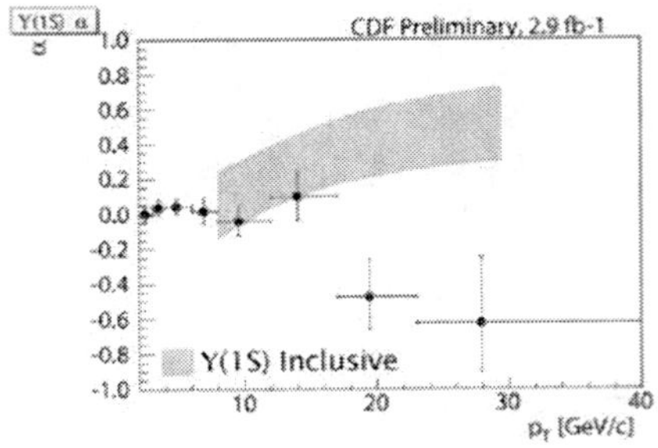

FIGURE 4. $\Upsilon(1S)$ polarization using 2.9 fb^{-1} of data. The $\Upsilon(1S)$ is longitudinal polarized at large $p_T(\Upsilon(1S))$, while the NRQCD predicts tranverse polarization in this regime.

The Tevatron accelerator continues to provide a high luminosity and these analyses can be expected to be updated with 10 fb^{-1}. This will allow for significantly improved measurements for those that are currently limited by statistical uncertainty. It also will allow to solve the found disagreements between CDF and DØ experiments, as in the Ω_b^- observation and the $\Upsilon(1S)$ polarization.

ACKNOWLEDGMENTS

The author thanks the Fermilab staffs and the technical staffs of the participating institutions for their vital contributions.

REFERENCES

1. D. Acosta *et al.* [CDF Collaboration], *Phys. Rev.* D **71**, 032001 (2005).
2. T. Aaltonen *et al.* [CDF Collaboration], *Phys.Rev.Lett.* **99**, 202001 (2007).
3. T. Aaltonen *et al.* [CDF Collaboration], *Phys.Rev.Lett.* **100**, 182002 (2008).
4. E.J. Eichten and C. Quigg, *Phys. Rev. D* **49**, 5845 (1994); W.K. Kwong and J.L. Rosner, *Phys. Rev. D* **44**, 212 (1991); S. Godfrey, *Phys. Rev. D* **70**, 054017 (2004).
5. I.F. Allison *et al.*, *Phys. Rev. Lett.* **94**, 172001 (2005).
6. A. Abulencia *et al.* [CDF Collaboration], *Phys. Rev. Lett* **98**, 132002 (2007).
7. T. Aaltonen *et al.* [CDF Collaboration], *Phys.Rev.Lett.* **103**, 152001 (2009).
8. T. Aaltonen *et al.* [CDF Collaboration], *Phys.Rev.Lett.* **100**, 082001 (2008).
9. E.J. Eichten, C.T. Hill, and C. Quigg, *Phys. Rev. Lett.* **71**, 4116 (1993); D. Ebert, V.O. Galkin, and R.N. Faustov, *Phys. Rev. D* **57**, 5663 (1998); S. Godfrey and R. Kokoski, *ibid.* **43**, 1679 (1991); A.F. Falk and T. Mehen, *ibid.* **53**, 231 (1996); M. Di Pierro and E. Eichten, *ibid.* **64**, 114004 (2001); A.M. Green *et al.*, *ibid.* **69**, 094505 (2004); W.A. Bardeen, E.J. Eichten, and C.T. Hill, *ibid.* **68**, 054024 (2003); P. Colangelo, F. De Fazio, and R. Ferrandes, *Nucl. Phys. B.*, Proc. Suppl. **163**, 177 (2007).
10. T. Aaltonen *et al.* [CDF Collaboration], *Phys.Rev.Lett.* **102**, 242002 (2009).
11. S.L. Zhu, *Phys. Lett. B* **625**, 212 (2005); F. Close and P. Page, *Phys. Lett. B* **628**, 215 (2005).
12. N.V. Drenska, R. Faccini, and A.D. Polosa, arXiv:hep-ph/0902.2803v1.
13. F.E. Close *et al.*, *Phys. Rev. D* **57**, 5653 (1998); F.E. Close and S. Godfrey, *Phys. Lett. B* **574**, 210 (2003).
14. T. Aaltonen *et al.* [CDF Collaboration], *Phys.Rev.D* **80**, 072003 (2009).
15. V.M. Abazov *et al.* [DØ Collaboration], *Phys.Rev.Lett.* **99**, 052001 (2007).
16. V.M. Abazov *et al.* [DØ Collaboration], *Phys.Rev.Lett.* **101**, 232002 (2008).
17. E. Jenkins, *Phys. Rev. D* **54**, 4515 (1996);**55**, R10 (1997); N. Mathur, R. Lewis, and R.M. Woloshyn, *ibid.* **66**, 014502 (2002); D. Ebert, R.N. Faustov and V.O. Galkin, *ibid.* **72**, 034026 (2005); M. Karliner, B. KerenZur, H.J. Lipkin, and J.L. Rosner, arXiv:0706.2163v5.
18. E. Jenkins, *Phys. Rev. D* **77**, 034012 (2008); R. Lewis and R.M. Woloshyn, *ibid.* **79**, 014502 (2009); D. Ebert, R.N. Faustov, and V.O. Galkin, *ibid.* **72**, 034026 (2005); M. Karliner, B. Keren-Zur, H.J. Lipkin, and J.L. Rosner, *Ann. Phys. (N.Y.)* **324**, 2 (2009); A. Valcarce, H. Garcilazo, and J. Vijande, *Eur. Phys. J. A* **37**, 217 (2008).
19. A. Abulencia *et al.* [CDF Collaboration], *Phys. Rev. Lett.* **99**, 132001 (2007).
20. http://www-cdf.fnal.gov/physics/new/bottom/090903.blessed-Upsilon1S-polarization/blessed_plots.html
21. V.M. Abazov *et al.* [DØ Collaboration], *Phys.Rev.Lett.* **101**, 182004 (2008).

B hadrons spectroscopy at the D0 experiment

Eduard De La Cruz Burelo[1]

Physics Department
Centro de Investigacion y de Estudios Avanzados del IPN
(CINVESTAV-IPN), México city, México.

Abstract. The latest results on B hadron spectroscopy from the DØ experiment at the Tevatron are presented. The mass measuremen of the B_c^- meson in the $B_c^- \to J/\psi\pi^-$ decay channel, the observation of excited B_s mesons, the first direct observation of the Ξ_b^- baryon, and the first observation of the Ω_b^- baryon.

Keywords: B hadrons
PACS: 14.20.Mr,13.30.Eg,13.85.Ni

INTRODUCTION

The study of B hadrons, particles in which the heaviest constituent quark is a b quark, is an important part of the rich physics program of the DØ [1] experiment at the Tevatron. DØ has measured many properties of B hadrons, such as lifetimes, masses, decay rates, CP violation phase, and mixing parameters, to name some of them. All these measurements are due to the fact that the Tevatron is an excellent place to study b-hadron physics, since all species (B^-, B^0, B_s, B_c^-, Λ_b, *etc.* [2]) are produced in large quantities. However, the level of background due to the total inelastic cross section in $p\bar{p}$ collisions compared to the cross section of $b\bar{b}$ is huge as well. This leads to the necessity for dedicated triggers to select a higher purity of b events, as well as the use of multivariate techniques to discriminate signal from background at the analysis level.

In this contribution, the observation of the B_c^- meson in a fully decay channel is reported, as well as the measurement of its mass. In addition, the first observations of excited B_s mesons, and the Ξ_b^- and Ω_b^- baryons are presented.

B_c MASS MEASUREMENT

The B_c^- is a a kind of special meson in which two heavy quarks, c and b, are combined to make a particle. Both quarks can decay weakly and in fact the lifetime of the B_c^- [2] is of a short nature, more a c-like meson lifetime than a b-like. This short lifetime makes the reconstruction of this meson experimentally challenging due to combinatorial background near to the collision point. In addition to its short lifetime, the B_c^- production rate is very low, making even more difficult to obtain sufficient statistics to measure its properties.

DØ have reconstructed this meson in the exclusive decay channel $B_c^- \to J/\psi\pi^-$, follow by $J/\psi \to \mu^+\mu^-$. The B_c^- selection criteria is optimized on data using $B^- \to J/\psi K^-$ decays, which have a similar topology to $B_c^- \to J/\psi\pi^-$. Figure 1 shows the invariant mass distribution of $J/\psi\pi^-$ decays reconstructed in the DØ experiments. In a dataset corresponding to 1.3 fb^{-1} of integrated luminosity DØ has observed [3] a B_c^- signal with a significance higher than five standard deviations above background, and has measured its mass to be $M(B_c^-) = 6300 \pm 14$ (stat) $\pm$ 5 (syst) MeV/c^2.

---

[1] On behalf of the DØ collaboration.
[2] Charge conjugation is always assumed.

12th International Conference on Meson-Nucleon Physics and the Structure of the Nucleon (MENU 2010)
AIP Conf. Proc. 1374, 573-576 (2011); doi: 10.1063/1.3647206

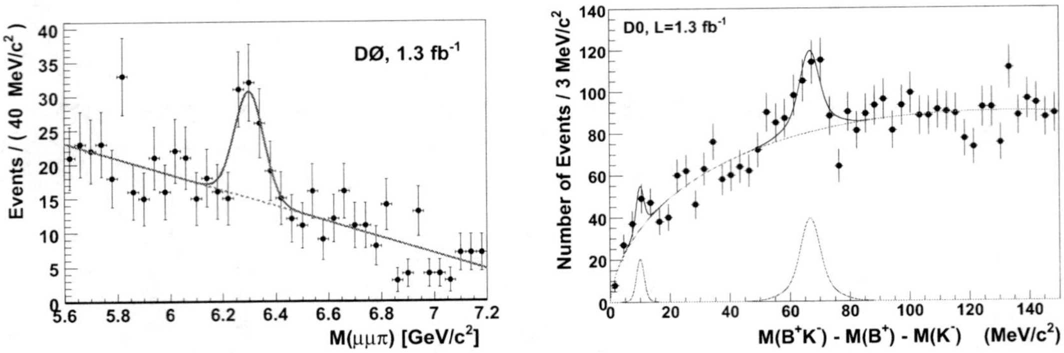

FIGURE 1. (Left) $J/\psi\pi$ invariant mass distribution of B_c^- candidates and a projection of the unbinned maximum likelihood fit to the distribution is shown overlaid. (Right) Invariant mass difference for the B_s^{**} candidates.

EXCITED B_s MESONS

Quark models predict the existence of four P-wave ($L = 1$) states in the $\bar{b}s$ system: two broad resonances (B_{s0}^* and $B_s^{'}$) and two narrow resonances (B_{s1} and B_{s2}^*) [4, 5]. The narrow B_s excited resonances are commonly called as B_s^{**}. These B_s^{**} resonances decay via D-wave processes ($L = 2$), and if the mass of the B_{sJ} ($J = 1,2$) is large enough, the main decay channel should be $B_{sJ} \to B^*K$, since the $B_s\pi$ channel is forbidden by isospin conservation.

The DØ experiment looks for the decays $B_s^{**} \to B^{*+}K^-$ followed by $B^{*+} \to B^+\gamma$ (γ is not detected) and $B^+ \to J/\psi K^+$. The invariant mass difference $M(BK) - M(B) - M(K)$ is shown in Fig. 1. A statistical significant signal associated to the B_{s2}^* is observed, and while there is an indication of the B_{s1}^* at low masses, more statistics is needed at DØ to declare observation. DØ finds [6] $M(B_{s2}^*) = 5839.6 \pm 1.1$ (stat) ± 0.7 (syst) GeV/c^2 in 1.3 fb^{-1} of integrated luminosity.

Ξ_b^- BARYON OBSERVATION

In the quark model the Ξ_b^- is formed by the combination of a b, d, and s quark, and it is expected to have $J^P = 1/2^+$ although I, J or P have yet to be measured. Evidence for the Ξ_b^- has been inferred from an excess of same sign $\Xi^\pm\ell^\pm$ events in jets which are interpreted as $\Xi_b^- \to \Xi^-\ell^-\bar{\nu}_\ell X$ [7]. From this decay mode, the average lifetime of the Ξ_b^- is measured to be $1.42^{+0.28}_{-0.24}$ ps [8]. These semileptonic decays of the Ξ_b^- did not allow for a mass measurement, but theoretical calculations predict the Ξ_b^- mass in the range $5.7 - 5.8$ GeV [9].

The DØ experiment searched for the Ξ_b^- baryon in the decay channel $\Xi_b^- \to J/\psi\Xi^-$, followed by $J/\psi \to \mu^+\mu^-$, and $\Xi^- \to \Lambda\pi^-$, where Λ decays to $p\pi^-$. DØ optimizes selection criteria based on signal Monte Carlo events and background from wrong-sign combination events $J/\psi(\Lambda\pi^+)$. Figure 2 shows the Ξ_b^- observation in DØ data. Different cross-checks such as trying to reconstruct the signal by using sideband events from J/ψ and Ξ^- signals, or from wrong-sign combinations ($J/\psi(\Lambda\pi^+)$), are performed to make sure that the observed signal is not due to artifacts of the analysis. In a data sample of 1.3 fb^{-1} of integrated luminosity DØ collaboration observes [10] 15.2 ± 4.4 Ξ_b^- events and measures the Ξ_b^- mass to be 5.774 ± 0.011 (stat.) ± 0.015 (syst.) GeV/c^2. DØ also determines its $\sigma \times \mathcal{B}$ relative to that of the Λ_b to be 0.28 ± 0.09 (stat.)$^{+0.09}_{-0.08}$ (syst.). The statistical significance of the signal was found to be 5.5σ.

Ω_b^- BARYON OBSERVATION

In the quark model the Ω_b^- is composed of a b quark and two s quarks. Nothing is experimentally known about this particle, but it is expected to have J$^P = 1/2^+$, a mass between 5.94 – 6.12 GeV/c^2 and a lifetime such that $0.55 < \tau(\Omega_b^-)/\tau(B^0) < 1.10$ [11].

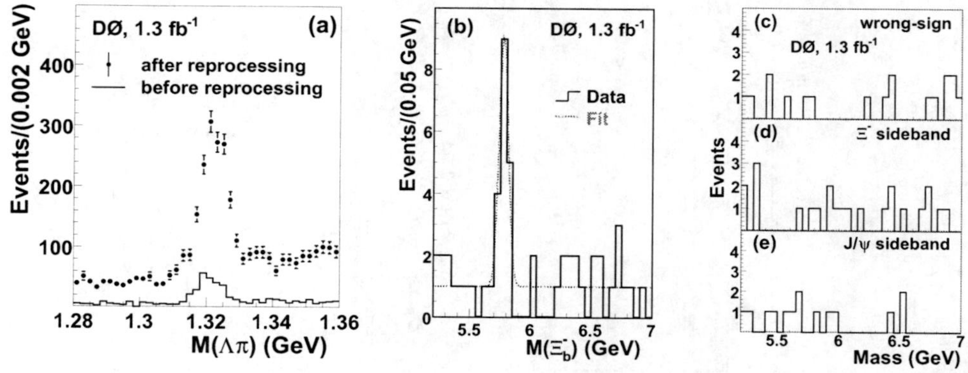

FIGURE 2. Ξ_b^- observation in DØ . (a) Invariant mass distributions of the $\Lambda\pi^-$ pair before and after data reprocessing. The reprocessing significantly increases the Ξ^- yield. (b) The $M(\Xi_b^-)$ distribution of the Ξ_b^- candidates after all selection criteria. The dotted curve is an unbinned likelihood fit to the model of a constant background plus a Gaussian signal. The $(\mu^+\mu^-)\Lambda\pi$ mass distributions for (c) the wrong-sign background, (d) the Ξ^- sideband, and (e) the J/ψ sideband events. The mass $M(J/\psi\Lambda\pi) - M(\mu^+\mu^-) + M_{\rm PDG}(J/\psi)$ is plotted for (c) and (d) while the mass $M(\mu^+\mu^-\Xi^-) - M(\Lambda\pi^-) + M_{\rm PDG}(\Xi^-)$ is plotted for (e).

The DØ have searched for the Ω_b^- in the decay channel $\Omega_b \to J/\psi\Omega^-$ followed by $J/\psi \to \mu^+\mu^-$, $\Omega^- \to \Lambda K^-$ and $\Lambda \to p\pi^-$. DØ reconstructs $\Omega_b^- \to J/\psi\Omega^-$ decays in the same data in which the Ξ_b^- is observed. A similar procedure to the reconstruction of $\Xi_b^- \to J/\psi\Xi^-$ decays is applied, but in the $\Omega^- \to \Lambda K^-$ reconstruction, kinematic variables associated with daughter particle momenta, vertices, and track qualities are combined using Boosted Decision Trees (BDT) [12, 13]. A blind optimization for the $J/\psi + \Omega^-$ combinations is performed by using signal Monte Carlo events compared to background from wrong-sign combinations ($J/\psi + (\Lambda(p\pi^-)K^+)$). From this optimization a $p_T(J/\psi\Omega^-) > 6$ GeV criterion is imposed, in addition to a requirement that the uncertainty on the boost-corrected decay length of the Ω_b^- candidates is less than 0.03 cm.

Figure 3(a) shows the invariant mass distribution of $J/\psi + \Omega^-$ combinations. The mass is corrected event per event by calculating $M(\Omega_b^-) = M(J/\psi\Omega^-) - M(\mu^+\mu^-) - M(\Lambda K^-) + \hat{M}(J/\psi) + \hat{M}(\Omega^-)$. Here $M(J/\psi\Omega^-)$, $M(\mu^+\mu^-)$, and $M(\Lambda K^-)$ are the reconstructed masses while $\hat{M}(J/\psi)$ and $\hat{M}(\Omega^-)$ are taken from Ref. [14]. This calculation improves the mass resolution of the MC Ω_b^- events from 0.080 GeV to 0.034 GeV.

By performing an unbinned likelihood fit of a Gaussian plus a flat background to data, DØ measures [15] the Ω_b^- mass to be $6.165 \pm 0.010\,(\text{stat.}) \pm 0.013\,(\text{syst.})$ GeV. The significance of the observed signal is 5.4σ, and the number of observed events is $17.8 \pm 4.9\,(\text{stat.}) \pm 0.8\,(\text{syst.})$. In addition DØ calculates the Ω_b^- production rate relative to that of the Ξ_b^-. By using the reported Ξ_b^- events and the observed Ω_b^- yield, DØ estimates

$$\mathscr{R} = \frac{f(b \to \Omega_b^-)}{f(b \to \Xi_b^-)} \cdot \frac{Br(\Omega_b^- \to J/\psi\,\Omega^-)}{Br(\Xi_b^- \to J/\psi\,\Xi^-)}$$

to be $\mathscr{R} = 0.80 \pm 0.32\,(\text{stat.})^{+0.14}_{-0.22}\,(\text{syst.})$. Here $f(b \to \Omega_b^-)$ and $f(b \to \Xi_b^-)$ are the fractions of b quarks that hadronize to form Ω_b^- and Ξ_b^-, respectively. The systematic uncertainty includes contributions from the signal yields as well as the efficiency ratio. Using a theoretical estimate for $\Gamma(\Omega_b^- \to J/\psi\Omega^-)/\Gamma(\Xi_b^- \to J/\psi\Xi^-) = 9.8$ [16], the central values of $\tau(\Xi_b^-) = 1.42^{+0.28}_{-0.24}$ ps [14], the $\mathscr{R}$ value above, and $\tau(\Omega_b^-)$ in the range of $0.83 - 1.67$ ps [11], we obtain $f(b \to \Omega_b^-)/f(b \to \Xi_b^-) \approx 0.07 - 0.14$.

DØ performs many different cross checks to validate the observation of the Ω_b^-. Figure 3(b) shows the same analysis procedure on control samples where no signal should be present. The analysis procedure is also applied to many different b-hadron decays in Monte Carlo events with more than ten times statistics than that observed on data for each of studied decays,and no signal is found. Tighter cuts in $p_T(\Omega_b^-)$ reduces background and increases the signal significance to more than 6σ. In addition, an independent cut based analysis (no BDT used) reproduces the signal but with lower statistical significance due to the better discrimination power of the BDT in the Ω^- selection.

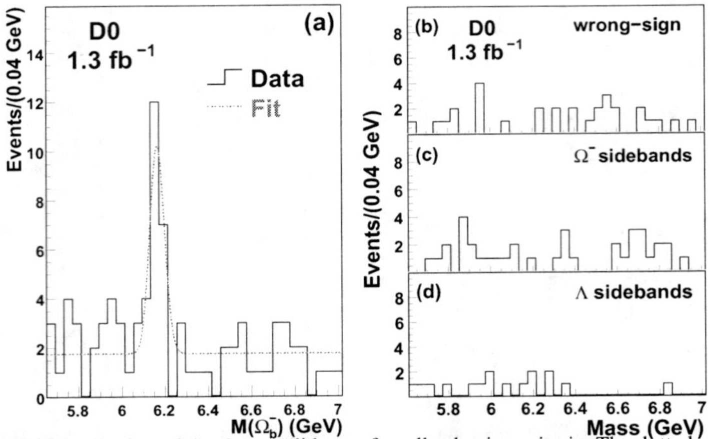

FIGURE 3. (a) The $M(\Omega_b^-)$ distribution of the Ω_b^- candidates after all selection criteria. The dotted curve is an unbinned likelihood fit to the model of a constant background plus a Gaussian signal. The mass distributions for the wrong-sign background (b), the Ω^- sideband events (c), and the Λ sideband events (d).

CONCLUSIONS

The DØ experiment is performing very exciting analyses on b-hadron spectroscopy with the data accumulated during Run II of the Tevatron. Many of these results are the first observations of particles like the excited B_s mesons, the Ξ_b^-, and the Ω_b^- baryon. In addition other properties of the b-hadrons such has masses and lifetimes are also measured, giving valuable information that can be compared directly to theoretical predictions.

REFERENCES

1. V.M. Abazov *et al.* (D0 Collaboration), Nucl. Instrum. Methods Phys. Res. A **565**, 463 (2006).
2. A. Abulencia et al. (CDF Collaboration), Phys. Rev. Lett. **97**, 012002 (2006).
3. V.M. Abazov *et al.* (D0 Collaboration), Phys. Rev. Lett. **101**, 012001 (2008).
4. M. Di Pierro and E. Eichten, Phys. Rev. D **64**, 114004 (2001); E. J. Eichten, C. T. Hill, and C. Quigg, Phys. Rev. Lett. **71**, 4116 (1993).
5. D. Ebert, V. O. Galkin, and R. N. Faustov, Phys. Rev. D **57**, 5663 (1998); Phys. Rev. D **59**, 019902(E) (1998).
6. 2008 V.M. Abazov *et al.* (D0 Collaboration), Phys. Rev. Lett. **100**, 082002 (2008).
7. J. Abdallah *et al.* (DELPHI Collaboration), Eur. Phys. J. **C44**, 299 (2005); D. Buskulic *et al.* (ALEPH Collaboration), Phys. Lett. B **384**, 449 (1996).
8. E. Barberio *et al.* (Heavy Flavor Averaging Group Collaboration), `arXiv:0704.3575`.
9. E. Jenkins, Phys. Rev. D **55**, R10 (1997); *ibid*, Phys. Rev. D **54**, 4515 (1996); N. Mathur, R. Lewis and R.M. Woloshyn, Phys. Rev. D **66**, 014502 (2002).
10. V.M. Abazov *et al.* (D0 Collaboration), Phys. Rev. Lett. **99**, 052001 (2007).
11. X. Liu *et al.*, Phys. Rev. D **77**, 014031 (2008); M. Karliner *et al.*, arXiv:0804.1575 (2008); E.E. Jenkins, Phys. Rev. D **77**, 034012 (2008); R. Roncaglia, D.B. Lichtenberg, and E. Predazzi, Phys. Rev. D **52**, 1722 (1995); N. Mathur, R. Lewis, and R.M. Woloshyn, Phys. Rev. D **66**, 014502 (2002); D. Ebert, R.N. Faustov, and V.O. Galkin, Phys. Rev. D **72**, 034026 (2005); T. Ito, M. Matsuda, and Y. Matsui, Prog. Theor. Phys. **99**, 271 (1998).
12. L. Breiman *et al.*, *Classification and Regression Trees*, (Wadsworth, Stamford, 1984).
13. A. Höcker *et al.*, arXiv:physics/0703039 [physics.data-an] (2007).
14. C. Amsler *et al.*, Physics Letters B **667**, 1 (2008).
15. V.M. Abazov *et al.* (D0 Collaboration), Phys. Rev. Lett. **101**, 232002 (2008).
16. H.-Y. Cheng, Phys. Rev. D **56**, 2799 (1997).

Charm Baryon Results from BaBar

Veronique Ziegler

SLAC National Accelerator Laboratory, 2575 Sand Hill Rd., Menlo Park, CA 94025

Abstract. We present experimental results from the BaBar experiment on charm baryon spectroscopy and production studies, including studies of excited cascades produced in charm baryon decays. We review the discovery of new decay modes of known states and searches for predicted states.

Keywords: charm baryons
PACS: 13.30.Eg,14.20.Jn,14.20.Lq

1. INTRODUCTION

Charm baryons are expected to exhibit a rich spectrum of states. Only a few of these states have been confirmed [1]. In this paper, we present selected searches for predicted states, and studies of charm baryon production and decays, carried out in the BaBar experiment. Note that throughout this paper, whenever a particle or decay mode is given, inclusion of the charge conjugate state is implied. The results described here were obtained from data samples collected in e^+e^- collisions near $\sqrt{s} = 10.58$ GeV by the BaBar detector [2] at the PEP-II asymmetric-energy e^+e^- storage rings.

2. OBSERVATION OF A CHARM BARYON DECAYING TO D^0p AT MASS NEAR 2.94 GeV/c^2

The $\Lambda_c(2880)^+$, which was first observed in its decay to $\Lambda_c\pi^+\pi^-$ [3], is one of the few singly-charmed baryons, along with the $\Xi_c(2815)$, $\Xi_c(2980)$, and $\Xi_c(3080)$, found above the Dp mass threshold. A study of the inclusive D^0p mass spectrum led to the discovery of the $\Lambda_c(2940)^+$, a new charm baryon decaying to D^0p [4]. The previously observed [3] decay $\Lambda_c(2880)^+ \to D^0p$ is also seen in the reconstructed D^0p invariant mass spectrum (Fig. 1). The mass and width values for the $\Lambda_c(2880)^+$ [$m = 2881.9 \pm 0.1(\text{stat.}) \pm 0.5(\text{syst.})$, $\Gamma = 5.8 \pm 1.5(\text{stat.}) \pm 1.1(\text{syst.})$] and the $\Lambda_c(2940)^+$ [$m = 2939.8 \pm 1.3(\text{stat.}) \pm 1.0(\text{syst.})$, $\Gamma = 17.5 \pm 5.2(\text{stat.}) \pm 5.9(\text{syst.})$] are obtained from the fit shown in Fig. 1. Furthermore, the absence of these signals in the corresponding D^+p mass distribution confirms their interpretation as Λ_c baryons. (See ref. [4] for further details.)

3. OBSERVATION OF THE Ω_c^* BARYON IN THE DECAY TO $\Omega_c^0\gamma$

The J^P = 3/2 *css* state, denoted as Ω_c^*, belongs to the family of singly-charmed baryons having zero orbital angular momentum. It was discovered by BaBar in its decay to the Ω_c^0 and a photon, with a significance of 5.2σ [5]. The measured mass difference between the Ω_c^* and the Ω_c^0, $\Delta M = 70.8 \pm 1.0(\text{stat.}) \pm 1.1(\text{syst.})$ MeV/c^2 is consistent with quark model predictions (see ref. [5] for details). With this observation of the Ω_c^* baryon, every predicted SU(4) ground-state singly-charmed baryon has been experimentally observed.

4. EVIDENCE FOR THE $\Xi_c(3055)^+$ AND $\Xi_c(3123)^+$ BARYONS

Both the BaBar and Belle collaborations have searched for ground-state doubly-charmed baryons decaying to the final state $\Lambda_c^+K^-\pi^+$ [6, 7]. These searches reveal no evidence for doubly-charmed states. However, an analysis carried by the Belle Collaboration using similar decay modes as those used in the search for double-charm baryons led to evidence for two excited charm-strange baryon states, $\Xi_c(2980)^{+,0}$ and $\Xi_c(3077)^{+,0}$, decaying to $\Lambda_c^+K_S\pi^-$ and

12th International Conference on Meson-Nucleon Physics and the Structure of the Nucleon (MENU 2010)
AIP Conf. Proc. 1374, 577-580 (2011); doi: 10.1063/1.3647207

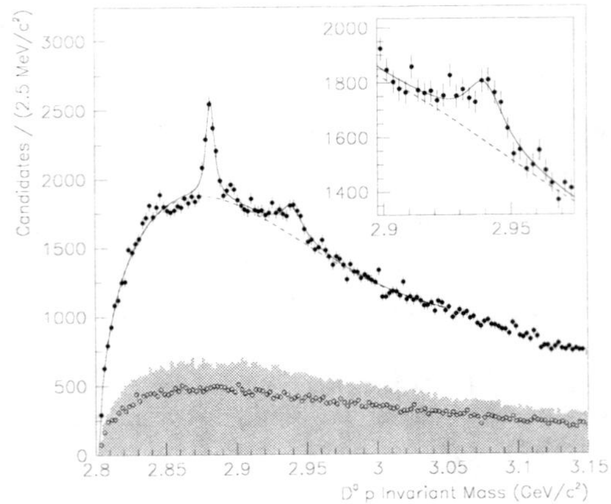

FIGURE 1. The solid points are the D^0p invariant mass distribution of the final sample. Also shown are (gray) the contribution from false D^0 candidates estimated from D^0 mass sidebands and (open points) the mass distribution from wrong-sign $\overline{D}^0p$ candidates. The solid curve is the result of a fit including the $\Lambda_c(2880)^+$ and $\Lambda_c(2940)^+$ states, each modeled by a relativistic Breit-Wigner lineshape convolved with a Gaussian resolution function. The dashed curve is the portion of that fit attributed to combinatorial background.

$\Lambda_c^+K^-\pi^+$ [7]. The $\Xi_c(2980)^{+,0}$ and $\Xi_c(3077)^{+,0}$ are identified as charm-strange states based on the measured masses, natural widths (which indicate strong decays), and charges of the members of the isospin doublet. Previously known excited Ξ_c baryons have been observed only in decays to a lower-mass Ξ_c baryon plus a pion or photon. In contrast, the $\Xi_c(2980)^{+,0}$ and $\Xi_c(3077)^{+,0}$ baryons are observed in decays in which the charm and strange quarks are in separate hadrons. The observed decay modes may have implications for the internal quark dynamics of these new states.

The BaBar Collaboration searched for decays to the three-body final states $\Lambda_c^+K^-\pi^+$ and $\Lambda_c^+K_S\pi^-$, the two-body final states $\Lambda_c^+K_S$ and $\Lambda_c^+K^-$, and the four-body final states $\Lambda_c^+K_S\pi^-\pi^+$ and $\Lambda_c^+K^-\pi^-\pi^+$ [8]. Significant signals are found only in decays to three-body final states (shown in Fig. 2). The existence of the states $\Xi_c(2980)^+$, $\Xi_c(3077)^+$, and $\Xi_c(3077)^0$ was confirmed, with an improvement over existing measurements of the $\Xi_c(2980)^+$ mass and width. For the three-body final states, searches for decays through intermediate resonant $\Sigma_c(2455)^{++}K^-$, $\Sigma_c(2455)^0K_S$, $\Sigma_c(2520)^{++}K^-$, and $\Sigma_c(2520)^0K_S$ states were performed, with the $\Sigma_c(2455)$ and $\Sigma_c(2520)$ baryons decaying exclusively to $\Lambda_c^+\pi$. In addition, evidence was seen for two additional new states $\Xi_c(3055)^+$ and $\Xi_c(3123)^+$ decaying through the intermediate-resonant states $\Sigma_c(2455)^{++}K^-$ and $\Sigma_c(2520)^{++}K^-$, respectively. (See ref. [8] for further details.)

5. BARYONIC DECAYS OF *B* MESONS

Baryonic decays of B mesons, which contain a heavy bottom quark and a light up or down anti-quark, provide a laboratory for a range of particle physics investigations such as decay rates and baryon production mechanisms, searches for exotic states and excited baryon resonances, and possibly new physics through flavor-changing neutral currents [9].

Substantial B meson decay to the ground- and first-excited Ξ_c states have been observed by the BaBar Collaboration [11]. In addition, BaBar carried out a detailed study of $\overline{B}^0 \to \Lambda_c^+\overline{p}$ and $B^- \to \Lambda_c^+\overline{p}\pi^-$ based on 383×10^6 $\Upsilon(4S) \to B\overline{B}$ decays recorded with the BABAR detector [10]. A branching ratio measurement $\mathscr{B}(B^- \to \Lambda_c^+\overline{p}\pi^-)/\mathscr{B}(\overline{B}^0 \to \Lambda_c^+\overline{p}) = 15.4\pm1.8\pm0.3$ was obtained. The resonant decays $B^- \to \Sigma_c(2455)^0\overline{p}$ and $B^- \to \Sigma_c(2800)^0\overline{p}$ were observed but no evidence for $B^- \to \Sigma_c(2520)^0\overline{p}$ was obtained. This is the first observation of the decay $B^- \to \Sigma_c(2800)^0\overline{p}$ and the measured mass of the observed excited Σ_c^0 state is $(2846\pm8\pm10)$ MeV/c^2. The $B^- \to \Lambda_c^+\overline{p}\pi^-$ system was used to measure the spin of the $\Sigma_c(2455)^0$. Consistency with spin-1/2 was obtained [10] (see Fig. 3).

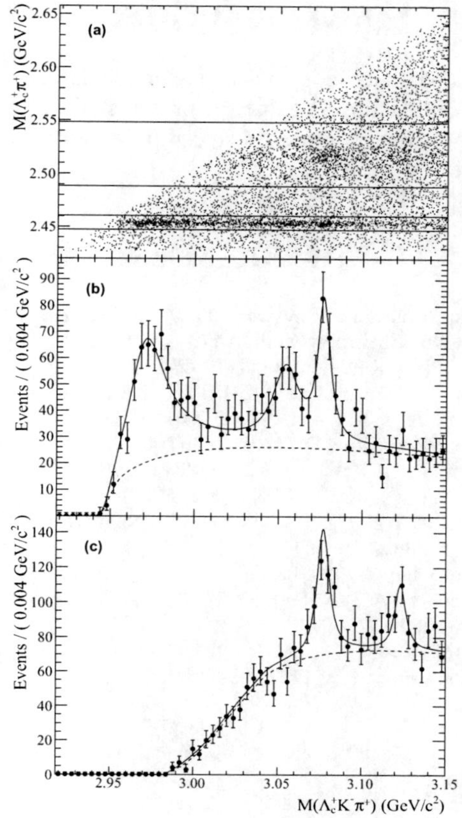

FIGURE 2. (a) Scatterplot of $M(\Lambda_c^+\pi^+)$ versus $M(\Lambda_c^+K^-\pi^+)$, and (b,c) projections of the $M(\Lambda_c^+K^-\pi^+)$ distribution for data (points with error bars) for all five reconstructed Λ_c^+ decay modes combined. The curves correspond to the two-dimensional fit results. The $M(\Lambda_c^+\pi^+)$ ranges are (b) within 3.0 natural widths of the $\Sigma_c(2455)^{++}$ mass, and (c) within 2.0 natural widths of the $\Sigma_c(2520)^{++}$ mass. These $M(\Lambda_c^+\pi^+)$ ranges are delineated by the horizontal lines on the scatterplot in (a). The projections of the combined background PDF components are illustrated by the dashed curves.

6. LIGHT QUARK SPECTROSCOPY

Charm baryons also provide a useful laboratory for the investigation of light quark spectroscopy, in particular that of cascade hyperons [13].

The $\Xi(1530)^0$ is observed in in $\Lambda_c^+ \to (\Xi^-\pi^+)K^+$ decay [15]. Spin information for the $\Xi(1530)$ is obtained using Legendre polynomial moments ($P_L(\cos\theta_{\Xi^-})$, where $\cos\theta_{\Xi^-}$ is the Ξ^- helicity cosine). The $\sqrt{10}P_2(\cos\theta_{\Xi^-})$ moment of the $(\Xi^-\pi^+)$ system invariant mass distribution for the Λ_c^+ signal region clearly favors spin 3/2, whereas the $7/\sqrt{2}P_4(\cos\theta_{\Xi^-})$ moment is consistent with being flat implying that spin 5/2 is ruled out. Based on an analysis of Schlein *et al.* [14] that showed $J^P = 3/2^+$ or $J^P = 5/2^-$ and claimed $J > 3/2$ not required, the present analysis by establishing $J = 3/2$ also establishes positive parity. A similar analysis of the decays $\Xi_c^0 \to \Omega^-K^+$ and $\Omega_c^0 \to \Omega^-\pi^+$ clearly established that the Ω^- has spin 3/2 [12].

Likewise, the properties of the $\Xi(1690)^0$ are extracted from a detailed isobar model analysis of the $\Lambda_c^+ \to \Lambda K_S K^+$ Dalitz plot [13]. The hypothesis that the $\Xi(1690)$ has spin 1/2 yields an excellent description of the data, whereas spin values of 3/2 and 5/2 result in poorer fit probabilities, and fail to reproduce the observed skewed $\Xi(1690)$ lineshape (see Fig. 4).

7. CONCLUSIONS

Although BaBar was primarily intended to be a B-meson factory, it is also an excellent charm hadron factory. Information on charm baryon baryon production is gained from the study of their production from B mesons, and measurements of charm baryon spins obtained from exclusive B decay processes. In addition charm baryon decays provide insight into cascade spectroscopy.

REFERENCES

1. K. Nakamura *et al.* (Particle Data Group), Journal of Physics G **37**, 075021 (2010).
2. B. Aubert *et al.* (BaBar Collaboration), Nucl. Instrum. Meth. **A479**, 1 (2002).
3. M. Artuso *et al.* (CLEO Collaboration), Phys. Rev. Lett. **86**, 4479 (2001).
4. B. Aubert *et al.* (BABAR Collaboration), Phys. Rev. Lett. **98**, 012001 (2007).
5. B. Aubert *et al.* (BABAR Collaboration), Phys. Rev. Lett. **97**, 232001 (2006).
6. B. Aubert *et al.* (BABAR Collaboration), Phys. Rev. D **74**, 011103 (2006).
7. R. Chistov *et al.* (BELLE Collaboration), Phys. Rev. Lett. **97**, 162001 (2006).
8. B. Aubert *et al.* (BABAR Collaboration), Phys. Rev. D **77**, 012002 (2008).
9. M. Misiak *et al.*, Phys. Rev. Lett. **98**, 022002 (2007); T. Becher and M. Neubert, Phys. Rev. Lett. **98**, 022003 (2007).
10. B. Aubert *et al.* (BABAR Collaboration), Phys. Rev. D **78**, 012003 (2008).
11. B. Aubert *et al.* (BABAR Collaboration), hep-ex/0607086.
12. B. Aubert *et al.* (BABAR Collaboration), Phys. Rev. Lett. **97**, 112001 (2006).
13. Veronique Ziegler, *Ph.D. Thesis*, SLAC-R-868 (2007).
14. P. Schlein *et al.*, Phys. Rev. Lett. **11**, 167 (1963).
15. B. Aubert *et al.* (BABAR Collaboration), Phys. Rev. D **78**, 034008 (2008).

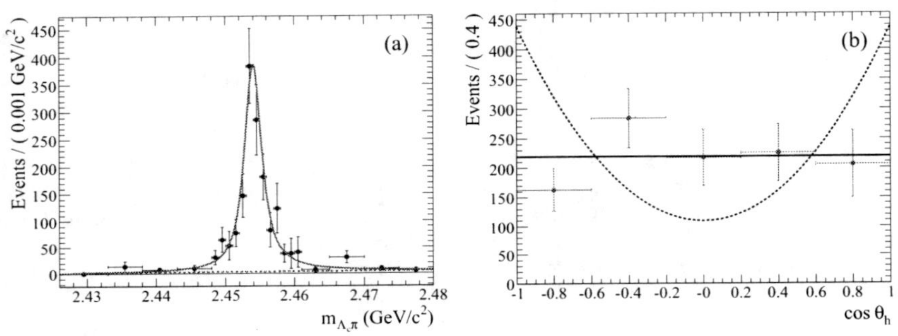

FIGURE 3. (a) The invariant mass distribution of reconstructed $\Sigma_c(2455)^0$ candidates in $\overline{B}^- \to \Sigma_c(2455)^0\overline{p}$ events. (b) The efficiency-corrected helicity angle distribution for $\Sigma_c(2455)^0$ candidates in (a). The curves for the spin-1/2 (solid line) and spin-3/2 (dashed line) hypotheses are overlaid.

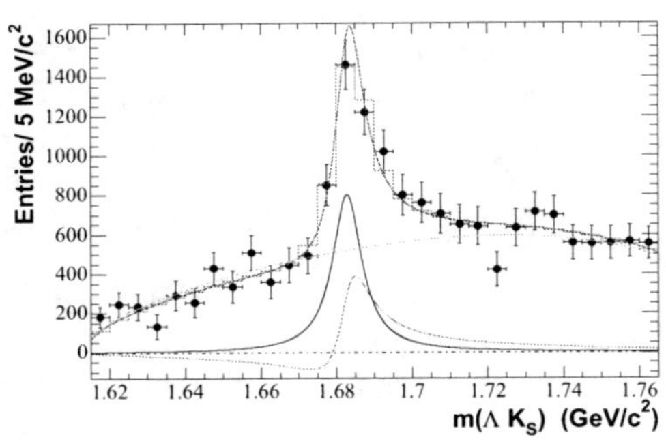

FIGURE 4. The $m(\Lambda K_S)$ projection of the efficiency-corrected, background-subtracted rectangular $\Lambda_c^+ \to \Lambda K_S K^+$ Dalitz plot in data. The details of the fit are described in ref. [13].

Bottom and charmed hadron spectroscopy from lattice QCD

Randy Lewis

Department of Physics and Astronomy, York University, Toronto, Ontario, Canada M3J 1P3

Abstract. A survey of recent lattice QCD simulations for the mass spectrum of bottom and charmed hadrons is presented.

Keywords: lattice QCD, bottom hadrons, charmed hadrons
PACS: 12.38.Gc, 14.20.Lq, 14.20.Mr, 14.40.Df, 14.40.Nd

INTRODUCTION

During the three years since the previous meson-nucleon (MENU) conference, there have been many lattice QCD computations of bottom and charmed hadron masses. These studies must handle challenges due to the physical quark masses being spread out relative to the natural lattice scales, λ_{QCD} and $1/a$ (the ultraviolet cutoff):

$$m_u \approx m_d \quad < \quad m_s \sim \Lambda_{\mathrm{QCD}} \quad < \quad m_c \quad < \quad \frac{1}{a} \quad < \quad m_b\,. \tag{1}$$

BOTTOM BARYONS

During the past three years, the CDF and D0 collaborations have published first observations of the Σ_b[1], Σ_b^*[1], Ξ_b[2, 3] and Ω_b[4, 5] baryons. The two results for the Ω_b are not consistent with one another.

Various lattice QCD groups have computed the mass spectrum using different b-quark actions (*NRQCD*[6, 7], *Fermilab*[8], *static*[9, 10, 11]), different light-quark actions (*nonperturbatively-tuned clover*[6], *improved staggered*[8], *chirally-improved*[9], *domain wall*[7, 10], *improved-staggered sea with domain wall valence*[11]) and different gauge actions (*Iwasaki*[6, 7, 10], *one-loop Symanzik/Lüscher-Weisz*[8, 9, 11]). Any action with the correct continuum limit is a valid approach, and these complementary studies provide valuable information about systematic uncertainties at present lattice spacings and volumes. Chiral extrapolations toward the physical u and d quark masses are also a source of systematic uncertainty, and the recent bottom baryon studies reach a minimum pion mass near 275 MeV[10], 290 MeV[8, 11], 331 MeV[7], 461 MeV[9] and 600 MeV[6].

Figure 1 conveys an impression of some results from these recent studies of bottom baryons, but of course the original publications provide many additional insights. For example, [8] provides a discussion of systematically-large splittings for (Λ_b, Ξ_b) versus $(\Sigma_b, \Xi_b', \Omega_b)$ that arise with a staggered lattice action due to the inability to separate $J^P = \frac{1}{2}^+$ and $\frac{3}{2}^+$ states using the spin projection operators.

It is clear from figure 1 that lattice simulations are consistent with the CDF measurement of the Ω_b mass as opposed to the D0 result. Lattice findings also agree with experiment for all other available masses, and lattice predictions are in place for comparison to future experimental discoveries.

Two related preprints have appeared after the MENU 2010 conference. Wagner and Wiese[12] have computed the baryon spectrum in the limit of a static bottom quark, where the twisted mass action was used for u,d quarks. Meinel[13] has produced a preprint that is focused on the mass of the Ω_{bbb}. Using NRQCD bottom quarks on dynamical lattice configurations, he arrives at

$$M_{\Omega_{bbb}} = 14.371 \pm 0.004_{stat} \pm 0.011_{syst} \pm 0.001_{exp}\ \mathrm{GeV} \tag{2}$$

where the third uncertainty arises from required experimental input. This mass with excusively heavy valence quarks can be of significant value for testing the content of theoretical models.

12th International Conference on Meson-Nucleon Physics and the Structure of the Nucleon (MENU 2010)
AIP Conf. Proc. 1374, 581-584 (2011); doi: 10.1063/1.3647208

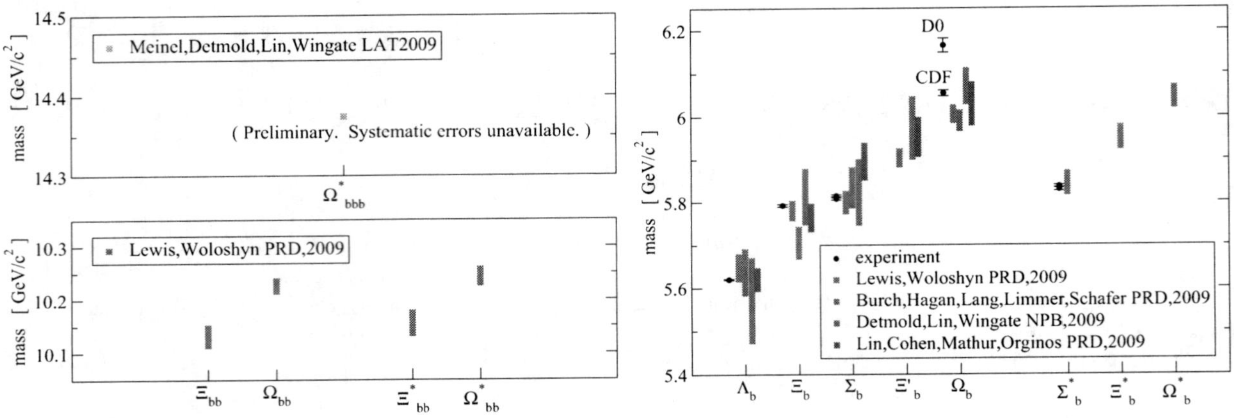

FIGURE 1. Bottom baryon spectrum from lattice QCD from [6, 7, 9, 10, 11]. Not shown are preliminary results from [7] for singly- and doubly-b baryons, valuable data from [8] for mass differences among singly- and doubly-b baryons, and the prediction for $M_{\Omega_{bbb}}$ given in equation (2).

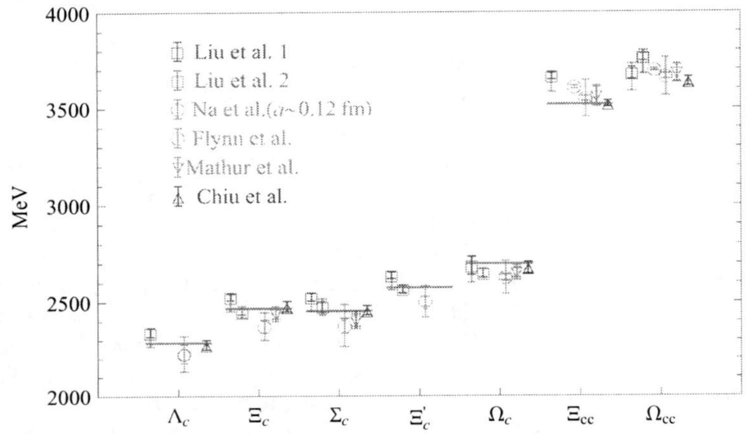

FIGURE 2. Charmed baryon spectrum from lattice QCD[8, 14] compared to older quenched results from [15, 16, 17]. Not shown are results in [8] for mass differences, including Ξ_{cc}^* and Ω_{cc}^*. This graph has been taken directly from [14].

CHARMED BARYONS

In contrast to m_b, the charmed quark mass is smaller than the standard lattice cutoff so neither NRQCD nor static quark actions are generally useful. Two lattice groups[8, 14] have recently used the Fermilab action for the charmed quark and gauge configurations containing staggered u,d,s sea quarks to obtain charmed baryon masses. Liu, Lin, Orginos and Walker-Loud[14] used domain wall valence u,d,s quarks while Na and Gottlieb[8] used the staggered action. The lightest u,d masses correspond to $m_\pi = 290$ MeV.

Some results are shown in figure 2, which was taken directly from [14]. It is interesting to see the consistency of the new results with the older simulations[15, 16, 17] which had relied on the quenched approximation.

BOTTOM MESONS

Recent lattice results for B, B_s and B_c mesons are compiled from [6, 18, 19, 20, 21] in figure 3 and compared to experimental data where available. Two groups[6, 21] used the NRQCD action for the bottom quark while the others[18, 19, 20] extrapolated from static quark simulations. The light quark action was nonperturbatively-improved

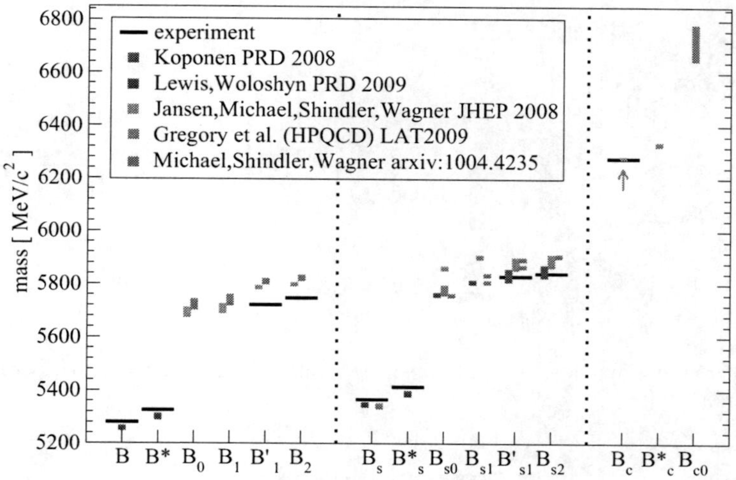

FIGURE 3. Bottom meson spectrum from lattice QCD[6, 18, 19, 20, 21].

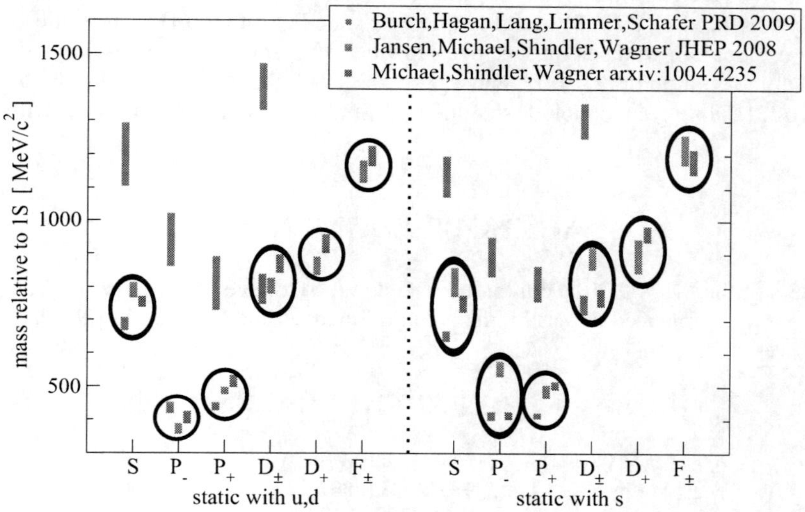

FIGURE 4. Meson spectrum of a static valence anti-quark with a u, d or s valence quark[9, 19, 20]. Black ovals are only to guide the eye in grouping relevant simulation results.

clover[6, 18], twisted mass[19, 20] or HISQ (staggered)[21].

Systematic deviations of lattice from experiment are observed in some cases. The spread among lattice results in figure 3 provides an impression of the systematic errors in current lattice simulations, but for detailed discussions see the original lattice papers[6, 18, 19, 20, 21].

It should be noted that the precise and accurate lattice result for the B_c mass, coming from the HPQCD collaboration, was obtained by tuning the standard QCD inputs to match the experimental masses of pion, kaon, η_c and Υ[21]. Their B_c^* mass stands as a precise prediction for future experiments.

Interesting work has also been reported for mesons in the limit of a static (infinitely-heavy) bottom quark[9, 19, 20]. This unphysical limit shares many qualitative features with the physical spectrum and has been used in combination with charmed meson masses to interpolate to the region of physical bottom mesons. As evident from figure 4, a long list of quantum numbers has been studied. The variance among different lattice determinations of some masses suggests the presence of significant systematic effects. Details can be found in the original articles[9, 19, 20].

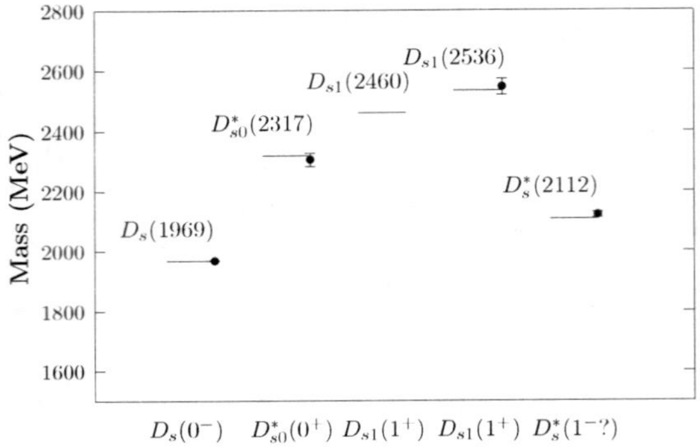

FIGURE 5. Charmed meson spectrum from a lattice QCD simulation (filled circles) compared to experiment (horizontal lines). This graph was taken directly from [22].

CHARMED MESONS

The spectrum of charmed mesons has not received much attention recently from the lattice community, and thus represents a future opportunity. Dong et al.[22] have studied the D_s spectrum, motivated in part by published speculations that the experimentally-observed $D_{s0}^*(2317)$ might not be a $c\bar{s}$ state but rather a DK molecule, a four-quark state or a threshold effect. They use domain wall sea quarks, overlap valence quarks and an Iwasaki gauge action with a lattice spacing of 0.08 fm, a lattice volume of $(2.7\text{ fm})^3$ and a pion mass as light as 331 MeV. Figure 5 is taken directly from their conference report[22]. Dong et al. conclude that their preliminary results are consistent with the $D_{s0}^*(2317)$ being a standard $c\bar{s}$ meson.

ACKNOWLEDGMENTS

Thanks to the organizers of the MENU 2010 conference, held at the College of William and Mary, for the opportunity to participate. This work was supported in part by the Natural Sciences and Engineering Research Council of Canada.

REFERENCES

1. CDF collaboration: T. Aaltonen et al., Phys. Rev. Lett. 99, 202001 (2007).
2. D0 collaboration: V. Abazov et al., Phys. Rev. Lett. 99, 052001 (2007).
3. CDF collaboration: T. Aaltonen et al., Phys. Rev. Lett. 99, 052002 (2007).
4. D0 collaboration: V. Abazov et al., Phys. Rev. Lett. 101, 232002 (2008).
5. CDF collaboration: T. Aaltonen et al., Phys. Rev. D80, 072003 (2009).
6. R. Lewis and R.M. Woloshyn, Phys. Rev. D79, 014502 (2009).
7. S. Meinel, W. Detmold, C.-J. D. Lin and M. Wingate, PoS(LAT2009)105 (2009).
8. H. Na, Ph.D. thesis, Indiana University (2008); H. Na and S. Gottlieb, PoS(LAT2008)119 (2008).
9. T. Burch, C. Hagen, C.B. Lang, M. Limmer and A. Schäfer, Phys. Rev. D79, 014504 (2009).
10. W. Detmold, C.-J. Lin and M. Wingate, Nucl. Phys. B818, 17 (2009).
11. H.-W. Lin, S.D. Cohen, N. Mathur and K. Orginos, Phys. Rev. D80, 054027 (2009).
12. M. Wagner and C. Weise, arXiv:1008.0653 (2010).
13. S. Meinel, arXiv:1008.3154 (2010).
14. L. Liu, H.-W. Lin, K. Orginos and A. Walker-Loud, Phys. Rev. D81, 094505 (2010).
15. N. Mathur, R. Lewis and R.M. Woloshyn, Phys. Rev. D66, 014502 (2002).
16. J.M. Flynn, F. Mescia and A.S.B. Tariq, JHEP 07, 066 (2003).
17. T.-W. Chiu and T.-H. Hsieh, Nucl. Phys. A755, 471 (2005).
18. J. Koponen, Phys. Rev. D78, 074509 (2007).
19. K. Jansen, C. Michael, A. Shindler and M. Wagner, JHEP 12, 058 (2008).
20. C. Michael, A. Shindler and M. Wagner, arXiv:1004.4235 (2010).
21. E.B. Gregory, C.T.H. Davies, E. Follana, E. Gamiz, I. Kendall, G.P. Lepage, H. Na, J. Shigemitsu, K.Y. Wong, arXiv:0911.2133 (2009).
22. S.J. Dong, A. Alexandru, T. Draper, K.F. Liu, A. Li, T. Streuer and J.B. Zhang, PoS(LAT2009)090 (2009).

LHCb Physics and 2010-11 prospects

Stefano Perazzini, on behalf of the LHCb Collaboration

INFN Bologna, Italy
stefano.perazzini@bo.infn.it

Abstract. LHCb is one of the four major experiments operating at the Large Hadron Collider, and is specifically dedicated to the measurement of CP-violation and rare decays in the beauty and charm quark sectors. By employing data from early LHC runs it is possible to assess the performance of the detector and to better understand the potential of the LHCb flavour programme. After a brief introduction of the motivations and of the relevant physics goals, the prospects about key CP-violation and rare decay measurements will be presented. Emphasis will be given to those topics where results with particular sensitivity to New Physics are expected during the 2010-11 run.

Keywords: Conference proceedings, High Energy Physics, CP violation, CKM matrix, B mesons
PACS: 12.15.Hh

INTRODUCTION

Both direct and indirect searches for New Physics are going to be pursued at the LHC. In the case of indirect measurements, New Physics particles might appear as virtual particles in loop and penguin diagrams. It can become possible in this way to observe new particles before finding them in direct searches. It also means, in principle, that indirect measurements can have access to mass scales beyond the reach of direct searches. Indirect measurements also allow to measure the phases of the new couplings, hence probing the flavour structure of physics beyond the Standard Model (SM).

Very schematically, we can consider two approaches for New Physics searches in heavy flavour physics. In a first approach we can look for flavour changing neutral currents (FCNC) transitions, which are forbidden at tree level in the SM. Such rare B and D decays which are suppressed in the SM can be significantly enhanced in models beyond the SM. In particular, in $b \to s$ transitions large effects might be observed.

The second approach in probing for NP is to make precise tests of the SM predictions of CP-violating phenomena. In the B_d^0 system this can be done through overconstraining the classical unitarity triangle. Although the current data are consistent with the SM, it is still open to corrections of the order 10% [1]. In this case, New Physics can be spotted out by comparing many different measurements, which can have contributions from New Physics or can be unaffected, thus leading to observable discrepancies. With this respect it is important to note that the Unitarity Triangle is not yet tightly constrained if one considers only measurements from tree decays [2]. In practice, a precise measurement of the CKM angle γ from tree decay processes in conjunction with the measurement of $|V_{ub}|$ provides a New Physics free reference point in the $\bar{\rho} - \bar{\eta}$ plane, which can be matched with the results achieved by using other measurements potentially sensitive to New Physics. Tests of the SM description of CPV are far weaker in the B_s^0 system and here the effects of large physics could be much more evident.

The LHCb detector [3] provides a series of unique opportunities for indirect searches. The key features of LHCb are given by excellent vertexing, particle identification and momentum resolution capabilities, as well as an efficient trigger and an acceptance particularly suited for detecting decays of beauty and charm hadrons. With only a few pb^{-1} of integrated luminosity so far recorded, LHCb has already collected a clean sample of B decays, including several exclusive decay modes. By the end of 2011 LHCb is expected to record order of 1 fb^{-1} of data.

KEY MEASUREMENTS

In this section we will present some of the top priority measurements that are planned for the 2010-11 run. More detailed information about some of the analyses can be found in the so-called LHCb roadmap document [7].

12th International Conference on Meson-Nucleon Physics and the Structure of the Nucleon (MENU 2010)
AIP Conf. Proc. 1374, 585-588 (2011); doi: 10.1063/1.3647209

Charm Physics

LHCb also has excellent potential for improving existing measurements and probing for new effects in charm physics. The charm cross-section is expected to be ~ 20 times higher than that for b-production, and the trigger and capabilities of the LHCb detector are well suited to exploiting this huge production rate.
The SM predicts that CP violation in the D system is tiny. For this reason any sizeable measurement of CP violation would be a clear indication of New Physics. Several measurements can probe CP violation in the D system. For example, employing about 100 pb^{-1} of data we can exploit a sample of 17 millions of tagged $D^0 \to K^{\pm}\pi^{\mp}$ decays and 1.3 millions of tagged $D^0 \to K^{\pm}K^{\mp}$ decays. These two-body decays can be used to probe CP violation in the mixing of the D^0 meson system. Furthermore, CP violation can be measured by means of a Dalitz analysis of the singly Cabibbo suppressed decay $D \to KK\pi$. Many millions of events are expected in only 100 pb^{-1} of data.

CKM angle γ from tree-level B decays

Transitions at the $b \to u$ tree level allow for a theoretically clean extraction of the γ angle, i.e. the CKM angle known with large uncertainty so far from direct measurements. The present knowledge on γ arising from Unitarity Triangle fits, assuming the validity of the SM, is $\gamma = (69 \pm 3)^\circ$ [2]. This is to be compared with the value measured from $B \to DK$ tree decays, which is $\gamma = (74 \pm 11)^\circ$ [2]. Apparently γ is not yet very constrained from direct measurements and experimental improvements are of paramount importance in the quest for New Physics in the flavour sector.
Two strategies for extracting γ can be exploited. The first is to make measurements involving B->DK decays. Many strategies exist - particularly powerful is the ADS+GLW method [4] [5], which proceeds through the measurement of the relative decay rates for two-body D^0 decays into $K\pi$, KK, $\pi\pi$, and the GGSZ method [6] which requires a Dalitz analysis of the three-body $D^0 \to K^0_S\pi\pi$ decay. The second strategy. only possible at hadron machines, and also very well suited to LHCb on account of its PID capabilities and excellent proper time resolution, involves the time dependent measurement of the $B^0_s \to D^{\mp}_s K^{\pm}$ decay, which provides access to $\gamma + \phi_s$ through the interference between the B^0_s mixing and the decay, where the B^0_s mixing phase ϕ_s can be measured directly thanks to the $B^0_s \to J/\psi\phi$ decay. This analysis requires the additional complication of the flavour tagging and of the proper time measurement. The combined sensitivity on γ from tree-level decays at LHCb is expected to be $\sigma_\gamma \simeq 7^\circ$ for 1 fb^{-1} of integrated luminosity.

CKM angle γ from penguins

The angle γ can also be measured from decays where contributions of loop topologies in the decay diagrams are not negligible. A discrepancy between measurements of γ with and without penguin diagrams involved in the decay would be a strong indication of a contribution from NP. One way to measure γ from such kind of decays is the analysis of the charmless charged two-body decays of B mesons. These decays can be subdivided into two main categories: decays into CP eigenstates, $B^0_d \to \pi^+\pi^-$ and $B^0_s \to K^+K^-$ characterized by time dependent CP asymmetry, and decays into flavor specific final states, $B^0_d \to K^+\pi^-$ and $B^0_s \to \pi^+K^-$, where the observable of interest is the direct CP asymmetry. Within the validity of the U-spin symmetry, from the evaluation of time-dependent and direct CP asymmetries of these channels it is possible to extract γ and also ϕ_s, using as an input the well known value of the B^0_d mixing phase ϕ_d [8]. The expected sensitivity on γ for this analysis is $\sigma_\gamma \simeq 7^\circ$ for 2 fb^{-1} of integrated luminosity.
From this family of decays it is also possible to measure various CP-averaged branching ratios and the direct CP asymmetries of the B baryon decays, such as $\Lambda_b \to p\pi^-$ and $\Lambda_b \to pK^-$. It will be also possible to observe for the first time the very rare decays $B^0_d \to K^+K^-$ and $B^0_s \to \pi^+\pi^-$.

B^0_s mixing phase from $B^0_s \to J/\psi\phi$

The decay $B^0_s \to J/\psi\phi$ is sensitive to the B^0_s mixing phase ϕ_s through the interference between the mixing and decay amplitudes. This decay is the B^0_s counterpart of the decay $B^0_d \to J/\psi K^0$ which measures the B^0_d mixing phase. The contribution from penguin diagrams is negligible and, in contrast with the B^0_d counterpart, the mixing phase is

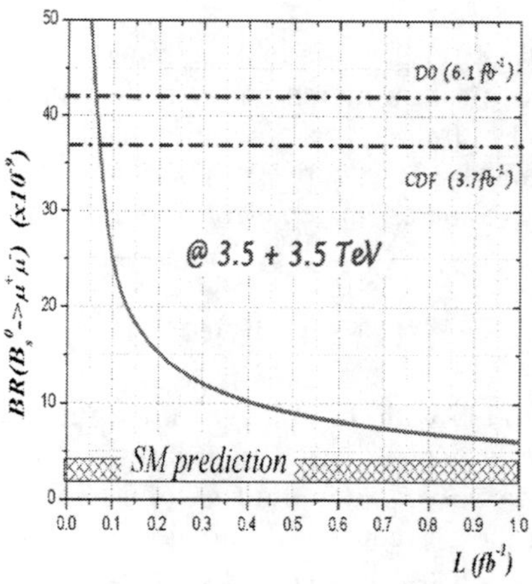

FIGURE 1. Exclusion limits at 90% CL for the $B_s^0 \rightarrow \mu^+\mu^-$ branching fraction versus integrated luminosity at LHCb.

expected to be very small in the SM, with a value of $\phi_s = \left(36.0^{+1.6}_{-2.0}\right)$ mrad [2]. New particles in box diagrams can easily modify the measured phase making it a promising measurement to search for New Physics. The current CDF sensitivity is $\sigma_{\phi_s} \simeq 0.5$ for 5.2 fb^{-1} of integrated luminosity [9].
Experimentally, this decay requires measuring the time-dependent decay rates of the B_s^0 and $\bar{B}_s^0$. Hence, the initial flavour of the B_s^0 needs to be tagged. From Monte Carlo simulations a mistag rate of 33% is obtained, with an effective tagging power of about 6%. Since the B_s^0 decays into two vector mesons, an angular analysis needs to be carried out to disentangle the contributions from the CP-even and CP-odd final states. In order to check the angular acceptance it is foreseen to use control channels as $B^+ \rightarrow J/\psi K^+$ and $B_d^0 \rightarrow J/\psi K^*$. About 60,000 untagged decays are expected to be triggered and offline-selected for an integrated luminosity of 1.0 fb^{-1}. The corresponding error on ϕ_s is $\sigma_{\phi_s} \simeq 0.07$.

Search for $B_s^0 \rightarrow \mu^+\mu^-$

The SM prediction of the branching fraction of the $B_s^0 \rightarrow \mu^+\mu^-$ decay is very small: $BR = (3.35 \pm 0.32) \times 10^{-9}$ [10]. In the Minimal Supersymmetric Standard Model (MSSM) the branching fraction is enhanced by the sixth power of $\tan\beta$. This makes this decay channel a very sensitive probe for New Physics. The present limit from CDF (using a data sample with an integrated luminosity of 3.7 fb^{-1}) is $BR < 3.6 \times 10^{-8}$ (90% CL) [11]. The selection parameters are divided into three uncorrelated sets, which are the invariant mass, the muon identification likelihood and the so-called geometrical likelihood. This procedure allows the calibration of the selection parameters with different control samples.
It is planned to use the kinematically similar decays $B_s^0 \rightarrow K^+K^-$ to extract the mass resolution from data. The simulation indicates a B_s^0 mass resolution of about 20 MeV.
Normalisation channels, such as $B_d^0 \rightarrow K\pi$ and $B^+ \rightarrow J/\psi K^+$, will be used to determine the branching ratio. The main contribution to the systematic uncertainty comes from the production ratio of B_s^0 over $B_{u,d}$, which is only known to $\sim 13\%$ [12]. Various strategies are under consideration for measuring this ratio at LHCb [13]. With 200 pb^{-1} the 90% C.L. upper limit for the branching ratio is expected to be 16×10^{-9} (see Fig. 1). Furthermore, with 1 fb^{-1} we could exclude branching ratios down to 7×10^{-9}, or observe a signal in case the branching ratio is 3.5 times the one in the SM.

Asymmetries in $B_d^0 \to \mu^+\mu^- K^*$ decays

Another interesting decay is $B_d^0 \to \mu^+\mu^- K^*$. It has a branching ratio of $BR \simeq 1.0 \times 10^{-6}$. The angular distributions of the decay products provide a rich source of information on the underlying physics. In the 2010-11 run, LHCb will focus on measuring the forward-backward asymmetry as a function of the q^2 of the two muons, AFB(q^2). The zero-crossing point of AFB is well predicted in SM. LHCb expects to have about 1,400 events selected with 1 fb^{-1} of data. In the most sensitive bin (16 GeV^2) this gives an estimated error on AFB of $\sigma_{\mathrm{AFB}} = 0.07$.

CONCLUSIONS

The LHCb experiment is proceeding smoothly with the analysis of the first data. The first clean samples of B decays have already been reconstructed, and the projected performances for the key measurements agree well so far with Monte Carlo expectations. Already with an integrated luminosity of 200 pb^{-1} we expect to provide very competitive measurements, in some cases surpassing the current world knowledge. With 1 fb^{-1} it will be hopefully possible to spot out physics beyond the SM, or at least narrow down the allowed parameter space dramatically.

REFERENCES

1. E. Barberio et al. [Heavy Flavor Averaging Group], arXiv:0808.1297v3 [hep-ex].
2. M. Bona et al., "Status of the unitarity triangle analysis," PoS E **PS-HEP2009**, 160 (2009).
3. A. A. Alves et al. [LHCb Collaboration], JINST 3 (2008) S08005.
4. D. Atwood, I. Dunietz and A. Soni, Phys. Rev. Lett. 78, 3257 (1997); D. Atwood, I. Dunietz and A. Soni, Phys. Rev. D 63, 036005 (2001).
5. M. Gronau and D. London, Phys. Lett. B 253, 483 (1991); M. Gronau and D. Wyler, Phys. Lett. B 265, 172 (1991).
6. Yu. Grossman, A. Soffer and J. Zupan, Phys. Rev. D 72, 031501 (2005).
7. B. Adeva et al. [The LHCb Collaboration], arXiv:0912.4179v2 [hep-ex].
8. R. Fleischer, Phys. Lett. B 459 (306) 1999.
9. L. Oakes, Proc. of FPCP 2010.
10. M. Blanke, A. Buras, D. Guardagnoli and C. Tarantino, arXiv:hep-ph/0604057v5.
11. CDF Collaboration, CDF Public Note 9892.
12. B. Adeva et al. [The LHCb Collaboration], arXiv:0912.4179v2 [hep-ex].
13. R, Fleischer, N. Serra and N. Tuning, arXiv:1004.3982 [hep-ph].

Session R: Symmetries, Parity Violation and Beyond the Standard Model

Study of charge symmetry breaking in dd collisions with WASA-at-COSY

Aleksandra Wrońska for the WASA-at-COSY Collaboration

Institute of Physics, Jagiellonian University, Cracow, Reymonta 4, 30-059 Cracow, Poland

Abstract. Charge symmetry is an approximate symmetry of the strong interaction. Studies of its breaking can yield information on the u and d quark mass difference. A theoretical collaboration is currently working on the description of charge symmetry breaking mechanisms for $dd \to \alpha\pi^0$ and $n\,p \to d\,\pi^0$ within Chiral Perturbation Theory, using the data from TRIUMF and IUCF. One of the items in the program of the WASA-at-COSY collaboration is to extend the data base for the $dd \to \alpha\pi^0$ reaction to higher energies, which would allow the extraction of the information on the p-wave. Status of the analysis of experimental data along with the preliminary results from the pilot run will be presented here.

Keywords: charge symmetry, pion production, quark mass difference
PACS: 11.30.Hv, 24.80.+y, 25.10.+s, 25.45.-z

PHYSICS MOTIVATION

Charge symmetry is a subset of the isospin symmetry, related to the invariance of a system under the rotation by 180° around the I_2-axis in the isospin space, which corresponds to the interchange of u and d quarks [1]. This symmetry is broken due to different masses and charges of those quarks.

Unlike in the case of the more general isospin symmetry, which is broken predominantly due to the electromagnetic effects introduced by the pion mass difference, for the charge symmetry breaking observables the electromagnetic effects and those resulting from the $up - down$ quark mass difference are typically of the same order. Thus, experimental studies of charge symmetry breaking offer a unique insight into the light quark mass difference, one of the fundamental parameters of the Standard Model.

The hitherto performed experiments addressing the topic belonged to the group of the *null experiments*, i.e. their goal was to determine observables for certain reactions, which vanish if charge symmetry holds. One was the TRIUMF experiment [2], which determined a non-zero forward-backward asymmetry in the reaction $n\,p \to d\,\pi^0$. The other was the IUCF measurement [3] of the cross-section of $d\,d \to \alpha\,\pi^0$ near threshold. This observable is a direct measure of the charge symmetry breaking matrix element.

A theoretical collaboration is currently working on a description of charge symmetry breaking mechanisms within Chiral Perturbation Theory, with a rigorous treatment of the few-body initial and final state interactions [4, 5, 6, 7, 8, 9]. The existing experimental data base is scant, still the results have to be reproduced by the theoretical calculations. One of the points achieved so far on the theory side is that the estimated total cross-section for $d\,d \to \alpha\,\pi^0$ near threshold is consistent in the order of magnitude with the experimental data of [3] with the assumption of pure s-wave production. The mechanism that dominates $n\,p \to d\,\pi^0$ forward-backward asymmetry is suppressed here due to specific selection rules in the $d+d$ collisions. However, with the onset of p-wave in the final state more graphs can contribute (πNN), thus increasing the cross-section, see [5]. In order to validate this calculation, dedicated experiments allowing the isolation the p-wave contribution are necessary. In addition, the approach should as well describe the related charge symmetry conserving channels, i.e. $d\,d \to ^3\mathrm{A}\;N\;\pi$ in order to prove the correct treatment of the dynamics of the four-nucleon system.

MEASUREMENTS WITH WASA-AT-COSY – PRELIMINARY RESULTS

One of the key items in the program of the WASA-at-COSY collaboration is the series of experiments allowing determination of the p-wave contribution in the reaction $d\,d \to \alpha\,\pi^0$ [10] and thus a deeper understanding of the physics behind the charge symmetry breaking. The total cross-section, estimated by s-wave phase-space scaling of IUCF data is of the order of 75 pb, while the total cross-section for $d+d$ collisions is about 80 mb, with a lot

12th International Conference on Meson-Nucleon Physics and the Structure of the Nucleon (MENU 2010)
AIP Conf. Proc. 1374, 591-593 (2011); doi: 10.1063/1.3647210

of breakup channels as well as those including pion production open. In order to successfully complete such a demanding measurement, the following prerequisites are necessary: high luminosity, very selective trigger, good methods of background suppression in the off-line analysis, good angular resolution. For the p-wave to be extracted unambiguously, a measurement with a polarized deuteron beam is needed.

So far, the WASA-at-COSY collaboration has performed two measurements at a beam momentum of 1.2 GeV/c from the planned series, one (of one week) focusing on the charge symmetry conserving channels, tuning of the apparatus, checking feasibility, estimating relevant cross-sections, the other (of two weeks) being a pilot run performed with unpolarized beam, for the measurement of the $d\,d \rightarrow \alpha\,\pi^0$ cross-section. Preliminary results of the analysis of data from the latter are presented here. The experiments were performed with the WASA facility installed at the COSY synchrotron located in Jülich, Germany. The setup is described in detail in [11]. Forward going α particles were registered in the Forward Detector of WASA, producing signals in several layers of the scintillation hodoscope, while the π^0 mesons were reconstructed from the $\gamma\gamma$ events registered in the CsI calorimeter of the Central Detector of WASA.

Calibration of energy deposits was performed using ^{3}He particles, much more abundant than the α particles. The $d\,d \rightarrow ^3$He $N\,\pi$ channels are the main background to the main reaction channel. Calibration procedures for the three layers of the scintillation hodoscope were performed using characteristic points in the $\Delta E_i - \Delta E_j$ plots (see fig. 1) and a comparison of the observed structures with the results of the Monte Carlo simulations.

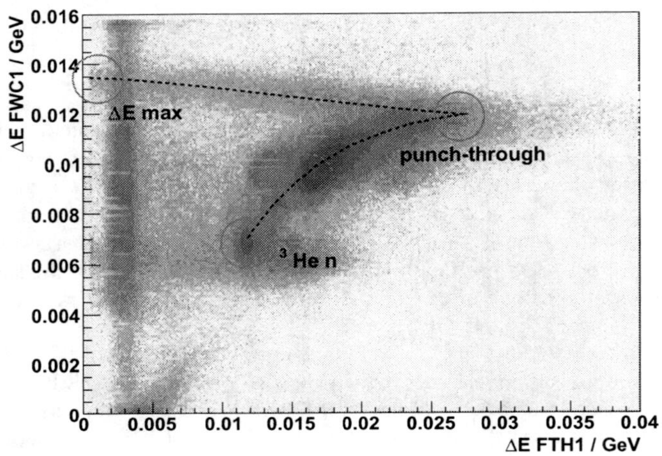

FIGURE 1. Structure stemming from the detected ^{3}He particles in the $\Delta E - \Delta E$ plot. Points used for calibration are marked with circles. For their description see the text.

The characteristic points were: location of the peak of ^{3}He stemming from the binary final state with a neutron, onset of the ^{3}He band formed by the nuclei produced with a neutron and a pion in the final state, the punch-through point, as well as the maximum energy loss. Energy calibration has to be very precise, offering the best achievable resolution. For this reason it was performed angle-dependent. With calibrated energy losses, the $\Delta E - \Delta E$ cut on the α band was determined by means of simulations. In the next step, the energy loss pattern in the hodoscope layers was used to reconstruct kinetic energy of the particles.

The reaction $d\,d \rightarrow ^3$He $n\,\pi^0$ was also used for the luminosity determination. The number of registered events of this process amounted to $5 \cdot 10^6$ in the whole data sample. Based on the cross-section ratio for the charge symmetry conserving and violating channels, the estimate for the expected number of the $d\,d \rightarrow \alpha\pi^0$ events is around $300 - 500$.

For efficient background suppression, apart from the $\Delta E - \Delta E$ cut on the α band, the differences between the two- and three-body final states were exploited. Those comprised cuts on the fixed center-of-mass momenta and back-to-back angular configuration of π^0 and α tracks in the center-of-mass system.

Fig. 2 shows missing mass spectra of the α particles, obtained after imposing the contraints listed above for one of the analysed angular bins, which is expected to contain close to a half of the total statistics, assuming s-wave production. In the left, open circles represent data obtained with the additional demand for the $\gamma\gamma$ invariant mass to exceed 0.1 GeV/c (signal), while filled circles are the spectrum obtained for smaller $\gamma\gamma$ invariant masses (background).

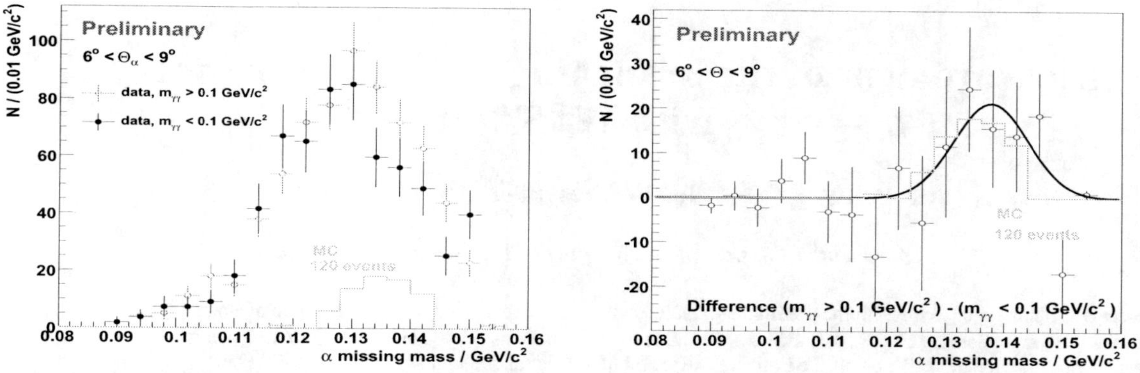

FIGURE 2. Left: Open circles represent data for which $m_{\gamma\gamma} > 0.1$ GeV/c (signal), filled ones are for smaller $\gamma\gamma$ invariant masses (background). **Right**: Difference of those two spectra superimposed on the simulated spectrum containing 120 events (green histogram).

In the right, the difference of those two spectra is superimposed on the simulated spectrum containing 120 events, for comparison and scale.

SUMMARY AND OUTLOOK

Measurements of charge symmetry violating observables offer a unique tool to study the impact of the light quark mass difference on hadron dynamics. A dedicated measurement of the $d\,d \to \alpha\pi^0$ total cross-section (vanishing in a charge symmetry conserving world) at an excess energy of 60 MeV has been performed using the WASA-at-COSY facility. Data analysis is in progress, current results indicating that refinements of the calibration procedures and more efficient background suppression methods are necessary. Preliminary results for the count rate are 2-3 times smaller than the value expected from the s-wave scaling of the IUCF data. Further experimental strategy will be developed when the current analysis has been completed. Extraction of the p-wave contribution requires a run with a polarized deuteron beam. Since the available polarized beam intensity is lower than this of unpolarized one, and the cross-section seems to undershoot the previously estimated value, the planned run will need to be much longer than the pilot run.

ACKNOWLEDGMENTS

This work is supported by the German Federal Ministry of Education and Research, the Polish Ministry of Science and Higher Education and Forschungszentrum Jülich.

REFERENCES

1. G. A. Miller *et al.*, *Phys. Rept.* **194**, 1 (1990).
2. A. K. Opper *et al.*, *Phys. Rev. Lett.* **91**, 212302 (2003).
3. E. J. Stephenson *et al.*, *Phys. Rev. Lett.* **91**, 142302 (2003).
4. U. van Kolck, J. A. Niskanen, G. A. Miller, *Phys. Lett. B* **493** 65 (2000).
5. A. Gårdestig *et al.*, *Phys. Rev. C* **69**, 044606 (2004).
6. A. Nogga *et al.*, *Phys. Lett. B* **639**, 465 (2006).
7. T.A. Lahde and G.A. Miller, *Phys. Rev. C* **75**, 055204 (2007), *Erratum: Phys. Rev. C* **77**, 019904 (2008).
8. A.C. Fonseca, R. Machleidt, G.A. Miller, *Phys. Rev. C* **80**, 027001 (2009).
9. D. R. Bolton, G. A. Miller, *Phys. Rev. C* **81**, 014001 (2010), A. Filin *et al.*, *Phys. Lett. B* **681** 423 (2009).
10. H.-H. Adam *et al.*, arXiv:nucl-ex/0411038.
11. A. Kupść *et al.*, *contribution to these proceedings.*

Measurement of the branching ratio of a rare decay $\eta \to \pi^0\gamma\gamma$ with WASA-at-COSY

Kavita Lalwani* for the WASA-at-COSY Collaboration

* *Department of Physics, Indian Institute of Technology Bombay, India*

Abstract. In this paper we present the preliminary results on the measurement of the branching ratio of a rare decay $\eta \to \pi^0\gamma\gamma$ with the WASA Detector at COSY. We have used a sample of 10^7 η mesons produced at the COSY ring using the pd $\to$ ^{3}He η reaction close to threshold. We detail the intricate extraction of the signal, which has about 360 $\pm$ 70 (stat) $\eta \to \pi^0\gamma\gamma$ events, from the overwhelming background channels for example $\eta \to 3\pi^0$, pd $\to$ ^{3}He $3\pi^0$ and pd $\to$ ^{3}He $2\pi^0$.

Keywords: Branching ratio, Pluto, WASA-at-COSY, $\eta \to \pi^0\gamma\gamma$
PACS: PACS: 14.40.Aq, 13.20.-v, 12.39.Fe

INTRODUCTION

The branching ratio for the decay $\eta \to \pi^0\gamma\gamma$ has fundamental implications for intermediate energy physics, where perturbative QCD breaks down. In this regime effective field theories are used to describe and reproduce experimental observables. One of the most successfull effective field theory is the Chiral Perturbation Theory (χPTh) which uses a momentum expansion and provides a good description of the strong and electroweak interactions of pseudoscalar mesons at low energy.

The uncertainties in χPTh calculations of the amplitude for the $\eta \to \pi^0\gamma\gamma$ transition is due to the fact that the tree-level contributions of the χPTh lagrangian of O(p^2) (lowest order) and O(p^4) (next-to-lowest order) are zero as neither π^0 and η can emit a photon. The pion and kaon loops at O(p^4) are greatly suppressed due to G-parity invariance and large mass of the kaons. The first sizable contribution to the $\eta \to \pi^0\gamma\gamma$ decay amplitudes comes from the O(p^6) but the coffecients involved are not precisely determined. One therefore must turn to models as the Vector Meson Dominance (VMD), the Nambu-Jona-Lasinio model (NJL) or the extended Nambu-Jona-Lasinio model (ENJL), to determine these parameters [1]

Similarly experimental results, by different collaborations, on the branching ratio of this channel also do not agree with each other as shown in table 1. This is primarily, because of significant background coming from other decay channels.

TABLE 1. Measurements of BR ($\eta \to \pi^0\gamma\gamma$).

Experiments	BR ($\eta \to \pi^0\gamma\gamma$)
GAMS-2000	$(7.1 \pm 1.4) \times 10^{-4}$ [2]
Crystal Ball	$(2.21 \pm 0.24_{stat} \pm 0.47_{syst}) \times 10^{-4}$ [3]
KLOE	$(8.4 \pm 2.7_{stat} + 1.4_{syst}) \times 10^{-5}$[4]

The KlOE results are preliminary and the Crystal Ball results is based on 500 events. The tenuous agreement between theory and existing measurements motivated us to measure the branching ratio of this rare decay $\eta \to \pi^0\gamma\gamma$ with WASA at COSY.

EXPERIMENT

The WASA detector facility is an internal experiment at the COoler SYnchrotron (COSY) in Juelich, Germany. WASA is a 4π multi detector system with capability of detecting both neutral as well as charged particles (figure 1). It consists of a forward part for measurement of scattered particles and a central part to measure the decay products of the mesons. COSY delivers phase space cooled protons and deuterons in the momentum range between 0.3 GeV/c to 3.7 GeV/c.

12th International Conference on Meson-Nucleon Physics and the Structure of the Nucleon (MENU 2010)
AIP Conf. Proc. 1374, 594-597 (2011); doi: 10.1063/1.3647211

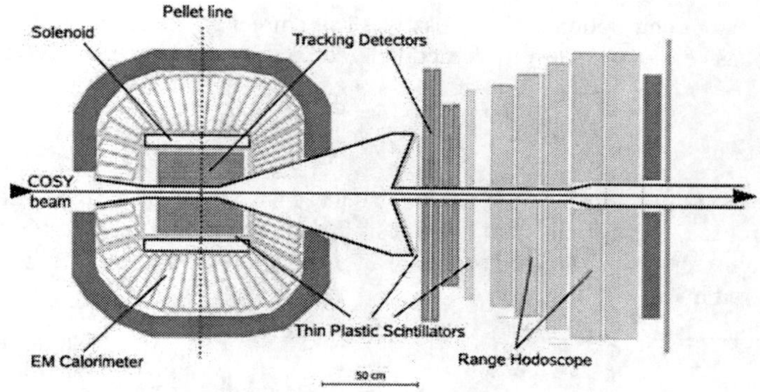

FIGURE 1. Cross section of the WASA detector.

ANALYSIS

Here we present the preliminary results of analysis of decay $\eta \rightarrow \pi^0\gamma\gamma$ with WASA. The data presented here has been obtained during the four week production run taken in September-November 2008, where η-meson were produced in the reaction pd $\rightarrow$ ^{3}He η, obtained by tagging ^{3}He in the Forward Detector, resulting in approximately 10^7 η's at beam kinetic energy T_{beam}= 1.0GeV. Figure 2 (a) shows Identified ^{3}He using ΔE-E method in the Forward Detector.

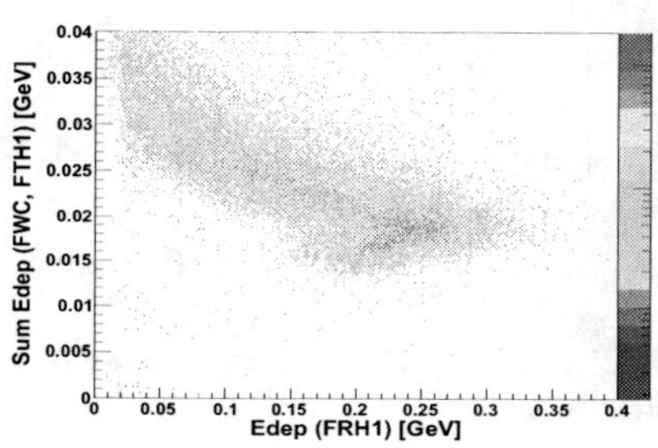

(a)

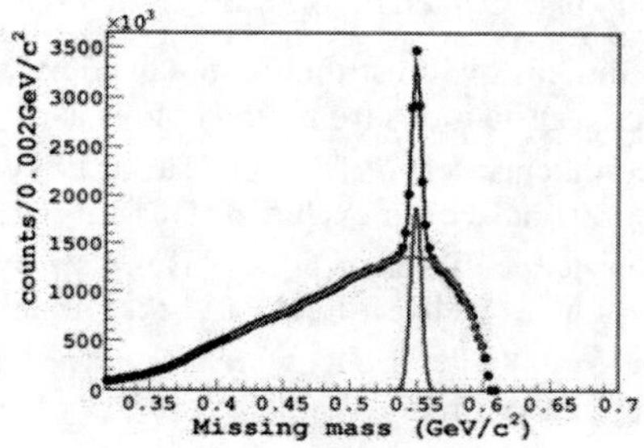

(b)

FIGURE 2. (a) Identified ^{3}He using ΔE- E - method in the Forward Detector (b) Missing mass of scattered ^{3}He for data.

For the selected events in which a helium was identified in the Forward Detector, the missing mass was calculated from the reconstructed helium four vector according to

$$MM(^3He) = \sqrt{(P_{beam} + P_{target} - P_{^3He})^2} \tag{1}$$

Figure 2 (b) shows the missing mass of scattered ^{3}He in the Forward Detector. It shows a pronounced peak at the η mass, which contains 10^7 η's.

Extraction of the branching ratio of $\eta \rightarrow \pi^0\gamma\gamma$, relies heavily on deep understanding of the background channels $\eta \rightarrow 3\pi^0$, double π^0, direct $3\pi^0$ (BR and cross sections : 32%, 1.5 μb, 0.108 μb respectively), since their contributions to the data sample exceeds that of the signal (BR ($\eta \rightarrow \pi^0\gamma\gamma$) $\sim 10^{-4}$). To select the $\eta \rightarrow \pi^0\gamma\gamma$ events, we demand for one charged track in the forward detector and four neutral clusters in the central detector with energy deposit more than 20 MeV for each cluster. In the central detector, to reconstruct π^0-mass from the four reconstructed photons, six possible combinations of photon pairs were made and the χ^2 minimization method was used to choose the best π^0 out of all six possible combinations. Figure 3 (left) shows the invariant mass of π^0. 3σ cut on the invariant mass of $\pi^0(\sigma = 11$ MeV) has been applied to select the $\pi^0\gamma\gamma$ events. A cut on the missing energy of the full event pd $\rightarrow^3$He $\eta(\rightarrow \pi^0\gamma\gamma)$ was applied in the range of -0.1 to 0.1 GeV to reduce contributions coming from six photons in the final state for example $\eta \rightarrow 3\pi^0$ and direct $3\pi^0$ background channels. Using these basic conditions signal to background ratio was 5.1×10^{-4}. The major contribution in this decay was from direct $2\pi^0$ production. To reduce this background channel, kinematic fitting has been used to test two reaction hypotheses needed in our analysis.

$$p\,d \rightarrow {}^3He\,\pi^0\,\pi^0 \qquad (2)$$

$$p\,d \rightarrow {}^3He\,\pi^0\gamma\gamma \qquad (3)$$

In this case four constraints on the energy and momentum balances have been used. In addition the invariant masses of the two π^0's have been used as constraints for hypothesis 2 and constraint on single π^0 mass has been used for hypothesis 3. Figure 3 (right) shows the invariant mass of 2γ versus invariant mass of $\pi^0\gamma\gamma$. We accept only those events which is above the red line.

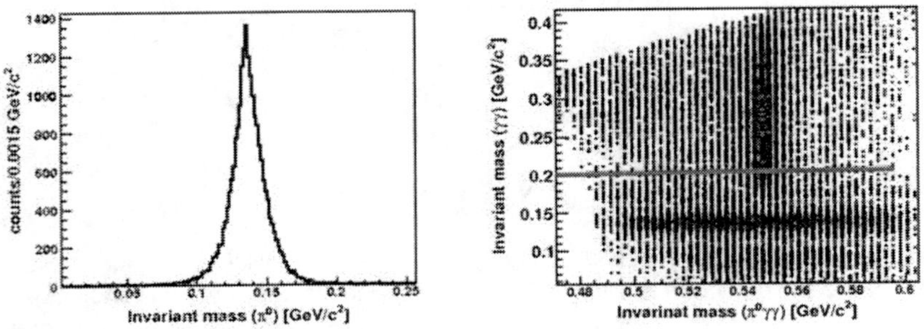

FIGURE 3. Invariant mass of π^0 (left) and Invariant mass of 2γ vs invariant mass of $\pi^0\gamma\gamma$ (right) for data.

To enhance signal to background ratio further, following conditions have been used:

- Cut on the two dimensional distribution obtained by plotting the larger mass of $\pi^0\gamma$ vs mass of $\pi^0\gamma\gamma$ has been applied to reduce contributions from direct pions.
- Condition on confidence level of pd $\rightarrow$ ^{3}He $\pi^0\pi^0$ hypothesis - Events having confidence level of double π^0 greater than 0.00001 have been excluded. This cut reduces the double π^0 events by 94%.
- Condition on confidence level of pd $\rightarrow$ ^{3}He $\pi^0\gamma\gamma$ hypothesis - Selecting CL ($\pi^0\gamma\gamma$) > 0.1 reduces 96 % of $\eta \rightarrow 3\pi^0$ background as well as direct $2\pi^0$ background and 98 % of direct $3\pi^0$ background.
- $m_{\gamma\gamma} > 0.200$ GeV/c^2 - In the events where a single π^0 is identified, an additional condition on the invariant mass of the remaining pair of photons requires that they has to be 4σ above m_{π^0} to reduce contributions from direct pions.
- $\theta_{\pi^0\gamma\gamma} < 57^0$ has been applied to reduce remaining background contributions.

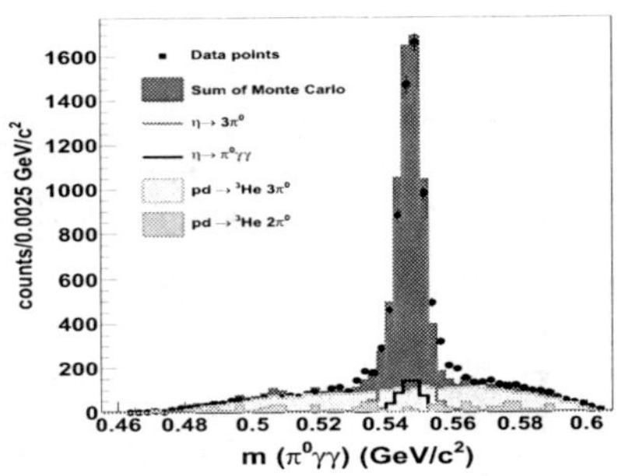

FIGURE 4. Invariant mass of $\pi^0\gamma\gamma$. Comparison of simulation and data.

After applying all selection criteria our Monte Carlo reproduces the experimental data and signal to background ratio is 0.12 as shown in figure 4.

To calculate the branching ratio of $\eta \rightarrow \pi^0\gamma\gamma$, we need to have η's from prompt decay channel for example $\eta \rightarrow 3\pi^0$ (32%) and $\eta \rightarrow 2\gamma$ (39%). We have analyzed $\eta \rightarrow 3\pi^0$ for the same data set. The $\eta \rightarrow 3\pi^0$ events were selected by asking one charged track in the forward detector and six neutral clusters in the central detector with energy deposit more than 20 MeV for each cluster. In the central detector, to select $3\pi^0$ candidates from the six reconstructed photons, fifteen possible combinations of two photon pairs were made and the χ^2 minimization method was used to choose the best π^0 out of all fifteen possible combinations. The only contributed background was direct production of $3\pi^0$. In addition to basic selection criteria, a cut on $\chi^2 < 15.0$ and missing energy of the full event pd $\rightarrow$ ^{3}He $\eta(\rightarrow 3\pi^0)$ in the range of -0.15 to 0.15 GeV have been applied to reduce direct $3\pi^0$ background.

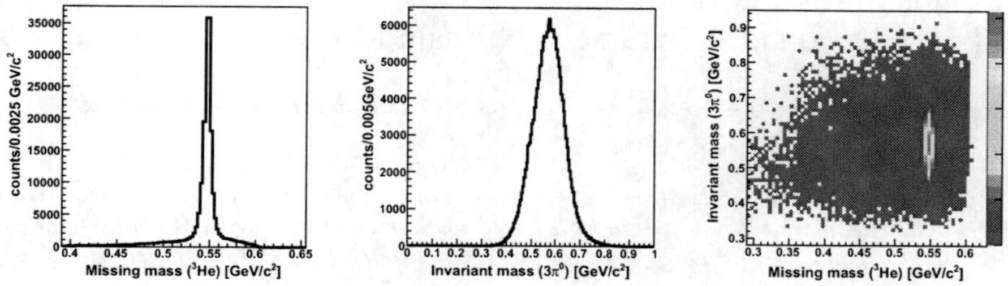

FIGURE 5. Missing mass of scattered ^{3}He (left), invariant mass of $3\pi^0$ (center) and invariant mass of $3\pi^0$ vs missing mass ^{3}He (right) for data.

Figure 5 shows missing mass of scattered ^{3}He (left), invariant mass of $3\pi^0$ (center) and invariant mass of $3\pi^0$ vs missing mass ^{3}He (right). To extract the signal, we have fitted the missing mass distribution by taking signal as Gaussian and background as a polynomial. Subtracting background contribution we have 3.2×10^6 $\eta \rightarrow 3\pi^0$ events.

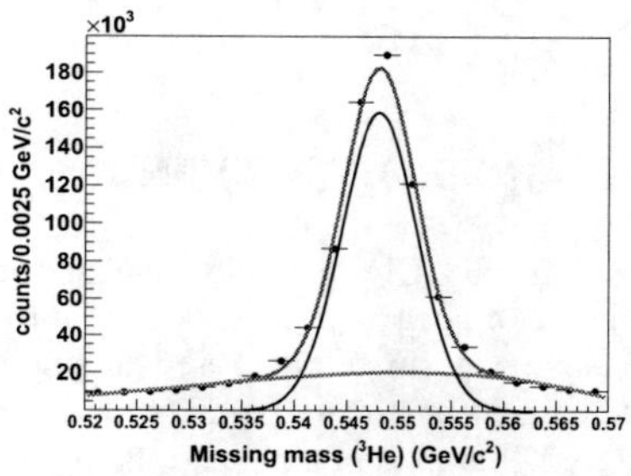

FIGURE 6. Missing mass of ^{3}He for data.

PRELIMINARY RESULTS

We have reconstructed $(3.2\pm0.38_{syst}) \times 10^6$ $\eta \rightarrow 3\pi^0$ events, here statistical error is negligible. The extracted $\eta \rightarrow \pi^0\gamma\gamma$ events from the present analysis is 360 $\pm$ 70 (stat) (figure 4). Evaluation of systematical errors inorder to extract the branching ratio are in progress.

ACKNOWLEDGMENTS

We acknowledge the financial support given by the DAAD-DST exchange program and support by the technical and administration staff at Forschungszentrum, Juelich.

REFERENCES

1. E. Oset et al, Phys. Rev. D **67**, 073013 (2003).
2. D. Alde et al. Z. Phys. C **25**, 225 (1984).
3. S. Prakhov et al., Phys. Rev. C **78**, 015206 (2008).
4. B. Di Micco et al., Acta Phys. Slov. **56**, 403 (2006).

Nucleon strangeness form factors and moments of PDF

Takumi Doi*, Mridupawan Deka†, Shao-Jing Dong**, Terrence Draper**, Keh-Fei Liu**, Devdatta Mankame**, Nilmani Mathur‡ and Thomas Streuer§

*Graduate School of Pure and Applied Sciences, University of Tsukuba, Tsukuba, Ibaraki 305-8571, Japan
email: doi@ribf.riken.jp
† Institute of Mathematical Sciences, Chennai- 6000113, India
** Department of Physics and Astronomy, University of Kentucky, Lexington KY 40506, USA
‡ Department of Theoretical Physics, Tata Institute of Fundamental Research, Mumbai 40005, India
§ Institute for Theoretical Physics, University of Regensburg, 93040 Regensburg, Germany

Abstract. The calculation of the nucleon strangeness form factors from $N_f = 2+1$ clover fermion lattice QCD is presented. Disconnected insertions are evaluated using the Z(4) stochastic method, along with unbiased subtractions from the hopping parameter expansion. We find that increasing the number of nucleon sources for each configuration improves the signal significantly. We obtain $G_M^s(0) = -0.017(25)(07)$, which is consistent with experimental values, and has an order of magnitude smaller error. Preliminary results for the strangeness contribution to the second moment of the parton distribution function are also presented.

PACS: 13.40.-f, 12.38.Gc, 14.20.Dh

INTRODUCTION

Understanding the structure of the nucleon from QCD has been one of the central issues in hadron physics. In particular, the strangeness content of the nucleon attracts a great deal of interest lately. It is also an ideal probe for the virtual sea quarks in the nucleon. Extensive experimental/theoretical studies indicate that the strangeness content varies depending on the quantum number carried by the $s\bar{s}$ pair: the scalar density is about 0–20% of that of up, down quarks, the quark spin is about -10 to 0% of the nucleon, and the momentum fraction is only a few percent of the nucleon. In general, the uncertainties in the strangeness matrix elements are quite large in both experiments and theories. Under these circumstances, it is desirable to provide the definitive quantitative results using lattice QCD.

The challenge in the lattice QCD calculation of strangeness matrix elements resides in the evaluation of the so-called disconnected insertion (DI). In fact, it requires the calculation of all-to-all propagators, which is prohibitively expensive compared to the connected insertion (CI). Consequently, there are only a few DI calculations [1, 2, 3], where the all-to-all propagators are stochastically estimated [4]. In this proceeding, we report the improvement of the calculation of all-to-all propagators using the stochastic method along with unbiased subtractions from the hopping parameter expansion [5], and the increment of the number of nucleon sources [6, 7]. We present the results for the strangeness contribution to the electromagnetic form factors [7] and the second moment of the nucleon. The preliminary result for the first moment of the nucleon is presented in Ref. [8].

FORMALISM AND SIMULATION PARAMETERS

We employ $N_f = 2+1$ dynamical configurations with nonperturbatively $\mathcal{O}(a)$ improved clover fermion and RG-improved gauge action generated by CP-PACS/JLQCD Collaborations [9]. We use $\beta = 1.83$ and $c_{sw} = 1.7610$ configurations with the lattice size of $L^3 \times T = 16^3 \times 32$, which corresponds to $(2\text{fm})^3$ box in physical spacial size with the lattice spacing of $a^{-1} = 1.625$ GeV [9]. For the hopping parameters of u,d quarks (κ_{ud}) and s quark (κ_s), we use $\kappa_{ud} = 0.13825, 0.13800$, and 0.13760, which correspond to $m_\pi = 0.60, 0.70$, and 0.84 GeV, respectively, and $\kappa_s = 0.13760$ is fixed. We perform the calculation only at the dynamical quark mass points, where 800 configurations are used for $\kappa_{ud} = 0.13760$, and 810 configurations for $\kappa_{ud} = 0.13800, 0.13825$.

12th International Conference on Meson-Nucleon Physics and the Structure of the Nucleon (MENU 2010)
AIP Conf. Proc. 1374, 598-601 (2011); doi: 10.1063/1.3647212

The nucleon matrix elements can be obtained through the calculation of 3pt function Π_J^{3pt} (as well as 2pt function Π^{2pt}), defined by

$$\Pi_J^{3pt}(\vec{p},t_2;\ \vec{q},t_1;\ \vec{p}'=\vec{p}-\vec{q},t_0) \;=\; \sum_{\vec{x}_2,\vec{x}_1} e^{-i\vec{p}\cdot(\vec{x}_2-\vec{x}_0)}\cdot e^{+i\vec{q}\cdot(\vec{x}_1-\vec{x}_0)}\langle 0|\mathrm{T}\left[\chi_N(\vec{x}_2,t_2)J(\vec{x}_1,t_1)\bar{\chi}_N(\vec{x}_0,t_0)\right]|0\rangle, \quad (1)$$

where χ_N is the nucleon interpolating field and J is the insertion operator. Since there is no strange quark as a valence quark in the nucleon, the 3pt is a DI which entails a multiplication of the nucleon 2pt correlator with the current quark loop. For the evaluation of the quark loop, we use the stochastic method [4], with Z(4) noises in color, spin and space-time indices. We generate independent noises for different configurations, in order to avoid possible auto-correlation. We use $N_{noise}=600$ noises for $\kappa_{ud}=0.13760, 0.13800$ and $N_{noise}=800$ for $\kappa_{ud}=0.13825$. To reduce fluctuations, the charge conjugation and γ_5-hermiticity (CH), and parity symmetry are used [6, 7]. We also perform unbiased subtractions [5] to reduce the off-diagonal contaminations to the variance. For subtraction operators, we employ those obtained through hopping parameter expansion (HPE) for the propagator M^{-1}, $\frac{1}{2\kappa}M^{-1}=\frac{1}{1+C}+\frac{1}{1+C}(\kappa D)\frac{1}{1+C}+\cdots$ where D denotes the Wilson-Dirac operator and C the clover term. We subtract up to order $(\kappa D)^4$ $((\kappa D)^3)$ term for the form factor (second moment) calculation, and observe that the statistical errors become about 50 (70) %, compared to the results without subtraction.

In the stochastic method, it is quite expensive to achieve a good signal to noise ratio (S/N) just by increasing N_{noise} because S/N improves with $\sqrt{N_{noise}}$. In view of this, we use many nucleon point sources N_{src} in the evaluation of the 2pt part for each configuration. Since the calculations of the loop part and 2pt part are independent of each other, this is expected to be an efficient way. We take $N_{src}=64$ for $\kappa_{ud}=0.13760$ and $N_{src}=82$ for $\kappa_{ud}=0.13800, 0.13825$, where locations of sources are taken so that they are separated in 4D-volume as much as possible. Details of the simulation setup are given in Ref. [7].

STRANGENESS ELECTROMAGNETIC FORM FACTORS

The formulas for Sachs electric (magnetic) form factors G_E^s (G_M^s) are given by

$$R_\mu^\pm(\Gamma_{\rm pol}^\pm) \equiv \frac{\mathrm{Tr}\left[\Gamma_{\rm pol}^\pm\cdot\Pi_{J_\mu}^{3pt}(\vec{0},t_2;\ \pm\vec{q},t_1;\ -\vec{q},t_0)\right]}{\mathrm{Tr}\left[\Gamma_e^\pm\cdot\Pi^{2pt}(\pm\vec{q},t_1;\ t_0)\right]}\cdot\frac{\mathrm{Tr}\left[\Gamma_e^\pm\cdot\Pi^{2pt}(\vec{0},t_1;\ t_0)\right]}{\mathrm{Tr}\left[\Gamma_e^\pm\cdot\Pi^{2pt}(\vec{0},t_2;\ t_0)\right]}, \quad (2)$$

$$G_E^s(Q^2)=\pm R_{\mu=4}^\pm(\Gamma_{\rm pol}^\pm=\Gamma_e^\pm), \qquad G_M^s(Q^2)=\mp\frac{E_N^q+m_N}{\varepsilon_{ijk}q_j}R_{\mu=i}^\pm(\Gamma_{\rm pol}^\pm=\Gamma_k^\pm), \quad (3)$$

where $J_\mu(x+\mu/2)=\frac{1}{2}\left[\bar{q}(x)(1-\gamma_\mu)U_\mu(x)q(x+\mu)-\bar{q}(x+\mu)(1+\gamma_\mu)U_\mu^\dagger(x)q(x)\right]$ is the point-split conserved vector operator, $\{i,j,k\}\neq 4$, $\Gamma_e^\pm\equiv(1\pm\gamma_4)/2$, $\Gamma_k^\pm\equiv(\pm i)/2\times(1\pm\gamma_4)\gamma_5\gamma_k$ and $E_N^q\equiv\sqrt{m_N^2+\vec{q}^2}$. The upper sign corresponds to the forward propagation ($t_2\gg t_1\gg t_0$), and the lower sign corresponds to the backward propagation ($t_2\ll t_1\ll t_0$).

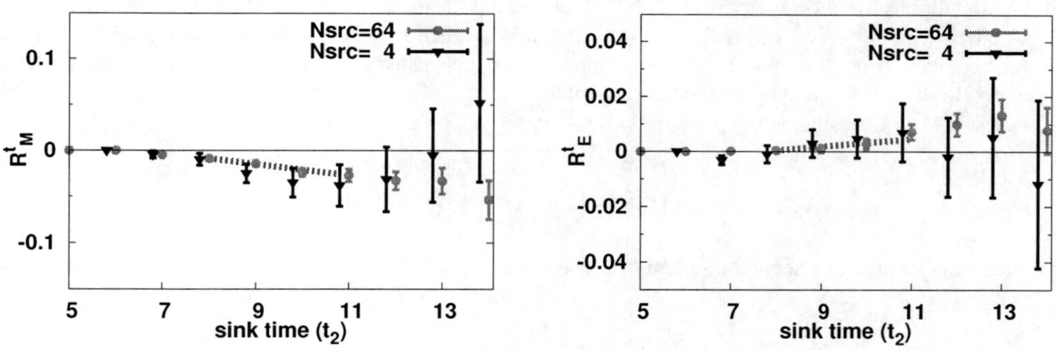

FIGURE 1. R_M^t (left) and R_E^t (right) with $\kappa_{ud}=0.13760$, $N_{src}=64$ (circles) and $N_{src}=4$ (triangles), plotted against the nucleon sink time t_2. The dashed line is the linear fit where the slope corresponds to the form factor.

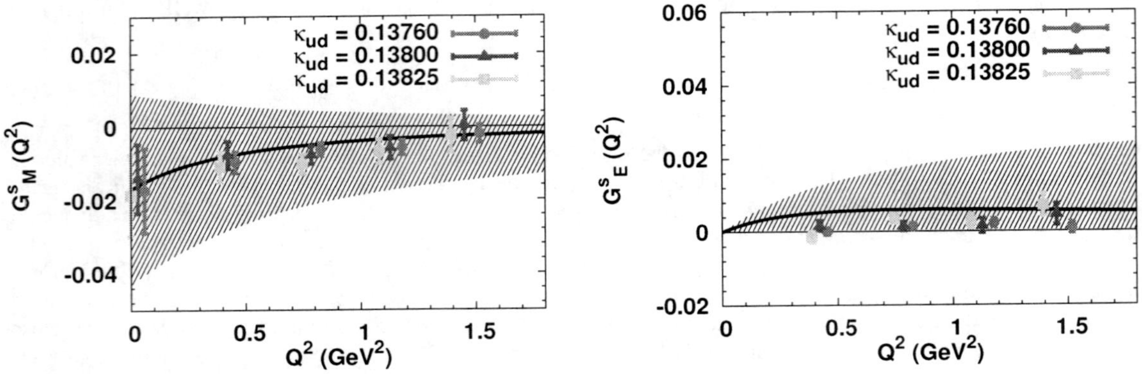

FIGURE 2. The chiral extrapolated results for $G_M^s(Q^2)$ (left) and $G_E^s(Q^2)$ (right) plotted with solid lines. Shaded regions represent the statistical and systematic error added in quadrature. Shown together are the lattice data for each κ_{ud}.

In Fig. 1, we plot typical figures for $R'_{M,E}$, where $R'_{M,E} \equiv \frac{1}{K^{\pm}_{M,E}} \sum_{t_1=t_0+t_s}^{t_2-t_s} R^{\pm}_{\mu}$ with $K^{\pm}_{M,E}$ being trivial kinematic factors in Eq. (3). Since $R'_{M,E} = \text{const.} + t_2 \times G^s_{M,E}$, the linear slope corresponds to the signal of the form factor. One can observe the significant S/N improvement by increasing N_{src}. In fact, the improvement is found to be nearly a factor of $\sqrt{N_{src}}$ (ideal improvement).

We then study the Q^2 dependence of the form factors. For the magnetic form factor, we employ the dipole form, $G_M^s(Q^2) = G_M^s(0)/(1+Q^2/\Lambda^2)^2$, where reasonable agreement with lattice data is observed. For the electric form factor, we employ $G_E^s(Q^2) = g_E^s \cdot Q^2/(1+Q^2/\Lambda^2)^2$, considering that $G_E^s(0) = 0$ from the vector current conservation, and the pole mass Λ is taken from the fit of magnetic form factor.

Finally, we perform the chiral extrapolation for the fitted parameters. Since our quark masses are relatively heavy, we consider only the leading dependence on m_K, which is obtained by heavy baryon chiral perturbation theory (HBχPT) [10]. The chiral extrapolated results are $G_M^s(0) = -0.017(25)$, $\Lambda a = 0.58(16)$, $\langle r_s^2 \rangle_M \equiv -6\frac{dG_M^s}{dQ^2}|_{Q^2=0} = -7.4(71)\times 10^{-3}\text{fm}^2$ and $g_E^s = 0.027(16)$ (or $\langle r_s^2 \rangle_E \equiv -6\frac{dG_E^s}{dQ^2}|_{Q^2=0} = -2.4(15)\times 10^{-3}\text{fm}^2$).

We examine the systematic uncertainties in the result of form factors. For the ambiguity of Q^2 dependence, we reanalyze the data using the monopole form, and obtain the results which are consistent with those from the dipole form. For the uncertainties in chiral extrapolation, we test two alternative extrapolations [7], and find that all results are consistent with each other. For the contamination from excited states, we employ the new projection operator [7] which eliminates the S_{11} state, and conclude that such contaminations are negligible.

Our final result for the magnetic moment is $G_M^s(0) = -0.017(25)(07)$, where the first error is statistical and the second is systematic from uncertainties of the Q^2 extrapolation and chiral extrapolation. We also obtain $\Lambda a = 0.58(16)(19)$ for dipole mass or $\tilde{\Lambda} a = 0.34(17)(11)$ for monopole mass, and $g_E^s = 0.027(16)(08)$. These lead to $G_M^s(Q^2) = -0.015(23)$, $G_E^s(Q^2) = 0.0022(19)$ at $Q^2 = 0.1\text{GeV}^2$, where error is obtained by quadrature from statistical and systematic errors. In Fig. 2, we plot $G_M^s(Q^2)$, $G_E^s(Q^2)$, where the shaded regions correspond to the square-summed error. Compared to the global analysis of the experimental data, e.g., $G_M^s(Q^2) = 0.29(21)$ and $G_E^s(Q^2) = -0.008(16)$ at $Q^2 = 0.1\text{GeV}^2$ [11], our results are consistent with them, with an order of magnitude smaller error [7].

SECOND MOMENT OF THE NUCLEON

The (asymmetry of) strangeness second moment of the nucleon $\langle x^2 \rangle_{s-\bar{s}} = \int_0^1 dx x^2 (s(x) - \bar{s}(x))$ can be obtained by

$$\frac{\mathrm{Tr}\left[\Gamma_e^{\pm} \cdot \Pi^{3\mathrm{pt}}_{T_{4ii}}(\pm\vec{p},t_2;\, \vec{0},t_1;\, \pm\vec{p},t_0)\right]}{\mathrm{Tr}\left[\Gamma_e^{\pm} \cdot \Pi^{2\mathrm{pt}}(\pm\vec{p},t_2;\, t_0)\right]} = \pm p_i^2 \cdot \langle x^2 \rangle_{s-\bar{s}}, \tag{4}$$

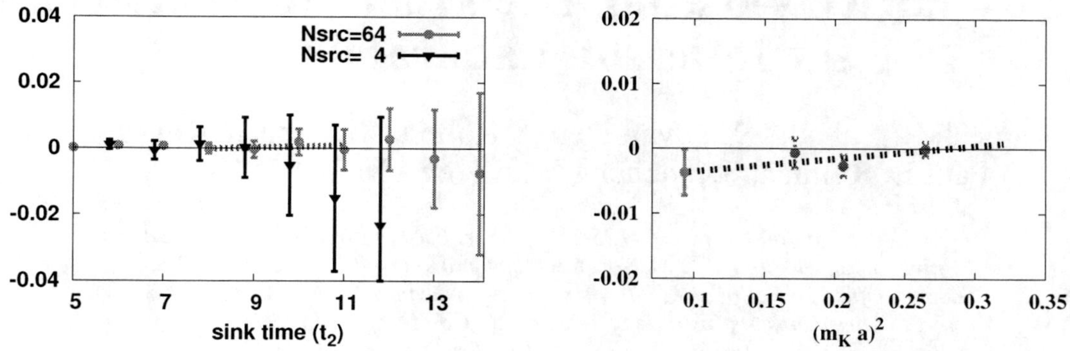

FIGURE 3. LEFT: The ratio of 3pt to 2pt with $\kappa_{ud} = 0.13760$, $N_{src} = 64$ (circles) and $N_{src} = 4$ (triangles), plotted against the nucleon sink time t_2. The dashed line is the linear fit where the slope corresponds to the second moment. RIGHT: The lattice bare results for the second moment at each valence quark mass κ_{ud} for the nucleon, plotted against $(m_K a)^2$. The dashed line corresponds to the linear chiral extrapolation, and the red point is the chiral extrapolated result.

with the three-index operator defined as

$$T_{4ii} \equiv -\frac{1}{3}\left[\bar{q}\gamma_4 \overleftrightarrow{D}_i \overleftrightarrow{D}_i q + \bar{q}\gamma_i \overleftrightarrow{D}_4 \overleftrightarrow{D}_i q + \bar{q}\gamma_i \overleftrightarrow{D}_i \overleftrightarrow{D}_4 q\right], \tag{5}$$

where $i \neq 4$, and the upper (lower) sign corresponds to the forward (backward) propagation as before.

In Fig. 3 (left), we plot the ratio of 3pt to 2pt for $\langle x^2 \rangle_{s-\bar{s}}$ in terms of t_2 for $\kappa_{ud} = 0.13760$, $\vec{p}^2 = (2\pi/La)^2$, where the summation of operator insertion time t_1 is taken as was done for the form factor analysis. Note that the linear slope corresponds to the signal for $\langle x^2 \rangle_{s-\bar{s}}$. One can clearly see that increasing N_{src} reduces the error bar significantly again (about a factor of $\sqrt{N_{src}}$, i.e., almost ideally). In Fig. 3 (right), we plot the bare value of the $\langle x^2 \rangle_{s-\bar{s}}$ in terms of $(m_K a)^2$, and perform the chiral extrapolation. We find that the result at each κ_{ud} and the chiral extrapolated result are basically consistent with zero within the error-bar. For the final quantitative result, it is necessary to take the renormalization factor into account. Systematic uncertainties have to be examined as well. The study along this line is in progress.

ACKNOWLEDGMENTS

We thank the CP-PACS/JLQCD Collaborations for their configurations. This work was supported in part by U.S. DOE grant DE-FG05-84ER40154. TD is supported in part by Grant-in-Aid for JSPS Fellows 21·5985. Research of NM is supported by Ramanujan Fellowship. The calculation was performed at Jefferson Lab, Fermilab and the University of Kentucky, partly using the Chroma Library [12].

REFERENCES

1. S.-J. Dong, K.-F. Liu and A.G. Williams, Phys. Rev. D **58**, 074504 (1998).
2. N. Mathur and S.-J. Dong, Nucl. Phys. Proc. Suppl. **94**, 311 (2001); *ibid.*, **119**, 401 (2003).
3. R. Lewis, W. Wilcox and R.M. Woloshyn, Phys. Rev. D **67**, 013003 (2003).
4. S.-J. Dong and K.-F. Liu, Phys. Lett. B **328**, 130 (1994).
5. C. Thron, S.-J. Dong, K.-F. Liu and H.P. Ying, Phys. Rev. D **57**, 1642 (1998).
6. M. Deka *et al.*, Phys. Rev. D **79**, 094502 (2009).
7. T. Doi *et al.* (χQCD Collab.), Phys. Rev. D **80**, 094503 (2009).
8. T. Doi *et al.* (χQCD Collab.), PoS (LAT2008), 163 (2008), T. Doi *et al.* (χQCD Collab.), PoS (LAT2009), 134 (2009).
9. T. Ishikawa *et al.*, PoS (LAT2006), 181 (2006); T. Ishikawa *et al.*, Phys. Rev. D **78**, 011502(R) (2008).
10. T.R. Hemmert *et al.*, Phys. Lett. B **437**, 184 (1998); *ibid.*, Phys. Rev. C **60**, 045501 (1999).
11. S.F. Pate, D.W. McKee and V. Papavassiliou, Phys. Rev. C **78**, 015207 (2008).
12. R.G. Edwards and B. Joó, Nucl. Phys. Proc. Suppl. **140**, 832 (2005); C. McClendon, `http://www.jlab.org/~edwards/qcdapi/reports/dslash_p4.pdf`

6 GeV Parity Violating Deep Inelastic Scattering at Jefferson Laboratory

Ramesh R. Subedi*, Xiaoyan Deng†, Robert Michaels**, Kai Pan‡, Paul E. Reimer§, Diancheng Wang¶ and Xiaochao Zheng¶

*George Washington University, 725 21st St, NW, Washington, DC 20052, USA
†University of Virginia, 382 McCormick Rd., Charlottesville, VA 22904, USA.
**Thomas Jefferson National Accelerator Facility, Newport News, VA 23606, USA.
‡Massachusetts Institute of Technology, Cambridge, MA 02139
§Physics Division, Argonne National Laboratory, Argonne, IL 60439, USA.
¶University of Virginia, 382 McCormick Rd.,Charlottesville, VA 22904, USA.

Abstract. The 6 GeV Parity Violating Deep Inelastic Scattering (PVDIS) experiment has measured a 10^{-4} level asymmetry through polarized electron scattering off a liquid deuterium target with a beam energy of 6 GeV. This experiment has a goal of measuring a combination of the product of the weak neutral couplings of the electron and the quark with a factor of six improvement in precision over world data. Precise data for the couplings are essential to search for physics beyond the Standard Model. The experiment took place in Hall A at Thomas Jefferson National Accelerator Facility (Jefferson Laboratory) and data collection was completed in the end of 2009. A highly specialized counting data acquisition system with an inherent particle identification was developed and utilized. We have taken data at two Q^2 points in order to possibly address the hadronic correction due to higher twist effects. An overview of the experiment will be presented.

Keywords: parity, asymmetry, Standard Model
PACS: 21.10.Hw, 25.30.-c

INTRODUCTION

Nature distinguishes between left and right. That is where parity violation experiments, like PVDIS, find their scope. The objective of the 6 GeV PVDIS experiment (E08-011) is to measure the coupling constant combination $2C_{2u} - C_{2d}$ with an unprecedented accuracy by measuring the PVDIS asymmetry better than 4% statistical accuracy using a polarized electron beam on an unpolarized liquid deuterium target. Here $C_{2q} = g_V^e g_A^q$, with $q = u, d$ indicating an up or a down quark, is a coupling constant, g_V^e is the electron vector coupling and g_A^q is the quark axial coupling. The coupling constant combination appears in the expression for the parity violating asymmetry (A_{PV}) for deep inelastic scattering (DIS) of electrons from a liquid deuterium target as shown below:

$$
\begin{aligned}
A_{PV} &= \frac{\sigma_+ - \sigma_-}{\sigma_+ + \sigma_-} \\
&= \left(\frac{3G_F Q^2}{\pi\alpha^2\sqrt{2}}\right) \frac{2C_{1u}[1+R_c(x)] - C_{1d}[1+R_s(x)] + Y(2C_{2u} - C_{2d})R_v(x)}{5 + R_s(x) + 4R_c(x)},
\end{aligned}
$$

where σ_+ and σ_- are the left-handed and right-handed electron scattering cross-sections, respectively, Q^2 is the negative of the four-momentum transfer squared, G_F is the Fermi

12th International Conference on Meson-Nucleon Physics and the Structure of the Nucleon (MENU 2010)
AIP Conf. Proc. 1374, 602-605 (2011); doi: 10.1063/1.3647213

weak coupling constant, α is the fine structure constant, Y is a kinematic factor, $R_{S,C}$ and R_V are related to sea- and valence-quark distribution functions, and x is the Bjorken scaling variable (for details see Ref. [1, 2, 3]). The magnitude of this asymmetry is approximately Q^2 times 100 parts per million.

The data taking of the experiment took place in experimental Hall A at Jefferson Laboratory and was completed in December 2009. The experiment was performed by using two high resolution spectrometers, with their standard settings [5], at two Q^2 points: the first Q^2 point being 1.1 (GeV/c)^2 with incident electron energy $E = 6.0$ GeV, scattered electron energy $E' = 3.66$ GeV, Bjorken $x = 0.25$, and scattered electron angle $\theta = 12.9°$, and the second Q^2 point being 1.9 (GeV/c)^2 with $E = 6.0$ GeV, $E' = 2.63$ GeV, $x = 0.30$, and $\theta = 20.0°$. Both Q^2 points used about 89% polarized electron beam about 90 – 115 μA on a 20 cm long deuterium target. The two Q^2 data points can be used to extract experimental results for the hadronic correction, $c(x)$, due to higher-twist effects

$$A_{pv}(x, Q^2) = A_{pv}(x)(1 + c(x)/Q^2),$$

whose value is believed to be impossible to get from the first principles [4]. The plan is to find asymmetry at low Q^2 point and then set a limit on higher twist there. Next, use that higher twist while measuring asymmetry at the high Q^2 point.

INSTRUMENTATION

The experiment utilized the standard Hall A instrumentation consisting of Compton and Moller polarimeters for measuring (and monitoring) beam polarization, target chamber consisting of a production target (liquid deuterium) along with other calibration targets, and two high resolution spectrometers. Each spectrometer had a detector package comprising of two vertical drift chambers for tracking, two scintillator planes for main triggers, and a set of particle identification detectors made up of a gas-cerenkov and two layers of lead-glass detectors. The tracking chambers were used only for some calibration runs at low event rate to check tracking efficiency, and they were turned off during production runs suspecting that they could not take a high event rate without being damaged. Lumi monitors, installed downstream of the target, were also used to monitor beam and target noise.

DEADTIME AND ASYMMETRY MEASUREMENT

This was the first time that a scaler based counting data acquisition (DAQ) system was implemented at Jefferson Lab for an experiment bearing an expected high event rate of about a mega hertz. The particle identification capability for identifying electrons and pions was included in the DAQ hardware. The main purpose of the counting DAQ was to take a high event rate data by keeping the deadtime negligible (<1%) with about $< 0.3\%$ accuracy so that it may not contribute anything significant to the accuracy in the physics asymmetry of the experiment.

For the deadtime measurement, two different resolution times of the electronic modules were used: 30 ns ("narrow path") and 100 ns ("wide path"). Subsequent events falling within the resolution time of the previous events cannot be recorded by the DAQ, causing a counting loss due to the deadtime.

As a specific method to study deadtime, a 10 kHz gate generator signal (called a tagger) was combined with all lead-glass (preshower and total shower) signals using analog summing modules producing a total input rate as R_i. The signal, after passing through a chain of electronics modules including discriminators that define both "narrow path" and "wide path", was then ANDed with the delayed tagger signal, giving R_o as the output. Now the fractional loss due to deadtime (D) is obtained as $D = 1-(1-p)R_o/R_i$, where p is a correction factor for a pileup effect: when the tagger signal follows closely a detector signal, the DAQ output caused by the detector signal would coincide with the tagger signal, causing a false count that should have been lost due to deadtimes. Fastbus TDCs were used to measure pileup effect, with the pileup factor p being measured by using a TDC spectrum.

In the deadtime analysis, we find that the "wide path" signals appear to be about 100 ns wide as expected, whereas the "narrow path" signals appear to be about $60-80$ ns wide instead of being 30 ns wide. The observation that the "narrow path" signal being unexpectedly wide is due to the reason that the input signal to the "narrow path" was $60-80$ ns wide by itself. The deadtime from the "wide path" boils down to a 0.1% value with a relative uncertainty of 1.2%.

The expected physics asymmetry of this experiment is about 90 ppm for the low Q^2 point and about 170 ppm for high Q^2 point. In the preliminary asymmetry analysis we find that asymmetry plots appear gaussian as expected, the asymmetry values also agree with expectation within statistical errors and flip sign under the flip of polarization direction by using a beam half wave plate. The asymmetry analysis is still in a very preliminary stage, since other ingredients (pileup, deadtime, polarization, etc.) that involve in revealing the accuracy of the asymmetry are being analyzed.

SUMMARY

The Jefferson Lab experiment $E08-011$ has measured the parity violating asymmetry in deep inelastic scattering from a liquid deuterium target. The preliminary result on deadtime analysis shows that the deadtime is well below one percent with a relative accuracy of about a percent. Very preliminary asymmetry analysis shows that the measured asymmetry has met expectation. Active data analysis is underway and final results could be out in about a year.

ACKNOWLEDGEMENT

This work is supported in part by the U. S. National Science Foundation and the U. S. Department of Energy (DOE), under Contract No. DE-AC02-06CH11357.
Notice: Authored by Jefferson Science Associates, LLC under U. S. DOE Contract No. DE-AC05-06OR23177. The U. S. Government retains a non-exclusive, paid-up, irrevocable, world-wide license to publish or reproduce this manuscript for U. S. Government purposes.

REFERENCES

1. R. Michaels, P. Reimer, and X.-C. Zheng, *$\vec{e}$-^2H Parity Violating Deep Inelastic Scattering at CEBAF at 6 GeV*, Jefferson Lab Hall A Proposal E08-011.
2. X. Zheng, *Parity Violation in Deep Inelastic Scattering at JLab 6 GeV, in PAVI 2006 May 16-20: From Parity Violation to Hadronic Structure and more... (Part III) (Springer, Milos, Greece, 2006)*
3. R. Subedi et al., *AIP proceedings of the 18th International Spin Physics Symposium (SPIN2008)*, 245-248, 2009.
4. K. S. Kumar, *Eur. Phys. J. A* **24**, s2, 191-195 (2005).
5. J. Alcorn, et al., *Nucl. Instr. and Meth.* **A 522**, 294 (2004).

Session S: Nonperturbative Models and Lattice QCD

Equivalence of light-front and covariant approaches in meson-baryon interactions

Chueng-Ryong Ji* and Wally Melnitchouk†

*Department of Physics, Box 8202, North Carolina State University, Raleigh, North Carolina 27695-8202
†Jefferson Lab, 12000 Jefferson Avenue, Newport News, Virginia 23606

Abstract. We demonstrate the equivalence of the light-front, equal-time and covariant formulations in meson-baryon interactions. In particular, we discuss the self-energy Σ of a nucleon dressed by pion loops with the pseudovector πNN coupling. It is shown that the leading nonanalytic behavior of Σ is equivalent whichever formulations are used for the derivation.

Keywords: Chiral Perturbation Theory, Non-Analytic Behaviors, Nucleon Self-Energy
PACS: 11.30.Rd, 12.39.Fe, 13.75.Gx

INTRODUCTION

From the approximate chiral symmetry of QCD it is known that the pion cloud of the nucleon plays a vital role in understanding the nucleon's long-range structure. It provides important corrections to static nucleon properties, such as the mass, magnetic moment and axial charge, and significantly influences its electric and magnetic charge distributions (for a review see Ref. [1]). This will also have important consequences for correctly implementing chiral extrapolation of lattice QCD data on parton distribution moments and subsequently on other quantities, such as generalized parton distributions (GPDs)[2], as well as on understanding the pion distributions in the nucleon in a well-defined theoretical framework. In particular, there has also been a great deal of interest in GPDs, in addition to the traditional studies of elastic form factors and parton distributions. As with the ordinary parton distribution functions, the physical interpretation of GPDs in terms of probability distributions is most natural in the light-front dynamics (LFD), and one knows that here too chiral corrections can be very important. It is timely, therefore, to address the question of how to provide a consistent derivation of the chiral corrections to all of these observables in LFD, and to explain in some detail the technical differences between these calculations in LFD and either in equal-time or covariant formulations.

Since it has been established from the chiral expansion in QCD that pion cloud contributions to moments of twist-two parton distributions of the nucleon have a leading nonanalytic (LNA) behavior determined by the infrared properties of chiral loops (characteristic of Goldstone boson loops in chiral perturbation theory[3]), the LNA behavior is model-independent [4, 5, 6], and places the physics of the pion cloud on a firm footing in QCD. In this presentation, we briefly summarize the distinguishing features of LFD and discuss our recent demonstration [7] of equivalence between LFD and covariant approaches for the interactions of nucleons with pions. We find that there is, in fact, no conflict between the results in equal-time, light-front or covariant frameworks, provided care is taken when performing loop integrations and if one compares consistently the same theories. The detailed analysis on the self-energy Σ of a nucleon dressed by a pion loop (with pseudovector or pseudoscalar πN interactions) can be found in Ref.[7].

In the next Section, Section II, we begin with a brief review on distinguishing features of LFD in comparison with other forms of Hamiltonian dynamics. In Section III, we present a specific example of a treacherous amplitude in LFD. In Section IV, we discuss the LNA behavior of the nucleon self-energy in the chiral limit and its equivalence, regardless of formulations used for the derivation. Conclusions follow in Section V.

DISTINGUISHING FEATURES OF LIGHT-FRONT DYNAMICS (LFD)

Among the three forms of Hamiltonian dynamics proposed by Dirac in 1949[8], i.e. instant ($x^0 = 0$), front ($x^+ = x^0 + x^3 = 0$), point ($x_\mu x^\mu = a^2 > 0, x^0 > 0$), the dimenson of the stability group which leaves the hypersurface of corresponding time invariant is the largest (i.e. seven out of the ten Poincare generators) in the front form which we call the LFD. Since the number of kinematic operators is maximum among the possible forms of Hamiltonian

12th International Conference on Meson-Nucleon Physics and the Structure of the Nucleon (MENU 2010)
AIP Conf. Proc. 1374, 609-612 (2011); doi: 10.1063/1.3647214

dynamics, to a certain extent LFD is like sweeping dirt to a corner so that it leaves the rest of space clean[9, 10]. More reasons for our sweeping analogy may be due to the energy-momentum dispersion relation in LFD is rational; i.e.

$$p^- = \frac{\vec{p}_\perp^2 + m^2}{p^+}, \tag{1}$$

where $p^- = p^0 - p^3$ is the light-front (LF) energy (the conjugate variable of the LF time $x^+ = x^0 + x^3$) and $p^+ = p^0 + p^3$ is the longitudinal LF momentum orthogonal to p^- for the particle with mass m (or $p^\mu p_\mu = m^2$). From Eq.(1), it is very clear that the signs of p^+ and p^- are correlated. Thus, for all the particles and antiparticles involved in the physical process, the LF longitudinal momentum cannot be negative,i.e. $p^+ \geq 0$ in the LFD. This leads to the simplicity of LF vacuum, i.e. the quantum fluctuation of the vacuum is suppressed in LFD, except the zero-mode participation when all the particles and antiparticles in the physical process have zero LF longitudinal momenta, i.e. $p^+ = 0$ for all individual constituents.

Although the simple vacuum except the zero-modes is a remarkable achievement in LFD, the apparent simplicity of the LF vacuum yields a difficulty in understanding the novel phenomena associated with nontrivial vacua such as the spontaneous symmetry breaking, Higgs mechanism, chiral symmetry breaking, axial anomaly, θ-vacuum, etc.. To understand these phenomena in LFD, it is very important to also look into the "dirt" piled in the corner in our sweeping analogy. This task, however, is not so trivial, and indeed is sometimes remarkably treacherous [7, 11, 12, 13, 14].

TYPICAL TREACHEROUS AMPLITUDE

To discuss a treacherous point involved in LFD, one may look into a simple example of two-dimensional loop amplitude for a particle of mass m:

$$I = \int d^2k \frac{1}{k^2 - m^2 + i\varepsilon}. \tag{2}$$

In the covariant approach, this amplitude can be computed by dimensional regulation and is given by

$$I = -i\pi \left(\frac{2}{2-d} - \log\pi - \gamma + \mathcal{O}(2-d) - \log\frac{m^2}{\mu^2} \right), \tag{3}$$

where μ is a mass parameter introduced to give the correct mass dimensions in d dimensions, $\gamma \approx 0.577$ is Euler's constant, and d goes to 2 in this example. The LNA behavior of this amplitude in the limit $m \to 0$ is thus given by $I_{\rm LNA} = i\pi \log m^2$. In time-ordered perturbation theory, it is straightforward to use Cauchy's integral theorem and compute the amplitude as follows:

$$I = \int dk^1 dk^0 \frac{1}{(k^0 + \omega_k - i\varepsilon)(k^0 - \omega_k + i\varepsilon)} = -2i\pi \int_0^{\Lambda\to\infty} dk^1 \frac{1}{\sqrt{(k^1)^2 + m^2}} = \lim_{\Lambda\to\infty} 2i\pi \log\frac{m}{2\Lambda}, \tag{4}$$

where $\omega_k = \sqrt{(k^1)^2 + m^2}$ is the on-mass-shell energy of the particle and Λ is the cutoff parameter in the momentum integration of k^1. The LNA behavior of this amplitude is identical to $I_{\rm LNA} = i\pi \log m^2$. However, the LFD calculation of this amplitude is not so simple due to a dramatic change of pole structure:

$$I = \frac{1}{2} \int dk^+ dk^- \frac{1}{k^+k^- - m^2 + i\varepsilon} = \frac{1}{2} \int \frac{dk^+}{k^+} \int dk^- \frac{1}{k^- - \frac{m^2}{k^+} + i\frac{\varepsilon}{k^+}}, \tag{5}$$

where the pole ($\frac{m^2}{k^+} - i\frac{\varepsilon}{k^+}$) of the LF energy k^- depends on the k^+ variable. Not only is there just one pole in the k^- plane but also the k^- pole is moving as k^+ varies. If one naively applies the Cauchy's integral theorem, then it would trivially result in $I = 0$, however, this is obviously the wrong result. The details of this treacherous point have been discussed in Ref.[13]. We may use the LF cylindrical coordinates ($k^+ = r\cos\phi$, $k^- = r\sin\phi$) to perform the k^+ and k^- integration as follows:

$$I = \frac{1}{2} \int_0^\infty dr\, r \int_0^{2\pi} d\phi \left(r^2 \sin\phi\cos\phi - m^2 + i\varepsilon\right)^{-1} = \lim_{R\to\infty} \left(i\pi \log\frac{2m^2}{R^2 e^{-i\pi/2}} + \mathcal{O}(1/R^4) \right), \tag{6}$$

where R is the cutoff parameter in the LF cylindrical coordinate integration. The result correctly yields $I_{\rm LNA} = i\pi \log m^2$, identical to the results of the covariant and equal-time calculations. It demonstrates the equivalence of the LF formalism to the equal time and covariant formulations.

LEADING NONANALYTIC (LNA) BEHAVIOR OF NUCLEON SELF-ENERGY

Paying attention to such treacherous points in LFD, we considered the lowest order πN interaction with a pseudovector(PV) coupling defined by the Lagrangian density [15, 16]

$$\mathscr{L} = \frac{f_{\pi NN}}{m_\pi}(\bar{\psi}_N \gamma^\mu \gamma_5 \vec{\tau}\, \psi_N)\cdot \partial_\mu \vec{\phi}_\pi\,, \tag{7}$$

where ψ_N and $\vec{\phi}_\pi$ are the nucleon and pion fields, $\vec{\tau}$ is the Pauli matrix operator in nucleon isospin space, m_π is the pion mass, and $f_{\pi NN}$ is the πNN PV coupling constant with $f^2_{\pi NN}/4\pi \approx 0.08$. The analogous pseudoscalar(PS) interaction is given in Ref. [7]. Often the PV coupling is expressed in terms of the PS coupling constant $g_{\pi NN}$, $\frac{g_{\pi NN}}{2M} = \frac{f_{\pi NN}}{m_\pi}$, where M is the nucleon mass, with $g^2_{\pi NN}/4\pi \approx 14.3$. Using the Goldberger-Treiman relation the πNN coupling can also be expressed in terms of the axial vector charge of the nucleon, g_A, $\frac{g_A}{f_\pi} = \frac{g_{\pi NN}}{M}$, where $g_A = 1.267$ and $f_\pi \approx 93$ MeV is the pion decay constant.

The self-energy (or mass shift) of a nucleon is given by[7]

$$\Sigma = -\frac{3ig_A^2}{2f_\pi^2}\int \frac{d^4k}{(2\pi)^4} M\left[\frac{1}{D_N} + \frac{m_\pi^2}{D_\pi D_N}\right], \tag{8}$$

where $D_\pi = k^2 - m_\pi^2 + i\varepsilon$ and $D_N = (p-k)^2 - M^2 + i\varepsilon$. Because the $1/D_N$ term in Eq. (8) is independent of m_π, the LNA behavior of Σ is solely determined by the $1/D_\pi D_N$ term.

The self-energy can be evaluated in a manifestly covariant manner using dimensional regularization to regulate the ultra-violet divergence in the integral as d goes to 4:

$$\begin{aligned}\Sigma_{\rm cov} = & -\frac{3g_A^2 M}{32\pi^2 f_\pi^2}\left\{\left(\gamma + \log\pi - \frac{2}{4-d} + \log\frac{M^2}{\mu^2}\right)(M^2+m_\pi^2) - M^2 - 2m_\pi^2\right. \\ & + \frac{m_\pi^3\sqrt{4M^2-m_\pi^2}}{M^2}\left(\tan^{-1}\frac{m_\pi}{\sqrt{4M^2-m_\pi^2}} + \tan^{-1}\frac{2M^2-m_\pi^2}{m_\pi\sqrt{4M^2-m_\pi^2}}\right) \\ & \left. + \frac{m_\pi^4}{2M^2}\log\frac{m_\pi^2}{M^2}\right\}.\end{aligned} \tag{9}$$

It is remarkable that a closed form exists for the complete result of the self-energy, even in the relativistic formulation. Taking into account the treacherous point of LFD, we find the self-energy also in LFD:

$$\begin{aligned}\Sigma_{\rm LF} = & -\frac{3g_A^2 M}{32\pi^2 f_\pi^2}\left\{\Lambda_\perp^2 + m_\pi^2\log\frac{M^2}{\Lambda_\perp^2} - M^2\log\left(1+\frac{\Lambda_\perp^2}{M^2}\right) - \Lambda_\perp^2\log\left(\frac{\Lambda_\perp^2+M^2}{R^2e^{-i\pi/2}}\right)\right. \\ & + \frac{m_\pi^3\sqrt{4M^2-m_\pi^2}}{M^2}\left(\tan^{-1}\frac{m_\pi}{\sqrt{4M^2-m_\pi^2}} + \tan^{-1}\frac{2M^2-m_\pi^2}{m_\pi\sqrt{4M^2-m_\pi^2}}\right) \\ & \left. + \frac{m_\pi^4}{2M^2}\log\frac{m_\pi^2}{M^2}\right\}\end{aligned} \tag{10}$$

where $\Lambda_\perp$ is the high momentum cutoff in $k_\perp$ integration and R is again the cutoff parameter in the LF cylindrical coordinate integration. The equivalent result can also be realized using time-ordered perturbation theory (TOPT), in which the pion and nucleon propagators are split up into their positive and negative energy poles. As shown in Ref.[7], the relative contributions naturally depend on the frame of reference, and we considered two commonly used examples, namely, the nucleon rest frame (RF) and the infinite momentum frame (IMF). Expanding Σ in powers of m_π/M and isolating the nonanalytic terms (namely, ones which are odd powers or logarithms of m_π), the LNA behavior as $m_\pi \to 0$ is given by

$$\Sigma^{\rm LNA}_{\rm cov} = \Sigma^{\rm LNA}_{\rm LF} = \Sigma^{\rm LNA}_{\rm TOPT(RF)} = \Sigma^{\rm LNA}_{\rm TOPT(IMF)} = -\frac{3g_A^2}{32\pi f_\pi^2}\left(m_\pi^3 + \frac{1}{2\pi}\frac{m_\pi^4}{M}\log m_\pi^2 + \mathscr{O}(m_\pi^5)\right), \tag{11}$$

where in addition to the $\mathscr{O}(m_\pi^3)$ term, which agrees with the established results from chiral perturbation theory [15], for completeness we have also kept the next order, $\mathscr{O}(m_\pi^4 \log m_\pi^2)$, nonanalytic term. This is not the complete contribution to this order, however, as there exists an $\mathscr{O}(m_\pi^4 \log m_\pi^2)$ term arising from diagrams with a pion loop accompanied by a Δ intermediate state [17]. In fact, since this contribution depends on $1/(M_\Delta - M)$, where M_Δ is the mass of the Δ, rather than on $1/M$ as in Eq. (11), it will give the next-to-leading contribution in the heavy baryon limit.

CONCLUSIONS

We have demonstrated the equivalence of the covariant, equal-time (both in the rest frame and the IMF), and LF formalisms in computing the renormalization of the bare nucleon by pion loops. As a specific example, we have focussed on the self-energy of the nucleon, which finds applications in computations of pion cloud corrections to deep inelastic structure functions, form factors, and other observables. While this equivalence is expected, it can be achieved only aftern taking into account the treacherous point as we discussed in this presentation.

There are a number of applications of the methodology developed here which will be important to pursue in future. As well as the LNA behavior of the vertex renormalization, Z_1, it is important to compute the LNA behavior of the moments of the twist-two parton distribution functions[18]. In particular, one will be able to explore within the LF formalism the results obtained by Drell, Levy and Yan [19] in the IMF, as well as the application to this problem of effective field theory [6, 20, 21]. It will also pave the way for a consistent interpretation of the physics of the pion cloud at the partonic level and for the development of realistic chiral models of hadron structure on the LF suitable for discussing form factors, parton distribution functions and GPDs.

ACKNOWLEDGMENTS

This work was based on the collaborations with B.L.G.Bakker, H.-M.Choi, M.A.DeWitt, A.W.Thomas and Y.Mishchenko. One of the authors (W.M.) was supported by the DOE contract No. DE-AC05-06OR23177, under which Jefferson Science Associates, LLC operates Jefferson Lab. We also acknowledge partial support from the Triangle Nuclear Theory fund and the NCSU theory group fund (DOE contract No. DE-FG02-03ER41260).

REFERENCES

1. A. W. Thomas, Adv. Nucl. Phys. **13**, 1 (1984).
2. For review see X. Ji, Ann. Rev. Nucl. Part. Sci. **54**, 413 (2004); A. V. Belitsky and A. V. Radyushkin, Phys. Rept. **418**, 1 (2005).
3. A. W. Thomas, W. Melnitchouk and F. M. Steffens, Phys. Rev. Lett. **85**, 2892 (2000).
4. W. Detmold, W. Melnitchouk, J. W. Negele, D. B. Renner and A. W. Thomas, Phys. Rev. Lett. **87**, 172001 (2001).
5. J.-W. Chen and X. Ji, Phys. Lett. B **523**, 107 (2001).
6. D. Arndt and M. J. Savage, Nucl. Phys. A **697**, 429 (2002).
7. C. R. Ji, W. Melnitchouk and A. W. Thomas, Phys. Rev. D **80**, 054018 (2009)
8. P. A. M. Dirac, Rev. Mod. Phys. 21 (1949) 392.
9. C.-R. Ji, IPPP/03/71, pp.86-89, Durham 2003; Nucl. Phys. A737 (2004) 255.
10. C.-R. Ji and B. L. G. Bakker, Few-Body Systems 36 (2005) 137.
11. C.-R. Ji and S.-J. Rey, Phys. Rev. D53 (1996) 5815.
12. B. L. G. Bakker and C. R. G. Ji, Phys. Rev. D **71** (2005) 053005.
13. B. L. G. Bakker, M. A. DeWitt, C.-R. Ji and Y. Mishchenko, Phys. Rev. D **72**, 076005 (2005).
14. A. Misra and S. Warawdekar, Phys. Rev. D **71**, 125011 (2005).
15. See, *e.g.*, V. Bernard, N. Kaiser and U. G. Meissner, Int. J. Mod. Phys. E **4**, 193 (1995).
16. T. Ericson and W. Weise, *Pions and Nuclei* (Clarendon Press, Oxford, 1988).
17. D. B. Leinweber, A. W. Thomas, K. Tsushima and S. V. Wright, Phys. Rev. D **61**, 074502 (2000).
18. M. Burkardt, C.-R. Ji, W. Melnitchouk and A. W. Thomas, in preparation.
19. S. D. Drell, D. J. Levy and T. M. Yan, Phys. Rev. D **1**, 1035 (1970).
20. J.-W. Chen and X. Ji, Phys. Rev. Lett. **87**, 152002 (2001) [Erratum-ibid. **88**, 249901 (2002)].
21. J.-W. Chen and X. Ji, Phys. Lett. B **523**, 73 (2001).

A five-dimensional effective model for excited light mesons

S.S. Afonin

Saint-Petersburg State University, St. Petersburg, Russia

Abstract. In the usual holographic approach to QCD, the meson spectrum is generated due to a non-trivial 5-dimensional background. We propose an alternative 5-dimensional scenario in which the spectrum emerges due to the coupling to a scalar field whose condensation is supposed to be dual to the formation of a gluon condensate and mimics the scale anomaly in QCD. The spectrum of model has a finite number of discrete states plus continuum and reveals a Regge-like behavior in the strong coupling regime.

Keywords: Extra-dimensional models, holographic approach
PACS: 11.25.Tq, 11.10.Kk, 11.25.Wx, 12.38.Cy

INTRODUCTION

In the given report, we will discuss an effective five-dimensional model for excited light mesons (based on our paper [1]) which represents an alternative to the standard bottom-up holographic approach to QCD [2]. The main difference consists in the mechanism for mass generation: Instead of mimicking the spectrum as Kaluza-Klein excitations of 5D fields living in a non-trivial 5D background we consider 5D fields in flat space coupled to a scalar field which acquires a non-zero vacuum expectation value (v.e.v.) dependent on the fifth coordinate that has a physical meaning of inverse energy scale. The scalar field is supposed to be dual to the gluonic field strength tensor squared. Such a construction mimics the scale anomaly in QCD and generates a discrete hadron spectrum with a finite number of states plus continuum.

We will regard our model as an effective field theory that is valid below some energy scale, say below 2.5 GeV, where the light mesons have been detected [3]. Staying within this viewpoint, we do not need the AdS metric in the UV limit as we do not perform matching to the UV asymptotics of QCD correlators and the simplest possibility will be considered — the flat metric. In addition, as is the case of many effective theories, our model is supposed to be applicable only on the tree level even in the strong coupling regime.

VACUUM SECTOR

Consider a scalar field $\varphi(x_\mu,z)$, $\mu=0,1,2,3$, living in five dimensions, where the 5th space-like coordinate z is the inverse energy scale, $0\leq z<\infty$. Let us assume that this field is dual to the gluonic field strength tensor squared $G_{\mu\nu}^2$. The latter is known to acquire a non-zero v.e.v. $\langle G_{\mu\nu}^2\rangle$ which causes the anomaly in the trace of the energy-momentum tensor [4] in QCD,

$$4\varepsilon_{\text{vac}}=\langle\Theta_\mu^\mu\rangle_{\text{n.p.}}=\frac{\beta(\alpha_s)}{4\alpha_s}\langle G_{\mu\nu}^2\rangle_{\text{n.p.}}+\mathcal{O}(\alpha)+\ldots. \quad (1)$$

where ε_{vac} denotes the vacuum energy, n.p. is the "nonperturbative part" and the term $\mathcal{O}(\alpha)$ is the contribution of quark polarization effects. We suppose that hadrons acquire observable masses due to interaction with the field φ. Taking the metric $\eta_{AB}=(1,-1,-1,-1,-1)$, the action describing the pure vacuum sector is postulated to be ($A=0,1,2,3,4$)

$$S_{\text{CSB}}=\int d^4xdz\left(\frac{1}{2}\partial_A\varphi\partial^A\varphi+\frac{1}{2}m^2\varphi^2-\frac{1}{4}\lambda\varphi^4\right). \quad (2)$$

By assumption, the self-interaction is weak, $\lambda m\ll 1$, hence, the semiclassical analysis may be applied.

The classical equation of motion is

$$\partial_\mu^2\varphi-\partial_z^2\varphi-\varphi(1-\varphi^2)=0. \quad (3)$$

12th International Conference on Meson-Nucleon Physics and the Structure of the Nucleon (MENU 2010)
AIP Conf. Proc. 1374, 613-616 (2011); doi: 10.1063/1.3647215

We also assume that the v.e.v. φ_{vac} does not depend on the usual space-time coordinates, $\varphi(x_\mu, z) = \varphi(z)$. Equation (3) has then a kink solution

$$\varphi_{\text{vac}} = \pm\tanh(z/\sqrt{2}). \tag{4}$$

We choose the plus sign. The solution (4) breaks the translational invariance along the z-direction thereby making different energy scales non-equivalent. The given effect is important at large enough z, *i.e.* at low energies; at high energies the effect becomes negligible. This construction mimics the scale anomaly in QCD and the dependence on the fifth coordinate in the solution φ_{vac} reflects the fact that the operator $G^2_{\mu\nu}$ has anomalous dimension.

Consider the particle-like excitations of the vacuum state. Substituting $\varphi = \varphi + \varepsilon$ into Eq. (3), retaining only linear in ε contributions and assuming $\varepsilon(x_\mu, z) = e^{ipx}\varepsilon(z)$, where $p^2 = M^2$ is the usual 4D momentum defining the physical mass M, we obtain the following equation for the discrete mass spectrum,

$$\left(-\partial_z^2 + 3\tanh^2(z/\sqrt{2}) - 1\right)\varepsilon_n = M_n^2\varepsilon_n. \tag{5}$$

This one-dimensional Schrödinger equation can be easily solved analytically, it is known to have two normalizable discrete states (we omit the normalization factors),

$$\varepsilon_0 = \frac{1}{\cosh^2(z/\sqrt{2})}, \qquad M_0^2 = 0; \tag{6}$$

$$\varepsilon_1 = \frac{\tanh(z/\sqrt{2})}{\cosh(z/\sqrt{2})}, \qquad M_1^2 = \frac{3}{2}. \tag{7}$$

The continuum begins at $p^2 = 2$. The zero-frequency mode ε_0 corresponds to the Goldstone boson of spontaneously broken translational invariance along the z-direction. This "dilaton" mode is located at high energies, $z \to 0$, where the scale invariance of the vacuum is restored asymptotically. An interesting feature of the model is the existence of a massive mode that is located near $z_0 = \sqrt{2}\,\text{arctanh}(1/\sqrt{2}) \approx 1.25$ and could be associated with a glueball according to the physical meaning of the field φ.

ADDING BOSONS

Let us embed a massless hadron H in the vacuum. The interaction with the field φ should generate a certain mass for the hadron. We consider the simplest situation: The relevant action is quadratic in H. We will look for the particle-like excitations assuming the following ansatz for a solution: $H(x_\mu, z) = H(x_\mu)H(z)$. Since the field H must be normalizable, we can then easily integrate over z and arrive at a 4D effective action. We will not introduce the couplings of bosons to the gradients of the scalar field φ since such vertices, which could play a role at high energies, are not important for the mass generation at low energies.

For simplicity, we analyse a scalar field Φ coupled to the vacuum field φ; the extension of the analysis below to the higher spin fields is straightforward. The action is

$$S_{\text{bos}} = \int d^4x dz \left(\frac{1}{2}\partial_A\Phi\partial^A\Phi - \frac{G}{2}\varphi^2\Phi^2\right). \tag{8}$$

Consider the particle-like excitations $\Phi(x_\mu, z) = e^{ipx}f(z)$, $p^2 = M^2$ over the background (4). The classical equation of motion represents a one-dimensional Schrödinger equation

$$\left(-\partial_z^2 + \frac{G}{\lambda}\tanh^2(z/\sqrt{2})\right)f_n = M_n^2 f_n, \tag{9}$$

which determines the discrete spectrum

$$M_n^2 = \frac{1}{2}\left[\sqrt{1+\frac{8G}{\lambda}}\left(n+\frac{1}{2}\right) - \left(n+\frac{1}{2}\right)^2 - \frac{1}{4}\right], \tag{10}$$

$$f_n = \cosh^{n-s}(z/\sqrt{2})F\left[-n, 2s+1-n, s+1-n, \frac{1-\tanh(z/\sqrt{2})}{2}\right] \tag{11}$$

where F is the hypergeometric function and

$$s = \frac{1}{2}\left(\sqrt{1+\frac{8G}{\lambda}}-1\right), \qquad n = 0,1,2,\ldots, \qquad n < s. \tag{12}$$

The continuum spectrum sets in at $n = s$. The states with $n > 0$ carry the same quantum numbers as the ground state, in the language of potential models they are referred to as radial excitations. These excitations emerge if $G/\lambda > 1$, the number of the radial excitations is controlled by the value of coupling G/λ to the field φ. An interesting feature of the obtained spectrum is that it is Regge-like, $M_n^2 \sim n$, in the strong coupling regime $G/\lambda \gg 1$. A numerical fit for the vector case (the experimental situation with the scalar mesons is ambiguous) is presented in Ref. [1].

ADDING FERMIONS

Consider massless fermions coupled to the scalar field φ. Following the standard procedure for introduction of fermions in 5D space, the simplest action is

$$S_{\text{ferm}} = \int d^4xdz\,(i\bar{\Psi}\Gamma^A\partial_A\Psi - h\varphi\bar{\Psi}\Psi)\,, \tag{13}$$

where ψ is a four component spinor and $\Gamma^\mu = \gamma^\mu$, $\Gamma^4 = -i\gamma^5$; here γ^μ, γ^5 represent the usual 4D Dirac matrices. Let us find the particle-like excitations over the background (4). With the factorization $\Psi_{L,R}(x_\mu,z) = e^{ipx}U_{L,R}(z)$ for the left and right components, $\gamma_5\Psi_{L,R} = \pm\Psi_{L,R}$, the relation for masses follows from the classical equation of motion,

$$\left(\pm\partial_z + \frac{h}{\sqrt{\lambda}}\tanh(z/\sqrt{2})\right)U_{L,R} = MU_{L,R}. \tag{14}$$

At $z \geq 0$, the equation (14) possesses a normalizable zero-mode solution describing a massless left-handed fermion,

$$M = 0, \qquad U_L = \cosh^{-\frac{\sqrt{2}h}{\sqrt{\lambda}}}(z/\sqrt{2}), \qquad U_R = 0. \tag{15}$$

This mode is localized near $z = 0$. If we considered the region $z < 0$ the situation would be opposite: The normalizable zero-mode solution describes a massless right-handed fermion. This feature is in agreement with the interpretation of negative-energy solutions of the Dirac equation as antiparticles. On the other hand, there is an asymptotic solution

$$z \to \infty: \qquad M = \frac{h}{\sqrt{\lambda}}, \qquad U_{L,R} = C_{L,R}. \tag{16}$$

Here $C_{L,R}$ are some constants.

The generation of mass for the fermion in question can be obtained directly by means of integration over z. With the help of the parametrization

$$\Psi(x_\mu,z) = s(z)\psi(x_\mu), \qquad \int_0^\infty s^2(z)dz = 1, \tag{17}$$

we can integrate over z (the zero-mode will not contribute) and arrive at an effective 4D Lagrangian

$$\mathscr{L}^{(4)}_{\text{ferm}} = \bar{\psi}(i\gamma^\mu\partial_\mu - \mathscr{M})\psi, \tag{18}$$

with

$$\mathscr{M} = \frac{h}{\sqrt{\lambda}}\int_0^\infty s^2(z)\tanh(z/\sqrt{2})dz. \tag{19}$$

We note that $\mathscr{M} < h/\sqrt{\lambda}$ and the principal contribution to the effective mass $\mathscr{M}$ comes from integration over large enough z, *i.e.* over the low-energy region.

Thus, the fermion described by Eq. (14) is massless and left handed. This fermion is localized only at high energies, at low energies its wave function is suppressed exponentially; this phenomenon could mimic confinement. In the limit of zero energy, $z \to \infty$, the massless fermion disappears and instead a massive fermion emerges that resembles a kink propagating like a particle. It is natural to associate this solution with a light quark in QCD which looks practically massless and left-handed if we probe it at high energies and acquires a constituent mass at low energies. After integration over the energy scales the zero-mode is lost and the model describes a massive fermion.

DISCUSSIONS AND OUTLOOK

The presented phenomenological model can serve as a basis for constructing various effective models for QCD. We have considered an application of the model to the calculation of boson masses and a coupling of fundamental quarks to a field mimicking the physical vacuum. The next natural applications of the proposed approach is the description of the baryon spectrum. A somewhat related question to be answered is the computation of dynamical information encoded in the correlation functions and finding restrictions from the QCD sum rules in the narrow-width approximation (see, *e.g.*, [5] for a review).

The model has a natural limitation — it should be regarded as an effective model valid below some scale. As long as the model has only one massive excitation that might be associated with the scalar glueball, this scale could be the mass of the second scalar glueball. Since this mass is expected at approximately the same scale as the onset of perturbative continuum in the light meson spectrum, about 2.5 GeV, the model is able to describe the whole discrete meson spectrum. As is the case of all effective models of QCD, the considered model has no well established relation to QCD itself, but rather mimics some features of the physics of strong interactions at low and intermediate energies.

The five-dimensional model of low-energy strong interactions studied here shares some general features with the bottom-up AdS/QCD models (see, *e.g.*, [2] and references in [6]). First of all, the 5th space dimension is introduced that has the physical meaning of inverse energy scale. The equations determining the mass spectrum of hadrons are reduced to the one-dimensional equations of the Schrödinger type. The holographic models, however, suffer from a large arbitrariness in the choice of background metric and boundary conditions on the holographic fields. As a consequence, any *ad hoc* spectrum can be obtained, in particular, the spectrum of the kind (10) [7]. The presented approach is much less arbitrary since the fifth coordinate is singled out dynamically and as a result one has a dynamical violation of the scale invariance while within the usual holographic models this is implemented "by hand". The considered approach represents thus a scheme for building low-energy models of QCD in which the dependence on the energy scale is included via the extension of usual space-time to the fifth "energetic" dimension; the latter can always be integrated out but we lose then the dynamical content of the theory. It turns out that the introduction of "energetic" space permits to calculate the spectrum of the 4D theory and perhaps other static properties of hadrons in an essentially semiclassical way; such a possibility was the starting hypothesis of the AdS/QCD conjecture.

ACKNOWLEDGMENTS

The work is supported by the Alexander von Humboldt Foundation and by RFBR, grant 09-02-00073-a and travel grant 10-02-09280-mob_z.

REFERENCES

1. S. S. Afonin, *Int. J. Mod. Phys. A*, to be published; arXiv:0908.0457 [hep-ph].
2. J. Erlich, E. Katz, D. T. Son, and M. A. Stephanov, *Phys. Rev. Lett.* **95**, 261602 (2005); G. F. de Teramond and S. J. Brodsky, *Phys. Rev. Lett.* **94**, 201601 (2005); L. Da Rold and A. Pomarol, *Nucl. Phys.* **B721**, 79 (2005); A. Karch, E. Katz, D. T. Son, and M. A. Stephanov, *Phys. Rev.* **D74**, 015005 (2006).
3. C. Amsler *et al.*, *Phys. Lett.* **B667**, 1 (2008).
4. J.S. Collins, L. Duncan ,and S.D. Joglekar, *Phys. Rev.* **D16**, 438 (1977).
5. S. S. Afonin, A. A. Andrianov, V. A. Andrianov and D. Espriu, *JHEP* **0404**, 039 (2004).
6. S. S. Afonin, arXiv:1001.3105 [hep-ph].
7. S. S. Afonin, *Phys. Lett.* **B675**, 54 (2009).

Baryons in and beyond the quark-diquark model

G. Eichmann*, R. Alkofer†, C. S. Fischer**,*, A. Krassnigg† and D. Nicmorus‡

*Institut für Kernphysik, Technische Universität Darmstadt, D-64289 Darmstadt, Germany
†Institut für Physik, Karl-Franzens-Universität Graz, A-8010 Graz, Austria
**Institut für Theoretische Physik, Universität Giessen, D-35392 Giessen, Germany
‡FIAS, Johann Wolfgang Goethe-Universität, D-60438 Frankfurt am Main, Germany

Abstract. We examine the nucleon's electromagnetic form factors in a Poincaré-covariant Faddeev framework. The three-quark core contributions to the form factors are obtained by employing a quark-diquark approximation. We implement the self-consistent solution for the quark-photon vertex from its inhomogeneous Bethe-Salpeter equation. We find that the resulting transverse parts which add to the Ball-Chiu vertex have no significant impact on nucleon magnetic moments. The current-quark mass evolution of the form factors agrees with results from lattice QCD.

Keywords: Nucleon, Electromagnetic form factors, Faddeev equation, Quark-diquark model
PACS: 11.10.St, 12.38.Lg, 13.40.Gp, 14.20.Dh

INTRODUCTION

Investigating the nucleon's electromagnetic structure continues to be an experimental challenge in contemporary hadron physics. High-precision electron scattering measurements have provided information on the nucleon's electromagnetic form factors as well as longitudinal vs. transverse momentum fractions and the distribution of spin and orbital angular momentum among its constituents; see [1, 2] for reviews. Approaching these issues from a theoretical perspective aims at an understanding of the hadrons' substructure in terms quarks and gluons, the fundamental degrees of freedom in Quantum Chromodynamics.

In a Dyson-Schwinger/bound-state equation approach (see [3–5] for reviews) baryons are obtained as solutions of the covariant Faddeev equation. It describes the binding of a baryon by iterated quark-quark correlations and thereby constitutes the three-body analogue of a quark-antiquark Bethe-Salpeter equation (BSE). The nucleon's Faddeev equation was recently solved in a numerical setup where the full covariant structure of the nucleon amplitude was implemented [6]. The quark-quark kernel was modeled by a rainbow-ladder interaction, i.e. a dressed gluon exchange between any two quarks, where the quark propagator was consistently determined from its Dyson-Schwinger equation (DSE). A rainbow-ladder truncation excludes the presence of pion-cloud corrections which have significant impact on the chiral structure of hadrons. It therefore reflects the properties of a hadronic 'quark core'.

The computation of electromagnetic form factors in this setup requires knowledge of another ingredient, namely the dressed quark-photon vertex. A frequently used truncation of the Faddeev equation is the quark-diquark model, where the binding of baryons is traced back to the interaction between quarks and effective scalar and axial-vector diquarks within a baryon [7, 8]. It was demonstrated in Ref. [6] that this simplification only introduces a $\sim 5\%$ change in the nucleon mass. Hence, as an intermediate step towards a fully self-consistent determination of electromagnetic form factors from the covariant Faddeev equation, we investigate the impact of the structure of the quark-photon vertex on nucleon form factors in the quark-diquark approach.

NUCLEON ELECTROMAGNETIC FORM FACTORS

The nucleon's electromagnetic current is expressed in terms of two form factors: the Dirac and Pauli form factors $F_1(Q^2)$ and $F_2(Q^2)$, or the Sachs form factors as their linear combinations $G_E = F_1 - \tau F_2$ and $G_M = F_1 + F_2$, with $\tau = Q^2/(4M_N^2)$. Their static values represent the proton and neutron charges $\lambda^{p,n} = (1,0)$, (anomalous) magnetic moments $\kappa^{p,n}$ and $\mu^{p,n}$, and electromagnetic radii $r_{1,2}^{p,n}$ and $r_{E,M}^{p,n}$:

$$\begin{array}{llll} F_1(0) = \lambda\,, & F_2(0) = \kappa\,, & r_1^2 = -6F_1'(0)\,, & \kappa r_2^2 = -6F_2'(0)\,, \\ G_E(0) = \lambda\,, & G_M(0) = \mu\,, & r_E^2 = -6G_E'(0)\,, & \mu r_M^2 = -6G_M'(0)\,. \end{array} \tag{1}$$

12th International Conference on Meson-Nucleon Physics and the Structure of the Nucleon (MENU 2010)
AIP Conf. Proc. 1374, 617-620 (2011); doi: 10.1063/1.3647216

Electromagnetic current conservation provides a construction principle for the current in a given model framework [9]. In the quark-diquark approach this construction involves quark and diquark impulse-approximation diagrams, but also a coupling to the exchanged quark between quark and diquark as well as seagull terms, i.e. the photon's coupling to the diquark amplitudes. Upon resolving the diquark-photon vertex into its constituents, all photon couplings are specified by two quantities: the quark-photon vertex and the seagull vertices. The details of this construction are described in Refs. [8, 10] for the nucleon and Ref. [11] for the Δ-baryon, respectively.

Herein we want to investigate the structure of the quark-photon vertex in the context of electromagnetic form factors. The most general form of the vertex can be expressed by a sum of the Ball-Chiu term [12] and a purely transverse contribution:

$$\Gamma^\mu_{\text{q}}(k,Q) = \Big[i\gamma^\mu \Sigma_A + 2k^\mu \left(i\not{k}\,\Delta_A + \Delta_B \right) \Big] + \Gamma^\mu_T(k,Q)\,. \tag{2}$$

The Ball-Chiu part involves the quantities

$$\Sigma_A = \frac{A(k_+^2)+A(k_-^2)}{2}\,, \quad \Delta_A = \frac{A(k_+^2)-A(k_-^2)}{k_+^2-k_-^2}\,, \quad \Delta_B = \frac{B(k_+^2)-B(k_-^2)}{k_+^2-k_-^2}\,, \tag{3}$$

where $k_\pm = k \pm Q/2$ are the in- and outgoing quark momenta and $A(k^2)$ and $B(k^2) = M(k^2)A(k^2)$ are the dressing functions of the quark propagator $S(k) = A^{-1}(-i\not{k}+M)/(k^2+M^2)$. Eq. (2) satisfies the Ward-Takahashi identity $Q^\mu \Gamma^\mu_{\text{q}}(k,Q) = S^{-1}(k_+) - S^{-1}(k_-)$. The transverse part can be written as

$$-i\Gamma^\mu_T = f_1\,\gamma^\mu_T + i f_2\,\gamma^\mu_T \not{Q} + \tfrac{i}{2} f_3\, k\!\cdot\! Q\left[\gamma^\mu_T, \not{k}\right] + f_4\,\gamma^\mu_t \not{k}_T \not{Q} + k^\mu_T \left(i f_5 + f_6\, k\!\cdot\! Q\, \not{Q} + f_7\, \not{k} + i f_8\, \not{k}_T \not{Q} \right), \tag{4}$$

where the $f_i(k^2, k\cdot Q, Q^2)$ are scalar dressing functions. We used the abbreviations $\gamma^\mu_T = T^{\mu\nu}_Q \gamma^\nu$, $k^\mu_T = T^{\mu\nu}_Q k^\nu$ and $\gamma^\mu_t = T^{\mu\nu}_{k_T} \gamma^\nu_T$, with $T^{\mu\nu}_Q = \delta^{\mu\nu} - \hat{Q}^\mu \hat{Q}^\nu$ being a transverse projector with respect to the photon momentum Q.

The quark-photon vertex is obtained from its inhomogeneous Bethe-Salpeter equation [13], given by

$$\left[\Gamma^\mu_{\text{q}}(k,Q) - Z_2\, i\gamma^\mu\right]_{\alpha\beta} = \frac{4}{3}\int_{p'} K_{\alpha\alpha'\beta'\beta}\left[S(k'_+)\,\Gamma^\mu_{\text{q}}(k',Q)\,S(k'_-)\right]_{\alpha'\beta'}\,, \tag{5}$$

where Z_2 is the quark renormalization constant and $K_{\alpha\alpha'\beta'\beta}$ the rainbow-ladder kernel that also appears in the quark DSE and the homogeneous meson and diquark BSEs. It reads

$$K_{\alpha\alpha'\beta\beta'} = Z_2^2\, \frac{4\pi\alpha(k^2)}{k^2}\, T^{\mu\nu}_k\, \gamma^\mu_{\alpha\alpha'}\, \gamma^\nu_{\beta\beta'}\,, \tag{6}$$

with k being the gluon momentum, and involves an effective interaction $\alpha(k^2)$ which defines the model input of the approach. We use the parametrization of Ref. [14] for $\alpha(k^2)$ whose infrared term is modeled by two parameters: an infrared scale Λ and a width parameter η. Adjusting the scale Λ to reproduce the experimental pion decay constant yields a good description of pseudoscalar-meson, vector-meson, nucleon and Δ ground states. Inflating this scale towards the chiral limit mimics the properties of an effective quark core: it generates hadron masses which are overestimated in the chiral region and must be dressed, for instance, by pionic corrections; see Refs. [11, 15] for a further discussion. These studies exhibited only a modest dependence on the width parameter η; i.e., the quoted observables are not sensitive to the details of the interaction in the infrared.

To solve the inhomogeneous BSE it is convenient to express the vertex in an orthonormal basis, for instance defined by the elements $\tau^\mu_i(k,Q) \in \{\gamma^\mu_t/\sqrt{2},\, \widehat{k_T}^\mu,\, \hat{Q}^\mu\} \times \{\mathbb{1},\, \widehat{\not{Q}},\, \widehat{\not{k}}_T,\, \widehat{\not{k}_T \not{Q}}\}$ which satisfy the orthogonality relation $\text{Tr}\{\bar{\tau}^\mu_i \tau^\mu_j\} = 4\delta_{ij}$. Since the inhomogeneous equations for the transverse and longitudinal parts decouple, it is sufficient to consider the eight transverse elements alone. The longitudinal result reproduces the longitudinal projection of the Ball-Chiu vertex and thereby the Ward-Takahashi identity, and purely longitudinal terms do not contribute to nucleon form factors because of current conservation: $Q^\mu J^\mu = 0$. The inhomogeneous BSE self-consistently generates a timelike vector-meson pole in the quark-photon vertex at $Q^2 = -m_\rho^2$ which significantly increases the charge radii of pseudoscalar and vector mesons [13, 16].

RESULTS AND DISCUSSION

Upon implementing the numerical solution for the quark-photon vertex according to Eqs. (5–6) in the nucleon form factor setup, one can quantify the impact of the purely transverse structure (4) of the vertex. In previous form factor

TABLE 1. Static electromagnetic properties of the nucleon at the physical pion mass $m_\pi = 140$ MeV. The magnetic moments $\mu^{p,n}$ are dimensionless and the squared charge radii $R^{p,n}_{E,M} = (r^{p,n}_{E,M} M_N)^2$ are given in (GeV fm)^2. The first row shows the results obtained with a Ball-Chiu vertex augmented by the ansatz (7); the second row those with the quark-photon vertex solution from the inhomogeneous BSE. The brackets indicate the sensitivity to the model parameter η. The last row collects experimental results [17, 18].

	μ^p	μ^n	R^p_E	R^n_E	R^p_M	R^n_M
BC + ρ	2.56(6)	−1.59(3)	0.55(2)	0.00(1)	0.46(1)	0.45(1)
Full vertex	2.52(12)	−1.59(7)	0.58(1)	0.01(1)	0.50(1)	0.49(1)
Exp.	2.79	−1.91	0.69(2)	−0.10	0.64(5)	0.67(2)

studies this transverse part was modeled by an effective $\rho-$meson contribution [10, 19]:

$$\Gamma^\mu_T = -\frac{1}{g_\rho} \frac{Q^2}{Q^2+m_\rho^2} e^{-g(Q^2/m_\rho^2)} \Gamma_{\rm vc}(k,Q)\,, \tag{7}$$

where $\Gamma_{\rm vc}(k,Q)$ is the calculated vector-meson amplitude (see Ref. [19] for details). In consistency with the behavior of the pion's charge radius, the Ball-Chiu part and Eq. (7) each roughly contribute $\sim 50\%$ to the nucleon's squared charge radii $(r^p_E)^2$ and $(r^{p,n}_M)^2$, whereas Eq. (7) cancels the Ball-Chiu piece in the neutron's electric charge radius $(r^n_E)^2$. On the other hand, this ansatz does not contribute to the nucleon magnetic moments due to the inherent suppression with Q^2. The resulting static electromagnetic properties allow for the interpretation as a quark core which agrees well with lattice data at larger quark masses where meson-cloud effects are no longer important.

It is visible from Table 1 that the full vertex-BSE solution does not significantly change this behavior. The resulting nucleon charge radii differ by $\lesssim 5\%$ from those obtained via Eq. (7). The fractions of μ^p and μ^n which come from the transverse vertex-BSE solution of Eq. (4) owe to the amplitudes f_2 and f_4. Those two substantially contribute to a quark's anomalous magnetic moment [20] where they roughly cancel the Ball-Chiu part. Nonetheless, they have only a small impact upon the nucleon magnetic moments which are still sufficiently well represented by the Ball-Chiu ansatz alone. Moreover, we have verified that the Q^2-evolution of all electromagnetic form factors is essentially identical for both choices of transverse contributions up to $Q^2 \sim 3$ GeV^2.

In Fig. 1 we show the current-mass evolution of the nucleon magnetic moments and charge radii. The left panels depict the dimensionless magnetic moments $\mu^{p,n}$ and squared charge radii $R^{p,n}_{E,M} := (r^{p,n}_{E,M} M_N)^2$ which exhibit a moderately rising behavior with the pion mass. In the remaining panels we show the isovector and isoscalar anomalous magnetic moments and isovector Dirac and Pauli radii as obtained from the form factor combinations $F_i^v = F_i^p - F_i^n$ and $F_i^s = F_i^p + F_i^n$, with $i = 1,2$. To allow for a sensible comparison of dimensionful charge radii and magnetic moments in static nuclear magnetons with lattice results we plot the following rescaled quantities:

$$\kappa_{\rm resc} = \kappa \frac{M_N^{\rm exp}}{M_N^{\rm Ref}}\,, \quad r^2_{\rm resc} = r^2 \left(\frac{M_N}{M_N^{\rm Ref}}\right)^2, \quad \text{with} \quad M_N^{\rm Ref}(m_\pi^2)^2 = M_0^2 + \left(\frac{3m_\pi}{2}\right)^2 (1+f(m_\pi^2))\,, \tag{8}$$

where κ, r and M_N are our calculated results and $M_N^{\rm Ref}$, with $f(m_\pi^2) = 0.77/(1+(m_\pi/0.65\,\text{GeV})^4)$ and $M_0 = 0.9$ GeV, is a reference mass which describes the lattice results for M_N of Refs. [21–26] reasonably well.

Fig. 1 shows a good agreement between our calculated magnetic moments and the lattice data. This is especially relevant at higher pion masses where the pion-cloud dressing effects of the nucleon are diminished. In combination with the observed missing chiral curvature, our results can be viewed to represent the quark core of the nucleon. The implementation of such chiral effects, together with a form factor determination beyond the quark-diquark model, remains a task for future investigations.

ACKNOWLEDGMENTS

We are grateful to M. Blank, I. C. Cloët, and R. Williams for discussions. This work was supported by the Austrian Science Fund FWF under Projects No. J3039-N16, P20592-N16 and P20496-N16; by the Helmholtz Young Investigator Grant VH-NG-332; by HIC for FAIR within the LOEWE program launched by the State of Hesse, GSI, BMBF and DESY; and in part by the European Union (HadronPhysics2 project "Study of strongly interacting matter").

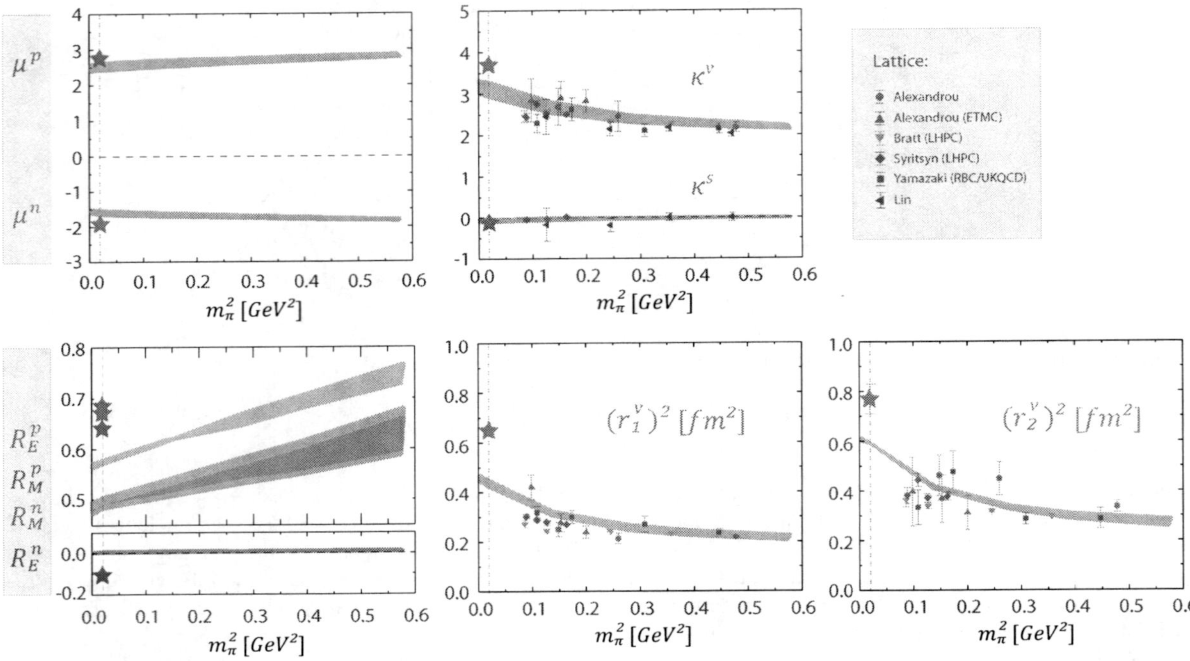

FIGURE 1. Static electromagnetic properties of the nucleon as functions of the squared pion mass. *Left panels:* dimensionless magnetic moments and squared charge radii in units of $(\text{GeV fm})^2$. *Middle and right panels:* isovector and isoscalar anomalous magnetic moments and squared isovector charge radii, rescaled as described in the text and compared to a selection of recent lattice results [21–26]. The bands show the dependence on the model parameter η; stars denote experimental values [17, 18].

REFERENCES

1. J. Arrington, C. D. Roberts, and J. M. Zanotti, *J. Phys.* **G34**, S23–S52 (2007).
2. C. F. Perdrisat, V. Punjabi, and M. Vanderhaeghen, *Prog. Part. Nucl. Phys.* **59**, 694–764 (2007).
3. R. Alkofer, and L. von Smekal, *Phys. Rept.* **353**, 281 (2001).
4. C. S. Fischer, *J. Phys.* **G32**, R253–R291 (2006).
5. C. D. Roberts, M. S. Bhagwat, A. Holl, and S. V. Wright, *Eur. Phys. J. ST* **140**, 53–116 (2007).
6. G. Eichmann, R. Alkofer, A. Krassnigg, and D. Nicmorus, *Phys. Rev. Lett.* **104**, 201601 (2010).
7. G. Hellstern, R. Alkofer, M. Oettel, and H. Reinhardt, *Nucl. Phys.* **A627**, 679–709 (1997).
8. G. Eichmann, A. Krassnigg, M. Schwinzerl, and R. Alkofer, *Annals Phys.* **323**, 2505–2553 (2008).
9. M. Oettel, M. Pichowsky, and L. von Smekal, *Eur. Phys. J.* **A8**, 251–281 (2000).
10. G. Eichmann, *Ph.D. Thesis, University of Graz* (2009), `0909.0703`.
11. D. Nicmorus, G. Eichmann, and R. Alkofer (2010), `1008.3184`.
12. J. S. Ball, and T.-W. Chiu, *Phys. Rev.* **D22**, 2542 (1980).
13. P. Maris, and P. C. Tandy, *Phys. Rev.* **C61**, 045202 (2000).
14. P. Maris, and P. C. Tandy, *Phys. Rev.* **C60**, 055214 (1999).
15. G. Eichmann, R. Alkofer, I. C. Cloet, A. Krassnigg, and C. D. Roberts, *Phys. Rev.* **C77**, 042202 (2008).
16. M. S. Bhagwat, and P. Maris, *Phys. Rev.* **C77**, 025203 (2008).
17. M. A. Belushkin, H. W. Hammer, and U. G. Meissner, *Phys. Rev.* **C75**, 035202 (2007).
18. K. Nakamura, *J. Phys.* **G37**, 075021 (2010).
19. G. Eichmann, I. C. Cloet, R. Alkofer, A. Krassnigg, and C. D. Roberts, *Phys. Rev.* **C79**, 012202 (2009).
20. L. Chang, Y.-X. Liu, and C. D. Roberts (2010), `1009.3458`.
21. C. Alexandrou, G. Koutsou, J. W. Negele, and A. Tsapalis, *Phys. Rev.* **D74**, 034508 (2006).
22. C. Alexandrou, et al., *PoS* **LATTICE2008**, 139 (2008).
23. S. N. Syritsyn, et al., *Phys. Rev.* **D81**, 034507 (2010).
24. T. Yamazaki, et al., *Phys. Rev.* **D79**, 114505 (2009).
25. J. D. Bratt, et al. (2010), `1001.3620`.
26. H.-W. Lin, S. D. Cohen, R. G. Edwards, K. Orginos, and D. G. Richards (2010), `1005.0799`.

A new approach for Delta form factors

C. Aubin[*,†] and K. Orginos[†,**]

[*] Dept. of Physics, Fordham University, Bronx NY
[†] Dept. of Physics, College of William and Mary, Williamsburg, VA
[**] Thomas Jefferson National Accelerator Facility, Newport News, VA

Abstract.
We discuss a new approach to reducing excited state contributions from two- and three-point correlation functions in lattice simulations. For the purposes of this talk, we focus on the $\Delta(1232)$ resonance and discuss how this new method reduces excited state contamination from two-point functions and mention how this will be applied to three-point functions to extract hadronic form factors.

PACS: 11.15.Ha,12.38.Gc,13.40.Gp,14.20.Gk

The calculation of the electromagnetic form factors for mesons and baryons is crucial to understanding the structure of hadronic states in QCD. However, for the $\Delta(1232)$ and other extremely short-lived resonances, they are notoriously difficult to both measure experimentally as well as calculate theoretically due to the complications that arise from the strong interactions. In the case of the $\Delta(1232)$ resonance, the form factors themselves are not currently experimentally accessible, although two of the form factors in the static limit are known (the charge) or measured to some degree (the magnetic dipole moment). For the nucleon, experimental results do exist for the form factors as a function of the momentum transfer. This makes a calculation of the nucleon form factors both a check of methodology as well as of QCD, and allows us to be confident that lattice results for the Δ form factors are reasonable.

The electromagnetic form factors of the Δ are encoded in the matrix element

$$\langle \Delta(p')|J_\mu|\Delta(p)\rangle = \bar{u}_\alpha(p')\Gamma_{\alpha\beta\mu}u_\beta(p) \tag{1}$$

where u_α is a Rarita-Schwinger vector-spinor describing the external Δ, and $J_\mu = \sum_q \bar{q}\gamma_\mu q$ is the vector current. The Lorentz structure of Γ is given, for example, in [1], and has four form factors $F^*_{1,2,3,4}(Q^2)$ that are functions of $Q^2 = -(p'-p)^2$ alone. These form factors give rise, in the limit $Q^2 \to 0$, to the electric charge, magnetic dipole moment, electric quadrupole moment, and magnetic octupole moment of the Δ.

Of these moments, the charge is of course known, and from the PDG [2], we have

$$\mu_{\Delta^{++}} = (5.6 \pm 1.9)\mu_N \,, \tag{2}$$

$$\mu_{\Delta^{+}} = (2.7 \pm 3.5)\mu_N \,, \tag{3}$$

where we have added all of the errors (including theoretical) in quadrature just to get an idea for how well these are determined experimentally. Thus, it is essential even for this simple quantity to have a well-determined lattice result. Some unquenched results were obtained using the form factor approach in [3] and using a background field technique [4], but there are several difficulties that arise in these different methods. For this work, we will focus on the difficulties with the form factor approach.

These are determined by calculating the 3-point correlator:

$$C^{3\text{pt}}(t_i,t,t_f,\mathbf{p}_i,\mathbf{p}_f) = FT\left[\langle 0|\chi(t_f,\mathbf{x}_f)J_\mu(t,\mathbf{0})\overline{\chi}(t_i,\mathbf{x}_i)|0\rangle\right] \,, \tag{4}$$

where FT is the Fourier Transform of the correlator, and χ is some appropriate interpolating operator for the Δ. In the large time limit $t_f \gg t \gg t_i$:

$$C^{3\text{pt}}(t_i,t,t_f,\mathbf{p}_i,\mathbf{p}_f) \to Z(\mathbf{p}_i,\mathbf{p}_f)e^{-E_f(t_f-t)}e^{-E_i(t-t_i)}\langle \Delta(p_f)|J_\mu(0)|\Delta(p_i)\rangle + \cdots \,. \tag{5}$$

Here we have schematically written this so that Z contains various overlap factors of the form $\langle 0|\chi|\Delta\rangle$ as well as other kinematic factors that are known. The dots denote contributions from excited states that are generally ignored.

12th International Conference on Meson-Nucleon Physics and the Structure of the Nucleon (MENU 2010)
AIP Conf. Proc. 1374, 621-624 (2011); doi: 10.1063/1.3647217

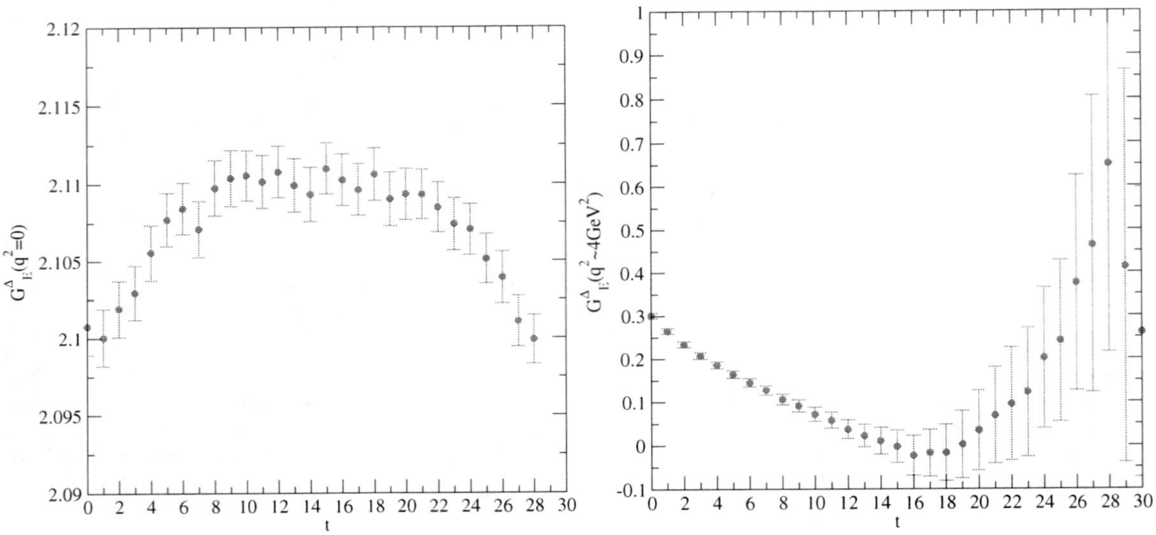

FIGURE 1. On the left is the matrix element G_E for the Δ^{++} evaluated at $q^2 = 0$. While there is significant excited state contamination, there is a noticeable plateau that gives (up to renormalization) the charge of the Δ. On the right is the same matrix element for $q^2 = 11(2\pi/a_sL)^2 \approx 4$ GeV^2, where there is no trustworthy signal.

The standard approach then is to note that the overlap factors and kinematics can be canceled by an appropriate ratio with 2-point correlators, and from these ratios we can extract the matrix element of interest. Note that additional complications arise in separating the Δ state and the $N-\pi$ state when the pion mass is below 300 MeV. Currently we are not near this regime, so it is not an issue for the current analysis.

The problem is that the contamination from excited states can be seen to be large, even for the simplest cases. Take here the Δ^{++} $E0$ form factor, $G_{E0}(q^2)$, which is a linear combination of the F_i^* form factors. In the limit $q^2 \to 0$, $G_{E0}(0) = +2$, the electric charge of the Δ^{++} in units of $|e|$. We show on the left of Fig. 1 the appropriate ratio of 3-to-2-point correlators to get this quantity using the Hadron Spectrum collaboration lattices of 2+1-flavor anisotropic Clover lattices [5] (here with a volume of $16^3 \times 128$ and a pion mass of roughly 390 MeV, and $a_t^{-1} \approx 5.5\,\text{GeV}, a_s/a_t \approx 3.5$). The source and sink are located at $t = 0, 28$, and we can see a plateau in the center where we could reliably extract the form factor, but there is significant contamination from excited states, seen from the deviation from the plateau.

This problem is more pronounced at higher momenta, and we show the same form factor at $q^2 = 11(2\pi/a_sL)^2 \approx 4$ GeV^2 on the right of Fig. 1. The location of a plateau is questionable at best for this plot.

We can extract the form factor using this standard approach and one gets reasonable results. For the $q^2 = 0$ point, we find the charge of the Δ^{++} is precisely twice that of the proton, and we can extract the renormalization factor for the vector current (note it is not conserved here because we are using a local current) from $G_{E0}(0)$, and we obtain

$$\frac{1}{Z_V} = 1.05(1) \text{ (statistical errors only).}$$

While this determination is trustworthy, extracting the higher momenta form factors is dangerous, due to the excited state contamination. So we would like to examine a new approach, which makes use of the Variational method to better extract states that contribute to a correlator. For now we will discuss this in the context of 2-point correlators for simplicity.

For a given state, there are many interpolating operators that could be used to calculate a two-point function, and one could form a matrix of correlators

$$C_{ij}(t) = \langle 0|O_i(t)\bar{O}_j(0)|0\rangle \ . \tag{6}$$

From this, by solving the generalized eigenvalue problem (GEVP)

$$C(t)\mathbf{x} = \lambda(t,t_0)C(t_0)\mathbf{x} \ , \tag{7}$$

one can show (see [6] and references therein) that the eigenvalues behave like

$$\lambda_i(t,t_0) \sim e^{-m_i(t-t_0)} + \cdots \ . \tag{8}$$

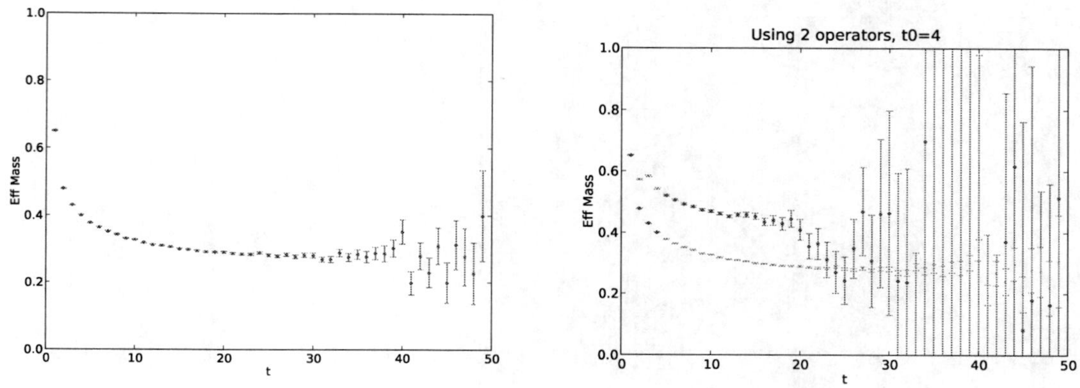

FIGURE 2. On the left we show the effective mass for a single Δ operator with noticeable excited state contamination before reaching a plateau. On the right is the effective masses for the two states extracted using the GEVP and two different local Δ operators.

The reference time t_0 is empirically chosen so that at that time, all n states (for an $n \times n$ system) would contribute to the correlator; no more, no less. For our purposes, we find that the choice of t_0 has little effect on our results, primarily because we are using a very small basis of operators.

If we restrict ourselves to local operators, there are two Δ interpolating fields we can use, and we can see that this has little effect to reduce contamination from excited states in the ground state, as shown by comparing the two plots in Fig. 2. On the left is a single operator effective mass, and we can see the excited state contamination before the plateau. On the right is the case with two operators, and solving the GEVP. The ground state is unchanged, and although this allows us to perhaps extract an excited state, this is not what we are currently interested in.

So we would like to find a way to kill off the excited state contribution in the ground state. This is where the Generalized Pencil-of-Function (GPoF) method comes in [7, 8]. In a quantum mechanical system, the important point is that if $O_\Delta(t)$ is an interpolating operator for the Δ, then so is

$$O_\Delta^\tau(t) \equiv e^{H\tau} O_\Delta(t) e^{-H\tau} = O_\Delta(t+\tau)\,, \tag{9}$$

and this new operator is linearly independent from the original operator.

So if we use $O_\Delta(t), O_\Delta^\tau(t)$ as our two operators, we can construct a correlator matrix using only a single correlator. This matrix is

$$\begin{aligned} \mathbf{C}(t) &= \begin{pmatrix} \langle O_\Delta(t)\bar{O}_\Delta(0)\rangle & \langle O_\Delta^\tau(t)\bar{O}_\Delta(0)\rangle \\ \langle O_\Delta(t)\bar{O}_\Delta^\tau(0)\rangle & \langle O_\Delta^\tau(t)\bar{O}_\Delta^\tau(0)\rangle \end{pmatrix} = \begin{pmatrix} \langle O_\Delta(t)\bar{O}_\Delta(0)\rangle & \langle O_\Delta(t+\tau)\bar{O}_\Delta(0)\rangle \\ \langle O_\Delta(t)\bar{O}_\Delta(-\tau)\rangle & \langle O_\Delta(t+\tau)\bar{O}_\Delta(-\tau)\rangle \end{pmatrix} \\ &= \begin{pmatrix} C(t) & C(t+\tau) \\ C(t+\tau) & C(t+2\tau) \end{pmatrix}. \end{aligned} \tag{10}$$

We can replicate this and use a set of operators $O_\Delta^{\tau,n}(t) = O_\Delta(t+n\tau)$, and make this correlator as large as we wish. It turns out that using multiple shifts does not give us any additional information, so we will ultimately use just a single shift and a $\tau = 4$. This would be something that is determined for each correlator one is interested in. Once this matrix is created then we follow through the same procedure as before with the GEVP.

Note that when the correlator basis grows, there is more of a likelihood of zero singular values, so we perform a singular value decomposition (SVD) cut on the correlator matrix to exclude states that have significantly smaller singular values than the largest by some cutoff, here chosen to be $\approx 10^{-3}$.

In Fig. 3 we show the same operator(s) as before, but now using the GPoF method (again, $\tau = 4$). On the left is one operator with one shift and the right uses both local operators and a single shift. We see the effective mass comes to a plateau much earlier than before, immediately after t_0 (which is also set to 4). Of course, the excited state(s) is(are) much noisier and we could not reliably extract information from that, but if one is only interested in the ground state, this is not an issue. However, one could apply the GPoF method to a large correlator basis to perhaps get cleaner signals for the lower lying states. On the right we perform the shift to the two operators, creating a 4×4 basis, but

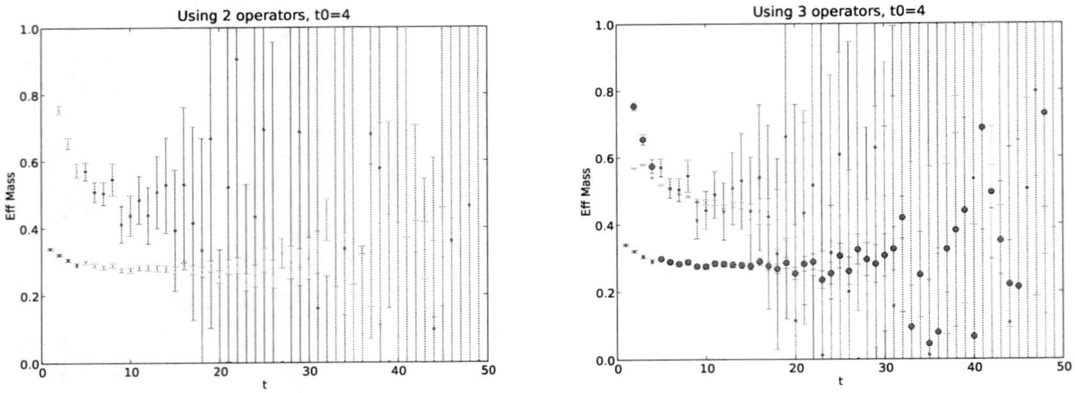

FIGURE 3. This is the same as in Fig. 2, but with a single shift of $\tau = 4$ for the operator(s) included.

after the SVD cut, this is reduced to three operators. In either case, we see that the ground state is unchanged, and a plateau is reached far sooner than without the use of the GPoF method.

For the unshifted operator if we perform a correlated fit to the correlator using an exponential in the range $t \in \{29, 40\}$, we get $a_t m_\Delta = 0.2770(50)$ with a confidence level of 78%. For the single shift of $\tau = 4$, we use the larger (and earlier) range $t \in \{5, 28\}$, and get $a_t m_\Delta = 0.2825(17)$ with a confidence level of 84%. Adding more shifts does not change this result, but it is amazing how well the signal improves with just this single shift.

It can be seen what is happening qualitatively, since the correlator $C(t+\tau)$ comes from the correlation with an operator that is τ time steps away from the other operator, and thus excited states are not going to contributed much to that, and even less to $C(t+2\tau)$. The ground state, however, is going to still contributed significantly to this correlator, so its signal, in a sense, is effectively enhanced.

In the three-point case, working through the correlator matrix, we would get

$$\mathbf{C}^{3-\mathrm{pt}}(t_i,t,t_f) = \begin{pmatrix} C^{3-\mathrm{pt}}(t_i,t,t_f) & C^{3-\mathrm{pt}}(t_i,t,t_f+\tau) \\ C^{3-\mathrm{pt}}(t_i,t+\tau,t_f+\tau) & C^{3-\mathrm{pt}}(t_i,t+\tau,t_f+2\tau) \end{pmatrix} . \tag{11}$$

So unlike the two-point function, we have to actually use three different sink locations, and thus generate a factor of three in the computational cost. However, if this allows a better determination of our ground state signal, this should be worth the additional propagator generation.

Once the two-point correlator matrix is diagonalized with the vectors $\mathbf{x}$, we form a matrix $V_{ij} = (\mathbf{x}_j)_i$ and then use that to diagonalize the three-point correlator

$$\mathbf{C}^{3-\mathrm{pt}}_{\mathrm{diag}}(t_i,t,t_f) = V^{-1}\mathbf{C}^{3-\mathrm{pt}}(t_i,t,t_f)V . \tag{12}$$

Currently, the analysis using this approach is under way and we hope to see a noticeable improvement as we did in the two-point case.

This work was partially supported by the US Department of Energy, under contract nos. DE-AC05-06OR23177 (JSA), DE-FG02-07ER41527, and DE-FG02-04ER41302; and by the Jeffress Memorial Trust, grant J-813.

REFERENCES

1. V. Pascalutsa, M. Vanderhaeghen, and S. N. Yang, *Phys. Rept.* **437**, 125–232 (2007).
2. K. Nakamura, *J. Phys.* **G37**, 075021 (2010).
3. C. Alexandrou, et al., *Phys. Rev.* **D79**, 014507 (2009).
4. C. Aubin, K. Orginos, V. Pascalutsa, and M. Vanderhaeghen, *Phys. Rev.* **D79**, 051502 (2009).
5. H.-W. Lin, et al., *Phys. Rev.* **D79**, 034502 (2009).
6. B. Blossier, M. Della Morte, G. von Hippel, T. Mendes, and R. Sommer, *JHEP* **04**, 094 (2009).
7. Y. Hua, and T. Sarkar, *IEEE Transactions on Antennas and Propagation* **37**, 229–234 (1989).
8. T. Sarkar, and O. Pereira, *IEEE Antennas and Propagation Magazine* **37**, 48–55 (1995).

Session T: Hadron Interactions

Calculation of Helium nuclei in quenched lattice QCD

T. Yamazaki*,†, Y. Kuramashi*,** and A. Ukawa*

*Center for Computational Sciences, University of Tsukuba, Tsukuba, Ibaraki 305-8577, Japan
†Center for Theoretical Studies, Nagoya University, Nagoya, Aichi 464-8602, Japan
**Graduate School of Pure and Applied Sciences, University of Tsukuba, Tsukuba, Ibaraki 305-8571, Japan

Abstract. We present results for the binding energies for ^{4}He and ^{3}He nuclei calculated in quenched lattice QCD at the lattice spacing of $a = 0.128$ fm with a heavy quark mass corresponding to $m_\pi = 0.8$ GeV. Enormous computational cost for the nucleus correlation functions is reduced by avoiding redundancy of equivalent contractions stemming from permutation symmetry of protons or neutrons in the nucleus and various other symmetries. To distinguish a bound state from an attractive scattering state, we investigate the volume dependence of the energy difference between the ground state of the nucleus channel and the free multi-nucleon state by changing the spatial extent of the lattice from 3.1 fm to 12.3 fm. A finite energy difference left in the infinite spatial volume limit leads to the conclusion that the measured ground states are bounded. It is also encouraging that the measured binding energies and the experimental ones show the same order of magnitude.

PACS: 11.15.Ha, 12.38.Aw, 12.38.-t 12.38.Gc

INTRODUCTION

The atomic nuclei have been historically treated as collections of protons and neutrons. The great success of the nuclear shell model since 1949 [1, 2], explaining the nuclear magic numbers and detailed spectroscopy, has established that protons and neutrons are very good effective degrees of freedom at the nuclear energy scale of a few MeV. Nonetheless, 60 years later, we know for certain that protons and neutrons are made of quarks and gluons whose laws are governed by QCD. It is a great challenge to quantitatively understand the structure and property of known nuclei based on the first principle of QCD. This direct approach will be more important and indispensable if we are to extract reliable predictions for experimentally unknown nuclei in the neutron rich regions of the nuclear chart. In this article we address the fundamental question in the research in this direction, namely the binding energies of nuclei.

Multi-baryon have been investigated by several studies in lattice QCD. *S*-wave nucleon-nucleon scattering was first studied in quenched QCD [3, 4]. This work was followed by a partially-quenched mixed action [5] and $N_f = 2+1$ anisotropic Wilson action [6] simulations. Extraction of nuclear force between two nucleons has been investigated in quenched and 2+1 flavor QCD [7, 8, 9]. All these studies assumed that the deuteron state is not bound for the heavy pion mass, $m_\pi \gtrsim 0.3$ GeV, employed in the calculations. NPLQCD Collaboration has tried a feasibility study of the three-baryon systems focusing on the quantum numbers of $\Xi^0\Xi^0 n$ and nnp channels. They found both the interactions to be repulsive [10, 11]. This means that the triton is unbound in the calculation.

So far no evidence supporting bound state formation even in a few-baryon systems has been observed in lattice QCD. In this work we focus on the helium nuclei especially ^{4}He, because it has much larger binding energy, $\Delta E = 28.3$ MeV, than that of the lighter nuclei. To calculate the ^{4}He channel, we might be able to observe the bound state of the nucleons. In this letter we present a calculation of the helium nuclei, ^{4}He and ^{3}He, in quenched lattice QCD using a heavy quark mass at a single lattice spacing. The results in this article have been already reported in other letter [12].

COMPUTATIONAL PROBLEMS FOR NUCLEUS

A major computational problem with multi-nucleon systems in lattice QCD is a factorially large number of Wick contractions of quark-antiquark fields required for evaluations of the nucleus correlation functions. A naive counting would give $(2N_p + N_n)!(2N_n + N_p)!$ for a nucleus composed of N_p protons and N_n neutrons, which quickly becomes prohibitively large beyond three-nucleon system, *e.g.,* 2880 for ^{3}He and 518400 for ^{4}He.

This number, however, contains equivalent contractions under the permutation symmetry in terms of the protons or the neutrons in the interpolating operator. We can reduce the computational cost by avoiding the redundancy. In case of the ^{4}He nucleus which consists of the same number of the protons and the neutrons, the isospin symmetry also helps us

12th International Conference on Meson-Nucleon Physics and the Structure of the Nucleon (MENU 2010)
AIP Conf. Proc. 1374, 627-630 (2011); doi: 10.1063/1.3647218

TABLE 1. Number of configurations ($N_{\rm conf}$), number of measurements on each configuration ($N_{\rm meas}$), acceptance rate in the HMC algorithm, pion mass (m_π) and nucleon mass (m_N).

L	$N_{\rm conf}$	$N_{\rm meas}$	accept.(%)	m_π [GeV]	m_N [GeV]
24	2500	2	93	0.8000(3)	1.619(2)
48	400	12	93	0.7999(4)	1.617(2)
96	200	12	68	0.8002(3)	1.617(2)

reduce the necessary contractions. After a scrutiny of the remaining equivalent contractions by a computer we find that only 1107 (93) contractions are required for the ^{4}He (^{3}He) nucleus correlation function. We have made a numerical test that the result with the reduced contractions reproduces the one with the full contractions on a configuration.

Another technique to save the computational cost is a modular construction of the nucleus correlation functions. We first make a block of three quark propagators where a nucleon operator with zero spatial momentum is constructed in the sink time slice. In this procedure we can incorporate the permutation symmetry of two up (down) quarks in a proton (neutron) sink operator. This is a simple trick to calculate 2^{N_N} contractions simultaneously. We also prepare several combinations of the two blocks which are useful for the construction of the nucleus correlators.

Another major problem is less known than the above one. The binding energy ΔE of the nucleus consisting of N_N nucleons with the mass m_N is very tiny compared with the mass M of the nucleus: $\Delta E/M \sim O(10^{-3})$ with $\Delta E = N_N m_N - M$. This causes a complicated situation that it is difficult to distinguish the physical binding energy from the energy shift due to the finite volume effect in the attractive scattering system [13]. One way to solve the problem is to investigate the volume dependence of the measured energy shift: In the attractive scattering system the energy shift is proportional to $1/L^3$ at the leading order in the $1/L$ expansion [13, 14], while the physical binding energy remains at a finite value at the infinite spatial volume limit.

If the volume is not enough large, it is difficult to distinguish a constant from a $1/L^3$ behavior of the energy shift. Thus, in our simulation we employ large volume as much as possible, and then choose three spatial extents corresponding to 3.1, 6.1 and 12.3 fm, which are much larger than those employed in current numerical simulations so as to provide sufficient room for the interacting multi-nucleon system.

SIMULATION PARAMETERS

We carry out calculations on quenched configurations generated with the Iwasaki gauge action [15] at $\beta = 2.416$ whose lattice spacing is $a = 0.128$ fm determined with $r_0 = 0.49$ fm as an input [16]. We employ the HMC algorithm with the Omelyan-Mryglod-Folk integrator [17, 18]. The step size is chosen to yield reasonable acceptance rate presented in Table 1. We take three lattice sizes, $L^3 \times T = 24^3 \times 64$, $48^3 \times 48$ and $96^3 \times 48$, to investigate the spatial volume dependence of the energy difference between the nucleus and the free multi-nucleon states. The physical spatial extents are 3.1, 6.1 and 12.3 fm, respectively.

We use the tadpole improved Wilson action with $c_{\rm SW} = 1.378$ [16]. Since it becomes harder to obtain a reasonable signal-to-noise ratio at lighter quark masses for the multi-nucleon system, we employ a heavy quark mass at $\kappa = 0.13482$ which gives $m_\pi = 0.8$ GeV for the pion mass and $m_N = 1.6$ GeV for the nucleon mass. Statistics is increased by repeating the measurement of the nucleus correlation functions with the source points in different time slices on each configuration. The numbers for the configurations and the measurements on each configuration are summarized in Table 1. We separate 100 trajectories between each measurement with $\tau = 1$ for the trajectory length. The errors are estimated by the jackknife analysis choosing 200 trajectories for the bin size.

The quark propagators are solved with the periodic boundary condition in all the spatial and temporal directions, and using the exponentially smeared source $Ae^{-B|\vec{x}|}$ after the Coulomb gauge fixing. On each volume we employ two sets of the smearing parameters: $(A,B) = (0.5,0.5)$ and $(0.5,0.1)$ for $L = 24$ and $(0.5,0.5)$ and $(1.0,0.4)$ for $L = 48$ and 96. Using the two sources, we confirm the consistency of the ground state energy of the nucleus channels. Hereafter the first and the second smearing parameter sets are referred to as "$S_{1,2}$", respectively.

The interpolating operator for the proton is defined as $p_\alpha = \varepsilon_{abc}([u_a]^t C\gamma_5 d_b)u_c^\alpha$ where $C = \gamma_4\gamma_2$ and α and a,b,c are the Dirac index and the color indices, respectively. The neutron operator n_α is obtained by replacing u_c^α by d_c^α in the proton operator. To save the computational cost we use the nonrelativistic quark operator, in which the Dirac index is restricted to upper two components. We employ simple interpolating operators for the ^{4}He [19] and ^{3}He [20] channels.

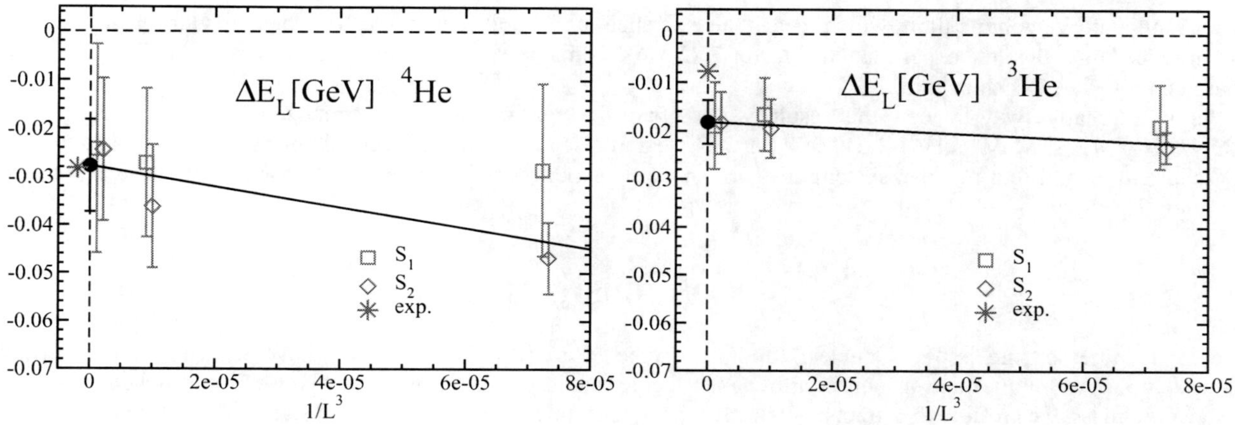

FIGURE 1. Spatial volume dependence of $-\Delta E_L = M - N_N m_N$ in GeV units for ^{4}He (left) and ^{3}He (right) nuclei with S_1 (open square) and S_2 (open diamond) sources. Statistical and systematic errors are added in quadrature. Diamond symbols are slightly shifted to positive direction in horizontal axis for clarity. Extrapolated results to the infinite spatial volume limit (filled circle) and experimental values (star) are also presented.

TABLE 2. Binding energies of ^{4}He and ^{3}He nuclei on each spatial volume. Extrapolated results to the infinite spatial volume limit are also presented. The first and second errors are statistical and systematic, respectively.

L	ΔE_L [MeV]			
	^{4}He(S_1)	^{4}He(S_2)	^{3}He(S_1)	^{3}He(S_2)
24	28(14)(11)	46.8(7.3)(1.6)	19.0(6.3)(6.0)	23.2(3.2)(0.5)
48	27(14)(05)	36(12)(04)	16.6(6.9)(3.2)	19.5(5.6)(2.3)
96	24(18)(12)	24(14)(03)	19.0(7.6)(4.9)	18.4(6.1)(1.9)
∞	27.7(7.8)(5.5)		18.2(3.5)(2.9)	

RESULTS

The volume dependence of the energy shift ΔE_L for the ^{4}He channel is plotted as a function of $1/L^3$ in the left panel of Fig. 1. Table 2 summarizes the numerical values of ΔE_L at three spatial volumes, where the statistical and systematic errors are presented in the first and second parentheses, respectively. In the figure the two errors are added in quadrature. In the following discussions we use the combined error. The results for the $S_{1,2}$ sources are consistent within the error bars. We observe little volume dependence for ΔE_L indicating a bound state, rather than the $1/L^3$ dependence expected for a scattering state, for the ground state in the ^{4}He channel.

The physical binding energy ΔE defined in the infinite spatial volume limit is extracted by a simultaneous fit of the data for the $S_{1,2}$ sources employing a fit function of $\Delta E + C/L^3$ with ΔE and C free parameters. The $1/L^3$ term is added to allow for contamination of scattering states. A systematic error is estimated from the difference of the central values of the fit results using the data with the different fit ranges in the determination of ΔE_L. The result for ΔE is 0.0180(62) in lattice units, which is 2.9 σ away from zero as shown in the left panel of Fig. 1. We also try a pure bound state fit allowing for an exponentially small finite size correction: ΔE and $\Delta E + C_1 e^{-C_2 L}$ with ΔE and $C_{1,2}$ free parameters. We find all the results are in agreement with reasonable values of χ^2.

Based on these analyses we conclude that the ground state of the measured four-nucleon system is bounded. An encouraging finding is that ΔE = 27.7(9.6) MeV with a^{-1} = 1.54 GeV agrees with the experimental value of 28.3 MeV. However, we do not intend to stress the consistency because our calculation is performed at the unphysically heavy pion mass, $m_\pi = 0.8$ GeV, and the electromagnetic interactions and the isospin symmetry breaking effects are neglected.

The results of the ^{3}He channel are shown in the right panel of Fig. 1 and Table 2. The statistical error is slightly smaller than the ones of the ^{4}He nucleus. The trend of the volume dependence is similar to the ^{4}He nucleus case. A simultaneous fit of the data for the $S_{1,2}$ sources with a fit function of $\Delta E + C/L^3$ yields a finite value of ΔE = 18.2(4.5)

MeV in the infinite volume limit, which means the existence of a bound state in the ^{3}He channel. Our result for ΔE is about twice larger than the experimental value of 7.72 MeV. A main reason for this difference could be the heavy pion mass employed in this calculation.

As an alternative way to view this result, we compare the binding energies normalized by the atomic number: $\Delta E/N_N = 6.9(2.4)$ MeV and 6.1(1.5) MeV for the ^{4}He and ^{3}He nuclei, respectively. At our unphysically heavy pion mass, the three and four nucleon system does not show the experimental feature that the binding is stronger for ^{4}He than for ^{3}He.

CONCLUSIONS

We have addressed the issue of nuclear binding for the ^{4}He and ^{3}He nuclei. We have shown that the current computational techniques and resources allow us to tackle this issue. Albeit in quenched QCD and for unphysically heavy pion mass, we are able to extract evidence for the bound state nature of the ground state and the binding energies for these nuclei.

A future direction of primary importance is to investigate the quark mass dependence of the binding energies of the nuclei. There are several model studies of the quark mass dependence of the nuclear binding energies [21] which suggest that the quark masses play an essential role in a quantitative understanding of the binding energies. Another important issue is development of a strategy to calculate nuclei with larger atomic numbers. The required number of the Wick contractions quickly diverges as the atomic number increases, even if the redundancies are removed with various symmetries. We leave it to future work.

ACKNOWLEDGMENTS

Numerical calculations for the present work have been carried out on the HA8000 cluster system at Information Technology Center of the University of Tokyo and on the PACS-CS computer under the "Interdisciplinary Computational Science Program" of Center for Computational Sciences, University of Tsukuba. We thank our colleagues in the PACS-CS Collaboration for helpful discussions and providing us the code used in this work. This work is supported in part by Grants-in-Aid for Scientific Research from the Ministry of Education, Culture, Sports, Science and Technology (Nos. 18104005, 18540250, 20105002, 21105501, 22244018).

REFERENCES

1. M. G. Mayer, *Phys. Rev.* **75**, 1969 (1949).
2. O. Haxel, J. H. D. Jensen, and H. E. Suess, *Phys. Rev.* **75**, 1766 (1949).
3. M. Fukugita, Y. Kuramashi, H. Mino, M. Okawa, and A. Ukawa, *Phys. Rev. Lett.* **73**, 2176–2179 (1994), hep-lat/9407012.
4. M. Fukugita, Y. Kuramashi, M. Okawa, H. Mino, and A. Ukawa, *Phys. Rev.* **D52**, 3003–3023 (1995), hep-lat/9501024.
5. S. R. Beane, P. F. Bedaque, K. Orginos, and M. J. Savage, *Phys. Rev. Lett.* **97**, 012001 (2006), hep-lat/0602010.
6. S. R. Beane, et al., *Phys. Rev.* **D81**, 054505 (2010), 0912.4243.
7. N. Ishii, S. Aoki, and T. Hatsuda, *Phys. Rev. Lett.* **99**, 022001 (2007), nucl-th/0611096.
8. S. Aoki, T. Hatsuda, and N. Ishii, *Prog. Theor. Phys.* **123**, 89–128 (2010), 0909.5585.
9. S. Aoki, T. Hatsuda, and N. Ishii, *Comput. Sci. Dis.* **1**, 015009 (2008), 0805.2462.
10. S. R. Beane, et al., *Phys. Rev.* **D80**, 074501 (2009), 0905.0466.
11. W. Detmold, *PoS* **LAT2009**, 008 (2009).
12. T. Yamazaki, Y. Kuramashi, and A. Ukawa, *Phys. Rev.* **D81**, 111504(R) (2010), 0912.1383.
13. M. Lüscher, *Commun. Math. Phys.* **105**, 153–188 (1986).
14. S. R. Beane, W. Detmold, and M. J. Savage, *Phys. Rev.* **D76**, 074507 (2007), 0707.1670.
15. Y. Iwasaki, ReportNo.UTHEP-118,1983(unpublished).
16. A. Ali Khan, et al., *Phys. Rev.* **D65**, 054505 (2002), hep-lat/0105015.
17. I. P. Omelyan, I. M. Mryglod, and R. Folk, *Comput. Phys. Commun.* **151**, 272–277 (2003).
18. T. Takaishi, and P. de Forcrand, *Phys. Rev.* **E73**, 036706 (2006), hep-lat/0505020.
19. J. E. Beam, *Phys. Rev.* **158**, 907–916 (1967).
20. M. Bolsterli, and E. Jezak, *Phys. Rev.* **135**, B510–B515 (1964).
21. V. V. Flambaum, and R. B. Wiringa, *Phys. Rev.* **C76**, 054002 (2007), 0709.0077.

Hadronic Interactions with lattice QCD

Martin J. Savage [1]

Department of Physics, University of Washington, Seattle, WA 98195-1560.

Abstract. I will describe recent progress toward calculating hadronic interactions with Lattice QCD.

Keywords: Lattice QCD, Nuclei
PACS: 21.10.-k,11.10.Hi,11.30.Rd,12.39.Fe,11.15.Ha

INTRODUCTION

It has been known for the past four decades that Quantum Chromodynamics (QCD), together with the electroweak interactions, underlies all of nuclear physics. Soon after the discovery of QCD it became apparent that the complexities of the theory at strong coupling would hinder analytic progress in understanding the properties of the simplest hadrons, let alone the simplest features of the nuclear forces. Wilson pointed the way to eventual direct quantitative confirmation of the origins of nuclear physics by formulating Lattice QCD [1], a regularization and non-perturbative definition of QCD that is suitable for the intensive computational demands of solving QCD in the infrared. Only in the last five years or so has Lattice QCD emerged from a long period of research and development —where only qualitative agreement between calculations and experiments could be claimed— to the present, where precise predictions for hadronic quantities are being made. In particular, presently, fully-dynamical calculations with near-exact chiral symmetry at finite lattice-spacing have become standard, with lattice volumes of spatial extent $L \gtrsim 2.5$ fm and with lattice spacings in the range $b \lesssim 0.12$ fm. It is still the norm that the light-quark masses, m_q, are larger than those of nature, with typical pion masses $m_\pi \sim 300$ MeV.

Nuclear physics is a vast, rich field, whose phenomenology has been explored for decades through intense experimental and theoretical effort. However, there is still little understanding of the connection to QCD and the basic building blocks of nature, quarks and gluons. For instance, as a first "benchmarking" step, Lattice QCD should post-dict the large nucleon-nucleon scattering lengths and the existence of the deuteron, the simplest nucleus. The connection between QCD and nuclear physics will be firmly established with Lattice QCD, and will allow for an exploration of how nuclei and nuclear interactions depend upon the fundamental parameters of nature. In particular, it is believed that an understanding of the fine-tunings that permeate nuclear physics will finally be translated into fine-tunings of the light-quark masses. While these issues are of great interest, and it is important to recover what is known experimentally to high precision, these goals are not the main objective of the Lattice QCD effort in nuclear physics. The primary reason for investing resources in this area is to be able to calculate physical quantities of importance that cannot be accessed experimentally, or which can be measured with only limited precision in the laboratory. Two important examples of how Lattice QCD calculations can impact nuclear physics are in the structure of nuclei and in the behavior of hadronic matter at densities beyond that of nuclear matter. Lattice QCD will be able to calculate the interactions of multiple nucleons, bound or unbound, in the same way that it can be used to determine the two-body scattering parameters. For instance, a calculation of the three-neutron interactions will be possible.

Extracting hadronic interactions from Lattice QCD calculations is more complicated than the determination of the spectrum of stable particles. This is encapsulated in the Maiani-Testa theorem [2], which states that S-matrix elements cannot be extracted from infinite-volume Euclidean-space Green functions except at kinematic thresholds. This is clearly problematic from the nuclear physics perspective, as a main motivation for pursuing Lattice QCD is to be able to compute nuclear reactions involving multiple nucleons. Of course, it is clear from the statement of this theorem how it can be evaded, Euclidean-space correlation functions are calculated at finite volume to extract S-matrix elements, the formulation of which was known for decades in the context of non-relativistic quantum mechanics [3] and extended to quantum field theory by Lüscher [4, 5]. The energy of two particles in a finite volume depends in a calculable way

[1] for the NPLQCD collaboration

12th International Conference on Meson-Nucleon Physics and the Structure of the Nucleon (MENU 2010)
AIP Conf. Proc. 1374, 631-634 (2011); doi: 10.1063/1.3647219

upon their elastic scattering amplitude and their masses for energies below the inelastic threshold.

MESON-MESON INTERACTIONS

Due to the chiral symmetry of QCD, $\pi\pi$ scattering at low energies is the simplest and best-understood of the hadron-hadron scattering processes. The scattering lengths for $\pi\pi$ scattering in the s-wave are uniquely predicted at LO in χ-PT [6]: $m_\pi a_{\pi\pi}^{I=0} = 0.1588$ and $m_\pi a_{\pi\pi}^{I=2} = -0.04537$ when m_π is set equal to the charged pion mass. While experiments do not directly provide stringent constraints on the scattering lengths, a determination of s-wave $\pi\pi$ scattering lengths using the Roy equations has reached a remarkable level of precision [7, 8]: $m_\pi a_{\pi\pi}^{I=0} = 0.220 \pm 0.005$ and $m_\pi a_{\pi\pi}^{I=2} = -0.0444 \pm 0.0010$. At present, Lattice QCD has computed $\pi\pi$ scattering only in the $I = 2$ channel with precision as the $I = 0$ channel contains disconnected diagrams. The only existing $n_f = 2+1$ Lattice QCD prediction of the $\pi^+\pi^+$ scattering length involves a mixed-action Lattice QCD scheme of domain-wall valence quarks on a rooted staggered sea [9]. The scattering length was computed at pion masses, $m_\pi \sim 290$ MeV, 350 MeV, 490 MeV and 590 MeV, and at a single lattice spacing, $b \sim 0.125$ fm and lattice size $L \sim 2.5$ fm [9]. The physical value of the scattering length was obtained using two-flavor mixed-action χ-PT which includes the effect of finite lattice-spacing artifacts to $\mathcal{O}(m_\pi^2 b^2)$ and $\mathcal{O}(b^4)$ [10]. The final result is: $m_\pi a_{\pi\pi}^{I=2} = -0.04330 \pm 0.00042$. Lattice QCD calculations at one or more smaller lattice spacings are required to verify and further refine this calculation. Recently, the ETM collaboration has performed a $n_f = 2$ calculation of the $\pi^+\pi^+$ scattering length at pion masses ranging from $m_\pi \sim 270$ MeV to 485 MeV, at two lattice spacings and in two lattice volumes. The result extrapolated to the physical pion mass is: $m_\pi a_{\pi\pi}^{I=2} = -0.04385 \pm 0.00028 \pm 0.00038$.

In Figure 1 (left panel) one sees that the $\pi^+\pi^+$ scattering length is consistent with the current algebra result up to pion masses that are expected to be at the edge of the chiral regime in the two-flavor sector. While in the two flavor theory one expects fairly good convergence of the chiral expansion and, moreover, one expects that the effective expansion parameter is small in the channel with maximal isospin, the lattice calculations clearly imply a cancellation between chiral logs and counterterms. However, as one sees in Figure 1 (right panel), the same phenomenon occurs in K^+K^+ where the chiral expansion is governed by the strange quark mass and is therefore expected to be much more slowly converging. The π^+K^+ scattering length exhibits similar behavior when $\mu_{K\pi}\, a_{K^+\pi^+}$ is plotted against $\mu_{K\pi}/\sqrt{f_K f_\pi}$. This cancellation between chiral logs and counterterms for the meson-meson scattering lengths is quite mysterious.

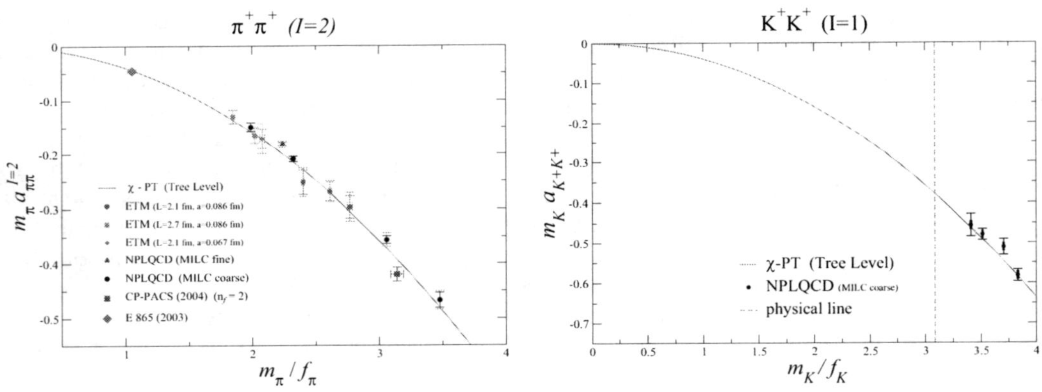

FIGURE 1. $m_\pi a_{\pi^+\pi^+}$ vs. m_π/f_π (left panel) and $m_k a_{K^+K^+}$ vs. m_k/f_k (right panel). The black points are the results of Lattice QCD calculations by the NPLQCD collaboration, the blue point (left panel) is from the CP-PACS collaboration, the purple, red and magenta points (left panel) are from the ETM collaboration. The red lines are the current algebra predictions.

BARYON-BARYON INTERACTIONS

The NPLQCD collaboration performed the first $n_f = 2+1$ QCD calculations of nucleon-nucleon interactions [11] and hyperon-nucleon [12] interactions at low-energies but with unphysical pion masses, and the nucleon-nucleon scattering lengths were found to be of natural size. Such calculations are far more challenging than those involving mesons due to the exponential degradation of the signal-to-noise at large times. A summary of all Lattice QCD calculations of NN scattering is shown in Figure 2. The results calculated on the anisotropic clover gauge configurations are consistent with those obtained with mixed-action Lattice QCD [11], and it is interesting to note that the results of quenched calculations [13] yield scattering lengths that are consistent within uncertainties with the fully-dynamical $n_f = 2+1$ values.

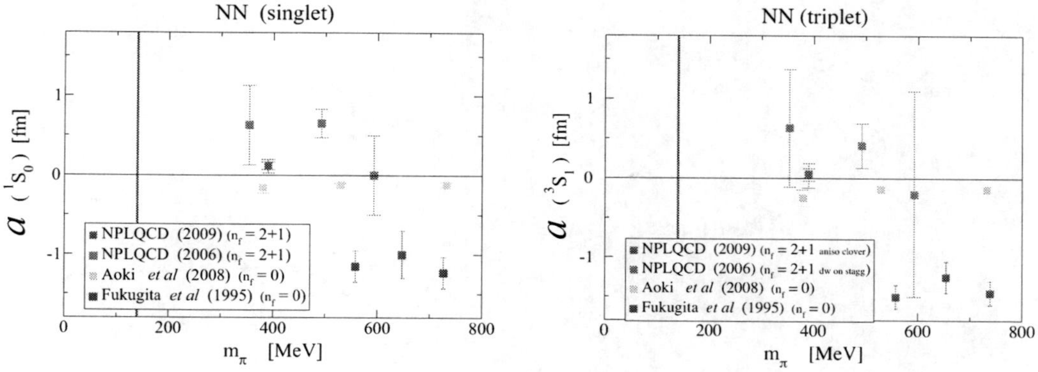

FIGURE 2. A compilation of the scattering lengths for NN scattering in the $^1S_0$ (left panel) and $^3S_1$ (right panel) calculated with Lattice QCD and with quenched Lattice QCD. The vertical lines correspond to the physical pion mass.

Figure 3 shows a summary of the high-statistics results for baryon-baryon interactions recently obtained on the $20^3 \times 128$ anisotropic clover gauge configurations with $m_\pi \sim 390$ MeV. The $\Lambda\Lambda$ channel is found to be negatively shifted in energy which may signal that the lowest-state is in fact a bound-state, but calculations in a larger volume are required before more definitive conclusions can be drawn. Calculations at this pion mass should be viewed to be only a proof of principle, and in order for lattice calculations to provide meaningful constraints on interactions at the physical pion mass through chiral extrapolation, the pion mass must be substantially reduced down toward the physical value ($m_\pi \sim 140$ MeV) while maintaining the integrity of the calculation (i.e. small enough lattice spacings and large enough volumes).

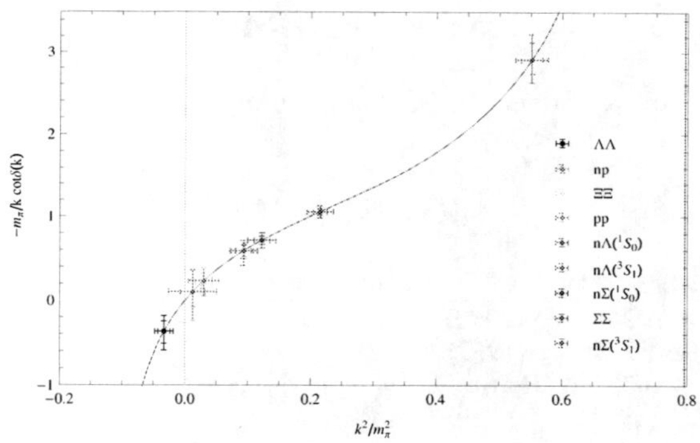

FIGURE 3. Baryon-baryon interactions extracted from calculations at $m_\pi \sim 390$ MeV [14].

I should also mention that there have been calculations of energy-dependent and source-dependent "potentials" that are constructed in such a way to reproduce the calculated phase-shifts at the energies of the states in the finite lattice volumes, for instance, see Ref. [16]. Currently unjustified assumptions are required in order for these objects to be compared with the modern nucleon-nucleon or hyperon-nucleon potentials that are used in the calculation of nuclear and hyper-nuclear structure and interactions, and I will discuss them no further.

MULTI-BARYON INTERACTIONS

The first significant steps towards the calculation of the properties of nuclei directly from QCD were taken by the NPLQCD collaboration [15] and the PACS-CS collaboration [17] during 2009. The NPLQCD collaboration performed a $n_f = 2+1$ calculation of the correlation function with the quantum numbers of the strangeness $s = -4$, baryon number $B = 3$ system, which is labeled as "$n\Xi^0\Xi^0$" for convenience, and also of the correlation function with the quantum numbers of the triton (or ^{3}He), at a pion mass of $m_\pi \sim 390$ MeV [15] as shown in fig. 4. Further, the PACS-CS collaboration performed a quenched calculation of the system with the quantum numbers of the triton and of the α-particle at a pion mass of $m_\pi \sim 800$ MeV [17] as shown in fig. 4.

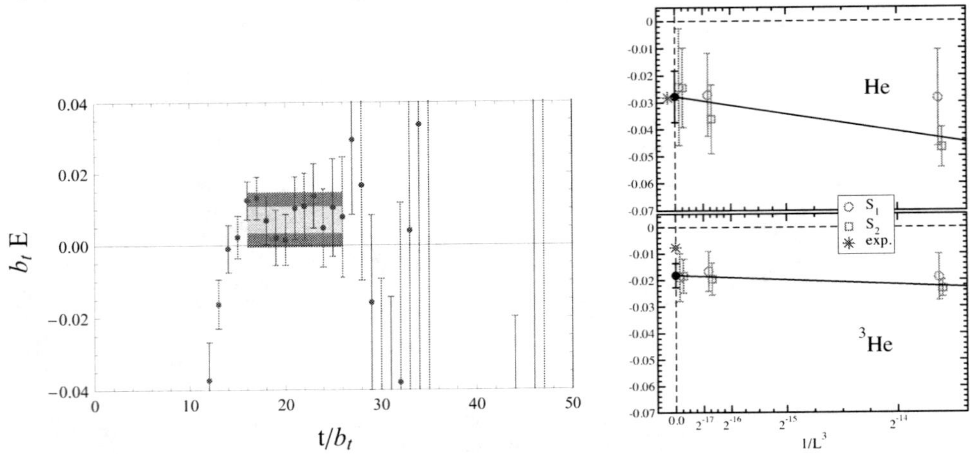

FIGURE 4. The left panel shows the effective mass plot in the triton channel resulting from the $n_f = 2+1$ calculation with $m_\pi \sim 390$ MeV after the subtraction of $3 \times M_N$. The right panel shows the quenched results for the binding energies (in lattice units) obtained by the PACS-CS collaboration in the channels with the quantum numbers of the triton and the α-particle [17] with $m_\pi \sim 800$ MeV.

The dream of being able to perform reliable calculations of the interactions among multiple nucleons and hyperons, and of the structure and reactions of light-nuclei, directly from QCD is starting to be realized. The path forward is clear, and the next decade will be a truly remarkable period for nuclear physics.

ACKNOWLEDGMENTS

I would like to thank all of the members of the NPLQCD collaboration for their contributions to this program.

REFERENCES

1. K. G. Wilson, Phys. Rev. D **10**, 2445 (1974).
2. L. Maiani and M. Testa, Phys. Lett. B **245**, 585 (1990).
3. K. Huang and C. N. Yang, Phys. Rev. **105**, 767 (1957).
4. M. Lüscher, Commun. Math. Phys. **105**, 153 (1986).
5. M. Lüscher, Nucl. Phys. B **354**, 531 (1991).
6. S. Weinberg, Phys. Rev. Lett. **17**, 616 (1966).
7. G. Colangelo, J. Gasser and H. Leutwyler, Nucl. Phys. B **603**, 125 (2001) [arXiv:hep-ph/0103088].
8. H. Leutwyler, PoS C **CONFINEMENT8**, 068 (2008) [arXiv:0812.4165 [hep-ph]].
9. S. R. Beane, T. C. Luu, K. Orginos, A. Parreno, M. J. Savage, A. Torok and A. Walker-Loud, Phys. Rev. D **77**, 014505 (2008) [arXiv:0706.3026 [hep-lat]].
10. J. W. Chen, D. O'Connell and A. Walker-Loud, Phys. Rev. D **75**, 054501 (2007) [arXiv:hep-lat/0611003].
11. S. R. Beane, P. F. Bedaque, K. Orginos and M. J. Savage, Phys. Rev. Lett. **97**, 012001 (2006) [arXiv:hep-lat/0602010].
12. S. R. Beane, P. F. Bedaque, T. C. Luu, K. Orginos, E. Pallante, A. Parreno and M. J. Savage [NPLQCD Collaboration], Nucl. Phys. A **794**, 62 (2007) [arXiv:hep-lat/0612026].
13. S. Aoki, T. Hatsuda and N. Ishii, Comput. Sci. Dis. **1**, 015009 (2008) [arXiv:0805.2462 [hep-ph]].
14. S. R. Beane *et al.*, arXiv:0912.4243 [hep-lat].
15. S. R. Beane *et al.*, Phys. Rev. D **80**, 074501 (2009) [arXiv:0905.0466 [hep-lat]].
16. T. Inoue *et al.* [HAL QCD collaboration], arXiv:1007.3559 [hep-lat].
17. T. Yamazaki, Y. Kuramashi, A. Ukawa (PACS-CS Collaboration), arXiv:0912.1383 [hep-lat].

Dependence on the quark masses of elastic phase shifts and light resonances within standard and unitarized ChPT

J. Nebreda and J.R. Peláez

Dept. Física Teórica II. Universidad Complutense, 28040, Madrid. Spain

Abstract. We study the dependence of the $\pi\pi$ scattering phase shifts on the light quark mass in both standard and unitarized SU(2) Chiral Perturbation Theory (ChPT) to one and two loops. We then use unitarized SU(3) ChPT to study the elastic $f_0(600)$, $\kappa(800)$, $\rho(770)$ and $K^*(892)$ resonances. The quark masses are varied up to values of interest for lattice studies. We find a very soft dependence on the light quark mass of the $\pi\pi$ phase shifts at one loop and lightly stronger at two loops and a good agreement with lattice results. The SU(3) analysis shows that the properties of the $\rho(770)$ and $K^*(892)$ depend smoothly on the quark mass whereas the scalar resonances present a non-analyticity at high quark masses. We also confirm the lattice assumption of quark mass independence of the vector two-meson coupling that, however, is violated for scalars.

PACS: 14.40.Cs, 12.39.Fe, 13.75.Lb

INTRODUCTION

Although QCD is well established as the theory of strong interactions, the hadronic realm is beyond the reach of perturbative calculations. In that regime, lattice methods are a useful tool to calculate QCD observables, but there are still few results at small quark masses. On the other hand, Chiral Perturbation Theory (ChPT) provides the quark mass dependence of the meson-meson scattering amplitudes at low energies. This allows us to increase the quark mass in the amplitudes in order to compare with lattice results. In this talk we first study the phase shift dependence on the averaged u and d quark mass, $\hat{m}$, in standard one and two-loop SU(2) ChPT[1]. This approach has the advantage of being completely model independent. Then we use the elastic Inverse Amplitude Method (IAM) to unitarize the amplitudes, which allows first, to calculate the phase shift quark mass dependence up to higher values of $\hat{m}$ and next, to generate poles on the second Riemann sheet and study their evolution. This technique was applied in [2] to calculate the $\hat{m}$ dependence of the $f_0(600)$ and $\rho(770)$ resonances. Here we extend this study to include the strange quark within a unitarized SU(3) ChPT formalism [3], so that we can also generate the $K^*(892)$ and $\kappa(800)$ resonances and vary both the light and the strange quark masses.

QUARK MASS DEPENDENCE IN STANDARD CHPT

We use the SU(2) scattering amplitudes in [1] and the LECs in [4] to study the dependence of the $\pi\pi$ phase shifts on $\hat{m}$. In Fig. 1 we show preliminary results for phases on different channels for different pion masses, to one and two loops (first and second row respectively). They are plotted as a function of the center of mass momentum -and not of the energy- in order to subtract the shift of the thresholds. Since the applicability of plain ChPT is limited to low energies, we do not plot phase shifts beyond $\sqrt{s} = 1$ GeV. In order to compare with lattice results [6], we rise M_π up to 420 MeV, although most likely the IAM is not reliable in that region. A Montecarlo gaussian sampling based on the errors of the LECs has been used to calculate the error bands.

We find that the dependence of the phase shifts on $\hat{m}$ is very soft at one loop and somehow stronger at two loops, specially for the I=2, J=2 channel, and that the results are compatible with those from lattice in the I=2, J=0 channel [6].

QUARK MASS DEPENDENCE IN UNITARIZED CHPT

We use now the IAM to unitarize our amplitudes. This procedure extends their applicability up to the resonance region and generates poles in the second Riemann sheet, which relate to resonances through the usual Breit-Wigner identification $\sqrt{s_{pole}} \equiv M - i\Gamma/2$. Furthermore, the coupling of the resonance to two mesons is given by the residue of the amplitude at the pole position.

12th International Conference on Meson-Nucleon Physics and the Structure of the Nucleon (MENU 2010)
AIP Conf. Proc. 1374, 635-638 (2011); doi: 10.1063/1.3647220

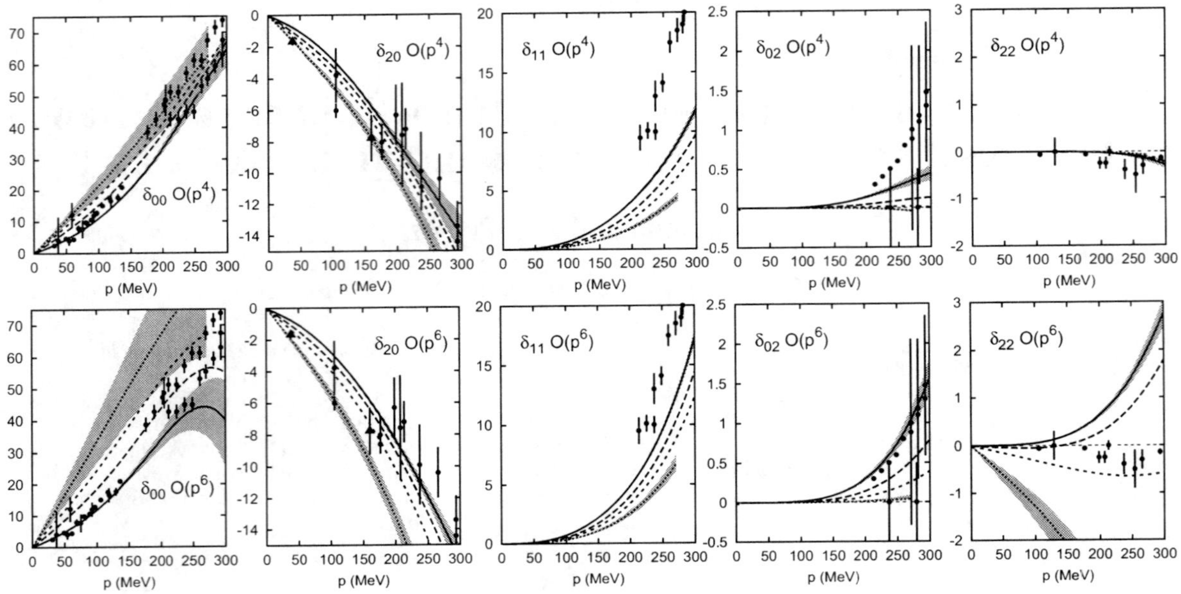

FIGURE 1. $\pi\pi$ phase shifts from standard ChPT up to one loop (first row) and to two loops (second row). Different lines stand for different pion masses: continuous, long dashed, short dashed and dotted for $M_\pi = 139.57$, 230, 300 and 420 MeV respectively. Because of applicability considerations, we do not plot phase shifts beyond $\sqrt{s} = 1$ GeV. Since the lines are too close to each other, we only show error bands for the lightest and heaviest masses. Experimental data (circles) come from [5]. Lattice results for the I=2, J=0 channel (triangles) come from [6].

Phase shifts. In Fig.2 we show preliminary results for $\pi\pi$ phase shifts (note that the one and two-loop IAM cannot be used for D-waves) for different pion masses. For the one loop analysis (first row) we used the LECs in [2] and found that the dependence is again quite soft, specially for the I=2, J=0 channel. For the two loop analysis (second row) we used some preliminary sets of LECs [7]. Once more the dependence on M_π is stronger at two loops than at one loop.

Resonance dependence on the light quark mass [3]. The vector resonances are well established $q\bar{q}$ states belonging to an SU(3) octet. In the three upper rows of Fig.3 we show the dependence of the $\rho(770)$ (first column) and $K^*(892)$ (second column) on the light quark mass. The results obtained for the former are very consistent with those in SU(2) [2] (dotted line) and the estimations for the two first coefficients of the M_ρ chiral expansion [8]. Both resonances behave very similarly: their masses increase smoothly, but much slower than M_π. As a consequence there is a strong phase space suppression which accounts by itself for the width decrease, without a dynamical effect through the couplings $g_{\rho\pi\pi}$ and $g_{K^*\pi K}$, that are remarkably constant, which is an assumption made in lattice studies of the $\rho(770)$ width [9]. On the other hand, the $f_0(600)$, or sigma, and the $\kappa(800)$ scalar mesons are still somewhat controversial since their huge width makes their experimental identification complicated. The third and fourth columns show their dependence on $\hat{m}$. As before, the results for $f_0(600)$ are in good agreement with [2]. The most prominent feature of the scalars behavior is the splitting of the mass into two branches. This happens when the associated pair of conjugated poles in the second Riemann sheet, which are approaching each other as the quark mass increases, join in a single pole below threshold to split again and remain in the real axis. Quantitatively, the growth of the $f_0(600)$ mass before the "splitting point" is much faster than that of the $\kappa(800)$. Their width decrease cannot be attributed to the phase space reduction now, as show by the thin lines, because their coupling to two mesons depend strongly on the quark mass.

Resonance dependence on the strange quark mass [3]. We will only vary the strange quark mass in the limited range $0.8 < m_s/m_{s\,\mathrm{phys}} < 1.4$ to ensure that the kaon does not become too heavy to spoil the ChPT convergence nor too light to require a coupled channel formalism. In the three lower rows of Fig. 3 we show the kaon (or strange quark) mass dependence of the ρ, $K^*(892)$, sigma and $\kappa(800)$. As it could be expected, the properties of the $\rho(770)$ and sigma non-strange resonances are almost independent of the strange quark mass within the range of study. Obviously, the $K^*(892)$ and the $\kappa(800)$ show a strong m_s dependence. As the kaon mass is made heavier, their masses grow much faster than they did when increasing the light quark mass, but still much slower than the kaon mass. As before,

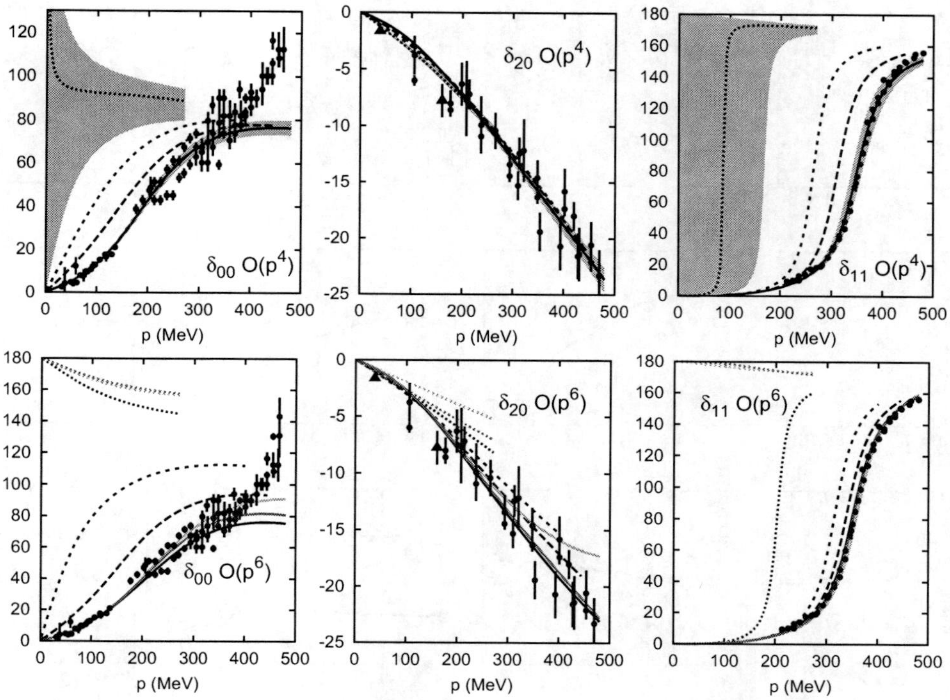

FIGURE 2. $\pi\pi$ phase shifts from unitarized ChPT up to one loop (first row) and to two loops (second row). The conventions are as in Fig. 2 except that, in the case of the two loop analysis, instead of error bands we show results for three different sets of LECs[7] (black, dark gray and light gray).

in the case of the vector resonance the width decreases almost exactly as it would be expected from phase space suppression only and its coupling to $K\pi$ is almost constant, while the scalar width decrease deviates significantly from that behavior, in agreement with $g_{K\pi K}$ depending quite strongly on the strange quark mass.

ACKNOWLEDGMENTS

Work partially supported by Spanish MICINN: FPA2007-29115-E, FPA2008-00592 and FIS2006-03438, U.Complutense/ Banco Santander grant PR34/07-15875-BSCH and UCM-BSCH GR58/08 910309 and the EU-Research Infrastructure Integrating Activity "Study of Strongly Interacting Matter" (HadronPhysics2, Grant n 227431) under the EU Seventh Framework Programme.

REFERENCES

1. J. Bijnens, G. Colangelo, G. Ecker, J. Gasser and M. E. Sainio, Phys. Lett. B **374**, 210 (1996).
2. C. Hanhart, J. R. Pelaez and G. Rios, Phys. Rev. Lett. **100**, 152001 (2008).
3. J. Nebreda and J. R. Pelaez., Phys. Rev. D **81**, 054035 (2010).
4. G. Colangelo, J. Gasser and H. Leutwyler, Nucl. Phys. B **603**, 125 (2001).
5. L. Rosselet *et al.*, Phys. Rev. D **15**, 574 (1977). S. Pislak *et al.* [BNL-E865 Collaboration], Phys. Rev. Lett. **87**, 221801 (2001) G. Grayer *et al.*, Nucl. Phys. B **75**, 189 (1974). S. D. Protopopescu *et al.*, Phys. Rev. D **7**, 1279 (1973). W. Hoogland *et al.*, Nucl. Phys. B **126**, 109 (1977). M. J. Losty *et al.*, Nucl. Phys. B **69**, 185 (1974). N. B. Durusoy *et al.*, Phys. Lett. B **45**, 517 (1973). P. Estabrooks and A. D. Martin, Nucl. Phys. B **79**, 301 (1974). B. Hyams *et al.*, Nucl. Phys. B **64**, 134 (1973). D. H. Cohen *et al.*, Phys. Rev. D **7**, 661 (1973). J. R. Batley *et al.* [NA48/2 Collaboration], Eur. Phys. J. C **54**, 411 (2008).
6. K. Sasaki and N. Ishizuka, Phys. Rev. D **78**, 014511 (2008).
7. J. R. Pelaez and G. Rios, in preparation.
8. P. C. Bruns and U.-G. Meißner, Eur. Phys. J. C **40** (2005) 97.
9. S. Aoki *et al.* [CP-PACS Collaboration], Phys. Rev. D **76**, 094506 (2007).

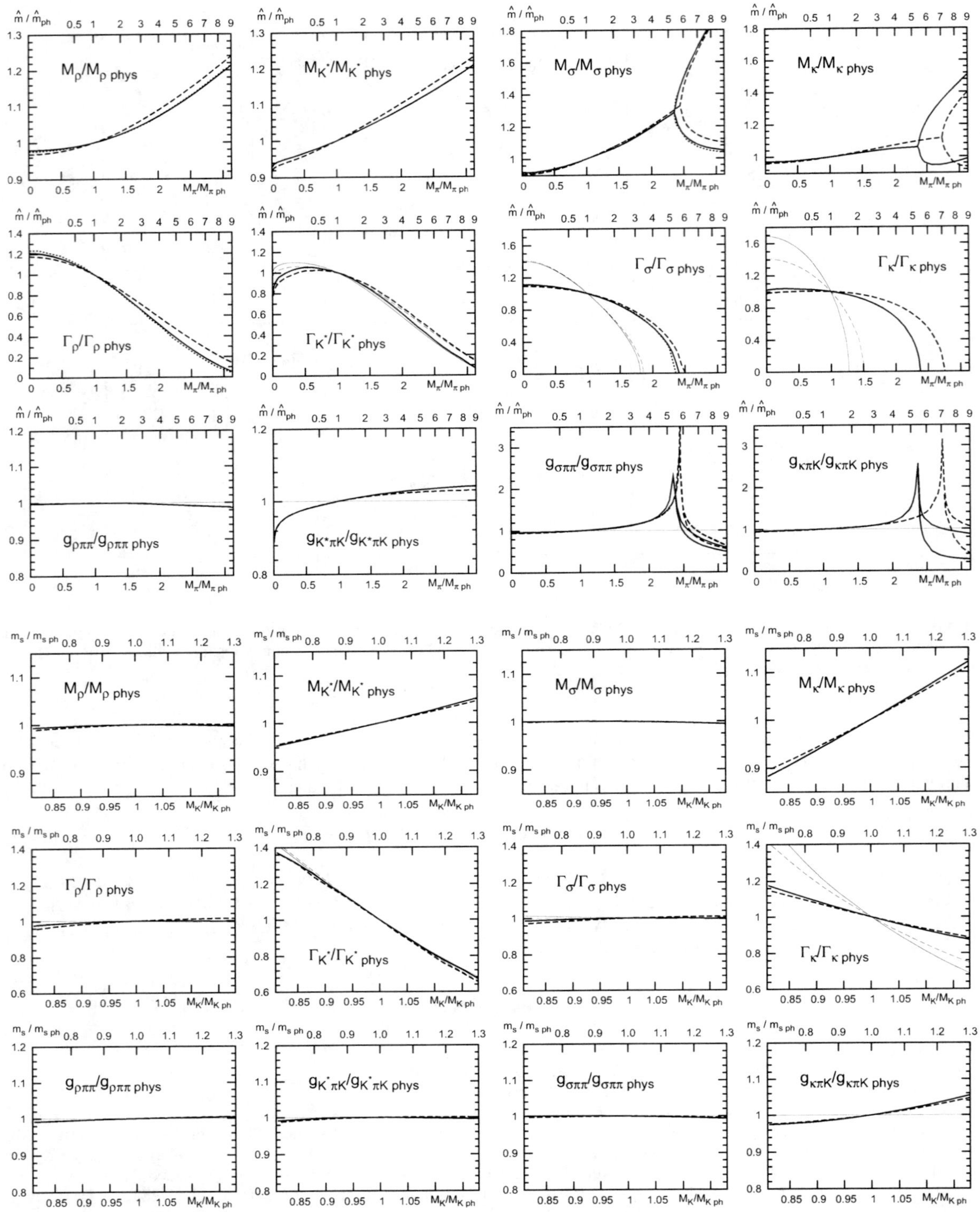

FIGURE 3. Dependence of the $\rho(770)$, $K^*(892)$, $f_0(600)$ and $\kappa(800)$ mass, width and coupling to two mesons with respect to $\hat{m}$ (horizontal upper scale) or M_π (horizontal lower scale) for the three upper rows and with respect to m_s (horizontal upper scale) or M_K (horizontal lower scale) for the three lower rows. Note that all quantities are normalized to their physical values. The thick continuous and dashed lines correspond to Fit I and Fit II in Ref. [3], respectively. The $\rho(770)$ and $f_0(600)$ dependence on the light quark mass is very compatible with that in [2] (dotted lines). Continuous (dashed) thin lines show the dependence of the widths from the change of phase space only, assuming a constant coupling of the resonances to two mesons, using Fit I (II).

Precision measurements of $\pi\pi$ scattering lengths at NA48/2

Andrea Bizzeti

Dipartimento di Fisica, Università di Modena e Reggio Emilia, via G. Campi 213/A, I-41100 Modena, Italy
Istituto Nazionale di Fisica Nucleare, Sezione di Firenze, via G. Sansone 1, I-50018 Sesto Fiorentino (FI), Italy

Abstract. Low energy $\pi\pi$ scattering has been studied with high precision by the NA48/2 experiment at the CERN SPS using large data samples of $K^\pm \to \pi^\pm\pi^0\pi^0$ and $K^\pm \to \pi^+\pi^-e^\pm\nu$ (K_{e4}) decays. The $\pi^0\pi^0$ invariant mass distribution in the $K^\pm \to \pi^\pm\pi^0\pi^0$ decay exhibits a Wigner cusp at the $\pi^+\pi^-$ threshold, from which the scattering lengths a_0 and a_2 are extracted. The measurement of a_0 and a_2 is also performed from K_{e4} decay, with similar precision. The combination of these measurement provides a stringent test of Chiral Perturbation Theory.

Keywords: Kaon decays; pion; scattering length
PACS: 13.20Eb; 13.75Lb

INTRODUCTION

The Chiral Perturbation Theory (ChPT), based on an approximate chiral symmetry of the QCD Lagrangian, can be used to study strong interactions at low energy. In this theory the value of the isospin 0 and 2 S-wave $\pi\pi$ scattering lengths a_0 and a_2 are directly connected with the pion mass m_{π^+} [1].

Under very general assumptions (analyticity, unitarity and crossing symmetries) it has been shown[2] that experimental data at higher energies provide a stringent constraint between a_0 and a_2 scattering lengths:

$$a_2 m_{\pi^+} = (-0.0444 \pm 0.0008) + 0.236\,(a_0 m_{\pi^+} - 0.22) - 0.61\,(a_0 m_{\pi^+} - 0.22)^2 - 9.9\,(a_0 m_{\pi^+} - 0.22)^3 \qquad (1)$$

Moreover, the standard ChPT provides an accurate prediction [3] of both a_0 and a_2: $a_0\,m_{\pi^+} = 0.220 \pm 0.005$; $a_2\,m_{\pi^+} = -0.044 \pm 0.010$.

Previous measurements of the $\pi\pi$ scattering lengths have been mostly performed in the semileptonic decay $K^\pm \to \pi^+\pi^-e^\pm\nu$ (K_{e4}), where the pions are the only strongly interacting particles in the final state. In the past years, only two experiments collected data samples large enough to study the properties of the K_{e4} decay, but not to reach the same precision as the theoretical predictions: the Geneva-Saclay S118 experiment at CERN analysed about 30,000 events[4]. the E865 experiment at BNL about 400,000[5]. A different approach was carried out by the DIRAC experiment at CERN, where the difference between the scattering lengths is extracted from the measured lifetime of the pionium atom. From an analysis of a partial data sample they obtain $|a_0 - a_2|\,m_{\pi^+} = 0.264^{+0.033}_{-0.020}$ [6].

Here we report two new measurements of the $\pi\pi$ scattering lengths by the NA48/2 collaboration.

THE NA48/2 EXPERIMENT

The NA48/2 experiment has been taking data in the years 2003 and 2004, detecting in-flight decays of charged kaons in two simultaneous oppositely charged beams of (60 ± 3) GeV/c momentum produced as secondary beams at CERN SPS. The charged particle reconstruction is provided by a magnetic spectrometer, consisting of a dipole magnet and four drift chambers, with a spatial resolution of 100 μm and a momentum resolution $\Delta p/p = (1.0 \oplus 0.044 p[\mathrm{GeV/c}])\%$.

The energy and position of photons and electrons are precisely measured by a Liquid Krypton electromagnetic calorimeter, consisting of a $27X_0$ almost homogeneous ionization chamber with high-granularity tower read-out: its energy resolution is $\Delta E/E = 3.2\%/\sqrt{E[GeV]} \oplus 9\%/E[GeV] \oplus 0.42\%$ and its spatial resolution about 1.5 mm.

A scintillator hodoscope for fast triggering and precise time measurement, muon and photon veto counters and an iron-scintillator hadron calorimeter complete the experimental apparatus, a detailed description of which can be found in Ref. [7].

Large data samples have been recorded, containing about $4 \cdot 10^9$ $K^\pm \to \pi^\pm\pi^+\pi^-$, 10^8 $K^\pm \to \pi^\pm\pi^0\pi^0$ and 10^6 $K^\pm \to \pi^+\pi^-e^\pm\nu$ (K_{e4}) well reconstructed decays.

12th International Conference on Meson-Nucleon Physics and the Structure of the Nucleon (MENU 2010)
AIP Conf. Proc. 1374, 639-642 (2011); doi: 10.1063/1.3647221

MEASUREMENT OF $K^{\pm} \to \pi^{\pm}\pi^{0}\pi^{0}$ DECAYS

A cusp-like anomaly in the $\pi^0\pi^0$ invariant mass (M_{00}) spectrum of the $K^{\pm} \to \pi^{\pm}\pi^{0}\pi^{0}$ ($K_{3\pi(\pm 00)}$) decays in the region $M_{00} \simeq 2m_{\pi^+}$ has been observed for the first time[8] in an analysis of a partial sample of $2.4 \cdot 10^7$ NA48/2 events.

The existence of this threshold anomaly, never observed in previous experiments, was originally predicted in 1961 by Budini and Fonda[9]. A recent paper by Cabibbo[10] proposed an interpretation of this anomaly, describing the $K_{3\pi(\pm 00)}$ decay amplitude in terms of the $\pi^+\pi^- \to \pi^0\pi^0$ amplitude at threshold.

The scattering lengths a_0 and a_2 are determined from the full NA48/2 data sample, fitting the M_{00} spectrum to theoretical predictions for different values of a_0 and a_2. Two theoretical approaches were used: a detailed model[11] proposed by Cabibbo and Isidori (CI) taking into account all $\pi\pi$ re-scattering processes at the one-loop and two-loop level and an effective field theory calculation[12] by Colangelo, Gasser, Kubis and Rusetsky (CGKR).

The final NA48/2 result[13] on the $\pi\pi$ scattering lengths from $K^{\pm} \to \pi^{\pm}\pi^{0}\pi^{0}$ decays is:

$$\begin{aligned} (a_0 - a_2)\, m_{\pi^+} &= 0.2571 \pm 0.0048_{\text{stat}} \pm 0.0025_{\text{syst}} \pm 0.0014_{\text{ext}} \\ a_2\, m_{\pi^+} &= -0.024 \pm 0.013_{\text{stat}} \pm 0.009_{\text{syst}} \pm 0.002_{\text{ext}} \end{aligned} \tag{2}$$

with a correlation coefficient $\rho = -0.774$.

By imposing the ChPT constraint[2] between a_0 and a_2 we obtain the very precise value of

$$(a_0 - a_2) m_{\pi^+} = 0.2633 \pm 0.0024_{\text{stat}} \pm 0.0014_{\text{syst}} \pm 0.0019_{\text{ext}} \tag{3}$$

MEASUREMENT OF $K^{\pm} \to \pi^{+}\pi^{-}e^{\pm}\nu$ (K_{e4}) DECAYS

The analysis of $K^{\pm} \to \pi^{+}\pi^{-}e^{\pm}\nu$ (K_{e4}) decays provides an independent and complementary method for the determination of $\pi\pi$ scattering lengths. The K_{e4} decay is a rare decay with a branching fraction of about $4 \cdot 10^{-5}$[14], whose amplitude depends on the difference $\delta = \delta_0 - \delta_1$ between the S- and P-wave phase shifts in isosinglet (I=0) $\pi\pi$ scattering. This phase δ can be measured as a function of the di-pion invariant mass $m_{\pi\pi}$.

The NA48/2 data sample contains 1.13×10^6 K_{e4} decays, selected from about 3.5×10^9 charged triggers using criteria described in [15]. The very small background contamination (0.6%), mainly due to $K^{\pm} \to \pi^{+}\pi^{-}\pi^{\pm}$ decays (followed by a $\pi \to e\nu$ decay or with a mis-identified pion) and to $K^{\pm} \to \pi^{\pm}\pi^{0}$ decays (with a Dalitz decay of the π^0 and a mis-identified electron), is estimated from "wrong-sign" combinations $\pi^{\pm}\pi^{\pm}e^{\mp}$.

The analysis is performed in the five independent Cabibbo-Maksymowicz variables[16] describing the K_{e4} decay (see Fig. 1(a)): the di-pion and di-lepton squared invariant masses $s_\pi = m^2_{\pi\pi}$ and $s_e = m^2_{e\nu}$; the angles θ_π (θ_e) in the di-pion (di-lepton) rest frame of $\pi^{\pm}$ ($e^{\pm}$) with respect to the $\pi\pi$ ($e\nu$) flight direction in the $K^{\pm}$ rest frame; and the angle φ between the $\pi\pi$ and $e\nu$ decay planes in the $K^{\pm}$ rest frame. The decay form factors and the phase shift difference δ are measured independently in each $m_{\pi\pi}$ bin by fitting the event distribution in the 4-dimensional space of the remaining Cabibbo-Maksymowicz variables. The fit results for the distribution of the events in each of the five Cabibbo-Maksymowicz variables are shown in Fig. 1(b-f), while the fit results for the phase δ are shown in Fig. 2, together with the measurements from previous experiments.

Numerical solutions[17, 18] of the Roy equations[19] can be used to predict the value of the phase δ as a function of $m_{\pi\pi}$ for any value of a_0 and a_2. The scattering lengths a_0 and a_2 are extracted performing a 2-parameter fit on the measured values of the phase δ near threshold. The resulting values of the scattering lengths are:

$$\begin{aligned} a_0\, m_{\pi^+} &= 0.2220 \pm 0.0128_{\text{stat}} \pm 0.0050_{\text{syst}} \pm 0.0037_{\text{th}} \\ a_2\, m_{\pi^+} &= -0.0432 \pm 0.0086_{\text{stat}} \pm 0.0034_{\text{syst}} \pm 0.0028_{\text{th}} \end{aligned} \tag{4}$$

with a large correlation coefficient $\rho = 0.969$. Effects from isospin symmetry breaking [20] ($m_{\pi^+} \neq m_{\pi^0}$, $m_u \neq m_d$) on the phase shift difference δ have been properly taken into account.

By imposing the CHPT constraint[2] between a_0 and a_2 and performing a one-parameter fit we obtain:

$$a_0 m_{\pi^+} = 0.2206 \pm 0.0049_{\text{stat}} \pm 0.0018_{\text{syst}} \pm 0.0064_{\text{th}}\ . \tag{5}$$

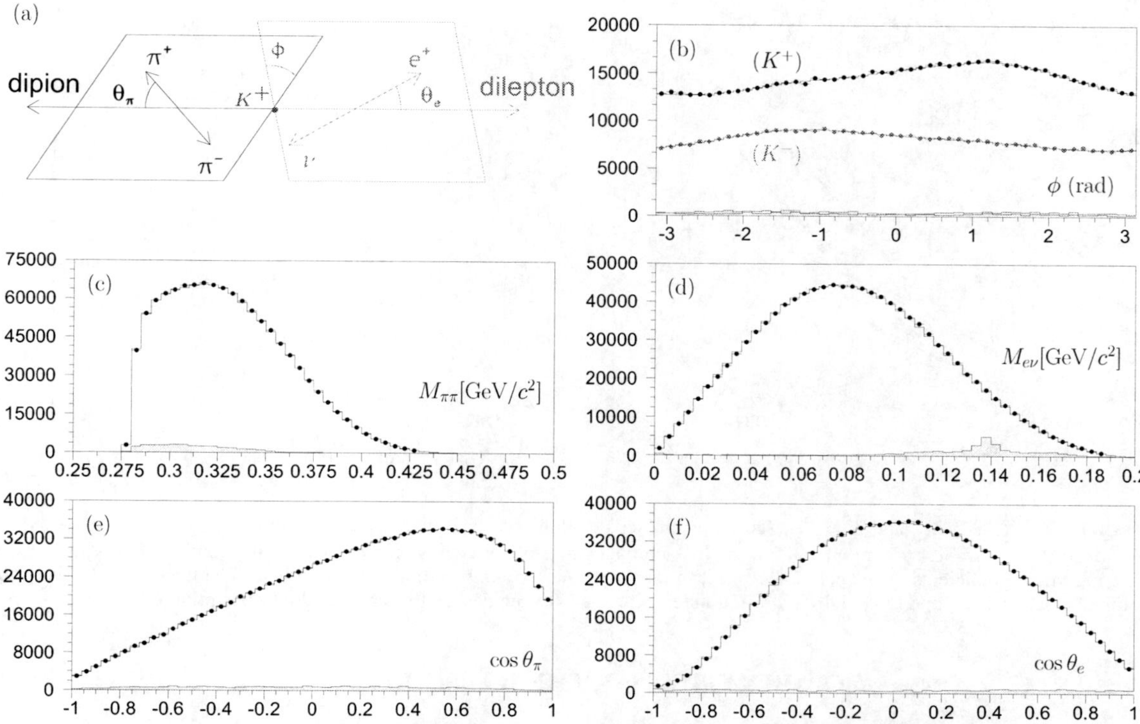

FIGURE 1. (a) Topology of the K_{e4} decay showing the angle definitions. (b-f) Distribution of the five Cabibbo-Maksymowicz variables. Data (after background subtraction) are shown as *symbols with error bars*, simulation using the fitted form factors values as *histograms*. The background contribution is superimposed in the *shaded area*, increased by a factor of 10 to be visible.

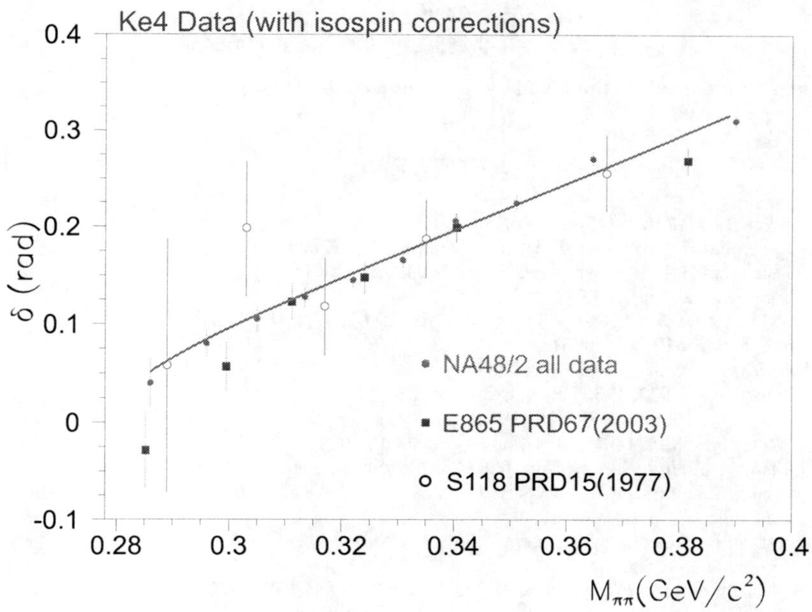

FIGURE 2. Phase shift (δ) measurements with isospin-breaking mass effects included for all K_{e4} experimental results: NA48/2 (solid circles), BNL E865 (solid squares) and S118 (open circles). The line corresponds to the 2-parameter fit of the NA48/2 data.

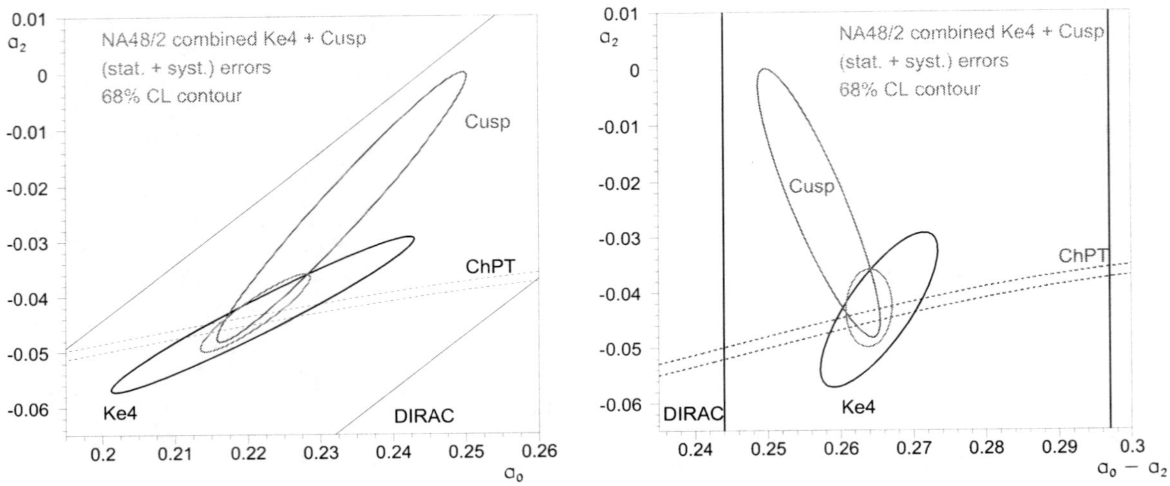

FIGURE 3. NA48/2 K_{e4} and $K_{3\pi(\pm 00)}$ (cusp) results as 2-parameter fits in the (a_0, a_2) plane (left) and in the $(a_0 - a_2, a_2)$ plane (right). The numbers are in units of $1/m_{\pi^+}$. In each plane the smaller contour corresponds to the combination of the results. The correlation coefficients of the combined result in the two planes are 0.91 and 0.28, respectively. The large band between the continuous lines correspond to the DIRAC result, the small band between dashed lines to the ChPT constraint[2].

COMBINATION OF THE NA48/2 RESULTS

The NA48/2 high-precision analyses of $K_{3\pi}$ and K_{e4} data provide complementary information on the scattering lengths, with independent experimental uncertainties and different theoretical inputs. They can be combined (see Fig. 3) to get an even more precise result:

$$\begin{aligned} a_0\, m_{\pi^+} &= 0.2210 \pm 0.0047_{\rm stat} \pm 0.0015_{\rm syst} \;; \\ a_2\, m_{\pi^+} &= -0.0429 \pm 0.0044_{\rm stat} \pm 0.0016_{\rm syst} \;; \\ (a_0 - a_2) m_{\pi^+} &= -0.2639 \pm 0.0020_{\rm stat} \pm 0.0004_{\rm syst} \;, \end{aligned} \tag{6}$$

in excellent agreement with the values predicted by the standard ChPT[3].

REFERENCES

1. S. Weinberg, *Phys. Rev. Lett.* **17**, 616 (1966).
2. G. Colangelo, J. Gasser and H. Leutwyler, *Phys. Rev. Lett.* **86**, 5008 (2001).
3. G. Colangelo, J. Gasser and H. Leutwyler, *Nucl. Phys. B* **603**, 125 (2001).
4. L. Rosselet *et al.*, *Phys. Rev. D* **15**, 574 (1977).
5. S. Pislak *et al.*, *Phys. Rev. Lett.* **87**, 221801 (2001); *Phys. Rev. D* **67**, 072004 (2003).
6. B. Adeva *et al.*, *Phys. Lett. B* **619**, 50 (2005).
7. V. Fanti *et al.*, *Nucl. Instr. Methods A* **574**, 433 (2007).
8. J.R. Batley *et al.*, *Phys. Lett. B* **633**, 173 (2006), `hep-ex/0511056`.
9. P. Budini and L. Fonda, *Phys. Rev. Lett.* **6**, 419 (1961).
10. N. Cabibbo, *Phys. Rev. Lett.* **93**, 121801 (2004), `hep-ph/0405001`.
11. N. Cabibbo and G. Isidori, *J. High Energy Phys.* **0503**, 21 (2005), `hep-ph/0502130`.
12. G. Colangelo, J. Gasser, B. Kubis and A. Rusetsky, *Phys. Lett. B* **638** 187 (2006), `hep-ph/0604084`.
13. J.R. Batley *et al.*, *Eur. Phys. J. C* **64**, 589 (2009).
14. C. Amsler *et al.* (Particle Data Group), *Phys. Lett. B* **667**, 1 (2008).
15. J.R. Batley *et al.*, *Eur. Phys. J. C* **54**, 411 (2008).
16. N. Cabibbo and A. Maksymowicz, *Phys. Rev.* **137**, B438 (1965); *ibidem* **168**, 1926 (1968).
17. B. Ananthanarayan, G. Colangelo, J. Gasser and H. Leutwyler, *Phys. Rep.* **353**, 207 (2001).
18. S. Descotes, N. Fuchs, L. Girlanda and J. Stern, *Eur. Phys. J. C* **24**, 469 (2002).
19. S. Roy, *Phys. Lett. B* **36**, 353 (1971).
20. G. Colangelo, J. Gasser and A. Rusetsky, *Eur. Phys. J. C* **59**, 777 (2009).

AUTHOR INDEX

N

O

P

Q

R

X

Y

Z